# CRYOCOOLERS 9

A publication of the International Cryocooler Conference

# CRYOCOOLERS 9

**Edited by**

# R. G. Ross, Jr.

Jet Propulsion Laboratory
California Institute of Technology
Pasadena, California

## Springer Science+Business Media, LLC

Library of Congress Cataloging-in-Publication Data

International Cryocooler Conference (9th : 1996 : Waterville Valley,
N.H.)
    Cryocoolers 9 / edited by R.G. Ross, Jr.
        p.   cm.
    At head of title: A publication of the International Cryocooler
Conference.
    "Proceedings of the 9th International Cryocooler Conference, held
June 25-27, 1996, in Waterville Valley, New Hampshire"--T.p. verso.
    Includes bibliographical references and indexes.
    ISBN 978-1-4613-7691-0      ISBN 978-1-4615-5869-9 (eBook)

    DOI 10.1007/978-1-4615-5869-9
    1. Low temperature engineering--Congresses.   I. Ross, R. G.
(Ronald Grierson), 1942-    . II. Title.
TP480.I46   1996
621.5'6--dc21
                                                        97-9269
                                                          CIP

Proceedings of the 9th International Cryocooler Conference, held June 25 – 27, 1996,
in Waterville Valley, New Hampshire

ISBN 978-1-4613-7691-0

© 1997 Springer Science+Business Media New York
Originally published by Plenum Press,New York in 1997

10 9 8 7 6 5 4 3 2 1

# Preface

The last two years have witnessed an explosion in interest in pulse tube cryocoolers following the achievement by TRW of high efficiency long-life pulse tube cryocoolers based on the flexure-bearing Stirling-cooler compressors from Oxford University, and have seen the initiation of development of long-life, low-cost cryocoolers for the emerging high temperature superconductor electronics market. Hydrogen sorption cryocoolers achieved their first operation in Space this year, and closed-cycle helium Joule-Thomson cryocoolers continue to make progress in promising long-life space applications in the 4 K temperature range. On the commercial front, Gifford-McMahon cryocoolers with rare earth regenerators are making great progress in opening up the 4 K market, and new closed-cycle J-T or throttle-cycle refrigerators are taking advantage of mixed refrigerant gases to achieve low-cost cryocooler systems in the 65 - 80 K temperature range. Tactical Stirling cryocoolers, now commonplace in the defense industry, continue to find application in a number of cost-constrained commercial applications and space missions, but are shrinking in numbers as the defense industry goes through a period of consolidation.

Building on the expanding stable of available cryocoolers, numerous new applications are being enabled; many of these involve infrared imaging systems, and high-temperature superconductors in the medical and communications fields. Application experiments, designed to explore, troubleshoot and resolve product integration issues, continue to occur on an ever widening front, particularly in the fields of infrared imaging and spectroscopy, gamma-ray spectroscopy, and high-temperature superconductor applications. An important lesson is that integrating cryogenic systems requires care and thoughtfulness in a broad range of engineering and scientific disciplines. In this regard, the vibration sensitivity of many of the infrared and medical imaging applications has led to the recognition that cryocooler-generated vibration and EMI is a critical performance parameter for these applications. In response, several of the application experiments involve the measurement of vibration and EMI susceptibility, and the development of advanced closed-loop active vibration control systems.

This book draws upon the work of many of the international experts in the field of cryocoolers, and is based on their contributions at the 9th International Cryocooler Conference, held in Waterville Valley, New Hampshire, in June 1996. The program of this conference consisted of 124 papers. Of these, 106 are published here in *Cryocoolers 9*. Although this is the ninth meeting of the conference, which has met every two years since 1980, the authors' works have only been made available to the public in hardcover book form since 1994. This book is the second hardcover volume of what we hope will be a series of professional texts for users and developers of cryocoolers. Prior to 1994, proceedings of the International Cryocooler Conference were published as informal reports by the particular government organization sponsoring the conference — typically a different organization for each conference. A listing of previous conference proceedings is presented in the Proceedings Index, at the rear of this book. Most of the previous proceedings were printed in limited quantity and are out of print at this time.

Because this book is designed to be an archival reference for users of cryocoolers as much as for developers of cryocoolers, extra effort has been made to provide a thorough Subject Index that covers the referenced cryocoolers by type and manufacturer's name, as well as by the scientific or engineering subject matter. Extensive referencing of test and measurement data is included in the Subject Index under a wide variety of performance topics. Examples include refrigeration performance data, complete cryocooler characterization test data, vibration and EMI measurements, and qualification and life test experience. Application and integration experience is also highlighted by specific index entries. To aide those attempting to locate a particular contributor's work, a separate Author Index is also provided, listing all authors and coauthors. Contributing organizations are listed in the Subject Index to assist in finding the work of a known institution, laboratory, or cryocooler manufacturer.

The content of the book is organized into 15 chapters by cryocooler type, starting with Stirling cryocoolers, pulse tube cryocoolers, and associated research. Next, Brayton, Joule-Thomson and sorption cryocoolers are covered in a progression of lowering temperatures. Gifford-McMahon cryocoolers and low-temperature regenerators in the 4 to 10 K range are covered next, followed by a glimpse into the future with miniature solid-state refrigerators receiving increased interest in the laboratory. The last three chapters deal with cryocooler integration technologies and experience to date in a number of representative applications. The articles in these last three chapters contain a wealth of information for the potential user of cryocoolers, as well as for the developer.

It is hoped that this book will serve as a valuable source of reference to all those faced with the challenges of taking advantage of the enabling physics of cryogenics temperatures. The expanding availability of low-cost, reliable cryocoolers is making major advances in a number of fields.

Ronald G. Ross, Jr.

*Jet Propulsion Laboratory*
*California Institute of Technology*

# Acknowledgments

The International Cryocooler Conference Board wishes to thank Creare Inc., which hosted the 9th ICC, and to express its deepest appreciation to the Conference Organizing Committee, whose members dedicated many hours to organizing and managing the conduct of the Conference. Members of the Organizing Committee and Board for the 9th ICC include:

**CONFERENCE CO-CHAIRS**
Walter Swift, *Creare Inc.*
Marko Stoyanof, *AF Phillips Lab*

**CONFERENCE ADMINISTRATOR**
Kathleen Cassedy, *Creare Inc.*

**PROGRAM CHAIRMAN**
Ralph Longsworth, *APD Cryogenics*

**CONFERENCE SECRETARY**
Jill Bruning, *Nichols Research Corp.*

**PUBLICATIONS**
Ron Ross, *Jet Propulsion Lab*

**TREASURER**
Ray Radebaugh, *NIST*

**PROGRAM COMMITTEE**
Alan Crunkleton, *Consultant*
Dean Johnson, *Jet Propulsion Lab*
Peter Jones, *Aerospace Corp.*
Peter Kittel, *NASA/ARC*
Martin Nisenoff, *NRL*
Dodd Stacy, *Creare Inc.*
Emanual Tward, *TRW*

**ADVISORY BOARD**
Stephen Castles, *NASA/GSFC*
Takasu Hashimoto, *Tokyo Univ.*
Chris Jewell, *ESA*
Peter Kerney, *CTI Cryogenics*
Martin Nisenoff, *NRL*
George Robinson, *NRC*
Joseph Smith, *MIT*
Michael Superczynski, *NSWC*
Klaus Timmerhaus, *U. of Colorado*
Jia Hua Xiao, *Chinese Acad. of Science*

In addition to the Committee and Board, key staff personnel made invaluable contributions to the preparations and conduct of the conference. Special recognition is due K. Alexander, D. Kametz, C. Kerney, W. Sixsmith, C. Stoyanof, and K. Swift.

# Contents

## Tactical/Commercial Stirling Cryocoolers                               97

## Stirling Cryocooler Research and Theory                               139

## Pulse-tube Cryocooler Developments                                    173

## Pulse Tube Cryocooler Configuration Investigations          261

## Pulse-tube Modeling and Diagnostic Measurements      327

## Generic Stirling/PT Components Development      403

# Brayton Cryocooler Developments                                     465

# J-T and Throttle-cycle Cryocooler Developments                      493

## Sorption Cryocooler Developments                                     567

## G-M Refrigerators and Low-Temperature Regenerators     607

## Advanced Refrigeration Cycles and Developments      681

## Cryocooler Vibration Characterization and Control      697

# Cryocooler Integration and Test Technologies     747

# Cryocooler Applications Experience     873

# An Overview of Air Force Phillips Laboratory Cryocooler Programs

**L. D. Crawford** and **C.M. Kalivoda**

U. S. Air Force Phillips Laboratory
Albuquerque, NM USA 87117

**D. S. Glaister**

The Aerospace Corporation
Albuquerque, NM USA 87117

## ABSTRACT

This paper presents an overview of the cryogenic refrigerator programs currently under development and test at the U. S. Air Force Phillips Laboratory. The vision statement for the Phillips Laboratory Cryogenic Focused Technology Area is to support the Department of Defense (DoD) space community as a center of excellence for developing and transitioning space cryogenic thermal management technologies. The primary customers for the Laboratory's programs are the Ballistic Missile Defense Organization and the USAF/Space and Missile Tracking System Program Office (formerly Brilliant Eyes). The SMTS program is concentrating on the near term Flight Demonstration System (1998 launch), while the BMDO organization emphasizes development of technology to support the far term Objective System launch (2002).

Phillips Laboratory's cryocooler development programs are generally categorized according to cooling requirement and operating temperature. This paper will describe a variety of Stirling, pulse tube, and Brayton cryocoolers currently under development to meet single stage cooling requirements at 60 K, 150 K, 35 K, and multi-load cooling requirements at 35/60 K. Phillips Laboratory is also developing several advanced cryogenic integration technologies that will result in the reduction in current cryogenic system integration penalties and design time. These technologies include the continued development of the Cryogenic Systems Integration Model (CSIM), 60 and 120 K thermal storage units and heat pipes, cryogenic straps, thermal switches, and development of a cryogenic integration bus.

## INTRODUCTION

The objective of the Phillips Laboratory cryocooler effort is to develop and demonstrate spacecraft cryogenic technologies required to meet future requirements for Air Force and Department of Defense (DoD) missions. Cryocoolers currently under development include Stirling cycle, pulse tube, and Brayton cycle machines that produce cooling in the 35 K through 150 K temperature range. Compared with state-of-the-art dewars and cryogenic radiators, mechanical cryocoolers offer space systems significant weight savings, performance improvements, and long life potential (greater than 5 years).

After a specific cryocooler is developed, the unit is usually subjected to acceptance, characterization, and endurance tests. Acceptance tests are performed to determine if the unit meets contractual specifications. Characterization is then performed to determine the operating envelope of the cryocooler. Endurance tests are then used to demonstrate the reliability of the machine; identify long term, life limiting failure mechanisms; and to identify long term performance degradation.

Components that significantly improve the efficiency, extend life, reduce mass, or limit induced vibration are developed and transitioned into next generation cryocooler designs. These technologies are typically developed under contractor sponsored In-house Research and Development efforts or from the Small Business and Innovative Research (SBIR) program.

Advanced, high pay-off cryogenic integration technologies are developed that reduce risk, complexity, mass, and volume of the cryogenic system. These technologies include thermal straps, thermal storage units for load leveling, thermal switches, and diode heat pipes. Utilization of these technologies ensures an optimum cryogenic thermal management system is developed that limits or eliminates operational constraints imposed on the spacecraft platform.

## REQUIREMENTS

Cryocooler technologies developed by the Phillips Laboratory are required for cooling electro-optical devices, infrared sensors, super cooled electronics, and superconducting devices. Department of Defense missions utilizing cryocoolers typically include missile warning, space and ground surveillance, mapping, and weather monitoring.

Spacecraft cryocooler requirements differ quite dramatically from tactical cryocooler applications due to the imposition of a requirement for long life and continuous operation. Operational lifetimes for spacecraft are usually in the 5-10 year range. With most Air Force spacecraft operating in orbits that preclude periodic maintenance, utilization of highly reliable components is critical. Current Air Force operational requirements result in the need for single and multi-load cooling for short, medium and long wavelength infrared sensors. In some instances, DoD spacecraft are limited in terms of allowable payload power, mass, or volume. This results in an additional requirement for highly efficient thermodynamic machines requiring minimal input power, lightweight structure, and packaged into a small volume as possible. Additional constraints such as vibration suppression or minimization is required to mitigate or eliminate induced vibration from affecting sensor operation. The following programs are categorized in order of temperature range to show how Phillips Laboratory has structured cryogenic technology development activities to satisfy DoD and Air Force mission requirements.

### 60 K Single Stage Cryocoolers

Cryocoolers under development in this category include a 60 K pulse tube, a 65 K single stage reverse Brayton machine, and single stage 60/65 K Stirling cryocoolers, the most mature of all spacecraft cryocoolers under development by the Air Force. The objective of these units is to provide approximately 2 watts of cooling at 60 or 65 K with an efficiency of less than 50 W/W (input power to the electronics divided by cooling load). It is desirable for these units to operate continuously in excess of 5 years and to limit induced vibrations to less than 0.1 Nrms.

The highest visibility cryocooler development program at Phillips Laboratory is the 60 K Protoflight Spacecraft Cryocooler (PSC) under development by the Hughes Aircraft Corporation and funded by BMDO (figure 1). This cryocooler is the most mature cryocooler in a long series of successful cryocoolers sponsored by the government and Hughes In-house Research and Development resources. Specific objectives of this program are to develop a unit requiring less than 100 watts of input power to the compressor motors (2 watt of cooling), life in excess of 7 years, and total system mass less than 33 kg. Unless otherwise noted in this paper, input power

**Figure 1.** Hughes 60 K Protoflight Spacecraft Cryocooler

refers to the power delivered to the compressor motor. In order to meet the objectives of this program, Hughes has incorporated a linear tangential flexure into the compressor. This innovation allows for improvement in radial stiffness and is expected to result in improved cryocooler reliability and induced vibration. Titanium has been added into the housing and piston assembly to increase unit efficiency and result in reduced system mass. Hughes will also demonstrate the adequacy of piston alignment techniques necessary for ensuring adequate sub-mil clearance of moving parts. At present time, delivery of the first protoflight unit is scheduled for November 1996, at which time the unit will be subjected to acceptance, characterization and endurance tests at Phillips Laboratory.

Air Force Science and Technology (S&T) funds are also contributing to the design and fabrication of a 65 K, 5 watt single stage reverse Brayton (SSRB) protoflight cryocooler under development by Creare, Inc. Brayton cycle technology offers the potential for dramatic reduction in induced vibration over Stirling or pulse tube cryocoolers due to the operation of the turbine at high speeds that results in higher frequency vibration modes. This particular program builds on the successes of a 65 K Engineering Demonstration Model (EDM) that is currently undergoing endurance tests at Phillips Laboratory[1]. Design of the protoflight unit has built upon the successes of the engineering unit by developing a miniaturized turboexpander and compressor. Miniaturization of these components results in a technical challenge to reduce component size without increasing efficiency, because parasitic losses are a larger percentage of the overall system when miniaturized. The SSRB cryocooler is currently a joint effort with the NASA Goddard Space Flight Center. The Air Force is providing funds to an existing NASA contract to Creare for the development of this unit and for engineering support for the endurance testing of the EDM unit (figure 2).

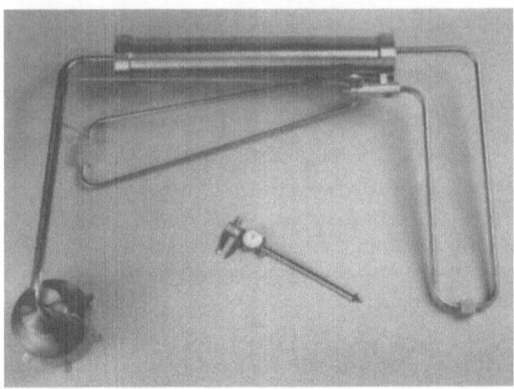

**Figure 2**. Creare 65 K Reverse Brayton Cryocooler Concept

Phillips Laboratory, BMDO, and the SMTS Program Office are co-sponsoring the development of a single stage pulse tube cryocooler that operates at 60 K. The purpose of developing pulse tube technology development program (referred to as the 35 K Pulse Tube contract) was to improve the performance and life of these units to match that of Stirling technology. TRW developed a new 10 cc compressor design that dramatically reduces the size, mass, and improves the overall efficiency of the cryocooler. The 6020 unit delivers 2 watts of cooling and requires only 77 watts of input power (to the compressor) while rejecting heat to 300 K (figure 3). The demonstrated efficiency of this unit is 38 W/W (down from 70 W/W for the 20 cc, 35 K pulse tube design) making it one of the most efficient cryocoolers available at 60 K (pulse tubes or Stirling). This particular pulse tube is nearing completion of characterization testing at JPL and will begin acceptance testing at Phillips Laboratory in July 96.

The final cryocooler in the 60 K family is a 2.25 watt, 65 K pulse tube (slight variant of the 2 W, 6020 unit discussed above) currently being fabricated by TRW and managed by Phillips Laboratory for the DOE/SNL Multispectral Thermal Imager (MTI) flight. Flight electronics developed for this machine are leveraged from the cryocooler electronics being developed for the NASA/AIRS instrument and managed by the Jet Propulsion Laboratory[2]. Delivery of the cryocooler and associated flight electronics is currently scheduled for April 1997.

Development of cryocoolers usually includes the development of flight quality electronics. Recent attempts by Phillips Laboratory to produce flight electronics have resulted in inflexible, ineffective, expensive, and outdated designs. In order to achieve a low cost and effective solution to this problem for users, PL has initiated an in-house effort to develop a standardized set of flight electronics applicable to Stirling and pulse tube cryocoolers. PL has assembled a multi-disciplinary, in-house team that will use a modified 8051 microcontroller incorporating feed forward vibration control algorithms. The initial phase of this effort is to demonstrate the feasibility of this concept by developing and evaluating breadboard hardware. Phase I of this effort is scheduled for completion in the November 1996 time frame. If successful, and approved for continuation, Phase II will focus on the development of brassboard hardware.

### 35 K Single Stage Cryocoolers

Phillips Laboratory, BMDO, and the SMTS program office sponsored the development of single stage pulse tube cryocoolers under the 35 K Pulse Tube contract with TRW. This activity includes the development of single stage pulse tube cryocoolers that deliver 0.3 and 0.9 Watt of cooling at 35 K for less than 100 and 200 watts of input power respectively. Performance and life of these cryocoolers are enhanced by the incorporation of several advanced technology components. The challenges of this technology effort are to implement precision alignment

**Figure 3.** TRW 60 K Pulse Tube Cryocooler

techniques and advanced components that will improve radial stiffness, eliminate piston/cylinder contact, minimize induced vibrations to below 0.1 Nrms, and result in improved efficiency (less than 50 W/W).

The goal of this effort is to improve pulse tube performance and lifetime to match the maturity of Stirling cryocoolers. TRW delivered three engineering models under this effort (2 units designed for 35 K and 1 unit designed for 60 K); all built with flight heritage in mind (details of the 60 K unit were discussed previously)[4]. The first 35 K machine (referred to as the 3585 unit) delivers 0.85 W of cooling at 35 K and requires 200 watts of input power while rejecting heat at 290 K. This pulse tube is currently in endurance testing at Phillips Laboratory. The 35 K Pulse Tube program also focused on minimizing weight and input power for the second (35 K) and third (60 K) pulse tubes. To accomplish this, TRW improved compressor performance and added a fixed regenerator. The compressor size on the 3503 and 6020 cryocoolers was reduced from 20 cc to 10 cc. The compact design and smaller 10 cc compressor results in a mass of only 12.1 kg for both or these cryocoolers. The second pulse tube delivers 0.3 W of cooling at 35 K while requiring only 83 Watts of input power and rejecting heat at 300 K. The 3503 machine recently completed characterization testing at JPL and is in acceptance testing at PL. The Jet Propulsion Lab's characterization of the 10 cc cryocoolers (35 and 60 K) are the subject of another paper presented at this conference. Once endurance testing is completed at Phillips Lab (FY98), these three classes of cryocoolers will be fully characterized in terms of both short and long term performance.

### 150 K Single Stage Cryocoolers

Phillips Laboratory, BMDO, and the SMTS Program Office also sponsored the 150 K Prototype Spacecraft Cryocooler program with TRW. The objective of this effort was to advance pulse tube technology available at 150 K to protoflight maturity. Under this contract, six miniature pulse tube cryocoolers have been developed for Phillips Lab. These light weight cryocoolers ( 2.0 Kg) provide 1.0 W of cooling, require only 9.4 watts input power, and reject heat at 300 K (figure 4).

There is additional performance to be gained by using an advanced cold head also developed under this contract with funding provided by NASA/GSFC. Results of the performance of the advanced cold head are the subject of another paper to be presented at this conference. The mini-pulse tube cryocoolers are designed for a lifetime that exceeds 7 years. Two of the mini-pulse tubes developed under this effort are currently in endurance test at TRW and have each accumulated over 13,500 hours.

As stated above, the mini-pulse tube cryocoolers were designed and fabricated to protoflight levels. There are several space flight experiments which will incorporate the mini-pulse tubes to meet cryogenic cooling requirements. One the mini pulse tubes was transferred to NASA to

**Figure 4.** 150 K Protoflight Spacecraft Cryocooler

support the LEISA instrument on an upcoming SSTI flight experiment. Two of the PL mini-pulse tubes are also being developed for an upcoming DOE/SNL experiment. The main emphasis of the Phillips Laboratory managed program for DOE/SNL was to develop flight electronics for these cryocoolers. Finally, mini-pulse tube cryocoolers are under consideration for several DoD missions. As an example, these cryocoolers are currently baselined for optics cooling on the SMTS Flight Demonstration System currently scheduled for launch in FY98.

### 35/60 K Two Stage Cryocoolers

In addition to single stage cooling, the Air Force is pursuing a joint cooling requirement at 35/60 K using pulse tube and Brayton cycles. The objective of this activity is to develop units that will simultaneously provide 0.4 watt at 35 K and 0.6 watt at 60 K for an input power less than 100 watts and life in excess of 5 years. The purpose of providing simultaneous dual load cooling is to use one active cryocooler to provide cooling to either two focal planes or one focal plane and its optics. There are currently three 35/60 K cryocoolers under development at PL. One is a near term Stirling cryocooler program with Ball Aerospace and two are longer term programs that are focusing on developing a Brayton cycle (Creare) and a co-axial pulse tube design (LMMS).

Phillips Laboratory/BMDO and the SMTS program office are jointly sponsoring a multi-load Stirling cryocooler under development with Ball Aerospace (figure 5). This program was initiated to provide technology for the AFSMC/Space and Missile Tracking System (SMTS) Flight Demonstration System in the event a 35 K sensor requirement was added. As with the 5 watt Creare Brayton program discussed earlier, this effort is joint with NASA/GSFC. The actual contract for this effort is administered through NASA. In this particular case, the Air Force cryocooler is leveraging technology developed for a NASA 30 K two stage cryocooler[3]. The Air Force is modifying the NASA compressor design to help double the cooling capacity. In addition, there is an effort underway to determine if a two or three stage cold finger design will satisfy contractual requirements. The cold finger features a fixed regenerator that improves life, efficiency, and reduces induced vibration. Additional performance improvement is realized by the incorporation of precision piston alignment techniques that eliminate piston/cylinder contact in the cryocooler. A protoflight cryocooler with associated flight electronics will be delivered in March 1997.

Phillips Laboratory/BMDO recently initiated the Miniature, Multi-load Reverse Brayton Cryocooler (MMRBC) program with Creare, Inc. The objective of this program is to develop a

**Figure 5.** Ball 35/60 K Cryocooler Conceptual Configuration

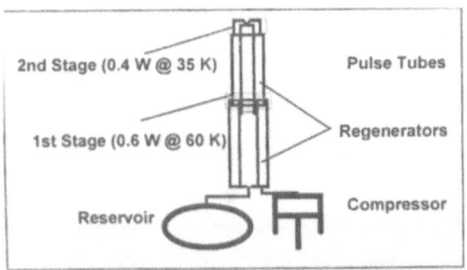

**Figure 6.** Co-axial Pulse Tube Cryocooler

dual load cooling requirement as stated above and also provide an additional 2 watt cooling load at 60 K. Total input power for either cooling requirement is not to exceed 125 watts and must demonstrate a operating life in excess of 5 years. Vibration of this unit is negligible due to the operation of the turbine at very high operating frequencies. In order to achieve these requirements, Creare plans to implement a cryogenic turboalternator that will result in increased efficiency of the unit. Current Brayton cycle machines rely on the working fluid to control the speed of the turbine. Addition of the turboalternator will allow the turbine speed to be controlled with minimum parasitic heat loss. Advanced materials will be used in the design of the unit which will result in additional motor efficiency improvements. A higher efficiency heat exchanger is also being incorporated into the design that is expected to dramatically reduce the mass of the system. A critical design review for this program is currently scheduled in the November 1996 time frame. If selected for continuation, a protoflight cryocooler with rack electronics should be delivered in November 1998.

The other far term 35/60 K program was initiated by Phillips Laboratory/BMDO in 1996 and is being developed by LMMS (figure 6). The objective of this cryocooler is identical to the MMRBC effort discussed above. This program proposes to improve state-of-the-art pulse tube cryocooler technology by incorporating several advanced technology components developed at PL, in-house, or by several Small Business and Innovative Research (SBIR) programs. These innovations include application of co-axial pulse tube configuration that enhances integration, incorporation of tangential flexures (Aerospace Corporation) for reduction in vibration and improved life, incorporation of a flexure cartridge (Peckham Engineering) that eases assembly time and increases lifetime, and incorporation of the etched foil regenerator (Ran Yaron) that results in improved system efficiency. The risk of this effort is considered high due the incorporation of numerous advanced concepts. If successful in the basic phase, delivery of a protoflight cryocooler with flight electronics will occur in October 1998. A Critical Design Review is scheduled for January 1997.

## CRYOGENIC INTEGRATION TECHNOLOGIES

The objective of this activity is to develop components that improve the integrated cryogenic system by reducing large system penalties and analytic errors. Technologies under development include the Cryogenic Systems Integration Model (CSIM), 60 and 120 K thermal storage units, oxygen and methane heat pipes, thermal switches, and flexible cryogenic joints and straps. The technical challenges to be addressed during component technology development includes management of a 2 phase liquid in a zero-G environment, poor capillary wicking of a cryogenic heat transport device, poor phase change material thermal conductivity, and a thermal switch that maintains low on thermal resistance and high off resistance.

The Cryogenic Systems Integration Model (CSIM) is an interactive, user friendly, PC Windows based software tool[5]. The objective is to develop an analysis tool that is capable of performing the integration of mechanical cryogenic systems. The goal of this effort was to develop an analysis tool that is available to industry and the government. This package includes

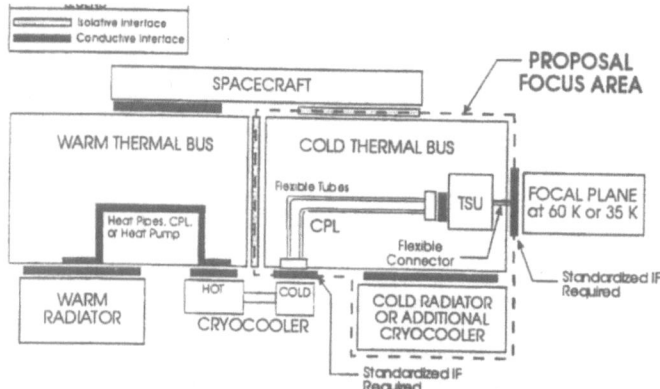

**Figure 7.** Cryogenic Integrated Thermal Bus

design algorithms for all cryogenic integration components and includes algorithms for heat load parasitics and heat rejection. The tool is also capable of accessing a database that includes materials, cryocoolers, and other components. Presently, CSIM is being distributed to industry and government personnel for evaluations against actual designs.

Phillips Laboratory recently initiated a contract with Swales & Associates for the development of a Cryogenic Integrated Thermal Bus (figure 7). The objective of this contract is to develop and integrated, lightweight systems testbed for linking a mechanical cryocooler with a cooled satellite component (sensor). The goal is to develop a standardized interfaces between the system and the component and to standardize the integration onto the spacecraft. This is a two phase program with phase I to be completed in CY96. The objective of phase I will be to survey potential user requirements and to develop the integration technology needs for the following phase II hardware development effort.

Phillips Laboratory has also awarded contracts through the Small Business and Innovative Research (SBIR) program to Jackson and Tull and Swales & Associates for the development of flexible diode heat pipes and thermal storage units (figure 8). Both of these efforts are focused on the design, development, and fabrication of devices that operate at 60 and 120 K.

**Figure 8.** 120 K Thermal Storage Unit

Currently, both of these devices are scheduled for zero-G evaluations onboard Space Shuttle missions. Delivery of the diode heat pipe will occur in October 1996 with flight experiment occurring in March 1997. Delivery of the thermal storage units will occur in May 1997 with a planned flight experiment that will occur in May 1998. Flight characterization of these units is required to determine that ability to manage multi-phase fluid in a zero-G environment and be able to correlate flight data with ground predictions.

## SUMMARY

The Air Force Phillips Laboratory is currently developing a broad spectrum of cryocoolers and integration components necessary for DoD missions. The implementation of cryocooler development has resulted in the successful teaming of government agencies through multiple sponsored programs. Spacecraft cryocooler technology has reached the point of maturity to be transitioned to several government programs. Applications of technology developed by Phillips Laboratory/BMDO is now being baselined on the AFSMC/SMTS spacecraft, MTI spacecraft, and the NRL sponsored HTSSE-II superconductivity flight experiment[6]. In-house evaluation has also demonstrated cryocoolers are capable of operating continuously for long periods of time. Continued technology development is required to continually improve efficiency and demonstrate operational lifetime in excess of 5 years that is required by the DoD spacecraft community.

## ACKNOWLEDGMENT

The work described in this paper was carried out by personnel from the U.S. Air Force Phillips Laboratory, Orion International Corporation, and Nichols Research Corporation. Technical effort was sponsored by the Ballistic Missile Defense Organization, the Space and Missile Tracking System Program Office, and DOE/Sandia National Laboratories. Additional technical effort for the characterization of pulse tube cryocoolers was provided by the Jet Propulsion Laboratory, California Institute of Technology, through an agreement with NASA. Administration and technical oversight of the Ball Aerospace 65 K cryocooler and Creare reverse Brayton contracts is provided by the NASA/Goddard Space Flight Center. We would also like to thank all of the contractors for their continuing efforts in developing enabling technologies that meet DoD mission needs.

## REFERENCES

1. Swift, W.L., "Single Stage Reverse Brayton Cryocooler: Performance of the Engineering Model", *Cryocoolers 8*, Plenum Press, New York (1995), pp. 499-506.

2. Ross, R.G., "JPL Cryocooler Development and Test Program Overview", *Cryocoolers 8*, Plenum Press, New York (1995), pp. 173-184.

3. Burt, W.W, and Chan, C.K, "Demonstration of a High Performance 35 K Pulse Tube Cryocooler", *Cryocoolers 8*, Plenum Press, New York (1995), pp. 313-319.

4. Sparr, L., et al, "NASA/GSFC Cryocooler Test Program Results for FY94", *Cryocoolers 8*, Plenum Press, New York (1995), pp. 221-232.

5. Donabedian, M., et al, "Cryogenic Systems Integration Model (CSIM)", *Cryocoolers 8*, Plenum Press, New York (1995), pp. 695-707.

6. Kawecki, T., "High Temperature Superconducting Space Experiment II (HTSSE II) overview and Preliminary Cryocooler Integration Experience", *Cryocoolers 8*, Plenum Press, New York (1995), pp. 893-900.

# A Case Study of a Successful Cryogenic Cooler Development Program

Stephen Castles
NASA, GSFC
Greenbelt, Md. USA

## ABSTRACT

NASA/Goddard Space Flight Center (GSFC) has been developing long life cryogenic coolers for use in space since the 1970s. Emphasis has been steadfastly maintained on the goal of developing cryogenic coolers that can operate in an unattended manner for 5 to 10 years with very high reliability. It has been the experience of Goddard cooler personnel that many institutions still do not have an appreciation of the difficulties involved in developing such a cooler. In this paper, a successful cooler development program will be presented as a case study of the process required to develop a highly reliable, long-life cryogenic cooler.

The case study covers the period 1987 through the present. It will cover the setting of the cooler requirements, the development of a program plan that lays out the resources and schedule for the overall program, the development of analytical tools used to model the most difficult technical aspects of the cooler, the hardware development and functional test cycle and the iteration of the cycle through a series of models, and the life testing. Lessons learned during the cooler development will be discussed.

## INTRODUCTION

The development of a long-life, closed cycle cooler is a major undertaking. Both government and private institutions have made numerous attempts to develop long-life coolers over the past 3 decades. Most of these efforts have failed. Some of the failures were caused, at least in part, by the large magnitude of the resources required to successfully develop a reliable cooler. In a number of instances, the participants in the development did not enter into the development with an adequate vision of the magnitude of the total required effort, either the resources required or the duration of the development. As a result, the effort was eventually abandoned.

A historical overview of a successful cooler development is presented herein as a case study to illustrate the resources required, the development schedule to be expected, and the typical cooler development process. The major technical issues confronted during the development process will be presented, along with how they were addressed. In addition, programmatic issues confronted during the development process will be discussed. It is hoped that this overview will provide some insight into the cost, schedule, and technical requirements for such a development.

Cryocoolers 9, Edited by R.G. Ross, Jr.
Plenum Press, New York, 1997

## HISTORICAL BACKGROUND

Long-life cryogenic coolers of various types have been under development for more than 3 decades. The first cooler to successfully complete a 5-year life test with no detectable degradation was the linear Stirling cycle cooler produced by Philips Laboratory, a division of North American Philips. NASA/Goddard began the development of this cooler in the late 1970s. The cooler used active magnetic bearings to maintain true clearance seals. It produced 5 W of cooling at 65 K. Unfortunately, it was larger, more massive, and more expensive than desirable for NASA payloads. NASA/Goddard terminated the program to develop a cooler with magnetic bearings in 1990, partly because of increasing success with the alternative cooler technology described below.

In the 1980s, Goddard began reviewing the Oxford University effort to develop a cooler for a British instrument (ISAMS) on the Upper Atmospheric Research Satellite. This small, linear Stirling cycle cooler used the "Oxford" flexure spring technology. By reducing the loading on the seals, this technology held the potential for enabling long-life coolers. This technology was commercialized by British Aerospace, and two coolers were flown on the ISAMS instrument.

In 1986, NASA/Goddard, working with the present NASA Headquarters Office of Space Access and Technology, began a program to develop a small, long-life cooler for the NASA Earth Observing System (EOS) program. Because of the relatively large cost of the instruments on this program, it was desirable to have a cooler with the highest possible reliability. The goal was to produce a cooler whose reliability would be limited primarily by the electronics that drive the cooler, and which could be treated by an instrument integrator in the same manner as any other aerospace component.

## EARLY STUDIES

After reviewing the available cooler options, Goddard decided to pursue several cooler technologies in parallel. One of these technologies was the flexure bearing-based linear Stirling cycle cooler. Goddard began this development effort with a program to study the most difficult technical aspects of the cooler, the aspects that might lead to insurmountable difficulties or that might result in an expensive cooler development if they were not well understood. After a review of all aspects of the flexure bearing-based cooler technology, the following items were identified for further study:

1. The dynamic motion of the flexure bearing-supported piston and displacer during operation of the cooler. The dynamic motion of the piston is affected by the structural modes of the piston when mounted on the flexures and by the gas forces acting on the piston. The same is true for the displacer, but the piston turned out to be the worst case.
2. The control of the residual vibration resulting from the linear motion of the piston and displacer.
3. The thermodynamic efficiency of the cooler.
4. The outgassing of materials within the cooler.

In 1987 Goddard began a program to develop both analytical and experimental tools to understand and control these four items. These tools will be summarized briefly . More details are available in the referenced papers.

Working with Swales and Associates and Fred Costello, Inc., Goddard personnel spent 2-years developing and verifying the analytical tools necessary to analyze the dynamic motion of a flexure bearing-supported piston. Experience has shown that this was a critical development. As expected, the analysis indicated that gas forces on the piston were the dominant forces. What was not expected was the critical nature of the alignment of the piston within the cylinder. The analysis showed that unless this alignment was maintained extremely well, the gas forces would drive the piston into the cylinder, particularly at the end of the compression stroke.[1] This analysis was provided to all of the cooler developers then under contract to the U. S. Government. Two of the contractors independently verified the analysis. Later, data from the radial position sensors on the Ball Aerospace "30 K" cooler was observed to match the qualitative predictions of the analysis.

The Goddard approach to controlling residual vibration was to attack the problem both experimentally and analytically. Because of the repetitive nature of the observed residual force from linear compressors and displacers, Goddard personnel decided to investigate the use of a feed-forward control system. Using an early linear Stirling cooler with an active counterbalancer, it was experimentally determined that a nonreal-time control system could provide excellent vibration control, literally driving the vibration levels into the noise floor of the test stand, less than 0.001 N. To limit the influence of external disturbances on the control system, a narrow-band control algorithm was created that could be tuned to specific harmonics. This algorithm was perfected and verified experimentally on a number of linear Stirling coolers.[2] It was then provided to cooler developers under contract to the U. S. Government. All but one of these companies have now adopted this control system for use on their cooler. It should be mentioned that further development of vibration control systems is still on-going in an attempt to reduce the cost of the electronics required to control coolers.[3,4]

The Goddard team also considered the thermodynamics of linear Stirling cycle coolers. It was decided very early not to attempt to reproduce the standard thermodynamic models of Stirling coolers. Instead, a more simple but still important aspect of the cooler thermodynamics was investigated. Specifically, Goddard, working with Swales and Associates, created numerical tools to analyze the conduction of heat from the working fluid to the cooler heat sink(s). This model incorporated both the thermal nodal analysis for the structure and the gaseous conduction within the cooler. The interesting feature was the ability of the model to account for the influence of the gas dynamics on the heat distribution within the cooler[1]. When this analysis was applied to several coolers, it was determined that certain coolers had significant thermodynamic losses resulting from inadequate heat exchange between the working fluid and the heat exchangers for the working fluid. Equally important, a number of issues were surfaced relevant to methods of interfacing coolers to external heat sinks.[5]

Finally, extensive outgassing analyses were run to determine the potential for long-term degradation in performance resulting from outgassing of materials within the cooler. These studies proved exceedingly valuable in evaluating the potential for cooler internal contamination.

## PROGRAM PLAN

When this background work was well underway, Goddard created a program plan to produce a multi-stage cooler to provide 0.3 watts of cooling at 30 K with no more than 75 watts of input power. A detailed program plan laid out the cost and schedule of the program, as well as the technical risks. The plan acknowledged the difficult technical nature of this undertaking by requiring multiple hardware iterations. The plan was presented to NASA Headquarters in 1990 and resulted in a 6-year cooler development program. An important aspect of the plan was its realism. This realism was made possible by the knowledge that Goddard had already obtained on the development program for the Philips magnetic bearing cooler mentioned earlier and by the analyses and tests described above that were in process at Goddard. By the time this plan was written, Goddard engineers, taken as a whole, had accumulated more than 75 years of experience working on mechanical cooler development.

Reviewing the original 1990 program plan, it is remarkable how accurately the plan predicted the actual development. For instance, the schedule presented in 1990 called for the delivery of the prototype cooler in July 1996, which is now expected to be the actual delivery date.

It is often difficult to maintain a major NASA research and development program for 6 years. The fact that the program described herein was adequately maintained is a tribute to the management of the Office of Space Access and Technology at NASA Headquarters. But it is also true that the careful planning performed by Goddard resulted in a well understood program that led to increased confidence in the ultimate outcome.

## COOLER DEVELOPMENT CONTRACT

In 1992, a contract was issued to Ball Aerospace for the development of a two-stage Stirling cycle cooler. The contract contained a basic phase and numerous options. This structure ensured maximum contractual flexibility. In addition, the contractual requirements were stated as functional requirements, leaving the particular choice of implementation to the contractor. Again, this type of contract provided Ball with maximum flexibility.

This flexible type of contract assumes that the contractor has personnel with sufficient experience to make use of the flexibility. Ball personnel did, indeed, have such experience. The lead technical person on the Ball cooler development contract had many years of experience working with linear Stirling cycle cooler technology.

The Ball design was based on a two-stage linear Stirling cycle cooler design that Ball licensed from Rutherford Appleton Laboratory (RAL). The design was significantly altered to include a number of new features. Particularly important to Goddard was the design of the clearance seals. Goddard insisted that the seals be totally noncontacting under all operating conditions. To accomplish this difficult goal, Ball redesigned the cooler alignment features and incorporated radial position sensors. These sensors can be used to determine the clearances at the seals while the cooler is fully operational. Perhaps equally important, they assist in the alignment of the piston and displacer. It is actually this very precise alignment that results in noncontacting seals during operation of the cooler. However, experience has shown that it is difficult to ensure non-contacting seals in a flexure-based cooler without radial position sensors.

In early 1994, Ball delivered the technology model of the cooler to Goddard. This cooler had a compressor with radial position sensors but, because of lack of funds, did not incorporate radial position sensors into the displacer. Tests on this cooler indicated that the compressor seals were indeed noncontacting and that the displacer seals were lightly loaded. The cooler also displayed excellent thermodynamic efficiency and low residual vibration.[6]

At this time, Ball and Goddard jointly decided to pursue an expander development to solve an ongoing problem with small Stirling cycle coolers. The problem is as follows. The outer wall of the cold finger must be very thin to reduce the parasitic heat load on the cold end of the cold finger. As a result, a small force applied transverse to the outer wall of the cold finger will push the outer wall into the moving displacer within it and cause rubbing contact between the outer wall and the displacer. The small cold fingers on small Stirling cycle coolers are typically so delicate that a force of 1 newton applied transversely to the cold finger will distort the outer wall sufficiently to cause touch contact between the outer wall of the cold finger and the moving displacer. A transverse force of this magnitude can easily result from integrating the cooler with a thermal strap and detector assembly.

To eliminate this problem and another problem discussed below, Ball proposed the use of a fixed regenerator.[7] The fixed regenerator occupies an annular region just inside the outer wall of the cold finger and is supported by a cylinder that surrounds the moving displacer. Tests have shown that even with more than 4 newtons applied transversely to the cold tip of the cold finger, the inner cylinder supporting the regenerator is not visibly distorted. The stack of screens that compose the regenerator distorts sufficiently to prevent the small transverse force from being transmitted to the inner cylinder.

The fixed regenerator has two other nice properties. First, the mass of the moving displacer is greatly reduced. The displacer becomes a thin wall "bottle." Since the mass is so small, g-forces have little effect on the displacer. That is, g-forces do not cause significant misalignment between the moving displacer and the cylinder surrounding it. Such a misalignment could easily cause rubbing contact between the displacer and the cylinder.

Also, because this "bottle" is so light, new cold finger configurations can be readily implemented without major impacts on the structures supporting the displacer. Changes to the cold finger, therefore, can be made without redesigning the entire expander assembly. This allows easy redesign of the cold finger to optimize the performance of the cooler at different temperatures and heat loads. This flexibility has allowed Ball to design multistage cold fingers to cool two or more detector assemblies operating at different temperatures.

After testing the technology demonstration model 30 K cooler at Goddard during the first 6 months of 1994, Goddard shipped the cooler back to Ball. Ball disassembled the cooler and integrated the compressor with the new expander. The new expander not only incorporated the fixed regenerator but also included radial position sensors. A series of tests demonstrated that the new fixed regenerator configuration had the same thermodynamic performance as the old moving regenerator configuration. I should mention that the technology demonstration model and this "engineering model" produced 0.4 watts of cooling at 30 K. The engineering model used a development model set of electronics but the input power with flight-like electronics was predicted to be 75 watts. That prediction subsequently has been verified.

Ball then fabricated the prototype model of the cooler in late 1994 and 1995. By September 1995, the cooler was ready for final assembly. This assembly was delayed to consider incorporating advances that were being investigated on a breadboard three-stage cold finger. This three-stage cold finger was being developed for the "35/60K" cooler being developed by the Air Force Phillips Laboratories. The breadboard coldfinger test demonstrated the importance of incorporating a gas flow ring to improve the flow of gas as it enters the regenerator.

The flow ring was added to the cold finger of the 30K prototype cooler and the cold finger assembly was begun in November 1995. It was thought that the final assembly of the cold finger would be straightforward. The cooler was assembled, and initial thermodynamic performance tests were run in February 1996. The thermodynamic performance was found to be extraordinary. The cooler produced 0.5 watts of cooling at 30 K for 75 watts of input power.

Unfortunately, the radial position sensors indicated that the cylinder that surrounds the moving displacer (the "bottle") was not machined correctly. The portion between the midstage to the cold-stage was offset from the center line of the upper portion of the cylinder. By a quirk of fate, the offset portion was also tilted so that the cylinder was in the correct location at the tip of the cold finger. That is, there appeared to be no offset at the tip of the cold finger, even though the cylinder was not machined correctly. This subtle machining error was not discovered during the initial inspection of the part. However, the resulting misalignment was clearly observed with the radial position sensors.

To correct this problem, the expander had to be disassembled, a new part had to be made, and then the cooler had to be reassembled. During this process, a number of mishaps and other small problems occurred that further delayed the final reassembly until May 1996. After reassembly, thermodynamic performance tests indicated that the improved alignment in the cold finger produced a further improvement in the performance of the cooler, particularly under high heat loads.

To complete this discussion on the cooler development, I should note that NASA Headquarters and Goddard did not have sufficient funding to develop flight electronics for the cooler. Fortunately, a flight program provided funding to design such electronics, and Ball worked on the electronics using their internal "IR&D" funds. With the designs that resulted from these efforts, Ball was able to produce a set of prototype electronics under the Goddard contract that was suitable for a space environment similar to that expected for satellites envisioned for the EOS program.

## QUALIFICATION AND LIFE TESTING OF THE COOLER

Ball is presently conducting a cooler qualification test program which includes a simulated launch vibration test; thermal cycling and operation in a vacuum; and an electromagnetic interference (EMI) test.

At the completion of this qualification test program, the prototype cooler will be delivered to Goddard. After initial functional tests, Goddard will perform a series of tests to simulate the use of the cooler with various combinations of radiative coolers. Such hybrid radiator/cooler configurations can greatly reduce the input power requirement and increase the cooling power available. The cooler will be operated with both a cold heat sink and with a heat intercept on the cold finger.

At the completion of these tests, a 5-year life test will be performed on the cooler. The majority of this test will be run under simulated on-orbit conditions. Specifically, the cooler will be operated in a vacuum, and a chiller will be used to control the temperature of the heat sink. A portion of the test may be conducted with the cooler operating with a reduced heat sink temperature. Throughout the life test, the cooler will be mounted on a 6-axis force dynamometer so that any change in the signature of the residual vibration can be detected. Also, periodically a well defined characteristic load curve will be measured to determine whether or not a degradation in the thermodynamic performance has occurred.

## CONCLUSION

At the time of this writing (early June 1996) the cooler is undergoing flight qualification testing. It is anticipated that the cooler will be delivered to Goddard in July 1996, which is the precise month predicted 6 years earlier when Goddard personnel presented the Program Plan for the development of the cooler to the Office of Space Access and Technology at NASA Headquarters. The cost of the cooler development was within budget, and the performance of the cooler exceeds the specified requirements in every respect.

This successful outcome has been the result of 10 years of concentrated and continuous effort by Goddard personnel. It has required an intense 6-year contractual effort by Ball Aerospace. Both Ball personnel and Goddard personnel had many years of experience working on cryogenic coolers prior to embarking on this effort. The successful outcome also required an enlightened management at NASA Headquarters with a long-term commitment to the development of a difficult new technology.

This is not the end of the story but just the beginning. The Air Force is now procuring a three-stage version of the cooler designed to cool one set of detectors to just over 35 K and a separate set of detectors to just over 60 K. Recently, the team developing HIRDLS, an instrument on the Earth Observing System (EOS) announced that HIRDLS will use the Ball cooler. Other flight instruments are also considering the use of the cooler. I look forward to seeing the scientific results from the powerful instruments that will make use of this versatile cooler.

## REFERENCES

1. Castles, S. H. et.al., "NASA/GSFC Cryocooler Development Program", 7th International Cryocooler Conference Proceedings, PL-CP--93-1001, (1992), p. 26.

2. Boyle, R. et. al., "Non-Real Time, Feed Forward Vibration Control System Development and Test Results", 7th International Cryocooler Conference Proceedings, PL-CP--93-1001, (1992), p. 805.

3. Boyle, R. et. al., "Flight Hardware Implementation of a Feed Forward Vibration Control System for Spaceflight Cryocoolers", *Cryocoolers 8*, Plenum Press, New York, (1995), p. 449.

4. James, E., et. al. "Investigation Into Vibration Issues on Sunpower M77 Cryocoolers", presented at the 9th International Cryocooler Conference, Waterville, N.H., (1996).

5. Sparr, L. et. al.,"Spaceflight Cryocooler Integration Issues and Techniques", to be published in *Advances in Cryogenic Engineering s*, (1996).

6. Sparr, L.et. al., *Cryocoolers 8*, Plenum Press, New York, (1995), p. 221.

7. Berry , D. et. al.,"System Test Performance Data for the Ball Two Stage Stirling Cycle Cooler", presented at the 9th International Cryocooler Conference, Waterville, N.H., (1996).

# The DOD Family of Linear Drive Coolers for Weapon Systems

**H. Dunmire and J. Shaffer**

U.S. Army Communication and Electronics Command,
Research, Engineering, and Development Center
Night Vision & Electronic Sensors Directorate
Fort Belvoir, VA 22060-5677

## ABSTRACT

The critical module of all second generation thermal imaging systems is the Standard Advanced Dewar Assembly (SADA). To meet the requirements of advanced Infrared (IR) Imaging systems of the 1990s and beyond the Department of Defense (DOD) has established a family concept for the SADAs. The family concept for second generation systems consists of the SADA I, SADA II, and SADA III. The SADA family addresses the needs of high performance systems (Comanche), high/medium performance systems (Horizontal Technology Integration (HTI) Second Generation Imaging Systems and the Improved Target Acquisition System (ITAS)) and compact class systems (Javelin). A SADA consists of the Focal Plane Array (FPA), dewar, Command & Control Electronics (C&CE), and the cryogenic cooler. In support of the SADA family DOD has also defined a family of linear coolers.

The purpose of this paper is to highlight the family of linear drive coolers that address the requirements of the SADA family and first generation applications. The linear drive cooler maintains the FPA at the desired operating temperature. The coolers are highly reliable, have low input power, a quick cooldown time, have low audible noise & vibration output, and are nuclear hardened.

This paper will (1) outline the family of coolers that the Department of Defense has standardized on, (2) highlight the characteristics of each cooler, (3) present the status of the various developmental & qualification efforts, (4) discuss the various second generation users and first generation systems, and (5) cover application and integration.

## INTRODUCTION

The Night Vision and Electronic Sensors Directorate (NVESD) has developed linear drive technology for the next generation of closed cycle cryogenic coolers. The increased reliability and improved performance characteristics of these coolers successfully addresses the shortcomings of the first generation rotary coolers and made it possible to develop an integrated Dewar/Cooler assembly for second generation thermal imaging applications. NVESD has defined a family of linear drive coolers that meets the needs of the second generation scanning and emerging staring focal plane arrays. The family of linear drive coolers consist of the 0.15 Watt, 0.35 Watt, 0.60 Watt, 0.70 Watt, 1.0 Watt and 1.75 Watt and are pictured below.

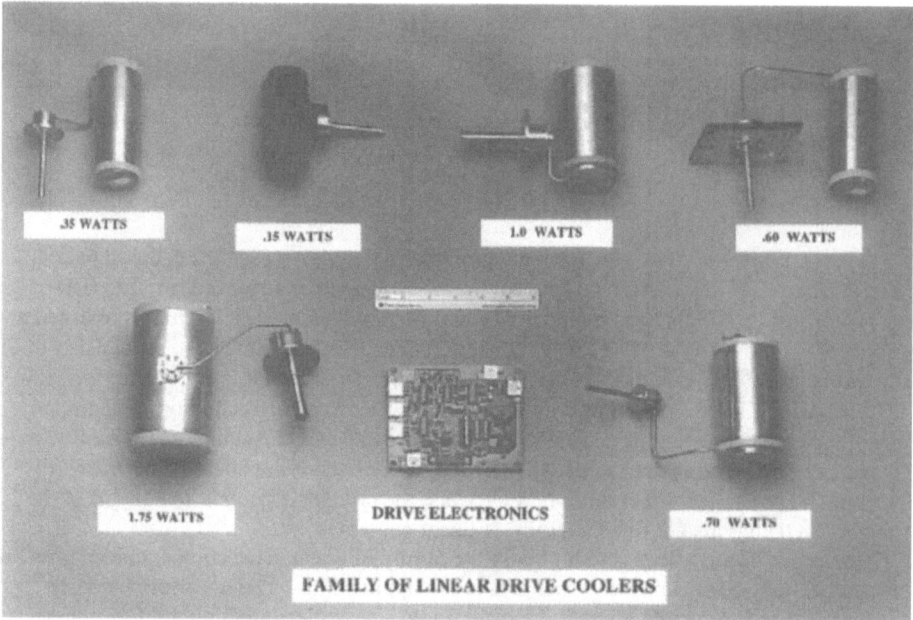

**Figure 1.** Linear Cooler Family

Given the problems (low reliability, poor shelf life, multi-axes vibration & torque, high vibration output levels, excessive acoustic noise, poor temperature stability of the detector array and unneeded power consumption) that were being experienced with rotary coolers in the mid 80's, linear drive cooler programs were initiated to address the failure modes that were being experienced on the coolers and resolve the other shortcomings and inherent limitations associated with the rotary design. Unlike rotary units, compression of the helium is achieved by a clearance seal piston coupled to a moving coil or magnet of the linear motor. The piston's motion can be dynamically balanced by the use of a dual opposed piston concept or a single piston with a vibration absorber system. With either approach no gears, bearings, connecting rods, wrist pins, crankshafts, lip seals or lubricants are required. Because the forces acting on the pistons are in the direction of its motion, the wear on the clearance seals is greatly decreased. Additionally, the linear coolers have temperature control circuitry to regulate the detector temperature about a predetermined set point. This control enables the cooler to operate at a reduced power level by limiting the stroke of the compressor after the desired operating temperature of the array has been achieved. This shortened stroke further reduces the wear, and noise. For these reasons, the linear drive coolers operational life exceeds 40 ') hours, the vibration output is 0.25-0.50 pounds force, they are inaudible at 15 meters, require low input power in the temperature control mode and provides a detector temperature stability of +/- 0.5K at any ambient temperature. Further, since the coolers are welded, the shelf life is in excessive of 10 years. Table 1 highlights the key parameters of these coolers:

**Table 1.** Family of Linear Cooler Key Parameters

| Requirements | 0.35 Watt | 0.60 Watt | 1.0 Watt | 1.75 Watt |
|---|---|---|---|---|
| Cooling Capacity @ 23°C | 0.35 Watt, @ 80 K | 0.60 Watt, @ 77 K | 1.0 Watt, @ 77 K | 1.75 Watt, @ 67 K |
| Cooldown time to 80 K @ 23°C (max) | 10 minutes, with 250 joules | 8.5 minutes, with 450 joules | 13 minutes, with 1440 joules | 6.5 minutes, with 1200 joules |
| Input Power (max) | 35 W | 40 W | 60 W | 94 W |
| Temp Regulation | ± 0.5 K | ± 0.5 K | ± 0.5 K | ± 0.5 K |
| Operating Voltage | 17-32 Vdc | 17-32 Vdc | 17-32 Vdc | 17-32 Vdc |
| Vibration Output (max) | 0.5 lbf | 0.5 lbf | 0.5 lbf | 0.5 lbf |
| Weight (max) | 2.5 lbs | 2.5 lbs | 4.2 lbs | 7.5 lbs |
| Reliability (min) | 4000 MTTF | 4000 MTTF | 4000 MTTF | 4000 MTTF |
| Operating Environment | -54°C to +71°C | -54°C to +71°C | -54°C to +71°C | -54°C to +71°C |

**0.15 Watt Linear Drive Cooler**

The 0.15 Watt Linear Cooler was developed and qualified for second generation, manportable IR applications. It is light weight, compact and in the temperature regulation mode its power consumption is very small. In order to maximize cooling performance and life the cooler is configured with as short of transfer tube as possible. The primary application is the Command and Launch Unit (CLU) of the Javelin anti-tank missile system. To achieve the cooldown time, weight and power requirements of this system, an integrated cooler/dewar approach is used. This configuration mounts the focal plane array directly to the end of the coldfinger significantly reducing thermal mass and heatload. During the Engineering and Manufacturing Development (EMD) phase of the Javelin program the Texas Instruments cooler/dewar assembly went through formal qualification testing per specification MIS-42704 and successfully passed all requirements. Under life test performed by Texas Instruments the coolers demonstrated a 5,526 hour Mean Time To Failure (MTTF) over a temperature profile identified in the specification [1].

### 0.35 Watt Linear Drive Cooler

The 0.35 Watt Linear Drive Cooler has been developed and qualified to address all the systems that are currently using either the 1/3 Watt integral or split rotary drive cooler. This cooler has been integrated and successfully tested on the TOW Night Sight, Remote Display System, Chemical Warfare Directional Detector System and the Night Observation Device, Long Range (NODLR). The cooler is also a component of the Integrated Dewar Assembly (IDA). With the inception of the long life linear drive coolers, NVESD initiated the IDA program for the first generation common module coolers and detector/dewars (60, 120 and 180 element linear arrays).

In addition to all the advantages that the linear drive cooler technology offers, the IDA provides the following extra benefits. Improved operational life over a basic linear drive cooler used in a slip on approach. Since the integrated approach allows more room for cooler performance degradation before it is unable to meet the specification requirements, the IDA life should be improved 20-30%. In qualification testing under thermal cycling (-32°C to +52°C) the IDA surpassed the 4000 hour Mean Time To Failure (MTTF) requirement and demonstrated a MTTF of 5,459 hours. Because the thermal mass to be cooled has been significantly reduced and the thermal interface losses between cooler and dewar eliminated, the operating temperature of the detector will be achieved significantly faster. A typical common module 120 element dewar and cooler assembly will take 10 minutes to cooldown to 77 K @ 23°C, the cooldown time of a 120 element IDA is about 2 minutes. Lastly, because of the reduce heat load imposed by the IDA, a single cooler (0.35 Watt) can cool all the first generation common module detector types regardless of size (60, 120 and 180) as opposed to the slip on approach where a 1/3 watt is needed for the 60 element detector/dewar and a 1 watt is required to cool either the 120 or 180 element detector dewar.

### 0.60 Watt Linear Drive Cooler

The 0.60 Watt Linear Drive Cooler is being developed for second generation, light weight compact systems and medium performance IR applications. It consists of the proven 0.35 watt linear compressor discussed above and uses a larger diameter coldfinger than the 0.35 watt (0.25 inch vs 0.196 inch) in order to increase cooling performance. The primary application of the 0.60 watt is the Standard Advanced Dewar Assembly III B (SADA III B). As previously described, the integrated approach offers quicker cooldowns, longer life and requires a lower capacity cooler to perform the necessary cooling which leads to lower input power, a smaller total package envelope and less weight. Because of the many significant thermodynamic and system advantages that the integrated approach offers over the modular cooler and dewar, it is the approach being used for the entire SADA family. Thus, the cooler's coldfinger becomes the coldwell of the dewar and the interface dimensions are controlled by the dewar to assure interchangeability of the dewar amongst all cooler manufacturers.

### 1.0 Watt Linear Drive Cooler

The One Watt Linear (OWL) Drive Cooler was developed and qualified to address the needs of the medium/high performance second generation thermal imaging systems that will use the SADA II. NVESD has performed various evaluations on the cooler and all testing has been successful. The Texas Instrument cooler has successfully passed the formal qualification test as identified in specification B2-A3165823. The reliability test is still continuing with the coolers having already demonstrated a 5,550 hour MTTF. Testing is scheduled to continue until the units have failed or the test facility is required for other testing. Additionally, under the NVESD SADA II program at Santa Barbara the OWL coolers have been integrated into units and the SADA/coolers have successful passed all the qualification requirements. Numerous OWL's have been successfully integrated into prototype and qualification hardware SADAs from various manufacturers (SBRC, TI and Lockheed Martin (LORAL)). The SADAs have also been

**Table 2.** Cooler Qualification Tests

| Test | Specification Requirement |
|------|---------------------------|
| Audible Noise | < 20 dB at 5 meters at 2000 Hz |
| Vibration Output | < 0.5 lbf peak |
| Operating Mode | Operate in any orientation |
| High Temperature | Seven 30 hour cycles, +71°C |
| Low Temperature | 24 hour soak, 6 hour operation at -54°C |
| Temperature Shock | -62°C to +71°C |
| Mechanical Vibration | MIL-STD-810, modified |
| Mechanical Shock | 250 g, 6 milliseconds |
| Salt Fog | MIL-STD-810 |
| Electromagnetic Compatibility | MIL-STD-461, modified |
| Interchangeability | Compressors/expanders/Dewars not selectively matched |
| Nuclear Survivability | US Army tank mounted FLIR, internal and Family of Light Helicopters, internal |
| Reliability | 4,000 hour MTTF |

installed into prototype second generation thermal imaging systems SAIRS, LOSAT test bed, SGTS, TI plate FLIR, ASTAMIDS, ITAS test bed, JSPD, and EMD systems (ITAS, IBAS and HTI CIV/TIS/CITV). The SADA/coolers have performed well in all the system applications and some of the early prototypes systems have extensive hours on the units. Further, the stand alone cooler has been mated with common module 120 & 180 element detector/dewars and is being used in several first generation thermal imaging systems.

Through a DOD Foreign Comparative Testing program with NVESD and PM-FLIR, AEG and Signaal are under contract to provide coolers to NVESD for qualification testing starting in the 3QFY96. Highlighted below are the qualification tests that the OWL coolers are subjected to. Although certain levels may differ for a specific capacity cooler the entire family of coolers must go through these various test in order to become qualified.

**1.75 Watt Linear Drive Cooler**

The 1.75 Watt Linear Drive Cooler is being developed to address the needs of the high performance second generation FLIR systems that will use the SADA Is or the medium/high performance systems that will use the SADA II but require a fast cooldown time that cannot be meet with an OWL. Prototype coolers have been successfully demonstrated and undergone evaluation testing at NVESD. Coolers from various manufacturers (Hughes, TI & AEG) have been integrated into SADA Is and provided for system applications.

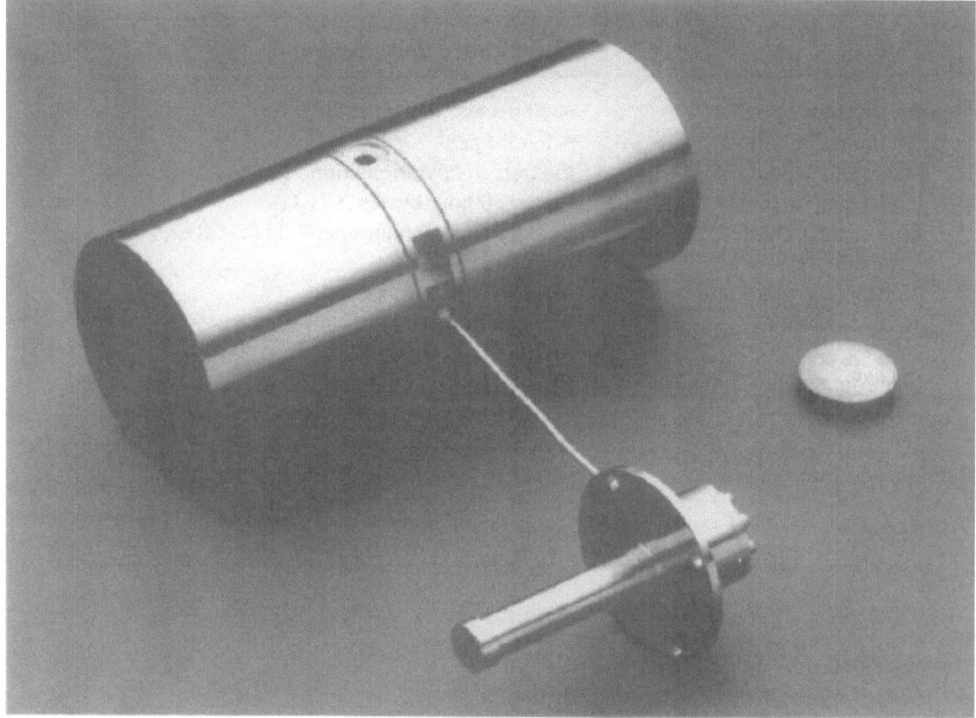

**Figure 2.** 1.75 Watt Linear Cooler

**APPLICATIONS**

As has been identified the family of coolers covers a board range of DOD applications, ranging from manportable to high performance aviation platforms. The operational, environmental, performance, and packaging requirements placed on the coolers in these applications are in most cases the most severe and harsh requirements placed on cryogenic coolers. Figure 2 covers most of the applications were these cooler are being used.

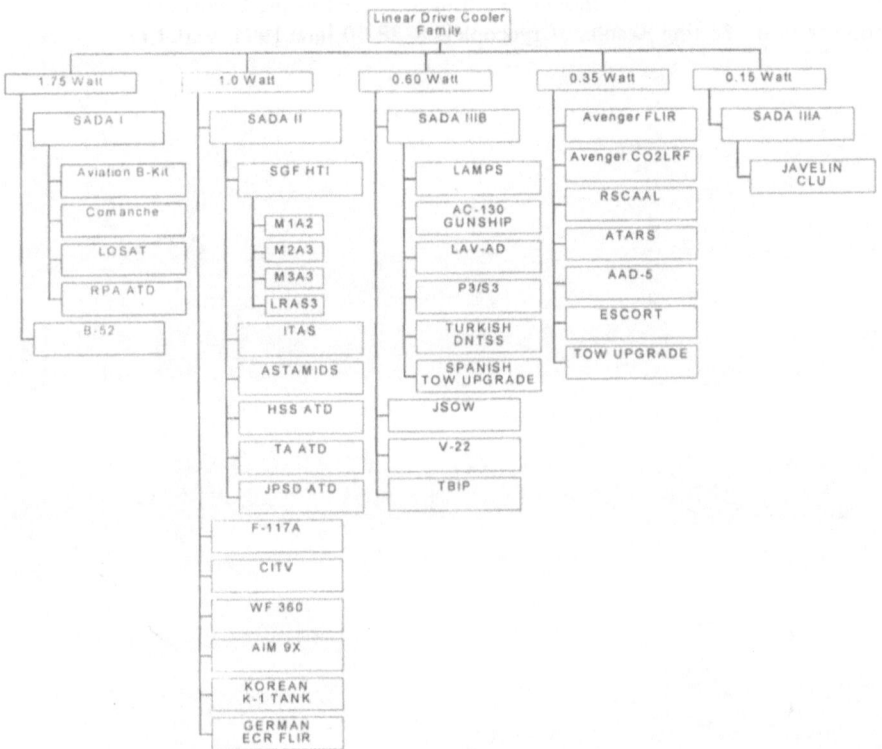

**Figure 3.** Linear Drive Cooler Weapon System Applications

## SUMMARY

A family of linear drive coolers has been defined by the Department of Defense that meets the requirements of second generation Infrared Imaging systems. This family covers almost all the requirements of the emerging second generation thermal imaging systems and also addresses the needs of the first generation applications. The coolers have demonstrated reliability in excess of the 4,000 MTTF requirement and met the environmental and system requirements. The coolers have been successfully interfaced in numerous SADAs highlighting interchangability amongst vendors, dewar designs, and coolers. The cooler/SADAs have been integrated into second and first generation systems and have gone through extensive evaluations and testing. The linear drive coolers are a significant improvement over the rotary drive coolers, however, as was the case with first generation common module coolers, NVESD is committed to the continual enhancement of the performance and operation of the linear drive units.

## REFERENCES

1. Rawlings, R.; Granger, C.; and Hinrichs, G., "Linear Drive Stirling Cryocoolers: Qualification and Life Testing Results", Cryocoolers 8, 28-30 June 1994, Vail, CO.

# And What about Cryogenic Refrigeration?

**M. Nisenoff**

Electronics Science and Technology Division
Naval Research Laboratory
Washington, DC 20375-5347 USA

**F. Patten and S. A. Wolf**

Defense Advanced Research Projects Agency
Arlington, VA 22231 USA

## INTRODUCTION

In attempting to transition any cryogenic electronic technology, such as superconductivity or cooled semiconductor chips, from the laboratory into the marketplace, one of the most frequently asked questions is " . . . . and what about cryogenic refrigeration?" A answer to this question that satisfies the end user is crucial if a new cryogenic electronic technology is to be introduced into the market place for, either, civilian or military applications. In the past, the cryogenics community has responded to the needs of the infrared sensor community.

Shortly after the discovery of superconductivity in materials with superconducting transition temperatures above the temperature of liquid nitrogen (at atmospheric pressures), the Defense Advanced Research Projects Agency (DARPA) of the US Department of Defense started a program to exploit the properties of superconducting at these "elevated" cryogenic temperatures in various system applications, both high power as well as electronic applications. From the beginning the importance of closed cycle cryogenic refrigeration technology was fully appreciated but the approach, during those years, was to use existing cryogenic refrigeration technology which had been developed primarily for the infrared detector community. As high temperature superconducting electronic technology matured, it was realized that cryocoolers with characteristics suitable for cooling infrared sensors would not be appropriate for superconducting or cooled semiconductor applications. Accordingly, in 1993, DARPA allocated funds from their High Temperature Superconductivity program to the development of cryogenic refrigeration systems which would be optimized for use with high temperature superconducting and cooled electronic applications. The Naval Research Laboratory was selected to serve as the agent for DARPA.

## GOALS OF THE PROGRAM

In formulating the DARPA program, it was decided to focus on those parameters which are the most crucial if cooled electronic technologies are to be introduced into operational systems, that is on a nominal cooling capacity at some operating temperature and on reliability and low cost. Specifically, the goals for the program were to develop cryocoolers with cooling capacities ranging from about 4 watts at 60 K (for a typical superconducting applications) at the low temperature limit, up to a cooling capacity of 50 watts at 150 K (for a cooled microprocessor applications) at the high temperature end. The prospective contractors were asked to explore candidate system applications and, from these inputs, formulate a new design concept for a cryogenic refrigerator which would cool the system of interest with emphasis on low cost and high reliability. The prospective contractor was free to vary all cryocooler characteristics, provided that the final design would be compatible with the requirements proposed by the system designers who might use the developed cryocooler. To permit verification of these design concepts, each contractor was required to provide to the Government three demonstration units for extended testing..

In the announcement soliciting proposals, the prospective contractors were told that there would be no military specifications or space specifications. These specifications, which have been applied to earlier cryocooler development, typically imposed constraints on the characteristics of a cryocooler intended for a specific applications. In the case of space cryocoolers, there are stringent requirements on shock and vibration and on weight and input power. For terrestrial "tactical" cryocoolers, there are constraints on weight, volume and dimensions, as these tactical coolers are frequently replacement units for existing systems which are being up-dated. The prospective vendors under the DARPA program were told that once the critical thermodynamic requirements of the candidate application were defined, the other physical and electrical parameters were negotiable, consistent with the system requirements and with low cost and high reliability..

### Low Cost

The only constraints imposed on the cryocooler development program by DARPA and NRL were low cost and high reliability. Both of these terms are relative. For space applications where extreme reliability is required, the cost of a 1 watt at 80 K cryocooler might be as much as US $ 500,000 while for terrestrial applications, such a military IR detector systems, a cryocooler with comparable cooling capacity sells for between US $ 5,000 and US $ 10,000 in small quantities. For a viable commercial applications, a goal of US $ 1,000 per unit (in quantities of 10,000 units per year) was, somewhat arbitrarily, selected. For more expensive systems (costing possibly US $ 20,000 or more), a cost target of 10 percent of the total system cost might be acceptable. In either cast, these costs targets are significantly below those that existing technology can offer.

### Reliability

The other targeted goal of the program was high reliability. An electronic systems must function, uninterrupted and without maintenance, for a period of about three years: the specific requirement, of course , will vary from application to application. In the solicitation, the term "Mean Time Before Failure (MTBF)" was used, while, in other cases, reliability is specified by other terms, for example, the system must be 95 percent reliable for a period of three years, There appears to be no generally accepted definition for reliability of cryocoolers. The precise requirement for these cryocoolers will have to respond to the needs and requirements of the end user and, thus, during the course of testing of these cryoocolers an appropriate definition and testing methodology will have to be formulated to determine whether these cryocoolers will meet the stated end user requirements

**Contract Awards**

A total of 15 proposals were received in response to the NRL Broad Agency Announcement (NRL BAA 05-93) and six were selected for contract award.. The number of awards was limited not by the quality of the proposals received but by the funds available to fully fund a given number of contracts. The successful awards went to:

APD Cryogenics
      Allentown, PA
Cryomech, Inc.
      Syracuse, NY
CTI Cryogenics
      Mansfield, MA
Hughes Aircraft Co.
      El Segued, CA
MMR Technologies, Inc.
      Mountain View, CA
Superconductor Technologies Inc.
      Santa Barbara ,CA

## CONCLUSION

The Defense Advanced Research Projects Agency has initiated a program to develop cryogenic refrigeration systems for high temperature superconducting electronic applications and for cooling conventional semiconductor electronic devices and chips to cryogenic temperatures. The goal of the program is to lower the cost of building such cryocoolers to approximately US$1,000 (or, possibly, ten percent of the cost of the system into which the cryocooler would be inserted) when built in quantities of 10,000 units per year, and to extend the "lifetime" for uninterrupted operation of such cryocoolers to more than three years. Under these contracts, the vendors were to formulate new design concepts for building cryocoolers to satisfy some candidate system requirements and to conform to the goals of this program of low cost and high reliability and to deliver three demonstration units built to these devices to the Government for verification and testing.

A total of six contracts were awarded in September 1994 for a nominal 24 month period of performance. The completion of these contracts and the associated deliveries of the three demonstration units to NRL are scheduled for September 1996 for extended testing at NRL. The description of many of these development activities are included in this volume, immediately following this manuscript.

# Prototype Spacecraft Cryocooler Progress

**K. D. Price, M. C. Barr,  and G. Kramer**

Hughes Aircraft Company
El Segundo, CA, USA  90245

## ABSTRACT

The Phillips Lab/Hughes Protoflight Spacecraft Cryocooler (PSC) is a space qualified, low vibration, single stage Stirling refrigerator that provides over 3.0 W refrigeration at 60 K and over 1.25 W at 35 K. All program performance requirements have been met or exceeded and data are presented in this paper.  The cooler design is derived from the Phillips Lab/Hughes 65K Standard Spacecraft Cryocooler (SSC), with significant improvements in the areas of residual vibration, weight, efficiency, mechanical robustness, and user interfaces. One unit has been built and tested. The compressor module uses a three finger tangential flexure suspension system, resulting in significant vibration reductions. The expander module retains the low vibration spiral flexure design utilized in the SSC. PSC mass has been reduced to 13 kg from the 19 kg mass of the SSC. Specific power referenced to the motor inputs for the 2 W/60 K operating point at 300 K rejection temperature has improved from 35 W/W (SSC) to less than 30 W/W (PSC). User interfaces have been improved by the addition of aluminum heat rejection flanges on the compressor and expander modules, which simplifies the end users heat rejection hardware and installation procedures.  A method to project cooler performance when integrated with a user's system specific thermal interfaces is also provided.

## INTRODUCTION AND MACHINE DESCRIPTION

The PSC, Figure 1, is an "Oxford" class Stirling cycle cryocooler, with compressor pistons and displacer suspended on flexures to maintain non-contacting clearance seal alignment.  The compressor has two opposed pistons for vibration balance working into one compression volume. The expander has a single stage displacer to produce refrigeration.  Vibration balance is accomplished with an internal balance mass driven in opposition. Key program requirements are compared to measured performance in Table 1.

The compressor housing is aluminum and the linear motors employ hermetically sealed coils and magnets to prevent release of volatile agents into refrigerating gas.  Compressor pistons are suspended on three finger tangential flexures, Figure 2. These replace spiral finger flexures used in previous compressors of this type.  The flexures were sized by Marquart using his published method[1].  Stress, stiffness, and rotation of the flexure during displacement were calculated by finite element analysis.  These results validated the accuracy of Marquart's method.  The compressor housing mounting and thermal interfaces are made at a flange located at the axial centerline.  One side of the flange is used for mounting load washers for an adaptive vibration control system and the other side is the heat rejection interface surface.

Cryocoolers 9, Edited by R.G. Ross, Jr.
Plenum Press, New York, 1997

**Figure 1.** PSC thermodynamic machine in random vibration test structure.

**Table 1.** Key PSC requirements and measured performance

| Specification | Requirement | Performance |
|---|---|---|
| Cooling capacity & temperature | 2.0 W at 60 K | 3.2 W at 60 K, max. |
| Power consumption at 300 K | < 75 W | 55 W (2 W at 60 K) |
| Residual vibration | 0.1 N rms | TBD |
| Penalty weight  (W + 0.3*Power) | 33 kg | 13 kg + 0.3 * 55 W = 29.5 kg |
| Cold tip side load capacity | 6 N, minimum | > 14 N |
| Temperature stability | < 0.1 K / 24 hours | < 0.1 K / 24 hours |

The expander housing is titanium for weight reduction over previously used steel.  Linear motors also employ hermetically sealed coils and magnets.  The expander module retained the spiral flexure previously used in similar designs because of their demonstrated low vibration characteristics and because incorporating tangential flexures proved difficult.  Since the PSC expander evolved from the SSC, incorporating tangential flexures required extensive modifications that were not warranted due to the low vibration levels already obtained.

Cold cylinder side load capacity is a robust 14 N or greater in all directions, a direct measurement enabled by the electrical isolation of the metal displacer from the cylinder and housing.  Side load is applied and measured with a calibrated force gage while contact is accurately detected by ohmic contact.  The expander housing has an aluminum heat rejection flange on the front face of the pressure housing.  The flange is inertia welded onto the titanium housing and the joint has been shown to be stronger than the parent materials.  The PSC retains the reliability heritage of the ISSC[2], a Hughes capital cooler derived from the SSC.  Two ISSCs have completed 2 1/2 years of continuous life test with no degradation and a third was successfully launched and operated on the Space Shuttle in 1995[3].

## PERFORMANCE

In this paper, the terms warm and cold interface temperature are used in place of heat rejection and refrigeration temperature in recognition that, in a real system, there are additional thermal

**Figure 2.** PSC compressor uses three finger tangential flexures.

linkages between these interfaces and the actual cooled article and heat sink. The effect of these linkages on cooler performance is discussed in the next section.

All data presented in this section was obtained with the PSC thermodynamic machine in a lab environment and with a vacuum dewar over the cold end. Waste heat was removed from the warm end thermal interfaces to fluid cooled plates attached to each module.

Coolant inlet temperature was defined as the warm interface temperature, that is, the temperature at which heat is removed from the cooler. The cold end temperature sensor and cooling load resistor were mounted to an instrument ring clamped to the cold tip. The temperature of the instrument ring is defined as the cold interface temperature. These definitions are used so it is clear that heat transport in and out of the cooler are through the actual thermal interfaces, which include the contact thermal resistance at the interface joint. For all tests, compressor and expander waste heat were rejected to essentially the same temperature, as coolant temperature rise through the system was less than 3 C. Motor power was measured at each data point, including both compressor piston motors, the displacer motor, and the balance mass motor. Also measured were motor drive voltage and current. Motor cable resistance losses have been subtracted from motor power consumption.

The thermodynamic performance map at 285 K warm interface temperature is shown in Figure 3. This map shows cooler motor power consumption against delivered cooling capacity at 35 K and 60 K cold interface temperatures.

Figure 4 shows the constant stroke load line with the PSC adjusted to deliver 2.0 W at 60 K (Nominal Load and Power) and at maximum stroke (Full Load and Power). This chart shows the effectiveness of refrigeration as low as 20 K.

## ADDING SYSTEM SPECIFIC EFFECTS

The goal of the test program was to characterize cryocooler performance relative to cold and warm thermal interface temperatures at the cooler. This approach isolates cooler performance from the user's cold and warm thermal linkage resistances. The cold thermal linkage is usually a compliant, thermally conductive strap connecting the cooler's cold tip to the user's thermal load. The warm thermal linkage connects the compressor and expander modules to the heat sink, typically a radiator on a space craft. Examples of heat sink linkages include thermally conductive structure, heat pipes, and capillary pumped loops.

The thermal resistance of these linkages vary from system to system, affecting overall thermal system performance. The following discussion combines the effect of the added thermal resistances with the cooler specific data given above to estimate cooler power consumption in a specific system. The level of analysis presented is intended to highlight the effects of the added elements and serve as an aid during conceptual level design, but is not recommended as a substitute for rigorous analysis during detail design.

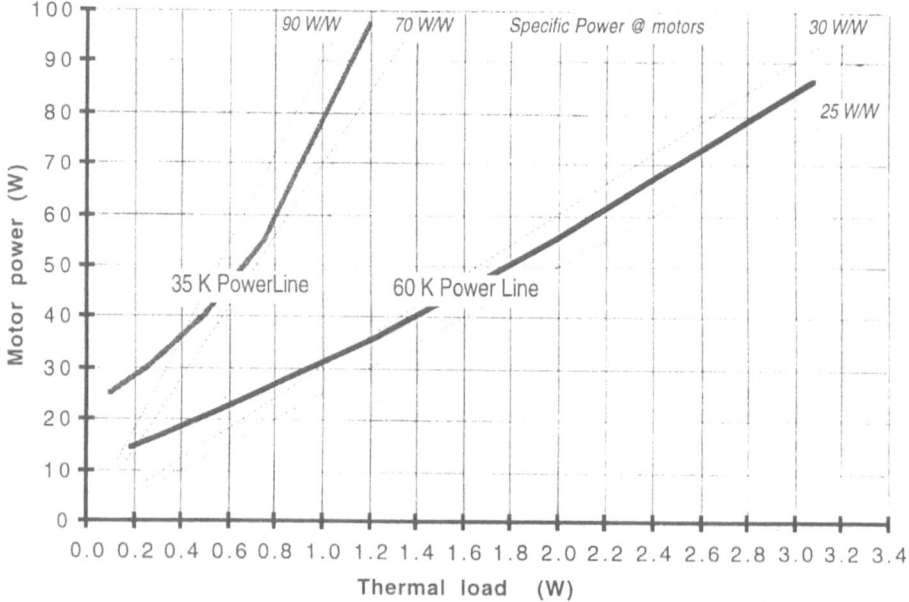

**Figure 3.** Thermodynamic performance map at 285 K warm interface temperature.

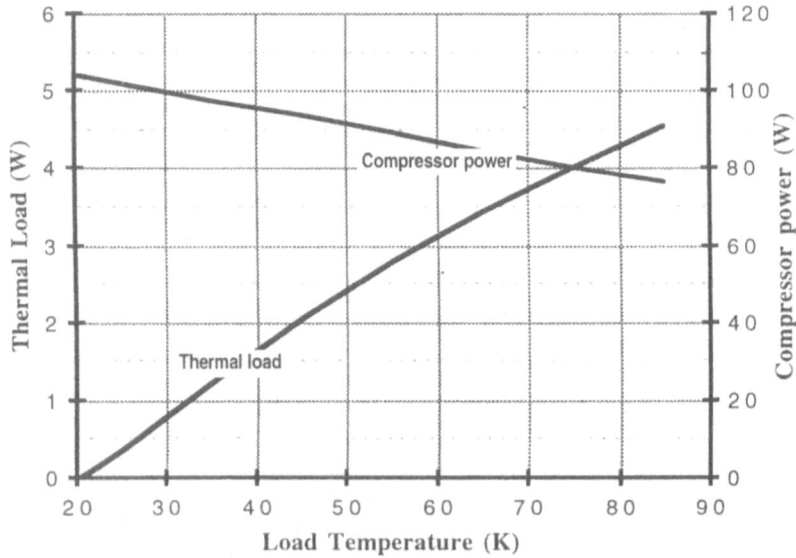

**Figure 4.** Cooling load vs. cold interface temperature at constant stroke, 285K rejection.

As shown for a Carnot cycle diagram, Figure 5, the two thermal linkage resistances decrease cooler efficiency and cooling capacity by increasing the temperature spread between warm and cold ends of the cooler. Adding a thermal linkage at the cold interface introduces a temperature difference between thermal load and cold tip, requiring the cold interface temperature to operate below load temperature. Adding a thermal linkage at the warm interface introduces temperature difference between cooler heat rejection surfaces and the heat sink, causing the warm thermal interface temperature to increase. Load and rejection temperature are fixed by system requirements, but from the cooler's point of view heat must be pumped over a larger temperature span, from $T_{ci}$ to $T_{wi}$. Therefore, more work must be performed to deliver the same cooling capacity.

Adjustment for added thermal linkages can proceed directly from performance charts, but interpolation and iteration are required. Alternatively, if performance data are reduced to equation form using empirical relations, solution may proceed analytically. Table 2 identifies five knowns and three unknowns that are typically involved in this analysis. Three equations are required to solve for the three unknowns.

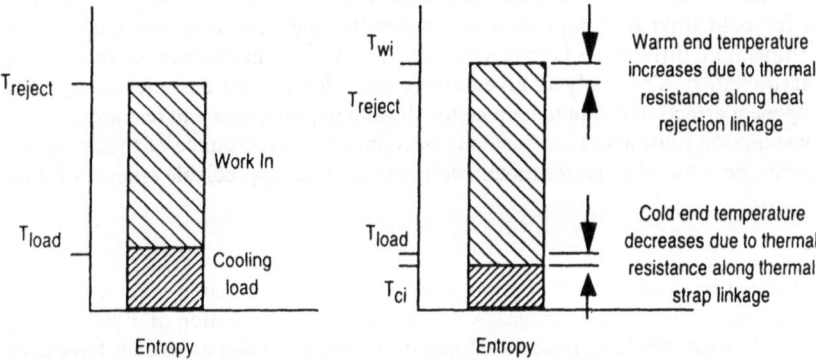

**Figure 5.** Adding thermal resistance between load and cold tip and between heat sink and the cooler's heat rejection surface increases temperature span over which heat must be pumped. Required power increases to maintain same cooling load.

**Table 2.** Cooler interface temperatures at warm and cold ends and cooler power consumption can be expressed in terms of load and rejection temperatures, cooling load, and warm and cold thermal linkage resistance.

| Knowns | | Unknowns | |
|---|---|---|---|
| $T_{load}$ | Temperature of thermal load, K | $T_{ci}$ | Cold thermal interface temperature, K |
| $Q_{load}$ | Cooling load at $T_{load}$, W | $T_{wi}$ | Compressor and expander warm thermal interface temperature, K |
| $R_{ci}$ | Cold interface resistance, K/W | $P_{motor}$ | Total motor input power, W |
| $T_{reject}$ | Temperature of heat sink, K | — | |
| $R_{wi}$ | Warm interface resistance, K/W | — | |

First note that specific power is a function of three variables:

$$P_{motor} / Q_{load} = SP(T_{wi}, T_{ci}, Q_{load}) \tag{1}$$

Then, power can be expressed as a function of specific power and cooling load.

$$P_{motor} = SP(T_{wi}, T_{ci}, Q_{load}) \times Q_{load} \tag{2}$$

Second, the temperature difference between cooling load, $T_{load}$, and cold thermal interface, $T_{ci}$, is a function of thermal load and cold linkage resistance:

$$(T_{load} - T_{ci}) = R_{ci} \times Q_{load} \tag{3}$$

Or, rewritten to isolate $T_{ci}$:

$$T_{ci} = T_{load} - R_{ci} \times Q_{load} \tag{3a}$$

Third, the temperature difference between cooler warm thermal interface, $T_{wi}$, and heat sink rejection temperature, $T_{reject}$, :

$$(T_{wi} - T_{reject}) = R_{wi} \times P_{motor}$$

(4)

Or, rewritten to isolate $T_{wi}$:

$$T_{wi} = T_{reject} + R_{wi} \times P_{motor}$$

(4a)

Equations 2, 3a, and 4a are typically non-linearly coupled and must be solved by iterative technique. For the level of analysis intended by this discussion, solution can be initiated by assuming the warm and cold interface temperatures are equal to the rejection and load temperatures. Proceed through the three equations in sequence using the data charts to determine power consumption until a suitable convergence is achieved, usually two to four cycles.

Ideally, the thermal linkage resistance should be as close to zero as possible. In practice, requirements for cold linkage compliance and warm linkage geometry and length results in appreciable temperature differences between the cooler's thermal interfaces with the cooling load and waste heat radiator. The analysis given above provides system designer's a quick aid in assessing component effects and should help better allocate requirements and resources.

Power consumption must also be adjusted to account for the resistance of motor cables. For practical purposes, only the compressor motor cables consume an appreciable amount of power.

## SUMMARY

The PSC has met or exceeded all program requirements and is capable of providing efficient refrigeration over the 25 K to 100 K range. A significant weight reduction of 30% was achieved relative to the predecessor SSC and residual vibrations in both drive and cross axes have decreased about 50%. Side load capacity was improved an order of magnitude while achieving a 25% improvement in efficiency. PSC reliability is at least as great as the ISSC machines that have already demonstrated over 30 months of continuous service with no performance degradation.

Stirling space coolers such as the PSC have now demonstrated efficiency, reliability, low vibrations and weight, and ruggedness to enable space applications throughout the 25 to 60 K regime. The superior efficiency of the Stirling machine with an active displacer produces a lower system weight penalty as a result of reduced power consumption and waste heat than a comparable pulse tube.

## ACKNOWLEDGEMENT

Program managed by Phillips Laboratory, Kirtland AFB, NM 87117-5320
Project Manager, John Reilly, Contract F29601-92-C-0134
Support from the BMDO and the SMTS Program is also gratefully acknowledged

## REFERENCES

1. Marquart, E., Radebaugh, R., "Design Optimization of Linear-Arm Flexure Bearings", *Cryocoolers 8*, Plenum Press, New York (1995), pp.293-304.

2. Wakagawa, J.M., Haque, H., and Price, K.D., "Improved Standard Spacecraft Cryocooler Life Test for Space-Based Infrared Surveillance", *Cryocoolers 8*, Plenum Press, New York (1995), pp. 69-76.

3. Sugimura, R.S., Russo, S.C., "Lessons Learned During the Integration Phase of the NASA IN-STEP Cryo System Experiment", *Cryocoolers 8*, Plenum Press, New York (1995), pp. 869-882.

# Hughes Aircraft Company SSC I & II Performance Mapping Results

**T. Roberts and J. Bruning**

Phillips Laboratory, USAF
Albuquerque, NM, USA 87117

## ABSTRACT

The results through March 1996 of profiling the performance of the Hughes Aircraft Company's SSC cryocooler family are presented. The similarities in this cooler family's operation are noted and the progress that this engineering design model program has made since inception is documented. Model estimates of this performance manifold are prepared using multivariate regression methods and the assumption of a smooth performance manifold. Anomalous results during the operation of SSC II during continuous operations are discussed with respect to future plans to profile these coolers in more detail.

## INTRODUCTION

The Hughes Aircraft Company Standard Spacecraft Coolers (SSC) design goals were to provide 2 watts of cold end cooling at 65 K, while rejecting heat to a 300 K temperature reservoir. Other major subsidiary goals which were assessed in mapping were [1]:

1. Have a specific power of 30 W/W.
2. Demonstrate active control of induced vibration to the cooler mount.
3. Demonstrate cold end temperature stability of .05 K/hour under constant load.
4. Demonstrate valid heat rejection interface to high temperature reservoir.
5. Operate in 28 Vdc ±5% range.
6. Operate with a heat sink temperature in the range of 250-300 K.

The SSC I cooler began acceptance testing in July 1995 and the SSC II in December 1995. The fabrication of SSC I was more than one year before that of SSC II, and the difference in the performance between these two Engineering Design Model (EDM) coolers reflects the progress and lessons learned which Hughes Aircraft experienced during this development program.

During the course of acceptance testing and performance mapping of "as delivered" performance these coolers supported qualitatively the design goals outlined above. However, tests also demonstrated several deficiencies which are being fully measured. The mapping results are intended for use in design refinement of the Hughes SSC Family of cryocoolers.

Cryocoolers 9, Edited by R.G. Ross, Jr.
Plenum Press, New York, 1997

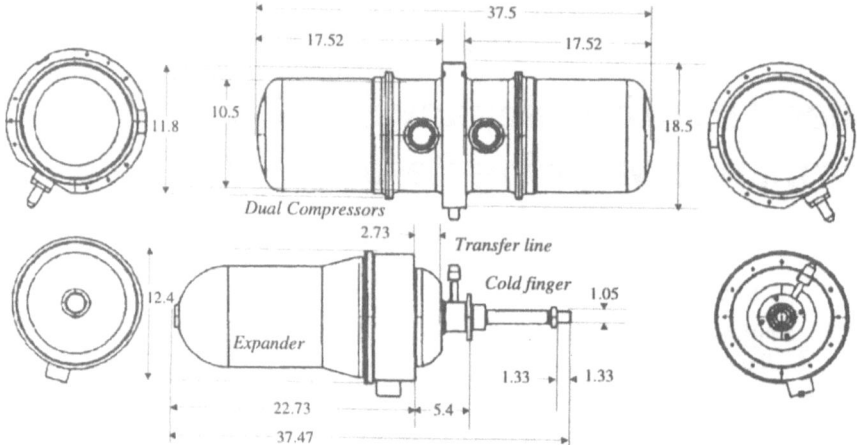

**Figure 1**. SSC Cooler Geometry (dimensions in cm, *courtesy Hughes Aircraft Co.*).

## DESCRIPTION OF THE COOLERS

The Hughes SSC family is characterized as a split Stirling cycle cooler which uses flexure bearings and clearance seals in the compressor and expander sections. The regenerator is attached to the expander piston. The expander is actively counterbalanced by a separately controlled dynamic mass. Additional sinusoid wave forms are added to the drive wave form to reduce induced vibration. The structural and thermal interfaces are located on the wide midsections of the compressor and expander. The two coolers are structurally similar except that SSC II has Rulon lining on the compressor pistons and the expander cylinder, while the SSC I is an all metal design with no lining of its moving parts. Hughes Aircraft has also documented the cold end brazing on SSC I as being marginally defective [1]. The coolers' casings are not hermetically sealed but have O-rings which are designed to support the nominal fill pressure of 32 atm over a five year endurance test. The coolers are controlled and powered from similar rack electronics panels. The control method used for this cooler family is essentially an open loop setting of mechanical parameters such as stroke length, phase angle, and operating frequency. The cold end temperature is controlled either by varying the cold end heat load or by manually changing the setting of these mechanical parameters. An overall diagram of the cooler geometry is shown in figure 1.

## ACCEPTANCE TESTING & PERFORMANCE MAPPING PLAN

The measurement of the thermodynamic performance of these two coolers can be described as having four segments:

1. Cool down and no load operation performance mapping.
2. Steady state loaded performance mapping.
3. Long term temperature stability at nominal design conditions.
4. Extended performance mapping by documenting performance dependence on mechanical and thermodynamic parameters such as stroke length or rejection temperature.

The first two items were planned to be conducted at various heat rejection temperatures using the nominal design point mechanical parameters:

32 atm gas fill pressure
9 mm compressor stroke
2 mm expander stroke
32.5 Hz operating frequency
75-78° phase angle.

The long term temperature stability test was conducted at the nominal design point described above with a cold end temperature of approximately 65 K and a rejection temperature of 300 K over a 72 hour period of continuous operation. Extended performance mapping seeks to extend the performance manifold shown by the cooler in mapping segments 1 to 3 (last page) by measuring the smoothness and limits of this manifold and by measuring performance at discrete intervals within the manifold.

## Instrumentation

The data acquisition (DAQ) equipment supporting this effort consisted of the items in Table 1. The coolers were internally instrumented with linear voltage displacement transducers (LVDT) to document piston position and a 5 Vdc potential between the cooler's piston and cylinders. This latter potential can be measured to monitor whether the moving elements are contacting the end stops or cylinder walls. The transfer line was instrumented with a pressure strain gage which allowed for the user to monitor fill pressure. The structural mount is instrumented with 6 Kistler load cells which indicate the induced vibration levels imparted to the mount.

## MAPPING RESULTS

### SSC I Acceptance Testing and Initial Operations

When SSC I was integrated into its stand in June 1995 it was 4 atm below its normal working fluid fill pressure of 32 atm. Upon initial start up, it was noted that the expander isolation trace indicated expander side contact during normal operations. This contact was eliminated by applying a permanent bending moment on the cold finger mount. During early mapping it was evident that the compressors were extremely sensitive to variations in rejection temperature and operating frequency. At reject temperatures greater than 280 K, the full compressor stroke of 9 mm could

**Table 1.** DAQ Equipment and Measured Properties

| Equipment Item | # Sensors | Measured Property | Accur. |
|---|---|---|---|
| Lakeshore 330 Cryo. Controller | 2 | Cold End Temps. | ±.1 K |
| Lakeshore 820 Cryo. Thermometer | 6 on SSC I | Cold End and Mount Temps. | ±.1 K |
| Fluke 2620A Multimeter (Thermocouple Measurements) | 18 SSC I 9 SSC II | Casing and Mount Temperatures | ±.7 K |
| Vahalla 2300/2300L | 5 (2 units/ stand) | Cooling Load, Exp., Balancer, & Comp. Powers | ±1% full scale |
| Lecroy 9304A Dig. Oscilloscope | 4 channels | Piston displacements | ±2% full scale |
| HP 35660A Dynamic Signal Analyzer | 2 channels | Induced vibration | N/A |
| Phillips PM3057 Analog Oscilloscope | 2 channels | Piston contact | N/A |
| HP 5334A Universal Counter | 1 channel | Piston contacts | N/A |

The Phillips PM3057 and HP 5334A were used to track qualitatively the touch contact events between the pistons and cylinder walls.

only be used without side piston contact when the operating frequency was limited to 31.8 to 31.9 Hz. Generally, compressor A would contact below that frequency and compressor B above that frequency. The general performance of the cooler was substantially below that reported previously by Hughes Aircraft [conversation with Mr. Pollack]; 1.06 watts cooling versus a previous high of 1.2 watts (at 300 K rejection temperature and 65 K cold end temperature). Due to the low fill pressure and frequency sensitivity, further mapping was discontinued until SSC I could be replenished with helium working fluid by Hughes Aircraft personnel.

## SSC I Performance Mapping

After replenishment to 33 atm working fluid pressure, performance mapping was continued. The same highly restricted operating frequency band was used as side contact was shown outside of this band even after replenishment. The mechanical parameters used in the following results were the stroke lengths and phase angle described above.

During unloaded cool down operations the SSC I demonstrated very uniform cooling paths as shown in figure 2.

The steady state performance envelope was measured with a series of load lines at various rejection temperatures, refer to figures 3 and 4. This data was then analyzed using multivariate regression procedures and a performance envelope was developed in polynomial form relating the cooling load and power draw to rejection and cold end temperatures. The polynomial developed for the power draw and cooling load functions are:

$$power = -368.7 - 0.2717T_{ct} + 0.0005688T_{ct}^2 + 2.829T_{rej} - 0.004595T_{rej}^2 \qquad [1]$$

$$cooling = -20.58 + 0.04591T_{ct} - 0.0008655T_{ct}^2$$
$$+ 0.1397T_{rej} - 0.0002522T_{rej}^2 - \frac{0.2785T_{ct}}{T_{rej} - T_{ct}} \qquad [2]$$

where $T_{rej}$ = rejection temperature and
$T_{ct}$ = cold end temperature

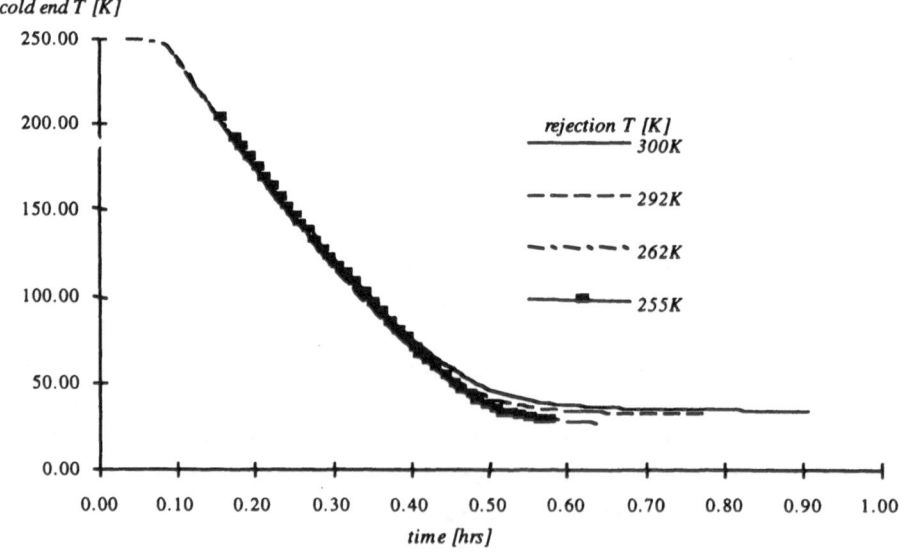

Figure 2. SSC I Cool down performance.

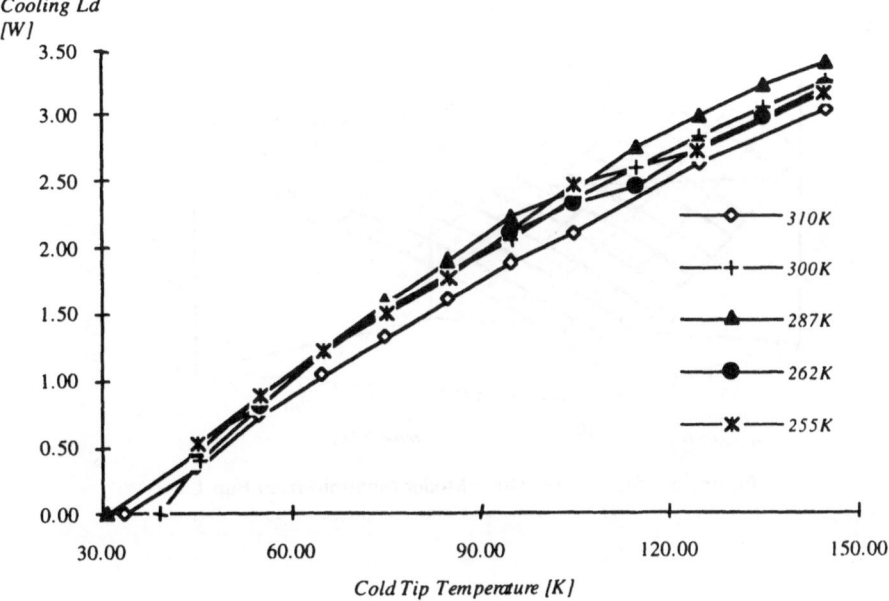

**Figure 3.** SSC I Cooling Load Performance.

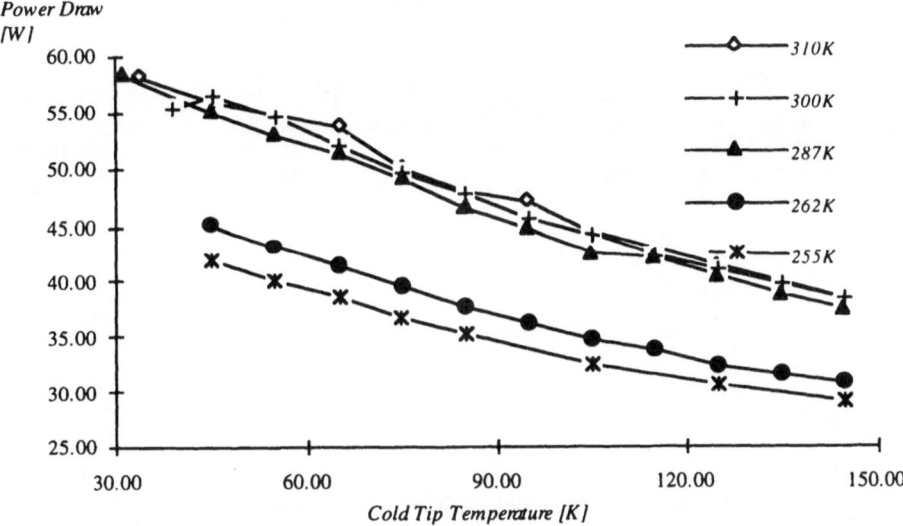

**Figure 4.** SSC I Power Draw Performance.

Figure 5 show the graphical results of the multivariate regression analysis for power draw. It should be noted that while the models in equations 1 and 2 have very high $R^2$ factors (.984 and .996 respectively) there is a considerable amount of deviation that occurs between the data and the model manifold as the rejection temperature goes above 300 K and the cold end temperature below 65 K. Gathering experimental data in this region was very difficult due to the strong tendency of the cooler to make compressor side contact.

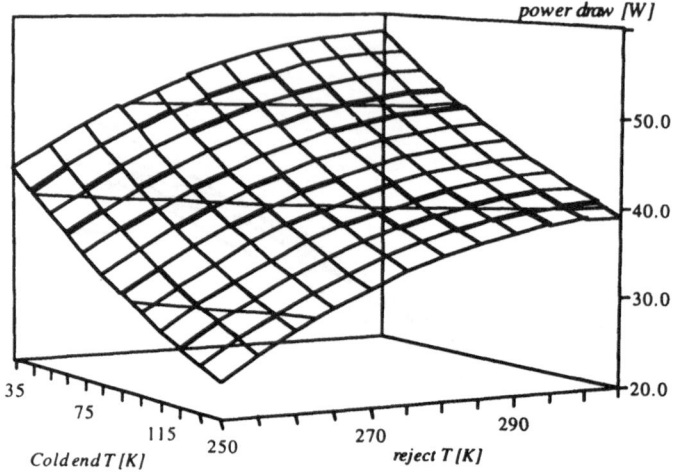

**Figure 5**. SSC I Power Draw Model Manifold from Eqn 1.

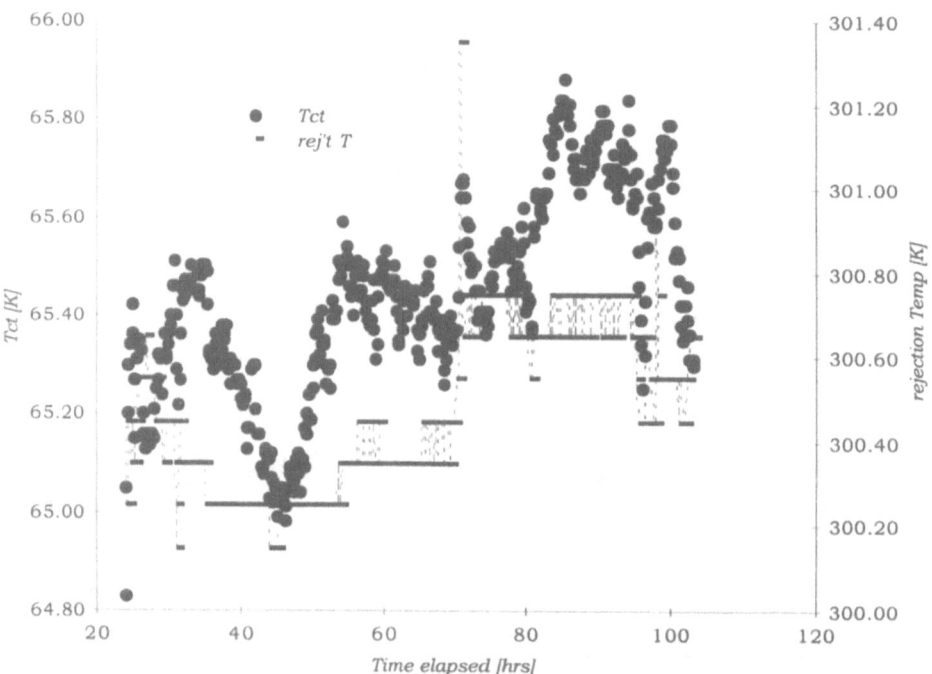

**Figure 6.** SSC I Temperature stability under constant load.

Figure 6 demonstrates the ability of the SSC I to maintain a steady cold end temperature under a constant cooling load of 1.21 watts. During the 80 hours in which this load was applied, the rejection temperature was with two exceptions maintained within a range of 300.2-300.8 K. These two exceptions occurred due to other vacuum chamber experiments coming on and off line on the common cooling fluid loop. During the steady portions of the 80 hours, from 35 to 70 hours or 75 to 95 hours for example, the cold end temperature drifted in two 0.5 K bands. At the end of these steady periods were two relatively sharp changes in rejection temperature, the first at 70 hours being 0.8 K and the second at 95 hours being 0.6 K in magnitude. The changes in rejection temperature apparently led to rapid changes in the thermodynamics of the expander's regenerator and cold end, as shown by the rapid 0.5 K shifts in cold end temperature. These cold end temperature spikes then exponentially decayed into the general drift pattern of the cold end.

## SSC II Performance Mapping

Performance mapping of SSC II began in January 1996 after replenishment of its working fluid to 33 atm. The same overall procedures and mechanical input parameters as in SSC I's mapping were used to develop a comparable performance manifold.

Uniformity of cool down paths at various rejection temperatures was demonstrated, as was noted before for SSC I. Figure 7 shows the various cool down paths measured.

The performance of SSC II at its nominal design point is shown on the load lines in figures 8 and 9. Notable is the downwards concave shape of the cooling load supported. The same regression analysis methods were used to model the cooling load performance and power draw manifolds, with equations 3 and 4 giving the functional results. Again, high $R^2$ factors were obtained in the multivariate analysis (.935 for the power draw manifold and .992 for the load manifold).

$$power = 144.4 - 0.3915T_{ct} + 0.001062T_{ct}^2 - 0.80161T_{rej} + 0.001909T_{rej}^2 \quad [3]$$

$$cooling = -20.92 + 0.07765T_{ct} - 0.0003195T_{ct}^2 + 1.193x10^{-6}T_{ct}^3$$
$$+ 0.1363T_{rej} - 0.0002361T_{rej}^2 - \frac{2.4703T_{ct}}{T_{rej} - T_{ct}} \quad [4]$$

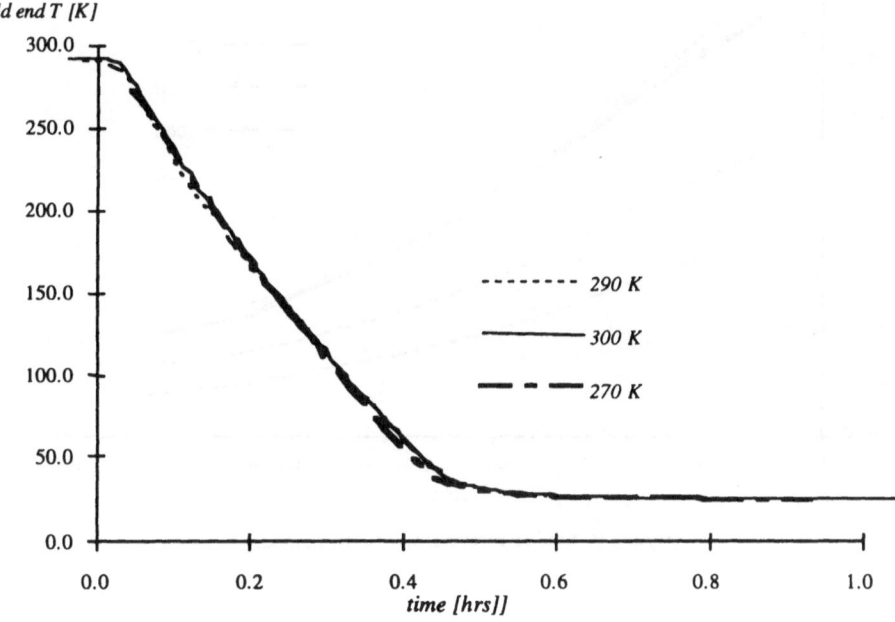

Figure 7. SSC II Cool down performance under no load.

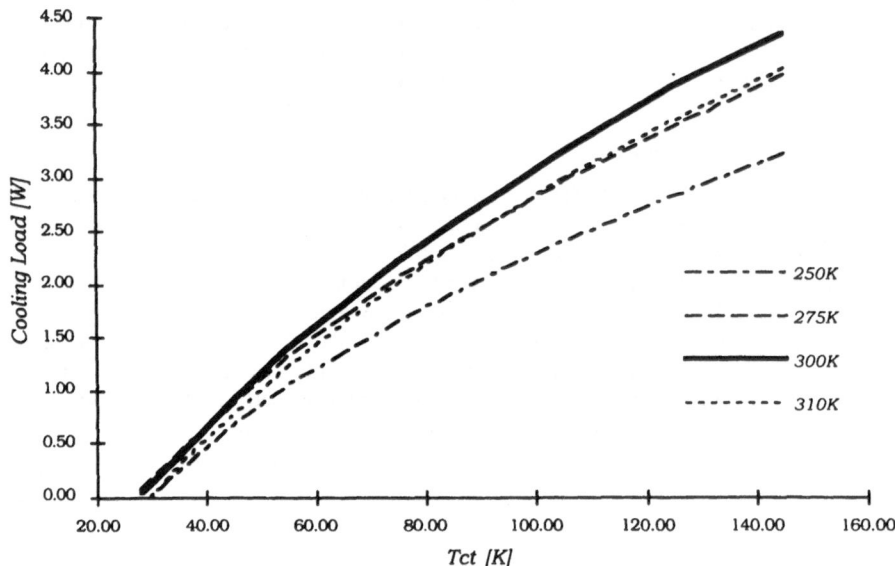

**Figure 8**. SSC II Cooling Load Performance

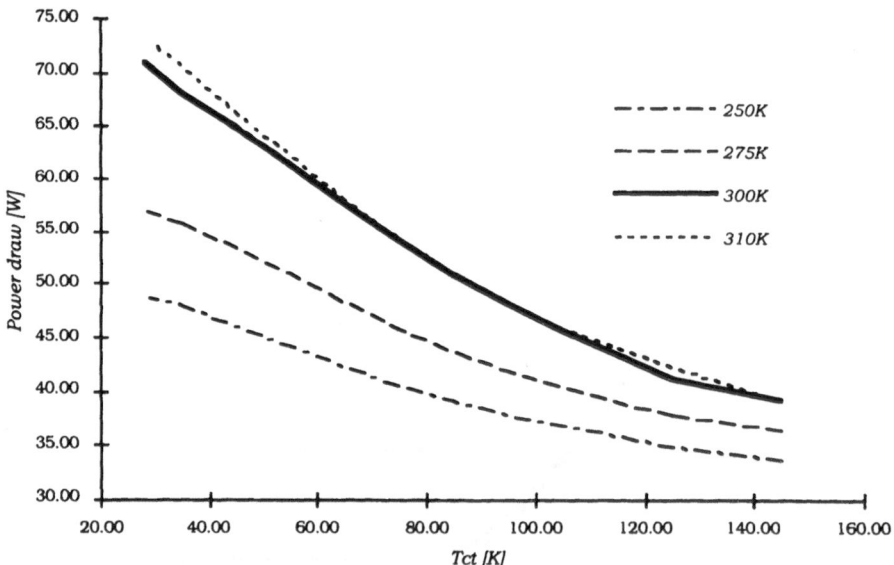

**Figure 9**. SSC II Power Draw Performance.

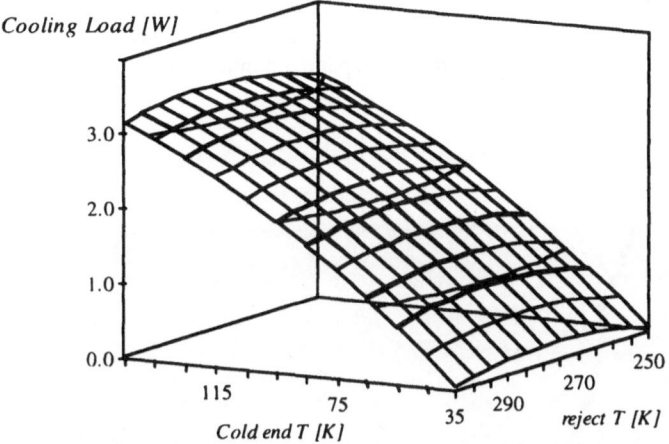

**Figure 10**. SSC II Cooling Load Model Manifold from Eqn 4.

## SSC II Anomalous Results

As the performance manifolds exhibited by SSC II up to 13 February 1996 did not demonstrate the ability to support a 2 watt load at 65 K cold end temperature, additional performance mapping was done concerning operating frequency and compressor stroke length. During this additional mapping the cooler demonstrated two significant shifts in performance. During the latter part of February 1996 the cooler began to have difficulty in maintaining a steady load in the 35-38 K cold end range. This was not evident at lower than 32 K nor higher than 40 K. Figure 11 shows a typical shedding of cooling load over a 3 hour period.

Mapping of the cooler's performance at a 10 mm stroke was initially quite successful. The cooler demonstrated on 4 March 1996 its ability to support the design load of 2 watts at a 65 K cold end temperature and 300 K rejection temperature. Additional data was gathered at 75, 85, and 95 K cold end temperatures. During this period of 10 mm compressor stroke operation, SSC II was allowed to run continuously, as opposed to nightly shutdowns. During the second day of continuous operation a stable no load temperature could not be obtained, with the cold end rising 1 K and power draw rising from 72 watts to 90 watts over an 8 hour period. One instance of compressor side contact was noted at ±15° around bottom dead center. The compressor stroke or rejection temperature was reduced nightly for three days and raised in the morning to repeat the same rise in no load cold end temperature and power draw. On the fourth day of continuous operation an attempt was made to extend the 4 March results downwards to 55 and 45 K cold end temperatures. Data was gathered at 55 K, but when the cold end had been stabilized at 45 K for approximately one hour, extensive expander side contact ±15° around top dead center started. This contact was operating frequency and cold end temperature insensitive in the ranges of 30-34 Hz and 45-65 K, but could be eliminated by reducing the expander stroke from 2 to 1.6 mm. The cooler was shut down and a cold hysteresis loop performed on the expander which showed signs of internal friction in a 0.9-1.2 mm expander displacement towards the cold end. After the cooler reached ambient a slow sweep of the expander showed that the contact disappeared over the whole sweep range. The cooler was started again and when the cold end reached 210 K expander contact began again. Ambient temperature hysteresis loops showed no sign of internal stiction.

Mechanically bending the cold finger was investigated as a possible method for eliminating the expander contact, as had been done with SSC I. A bending moment eliminated the contact until a normal operating drive wave form began, whereupon expander contact was shown again.

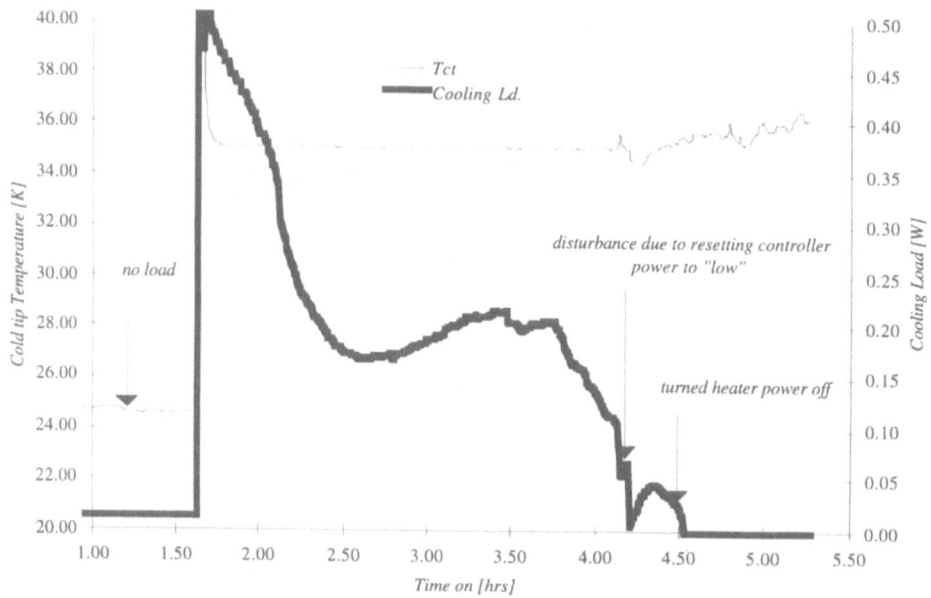

**Figure 11**. SSC II Anomalous Loss of Cooling Ability at 35 K.

## CONCLUSIONS

Review of these results leads towards several hypotheses. First, the cool down path uniformity and the insensitivity of the SSC family's cooling load manifold show the same dependence on reject temperature. Essentially, any one cooling path will follow the actual manifolds modeled by equations 1 through 4. Second, the transient performance of this family indicates that there exists a relatively quick thermodynamic transfer function between changes in rejection temperature and cold end performance, but a relatively slow transfer function between changes in rejection temperature and overall power draw. Third, the assumption of performance manifold smoothness appears to be justified and is helpful in showing family performance parallels. This common behavior will be the subject of further mapping across a wider range of mechanical parameters.

The SSC I will remain at Phillips Laboratory to continue its performance mapping program. SSC II will be returned to Hughes Aircraft in order to eliminate the cause of the expander contact. Working fluid analysis and gas purging will be used to eliminate contamination as a possible cause for contact. If needed, the expander will be opened and examined for misalignment and evidence of wear surfaces on the clearance seal components. Upon return to Phillips Laboratory, Acceptance Testing and Performance Mapping will be restarted for SSC II. The feedback of performance data to Hughes Aircraft will continue during these continued performance mapping activities.

## REFERENCES

1. 65 Degree K Standard Spacecraft Cryocooler Program Final Report, Contract # F29601-89-C-0082, Hughes Aircraft Company, Electro-Optical Systems, El Segundo CA; November 1995

# Development and Demonstration of the Creare 65 K Standard Spacecraft Cryocooler

W. D. Stacy

Creare Incorporated
Hanover, NH, USA 03755

T. Pilson, A. Gilbert and J. Bruning

USAF Phillips Laboratory
Kirtland AFB, New Mexico 87117

## ABSTRACT

In the late 1980s, both the Air Force and the Strategic Defense Initiative Organization (now Ballistic Missile Defense Organization) recognized the need for a long-life space cryocooler capable of handling small cooling loads in the mid-cryogenic temperature range of 50 – 80 Kelvin. This class of cryocooler had a multitude of system-level applications. Potential uses included cryogenic cooling of chilled electronics/CMOS, MWIR sensors, missile seeker precooling, health monitoring, and high temperature superconducting devices.

In response to these needs, the DoD initiated the Standard Spacecraft Cryocooler (SSC) development and demonstration program. As part of the SSC initiative, Creare developed an engineering model of a novel diaphragm flexure Stirling cryocooler designed to provide 2 watts of cooling capacity at 65 Kelvin. Design, fabrication, and initial performance tests have been completed. The unit displayed stable performance at 65 K over an 1,150 hour continuous burn-in test at Creare. Further performance mapping tests continued at Phillips Laboratory to fully characterize the cooler's operational capabilities. The unit is now in endurance test and has accumulated over 2,800 additional hours of operation at various loads, cold end, and heat rejection temperatures.

This paper presents the results of the design, development, and demonstration test activities of the 65 K SSC. Performance characterization test results are analyzed, including suggested design changes for improved performance and reliability.

## INTRODUCTION

Design, construction, and shakedown testing of the diaphragm 65 K SSC have been previously reported [1], [2]. The diaphragm 65 K SSC engineering model was developed to prove and demonstrate a technology alternative to the Oxford heritage flexure bearing cryocoolers currently under development by numerous sources. In-house performance and burn-in testing have been completed, and the engineering model unit has been transferred to USAF Phillips Laboratory for verification and life testing. This paper describes completion of electronic controls development and presents results for performance, environmental verification, and endurance testing to date.

## CONTROLS DEVELOPMENT

The original engineering model electronic power and control module, shown in Figure 1 and previously described [2], employed a capacitive discharge architecture with integrating control on stroke amplitude. Stability of this control scheme proved inadequate when driving the back-to-back compressor pair at design stroke. An alternate, rack-mounted drive and control system, shown in Figure 2, was subsequently developed, based upon linear power amplifiers and DSP controllers. Control feedback loops were closed on compressor and expander diaphragm position signals from Kaman inductance probes and on motor flux sensing coils in the compressor motor stators. This system provided adequate stability to enable unattended burn-in testing and parametric performance sensitivity testing. Effective bandwidth of the controls, however, was limited by the position sensor signal-to-noise ratio, limiting the ability to suppress higher harmonic content in the compressor position waveforms as shown in Figure 3. Compressor pressure ratio and effectiveness of vibration cancellation consequently fell short of design. The narrow band controller on the expander drive motor successfully limited total harmonic content in the expander position waveform to less than 1%.

**Figure 1.** Diaphragm 65 K standard spacecraft cryocooler engineering model.

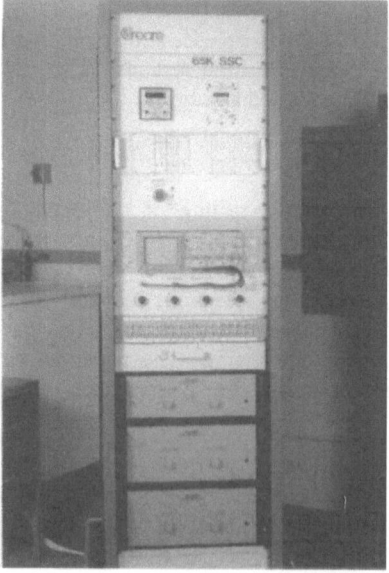

**Figure 2.** 65 K SSC electronics rack.

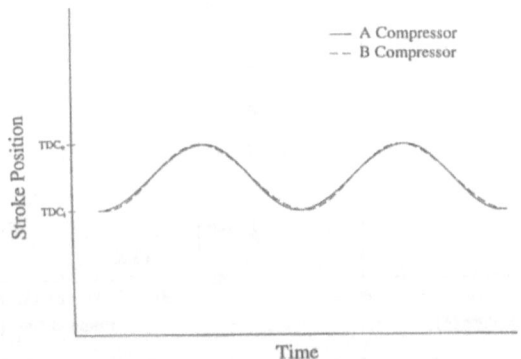

**Figure 3.** 65 K SSC compressor A and B position waveforms.

Successive generations of diaphragm Stirling cryocoolers, such as the two stage 30 K engineering model developed for NASA, have eliminated the compressor instability tendency by pneumatically isolating the compressors from one another.

## TESTING AT CREARE

**Initial Burn-in Test.** An initial 1,150 hour continuous test was conducted in order to verify long-term control stability and efficacy of monitoring equipment. Thermocouples embedded in the motor stator coils were wired to a multi-channel temperature alarm unit which would open relays powering the linear amplifiers if any motor were to exceed its 50°C set-point temperature. The primary challenge in providing reliable, long-term operation was managing the effects of calibration drift of the position sensors providing control system feedback signal. The rack mounted signal conditioning circuitry for the inductive position sensors proved highly sensitive to temperature of the air in the laboratory, due to the very high gain required by the metal sheathed probes. Calibration drift was successfully managed by a combination of stroke clearance margin and programming an automatic recalibration routine in the control system software. At regular time intervals, ultimately set at 8 hours, the control computer would automatically stop the cryocooler and recalibrate the position probes at top and bottom dead center by briefly energizing the motors in a DC mode in both directions sequentially. The computer would then automatically restart the cryocooler at the preexisting stroke percentage settings. The automatic recalibration sequence required approximately 50 seconds to execute, during which time the cold tip temperature would rise slightly less than 1°C.

The continuous test period was utilized for tuning of control parameters and optimizing performance through parametric sensitivity tests. Speed, charge pressure, phase angle, and expander and compressor stroke were varied during this testing. A Lakeshore 320 cryogenic temperature controller powered a load heater mounted to the cold tip to maintain temperature read by a silicon diode on the cold tip to better than the 0.1 K resolution of the readout. More than 1,000 hours of the test were at 65 K or lower. Operation was trouble-free throughout the test.

**Performance Verification Test.** Speed, charge pressure, and phase angle for performance testing were set at values which optimized net cooling capacity at 65 K. The cooler produced 1.0 watts of net cooling at 65 K, with 47.6 watts of net power input, measured at the motor leads. Gross motor input power was approximately twice the net power. The difference corresponds to the reactive power which cycles between the motors and the drive electronics in a non-resonant valveless compressor. The engineering model electronics architecture did not address regeneration of the reactive component of the motor power, a requirement for future high efficiency systems. Steady state cooling capacity as a function of cold end temperature is plotted in Figure 4. Cooldown required approximately 4 hours, as shown in Figure 5.

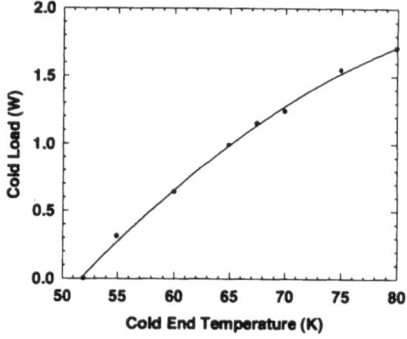

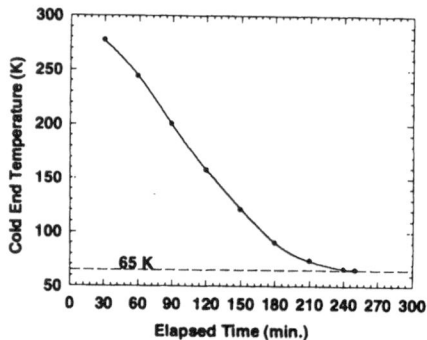

**Figure 4.** 65 K SSC load curve for nominal
design point stroke/phase/charge.

**Figure 5.** Cooldown curve.

A test of cooling capacity at 65 K versus expander stroke was conducted to assess the impact of an error in axial alignment induced by weld distortion during final machine assembly. This misalignment resulted in limiting the maximum available expander stroke to 78% of the design stroke. Results of this test are plotted in Figure 6 and indicate potential for capacity improvement with correct alignment.

A final continuous burn-in test run of 300 hours at 65 K was conducted prior to shipping the unit to Phillips Laboratory for performance characterization and endurance testing. This was to confirm that final wire harness bundling and several cycles of system breakdown and setup had not introduced faults and to minimize the likelihood of problems in transferring the system to operational status at Phillips Laboratory. The 300 hour run was completed without problems.

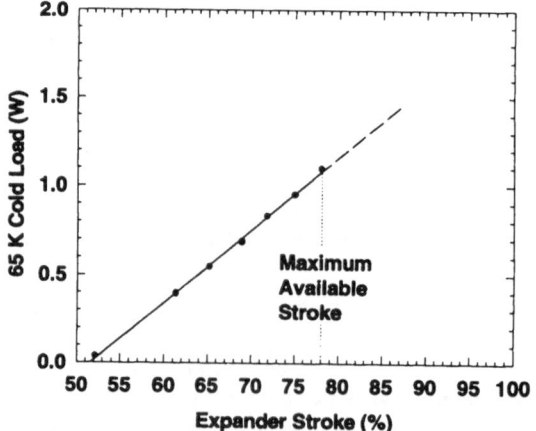

**Figure 6.** Net cooling vs. expander stroke.

## TESTING AT PHILLIPS LABORATORY

The engineering model 65 K SSC was shipped to Phillips Laboratory in February, 1994 where the cooler began its formal performance characterization test program. Since then, the cooler has undergone an extensive battery of tests as part of the acceptance and performance mapping phases of the overall test effort. The following tests were conducted on the SSC:

### Acceptance Testing Phase

a. Post-shipping receipt and inspections
b. Functional checkout
c. Cooldown/warmup
d. No-load low temperature test
e. Nominal load line
f. Low temperature stability
g. Input power evaluation

### Performance Mapping Phase

a. Stroke variation
b. Phase angle variation
c. Fill pressure variation
d. "New nominal" load line determination
e. Heat rejection temperature sensitivity
f. Thermal cycle (cold soak)

The results of all performance characterization test activities at Phillips Laboratory are described in detail in the test program's interim technical report [3]. The "as-delivered" cooler had a small, warm-end helium leak at a defective instrument penetration weld. The leak resulted in a loss of charge pressure of approximately 5 KPa/day. In order to test the cooler over extended periods, a regulated helium supply bottle was plumbed to the unit's internal reservoir volume to maintain charge pressure. The cooler was also periodically evacuated and recharged during the testing period. The SSC test setup at Phillips Laboratory is shown in Figure 7.

**Figure 7.** 65 K SSC test setup at Phillips Laboratory.

Figure 8 shows a composite of the load lines generated at various heat rejection temperatures between 270 K and 310 K. During load line testing, net cooling capacity was observed to diminish after several days of testing following helium purge and refill on the cooler. Upon recharge, the performance returned again to nominal conditions. The family of load line curves shows some slight shifting due to this effect. The extent of the shift varied with the length of time since the previous purge and refill operation. It remains to be determined whether this apparent performance shift is due to contamination effects or to cooler mechanical operation. If contamination is the cause, Phillips Laboratory will attempt to ascertain the nature of the contamination, whether from internal machine out-gassing, cryogenic gettering of make-up gas impurities, or back diffusion of environmental air through the leak against the relatively low 0.16 MPa charge pressure.

The results of displacer phase angle sensitivity testing are illustrated in Figure 9. While running the cooler at maximum stroke and nominal operating conditions (Trej = 300 K and 0.14 MPa fill pressure), the phase angle was varied to determine maximum cooling capacity for various target cold end temperatures. The phase angle yielding the best cold load at the design point 65 K temperature was approximately 50°. However, a 60° phase angle used while cooling down between 295 K and 150 K results in the fastest cooldown time to steady state at 65 K.

Fill pressure sensitivity was also tested in a similar manner. For each of three target cold end temperatures (55 K, 65 K, and 77K), the fill pressure was varied between 0.130 and 0.165 MPa. The maximum cooler capacity for 65 K cold end operation was obtained with an internal charge pressure of 0.14 MPa. Figure 10 shows the results of these tests and illustrates the decline in optimum cooler fill pressures with decreasing target cold end temperature.

Phillips Laboratory also observed slightly less net motor input power required to run the cooler at the nominal design point conditions. Creare observed 47.6 W net input power for 65 K/1 W operation while Phillips Laboratory measured approximately 40.2 W net input power. Net motor input power is an indirectly measured quantity, derived by integrating the product of the instantaneous voltage and current at the motor leads over one cycle of operation. Phillips Laboratory and Creare used different, in house data acquisition systems with different data reduction software to measure net motor power. Phillips Laboratory will investigate the measured power draw discrepancy before conducting more long-term endurance tests on the SSC.

Cooldown time (<5 hours), warmup time, cold end temperature stability (<0.1 K), and no-load low temperature test results (<45 K with no heat load applied) all correlated well with Creare's initial performance test findings. Phillips Laboratory also subjected the SSC to a 24-hour, non-operational cold soak at 250 K. Load lines run immediately afterwards indicated no change in performance due to thermal cycle test conditions.

**Endurance Testing Phase**

After investigating the cooler's internal gas charge history, lower net input power at design point conditions and its effects, Phillips Laboratory will place the 65 K SSC into its endurance test phase. Endurance characterization is conducted as a long-term, near continuous duty-cycle test meant to surface any wearout, drift, or fatigue-type failure mechanisms inherent in the design. To date, the SSC has accumulated over 4,300 hours of intermittent operation while conducting various types of tests. The endurance test is scheduled to accumulate up to 80,000 hours of total cooler operational time.

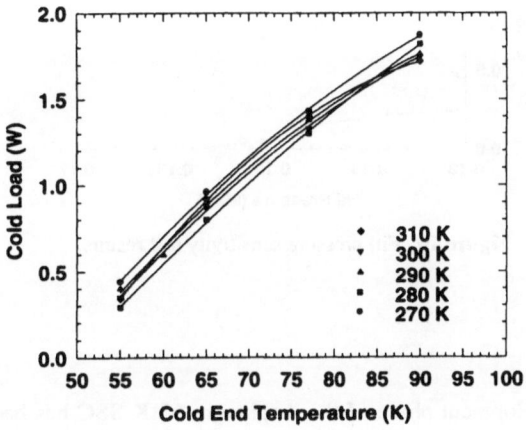

**Figure 8.** Composite plot of load lines at various reject temperatures.

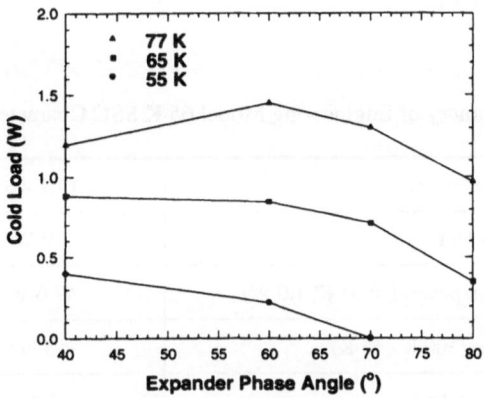

**Figure 9.** Phase angle sensitivity test results.

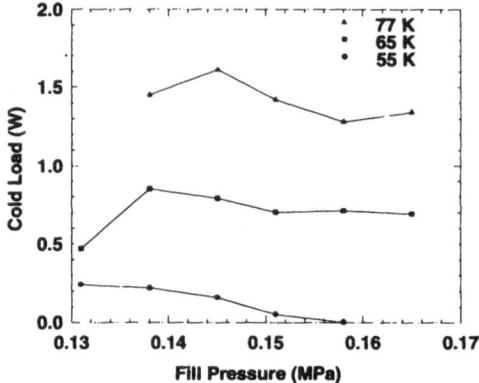

**Figure 10.** Fill pressure sensitivity test results.

## CONCLUSIONS

The engineering development phase of the diaphragm 65 K SSC has been completed. Key parameters at this development phase are summarized in Table 1. The machine has proven robust and mechanically trouble-free through shakedown and environmental testing and more than 4,300 hours of operation. Experience to date continues to support analytical predictions of 10 year life at > 95% reliability. Lessons learned and reflected in subsequent machines include improved diaphragm manufacturing and axial alignment techniques and elimination of coupled system instabilities.

**Table 1.** Summary of Engineering Model 65 K SSC Characteristics.

| | |
|---|---|
| Mechanical cooler weight | 12.1 Kg |
| Net cooling @ 65 K | 1.0 W |
| Net motor input power (@ 65 K, 1.0 W) | 47.6 W |
| Cooldown time (300 K - 65 K) | < 5 hrs |
| Temperature stability | < 0.1 K |
| Safe cold tip normal load (running) | > 2 Kg |
| Heat sinking method | Conduction interface |

## ACKNOWLEDGMENTS

This work was funded by BMDO and administered by USAF/Phillips Laboratory.

## REFERENCES

[1] Stacy, W.D.; *A 10 Year Life 65 K Stirling Cryocooler for Satellite Sensor Applications;* Presented at 6th Inter-Agency Cryocooler Conf., Plymouth MA, Oct. 24, 1990.

[2] Stacy, W.D., McCormick, J.A., and Valenzuela, J.; *Development and Demonstration of a Diaphragm Stirling 65 K Standard Spacecraft Cryocooler;* Presented at the 7th Int. Cryocooler Conf., Santa Fe, NM, Nov. 16-18, 1992.

[3] Bruning, J.A., Gilbert, A., and Pilson, T.; *Creare 65 Kelvin Standard Spacecraft Cryocooler Acceptance and Performance Characterization Test Interim Report;* Draft PL TR, Kirtland AFB, NM, June 1996.

# Test Results for the Ball Single-Stage Advanced Flight Prototype Cryocooler

W. J. Horsley, D.W. Simmons, and J. A. Wells

Ball Aerospace & Technologies Corp.
Box 1062
Boulder, Colorado 80306, USA

## ABSTRACT

This paper describes performance test results for the next generation Ball Single-Stage Advanced Flight Prototype split Stirling cycle cryocooler.

The cooler features smooth operation over a frequency range of 30 - 50 Hz, vibration cancellation using force transducers at the mounting interfaces and thermal interfacing of the cooler structure allowing both vibration control and mounting to vacuum tight structures.

The cryocooler incorporates the basic Oxford heritage cryocooler design acquired by Ball from Rutherford Appleton Laboratory. Two opposed compressors are connected to a momentum compensated single-stage integral displacer/regenerator coldhead.

Data on the thermal performance and vibration control are presented.

## INTRODUCTION

This paper describes performance test results for the Ball Single-Stage Advanced Flight Prototype split Stirling cycle cryocooler.

## ADVANCED FLIGHT PROTOTYPE CONFIGURATION

An Advanced Flight Prototype compressor pair driving a single-stage integral displacer/regenerator cold head with momentum compensation was fabricated and tested using Ball IR&D funds. This cooler maintains the Rutherford Appleton Laboratory design and fabrication heritage with the following features:

- Non-contacting moving parts with concentricities maintained by diaphragm springs
- Tight control on piston and displacer gaps
- Capacitive position sensors

Several enhancements to the basic design have resulted in a cooler capable of stable operation over an extended frequency range of 30 - 50 Hertz, enhanced mounting interfaces which allow vacuum tight mounting combined with thermal interfacing to an external heat sink and vibration control via force sensitive interface washers. Also featured are a wider operating temperature range, improved producibility, and reduced contaminants due to higher bakeout temperature. The displacer has a stronger motor which, when shorted, provides an effective launch lock. The compressors and displacer both meet random vibration of over 14 grms and sine vibra-

**Figure 1.** Ball Single-Stage Advanced Flight Prototype Cryocooler

tions of over 15 G from 10 to 100 Hertz. The moving regenerator is produced from new materials resulting in better long-term stability and totally non-contacting operation in the vertical position. The cooler electronics have been developed by Ball and include:
- Efficient PWM compressor drive
- Automatic computer control of temperature, stroke, phase, and vibration cancellation
- Control of EMI

The mechanical cooler is shown in **Figure 1**.

## ADVANCED FLIGHT PROTOTYPE TESTING

The following tests were conducted:
- Heat lift and efficiency test data on the single-stage cooler.
- Self-induced vibration control

## HEAT LIFT AND EFFICIENCY

The Ball Advanced Flight Prototype Cooler's nominal operating frequency is 45 Hz. We measured load lines for this cooler at a nominal 3.5 mm displacer stroke with compressor strokes varying from 5 to 8 mm. The cooler was provided with interface plates controlling the temperatures of the compressor and displacer bases. The measured power (included in the efficiency calculation) was the input power to both compressors and the displacer. The heat load was found through a direct voltage-times-current measurement to a load resistor on the cold tip. The data is also shown using the format provided by Ron Ross (personal communication), Jet Propulsion Lab.

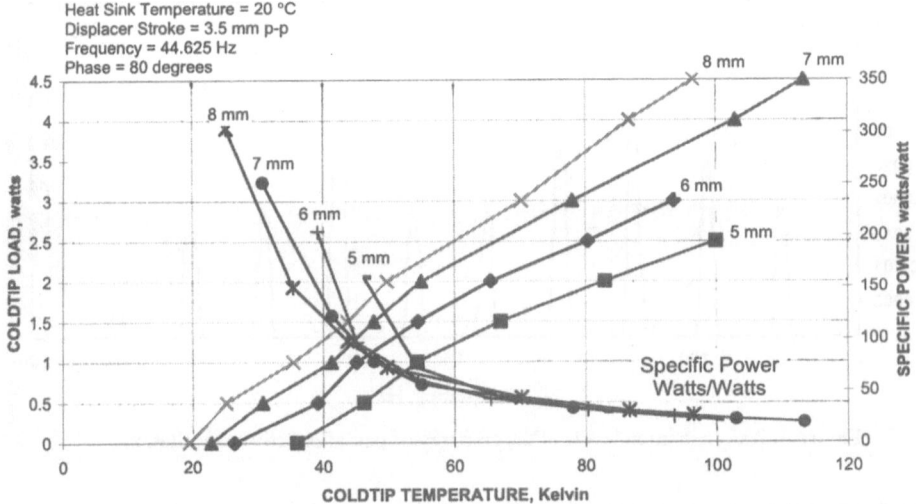

**Figure 2.** Thermal Performance Data

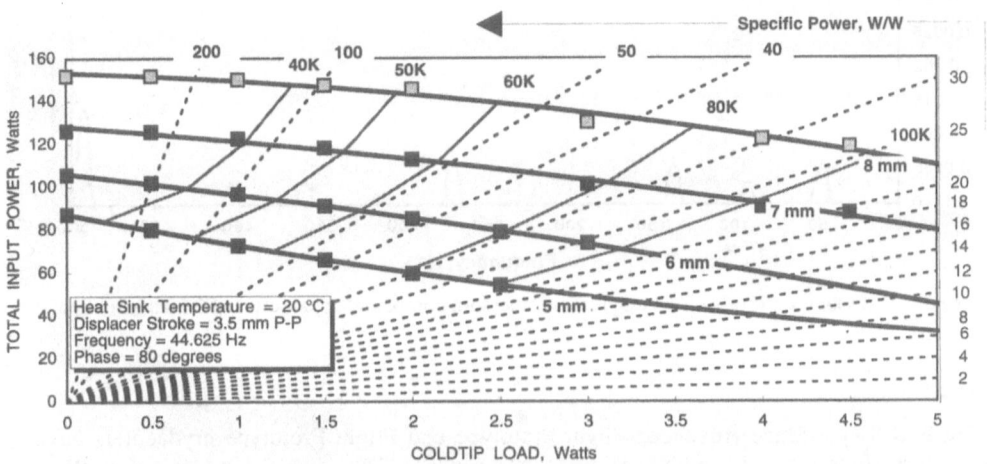

**Figure 3.** Ross Thermal Performance Sensitivity Plot

## AUTOMATIC VIBRATION CANCELLATION

Low induced vibration levels are achieved through the use of digital algorithms obtaining input from force transducers incorporated into mounting washers (Kistler Type 9011A) and outputting equal and opposite forces as perturbations to the basic PWM drives. Routines for the automatic control of the fundamental and first 9 harmonic axial vibrations are incorporated into the cooler automatic control software. This allows the vibration to be canceled to the millipounds force level while the cooler is operating under normal conditions. Canceled vibration records are shown in **Figures 4** and **5**.

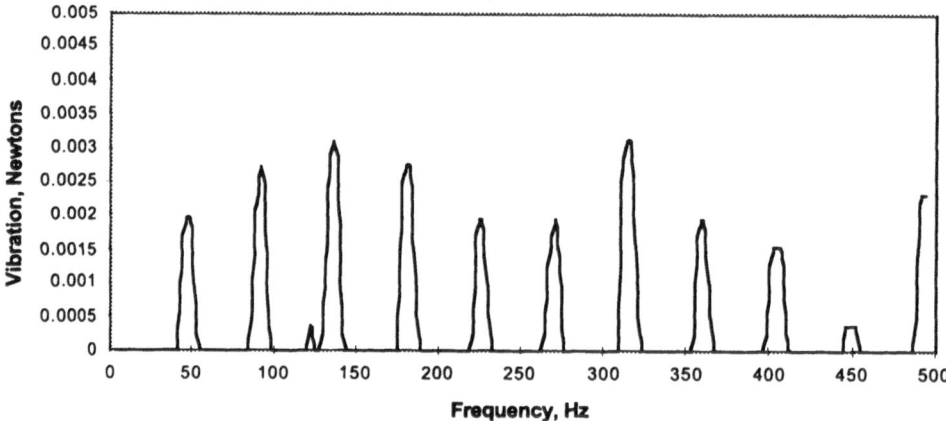

**Figure 4.** Axial vibration levels of the compressor pair after active cancellation.

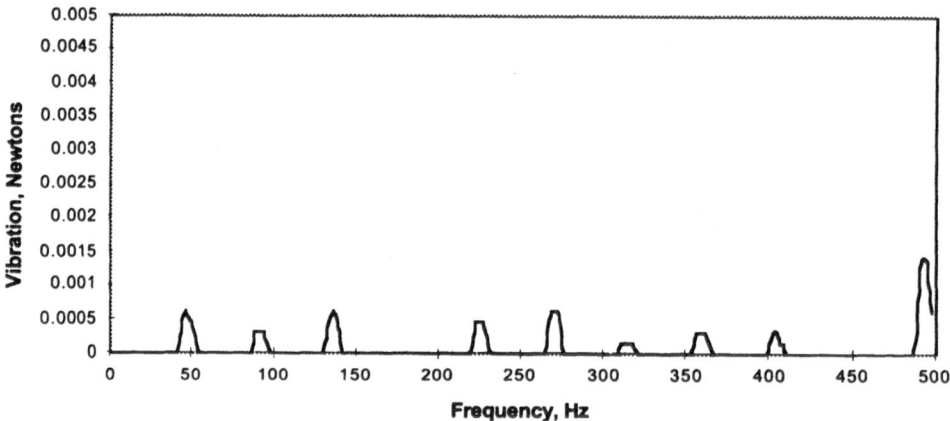

**Figure 5.** Axial vibration levels of the displacer momentum compensator pair after active cancellation.

## CONCLUSION

The Ball Single-Stage Advanced Flight Prototype and Flight Prototype cryocoolers have been extensively tested and exhibit exceptional performance. The advanced model has excellent heat lift and efficiency. It exhibits self-induced vibration levels of less than one millipound force, operates over a temperature range of ±60 °C and meets EOS conducted and radiated emission requirements. The initial Flight Prototype cooler has been operating since the last ICC conference and now has over 12000 hours of life testing with no degradation.

## REFERENCES

1.  The author can be reached at bhorsley@ball.com.

# The Batch Manufacture of Stirling-Cycle Coolers for Space Applications Including Test, Qualification, and Integration Issues

B.G. Jones[1], S.R. Scull[1], and C.I. Jewell[2]

1 Matra Marconi Space (UK) Ltd
P.O. Box 16, Filton, Bristol
UK, BS12 7YB
2 European Space Research and Technology Centre
2200 AG, Noordwijk, The Netherlands

## ABSTRACT

The Cryogenic Systems Group in Matra Marconi Space (UK) Ltd has been continuously involved in the development of Stirling Cycle coolers since 1986. The work started out under British Aerospace Space Systems Ltd which was acquired by MMS in 1994. A 50-80K Stirling cycle cooler has been developed with a heat lift capability of 1700 mW at 80K and 1100mW at 65K for a single compressor, single displacer unit. For most applications two coolers will be mounted in mechanical opposition for vibration cancellation purposes and therefore doubles the heat lift capability potential. This 50-80K cooler is a direct design derivative of the existing 80K cooler, formerly known as the BAe 80K cooler. The design changes incorporated were minimal and therefore achieves very strong linkage to the original Oxford 80K cooler for ISAMS. Fifteen of these coolers have been manufactured as a single batch quantity for various earth observation space instruments. The programme requirements are briefly presented, and the experience gained in manufacture and acceptance testing is discussed. During 1995 a cooler from this batch was allocated to a generic qualification test programme, including flight vibration, thermal vacuum and EMC testing before being put onto long term life testing. The data from this test programme is presented, including an update of the life test.

All of these coolers with the exception of the qualification cooler were supplied into instrument flight programmes. The experience gained and issues arising from the requirement for MMS to integrate some of these coolers into flight bracket systems is discussed. These issues include factors such as end load limitations on the displacer cold finger.

Cryocoolers 9, Edited by R.G. Ross, Jr.
Plenum Press, New York, 1997

## INTRODUCTION

During the past two years, MMS have manufactured a single batch consisting of fifteen 50-80K Stirling cycle coolers. These are of a common design standard and are being provided into a range of space programmes. This design is very closely based upon the previous MMS 80K cooler[1] (was the BAe 80K cooler), which in turn is a direct derivative of the Oxford 80K cooler. The design has evolved through a conservative approach to improvements during earlier 80K cooler manufacture (4 batches totalling 23 coolers in six years) with design changes being incorporated based upon lessons learnt, and reliability improvements. This approach has allowed the life test history of these earlier coolers to be directly applicable to the 50-80K cooler from this recent batch manufacturing programme.

Figure (1) shows a photograph of the 50-80K cooler. The design improvements necessary to improve heat lift capacity and efficiency include:

- increased swept volume (larger piston) 20mm $\varnothing$ x 9mm
- improved motor design, configuration and materials
- cold finger sizing (length and internal detail changes)

The design and specification for this cooler are presented in greater detail in reference 2. The basic heat lift specification is 1700 mW at 80K and 1100 mW at 65K for mechanical cooler input power of less than 50 W.

This approach of manufacturing a batch quantity of coolers has allowed units to be available to space instrument programmes in a shorter time than would otherwise be possible. In the majority of cases manufacture had advanced to the point where all component parts had been received when flight requirements were confirmed. It has been necessary for MMS to work closely with instrument designers in order for their system design to be optimal to accommodate this existing cooler. In some cases, it has been necessary to change some details of the cooler design and these include interconnecting transfer line lengths, launch support tube, cold tip interface details and specific EMC shielding requirements. The EMC shielding has evolved to become a standard design for a current further batch of coolers.

Flight applications for those coolers include:

- MIPAS instrument on the European ENVISAT platform scheduled for launch in 1999.
- ODIN micro-satellite for Swedish Space.
- MOPITT instrument for COM-DEV/Canadian Space Agency on the USA's EOS platform.
- AATSR instrument on the European ENVISAT platform
- HELIOS 2 French military application for Aerospatiale.

Generally the heat lift requirements for many of these applications were significantly less than the cooler capacity.

One cooler (No. 4, the fourth unit manufactured from this batch)was taken as a qualification unit. The test programme included heat lift performance tests, flight vibration, mechanical shock, thermal vacuum, leak tests, EMC tests and life testing. This qualification programme is presented briefly within.

Figure (2) shows a schematic representation of the cooler

## FLIGHT PROGRAMMES AND REQUIREMENTS

The Michelson Interferometer for Passive Atmospheric Sounding (MIPAS) instrument being developed by Daimler Benz Aerospace of Munich in Germany uses two 50-80K coolers mounted in back to back configuration for vibration cancellation purposes. This programme required two flight coolers and two flight spare units. These are shown in their configured form on the flight bracket in figure (3). MMS were responsible for cooler integration onto the bracket which was free issued by the customer. Vibration cancellation electronics in this case are being

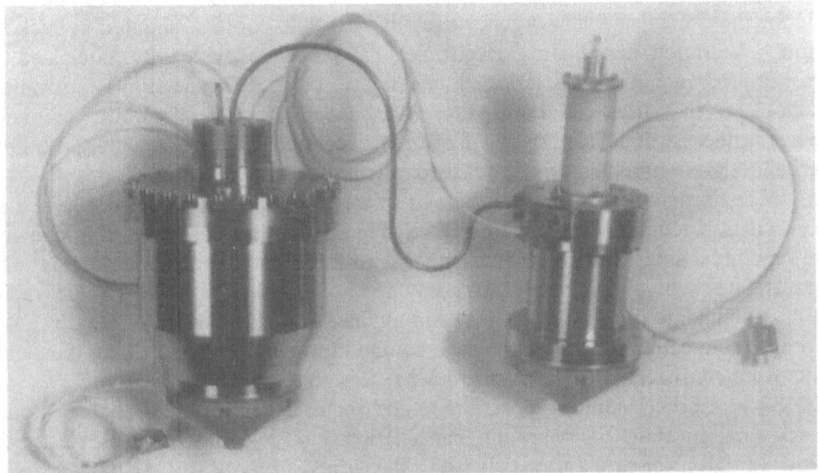

**Figure 1.** Photograph of MMS 50-80K cooler.

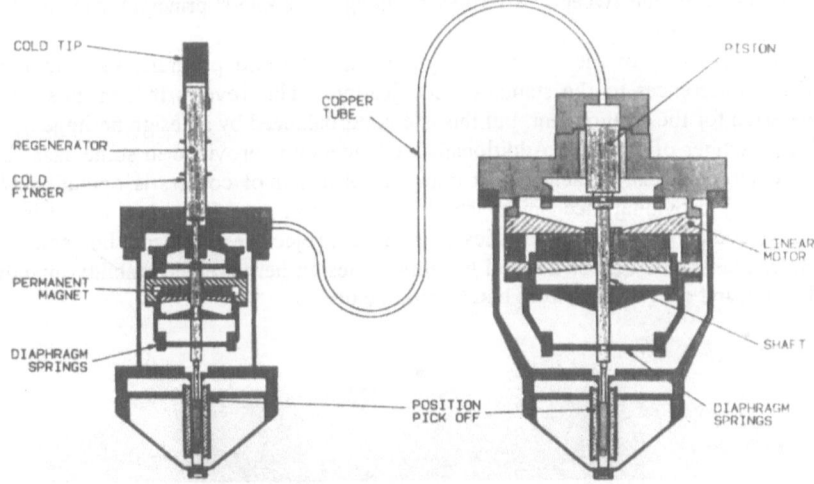

**Figure 2.** Schematic representation of the MMS 50-80K cooler

developed by Rutherford Appleton Laboratory (RAL) whose system uses acceleration feedback. Issues specific to the MIPAS programme included EMC shielding, a longer interconnecting transfer tube, and a flexible interconnecting thermal link (developed for MMS by Rutherford Appleton Laboratory (RAL)) with adequate thermal performance whilst not imparting side loads greater than 0.4N to the cold finger.

The Advance Along Track Scanning Radiometer (AATSR) being developed by MMS and supported by RAL uses two 50-80K coolers mounted in back to back configuration. The electronics in this case only provide vibration cancellation at the fundamental frequency. This programme required two flight coolers and one flight spare. AATSR is the third generation spectrometer instrument whose objective is to monitor global sea surface temperatures which may lead to climate changes. This instrument has 4 infrared channels, and 3 visible channels. With the exception of interconnecting transfer tube shape configuration, the mechanical requirements were consistent with the standard MMS design.

The Instrument for Measurements of Pollution In The Troposphere (MOPITT) developed by COM-DEV for the Canadian Space Agency uses two 50-80K coolers mounted in back to back configuration. Vibration cancellation electronics have been provided by Lockheed[3]. This programme required two flight coolers and one flight spare. The MOPITT instrument is a infrared radiometer that uses gas correlation spectroscopy to detect carbon monoxide and methane gases in the Earth's Troposhere. With the exception of EMC shielding requirements, the mechanical design requirements were consistent with the standard MMS design.

The ODIN instrument developed by the Swedish Space Corporation does not have a specific vibration cancellation requirement, and, therefore, operates with a single 50-80K cooler. This programme required a single flight cooler together with full flight standard electronics. ODIN is a scientific satellite with a combined astronomy/aeronomy mission. The cooled radiometer instrument has submillimetric (420-575 GHz) and millimetric (119 GHz) wavelengths. With the exception of interconnecting transfer tube shape configuration and EMC shielding the mechanical requirements were consistent with the standard MMS design.

HELIOS 2 is a French military programme and requires two 50-80K coolers mounted in back to back configuration for vibration cancellation. The cooler control electronics for achieving low vibration (circa 0.2N rms) is being developed for MMS by Sextant Avionique of France. Particular requirements include EMC shielding, interconnecting transfer pipe configuration, cold tip interfacing. Additionally, the launch vibration requirement is particularly stringent, requiring 20g sine sweep (5 to 100 Hz) along the coolers' principal axis and 50g sine sweep along the cross axes.

Table 1 gives a summary of the principal requirements of these programmes with particular regard to major differences to the standard specification. This reveals that in most cases the cooler is oversized for the requirement, but this is counterbalanced by a design heritage applicable to a significant number of coolers. Additionally, redundancy is provided in some cases because the heat lift requirement can be achieved if only one of a pair of coolers is operating with the other cooler simply providing mechanical vibration compensation. It is clear that it is unlikely that a standard cooler design fully satisfies any single project requirement, but some project requirements can be traded off for the real benefits of design heritage and stability, manufacture and test schedule, and cost savings from batch manufacture.

**Figure 3.** MIPAS Instrument Coolers Installation

**Table 1.** Summary of principal performance requirements

| PROJECT | PERFORMANCE REQUIREMENT |
|---------|------------------------|
| MIPAS | • Nominal heat lift 300mW at 65K for 19W input power<br>• Vibration disturbance (pair) 0.6N peak at fundamental freq.<br>• Mechanical and Thermal environment very similar to MMS product specification |
| AATSR | • Nominal heat lift 400mW at 70K (actual load expected to be circa 100mW)<br>• Vibration disturbance typically 1.0N (TBC)<br>• Mechanical and thermal environment as MMS product specification |
| MOPITT | • Nominal heat lift 1000mW at 88K for 23W input power and 800mW at 64K for 37W input power<br>• Vibration disturbance (pair) <0.2N.Rms<br>• Launch vibration<br>    Random  0.08g$^2$/Hz (50-800 Hz)<br>    Sine     2g peak (5 to 50 Hz) |
| ODIN | • Nominal heat lift 1200mW at 80K for 40W input power<br>• Vibration disturbance not specified, single cooler used with no vibration compensation |
| HELIOS 2 | • Nominal heat lift is as per the MMS product specification<br>• Vibration disturbance <0.2N.Rms<br>• Launch vibration<br>    Random 0.3g$^2$/Hz (80-600Hz)<br>    Sine     20g peak along  cooler axis<br>           50g peak along cooler X-axes (5-100Hz) |
| MMS product specification | • Heat lift; 1700 mW at 80K, 1100 mW at 65K, for 50W input<br>• Vibration disturbance is dependent upon electronics type<br>• Launch vibration<br>    Random 0.24g$^2$/Hz (80-1200Hz)<br>    Sine     15g peak (5-80Hz) |

## ACCEPTANCE TEST PROGRAMME

Each cooler was subject to an acceptance test programme consisting of the following in the sequence shown:
- proof pressure and leak detection
- vacuum bakeout and residual gas analysis followed by helium fill and purging
- stiction tests to verify clearance seals at 20° and +55°C
- seal the cooler with standard gas pressure charge
- initial performance heat lift tests
- 100 hour initial life test
- flight vibration test
- thermal vacuum test
- definitive helium leak test
- heat lift test
- 400 hour life test
- final heat lift performance and characterisation

This sequence was adherred to for each cooler, but specific environmental conditions and/or performance tests points could differ for specific projects. The original intent was to implement a standard test programme to maximise flexibility for cooler allocation, and minimise cost and schedule impacts. The reasonable compromise achieved in most cases was to finalise cooler allocation to projects at the point flight vibration testing was initiated.

Table 2 summarises the heat lift performance results achieved for each cooler. Test points do differ because they were driven by specific contract requirements. The variation in performance seen is generally attributable to differences in the actual clearance seals achieved, heat exchange matrix, and repeatability of test installations. The results represent a reasonable spread of performance for batch manufacture. Clearly, where there is a strong need for best performance, detailed investigations can be performed on the poorer performing coolers and these could be reworked to achieve better performance. However this approach would not be consistent with batch manufacture and the general specification is set at a point where an acceptable yield for performance can be achieved, thus realising an acceptable cost/schedule baseline.

**Table 2.** Summary of heat lift performance achieved in thermal vacuum conditions

| COOLER No. | PROJECT | HEAT LOAD (mW) | TEMPERATURE (K) | POWER (W) |
|---|---|---|---|---|
| 1 | MOPITT | 1700 | 77.9 | 50 |
|   |        | 950  | 57.9 | 50 |
| 2 | AATSR | 1750 | 94.1 | 50 |
|   |       | 1100 | 77.1 | 46.6 |
| 3 | MOPITT | 1700 | 81.5 | 50 |
|   |        | 950  | 63.1 | 50 |
| 4 | MMS Qualification | 1750 | 80.3 | 50 |
|   |                   | 1100 | 65.3 | 50 |
| 5 | MIPAS | 850 | 63.1 | 40 |
|   |       | 300 | 63.3 | 18.5 |
| 6 | MIPAS | 850 | 68.2 | 40 |
|   |       | 300 | 64.7 | 20 |
| 7 | AATSR | 1750 | 76.9 | 50 |
|   |       | 1100 | 62.3 | 50 |
| 8 | AATSR | 1750 | 73.2 | 48.1 |
|   |       | 1100 | 58.2 | 50 |
| 9 | ODIN | 1750 | 77.7 | 47.4 |
|   |      | 1100 | 61.9 | 50 |
|   |      | 1200 | 71.8 | 35 |
| 10 | MOPITT | 1700 | 80.1 | 46.8 |
|    |        | 950  | 60.7 | 50 |
| 11 | Spare Unit Test Programme not started | | | |
| 12 | MIPAS | 850 | 70 | 27.3 |
|    |       | 300 | 65 | 15.7 |
| 13 | MIPAS | 850 | 70.0 | 28.8 |
|    |       | 300 | 65.0 | 17.3 |
| 14 | HELIOS 2 | Test programme starting and | | |
| 15 | HELIOS 2 | data not yet available | | |

## QUALIFICATION

The fourth cooler (No. 4) from this batch was selected as a qualification unit for the design. It was subjected to all the normal acceptance tests applicable to the batch (with the exception of flight vibration and thermal vacuum tests) prior to starting the qualification test programme. The qualification tests consisted of a flight launch vibration test, mechanical shock, thermal vacuum test, definitive leak test, EMC tests with laboratory drive electronics and finally a long term life test. These results are presented briefly below.

### Flight Vibration

The flight vibration test consisted of the following sine and random vibration levels applied in turn to each of the three principal cooler axes. The cooler was powered during these tests and zero position demanded for launch locking.

- Sine vibration of 1 sweep up and 1 sweep down in each axis:
  - 5 - 17.1 Hz constant displacement of 25.4mm peak to peak
  - 17.1 - 100Hz constant acceleration of 15g peak.
- Random vibration applied for 2 minutes per axis:
  - 20 - 80Hz increasing at 3db/octave
  - 80 - 1200Hz  0.24 $g^2$/Hz
  - 1200 - 2000 Hz decreasing at 6db/octave.

A schematic representation of the test installation fixturisation is shown in figure (4). Low level sine sweeps were performed to 2000 Hz after each individual test run in order to confirm that no mechanical deterioration had occurred. No changes were observed in this resonance response. During sinusoidal vibration along the cooler axis the launch locking power requirements measured were 4.0 watts for the compressor and 4.0 watts for the displacer, peaking at this value at 95 Hz. Stiction tests were performed and confirmed the adequacy of the clearance seals.

### Launch Mechanical Shock

The test configuration was the same as that shown in figure (4) but with the control and safety accelerometers removed. A three axis control accelerometer was placed on the base of the fixture. The compressor and displacer mechanisms were powered during the shock tests to minimise piston and regenerator excursions.

The shock level applied was :
- 60g amplitude
- ½ msec duration
- ½ sine
- 1 shock in each direction (+/-) of the three principal cooler axes.

The cooler survived this shock with no change in the mechanical characteristics, and a stiction test verified the adequacy of the clearance seals.

### Thermal Vacuum Tests

The test configuration for the thermal vacuum tests was as shown in figure (5). The support rig used was the same as that used for vibration and shock testing.The temperature was controlled from T1 on the compressor.

Figure (6) shows the temperature cycles performed and identifies the specific heat lift test points at -20°, +20°C and +50°. Stiction tests based upon current measurements were performed at each temperature extreme (-20°C and +50°C). This verified the adequacy of the clearance seals.

The thermal vacuum heat lift performance achieved for compressor input power of 50W is shown in table (3).

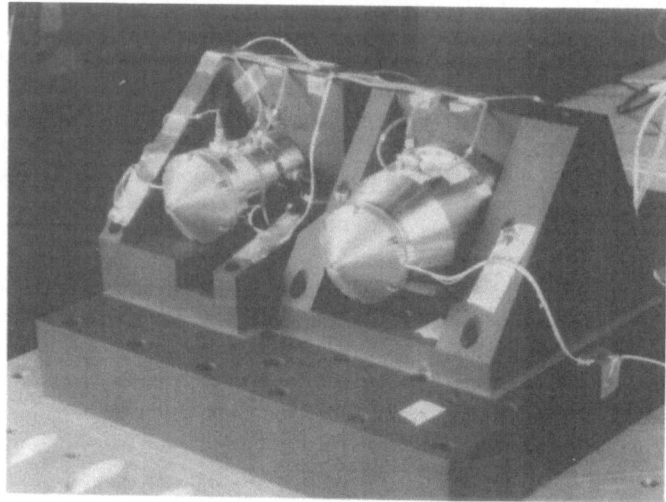

**Figure 4.** Vibration test configuration with accelerometer positions.

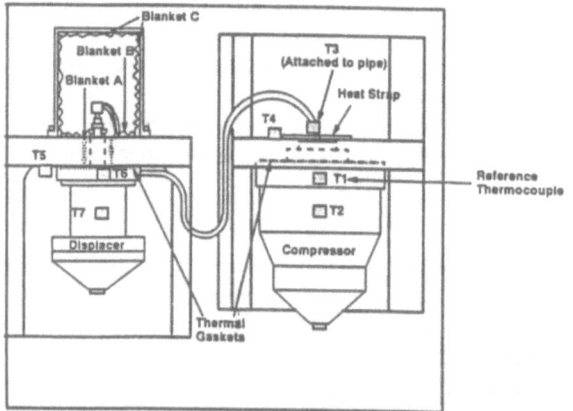

**Figure 5.** Test configuration for thermal vacuum testing.

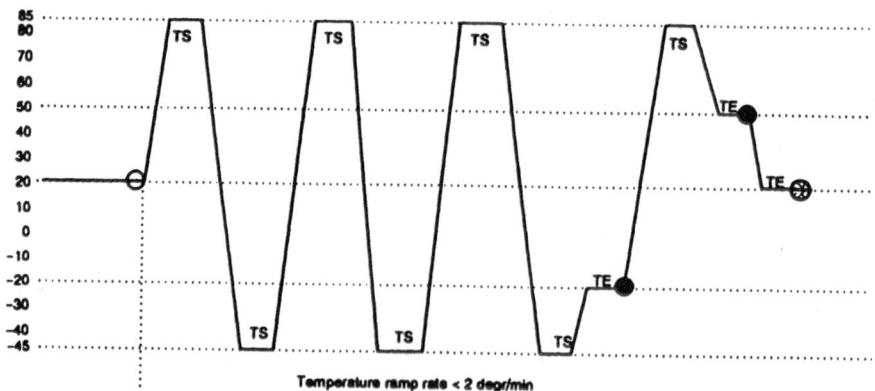

**Figure 6.** Thermal vacuum test temperature cycle profile.

**Table 3**. Cooler heat lift performance

| Heat Load (mW) | Cold Tip Temperature (K) | | |
|---|---|---|---|
| | -20°C Test | +20°C Test | +50°C Test |
| 0 | | 46.7 | |
| 400 | | 54.1 | |
| 800 | 58.8 | 61.9 | 67.8 |
| 1100 | 64.7 | 68.2 | 75.0 |
| 1600 | | 80.1 | |
| 1750 | 78.8 | 84.2 | 92.7 |
| 2000 | | 91.4 | |

A review of the test data showed that the compressor head temperature was nominally 23°C hotter than the controlled interface temperature. Correcting the heat lift performance data for K/°C confirms a heat lift performance at 20°C of:

    80.3K for 1750 mW
    65.3K for 1100 mW.

A similar correction is applicable at -20°C and +50°C.

## Leak Test

The helium leak rate of the cooler was measured to be $0.4 \times 10^{-9}$ mbar.lit/sec. (Specification $1 \times 10^{-7}$ mbar.lit/sec.)

## EMC Tests

EMC tests were performed against the requirements of MIL STD 461 and 462. Note that laboratory cooler drive electronics were used with linear amplifiers. This approach was considered compatible with verifying the mechanical cooler performance. The laboratory electronics were housed outside the EMC test chamber. The cooler was fitted with a local vacuum enclosure to allow for nominal operation during EMC testing and compressor input power of 50 watt.

The EMC tests consisted of the following:

- conducted narrow band, common mode emissions on control and signal leads. (20Hz to 50 M Hz)
- conducted, narrow band susceptibility (20 Hz to 50 MHz).
- radiated narrow band E field emissions. (10kHz to 1GHz).
- radiated broad band E field emissions (10kHz to 1GHz.)
- radiated H field emission
- radiated narrow band E field susceptibility (14 kHz to 1GHz).
- radiated broad band susceptibility (Electrostatic discharge).
- radiated susceptibility - magnetic (30 Hz to 50 Hz).

The performance achieved for all the above tests were within the requirements of MIL STD 461 and 462.

## Life Testing

After completion of the test programme, the cooler was put on to life test. The cooler is controlled by laboratory electronics and tested under general laboratory conditions with a local vacuum enclosure around the cold finger.

The test conditions are:

- Frequency:                44Hz
- Relative Phase:           63. deg
- Compressor input power :  50W
- Displacer input power:    1W

- Heat Load :                                    1750mW
- Head Temperature :                        34°C  (nominal)
- The cold tip temperature achieved: 83.3K.(nominal)

The operational hours achieved to date is 7,000 hours, without degradation in performance.

## DISCUSSION

A single stage Stirling cycle cooler has been demonstrated to meet the performance requirements of a range of space instruments for an application range of circa 50 to 80K. Furthermore it has been qualified by test to achieve an overall performance level that encompasses these broadranging requirements.  The batch manufacture approach is clearly beneficial to schedule, cost and design stability.

## ACKNOWLEDGEMENTS

The authors wish to acknowledge the support of Daimler Benz Aerospace, Com Dev Atlantic, Swedish Space Corporation and Aerospatiale whose flight instruments will incorporate these coolers.  Acknowledgement is also due to workers at RAL and MMS who have supported these activities.

## REFERENCES

1.Scull, S.R., et al.  Pre-Qualification Testing of an 80K Stirling Cycle Cooler.  Proceedings of the European Symposium on Space Thermal Control, Florence, October 1991.

2.Jones, B.G.,  Development for Space Use of BAe's Improved Single Stage Stirling Cycle Cooler for Applications in the Range 50-80K.  Cryocoolers 8, Plenum Press, New York (1995), pp.1-11.

3.Cook, E.L. et al.  MOPITT Stirling Cycle Cooler Vibration Performance Results. Cryocoolers 9, Plenum Press, New York.

# System Test Performance for the Ball Two-Stage Stirling-Cycle Cryocooler

D. Berry, H. Carrington, W.J. Gully, M. Luebbert, and M. Hubbard

Ball Aerospace & Technologies Corp.
Box 1062
Boulder, Colorado 80306, USA

## ABSTRACT

In this paper we will present the current status of the ongoing development of a Stirling cycle cryocooler specifically tailored for space applications. We are putting the final version of our cryocooler through standard qualification tests and will report on the results we have obtained to date.

## INTRODUCTION

Our Stirling cycle cryocooler is a derivative of the original "Oxford" style Stirling cryocooler, which was a significant improvement in linear cryocooler technology. In these cryocoolers diaphragm springs support the oscillating armature, which greatly increases the .operating lifetime. We were funded by NASA Goddard Space Flight Center to improve the technology where we could and to adapt the cryocooler specifically for space applications. We focused on optimizing the thermal performance of a two-stage cryocooler, on achieving and verifying non-contacting operation of the close tolerance seals, and on developing an efficient and reliable electronic controller.

At the beginning of our program we concentrated on achieving thermal performance with a development breadboard cold head.[1] We subsequently introduced a fixed regenerator version of the displacer to solve some of the fundamental system problems we were encountering.[2] We recently completed the final "flight" version of the cryocooler and are currently conducting a series of tests to demonstrate that our design meets its goals. This will complete our two-stage development activity. We are continuing to adapt the technology to other cooling requirements in the form of single-stage and three-stage versions of this basic cryocooler.

## CRYOCOOLER SYSTEM DESCRIPTION

The complete cryocooler system is shown in **Figure 1** and is shown schematically in **Figure 2**. The cryocooler has three components: the compressor, the displacer, and the electronics package.

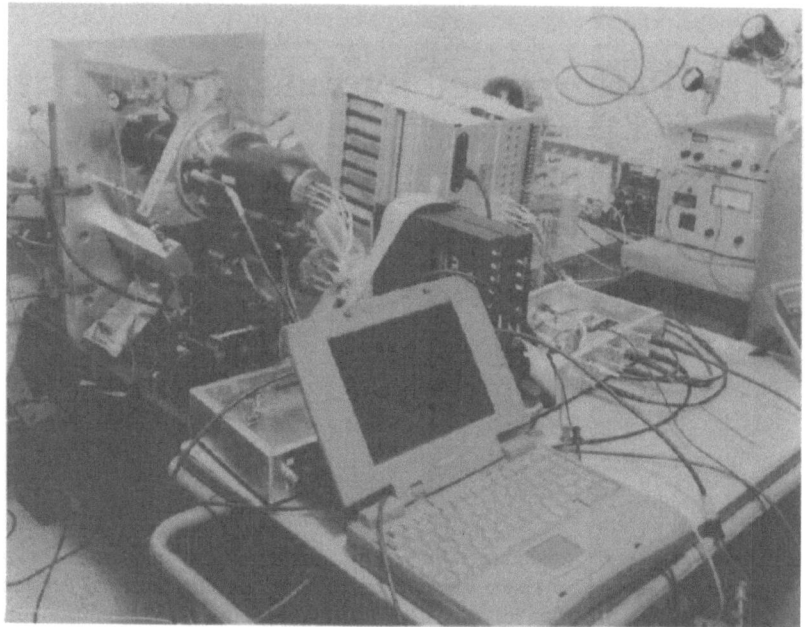

**Figure 1.** The Ball two-stage Stirling cryocooler is undergoing qualification testing.

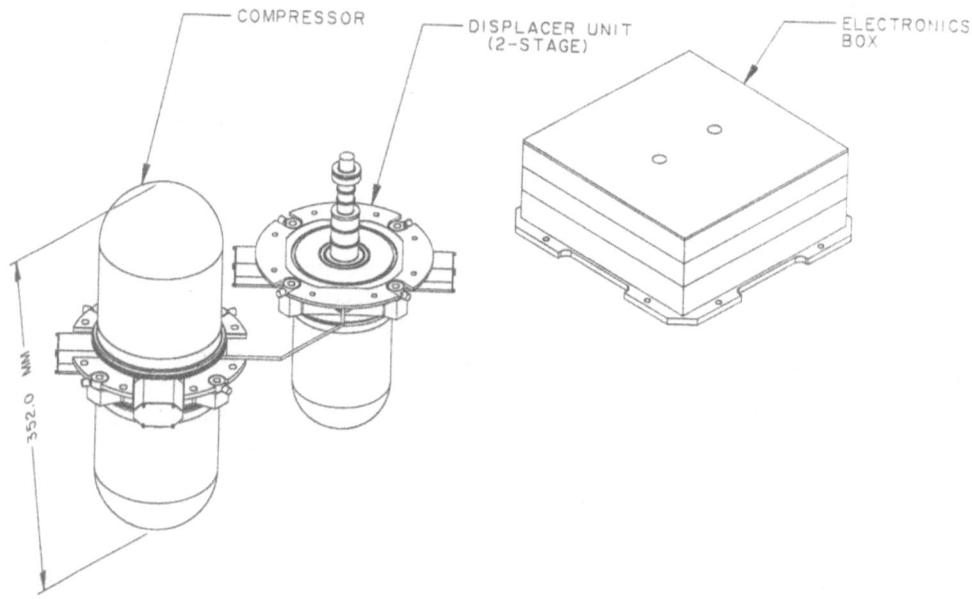

**Figure 2.** The cryocooler consists of three primary components: the twin compressor, the displacer and counterweight assembly, and the electronics

The compressor is essentially identical to the unit we built for the initial phase of the program. It has two opposed armatures, each comprised of a piston shaft assembly suspended on conventional "Oxford" springs and driven by a high efficiency moving coil motor. The basic unit is 35 cm long and weighs 7.3 kg. It has a total swept volume of 5 cubic centimeters, operates between 30 and 45 Hz, and is designed to operate at power levels up to 100 watts. The two armatures mount on a central housing that provides the interface to the mounting structure. It is designed to drop through a hole in a surface from one side. To enhance thermal heat transfer to this surface, the aluminum central housing has four broad integral thermal flanges that provide ample contact area. Mechanical support is provided by four large lugs that mount to load cells spaced about the perimeter of the housing between the thermal flanges. These load cells are the sensor element for our closed loop axial vibration control.

The compressor shown in Figure 1 also has a special set of sensors used to monitor the radial spacing of the clearance seals in the cryocooler. These sensors require a special hermetic interface flange located on the rear of the housing that extend the length of the entire assembly by 20% and increase the weight by 0.8 kg. As we have previously reported,[1] we have used these sensors to study the dynamics of the moving armature in full power operation. These sensors are designed to be removed after the cryocooler has been aligned and initially tested but before delivery of the unit to the customer.

The displacer, illustrated in **Figure 3**, is the new component in our system. We have combined the cold finger design developed on the breadboard with a mature mechanism meant to address other environmental factors. The displacer now has an integral counterbalance in the dome behind the displacer drive mechanism for axial vibration cancellation. Its mechanical and thermal interface to the support surface is similar to that of the compressor, for it also has an aluminum body with thermal tabs and lugs for mounting the assembly on load cells at the interface. The version of the displacer shown in Figure 3 weighs 3.35 kg.

We have included internal lateral position sensors in the displacer for the first time. We use these for the displacer alignment and for monitoring the close tolerance gaps as will be described later. The sensors will remain in this development cryocooler during its life test, but we plan to remove the sensor suite after the alignment and initial testing of the displacer for subsequent units.

The cold head is unusual for a cryocooler of this size in that its regenerator is not carried in the displacer but is annular, stationary, and outside the displacer piston. We have illustrated this arrangement with the exploded view shown in Figure 3. We evolved to this design to avoid many serious system problems arising from the cantilevered weight of the regenerator. This has been discussed more fully in reference 2.

The change to a fixed regenerator design has had a neutral effect on the cryocooler performance thermally, and we have been content to maintain our original performance. The

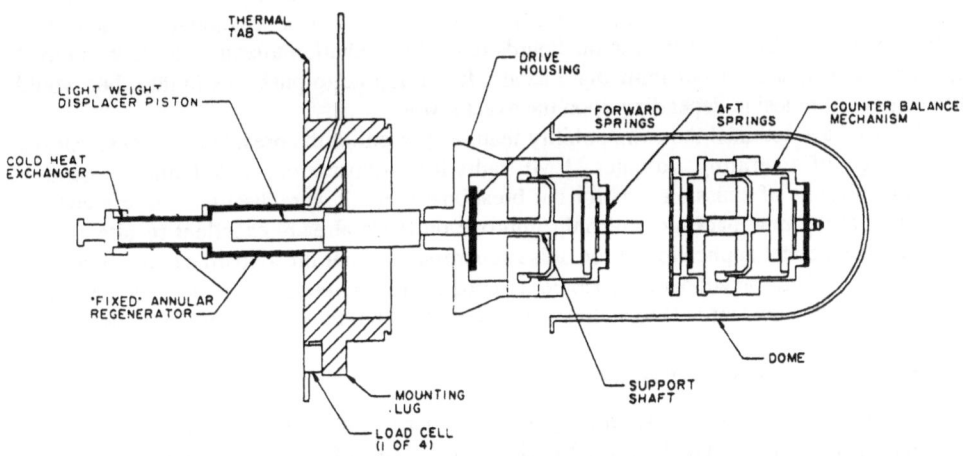

**Figure 3.** The displacer assembly has a lightweight piston operating within a stationary regenerator.

design adds extra heat conduction into the cold space but apparently offsets this with better heat exchange at both ends of the regenerator. As a practical matter, it has been quite important to understand the change in the flow geometry brought about by the new design.

The new approach has many benefits. We were able to reduce the moving mass to 40% of the value projected for the equivalent displacer with a moving regenerator design. This leads to a lower output vibration. It also reduces the sag on the springs and the bending in the shaft, so there is less droop when the unit is on its side. We were also able to lock the displacer with a smaller motor and to support the assembly with a more compact housing. Finally, the fixed regenerator obviously stiffens the coldfinger, which helps with integration.

Our new electronics package is shown in the cryocooler diagram in Figure 2. We have streamlined our original breadboard circuit for flight while retaining all principal functions. We have reconfigured the three components of our breadboard, the PWM motor driver box, the computer and digital interface box, and the analog sensor box into three equivalent boards in a standard Ball flight electronics package. The final package, as shown in Figure 2, has an outline of 9x23x23 cm and a weight of 3.9 kg. The shielded cable adds approximately 0.4 kg per meter of cable to the weight of the system.

The control electronics provides several autonomous control modes of operation of the cryocooler. In stroke control mode, it will maintain the amplitudes and phases of all the moving components as commanded by the user. In temperature control mode, it will adjust the strokes to provide a constant cold tip temperature. In vibration control, which can be added to either operating mode, the cryocooler adjusts the motor drive to minimize vibration as measured in the mount sensors. The electronics also provides status telemetry to the user over a RS 422 serial interface. At present, we maintain this interface with a simple program running in a laptop computer with the appropriate interface card. The program provides a menu of operating modes, displays the status, and logs all of the operating data.

Our breadboard circuit was developed to provide a natural interface to the cryocooler. As we converted from development to flight we focused on reliability, quiescent power, heat sinking, temperature operating range, and radiation survivability. By choosing parts carefully and paring them wherever possible, we were able to cut the quiescent power of our original breadboard circuit in half. With fewer parts we were able to achieve a nominal reliability of 92% for 5 years, and we plan a revision to boost that number to 95% after we gain more operating experience with the present unit. Our thermal management builds upon the features of the Ball flight electronics package, which has thermal lugs around the inside of each slice in the package. We gave each printed circuit board heavy thermal planes, and mounted the heat producing components around the perimeter. We have demonstrated performance with the electronics from -20 °C to +30 °C in air and will be testing the electronics in vacuum later in our test sequence. Finally, even though the circuit is built with commercial parts, we can upgrade to "S" and "Mil-Std 883B" parts without any change to the boards to attain a radiation tolerance of 20 kRad total dose, which is adequate for our immediate needs. It is possible to push this higher, but would require degrees of redesign depending upon the exact levels.

Finally, we have retained the compatibility features present in the breadboard. These consist of wide range of operating voltages, EMI radiation suppression, and launch vibration survivability. Results of emission tests for our breadboard circuit were discussed in reference 1. We believe that the new package will only improve on these already excellent results. A test confirming the practical application of the noise suppression of our breadboard can be found in a report on the use of our breadboard cryocooler to cool a sensitive germanium radiation detector.[3] We have added a few new circuits to address susceptibilities, but these are as of yet untested.

## QUALIFICATION TEST PROGRAM

The qualification test program for the cryocooler is derived from the General Environmental Verification Specification issued by NASA GSFC. We list in **Table 1** the test program we have constructed for our cryocooler. We began the tests this month, and our current status is shown.

**Table 1.** Table of qualification tests for Phase IV of the NASA GSFC cryocooler program

| Test | Status |
|------|--------|
| Thermal performance | Complete |
| Non-contact | In process |
| EMI | July |
| Vibration out | In process |
| Shaker | In process |
| Go to weld | |
| Thermal vacuum | July |

## THERMAL PERFORMANCE

The heat lift of our cryocooler system as shown in Figure 2 is displayed in **Figure 4**. The cryocooler is operating at 32 Hz and at fixed pressure but has not yet been welded shut. The cryocooler is operating in air at an ambient temperature of approximately 25 °C. Data at two fixed strokes are plotted to facilitate interpolation to other operating conditions. In our cryocooler, 90% of attainable stroke is the nominal "full stroke" of the cryocooler. We show on the vertical axis the power delivered to the compressor because it is directly related to efficiency.

To relate this to the total power that would be required from the dc power supply, we select the data point at 90% stroke, where the cryocooler is carrying 0.5 watts at 31 K for 53 watts of input power, for this sample calculation. We first add a 10% "power tax" to the compressor power to account for the efficiency of the PWM motor drivers, whose efficiency varies from 90% to 95%, depending upon the operating conditions. We then add a quiescent power "overhead" of 14 watts for the circuitry in the box, which obviously is present even when the compressor is not running. Finally, we add approximately 2 watts to power the displacer and its counterweight. This power primarily goes to oppose pneumatic forces on the displacer.

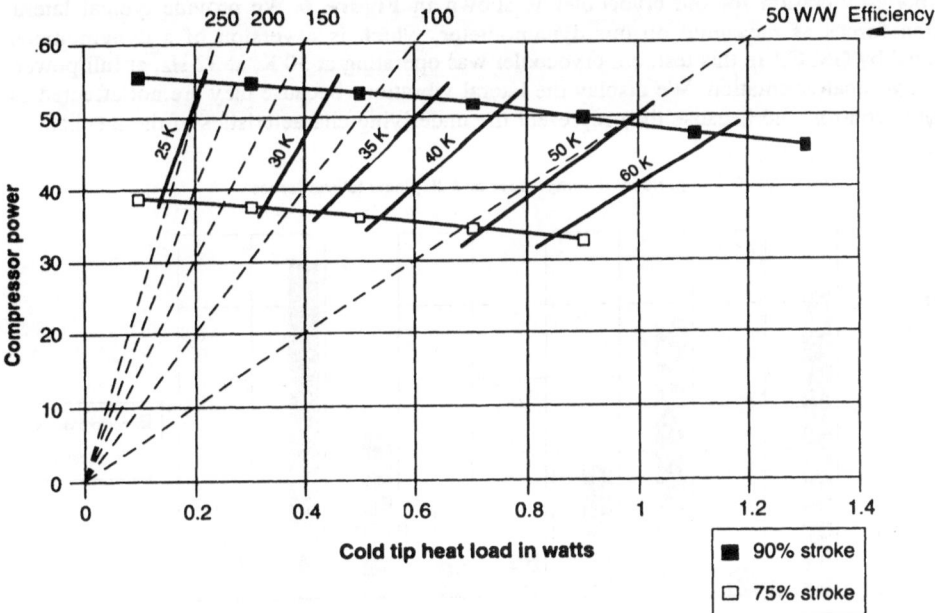

**Figure 4.** Performance tests results for the two-stage cryocooler at 75% and 90% stroke. The total input power from the dc supply would be 100% of the compressor power plus 15 watts.

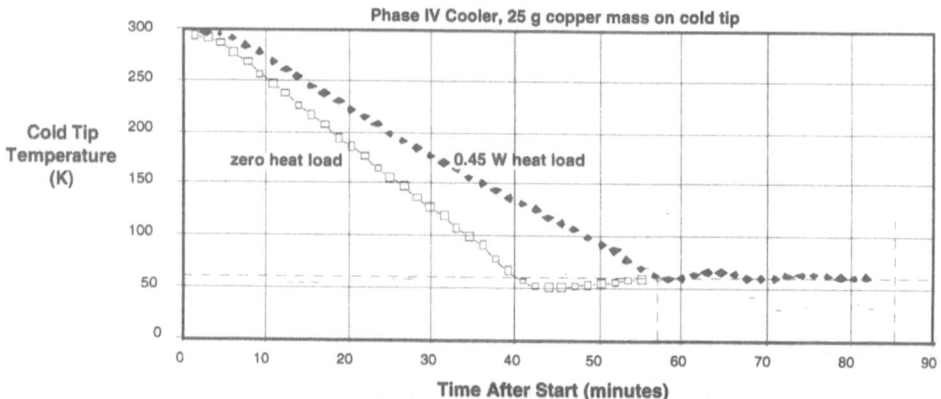

**Figure 5.** Cooldown performance under various heat loads and
the beginning of temperature control at 60 K

Consequently, for our sample data point the total power required to provide 0.5 watts at 31 K
is approximately 75 watts input power.

Another measure of our performance is our cooldown time under different applied heat loads.
The data in **Figure 5** (which was taken with this cryocooler, but before the start of this particular
qualification test schedule) show that we cool to below 60 K in less than an hour even when
carrying a heat load. The data also illustrates the transient response as our cryocooler locks on to
60 K in temperature control using stroke modification. The temperature is controlled to the
resolution of our sensor measurement, which is 0.1 K.

## VIBRATION OUT

Exported vibration for our cryocooler is shown in **Figure 6.** We provide typical lateral
vibration forces as measured on our dynamometer, which is a version of a dynamometer
developed by GSFC.[4] In this test, our cryocooler was operating at 30 K, at 32 Hz, at full power,
and in a vertical orientation. We display the lateral vibrations because they are not affected by
electronic control, and because they represent the underlying characteristics of the mechanical

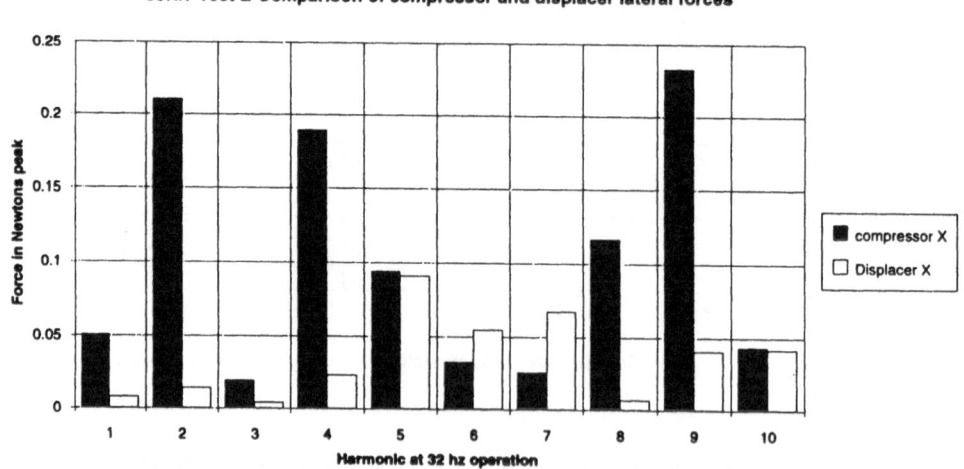

**Figure 6.** The cryocooler meets its lateral vibration requirement of 234 millinewtons
at each harmonic for 8 harmonics. The displacer is significantly quieter than the compressor.

cryocooler. These levels meet the 0.05 lb requirements of our present program. As expected, the vibration from the fixed regenerator displacer is substantially lower than that of the compressor and in some cases is so low as to be comparable to the cross coupling in our dynamometer.

Even though we have demonstrated vibration control on our initial cryocooler[1] several years ago, we have not yet demonstrated combined temperature and vibration control in our flight system. We are operating this more complex system on a breadboard cryocooler and are in the process of transferring the code into our flight electronics for further testing.

## NON-CONTACTING OPERATION

Non-contacting operation has been a high priority for this program since its inception. We have included sensors in our cryocooler for monitoring critical clearances as it operates at full power. We have discussed the operation of the sensors in the compressor in reference 1. This is the first time that we have used the sensors in an operating displacer, and we will give a brief overview of our findings to date. We are just beginning these measurements, and the interpretations given below are certainly preliminary.

The scheme for monitoring the non-contacting operation of the displacer cold tip is shown in **Figure 7**. It is clearly difficult to locate a sensor at the cold tip, so we have to extrapolate its position. To determine if we contact, we compare the lateral deflection with the diametric clearance available. We confirmed the expected clearance (from the dimensions of the parts) by pushing the displacer tip to the wall when the tip was accessible and noting the limiting size of the signal.

We are just beginning a systematic study for the displacer and already find that the signals must be interpreted with care. The drift in the electronics precludes an absolute calibration, so we must rely on the location of the tip relative to its static location. The veracity of this approach was verified on the phase II compressor.

We show a Lissajous figure representing the motion of the cold tip in **Figure 8** in full power operation in a vertical orientation. It indicates that the cold tip is operating well within its limits, which would be a circle whose diameter is the size of the axes shown. We find that it primarily vibrates in one lateral direction, as had our compressors, and that it is phased to the axial motion of the displacer. This is in qualitative agreement with the predictions of our model that predicts lateral disturbances based pneumatic forces that are keyed to residual misalignments of the components.

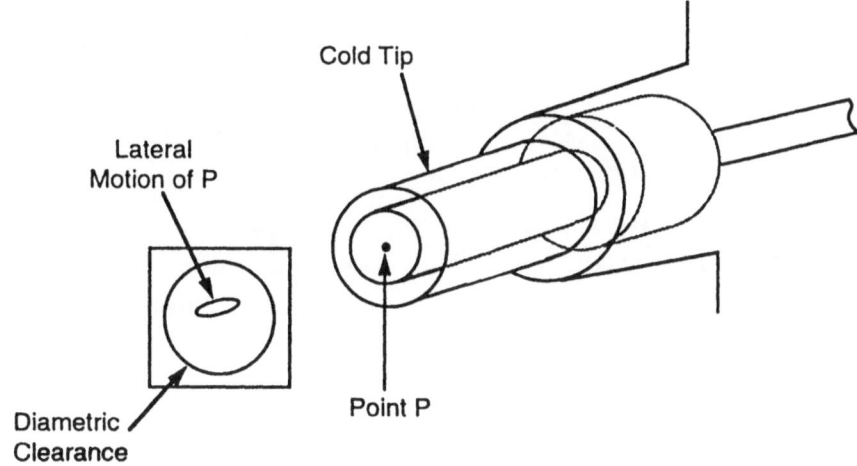

**Figure 7.** The location of the cold tip is compared to the diametric clearance to determine if the cryocooler is non-contacting.

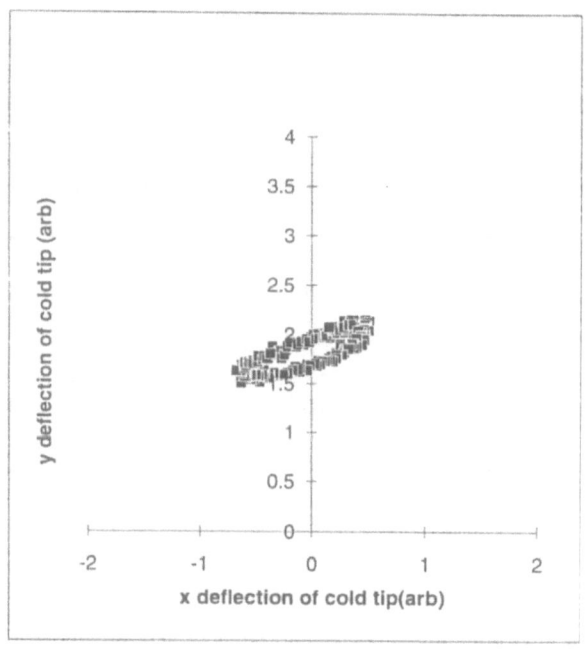

**Figure 8.** The displacer excursion is shown to be approximately 30% of its clearance.

Vibration tests have just been performed. This is the second test for the compressor,[1] but represents the first test for the displacer and the electronics. The tests included random vibrations at 14.1 g (integrated PSD) along and perpendicular to the axis of the cryocooler. We also performed swept sine excitations of up to 15 g, at 4 octaves/minute from 5 to 100 Hz in both directions. The electronics were shaken at all levels in all three orientations. The cryocooler has passed the test.

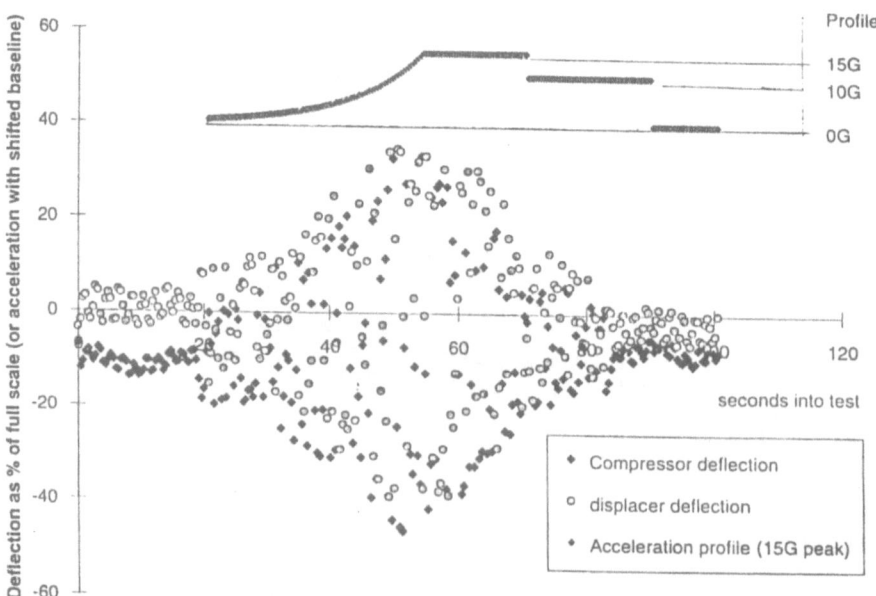

**Figure 9.** Initial results show that the passive launch locks hold both the compressor and displacer armatures to half stroke during a 15-g swept sine vibration.

We show in **Figure 9** the performance of a particular launch lock feature of our cryocooler. In the figure we show the record of the axial position sensors in a compressor and in the displacer during the 15-g sine test. The armatures have low frequency mechanical resonances and are quite susceptible to a coherent sine input along their axis. Our electronics box includes a latching relay that passively shorts the coils of each of our motors, which puts a magnetic brake on the armature and locks it to the housing. Neither compressor nor displacer exceeds half stroke under the worst conditions of the test. The excursions under random vibration and all vibrations at cross axis are even less.

The cryocooler operated normally after the vibration. We are currently trying to determine if it still meets its original performance and to determine if we have had any shift in alignment. We are scheduled to open the cryocooler for inspection in preparation of its final weld in preparation for our thermal vacuum test.

## CONCLUSION

We have begun our qualification tests on our two-stage Stirling cycle cryocooler. When we finish, the cryocooler will be delivered to the Goddard Space Flight Center for continued testing, including an extended life test.

## ACKNOWLEDGMENTS

I wish to thank David J. Taylor and Lee Webb for their extraordinary efforts in support of this work.

## REFERENCES

1. "Multistage Cryocooler for Space Applications," p. 93, Proceedings of the eighth international Cryocooler conference, Plenum Publishing, 1994.

2. "Two-Stage Cryogenic Refrigerator for High Reliability Applications," Proceedings of the Cryogenics Engineering Conference, to be published in ACE, Plenum Publishing, 1995.

3. "Germanium gamma-ray detector operation with a flight like mechanical cryocooler," Proceedings of the Cryogenics Space Workshop, to be published in Cryogenics.

4. "Structural and Thermal Interface Characteristics of Stirling Cycle Cryocoolers for Space Applications," ACE 37, pp. 1063-1068, Plenum Publishing, 1991.

# Improvements to the Cooling Power of a Space Qualified Two-Stage Stirling Cycle Cooler

**T. W. Bradshaw[1], A. H. Orlowska[1], C. Jewell[2], B. G. Jones[3] and S. Scull[3]**

[1]Rutherford Appleton Laboratory, Chilton, Didcot, UK, OX11OQX
[2]European Space Research and Technology Centre, PO Box 299, 2200 AG
Noordwijk, Netherlands
[3]Matra Marconi Space UK Limited, PO Box 16, Filton, Bristol, UK, BS12 7YB

## ABSTRACT

A long life two stage cooler has been developed at the Rutherford Appleton Laboratory (RAL) for space purposes. This cooler has been qualified for space use by Matra Marconi Space Systems (MMS). This cooler is used for shield cooling and for pre-cooling a Joule Thomson stage for the production of temperatures below 4K. The cooling power at 4K can be increased by reducing the pre-cooler temperature. This is a critical technology for the Far Infra Red Space Telescope (FIRST) and improvements to the performance of the pre-cooler are being sought.

Recent measurements at JPL on single stage coolers show that performance improvements can be made by actively cooling part of the cold finger of a Stirling cycle cooler. On FIRST, extra cooling is available from radiators. Cooling the mid stage of the two-stage cooler from around 130K to 100K improved the cooling power at 20K from 140mW to 190mW. There was no benefit in cooling the cold finger of the upper stage.

An experimental programme has been conducted to look at the effect of different regenerator materials on the cooling power of the cooler. In order to maintain the clearance seals, the cooler operates at a relatively high frequency, 30-35 Hz. and pressure drop losses are an important consideration. The best cooler performance was obtained with a combination of lead wire and Er3Ni.

## INTRODUCTION

Closed cycle coolers are being considered for use on FIRST [1]. This telescope has focal plane instruments that require cooling to 4K and below. The cooler that is being studied for this instrument is based on the RAL 4K cooler which is being industrialised by MMS. The RAL 4K cooler consists of a two-stage 4K JT cooler pre-cooled by a two-stage Stirling cycle cooler[2]. Performance improvements to the 4K cooler are being sought. One way of achieving this with the current design is to reduce the temperature of the Stirling cycle pre-cooler. This has been done by improving the regenerator in the upper stage of the cooler and by investigating the use of a heat intercept to take advantage of the radiative cooling available on FIRST. These measurements are similar to those performed at JPL on a single stage cooler[3].

Cryocoolers 9, Edited by R.G. Ross, Jr.
Plenum Press, New York, 1997

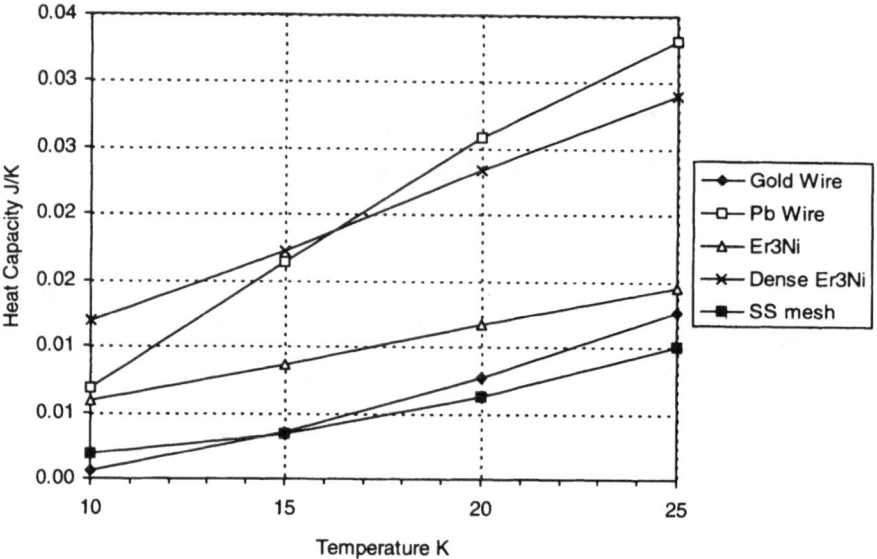

**Figure 1.** Heat capacity of the regenerator matrices

## REGENERATOR OPTIMISATION

The RAL two stage cooler consists of two opposed free piston compressors connected to a two-stage displacer unit[4]. In this type of cooler the regenerators are mounted inside the moving displacer. Several new regenerator materials have been tested in the cold end of the two stage cooler. For each regenerator a load line of cooling power against temperature in the range of base temperature up to 35 K was recorded with constant piston strokes (8 mm), displacer stroke (3 mm), phase between compressor and displacer (50°) and frequency (30 Hz). The maximum input power into the compressors under these conditions was found to be 42 W per compressor (when the laboratory temperature was 32°C), but under normal laboratory conditions (24 °C) was about 35 W per compressor. The fill pressure was 8 bar gauge for all tests. An additional test with the $Er_3Ni$ and lead wire regenerator was performed at 7 bar gauge to investigate the effect of fill pressure. The separation between the compressors and displacer in all tests was 170 mm.

### Regenerator containment

The standard regenerator that we use in the cooler consists of a stack of gauze discs held in place by a circlip. The particulate form of other candidate materials requires a different means of containment. Thin walled, low thermal conductivity titanium alloy, cylinders, 5 mm long, sealed at both ends with gauzes glued to the cylinder lip were used to hold the wire, ribbon and sphere materials selected. These were a sliding fit into the displacer enabling the regenerator material to be changed from the cold end, without disassembly of the cooler. Two stainless steel mesh discs were added on top of the sample holder as an additional precaution against leakage of the regenerator material. This arrangement was found to have no effect on the parasitic losses in the displacer since it only occupies a very short region of the cold displacer, which itself is only one of the contributing parts of the total heat load.

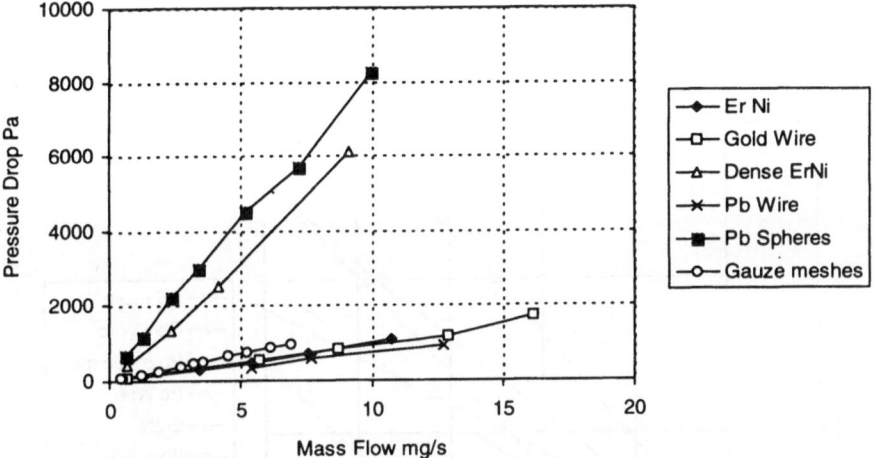

**Figure 2.** Pressure drop at room temperature through regenerators

## Regenerator materials

The following materials were tested:

1. The existing regenerator matrix, consisting of 250 and 325 mesh phosphor bronze discs and 400 mesh stainless steel discs with 6 lead plated 250 mesh Phosphor bronze discs at the cold end.
2. As above but with the lead plated discs removed and replaced with 400 mesh stainless steel mesh.
3. With the top 9 mm of discs removed and the sample holder filled with gold wire.
4. As above but with lead wire
5. As above but with amorphous ribbon of $E_3Ni$ [5].
6. As above but with more densely packed $E_3Ni$ ribbon.
7. As above but packed with lead wire and $E_3Ni$ ribbon

The heat capacities as a function of temperature of the contents of the sample cylinder are shown in Fig. 1. The pressure drop through the sample holders was measured with steady state helium flow before integration into the cooler. These are shown in Fig. 2 together with results from 250# gauze discs for comparison.

## RESULTS

### Performance

The load lines obtained with the different regenerators at temperatures between base temperature - 20 K and at higher temperatures are shown in Fig. 3 and Fig. 4.

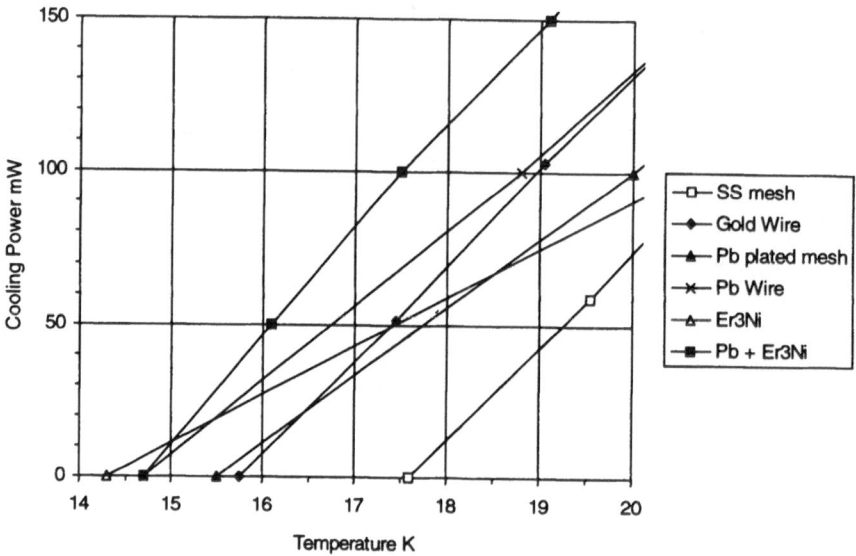

**Figure 3.** Cooler performance with various regenerators at low temperatures.

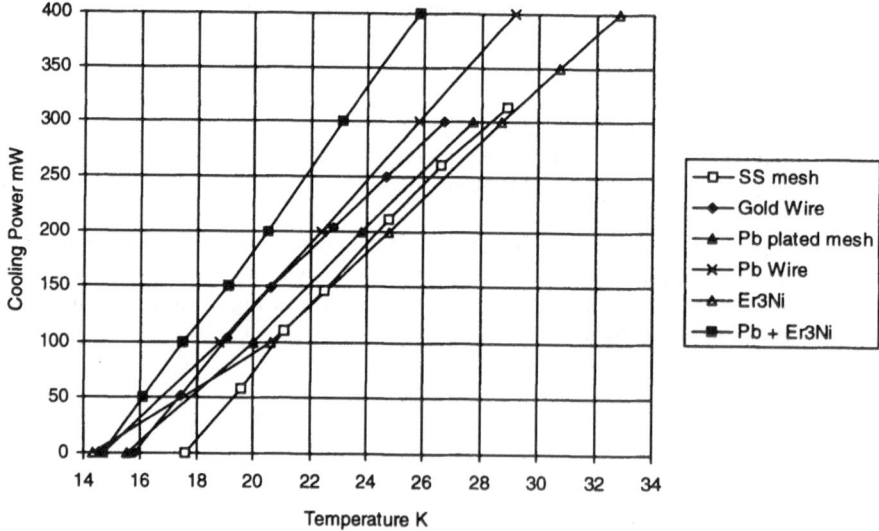

**Figure 4.** Cooler performance with various regenerators at high temperatures.

The performance of the base line cooler with a few lead plated discs is shown in comparison. The effect of this small amount of lead is dramatic, as shown by the reduced heat lift obtained when the discs were replaced with stainless steel. Further enhancements of the heat capacity, by the use of gold wire or lead wire led to still higher heat lift.

The first load line obtained with the $Er_3Ni$ was disappointing. The heat lift was reduced at all temperatures above 16 K. However, the cooler reached a base temperature below 15 K indicating that there was some advantage at the lowest temperatures. Due to packing difficulties the amount of $Er_3Ni$ in the sample can was limited. A further sample was made up with considerably more $Er_3Ni$, by packing it in layers between gauze discs. Unfortunately this led to a very high pressure drop through the matrix (see Fig. 2) and the cooler would not cool below 20 K.

A third attempt to make use of the $Er_3Ni$ ribbon was made by using it together with the lead wire. This matrix gave the best performance of all at all temperatures.

The effect of the variable heat capacity with temperature can be seen by listing the regenerators in order of performance at 30 K and at 14 K (extrapolating where necessary) as shown in Table 1. The change in position of the $Er_3Ni$ from worst at 30 K to best at 14 K is particularly marked.

**Table 1** Relative performance of regenerators at 30 K and 14 K.

| 30 K order of performance | 14 K order of performance |
|---|---|
| lead wire + $Er_3Ni$ | $Er_3Ni$ |
| lead wire | lead wire + $Er_3Ni$ |
| gold wire | lead wire |
| lead plated mesh | lead plate mesh |
| stainless steel mesh | gold wire |
| $Er_3Ni$ | stainless steel |

best ↓ worst

## HEAT INTERCEPT MEASUREMENTS

An investigation was made into the effect of actively cooling parts of the two-stage cooler in order to enhance its performance. On FIRST cooling could be available from a radiator but a study was required to assess the best way of using this. There are various places on the two stage cooler where the cold stage can be attached on one of the stiffening rings on the upper cold finger or on the mid-stage of the cooler. These are denoted by positions A and B in Figure 5. In order to attach to one of the stiffening rings on the pre-cooler a clamp was manufactured. It is important that the intercept does not distort the thin wall of the cold finger as the displacer is a close fit in this item. At high clamping forces some deformation of the cold finger was observed and the clamp force was kept below this level. In an actual system the heat intercept point would be brazed on to the stiffening ring before final machining of the cold finger and the heat transfer at this point would be improved.

The cooler used for all of these measurements had a lead wire/ $Er_3Ni$ regenerator as described above. The distance between the compressors and the displacer was increased to 0.5m to simulate the arrangement to be used on FIRST.

The temperature of the intercept was varied by changing the power to the heater on the top of the GM machine. The power through the intercept was calculated from the temperature difference across a brass shunt installed between the thermal link to the GM machine and the flexible copper braid used to connect the shunt to the intercept. This was wrapped in MLI and hung loosely inside the radiation shields. The radiation shields were attached to the GM cooler at one end and were at the temperature of the GM machine.

The performance of the cooler was measured under three conditions : With the heat intercept on the stiffening ring, with the intercept on the mid-stage and with the intercept disconnected.

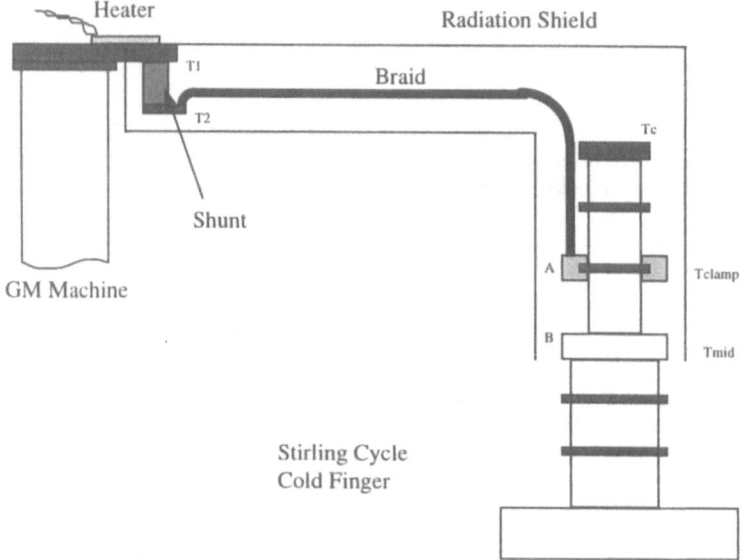

**Figure 5.** The experimental arrangement.

## EXPERIMENTAL ARRANGEMENTS

### General operating conditions

In order to obtain reproducible results the following operating conditions were established:

**Table 2.** Cooler parameters for the heat intercept measurements.

| | |
|---|---|
| Compressor strokes | 7.4 mm |
| Displacer stroke | 3.4 mm |
| Phase | 55 degrees |
| Fill pressure | 10 Bar absolute |
| Frequency | 31.8 Hz |

## RESULTS

### Heat intercept on the first stiffening ring

The cooling power of the cooler as a function of the temperature of the heat intercept for cold end temperatures of 20, 25 and 30 K is shown in Figure 6. There is little improvement to the performance of the cooler as the clamp temperature is reduced. The reason for this became clearer in the next set of measurements.

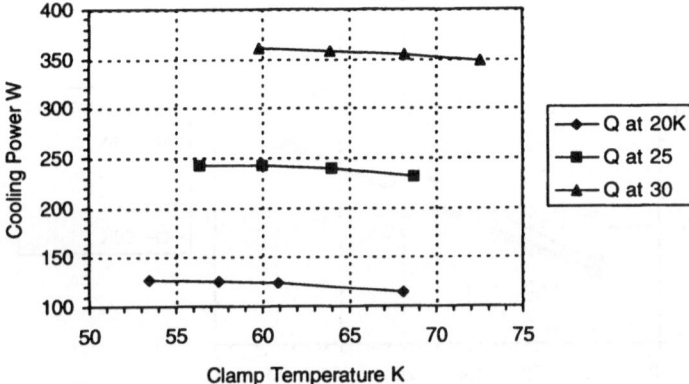

**Figure 6.** The cooling power of the cooler vs heat intercept temperature.

The power through the heat intercept is shown in Figure 7. There is some scatter on this data as the calculation relies on the difference between two fairly close temperatures. There are large thermal time constants in the system, longer than the stability in the cooling power of the GM machine.

### With no heat intercept on the cooler

For these measurement the clamp with the platinum resistance thermometer was left on the first stiffening ring of the upper stage of the cooler but the braids connecting the intercept to the GM machine were disconnected.

The load lines for the cooler in the 20 - 30 K region are shown in Fig. 8. This figure shows the results from this set of measurements without the heat intercept and the results from the previous set with the intercept. The relative change in performance of the cooler can therefore be seen. The key gives the average temperature of the heat intercept during the series of measurements. This value was not constant during these measurements but only varied by typically 3K.

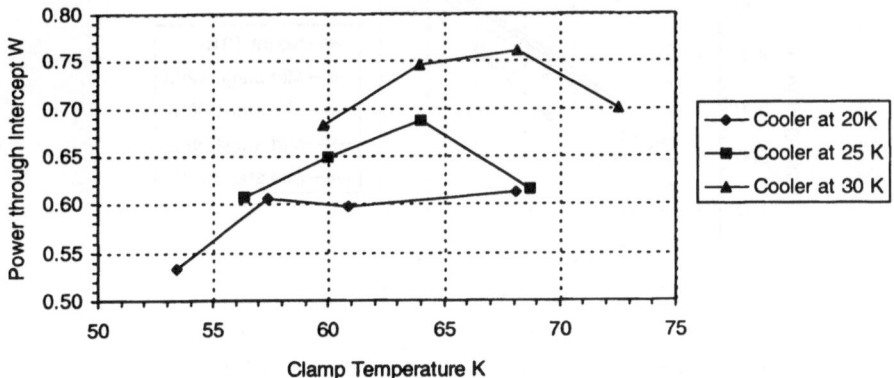

**Figure 7** The power through the intercept

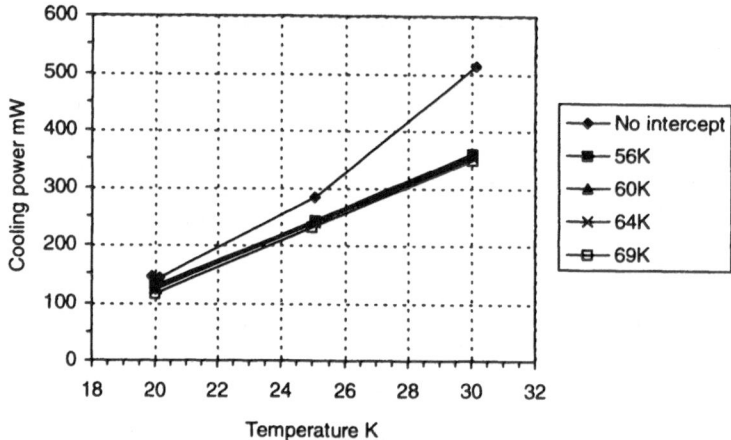

**Figure 8** The cooler load line with and without the intercept

The cooling power of the cooler is worse with the heat intercept than without it. The temperature of the clamp on the first stiffening ring was measured and was found to be 84K with the cooler cold end at 20K. The heat intercept therefore makes little difference to the temperature at that point. It is unlikely that the temperature gradient within the regenerator or along the displacer is therefore significantly affected by the intercept. The total power to the compressors was reasonably constant during these measurements at 85 +/- 1 W.

The residual power through the (disconnected braid) was always less than 100mW and represents errors present in the measuring system and small parasitics on the braid connection.

It should be noted that the normal operating temperature of the mid-stage of the cooler is approximately 150K. With the radiation shields surrounding the cooler and connected to the GM machine the mid stage temperature was cooled to 130 K.

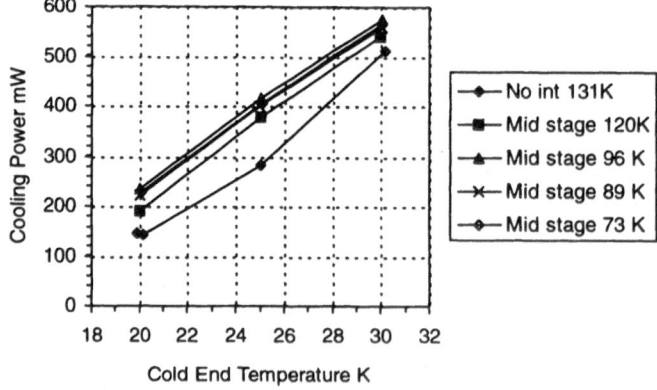

**Figure 9.** The cooling power with the mid stage cooled

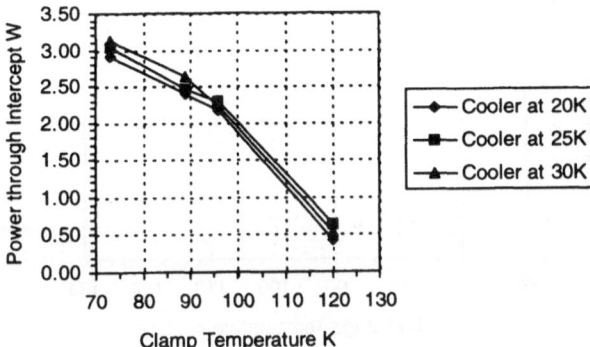

**Figure 10.** The power through the intercept as a function of mid stage temperature.

## With the intercept on the Mid Stage of the cooler

For these measurements the braid was connected to the mid stage of the cooler. The clamp ring for the intercept was left on the first stiffening ring of the upper stage of the cold finger to enable temperature measurements to be made at that point.

The temperature of the GM machine was varied so that the mid stage of the cooler was cooled to temperatures below 120K. The cooling power of the upper stage of the cooler was then measured at 20, 25 and 30K. The cooling power of the cooler at these temperatures for various mid-stage temperatures is shown in Figure 9. There was little benefit in cooling the mid-stage much below 100K. Also shown on this figure is the load line with no heat intercept. Under these conditions the mid-stage temperature stabilised at 131 K. The values of the mid stage temperature given in the key are the average values during the load line measurement. The mid stage temperature varied by +/- 3K around this value. The thermal time constant in the system were such that it was not possible to maintain the mid stage at a fixed temperature, rather the measurements were made at fixed power to the GM cooler. The power through the intercept as a function of mid-stage temperature is shown in Fig.10. There is less scatter in this data than on earlier measurements as the power through the shunt is larger and the measurement therefore more precise. The power through the shunt rises as the mid stage temperature is reduced. At the low temperature end the power is rather high rising to 2 W at 100K. The input power reduces as the mid stage temperature is reduced (the data was taken with the cold end at 20K). There is an 8 W reduction in input power in cooling the mid stage from 120K to 70K as shown in Fig. 11.

## CONCLUSIONS

Some methods of improving the cooling power of a two-stage cooler have been investigated. Modifications to the regenerator with the addition of $Er_3Ni$ have increased the cooling power of the cooler from 100mW to 180mW at 20K.

On a spacecraft where there is extra radiative cooling available, performance improvements to a two-stage cooler can be made by using this to actively cool the mid-stage of the cooler. Reducing the temperature of the mid-stage of the RAL cooler from 130K to 100K improves the cooling power at 20K from 140mW to 220mW, a useful increase of 80mW. There was no performance improvement cooling much below this temperature.

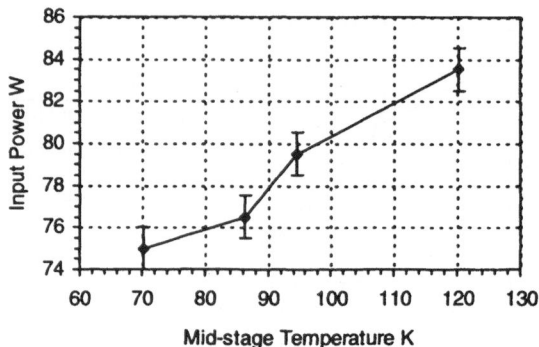

**Figure 11.** The Input power as a function of mid-stage temperature.

## ACKNOWLEDGEMENTS

We would like to acknowledge the generous assistance given in the preparation of the materials by Professor R Cywinski at the University of St. Andrews. The assistance of G Brown, J Hieatt and R Wolfenden is gratefully acknowledged.

## REFERENCES

1.  Beckwith, S. et al., *FIRST-Far Infra-Red and Submillimetre Space Telescope*, SCI(93)6, ESA, (1993).

2.  Bradshaw, T. W. and Orlowska, A. H., "A Closed Cycle 4 K Mechanical Cooler for Space Applications", Proceedings of the fourth European Symposium on space Environmental and Control systems, Florence Italy, 21-24th October 1991, published in ESA SP-324/ISBN 92-9092-138-2.

3.  Ross, R jnr., Private Communication, (1995).

4.  Orlowska, A. H., Bradshaw, T. W. and Hieatt, J., "Closed Cycle Coolers for Temperatures below 30 K", Cryogenics vol. 30, (1990), pp. 246-248.

5.  Prepared by Professor R Cywinski, Department of Physics, University of St Andrews

# Design and Development of a 20K Stirling-Cycle Cooler for FIRST

S.R. Scull[1], B.G. Jones[1], T.W. Bradshaw[2],
A.H. Orlowska[2] and C.I. Jewell[3]

[1]Matra Marconi Space, Filton, Bristol, England.
[2]Rutherford Appleton Laboratory, Chilton, Oxfordshire, England.
[3]ESA - ESTEC, Noordwijk, The Netherlands.

## ABSTRACT

The Far Infra-Red Space Telescope (FIRST) is the 4th Cornerstone mission of the European Space Agency (ESA) 'Horizon 2000' space science programme and is designed to offer observational opportunities in the last poorly explored parts of the electromagnetic spectrum, the sub-millimetre and far infra-red. The cryogenic sub-system required for instrument cooling is based upon the use of mechanical coolers to achieve cooling of shields at approximately 65K, 20K and 4K. A further cooler capable of cooling to 0.1K is to be provided by the scientific instrument.

This paper describes the design and development of a two stage Stirling cycle cooler to meet the 20K requirements of the FIRST mission. This cooler is used both as a stand alone cooler and as a pre-cooler for a closed cycle Joule-Thomson 4K Cooler.

A cooler of this type has previously been development by RAL and subject to a pre-qualification test programme by MMS under ESA contract, but the criticality of the cryogenic cooling sub-system to the mission is such that ESA have identified the need to prove and qualify the cooler against the FIRST specific requirements before the start of the spacecraft development.

The environmental and performance requirements of the cooler are challenging and the FIRST 20K Cooler includes many features to enhance performance over previous designs.

In making changes due regard has been taken of the need to maintain the heritage gained on other MMS coolers and the cooler has been designed with a high degree of modularity with respect to other MMS Stirling coolers that are in batch manufacture.

In addition to the design of the cooler, this paper describes the design verification process required to ensure that all of the FIRST requirements have been addressed.

## INTRODUCTION

In December 1994 the ESA initiated a programme of work at MMS to qualify the critical cryocooler technology required for the FIRST mission. This includes a 20K two stage Stirling cycle cooler, a 4K three stage Stirling Joule Thomson cooler, and key design issues associated with the low vibration drive electronics (LVDE) for cooler control. It is required for all the activities to be complete by the end of 1997 to ensure that no cryocooler technology issues could prevent a cryocooler based solution for FIRST from proceeding.

The specific schedule needs for the 20K cooler qualification presented herein is to complete qualification by February 1997. All design activities have been completed and cooler manufacture is well advanced. Two coolers to a full flight standard will be built and qualification tested.

Presented herein are; the key requirements and design drivers, overall cooler description, cooler design features and verification approach.

## KEY REQUIREMENTS AND DESIGN DRIVERS

### Cooler Configuration Requirements

The cooler development is to be based heavily upon the heritage of the 'Oxford' Stirling cooler technology and the two-stage cooler technology developed at RAL[1] and subsequently transferred to industry by MMS[2] and subject to a pre-qualification test programme by MMS under ESA contract. The cooler is thus to comprise two compressors and a two stage displacer complete with integral momentum balancer unit to limit the levels of exported vibration.

The cooler is to have a transfer tube length of $\geq 500$mm to ease integration issues and to be designed and flight qualified as a complete cooler unit that includes support structure, cold finger launch support tube and temperature and force sensors.

The cooler is to be configured in such a way that it could be used either as a stand alone unit or as a pre-cooler for a closed cycle J-T 4K Cooler of the type developed by RAL[3]. It should also be configured to interface directly with a space radiator without the need for further flight brackets.

A Cooler drive electronics is to be developed to a flight design and built and tested to development model stage using commercial components. It is intended that this electronics unit will be subjected to final design and qualification later in the FIRST programme.

The main performance requirements of the cooler are summarised in the following sections.

### Cooler Heat Lift Performance

The primary heatlift requirement is to achieve $\geq 80$mW of heatlift at 20K with 500mW of heat being lifted at the cold finger mid-stage with the mounting interfaces of the cooler at a temperature of $\geq 300$K. This heatlift requirement is to be achieved with the cold finger launch support tube and the flight temperature sensors in place. The total power input into the mechanical cooler is not to exceed 85W to achieve this performance. In addition, with the input power increased to 105W the cooler is to be able to lift 120mW of heat at 20K.

### Cooler Exported Vibration

The level of out of balance force measured at the displacer support structure interface is not to exceed 0.04N Rms in any of the 3-axes for the cooler fundamental operating frequency and its' harmonics up to a frequency of 250Hz.

The out of balance measured at the compressor support structure is not to exceed 0.2N Rms under the same conditions.

### Launch Vibration Environment

The levels of launch vibration that the cooler must survive are particularly challenging and are well in excess of levels to which coolers of this type have been previously qualified.

The cooler must be capable of withstanding qualifications vibration levels of 30g peak sinusoidal and 21gRms random. This can be compared with present cooler qualification limits of 15g peak sinusoidal and 19g Rms random.

**Thermal Environment**

For qualification test purposes the cooler must be capable of operation over the range -30°C to +50°C.

**Cold Finger 'Off-Cooler' Parasitic Head Load**

It is anticipated that the FIRST cryocooler sub-assembly will include redundant coolers in order to meet reliability requirements and thus the parasitic load on the cold bus bar of any 'off-cooler' is of great significance. In order to minimise this heat load the mid stages (150K) of the cooler are to be linked together. The 'Off-Cooler' parasitic head load requirement is thus specified as <90mW with the second stage held at 20K and the mid-stage at 150K.

**Cold Finger Side Load**

In order to ease cooler installation requirements a requirement has been placed that the cooler must be capable of operation with a side load of 0.7N applied to the cold finger.

**Radiation Requirements**

The radiation requirement is greater than 100K Rad.


**COOLER OVERALL DESCRIPTION**

**Overall Cooler Assembly**

The overall cooler installation is shown in Figure (1). The compressors which are based upon the MMS 50-80K cooler[4] design are supported relative to each other by a tubular type structure. This tubular structure is itself supported onto the main compressors support bracket via four piezo-electric force transducers. These force transducers are used to provide a measure of the vibration disturbance from the balanced compressors for active vibration suppression purposes using the LVDE. Waste heat from the compressors is transported to the base of the flight bracket via thermal straps designed to achieve optimum conductivity whilst not impacting the sensitivity of the force transducers.

The displacer is connected to the compressors via a 500mm copper tube. The support structure for the displacer with its integral vibration compensator mechanism consists of a tubular structure. As for the compressors assembly, the displacer/compensator is supported to the structure via force transducers, and thermal straps carry waste heat to the base of this structure. The two stage cold finger is not visible in this drawing since it is surrounded by the launch support tube.

The cooler structure is designed to interface directly with the spacecraft radiator, and heat lift performance assumes the operating temperature environment to be at the compressors/displacer mounting interface not the base of the flight brackets.

**Compressor Assembly**

The compressor assembly is shown in Figure (2). It is based upon the MMS 50-80K cooler compressor with mostly minor changes in order to ease integration. Detail changes include material changes to improve thermal conductivity, fully enclosed position sensor electronics for EMC improvements, electrical feedthroughs and pumpdown ports taken out at the side. More significantly the piston sizing is increased to 22mm diameter.

**Displacer/Momentum Balancer Assembly**

Figure (3) shows the displacer/momentum balancer assembly. The mechanism of both the displacer and the compensator is based upon the displacer mechanism of the MMS 50-80K cooler. The cold finger is a two stage design in order to achieve temperatures as low as 20K.

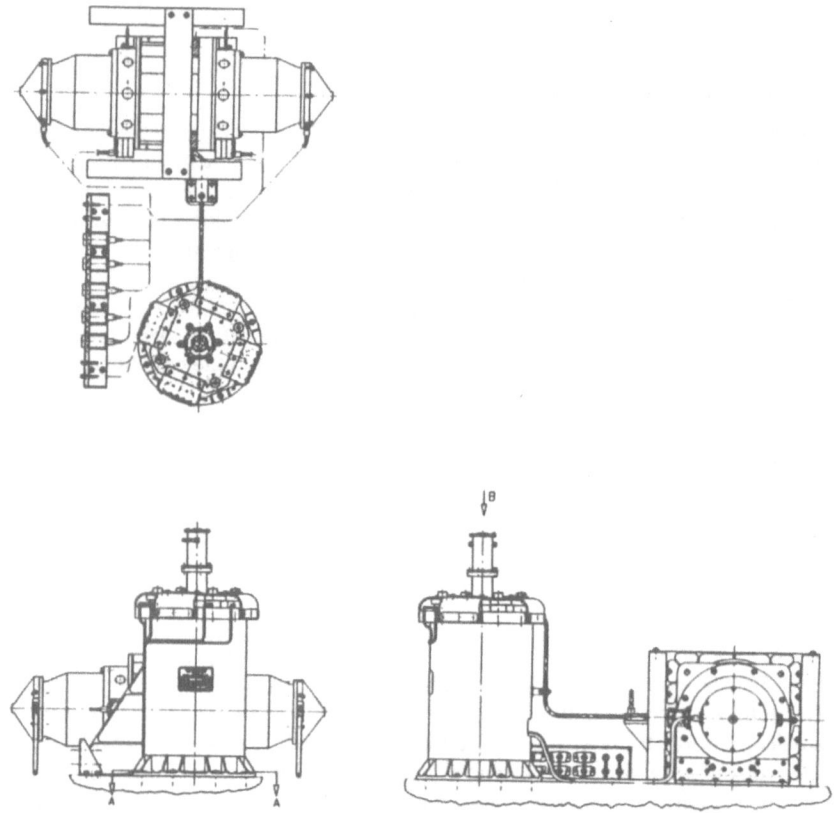

**Figure 1** Overall 20K cooler installation

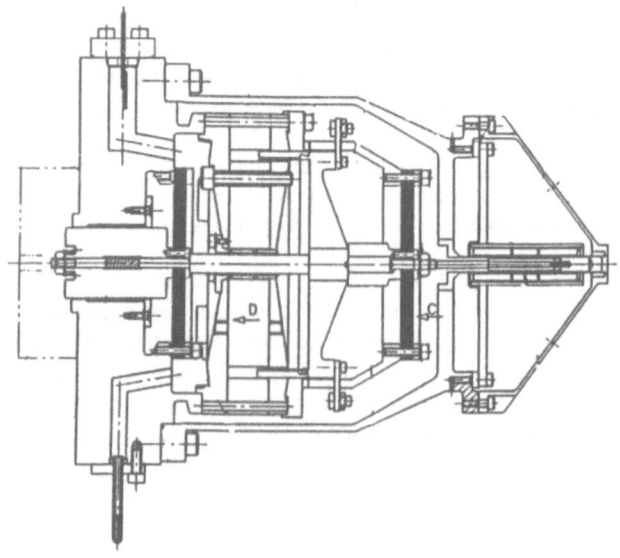

**Figure 2** Compressor Assembly

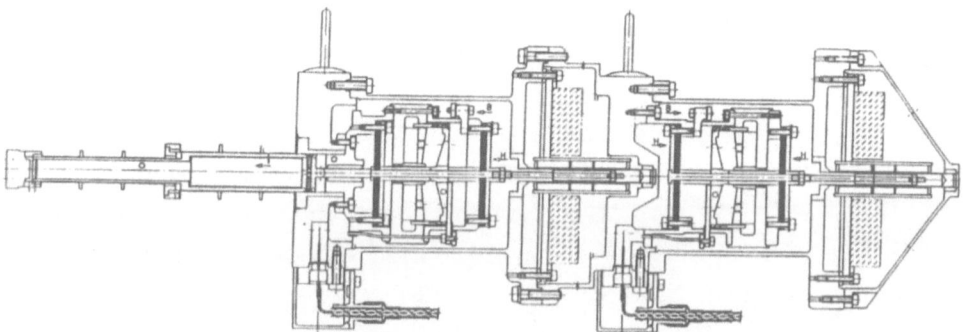

**Figure 3** Displacer/Momentum compensator assembly

## COOLER DESIGN FEATURES

### Cooler Geometry and Sizing

MMS have been supported in this work by RAL who were given the specific task of modelling the cooler performance and to identify the optimum geometry requirements for this application. This confirmed a compressor piston diameter requirement of 22mm, and the displacer cold finger sizing. Additionally, RAL under a separate contract from ESA have investigated regenerator materials for this 20K application[5], and the results of that programme have been incorporated into this cooler design.

### Launch Support Tube Design

The launch support tube design shown in Figure (4) provides functions beyond the ability to survive launch vibration loads. These additional functions include direct structural support to the first stage of the cold finger, which is necessary in order for the second stage to withstand side loads of 0.7N and not generate internal rubbing contact between the regenerator and the thin wall tube. Without this additional support, side load tolerance would be unacceptably low at about 0.2N. The first stage launch support tube is of GFRP structure and assembled initially with clearance around the second stage; this gap is then filled with a glue/epoxy thus achieving support without risk of inducing distortion. The thermal parasitic this induces is allowed for within the design.

The second stage support tube is metallic and therefore since it is also thermally bonded to the 1st stage provides a 150K shield around the 20K cold finger. Connecting to the 150K stage for cooling is also eased since it can be achieved at any point along the length of the upper support tube.

### Force Transducer/Thermal Strap Issues

Figure (5) shows the compressors assembly mounted to their common bracket. This shows that heat flow from the compressors is managed independently from the support structure. The heat from the compressors is taken to the radiator/support plate directly via thermal straps. The thermal straps in turn consist of a loose laminate construction optimised for maximum conductivity and minimum stiffness. The compressor head volume is shrouded in copper in order to achieve the best thermal conductivity from the compressor head to the heat straps. The common support structure for the compressors is supported on force transducers to measure the vibration force. The force transducers have been reviewed for their ability to be fully qualified for FIRST at a later date.

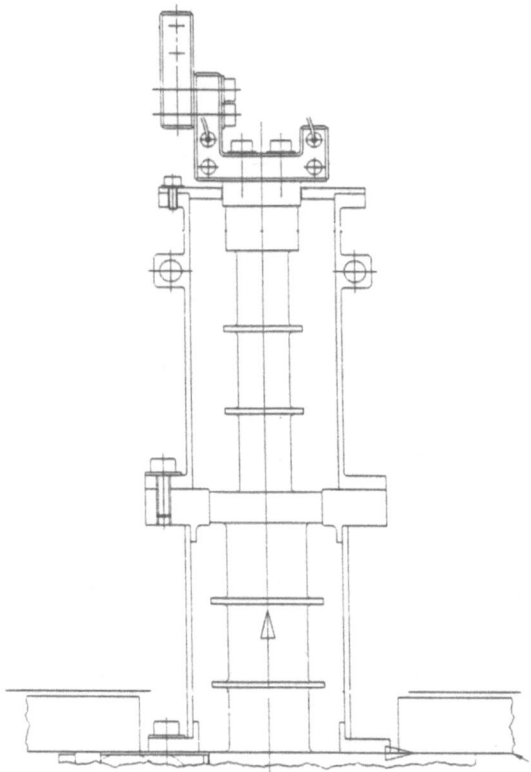

**Figure (4)** Launch support tube assembly

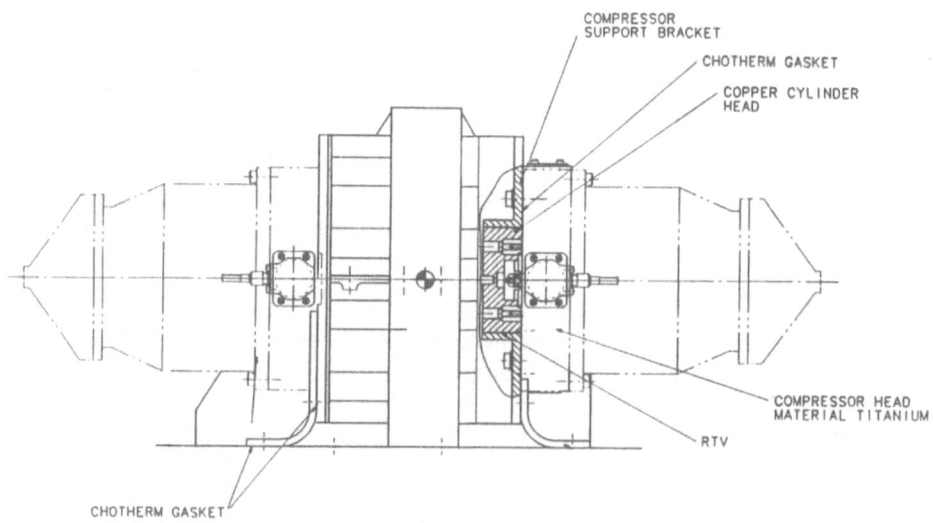

**Figure (5)** Compressors assembly and heat straps

**Exported Vibration**

The need for exported vibration of less than 0.2NRms for compressors and 0.04NRms at the displacer was a major design driver. This confirmed the need for low vibration cooler drive electronics (LVDE), head to head compressors, and an integrated vibration compensator for the displacer assembly.

An accelerometer based vibration sensing system was traded-off against force transducers and the latter chosen. The major driver for this selection is the need for the cooler to interface directly with a radiator or baseplate and not impact the spacecraft design. Accelerometers require resilience in the system in order to have sufficient sensitivity, and this may not necessarily be the case for FIRST.

**Low Vibration Drive Electronics (LVDE) Development**

As part of this cooler development MMS have placed a contract with Dornier to develop vibration cancellation electronics. The LVDE is required to be compatible with controlling the MMS 50-80K single stage Stirling cooler, this 20K cooler, and the 4K cooler being developed for FIRST. This development achieves a flight standard design, but for the purposes of this work will be manufactured using commercial components. It is intended to fully qualify this unit at a future date in the FIRST programme. Meanwhile three units of this type will be available to support the 20K cooler qualification programme allowing vibration and temperature control issues to be verified.

## COOLER QUALIFICATION TEST PROGRAMME

Two coolers are being manufactured to a full flight design, and will be subject to a full qualification programme against the FIRST requirements. Prior to commencing the qualification tests an acceptance test programme will be performed.

Acceptance level testing will include the following:
- Proof pressure and leak testing
- Vacuum bakeout and gas analysis
- Helium fill and purge
- Stiction tests to verify clearance seals
- Performance optimisation
- 100 hour life test
- Performance test
- Crimp and final sealing of helium volume
- Performance test with LVDE including LVDE verification
- 400 hours life test
- Performance test

Upon successful completion of the acceptance testing the formal qualification programme will start. This will consist of:
- Detailed performance characterisation
- Vibration cancellation verification
- Vertical and horizontal operation
- Flight vibration tests
- Mechanical shock test
- Thermal vacuum tests
- Definitive leak test
- EMC tests with LVDE electronics
- Full performance test

- Cold finger parasitic heat load test
- Life testing

Full qualification will be demonstrated against the FIRST requirements based upon the outputs from the above test programme and detailed analysis activities performed in support of the design process.

## SCHEDULE

This programme started in December 1994, and other key dates are as given below:
- Complete detail design (September 1995)
- Complete coolers assembly (July 1996) (not including flight bracketry)
- LVDE available (August 1996)
- Complete full cooler assembly (September 1996) (including brackets, etc).
- Tests with LVDE (September 1996)
- Commence qualification (September 1996)
- Complete qualification (February 1997).

## DISCUSSION

Completion of this programme of work will ensure that a fully qualified 20K cooler will be available to support the 'FIRST' mission. Since the coolers are being manufactured to a full flight standard, it is possible that with minimal refurbishment (force transducer replacement for example) these units could be available as flight spares for the flight programme. It is also desirable that these coolers are put on permanent life test or even disassembled at some stage to allow examination of the internal condition of the hardware. This will be reviewed by MMS and ESA.

Having qualified this 20K cooler for FIRST which has particularly challenging requirements in several key areas, this cooler design could be considered qualified for a number of similar space based missions.

## ACKNOWLEDGEMENTS

The authors wish to acknowledge the support of other workers at ESA, RAL and MMS who have supported this work. Thanks are due to Dornier for their efforts in developing the LVDE.

## REFERENCES

1. Bradshaw T.W., Orlowska A. Closed Cycle Cooler for Temperatures Below 30K. Cryogenics volume 30, 1990 (pages 246 to 248).

2. Jones, B.G. et al. Long Life Stirling Cycle Cooler Developments for the space application range of 20K to 80K. Proceedings of the 7th International Cryocooler Conference, Santa Fe, New Mexico, USA, November 1990.

3. Bradshaw, T.W. et al. A Closed Cycle 4K Mechanical Cooler for Space Applications. Proceedings of the 4th European Symposium on Space thermal Control. Florence, Italy, 1990.

4. Jones, B.G. et al. The Batch Manufacture of Stirling Cycle Coolers for Space Applications Including Test, Qualification and Integration Issues. Proceedings of the 9th International Cryocooler Conference, Waterville Valley, New Hampshire, USA, June 1996.

5. Bradshaw, T.W. et al. Improvements to the Cooling Power of a Space Qualified Two Stage Stirling Cycle Cooler. Proceedings of the 9th International Cryocooler Conference, Waterville Valley, New Hampshire, USA, June 1996.

# Path to Low Cost and High Reliability Stirling Coolers

V. Loung, A. O'Baid, and S. Harper

Superconductor Technologies Inc. (STI)
Santa Barbara, CA, USA

## ABSTRACT

Superconductor Technologies Inc. (STI) has developed a novel Stirling cycle refrigerator to meet the cryogenic cooling requirements demanded for high temperature superconductor (HTS) electronics applications. This free piston, gas bearing, external motor coil Stirling cooler shows high efficiency (>4W lift at 77K with 130W input power), ruggedness, and the potential for low cost manufacturing and long life. It is now being manufactured in small quantities in a pilot production facility at STI while also undergoing reliability testing and development. In addition, manufacturing is being transferred to a volume production facility.

## INTRODUCTION

HTS suppliers need a cryocooler for an emerging cryoelectronics industry. Present cryocoolers are not attractive to potential HTS users because they do not satisfy user expectations and requirements of size, input power, efficiency, lift capacity, cost, reliability, and operation at high and low ambient temperatures. In addition, potential users have expressed skepticism because of the added complexity and costs of cooling. However, HTS filters offer significant performance advantages for wireless communications applications which conventional technology cannot provide. Cryocooling must be provided reliably and inexpensively to exploit this opportunity for HTS.

The cryocooler mated to a dewar with HTS circuitry must be accepted as a transparent subsystem of the HTS system for product acceptance. Wireless system providers expect systems capable of unattended operation for years in an outdoor environment subjected to rain, snow, hail, fog, salt, dust, high winds, lightning, and thunderstorms. HTS users/customers must accept the improved performance, form and fit of HTS as an acceptable trade-off for the added complexity of using a cryocooler, dewar, drive and control electronics. The cryocooler development program at STI was to develop a low cost, reliable cryocooler sufficient for HTS wireless applications.

At the onset, STI analyzed available cooler technology and concluded that there was no available cryocooler which would satisfy customer/user requirements and expectations. Therefore, this would delay customer acceptance of HTS as a viable solution and delay product acceptance. STI concluded that a free piston Stirling cryocooler potentially could provide an attractive solution for HTS applications. It is efficient, compact, quiet, and provides high specific cooling per unit weight. It has great promise to withstand extreme operating and storage environmental conditions including electromagnetic pulse (EMP) conditions, can potentially be produced at low cost and operate reliably in the wireless application.

STI contracted with Sunpower Inc. of Athens, Ohio, to develop and provide a proof of concept cryocooler for HTS applications. Proof of concept units were delivered to STI in 1993. This paper describes the second phase of the cryocooler program at STI to build on the Sunpower proof of concept coolers and develop and produce a low cost and reliable cryocooler to meet HTS system requirements. A system approach was taken in the design of the cryocooler to achieve low cost and high reliability. This required a concurrent engineering effort with input from cryopackaging, cryocooler drive electronics, systems control, and microwave engineering.

Over the past five years, STI has been developing Stirling cycle coolers to support its internal needs for HTS. Given the requirements for low cost and high reliability cryocoolers, STI has determined that other market opportunities for cryocoolers exist if the cost of entry is low enough. Identified markets are general HTS applications and cold-computing. In the HTS marketplace, if a highly reliable cryocooler costs under $5,000, the entire market can likely be captured. The cost target for cold-computing is significantly lower, but a cooler with a cost of $1,000 will secure a significant portion of the market. It is clearly evident that as the cost of cryocoolers decreases, the market opportunities increase dramatically.

STI has chosen to focus on Stirling cycle coolers, as they are lightweight, small and provide efficient use of power. Stirling provides these features over a temperature range which HTS can effectively use (~75K) and also provides adequate lift for cold-computing (e.g., ~35W @ 200K). STI has developed a Stirling cycle cooler that provides the proper lift, meets the input power requirements and is designed for high reliability. The next steps in the path to low cost and high reliability are: (a) proof of reliability; (b) reduction in complexity; (c) standardization; (d) reduction in the cost of components through use of alternative materials, and (e) manufacturing methods like injection molding, die casting, stamping, etc. When these elements are combined, STI will be able to serve a large market with a need for cryocoolers.

## WIRELESS SYSTEM & REQUIREMENTS

### Wireless System

The first application targeted was HTS filters for cellular communications. However, over the course of this program the scope was widened to include filters for Personal Communication Services (PCS), HTS defense electronics applications, and finally cold computing. PCS applications required higher cooling load than initial cellular applications because of the use of active Low Noise Amplifiers (LNAs) on the cold stage. The defense electronics applications require faster cooldown times. A cryocooler size was chosen to meet all these applications requirements.

Fig. 1 shows a first generation standard product HTS wireless filter system consisting of the STI integral, single stage Stirling cryocooler (right front of the open unit in the figure) mated to a permanently evacuated dewar (right rear of the figure) containing the HTS filters. The RF connections in and out of the system connect to the dewar at the rear of the enclosure. The power, control and telemetry cables are also connected at the rear of the enclosure. The RF bypass and lightning protectors are connected at the rear of the unit. The drive and control electronics for the cryocooler and the HTS system control and diagnostics electronics fit in the left half of the system(left side of the open system in the figure). Therefore, the entire HTS system, of which the cryocooler is the largest component, fit very well in a standard 19" electronics rack (20" long x 17" wide x 6 31/32" high). The enclosure is designed to be installed in a electronics rack assembly with the cooling air flow moving from the front to the rear of the unit. All access is at the rear of the unit.

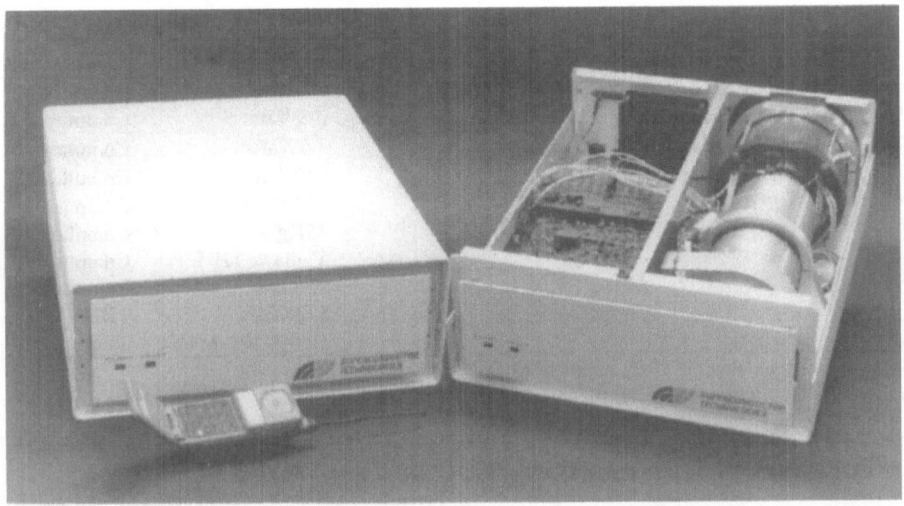

**Figure 1.** Cellular base station front end receiver subsystem.

At high ambient temperatures, the dewar radiation load to the cooler increases and the lift of the cryocooler decreases. Maintaining the lowest possible reject temperature at high ambient operating conditions is important to minimize cryocooler size. Implementing an efficient systems thermal management approach for heat rejection was necessary. The enclosure was designed for operation in high (60°C) and low (-10°C) ambient temperatures and all relative humidities.

### Cryocooler Requirements

The cryogenic cooling industry has been supported either by the semiconductor processing community (Gifford McMahon cycle cryocoolers) or by the infrared detection and imaging community (Stirling cycle cryocoolers). However, these industries imposed requirements on cooler manufacturers specific to their applications and these resulting parameters impacted cost and reliability. The infrared industry requires a cryocooler which is compact, light weight, efficient and generally less than 1 watt of lift. Because of microphonics concerns, these cryocoolers are required to have very low residual vibration levels. In semiconductor processing, high lift at low temperature (~12K) is required to pump hydrogen and other gases. Size, weight and input power are not severe constraining factors. Scheduled maintenance is acceptable.

The targeted HTS wireless applications (having a maximum of 6 filter pairs with LNAs) demanded a reliable, low cost cooler with intermediate lift (4W) at "reasonable" input levels (less than 150W), at reasonable coldfinger vibrations. A summary of the application design specifications and present compliance is in Table 1. The worst case lift requirement with an optimized dewar and coaxial cryocable design is 2.5 watts. Therefore a nominal 4W cooler design was specified. Although weight was not a critical item, it was decided to minimize the cryocooler weight to be less than 5 Kg. Total system input power had to be less than 200W including the electronics and the cooling fan power. Therefore the cryocooler input power had to be less than ~150W. Acceptable coldfinger vibration levels are modest and have allowed inexpensive methods for balancing the cryocooler.

**Table 1.** Cryocooler Design Goals.

| Parameter | Design Goal | Compliance |
|---|---|---|
| Operating Temperature Range | 50-100K | Complies |
| Lift at 50-100K (23°C ambient temp.) | 0-5Watts | Complies |
| Lift at 77K (23°C ambient temp.) | 3.5Watts | Complies |
| Lift at 77K (60°C ambient temp.) | 2.7Watts | Complies |
| Input Power | <150Watts | Complies |
| Weight | <5Kg | Complies* |
| Size | 5" dia. x 12" L | Complies** |
| Cooldown Time | <5 mins. | Complies |
| Cost | $1,000 | TBD |
| Reliability | 40,000 hrs. MTTF | TBD |

\* 2.25Kg w/o fan & balancer; 4.5Kg complete w/ fan & balancer

\*\* Complete assy. is 4.5 dia. x 12.5" L

## LOW COST CRYOCOOLER DESIGN

Attention was paid to meeting application requirements while minimizing ultimate cryocooler manufacturing cost from the outset of this cryocooler development effort. Fig. 2 shows the final cryocooler and dewar subassembly. Fig. 3 shows a drawing of the cooler. It shows the cold end, the motor end with an external coil, the dynamic balancer and a cooling fan.

**Figure 2.** Wireless cryocooler/dewar subassembly.

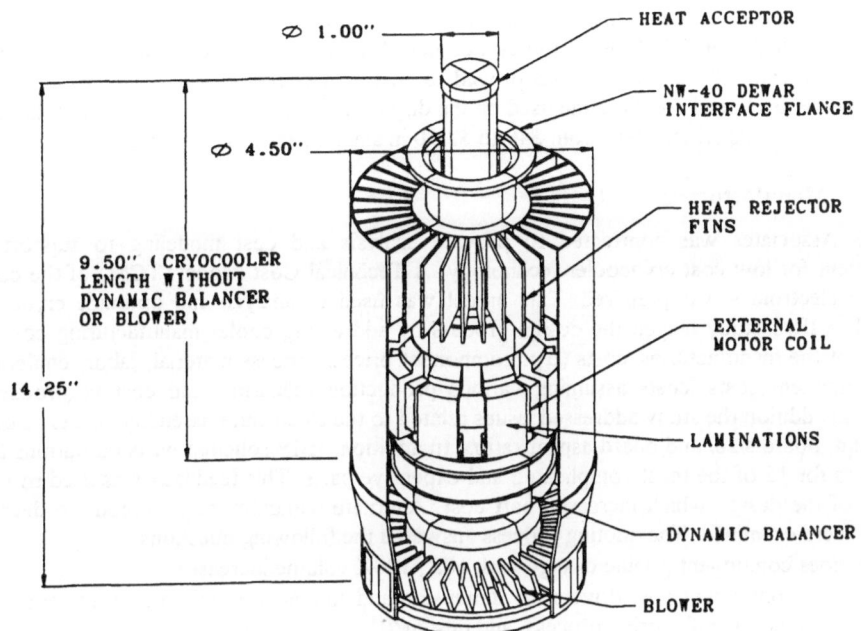

**Figure 3.** Drawing of HTS cryocooler.

The design approach was to establish very aggressive cost goals for the cryocooler as a percentage of the overall system cost. The cooler was further divided into 5 cost areas: motor, piston/cold end, electronics/miscellaneous hardware, assembly and test, and overhead. To ensure low cost and high reliability, the following design steps were taken:

1.  Gas bearings are used for contacting surfaces to eliminate friction. These surfaces are coated with high performance low friction materials to provide repeated starts and stops without wear to the contacting surfaces when the bearing is not activated. The piston and bore are made from the same material to accommodate operation over a wide range of ambient temperatures.

2.  The motor has a low cost external coil. This eliminates electrical feed-thrus in the pressure wall. It reduces the contaminants due to electrical insulation varnishes and improves motor cooling. The penalty is about 5 to 7% increased motor losses. This trade-off is acceptable considering the reliability benefits and the substantial decrease in final assembly time. This feature proves very valuable in optimizing the cryocooler for outdoor applications. It provides the capability to change the cryocooler operating voltage very quickly after the unit has been processed and sealed. This is a desirable option when optimizing the power consumption of an antenna mount unit. An optimum cryocooler drive voltage has to be selected to minimize the cost and heat generation of a  secondary supply and the $I^2R$ losses of the power line supplying the system.

3.  Assembly and alignment time have been minimized. The displacer, piston bore, piston and moving magnet assembly are very easily assembled in the body of the cooler and welded closed. Through the helium fill port, the unit is vacuum baked back filled with helium and hermetically sealed.

4.  The cooler housing is a vacuum brazed and welded metal can with no possibility for helium leakage through ceramic or glass to metal seals.

5.  Free piston designs with no position feedback sensors are prone to overstroking. This situation is cost effectively handled by the controller which monitors the motor and reduces drive current to prevent overstroking.

6.   A dynamic balancer, consisting of a mass located between two sets of the flexures is used to reduce the residual vibration level of the cryocooler. This mass spring assembly is tuned to the operating frequency of the cooler. It has proved to be a cost effective solution for balancing the cooler. The flexures used are the same used on the displacer and it has been possible to reduce the piece part cost of this critical item from $40 to $2.50 in small quantities (500 pieces).

## Design for Manufacturing

IBIS Associates was contracted to provide analysis and cost modeling to support the development for low cost cryocooler technology. A Technical Cost Model (TCM) of the cooler and drive electronics was prepared. This model was used to analyze and optimize cryocooler design. The IBIS study helped the design process by addressing cooler manufacturing cost, the elements of the manufacturing costs (e.g. component prices, process material, labor, equipment, etc.), component costs, costs assuming various production scenarios, and cost vs production volume. In addition the study addressed issues related to the electronics assembly: costs vs board layer count, board size, and board aspect ratio. In addition, IBIS solicited parts quotations from 20 vendors for 15 of the most complicated and expensive parts. This feedback was used to focus on areas of the design which increased part cost. Alternate manufacturing methods to decrease part cost were examined. The quoting process answered the following questions:

- How does component pricing change with 10X, 100X volume increase?
- What are component cost drivers for dimension and tolerance requirements, material cost, labor content, manufacturing process, and design?
- How can component prices be lowered?
- Is there a trend for low cost suppliers?

Cost goals were established. The initial goals for the cryocooler were: motor, piston/cold end, miscellaneous parts, assembly & test, overhead each $175.

Though simple this mechanism guided the development and established clear goals for engineering. The cost goals improved decision making because they demonstrated how design impacts overall costs. For instance, it was decided that final assembly and alignment had to be completed in 15 minutes or less. This implied that "C" clips rather than screws had to be incorporated in some areas of assembly. The dynamic balancer must not require additional tuning after it is installed on the cryocooler. These and other similar guidelines were followed during the design phase. The study was performed during the prototype phase of the design and indicated that component costs were the dominant cost factor. Much progress has been made in lowering these costs. The design for low cost manufacture focused on using parts and material sizes close to commercially available dimensions, taking advantage of vacuum batch braze processing, minimizing sub-assembly and final assembly time, and increasing assembly tolerances where possible.

## Cold End Design

The cold end sub-assembly (Fig. 4) is batch vacuum brazed. This has proved a very successful and cost effective method of assembling the entire cold end. Assembly of the heat exchanger is completed at this step. To obtain a low cost coldfinger tube, a standard mill size seamless thin wall stainless steel tube is used. The tube is then re-drawn to final dimensions and swaged at one end to accommodate brazing and easy removal from the brazing fixture.

The internal and external heat exchanger fins were sized to accommodate a size commonly fabricated for other industries such as air conditioning and refrigeration. Several fabrication methods such as photo chemical etch, water jet, laser, and EDM were investigated to reduce manufacturing costs. However, it was concluded that method of shearing and forming is the most cost effective method of manufacture.

The long life flexure spring used on the dynamic balancer and the displacer was a costly component during prototype fabrication. Much effort was devoted to this part to substantially reduce the cost. Several manufacturing methods such as wire EDM, water jet, laser and stamping

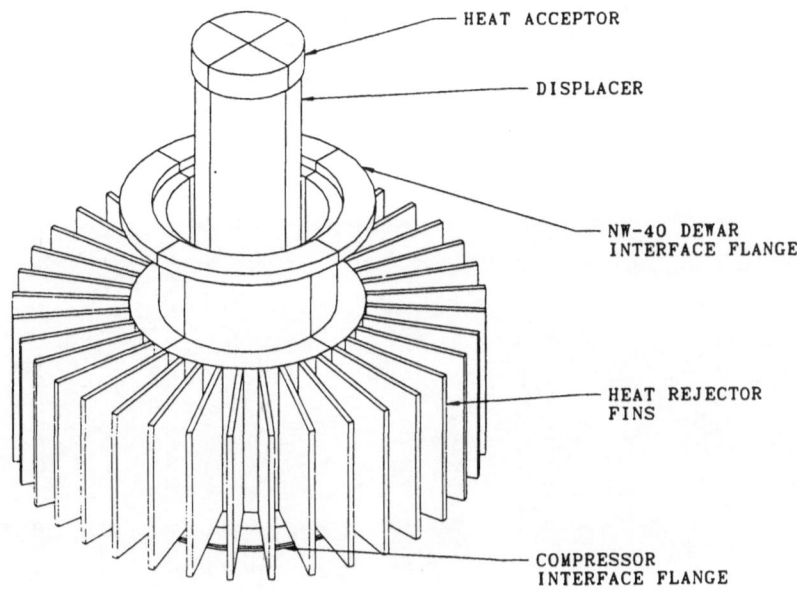

HEAT ACCEPTOR

DISPLACER

NW-40 DEWAR
INTERFACE FLANGE

HEAT REJECTOR
FINS

COMPRESSOR
INTERFACE FLANGE

**Figure 4.** Cryocooler brazed assembly.

were investigated. By changing the manufacturing methods, it has been possible to reduce piece part costs from $40 per spring to $2.50 per spring in small quantities. When volume can justify the cost for proper hard tooling, the part cost can be decreased to less than $1. These costs are consistent with the production costs of similar parts for the high volume pneumatic industry.

Similar examinations are continuing on cryocooler parts. An industry is identified where a similar part is presently manufactured in high volume. Technical discussions are then initiated with that vendor to discuss ways in which costs can be designed out of the cryocooler part.

**Measured Performance**

To date a total of 25 units have been built and tested. The units were built in three lots. The last lot (Fig. 5) consists of 8 air cooled units and one unit without the external fins. Alternative methods of heat rejection will be installed on the heat rejector of this unit for a sealed outdoor antenna mount application. The units have valves on the helium fill tube and the test dewar enclosing the coldfinger.

Cooldown and steady state performance tests were performed on each of the 25 coolers. The last lot of coolers (which were more tightly controlled during manufacture) showed the closest spread in the data. The spread in performance with the first lot was about 1W. The last lot exhibited a spread between the best and the worst performers of about 0.5W.

The transient cooling at room temperature ambient with different thermal capacities are shown on Fig. 6. From the two measured results, an average transient cooling lift of 14W is calculated. This result while not pertinent to base station and antenna applications, provided good performance for another HTS application requiring fast cooldown. Fig. 7 shows that the cooldown time does not increase substantially at high reject temperatures. With an additional thermal capacity of 15,000J, there is approximately a 5 minute increase in cooldown time from a rejecter temperature of 40°C to 70°C.

Fig. 8 shows the performance of an 'average' cooler from the last batch. This figure shows the decrease in cooler lift at higher rejecter temperature. It demonstrates that the cooler will still satisfy the requirements for the worst case PCS applications (i.e., most filter pairs with LNAs). In addition to in-house reliability testing and testing with cellular, PCS and defense electronics systems, a series of accelerated life tests was successfully completed recently by a large wireless equipment manufacturer. These tests involved repeated temperature cycling from -10°C to +60°C while operating, thermal shock from -40°C to +85°C, with humidity from 0 to 95%.

**Figure 5.** Batch 3, low cost/high reliability cryocoolers.

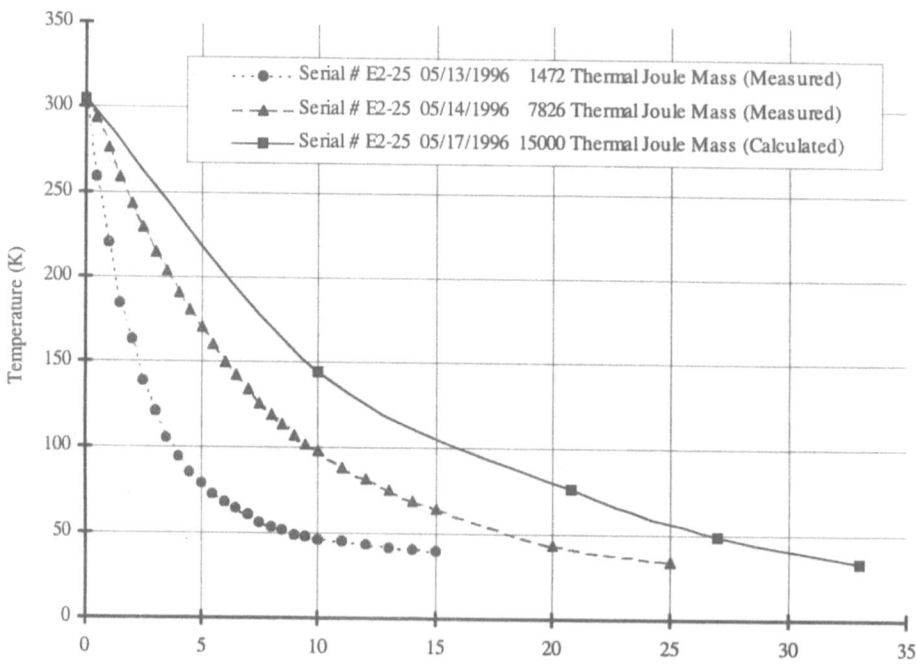

**Figure 6.** Stirling cryocooler cooldown performance from room temperature.

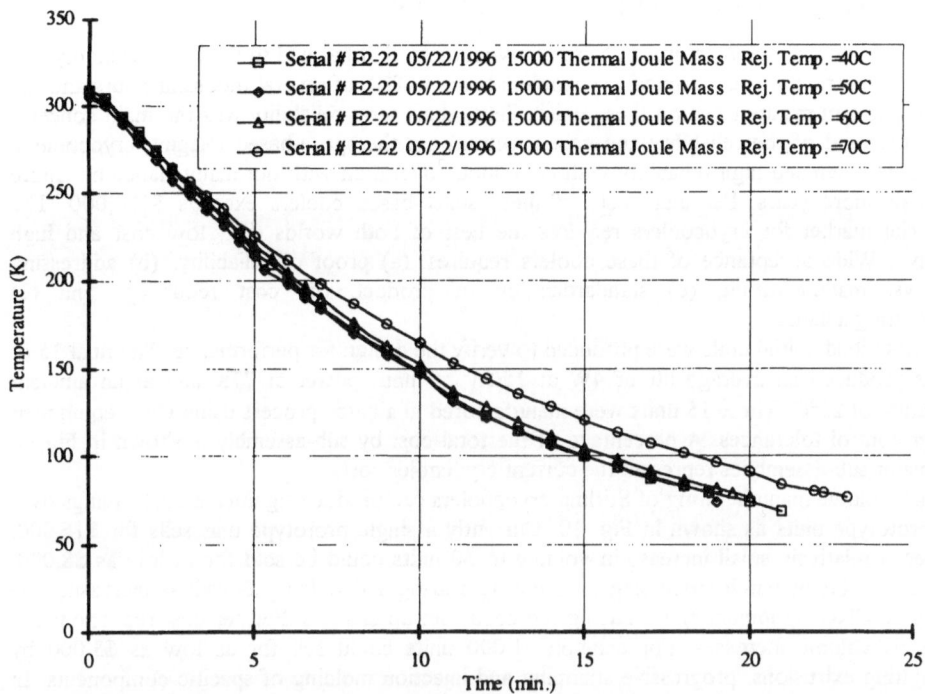

**Figure 7.** Stirling cryocooler cooldown performance at high ambient temperature.

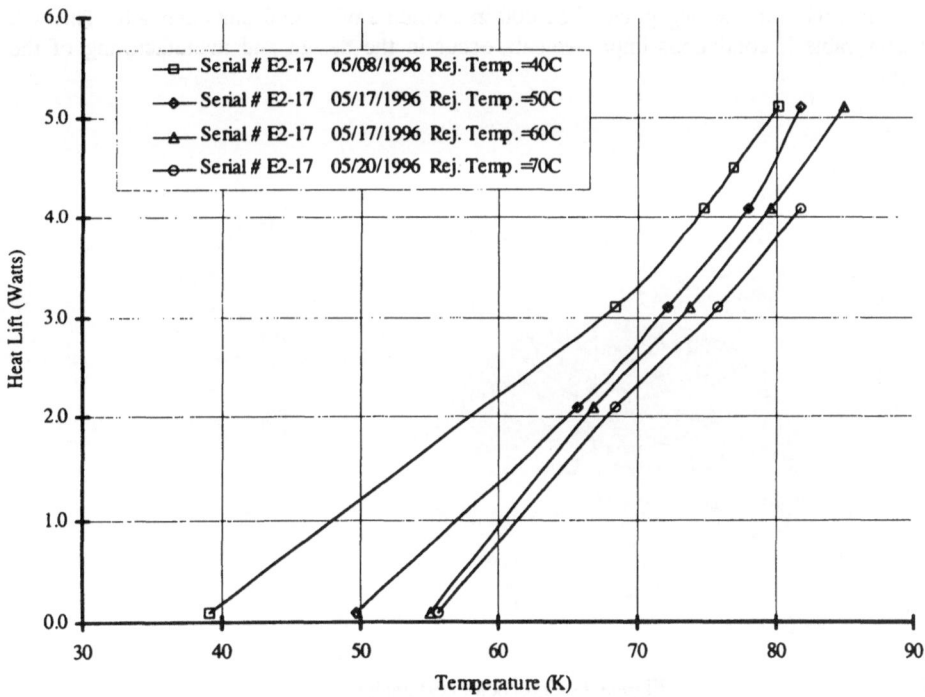

**Figure 8.** Stirling cryocooler steady state performance.

## MANUFACTURING OF STIRLING CRYOCOOLERS

As a commercial manufacturer of Stirling cryocoolers, STI has had to focus on reliability and costs. In the past, "mil-spec" tactical cryocoolers had a MTBF of several thousand hours and the ability to perform maintenance on these units. Thus, short term reliability was the main concern. At the other end of the reliability and cost spectrum are the space-based imaging cryocoolers. These coolers demand high reliability with continuous operation with out maintenance or failure for five or more years. Per unit costs of these space-based coolers exceeds $300,000. The commercial market for cryocoolers requires the best of both worlds (i.e., low cost and high reliability. Wide acceptance of these coolers requires: (a) proof of reliability, (b) addressing design vs. manufacturing, (c) standardization of product, (d) cost reduction, and (e) manufacturing alliance.

As described, initial units were produced to verify the design for performance. The final 15 of 25 units produced an average lift of 4W at 150W of input power at 77K and at an ambient temperature of 23°C. These 15 units were manufactured in a batch process using CNC equipment to better control tolerances. A percentage of the total cost by sub-assembly is shown in Fig. 9. These major sub-assemblies represent the current cryocooler costs.

Small volume manufacturing of Stirling cryocoolers can produce significant cost savings over single prototype units as shown in Fig. 10. Currently a single prototype unit sells for $15,000. However, a relatively small increase in volume to 50 units could be sold for as low as $8,000. This is achievable by batch processing units during brazing and welding as well as increasing the run time on CNC equipment and reducing the costly set up charges. Further cost reductions are possible as volume increases. For example, 1,000 units could sell for as low as $3,000 by incorporating extrusions, progressive stamping and injection molding of specific components. In addition to component cost reduction, STI has also formed a Joint Venture to assemble cryocoolers in volume production. The production occurs in a work center format which allows parallel processing of sub-assemblies. The goal is to manufacture cryocoolers for HTS and other commercial markets for a selling price of $1,000 in a volume of 10,000 units annually. This will only be achievable if continuous improvements occur in the design and manufacturing of the cryocoolers.

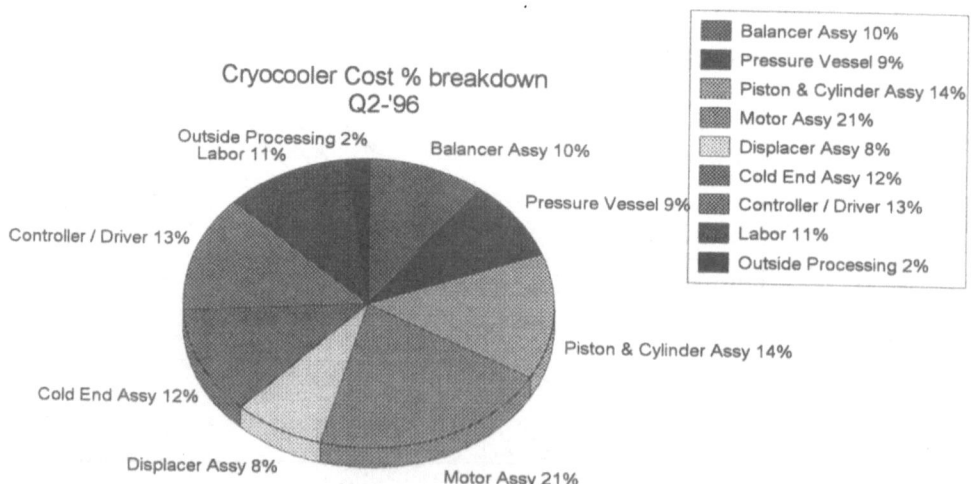

**Figure 9.** Cryocooler cost analysis.

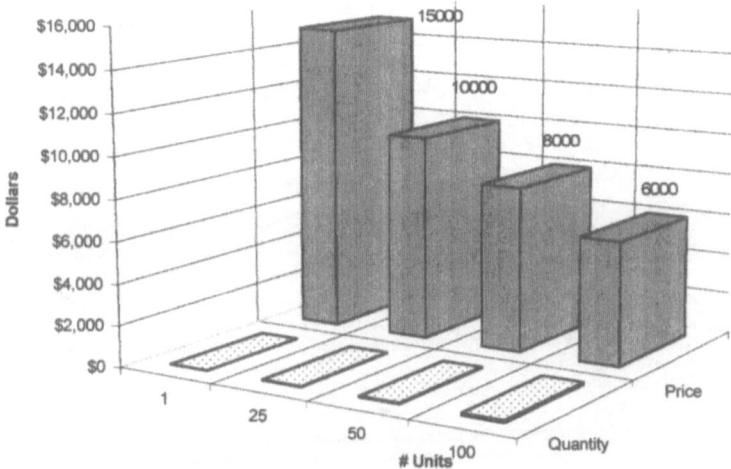

**Figure 10.** Cryocooler price vs. quantity.

## CRYOCOOLER RELIABILITY

STI has developed a cryocooler reliability program that integrates the design/development, manufacturing/production and the quality/reliability teams together to produce a continuous improvement philosophy. For any cooler to be "reliable" it must function under the expected range of environmental conditions, for an expected lifetime, as well as meet all safety requirements. The reliability goal of the cryocooler is to maintain a cold end temperature of 77K with a 4W heat load at room temperature and have a MTTF of 40,000 hours. Cryocooler reliability is best assured when it is designed into the product.

The STI cryocooler has many features incorporated into the design that make it inherently reliable. The free piston integral Stirling design utilizing gas bearings to reduce frictional wear is key to the long life of this cryocooler. Also having an external coil motor eliminates the need for hermetic feed-thrus while reducing the potential risk of helium contamination. The tolerances associated with the reliable gas bearing design are not over stringent on current machining capability. The cooler design also has allowed for sub-assemblies to be quickly and easily assembled. This simplistic approach makes cooler assembly not only reliable but reproducible as well. Helium gas contamination over time was considered to be one of the biggest threats to maintenance free long life of this cryocooler. For this reason the motor coil was moved external of the helium space. Bake out procedures were developed to monitor and reduce the potential risks of gas contamination.

Cryocooler life testing is an important part of STI's reliability program. Step 1 of our 3 step approach is a Failure Reporting And Corrective Actions System (FRACAS--see Fig. 11). Which is a positive feedback system to design and manufacturing of failures during the development phase. Step 2 is a Failure Modes Effects Analysis (FMEA) system which encompasses both the process of assembly as well as cooler design criteria. This pre-production reliability tool identifies critical characteristics on the drawings and process steps. It is also used to develop the Process Control Plan and Process Travelers. Lastly, the Accelerated Life Tests (ALTs) are environmental stress screen tests used to thermally shock, temperature and humidity cycle as well as vibration test the cryocoolers on HTS systems. STI's wireless communication system (with cryocooler) has passed ALT testing by Motorola. This is a 1000 hour test that is an environmental stress screen. The many iterations of this type of test prove the HTS system to be reliable under extreme conditions.

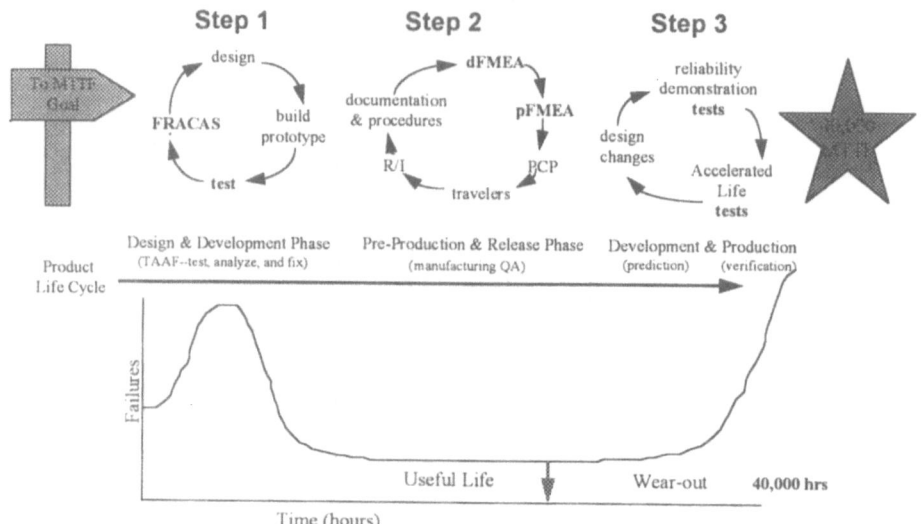

**Figure 11.** Cryocooler reliability program.

## SUMMARY AND CONCLUSIONS

STI embarked on a vertically integrated approach to design and develop HTS filter systems for cellular and PCS applications. This approach hastened the development of the cryocooler to meet application requirements. It has provided quick determination of a cryocooler configuration and prioritization at the system level the first order and second order issues to be addressed. It has provided valuable feedback on the areas to cut costs.

Recent developments indicate that very long life and high reliability can be expected from these machines at a cost consistent with the wireless system cost requirements. In addition, their efficiency, cost, and long life potential indicate the potential of these Stirling coolers to enable cold computing applications. Further manufacturing development and needed comprehensive life and reliability tests are continuing at STI.

## ACKNOWLEDGMENTS

The authors wish to acknowledge and thank the many individuals who have contributed to the success of this program. These include but are not limited to Betsy Adams (IBIS), Mah Yuen Ann, David Chase, Dennis Ebejer, David Gedeon, Jim Gibson, Randy Gilligan, Tim James, Wally Kunimoto, Bob Hammond, and Adam Singer (IBIS). Funding for this work was provided by ARPA under a contract monitored by the Naval Research Laboratory (N00014-94-C-2213).

## REFERENCES

1.   Raihn, K., Fenzi, N., Hey-Shipton, G., Saito, E., Loung, V., Aidnik, D., " Adaptive High Temperature Superconducting Filters for Interference Rejection", *IEEE Transactions on Microwave Theory and Techniques,* vol. 44, no. 7 (1996), pp. 1374-1381.

2.   Nisenoff, M., "Cryocoolers and High -Temperature Superconductors: Advancing toward Commercial Applications", *Cryocoolers 8,* Plenum Press, New York (1995), pp. 913-917.

3.   Redlich, R.W. and Berchowitz, D.M., "Linear Dynamics of Free-Piston Stirling Engines", *Proceedings of Institute of Mechanical Engineers,* vol. 199, no. A3 (1985), pp. 203-213.

# Miniature Long Life Tactical Stirling Cryocoolers

C. R. Aubon and N. R. Peters

The Hymatic Engineering Company Limited
Redditch, Worcestershire, United Kingdom B98 9HJ

## ABSTRACT

This paper describes the development and productionisation of a miniature Stirling cycle cryocooler for both military and industrial applications. The design is described and the features which overcome common Stirling cycle cryocooler failure mechanisms are highlighted. Performance, life and environmental test data gathered from a pre-production batch of coolers is presented. Issues arising from the integration of linear, monobloc, integrated Stirling cycle coolers into infra-red and other systems are described. The paper concludes that long life, low cost Stirling cycle coolers can be designed in a form suitable for volume production.

## INTRODUCTION

One of the major drivers in the selection of a cooling technology for cryogenic devices is the requirement to minimise the logistical burden associated with the cooler. Consequently, in recent years Stirling cycle coolers (which simply require an electrical power supply in order to operate) have become the preferred technology in many cryogenic cooling applications.

### Background

Hymatic has been one of the major developers and suppliers of Joule-Thomson (JT) cryostats for over 40 years. This experience has been applied to the development and productionisation of a miniature, linear, monobloc, Stirling cycle cryocooler for military and industrial applications.

The objective of this development programme has been to minimise input power and size whilst providing sufficient cooling power to operate with relatively high heat load detectors over a normal military temperature range.

### University of Oxford Cryocooler

Hymatic's linear cryocooler design has evolved from a space-qualified design previously built by The University of Oxford with whom Hymatic is teamed. A fundamental principle of the programme has been to retain the features of the Oxford design which assure long life whilst using Hymatic's production expertise to generate a product which can be manufactured in large quantities at low cost. It is clear that there is considerable commonality between the manufacturing requirements for JT and Stirling cycle coolers such as precision mechanical assembly, clean processing, and high integrity joints.

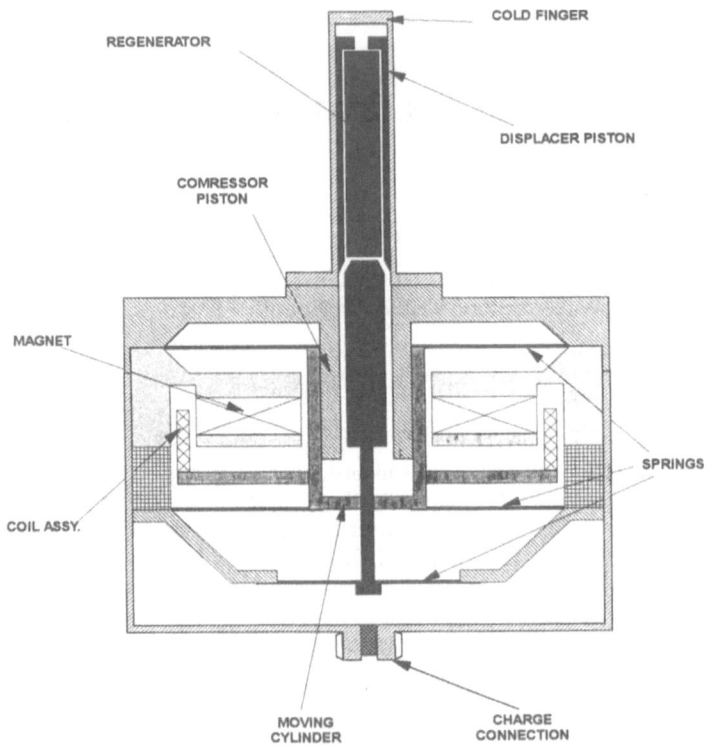

**Figure 1.** Arrangement of design.

## DESCRIPTION OF DESIGN

The layout of the cryocooler is shown schematically in figure 1. This cooler has been designed to provide nominally 200mW of cooling power at a refrigeration temperature of 85K over a range of rejection temperatures from -40°C to +70°C. The technology employed is scaleable and prototype coolers in a range up to 5W refrigeration at 80K have been assembled by Hymatic.

A monobloc configuration was chosen in order to maximise efficiency and refrigeration power within a compact assembly. A single compressor was selected to minimise component count and overall cost. Passive and active balancers have been engineered as a lower cost and modular approach to vibration reduction.

### Pressure Vessel

The working parts of the cryocooler are enclosed in a stainless steel pressure vessel which is hermetically sealed with a 20 bar helium charge. An integral part of the pressure vessel is the cold finger - a very thin walled stainless steel tube which has a copper plug electron beam (eb) brazed into its end. The external face of this plug is the spot where cryogenic refrigeration is available.

### Compressor

Inside the main body of the pressure vessel is a compressor. This consists of a moving cylinder and coil assembly, and a fixed piston. The cylinder is supported by a flexure bearing (electrochemically etched suspension spring) at each end. The suspension springs hold the cylinder clear of the piston and constrain it to move along its axis; this ensures that a truly non-contact clearance seal is maintained under all conditions.

## Linear Motor

The cylinder drive coil is mounted concentrically with the cylinder and, when assembled into the cryocooler, the coil is positioned in the field of a toroidal permanent magnet. The magnet's field is directed around a soft iron magnetic circuit and concentrated in the working gap such that the turns of the coil cut the flux. The motive force for the compressor is generated by an alternating current in the coil.

## Displacer

In order to generate refrigeration at the cold end the gas is shuttled from the compression space to the cold end by a displacer piston. This is supported at one end by a flexure bearing and is not mechanically linked to the compressor. The displacer is designed such that out of balance pressure forces pneumatically drive it with the correct phase and stroke relationship.

## Regenerator

The movement of the displacer causes the gas to flow over a stack of gauze disks, which is located inside the displacer piston. This stack of disks constitutes a regenerative heat exchanger. Heat is removed from the gas as it flows to the cold end and is returned to the gas as it flows back to the compression space.

## ADVANTAGES OF THIS DESIGN

In order to understand the advantages of the design it is useful to consider the common failure modes of other Stirling cryocooler designs.

### Common Failure Mechanisms of Stirling Cycle Cryocoolers

These fall into the following categories.

**Wear.** This is caused by rubbing contact between moving parts. Since lubricants contaminate the system and dry bearings inevitably wear, long term reliability can be achieved only by eliminating rubbing contact.

**Charge Gas Leakage.** Sealing of the high pressure helium charge gas used in Stirling cycle cryocoolers is a difficult technological problem. This is a particular problem when the integration with the cooled component necessitates the use of a demountable seal for the cold finger to body joint.

**Contamination.** For long term operation of a Stirling cycle cryocooler without deterioration it is essential to eliminate condensable vapours from the system. These vapours, when present, will condense or freeze at cryogenic temperatures. This can cause restrictions in gas passages and undesirable rubbing of moving parts.

### Design Features Which Overcome the Common Failure Mechanisms

Hymatic's cryocooler addresses each of the above-mentioned failure mechanisms.

**Wear.** The compressor in Hymatic's cooler achieves a dynamic pressure seal by the maintenance of very small clearances between the moving parts.

The cooler utilises photochemically etched stainless steel suspension springs. These provide a large axial movement whilst maintaining very high radial stiffness. The springs enable the elimination of rubbing contact. Careful design guards against the effect of environmental vibration and ensures that the flat spiral springs are subjected to stress levels lower than their high cycle fatigue limit.

**Charge Gas Leakage.** Hymatic coolers have three static pressure seals which are hermetically joined using eb welding. The final seal, which is made after the cooler has been charged, is produced using a metal to metal (cold weld) seal. This sealing technology has been well proven by Hymatic on long life bottle systems with pressures up to 600 bar. Measured leak rates of less than $0.4 \times 10^{-10}$ mbar litres per second are routinely achieved for cryocoolers.

**Contamination.** The cryocooler is charged with very pure helium. However, the main potential for contamination arises from the outgassing of vapour from the materials of which the cryocooler is constructed. The major potential contaminant is water vapour. The effect is minimised by the selection of constructional materials which allow the whole cooler to be baked out at high temperature in a vacuum prior to filling with charge gas.

Control of the clean precision assembly processes within a purpose-built production clean room facility, together with careful material selection, minimise the risk of contamination. The cleaning strategy is based on Hymatic's long experience with cleaning pure air components for JT coolers.

## DESIGN OF A PRACTICAL LONG LIFE TACTICAL CRYOCOOLER

It is well understood that the design and manufacture of a new Stirling cycle cooler is not an inconsiderable task. Hymatic was aware from the start of the programme that there are many potential pitfalls.

### Fundamentals

From the outset it has been an essential aim to produce a design which is suited to volume production. For this reason a multi-disciplinary team including Oxford University engineers as well as design, development and manufacturing personnel, production engineers, and quality engineers has worked together on the design and the development process throughout the programme. A decision was quickly taken to manufacture a pre-production batch of 50 coolers concurrently with the 15 engineering development models.

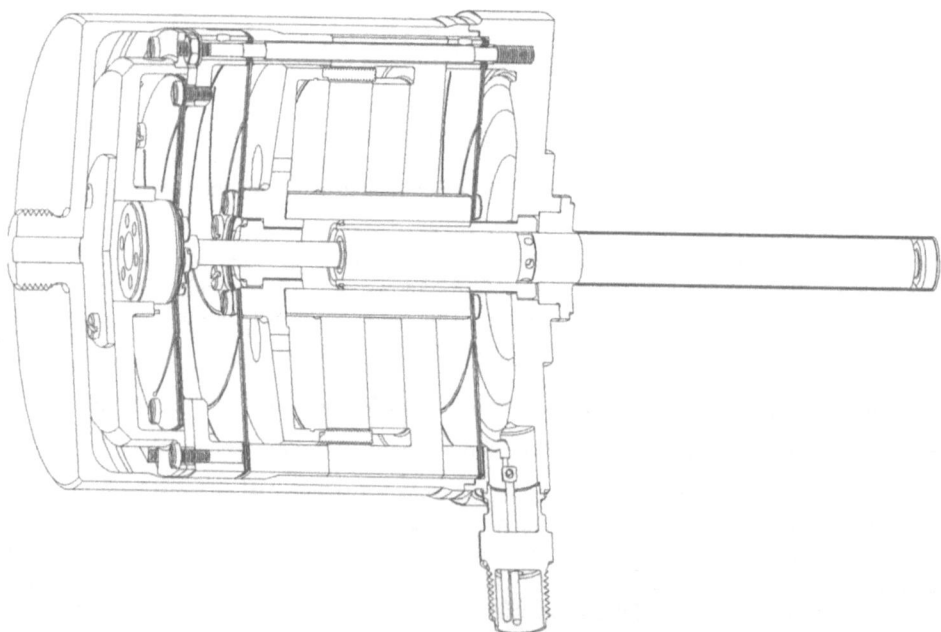

**Figure 2.** Indicating the level of complexity.

## Design for Volume Manufacture

**Simplicity of Design.** Figure 2 shows that a low component count and simple construction have been achieved. This enables the manufacturing process to be made inexpensive.

**'In Process' Checking.** Particular attention has been given to 'in process' checking which ensures that problems are spotted early in the build process enabling rectification. Once a cooler reaches completion, there is low probability that the performance measured during the final acceptance test will be below specification.

**Component Assessment.** Each component of the cooler has been individually assessed for manufacturability and cost.

A particular example is the displacer piston which is manufactured from a glass-filled engineering thermoplastic. The displacer is designed to run inside the precision bore of the cold finger with no contact permitted except for a light rub at the warm end seal. Difficulty in maintaining the very rigorous straightness tolerance for the component was experienced during machining of the plastic. In order to improve the manufacturability of the component (and hence minimise cost) several parallel activities were undertaken.

A thorough investigation into alternative materials was carried out. The Rubber & Plastic Research Association were charged with carrying out a comprehensive assessment of the thermoplastic raw material. Alternative machining processes were investigated. Changes to the design of the component were researched with particular regard to the influence of the change on the thermodynamic performance of the cooler.

The final solution drew on the results of all of these activities and the outcome is a design which is considerably easier to manufacture than the original.

Figure 3 is a scatter graph showing the effect of adding a 60μm deep by 20mm long relief at the cold end of the displacer in order to mitigate the straightness requirement. This was difficult to model thermodynamically as the effect on performance depended on several physical processes which were not well defined (such as the regenerative heat exchange which occurs between gas which leaks past the displacer warm end seal and the cold finger wall). Because of the nature of the cooler assembly it is not possible to test a single cooler with several displacers, hence, a number of coolers were evaluated with each type of displacer.

**Operator Training.** Early in the programme production assembly personnel began working alongside the engineering technicians who were building the first coolers. The whole pre-production batch of coolers has been manufactured by these personnel.

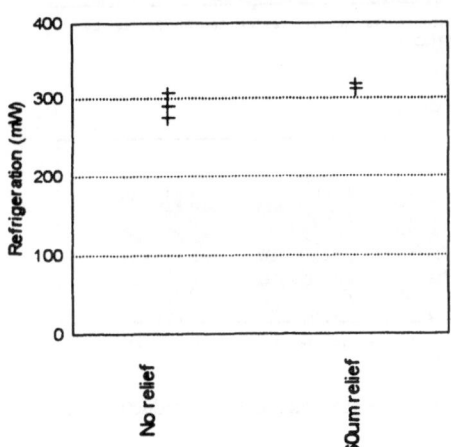

**Figure 3.** Displacer relief scatter graph.

## Design for Ruggedness

Hymatic has many years of experience of designing precision mechanical systems for military environments. Specific features were added to the basic Oxford cooler layout in order to improve the fitness for use in high vibration and extended ambient temperature environments.

## DATA GATHERED FROM PREPRODUCTION BATCH

A preproduction batch of 50 sets of components has been manufactured; 23 of these have completed assembly.

### Performance Bar Charts for Engineering and Production Batches

Figure 4 shows the cooler performance trend. Blank entries indicate coolers which have been rejected during the build process before initial performance testing.

This data validated the design model enabling the generation of performance data that can be offered on a production basis at an acceptable yield (figure 5).

### Operating Frequency

Each cooler is optimised during the build process by varying charge pressure, displacer damping, and drive frequency. Figure 6 indicates the variation in optimum drive frequency for engineering and pre-production batches. The variation in performance between the optimum drive frequency and a drive frequency of 74Hz is being monitored. This is typically no more than 10mW at 80K with 6W input power to the cooler and an ambient of 20°C.

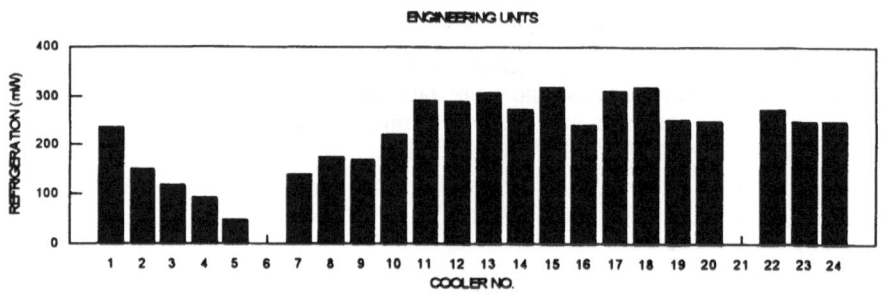

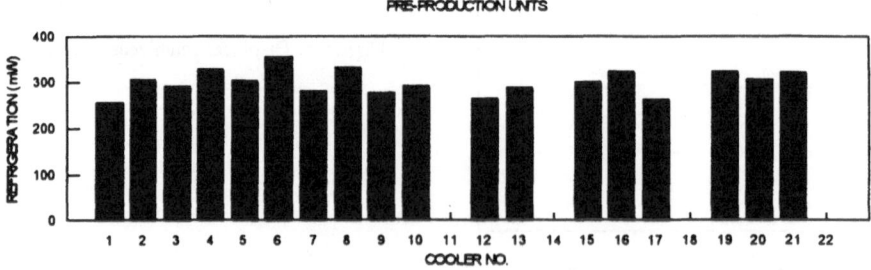

**Figure 4.** Cooler performance at 80K, 6W input, and 20°C - engineering and pre-production units

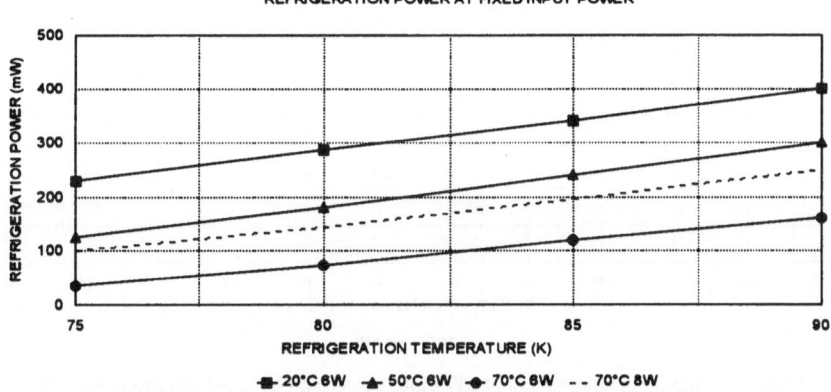

**Figure 5.** Performance data offered on a production basis.

## LIFE AND ENVIRONMENTAL TEST PROGRAMME

**Life.** A sample of the nominally 200mW cryocooler described in this document has recently begun life test. To date this cooler has operated for approximately 1,750 hours at a refrigeration temperature of 77K.

The technology utilised in the 200mW cryocooler has been scaled from a larger 500mW prototype design. Hymatic produced 10 of these 500mW coolers. Two coolers are currently running on an extended life test and to date have achieved 15,700 and 16,600 hours of operation at 80K.

**Operating Temperature Range.** Samples of the 200mW cooler have been tested at -40°C and +70°C rejection temperature. These have met the performance criterion specified in figure 7.

## SYSTEM INTEGRATION

There are many potential problems associated with the integration of the cooled component and the cryocooler. Hymatic's strategy has been to supply coolers to possible users for integration in order to identify and resolve these problems.

Figure 8 shows a typical infra-red detector encapsulation with cryocooler and control electronics.

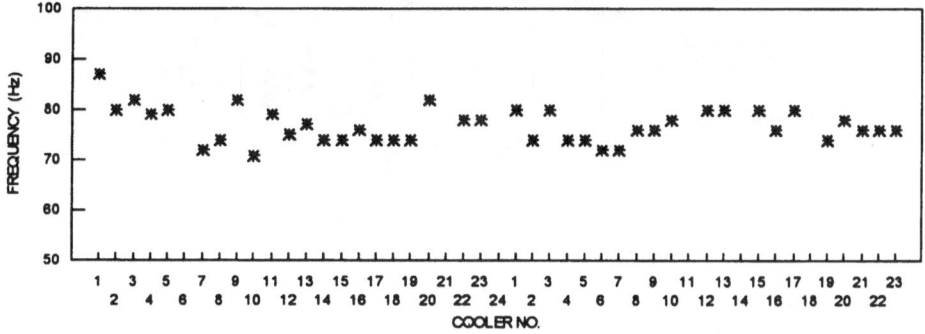

**Figure 6.** Variation of optimum drive frequency for engineering and pre-production batches.

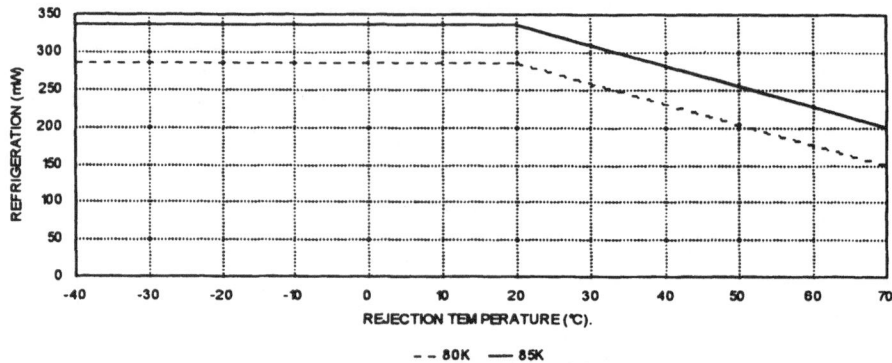

**Figure 7.** Cooler performance requirement at 6W input power and 80K refrigeration temperature.


**Exported Vibration.** Operation of the cooler requires the reciprocation of a compressor cylinder and a displacer piston. The mass and stroke of the displacer are considerably smaller than those of the compressor and hence the generated vibration is dominated by the acceleration of the compressor cylinder.

The vibration primarily occurs at the cooler drive frequency with smaller components at harmonic frequencies.

Hymatic's strategy has been to minimise the vibration level at the fundamental drive frequency by developing active and passive vibration dampers. These consist of spring - mass assemblies which are designed to be resonant at the cryocooler drive frequency. When attached to the cryocooler the vibration generated by the cooler is balanced by motion induced in the absorber. The

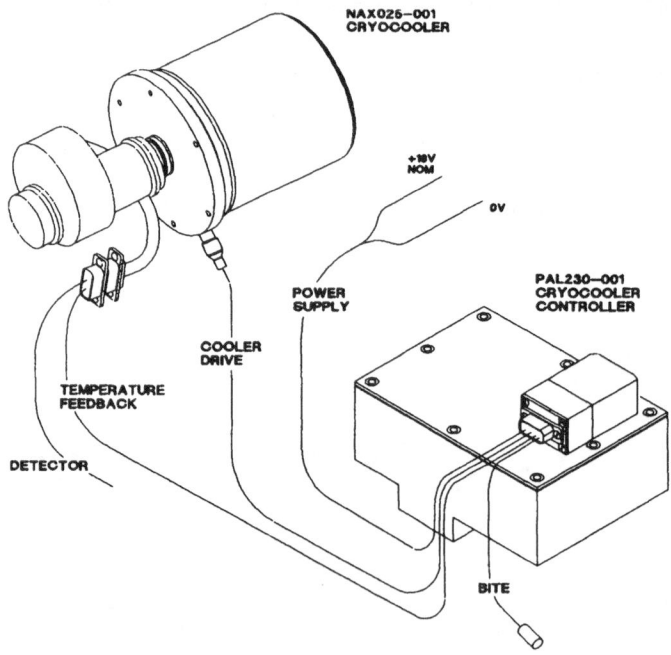

**Figure 8.** Typical system.

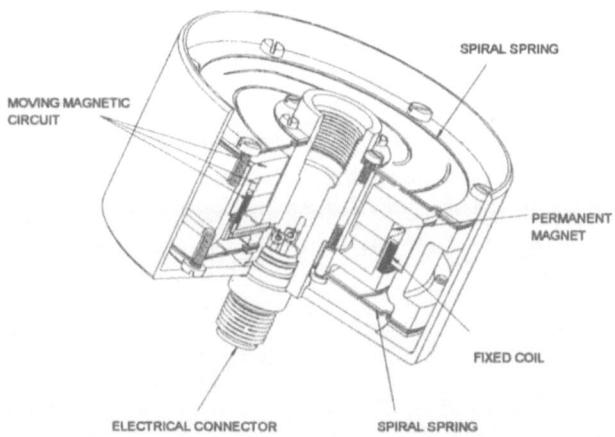

**Figure 9.** Active vibration absorber cut-away.

moving mass in the active absorber is driven by a small linear motor in order to optimise the phase and stroke of the balancer. Figure 9 shows the layout of the active absorber.

Coolers are tested on a representative base plate and mounting bracket. The mass of the cooler, mounting bracket, test dewar, accelerometer and base plate is 4.256kg. The mass of the passive balancer is 73g and the mass of the active balancer is 137g. The accelerometer is used to measure the vibration in the axis of the compressor reciprocation. Acceleration generated in the orthogonal axes is negligible. Figure 10 shows typical vibration spectra measured on a cooler with and without balancing.

The simplicity and absence of close tolerance parts in the absorbers make this strategy a more economic approach than the alternative approach of using twin balanced compressors.

**Controllers.** The cryocooler compressor cylinder is driven by the application of a sinusoidal voltage drive waveform to the linear motor coil.

For the purpose of evaluation two standards of controller have been produced. A military grade switching controller has been designed to operate from a standard 9 to 16V battery supply. By using a switching converter design efficiencies of 75% are achieved. A commercial grade linear amplifier has also been designed. This provides suitable drive waveforms for the cryocooler and active balancer.

Both types modulate the cryocooler drive waveform amplitude in order to control the refrigeration temperature based on feedback from a silicon diode thermometer. Proportional (P) control and Proportional + Integral (PI) control have both been evaluated.

**Cooled Component Interface.** The cryocooler has been designed for integration with the cooled component such that the cold finger forms the inner wall of the dewar. The cold finger features a weld flange which enables the dewar outer housing to be laser or eb welded to the cryocooler.

Mounting of the cooled component to the end of the cold finger has been carefully considered. Hymatic has collaborated with potential end users to develop successful mounting processes.

**EMI Considerations.** Hymatic were conscious of the need to minimise the effects of electromagnetic interference. The approach adopted has been to integrate coolers with typical infrared detectors and analyse the noise on the output of the detectors. Satisfactorily low levels have been achieved.

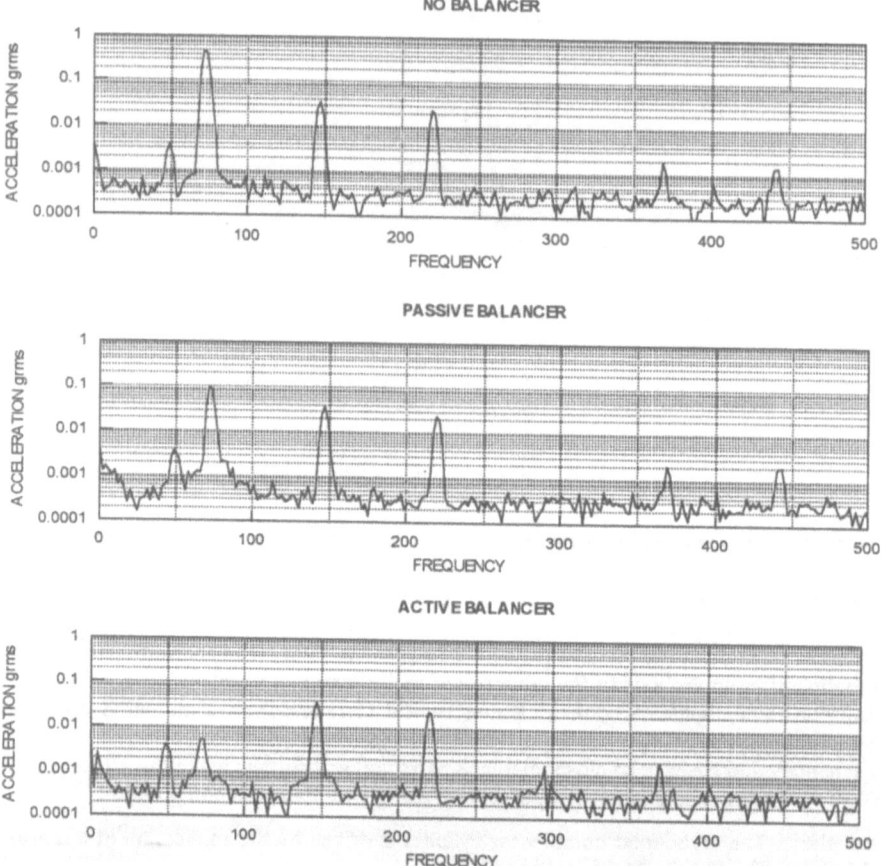

**Figure 10.** Vibration spectra with 6W input power.

## CONCLUSIONS

By their very nature, Stirling coolers are devices in which product and process development must occur simultaneously. This demands the use of a multi-disciplined team and influenced Hymatic's decision to launch a pre-production batch of 50 samples in order to test the production methods and tooling and also to train assembly personnel. This approach has exposed several problems that might otherwise have only emerged at the start of true production.

Hardware is now built by Production Assembly staff using production methods and tooling. With half of the pre-production batch complete, it is evident that the aims of producibility have been achieved.

Environmental and life testing on this 200mW model and a related design indicate that the performance and operating lives over 15,000 hours can be achieved .

Component cleaning and the control of the assembly process has eliminated contamination.

No cooler failures have occurred after final charge and weld.

The design and manufacturing approach adopted by Hymatic has resulted in an efficient, small, affordable, Stirling cryocooler with an expected operating life beyond 20,000 hours.

# Experimental and Predicted Performance of the BEI Mini-Linear Cooler

D.T. Kuo, A.S. Loc and S.W.K. Yuan

BEI Sensors and Systems Company
Sylmar, CA 91342

## ABSTRACT

BEI is a manufacturer of Stirling-cycle cryocoolers based on the concept of clearance seal, pneumatically driven displacer and linear drive motors. Several cooler models are available covering refrigeration requirements ranging from 150 mW to 5.0 Watts of cooling at 78 K. This paper describes a computer simulation model, herein referred to as Hybrid Refrigeration Model (HRM), that was developed for the design of BEI coolers. The BEI computer model uses a novel technique that greatly reduces computation time without compromise in accuracy.

The hybrid model is used at BEI for the design of both Stirling and Pulse-Tube refrigerators. The model is similar to a third order model described in the literature as the Stirling Refrigerator Performance Model (SRPM), that has been extensively validated against various Stirling and Pulse Tube designs from different manufacturers. Although very accurate in its prediction, the SRPM suffers a major shortcoming in its long run time. The Hybrid Refrigeration Model, on the other hand, uses a third order approach in calculating mass flows and pressure drops within the refrigerator while the losses are imposed using the second order analysis. This approach has provided an accurate tool for performance prediction with fast turnaround. An example of the use of this computer model is provided in this paper with the parameters of the BEI Mini-linear cooler. The performance of the cooler is discussed, in particular, the effects of the regenerator material, configuration and operating conditions.

## INTRODUCTION

The external view of the BEI Mini-linear Cooler is shown in Figure 1. The cooler comprises of a compressor that is pneumatically coupled to its expander by way of a flexible transfer tube. The entire unit is very compact, with the compressor housing measured 4 inches in length and 1.365 inches in diameter. The cooler can be viewed as having three distinct volumes to aid the understanding of its operation: 1) working volume, 2) crankcase volume, and 3) gas spring volume. The compressor functions as a pressure oscillator causing a rise and fall of the working volume pressure. As the working volume pressure approaches its peak value, this pressure pushes the displacer towards the warm end of the expander assembly. When the

Cryocoolers 9, Edited by R.G. Ross, Jr.
Plenum Press, New York, 1997

119

working pressure approaches its minimum point, the gas spring pushes the displacer toward the cold end.

The compressor generates the pressure oscillations from a pair of pistons driven in-line, and in opposite direction within a common cylinder. This configuration, known as "twin-opposed pistons," provides a well-balanced machine with minimum vibration. The piston also functions as a bearing to support the weight of the moving mass.

Actuation force for each piston comes from a linear motor. This motor consists of a circumferentially wound coil assembly that is free to move in the axial direction within an annular air gap. A magnetic field is established with magnets and back-iron to focus the magnetic flux density in the radial direction across the air gap. The actuation force is the product of this flux density and the current flowing in the coil assembly.

Mechanical springs are used to maintain position of the coil assembly centered relative to the magnets. The springs are made stiff enough to minimize static deflections in 1-g gravitational field. Another important feature of the spring is the manner which it is used to achieve resonance for the spring-mass assembly to be equal to the drive frequency. Driving at resonance results in the best motor efficiency.

The expander has a displacer assembly with a moving regenerator. The regenerator consists of a matrix (porous medium) that absorbs heat from and releases heat to the working fluid as it passes through. The cold end of the regenerator is at cryogenic temperature and the warm end can get quite hot. Therefore, an efficient regenerator must also have low axial thermal conductivity.

Dynamic clearance seals are used in three places to separate the volumes. A seal on the piston separates the crankcase volume from the working volume in the compressor. In the expander, a seal on the plunger separates the working volume from the gas spring volume. The regenerator seal forces the working fluid to pass through the regenerator. This seal also functions as the bearing for the moving regenerator assembly.

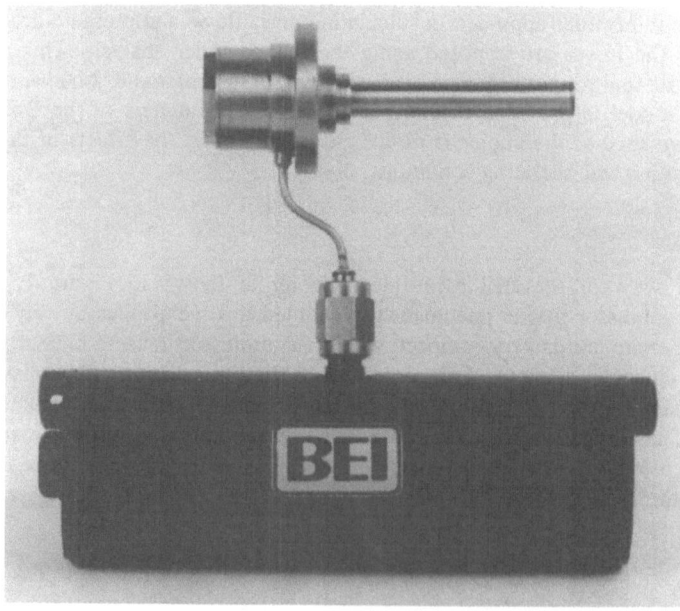

**Figure 1.** An external view of the BEI Mini-linear Cooler.

Retention of the working fluid over extended period dictates the use of electron-beam welding and brazing techniques. However to facilitate assembly, two static metal seals are used: a conical seal at the joint between the compressor and the transfer line assembly, and a C-seal between the expander housing and the warm end-cap.

## Computer Modeling

The modeling of cryocoolers ranges in degree of complexity and the resulted accuracy also varies accordingly. The First Order Analysis is an ideal cycle with no losses. By imposing losses on the ideal cycle, the Second Order is far more realistic. The Third Order Analysis, on the other hand, breaks down the cryocooler into a large number of nodes and the equations of conservation of energy, mass, and momentum are solved at each node until the solution converges. The number of nodes in each section depends on the value of the state variables. For examples, more nodes are required in the regenerator because of the large temperature difference and large pressure drop in the axial direction. Equation of states and empirical equations for pressure drop and heat transfer are also used. Among all the Third Order Models in the literature, only the Stirling Refrigerator Performance Model (SRPM) has been validated extensively against various refrigerators in the literature. They include the Lucas-Lockheed 60K unit[1], the NASA/Philips Magnetic Bearing unit[2], the Oxford refrigerators[3], and the Astronomic Infrared Sounders (AIRS) units A, B, and C. A detailed description of the model can be found in Reference 4. SRPM was also found to give accurate prediction on the performance of Pulse Tube coolers (including a blind test, Ref. 5 and 6), and is very useful in the design of Stirling refrigerators[7]. The only shortcoming of the above model is long runtime.

## The BEI Model

At BEI, a new Hybrid Model has been developed. This model is third order in fluid dynamics and second order in heat transfer. The heat transfer in a third order model requires long runtime to converge because of the finite heat capacity, whereas the dynamics only take a few cycles to reach steady state. The Third Order Model is also known for its accuracy in predicting PV loops in both the compressor and the displacer (Reference 1). The compressor PV loop allows one to calculate the input power, where the expander PV loop gives the total gross refrigeration of the cooler. Knowing the gross refrigeration, the BEI model imposes the losses as in the Second Order approach. Losses incorporated include, regenerator loss, regenerator matrix conduction loss, static losses, gas conduction loss, pumping loss, shuttle loss, blow-by loss, radiation loss, and friction loss. Since the pressure drop loss is already accounted for in the expander PV loop, it is not considered separately. The resulted model does not give up much accuracy as compared to the Third Order Model, but has a much faster turn-around time.

## RESULTS AND DISCUSSION

The normal operating conditions of the BEI Mini-linear Cooler is listed in Table 1.

**Table 1.** Normal operating conditions of the BEI Mini-linear Cooler.

| Charge Pressure | 400 PSIA |
|---|---|
| Frequency | 60 Hz |
| Max. Displacer Stroke | 0.1 inch |
| Max. Compressor Stroke | 0.44 inch |

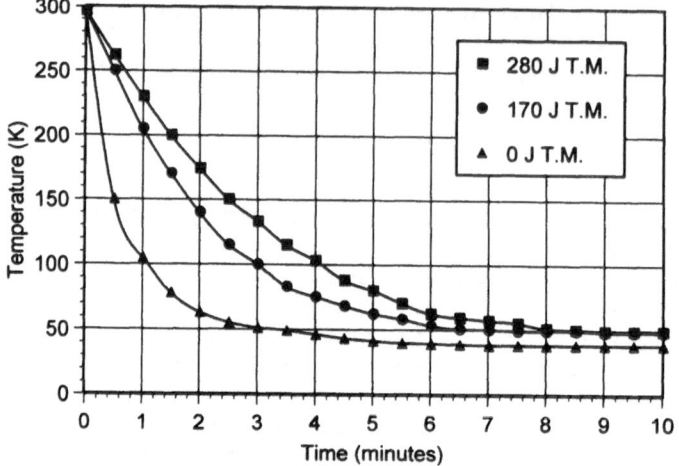

**Figure 2.** Cooldown characteristics of the BEI Mini-linear Cooler.

Figure 2 shows the cooldown characteristics of the cooler subject to various thermal masses (T.M.). Typical cooldown time to 80 K with a 250J thermal mass is around four minutes.

The modeling of the BEI Mini-linear Stirling Cooler which has a pneumatically driven displacer is far more challenging than that of a cooler with a motor-driven displacer. In the latter case, the size of the PV loop in the expansion space (which represents the gross cooling) is set by the stroke of the displacer motor, the compressor pressure wave and the phase angle between the compressor and the displacer motors. With a pneumatically driven displacer, the expansion space PV loop is also a function of the frictional force acting against the motion of the displacer. This friction term (f) was found in the present work to be a function of the compressor stroke (X), the coldtip temperature ($T_E$), and the ambient temperature ($T_A$).

$$f \propto T_A \left( \frac{X}{T_E} \right)^2 \tag{1}$$

The performance of the BEI Mini-linear Cooler (with a one-inch transfer line) at various ambient temperatures as predicted by the Hybrid Computer Model (HCM) is compared to the experimental data in Figure 3. The ambient temperature ranges from room temperature to 358 K.

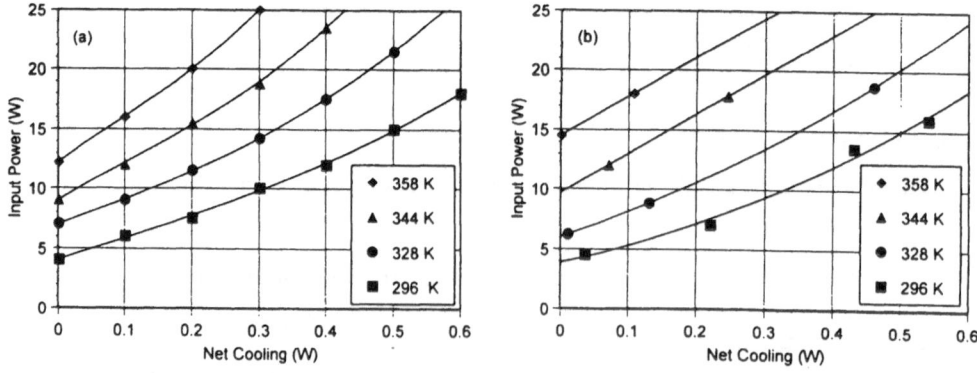

**Figure 3.** Comparison of the experimental performance of the BEI Mini-linear Cooler to the prediction of the Hybrid Computer Model; a) experimental data, b) prediction.

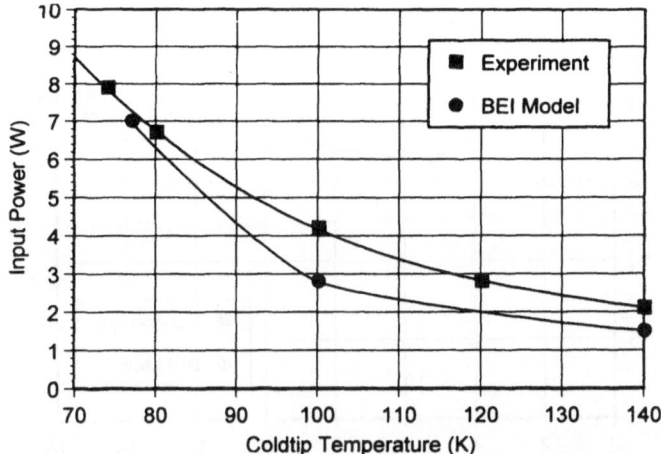

**Figure 4.** Input power (for 0.2 W refrigeration) as a function of coldtip temperature.

As the ambient temperature increases, the performance of the cooler degrades due to the limitation of heat rejection at the compressor and the increase of heat losses from the warm end of the coldfinger to the cold end. As a result, the input power required to produce the same cooling is higher at an elevated ambient temperature. By comparing Figures 3a and 3b, one can see that good agreement is found between the experimental data and HCM prediction.

Figure 4 depicts the input power of the Mini-linear Cooler (to produce 200 mW of cooling) as a function of the coldtip temperature. The input power increases exponentially as the coldtip temperature is decreased which is the characteristic of all Stirling coolers.

The effect of transfer-line length on the performance of the BEI Mini-linear Cooler is presented in Figure 5. A finite transfer-line length is required for most practical applications, which also acts as a buffer to isolate the coldtip interface from the vibration of the compressor. As the length of the transfer line increases, the dead volume in the system also increases accordingly. This results in a reduction of the pressure ratio, which in turn reduces the refrigeration power of the cooler. The prediction of the Hybrid Computer Model is also included in the figure. However, more experimental data is required to quantify the validation.

The HCM model is extremely useful for the design of mechanical coolers. Figure 6 shows the predicted input power of the cooler at 23 C ambient temperature, with a cooling load of 150 mW at 78 K as a function of regenerator length. A minimum input power is predicted by the model as too short a regenerator leads to large regenerator losses and too long a regenerator results in a large dead volume which translates into small pressure ratio and reduced cooling. The baseline length of the regenerator can be extended to enhance performance and this was verified experimentally.

BEI is in the process of developing an etched-foil regenerator to enhance the performance of the Stirling refrigerators and Pulse Tube coolers. The effect of using etched-foil regenerators and titanium coldfingers as predicted by the BEI Hybrid Computer Model is presented in Table 2. The pressure drop and heat transfer in the etched-foil regenerator is modeled as transport in parallel plates. By using a titanium coldfinger, the static conduction loss is reduced tremendously, resulting in a more efficient refrigerator (by 23%). The combination of enhanced heat transfer and reduced pressure drop in the etched-foil regenerator result in further improvement in performance (by 15%). It is noteworthy that one should be cautious in the application of etched-foil regenerators in pneumatically driven displacers, as the reduction in pressure drop across the regenerator might influence the dynamics of the displacer motion.

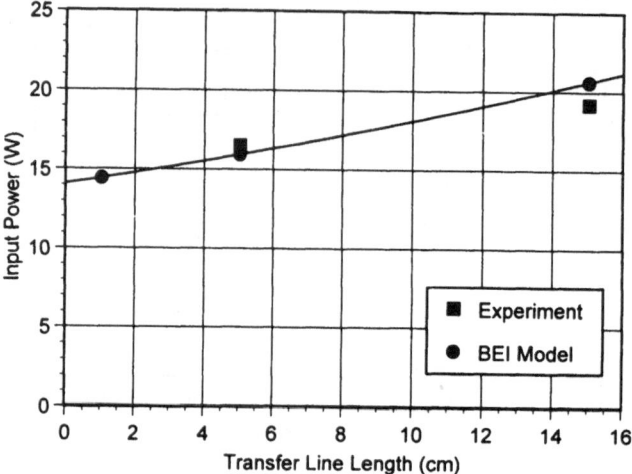

**Figure 5.** Input power (for 0.55 W refrigeration) as a function of transfer-line length.

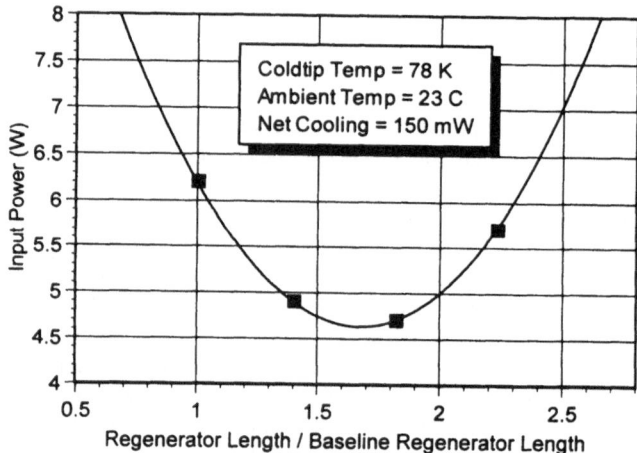

**Figure 6.** Input power as a function of regenerator length.

**Table 2.** Predicted performance of the BEI Mini-linear cooler.

| Conditions | Baseline | Baseline + Titanium Coldfinger | Baseline + Titanium Coldfinger + Etched Foil Regenerator |
|---|---|---|---|
| Ambient Temperature | 50 C | 50 C | 50 C |
| Coldtip Temperature | 73 K | 73 K | 73 K |
| Refrigeration Load | 150 mW | 150 mW | 150 mW |
| Input Power | 13 W | 10.0 W | 8.5 W |

## CONCLUSIONS

The performance of the BEI Mini-linear Cooler is discussed in this paper along with the performance predicted by a Hybrid Computer Model. This model is third order in fluid dynamics and second order in heat transfer. Since the momentum equation converges much faster than the energy equation, the resulted model is fast with short turn-around time. With accurate predictions of the compressor PV-loop and the expander PV-loop (gross refrigeration) from the third order approach, one can thus calculate the input power (from the compressor PV-loop) and the net cooling (by subtracting the second order losses from the expander PV-loop). Good agreement was found between the prediction and experimental data. The frictional force acting against the motion of a pneumatically driven displacer was found to be directly proportional to the ambient temperature and proportional to the square of compressor stroke over the coldtip temperature. The model is invaluable in optimizing the performance of Stirling refrigerators. BEI is in the process of constructing a hybrid computer model for Pulse Tube coolers.

## REFERENCES

1) Yuan, S.W.K., Spradley, I.E., Yang, P.M., and Nast, T.C., "Computer Simulation Model for Lucas Stirling Refrigerators," *Cryogenics*, vol. 32, (1992) p.143.

2) Yuan, S.W.K., and Spradley, I.E., "Validation of the Stirling Refrigerator Performance Model Against the Philips/NASA Magnetic Bearing Refrigerator," *Proc. 7th Int. Cryocooler Conf.*, vol. 1, Phillips Lab, USA (1993) p.280.

3) Yuan, S.W.K., Spradley,I.E., and Foster, W.G., "Validation of the Stirling Refrigerator Performance Model Against the Oxford Refrigerator," *Advances in Cryogenic Engineering*, vol. 39, (1994) p.1359.

4) Yuan, S.W.K., and Spradley, I.E., "A Third Order Computer Model for Stirling Refrigerators," *Advances in Cryogenic Engineering*, vol. 37 (1992) p.1055.

5) Yuan, S.W.K., and Radebaugh, R., "A Blind Test on the Pulse Tube Refrigerator Model", to be published in *Advances in Cryogenic Engineering*, 1996.

6) Yuan, S.W.K., "Validation of the Pulse Tube Refrigerator Model Against a Lockheed Built Pulse Tube Cooler", to be published in *Cryogenics*, 1996.

7) Yuan, S.W.K., Naes, L.G. and Nast, T.C., "Prediction of Natural Frequency of the NASA 80 K Cooler by the Stirling Refrigerator Performance Model", *Cryogenics*, vol. 34, No.5, (1994) p.383.

# Space Qualification Test Plan Development, Implementation, and Results for the STRV-2 1.0-watt Tactical Cryocooler

**K. S. Moser**

Nichols Research Corporation
Albuquerque, NM, USA, 87106


**T. P. Roberts**

United States Air Force, Phillips Laboratory
Kirtland AFB, NM, USA  87117


**R. M. Rawlings**

Texas Instrument Corp.
Dallas, Texas, USA 75243

## ABSTRACT

The BMDO Materials and Structures' STRV-2 (Space Test Research Vehicle #2) is the primary set of experiments which are manifested for flight on the Space Test Program's TSX-5 (Tri-Service eXperiment # 5) mission in July 1998. The Texas Instrument's (TI) 1.0 watt tactical cryocooler was selected to cool the focal plane on the STRV-2 MWIR. A rigorous series of flight qualification tests were developed and conducted on a set of four of these cryocoolers (and associated control electronics cards) in order to characterize cooler performance under the stressing environments of launch and space operations. This paper discusses the development of the test plan, the test stand and instrumentation, and environmental controls. Specific data from the test series (induced vibration, thermal performance mapping, EMI/EMC, random vibration and thermal cycling) are presented, and compared with mission performance requirements. Factory test data (which is designed to qualify the cooler for ground-based tactical systems) relevant to the space qualification effort are also presented. The ability of the Texas Instrument's cooler to perform under flight conditions for limited duration missions is confirmed by these test articles' robust behavior during the characterization and environmental test program.

Cryocoolers 9, Edited by R.G. Ross, Jr.
Plenum Press, New York, 1997

## INTRODUCTION

The cooler will provide 625 milliwatts of cooling at 80K for the Mid-Wave InfraRed (MWIR) experiment on the STRV-2 experiment module which is manifested for flight aboard the Space Test Program's TSX-5 (Tri-Service eXperiment) spacecraft. Initial launch capability for TSX-5 is planned for 3rd calendar quarter 1998. Nominal insertion will provide a 460 X 1750 nmi orbit at 70° inclination. Four coolers were procured for the STRV-2 program and assigned particular functions: 1) an engineering unit for use in MWIR development work, 2) a life-test cooler, 3) a primary flight cooler, and 4) a spare flight cooler. Following relevant testing, three of the four coolers were delivered to the MWIR experimenter (engineering unit: Jan. 1995; flight unit: Jan. 1996; flight spare: May 1996), with the life-test cooler remaining under test at the Air Force Phillips Lab's (AFPL) Cryocooler Testing and Characterization Facility (PL/VTPT) at Kirtland Air Force Base.

## COOLER DESCRIPTION

Figure 1. depicts the STRV-2 version of the Texas Instruments 1.0 watt linear cooler, drive electronics and typical cooling performance. The STRV-2 configuration is virtually identical to the cooler which TI delivers to its tactical customers with the following exceptions: The STRV-2 configuration has the opposing compressors wired out separately, it has a custom-designed transfer-tube geometry, the cold-finger side of the expander flange has a special surface finish, and the drive electronics were delivered without conformal coating. Moser, Das and Obal[1] provide a detailed review of the decision process which resulted in the selection of this cooler for STRV-2 and the rationale for STRV2-specific modification to the TI cooler. The cooler requires conditioned power between 17 and 32 volts. The conductively cooled drive electronics dissipate approximately 20% of the total power used; the compressor uses 60%, and the expander dissipates the remaining 20%.

## FLIGHT ACCEPTANCE TESTING

The flight acceptance test flow for each of the four coolers is depicted in Figure 2. There are two primary purposes for these test: 1) to verify the coolers' acceptability for flight, and 2) to characterize the cryocooler so that its interfaces with the STRV-2 module could be properly designed. As will be shown later, the coolers successfully passed the environmental screening and the cooler characterization helped define the MWIR experiment power conditioning circuitry design and the cooler/detector operational profile. Likewise, factory testing at Texas Instruments which supplements the test flow shown in figure 1. will be discussed separately.

### Induced Vibration Characterization

Induced vibration characterization was conducted by Johnson and Collins[2] at JPL on a six axes dynamometer. JPL's 6 degree of freedom dynamometer has a force sensitivity of 0.005 N and a full-scale range of 445 N between 10 and 500 Hz. The STRV-2 cooler configuration has collinear expander and compressor axes. Following convention, this axis is designated as the z-axis. The x and y axes are radial and arbitrary. The JPL data acquisition system simultaneously collects the x,y and z force data and the moment about the cooler's z-axis.

For the test, in order to simulate the coolers physical interface to the MWIR experiment, the expander and compressor bodies were rigidly linked by an interface fixture. This fixture, in turn is mounted to the dynamometer. Figure 3. reports the typical vibration produced by the coolers. Because the 2.0 N peak force is higher than desired (but not than expected), the MWIR

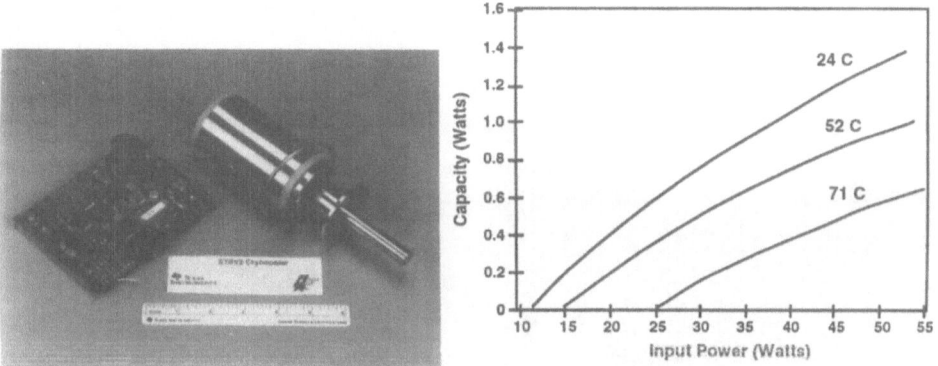

**Figure 1.** TI 1-watt linear tactical cryocooler for STRV-2 and typical cooling performance.

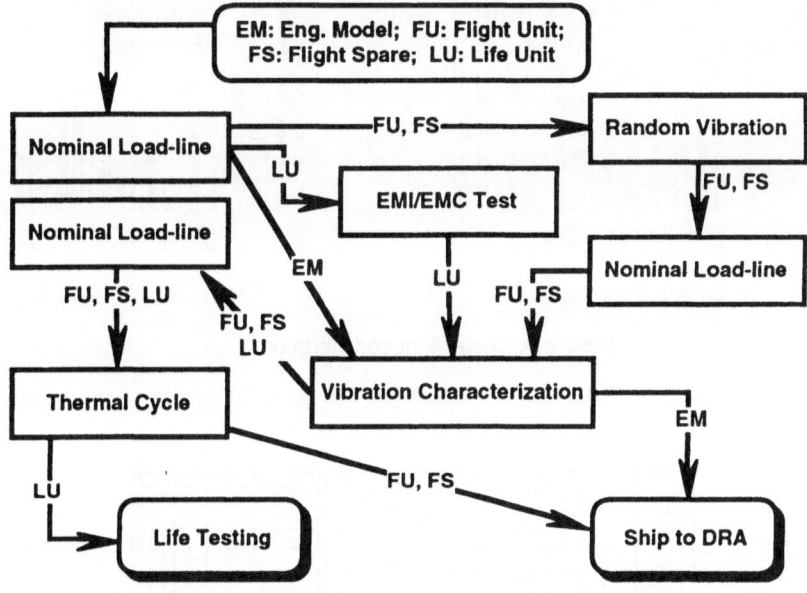

**Figure 2.** STRV-2 cooler flight acceptance test flow.

experiment incorporates a copper mass as part of it focal plane assembly. The copper will provide sufficient thermal capacitance to allow shutting down the cooler during image acquisition (10 to 20 seconds).

### Random Vibration Testing

Random vibration testing was only conducted on the primary flight unit and flight spare. Both of these coolers were subjected to the spectrum shown in figure 4., though random vibration testing of the flight article was conducted by Roberts and Tomlinson[3] and the flight spare was tested by Johnson, et. al.[4] at JPL. The coolers and electronics were tested for 3 minutes in each

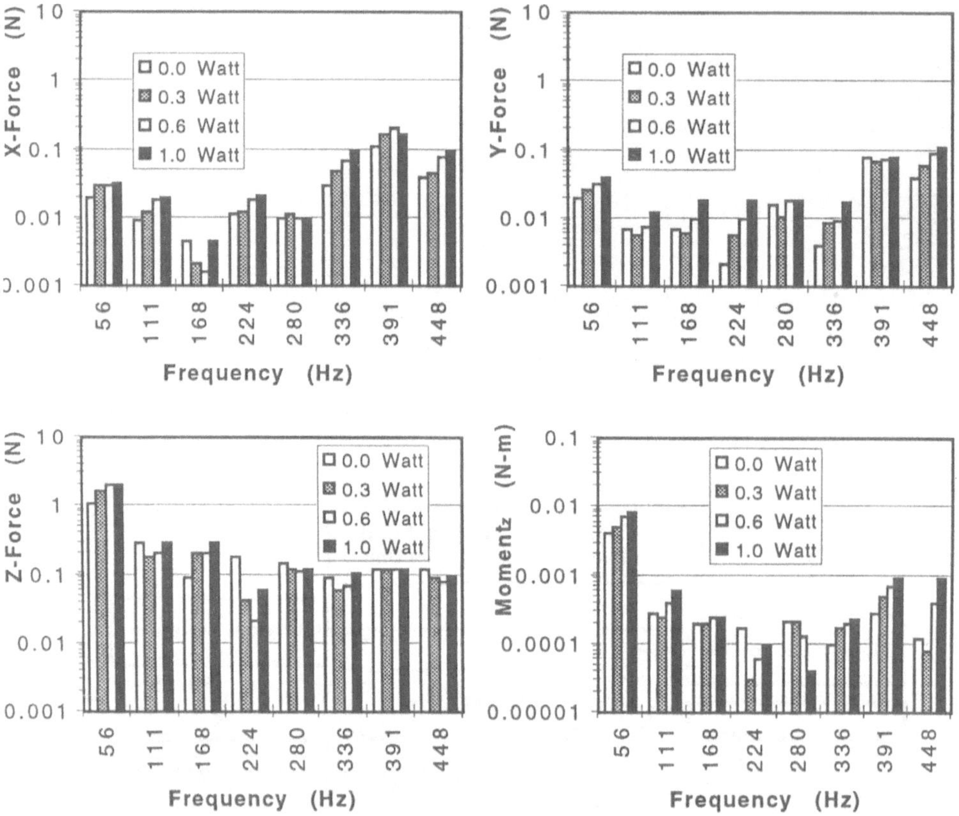

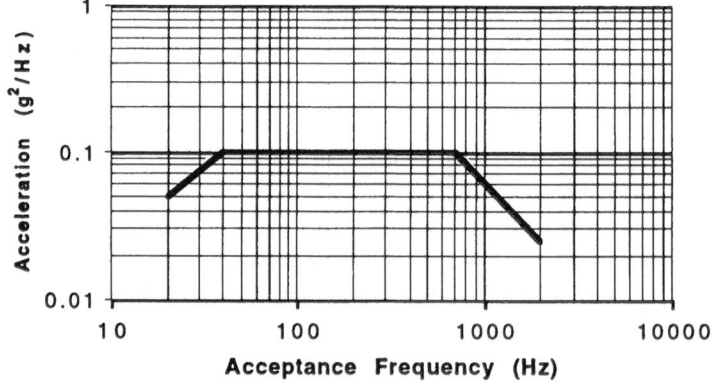

**Figure 3.** Cooler-induced vibration

**Figure 4.** Cooler component-level random vibration spectrum.

of 3 axes. It is worth noting that factory qualification testing included 4 grms random vibration testing on an operating cooler for durations of 1-hour in each of 3 axes. Following the random vibration tests, the flight and spare coolers' pre-test and post-test thermal performance were compared (see figure 5) to verify acceptance.  The apparent improvement in performance following random vibration is within instrument error.

### Electromagnetic Compatibility Testing/Characterization

EMIEMC tests were conducted at JPL during the last week of July 1995 and are reported by Johnson, et. al.[5]  EMI testing was conducted on the life-test cooler in accordance with the following MIL-STD-461 tests:  CE01/03, CS01/02, RE01/02, and RS03 per STRV-2 program requirements.  At the time that these tests were conducted, the power conditioning circuitry for the MWIR experiment (which provides bus power to the cooler) was not defined. Hence, testing was done with the cryocooler drive electronics (CDE) directly attached to simulated raw bus power.  This unfiltered power represents a much harsher environment than the CDE will ever be subjected to in flight where the CDE board will be buffered from the powerbus by MWIR Power Conditioning Unit (PCU).  In fact, the early EMI test results facilitated the design of a PCU which mitigates the susceptibility and emission issues revealed during the test sequence.  The following discussion of the test result only addresses those areas where the cooler had susceptibility or emission issues.

**Susceptibility.**  The most significant issue uncovered by the testing is that the cooler drive electronics are susceptible to high frequency voltage transients on the power line.  Susceptibility begins at about 900 khz at a 150 mV (peak-to-peak) amplitude.  At this point the cold-tip set point begins to change.  The degree of change in both set point and set point stability appeared to get worse as line voltage ripple frequency was increased beyond 900 khz.  This susceptibility was easily mitigated by incorporating a power converter unit (in the MWIR PCU) which filters line noise.

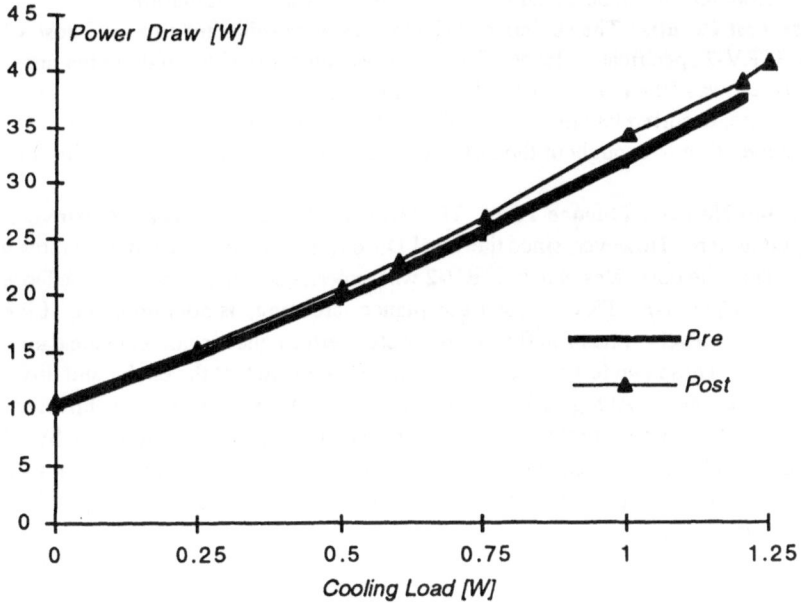

**Figure 5.** Flight unit performance before and after vibration test.

**Conducted Emissions.** The CDE exceeds the MIL-STD-461C levels for both current ripple and voltage ripple at high frequency (> 1 Mhz) by about 20 dB. The emission below 1 Mhz were within spec. These high frequency exceedances will be reduced to spec-levels by filters in the MWIR PCU.

**Radiated Emissions.** RE01: The cooler emits radiated peaks at the drive frequency of 56 hz and its harmonics. Specification exceedances were seen at harmonics to about 2000 hz. The worstcase specification exceedance was by about 10 dB-picoteslas at 56 hz. (102 dB-picoteslas measured, 92 allowed). RE02: The cooler passed the RE02 broadband test, but had a minor exceedance during the RE02 narrowband test at 46 Mhz. At this frequency the cooler produced a narrow spike of magnitude 47 dB μV/m vs. the spec value of 35 dB μV/m. Although the level is out-of-spec, the measured field strength is only 0.2 mV/m which is extremely benign when compared with the 10 V/m field which all other STRV-2 instruments are required to tolerate. In fact, there is a 93 dB margin of safety between the cooler emission at 46 Mhz, and what other STRV-2 equipment will be designed to tolerate.

## TEXAS INSTRUMENTS FACTORY TESTING

### Radiation Testing

The most relevant factory testing to the flight acceptance profiling was the radiation dosage testing conducted on two 1.0 watt coolers and drive boards with the same configuration as for STRV-2. Testing was accomplished at three separate facilities: Total Dose testing at the Texas Instrument Expressway facility in Dallas using a Nordion Cobalt-60 Gammacell 220 Irradiator; Dose Rate testing at the White Sands Missile Range Fast Burst Reactor; and Neutron Fluence testing at the Defense Nuclear Agency's Aurora Test Facility.

During the course of its one-year mission, the cooler and drive board will receive approximately 16krads total dose (this spec-level has a built-in Radiation Design Margin [RDM] of 2). The exact levels to which the cryocoolers/electronics were tested is sensitive and cannot be reported here; however, it can be stated that the board was tested with margin.

**Total Dose Test Results.** The coolers and electronics were subjected to a total dose which exceeded the STRV-2 specification level. The cooler was operational throughout the test. Dose rate during this test was 10.8 rad (Si)/sec. The only measurable effect, a 0.45% (5 millivolt) shift in set-point voltage, occurred late in the test. This voltage shift corresponds to about a 3K shift in set-point temperature. Throughout the test, the cooler successfully performed within required values.

**Dose Rate and Neutron Fluence Tests.** The levels for these two tests are sensitive and cannot be reported here. However, since the Total Dose test was conducted at dose rates which significantly exceed the dose rates which STRV-2 will experience, the omission of the Dose Rate test data, for the purposes of STRV-2 cooler acceptance screening, is not important. Likewise, the effects of various levels of neutron fluence on cooler performance is not requisite data for STRV-2 screening. These two test are mentioned only to indicate that the cooler and drive electronics are quite robust. At higher levels, in both of these tests, the cooler manifested a shift in set point temperature similar to that reported in the Total Dose testing. Additionally, at higher dose rates, both test units experienced a momentary current drop-out (less than two seconds). Nevertheless, at all test levels, the cooler continued to operate within specified performance requirements.

## MEASUREMENT OF THERMODYNAMIC PERFORMANCE

Thermodynamic performance was measured at the Air Force Phillips Lab's Cryocooler Characterization Facility (PL/VTPT). The cooler was integrated into a 24" vacuum chamber using the temperature and power sensors indicated in Figure 6. The thermocouples were used primarily to monitor thermal mount and cooler casing temperatures, while the silicon diode was used to monitor cold end temperatures. The TI PID control system also uses another silicon diode to implement a feedback control algorithm which alters the cooler's mechanical operation in order to keep the cold end at the desired temperature. The cooler's performance is monitored by a data acquisition (DAQ) program which polls (via GPIB network) the instrumentation listed in Table 1 on a 10-15 second data gathering cycle during transient and 1 to 10 minute cycle during steady state operations.

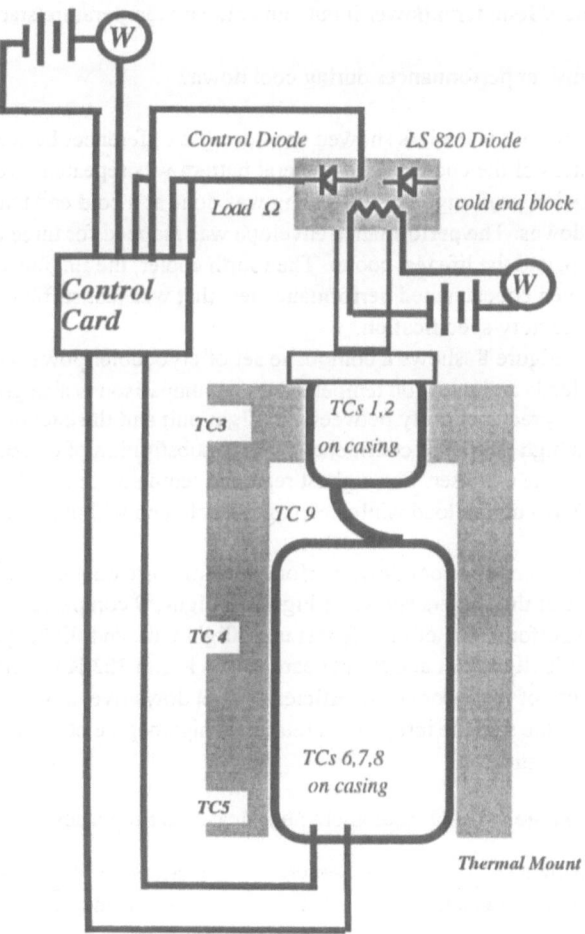

**Figure 6.** Temperature sensor placement.

The 24" chambers are equipped with turbo vacuum pumps capable of maintaining the chamber vacuum in the $10^{-6}$ torr neighborhood during sustained operations. The common vacuum conductance loop allows for some transient cryopumping to occur when other chambers are taken on or off the conductance loop. Each cooler thermal mount is cooled with a fluid cooling line maintained at a stable temperature by a fluid chiller serving the 4 chamber cooling loop. The amount of cooling fluid delivered to each chamber is controlled by proportional gate valves and preconditioning heating elements adjust the fluid temperature entering the thermal mount. Figure 7 shows the test chamber and a view of the instrumented cooler and card interfaced with the fluid chiller system.

### Acceptance and Characterization Profiles

The primary mapping comparisons between these coolers can be made with the following criteria:

1. What are their power input dependencies on cooling load and rejection temperature?
2. What are the coolers' long term power input and cold end temperature stability characteristics?
3. What are their transient performances during cool down?

The acceptance profiles of the coolers showed considerable differences between the nominal design point performances of the coolers. This general pattern was repeated through the subsequent characterization profiling. All the testing was done at a cold end temperature of 80 K, except during cool downs. The performance envelope was mapped for three coolers: the flight cooler, the flight backup, and the life test cooler. The fourth cooler, the Engineering Model, received a schedule-driven concatenated performance test that was just sufficient to verify that it was functioning within factory specification.

**Load Line Results.** Figure 8 shows a composite set of cryocooler power consumption for a given range of cooling loads and rejection temperatures. A comparison is also given with the mission profile unit. The great similarity between the flight unit and the backup unit's performance provides a high degree of confidence that the substitution of one unit for the other can be done if needed by the end user. The highest rejection temperature load line, 323 K, could not extend above a 1.15 W cooling load while keeping the cold end within the required 79-81 K range.

**Cool Down Performance.** The cool down performance of these coolers is closely related to the relative performance of the coolers shown in Figure 8. Figure 9 compares the 296 K rejection temperature cool down performance of the life test unit, flight unit, and flight spare. Similar cool down performances of the flight unit and flight spare at 255 K and 323 K rejection temperatures were obtained. The ability of these coolers to efficiently cool down over a wide range of rejection temperatures should give the satellite integrator a relatively high degree of flexibility concerning transient temperature operations.

**Table 1.** DAQ Equipment and Measured Properties

| Equipment Item | # Sensors | Measured Property | Accur. |
|---|---|---|---|
| Lakeshore 820 Cryo. Thermometer | 1 | Cold End Temp. | ±.1 K |
| Fluke 2620A Multimeter (Thermocouple Measurements) | 9 | Casing and Mount Temperatures | ±.7 K |
| Vahalla 2300/2300L | 2 | Cooling Load & Cooler Power | ±1% full scale values |

**Figure 7.** 24" Thermal Vacuum Chamber and instrumented cooler.

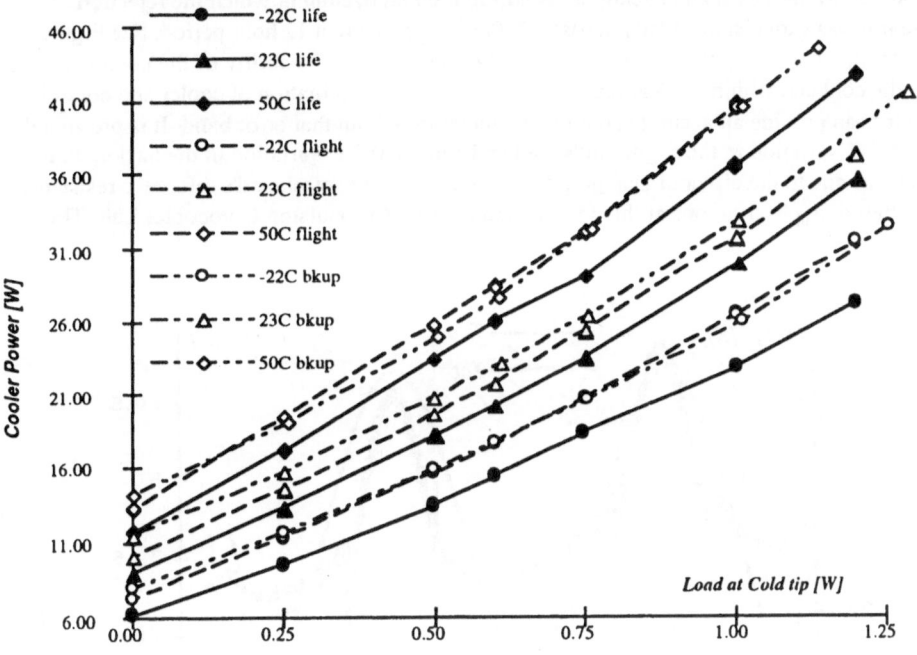

**Figure 8.** Power draw as function of reject temperature and cooling load.

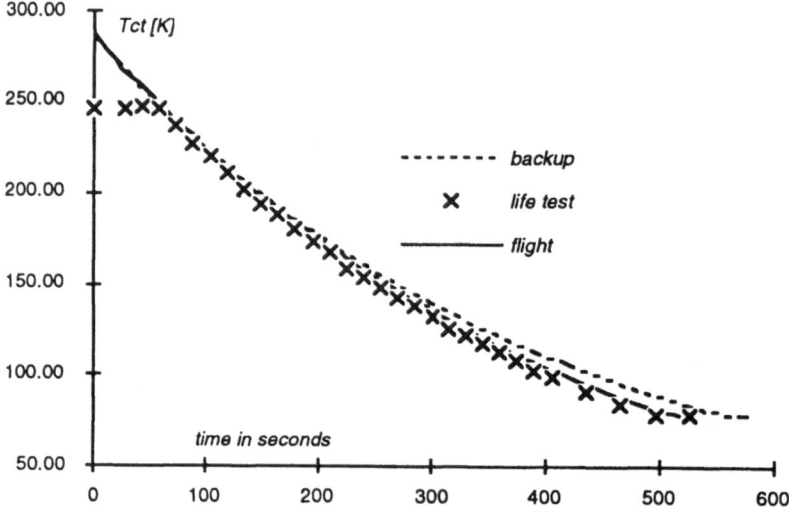

**Figure 9.** Cool down performance at 296 K reject temperature.

**24 Hour Stability and Thermal Cycling.** The ability of the cooler system to provide a stable cold tip temperature under a constant load of 0.6W at a rejection temperature of 296K was demonstrated for the flight and backup units during thermal cycling in which the rejection temperature was varied sinusoidally across a 230-340 K range in a 12 hour period. See Figure 10 for the stability shown by the flight unit over a 24 hour period. The ability of the cooler system to keep the cold end within a 1.5 K range indicates that the combination of cooler and control electronics can provide an accurate cold end temperature within that error band. It is presumed that the 1.5 K variation in the flight unit's cold end and the 0.5 K variation in the backup unit's cold end was due to the effect of changing temperature on the control card's reference resistance circuits, based on observations at this facility and at the Jet Propulsion Cryocooler Lab. The

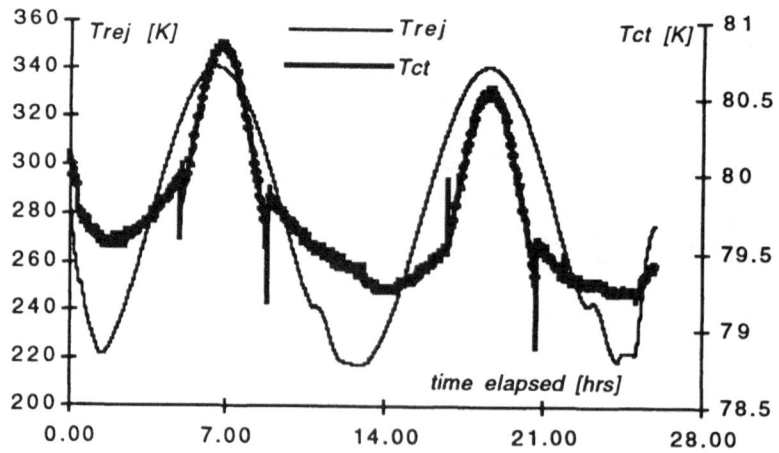

**Figure 10.** Cold tip stability of flight unit during thermal cycling.

difference in these bands for the flight unit and for the backup unit would therefore be due to the greater amplitude of the flight unit's rejection trace (about 10 K) and the interaction of the card's temperature with cold tip temperature control resistors.

## LONG TERM ENDURANCE PROFILING

The typical orbital operation consists of cooler turn-on and operation for approximately 45 minutes prior to an MWIR data collection opportunity. This provides ample time under worstcase conditions for the cooler to bring the focal plane to temperature. Seconds prior to the MWIR data-take, the cooler will be shutdown in order to reduce image jitter. The cooler will remain quiescent until it is turned on in preparation for another MWIR operation (up to two per day). Thus, for a 1-year mission, the cooler is nominally expected to see less than 600 hours of on-orbit operations, and a similar number of activations. Ground hours on each of the coolers will be limited to well under 400 hours. Hence, the nominal total operating life for the cooler in support of the STRV-2 mission is less than 1000 hours. The cooler is rated by Texas Instruments for a 4000 hour MTTF with a 1.0 watt load at 80K. In establishing the 4000 hour MTTF, TI deems the cooler to have failed when it is no longer able to cool a 1.0 watt load to 80K. Failure criteria for the STRV-2 endurance test is somewhat more generous as its nominal load is 625 milliwatts. Since the common failure mode in this family of coolers is graceful degradation, the cooler can be expected to provide required cooling for more than 4000 hours. Nevertheless, for the purposes of STRV-2 success criteria, the cooler has been derated to 2000 MTTF.

Endurance profiling at AFPL/VTPT for this cooler family is being made at the nominal mission cooling load of 0.6 watts. Additionally, both of the flight and back up cooler control cards were subject to 300 hour burn-in testing using a RCL circuit to simulate the cooler. During the burn in tests the cards exhibited very stable power draw and thermal characteristics. During the endurance profiling the rejection temperature was allowed to vary over the general mission rejection range of 250-325 K. The results of this operating regimen is shown in Figure 11 which shows the variation in the Coefficient of Performance (COP) with rejection temperature during a one week period of operation with 3 two hour shutdowns (15-21 April 1996).

The most notable aspect of this operating performance envelope is the width of the performance envelope. This is largely due to the short term degradation of performance over a 24-36 hour period. Immediately after a 2 hour shutdown and cool down from ambient, the COP will rise by 0.001-0.002, which is approximately the width of the performance band. This short-term degradation was qualitatively predicted prior to the test based on a contamination sublimation mechanism. That this short term effect is due to contamination sublimation is further confirmed when the effect "anneals out" after the cold finger is allowed to warm. Though this contamination-induced performance loss may be critical to scenarios requiring continuous cooling over long periods, the STRV-2 operational scenario consists of short-term operations followed by long quiescent period which reset the cooler's COP.

Beside this short term phenomena, no long term degradation has been noted during approximately 1200 hours of semi-continuous operation (46-70 hours on, 2 hours off).

## CONCLUSIONS

Although the Texas Instruments 1-watt split linear cryocooler was designed for use in terrestrial tactical systems, the successful flight acceptance tests and supplementary factory tests conducted on the cooler have demonstrated that the cooler is qualified for use on limited duration experimental space missions generally, and for the STRV-2 mission specifically.

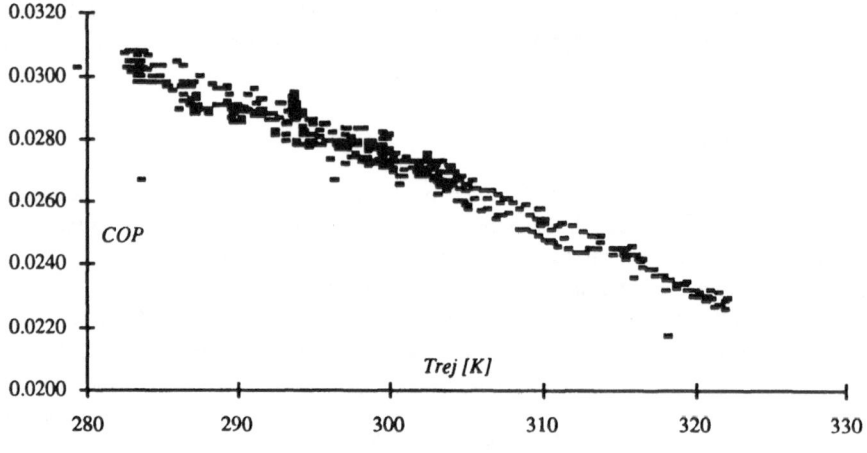

**Figure 11.** COP variation during endurance profiling.

## ACKNOWLEDGMENTS

The authors express gratitude to Dr. John Stubstad, the BMDO Materials and Structures Program Manager, for his sustaining support, timely guidance and encouragement; and to the UK/MOD for their assistance in defining flight requirements for the STRV-2 cooler.

## REFERENCES

1. Moser, K.S., Das, A., and Obal, M.W., "The Qualification and Use of Miniature Tactical Cryocoolers for Space Applications", *Cryocoolers 9*, Plenum Press, New York (1996).
2. Johnson, D.L., Collins, S.A., Heun, M.K., and Ross, R.G. Jr., *Texas Instruments 1-Watt Cryocooler Performance Characterization*, JPL D-13623, Jet Propulsion Laboratory, Pasadena (1996).
3. Roberts, T., and Tomlinson, B., *STRV-2 Flight and Backup Units Acceptance and Performance Characterization Mapping Results- Interim Report*, Phillips Laboratory, Kirtland AFB (1996)
4. Johnson, Ibid.
5. Johnson, Ibid.

# Reduction of Surface Heat Pumping Effect in Split-Stirling Cryocoolers

A. A. J. Benschop, F. C. v. Wordragen and P.C. Bruins

Signaal-USFA
Eindhoven, The Netherlands

## ABSTRACT

Many applications of Stirling cryocoolers require space-separation of oscillator and cold finger. Commonly a split pipe with a small volume is used to accomplish this spatial separation. The separated mounting may be necessary for purposes of positioning or remote removal of abundant compressor heat.

Even in a split Stirling cooler, however, the warm end of the cold finger is inevitably found in the temperature-critical part of the application. In order to prevent conductive heat leak to the cold side of the finger, it is important to limit the temperature of the warm end. Therefore, it is important to either assure sufficient heat rejection from the warm end or to prevent the flowing of excess heat towards the warm end.

In a theoretical analysis of enthalpy flows in a split-Stirling cryocooler it is shown that an important contribution to the heat flow towards the warm end stems from surface heat pumping in the split pipe. A reduction of the enthalpy flow associated with surface heat pumping thus has the desired effect of decreasing the total heat load of the cold spot. A formulation is discussed that can be used to estimate the excess heat that has to be discarded at the warm end due to heat pumping in the split pipe.

Based on experiences obtained in the modeling of pulse-tube cryocoolers, a patent-pending solution is proposed to reduce the heat-flow through the split pipe by establishing a critical temperature gradient over the pipe. Experimental verification of the concept is given in measurements performed on Signaal-USFA 80K, ½ watt and 1 watt linear coolers.

## INTRODUCTION

USFA delivers a complete range of linear split Stirling coolers with cooling performances between 0.25 and 2.0 Watts, operating at ambient temperatures between -52°C and 71°C. A schematic representation of a linear split Stirling cooler is given in Figure 1.

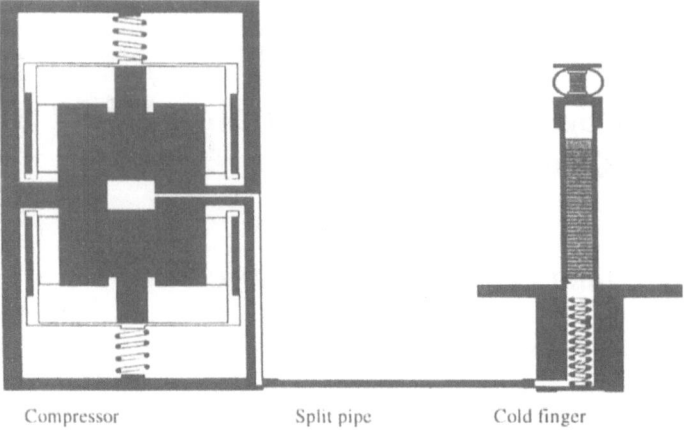

Compressor                    Split pipe                    Cold finger

**Figure 1**. Schematic view of split Stirling cooler

In a split Stirling cooler, the compressor and cold finger are spatially separated by means of a split pipe. The split pipe is a thin-walled stainless steel tube, which is easy to bend and therefore offers relatively free positioning of the compressor and the cold finger in an application. The compressor and cold finger can be placed up to 300 mm apart, and may have their axes in different directions for mechanical and noise-reduction purposes.

An important issue in the application of cryocoolers is the heat-sinking of both the compressor and the warm end of the cold finger. In the cooler, heat is produced as the electrical input energy is transformed to useful cooling power. This excess heat should be radiated or conducted through the housing of the system to the surroundings. A temperature rise of the cold finger has a dramatic effect on the performance of the cooler, because the increased conduction adds to the heat load of the cold spot. Especially in applications where the cooler is operated at high ambient temperatures this increased heat load is often problematic. In the specification of the cooler it is therefore stated that both the compressor and the warm end of the cold finger should be heatsinked in such a way that the skin temperature of the cooler does not exceed the ambient temperature by more than 10°C.

In many applications of split Stirling coolers, heat-sinking of the warm end is most problematic due to mechanical and spatial constraints. The amount of heat that has to be removed at the warm end can be reduced by introducing a *heatstop* in the split pipe. The heatstop prevents the flow of abundant compressor heat towards the warm end by eliminating surface heat pumping which occurs in the split pipe.

## ENERGY BALANCE OF THE COOLER

In order to determine the heat load on the warm end of the cryocooler, an analysis is made of the energy balance of an operating cooler. In Fig. 2, the energy exchange with the surroundings is depicted (neglecting radiation from the cooler surface). The total power input into the system is the sum of electrical input to the compressor and generated cooling power. It is clear from Fig. 2 that in a steady state situation the same amount of energy has to be removed via the two (parallel) heat sinks of both compressor and cold finger.

### Measurement of the energy balance

To determine the distribution of the heat flow over the two parallel heat sinks, a copper H-profile is used. As visualized in Fig. 3, the oscillator heat sink is connected to a temperature controlled test bed. The cold finger heat sink is mounted on the copper H-profile. A thermocouple bridge is used to measure the temperature gradient in the narrow part of the H-profile.

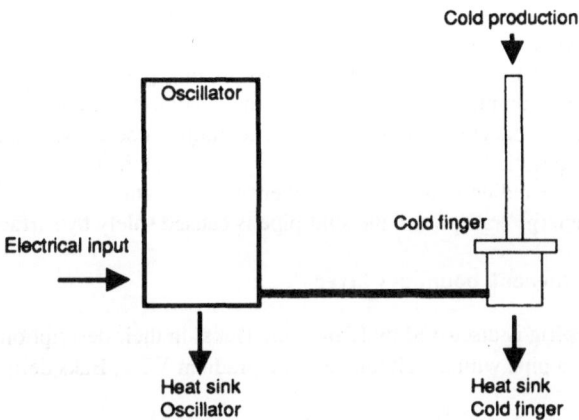

**Figure 2**. Energy balance of the complete cooler

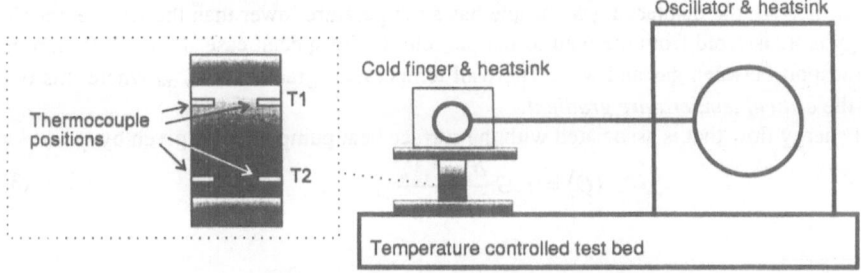

**Figure 3**. Set-up for measuring heat removal

With the described setup, the heat flow through the H-profile can be found from the temperature difference between the thermocouples as:

$$\dot{Q} = \lambda \cdot A \cdot \frac{dT}{dx} \tag{1}$$

with $\lambda$ the heat conductivity and A the cross sectional area of the copper profile, respectively.

It is found that without proper heatsinking (radiation only), at an input power of 55 watts the temperature of the warm end rises to 77° C (at an ambient temperature of 20° C). At 71° ambient, the warm end reaches temperatures up to 138° C.

Measurements with the described setup reveal that for a UP 7050 cooler (600 mW cooling power, Ø 5mm cold finger), 30% of the compressor input power has to be removed at the warm end of the cold finger. The ratio remains constant over a large range of input powers. The energy which is generated at the warm end because of the thermodynamic stirling cycle is in the order of a few watts (the undisturbed cooling power times the temperature difference). Therefore it is concluded that there is a transport of energy through the split pipe towards the cold finger.

## ENERGY TRANSPORT THROUGH THE SPLIT PIPE

If the cooler is operated at its maximum input power of 60 W, the heat flow through the split pipe is in the order of 15 W. Conduction through the thin-walled split pipe can not account for this heat flow. Obviously, the oscillating pressure and movement of gas particles inside the split pipe provide another energy transport mechanism.

In literature, energy transport in a pipe with an oscillating gas flow is described with two mechanisms. Surface heat pumping is used to explain the generation of cooling power in the basic pulse tube[1] and in thermoacoustic coolers[2]. In orifice pulse tube coolers, adiabatic enthalpy transport[3,4,5] is held responsible for the transport of energy through the tube.

Because the split pipe radius is less than the thermal penetration depth of approximately 1 mm, it is believed that energy transport in the split pipe is caused solely by surface heat pumping.

### Enthalpy transport in thermal boundary layer

**Surface heat pumping** is discussed by Liang[3] and Baks[4] in their description of the basic pulse tube. Considering a pipe with a wall temperature gradient $\nabla T_w$, Baks defines the factor G as:

$$G = 1 - \nabla T_w . \frac{\Delta x}{\Delta T} \qquad (2)$$

In this description, $\Delta x$ is the distance that a gas particle travels during a cycle, and $\Delta T$ is the temperature change it experiences. If the ratio $\frac{\Delta T}{\Delta x} > \nabla T_w$ or G<0, energy is transferred from the gas to the wall. If G>0, the displaced gas particle has a temperature lower than the wall temperature, and energy is transferred from the wall to the particle. In the special case where G=0 there is no energy transport between gas and wall. The wall temperature gradient $\nabla T_{w\ crit}$ where this occurs is called the *critical temperature gradient*.

The total energy flow that is associated with the surface heat pump effect is given by

$$\langle \dot{Q} \rangle = f.G. \frac{\delta_t . \tilde{p}^2 . V_b}{\gamma . a . p_0} \qquad (3)$$

Where:

$\delta_t = \sqrt{\dfrac{k}{\rho_0 . c_p . \omega}}$ thermal boundary layer thickness $\qquad \gamma = \dfrac{c_p}{c_v} = 1.66 \qquad$ for helium

$G = 1 - \nabla T_w \dfrac{\Delta x}{\Delta T}$ $\qquad\qquad\qquad\qquad\qquad V_b = \frac{\pi}{4}.D^2.L \qquad$ tube volume

$a$ = tube radius $\qquad\qquad\qquad\qquad\qquad\qquad\qquad f$ = frequency

This formulation reveals once more that for G<0, the heat flow has a negative sign and thus heat is pumped from the cold to the warm end of the tube. For G>0 heat is pumped towards the cold end, and for G=0 the heat pump effect is zero.

## REDUCTION OF THE ENERGY TRANSPORT

At start up, there is no temperature gradient over the split pipe, and thus surface heat pumping causes a transport of energy towards the warm end of the cold finger. This heat flow causes a temperature rise of the warm end. The temperature will increase until an equilibrium is reached between the supplied heat and the heat that is rejected from the warm end by means of conduction or radiation. From Eq.'s (2) and (3) it may be stated that the heat flow towards the warm end reduces if the temperature of the warm end increases. This leads to the following conclusions:

- if the warm end is not heat sinked, i.e. there is no conductive or radiative removal of energy from the warm end, the temperature of the warm end will increase until the heat flow through the split pipe is zero. This implies that the temperature gradient over the split pipe is just the critical temperature gradient.
- if a perfect heat sink is applied, the warm end will be exactly at the ambient temperature, which is the lowest temperature achievable. In this case, there is maximum energy transport from the compressor to the cold finger.

**Table 1.** typical split pipe skin temperatures

| with heatsink | T _osc | T$_1$: exit oscillator | T$_2$: before regen. | T$_3$: after regen. | T$_4$: warm end |
|---|---|---|---|---|---|
| Normal situation | 30.1 | 41 | -- | 54 | 33 |
| + Regenerator | 35.6 | 66 | 80 | 17 | 24 |
| without heatsink | | | | | |
| Normal situation | -- | 53 | -- | 75 | 77 |
| + Regenerator | -- | 71 | 89 | 25 | 44 |

## GAIN IN COOLER PERFORMANCE

In addition to the described beneficial effect of the lower warm end temperature, the introduction of a regenerative matrix in the split pipe has the disadvantage of obstructing the gas flow through the split pipe. In practice, the regenerator used for the heatstop is thus a tradeoff between a dense matrix (good regenerative properties) and a stack with relatively large maze openings (low pressure drop). The optimal gauze characteristics can be different for different types of cooler. In general, the effect of the obstruction is of key importance in high-capacity (> 1 W) coolers. These coolers have a large massflow through the split pipe, and the impact of a decrease of the pressure wave in the warm end leads to a relatively large reduction in cooling power. The 250 mW cooler range is less sensitive to a decrease in pressure wave. Moreover, there is less massflow through the split pipe, and thus the pressure loss over a given obstruction is smaller. Therefore, the optimal matrix characteristics and regenerator dimensions will differ from coolers in the higher capacity range.

### Measurements on a UP 7050 cooler

In order to verify the theoretical predictions, several heatstops have been tested in the split pipe of a UP 7050 cooler. The UP 7050 has a Ø 5mm cold finger, which is the smallest available at USFA. For this type of cooler, a reduction of the pressure wave with 1 bar reduces the cooling power with approximately 150 mW. The temperature of the warm end influences the cooling performance with a slope of typically 6 mW·K$^{-1}$.

As table 2 reveals, the warm end temperature and the heat flow to the warm end reduce as the length of the heatstop increases. However, a longer heatstop implies a larger pressure drop over the heatstop, and therefore a reduction in performance is found for the long heatstops.

**Table 2.** Performance of a UP 7050 with different heatstops

| heatstop gauze, [mm*mm] | sink | input power [W] | temp. warm end [°C] | power to warm end [W] | Qe @ 80K [mW] |
|---|---|---|---|---|---|
| *ambient 20 °C* | | | | | |
| none | yes | 40 | 41.0 | 12.0 | 625 |
| medium, 25*Ø4 | yes | 40 | 23.9 | 1.30 | 395 |
| medium, 12.5*Ø4 | yes | 40 | 26.1 | 1.16 | 620 |
| medium, 8 * Ø4 | yes | 40 | 27.8 | 3.48 | 630 |
| medium, 4 * Ø4 | yes | 40 | 30.4 | 5.07 | 630 |
| none | no | 40 | 76.8 | #### | 385 |
| medium, 12.5*Ø4 | no | 40 | 43.7 | #### | 530 |
| *ambient 70 °C* | | | | | |
| none | yes | 60 | 85.2 | 10.00 | 245 |
| medium, 12.5*Ø4 | yes | 60 | 76.4 | 1.16 | 400 |

### Blocking the enthalpy flow with a regenerator

In general, it can be stated that an ideal regenerator imposes locally isothermal conditions on the gas flow. In an ideal regenerator, the enthalpy flow is:

$$H = \rho_m \cdot c_p \left\langle \tilde{T} \tilde{u} \right\rangle \tag{4}$$

Because there is no temperature fluctuation, the enthalpy transport through an ideal regenerator is zero. Thus the perfect regenerator acts as a barrage to the energy flow through the split pipe. In a non-ideal regenerator heat will leak from the warm end to the cold end due to conduction (via the matrix and the outside wall) and due to regenerator inefficiency.

After start up the temperature at the compressor side of the heat stop will increase. This temperature raise continues until an equilibrium is reached between the energy flow through the split pipe and heat leak through the regenerator.

Measurements on the actual skin temperature of a split pipe (Fig. 4) confirm the energy flow hypothesis. In the standard configuration, the temperature of the warm end is at the value where the heat flow is balanced by heatsinking at the cold finger side. With a regenerator mounted in the split pipe, several effects are noticed:

- heating up of the left (oscillator-) side of the regenerator. This increases the temperature gradient over the split pipe and thus reduces the energy flow from the compressor to the cold finger.
- cooling down of the cold finger side of the regenerator due to the surface heat pumping in the remaining tube between heatstop and cold finger.
- an increase of the oscillator temperature, because the reduced energy flow implies a bigger heat load for the oscillator heat sink.
- a decrease in the energy flow through the heat sink of the cold finger

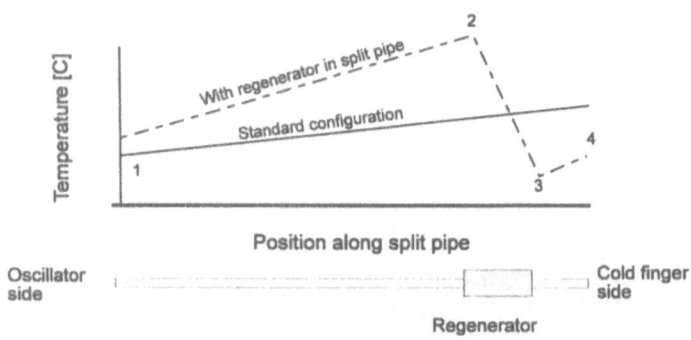

**Figure 4**. Temperature profiles over the split pipe

Typical values for the measured temperatures are given in table 1. It is shown experimentally that at higher ambient temperatures, the temperature gradient remains unchanged. Furthermore, if the regenerator is positioned closer to the oscillator, the temperature of its cold finger side decreases even further, even to temperatures below 0 °C. It may be concluded that the heat stop should be positioned as close as possible to the warm end of the cold finger. Furthermore, there is no use in removing heat from the heat stop. Heat sinking of the regenerator increases the surface heat pumping, due to a reduction of the temperature gradient over the split pipe.

**Table 3.** Performance of a UP 7088 with different heatstops

| heatstop gauze, [mm] | sink | input power [W] | gas temp. warm end [°C] | Qe @ 80K [mW] |
|---|---|---|---|---|
| *ambient 20 °C* | | | | |
| none | yes | 55 | 43.8 | 1820 |
| coarse, 6 mm | yes | 55 | 34.0 | 1800 |
| medium, 6 mm | yes | 55 | 30.3 | 1730 |
| fine, 6 mm | yes | 55 | 29.0 | 1560 |
| coarse, 9 mm | yes | 55 | 33.3 | 1720 |
| medium, 9 mm | yes | 55 | 29.3 | 1565 |
| fine, 9 mm | yes | 55 | 28.0 | 1400 |
| none | no | 55 | 84.9 | 1350 |
| coarse, 6 mm | no | 55 | 73.3 | 1560 |
| *ambient 70 °C* | | | | |
| none | yes | 55 | 92.3 | 1100 |
| coarse, 6 mm | yes | 55 | 86.0 | 1100 |
| none | no | 55 | 120.6 | 960 |
| coarse, 6 mm | no | 55 | 100.3 | 1050 |

## Measurements on a UP 7088 cooler

The cold finger of the UP 7088 has a diameter of 10 millimeter. The large finger is more sensitive to variations in the pressure wave, typically the cooling power is reduced with about 400 mW/bar if the pressure wave amplitude reduces. Therefore it is expected that the optimum heatstop has a low pressure drop, and inevitably non-optimal regenerative properties. Measurements are performed (table 3) with short (6mm and 9mm) regenerators, with different gauzes (fine, medium, coarse).

In fig. 5, the effect of the heatstop is demonstrated in a registration of the cooling performance and warm end temperature against the time after startup. During the registration, no heat sink is applied to the warm end of the cold finger.

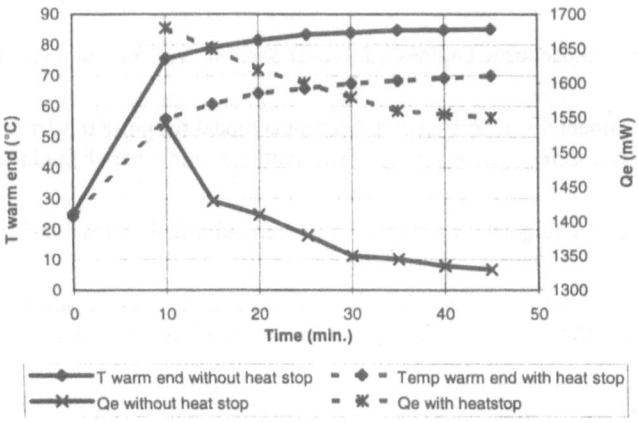

**Figure 5.** Cooling power and temperature of the warm end of the cold finger, registrated during cooling down

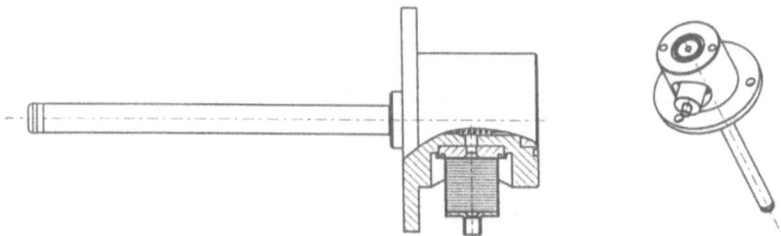

**Figure 6**. Implementation of the heatstop in a UP 7080

## Implementation

The implementation of the heatstop in a UP 7080 cooler is illustrated in fig. 6. Note that the compressor side of the heatstop makes no thermal contact with the warm end of the cold finger, except via the regenerator wall. As mentioned before, this is essential because of the required temperature gradient over the splitpipe.

## CONCLUSIONS

From the presented work, it may be concluded that

- The energy balance of an operating split Stirling cooler shows that an energy transport exists from the compressor to the cold finger, due to the oscillating gas flow in the split pipe.
- The energy transport is mainly caused by surface heat pumping in the split pipe.
- A patent-pending solution is proposed that reduces the energy flow by inserting a regenerator in the split pipe.
- The optimum size of the *heat stop* strongly depends on cooler characteristics.
- The application of the heat stop in Signaal USFA coolers leads to a considerable increase in performance. The increase in performance is most visible in applications functioning in high temperature surroundings, and with non-adequate heat sinking possibilities.

## REFERENCES

1. Gifford, W.E. and Longsworth, R.C., "Surface heat pumping", *Int.Adv.Cryo.Eng.*, vol.11, (1966), pp. 171-179.

2. Swift, G.W., "Thermoacoustic engines", *J.Acoust.Soc.Am.*, vol. 84, no. 4 (1988), pp. 1145-1180

3. Liang, J., "Experimental verification of a theoretical model for pulse tube refrigeration", *Chinese academy of sciences, Beijing*, (1993), submitted to Institue of Engineering Thermo-physics

4. Baks, M.J.A., "Eenvoudig analytisch model voor een pulsbuis-koelmachine", *Technical university of Eindhoven*, (1990)

5. Radebaugh, R., Zimmerman, J., Smith, D.R. and Louie, B.A., "A comparison of three types of pulse tube refrigirators: new methods for reaching 60 K", *Adv.Cryog.Eng.*, vol. 31, (1986), pp. 779-789.

# A Stirling Cycle Analysis with Gas-Wall Heat Transfer in Compressor and Expander

**J. S. Park and H.-M. Chang**

Hong Ik University
Department of Mechanical Engineering
Seoul, 121-791, Korea

## ABSTRACT

A cycle analysis that includes the gas-wall heat transfer in compressors and expanders of Stirling cryocoolers is presented. The gas-wall heat transfer in the working spaces is quite different from the conventional convective heat transfer, since the pressure of the working gas inside the spaces oscillates and the heat transfer may not be in phase with the temperature difference between the gas and the wall. Several experimental and theoretical expressions have been published to date to estimate the amount of the gas-wall heat transfer and the so-called hysteresis loss for a gas spring. While most of the expressions for the gas spring are not directly applicable to the working spaces of Stirling coolers, some might be useful in predicting its effect on refrigeration.

The conventional adiabatic analysis for Stirling cycle is generalized by adding the heat transfer terms to the energy balance equations for the two working spaces. Of the several existing heat transfer relations for the gas spring, three different relations are used and compared. The results are verified by observing that the cycle asymptotically approaches the Schmidt isothermal limit for very small Peclet numbers and the Finkelstein adiabatic limit for very large Peclet numbers. It is found that the effect of the gas-wall heat transfer on the refrigeration would not be significant except for low-speed miniature coolers. The analyses are repeated for various frequencies, the bore/stroke ratios, the dead volumes and the phase angles between the two pistons, and their effects on the cooler performance are discussed.

## NOMENCLATURE

$A$      Heat transfer area
$COP$    Coefficient of performance
$D_h$      Hydraulic diameter
$H$      Characteristic length of piston
$f$       Frequency
$K$      Coefficient defined by Eq.(6) or (7)
$k$       Thermal conductivity

| | |
|---|---|
| $L$ | Stroke of piston |
| $m$ | Mass |
| $Nu_c$ | Complex Nusselt number |
| $Nu_I$ | Imaginary part of $Nu_c$ |
| $Nu_R$ | Real part of $Nu_c$ |
| $P$ | Pressure |
| $Pe$ | Peclet number |
| $Pr$ | Prandtl number |
| $Q$ | Heat transfer for a cycle |
| $\dot{Q}$ | Heat transfer rate |
| $R$ | Gas constant |
| $Re$ | Reynolds number |
| $T$ | Temperature |
| $t$ | Time |
| $V$ | Volume |
| $W$ | Work for a cycle |
| $z$ | Coefficient defined by Eq.(3) |

## Greek letters

| | |
|---|---|
| $\alpha$ | Thermal diffusivity |
| $\phi$ | Phase angle advance |
| $\gamma$ | Specific heat ratio |
| $\lambda$ | Coefficient defined by Eq.(8) |
| $v$ | Kinematic viscosity |
| $\zeta$ | Coefficient defined by Eq.(8) |

## Subscripts

| | |
|---|---|
| $C$ | Compressor |
| $E$ | Expander |
| $H$ | Warm heat exchanger |
| $L$ | Cold heat exchanger |
| $R$ | Regenerator |
| *wall* | Wall temperature |

## INTRODUCTION

One of the most difficult problems in Stirling cycle analysis is to estimate the gas-wall heat transfer in working spaces. Concerning the heat transfer, there are two limiting analysis models - the isothermal (Schmidt) model and the adiabatic (Finkelstein) model.

In the isothermal analysis, the gas-wall heat transfer in the compressor and in the expander is assumed to be quite rapid so that the gas temperatures in the spaces are constant. The isothermal heat transfer represents a reversible process and always results in the maximum efficiency of the cycle. This model is valid for an extremely low speed operation or when extremely small wall-to-wall distances in the working spaces are involved.

In the adiabatic analysis, the heat transfer is neglected in the working spaces and an isothermal heat exchanger is placed between the working space and the regenerator. This model is valid when the work transfer is much dominant over the heat transfer in the compression and the expansion spaces. It is well-known that the most of the operating conditions for the real Stirling machines are closer to the adiabatic model than the isothermal one.

However, as far as the authors know, the effect of the gas-wall heat transfer in the working spaces on the cycle performance has not been clearly elucidated. One reason is that the gas-wall heat transfer is quite different from the conventional convective heat transfer, as the pressure of the working gas inside the spaces oscillates and the heat transfer may not be in phase with the temperature difference between the gas and the wall. Several experimental and theoretical expressions have been published to date to predict the gas-wall heat transfer and the so-called hysteresis loss for a gas spring. While most of the expressions for the gas spring are not directly applicable to the working spaces of Stirling coolers, some might be useful in predicting its effect on refrigeration.

In this paper, some of the published heat transfer models are incorporated into the first order analysis of Stirling cycle in order to predict how the heat transfer process affects the performance of the cycle. The heat transfer relations to be used here are introduced first. Then the method of the new Stirling cycle analysis is briefly described. Finally, the physical interpretations of the results are presented.

## GAS-WALL HEAT TRANSFER MODEL

A number of studies have been performed concerning the heat transfer of an oscillating fluid under oscillating pressures. However, most of the results are not directly applicable to the Stirling cycle analysis, since the situations for the fluid and the heat flow are not the same. On the other hand, some of the results are useful for a rough estimation of the performance of Stirling cycle. Three different functional relationships for the heat transfer in gas spring are introduced, and the results cycle analysis with the heat transfer are compared with the conventional case of the steady flow inside tubes.

### Lee-Smith Relation

The very first and the simplest attempt to relate the heat transfer rate with the gas temperature for the gas spring was performed by Lee and Smith. They analytically solved an energy equation without the convection terms to derive an expression for the wall heat flux when the pressure oscillation is first-order harmonic. Their result is given by

$$\dot{Q}(t) = A(t) \frac{k}{D_h} \left[ Nu_R (T_{wall} - T(t)) + \frac{Nu_I}{2\pi f} \frac{dT}{dt} \right] \tag{1}$$

where the heat transfer is expressed as a function of the rate of change of the gas temperature as well as the temperature difference between the wall and the gas residing outside of the boundary layer. In Eq.(1), $Nu_R$ and $Nu_I$ are the real and the imaginary parts of a complex Nusselt number, $Nu_c$, respectively. In their relation $Nu_c$ is defined by

$$Nu_c = \sqrt{\frac{\pi}{2} Pe \frac{D_h}{L_s}} \frac{(1+i)\tanh z}{1 - \tanh z / z} \tag{2}$$

where

$$z = (1+i) \sqrt{\frac{\pi}{32} Pe \frac{D_h}{L_s}} \quad \text{and} \quad Pe = \frac{2\pi f L_s D_h}{\alpha} \tag{3}$$

When the Peclet number is much smaller than 1, $Nu_R$ is much greater than $Nu_I$ and the second term in Eq.(1) becomes negligible so that the heat transfer is in phase with the temperature difference. As the Peclet number increases, $Nu_R$ and $Nu_I$ tend to become equal in magnitude. It should be noted that the Lee-Smith method is based on a simplified conduction analysis that neglects any convection effect.

## Kornhauser-Smith Relation

Kornhauser and Smith performed a series of gas-spring experiments for various gases and operating conditions. From their experimental data, a simple functional relationship for the complex Nusselt number for the case of relatively large Peclet numbers was presented.

$$Nu_R = Nu_I = 0.98\left(Pe\frac{D_h}{L_s}\right)^{0.59} \qquad \text{for } Pe \geq 100 \tag{4}$$

It is noted that Eq. (4) is a quite accurate experimental expression for gas spring and predicts the same behavior as Lee-Smith's for large Pe's.

## Jeong-Smith Relation

Jeong and Smith has obtained an approximate numerical solution for the fluid flow and the temperature in a two-dimensional (rectangular) gas spring. The heat transfer rate at the wall was expressed as

$$\dot{Q}(t) = A(t)\frac{k}{H}\left[K_S(T_{wall} - T(t)) + K_T\frac{H^2}{k}\frac{dP}{dt}\right] \tag{5}$$

where

$$K_S = \frac{6H}{\lambda}(1-\zeta)^2 \bigg/ \left\{ (5 - 16\zeta + 9\zeta^2) + (-19 + 104\zeta - 144\zeta^2 + 64\zeta^3 - 5\zeta^4)\left(\frac{6H}{\lambda}\right) \right\} \tag{6}$$

$$K_T = \frac{\lambda}{H}\frac{(1-\zeta)}{(5-\zeta)} \bigg/ \left\{ \frac{(-5 + 16\zeta - 3\zeta^2) + (25 - 140\zeta + 162\zeta^2 - 52\zeta^3 + 5\zeta^4)\left(\frac{6H}{\lambda}\right)}{(5 - 16\zeta + 9\zeta^2) + (-19 + 104\zeta - 144\zeta^2 + 64\zeta^3 - 5\zeta^4)\left(\frac{6H}{\lambda}\right)} \right\} \tag{7}$$

$$\zeta \equiv e^{-\frac{H}{\lambda}} \quad \text{and} \quad \lambda(t) = \sqrt{\frac{3kT_{wall}}{\pi f p(t)}\frac{\gamma - 1}{\gamma}} \tag{8}$$

In Eq. (8), $\lambda$ represents the thickness of the thermal boundary layer on the side-wall of the cylinder. This relation is different from Lee-Smith's in that the convection effect is included and also in that the heat transfer is a function of the temperature difference and the rate of change of pressure instead of the rate of change of temperature. This solution is approximate because it is valid only when the stroke is much smaller than the total length of gas spring such that the heat is transferred only through the side wall of the cylinder.

## McAdams Relation

The three functional relations just mentioned are derived from the heat transfer models for an oscillating gas spring. For the purpose of comparison, a well-known convective heat transfer relation by McAdams is also introduced here.

$$\dot{Q}(t) = A(t)\frac{k}{D_h}Nu(T_{wall} - T(t)) \tag{9}$$

where

$$Nu = 0.023\,Re^{0.8}\,Pr^{1/3} \qquad \text{for } Re \geq 3000 \text{ and } Re = \frac{2\pi f L_s D_h}{v} \tag{10}$$

Obviously, Eq. (10) is most accurate for steady, fully developed, and turbulent flows in a smooth round tube.

## METHOD OF CYCLE ANALYSIS

To illustrate the effects of the heat transfer in the working spaces, the simplest Stirling refrigerator is considered in this study. Figure 1 shows schematically a two-piston refrigerator with typical temperature distribution. For the analysis, it is assumed that the regenerator and the two heat exchangers are perfect without pressure drops and that the working fluid is the helium which behaves like an ideal gas. The direction of heat and work in the heat exchangers and in the compressor and the expander is shown in Figure 1.

The ideal adiabatic analysis for Stirling machines is modified by adding a heat transfer term in the energy balance equation for the compressor and the expander. By combining the mass balance and the energy balance equations, the rate of change of mass in the working spaces can be written as

$$\frac{dm}{dt} = \frac{1}{RT^*}\left\{ p\frac{dV}{dt} + \frac{V}{\gamma}\frac{dp}{dt} - \frac{\gamma-1}{\gamma}\dot{Q}\left(T, \frac{dp}{dt}, \frac{dT}{dt}\right) \right\} \tag{11}$$

where $T^*$ is dependent on the direction of flow, and can be defined as the temperature of the gas in the working space if the gas flows out and as the temperature of the adjacent heat exchanger if the gas flows in. The heat transfer rate $\dot{Q}$ is expressed in terms of the gas temperature and the rate of change of temperature or pressure.

For the two heat exchangers and the regenerator, the mass is directly proportional to the pressure, since the temperature does not vary. Once the sum of the five mass change rates is set to zero, the differential equation can be rearranged via the equation of state for an ideal gas such that the pressure change rate is the only unknown. The pressure is integrated while confirming the direction of flow in the working volumes iteratively until a cyclic steady-state is reached.

The final procedure in the analysis is to calculate the heat and work for a cycle in the working spaces and the heat exchangers by integrating the energy balance equations. The coefficient of performance(COP) for refrigeration is obtained by

$$COP = \frac{Q_L + Q_E}{W_C - W_E} \tag{12}$$

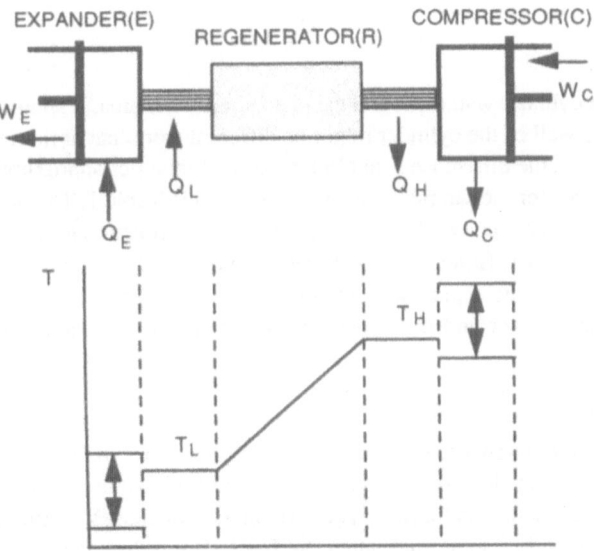

Figure 1. Schematic of two-piston Stirling refrigerator and typical temperature distribution

Table 1 Specifications of simple Stirling cycle for sample calculation

| | | | |
|---|---|---|---|
| working fluid | | He | helium |
| total mass | | $m$ | 5 g |
| warm temperature | | $T_H$ | 300 K |
| cold temperature | | $T_L$ | 150 K |
| compressor | swept volume | $V_{CM}$ | 1.5 liter |
| | clearance volume | $V_{CC}$ | 0.075 liter |
| expander | swept volume | $V_{EM}$ | 0.75 liter |
| | clearance volume | $V_{EC}$ | 0.0375 liter |
| bore/stroke | | $D/L_s$ | 0.55 |
| phase angle advance of compressor | | $\phi$ | $\pi/2$ rad |
| dead volume ratio | | | 0.3 |

In this study, the 4th-order Runge-Kutta method is used to numerically solve the differential equations. For an appropriate initial value of the pressure, the cyclic steady-state was reached after three or four cycles of iteration.

## RESULTS AND DISCUSSION

While the analysis method described above could be applied to the general Stirling cycle, the results are presented here for typical values of design parameters for illustrative purpose. In this study, the volumes of the compressor and the expander have sinusoidal variations given by

$$V_C(t) = V_{CC} + \frac{V_{CM}}{2}\left\{1 + \cos(2\pi f t - \phi)\right\}$$
(13)

and

$$V_E(t) = V_{EC} + \frac{V_{EM}}{2}\left\{1 + \cos(2\pi f t)\right\}$$
(14)

respectively. In the working spaces, the total area for the gas-wall heat transfer varies with time and are given by

$$A(t) = \pi D\left(\frac{D}{2} + L(t)\right)$$
(15)

which is the sum of the cylinder wall areas and the piston head wall area. It should be noted that the heat transfer in the side wall of the cylinder might be different from that at the top and the bottom walls in general. However, the difference is neglected for the first order approximation.

The basic specifications for the sample calculation are given in Table 1. The dead volume ratio is defined as the total volume of the two heat exchangers plus the regenerator divided by the swept volume of the compressor. It is further assumed that the three dead spaces have the same volume and that the temperature in the regenerator varies linearly along the axial direction. The wall temperatures of the compressor and the expander are taken to be the adjacent heat exchanger temperatures for simplicity.

To validate the analysis method, the pressure variation over a cycle is plotted for various frequencies in Figure 2 including the two limiting cases - isothermal and adiabatic. In the present analysis, the Jeong-Smith relation was used to obtain the heat transfer in the working spaces. It can be clearly observed that the cycle approaches the isothermal limit as the frequency decreases, and that the cycle approaches the adiabatic limit as the frequency increases. In this specific case, the cycle can be considered to be isothermal when the frequency is less than 0.001 Hz, and to be adiabatic when the frequency is greater than 10 Hz. Generally speaking, the dimensionless Nusselt

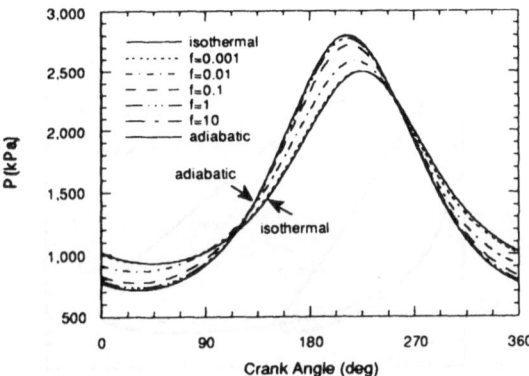

Figure 2.  Pressure variation over a cycle for various frequencies for Jeong-Smith model

numbers are functions of Peclet numbers. But it does not imply that the behavior described above is wholly determined by the Peclet numbers, since the heat transfer is dependent upon the operation speed and the heat transfer area as well as the heat transfer coefficient.

Figure 3 shows pressure-volume diagrams of the compressor and the expander for the four heat transfer models when the frequency is set at 0.1 Hz. For this operating condition, the cycles for the two gas-spring heat transfer models (Lee-Smith and Jeong-Smith) are in close agreement. Since the Kornhauser-Smith model is most accurate for large Peclet numbers at high frequencies, a slightly different result is observed. The simple convective heat transfer by McAdams, however, over-estimates the true value since the amplitude of the pressure oscillation is smaller. A more detailed comparison is made in Figure 4. The coefficient of performance in refrigeration was calculated for the four heat transfer models at various frequencies. The dotted curves at low frequencies for the Kornhauser-Smith model and the McAdams model imply that the operating conditions are outside of the applicable ranges. In all cases, as the frequency increases, COP approaches the value of about 0.49, which is COP of an ideal adiabatic cycle. For the Lee-Smith and Jeong-Smith models, as the frequency decreases, COP approaches 1, which is COP of an ideal isothermal or a reversible cycle at the given temperatures. At low frequencies, the higher COP values of the Lee-Smith model compared to the Jeong-Smith model implies correspondingly higher heat transfer rate.

Figures 5, 6, and 7 show the effects of three significant parameters - the phase angle advance of the compressor, the dead volume ratio, and the bore-to-stroke ratio - on the COP of refrigeration, respectively. As the value of the phase angle advance increases, the COP increases, but retains the same value at very low frequencies because it is the reversible asymptote. A similar behavior is observed for the dead volume ratios. As the dead volume ratio increases, the amplitude of the pressure oscillation gets smaller and more entropy generation occurs due to mixing in the working spaces and the irreversible heat transfer in the heat exchangers. On the other hand, the effect of the bore-to-stroke ratio is more complex. Generally, the hydraulic diameter or the characteristic length of the working spaces gets smaller as the ratio increases or decreases from unity. However, for small values of the ratio (the cases of "long" cylinders), the heat transfer area shows high degree of variability, while for large values (the cases of "wide" cylinders), the area does not vary much. Therefore, for large bore-to-stroke ratios, the amount of the overall heat transfer and the COP gets larger.

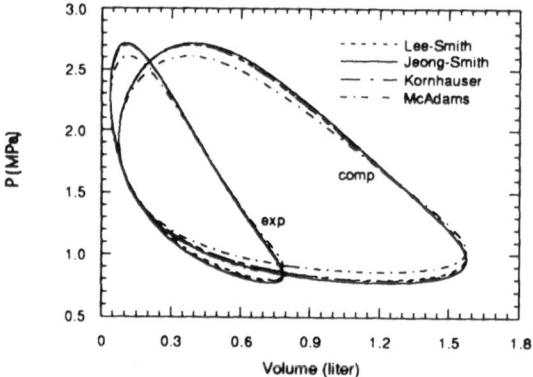

Figure 3.  Pressure-volume diagram for four different heat transfer models (f=0.1 Hz)

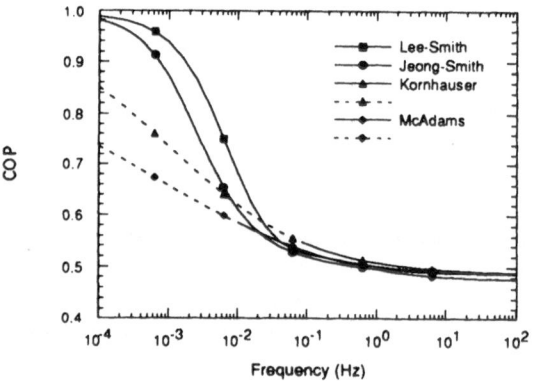

Figure 4.  COP vs. frequency for four different heat transfer models

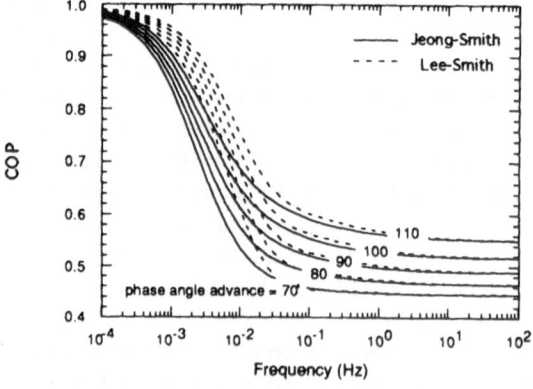

Figure 5.  COP vs. frequency for various values of the compressor phase angle advance

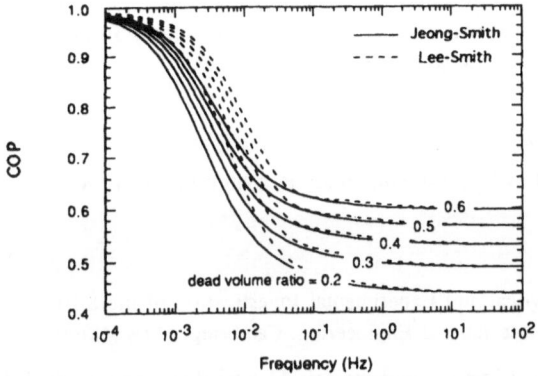

Figure 6.  COP vs. frequency for various dead volume ratios

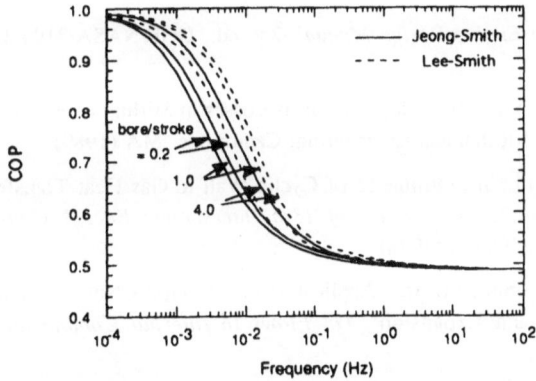

Figure 7.  COP vs. frequency for various bore-to-stroke ratios

Finally, it should be mentioned that from the definition of COP, Eq. (12), the heat transfer at the expander can be considered to be the refrigeration in addition to the heat transfer at the cold heat exchanger. In gas-springs, the net gas-to-wall heat transfer is always a dissipation loss or the so-called hysteresis loss, as it is identical to the difference between the (compression) work-in and the (expansion or recovered) work-out. However, the authors think that the issue of whether the heat from the expander is refrigeration or not depends upon one's point of view. If only the heat at the cold heat exchanger is considered to be refrigeration, then the results could be different.

## CONCLUSION

A cycle analysis that includes the gas-wall heat transfer in compressors and expanders of Stirling cryocoolers is presented along with previously reported heat transfer models obtained from the gas-spring analyses and experiments. The conventional adiabatic analysis for Stirling cycle is generalized by adding the heat transfer terms to the energy balance equations for the two working spaces. It is found that the effect of the gas-wall heat transfer would not be significant except for low-speed miniature coolers. The analyses are repeated for various operation frequencies, the bore-to-stroke ratios, the dead volumes and the phase angle advances, and their effects on the cooler performance are presented. It can be also mentioned that Jeong-Smith model seems to be most effective in predicting the heat transfer in the working spaces.

## ACKNOWLEDGMENT

This work is supported by Hong Ik University Research Fund - 1996.

## REFERENCES

1.  Rios, P.A., "An Analytical and Experimental Investigation of the Stirling Cycle", Ph.D. Thesis, M.I.T., Department of Mechanical Engineering, Cambridge, MA (1969)

2.  Walker, G., *Cryocoolers, Part 1: Fundamentals*, Plenum Press, New York (1983)

3.  Urieli, I. and Berchowitz, D.M., *Stirling Cycle Engine Analysis*, Adam Hilger., Bristol (1984)

4.  West, P.A., *Principles and Applications of Stirling Engines*, Van Nostrand Reinhold, New York (1986)

5.  Martini, W.R., *Stirling Engine Design Manual*, 2nd ed., DOE/NASA/3194-1, NASA CR-168088 (1983) pp.87-92

6.  Wang, A.C., "Evaluation of Gas Spring Hysteresis Losses in Stirling Cryocooler", M.S.M.E. Thesis, M.I.T., Department of Mechanical Engineering, Cambridge, MA (1989)

7.  Lee, K.P. and Smith, J.L. Jr., "Influence of Cyclic Wall-to-Gas Heat Transfer in the Cylinder of Valved Hot-Gas Engine", *Proceedings of 13th Intersociety Energy Conversion Engineering Conference*, Paper No. 809338 (1978)

8.  Kornhauser, A.A. and Smith, J.L. Jr., "Application of a Complex Nusselt Number to Heat Transfer During Compression and Expansion", *On Flows in Internal Combustion Engine-IV*, ASME (1988) pp.1-8

9.  Jeong, E.S. and Smith, J.L. Jr., "An Analytical Model of Heat Transfer with Oscillating Pressure", ASME HDT-Vol.204, Book No. H00762-1992 (1992) pp.97-104

10. Rohsenow, W. M. and Choi, H., *Heat, Mass, and Momentum Transfer*, Prentice-Hall, Englewood Cliffs (1961) pp.192-193

# Cyclic Simulation of Stirling Cryogenerator with Two-Component Two-Phase Fluid

**K.P. Patwardhan and S.L. Bapat**

Department of Mechanical Engineering
Indian Institute of Technology, Powai
Bombay, India 400076

## ABSTRACT

A Stirling cycle cryogenerator operates on a closed regenerative thermodynamic cycle with compression and expansion of the working fluid occurring at different temperature levels. Generally, hydrogen or helium is used as a single-component working fluid in the cryogenerators used for liquefaction of nitrogen. The use of a two-component working fluid, one condensing at the required refrigerating temperature, can improve the performance of the cryogenerator.

If maximum cycle pressure in the cryogenerator is kept the same as when using a single-component working fluid, the pressure varies over a larger range. This provides an increase in the available refrigerating effect, but at a correspondingly higher input power.

Thus, a particular cryogenerator can provide a larger liquefaction capacity when working with a two-component two-phase working fluid. Alternately, for a fixed liquefaction capacity, the maximum cycle pressure can be kept lower than the maximum cycle pressure with a single-component working fluid. The change over to a two-component fluid does not involve any change in the cryogenerator configuration.

In this paper a computer simulation using second order cyclic analysis has been carried out for a PLN-106 cryogenerator used for the liquefaction of nitrogen at a capacity of 6-6.5 dm³ h⁻¹. The results using a single-component working fluid are compared with those for a two-component working fluid with the same maximum cycle pressure.

## INTRODUCTION

The ideal Stirling cycle has two isothermal and two constant-volume processes. Stirling cycle machines can be used as engines or as cryogenerators. Cryogenerators can be used as liquefiers for gases that can be condensed at the cold space temperature. Normally, the working fluids used in Stirling cycle machines are those that do not condense at any temperature occurring in the system. Hydrogen or helium are preferred in Stirling cycle systems because of their thermodynamic and transport properties.

Walker[1] has discussed the use of two-component two-phase working fluids in Stirling cryocoolers. The analysis is based on Schmidt's method[2] for Stirling cycles, which considers sinusoidal variation of the working spaces and also void volumes in the heat exchangers. The details regarding working fluids, operating temperatures, and configurations are given only in nondimensional forms. Walker[1] has also assumed that the concentration of the working fluids is not affected by phase change of the working fluid. He concludes that the performance of the system will improve if a two-component two-phase working fluid is used. Metwalli and Walker[3] have discussed the use of chemically reactive working fluids in Stirling engines. They indicate that a substantial increase in net cycle work can be obtained without any penalty in size, weight or cost of the engine.

The analysis of the ideal Stirling cycle as presented by Schmidt[2], was used by Walker[1] for a cryocooler, and by Metwalli and Walker[3] for an engine. Further, Martini[4] analyzed the cycle for realistic losses while keeping the basic assumptions the same as Schmidt's analysis. Walker et. al[5] have applied Martini's analysis to a PPG-102 for a single-phase working fluid and compared the experimental and analytical results. The compression and expansion processes are considered to be isothermal in Martini's analysis; however, in practice the compression process tends to be adiabatic, and the expansion process approaches isothermal. Atrey et. al[6] simulated Stirling cryocoolers considering the isothermal expansion and adiabatic compression process. The analysis, when applied to the PPG-102 cryogenerator, shows a better matching with the experimental results of Walker, thus indicating the validity/usefulness of assuming isothermal expansion, and adiabatic compression.

The present paper extends the above analysis for a two-component two-phase working fluid in a Stirling cycle PLN-106 cryogenerator for liquefaction of Nitrogen.

## TWO-COMPONENT TWO-PHASE WORKING FLUID

The working fluid is assumed to be a mixture of two chemically non-reacting components. One component is assumed to behave as a perfect gas through the complete temperature and pressure range involved in the cycle; this fluid is hereafter referred to as the carrier gas. The other component, which can be in liquid phase at the cold space temperature, is referred to as the condensable fluid.

## ASSUMPTIONS

The important assumptions considered in the present analysis are as follows:
1) The movements of piston and displacer are sinusoidal.
2) The carrier gas behaves as a perfect gas.
3) The vapor of the condensable fluid also behaves as a perfect gas.
4) The expansion and compression processes are isothermal and adiabatic, respectively.
5) The condensable fluid fraction in the liquid phase occupies a negligible portion of the regenerator void volume.
6) The pressure is the sum of the partial pressure of the carrier gas and the condensable fluid, and remains the same in all the components at any instant.
7) The quasi-steady state is attained so that the volumes and pressures are subject only to cyclic variations.
8) The molar concentration of the condensable fluid in the vapor phase is maximum when the expansion space is maximum.
9) Only that mass of the condensable fluid that is in the maximum expansion space passes through the phase-change process; and when in the liquid phase, it does not contribute to the total pressure. Apart from temperature and pressure variation, this will also cause variation in composition with crank angle.

## ANALYSIS

The analysis starts with calculations of volume variations of working spaces with crank angle. The cycle is considered to be split into 12 equal intervals of 30° each. For the two-component working fluid, the total pressure is the summation of the partial pressures of the two working-fluid components.  The maximum contribution of the condensable fluid to the total pressure will be at the crank angle when all the condensable fluid is in the vapor phase.  This will result in a maximum concentration of condensable fluid and will occur when the expansion space volume is maximum.  When the expansion space volume decreases, some condensable fluid will condense at the cold end of the regenerator.  This fraction of condensable fluid that is in the liquid phase does not contribute to the partial pressure.  Simultaneously, the volume of liquid is considered to be negligible as compared to the regenerator void volume.  Thus, with a two-component two-phase working fluid, the volume and pressure variation with crack angle is also accompanied by a variation in the vapor phase concentration.

The cycle starts from the position where the expansion space is maximum; the pressure, temperature, and concentration then vary with crank angle.  The Gauss-Siedel iterative technique discussed by Atrey et al. is used for the calculations, which are done initially assuming one mole of the gas mixture.  The mean pressure in the cycle $P_m$, is therefore given by

$$P_m = \sum_{i=1}^{i=12} P(i)/12 \tag{1}$$

The exact mass of working fluids depends on the average pressure, $P_{avg}$, desired in the system.  The ratio of $P_{avg}$ to $P_m$ gives the number of moles.  Using this, the actual pressures over the complete cycle are recalculated.

## MASS FLOW RATES

To calculate the mass flow rates in the expansion and compression space, the mass fractions of working fluids present in these spaces are calculated for each interval.  Using the difference in mass fractions between the interval under consideration and the next interval, together with the molecular weight of the working fluid and the speed of cryogenerator, one obtains the mass flow rates for the expansion space WES(I) and compression space WCS(I), for the $I^{th}$ interval.

The procedure is then repeated until the mass flow rates in successive cycles do not vary beyond a specified convergence limit.

## IDEAL POWER INPUT

Once the pressure variable for the complete cycle is known, the ideal power input is calculated.  The algebraic sum of the product of pressure and volume difference for each interval gives the total power input:

$$\text{Total power input} = \int P(i) \ dV_E(i) \tag{2}$$

where
$$V_T(I) = V_C(I) + V_D + V_E(I) \tag{3}$$

## ACTUAL POWER REQUIREMENTS

The actual power is greater than the ideal due to: 1) the pressure drop caused by flow friction in the cooler, regenerator, and the condenser, and 2) the transmission or mechanical losses.

The pressure drop in the cooler, regenerator and the condenser can be calculated using mass flow rates WCS(I), WRS(I) and WES(I), respectively.  Standard formulae available in the literature have been used.  The power requirement to compensate for the pressure drop is added for the complete cycle to determine the power requirement.

The mechanical efficiency takes various transmission losses into consideration.  Atrey et al, have assumed the mechanical efficiency to be 80% for the PLN-106 machine.  The same is assumed here so that a proper comparison can be made.

## IDEAL REFRIGERATING EFFECT

The ideal refrigerating effect over the complete cycle is

$$R.E. = \oint P(i)\, dV_E \tag{4}$$

This expression is valid only when no phase change occurs in the expansion space, i.e., when all the working fluid enters and remains in the vapor phase. The increase in the available refrigerating effect due to phase change is considered below.

## ADDITIONAL REFRIGERATING EFFECT

The process of expansion is such that when the displacer moves from $V_E = 0$ to $V_E = V_{Emax}$, the fluid that is condensed at the cold end of regenerator is drawn into the expansion space and evaporates due to the continuous decrease in pressure during the expansion stroke. The latent heat of vaporization at the expansion space temperature is withdrawn from the nitrogen as it condenses on the condenser head. The mass of evaporating fluid in the expansion space is the mass of condensable fluid in the vapor phase in the expansion space at the end of the expansion process. Thus, this additional refrigerating effect is available per cycle.

## ACTUAL REFRIGERATING EFFECT AVAILABLE

There are some losses that result in reducing the available refrigeration effect. These are:

1.  Regenerator ineffectiveness
2.  Shuttle heat conduction losses
3.  Temperature swing
4.  Pumping action
5.  Instantaneous pressure drop
6.  Conduction through solid components

## LOSS DUE TO REGENERATOR INEFFECTIVENESS

Atrey et. al[6] have used Miyabe's method[7] for the calculation of regenerative effectiveness. In the case of a two-component two-phase fluid, the latent heat of condensation of the condensable component fluid adds up to the heat capacity of the hot gas and is considered in calculating the regenerative effectiveness. This forms a major deviation in the calculation of losses due to regenerator ineffectiveness when compared with a single-component working fluid where no phase change occurs.

Further, the surface area of the regenerative mesh is large and the volume of the condensate is small, thus the process of condensation can be considered as dropwise condensation. Very high values of heat transfer coefficients (up to 80,000 W/m²-K) have been reported in the literature for dropwise condensation. Generally, with an increase in hot gas heat capacity, an increase in the required surface area of the regenerator (increase in size of the regenerator) can be expected for the same effectiveness. Due to such a high heat transfer coefficient, a change in regenerator size can be avoided without affecting the regenerative effectiveness to any appreciable extent.

## OTHER LOSSES

The other refrigerating effect losses are calculated exactly along the lines mentioned by Atrey et. al[6] using two-component fluid properties, i.e., specific heat of the working fluid, viscosity of the working fluid, and thermal conductivity of the working fluid.

## RESULTS AND DISCUSSION

The present analysis has been applied to a PLN-106 cryogenerator. The mechanical/transmission efficiency of the drive system is assumed to be 80%. The results are obtained for

**Table 1.** Comparison of the performance of single-component and compound working fluids

| Quantity | Pure Working Fluid | Compound Working Fluid |
|---|---|---|
| Ideal Power (W) | 6846.06 | 6842.312 |
| Power loss in regenerator (W) | 169.33 | 298.5031 |
| Power loss in cooler (W) | 68.889 | 100.3597 |
| Power Loss in condenser (W) | 4.6884 | 11.51849 |
| Total power loss (W) | 242.93 | 410.3813 |
| Mechanical loss (W) | 1772.249 | 1813.173 |
| Net power (W) | 8861.244 | 9065.866 |
| Ideal refrigerating effect (W) | 1968.105 | 1969.406 |
| Loss due to regenerative ineffectiveness (W) | 371.245 | 602.69 |
| Pv loss (W) | 126.1101 | 125.3382 |
| Temperature swing loss (W) | 493.168 | 557.89 |
| Shuttle conduction loss (W) | 28.57 | 28.55 |
| Axial conduction loss (W) | 18.406 | 18.411 |
| Pumping loss (W) | 81.504 | 202.99 |
| Extra refrigerating effect (W) | 0.000 | 546.835 |
| Net refrigerating effect (W) | 849.57 | 980.3546 |
| Partial pressure of condensing component (bar) | 0.00 | 1.013196 |
| Molar concentration of condensing component | 0.000 | 0.06021 |
| Average pressure (bar) | 25 | 25 |
| C.O.P. | 0.09587 | 0.1081369 |

conditions when the partial pressure of the condensable fluid in the maximum expansion space volume is 1.01325 bar. The maximum pressure in the cycle with the single-component and two-component two-phase fluids are maintained the same. For a given maximum pressure, the refrigerating effect is more with the two-component working fluid, but at the same time the work input is also larger. This indicates that the same cryogenerator can be used to give a larger liquefaction rate. This occurs without any change in the cryogenerator, and without any change in safety factors, as the maximum pressure is kept unchanged. The results of the loss analysis are given in Table 1 for the single-component and two-component working fluids. It can be observed that net refrigerating capacity increases from 849.57 W to 980.3546 W, showing an increase of about 15.39%. As the cold end temperature is kept fixed, the comparison can be made for single-component and two-component systems on the basis of C.O.P. The C.O.P. changes from 0.09587 for the single-component fluid, to 0.10813 for the two-component fluid. Further, the analysis is repeated for the two-component fluid to obtain the same net refrigeration effect as when using the single-component fluid. It is observed that the maximum pressure drops from 25 bar to 11.3 bar. The reduction in maximum pressure will allow for a longer running time for the system. This will also ensure that the temperature after compression will reduce. A small benefit in the form of slightly improved regenerator effectiveness is also obtained; this reduces the regenerative ineffectiveness loss. As the hot-end temperature reduces regenerator contamination, regenerator cleaning should be required only after a larger number of continuous working hours. The results for this case are given in Table 2.

## CONCLUSION

The present paper uses the Stirling-cycle second-order cyclic analysis to analyze the PLN-106 cryogenerator with a single-component working fluid, and then compares these results with

**Table 2.** Comparison of the performance of a single-component and compound working fluid for the same refrigerating effect.

| Quantity | Pure Working Fluid | Compound Working Fluid |
|---|---|---|
| Ideal Power (W) | 6846.06 | 3083.7 |
| Power loss in regenerator (W) | 169.3573 | 264.9394 |
| Power loss in cooler (W) | 68.889 | 78.5352 |
| Power Loss in condenser (W) | 4.6884 | 10.344 |
| Total power loss (W) | 242.93 | 353.8189 |
| Mechanical loss (W) | 1772.249 | 859.3796 |
| Net power (W) | 8861.244 | 4296.8998 |
| Ideal refrigerating effect (W) | 1968.105 | 888.34342 |
| Loss due to regenerative ineffectiveness (W) | 371.245 | 56.15089 |
| Pv loss (W) | 126.1101 | 124.5 |
| Temperature swing loss (W) | 493.168 | 255.4955 |
| Shuttle conduction loss (W) | 28.57 | 28.52695 |
| Axial conduction loss (W) | 18.406 | 18.4019 |
| Pumping loss (W) | 81.504 | 111.5753 |
| Extra refrigerating effect (W) | 0.000 | 547.4005 |
| Net refrigerating effect (W) | 849.57 | 841.0081 |
| Partial pressure of condensing component (bar) | 0.00 | 1.01322 |
| Molar concentration of condensing component | 0.000 | 0.133 |
| Average pressure (bar) | 25 | 11.3028 |
| C.O.P. | 0.09587 | 0.195 |

the results for a two-component two-phase working fluid. The comparison indicates that a simple change from single-component to two-component two-phase fluid can be utilized to both increase the liquefaction capacity of the cryogenerator under identical operating conditions of maximum system pressure, or to obtain the same liquefaction capacity at a lower maximum pressure.

## REFERENCES

1. Walker, G., "Stirling Cycle Cooling Engine with Two-phase Two-component Working Fluid," *Cryogenics*, Vol. 14, No. 2 (August 1974), pp. 459-462.

2. Schmidt, G., "Theorie der Lehmannschen Calorischen Maschine," *Z. Ver Dtesch Ing.*, Vol. 15, No. 1 (1871).

3. Metwalli, M.M. and Walker, G., "Stirling Engines with a Chemically Reactive Working Fluid — Some Thermodynamic Effects," *Journal of Engineering for Power, Transactions of ASME* (April 1977), pp. 284-287.

4. Martini, W., "Design Manual for Stirling Engines," NASA CR 135382, NASA Lewis Research Center, Cleveland, Ohio (1978).

5. Walker, G., Weiss, M., Fauvel, R., and Reader, G., "Microcomputer Simulation of Stirling Cryocoolers," *Cryogenics*, Vol. 29 (1989), pp. 846-849.

6. Atrey, M.D., Bapat, S.L., Narayankhedkar, K.G., "Cyclic Simulation of Stirling Cryocoolers," *Cryogenics*, Vol. 30 (April 1990), pp. 341-347.

7. Miyabe, H., Takahashi, S., and Hamaguchi, K., "An Approach to Stirling Engine Regenerator Matrix Using Packs of Wire Gauzes," *Proc. 17th IECEC*, IEEE, New York, pp. 1833-1944.

# Cryocooler Transient Performance Modeling

T. Roberts

Phillips Laboratory, USAF
Albuquerque, NM, USA 87117

## ABSTRACT

The performance of cryogenic coolers during intermittent operations is dominated by the transient terms of the thermal energy equation and by the changes of the cooler's operating efficiency as the cold end temperature changes. This is in contrast to static operating conditions in which these effects are not relevant and for which most cooler operating data is available. In order to support the integration design effort of the STRV-2 satellite system, a study of the transient cool down and warm up performance of the Texas Instruments 1 Watt Cryocooler and Electronics Control Card was performed. The objective of this study was to model the cool down and warm up behavior of the cooler which is supporting a thermally loaded focal plane of 0 to 0.6 Watts.

The form of the model was derived from the fundamental thermodynamic and heat transfer principles operating within the combined cooler, cold end block, and heat rejection interface system. The predictions of the cool down and warm up models based on fundamental principles are presented and are compared to the observed performance envelopes at various loads and heat rejection temperatures. A multivariable regression is then performed in order to fit the model to the observed data for both warm up and cool down.

The conclusion is drawn that while a model formed from fundamental principles of the loaded warm up can be accurately based on heat transfer laws, such a model is not yet available for a loaded cool down. This is largely due to the lack of instrumentation to measure properties within a practical working cooler. Finally, equations correlating rejection temperature, cold end temperature, cold end load, and time are presented for cool down and warm up performances.

## NOMENCLATURE

| Roman Symbols | | Subscripts | |
|---|---|---|---|
| $A$ | area | $c$ | compressor |
| $c$ | specific heat | $ct$ | cold end or tip |
| $C$ | arbitrary constant | $D$ | dead space |
| $d$ | diameter | $e$ | expansion space |
| $k$ | thermal conductivity | $eff$ | effective |
| $L$ | length of displacer | $g$ | expander clearance |
| $m$ | mass | $ld$ | load |
| $P$ | pressure | $net$ | net refrigeration |
| $q$ | heat | $rej$ | rejection |
| $S$ | $2X\tau/(\tau+1)$ | $1,2$ | # of constant, $C$ |
| $t$ | time | | |
| $T$ | temperature [abs.] | | |
| $V$ | volume | | |
| $x$ | clearance seal | | |
| $X, X_o$ | $V_D/V_e$ at $T_c, T_D$ respectively | | |

| Greek Symbols | | | |
|---|---|---|---|
| $\alpha$ | expansion to compressor space phase angle | $\tau$ | $T_c/T_e$ |
| $\Delta$ | change in... | $\varepsilon$ | regenerator ineffectiveness |
| $\kappa$ | $V_c/V_e$ | $\nu$ | rotational frequency |
| $\theta$ | $arctan\left[\dfrac{\kappa Sin(\alpha)}{\tau - 1 + \kappa Cos(\alpha)}\right]$ | | |
| $\delta$ | $\dfrac{\sqrt{(\tau-1)^2 + X^2 + 2(\tau-1)XCos(\alpha)}}{\tau + \kappa + 2(X_o + S)}$ | | |

## INTRODUCTION

The STRV-2 flight experiment imposes a very transient mission profile on its supporting cryocooler. Present estimates of the mission life indicate that the cooler will be on approximately 25% of the one year mission. During this time the cooler will be operated in a principally transient mode, with the cooler cycling on and off during the period of an orbit whose length is approximately one or two hours. The heat rejection reservoir may periodically swing from 250 K to 325 K, with occasional excursions above that level. A model for such transiently loaded cool downs, verified by experimental results, provides the most reliable planning tool available to estimate a cooler's ability to support this mission profile. Cool down at a rejection temperature of 339 K was deemed of highest interest to the satellite integration effort, as:

1. at rejection temperatures higher than that the cooler would have internal temperatures above its operating limits, and

2. lower rejection temperatures had faster cool down rates, so that 339 K could be used as an engineering worst case scenario.

## DESCRIPTION OF THE COOLER

The cooler supporting the experiment is the Texas Instruments (TI) 1 Watt cryocooler. TI developed this cryocooler as part of an Army Night Vision Laboratory program for airborne night vision systems and the M1 Abrams tank. It is a split Stirling cycle cooler capable of 1W at 80 K cold end temperature. The power budget available to the cooler from the spacecraft bus is not yet known, but performance mapping indicates that the cooler will require about 25 watts of continuous power during operation. The Focal Plane Assembly (FPA) will include a thermal storage device which will be capable of maintaining stable FPA temperature (with the cooler shut down) during an Medium Wave Infra Red measurement (approximately 10 to 20 seconds).

The cooler system is comprised of the cooler unit and its electronics control card. The card uses a voltage drop across an externally adjusted resistance in to comparison to the voltage drop across a silicon diode in maintaining the cold end at a desired temperature set point with a feedback control algorithm. If the cold end is more than approximately 3 K from its set point, the card either turns on the cooler at maximum stroke or effectively turns off the cooler, as appropriate. The operator does not set any mechanical parameters, such as operating frequency, stroke length, or phase angle.

## MEASUREMENT OF COOL DOWN PERFORMANCE

The cooler was integrated into 24" vacuum chambers with an aluminum thermal and structural mount. The thermal interfaces for the compressor and expander are usually within 1 K of being isothermal, as the aluminum mount's halves grasping the entire circumference of the cooler's sides are relatively massive. The mount itself rests on a 9" flat copper plate which contains a fluid cooling loop. The thermal gradients within the mount are consequently established by the fluid velocity and temperature in the cooling loop and the amount of heat generated by the operational cooler. During any cool down the rejection temperature of the cooler generally rises by approximately 5 K as the cooling loop is unable to react to the practical step function in heat flux produced by the cooler being switched on at several hundred Kelvin above an 80 K set point.

The cooler's performance is monitored by a data acquisition (DAQ) program which polls (via GPIB network) the instrumentation listed in Table 1 on a 10-15 second data gathering cycle. The placement of the temperature sensors is shown in Figure 2.

The resulting cool down paths are shown in Figure 3. Similar paths result from cool downs at different temperatures, with lower reject temperatures leading towards faster cool down times and steeper paths. For the purposes of this program, the 339 K rejection temperature case was the principal scenario of interest.

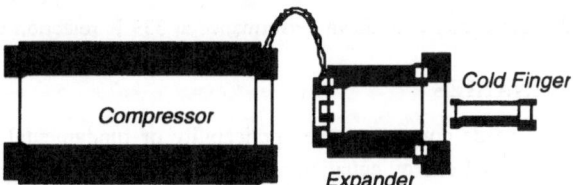

**Figure 1.** TI 1 watt cryocooler.

**Table 1**. DAQ Equipment and Measured Properties

| Equipment Item | # Sensors | Measured Property | Accur. |
|---|---|---|---|
| Lakeshore 820 Cryo. Thermometer | 1 | Cold End Temp. | ±.1 K |
| Fluke 2620A Multimeter (Thermocouple Measurements) | 9 | Casing and Mount Temperatures | ±.7 K |
| Vahalla 2300/2300L | 2 | Cooling Load & Cooler Power | ±1% full scale values |

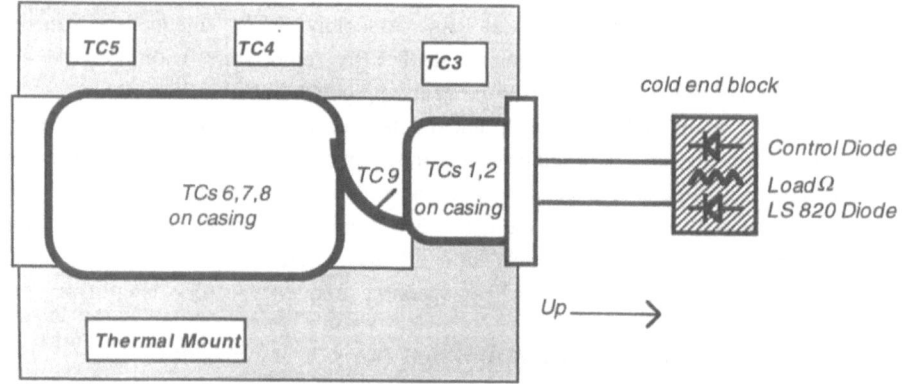

**Figure 2**. Temperature sensor placement.

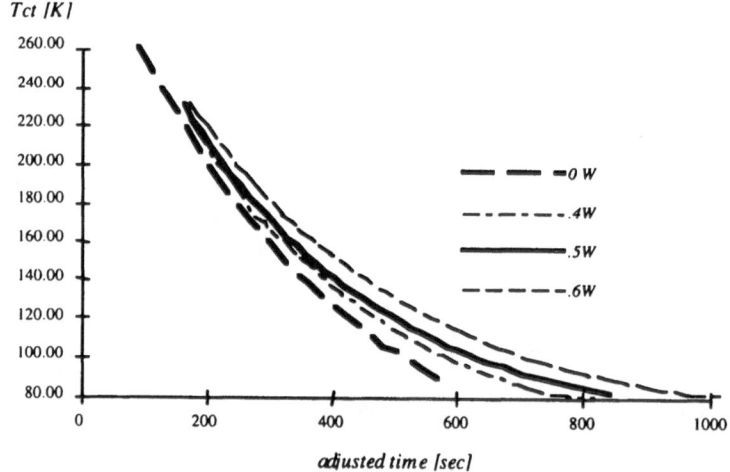

**Figure 3**. STRV-2 flight cooler cool down performance at 339 K rejection temperature.

## MODELING PRELIMINARIES

Assumptions have to be made concerning the applicability of fundamental principles to this study. We assume that the thermal energy equation and Fourier heat transfer law are relevant, but that the effects of radiation and external convective transfer are minimal. The behavior of the cooler working fluid should be in accordance with the Navier Stokes equations. This problem will be modeled in one spatial dimension. Finally, the assumption that the performance manifold of the cooler is smooth should be made unless observed otherwise. In the region of the manifold being modeled, we also do not expect to find any performance manifold boundaries (limits to how the cooler can be operated) other than the 80 K cold end set point.

## Initial Efforts

Given a three dimensioned data set comprised of cold end temperatures, elapsed time, and rejection temperatures, it is mathematically feasible to construct an exact fit of the experimental data using either interpolating polynomials or Fourier series expansions. A less ambitious method would be to use multivariate regression analysis to develop a smaller model of statistically significant linear components. These methods ignore three modeling goals to a greater or less extent:

1. A model should explain the data set, as well as represent it.
2. A model should allow for reasonable extrapolation as well as interpolation.
3. A model should efficiently use computational resources.

Deficiently addressing goal 2 is especially grave for this particular program as the intent of the model is to support extrapolation downwards in rejection temperature. Addressing goal 1 also serves to bring to the foreground one important feature of the thermal sciences: usually the solutions to heat transfer and thermodynamic problems are transcendental functions. Approximation of transcendental functions with polynomials in particular is fraught with errors.

## Eliminating the Obvious

Sorting through the above considerations and assumptions, and realizing that reference 2 provides the basis for a successful model of the transient off cooler warm up, it may be contemplated that a similar exponential model could be curve fitted to the cool down performance paths. Such a model would take the form of:

$$T_{CT} = C_1 + Exp\left(-tC_2 {}^{6.73}\!\!\big/\!(mc+.49)\right)\!\left[T_{CT,t=0} - C_1\right] \qquad [1]$$

The modification to the time constant, $6.73/(mc+.49)$, is a correcting factor for changes in cold end load block thermal mass. The *effective* thermal mass from the off state conduction calculations turned out to be 6.73 J/K, whereas the cold block physically was calculated to have only 6.24 J/K. As expected some portion of the cold finger acts as part of the effective thermal mass from which the cooler draws heat.

Curve fitting the data using steepest gradient methods to reduce the square of the errors gives the coefficients in Table 2 and the curve fits in Figures 4 and 5. As this model is that of one dimensional cooling with a set boundary temperature, $C_2$ is the inverse time constant and $C_1$ the eventual temperature to which the cooler would cool to without interference from temperature control algorithms, changes in thermodynamic performance, and other physical constraints.

**Table 2.** Exponential curve fit coefficients

| Load  | 0.4W     | 0.5W     | 0.6W     |
|-------|----------|----------|----------|
| $C_1$ | 40.92    | 48.46    | 59.73    |
| $C_2$ | 0.002704 | 0.002704 | 0.002704 |

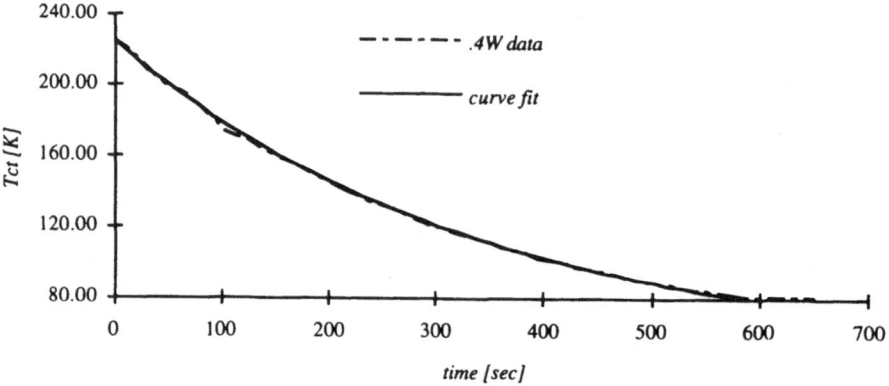

**Figure 4.** Simple exponential curve fitting of 0.4 W cool down.

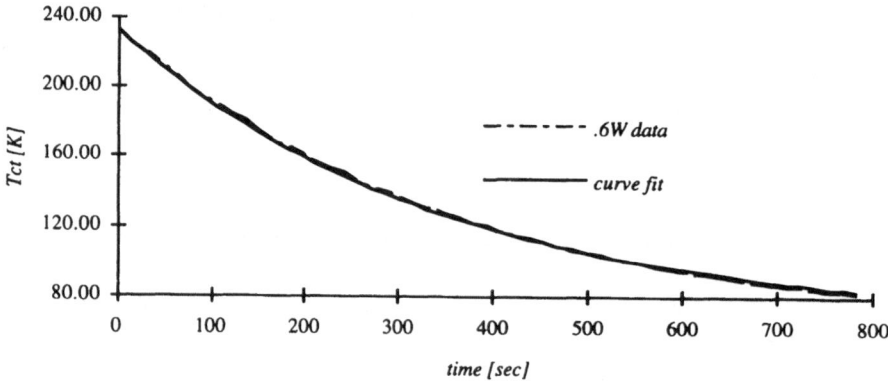

**Figure 5.** Simple exponential curve fitting of 0.6 W cool down.

As can be seen from the curve fits, accuracy is qualitatively acceptable. This model, however, explains *nothing* about the operation of the cooler. It instead says that it is analogous to a cooling lump of metal affixed to a mysterious heat sink. This hypothetical heat sink gets progressively colder to reflect reductions in cooling load and rejection temperatures until it sinks significantly below absolute zero.

## Resorting to Fundamentals

Following the lead of Radebaugh [1], there are several heat fluxes operating on the cold end of the cooler during operation. Equation 2 gives a summary of this situation.

$$q'_{refrigeration\,,net} = q'_{refrigeration\,,ttl} - q'_{conduction} - q'_{shuttle} - q'_{regenerator} - q'_{\Delta P} \qquad [2]$$

Essentially, the net refrigeration is the total refrigeration less the sum of the system irreversibilities. As a first order approximation, the assumption is made that the steady state equations used by Radebaugh [1] are sufficiently accurate to describe the cooler's operatation. The terms on the right hand side of equation 2 can then be examined for their dependencies on properties which can be measured during cooler operation or estimated. Total refrigeration can be described:

$$q'_{refrigeration,ttl} = \frac{P_{max}V_{ttl}\pi\delta Sin[\theta]\sqrt{1-\delta}}{\sqrt{1+\delta}(2+X)\left(1+\sqrt{1-\delta^2}\right)^2} \qquad [3]$$

Parasitic conduction and working fluid shuttling gains can be described, assuming the cold end temperature and expansion space temperature are equal:

$$q'_{conduction} = \frac{kA_c}{L}(T_c - T_{ct})$$

$$q'_{shuttle} = \frac{0.186s^2dk_g\pi}{t_gL}(T_c - T_{ct}) \qquad [4,5]$$

Regenerator effectiveness can be described:

$$q'_{regenerator} = m'_{average}C_p\frac{1-\varepsilon}{2}(T_c - T_{ct}) \qquad [6]$$

The effect of pressure drop from the compressor to expansion space is:

$$q'_{\Delta P} = 2q'_{refrigeration.ttl}\frac{\left(P_{max}/P_{min} + 1\right)\Delta P}{\left(P_{max}/P_{min} - 1\right)P_{average}} \qquad [7]$$

Reviewing equations 3 through 7, there are many physical parameters of the cooler which are constant or steady under these operating conditions, but they are also unknown or unmeasureable with the acquisition system at hand. The major exception is $T_c$, which can be easily estimated from the temperature of the transfer line at thermocouple 9 (Figure 2). Beginning most cool downs the cooler was isothermal at 5 K below the target rejection temperature, at 334 K in this study's case. The rejection temperature would rise rapidly to 339 K after turning on the cooler while the transfer line climbed exponentially in temperature to about 10 K above the rejection temperature. As this transfer line is well insulated and accessible to thermal fluxes from the compressor, it would be reasonable to use the temperature of thermocouple 9 as a guide for the profile of $T_c$. Substituting this exponential profile for $T_c$ into equations 4 through 6, and equations 3 through 7 into equation 2, and simplifying the unknown and unmeasured parameters into general coefficients, equation 3 can be restated as an ordinary differential equation using the Fourier heat conduction law operating on the cold end.

$$q'_{refrigeration,net} = mc\Big|_{eff}\frac{dT_{ct}}{dt} + q'_{load} = C_1 - C_2(T_c - T_{ct}) \qquad [8]$$

The solution of the differential equation is:

$$T_{ct}(t) = (q'_{ld} - C_1)\Big/C_2 + T_c + Exp\left[{tC_2}\Big/{mc|_{eff}}\right]\left((C_1 - q'_{ld})\Big/C_2 + T_{ct,t=0} - T_c\right)$$                    [9]

where $q'_{ld}$ = cooling load

During the experimental work, $T_{rej}$ and $T_{ct,t=0}$ are roughly comparable within 4 K. Due to the rapid initial cooling rate of this system, the interchange of the two values does not affect the model's accuracy drastically.

This solution can be curve fitted using steepest gradient reduction of the square of the errors to provide the coefficients:

$$C_1 = -6.0956 + 2.0470 q'_{ld}$$                    [10]

and $C_2 = -.019$.

The graphical results are in Figure 6. The generalized cool down manifold is shown in Figure 7.

## CONCLUSIONS

The inclusion of the major external thermodynamic parameters affecting the cooler system in equation 9 represents a significant advance over the simple model of equation 1. It also allows for a

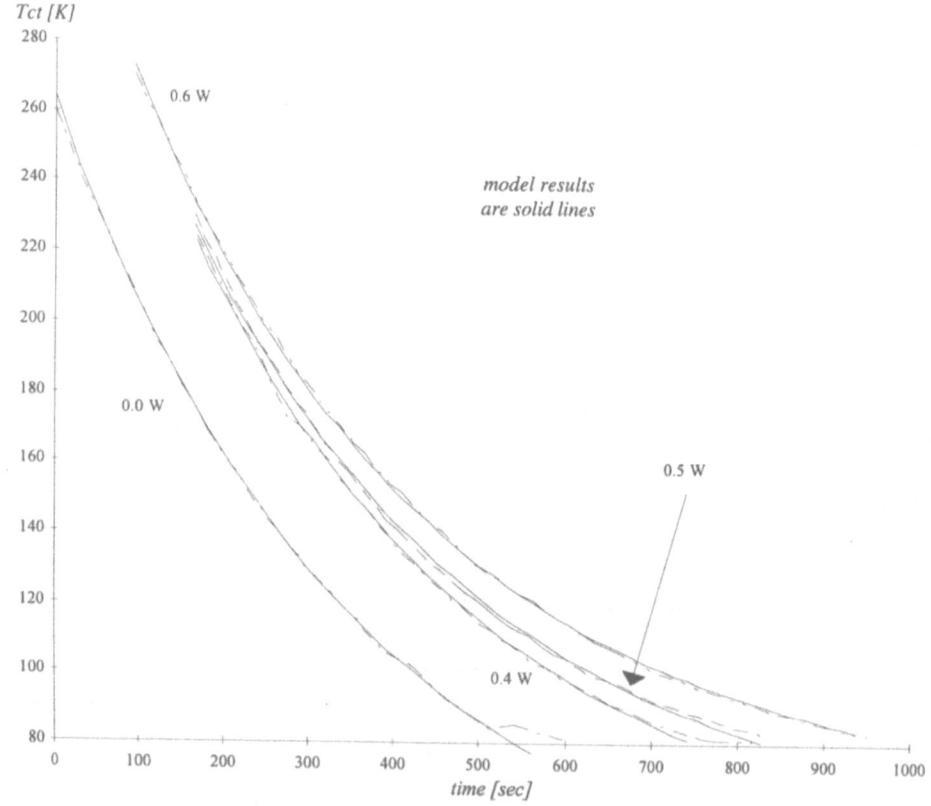

**Figure 6:** Model results compared to experimental data.

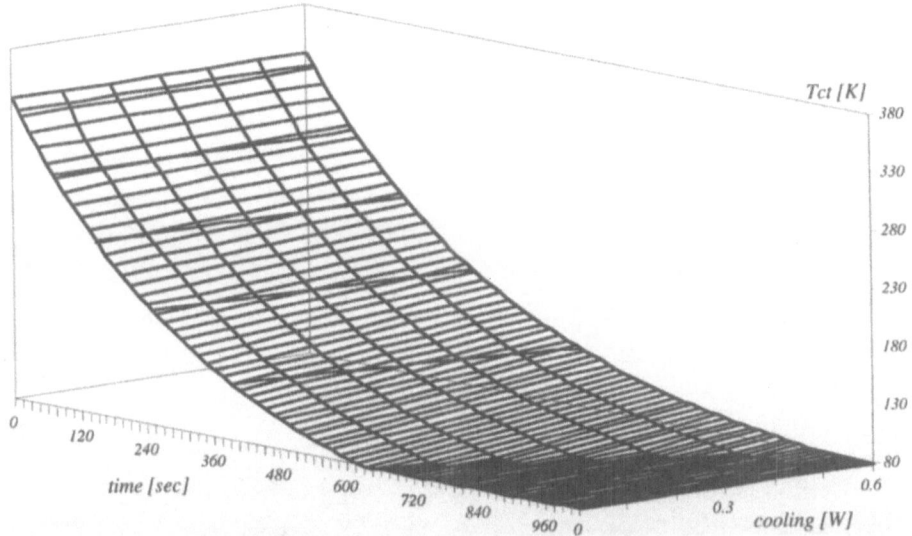

**Figure 7**. General model cooling manifold.

realistic extrapolation of the model to the rest of the rejection temperature envelope in order to allow for the satellite integration effort to more accurately estimate the loaded cool down rate of the focal plane under a variety of scenarios.

Gathering the unknown or unmeasured thermodynamic entities in equations 3 through 7 would allow for a true foundation of this model from fundamental principles. Additional instrumentation of the operating frequency and effective phase angle and piston displacements would be required. Given this more thorough approach to the model's foundations, a true verification of the derivation of equation 9 would be possible, and any inconsistencies would then be corrected. For the purposes of supporting the STRV-2 satellite integration effort, such basic research might not be an effective use of program resources, and reliance on the general form of equation 9 and curve fitting methods seems as reliable and a more efficient method to obtain a predictive performance model of loaded cooler cold end temperatures.

## REFERENCES

1. Radebaugh, R. "Low Capacity Cryogenic Refrigerators", in Low Capacity Cryogenic Refrigerators , Radebaugh, R. and Walker, G. eds. , UCLA, Los Angeles, CA, Feb. 1996

2. Horsley, W.J., et al, "Test Results for the Single-Stage Ball Flight Prototype Cooler", in Ross, R. ed., Cryocoolers 8, Plenum Press, NY, 1995; pp. 23-33

# New Mid-Size High Efficiency Pulse Tube Coolers

**W.W. Burt and C.K. Chan**

TRW Space & Technology Division
Redondo Beach, CA 90278

## ABSTRACT

TRW has developed, tested, and delivered a mid-capacity pair of 60 Kelvin and 35 Kelvin pulse tube cryocoolers for space applications. These coolers are the highest efficiency pulse tube coolers yet reported with efficiency exceeding common Stirling cycles. The coolers are a mechanically integral configuration designed for 2W at 60K and 300 mW at 35K. At the operating points the input power to the compressor is 76 W and 82 W respectively with a 300K sink temperature. The coolers are mechanically balanced using two head-to-head Oxford style flexure compressors in a common, hermetically sealed housing.

These coolers were developed for space applications where long life, light weight and high efficiency are of premium importance. These inherently long life coolers were found to operate with low vibration, good temperature stability, high day to day reproducibility, and to provide a large performance capability for off-design operating points. The coolers' development, operating features, performance and operating history are reported.

## INTRODUCTION

Since the late 1980s TRW has reduced to practice the technology of pulse tube coolers, producing the world's first miniature pulse tube cooler[1] and the world's first 35 Kelvin (K) high-capacity pulse tube cooler[2]. Designed for space operation, both products make use of the inherently high reliability of the passive pulse tube cooler and both represent milestones in efficiency for their time. This paper reports on the latest evolution of the pulse tube cooler, which further improves efficiency and has been sized into a new intermediate capacity for use on long-life space applications.

The attractiveness of the pulse tube cooler rests in its simplicity. Since it has no moving parts in the cold producing section, it is inherently reliable and producible. The pulse tube is a regenerative cycle cooler that operates similarly to a Stirling cryocooler but uses a passive expander and its operation has been described elsewhere[3]. To productize the pulse tube cold head, it must be mated with a high reliability room-temperature compressor. For space use, this is usually the Oxford-style non-contacting flexure bearing compressor, which is described elsewhere[4]. The coolers described in this article use a TRW designed Oxford-style compressor mated to the TRW-designed pulse tube cold heads.

The two coolers reported here were developed from April 1994 to October 1995. One cooler was designed for 2 W of cooling at 60 K while rejecting waste heat to 300 K and the other cooler

Cryocoolers 9, Edited by R.G. Ross, Jr.
Plenum Press, New York, 1997

was designed for 300 mW at 35 K rejecting to 300 K. For each cooler power efficiency was to be maximized. These units, shown in Figure 1, are designated the 60 K cooler and the 3503 cooler, respectively.

The coolers look similar because they use the same compressor design. The cold heads are slightly different and tuned for the respective operating temperatures. Designed as brassboard models, the coolers have the same form, fit, and function of a flight cooler. The coolers were tested to qualify them for space flight; the results of these tests are the primary content of this article. The coolers were delivered to the Air Force at JPL in the summer and fall of 1995 and they have since been intensively characterized at JPL[5]. The coolers are noteworthy for their efficiency. The 60 K cooler delivers 2 W at 60 K for 78 W into the compressor. The 3503 cooler delivers 300 mW at 35 K for 82 W into the compressor. Both powers are for heat sink temperatures of 300K. The 60 K cooler is particularly impressive and exceeds the performance of reported Stirling coolers at this temperature, as shown in

Figure 2. A new unit is currently under fabrication for the Multispectral Thermal Imager space flight in 1998.

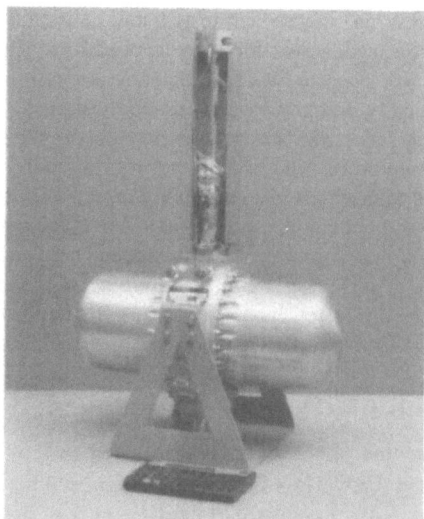

**Figure 1.** The 60K (left) and 3503 pulse tube cryocoolers.

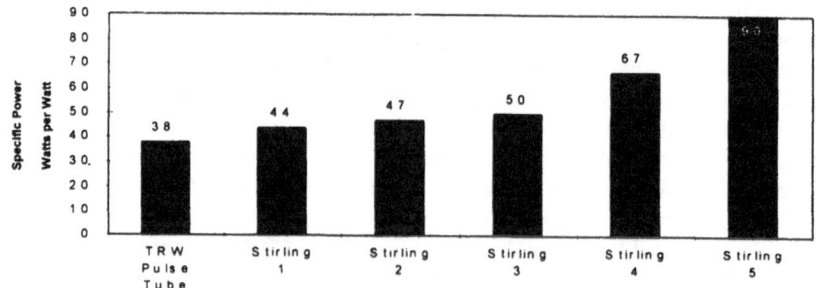

**Figure 2.** The TRW 60K pulse tube cooler exceeds the reported efficiency of aerospace Stirling coolers. ( Stirling data reference[2]: *Cryocoolers 8).*

## COOLER DESCRIPTION

The coolers described in this article are scaled-down models of the robust 35 K pulse tube cooler[2]. The new cooler compressor's swept volume is one half that of the 35K cooler. Except for size, the new coolers are largely identical in configuration to the larger 35K cooler.

Figure 3 shows outline and cutaway schematic views of the coolers. The cold head represents an improved performance design over that developed for the 35 K coolers. Mechanical and thermal rejection interfaces to the coolers occur via surface flats with the tapped inserts seen on each side of the centerplate. User attachment to the cold block is also via bolted fasteners into tapped inserts.

The cooler design seen in Figure 1 is referred to as an integral configuration because the cold head is mechanically fastened to the compressor (although it is thermodynamically split by a flow channel from the compressor). Under a separate program, TRW has also developed a similar sized cooler that has the cold head and compressor mechanically separated (split) and it is reported upon elsewhere[6]. That unit uses a functionally similar but non-interchangeable compressor design where great attention was paid to lightweight the compressor components.

The cross-section schematic in

Figure 3 shows that the 10 cc compressor actually contains two flexure compressor half assemblies, each 5 cc in swept volume, which are bolted together head to head. The in-phase operation of these two assemblies provides for inherent momentum cancellation and minimal vibration output. Note the location of the cold finger at the center of the cold head.  This is characteristic of the TRW pulse tubes in contrast to historical Stirling cold finger geometry where displacer mechanics are a design factor. User access to the cold finger is typically through the perpendicular leg of a vacuum dewar tee section provided as support equipment. Characteristics of the coolers are summarized in Table 1.

Both coolers are controlled using associated rack electronics that feature a motor controller, diagnostics, and a commercial linear power amplifier. This program made no effort to develop flight electronics and all of the reported power readings are as measured at the compressor.

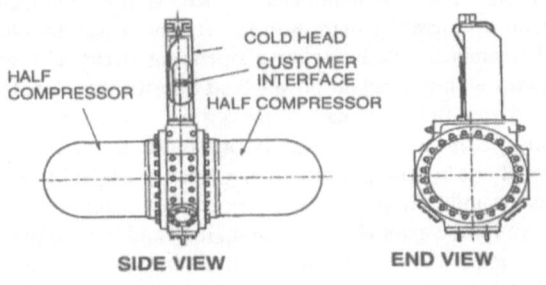

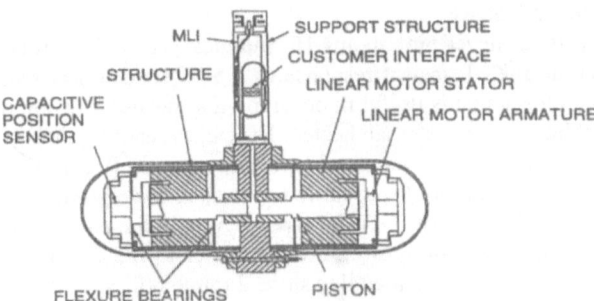

**Figure 3**. Outline and cutaway schematic views of the pulse tube cryocoolers. Outside end cap diameter is 130 mm for reference.

**Table 1.** Pulse Tube Cooler Characteristics.

| Housing materials | Aluminum alloy compressor; titanium alloy cold head |
|---|---|
| Design safety factor | 2.5 |
| Working fluid | Dry helium |
| Housing seals | Hermetic, helium leak tight |
| Transducers | Capacitive position sensor each motor, accelerometer vibration sensor |
| Flexure springs | BeCu Oxford heritage |
| Frequency, nominal operating range | 43-48 Hz |
| Fill pressure | 250 psig at 70°F |
| Rated input power | 150 W, 75 W per side |
| Weight | 12.0 kg (60K) |
| | 12.1 kg (3503) |
| Size (maximum length x width x height) | 341 mm x 200 mm x 490 mm (60 K) |
| | 341 mm x 200 mm x 498 mm (3503) |

## MEASURED 60 K COOLER PERFORMANCE

Characteristic load lines for the 60 K cooler are captured in Figure 4 with pertinent parameters identified. Much literature in the cooler community reports on cooler operating points but often neglects to identify all five inter-related parameters: cooling load Q, compressor power P, cold block temperature Tc, heat reject temperature Th, and operating stroke S. These five parameters should be universally quoted for any cooler performance data to make valid comparisons between coolers and applications. The last parameter, stroke, is not redundant with power, because this defines the amount of operating margin in the cooler for use by electronic control loops, for physical end stop clearance, and for relative operating stress. However, a constant stroke load line almost coincides with a constant power load line for these coolers. All thermal data in this article are measured with the cold head in the vertical (up) orientation relative to the gravitational field. Operating performance is slightly sensitive to orientation because of buoyancy (absent in zero g) of the working fluid in the open pulse tube interior. In a 1-g horizontal orientation, such as ground testing, this results in approximately 60.75 K compared to the 60.0 K vertical measurement for the same 2 Watt cooling load and input power.

Sensitivity to heat reject temperature is displayed in Figure 5 at constant 60 K source temperature and at constant 76 W into the compressor. At 60K, this line has a slope of about -20 mW/K of reject temperature change. The cooldown profile of the cooler with no applied load or attached mass is presented in Figure 6.

Nominal cooldown is achieved in about 15 minutes. At 65% stroke cooldown the temperature bottoms out at 35K. Transient response to a 5 W sudden heat reduction at 65K is shown in Figure 7. This information is useful in developing a transfer function relationship for a temperature loop controlled by an external heater. Efficiency and cooling robustness of the cooler are displayed in Figure 8 as a function of input power at 60 K cold tip and 303K reject temperature. Efficiency as measured by the specific power is relatively flat above about 70W of input power and increases sharply below 50W of input power as the parasitic losses become an increasing fraction of the gross cooling power. As the power to the compressor increases, the refrigeration capacity as measured by the load line slope through 60K steadily increases.

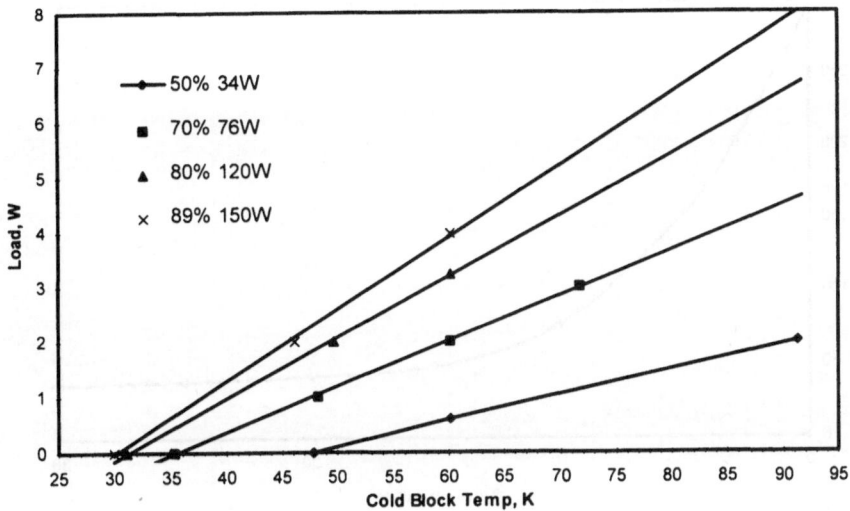

**Figure 4.** Load lines of variable power at 298 K reject temperature as measured on the 60K cooler.

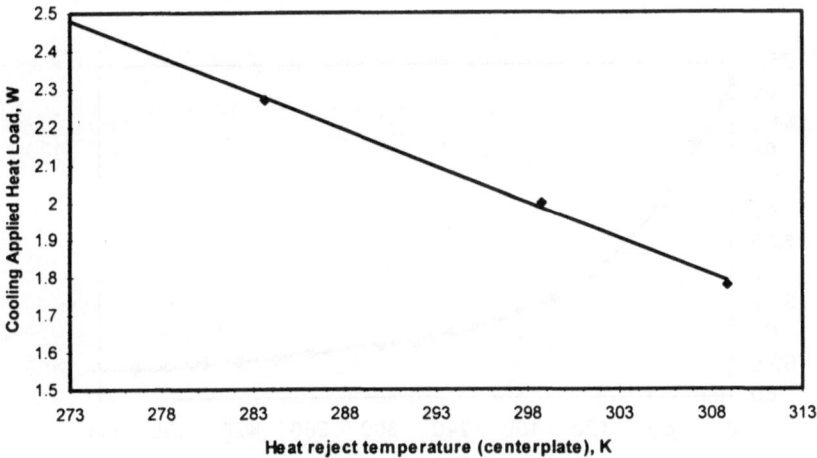

**Figure 5.** 60K cooler's sensitivity to cooling loads at 60K relative to heat rejection temperature for 76 Watts of input power at 44 Hz drive frequency.

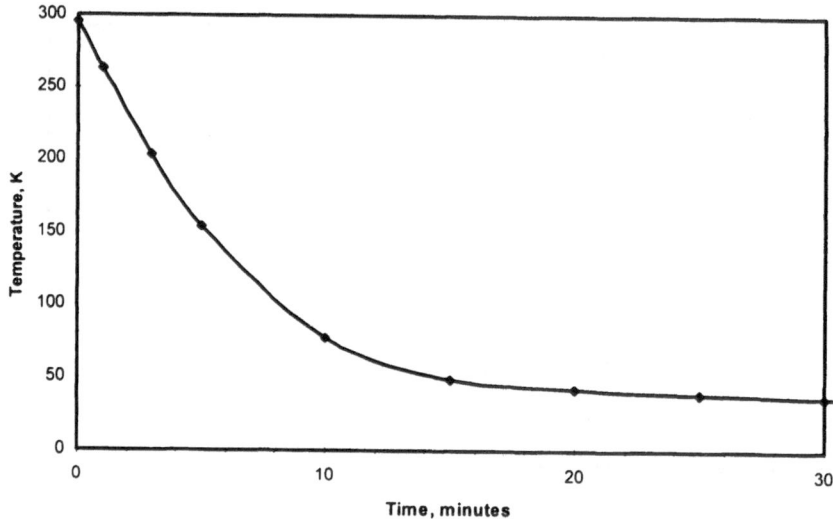

**Figure 6.** Cooldown profile of the 60K cooler with no applied load , 65% stroke, and no added cold mass. Final steady state temperature is 35K.

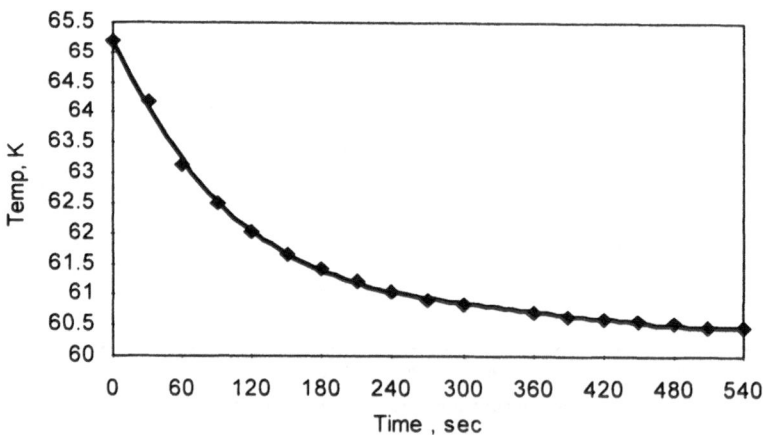

**Figure 7.**   Transient open loop cooler response to sudden 0.5 W cooling load reduction at 65 K. Constant reject temperature at 298 K, constant 92 W input power, no added mass.

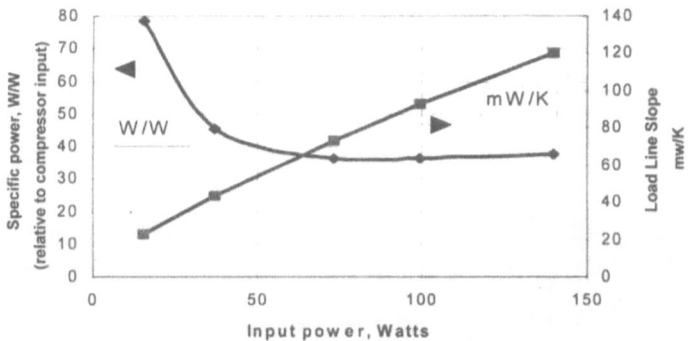

**Figure 8.** 60 K cooler efficiency and cooling capacity at 60K rejecting to 303 K.

The cooler is inherently quiet by its head-to-head compressor design and the pulse tube has no moving parts that produce vibration. Its operation is not perceptible by touch. Nonetheless, most space applications strive for ultra quiet operation. In coolers of the Oxford type, further mitigation of residual vibration from the compressor is accomplished by using noise cancellation methods[7] to minimize total harmonic distortion. Vibration output of the 60 K cooler is shown in Figure 9 where no harmonic cancellation is used and in Figure 10 where the drive waveform was optimized for harmonic suppression up to 1 kHz. The vibration data are measured using a Kistler commercial force dynamometer transducer to which the cooler is rigidly mounted. Cancellation provides up to a 10x reduction in some harmonics of the piston *drive* axis levels, such as the sixth harmonic, but has only a 2 to 3x reduction on the overall root sum square (RSS) three-axis vector sum of its peak components. In the case of the sixth harmonic, its cancellation results in a growth of the fifth harmonic. This diminishing return is a consequence of the algorithm, single (drive) axis control, and system response. The user must assess the dollar costs versus system benefits of implementing waveform suppression capability for a factor of 2-3 improvement.

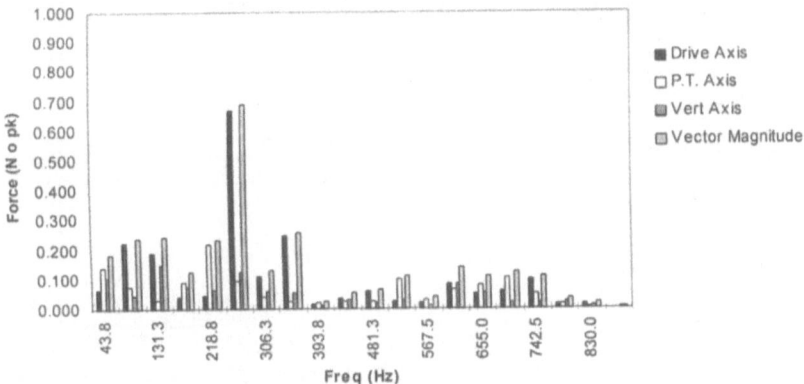

**Figure 9.** 60K cooler raw jitter output WITHOUT harmonic waveform suppression measured at 78 W input at 60K cold block and 43.8 Hz sinusoidal drive frequency.

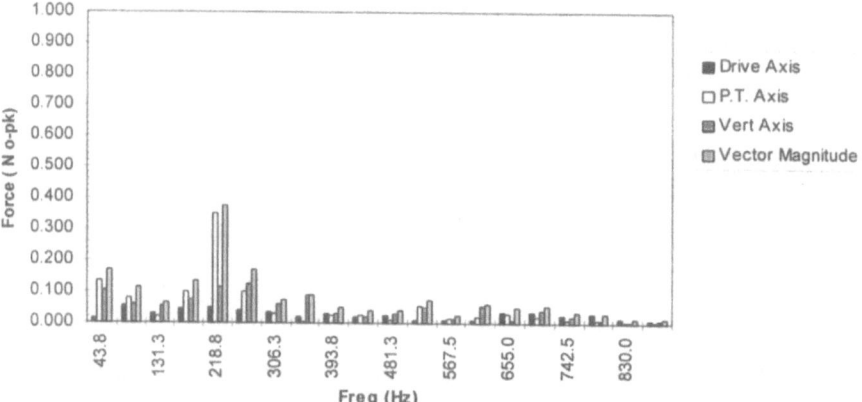

**Figure 10.** 60K cooler jitter output WITH harmonic waveform suppression applied as measured at 78 W input power at 43.8 Hz drive frequency and 60K cold block. The 6th harmonic is reduced by over 10x. Drive axis (black bar) levels are all reduced to the noise floor thresholds.

## MEASURED 3503 COOLER PERFORMANCE

The 3503 cooler performance is very similar to the 60 K with the primary exception being enhanced 35 K performance. Load lines for the 3503 cooler are displayed in Figure 11. Heat sink sensitivity at constant load and temperature is plotted in Figure 12. Vibration output characteristics are nearly identical to the 60 K and are not reproduced here.

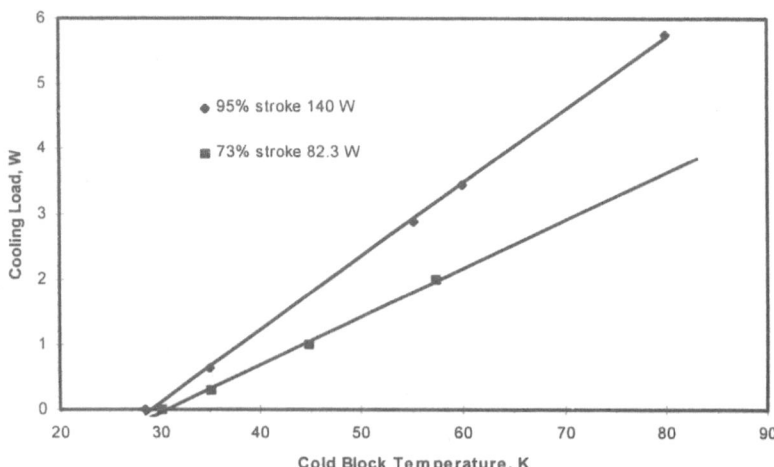

**Figure 11.** Measured load lines for the 3503 cooler.

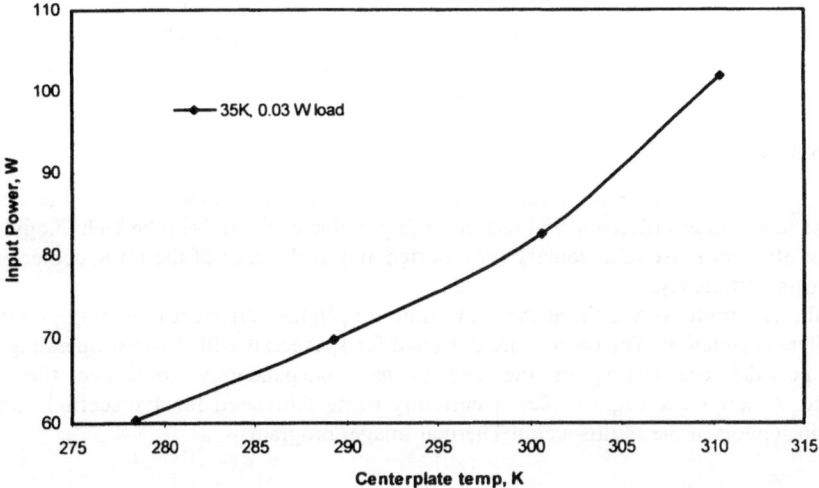

**Figure 12.** 3503 cooler sensitivity to sink temperature at constant load and source temperature.

## RANDOM VIBRATION

Space flight hardware is exposed to severe vibration environments during launch. The 60 K cooler was tested against the specified levels of Figure 13. Using NASTRAN and ANSYS finite element analysis in both beam and shell models, the cooler was designed to withstand these and higher loads.

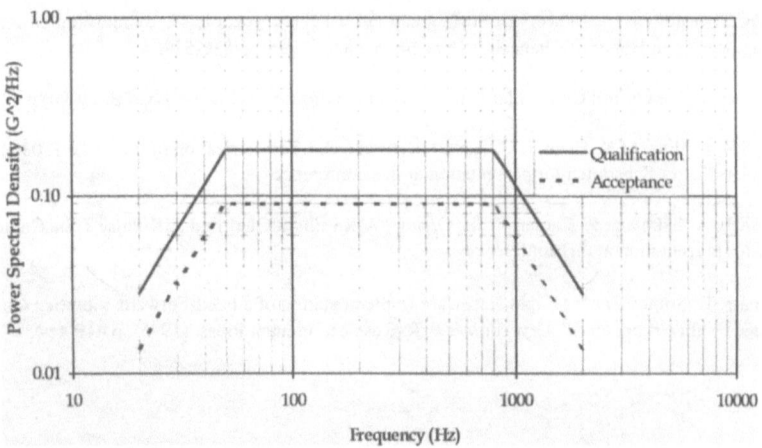

**Figure 13.** The random vibration qualification test level for the 60K cooler was 14.1 grms.

The cooler compressor and cold head components were successfully tested to the qualification loads of 14.1 grms where the plateau portion of the spectrum has a value of 0.16 $g^2$/Hz. The components were exposed separately to the specified levels in each of three orthogonal axes for 1 minute duration in each axis.

## CONCLUSIONS

The two coolers reported upon in this article represent a noteworthy advancement in state-of-the-art pulse tube cooler efficiency and reduction to practice of the pulse tube technology. They are the most efficient pulse tube coolers yet reported and in the case of the 60 K cooler exceed reported Stirling efficiency.

While the two units were built at the same time for slightly different operating points they show excellent correlation. The coolers are designed for spacecraft with 10-year operating lives. Extensive post-delivery testing of the coolers has independently confirmed the cooler performance. A new 60 K flight cooler is currently being fabricated for a spaceflight mission circa 1998 in support of the Multispectral Thermal Imager program.

## ACKNOWLEDGMENTS

The work reported on was sponsored by the United States Air Force, Air Force Systems Command, Phillips Laboratory, Kirtland AFB, New Mexico, 87117-5776.

## REFERENCES

[1]  E. Tward, C.K. Chan, J. Raab, R. Orsini, "Miniature Long-Life Space-Qualified Pulse Tube Cooler," *SAE Technical Paper* #941622, SAE International, Warrendale, PA (1994).

[2]  W.W. Burt, C.K. Chan, "Demonstration of a High Performance 35K Pulse Tube Cryocooler," *Cryocoolers 8*, R. Ross ed., Plenum Press, (1995), p 313-319.

[3]  C.K. Chan, "AC Thermodynamics Theory of Stirling and Pulse Tube Cryocoolers", Symposium on Thermal Science and Engineering in Honor of Chancellor Chang-Lin-Tien, (1995), p 531-539.

[4]  G. Davey, "Review of the Oxford Cryocooler," *Advances in Cryogenic Engineering*, v35B, (1990) p 1423-1430.

[5]  D.L. Johnson, S.A. Collins, M.K. Heun, R.G. Ross, "Performance Characterization of the TRW 3503 and 60 K Pulse Tube Coolers," JPL. Scheduled for presentation at this conference.

[6]  C.K. Chan, J. Raab, A. Eskovitz, R. Carden III, R. Orsini, "AIRS Flight Qualified 55K Pulse Tube Cooler," TRW Inc. Scheduled for presentation at this conference.

[7]  R. Boyle, L. Sparr, T. Gruner, *et al.*, "Flight Hardware Implementation of a Feed-Forward Vibration Control System for Space Flight Cryocoolers," *Cryocoolers 8*, R. Ross ed., Plenum Press, (1995), p 449-454.

# Performance Characterization of the TRW 3503 and 6020 Pulse Tube Coolers

D. L. Johnson, S. A. Collins, M. K. Heun, and R.G. Ross, Jr.

Jet Propulsion Laboratory
California Institute of Technology
Pasadena CA 91109

C. Kalivoda

Phillips Laboratory
Kirtland Air Force Base
Albuquerque, NM

## ABSTRACT

The Jet Propulsion Laboratory, under joint Ballistic Missile Defense Organization (BMDO)/Air Force Phillips Laboratory and NASA/EOS Atmospheric Infrared Sounder (AIRS) sponsorship, has conducted extensive characterization testing of the TRW Model 3503 and Model 6020 pulse tube cryocoolers. These coolers, built under BMDO/AFPL sponsorship, share a common design that utilizes a single-stage pulse tube integrally mounted onto 10-cc common compression space compressors, and are distinguishable by slight differences in the pulse tube designs which optimized cooler performance for operation at either 35 K or 60 K. The coolers were characterized over a range of heak rejection temperatures and cooler operating parameters (compressor stroke, piston offset, and drive frequency) to understand their effects on cooler thermal performance, cooler-generated vibration and cold block motion, and cooler-generated EMI. Pulse tube parasitic conduction as a function of cold block temperature has been studied for a non-operating cooler; the results show a strong angular dependence relative to gravity. The results of the parametric studies are presented.

## INTRODUCTION

The Air Force Phillips Laboratory has been developing pulse tube technology since 1992 when it awarded the 35 K Pulse Tube contract to TRW for development of a 1 W at 35 K pulse tube cooler using a 20-cc compressor. The overall objective was to improve pulse tube cryocoolers to be competitive with Stirling cryocoolers. TRW has significantly advanced the state-of-the-art of this technology in terms of performance, efficiency, and weight, thus enabling pulse tube cryocoolers to be competitive with the Stirling coolers. Phillips Laboratory and the Jet Propulsion Laboratory (JPL) have teamed to fully characterize the coolers' performance.

Cryocoolers 9, Edited by R.G. Ross, Jr.
Plenum Press, New York, 1997

JPL has recently completed extensive performance characterization of the TRW 3503 and TRW 6020 pulse tube coolers delivered under a Phillips Laboratory contract[1]. Characterization tests include varying environmental and cooler operating parameters to determine their effects on performance, off-state parasitics, self-induced vibration and EMI. The 3503 cooler provides a nominal 0.3 watts of cooling at 35 K with 80 watts of input power to the cooler; the 6020 provides a nominal 2.0 watts of cooling at 60 K with 80 watts input power to the cooler.

The TRW 3503 and 6020 pulse tube cryocoolers are very similar in design. The compressor design consists of two 5-cc back to back flexure-bearing compressors compressing the helium working fluid in a common compression space. The pulse tube is integrally mounted to the center plate of the compressor. The pulse tube cold head designs are also very similar; with only slight differences between the cold heads to optimize the cooler performance for either 35-K or 60-K operation, respectively. The two 10-cc coolers are shown side-by-side together with their laboratory drive electronics in Fig. 1.

The coolers are driven with low-distortion audio amplifiers and sinusoidal voltage waveforms. While there is no active temperature and vibration control for these cryocoolers in the present electronics, a separate effort is underway to develop this capability for space flight application. The TRW model 6020 cryocooler, with slightly modified interfaces, has been selected for the Multispectral Thermal Imager space mission in 1998.

The characterization testing of these 10-cc coolers has provided very informative and timely performance results for the JPL Atmospheric Infrared Sounder (AIRS) instrument[2]. The AIRS flight cooler in development by TRW also uses a 10-cc compressor and has similar thermal performance requirements[3]. There are several significant design differences in the AIRS cooler which has lengthened the development effort of this cooler. With the build of the AIRS engineering model instrument in progress, the TRW 3503 and 6020 cooler characterization results have helped finalize the design of the AIRS cooler and the planned thermal vacuum tests of the integrated AIRS engineering model instrument

## REFRIGERATION PERFORMANCE

### Thermal Performance

Separately, each cooler was mounted in a thermal-vacuum chamber on thin-walled stainless steel standoffs to minimize thermal conduction to the chamber. Copper heatsink plates with

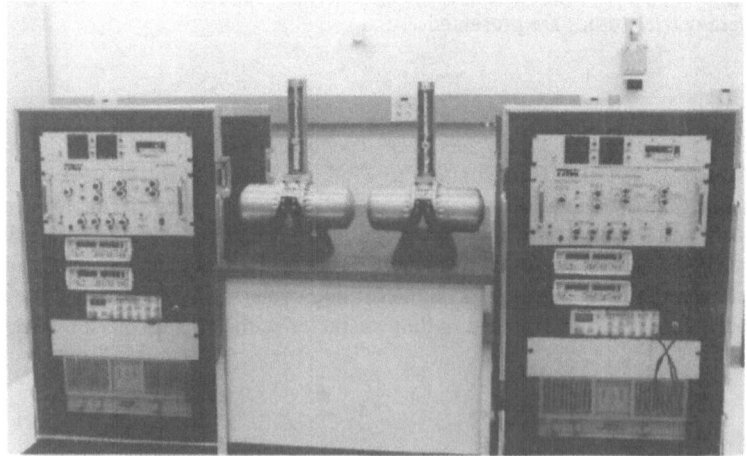

**Figure 1.** The TRW 3503 (left) and TRW 6020 (right) pulse tube coolers.

temperature controlled fluid loops were attached to opposing sides of the compressor center plate to simulate the heat rejection in a spacecraft implementation. The two coolers were comprehensively tested over similar coldblock temperature ranges (35 K to 100 K) with similar variations in the operational and environmental operating parameters — compressor piston stroke, piston position offset, drive frequency, and heat sink temperature — to provide performance sensitivity measurements with respect to each of these parameters.

Thermal performance measurements of the pulse tube coolers are plotted on multi-variable plots devised by JPL to describe the cooler thermal performance dependence on the cooler input power, coldtip load, cold block temperature, specific power, and other operational parameters.

**Piston offset.** Piston offset is a shifting of the piston neutral position towards or away from the compression end of the cylinder. A positive offset moves the pistons towards the compression end of the cylinder, and has the effect of reducing head room and cooler dead volume. This offset affects the peak-to-peak pressure ratio for a given stroke, and becomes increasingly important at lower compressor strokes. This parameter was examined first. Each cooler was operated at the nominal 11-mm operating stroke, and cooler performance was measured as a function of cold block load for different dc offset positions for the pistons; the optimal offset was used throughout the remainder of the thermal performance tests. The 3503 cooler was found to operate more efficiently with a +1-mm offset over the 35-K to 100-K temperature span, whereas the 6020 cooler was operated at zero offset since its thermal performance was rather insensitive to piston offset. Figures 2 and 3 show the sensitivity of the thermal performance to piston offset.

**Compressor stroke.** Figures 4 and 5 show the overall thermal performance of the respective 3503 and 6020 coolers as a function of compressor stroke. The nominal operating stroke for both coolers was 11-mm (about 70% of full stroke). As can be seen in comparing the graphs, the 3503 cooler, optimized for 35 K operation, is able to reach lower cold block temperatures although requiring higher cooler input power at a given stroke than the 6020 cooler. The 3503 cooler was able to provide 350 mW of cooling at 35 K for an input power of 105 watts, whereas for the same input power the 6020 cooler was unable to cool below 37 K. At 60 K, both coolers are able to produce 2 watts of cooling with approximately 80 watts of cooler input power. Above 60 K the 6020 performs more efficiently for a given cold block load and temperature. As can be seen from

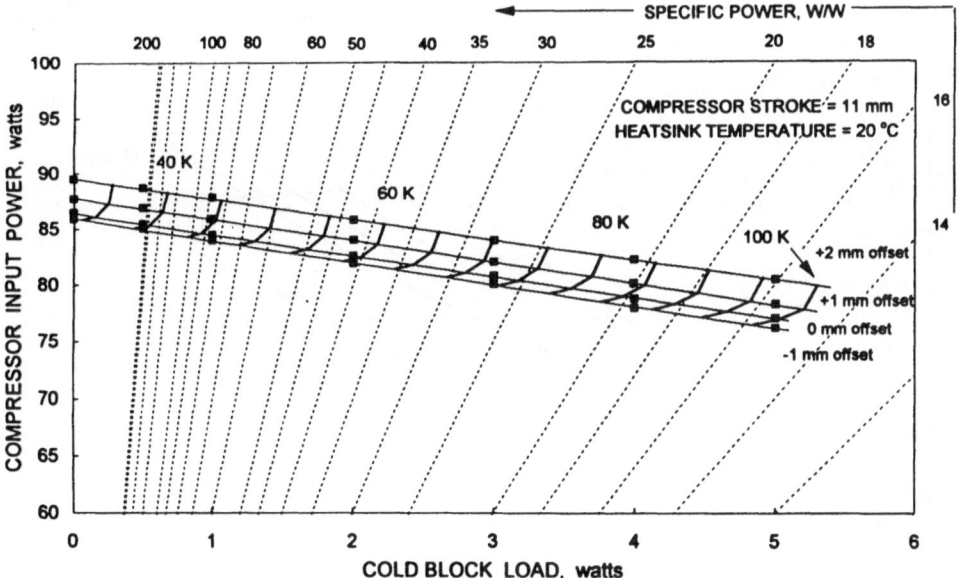

**Figure 2.** Sensitivity of the 3503 cooler thermal performance to piston offset.

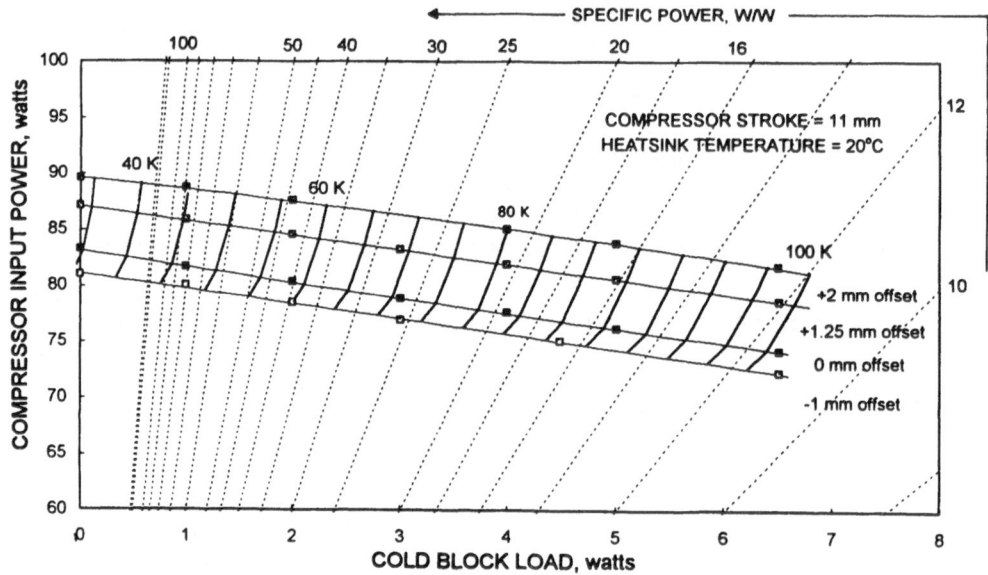

**Figure 3.** Sensitivity of the 6020 cooler thermal performance to piston offset.

the figures, cooler efficiency is a strong function of the stroke at the lowest temperatures, with improved efficiencies for higher strokes. Around 60 K, cooler efficiencies are insensitive to wide stroke variations.

**Drive frequency.** Cooler performance for each cooler was measured at the 11-mm operating stroke for several different drive frequencies ranging between 40 Hz and 48 Hz to demonstrate the performance sensitivity to cooler drive frequency (See Figs. 6 and 7). Nominal operating drive frequencies are 46 Hz for the 3503 cooler, and 44 Hz for the 6020 cooler. Note that at the

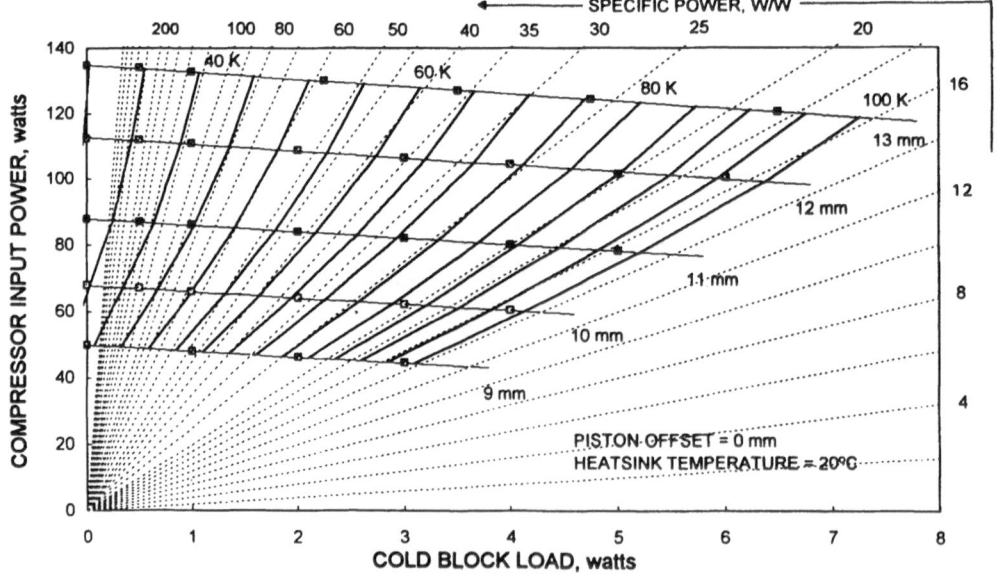

**Figure 4.** Sensitivity of the 3503 cooler thermal performance to compressor stroke.

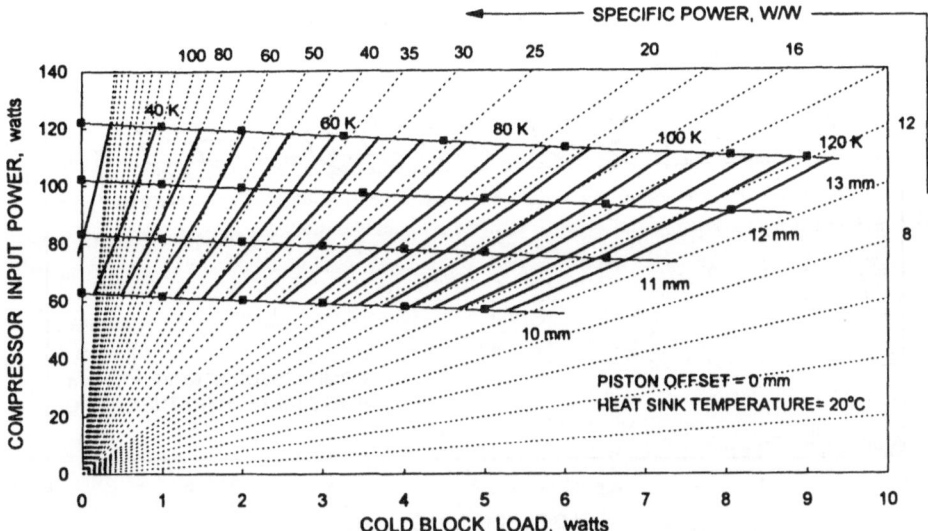

**Figure 5.** Sensitivity of the 6020 cooler thermal performance to compressor stroke.

suggested drive frequency for each cooler the cooling capacity and the specific power are optimum for temperatures below 60 K; above 60 K the efficiency of each cooler improves with slightly lower drive frequencies.

**Heat sink variation.** Figure 8 depicts the change in thermal performance of the 3503 cooler due to a change in heat sink temperture from 0°C to 40°C for a constant 11-mm stroke. There is a very significant improvement in efficiency, both with improved cooling capacity and with decreased input power by operating at the lower heat sink temperature. At higher tempratures, efficiency continues to improve with lower sink temperatures. Figure 9 demonstrates the shift in

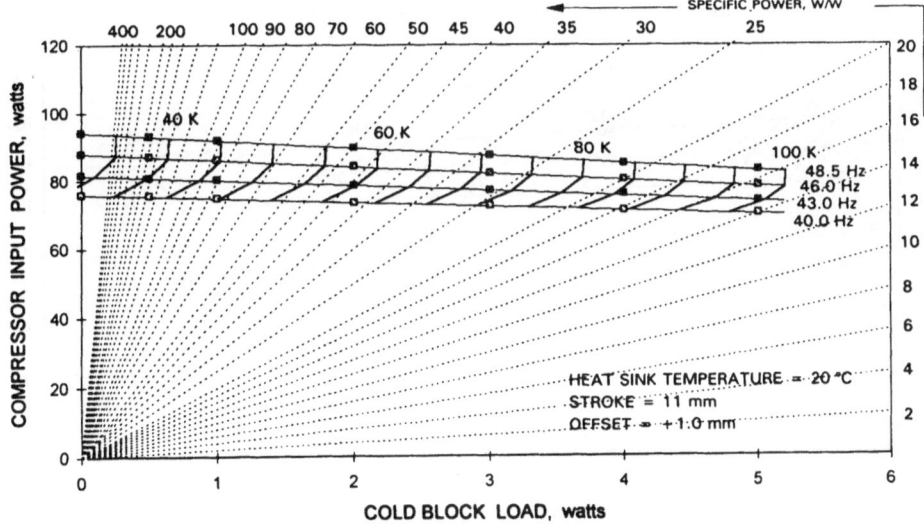

**Figure 6.** Sensitivity of the 3503 cooler thermal performance to drive frequency.

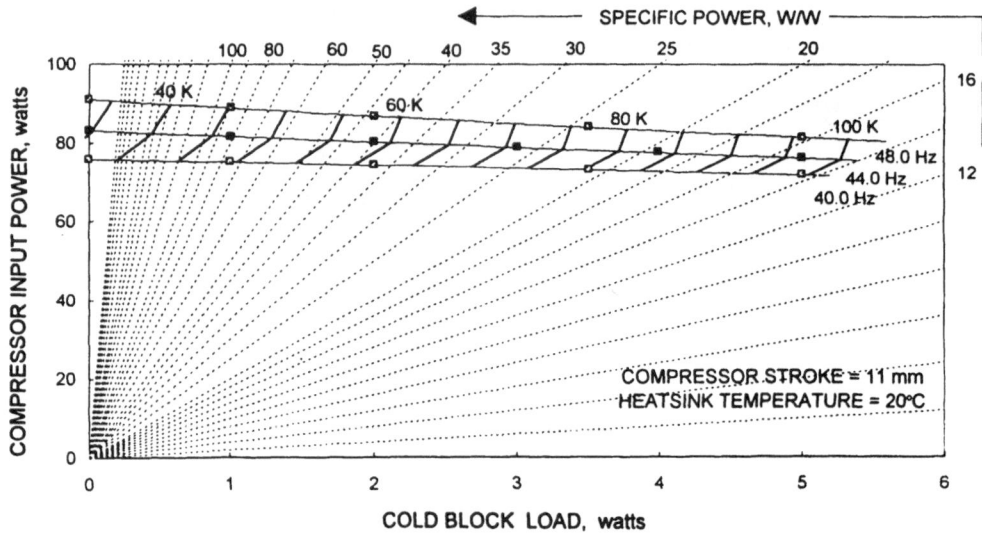

**Figure 7.** Sensitivity of the 6020 cooler thermal performance to drive frequency.

isotherms between the 11-mm and 13-mm stroke conditions for the heat sink variation between
20°C and 0°C. The 20°C change in sink temperature can result in as much as a 40 watt change in
input power for the same cooling load and coldblock temperature.   Similar performance
improvements with reduced heat sink temperatures were observed with the 6020 cooler but are
not presented here. Heat sink temperatures were monitored at the copper heat sink plates bolted
to the compressor center plate.   Prior to operating the cooler at either 0°C or 40°C heat sink
temperature, each cooler was driven at full stroke at very low frequencies (0.002 Hz to 0.005 Hz)
to make stiction measurements to insure there was no rubbing of the piston due to differential
thermal contraction mismatches.

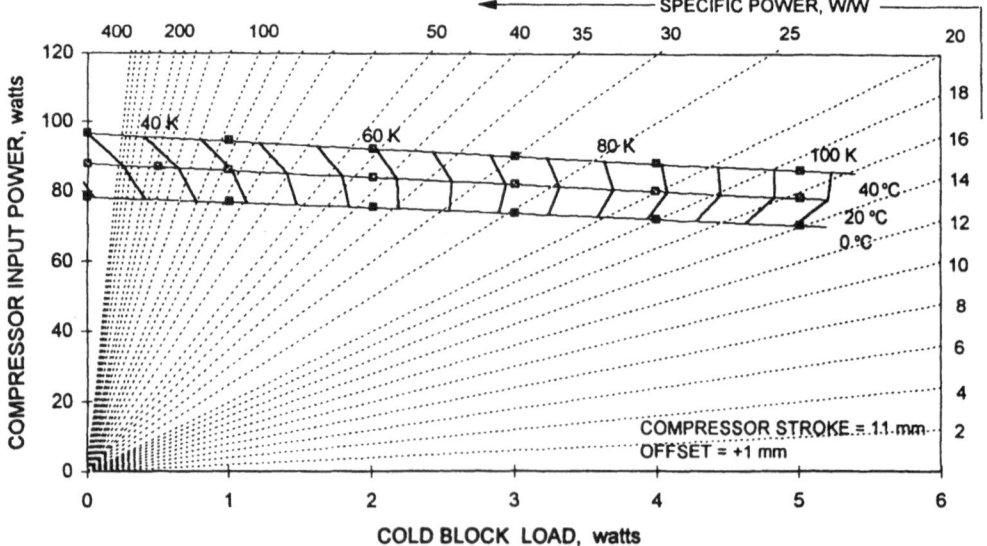

**Figure 8.** Sensitivity of the 3503 cooler thermal performance to heat sink temperature.

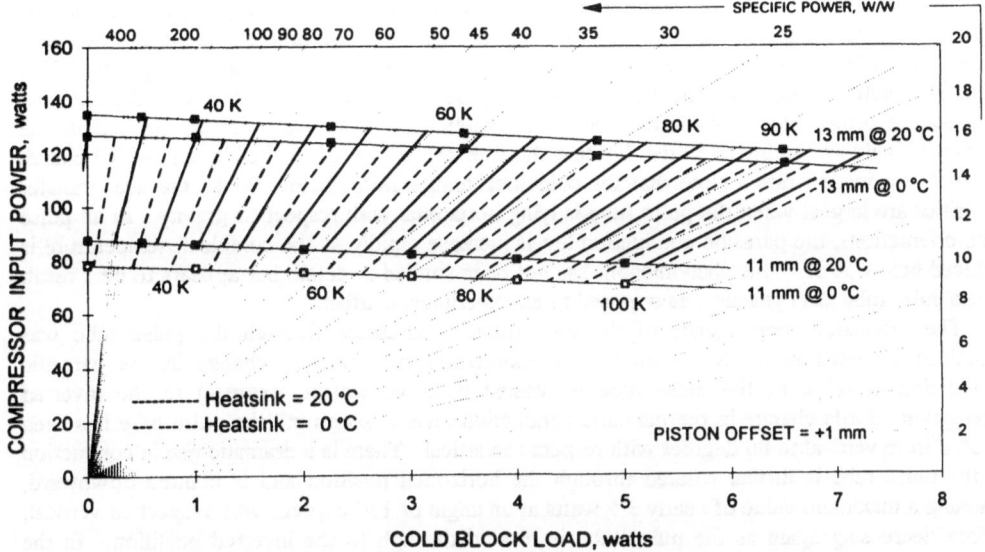

**Figure 9.** Sensitivity of the 3503 cooler specific power to heat sink temperatures.

## Cooler Efficiency

The figure of merit for cryocoolers, the thermodynamic coefficient of performance (COP), is defined as the ratio of the net cooling power to the net applied input power (input electrical power - $i^2R$ heating of the coil), and is expressed as a percentage of the ideal Carnot COP. The 3503 cooler, operating with an 11-mm stroke and 20 °C heat sink, ran at about 2 % of Carnot efficiency at 35 K, increasing to 4% at 35 K by reducing the heat sink temperature to 0°C. It also operated more efficiently with higher compressor strokes. The 6020 cooler operated with a Carnot efficiency near 12% at 60 K. Similar levels of Carnot efficiency were observed at 60 K for the 3503 cooler.

Motor efficiencies ranging between 80% to 88% were measured for the coolers under the various operating conditions. Motor efficiencies are dependent on the $i^2R$ losses within the coil. In general, motor efficiencies were found to increase with decreasing stroke, increasing cold block temperature for a given stroke, decreasing drive frequency, and increasing heat sink temperature. Motor efficiencies were highest at the optimal dc piston offset condition.

The measured power factors for the motors of the coolers range from 0.8 to as high as 0.96. The power factor is defined as the ratio of the true rms input power to the product of the measured true rms voltage and true rms current. The observed power factor followed the trends of the motor efficiency, that is, the power factor improved with decreasing stroke, decreasing drive frequency, and increasing heat sink temperature.

## OFF-STATE CONDUCTION

An important consideration in the use of redundant coolers is the parasitic heat load placed on operating coolers by the non-operating or standby coolers. Because the parasitic load can be a substantial fraction of the available cooling power at cryogenic temperatures, accurate data on these off-state loads are essential.

Heat transfer parasitics through the pulse tube of the non-operating pulse tube cryocooler were measured for both the 3503 and 6020 cryocoolers using an equilibrium heat conduction measurement technique[4]. In the test configuration, the pulse tube is enclosed in a vacuum housing together with the coldfinger of a second cryocooler that is used to cool the cold block of the pulse tube cooler down to typical flight-operating temperatures. An absolute heat-flow transducer measures the resulting heat flow passing through the pulse tube.

The measured parasitic conduction levels were very similar for the two pulse tube coolers. Figure 10 shows the measured parasitic heat flow through the pulse tube of the 6020 cooler as a function of coldblock temperature for three pulse tube orientations with respect to gravity. The most favorable position for the pulse tube is to be positioned vertically, with the ambient temperature orifice block positioned above the coldblock. In this orientation the helium gas becomes highly and stably stratified, limiting the free convection of the helium gas which causes the high conduction rate. Note that for coldblock temperatures below 90 K, the heat transfer parasitics are largest when the pulse tube is lying horizontal with respect to gravity. In all pulse tube orientations, the parasitic conduction levels increase rapidly as the coldblock temperature is reduced below 50 K. This phenomenom has not been studied in detail, but appears to be a result of the pulse tube gas dynamics, as opposed to an experimental affect.

The orientation dependence of the heat transfer parasitics through the pulse tube was examined in detail at 60 K. Figure 11 demonstrates the dramatic change in the parasitic conduction at 60-K as the pulse tube is rotated from vertical (0 degrees) to the inverted orientation. Little change in the parasitic conduction level is seen until the pulse tube has been rotated from vertical to 80 degrees with respect to vertical. There is a dramatic rise in conduction as the pulse tube is further rotated through the horizontal position and is pointed downward, reaching a maximum value of nearly 3.5 watts at an angle of 130 degrees with respect to vertical, before decreasing again as the pulse tube is rotated through to the inverted postition. In the inverted orientation the pulse tube is observed to have a conductance level twice that of the vertical orientation.

## GENERATED VIBRATION

The measurements of the vibration generated by the pulse tube coolers were conducted in the JPL cryocooler vibration characterization facility using a special-purpose six degree of freedom dynamometer[5]. The TRW coolers were mounted with the compressor aligned in the vertical direction. This facilitated the gathering of data for the moment about the piston axis along with the force vectors in the x, y, and z directions.

## Compressor Vibration

Vibration measurements were made as a function of such parameters as the drive frequency, the compressor stroke, the piston offset, and the coldblock temperature. The back-to-back compressor configuration facilitated the minimization of the vibration along the piston axis at the

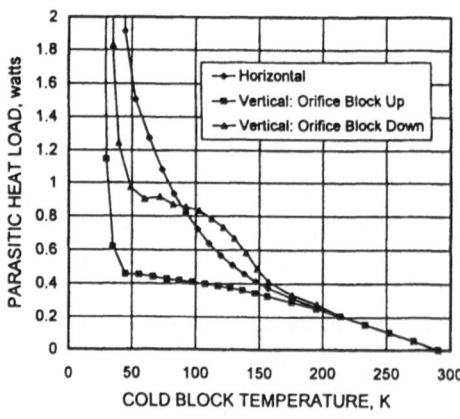

**Figure 10.** Heat flow parasitics of the 6020 cooler as function of cold block temperature.

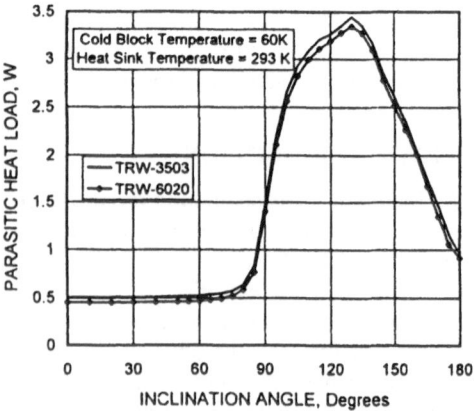

**Figure 11.** Heat flow parasitics as a function of inclination angle (0° refers to vertical orientation).

fundamental harmonic by trimming the stroke of one of the two compressors. There was no provision available with the laboratory drive electronics for multi-harmonic vibration cancellation.

A sample of the vibration levels measured for various piston strokes are shown for the two coolers in Figs. 12 and 13. The z-axis is the vertical axis measured along the piston axis, the x-axis is pointed along the pulse tube. Although identical in build, the vibration signatures from the two 10-cc compressors are significantly different. The measured vibration levels for the 3503 cooler were typically between 0.1 N and 1.0 N, the larger value being for a harmonic in the piston axis. Vibration levels typically fell off with increasing harmonics. The vibration levels for the first few harmonics for the 6020 cooler were similar to the 3503 cooler in all axes, however at the higher harmonics the the 6020 vibration levels did not fall, but rather were at or above the vibrational levels of the first few harmonics. The 6020 vibration levels could be reduced by running the cooler at 46 Hz.

## Coldblock Motion

The motion of the coldblock was also measured as a function of several operating parameters, such as drive frequency, piston stroke, etc. This motion was measured while the compressors were mounted on the dynamometer, using a set of three miniature cryogenically coolable accelerometers mounted on the coldblock and aligned to the same axes as the force sensors. Measured coldblock motions in the pulse tube direction were on the order of 1 to 2 $\mu$m, and an order of magnitude lower in the pulse tube transverse directions. These motions were rather insensitive to changes in operating parameters. However it was possible to excite the transverse bending mode of the 6020 pulse tube around 264 Hz

## EMI/EMC

Measurements of the radiated electric and radiated magnetic field emissions were made on the 3503 cooler operating at a nominal 80-W input power to the compressor (11-mm stroke). The cooler was placed in an RF-shielded room and grounded to a copper laminated table; the

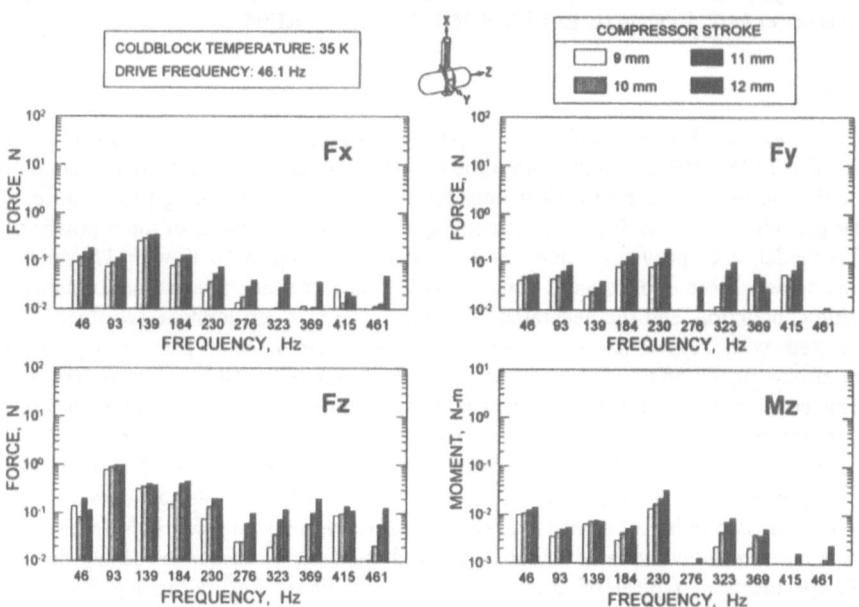

**Figure 12.** Sensitivity of the 3503 cooler-generated vibration to compressor stroke.

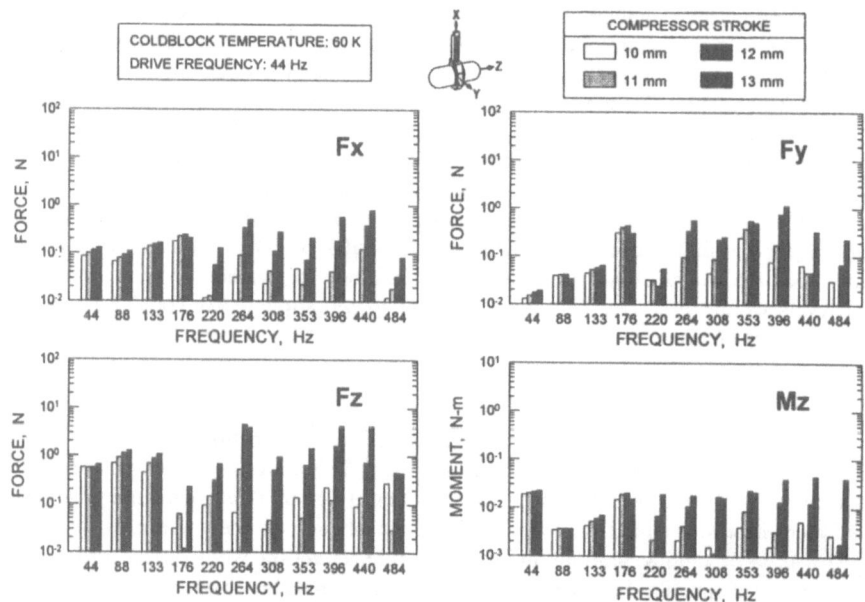

**Figure 13.** Sensitivity of the 6020 cooler-generated vibration to compressor stroke.

cooler electronics were placed in an adjacent room with the cabling fed through a bulkhead in the wall. The cabling was sheathed in aluminum foil and grounded to the copper table top to minimize any contributing radiation. Measured levels for the radiated AC electric field emissions were well below the specification limits of Mil-STD 461C RE02. The AC magnetic field emissions were measured at distances of 7 cm and 1 meter from the cooler. Figure 14 depicts the radiated magnetic emissions measured at 7 cm from the 3503 cooler. Typical of linear voice coil compressor designs, the radiated magnetic emissions of the first few harmonics exceeded the specifications under MIL-STD 461C RE01 and MIL-STD 461C RE04.

## SUMMARY

The effort described herein was a unique opportunity to test two pulse tube coolers from the same build. The TRW 3503 and 6020 pulse tube coolers were constructed using the same 10-cc compressor design, but optimized for their intended 35-K or 60-K operating temperature. The 3503 cooler provided a nominal 0.3 watts of cooling at 35 K with 80 watts of input power to the cooler. The model 6020 provides a nominal 2.0 watts of cooling at 60 K with 77 watts input power to the cooler. Both coolers were characterized in terms of their thermal performance, off-state parasitic conduction, and the generated vibration. Thermal performance sensitivity studies were conducted with respect to key operational and environmental parameters, including compressor stroke, piston offset, drive frequency, and heat sink temperature. The pulse tube off-state parasitic conduction was found to be very sensitive to the pulse tube orientation with respect to gravity. Measurements of the parasitic conduction were made both as a function of coldblock temperature for several fixed pulse tube inclination angles, and also at a fixed temperature for inclination angles between 0° and 180° with respect to gravity. Cooler-generated vibration measurements provided quite different vibration signatures between the two pulse tube coolers. EMI measurements made with the 3503 cooler showed magnetic emissions exceeding the MIL-STD 461C specification, typical of compressors utilizing linear flexure-bearing linear motor designs.

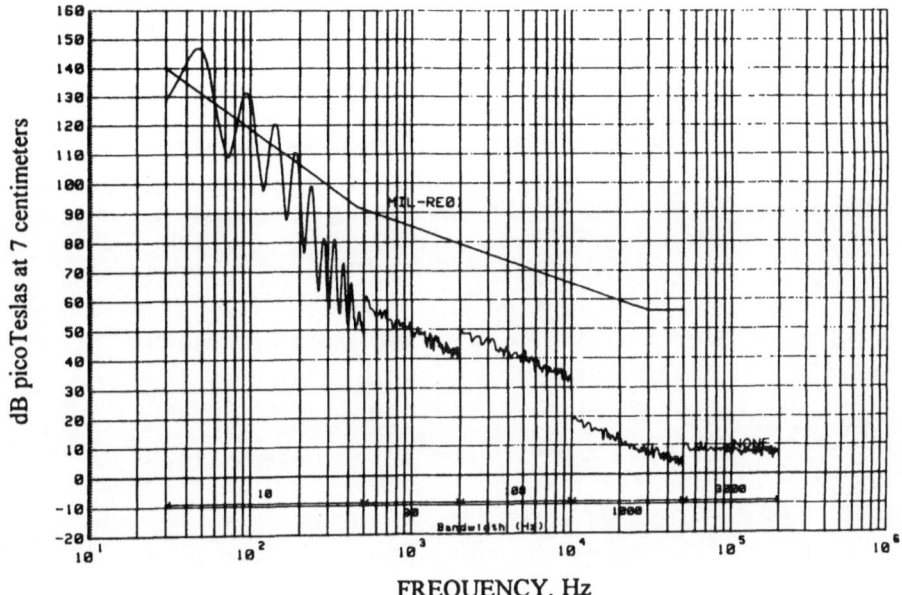

**Figure 14.** Radiated magnetic emissions measured at 7 cm from the TRW 3503 cooler operating with a nominal 80 W input power to the cooler.

## ACKNOWLEDGEMENTS

The work described in this paper was carried out by the Jet Propulsion Laboratory, California Institute of Technology, for the Air Force Phillips Laboratory, Albuquerque, New Mexico. The work was sponsored by the Ballistic Missile Defense Organization under a contract with the National Aeronautics and Space Administration. Particular credit is due S. Leland for his assistance and support with many of the test setups.

## REFERENCES

1.  Burt, W.W. and Chan, C.K.,"New Mid-size High Efficiency Pulse Tube Coolers", <u>Proceedings of the 9th International Cryocooler Conference</u>, Waterville Valley, New Hampshire, June 1996, Plenum Publishing Corp., New York.

2.  Ross, R.G., Jr. and Green, K.E., "AIRS Cryocooler System Design and Development," <u>Proceedings of the 9th International Cryocooler Conference</u>, Waterville Valley, New Hampshire, June, 1996, Plenum Publishing Corp., New York.

3.  Chan, C.K., Raab, J., Eskovitz, A., Carden, R., III, and Orsini, R., "AIRS Flight Qualified 55K Pulse Tube Cooler," <u>Proceedings of the 9th International Cryocooler Conference</u>, Waterville Valley, New Hampshire, June 1996, Plenum Publishing Corp., New York.

4.  Kotsubo, V, Johnson, D.L., and Ross, R.G., Jr., "Cold-tip Off-state Conduction Loss of Miniature Stirling Cycle Cryocoolers, *Advances in Cryogenic Engineering*, Vol.37B, Plenum Press, New York (1991), pp. 1037-1043.

5   Ross, R.G., Jr., Johnson, D.L., and Kotsubo, V., "Vibration Characterization and Control of Miniature Stirling-cycle Cryocoolers for Space Application," *Advances in Cryogenic Engineering*," Vol. 37B, Plenum Press, New York (1991), pp.1019-1027.

# Performance of the AIRS Pulse Tube Engineering Model Cryocooler

C. K. Chan, C. Carlson, R. Colbert, T. Nguyen, J. Raab, and M. Waterman

TRW Space & Technology Division
Redondo Beach, CA 90278, United States of America

## ABSTRACT

TRW has designed, analyzed, built and tested an engineering model (EM) pulse tube cryocooler for the NASA Atmospheric Infrared Sounder (AIRS) instrument. The cooler provides 1.63 W cooling at 55 K while rejecting to 300 K for a mission life of 50,000 hours in space. In the AIRS instrument, two of these split configuration pulse tube coolers are integrated into a structure that supports the cold head vacuum vessel, the compressors, and the heat rejection system to provide a standby redundant cooler system. The key components of the EM cryocooler are a dual-opposed balanced compressor, an efficient pulse tube cold head, and efficient high-reliability drive electronics. This paper presents the cooler performance test data, delineates system weight, and describes input power, envelope, and vibration and temperature control.

## INTRODUCTION

The AIRS pulse tube cryocooler features mechanical and electrical design that achieves major improvements in reliability, efficiency, weight, vibration and ease of interface for the instrument. It consists of three subsystems: the passive, exceptionally reliable, high-efficiency cold head, a heritage flexure-bearing balanced compressor, and control electronics. In comparison to a Stirling cooler, the pulse tube cooler eliminates mechanical complexity in the cold head and the resultant reduced electronics complexity. Figure 1 shows the subsystems.

The three separate mechanical and thermal interfaces and electrical interfaces, are provided by the cooler system. Mounting interfaces through which heat can be conducted are provided for each of the compressors, cold heads, and electronics control unit. Command and control and 28-volt power are provided by the AIRS instrument system to the electronics control system which in turn powers the compressor, and provides excitation and sensor signals to the compressor and cold head. Only redundant thermometer leads are required on the cold head and their leads are routed to the compressor electrical interface.

## DESIGN

Cooler longevity and reliability are inherent in the design using (1) the completely passive pulse tube cold head containing no moving parts (2) S-level electronics parts where possible (3) a mechanical compressor design that incorporates no wearing parts; (4) flexure springs operating

Cryocoolers 9, Edited by R.G. Ross, Jr.
Plenum Press, New York, 1997

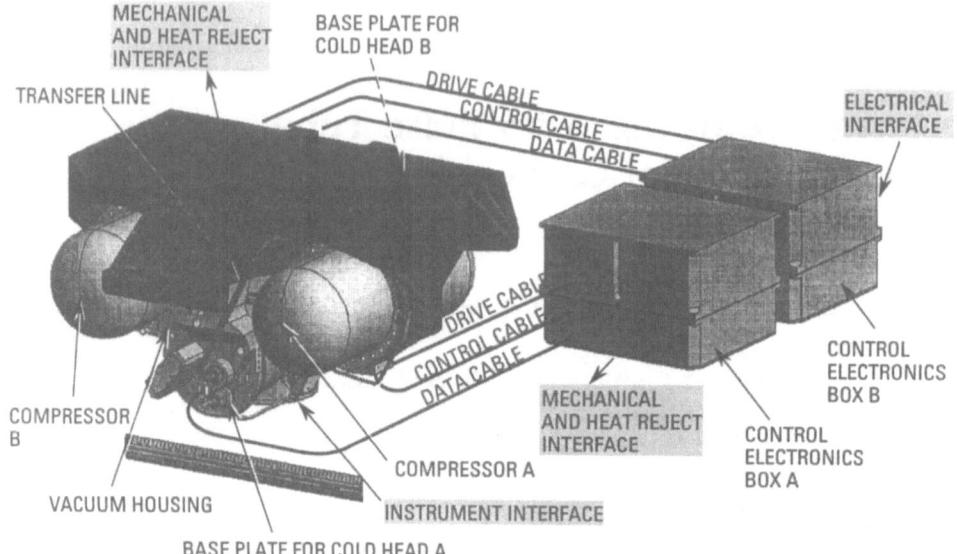

**Figure 1.** AIRS cooler subsystems.

well below fatigue failure limits; (5) hermetic sealing techniques to contain the helium working fluid well beyond the required life; and (6) contamination control techniques during manufacture.

Cooler electronics are powered by the spacecraft and provide 44.625 Hz drive electrical power to the identical linear motors in the head to head compressor. The two compressor linear motors oscillate noncontacting close fitting pistons in cylinders to produce a 44.625 Hz pressure wave and mass flow, resulting in delivery of the compressor shaft power (or PV work) to the cold head. For maximum efficiency, the compressor is operated at its resonant frequency determined by the piston moving mass on the compressed gas spring. Adaptive active vibration control on one of the motorsoperates continually to reduce vibration to levels below the requirement by electronically controlling the compressor stroke. In the pulse tube cold head, the compressor-generated pressure wave causes cooling at the cold block by pumping heat up the regenerator (reference 1).

The cryocooler drive electronics (CDE) control the cold tip temperature and the vibration produced by the cooler using a cold tip temperature sensor and accelerometer mounted on the compressor centerplate. Following a power-on command, the AIRS cooler operates to the default maximum compressor stroke command. During startup, the soft start cooler circuitry ramps and controls the stroke to the required levels.

Cooldown to 55 K has been measured to occur in less than 65 minutes (Figure 2). At the 55 K steady-state temperature sensed by a platinum resistance thermometer (chosen for radiation hardness), internal control loops maintain the cooler cold temperature by controlling the compressor stroke. Self-induced vibration from the compressor is controlled to levels below 0.22N using closed loop and adaptive feedback from the accelerometer to one of the compressor motors. The robust vibration control system learns transfer functions either during cool down or if required in steady state operation. The electronics drive also provides for a set-and-forget mode in which the active control is disabled and preset waveforms and active stroke control maintain a low vibration output. No vibration control is required on the cold head since it contains no moving masses.

The electronics drive provides the user option of disabling or resetting the temperature controller set point or redefining the fixed stroke windows up to the maximum limits, which are electronically established to shut down the cooler in case of overstroke. The user has the option of choosing either automatic or commanded soft restart following an overstroke shutdown.

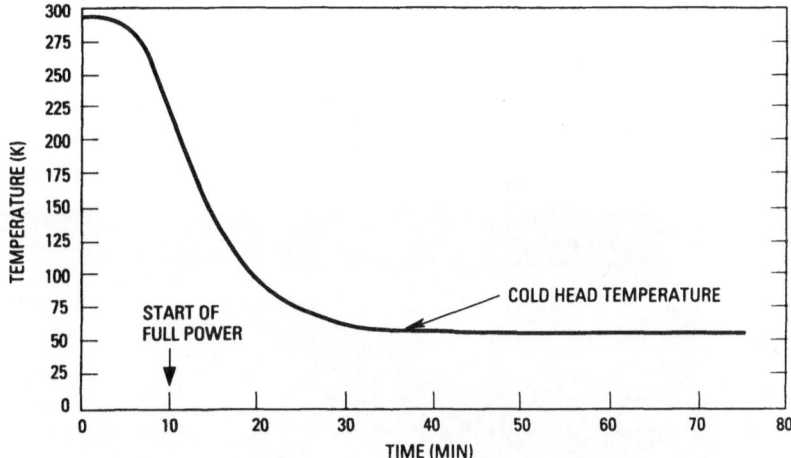

**Figure 2.** Cooldown curve.

## Mechanical Interfaces

The mechanical compressor centerplate serves as the structural/thermal interface to the cooler system structure. The cold head mounting plate provides for equally spaced threaded holes for mounting the cold head to the vacuum housing. The mounting plate also provides for attachment of an aluminum structure to remove heat from the cold head. The vacuum housing around the cold finger allows cooler testing on the ground. The copper cold block has a threaded hole for attachment to a flexible conductive strap for removal of heat from the cooled payload. The electronics control system is housed in a standard slice subassembly, which provides a mounting surface for attachment to a conductive plate on the spacecraft.

## Thermal Interfaces

The heat from the compressor is conducted through the structural conducting plate on which it is mounted. All of the heat generated in the cold head is collected at the flange and can be removed through the aluminum structure to the center plate. Using a finite element code, the temperature distribution from the baseplate to the compressor center plate as well as the heat flux distribution on the mounting plate were computed (Figure 3).

## System Weight

The EM compressor and the cold head subsystem weigh 8.65 kg, each electronic subsystem weights 4.33 kg, and the supporting structure and vacuum vessel that hold both coolers weighs 9.04 kg. Total system weight including two sets of coolers and electronics is 35.10 kg.

## Input Power

The EM compressor was connected to the prototype cold head without the structure. The bus power to the electronics measures 112.2 W with 19.4 W being consumed by the drive and the control electronics. With the remaining 92.8 W of power input into the compressor, the sytem has a cooling capacity of 1.63 W at 55 K when the heat reject temperature at the vacuum flange is 320 K. The load line for the 112.2 W bus power is shown in Figure 4.

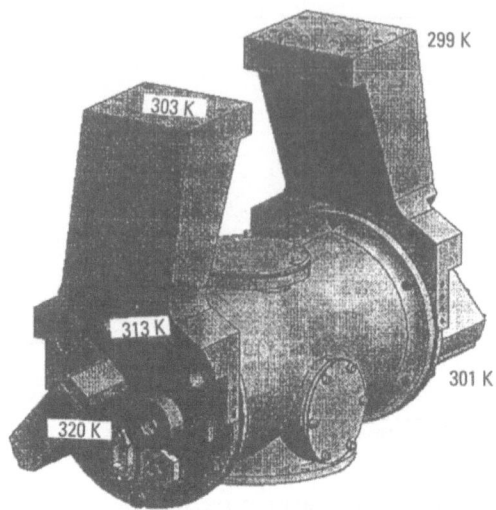

(a) Temperature Distribution

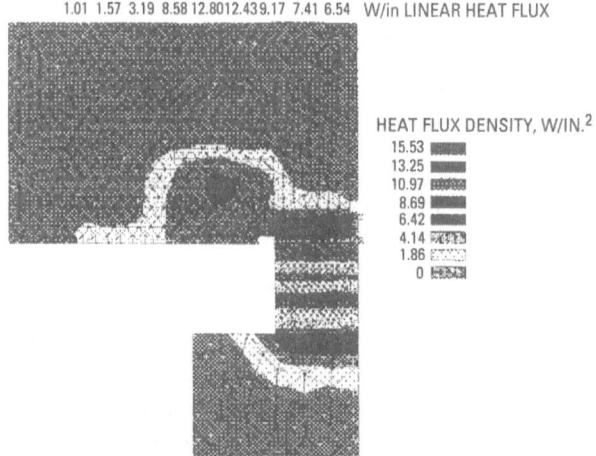

(b) Heat Flux Density and Linear Heat Flux on Structure

**Figure 3.** Code thermal calculation.

## Temperature Stability

The cold head temperatures during the EM acceptance test, were monitored over a short period (<25 minutes) and over a long period (<150 minutes) (Figure 5).  The cooler temperature is stable within the required limits.

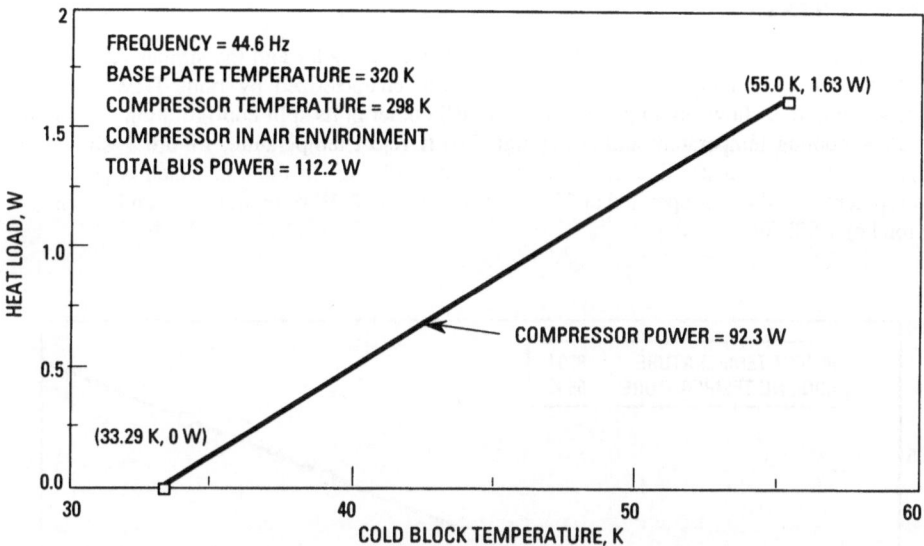

**Figure 4.** Measured AIRS EM cooler load line.

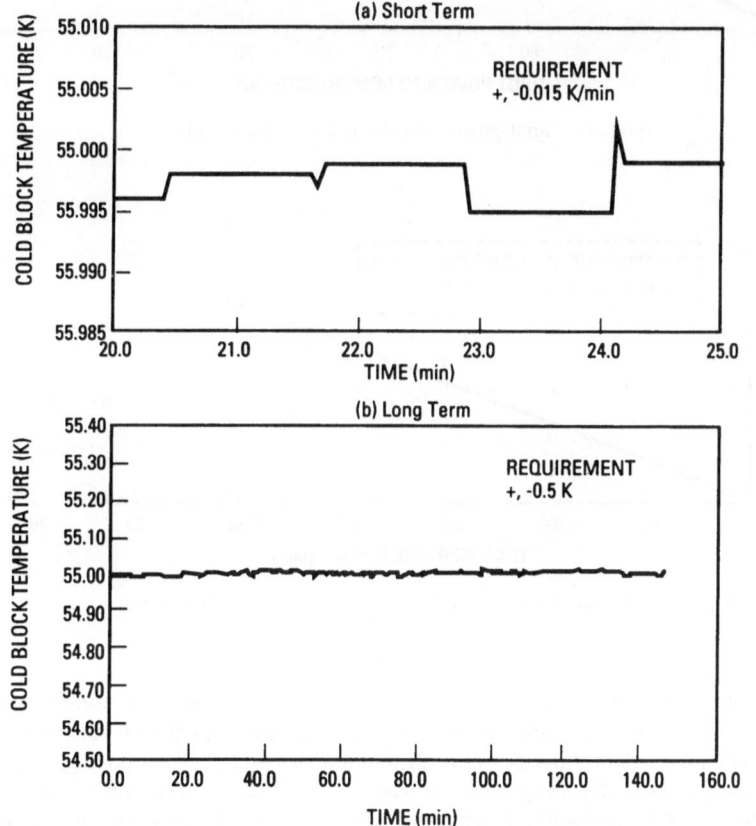

**Figure 5.** Temperature stability.

## SENSITIVITY TESTS

In addition to the operating point, the AIRS split pulse tube cooler and the similar AF60K and AF3503K (reference 3) integral coolers have been well characterized by both TRW and JPL (reference 4). Figure 6 shows input power for the AIRS cooler in its split configuration. Data was taken at 55 K cooling temperature and at the high 320 K reject temperature. Figure 7 shows the penalty paid for various transfer line lengths on a similar coldhead. In the test, which is performed as a development test, the unit operated at 55 K with a constant 80 W of input power and a constant cooling load of 1.522 W.

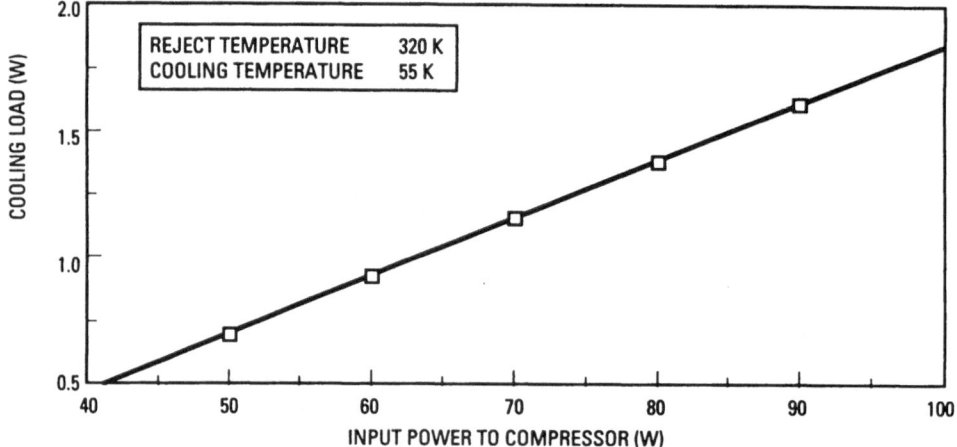

**Figure 6.** Input power sensitivity to cooling load.

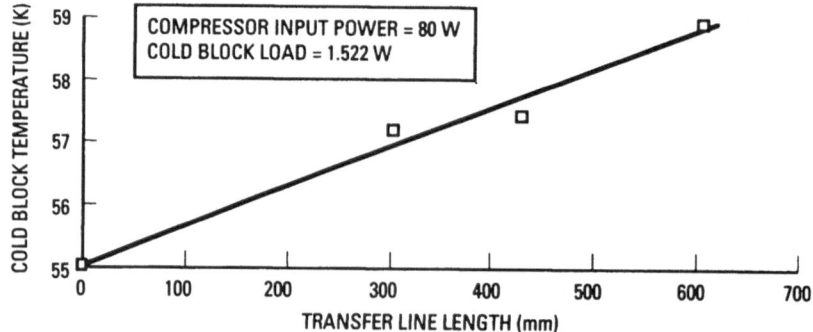

**Figure 7.** Cooling temperature sensitivity to transfer line length.

Since the split configuration can accommodate different cold head and compressor reject temperatures, Figure 8 shows the performance sensitivity measured in another development test for a similar cold head with constant input power to the compressor and constant cooling load. For a nominal 80 W input power to the compressor, reducing the cold head reject temperature from 310 to 281 K resulted in a lowered cooling temperature, from 57 to 55 K, and improved the load line slope, from 76 to 86 mW/K.

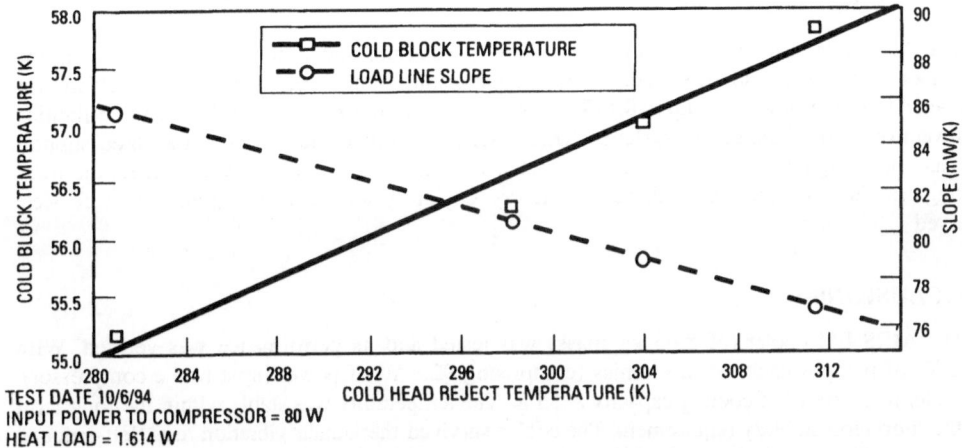

TEST DATE 10/6/94
INPUT POWER TO COMPRESSOR = 80 W
HEAT LOAD = 1.614 W

**Figure 8.** Cooler performance sensitivity to heat reject temperature.

## Compressor Self-Induced Vibration

Redundant accelerometers mounted on the centerplate of the compressor measure the vibration in the direction of the piston motion. This produces the error signal used by the control electronics to adaptively correct the waveform and reduce compressor vibration. Figure 9 gives vibration of the AIRS cooler in three axes as measured with a Kistler dynamometer (force measurement) with the compressor mounted on a 70-ton seismic platform. Measurement was taken with the first eight harmonics controlled. The vibration performance is below the requirement of 0.22 N for frequencies less than 150 Hz and 0.44 N for frequencies greater than 150 Hz. The pulse tube cold head contains no moving parts. However, the internal oscillating pressure causes a small vibration (which is much less than the requirement of 0.11 N) depending on how the cold head is mounted.

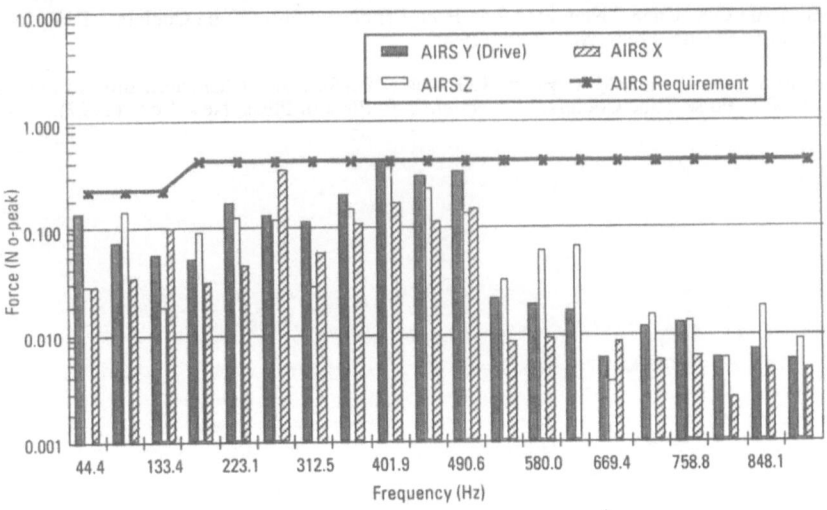

**Figure 9.** Three-axis self-induced vibration force.

## LAUNCH VIBRATION TEST

The AIRS EM cryocooler assembly with support structure and a non-contact snubber at the cold block underwent random vibration qualification testing in all three axes. The test effort was conducted to protoflight levels up to 8.1 Grms (gravity force root mean square) with force limiting to prevent overtest of the cooler system. This system monitored the forces at the three support structure mounting points, then automatically notched the input acceleration levels to meet the force limiting criteria. No structural failures of the cryocooler system and its support structure were observed.

## CONCLUSIONS

The AIRS EM cooler (of 8.65 kg mass) was tested and its performance was verified. With 112.2 W of bus power to the electronics (comprising 92.8 W of power input to the compressor), the cooler had 1.63 W of cooling capacity at 55 K. The temperature was stable within the long term and the short term stability requirement. The cooler survived the launch vibration test. Self-induced compressor vibration was less than the required level of 0.22 N for frequencies less than 150 Hz. Cold head vibration was less than the required level and its motion was less than ±8 μm (reference 2). Reliability analysis predicted an operative life of more than 50,000 hours for the EM cooler.

## ACKNOWLEGEMENT

This work was sponsored under contract by Lockheed Martin IR Imaging System, Inc., for the JPL AIRS Instrument Program.

## REFERENCES

1   C.K. Chan, "AC Thermodynamics Theory of Stirling and Pulse Tube Cryocoolers," *Symposium on Thermal Science and Engineering in Honor of Chancellor Chang-Lin-Tien* (1995), p 531-539.

2   C.K. Chan, J. Raab, A. Eskovitz, R. Carden III, R. Orsini, "AIRS Flight Qualified 55 K Pulse Tube Cooler," TRW Inc. *Cryocoolers 9*, Plenum Press, New York (1997).

3   W.W. Burt and C.K. Chan, "New Mid Size High Efficiency Pulse Tube Coolers," TRW Inc. *Cryocoolers 9*, Plenum Press, New York (1997).

4   D.L. Johnson, S.A. Collins, M.K, Heun, R.G. Ross, "Performance Characterization of the TRW 3503 and 60 K Pulse Tube Coolers," *Cryocoolers 9*, Plenum Press, New York (1997).

# Advanced Pulse Tube Cold Head Development

C. K. Chan, C. Jaco, and T. Nguyen

TRW Space & Technology Division
Redondo Beach, CA 90278, United States of America

## INTRODUCTION

The TRW miniature pulse tube cooler uses a 1cc swept volume compressor and a pulse tube cold head, developed in 1991, to provide long life cooling in space. The initial design objective of 280mW cooling at 73K was successfully achieved by operating the compressor at 95% of maximum stroke with 20W of electrical power input to the compressor. At 65K, the cold head has 100mW cooling power. The challenge ws to achieve a larger cooling load at 65K. Our advanced cold head is compatible with the compressor of 1cc swept volume and meets the objectives of maximizing the cooling power at 65K and minimizing the compressor input power.

Our approach to improving the cold head was to first thermodynamically design and analyze the new cold head and determine the size and the geometry of each component to give optimal net cooling at 65K with minimal swept volume and shaft power from the compressor. The component size and geometry was then incorporated into the mechanical design and the cold head fabricated. The performance of the new design was first tested with a stainless steel pulse tube and consequently with a flight-like titanium pulse tube cold head. Measured test parameters include cooling loads, cooling temperatures, heat reject temperatures, input power, operating frequencies, strokes and pressures.

### Thermodynamic Design and Analysis

The pulse tube cooler includes two major subsystems: a compressor and a cold head which contains an aftercooler, a regenerator, a cold- end heat exchanger, a pulse tube, a hot-end heat exchanger, a flow restrictor, and a reservoir (Figure 1). A bypass line that connects the aftercooler to the hot end heat exchanger enhances the efficiency. The system is filled with pressurized helium gas. In operation[2], the compressor generates an oscillating pressure wave and an out-of-phase oscillating mass flow. The dot product of the alternating pressure and mass flow vector produced by the compressor is the flowing PV work (or shaft power) which causes the regenerator to pump heat from the cooled load and the cold-end heat exchanger at temperature, TLOAD, to the aftercooler temperature TA, where heat is rejected to the heat sink at temperature TREJ. In this integral configuration the same aftercooler is used to pre-cool the compressed gas, whose temperature was raised to TCO in the compression chamber. Meanwhile, the PV work travels down the pulse tube, where it is rejected as heat to the heat sink at the hot-end heat exchanger.

Improving the regenerator makes the most impact on the cold head efficiency. As shown in Figure 2, the regenerator loss is affected by the effectiveness of the heat transfer process (or the

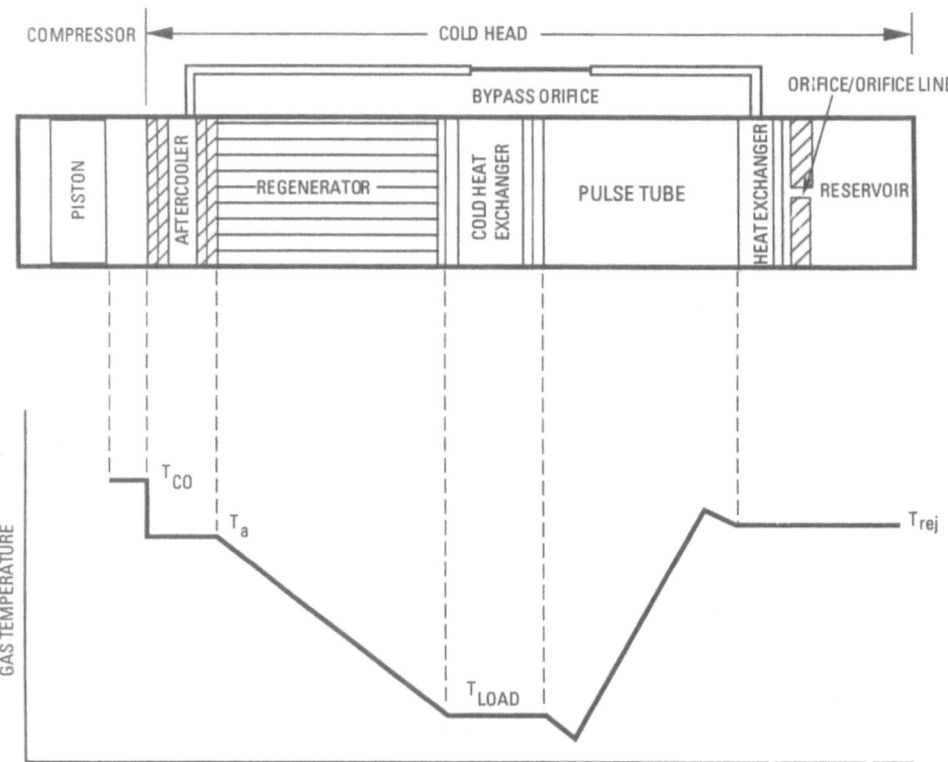

**Figure 1.** Miniature pulse tube cooler.

regenerator loss, Qr), the pressure drop loss, Qp, and the heat conducted through the regenerator matrix, Qcr. The regenerator geometry and the matrix material have significant effects on the mass flow distribution and hence the heat transfer process and the pressure drop. Other heat exchange

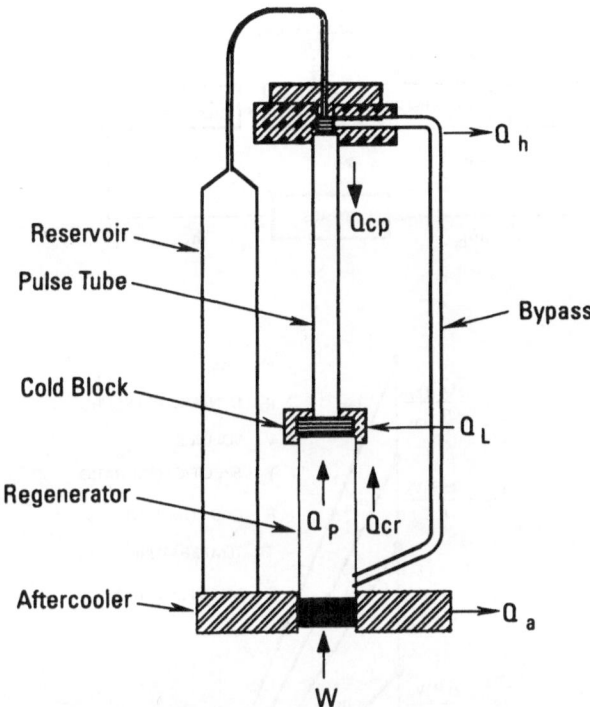

**Figure 2.** Energy loss terms in pulse tube cooler.

components also play an important role in cold head efficiency. The pulse tube is a thin-walled tube of a low thermal conductivity material where conduction heat loss is Qcp. The pulse tube contains copper heat exchangers, which are thermally connected to copper blocks, at both cold and hot ends. The heat transfer at these heat exchangers cannot be effective unless the mass flow has the correct phase relationship with the temperature. The phase angle in each component has to be within a certain range, so that heat from the cooling load is absorbed by the gas through the cold end heat exchanger and at the aftercooler. The phase angle between the alternating pressure and mass flow of the gas is fine tuned to maximize the power of the cooler by choosing the proper size and geometry of the cold head components.

Our design approach to the pulse tube cold head involves two stages. The first stage is the use of the TRW Pulse Tube Design Code to divide the cold head into functional components, such as regenerator, pulse tube (Figure 3). The oscillating mass flow and pressure form one component to the other are represented by vectors. The construction of the vector or the phasor diagram as shown in Figure 3, together with the heat transfer analysis of the regenerator and the heat exchanger, determines the loss term in the cold head and hence the net cooling power for a set of component geometries and dimensions. Table 1 shows gross cooling power and the energy loss term computed from the design code for the 91%, 95% and 98% swept volume cases. The net cooling power is the gross cooling power less the loss terms. The shaft power is determined in the second stage using the TRW performance code[3] where the differential equations of mass, momentum and energy are solved. The code gives the time dependent thermodynamic properties at each nodal point in the system. These include pressure, mass flow, temperature, density, enthalpy and entropy. The performance code also computes the net cooling and the energy loss terms, shown in Table 2. Comparing the results in Table 1 with Table 2, the design code predicts a net cooling power of 0.36W at 65K while the performance code predicts a higher value of 0.52W using 19.8W of shaft power with 98% of swept volume.

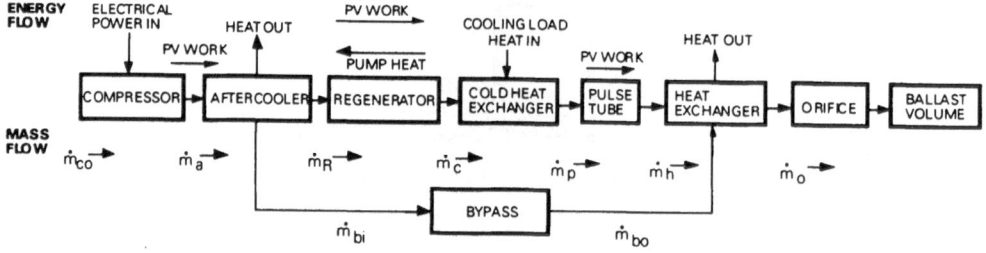

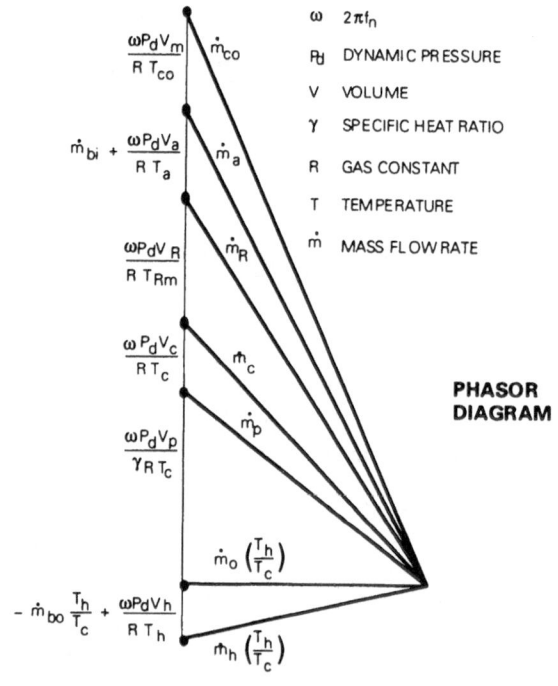

**Figure 3.** Phasor diagram of pressure and mass flowrate vectors.

**Table 1.** Energy Balance, Net Cooling Power and Compressor Swept Volume Prediction from Design Code

| ENERGY BALANCE | | | | |
|---|---|---|---|---|
| $Qg - (Qr + Qcr + Qcp + Qp) = Q_L$ | | | | |
| Term | Symbol | Case 1 | Case 2 | Case 3 |
| Gross Cooling Power, W | Qg | 2.225 | 2.601 | 3.039 |
| Regenerator Loss, W | Qr | 0.853 | 0.980 | 1.074 |
| Conduction Loss, Regenerator, W | Qcr | 0.160 | 0.160 | 0.160 |
| Conduction Loss, Pulse Tube, W | Qcp | 0.044 | 0.044 | 0.044 |
| Pressure Loss, W | Qp | 0.819 | 1.054 | 1.392 |
| Net Cooling Power, W | QL | 0.349 | 0.363 | 0.369 |
| Compressor Swept Volume CM³ | | 0.908 | 0.946 | 0.974 |

**Table 2.** Energy Balance, Net Cooling Power and Compressor PV Work Prediction for Performance Code

| ENERGY BALANCE $Qg - (Qr + Qcr + Qcp + Qp) = Q_L$ | | | | |
|---|---|---|---|---|
| Term | Symbol | Case 1 | Case 2 | Case 3 |
| Gross Cooling Power, W | Qg | 2.531 | 2.768 | 2.952 |
| Regenerator Loss, W | Qr | 1.04 | 1.089 | 1.124 |
| Conduction Loss, Regenerator, W | Qcr | 0.160 | 0.160 | 0.160 |
| Conduction Loss, Pulse Tube, W | Qcp | 0.044 | 0.044 | 0.044 |
| Pressure Loss, W | Qp | 0.967 | 1.045 | 1.104 |
| Net Cooling Power, W | QL | 0.32 | 0.43 | 0.520 |
| PV Work, W | W | 17.2 | 18.68 | 19.8 |

## Mechanical Design

The flexure bearing compressor is driven by a moving coil linear motor supported by flexure springs that provide drive current and maintain alignment for the attached piston. The piston oscillates and compresses gas into the pulse tube cold head. A narrow clearance between the cylinder and piston seals the compression space. The seal loss in the laboratory compressor used for the present tests has not been determined. The compressor operates at frequencies of 40 to 60 Hz, a range chosen as a compromise between a realizable and an efficient pressure-wave generating resonant long-life compressor. It is desirable to operate the compressor at a mechanical resônance that coincides with the frequency that yields maximum net cooling at the piston-coil assembly mass and the sum of the stiffness of the gas and mechanical spring. No attempt has been made at present to optimize the compressor dynamics for the new cold head.

Two cold heads were fabricated; one used stainless steel and the other used titanium alloy. The stainless steel unit was used to verify the regenerator design. The thermal performance tests whose results are quoted were performed with the titanium unit.

## Testing

The thermal performance of the cooler was measured using the test configuration shown in Figure 4. The pulse tube cooler was supported at its base plate by a structure with the cold head contained within a vacuum housing. To minimize the radiation heat leak, the cold head was loosely wrapped with ten layers of doubly aluminized Kapton isolation. The vacuum housing was evacuated to the range of 2 x 10-5 torr. The heat at the warm end heat exchanger was conducted through the aluminum surge tank to the base plate where the aftercooler was located. The heat at the baseplate was removed by a fluid loop, the inlet temperature of which was controlled by a chiller. The temperature at the warm-end heat exchanger was 5 to 10K higher than the aftercooler temperature, depending on the temperature drops across the surge tank and associated thermal joint and the amount of power applied to the cold finger heater.

Two redundant silicon diodes, one of which is NIST calibrated, were put on the cold finger to measure cooling temperature. Another diode was located at the aftercooler base plate. The diodes were read by a NIST calibrated electronics box and recorded in a data aquisition system.

The ac power to the compressor was calculated by integrating the measured ac current and voltage in the computer. This power was also independently measured by a power meter containing a four quadrant multiplier. The I2R loss in the compressor was also measured through the A-D card and the four quadrant multiplier by knowing the current and the dc resistance. The heat load at the cold block was produced by a resistance heater. The amount of the heat load was determined by the product of the dc voltage and the current.

Figure 5 shows the measured load and test data for 22.17W input power (200mW at 65K) as well as maximum cooling performance of 374mW at 65K. At the maximum measured stroke (98%), compressor efficiency was only 72.6% of the input (33.67W) power due to the fact that the compressor is not tuned for maximum efficiency. The 24.45W of shaft power includes the unknown seal loss for this compressor, thus providing an upper bound on the shaft power required from a flight compressor to achieve this cooling performance. Table 3 summarizes test data taken at the 55 Hz operating frequency and 500 psig operating pressure.

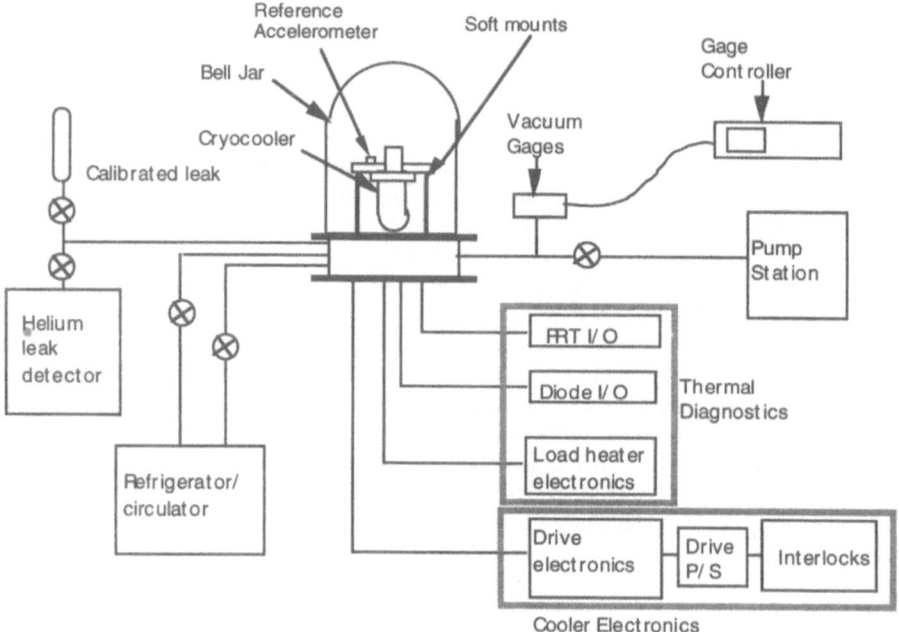

**Figure 4.** Cooler test configuration.

Figure 6 gives the cooling power at 65K as a function of compressor input power and the corresponding shaft power for 300K reject temperature. Figure 7 gives the effect of heat reject temperature on the net cooling power at 65 and 105K. This cold head has a maximum cooling capability of 500mW at 65K when the heat reject is 273K. The static conduction heat loss of the cold head at 65K was measured to be 280mW.

## CONCLUSION

A flight prototype cold head was designed, built, and tested with a 1cc laboratory compressor. Our test results demonstrated that this new cold head has 3.7 times the cooling capacity at 65K of the previous design with a parasitic heat loss is half that of the current flight unit.

**Table 3.** Performance Test Summary

| | Stroke, % of 1cc | Power, W | Shaft Power, W | Reject Temp., K | Applied Heat Load, mW | Cold Head Temp., K |
|---|---|---|---|---|---|---|
| Constant Stroke | 98% | 32.70 | 24.11 | 300.02 | 0 | 54.27 |
| | 98% | 33.69 | 24.45 | 300.00 | 374 | 65.08 |
| | 98% | 36.95 | 25.18 | 300.02 | 1707 | 105.03 |
| Constant Power | 98% | 32.70 | 24.11 | 300.02 | 0 | 54.27 |
| | 97% | 32.70 | 23.92 | 299.99 | 361 | 65.05 |
| | 93% | 32.70 | 22.86 | 300.00 | 1573 | 105.03 |
| | 89% | 32.71 | 21.64 | 299.96 | 2668 | 150.04 |
| Constant Power | 81.5 | 22.18 | 17.06 | 299.98 | `0 | 57.14 |
| | 80.7 | 22.20 | 16.78 | 300.00 | 200 | 65.03 |
| | 77 | 22.17 | 16.03 | 300.02 | 1109 | 105.05 |
| | 74 | 22.18 | 15.21 | 299.97 | 1928 | 150.01 |
| Variable Heat Reject Temperature | 98% | 34.67 | 24.69 | 310.00 | 325 | 65.03 |
| | 98% | 33.69 | 24.45 | 300.00 | 374 | 65.08 |
| | 98% | 31.62 | 23.81 | 272.99 | 500.6 | 65.08 |
| | 98% | 38.41 | 25.61 | 310.11 | 1680 | 105.09 |
| | 98% | 36.95 | 25.18 | 300.02 | 1707 | 105.03 |
| | 98% | 34.51 | 24.93 | 273.02 | 1832 | 105.03 |
| | 92% | 29.98 | 22.04 | 273.07 | 1649 | 104.99 |
| Variable Cooling Power | 98% | 33.69 | 24.45 | 300.00 | 374 | 65.08 |
| | 91% | 28.29 | 20.99 | 300.03 | 300 | 65.01 |
| | 86% | 25.07 | 18.79 | 299.99 | 250 | 65.00 |
| | 81% | 22.17 | 16.81 | 299.98 | 200 | 65.01 |

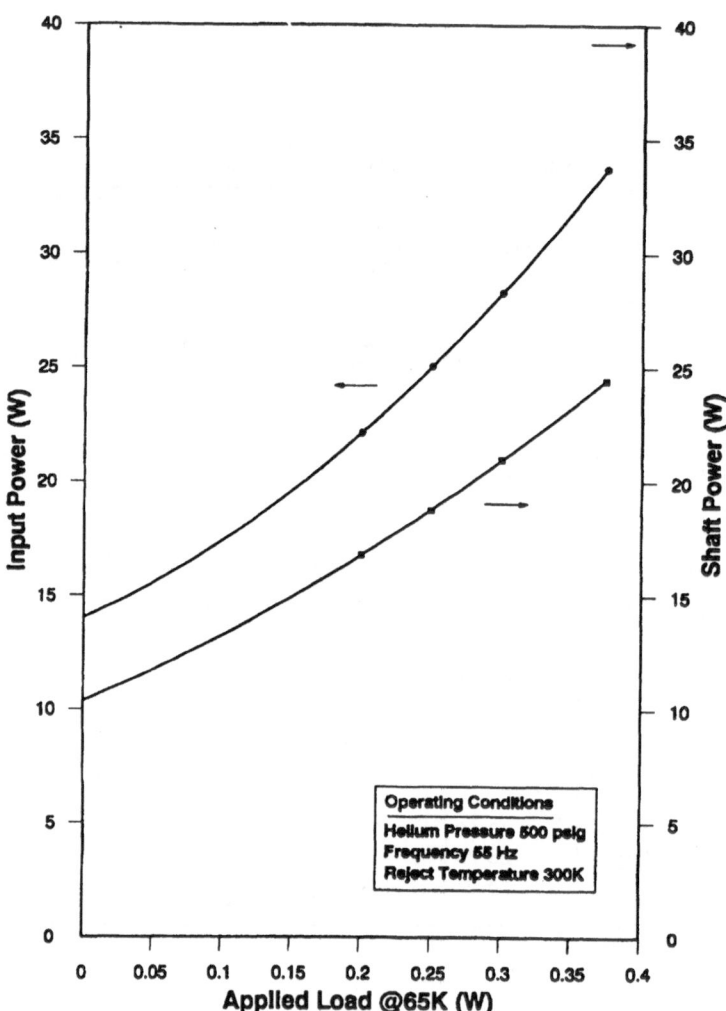

**Figure 5.** Heat load curves for constant (98%) stroke and constant compressor power.

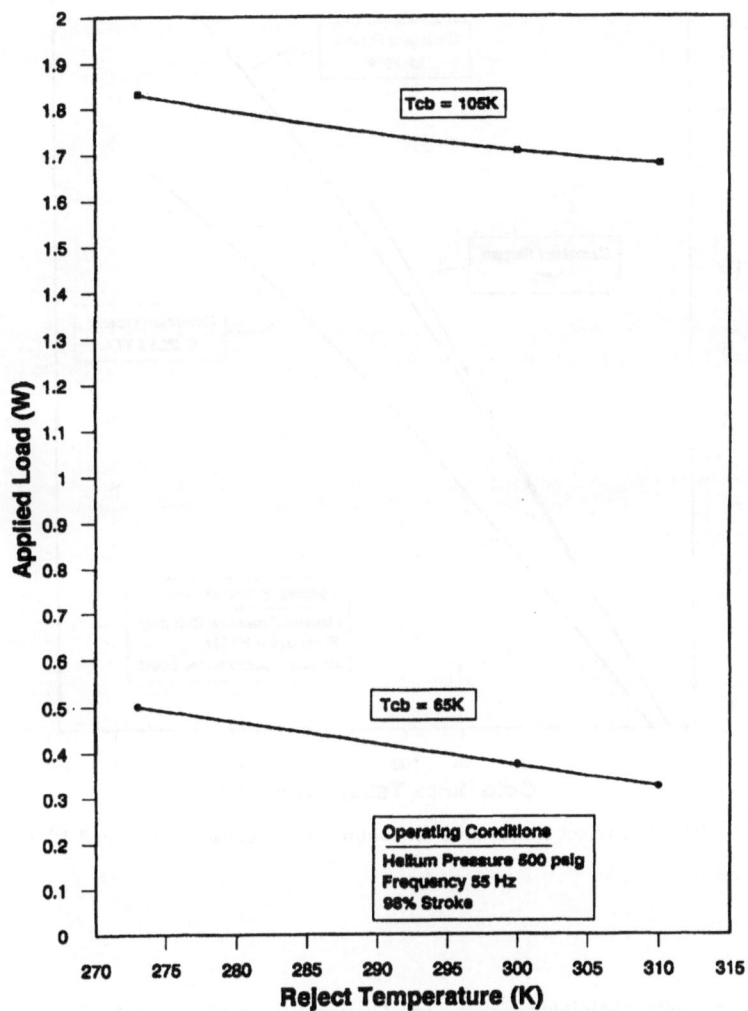

**Figure 6.** Compressor power and shaft power for various cooling power at 65 K.

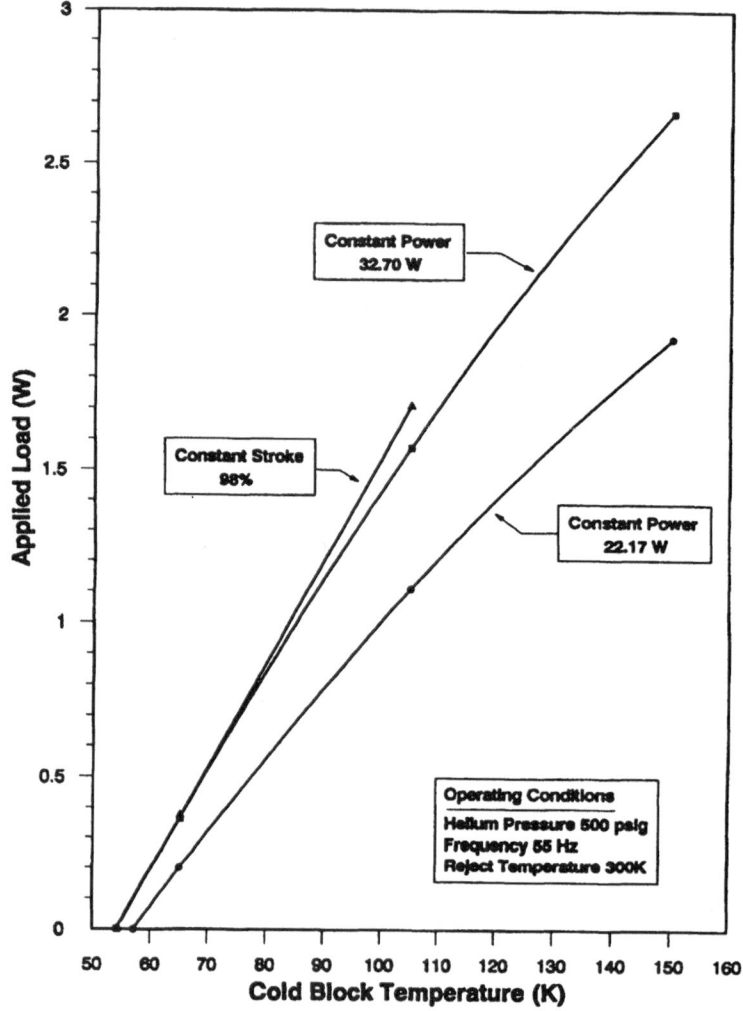

**Figure 7.** Effect of reject temperature on cooling power capacity at 65 and 105K.

## REFERENCES

1. Chan, C.K., et al, 1993. "Miniature Pulse Tube Cooler," *Proceedings of the 7th International Cryocooler Conference*, November 1992. PL-CP-93-1001, Phillips Laboratory, Air Force Material Command, Kirtland AFB, pp. 113-124.
2. Harpole, G.M., and Chan, C.K., 1991. "Pulse Tube Cooler Modeling," *Proceedings of the 6th International Cryocooler Conference*, pp. 91
3. Chan, C.K., 1995. "AC Thermodynamics Theory of Stirling and Pulse Tube Cryocoolers," *Symposium on Thermal Science and Engineering in Honor of Chancellor Chang-Lin-Tien*, University of Illinois, Urbana-Champaign, pp. 531-539.

# 50-80 K Pulse Tube Cryocooler Development

L. Duband, A. Ravex [1], T. Bradshaw, A. Orlowska [2]
C. Jewell [3] and B. Jones [4]

[1] Service des Basses Températures /DRFMC/CEA
17 rue des Martyrs - Grenoble 38054 Cedex 9 - FRANCE
[2] Rutherford Appleton Laboratory, Chilton, Didcot, UK, OX11OQX
[3] European Space Research and Technology Centre, PO Box 299, 2200 AG -
Noordwijk, Netherlands
[4] Matra Marconi Space UK Limited, PO Box 16, Filton, Bristol, UK, BS12 7YB

## ABSTRACT

A strong effort is being made world wide in developing miniature cryocoolers for multi-year space missions. These cryocoolers must satisfy strong requirements among which reliability, minimal exported vibration and efficient thermal performance are the most critical. Many cryocooler designs are under development. These designs include Stirling, Sorption and particularly Pulse Tube coolers. While this last technology is not new, major improvements have been made recently that promise significant benefits. Pulse tube coolers are very similar to Stirling coolers but they do not require any moving displacer in the cold finger and consequently no motorization, no synchronisation electronics, no bearings and clearance seals. This is very interesting in terms of mechanical simplicity, reliability, reduction of induced vibrations and electromagnetic noises. Performances comparable with those of Stirling coolers are achievable in terms of size, weight, cooling power, cooling temperature and efficiency.

In the framework of an ESA contract we are developing for space purposes a pulse tube cryocooler associated with a diaphragm springs oscillator. An improved version of the oscillator has been developed by the Rutherford Appleton Laboratory. Preliminary prototypes of the pulse tube cooler have been designed and manufactured, and are presently being characterised.

## INTRODUCTION / HISTORY

Since the early 1970's instruments on spacecraft have continued to demand higher performances which have entailed the development of such long life cryogenic cooling options as mechanical refrigerators. The applications associated with the use of such mechanical refrigerators operating in the temperature range 50-80K are numerous and well known (e.g. scientific and earth observational detector cooling).

To fulfil these requirements a collaborative European effort has successfully developed a split single stage 50-80K Stirling Cycle cooler which is now available from MMS-B as an "off the shelf" commercial product. Such coolers are based on the Oxford/RAL design[1, 2] and use the

European technology of linear motors and clearance seals. While these coolers are capable of fulfilling most of the requirements for future multi-year space missions in the temperature range of 50K to 80K, some system and instrument level issues have arisen which need to be addressed. These issues can be summarised as the following :

- Reduction of vibration at the detector
- EMC issues
- Assembly issues such as maximum side load allowable at cold finger tip.
- launch lock and power requirements during ascent
- orientation of coolers during on ground testing
- connection to detectors and possible need for electrical isolation
- reduction of parasitic heat load for coolers in cold redundancy etc.

Although solvable for the present MMS Stirling cooler it is obvious that a more fundamental solution to some of these issues would be available by developing a pulse tube technology[3] where the only moving part in the "displacer" is the gas. Indeed with the added benefit of a reduction in the number of mechanisms and hence an increase in the overall reliability of the cooler many R&D programmes have been instigated with a general trend towards pulse tubes obtaining thermal performances equivalent to that of the present Stirling coolers.

Hence in 1994 ESA issued a competitive request for quotation (RFQ) to develop a so called "2nd generation" space cryocooler based on pulse tube technology and Oxford/RAL type compressors. This RFQ was won by a collaboration composed of CEA-SBT, RAL and MMS-B, and some of the goals can be summarised as follows:

- a user friendly "plug-in" design
- a cooler tested to a conservative mechanical specification (see table 1) in order to avoid any delta mechanical design activities.
- Exported vibration at the cold tip of <0.01 N rms in any direction in a frequency bandwidth including the fundamental operating frequency and its first six harmonics.
- Compressors situated at >500mm from the pulse tube etc.

Activities associated with the ESA gamma ray satellite INTEGRAL are also on-going at the CEA-SBT, but this paper will outline some of the work related to the ESA technology development programme which has longer term goals.

**Table 1**. Mechanical Environment (Acceptance Levels)

| Sinusoidal Vibration (each orthogonal axis) | |
| --- | --- |
| Frequency (Hz) | Level |
| 5 - 20.5 | 12 mm peak |
| 20.5 - 60 | 20 g peak |
| 60 - 100 | 8 g peak |
| Sweep Rate | 2 octaves/minute, 1 sweep up |
| | |
| Random Vibration (any direction) | |
| 20 | 0.016 g2/Hz |
| 20 - 80 | +6 dB/oct. |
| 80 - 500 | 0.26 g2/Hz |
| 500 - 2000 | -3 dB/oct. |
| 2000 | 0.065 g2/Hz |
| Composite | 17.0 grms |
| | |
| Duration: | 2.5 min/axis |

## PULSE TUBE TECHNOLOGY

In 1963 Gifford and Longsworth[4] discovered a new refrigeration technique which they called Pulse Tube refrigeration. This technique has been extensively described and we will just give here a brief description. The Pulse Tube refrigerator in its present evolution is a variation of the Stirling refrigerator, with the difference that the moving mechanical displacer is eliminated. The thermodynamic cycle is very similar: in both cryocoolers a compressor associated with a distributor, or a pressure oscillator, generates an oscillating pressure wave and a mass flow rate in a cold head which contains a regenerator. The cooling effect is obtained by the proper phase shift at the cold end between the mass flow rate and the pressure oscillation. The coolers differ by the mechanism used to control this phase shift: • in a Stirling cryocooler, this phase shift is set by the relative motion of the cold finger displacer and the piston, • in a Pulse Tube the phase shift is adjusted by one or two impedances.

The absence of any moving components at low temperature in the Pulse Tube concept offers the potential to develop reliable cryocoolers. This feature leads to other benefits such as:

- • no motorisation and thus no EMI
- • no mechanical source of vibration in the cold head
- • no clearance seal (piston / cylinder)
- • simplicity
- • lower mass
- • lower cost

Moreover Pulse Tube cryocoolers may be powered by the same compressor or pressure oscillator used in Stirling coolers and thus can benefit from all the developments presently made on flexure bearings and hydrodynamic gas bearings for the space Stirling coolers.

Several configurations of the Pulse Tube cryocooler exist. These configurations can be classified under three main types: basic, orifice and double inlet (the thermoacoustic pulse tube, out of the scope of the present work, can also be defined as a fourth configuration). Each configurations differ by the state of the impedances V1 and V2 as shown on the following figure and table.

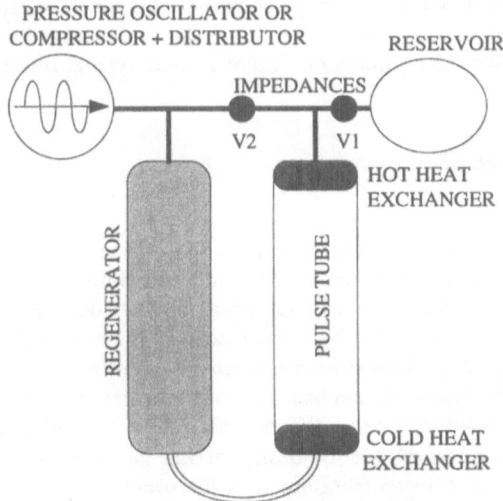

**Figure 1**. Schematic of a pulse tube refrigerator

**Table 2**. Evolution of performance with configurations

| Type | $V_1 - V_2$ | Year | Authors | Temperature |
|---|---|---|---|---|
| Basic | 0 - 0 | 1963 | Gifford / Longsworth[4] | ≈ 124 K |
| Orifice | 1 - 0 | 1984 | Mikulin[5] / Radebaugh[6] | 120 K / 60 K |
| Double inlet | 1 - 1 | 1990 | Zhu[7] | ≈ 42 K |
| Double inlet | 1 - 1 | 1993 | CEA-SBT[8] | 28 K |

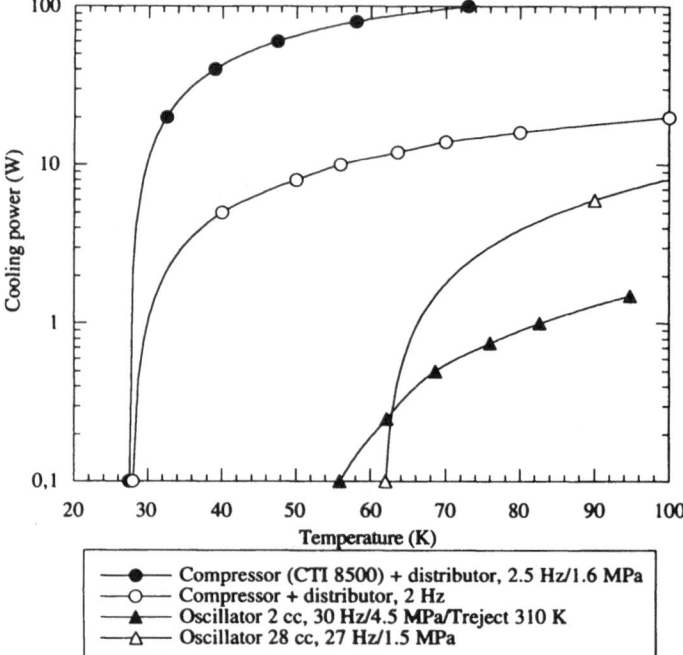

**Figure 2**. Performance of some typical pulse tube coolers developed at CEA-SBT

We commonly distinguish two families of pulse tube cooler, the large capacity/low frequency one operated using a compressor associated with a distributor and the high frequency one operated with a pressure oscillator. We have plotted on Figure 2 some typical results obtained on Pulse Tube coolers at CEA-SBT.

## ESA PULSE TUBE COOLER

### Sizing of the Pulse Tube

An analytical model describing the behaviour of the orifice Pulse Tube cooler has been developed by Storch and Radebaugh[9] to gain a better understanding of the refrigeration process. Based on this work we have developed our own numerical model to describe the double inlet Pulse Tube[10]. This model continues to evolve but predicted results are found to be in fairly good agreement with experimental data. This software is already an efficient tool for the design and optimisation of pulse tube refrigerators and has been used for the present work to size the first prototype. It permits to scan automatically different diameters, lengths, pressure, frequencies, etc... for the pulse tube and to search for the design giving the best cooling power at a given temperature or the lowest ultimate temperature. For the present study over 3000 pulse tube geometries were evaluated. The input parameters and goal are indicated in table 3.

**Table 3**. Input parameters

| | |
|---|---|
| Oscillator swept volume: | $6.84 \ cm^3$ |
| Mean pressure: | 1.2 to 1.5 MPa |
| Frequency: | 40 to 60 Hz |
| Goal: | maximise cooling power at 80 K |

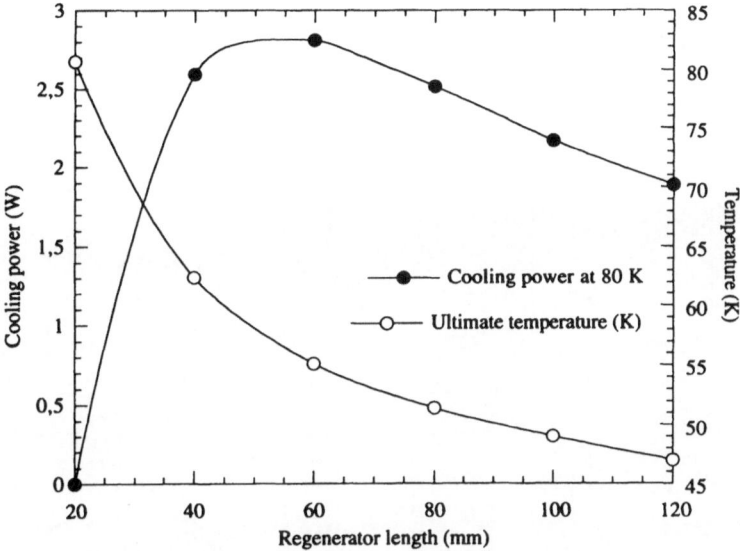

**Figure 3**. Example of output (1.2 MPa / 50 Hz)

An example of result is shown on figure 3. On this figure we have displayed the cooling power at 80 K as well as the ultimate temperature for a given set of operating conditions. It can be seen that a compromise has to be made between the ultimate temperature and the cooling power. It should also be noted that these results correspond to the gas temperature, the efficiency of the cold and hot heat exchangers will probably reduce these values. Nevertheless they indicate the influence of the main parameters and allow for optimisation.

A pulse tube cooler can be arranged in various architectures such as in-line, co-axial and U shaped. Most of these arrangements have been tested during previous projects. With respect to the requirements given by ESA a U shaped arrangement has been chosen for the present work.

## Description of the Pulse Tube

The first pulse tube prototype was built using standard materials and processes, i.e. copper hot head, stainless steel tubes and brazing or TIG welding. However in preparation of the second generation cooler the design has been conceived to be consistent with industrial manufacture and has been refined according to the main specifications as described hereafter.

**Reduction of Mass and Compactness** - The top head (hot heat exchanger level) of the pulse tube cooler, where most of the mass is concentrated, is made of copper. This material, easy to machine, chosen for its thermal performance and its reasonable cost will be replaced by titanium (T60) which provides reasonably good thermal performance and has a much lower density ($\approx 4.5$ g.cm$^{-3}$ to be compared with 8.9 g.cm$^{-3}$ for copper). The ballast volume as well, now included in the top head as shown on the schematic (Figure 4) will be made out of titanium.

**Integration of the Impedances and Ballast Volume within the Pulse Tube** - Schematically a pulse tube cooler, if we ignore the pressure oscillator, consists of 4 main elements: the tube and regenerator, the hot and cold heat exchangers, the injection lines that comprise the impedances, and the ballast volume. For laboratory purposes the impedances (valves, capillaries, diaphragms,...) and ballast volume are usually separated from the other components. For this new definition we have worked on a "plug in" type design which as much as possible is user friendly and integrates the above elements. Of course the design has kept in mind that the pulse tube cooler must survive the mechanical environment previously specified.

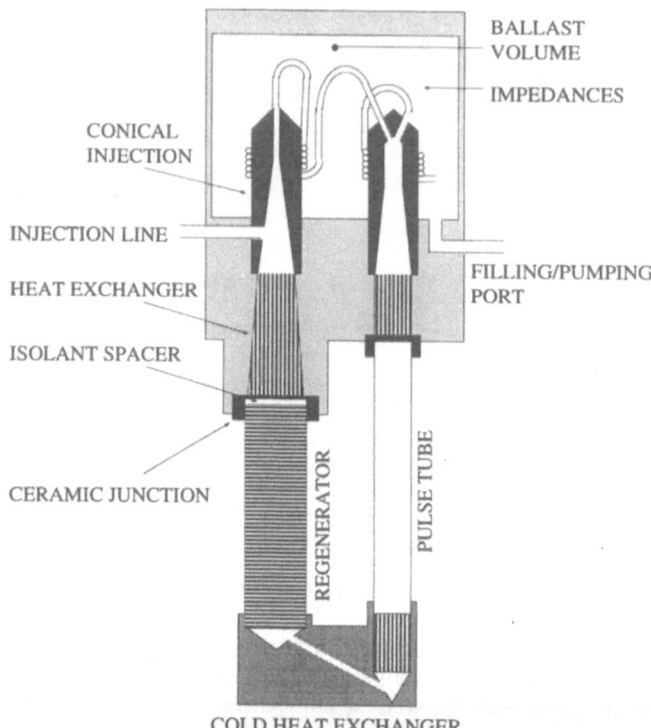

BALLAST VOLUME

IMPEDANCES

CONICAL INJECTION

INJECTION LINE

HEAT EXCHANGER

FILLING/PUMPING PORT

ISOLANT SPACER

CERAMIC JUNCTION

REGENERATOR

PULSE TUBE

COLD HEAT EXCHANGER

**Figure 4**. Schematic of the ESA pulse tube cooler

As of today the experiments done on various pulse tube coolers at CEA-SBT show that the use of capillaries or diaphragms for the impedances is beneficial to the thermal performance. Moreover they are easy to adjust and integrate. We have thus chosen this technology for the present work. As shown on figure 4 the two capillaries are mounted on the top head of the pulse tube cooler inside the ballast volume. The first capillary (J1) connects the output of the pulse tube to the ballast volume, the second one (J2) connects the input of the regenerator to the output of the pulse tube. Each capillary is wounded around a stainless steel connector and is soldered to it. These connectors which include a conical injection designed to reduce the jet effect are removable to ease the mounting and adjustment of the impedances.

The flat surface on the top cover is designed to ease the mounting of the cooler on various structure (thermal radiator,...).

**Electrical Isolation of the Cold Finger, i.e. Regenerator and Pulse Tube** - The electrical isolation for coolers is usually done after the cold finger, as part of a flexible connection between the cold finger and the focal plane/detectors. Unfortunately with the conflicting requirements of a good thermal conductor and poor electrical conductor, the impact on performance is quite significant. For this reason we have looked at a solution where the electrical isolation is implemented in the pulse tube itself. As mentioned previously the top head of the cooler is made out of titanium and so are the regenerator and pulse tube tubing. The solution we have studied is to electrically isolate this tubes with either the top or cold head of the pulse tube using a specific insulating ceramic junction. These two possible locations have advantages and drawbacks:
• at the cold heat exchanger level, the mechanical oscillations during launch and vibration testing should be fairly small, but the junction will be subjected to many thermal cycling.
• at the hot heat exchanger level (top head), the temperature will remain fairly constant, but most of the mechanical stresses will be concentrated at the junction.

Other considerations such as the need to keep the mass of the cold heat exchanger part as low as possible (mechanical aspects), the assembling order and the geometry lead us to choose to implement the electrical isolation at the top head level. This junction must mechanically link the titanium tubes to the titanium top head, provide electrical isolation and leak tightness. The junction consists of an alumina ring inserted between the titanium parts; the three resulting pieces are brazed together using silver alloys. The size and geometry of the ring is made such that it should guarantee the electrical isolation and provide mechanical strength. After the brazing process is done, an electrically isolant spacer is first inserted to prevent direct contact between the stainless steel meshes and the titanium top head.

**Reduction of the Parasitic Load of the Non Operating Cooler** - In the case of the use of multiple coolers, during operation any non operating cooler contributes to the total parasitic heat load. Thus it is important to minimise this contribution. which is due to two factors, the ratio area over length of the tubes (the stack of meshes do not contribute significantly to the load) and the thermal conductivity of the material(s) used. The tubes used are thin walled tubes. For mechanical and machining reasons their wall thickness is pretty much at the minimum allowable (100 μm). Thus to further reduce the parasitic load stainless steel will be replaced by titanium Ta6V which has a thermal conductivity roughly two times lower in this temperature range.

## PRESSURE OSCILLATOR

The pulse tube cooler is associated with a diaphragm spring type pressure oscillator originally developed by RAL and Oxford University for an atmospheric physics experiment[1]. For the present work its design has been refined to reduce the Joule heating losses and increase the motor force.

### Compressor Drive Motor

The linear drive motor in the RAL compressors is a simple loudspeaker-type mechanism. A permanent magnet provides a radial field across the windings of a coil (see Figure 5). Soft iron pole pieces are used to contain the magnetic field. If an alternating current is applied to the coil windings the mechanism moves backwards and forwards. At resonance, neglecting any inductive effects, the losses in the coil will be those from Joule heating. If the length of the coil, $L_c$, is greater than the effective depth of the pole piece (2 x the actual length, $L_p$, to account for fringing) then the force, F, can be estimated from:

$$F = B_a i L_w 2 \frac{L_p}{L_c} \qquad (1)$$

where $L_w$ is the length of the wire in the drive coil, $B_a$ the field between the pole pieces in the magnetic circuit and i the current. The resistance, R, of the coil is

$$R = \frac{\rho L_w}{f d^2} \qquad (2)$$

where f is the filling factor of the coil, d is the wire thickness and $\rho$ is the resistivity of the wire. The losses in the coil are therefore:

$$i^2 R = \left(\frac{F}{B_a}\right)^2 \left(\frac{L_c}{L_p}\right)^2 \frac{\rho}{4 L_w f d^2} \qquad (3)$$

The Joule heating losses are inversely proportional to the filling factor of the coil in the magnetic circuit. One way of reducing these losses is to use square wire in the drive coil. In the coil (excluding the gap between the coil and the magnetic circuit) we achieve fill factors of 0.87 for the square wire coil vs. 0.7 for conventional round wire.

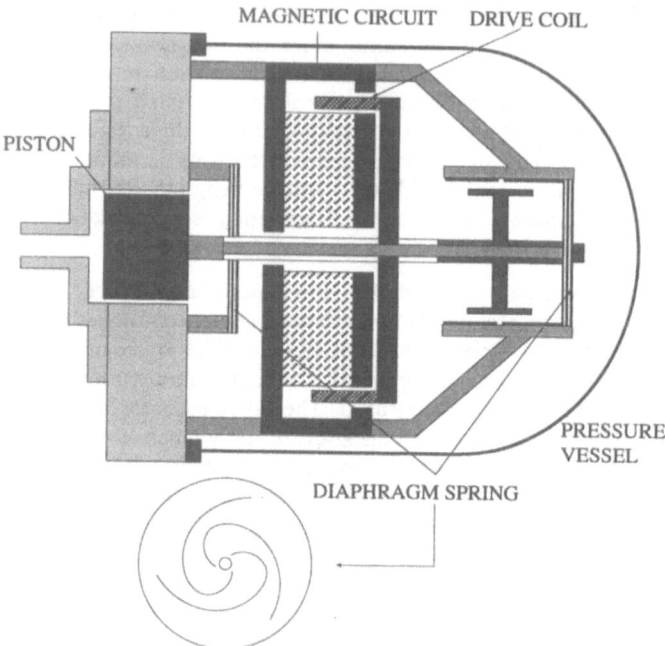

**Figure 5.** A Schematic diagram of the RAL compressor

The use of silver wire in the drive coil in the place of copper yields a further modest improvement in the $i^2R$ losses. Silver has a resistivity of $1.6 \times 10^{-8}\Omega$m while copper is $1.7 \times 10^{-8}\Omega$m. The effect of using silver wire is to reduce the $i^2R$ losses by approximately 0.5W in each compressor when running at 30 W input.

The motor force can be increased by optimising the magnetic field between the pole pieces. In our magnetic circuit we use NdFeB which gives a radial field across the windings of 1.0 Tesla. With double magnet configuration[11] consisting of an annular outer magnet we achieve 1.2 Tesla. This gives a useful increase in motor force of 20%. Figure 6 shows the results of modelling the double magnet configuration. The magnetic circuit is radialy symmetric around the z-axis. The calculated peak field of 1.3 Tesla is to be compared with the measured value of 1.2 Tesla.

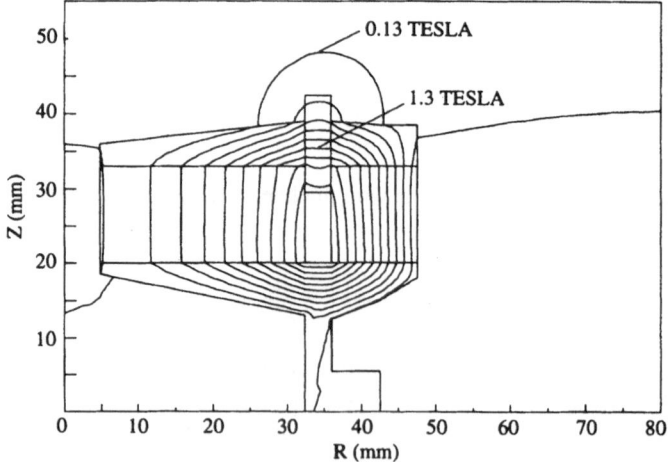

**Figure 6.** Computer modelling of the compressor drive motor.

## CONCLUSION

In the framework of an ESA contract we are developing for space purposes a pulse tube cryocooler associated with a diaphragm spring type oscillator. An improved version of the oscillator has been developed by the Rutherford Appleton Laboratory. Preliminary prototypes of the pulse tube cooler have been designed and manufactured, and are presently being characterised. In its final design this pulse tube cooler includes new features such as the electrical isolation of the cold finger with respect to the rest of the cooler and the integration of all the elements.

## REFERENCES

1    Werret S.T. & al, "Development of a Small Stirling Cycle Cooler for Spaceflight Applications", Adv. Cryo. Eng. (1986) 31, p 791

2    Bradshaw T.W. & al, "Performance of the Oxford Miniature Stirling Cycle Refrigerator" Adv. Cryo. Eng. (1986) 31, p 801

3    Chan C.K., Tward E. and Burt W.W., "Overview of Cryocooler Technologies for Space-Based Electronics and Sensors" Adv. Cryo. Eng. (1990) 35B, p 1239

4    Gifford W.E. and Longsworth R.C., "Pulse Tube Refrigeration", Trans. ASME J. Eng. Ind. (1964) 63, p. 264.

5    Mikulin E.I., Tarasov; A.A. and Shkrebyonock M.P., "Low Temperature Expansion Pulse Tubes", Adv. Cryo. Eng. (1984) 29, p.629

6    Radebaugh R., Zimmermann J. Smith D.R., Louie B., "A Comparison of Three Types of Pulse Tube Refrigerators", Adv. Cryo. Eng. (1986) 31, p. 779.

7    Zhu S.W., Wu. P.Y. and Chen Z.Q., "A Single Stage Double Inlet Pulse Tube Refrigerator Capable of Reaching 42 K", Cryogenics (1990) 30, p. 257.

8    Liang J., "Development and Experimental Verification of a Theoretical Model for Pulse Tube Refrigeration", Thesis (1993), Grenoble.

9    Storch P.J. and Radebaugh R., "Development and Experimental Test of an Analytical Model of the Orifice Pulse Tube Refrigerator", Adv. Cryo. Eng. (1988) 33, p. 851.

10   A. Ravex & al, "Pulse Tube Cooler Development at CEA-SBT", Proceedings of the 19th International Congress of Refrigeration, La Hague, The Netherlands (1995) Vol. IIIb, p. 1209

11   UK Patent Application No. 9504852.6

# 80 K Miniature Pulse Tube Refrigerator Performance

**M. David and J-C. Maréchal***

Air Liquide-DTA
BP 15, 38360 Sassenage, France
* Ecole Normale Supérieure, Département de Physique
24 Rue Lhomond, 75005 Paris, France

## ABSTRACT

Air Liquide, in collaboration with Ecole Normale Supérieure in Paris, has previously developed the theory of an ideal Orifice Pulse Tube Refrigerator, later extended to a complete Double Inlet Pulse Tube. This analytical model permits design of any PTR from only the knowledge of the oscillator characteristics (swept volume, average pressure and working frequency).

One of the first applications of this theory was the conception of a 80K miniature PTR devoted to high Tc superconductors and IR detectors. A 48K limit temperature with 1W available power at 82K were achieved. The 2 cm³ swept volume oscillator is a standard one developped for Stirling machines.

The exhaustive results of such PTR, exhibited during the Paris Show 95 at Le Bourget, are presented and compared to the predictions.

## INTRODUCTION

After the invention of the Orifice Pulse Tube by Mikulin[1], greatly improved by Radebaugh[2] in 1986, and the possibilty to add a second orifice[3], proposed by Zhu in 1989 (Fig. 1), the only industrial PTR was those developped by TRW[4] for space applications. The present work shows that it is now possible to dispose of miniature industrial PTR, ie reproducible, reliable and using standard oscillators, qualified for defense programs.

An « ideal » PTR is defined with the usual following assumptions, as justified in a previous paper[5] : ideal gas, uniform pressure along the pulse tube, one-dimensional laminar flow, isentropical transformations and perfect heat exchangers. For a miniature PTR driven by a piston oscillator, the gas pressure in the tube can be well approximated by a sinusoid with small amplitudes. In this section, the useful terms for the optimization of the system are derived from our analytical model[5,6] by taking into account the foregoing conditions.

Cryocoolers 9, Edited by R.G. Ross, Jr.
Plenum Press, New York, 1997

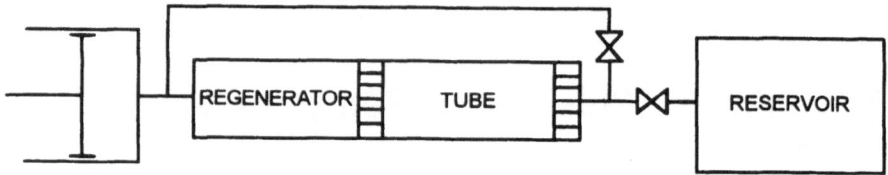

**Figure 1.** Double Inlet Pulse Tube Refrigerator

## MAIN LOSSES IN A MINIATURE PULSE TUBE REFRIGERATOR

We propose to discuss the variation of each term of the energy balance at the cold part, reminded hereafter, when the sizes and volumes are reduced.

$$q_c = q_{cth} - q_{adia} - q_{reg} - q_{cond} \tag{1}$$

where $q_c$ is the available cooling power, $q_{cth}$, the theoretical cooling power, $q_{adia}$ the losses due to the non-adiabatic process, $q_{reg}$ the regenerator losses and $q_{cond}$ the conduction losses.

In the ideal conditions, the cooling power provided by the gas in a Double Inlet Pulse Tube Refrigerator is equal to the heat averaged over a cycle, $<q>_{ath}$, evacuated to the hot end :

$$<q>_{ath} = - <q>_{cth} = \frac{r T_a P \varepsilon^2 k}{2 R_o} \tag{2}$$

where P is the average pressure, r the gas constant, $T_a$ the hot end temperature, $\varepsilon$ the ratio of the half of the pressure amplitude by P and $R_0$ the orifice impedance. k is a relation between the different impedances of the system and is given by :

$$k = (1 + \frac{T_a R_R}{T_c R_1})^{-1} \tag{3}$$

where $R_1$ is the second inlet impedance, $R_R$, the regenerator one and $T_c$ the cold end temperature.

For a miniature PTR, $\varepsilon$ can be deduced from an electrical analogy[7] :

$$\varepsilon = \frac{\frac{T_c}{T_a} R_0 C_0 \omega}{k \left\{ \left[ \frac{T_c}{T_a} R_0 (C_T + C_R) \omega \right]^2 + 1 \right\}^{\frac{1}{2}}} \tag{4}$$

where $C_T$ and $C_R$ are respectively the tube and regenerator « capacity »[7] and $\omega$ the angular frequency. The dead volume capacity, $C_{dv}$, is function of the temperature, T, and of the compression mode (adiabatical or isotherm) :

$$C_{dv} = \frac{'\alpha \; V_{dv}}{rT} \tag{5}$$

where $\alpha$ is equal to 1 for isothermal transformation and to 3/5 in adiabatical conditions, and where $V_{dv}$ is the dead volume value. As all the capacities are in parallel, each $C_{dv}$ can be added to $C_T + C_R$ in the equation (4). Therefore, a dead volume can dramatically reduced the $\epsilon$ value.

The adiabatic conditions, and then the $q_{adia}$ value, relies on the smallness of the thermal penetration compared with the tube diameter.

One of the major difficulties in a miniature PTR is to prevent gas jets from the tube ends to the tube which can achieve a few 10 meters per second. To overcome this problem and then to reduce the heat exchanges between gas and the tube wall and between the different gas parcels, the flow-straighteners must be well studied. In fact, the maximum cooling power due to the hysteresis effect is obtained with a flat velocity profile[5] (Fig. 2).

The conduction losses, $q_{cond}$ are minimized by using very thin stainless steel or titanium tube. In order to minimize the $q_{reg}$ losses, the regenerator sizes must be a compromise between large heat transfer coefficient and low pressure drop, low dead volumes, low longitudinal conductivity and large matrix heat capacity.

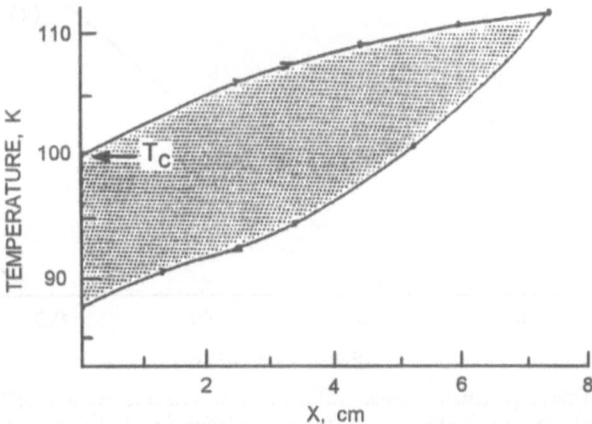

**Figure 2.** Visualization of the hysteresis effect. After travelling back and forth into the tube, the temperature of an element of gas has decreased significantly, due to the orifices-reservoir system (the data of this experiment are given in Ref 5 or 6).

## EXPERIMENTAL RESULTS

The data presented hereafter shows clearly the influence of the dead volumes, of the orifices and of flow-straighteners on a miniature PTR performance.

The first results of a similar PTR were given in a previous paper[8]. In this time, a 76K limit temperature was achieved with a 250 mW available power at 90K. The 1 cm³ pulse tube was in-line connected to the regenerator and the gas pressure supplied by a 2cm³ swept volume oscillator (see Figure 3, curve 1).

A few months later, with the same oscillator, the same regenerator connected to the same pulse tube in U shape configuration and the same distance between the oscillator and the cold finger, the performances were dramatically improved (see Figure 3, curve 2) :   48K limit temperature and 1W at 82K !

This evolution was mainly due to three changes :
. the amount of the dead volumes was reduced and their location and form precisely calculated,
. the design of the orifices sytem was modified,
. the conception of new flow straighteners have permitted to take into account the U shape.

This PTR is currently under tests with Infra-Red detectors.

The influence of the dead volumes is well predicted by our analytical model. On the other hand, the calculation of the orifices design and of the flow straighteners on the available power requests an other method which is currently studied.

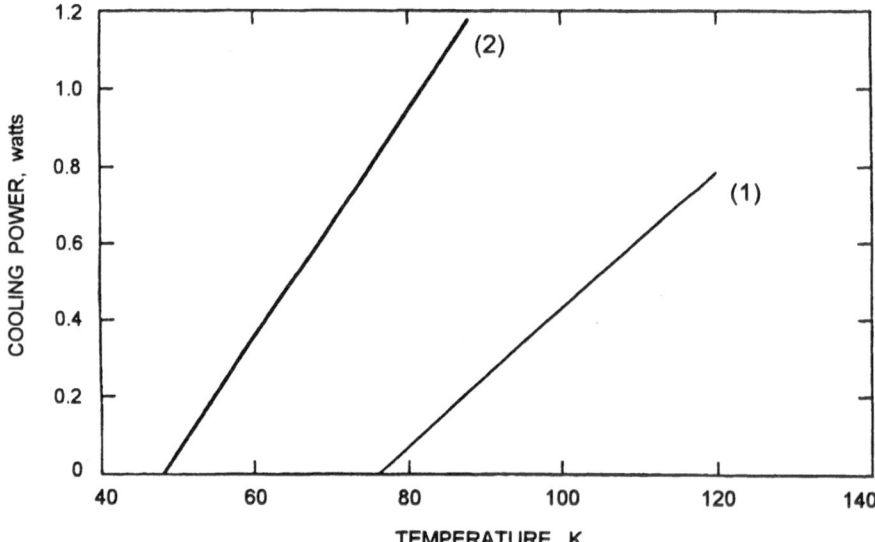

**Figure 3.** Influence of three parameters (dead volumes, flow straighteners and orifices design) on miniature PTR performances. The curves (1) and (2) were obtained with the same regenerator, pulse tube and oscillator and with the same distance between the oscillator and the cold finger. The

dramatic improvement between the two experiments was only due to the evolution of these 3 parameters.

For SQUIDS applications, we have developped and sold a PTR with a smaller oscillator (see Fig. 4). With a 20 cm length flexible between the oscillator and the cold finger, it provides 500 mW at 77K. When the oscillator-cold finger distance is increased up to 1 m, the available cryogenic power at the same temperature is still 200mW. This long length cable is useful both to avoid electomagnetic pertubation from the oscillator and to easily move around the patient. In this case, the application will be to realize electro and magneto-cardiogram of the foetus.

The ratio between the two available powers is in good agreement with the predictions calculated from Eq.(2) and (4) : at 77 K, $\varepsilon$ is reduced by a factor 1.6 due to the dead volume of the additional flexible tube and therefore $q_c$ is reduced by a factor 2.5.

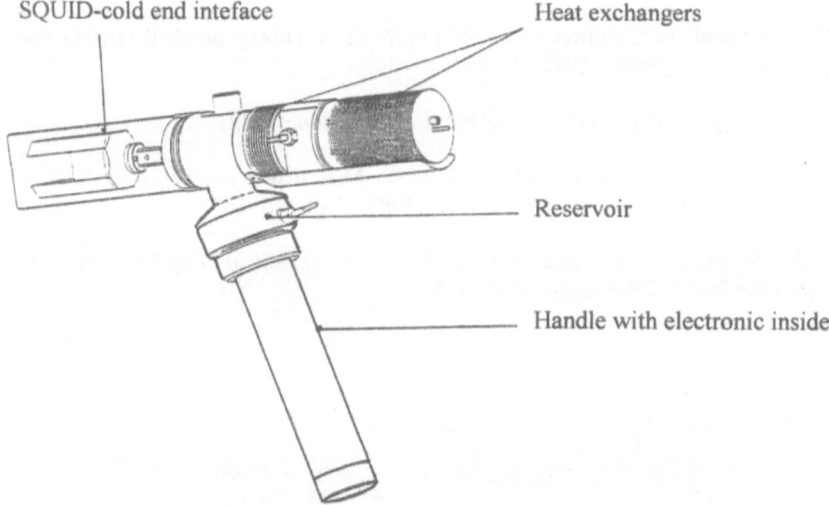

SQUID-cold end inteface

Heat exchangers

Reservoir

Handle with electronic inside

**Figure 4.** Portable miniature Pulse Tube Refrigerator for SQUID applications

## CONCLUSION

High performances of a miniature Pulse Tube Refrigerator can only be achieved if the gas works in ideal conditions, as already defined (see ref 5 or 6) : adiabatical transformations, flat velocity profile and perfect heat exchangers. Moreover, the dead volumes sizes and locations are dramatically important. For small sinusoidal amplitudes, the available power at a given temperature can be well approximated by using the electrical analogy previously described (see ref 7). However, the design of the global system, including dead volumes, flow straighteners and heat exchangers also requests an important know-how. The PTR here described are available for industrial applications in 100mW to 1W cooling power range at 80K.

## REFERENCES

1. Mikulin, E.I., Tarasov, A.A and Shrkrebyonock, M.P., « Low Temperature Expansion Pulse Tubes », Advances in Cryogenic Engineering, Vol. 12 p. 608, Plenum Press, New York, (1984) p. 629.

2. Radebaugh, R. « Pulse tube refrigeration : a new type of cryocooler », Proceedings of the 18 th International Conference on Low Temperature Physics (Kyoto, Japan 1987), J Appl Phys (1987) 26 p. 2016.

3. Zhu, S., Wu, P. and Chen, Z., Double Inlet Pulse Tube Refrigerator: « an important improvement », Cryogenics (1989) p. 1191.

4. Chan, C.K., Jaco, J.C., Raab J., Tward E. and Waterman, M., « Miniature Pulse Tube Cooler », Proceedings of Cryocoolers 7 conference, Santa Fe (1990), p.113.

5. David, M., Maréchal, J-C., Simon, Y. and Guilpin, C., « Theory of Ideal Orifice Pulse Tube Refrigerator », Cryogenics, (1993), Vol 33, p. 154.

6. David, M., « Réfrigération par Tube à Gaz Pulsé », PhD (June 1992).

7. Maréchal, J-C., Pety J., Simon Y. and David, M., « Analytical Comparison of Different Types of Pulse Tubes Refrigerators », Cryogenics, (1994), Vol 34p. 163.

8. David, M., Maréchal, J-C. and Simon, Y., SAE technical Paper, 24th ICES, Friedrischafen, June 20-23, 1994, paper n°941527.

# Development of a Low-Cost Cryocooler for HTS Applications

**S.C. Russo and G.R. Pruitt**

Hughes Aircraft Company
Electro-Optical Systems
El Segundo, CA 90245

## ABSTRACT

Cryogenic refrigeration is an essential and enabling technology for the commercialization of high temperature superconducting (HTS) devices. A reliable, low-cost, low-vibration cryocooler, designed for efficient integration with HTS devices must be commercially available before HTS electronic technology can be introduced into the market place. To satisfy the need of HTS applications, the Low Cost Cryocooler (LCC) Program has been initiated. The LCC design is based on a combination of new technology and upgraded existing commercial designs. The goal is to hasten the insertion of HTS technology into the military and commercial communities by proving that cryocooler technology exists for the right price and demonstrating that this technology is low risk and reliable.

The LCC Program is sponsored by the Advanced Research Projects Agency (ARPA) and the Naval Research Laboratory (NRL). This program will demonstrate a cryocooler operating in the 60K to 80K range (4 W at 70K) and satisfying the requirements of a number of HTS applications. The LCC design incorporates a unique compact and mechanically robust version of a pulse tube expander that offers cost, structural and thermal management advantages over other type expanders. The LCC compressor is an upgraded design of a linear twin-opposed piston design embodied in Hughes tactical and commercial coolers already in production. When integrated with the HTS device or other similar devices, this combination provides the system designer with a standard "plug-in" module.

This paper describes the overall LCC system, its design, performance and manufacturing philosophy.

## INTRODUCTION

The LCC technical approach builds on proven technology. The pulse tube expander design is based on technology developed and demonstrated on Hughes IR&D. The compressor is a modified linear compressor similar to models now in production as shown in Figure 1. The integrated design results in a low cost, long life production cooler.

## DESIGN CONCEPT

The LCC design approach is to provide a cryocooler with high performance at low cost that is flexible in design and integration capability, thus providing a cryocooler usable in a number of applications. In order to make the cryocooler immediately useful for integration with HTS devices Hughes is consulting with a users group of potential cryocooler users with similar cooling requirements. The HTS applications considered include filters, which when combined with a suitable low noise amplifier, can provide an extremely low noise frequency selection amplifier for advanced radar systems and for new communications systems. The LCC demonstration will show the validity of using a balanced cooler for HTS applications, which have demanding low vibration requirements.

For the intended applications, coefficient of performance (COP) is not an all important issue in designing the LCC. Other more important performance drivers are listed in order of importance: (a) low manufacturing cost; (b) reliability and life (MTBF); (c) minimum cold end vibration output; (d) mechanical ruggedness and tolerance to operate in potentially high stress environments; (e) weight and size; (f) suitability of expander cold finger integration into electronic system. These features were compared and the study indicated a preference of pulse tube over moving piston expanders in this application, primarily because of the superiority of pulse tube expanders in cost, reliability, vibration generation and ruggedness.

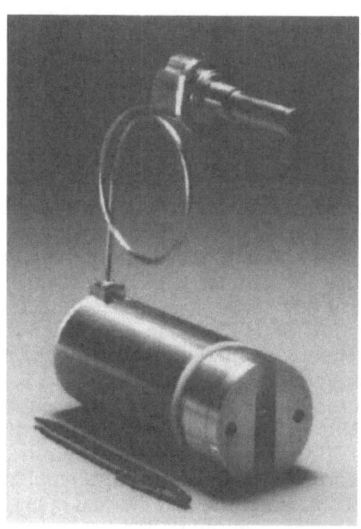

**Figure 1. Hughes commercial cooler.** This commercial split Stirling cooler, used for IR cameras and small instruments, is the basis for the compressor design on the Low Cost Cryocooler Program.

The low cost manufacturing philosophy began with a concurrent engineering design process. The focus on manufacturability began at the conceptual design phase. At this phase we concentrated on ease of assembly, reduction in number of piece parts, and use of self jigging assemblies. Ensuring that individual parts can be assembled with minimum handling pays off even in low volume production, and will pay great dividends in high volume. Reduction in parts count will pay off in procurement, parts handling prior to reaching the assembly area and after arrival on the assembly line.

## USER GROUP REQUIREMENTS

The LCC design process started with user needs. In order to develop this information, Hughes has collaborated with the user group identified in Table 1. This group contributed by helping to establish design requirements and participating in technical reviews. As a result of this cooperative effort the LCC design specification summarized in Table 2 was developed.

**Table 1.  User Group and Targeted Cryogenic Electronic Applications.** The LCC design process started with user needs.

| Markets | Applications | User Group |
|---|---|---|
| Communications<br>• Cellular Stations<br>• Satellites<br>• Aircraft | • Filters/Filter Banks<br>• Resonators | • Conductus<br>• DuPont<br>• Illinois Superconductor<br>• Superconducting Core Technology |
| Radar<br>• Ground Stations<br>• Aircraft<br>• Shipboard | • Filters/Filter Banks<br>• Resonators | • Tektronic Inc.<br>• West Science and Tech. Center |
| Computers/Workstation | • CMOS ICs<br>• MCMs | |

** LCC cooling load can be scaled up for this application

**Table 2.  LCC Design Requirements.** The cryocooler is designed to meet these requirements.

| Parameter | Requirement |
|---|---|
| Load Temperature | 70K |
| Cooling Capacity | 4 Watt |
| Life | > 3 Years |
| Residual Vibration | <0.1 N, Expander/<2.0 N, Compressor |
| Cooldown Time to 70K | 0.1 min/Grm Cu |
| Operating Frequency | ≈ 30 Hz |
| Heat Sink Temperature | 300 ± 20K |
| Envelope | 1' x 1' x 1' |
| Weight: Cooler/Electronics | <11 Kg (24.2 lb)/<7 Kg (15.4 lb) |
| Power Form | 110 vac/60 Hz |
| Input Power | <200 Watts |
| Large Quantity Production Cost | <$2000 Unit |

## COOLER SYSTEM DESCRIPTION

### General Background

A focus area of Hughes, IR&D sponsored research, has been the advancement of the state of the art of pulse tube expanders, and has succeeded in producing both linear and concentric pulse tubes exhibiting competitive performance. In addition, a more compact and mechanically robust version of concentric pulse tube expanders has been designed. The newer design, which has been implemented on the LCC program, is mechanically more robust and compact than other types of expanders. It offers structural and thermal management advantages for integration of the cooler with the users thermal load. The basic concept combines a long life, state of the art twin opposed piston linear compressor, with the new pulse tube expander.

The integrated LCC system is depicted in Figure 2. This arrangement, for cooling a small HTS device, has evolved from discussions between one potential user (DuPont) and Hughes. Shown inside a vacuum dewar is DuPont's equipment. In order to cool the high temperature superconducting (HTS) based oscillator package, the LCC provides cooling up to 4.0 watt at 70K. For other users, this same system is capable of producing 1.5 watts at 60K and more than 6 watts at 120K. The selected refrigerator concept and the heat lifting capacities and load temperatures can be readily adapted to different thermal envelopes as well as to different user thermal loads and cold end temperatures.

In Figure 2, the LCC expander is shown protruding into the cryo envelope containing the cooled electronic device. The expander cold tip is provided with a suitable thermal interface for easy attachment to the users equipment and radiation shields (not shown in the Figure) are used for thermal parasitics control. The cooler portion being constructed under this program ends at the interface flange, which is an integral part of the expander. As can be seen from the drawing, a rather simple and compact configuration has evolved.

In the configuration shown in Figure 2, the expander base is mounted to the flange of the dewar. The (HTS) electronic device and its radiation shield are then mounted to the cold finger flange as a module. The cover of both the dewar and the internal radiation shield are removable

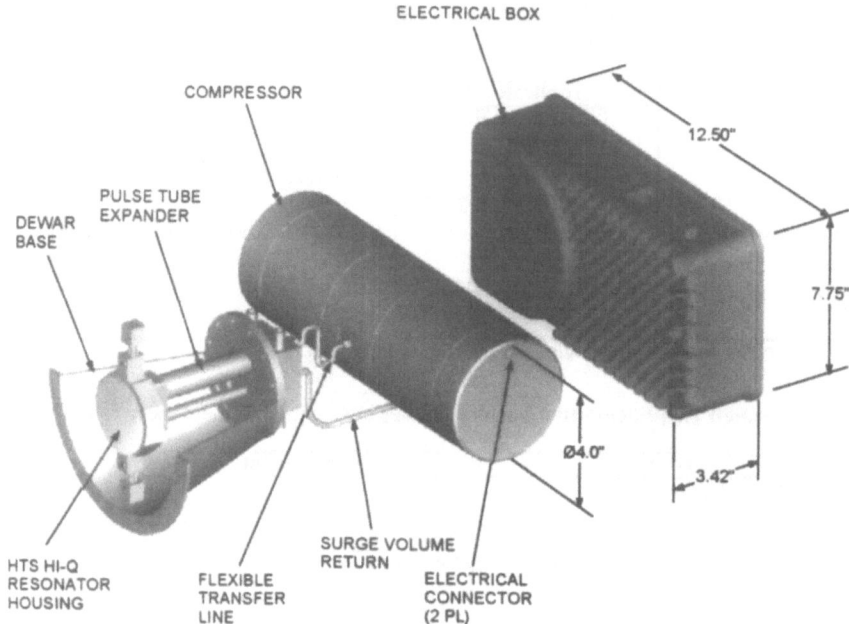

**Figure 2. Integrated cooler configuration.** The combination of a linear compressor and parallel pulse tube offer a simple package with the flexibility to satisfy many users.

to allow easy access to the equipment. Multi Layer Insulation (MLI) can be used to further reduce radiative parasitic loads, if needed. The shape of the radiation shield and dewar facilitate low cost MLI wrapping. Needed electrical connectors and vacuum fittings penetrate the vacuum dewar base (not shown in the Figure).

The linear compressor is balanced to the first order, thus benefiting users sensitive to vibration. Since the configuration is also split, the compressor can be isolated on a flexible mount further lowering the vibration propagated to the system. The pulse tube expander does not contain any moving parts that can cause vibration, and is inexpensive to make and highly immune to failure for the same reason. The LCC expander should be easier to integrate than traditional linear pulse tube expanders, since the cold station is located at an end rather then in the middle of the expander.

The LCC operating parameters, power requirement and weight are summarized in Table 3.

**Table 3. LCC Operating Parameter, Power Requirement, and Mass.**

| Operating Parameters | Nominal | Allowable Ranges |
|---|---|---|
| Speed | 30 Hz | 25-35 Hz |
| Charge pressure | 30 atm | Up to 40 atm |
| Rejection temperature | 300 K | 280 K - 320 K (Air temperature) $T_{reject}$ + 20 C |
| Power Requirement | | |
| Compressor | 162 | |
| Electronics | 38 | |
| Total | 200 W | |
| Mass/Weight | | |
| Expander | 1.7 (3.7) Kg (lb) | |
| Compressor | 9.4 (20.6) Kg (lb) | |
| Electronics | 6.4 (14.1) Kg (lb) | |
| Total | 17.5 (38.4) Kg (lb) | |

## Pulse Tube Expander

As shown in Figure 3 the LCC utilizes a "parallel" pulse tube expander where the pulse tube is folded back parallel to the regenerator tube. This has the advantage of significantly reducing the expander length and of essentially combining the equal temperature inlet and rejection heat exchangers into an integrated unit. As opposed to the linear pulse tube expander, this configuration embodies a single warm end with the cold finger protruding into the dewar space thus allowing easier attachment of the thermal load. The "parallel" pulse tube expander meets the special needs of the user group. Although the "parallel" expander differs substantially in configuration, it utilizes the same basic functional components and the thermodynamic process as the conventional linear pulse tube.

The calculated LCC load curve is shown in Figure 4. As shown the expander is expected to generate more than 2.0 watts of cooling at 60K, 4 watts at 70K and more than 7 watts at 100K. The calculated input power is estimated to be 200 watts at the 4 watt, 70K design point. As much as 30 per cent lower power would be required at lower operating frequencies near 15 Hz, but at the expense of a larger, heavier and more costly compressor module.

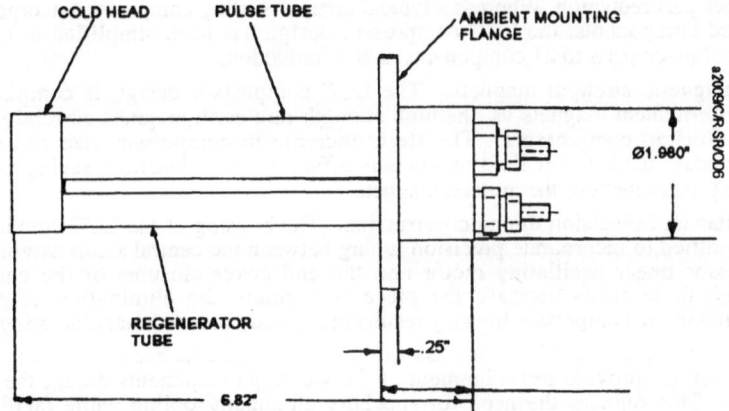

**Figure 3. Hughes parallel pulse tube.** The parallel pulse tube has been sized to meet the customer's heat lifting requirement.

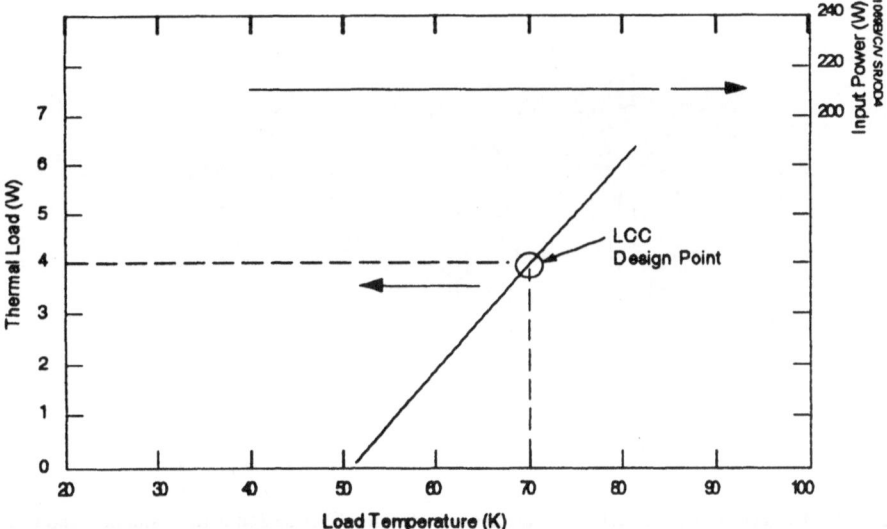

**Figure 4. Calculated load curve.** The LCC is designed to provide 4 watts of cooling at 70K.

## Compressor

To attain the cost objectives of the Low Cost Cooler (LCC) program, it was key that the design of the LCC compressor embrace producibility without compromising the reliability requirements of continuously operated HTS systems. To achieve these objectives, the LCC compressor baselined the design methodology demonstrated on tactical Stirling cycle compressors depicted in Figure 5. The baselined Stirling cycle compressor design offers reasonable manufacturing costs associated with a design compatible with a manufacturing facility with a demonstrated throughput of 10,000 coolers per year while incorporating a mechanical design with a demonstrated wearout life in excess of 15,000 hours. The LCC compressor built on this proven design to achieve greater affordability and to address life limiting features for improved operational life.

To address producibility of the LCC compressor design, a team of suppliers, customers, manufacturing personnel and design engineers reviewed the baseline design to simplify the design. This review process resulted in a LCC compressor design tailored for ease of parts manufacture and component assembly. Key areas leading to a more producible design include:

- total piece part reduction. Whereas a typical tactical Stirling compressor incorporates ~25 machined components, the LCC compressor design has been simplified to reduce the number of piece parts to 21 components, a 20% reduction.

- lower magnetic strength magnets. The LCC compressor design is compatible with molded permanent magnets vs. the high strength rare earth magnets currently in use in tactical Stirling compressors. The slight increase in compressor size and weight to accommodate these lower field magnets is offset by the substantial savings associated with the procurement of the molded magnets

- use of standard precision tubing construction. The housing of the LCC compressor has been modified to incorporate precision tubing between the central stator assembly of the compressor linear oscillating motor and the end cover closures of the compressor. Although these tubes increase the piece part count, the elimination of deep bore machining in the compressor housing results in a greatly reduced manufacturing cost;

- self-jigging alignment. The piston assemblies, which are attached to the compressor end cover closures, provide self-alignment of the wearing components during the assembly process. This reduces the need for specialty alignment tooling while facilitating the assembly process of the compressor.

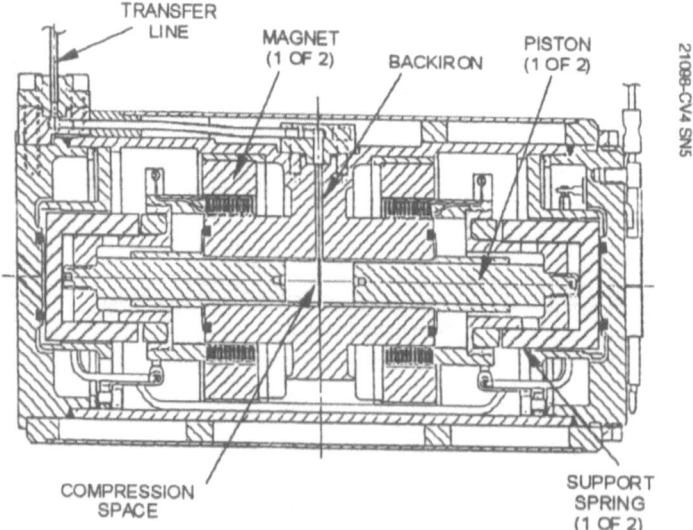

**Figure 5. Current twin-piston linear compressor design.** This design is in production and forms the basis for the LCC compression design.

These design and manufacturing changes result in a projected savings of 51% reduction in manufacturing steps, 68% reduction in touch assembly labor and 29% reduction in assembly and test cycle time.  Overall, these changes reduce the projected manufacturing cost of the compressor by more than 50% when compared with traditional tactical Stirling cooler compressor assemblies.

In addition to manufacturability, the LCC compressor must be compatible with operational life requirements of HTS systems. The minimum acceptable average wearout life is required to exceed three years or 26,300 hours. Wear test data amassed on tactical linear coolers have demonstrated wearout in ~16,000 hours[1]; however, the wear data indicated greatest wear at the interface of the compressor piston with the entrance to the mating sleeve. Wear rates measured at locations other than this interface were consistent with wearout life in excess of 36,000 hours.

The LCC compressor has taken into account this experience to eliminate the wear interface of the compressor piston with the entrance of the mating sleeve. Additionally, the geometry of the LCC compressor piston has been tailored to reduce the side pressure x velocity product that establishes wear rate of the compressor piston material. The result is a projected wearout life of the LCC compressor piston of 5.3 years or ~46,000 hours. Based on these projections, the cost and life requirements for the LCC cooler are considered achievable.

## CONTROL ELECTRONICS

The LCC control electronics draws 110Vac wall power and generates a 30 Hz stepped-square wave to power drive the compressor. The magnitude of the stepped-square wave power is controlled by a temperature sensor located at the expander. The stepped-square wave drive power is chosen to keep the electronics simple and low cost. The power electronic box (see Figure 2) is comprised of an un-regulated power supply, a buck regulator, a temperature control interface circuit, and a 30 Hz chopper. The function of the un-regulated power supply is to convert the 60 Hz AC wall power into a DC power. Since the 110Vac power varies from 85Vac to 135Vac, a buck regulator is used to regulate the voltage coming out from the un-regulated power supply. Cold end temperature control (±1K) is provided via a feedback control loop that is tied to a temperature sensor at the expander cold zone. The temperature control interface circuit translates the cold end temperature into a control voltage for the buck regulator. Therefore, the buck regulator output voltage is a function of the cooler temperature. The 30 Hz chopper is used to generate the stepped-square wave AC power to drive the LCC compressor motor. Industrial grade rather than commercial grade electronic components are used in the power electronics to ensure the reliability of the electronics operating up to an ambient temperature of +47°C. The electronics box is designed for still air convection cooling.

Compressor vibration is expected to be maintained below the 2 Newton's design requirement with balanced opposing displacers. Precision vibration control could be achieved (at additional cost) via feedback sensors that monitor the motion of the two moving pistons. Differences in piston motion drive signals to the two linear motors are used to drive these motion differences towards zero. This servo approach has been incorporated into space cooler designs developed at Hughes with a demonstrated vibration imbalance control to less than 0.1 newtons.

## USER INTERFACES

Three interfaces must be provided by the user: Cryo, Vacuum, and Thermal Management.

### Cryo Interface

Using the DuPont High-Q Resonator as an example of a user, one finds that it is desirable to mount the equipment to a flange on the cold end of the expander as shown previously in Figure 2 thus eliminating any support parasitics. The large diameter of the pulse tube and regenerator can support the anticipated load under most conditions. Therefore, inefficient flexible straps which are needed to protect the more delicate cold fingers of the conventional Stirling Cycle expanders are not required here. A radiation shield is used to lower the radiative parasitic load on the DuPont system. The base of the radiation shield is also mounted to the cold tip. Even with gold plating of the shield, (emissivity e = 0.02) the radiative load from the 330K surroundings to the cooler would equal the total 0.5 watts parasitic allocation. Mounting hardware and penetrations for leads and the required adjusting screws would create a higher than desired load. Thus, the use of a simple MLI blanket over the radiation shield can lower the radiative load to under 0.2 watts.

### Vacuum Interface

For overall operating efficiency, the dewar must be evacuated. The cooler, itself, is leak free, and can be welded to the dewar base if needed. Vacuum tight connectors can be incorporated into the dewar for the customer's equipment, 'and for a feedback temperature sensor, if required. Thus, the use of a vacuum pump or a permanent vacuum maintained by getters will be a required part of the user's hardware.

### Thermal Management/Heat Rejection

In order for the cooler to operate efficiently, the heat of compression and compressor motor dissipation must be managed. The heat load at the electronics, compressor, and expander are about 38, 60 and 102 watts, respectively. The electronics is box designed for natural convection cooling but forced convection cooling is required for the compressor and water cooling is required for the expander. Water coolant passages are provided in the expander base.

## MANUFACTURABILITY

The primary LCC features contributing to the low cost design are the incorporation of the pulse tube expander and a low part count compressor design. This design significantly reduces total system parts count. This reduced piece part count results in minimum assembly labor. The pulse tube consists of four basic parts, the cold expansion section, regenerator section, warm base, and cold end. The compressor design is also a significant cost improvement over existing long life designs. Existing designs use ultra close tolerance clearance seals and resonant springs. The LCC design uses compression space seals developed for Hughes' low cost commercial coolers. These seals use a proprietary process which eliminates the need for close tolerance secondary machining of the seal. Current long life seals are machined to very close clearances which require custom honing and maintaining of matched sets for the piston and cylinder. The proposed design does not require secondary machining or the maintenance of matched sets of pistons and cylinders.

The compressor design uses a resonant spring sized from Hughes' low cost commercial cooler compressors. These springs consist of a single component per piston as opposed to the laminated stacks of several springs used in some long life designs.

Motor coils consist of molded plastic assemblies which incorporate many of the normal coil support structures which are typically made of metal parts that are bonded or held together with screws. The coils consist of a single assembly per side which incorporate the copper motor winding, piston support member and the spring attachment in one molded assembly.

The cost model employed is based on manufacturing costs from vendor quotes and comparison with existing product cost including touch labor, support labor, and burdening. The projected cost per LCC system is approximately $2,000 in production quantities of 10,000 units or more. The projected unit cost is illustrated in Figure 6.

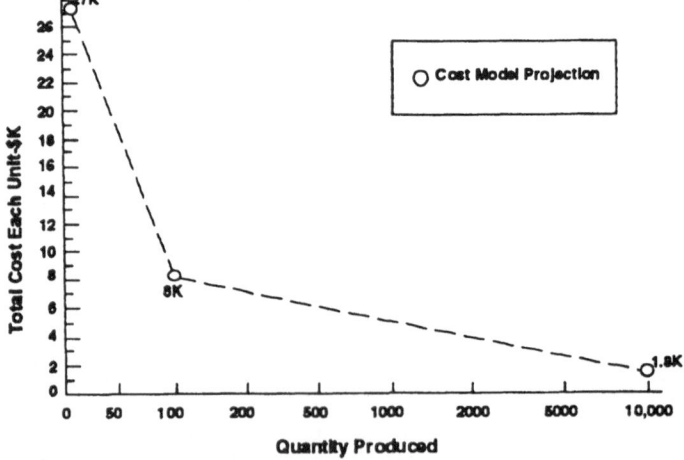

**Figure 6. LCC cost model.** Projected system cost is approximately $2,000/unit based on minimum quantities of 10,000 units.

## SUMMARY

Hughes is developing a low cost long life commercial cooler to satisfy the requirements of near term HTS applications. With the participation of a representative HTS user group, the LCC design specification has been tailored to meet the needs of several users. The cooler design combines the pulse tube technology with proven production expertise to provide minimum three year life cooling of 2 watt at 70K. The LCC can operate off of 110VAC wall power providing cooling meeting the cooling requirements for several applications. Three LCC systems are currently being assembled on the existing program and are scheduled to be delivered to the Naval Research Laboratory by September 1996.

## ACKNOWLEDGMENTS

Program managed by Naval Research Laboratory, Washington D.C. 20375-5320

Program Manager, Dr. Martin Nisenoff, Naval Research Laboratory, Contract No. N0001494C2128.

Support from LCC user group: Conductus, Dupont, Illinois Superconductor, Superconducting Core Technology, Tecktronic Inc., and West Science Tech. Center.

## REFERENCES

1.   "Hughes Long Life Linear Stirlings: A Status Report", Pruitt, G.R., 8th International Cryocooler Conference, 28-30 June 1994.

# An Experimental Investigation of the Pulse Tube Refrigerator

**D.Y.Koh, S.J.Park, S.J.Lee, H.K.Yeom, Y.J.Hong, and S.K.Jeong***

Thermal and Fluid Systems Department
Korea Institute of Machinery and Materials
Daejeon 305-600, Korea

*Department of Mechanical Engineering
Korea Advanced Institute of Science and Technology
Daejeon 305-701, Korea

## ABSTRACT

The experimental results of the pulse tube refrigerator are presented in this paper. The pulse tube refrigerator, which has no moving parts at its cold section, is attractive in obtaining higher reliability, simpler construction, and lower vibration than any other small refrigerators. The objectives of this study are to develop the design technology of pulse tube refrigerator and acquire its application technique. As a preliminary test, the refrigeration performance of the basic, orifice, and double inlet pulse tube refrigerators were investigated. The lowest temperature obtained in the single-stage pulse tube refrigerator was 34.4K and the refrigeration capacity at the optimum operation condition was 23W at 80K. The lowest temperature of the second stage cold head in the two-stage pulse tube refrigerator was 18.3K and the refrigeration capacities at the optimum condition were 0.45W at 20K and 20W at 80K, respectively.

## INTRODUCTION

The pulse tube refrigerator was first described by W.E.Gifford and R.C.Longsworth[1] in 1964. This type of the pulse tube refrigerator is now called as the basic pulse tube refrigerator. The practical development, however, did not proceed until E.I.Mikulin et al.[2] who investigated the "Low-Temperature expansion pulse tube refrigerator" in 1984. The performance of this pulse tube refrigerator has been greatly improved by introducing an orifice and a buffer volume added to the hot end of the pulse tube. This type of the pulse tube refrigerator, which is called as the orifice pulse tube refrigerator, was modified by R.Radebaugh et al.[3] in 1986, and reached the lowest temperature of 60K. In 1990, the double inlet pulse tube refrigerator, in which a bypass tube is connected between a pressure wave generator and the hot end of the pulse tube, was suggested by S.Zhu et al.[4,5]. The refrigeration power per unit mass flow rate through the regenerator was greatly increased in the double inlet pulse tube refrigerator. Many studies of the multiple-stage pulse tube refrigerator have been performed to achieve much lower temperature. Y.Zhou et al.[6]

and E.Tward et al.[7] achieved the lowest temperatures of 31K and 26K, respectively, by the orifice type two-stage pulse tube refrigerators. Y.Matsubara et al.[8] investigated the performance of the pulse tube refrigerator below 15K by using a two-stage Gifford-McMahon refrigerator to precool the hot end of the final stage regenerator. Recently, Y.Matsubara and J.L.Gao succeeded in achieving the lowest temperature of 3.6K by a three-stage pulse tube refrigerator[9].

In this paper, firstly we report on the performance test data of the single-stage basic, orifice, and double inlet pulse tube refrigerator. And then, the experimental results of the two-stage pulse tube refrigerator are described.

## EXPERIMENT DESCRIPTION

### Single-stage pulse tube refrigerator

Experimental apparatus of the single-stage pulse tube refrigerator is shown in Fig. 1. The refrigerator consists of a compressor, heat exchanger, regenerator, pulse tube, orifice(adjustable needle valve), double inlet valve, reservoir and vacuum chamber. The pressure oscillation is generated by using a commercial helium compressor for Gifford-McMahon refrigerator and solenoid valves with rotary timer for adjusting operating cycle frequency. The helium compressor has an electrical input power of 2.2kW.

The regenerator and the pulse tube are made of thin-walled stainless steel tubes with 35mm i.d. × 100mm length and 21mm i.d. × 203mm length, respectively. Volume ratio of the regenerator to the pulse tube is 1.33. The regenerator matrix consists of a stack of about 1,000 bronze screens of 250 mesh. The flow straightners at both ends of the pulse tube are packed with 50 stainless steel screens of 200 mesh. The hot end heat exchangers of the regenerator and the pulse tube are cooled by circulating cooling water. The volume of the reservoir is 1,000cm³. The adjustable needle valves($C_v$=0.03 at full-open, 9 turn-open) are used as the orifice and double inlet valve.

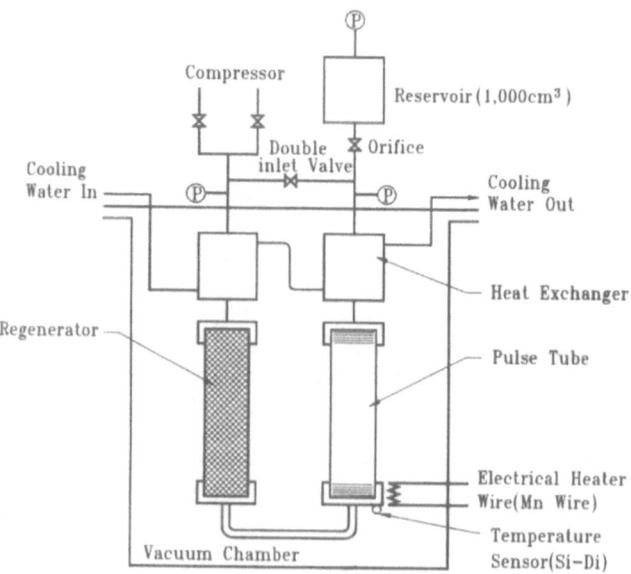

**Figure 1.** Experimental apparatus of the single-stage pulse tube refrigerator.

Strain gauge type pressure transducers are used to monitor the pressure oscillations at the hot ends of the regenerator and the pulse tube and the reservoir. The cold head temperature of the pulse tube is measured by a silicon-diode thermometer. A manganin resistance heater is provided at the cold end of the pulse tube for the refrigeration capacity measurement. After attaching all sensors, the cold end of the pulse tube is wrapped by multi-layer insulation(MLI) for radiation shield, and the apparatus of the pulse tube refrigerator except the component of the room temperature region is connected to the vacuum flange. During the experiment, a vacuum chamber is connected to the high vacuum pump under a pressure of $10^{-5}$ Torr. High vacuum pump system consists of a rotary roughing pump, a diffusion pump and vacuum gauges.

After the regenerator of the pulse tube refrigerator is cleaned by evacuating and purging with clean high pressure helium gas, the pulse tube refrigerator is connected to the compressor. Then, the system is charged up to $15 kg_f/cm^2$. In the experiment, the test apparatus acts as a basic pulse tube refrigerator if the orifice and the double inlet valve are closed, and the orifice pulse tube refrigerator if the orifice is open and double inlet valve is closed, finally the double inlet pulse tube refrigerator if the orifice and the double inlet valve are open.

## Two-stage pulse tube refrigerator

Fig. 2 shows the experimental apparatus of the two-stage pulse tube refrigerator. In general, the hot end of the second pulse tube is cooled by the first refrigeration stage. In this paper, however, the hot ends of the first and second pulse tubes are directly cooled by the air at the ambient temperature. This type of the multi-stage pulse tube refrigerator has been suggested by Y.Matsubara and has shown good results[8,9].

The regenerators and the pulse tubes are made of thin-walled stainless steel tube. The first and second regenerator are 35mm i.d. × 100mm length and 35mm i.d. × 80mm length with a stack of 1,000 bronze screens of 250 mesh and lead spheres($\phi$0.2mm - 0.3mm), respectively. The first and the second pulse tube are 21mm i.d. × 203mm length and 17.8mm i.d. × 200mm length, respectively. The volumes of the reservoirs are 1,000cm$^3$. The adjustable needle valves ($C_v$=0.03)

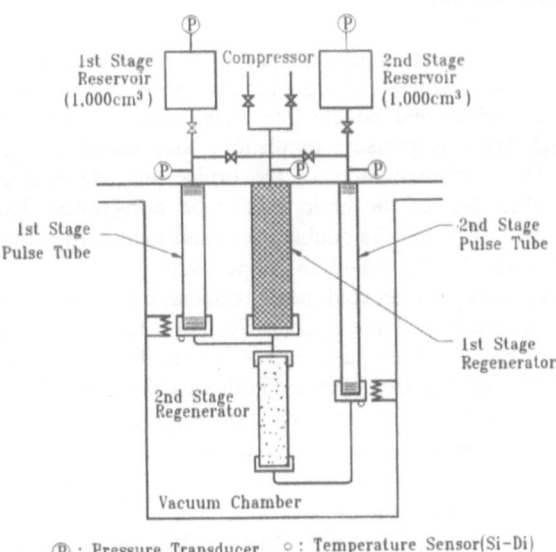

Figure 2. Experimental apparatus of the two-stage pulse tube refrigerator.

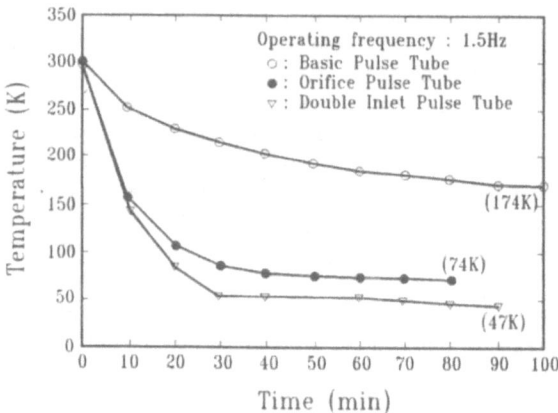

**Figure 3.**   Cooldown characteristics of the single-stage pulse tube refrigerator.

are used as the orifice and double inlet valve. Pressure transducers, thermometers and heater wires for the single-stage pulse tube refrigerator are again used for the two-stage pulse tube refrigerator. Pressures are measured at the hot ends of the first regenerator and the pulse tubes and the reservoirs. Temperatures are monitored at the cold ends of the first and the second pulse tube. Manganin resistance heaters are provided at the cold ends of the first and second pulse tube. Each stage has a double inlet line provided with adjustable needle valves. Charged pressure of the two-stage pulse tube refrigerator system is 15kg₀/cm². The same experimental method has been applied to the two-stage pulse tube refrigerator as the single-stage pulse tube refrigerator.

## EXPERIMENTAL RESULTS

### Single-stage pulse tube refrigerator

As a preliminary test, the refrigeration performance of the single-stage pulse tube refrigerator was investigated on the basic, orifice and double inlet type. Fig. 3 shows the cooldown characteristics of the basic, orifice and double inlet pulse tube refrigerators at the operating frequency of 1.5 Hz. High and low pressure oscillations have varied 19 - 21kg₀/cm² and 8 - 10kg₀/cm², respectively. The cooldown rates of the orifice and the double inlet pulse tube refrigerators were faster than that of the basic pulse tube refrigerator. The lowest no load temperature of the basic, the orifice and the double inlet pulse tube refrigerators at the operating frequency of 1.5Hz was 174.1K, 74.6K, and 47.5K, respectively.

Fig. 4 shows the variations of the cold head temperatures with the various operating frequency for the basic, the orifice and the double inlet pulse tube refrigerator. These results correspond to the optimized needle valve adjustments leading to the lowest cold end temperature. The lowest temperature of 34.4K has been obtained in the double inlet pulse tube refrigerator at the operating frequency of 2.5Hz. The lowest no load temperature of the orifice pulse tube refrigerator was 59.1K at the operating frequency of 2.5Hz. This figure shows that the optimal frequency for the lowest no load temperature decreased according to the order of the basic, the orifice and the double inlet pulse tube refrigerator.

Fig. 5 shows the behavior of the refrigeration capacity with the operating frequency for the orifice and the double inlet pulse tube refrigerator. Refrigeration capacity increases with the operating frequency in the orifice pulse tube refrigerator and has a maximum at the operating

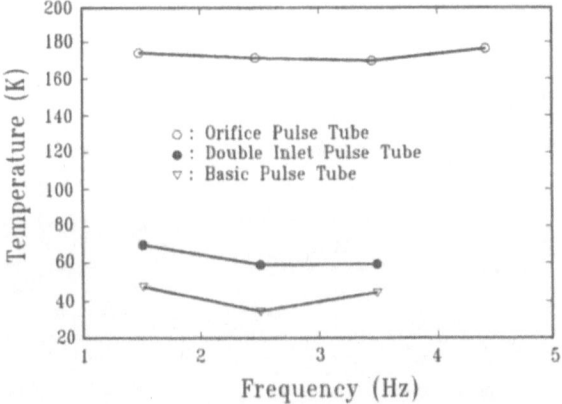

**Figure 4.**  The variations of the cold head temperature with operating frequency for the single-stage pulse tube refrigerator.

frequency of 2.5Hz in the double inlet pulse tube refrigerator. Maximum refrigeration capacities of the orifice and the double inlet pulse tube refrigerator are 8W and 23W at the cold end temperature of 80K, respectively.

## Two-stage pulse tube refrigerator

Fig. 6 shows the variations of no load temperature at the cold end of the second stage as a function of the first and the second orifice valve opening with the first and the second double inlet valve closed. In the two-stage orifice pulse tube refrigerator, the optimal condition leading to the lowest no load temperature of the second stage cold end is not related to the opening rate of the first stage orifice valve. The lowest no load temperature of the second stage cold end was achieved when the second stage orifice valve was 2 turn-open($C_v$=0.005). Furthermore, even though the lowest no load temperature of the second stage cold end usually decreased regarding

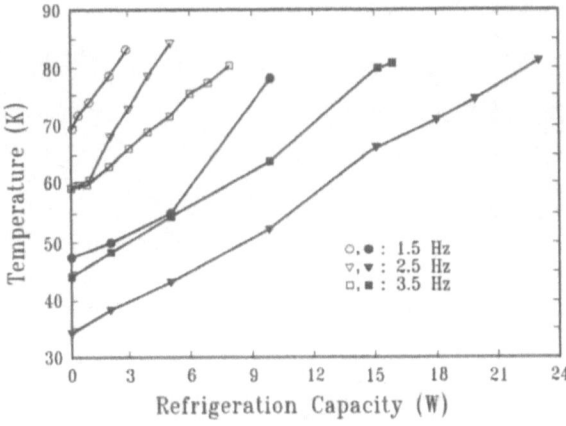

**Figure 5.**  Refrigeration capacity with the operating frequency for the single-stage pulse tube refrigerator.

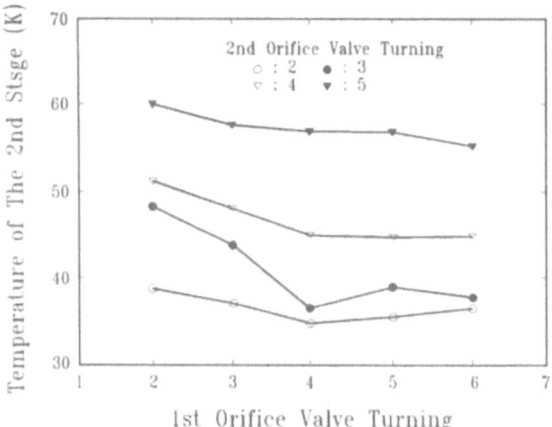

**Figure 6.** The variations of no load temperature with orifice valve opening in the two-stage pulse tube refrigerator.

the opening of the first stage orifice valve, there was a minimum at the first stage orifice valve of 4 turn-open($C_v$=0.012) when the second stage orifice valve was 2 turn-open and 3 turn-open ($C_v$=0.008). In this case, the lowest no load temperature achieved was 34.8K.

In Fig. 7, the relationship between the refrigeration capacities of the second stage cold end and the temperatures of the first and second stage cold ends are displayed for the two conditions. Firstly(case A), the first, the second orifice and the second double inlet valve are 3 turn, 3 turn and 4 turn-open, respectively. Secondly(case B), the first, second orifice and the second double inlet valve are 5 turn($C_v$=0.015), 3 turn and 4 turn-open, respectively. The lowest no load temperature and the refrigeration capacity of the second stage cold end in the case A are 18.9K and 0.45W, 20W at 20K, 80K, respectively. The lowest temperature and refrigeration capacity of

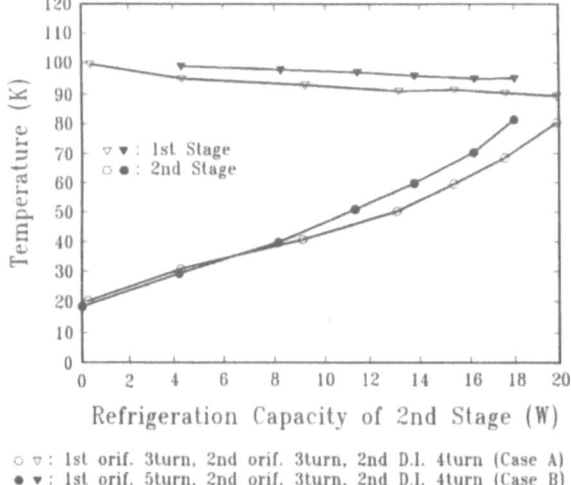

**Figure 7.** The relation between the refrigeration capacity of the second stage cold end and the temperature of the first and second stage cold ends.

the second stage cold end in the case B are 18.3K and 18W at 80K, respectively. Temperature of the first stage cold end is decreased by increasing the temperature of the second stage cold end with the refrigeration load.

## CONCLUSION

A single-stage and a two-stage pulse tube refrigerators were designed, fabricated and tested. The following conclusions are drawn from the experimental results.

(1) In the single-stage pulse tube refrigerator, the cooldown characteristics at the operating frequency of 1.5 Hz and the variations of the cold head temperature with the operating frequency were evaluated for the basic, the orifice and the double inlet pulse tube refrigerator. The cooldown rates of the orifice and the double inlet pulse tube refrigerators are faster than that of the basic pulse tube refrigerator. The lowest no load temperature of the orifice and the double inlet pulse tube refrigerator reached 59.1K and 34.4K at the operating frequency of 2.5Hz, respectively. The refrigeration capacity of the double inlet pulse tube refrigerator is 23W at the cold head temperature of 80K.

(2) In the two-stage orifice pulse tube refrigerator with the first and the second double inlet valve closed, the optimal condition is not related to the opening rate of the first stage orifice valve. The lowest no load temperature achieved was 34.8K at the operating frequency of 2.5Hz when the first and second stage orifice valves are 4 turn-open and 2 turn-open, respectively.

(3) In the two-stage pulse tube refrigerator with the second stage double inlet valve opened 4 turn, the lowest no load temperature of the second stage cold end was 18.3K, the refrigeration capacity of the second stage cold end were 0.45W and 20W at 20K and 80K, respectively, at the optimum operating conditions.

## REFERENCES

1. Gifford, W.E. et al., "Pulse-tube refrigeration", ASME paper No.63-WA-290 presented at Winter Annual Meeting of the American Society of Mechanical Engineers,(1963), pp.17-22.
2. Mikulin, E.I., "Low Temperature Expansion Pulse Tubes", Advances in Cryogenic Engineering, Vol.29, Plenum Press, New York, (1984), pp.629-637.
3. Radebaugh, R. et al., "A comparison of Three Types of Pulse Tube Refrigerators : New Methods for Reaching 60K", Advances in Cryogenic Engineering, Vol.31, Plenum Press, New York, (1986), pp.779-789.
4. Zhu, S. et al., "Double inlet Pulse Tube Refrigerators : An Important improvement", Cryogenics, Vol.30, (1990), pp.514-520.
5. Zhu, S. et al., "A Single Stage Double Inlet Pulse Tube Refrigerator Capable of Reaching   42K", Cryogenics, Vol.30, Sept. Supplement, (1990), pp.257-261.
6. Zhou, Y. et al., "Two-Stage Pules Tube Refrigerator", Proceedings of the 5th International Cryocooler conference, (1988), pp.137-144.
7. Tward, E. et al., "Pulse Tube Refrigerator Performance", Advances in Cryogenic Engineering, Vol.35, Plenum Press, New York, (1990), pp.1207-1212.
8. Matsubara, Y. et al., "An Experimental and Analytical Investigation of 4K Pulse Tube Refrigerator", Proceedings of the 7th International Cryocooler Conference PL-CP-93-1001, Part 1, (1993), pp.166-186.
9. Matsubara, Y. and Gao, J. L., "Multi-stage Pulse Tube Refrigerator for Temperatures below 4K", Proceedings of the 8th International Cryocooler Conference, (1995), pp.345-352.

# Pulse-Tube Refrigerator and Nitrogen Liquefier with Active Buffer System

**Y. Kakimi, S. W. Zhu, T. Ishige, and K. Fujioka**

Tsukuba Laboratory, DAIDO HOXAN Inc.
3-16-2, Ninomiya, Tsukuba, Ibaraki 305, Japan

**Y. Matsubara**

Atomic Energy Research Institute, Nihon University
7-24-1, Narashinodai, Funabashi, Chiba 274, Japan

## ABSTRACT

An experimental study has been carried out to determine the effects on cooling capacity of change in valve timing, frequency, buffer volume, and valve flow coefficients (Cv values) in an active buffer pulse-tube refrigerator. A cooling capacity of 164 W at 80 K is achieved with an input power of 3.63 kW. The percent carnot is 12 %. The original straight type pulse-tube refrigerator is modified to a U-type refrigerator for liquefying nitrogen gas. A liquefying rate of 1.5 liters per hour is achieved.

## INTRODUCTION

From the viewpoint of simplicity and reliability, GM type pulse tube refrigerators have great potential for some applications. The problem with most GM type pulse tube refrigerators is their low efficiency. To increase efficiency, an active buffer pulse-tube refrigerator was introduced [1]. It is a GM type pulse tube refrigerator with three buffers connected at the hot end of the pulse tube through on/off valves, and is highly efficient. A percent carnot of 11 % and a cooling capacity of 160 W were achieved at 80 K with a prototype refrigerator.

This paper describes in detail how changes in valve timing, frequency, volume of buffers, and valve flow coefficients (Cv values) affect the cooling capacity of a pulse tube refrigerator with an active buffer system. A cooling capacity of 164 W was achieved at 80 K with an input power of 3.63 kW. The percent carnot was 12 %.

In the first application of the active buffer pulse tube refrigerator, the refrigerator was modified into a U-type refrigerator as a nitrogen liquefier. The liquefying rate was 1.5 liters per hour.

## MECHANISM

Figure 1 shows the schematic structure of the active buffer pulse tube refrigerator with three buffers [1]. It consists of a regenerator; a heater; a pulse tube; a cooler; valves $V_H$, $V_L$, $V_1$, $V_2$, $V_3$; and buffers 1, 2, and 3. Two or more than tree buffers can be used for the active buffer system. Valve $V_H$ is connected to the high-pressure output of a compressor, and $V_L$ is connected to the low-pressure input. Unlike the tube expander [2], the active buffer pulse tube refrigerator has a regenerator. The

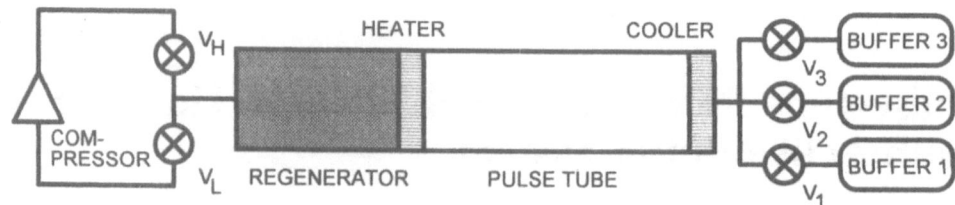

**Figure 1.** Active buffer pulse tube refrigerator with three buffers

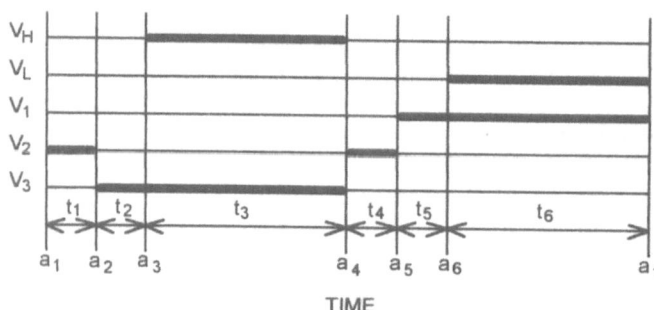

**Figure 2.** Valve timing chart

basic working mechanism of gas in the pulse tube and in the buffers is similar to that in an expander. The function of the buffers is different from that in orifice or double inlet pulse tube refrigerators. The gas flow between the pulse tube and the buffers is actively controlled by opening and closing valves $V_1$, $V_2$, and $V_3$, it is called an active buffer system. Displacement, expansion, and compression of gas in the refrigerator are controlled not only by the compressor.

Figure 2 shows a typical valve timing chart for all valves in a single cycle. The bold lines indicate the opening period. $P_1$, $P_2$, $P_3$, $P_H$, and $P_L$, the pressures in buffer 1, buffer 2, buffer 3, the high-pressure output of the compressor, and the low-pressure input of the compressor, respectively, have the following relations: $P_3$ is slightly lower than $P_H$, $P_1$ is slightly higher than $P_L$, and $P_2$ is moderate. The process of opening and closing valves and gas flow is as follows. After $V_L$ and $V_1$ close and $V_2$ opens at time $a_1$, the pressure in the pulse tube increases to near $P_2$ due to gas flow from buffer 2. After $V_2$ closes and $V_3$ opens at time $a_2$, the pressure in the pulse tube increases to near $P_3$ due to gas flow from buffer 3. After $V_H$ opens at time $a_3$, the gas flows from the compressor into the pulse tube through the regenerator, and the gas in the pulse tube from buffer 3 flows back to buffer 3. After $V_H$ and $V_3$ close and $V_2$ opens at time $a_4$, the pressure in the pulse tube decreases to near $P_2$ due to gas flow into buffer 2. After $V_2$ closes and $V_1$ opens at time $a_5$, the pressure in the pulse tube decreases to near $P_1$ due to gas flow into buffer 1. After $V_L$ opens at time $a_6$, the gas flows from the pulse tube to the compressor through the regenerator, and the gas in buffer 1 from the pulse tube flows back to the pulse tube. Pressure differences across valve $V_H$ at $a_3$ and across valve $V_L$ at $a_6$ are small. Thus, loss caused by the pressure difference is small.

When the pressure in the pulse tube and in the regenerator increases from $a_1$ to $a_3$, the gas in the pulse tube is compressed and flows toward the hot end of the regenerator through the cold end of the regenerator. When the pressure in the pulse tube and in the regenerator decreases from $a_4$ to $a_6$, the gas in the regenerator expands and flows toward the hot end of the pulse tube through the cold end of the regenerator. Thus, the regenerator loss is decreased.

## RESULTS OF EXPERIMENTS ON STRAIGHT TYPE PULSE TUBE REFRIGERATOR

An experimental apparatus comprising a single stage active buffer pulse tube refrigerator with three buffers was made. The pulse tube was a stainless steel tube of inner diameter 49 mm, wall thickness 0.5 mm, and length 202 mm. Its volume was 0.38 liters. Its regenerator comprised a stainless steel tube of inner diameter 54 mm, wall thickness 0.5 mm, and length 94 mm, filled with 200-mesh stainless screens of packing factor 0.27. An electric heater was installed in its heater to measure the cooling capacity. The structure of the pulse tube, the heater, and the regenerator was straight. The refrigeration temperature was measured by a Pt-Co thermometer. All valves were electric on/off valves. Timing of opening and closing valves was controlled by a personal computer. The refrigerator was in a vacuum vessel. Multi layer insulation was installed to decrease room temperature radiation. The compressor was a water-cooled GM refrigerator compressor with a standard power of 3.3 kW.

Experiments were performed to investigate the effects on cooling capacity of changes in valve timing, frequency, volume of buffers, and $C_V$ values of valves.

Two parameters were used to control the valve timing. One was the ratio of positive half period to negative half period. This is called half period time ratio. A typical half period time ratio used was (45:55). This occurs when the ratio of positive half period ($t_1+t_2+t_3$) to negative half period ($t_4+t_5+t_6$) is 45%:55%. The definitions of $t_1$, $t_2$, $t_3$, $t_4$, $t_5$, and $t_6$ are indicated in Figure 2. The other is the ratio of $t_1$ to $t_2$ to $t_3$, which is called valve timing ratio. A typical value used was (1:1:5). The same valve timing ratio is used for the negative half period.

Figure 3 shows the effect on cooling capacity of changes in half period time ratio of (45:55), (50:50), (55:45), and (64:36). The valve timing ratio was (1:1:5). Frequency was 3 Hz. Volume of each buffer was 4.4 liters. The $C_V$ values of $V_H$, $V_L$, $V_1$, $V_2$, and $V_3$ were 0.4, 0.4, 0.35, 0.15, and 0.35, respectively. The cooling capacity increased with decreasing positive half period above 40 K. The minimum temperature of 28 K was achieved at a half period time ratio of (55:45). This is the lowest temperature achieved with this refrigerator. The condition for large cooling capacity was not the same as that for minimum temperature.

Figure 4 shows the effect on the cooling capacity of changes in the valve timing ratio of (1:1:4), (1:1:5), and (1:1:7). The half period time ratio was (45:55). Frequency was 1.9 Hz. Other parameters were the same as in Figure 3. The cooling capacity for the valve timing ratio of (1:1:7) at 80 K was 15 W larger than that for (1:1:4). The largest cooling capacity of 120 W at 80 K was achieved at a valve timing ratio of (1:1:7). The condition for large cooling capacity was not the same as that for minimum temperature.

Figure 5 shows the effect on cooling capacity of changes in frequency from 1.5 to 4 Hz. The half period time ratio was (50:50). The valve timing ratio was (1:1:5). The volume of buffers and $C_V$ values were the same as those of Figure 3. The largest cooling capacity was achieved at 1.9 Hz. The minimum temperature was achieved at 3 Hz. The condition for large cooling capacity was not the same as that for minimum temperature.

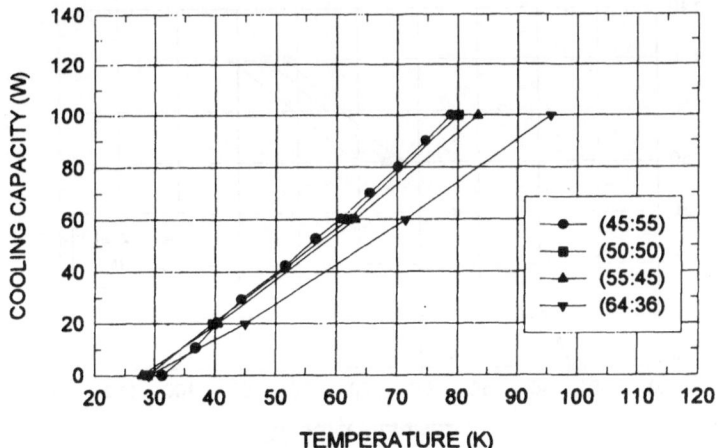

**Figure 3.** Cooling capacity as a function of half period time ratio

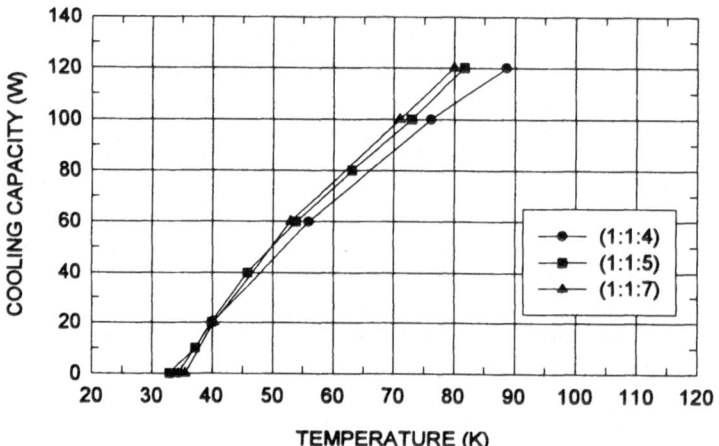

**Figure 4.** Cooling capacity as a function of valve timing ratio

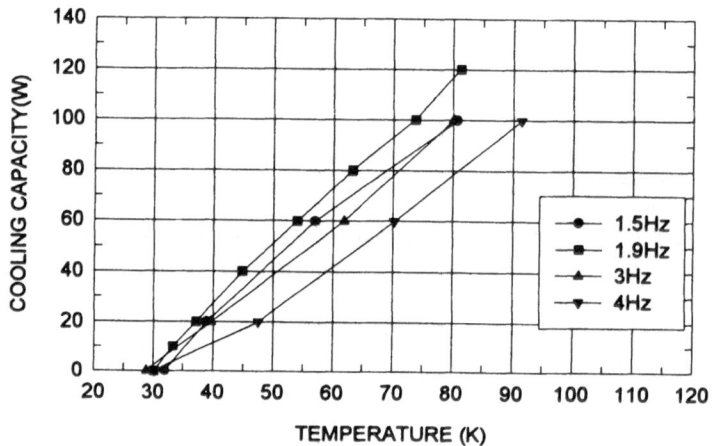

**Figure 5.** Cooling capacity as a function of frequency

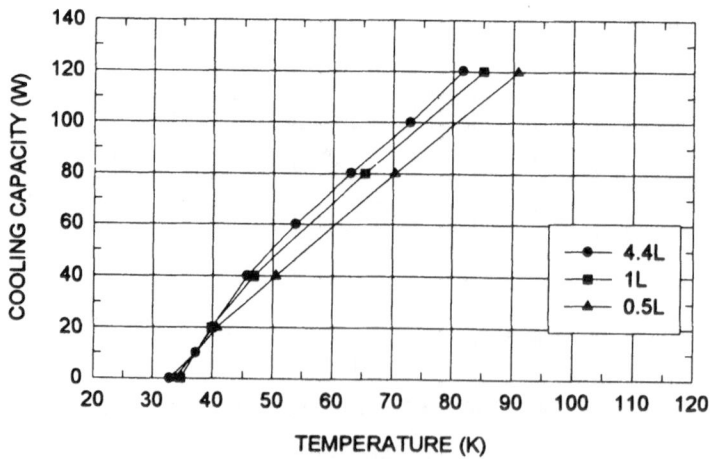

**Figure 6.** Cooling capacity as a function of volume of buffers

Figure 6 shows the effect on cooling capacity of changes in the volume of buffers from 0.5 to 4.4 liters. The half period time ratio was (45:55). The valve timing ratio was (1:1:5). Frequency was 1.9 Hz. The $C_V$ values were the same as those of Figure 3. The cooling capacity increased with increasing the volume of buffers. The difference in cooling capacity at 80 K between 0.5 liters and 1 liter was about 10 W. However, the difference between 1 liter and 4.4 liters was smaller.

To increase its cooling power, valves $V_H$, $V_L$, and $V_2$, that cause large pressure drops, were changed to different valves with larger $C_V$ values. The $C_V$ values of $V_H$, $V_L$, $V_1$, $V_2$, and $V_3$ were 1.24, 1.24, 0.35, 0.3, and 0.35, respectively. The half period time ratio, frequency, and volume of buffers were the same as those of Figure 4. Figures 7 and 8 show the effect on the cooling capacity and percent carnot of changes in valve timing ratio of (1:1:5), (1:0.5:5.5), and (1:0:6), respectively. The cooling capacity for (1:1:5) at 80 K was over 20 W larger than that with valves of small $C_V$ values as shown in Figure 4. The largest cooling capacity of 164 W at 80 K was achieved with an input power of 3.63 kW for (1:0:6). The percent carnot was 12 %. Figure 9 shows equivalent PV diagrams [3] for the gas in the cold end and the hot end of the pulse tube. These diagrams were analyzed from a single cycle of measured pressure waves in the pulse tube and in the buffers at 80 K. The length of the pulse tube is normalized such that the cold end is 0 and the hot end is 1. The shape of the equivalent PV diagram is changed by the valve timing ratio.

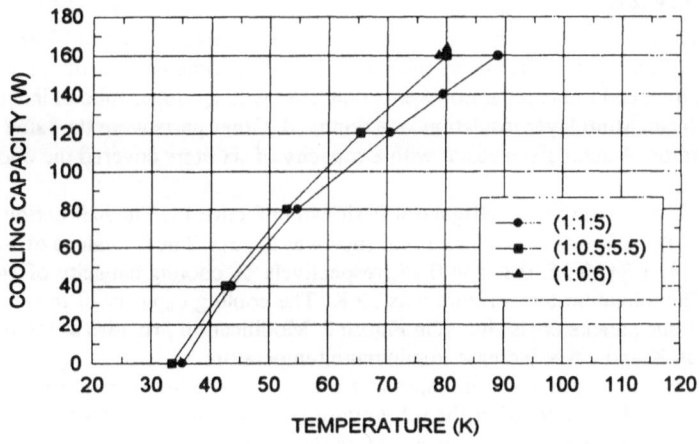

**Figure 7.** Cooling capacity as a function of valve timing ratio

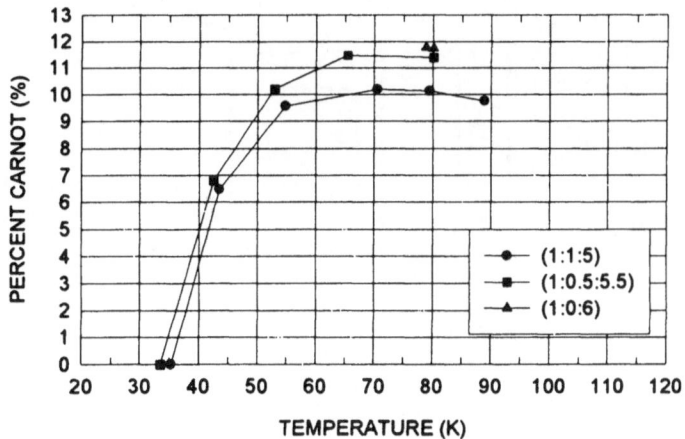

**Figure 8.** Percent carnot at 80 K

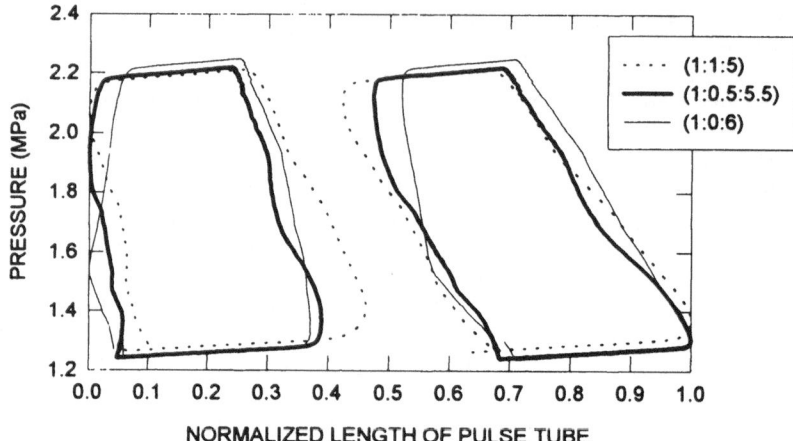

**Figure 9.** Equivalent PV diagrams at 80 K

## NITROGEN LIQUEFIER

The straight type refrigerator was modified to the U-type refrigerator to liquefy nitrogen gas as the first application of an active buffer pulse tube refrigerator. Its schematic structure is shown in Figure 10. The pulse tube and the regenerator were connected with a copper tube of inner diameter 7.9 mm and length 16 cm. Multi layer insulation was removed. Other parts were the same as for the straight type refrigerator. A stainless cryostat with a capacity of six liters covered the cooling parts of the refrigerator.

The cooling capacity of the U-type refrigerator is shown in Figure 11. The half period ratio was (45:55). The valve timing ratio was (1:0.5:5.5). Frequency was 1.9 Hz. The $C_V$ values of $V_H$, $V_L$, $V_1$, $V_2$, and $V_3$ were 1.24, 1.24, 0.35, 0.3, and 0.35, respectively. A cooling capacity of 140 W was achieved at 78 K. The minimum temperature was 39 K. The cooling capacity of the straight type refrigerator with the same parameters is shown in Figure 7. Modification provides a 9 % decrease in cooling capacity at 80 K and a 5 K increase in minimum temperature.

Nitrogen gas at room temperature was input at a pressure 2 kPa higher than the atmosphere. Nitrogen gas was liquefied 15 minutes after the refrigerator started. The temperature was kept at 75 K during liquefying. The liquefying rate was 1.5 liters per hour, which corresponds to a cooling capacity of 143 W.

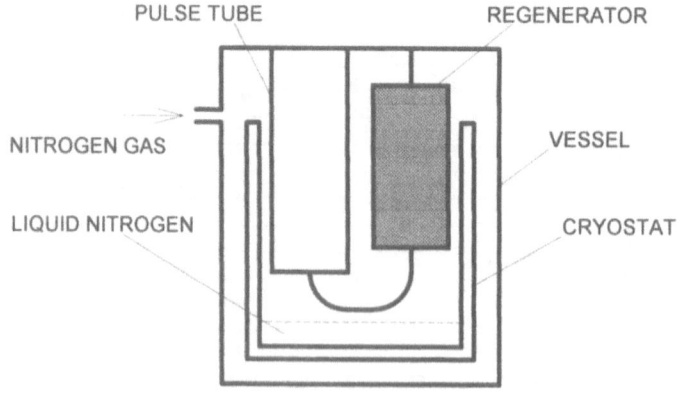

**Figure 10.** Nitrogen liquefier

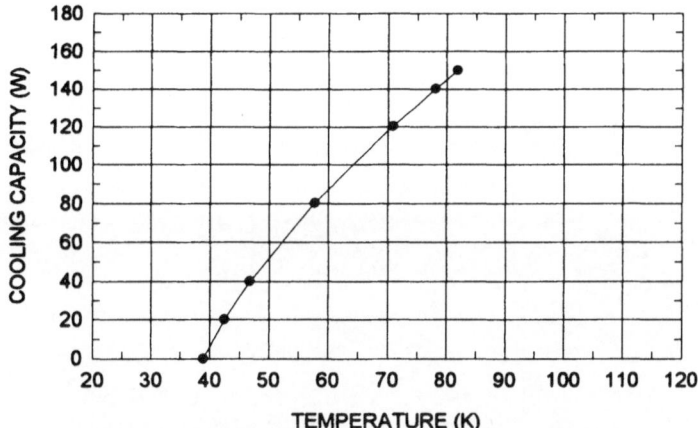

**Figure 11.** Cooling capacity of U-type pulse tube refrigerator

## CONCLUSION

This paper presented the effects on cooling capacity of the active buffer pulse tube refrigerator as a function of valve timing, frequency, volume of each buffer, and valve flow coefficients ($C_V$ values). A maximum cooling capacity of 164 W was achieved at 80 K with an input power of 3.63 kW. The percent carnot was 12 %. The minimum temperature achieved was 28 K. Nitrogen liquefier was made by modifying the straight type pulse tube refrigerator into a U-type refrigerator. A liquefying rate of 1.5 liters per hour was achieved.

## REFERENCES

1.  Zhu, S. W., Kakimi, Y., Fujioka, K., and Matsubara, Y.,"Active-Buffer Pulse-Tube Refrigerator", Presented at ICEC16, (1996).
2.  Zhu, S., and Matsubara, Y., "Proposal for a Tube Expander", *Cryogenics*, vol. 36, no. 6 (1996), pp. 403-408.
3.  Matsubara, Y., Gao, J. L., Tanida, K., Hiresaki, Y., and Kaneko, M., "An Experimental and Analytical Investigation of 4 K Pulse Tube Refrigerator", *Proceedings of the 7th International Cryocooler Conference*, (1993), pp. 166-186.

# Two-Stage Double-Inlet Pulse Tube Refrigerator down to 10 K

**S. Wild and L.R. Oellrich**

Institut für Technische Thermodynamik und Kältetechnik
Universität Karlsruhe (TH), D-76128 Karlsruhe, Germany

**A. Hofmann**

Institut für Technische Physik
Forschungszentrum Karlsruhe, D-76021 Karlsruhe, Germany

## ABSTRACT

A two-stage pulse tube refrigerator, driven by a 6 kW compressor equipped with magnetic valves, has been developed with the objective of evaluating how well pulse tube technology can compete with comparable Gifford-McMahon refrigerators.

The first stage, when operated separately as a double-inlet system, achieved more than 30 W at 80 K; the no-load temperature was 26 K. The second stage is of similar construction and is fed by a portion of the gas flow from the cold end of the first-stage regenerator. The second stage is provided with a cold phase shifter, whereby the orifice, bypass, and buffer volume are integrated into a single copper block that is in direct thermal contact with the first-stage cold end. The cold valves are actuated by a cardanic rod system with vacuum-sealed feedthrough. This enables adjusting the optimum phase shift of the second stage without interruption of operation.

First measurements of the two-stage double-inlet system showed a minimum no-load temperature of 15 K and a cooling power of 1 W at 19 K. Stepwise modification of both stages has improved the performance to a typical value of 11 K minimum temperature, and 20 K for 10 W and 2 W heat loads at first and second stage, respectively. Additional future improvement is expected.

## INTRODUCTION

The pulse tube cryocooler seems to be an attractive alternative to Gifford-McMahon (GM) refrigerators over a wide range of temperatures down to below 4K. Rapid development is underway throughout the world. In spite of very spectacular results, as for instance the different 4K refrigerators developed by Matsubara and co-workers [1,2], there are a lot of unanswered questions in the field of pulse tube refrigeration.

Optimization of multi-staged systems is just beginning, and achievement of the lowest temperatures has been the main impetus up until now. The best results have been obtained with the Matsubara-type system where all three stages are equipped with active phase shifters consisting of magnetic valves operated at ambient temperature.

Cryocoolers 9, Edited by R.G. Ross, Jr.
Plenum Press, New York, 1997

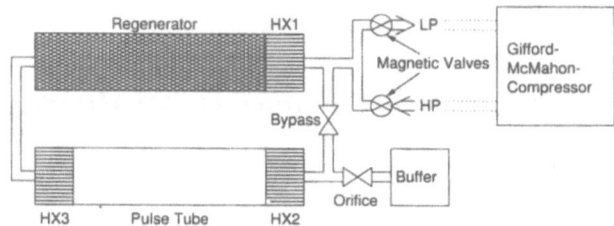

**Figure 1.** Schematic of the single-stage pulse tube refrigerator.

In the present work, an alternative of using passive phase shifters at both stages is investigated. The first-stage phase shifter is at ambient temperature, and that of the second stage is at an intermediate temperature. The aim is to achieve refrigeration powers of 2 to 5 W at 20 K, and about 10 W of additional power at a first-stage temperature of 80 K.

An assembly of components such as different pulse tubes, regenerators, and heat exchangers — all equipped with Kapton sealed flanges — has been prepared. This allows various configurations to be analyzed with minimal interruption for modifications. The pressure wave controller is also of high flexibility. It enables operation of different configurations with active (four-valve) and passive phase shifters. Experimental results obtained with different single- and two-stage configurations are described in this paper.

## SINGLE-STAGE PULSE TUBE REFRIGERATION

Work was started on pulse tube refrigerators (PTR) with the objective of building up a flexible system for detailed analysis and testing. For such a system it proved advantageous to work with a Gifford-McMahon compressor and valves for the pressure wave generation. A variable-valve unit with adjustable frequency and phase shift over a wide range has been obtained by using magnetic valves controlled by a PC with a digital I/O card. The complete valve unit, including buffer volume, orifice, and bypass valve, is mounted on a valve board. This enables operation of the PTR in basic, orifice, and double-inlet mode, as well as in the four-valve mode. All measurements presented in this paper have been made for the double-inlet mode. Further investigation will address improvement of our system with the four-valve mode. Figure 1 shows a schematic drawing of the single-stage system without the valve unit for the four-valve mode.

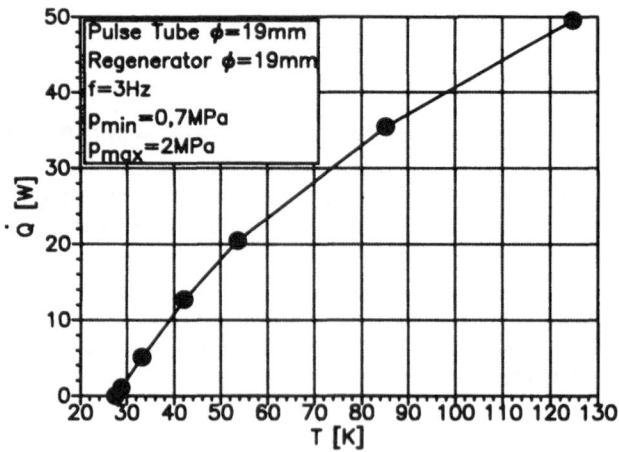

**Figure 2.** Cooling power of the single-stage pulse tube refrigerator.

The regenerator of the single-stage system is made of stainless steel tube with an inner diameter of 19 mm, a wall thickness of 0.5 mm, and a length of 230 mm. It is filled with 1860 stainless steel wire screens of 200 mesh.

We tried several pulse tubes, each 200 mm long, to find the optimum performance of the single-stage refrigerator. Best results were obtained with a pulse tube of 19 mm inner diameter; the wall thickness is 0.5 mm. At the optimum frequency of 3 Hz, and a mean pressure of 1.35 MPa, a minimum no-load temperature of 26 K was achieved. In this operating condition a cooling capacity of 32 W at 80 K was achieved. Figure 2 shows the cooling power of the single-stage unit plotted against the temperature at the cold end.

## TWO-STAGE PULSE TUBE REFRIGERATOR

### First Configuration

To start with, the unmodified single-stage pulse tube refrigerator was used as the first stage for the two-stage device. A schematic of the two-stage cooler is shown in Fig. 3. The two stages are connected by a copper plate with suitable boreholes for the gas flow. Both valve seats, for orifice and bypass, and the buffer volume ($V = 130 \text{cm}^3$), are integrated into the copper block. The valves are adjustable during the experiment from outside the vacuum vessel by a cardanic rod system with vacuum-sealed feedthrough. Later those valves may be replaced by fixed impedances, but the variable system proves to be very advantageous in that it allows optimum adjustment to be done without interruption of the running experiment.

The second-stage components are made of stainless steel tubes with inner diameters of 11 mm (regenerator) and 10 mm (pulse tube), both with a wall thickness of 0.5 mm. The lengths of the regenerator and pulse tube are 150 mm and 125 mm, respectively. The second stage regenerator is filled with 0.2 mm-diameter lead shot.

For this first two-stage unit, it was anticipated that the first stage would not work optimally and the regenerator would likely be overloaded. However, the configuration was expected to deliver preliminary information useful for improving the design.

Figure 4 shows the cooling map for this initial two-stage system. The data are presented for the optimum frequency of 1.9 Hz, a low pressure of 0.7 MPa, and a high pressure of 2 MPa. The first stage was driven with a heat load of 0 W, 5 W and 10 W, and the second stage with a heat load of 0 W, 1 W and 2 W. A minimum no-load temperature of 15 K at the second stage was reached. For this condition, the temperature of the first stage was 56 K. Cool down from room temperature to 15 K needed about 6 hours. It is remarkable that the no-load temperature of the second stage stays below 19 K when a heat load of 10 W is applied to the first stage.

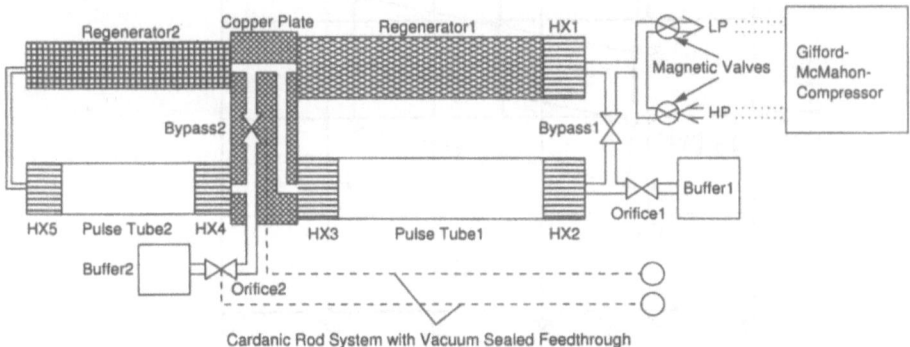

**Figure 3.** Schematic of the two-stage pulse tube refrigerator.

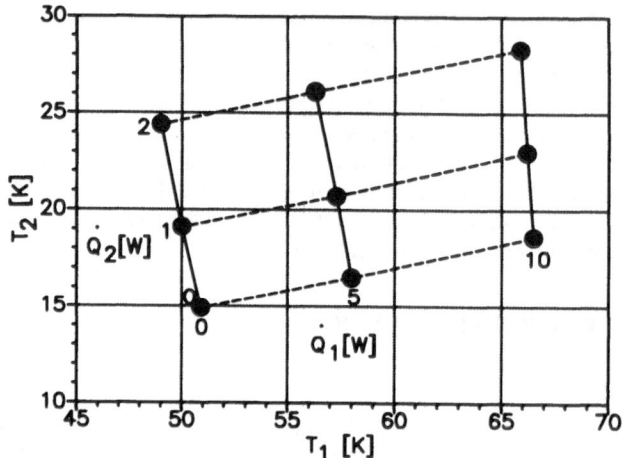

**Figure 4.** Cooling map of the initial configuration of the two-stage refrigerator.

An analysis of the temperature distribution and pressure amplitude in the first stage showed that the first-stage regenerator was considerably overloaded.

## Second Configuration

In the next phase of experimentation, the regenerator of the first stage was increased to an inner diameter of 24 mm, with a length and wall thickness of 230 mm and 0.5 mm, respectively. For the matrix, 200 mesh stainless steel wire screens were used.

The improvement of the two-stage pulse tube refrigerator is shown in Fig. 5. The main difference, compared to Fig. 4, is the higher cooling power at the second stage; it increased to 4 W. With no load on both stages, the minimum temperature at the second stage decreased to 14 K. In this operating mode the minimum temperature of the first stage decreased to 46 K ($f=1.9$ Hz; $P_{min}=0.7$ MPa; $p_{max}=2$ MPa). Also, the time for cool down to 14 K was reduced from 6 hours to about 3 hours. In summary, the cooling power of the system was almost doubled with the new regenerator at the first stage.

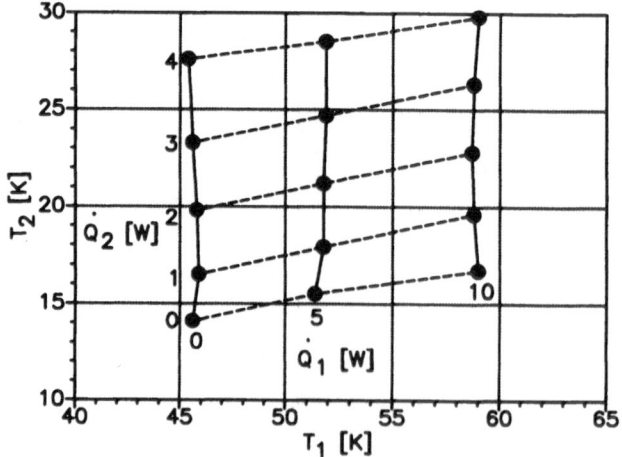

**Figure 5.** Cooling map of the two-stage refrigerator after the first improvement.

**Table 1.** Data for the two-stage double-inlet pulse tube refrigerator.

| | Pulse Tube Ø$_i$ x l x wall [mm] | Regenerator Ø$_i$ x l x wall [mm] | Regenerator Matrix |
|---|---|---|---|
| 1st Stage | 19 x 200 x 0.5 | 24 x 230 x 0.5 | SS-Mesh (No. 200) |
| 2nd Stage (config. 2) | 10 x 125 x 0.5 | 11 x 150 x 0.5 | Lead Shot (0.2mm) |
| 2nd Stage (config. 3) | 11 x 150 x 0.5 | 19 x 180 x 0.5 | Lead Shot (0.2mm) |

**Third Configuration**

Detailed analysis of the above test results suggested that now the second-stage regenerator had become the limiting component. In the third phase of the experiment the second-stage regenerator was replaced by a unit with 19 mm I.D. and 180 mm length. The first stage was not modified. The design parameters of both stages for configurations 2 and 3 are given in Table 1. This latest change again led to an appreciable enhancement of refrigeration power. With this two-stage pulse tube refrigerator, a no-load temperature of 11 K was obtained. The optimum frequency of the new system increased to 2 Hz, with a low pressure of 0.7 MPa and a high pressure of 2 MPa. To reach temperatures in the range of 15 K, a cool down time of 2.5 hours was necessary. The complete cooldown from 300 K to 11 K took about 4 hours.

Figure 6 shows the cooling map of the 11 K cooler. It can be seen that the temperature of the first stage decreases with increasing cooling power at the second stage. This might be caused by the higher temperature at the second stage, with a lower mass flow into this stage. As a result, the pressure amplitude and the cooling power in the first stage increases. This may affect the decrease of temperature. The same effect is seen in Fig. 4. This effect is also seen when the system is operated with a higher heat load at the first stage.

With respect to the initial objectives, the 10-W load line should be emphasized. It yields a working point with 10 W at 77 K, and additionally, 2 W at 20 K. This result is at the low end of the specified range. We expect that further improvement is possible; respective experiments are in progress.

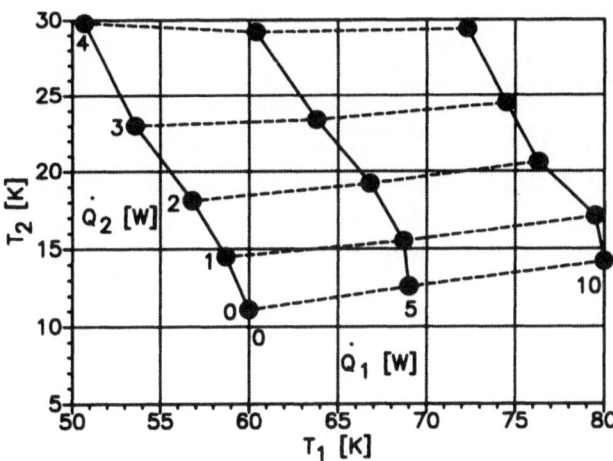

**Figure 6.** Cooling map of the two-stage refrigerator after the second improvement.

## CONCLUSION

A modular pulse tube system has been manufactured and tested successfully. With a single-stage pulse tube refrigerator, a minimum temperature of 26 K and a cooling power of about 32 W at 80 K was achieved at optimum operating conditions. When a second stage was added, stepwise improvement was achieved by systematic modification of components. Results are given for three different steps. The final system achieved a no-load temperature of about 11 K. At 20 K, a maximum cooling power of 2.5 W was available with no heat load at the first stage. With a heat load of 10 W at the first stage, the cooling power of the second stage decreased to about 2 W. Further enhancement of the cooling power seems likely.

## ACKNOWLEDGMENT

We are grateful to Mr. B Vogeley for his skilled design of the pulse tube refrigerator and useful discussions about improvement of the system. The present work is supported by the German Ministry of Science and Technology (BMBF) under contract number 13N65933.

## REFERENCES

1.  J.L Gao and Y. Matsubara, "Experimental Investigation of 4K Pulse Tube Refrigerator," *Cryogenics*, Vol. 34, No. 1 (1994), p. 25.

2.  Y. Matsubara and J.L. Gao, "Novel Configuration of Three-Stage Pulse Tube Refrigerator for temperatures below 4K," *Cryogenics*, Vol. 34, No. 4 (1994), p. 259.

# Early Pulse Tube Refrigerator Developments

**R. C. Longsworth**

ADP Cryogenics
Allentown, PA, USA

## ABSTRACT

This paper discusses some of the previously unpublished research work done on experimental valved pulse tubes that were built and tested at Syracuse U. by W. E. Gifford and myself in the mid 1960's. These included multi-tube, two stage, and four stage designs. The four stage unit had a novel valve design and had a no load temperature of about 38 K.

## INTRODUCTION

Pulse tube refrigerators have their origin in an observation that W. E. Gifford made while working on a compressor when he was at Arthur D. Little Inc. in the late 1950's. He noticed that a tube which branched from the high pressure line and was closed by a valve was hotter by the valve than at the branch. He recognized that there was a heat pumping mechanism that resulted from pressure pulses in the line. At this time a number of universities in the U. S. were recruiting engineers from industry in response to the successful launching of Sputnik by the USSR. W. E. Gifford, who I came to know as Prof. Gifford, joined the faculty at Syracuse U. and took over the Cryogenics Lab in the Chemistry building. One of his first proposals was to the Advanced Research Projects Agency to study pulse tube refrigeration. I was offered the position of graduate research assistant which I accepted at the time he was awarded the contract and thus had the opportunity to explore with him this new phenomena.

## FIRST PULSE TUBE

Gifford envisioned a pulse tube refrigerator [1] consisting of a compressor, a switch valve, a regenerator, and a tube with a cold end heat exchanger / flow smoother and a warm end with a heat transfer mesh to transfer heat from the gas to ambient temperature. The gas pressure would have to change enough so the gas entering the cold end would flow into the mesh at the warm end where it would be cooled. The Cryogenics Lab had a Collins helium liquefier which had ceased to function as a liquefier but was a source of compressed helium for the first pulse tube tests. My box of notes which survived from this early work indicates that the first experimental unit was designed to operate with a helium flow rate of 16.5 SL/s at 1.5 / .1 MPa. and a pulse rate of 3.3 Hz. Figure 1 is a copy of the original drawing that was given to the lab machinist to build, and Figure 2 is a photograph of the unit. The warm end of the pulse tube has a roll of Cu screen soldered to a water cooled plate which could be moved up and down in the tube. It was

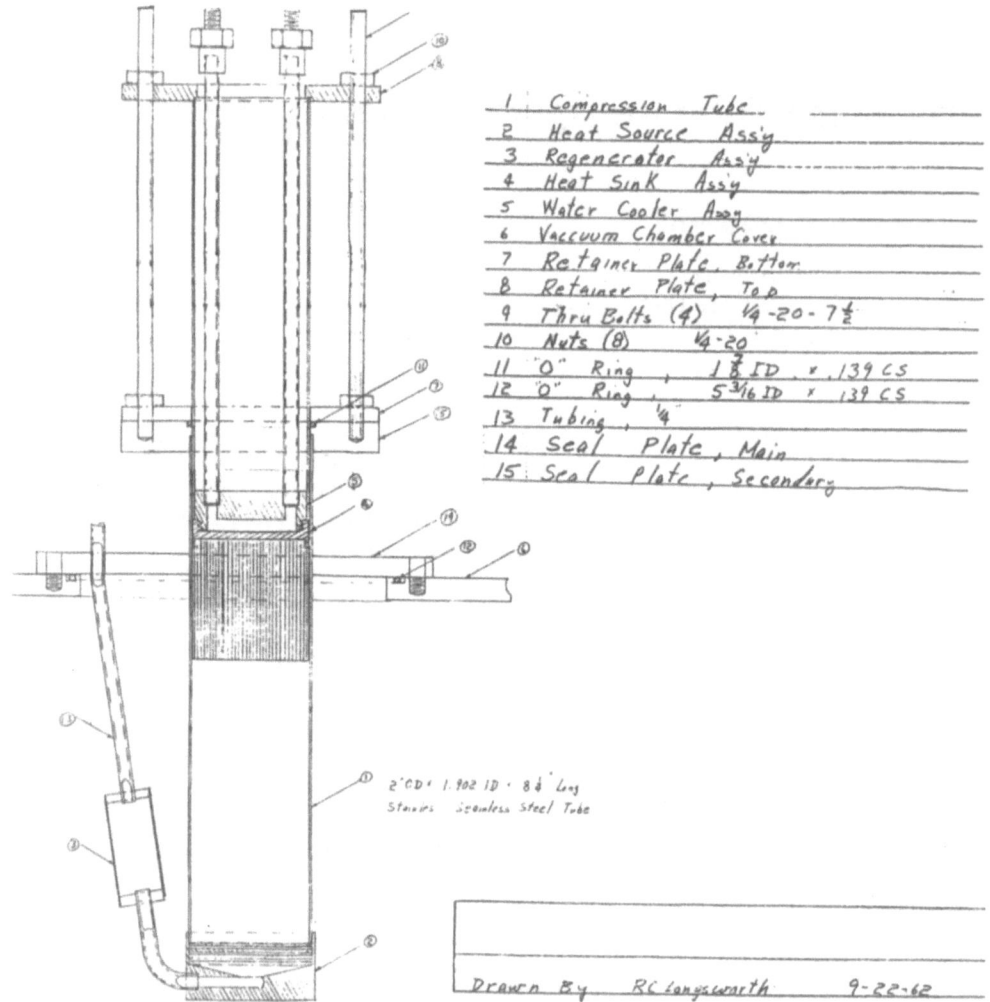

Figure 1. Drawing used to build the first pulse tube dated 9-22-62.

designed to occupy 30 % or more of the volume. It cooled from 278 K to 236 K, a beginning, but not very impressive considering the power input.

In parallel with work on the pulse tube we also worked on development of his Gas Balancing refrigerator concept [2]. The rotary valve that was designed for the new GM type refrigerator was also used for the Pulse Tube. It consisted of a filled Teflon disc rotating on a hardened steel plate driven by a variable speed motor. We also got Carrier Corp. to give us some small air conditioning compressors and started developing means to cool them while compressing helium and remove the oil. These proved to be more convenient than the large Collins compressor.

**Figure 2.**  Photograph of the first pulse tube refrigerator.

## ENCAPSULATED PULSE TUBES

Gifford then became enamored with the idea of building both the GM and pulse tube expanders in plastic tubes which were then inserted into a SS sleeve that had few joints that might leak.  This concept was used to make pulse tubes with up to four stages.  It provided a lot of flexibility in changing pulse tube and regenerator parameters.  Test results with one and two stage units of this design are described in the first pulse tube paper [3].  Temperatures with helium of 199 K for the one stage unit and 175 K for the two stage unit with cooling water at 289 K were reported.  A photograph of the four stage unit that was built is shown in Figure 3.  A cross section of one of the stages is shown in Figure 4.

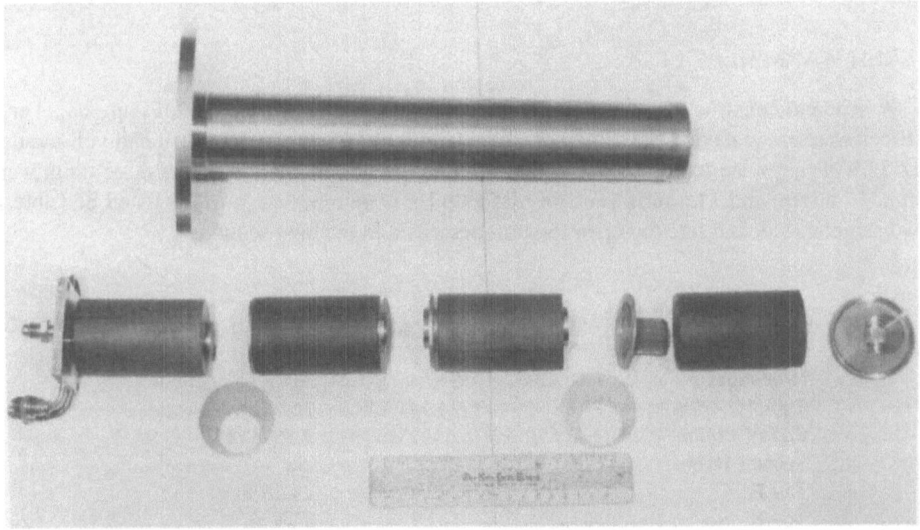

**Figure 3.**  Photograph of the four stage encapsulated pulse tube made in 1963.

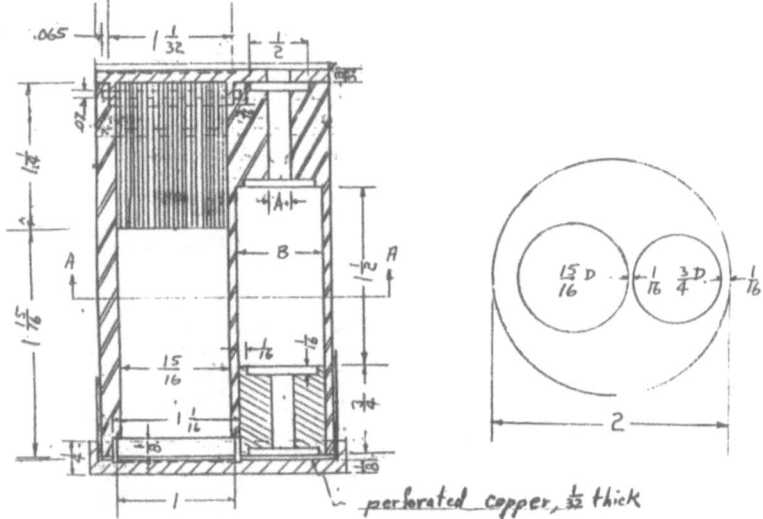

**Figure 4.** Drawing of the last stage of the four stage encapsulated pulse tube.

## CONCENTRIC PULSE TUBE

The next pulse tube that we built was a concentric design as shown in Figure 5, [4]. With He at 1.63 / .14 MPa and a pulse rate of .68 Hz it cooled to 212 K from 280 K and did almost as well with air. The poor performance was attributed to thermal interaction with the regenerator so this design was not pursued further. This unit was significant because one of the tests that we ran with this unit was to measure temperature and cooling rate as a function of pressure ratio. It was expected that it would warm up if the pressure ratio was insufficient to move gas from the cold end to the warm end during each pulse. To our surprise the unit stayed cold even when the pressure ratio was much less than that required to have some gas move the entire length of the tube. The concept of surface heat pumping was then postulated to explain the effect [5].

## SINTERED WARM HEAT SINK

A different construction of the warm end heat exchanger was tried next, Figure 6a. Three pulse tubes were made having the same open tube length of 165 mm and a warm end volume that was 22.5 % of the pulse tube plus warm end volume. Inside diameters of the three units were 24.1 mm, 17.8 mm, and 14.6 mm. Test results with He were reported in [6] as listed in Table 1. The two stage unit looked like the open top unit described in the next section.

**Table 1** Test Results With Sintered Warm Heat Sink Pulse Tubes

| No. Stages | 1 | 2 | 3 |
|---|---|---|---|
| Tube ID - mm | 17.8 | 24.1 / 17.8 | 24.1 / 17.8 / 14.6 |
| Ph / Pl - MPa | 2.41 / .33 | 2.41 / .14 | 2.35 / .48 |
| Speed - Hz | 1.27 | .75 | .58 |
| Ta - K | 282 | 279.4 | 287.8 |
| Tc - K | 167.2 | 123.3 | 84.4 |

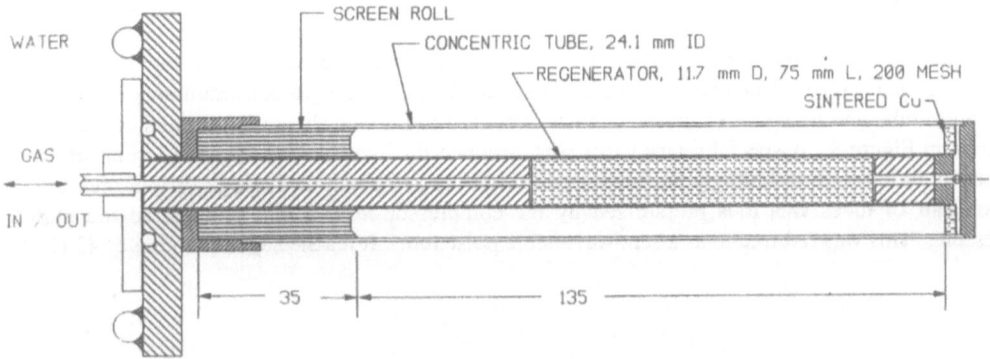

**Figure 5.** Concentric Pulse-Tube

## OPEN TOP PULSE TUBES

With the concept of surface heat pumping in mind an open top design as shown in Figure 6b was tried. The first unit that was built had a tube with an ID of 17.8 mm, a length (below the top) of 216 mm and a top with a volume that was 17.5% of the total volume. It cooled from 295 K to 144 K at 2.35/.48 MPa and .93 Hz with He. The top was then replaced with a top that was lined with sintered metal as shown in Figure 6c. It only cooled to 171 K so this design was not pursued further. At this point I would have preferred to continue exploring different design concepts; however the faculty said that would not be acceptable for a dissertation, so I focused on understanding the heat pumping mechanism in the open top type pulse tube. I built a tube with SS tabs spaced along the tube having thermocouples on the ends and differential thermocouples between the tip and base. These enabled me to measure the heat flowing in and out of the tube wall along the wall and correlate it with the total heat pumping rate [7]. Other correlation's between tube geometry and operating variables for air and He were reported in [8].

Two of the single stage pulse tubes that had been tested were assembled as a two stage unit as shown in Figure 7 and produced a minimum temperature of 95 K.

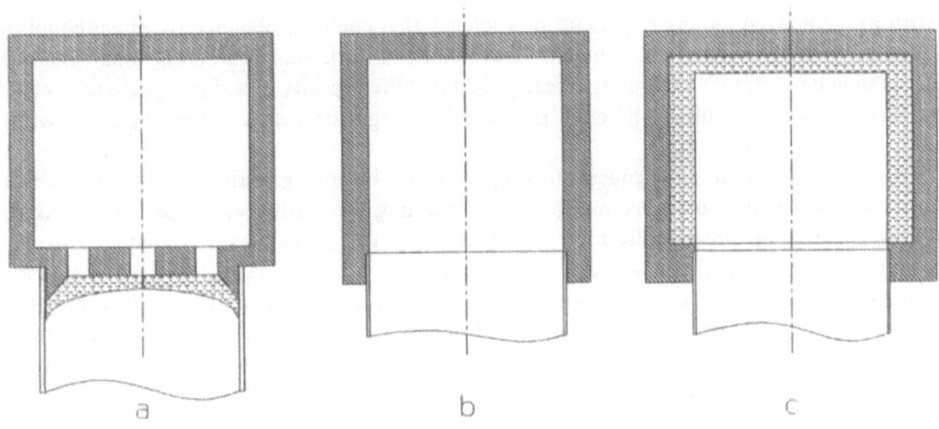

**Figure 6.** Three different types of warm end heat sinks.

## FOUR STAGE PULSE TUBE

The Air force funded a program to have Airesearch build a He compressor to supply gas to a four stage pulse tube of the open top design. In order to reduce the gas consumption we divided the pulse tube into two pairs of tubes, which would draw about the same amount of gas, as shown in Figure 8. A special rotary valve was designed that equalized the pressure in all of the tubes at an average of the high and low pressures before switching to the high or low pressure. Each pair of tubes was thus pressurized by the compressor from a mid pressure to the high pressure. This was referred to as a semi-reversible pulse tube. It reached a temperature of 43 K.

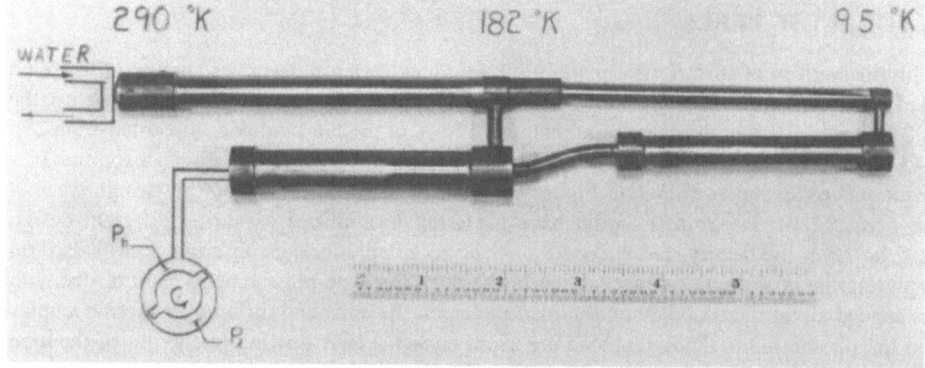

**Figure 7**. Two stage open top pulse tube

## OTHER DESIGNS

Gifford recognized the improvement in efficiency that could be obtained by direct coupling of the pulse tube to a compressor piston. Development work was done along this line with another graduate student [9]. It was also recognized that the capacity of a single pulse tube could be increased by having multiple tubes in parallel. A design for such a pulse tube is shown in Figure 9 but it was never built.

In response to a number of requests to supply his gas balancing refrigerators W. E. Gifford organized Cryomech Inc. and gave me the job of building the hardware. One of our products was the single stage air driven pulse tube shown in Figure 10. It has the distinction of being the first commercial offering of a pulse tube refrigerator but it was not a financial success. Perhaps it was this experience that led me to pursue refrigerator designs that relied on solid pistons rather than gas pistons in order to support my family. If one looks at the patent I received shortly after leaving Syracuse [10] it is seen that by removing the displacer in figure 1 one has a single orifice pulse tube which was discovered about 10 years later.

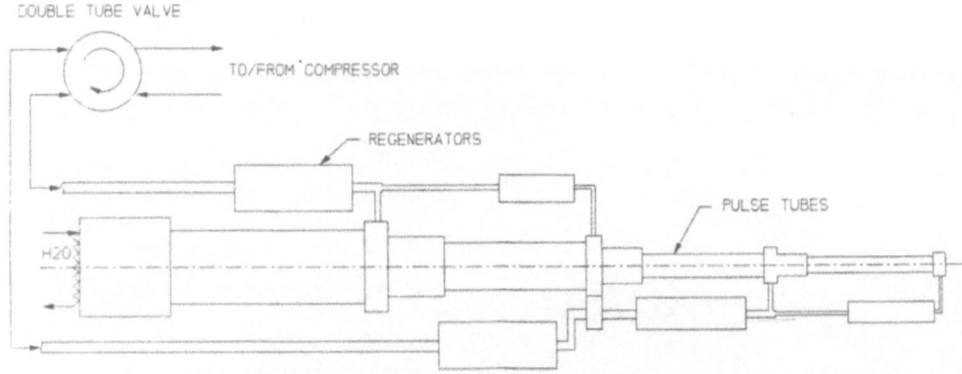

**Figure 8**. Four stage semi-reversible pulse tube

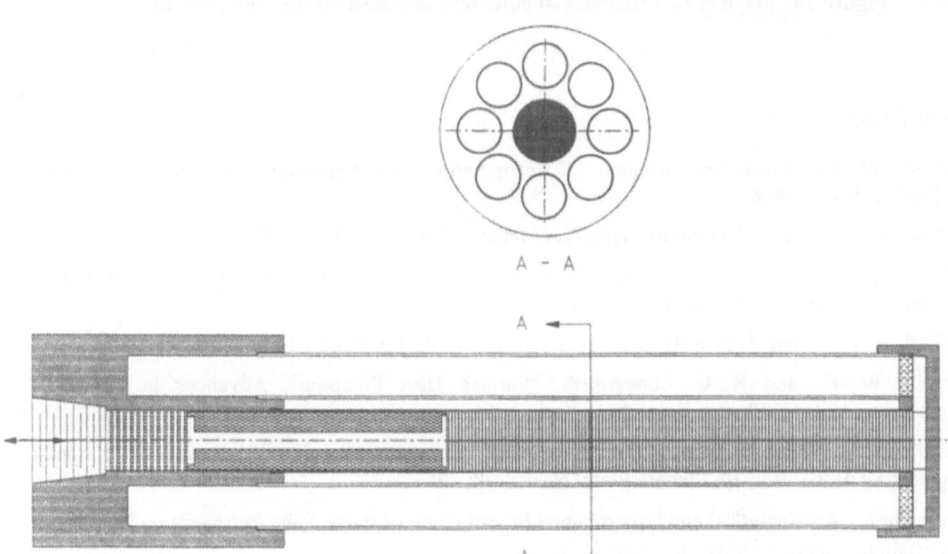

**Figure 9**. Multi-tube pulse tube concept.

**Figure 10**. Photo of first commercial pulse tube designed for use with shop air.

## REFERENCES

1.  Gifford, W. E., "Pulse-Tube Method of Refrigeration and Apparatus Therefore", US Patent 3,237,421, March 1966.

2.  Gifford, W. E., "Gas Balancing Refrigeration Method", US Patent 3,119,237 Jan 1964.

3.  Gifford, W. E., and R. C. Longsworth, "Pulse-Tube Refrigeration Progress", Trans. of the ASME, Journal of Engineering for Industry; August 1964.

4.  Gifford, W. E., "Pulse-Tube Refrigeration" Final Report, ARPA Project 1620.1002, Sept. 1964.

5.  Gifford, W. E., and R. C. Longsworth, "Surface Heat Pumping", Advances in Cryogenic Engineering, Vol. 11; Plenum Press/ New York; 1966.

6.  Gifford, W. E., and R. C. Longsworth, "Pulse-Tube refrigeration Progress", Advances in Cryogenic Engineering, R. C., Vol. 10; Plenum Press/ New York; 1965.

7.  Longsworth, "An Analytical and Experimental Investigation of Pulse-Tube Refrigeration" -- Ph. D. Dissertation, Syracuse University; June 1966.

8.  Longsworth, R. C., "An Experimental Investigation of Pulse Tube Refrigeration Heat Pumping Rates", Advances in Cryogenic Engineering, Vol. 12; Plenum Press/ New York; 1967.

9.  Gifford, W. E., and G. H. Kyanka, "Reversible Pulse-Tube Refrigeration", Advances in Cryogenic Engineering, Vol. 12; Plenum Press/ New York; 1967.

10. Longsworth, R. C., ""Refrigeration Method and Apparatus", US Patent #3,620,029  Nov 1971

# Phase Shift Effect of the Long Neck Tube for the Pulse Tube Refrigerator

S. W. Zhu*, S. L. Zhou**, N. Yoshimura** and Y. Matsubara**

*DAIDO HOXAN Inc., Tsukuba, Ibaraki, Japan
**Atomic Energy Research Institute, Nihon University, Chiba, Japan

## ABSTRACT

This paper discusses the effect of a long neck tube inserted between the reservoir and the pulse tube hot end. To improve pulse tube performance, the pulse tube enthalpy flow divided by the regenerator mass flow needs to be maximized. From the view point of the equivalent PV diagram at the cold end of the pulse tube, the optimum performance will be given by the PV shape of the idealized Stirling cycle. However, in the case of the orifice pulse tube, the phase shift effect generated by the orifice and reservoir is not sufficient. The double inlet pulse tube can approach the PV shape of the idealized Stirling cycle, but this configuration requires additional work input. The computer simulation developed in this study indicates that a long neck tube with a reservoir, instead of the orifice and the double inlet, gives an improved PV diagram with no additional work.

As a verifying experiment, a neck of 4 mm diameter and 2 m length was inserted between a two-liter reservoir and a pulse tube of 19 mm diameter and 120 mm length. The indicated optimum frequency was about 30 Hz, and a pressure ratio of 1.28 was achieved at the cold end of the pulse tube. The results from the experimental study are discussed together with the numerical simulations.

## INTRODUCTION

The lowest achievable temperature of a double inlet pulse tube refrigerator is lower than that of an orifice pulse tube refrigerator. The main reason is the equivalent PV work[1] at the cold end of the pulse tube approaches that of an idealized Stirling cycle. However, additional input work is required. If we connect a long tube between the hot end of the pulse tube and the buffer volume, the long tube will be a phase shifter[2,3,4]. We call this long tube the long neck tube.

This paper describes the numerical analysis of a pulse tube refrigerator with a long neck tube. The analyses explore the phase shift effect of the long neck tube including the effect of the diameter and length of the long neck tube on the phase angle between the pressure wave and mass flow rate, the shape of the equivalent PV diagram, the cooling capacity, and the pressure ratio.

To verify the phase shift effect, a simple experiment has been conducted using a long neck tube directly connected to a compressor. With the compressor piston head directly attached to the gas piston head of the pulse tube, the PV diagram of the compressor volume becomes the equivalent PV diagram of the gas within the pulse tube. This experiment was used to support the computer simulations.

Cryocoolers 9, Edited by R.G. Ross, Jr.
Plenum Press, New York, 1997

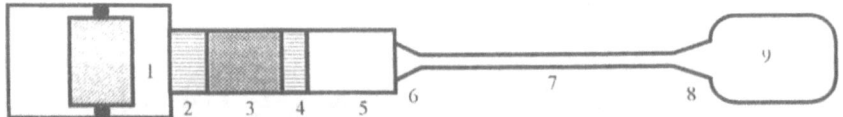

**Figure 1.** Pulse tube refrigerator with long neck tube, 1. Compressor, 2. Cooler (Ex1), 3. Regenerator, 4. Cryo Load Attachment (Ex2), 5. Pulse tube, 6. Gas smoother, 7. Neck tube, 8. Gas smoother, 9. Buffer

## THE STRUCTURE AND MECHANISM

Figure 1 shows a schematic of the long neck tube pulse tube refrigerator. It includes a compressor, a cooler (Ex1), a regenerator, a heater to represent the cryogenic load (Ex2), a pulse tube, a neck tube with gas smoother at both ends, and a buffer.

In the orifice pulse tube refrigerator, the expansion work of the pulse tube is simply dissipated by the orifice. If we connect a long neck tube as shown in Figure 1, the expansion work of the pulse tube will cause the gas in the neck tube to oscillate. The long neck tube becomes a resonator similar to that in a thermal acoustic engine. In the thermal acoustic engine, there is a standing wave and a progressive wave at the end of the resonator. A similar condition will exist at the hot end of the pulse tube in Figure 1. The standing wave and the progressive wave are generated by the long neck tube. The standing wave is necessary to change the phase between the mass and the pressure wave. The progressive wave is also necessary, because the expansion work must be changed to heat. The standing wave comes from the oscillation of the gas in the long neck tube; the progressive wave is caused by the loss along the long neck tube.

## NUMERICAL RESULTS

### Numerical Method

The numerical method used to model the neck tube pulse tube refrigerator was adapted from that previously used to model thermal acoustic engines[5]. In the program, one dimensional equations for the momentum, continuity, and energy of the gas and the matrix are considered, and sinusoidal displacement of the compressor piston is assumed. A finite difference method was used to solve the equations combined with the ideal gas state equation.

### Pressure Amplitude and Work Flow

The calculation conditions are shown in Table 1 for a compressor swept volume of 60 cm³, a buffer volume of 2 liters, and an operating frequency of 30 Hz. The refrigeration temperature is 80 K, while the temperature is assumed to be a constant 300 K for the regenerator, the pulse tube hot ends, the long neck tube, and the buffer. The diameter and length of the neck tube are treated parametrically.

Figure 2 shows the maximum, minimum and average pressure. The length from Ex1 to the hot end of the pulse tube, the length of the long neck tube including the two gas smoothers, and

**Table 1.** The parameters of the pulse tube refrigerator

|  | Length, $cm$ | Cross sectional area, $cm^2$ | Heat transfer area per unit length, $m^2 / m$ |
|---|---|---|---|
| Cooler | 1 | 3.58 | 2.87 |
| Regenerator | 4 | 10.1 | 50.9 |
| Heater | 1 | 1.13 | 4.54 |
| Pulse tube | 12 | 2.84 | 0.0597 |
| Buffer | 80 | 25.0 | 0.177 |

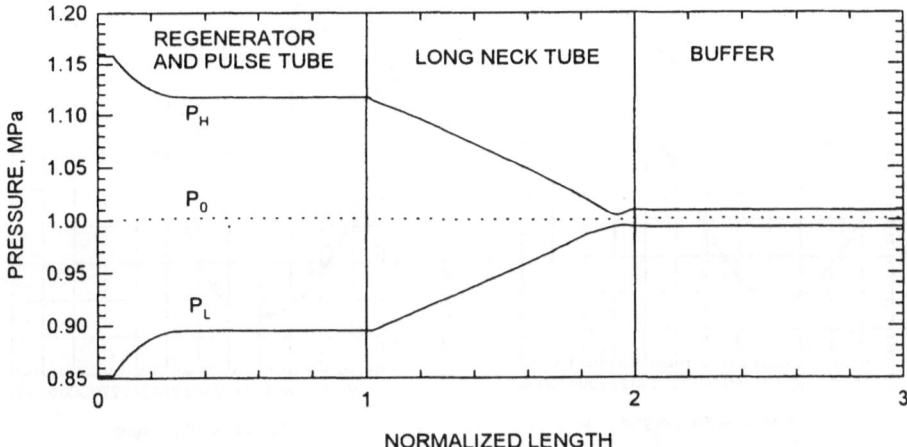

**Figure 2.** Variation of pressure amplitude from compressor to buffer

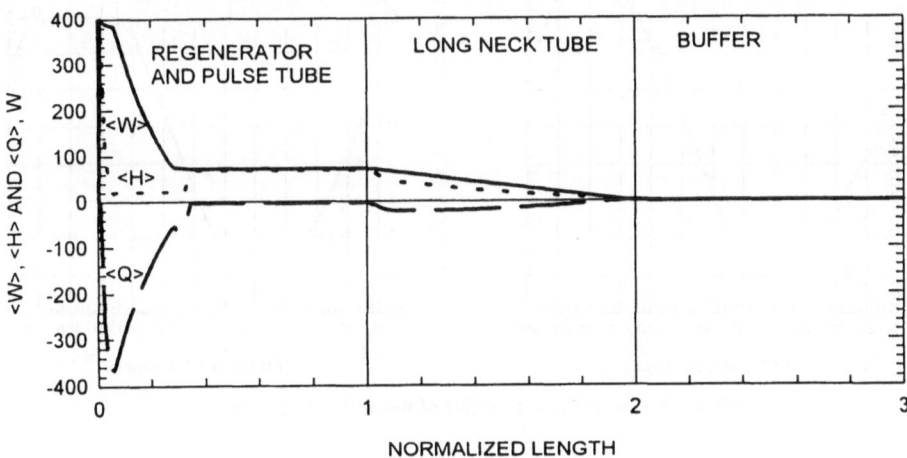

**Figure 3.** Work flow, enthalpy flow and heat flow from compressor to buffer

the length of the buffer, are each normalized to one. The pressure amplitude decreases slightly due to the pressure drop in the regenerator, then stays constant in the pulse tube, and finally decreases again along the neck tube. This decrease of pressure amplitude generates the work flow within the pulse tube.

Figure 3 shows the work flow $<W>$, the enthalpy flow $<H>$, and the heat flow $<Q>$ for the case of a 5-m long neck tube of 6 mm diameter. Each length is normalized to one as shown in Figure 2. In the figure, a positive value indicates flow toward the right, and a negative value indicates flow toward the left. The pattern of the energy flow from Ex1 to the hot end of the pulse tube is similar to that for the basic pulse tube refrigerator. The work flow within the neck tube decreases to zero at the end connected to the buffer. The enthalpy flow is slightly lower than the work flow in the neck tube, and it approaches zero at the end connected to the buffer. The heat flow is almost zero near the hot end of the pulse tube, and it decreases to negative along the long neck tube, which becomes zero at the end connected to the buffer. Note that the negative heat flow within the neck tube originates by the isothermal neck tube wall.

The numerical simulation predicts that 400 W of adiabatic compressor work is required for a cooling capacity of 45 W at 80 K.

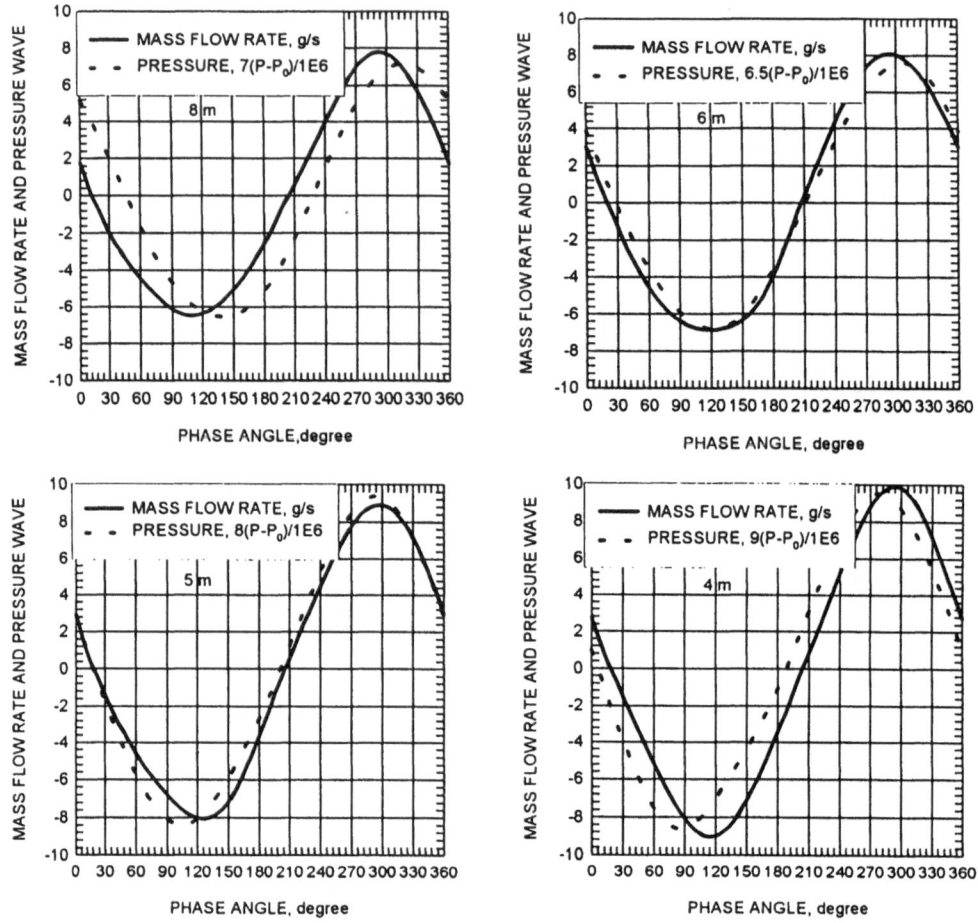

**Figure 4.** The phase shift effect of the neck tube length

## The Effect of Diameter and Length

Figure 4 shows the phase shift effect of the neck tube length in the case of the 6-mm tube diameter. The phase angle between the mass flow rate and the pressure wave at the cold end of the pulse tube changes with changing of the neck tube length. The pressure wave shows a delay in phase angle with respect to mass flow with an 8-m neck tube, is almost in phase with the mass flow with the 6-m neck tube, leads by a small phase angle with the 5-m neck tube, and leads by a large phase angle in the mass flow rate with the 4-m neck tube. This figure shows that, if the length of the long neck tube is reduced from 8 meters to 4 meters, the phase angle between the pressure wave and the mass flow rate will change progressively from the orifice condition, to the double inlet condition, and to the Stirling-refrigerator condition.

Figure 5 shows the PV diagram predicted at both ends of a pulse tube as a function of various diameters and lengths of the neck tube. Figure 5(a) is obtained by replacing the neck tube with an orifice that behaves according to the valve equation. The parameter numbers represent the amount of friction of the orifice. Figures 5(b) ~ (d) show the phase shift effect of long neck tubes having tube diameters of 3 to 6 mm, and lengths of 0.4 to 8 meters.

In the case of a tube diameter of 6 mm with 8-m length, Figure 5(d), the PV diagram at the cold end of the pulse tube has a shape similar to that of the orifice pulse tube refrigerator. With a reduction in the length to 5 meters, the shape becomes that of a double inlet pulse tube, or Stirling refrigerator. This tendency is also evident in the case of the 4-mm neck tube with a

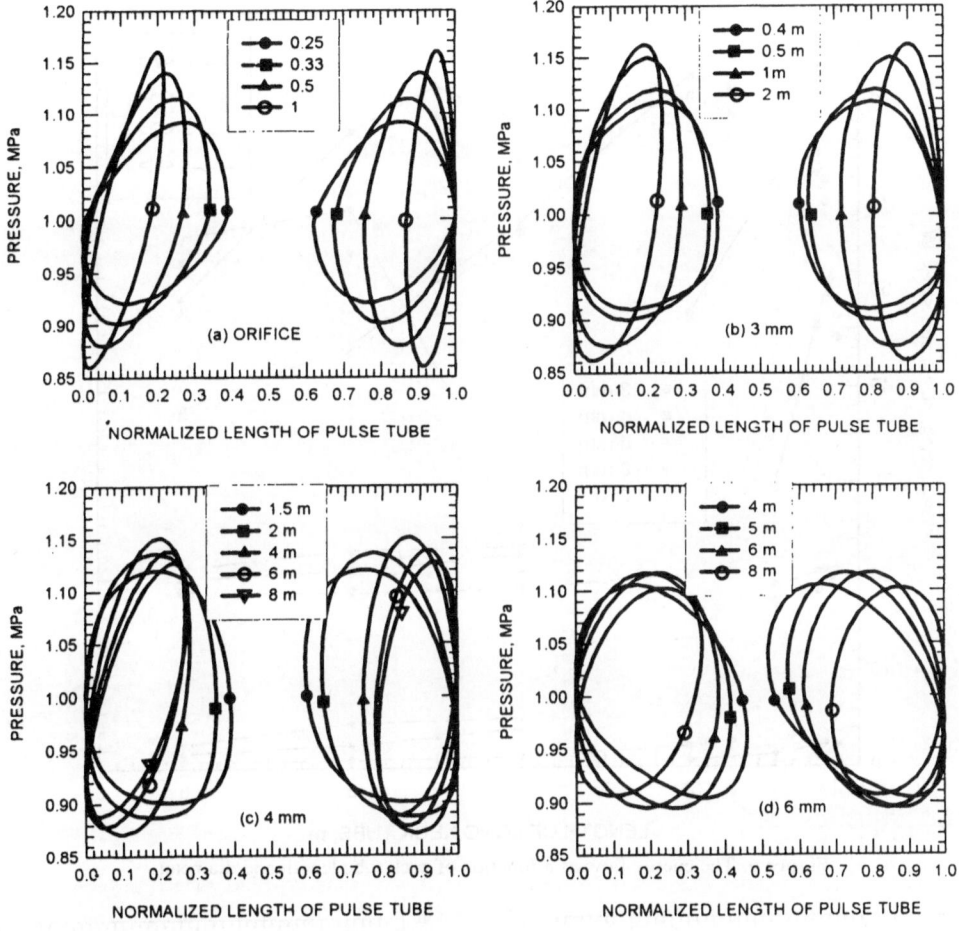

**Figure 5.** The phase shift effect of the diameter of the long neck tube

length shorter than 4 meters. The 3-mm neck tube does not show this tendency, even at a reduced length of 0.4 meters. These results indicate that the long neck tube gives a better phase shift effect than the orifice, and that there exists an optimum neck tube size. This additional phase shift effect comes from the momentum of the gas within the neck tube; thus, wider and longer neck tubes provide a larger phase shift effect.

The calculated results of the work flow at the cold end of the pulse tube $<W>_P$, the regenerator $<H>_R$, and the heat flow at the cold end of the pulse tube $<Q>_P = <H>_P - <W>_P$ are obtained as shown in Figure 6. This figure shows that the enthalpy flow through the regenerator and the heat flow at the cold end of the pulse tube are not changed significantly by the change of diameter and length of the long neck tube, but the work flow and the enthalpy flow in the pulse tube are increased significantly. The increase in the cooling capacity for the long neck tube seems to come from the changing of the PV diagram, and not from a decrease of the regenerator loss.

Figure 7 shows the efficiency (percent Carnot) based on the adiabatic compressor work. The parameter denoted ORIF signifies the case of the orifice pulse tube. For each diameter of the neck tube, there is an optimum length with which the efficiency nears a maximum. When the diameter of the neck tube decreases from 8 mm to 4 mm, the maximum efficiency increases. The minimum is the orifice case. Thus, there is an optimum length and diameter for which the cooling power is maximized. This figure indicates the optimum neck tube is around 2 meters

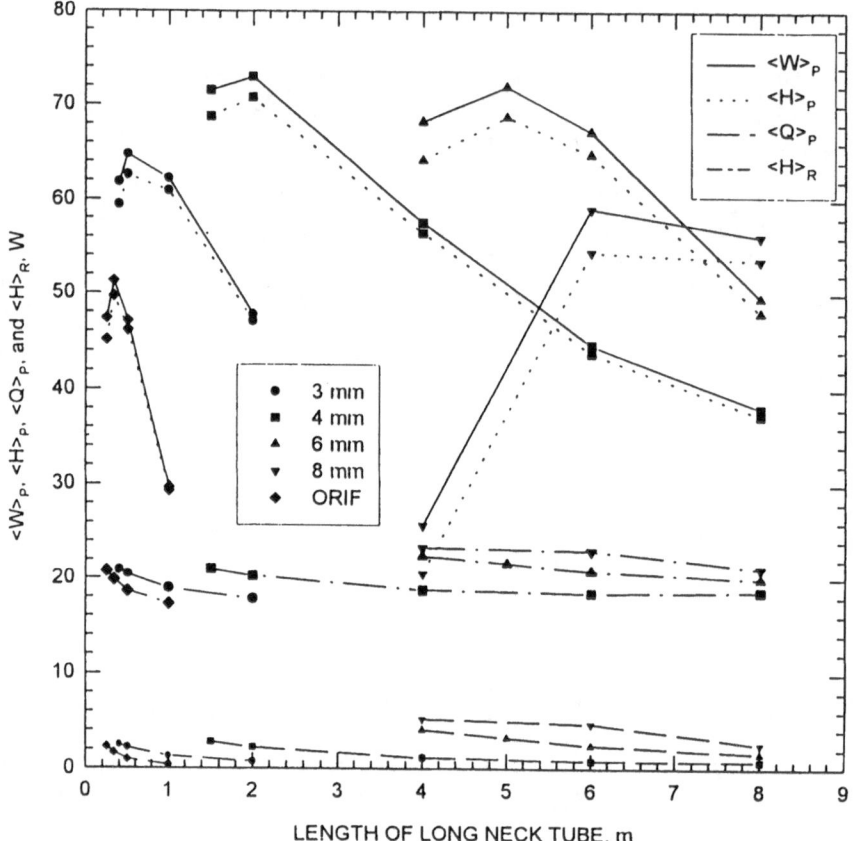

**Figure 6.** The energy flow as a function of neck tube length and diameter

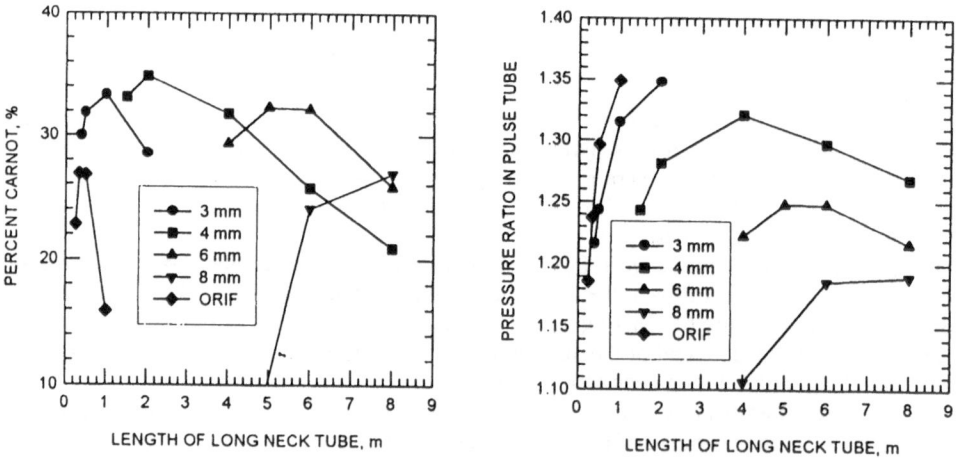

**Figure 7.** Percent Carnot based on adiabatic
compression work

**Figure 8.** Pressure ratio within the pulse tube

long with a 4-mm diameter; it also indicates the additional phase shift is not adequate.

Figure 8 shows that the pressure ratio changes with change of the diameter and length of the long neck tube. The pressure ratio will increase when the diameter decreases. The orifice condition is the highest pressure ratio condition.

**Table 2.**   The dimensions of the large-scale pulse tube refrigerator

|            | Length, $cm$ | Cross sectional area, $cm^2$ | Heat transfer area per unit length. $m^2 / m$ |
|------------|------|--------|--------|
| Cooler     | 1    | 10.14  | 10.14  |
| Regenerator| 4    | 40.45  | 203.8  |
| Heater     | 0.5  | 10.14  | 10.14  |
| Pulse tube | 12   | 11.95  | 0.1225 |
| Buffer     | 80   | 100.0  | 0.7088 |

**Table 3.**  The calculated result of a large scale pulse tube refrigerator.

| Length of long neck tube, m    | 8     | 6     | 4     |
|--------------------------------|-------|-------|-------|
| Input power, W                 | 1568. | 1949. | 2574. |
| Cooling capacity, W            | 143.5 | 232.0 | 354.1 |
| Percent Carnot, %              | 25.2  | 32.7  | 37.8  |
| Pressure ratio in the pulse tube | 1.39 | 1.45 | 1.48 |

## The Pressure Ratio Limitation

For any kind of pulse tube refrigerator, the displacement of the cold gas and hot gas at the ends of the pulse tube must not be too large; similarly, the gas piston within the pulse tube must not be too short. This is very important in order to decrease the shuttle heat loss within the pulse tube; the shuttle heat loss is the heat flow toward the cold end of the pulse tube caused by the heat transfer between the gas and the pulse tube wall, which is denoted as $<Q>_P$ in Figure 6. From Figure 5 we can find that in most cases the displacement of the cold gas is reasonable. It is noted from Figure 7 and 8 that the pressure ratio where the efficiency maximizes is around 1.3 or less.

To verify the size effect, another calculation was made whereby the input parameters were changed as shown in Table 2, and the compressor swept volume was changed to 300 cm³. For this case the operating frequency is 30 Hz, the cold end temperature is 80 K, and the temperature of the regenerator, the pulse tube hot ends, the long neck tube, and the buffer is 300 K.

The calculation has been done only for the case of an 8-mm neck tube diameter. Table 3 shows calculated results for the input power, cooling capacity, pressure ratio, and efficiency. Figure 9 shows the equivalent PV diagram within the pulse tube. Note that the 4-m long neck tube gives the optimum operating conditions with this 8-mm diameter neck tube. Compared to Figure 8, the pressure ratio is increased to nearly 1.5; the efficiency also slightly increased.

To assess the effect of the operating frequency, the frequency was changed to 50 Hz, with the same parameters as listed in Table 1, and with a 6-mm diameter neck tube. Figure 10 shows the equivalent PV diagram within the pulse tube. The pressure ratio, input power, and the cooling capacity are shown in Table 4. The optimum pressure ratio increased in comparison to Figure 9; however, the total performance (percent Carnot) decreased. The main reason for this degradation is hypothesized to be the increased enthalpy flow through the regenerator.

## EXPERIMENT

The phase shift effect of the long neck tube predicted by the computer simulation was confirmed by the test setup shown in Figure 11. A 2.1-m long neck tube of 4 mm diameter was connected to a 1-liter buffer at one end. The other end of the neck tube was connected directly to the compressor. Figure 12(a) shows PV diagrams obtained at the compressor volume while changing the frequency from 10 to 40 Hz. Replacing the neck tube with a 3 mm-diameter one, 1.4 m in length, led to the PV diagrams shown in Figure 12(b). The horizontal axis of these

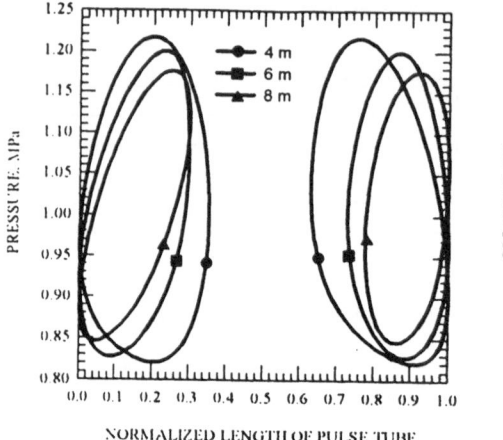

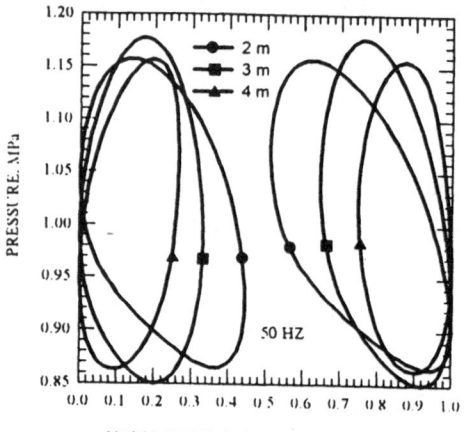

**Figure 9.** The phase shift effect of the large scale pulse tube

**Figure 10.** Equivalent PV diagram of 50 Hz operation

**Table 4.** Calculated results for the 50-Hz pulse tube refrigerator

| Length of long neck tube, m | 4 | 3 | 2 |
|---|---|---|---|
| Input power, W | 684.3 | 884.8 | 998.4 |
| Cooling capacity, W | 60.4 | 98.8 | 104.6 |
| Regenerator enthalpy flow | 36.6 | 39.0 | 43.8 |
| Percent Carnot, % | 24.3 | 30.7 | 28.8 |
| Pressure ratio in the pulse tube | 1.34 | 1.38 | 1.33 |

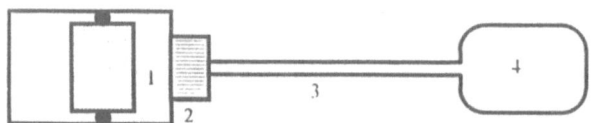

**Figure 11.** Schematic of the test setup, 1. Compressor, 2. Cooler (Ex1), 3. Long neck tube, 4. Buffer

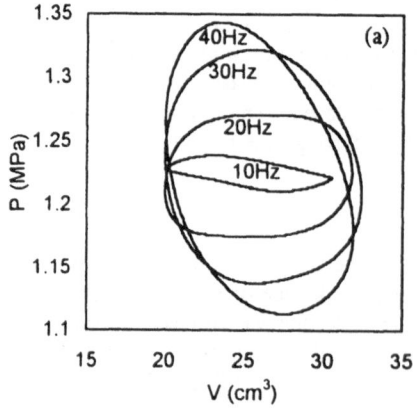

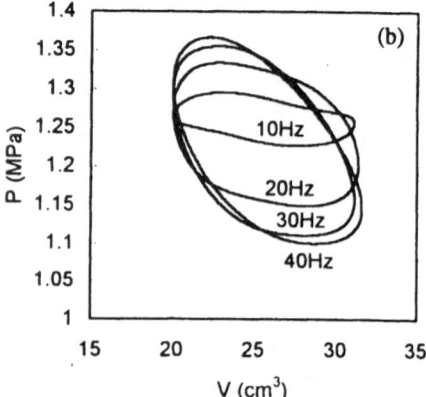

**Figure 12.** Experimental PV diagrams: (a) 4-mm neck tube, (b) 3-mm neck tube

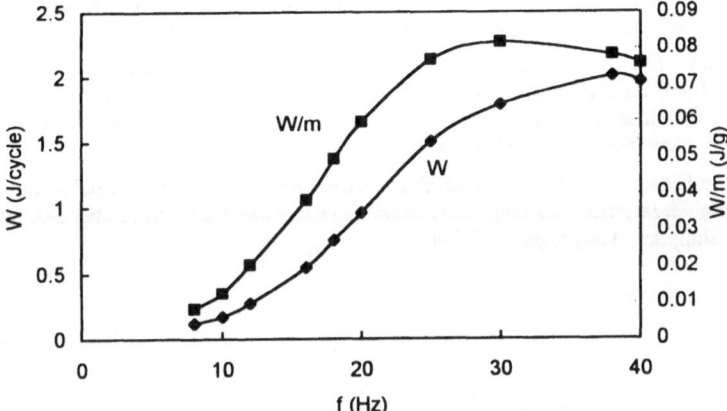

Figure 13. PV work and work per swept mass in the compressor volume

diagrams is the compressor swept volume. Thus, the reverse PV diagrams in Figure 12 represent the equivalent PV diagrams within the pulse tube as shown in Figure 5.

In comparison with Figure 12(a) and (b), the 3-mm neck tube does not show the double inlet effect; however, the 4-mm neck tube operated at 20 Hz provides the double inlet effect. Figure 13 shows the calculated results of PV work, W and W/m, obtained from Figure 12(a). Here m is the mass calculated from the difference of maximum and minimum mass within the PV diagram; this mass of gas represents that passing through the regenerator cold end of the actual pulse tube refrigerator. Thus, the maximum W/m gives the optimum operating condition of this neck tube. In this example, 30 Hz operation gives the optimum condition. The computer simulation described above, as shown in Figure 6 or 7, is verified by these experimental results.

## CONCLUSION

A computer simulation of the long neck tube refrigerator has been developed. The phase shift effect of a neck size ranging from 3 to 8 mm in diameter and from 0.4 to 8 meters in length has been demonstrated by means of equivalent PV diagrams. The work flow <W> through the system is broken down into the enthalpy flow <H> and the heat flow <Q>. With these results, the thermodynamic efficiency based on the adiabatic compressor work has been predicted. The results indicate the flowing characteristic performance:

(1) A long neck tube can generate a larger phase shift effect than an orifice. This results in an increased work flow per mass flow within the pulse tube.

(2) There exists an optimum neck tube size for a fixed operating frequency.

(3) The optimum pressure ratio is limited. However, a larger volume pulse tube or higher operating frequency increases the optimum pressure ratio.

(4) A large scale pulse tube gives better results than a small scale.

## REFERENCES

1. Matsubara, Y., Gao, J. L., Tanida, K., Hiresaki, Y. and Kaneko, M., "An Experimental and Analysis Investigation of 4 K Pulse Tube Refrigerator", *7th International Cryocooler Conference Proceedings*, Air Force Phillips Laboratory Report PL-CP--93-1001, 1993, Kirtland Air Force Base, NM, pp. 166-186.

2. Gu, U. and Timmerhaus, K. D., "Damping of Thermal Acoustic Oscillations in Hydrogen Systems", *Advances in Cryogenic Engineering*, Vol. 37A (1991), Plenum Press, New York, pp. 265-273.

3.   Kanao, K., Watanabe, N. and Kanazawa, Y., "A Miniature Pulse Tube Refrigerator for Tempera-
      ture below 100 K", ICEC Supplement, *Cryogenics*, Vol. 34 (1994), pp. 167-170.

4.   Godshalk, K.M., Jin, C., Kwong, Y.K., Hershberg, E.L., Swift, G.W. and Radebaugh, R., "Char-
      acterization of 350 Hz Thermoacoustic Driven Orifice Pulse Tube Refrigerator with Measurements
      of the Phase of the Mass Flow and Pressure", *Advances in Cryogenic Engineering*, Vol. 41B (1996),
      Plenum Press, New York, pp. 1411-1418.

5.   Zhu, S. W. and Matsubara, Y. "Theoretical and Experimental Study of Thermal Acoustic Engine",
      *Proceedings of 7th International Conference of Stirling Cycle Machines*, November 5-8, 1995, Waseda
      University, Shinjuku, Tokyo, pp. 579-584.

# Experimental Study on the Pulse Tube Refrigerator with Two Relief Valves

Y. Hagiwara, S. Yatuzuka and S. Ito

Advanced Mobile Telecommunication Technology, Inc.
Aichi-ken, 470-01 Japan

## ABSTRACT

In this paper we report on the performance of a pulse tube refrigerator with two relief valves. Pulse tube refrigerators have no moving parts in the cold end so there is the possibility of achieving long life. However, the pulse tube refrigerator has not achieved enough cooling power. One candidate approach to increasing the cooling power without increasing the size, operating frequency, or the cost is to more precisely control the pattern of the compression (expansion) and the travelling wave.

To this end, we have adopted two relief valves, instead of an orifice, to control the patterns of the compression (expansion) and of the travelling wave. The refrigerator consists of a piston, a regenerator, a pulse tube, a buffer tank, and two relief valves. The two relief valves are placed between the hot end of pulse tube and the buffer tank. The two relief valves divide the compression (expansion) time and the traveling time during which the gas moves at constant pressure. The two relief valves open when the pressure achieves a fixed value and make the travelling wave into a square wave. Thus, the PV diagram becomes similar to that of the Ericson cycle.

In our test apparatus, the pulse tube refrigerator with two relief valves achieved a minimum temperature of 69 K. The maximum cooling power measured at 90 K was 4.7 W, with an indicated work of 204 W — this corresponds to a 5.3% Carnot efficiency. When using an orifice instead of the two relief valves, the pulse tube refrigerator achieved a 74 K lower temperature limit; the maximum cooling power measured at 90 K was 3.6 W, with an indicated work of 209 W — this corresponds to a 4.0% Carnot efficiency.

## INTRODUCTION

In recent years, many reports on pulse tube refrigerators have been published.[1,2] In those reports, the pressure and displacement of the gas in the pulse tube are assumed to oscillate sinusoidally; little attention has been given to other patterns of oscillation. Added to this, it is important to note that the pulse tube refrigerator has not achieved sufficient cooling power. Thus the effect of patterns of oscillations on the cooling power is an interesting problem.

In this work, we propose a more precise method of controlling the oscillatory wave shape, and evaluate it via experiments.

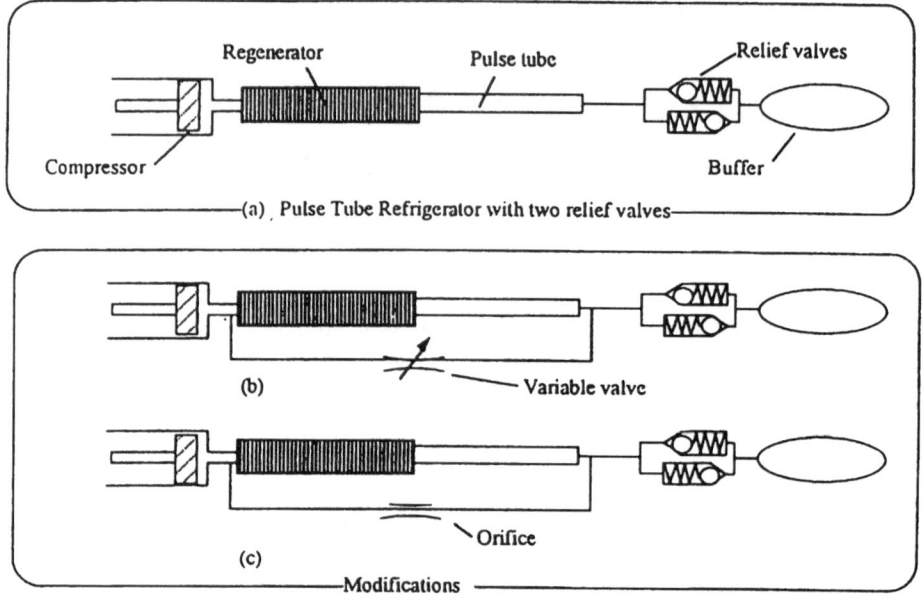

**Figure 1.** Three types of pulse tube refrigerators with two relief valves.

## REFRIGERATORS WITH TWO RELIEF VALVES

In this study we propose three types of refrigerators as noted in Figure 1. The key feature of the three refrigerators is the use of relief valves and a reciprocating compressor. The reciprocating compressor is used due to its high volumetric efficiency. The relief valves are used to more precisely control the pressure and travelling wave.

The combination of relief valves and a reciprocating compressor is a simple and well understood concept. A GM compressor could not be used for this case because the compressor needs an additional valve for controlling the operating pressure as well as the displacement of the gas. The use of many valves causes increasing energy consumption and size. In small refrigerators, the energy consumption of the active device using a rotary valve could not be neglected.

The three types of refrigerators with relief valves used in this study are referred to as types (a), (b), and (c). The first, type (a), is the simplest type and examines the effect of only the relief valves. The second, type (b), is equipped with a variable valve in addition to the two relief valves. The variable valve is controlled to open when the fluid is compressed and expanded in order to decrease shuttle loss and work-flow loss. The work-flow loss is caused by the flow from the double inlet into the buffer tank. The shuttle loss results from the large displacement of fluid in the orifice pulse tube. The third configuration, type (c), is equipped with an orifice instead of the variable valve. In this case, the work-flow loss is larger than type (b).

Next, we describe the justification for expecting an increase in cooling power and its dependence on the relief valves and orifice. In this regard, consider a parcel of fluid as shown in Figure 2. To simplify the discussion about the difference of heat flows we assume that all spaces in the regenerator are maintained isothermal, and the temperature gradient is equal to zero.

In the case of conventional double-inlet and orifice pulse tube refrigerators, the gas parcel undergoes sinusoidal pressure and movement. The parcel is thus compressed (expanded) while moving. During compression, the heat is transferred from the parcel to all surfaces that the parcel passes by.

In the case of the pulse tube refrigerator with relief valves, the parcel is compressed at a fixed position near the end of its displacement. Therefore, the travelling distance of heat is larger than with sinusoidal operation, and the heat flow increases.

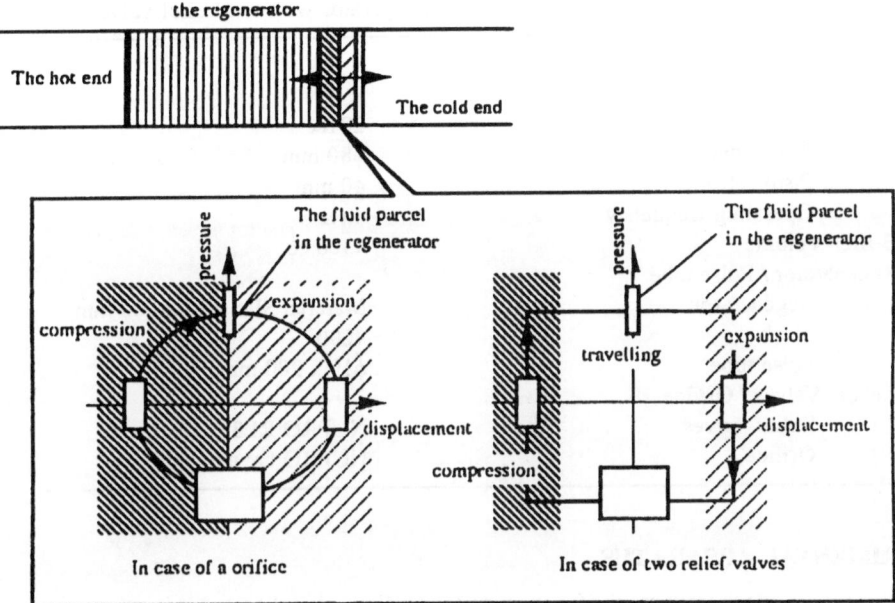

**Figure 2.** Pressure/displacement paths of a fluid parcel for the two cycle types: (a) the parcel undergoes sinusoidal pressure and movement; (b) the parcel is compressed at a fixed position near the end of its displacement, and during movement, the parcel maintains a constant pressure.

In the case of sinusoidal operation with an orifice, pressure in the regenerator and displacement of the parcel are expressed as follows

$$p = p_m + \Delta p \sin \theta \tag{1}$$

$$x = \Delta L \cos \theta \tag{2}$$

where $p$ is the pressure in the regenerator, $p_m$ is the mean pressure, $\Delta p$ is the amplitude of the pressure, $\theta$ is the phase angle, $x$ is the displacement of the parcel, and $\Delta L$ is the amplitude of displacement of the gas parcel. Thus, heat flow $\dot{q}_a$ is expressed as follows

$$\dot{q}_a = \pi f \, \Delta L \, \Delta p \, A \tag{3}$$

where $f$ is the operating frequency and $A$ is the cross-sectional area of the regenerator.

In contrast, the heat flow of the refrigerator with two relief valves, as shown in Figure 2, is expressed as follows

$$\dot{q}_b = 4 f \, \Delta L \, \Delta p \, A \tag{4}$$

Therefore, the ratio $\dot{q}_b / \dot{q}_a$ is as follows

$$\frac{\dot{q}_b}{\dot{q}_a} = \frac{4 f \, \Delta L \, \Delta p \, A}{\pi f \, \Delta L \, \Delta p \, A} = 1.27 \tag{5}$$

The refrigerator with two relief valves is thus expected to increase cooling power by 27%.

**Table 1.** Specification of a pulse tube refrigerator with two relief valves

| | |
|---|---|
| TYPE (a) | |
| Compressor | |
|     Swept volume | 201cc |
|     Piston diameter | φ80 mm |
|     Piston stroke | 40 mm |
|     Operating frequency | 4 Hz |
| Buffer Volume | 1300 cc |
| Regenerator / Pulse tube | |
|     Regenerator | Material SUS φ32 x 200 mm |
|     Mesh | #200 |
|     Pulse tube | φ19 x 200 mm |
| Relief  Valves / Orifice | |
|     Relief valves | NUPRO RL3 |
|     Orifice | NUPRO Cv max 0.03 |

## EXPERIMENTAL APPARATUS

In this report, we present results for the type (a) configuration, because it is the most suitable to examine the effect of the relief valves and is the simplest structure. Table 1 presents the construction parameters for the prototype pulse tube refrigerator with two relief valves. This apparatus allows the relief valves to be easily replaced with an orifice without changing other components. Therefore, we can compare the relief valves with the orifice.

## RESULTS

First, we demonstrate the function of the relief valves. The measurement pressures and the piston stroke at 4 Hz are shown in Figure 3. The pressures in the pulse tube are measured at the hot end of the pulse tube, at the hot end of the regenerator, at the buffer tank, and at the compressor. The top plot in Figure 3 shows the normalized piston stroke. The displacement of the piston is controlled to maintain a sinusoidal wave. The middle plot in Figure 3 shows the pressure in the buffer tank. We find that the pressure in the buffer tank changes intermittently. Therefore, the flow between the buffer tank and the pulse tube occurs intermittently. The bottom plot in Figure 3 shows the pressure in the compressor, in the hot end of the pulse tube, and in the hot end of the regenerator. The pressure at the hot end of the pulse tube and at the regenerator are nearly the same.

Figure 4 shows the relation between the pressure in the pulse tube and the fluctuation of mass in the buffer tank. The fluctuation of mass corresponds to the travelling wave at the hot end of the pulse tube. The fluctuation of mass $\Delta m$ is calculated from the pressure in the buffer tank as

$$\Delta m = \frac{pV_b}{RT_b} - \frac{P_m V_b}{RT_b} \tag{6}$$

where $V_b$ is the volume of the buffer tank, R is the gas constant, and $T_b$ is the temperature in the buffer tank. The thick line represents the case of the relief valves; the thin line represents the case of the orifice. We find there is a travelling time at constant pressure, a compression time, and an expansion time.

With the orifice, on the other hand, one cannot easily divide the time between the travelling time and the compression time. The experimental results discussed above show that the relief valves work as expected.

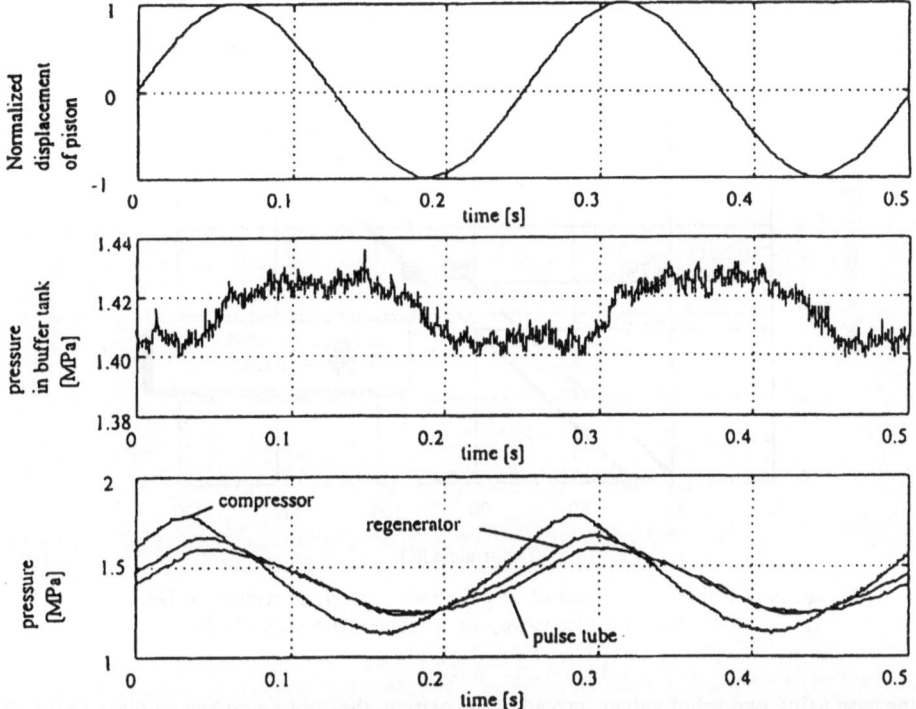

**Figure 3.** The measured pressures and piston stroke in the refrigerator with two relief valves at 4 Hz.

Next, Figure 5 describes the observed differences in cooling powers and minimum temperatures. The pulse tube refrigerator with two relief valves achieved a 69 K minimum temperature. The maximum cooling power measured at 90 K was 4.7 W, with an indicated work of 204 W; this corresponds to a 5.3% Carnot efficiency. In the case using an orifice instead of the relief valves, the pulse tube refrigerator achieved a 74 K minimum temperature, and the maximum cooling power at 90 K was 3.6 W, with an indicated work of 209 W; this corresponds to a 4.0% Carnot efficiency.

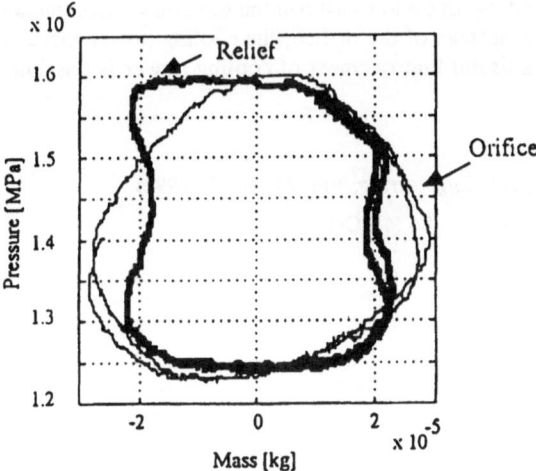

**Figure 4.** Pressure in the pulse tube versus the fluctuation of mass in the buffer tank.

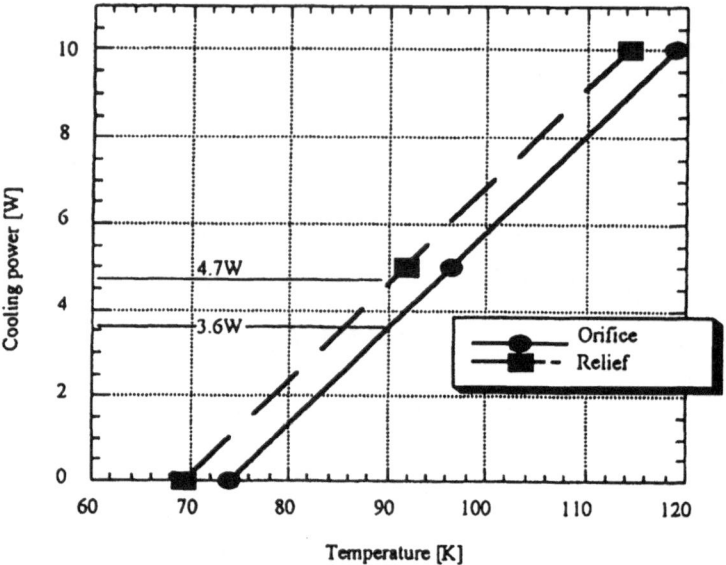

**Figure 5.** Cooling performance of the pulse tube refrigerator with two relief valves and an orifice valve; the operating frequency is 4 Hz.

In the case using two relief valves instead of the orifice, the cooling power increased by 31% for the same input power at 90 K. This result is larger than our prediction. Regarding the difference between the experiment and our prediction, there is room for further investigation. The above results show the feasibility of implementing relief valves. The next step is to consider other factors that increase the cooling power, or to analyze a more detailed model.

## CONCLUSION

In this work, two relief valves were introduced instead of an orifice so as to more precisely control the patterns of the compression (expansion) and the travelling wave. These relief valves were implemented in three types of refrigerators.

Our experimental results show that the two relief valves divide the time between the compression time and the travelling time and transform the travelling wave into a square wave. In the case using the relief valves instead of the orifice, the cooling power increased by 31% at 90 K. This result indicates that a useful improvement of cooling power is possible.

## REFERENCES

1.   Akira Tominaga, *Cryogenic Engineering*, Vol. 27, No. 7 (1992).

2.   G.W. Swift, *J. Acost. Soc. Am.*, 84 (4), October 1988.

# Experiments on the Effects of Pulse Tube Geometry on PTR Performance

**C. S. Kirkconnell, S. C. Soloski, and K. D. Price**

Hughes Aircraft Co.
El Segundo, California, U.S.A., 90245

## ABSTRACT.

Commercial and defense-related demand for reliable, low cost cryocoolers continues to drive the development of pulse tube cryogenic refrigerators. While related to the Stirling cryocooler, the Pulse Tube Refrigerator (PTR) has the distinct advantage over the Stirling of having no moving parts in the expander, yielding the potential for great improvement in the areas of reliability, cost, and vibration minimization. Though the fundamental science of pulse tube refrigeration is well understood, more accurate analytic and numerical modeling tools are needed to facilitate the development of different and higher efficiency PTRs. At the present time, one of the primary areas of uncertainty in pulse tube cryocooler modeling is the calculation of the refrigeration losses due to dissipative mechanisms occurring within the pulse tube itself. The purpose of the experiments described herein is to provide insight into how the volume and aspect ratio of the pulse tube influence both the performance of the PTR and the magnitude of the pulse tube losses. To accomplish that task, a modular PTR was designed such that the pulse tube component could be changed independently of the other components in the system (compressor, heat exchangers, etc.). This facilitated a series of parametric tests on distinct PTRs where the only design variables were those related to the geometry of the pulse tube component. The PTR performance was shown to be relatively insensitive to aspect ratio and sensitive to volume over the range of pulse tubes tested.

## INTRODUCTION

The elimination of moving parts in the cryogenic region of the pulse tube refrigerator (PTR) makes it an attractive alternative to other regenerative cryocoolers (such as the Stirling, Vuilleumier, and Gifford-McMahon). That is especially true for those applications where cost, reliability, and vibration at the cold station are primary areas of concern. For this reason PTRs are being proposed increasingly to meet the demands of both the commercial and military markets. The early history of pulse tube refrigeration has been well documented, beginning with the work of Gifford and Longsworth[1] on the basic pulse tube in the 1960's to the addition of an orifice by Mikulin[2] in 1984. Variations on the orifice pulse tube refrigerator theme, such as the double-inlet PTR[3] and the double-inlet reversible PTR[4], have appeared in recent years, and some researchers have observed significantly improved performance using these modified designs. Recently published experimental results reveal PTRs to be approaching the more-established Stirling devices in thermodynamic efficiency, further spurring interest in these types of

Cryocoolers 9, Edited by R.G. Ross, Jr.
Plenum Press, New York, 1997

cryocoolers. For example, Burt and Chan report obtaining 2.0 W net cooling at 60 K with compressor electrical input power of 88.6 W (44.3 W/W) in a paper published in 1995[5]. This is only 43% less efficient than the Hughes PSC Stirling cryocooler, one of the most efficient Stirling cryocoolers on the market at these temperatures, which delivers 3.0 W net refrigeration at 60 K with 76.3 W of electrical input power[6] (25.4 W/W).

The first thermodynamic model of a PTR was a one-dimensional enthalpy flow model of the pulse tube component, described by Storch and Radebaugh[7] in 1987. Subsequent enhancements to this model eventually led to the more-detailed SRZ model, published in a 1991 NIST Technical Note[8]. Though qualitatively informative, the SRZ model tends to over-predict observed pulse tube heat transfer rates by 3 to 5 times, presumably due to the simplifying assumptions inherent in the model (small pressure fluctuations, no axial mixing, etc.). Ensuing efforts to improve on the accuracy of the SRZ model have met with some success[9,10], usually over limited ranges of operation, but a general PTR model which is quantitatively accurate over a wide range of sizes and operating points has yet to be published. It is suggested that this is largely due to a lack of understanding about the dissipative losses which occur inside the pulse tube component caused by non-ideal flow characteristics (referred to herein as "pulse tube losses"). An example of this type of refrigeration loss, termed "enthalpy streaming", is discussed in a recent paper by Lee, et al.[11].

The goal of the present effort is to experimentally investigate the role of the relative size and aspect ratio of the pulse tube in the performance of a PTR. Five different PTRs were tested with only the geometry of the pulse tube component varying between refrigerators. This approach was taken to isolate the role of the pulse tube from that of the other system components. Using one pulse tube as a reference, two distinct parametric tests were performed with three data points per test: i) constant volume, variable aspect ratio, ii) constant length, variable volume. The results contained herein reveal the PTR to be relatively insensitive to changes in the aspect ratio over the range investigated, but significant changes in performance were observed when the volume of the pulse tube was varied.

This study was funded by Hughes Aircraft Co. Internal Research and Development (IR&D), and the results have been incorporated into our analytical tools. These tools are used in the design and development of all of our Pulse Tube Refrigerators including those for the Low Cost Cryocooler (LCC) Program.

## LIST OF SYMBOLS

| Symbol | Description | Symbol | Description |
|--------|-------------|--------|-------------|
| D | pulse tube diameter | T | temperature |
| f | compressor frequency | V | volume |
| $\phi$ | porosity | Vs | compressor swept volume |
| $\Gamma$ | pulse tube aspect ratio (L / D) | < > | quantity in is brackets time-averaged & steady-state |
| L | pulse tube length | | |
| P | thermodynamic pressure | | |
| $P_0$ | charge pressure | Subscripts | Definition |
| $\dot{Q}_{cond}$ | conduction loss | chx | cold heat exchanger |
| | | min | minimum |
| $\dot{Q}_g$ | gross refrigeration | opt | optimum |
| $\dot{Q}_{net}$ | net refrigeration | pt | pulse tube |
| $\dot{Q}_p$ | sum of parasitic losses | reg | regenerator |
| $\dot{Q}_{pt\,loss}$ | pulse tube loss | | |
| $\dot{Q}_{rad}$ | radiation loss | | |
| $\dot{Q}_{reg\,loss}$ | regenerator inefficiency loss | | |
| t | time | | |

## EXPERIMENTAL SETUP

### Pulse Tube Refrigerator Design

A linear pulse tube expander, shown schematically in Figure 1, was designed so that the pulse tube component could be changed with minimal impact on the other components in the system. To reduce costs, the cold seal was eliminated by designing the housing for the regenerator-cold heat exchanger (HX)-pulse tube section as a single machined part with a step in the diameter between the cold HX and the pulse tube. The cold HX is a copper insert filled with medium mesh copper screen press-fit into the thin-walled stainless steel housing. To minimize uncertainty associated with regenerator performance, the regenerator is of a similar design (geometry and matrix) to those used in the Hughes ISSC and PSC Stirling cryocoolers. It is composed of fine mesh stainless steel screen contained within the inside diameter of the large end of the center section housing. In the usual fashion, the pulse tube is simply a hollow cylindrical tube. The design of the regenerator and the cold HX are the same for all five sub-assemblies, and the measured porosities, $\phi$, ( $0.64 < \phi_{reg} < 0.67$; $0.57 < \phi_{chx} < 0.63$) reveal them to be functionally identical. Common inlet and rejection heat exchangers, both of water-cooled, packed copper mesh design, were used for all five assemblies. Thus the goal of assembling five virtually identical expanders with variation only in the geometry of the pulse tube was achieved.

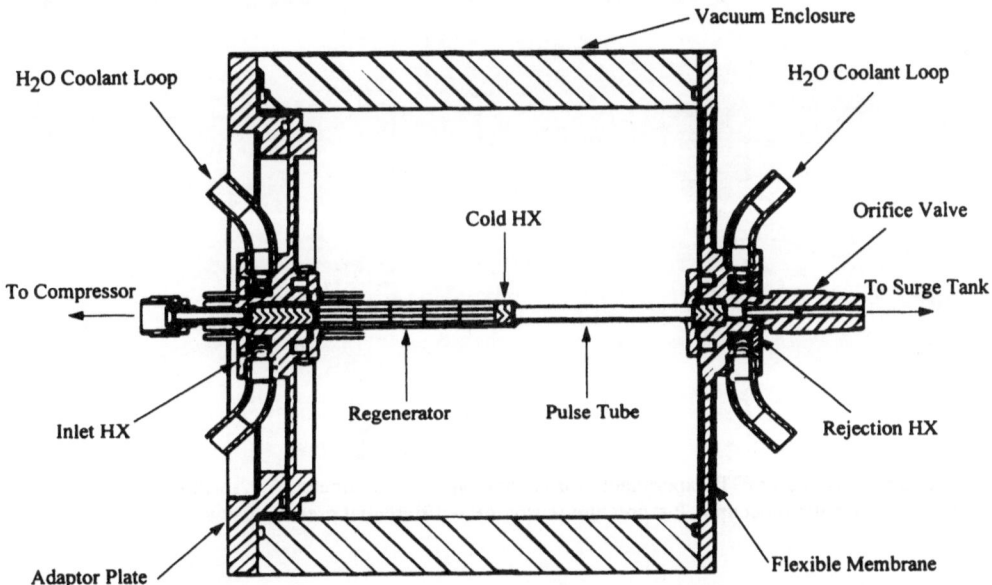

Figure 1. Layout/schematic of modular pulse tube expander.

The expanders for these experiments were coupled by a short (approximately 5 cm) transfer line to a Hughes Aircraft Co. Condor compressor: a dual-piston, linear, positive displacement device which produces an essentially sinusoidal pressure wave. At maximum stroke the swept volume of the compressor is less than 5 cc (small compressor), and the drive electronics utilized allowed the variation of the swept volume so that the input P-V (pressure-volume) power could be controlled. Within the structural limits of the compressor, the charge pressure could also be varied. By varying both the charge pressure and swept volume, the pressure ratio could be adjusted.

## Instrumentation

Figure 2 illustrates the extensive instrumentation of the test setup. The temperatures of each of the heat rejection end plates are regulated with a constant-temperature circulating bath and associated valving, while the corresponding temperatures are monitored with type-T thermocouples placed on the copper plates. Since both the flow rate and temperature of the cooling water are also monitored, calorimetric measurements at each of the heat exchangers are possible. Both temperature and heater power at the cold station are governed by a Lakeshore temperature controller using silicon diode temperature sensors and a resistive heater. Pressure waves in the gas are measured at various locations in the PTR using fast response strain gauges. A differential pressure transducer between the orifice and the surge tank permits an accurate estimate of the mass flow rate through the orifice valve. Together with a co-located pressure sensor, the use of an anemometer in the transfer line gas stream provides a direct measurement of the phase angle between the mass flow rate and the pressure at that particular location. The anemometer is not calibrated for this range of operation, so quantitative velocity measurements are not obtainable at present. This level of instrumentation is provided so that the individual effects to operating parameters such as pressure ratios and heat exchanger temperatures can be observed in addition to monitoring the top-level performance parameters such as load temperature and net refrigeration.

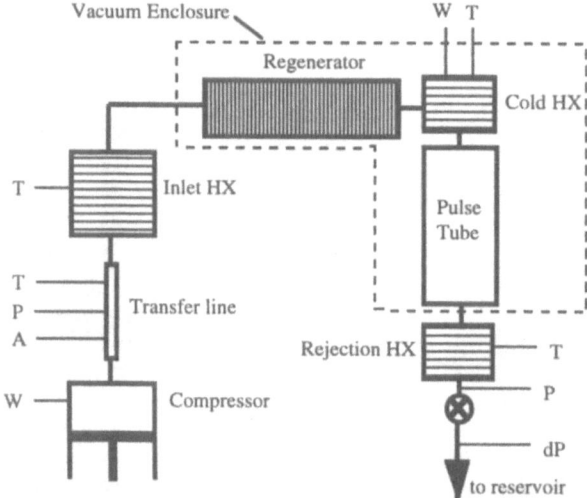

Figure 2. Modular PTR experiment instrumentation. W = wattmeter, A = anemometer, T = thermocouple, P = pressure sensor, dP = differential pressure gauge.

# RESULTS

## Optimum Operational Settings

The five different pulse tube geometries and performance characteristics are summarized in Table 1. For each of these pulse tubes, a series of tests were performed to characterize the PTR performance as a function of operating frequency and, in the case of the first pulse tube tested (PT1), charge pressure. The load curves in Figure 3 illustrate the results of the frequency test for PT1, which is typical of the other four pulse tubes, and Figure 4 summarizes the results of the charge pressure test, also for PT1. With the minor exception of PT4, which performed considerably worse than the other four pulse tubes, the optimum operating frequency was found to be largely independent of the pulse tube geometry. Though load curves were not obtained as a function of charge pressure for the other four pulse tubes, the influence of the charge pressure on the no-load operating temperature was examined, and little change was observed between 30 atm and 35 atm. The same type of "flat" behavior was also observed with regards to the valve setting. In general, the overall performance was seen to be relatively insensitive to changes in the frequency, charge pressure, and valve setting in the vicinity of the optimum operational settings for those parameters.

Table 1. Pulse Tube Geometry and Performance Data

|       | $V_{pt}$ / $V_{PT1}$ | $\Gamma$ | $P_{0,opt}$ [atm] | $f_{opt}$ [Hz] | $T_{min}$ [K] |
|-------|-------|------|------|------|------|
| PT1   | 1.00  | 13.0 | 30   | 38   | 75.2 |
| PT2   | 1.00  | 10.0 | 34   | 33   | 77.8 |
| PT3   | 1.00  | 16.5 | 33   | 34   | 78.5 |
| PT4   | 0.80  | 14.7 | 35   | 42   | 85.0 |
| PT5   | 1.17  | 12.1 | 33   | 33   | 73.2 |

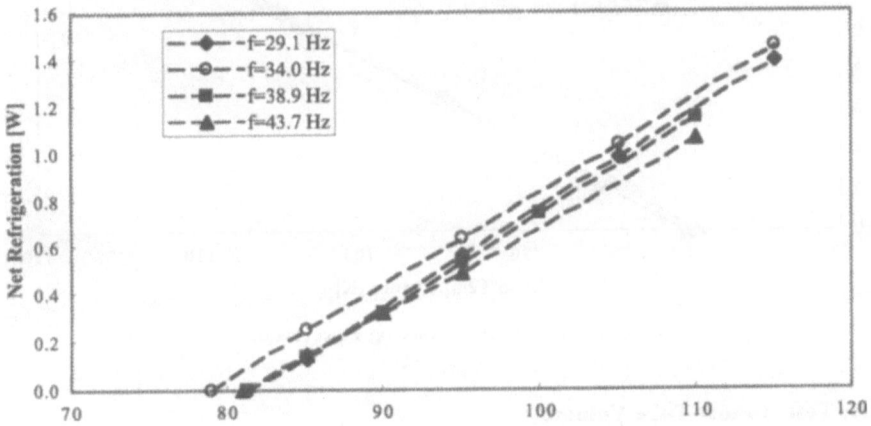

Figure 3. Load curves for varying operational frequencies. Data for PT1.
Charge pressure = 30.5±0.1 atm. Input P V power = 33±1 watt.

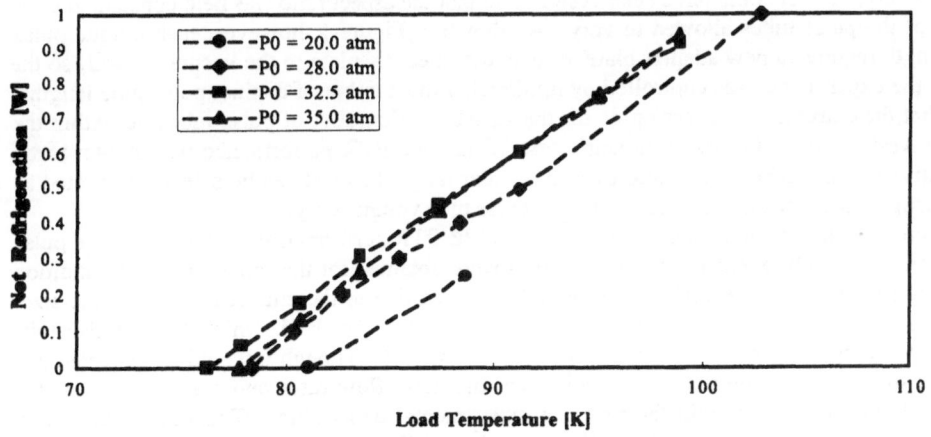

Figure 4. Load curves for varying charge pressure. Data for PT1.
f = 34 Hz. Input P-V power = 33±1 watt.

## Parametric Test 1: Pulse Tube Aspect Ratio

The first three pulse tubes in Table I (PT1, PT2, and PT3) comprise a set of constant volume, variable aspect ratio pulse tubes. The values for aspect ratio used here vary from 10.0 for PT2 to 16.5 for PT3, and these values are typical of those reported in the literature for single-stage orifice PTRs. Shown in Figure 5 are the load curves for the three configurations at virtually the identical operating point (f = 34.0 ± 0.0 Hz, $P_0$ = 30.5 ± 0.2 atm, Vs constant to within 3%). The overall PTR performance is essentially independent of the aspect ratio over the range of values investigated. Furthermore, the data acquisition system reveals that the pressure ratios, heat exchanger temperatures, and orifice flow rate are also largely unaffected by the changes in aspect ratio. As discussed later, this indicates that the pulse tube losses may not be highly dependent on the aspect ratio for the values considered.

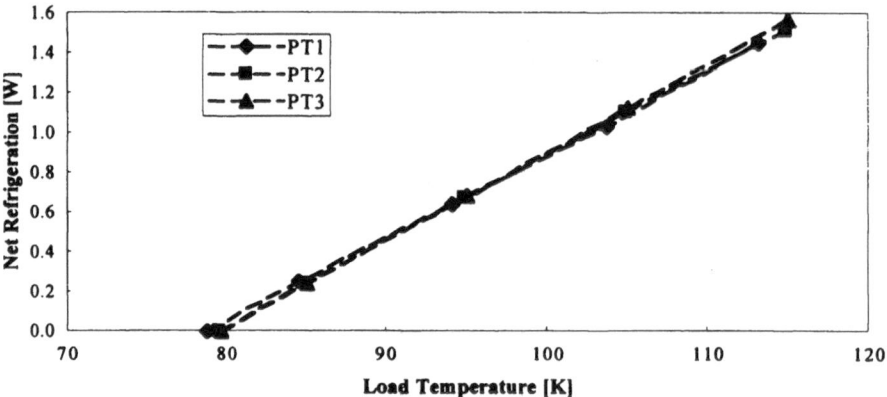

Figure 5. Load curves for varying aspect ratios.

## Parametric Test 2: Pulse Tube Volume

PT1, PT4, and PT5 define a set of constant length, variable volume pulse tubes. The length was held constant to model a hypothetical design trade which may arise in the development of a concentric pulse tube expander. Assuming the optimum regenerator length has been identified, it may be deemed desirable to design a pulse tube of the same length for manufacturing reasons. Alternatively, the test could have been designed so that the aspect ratio was held constant and the length of the pulse tubes allowed to vary. As shown in Figure 1, however, each unique pulse tube length requires a new adapter plate to mate the inlet HX plate to the vacuum dewar, so the cost of the experiments was controlled by minimizing the number of distinct pulse tube lengths. Note that the extremes in aspect ratios for the variable-volume set of pulse tubes lie within the range investigated in the first parametric test. Since the PTR performance was shown to be insensitive to pulse tube aspect ratio over that wider range, this test can be safely interpreted as an investigation of the effects of changing the pulse tube volume only.

Figure 6 clearly demonstrates a dependence of the PTR performance on the size of the pulse tube. As with Figure 5, each of the load curves were obtained for the same operating condition and piston stroke, thus the variation observed is indeed due to the changes in pulse tube size. The most efficient pulse tube tested is the largest pulse tube, PT5, which has a pulse tube volume-to-compressor swept volume ratio, $(V_{pt}/V_s)$, of 0.46. Though it would be expected that changes to the pulse tube volume would affect the mass flow rates and pressure waves, the measured changes were within the range of experimental uncertainty. This is not surprising, however, for the difference in volumes between the smallest and largest pulse tubes represents less than a 5% change in the total expander void volume. The significance of these results with regards to the pulse tube losses is discussed in the next section.

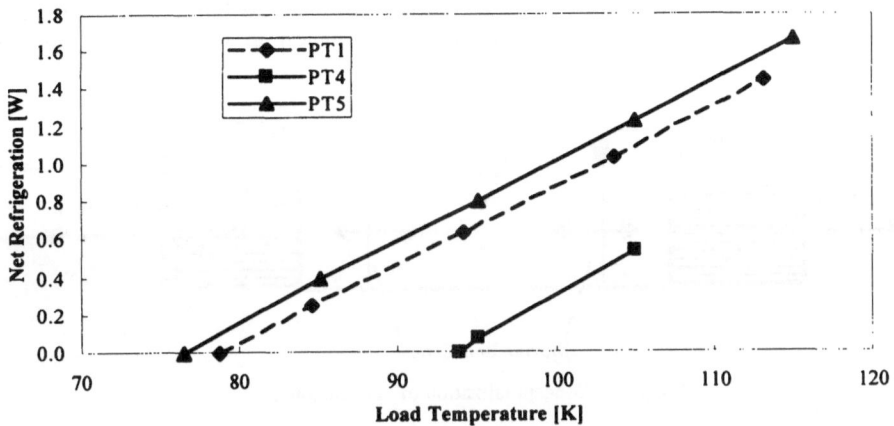

Figure 6. Load curves for varying pulse tube volumes.

## DISCUSSION

Consider the control volume partitioning of the cold heat exchanger - pulse tube - rejection heat exchanger portion of the PTR as illustrated in Figure 7. In addition, assume that the pulse tube flow field is ideal, where "ideal" is defined as one-dimensional and free of axial mixing. Under these assumptions, a gas piston can be identified which is a constant-mass control volume whose limits of travel are defined by the heat exchangers at either end of the pulse tube. The heat exchanger control volumes each have one fixed boundary and one moving boundary, the moving boundary being that in contact with the gas piston. By defining the control volumes in this fashion, the gross refrigeration, $\langle \dot{Q}_g \rangle$, can be calculated from the P-V power absorbed by the dynamic cold heat exchanger volume:

$$\langle \dot{Q}_g \rangle = f \oint P_{chx} dV_{chx} .$$  (1)

This equation reveals the origins of the term "gas piston, " for the constant-mass control volume inside the tube serves the same purpose as the physical expander piston in a Stirling cryocooler, causing P-V power to be absorbed by the cold gas. Since $P_{chx}(t)$ and $V_{chx}(t)$ can be calculated fairly accurately using a control volume mass conservation model of the PTR, the gross refrigeration can also be determined with some confidence by virtue of equation 1. Then the net refrigeration can be determined by utilizing a first law energy balance on the cold heat exchanger volume, i.e.,

$$\langle \dot{Q}_{net} \rangle = \langle \dot{Q}_g \rangle - \langle \dot{Q}_{parasitics} \rangle$$

$$\langle \dot{Q}_{parasitics} \rangle = \langle \dot{Q}_{reg\_loss} \rangle + \langle \dot{Q}_{rad} \rangle + \langle \dot{Q}_{cond} \rangle + \langle \dot{Q}_{pt\_loss} \rangle$$  (2)

Conduction, $\langle \dot{Q}_{cond} \rangle$, and radiation, $\langle \dot{Q}_{rad} \rangle$, losses can be calculated relatively accurately from first principles, and methods have been developed for calculating regenerator inefficiency loss, $\langle \dot{Q}_{reg\_loss} \rangle$, (NIST's REGEN program, for example). The magnitude of the "pulse tube loss", $\langle \dot{Q}_{pt\_loss} \rangle$, which is defined here as the sum of the refrigeration losses that occur in the pulse tube due to deviations in the real flow field from the ideal case described above, remains as an area of uncertainty in the calculation of the net refrigeration. The goal of these experiments is therefore to provide some qualitative and quantitative information on how this pulse tube loss varies with pulse tube design.

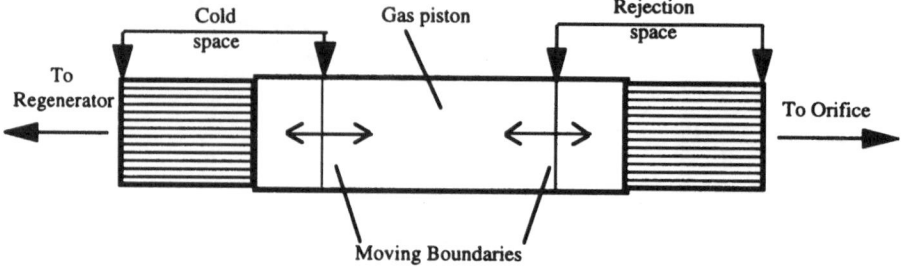

Figure 7. Conceptualization of "gas piston".

The first parametric test considers the effect of the pulse tube aspect ratio on the behavior of the PTR. In the limiting case where the aspect ratio approaches infinity for a finite volume, the tube diameter goes to zero and the flow impedance becomes so large that it eventually chokes the flow at an infinitesimal flow rate, effectively reducing the net enthalpy flux through the tube to zero. For the opposite case where the aspect ratio, hence the length, goes to zero for a finite volume, conduction through the gas and the tube walls from the rejection HX to the cold HX eventually dominates the desired convective flux in the opposite direction. Between these two extremes lies a range which encompasses the optimum pulse tube aspect ratios for practical PTR designs. As stated previously, the aspect ratios tested in the present effort clearly demonstrate that the PTR performance is not highly dependent within the range tested. Furthermore, since the pressure waves, mass flow rates, and temperatures are basically constant between cases, the gross refrigeration, the regenerator losses (conduction and enthalpy flux), and the radiation loss are also constant. The calculated change in the pulse tube conduction parasitic loss due to the variations in pulse tube geometry is less than 0.02 W, which is negligible. Therefore, since the same operating points yield virtually the same net cooling rate for all three PTRs, the pulse tube flow loss must also be essentially constant for the 3 different pulse tubes by virtue of equation 2.

As with the aspect ratio, the determination of the optimum pulse tube volume involves a trade-off between competing effects. Reducing the volume of the pulse tube, or any expander component for that matter, increases the amplitude of the pressure wave in the cold HX and, in the absence of a dramatic change in the phase angle between the volume and pressure, increases the amount of gross refrigeration (equation 1). However, as has been known for some time[1,9], a practical pulse tube must be of sufficient size so that a portion of the gas, in the absence of axial mixing, never leaves the pulse tube. This is the "gas piston" identified in Figure 7. It is hypothesized here that reducing the size of the pulse tube increases the refrigeration losses associated with non-ideal pulse tube flow characteristics such as axial mixing and secondary flows[11]. The purpose of this test was to analyze that hypothesis by observing and characterizing how the PTR performance varies with pulse tube volume.

A numerical analysis based upon the predictions a of well-correlated PTR model[12] supports the theoretical prediction of increasing gross refrigeration capacities as the volume of the pulse tubes decreases, but as illustrated in Figure 6, the net refrigeration capacities actually decrease along with the pulse tube volume. Therefore, by virtue of equation (2), the parasitic losses must increase as the size of the pulse tube decreases. Since the radiation, regenerator, and conduction losses are essentially constant, the increased losses must be due to the changes in the behavior of the pulse tube flow field. Preliminary analysis of the data reveals the pulse tube losses for the largest and smallest pulse tubes to be roughly 35% and 65%, respectively, of the gross refrigeration. A study is underway to more carefully quantify these losses and to seek correlations between the magnitude of the pulse tube losses and the dimensionless variables which govern the flow field (aspect ratio, Womersley number, etc.).

## CONCLUSIONS

A series of experiments have been conducted on a modular PTR design in which only the geometry of the pulse tube component was allowed to vary between test items. The influences of the charge pressure, frequency, and valve setting on the PTR performance were investigated. Furthermore, parametric tests were conducted to investigate the role of the aspect ratio and volume of the pulse tube on the overall PTR performance and on the magnitude of the refrigeration losses due to non-ideal pulse tube flow characteristics. The results were as follows:

- the PTR performance was found to be relatively insensitive to changes in the operational parameters in the vicinity of the optimum operating point;
- the PTR performance and pulse tube losses were virtually independent of pulse tube aspect ratio over the range investigated ($10.0 < \Gamma\ 16.5$);
- the best PTR performance characteristics and minimum pulse tube losses were observed for the largest pulse tube ($V_{pt} / V_s = 0.46$).

Since the PTR with the largest pulse tube proved to be the most efficient, it would desirable to continue these experiments with even larger pulse tubes to determine the optimum ($V_{pt} / V_s$) ratio for this particular design.

## ACKNOWLEDGMENTS

The authors would also like to thank the rest of the Hughes "Pulse Tube Team" for their valuable assistance: S. Russo - Product Line Manager, A. Rattray - Design Engineer, W. Croft - Test Engineer, T. Pollack - Test Engineer. This work would not have been possible without them.

## REFERENCES

1. Radebaugh, R. C., "A Review of Pulse Tube Refrigeration", *Advances in Cryogenic Engineering*, vol. 35, Plenum Press, New York (1990), pp. 1191-1205.

2. Kirkconnell, C. S., *Numerical Analysis of the Mass Flow and Thermal Behavior in High-Frequency Pulse Tubes*, Doctoral Thesis, Georgia Institute of Technology, Atlanta, (1995).

3. Zhu, S., P. Wu, and Z. Chen, "A Single Stage Double Inlet Pulse Tube Refrigerator Capable of Reaching 42 K", *Cryogenics*, vol. 30 (supplement), (1990), pp. 257-261.

4. Wang, P. Wu, and Z. Chen, "Theoretical and Experimental Studies of a Double-Inlet Reversible Pulse Tube Refrigerator", *Cryogenics*, vol. 33, (1993), p. 648.

5. Burt, W. W. and C. K Chan, "Demonstration of a High Performance 35 K Pulse Tube Cryocooler", *Cryocoolers 8*, Plenum Press, New York, (1995), pp. 313-319.

6. Price, K. D., Hughes PSC Performance Data, Personal Communication, (1996).

7. Storch, P. J. and R. C. Radebaugh, "Development and Experimental Test of an Analytical Model of the Orifice Pulse Tube Refrigerator", *Advances in Cryogenic Engineering*, vol. 33, Plenum Press, New York, (1987), p. 851.

8. Storch, P. J., R. C. Radebaugh, and J. E. Zimmerman, "Analytical Model for the Refrigeration Power of the Orifice Pulse Tube Refrigerator", *NIST Technical Note 1343*, Boulder, CO, (1991).

9. Wu, P. and S. Zhu, "Mechanism and Numerical Analysis of Orifice Pulse Tube Refrigerator With a Valvless Compressor", *Cryogenics and Refrigeration - Proceedings of International Conference*, May 22-26, 1989, International Academic Publishers, (1989), pp. 85-90.

10. Baks, M. J. A. et al., "Experimental Verification of an Analytical Model for Orifice Pulse Tube Refrigeration", *Cryogenics*, vol. 30, (1990), pp. 947-951.

11. Lee, J. M. et al., "Steady Secondary Momentum and Enthalpy Streaming in the Pulse Tube Refrigerator", *Cryocoolers 8*, Plenum Press, New York, (1995), pp. 359-369.

# An Experimental Investigation of a Single-Stage Two-Pulse-Tube Refrigerator

J. Yuan and J.M. Pfotenhauer

Applied Superconductivity Center
University of Wisconsin - Madison

## ABSTRACT

A single stage orifice pulse tube cryocooler incorporating one compressor and regenerator, but two pulse tubes has been constructed and tested. This development is directed toward use as cooling for superconducting current leads where the voltage insulation between the two leads is critical. The purpose of the present study is to investigate the thermal stability of parallel oscillating flows and compare the performance of a specific pulse tube in a two-tube configuration with that of the same pulse tube working in the standard single tube configuration. Performance is characterized by the system cool-down speed and the minimum cold end temperature. Three pulse tube specimens named I, II, and III, have been assembled and tested individually and in various combinations. The experimental results reveal the absence of any thermal stability problem in the two pulse tube configuration. However, the performance of a specific pulse tube is degraded in the two-tube configuration as compared to its individual use.

## INTRODUCTION

The pulse tube refrigerator has significant advantages over other cryocoolers due to the elimination of moving parts at low temperature and the relatively low working pressure.[1,2] Therefore it has great potential to be more reliable, simpler to construct and cheaper than the Stiring or GM cryocooler. With recent advances [3,4] in pulse tube refrigerators, the attention to the application of this kind of device has been increasing rapidly. In the applications such as the cooling of superconducting current leads researchers are seeking the possibility of using one pulse tube cryocooler with twin cold heads to cool leads in which the voltage insulation between two leads is crucial. [5] The multiple pulse tube configuration provides additional potential for future use, as for example in cooling multiple computer chips by using one cryocooler (one compressor, one regenerator) with multiple cold fingers. However, in such configurations, a new variety of possible problems must be addressed such as the thermal stability of parallel oscillating flows, and entropy production of mixing flows.

In the present paper, initial results are provided from an investigation with a single stage pulse tube refrigerator driving two tubes. The following sections describe the design, configuration, and performance of this kind of cooler.

## EXPERIMENT DESCRIPTION

### Selection of Specimens and Configurations

The experimental configurations are designed to study the performance of a pulse tube refrigerator with multiple cold heads. Three pulse tube specimens are used in our investigation, the dimensions of which are given in Table 1. Specimens I and II are identical in all dimensions,

Cryocoolers 9, Edited by R.G. Ross, Jr.
Plenum Press, New York, 1997

**Table** 1. Test specimens

| specimen No. | length (mm) | O.D. (mm) | wall thickness (mm) | volume (cm³) |
|:---:|:---:|:---:|:---:|:---:|
| I | 142 | 22.225 | 0.89 | 46.6 |
| II | 142 | 22.225 | 0.89 | 46.6 |
| III | 142 | 31.75 | 0.89 | 100.2 |

while specimen III has the same length as the other two, but the total volume is approximately doubled due to a larger diameter. Three same designed cold end heat exchangers filled with 100 mesh copper screens are soldered with three pulse tubes, respectively. In order to compare the overall system performance, three different pulse tube configurations have been accommodated. In the first case, a single pulse tube with both orifice and double-inlet valve configurations is used. This configuration is used as the reference for comparison with the other arrangements. All three specimens have been tested in this configuration. The second arrangement uses specimen I and II as twin tubes with two different orifice configurations. In case (a), each pulse tube has its own orifice while in case (b), the twin pulse tubes share one orifice. In the third configuration specimen I and III are combined. These configurations allow us to study the effect of combining two pulse tubes, the role of the orifice, and the thermal stability. The investigation of each of these issues will be further discussed below.

## Test Apparatus Description

The arrangement of the instrumentation and primary components for version (a) of the twin pulse tube test is illustrated schematically in Fig. 1. Here, the regenerator, two pulse tubes and five heat exchangers are contained within a chamber which is evacuated to a pressure of $1 \times 10^{-5}$ - $2 \times 10^{-5}$ torr, while the compressor, reservoir and all valves remain in the room environment. In order to reduce the thermal radiation loss at low temperature, more than 10 layers of aluminized Mylar are wrapped around the cold parts. A valveless and dry-lubricated compressor with swept volume of 289.6 cm³ is used in all the experiments. The frequency of the compressor can be adjusted from 6 Hz up to 12 Hz.

The regenerator is fabricated from a thin walled stainless steel tube which is 200 mm in length, 40 mm in diameter and 0.75 mm wall thickness. More than 2000 pieces of 400 mesh stainless steel screens are packed inside the regenerator tube. The porosity of the regenerator is about 67%. The regenerator and pulse tubes are arranged side by side but connected at the cold end heat exchangers. Each of the pulse tubes are equipped at the hot end with water-cooled heat

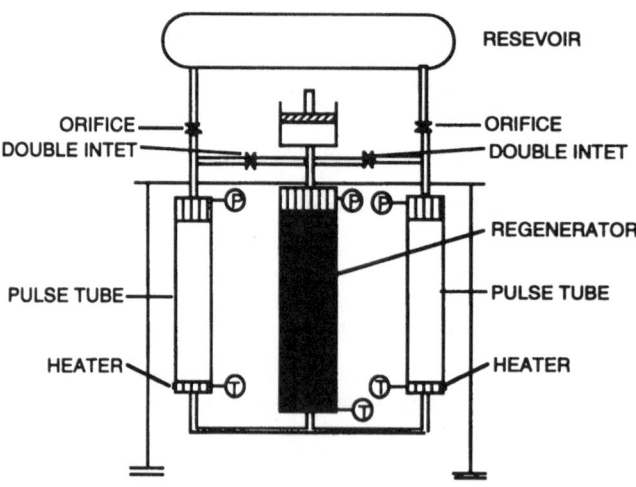

**Figure** 1. Schematic of test apparatus

exchangers. While the design for these heat exchangers are identical for each of the pulse tubes, it may be possible that small differences arose during the fabrication process. The heat exchangers at both ends of pulse tubes are filled with copper screens of 100 mesh to provided for both a large heat exchange surface area and flow straightening. The orifice and double-inlet vales are needle vales that can be adjusted over a wide range, and the volume of the reservoir for all experiments is 3785 cm³.

A 10 watt heater is placed at the cold end of each pulse tubes to measure the cooling capacity. Three separate piezoresistive pressure transducers which allow measurement of the absolute pressure oscillation are mounted on the hot ends of the pulse tubes and regenerator, respectively. Temperatures at different points of the system are measured with silicon diodes. The working medium used in all cases is helium gas.

## ANALYSIS

The following analysis of the two-pulse-tube arrangement is based on the enthalpy flow model of Radebaugh [6]. The energy flows for the two pulse tubes are displayed in Figure 2. Here $<H_1>$ and $<H_2>$ are the respective net enthalpy flows of pulse tube 1 and 2 integrated over one period of oscillation, and $<Q_{c1}>$ and $<Q_{c2}>$ are the respective refrigeration powers available at temperatures $T_1$ and $T_2$. In the case of zero heat load, the cold ends of both pulse tubes will reach their minimum temperatures. Using the first law of thermodynamics for an open system, it follows that the general energy balance for the cold ends of each pulse tube without heat loads is given by

$$(1)$$

and

$$<H_2> = \Sigma <Q_{2\_loss}> \tag{2}$$

where $\Sigma <Q_{1\_loss}>$ and $\Sigma <Q_{2\_loss}>$ are the total loss terms at the cold end of pulse tube 1 and pulse 2, respectively. From Equation (1) and (2), it can be seen that the cold end temperature of each pulse tube depends on the net enthalpy flow in the pulse tube. The time averaged enthalpy flow for an ideal gas is given by

$$<\dot{H}> = (\frac{C_p}{R\tau}) \oint \Phi P dt \tag{3}$$

where R is the gas constant, $\Phi$ is the volumetric flow rate, and P is the pressure in pulse tube. To a good approximation, the pressure is taken to be uniform within both pulse tubes. Therefore from equation (3) it follows that the minimum temperature of the cold end of each pulse tube depends on the volumetric flow rate and the phase angle between that flow rate and the pressure oscillation at the cold end. If the net enthalpy flow at pulse tube 1 is different from that of pulse tube 2, different cold end temperature will be produced.

Furthermore, one can expect that the cold end temperature of the single regenerator will be at an intermediate temperature between that of the cold ends of pulse tube 1 and the pulse tube 2 due to the mixing during half of the cycle of the gas flows with two different temperatures. The effect of mixing two gas streams with differing temperatures is the production of entropy which

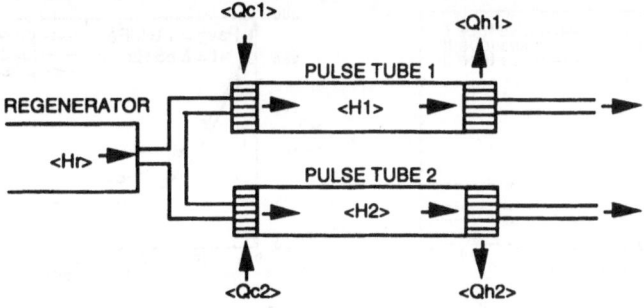

**Figure** 2. Enthalpy flows in the two pulse tube refrigerator

introduces an extra loss term for the system and will degrade system performance. In order to obtain the best performance, two pulse tubes of identical individual performance should be used in the system.

## EXPERIMENTAL RESULTS AND DISCUSSIONS

### Single Pulse Tube Configuration Tests

Prior to the two-pulse-tube experiments, and as a performance reference for the other arrangements, three pulse tube specimens I, II and III have been tested individually using the same compressor, regenerator and hot end heat exchanger, and under same conditions of frequency and pressure. More specifically, these tests have been conducted using an average pressure of 1.0 MPa, a pressure ratio of 1.6, and an operating frequency of 6.85 Hz. The influence of the double-inlet valve opening on the cold end temperature for three specimens has also been investigated. Table 2 shows the minimum temperature attained for the three specimens with and without opening of double-inlet valves. It should be noted that these operating conditions (and hot end heat exchanger configuration) are not necessarily optimized for all three pulse tubes. The cool-down curves of the three pulse tubes are displayed in Fig. 3. From the table and the figure, several interesting points can be observed. First, one may observe that although specimens I and II have the same dimensions, the performances of these two pulse tube are quite different. The minimum temperature of 64 K was achieved by pulse tube I while the minimum temperature of 110 K was achieved by pulse tube II. The exact reason for this performance difference is unknown. Possible reason could be the slight difference between the cold end heat exchangers. Second, for all specimens, use of the double-inlet valves benefit the system performance. Finally, note that the minimum temperature achieved by specimen III is lower than specimen I but higher than specimen II.

### Two Pulse Tube Configuration Tests

Two different combinations of pulse tubes have been tested. In the first case, specimen I and II are combined together, while in the second case pulse tubes I and III are arranged together. All experiments are carried out under the same conditions of frequency, average pressure, and valve opening as in the single pulse tube tests. Due to the volume change, the pressure ratio is slightly decreased to approximately 1.5. In all tests, the cool-down speed and minimum temperatures are gathered for each pulse tube. Table 3 displays the final cold end temperature by each pulse tube and by the regenerator in the two cases.

**Table 2.** Minimum temperature achieved by pulse tube I, II, and III

| specimen No. | Tmin (without double inlet) K | Tmin (with double inlet) K |
|:---:|:---:|:---:|
| I | 76 | 64 |
| II | 115 | 110 |
| III | 90 | 85 |

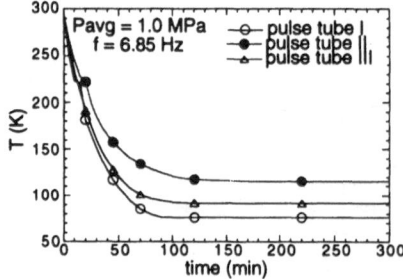

**Figure 3.** Cool-down curve for three pulse tubes without double inlet

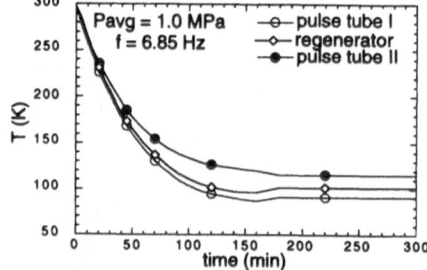

**Figure 4.** Typical cool-down curves for two pulse tube configuration

**Table** 3. The minimum temperature of each pulse tube in two cases

| specimen No. | Tmin (without double inlet) K | |
|---|---|---|
| | case 1 | case 2 |
| I | 82 | 100 |
| II | 125 | n/a |
| III | n/a | 118 |
| regenerator | 90 | 110 |

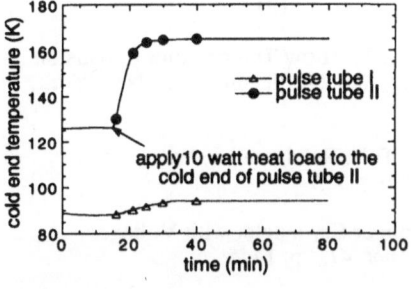

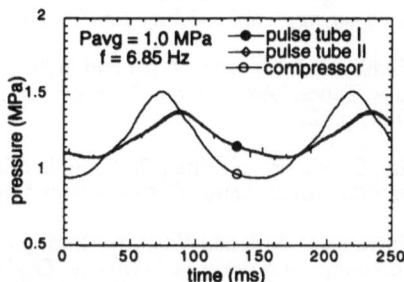

**Figure** 5. Temperature vs. time after the heater on the cold end of pulse tube II is on

**Figure** 6. Typical pressure oscillations in two pulse tubes and compressor

The results in Table 3 confirm the above analysis, displaying that different stable cold end temperatures can be achieved by the different pulse tubes. As mentioned earlier, this suggests that the net enthalpy flows in the two pulse tubes are different. Further, the tests suggest that the thermal coupling between the two pulse tubes is weak. Comparing the performance of each individual pulse tube in two-pulse-tube configuration with that of the same tube in the single configuration, one finds that the performance is somewhat degraded. Consideration of losses suggests that even though the regenerator loss is shared by the two pulse tubes in the two pulse tube configuration, the added loss due to gas mixing overwhelms any advantage gained.

To test the coupling effect between the two tubes, two different experiments are carried out. In the first experiment, a 10 watt heat load is applied to pulse tube II while observing the temperature response of pulse tube I. As can be seen in Figure 5, the temperature of pulse tube I increases by 5 K while the temperature of pulse tube II increases by 40 K. In the second experiment, the cold end of pulse tube I is forced to a higher temperature than that of pulse tube II by closing the orifice of pulse tube I. When a 30 K temperature difference is established, the orifice of pulse tube I is re-opened to its earlier position and in less than 30 minutes, pulse tube I reaches its original stable and colder temperature as compared with pulse tube II. Similar results are obtained for pulse tube I / pulse tube III arrangement. These experiments verify that the thermal coupling between the two pulse tubes is weak.

The pressure oscillations have also been gathered during the above experiments. Fig. 6 displays the pressure oscillations in the compressor and the two pulse tubes. Note that although the cold end temperatures are different for the two tubes, the pressure waves in both tubes are the same.

## CONCLUSIONS

A single stage pulse tube refrigerator capable of driving two pulse tubes has been constructed and tested. The test results demonstrate that the relative performance of the individual pulse tubes in the two-pulse-tube configuration is the same as their relative performance in a single pulse tube configuration, although the overall performances are degraded. The results also show that it is possible to establish two different temperatures at the ends of the two pulse tubes, and that the pulse tubes demonstrate only a weak thermal coupling. Finally the results suggest that losses associated with the gas mixing overwhelm any advantages gained by sharing the regenerator losses between the two pulse tubes. These observations further suggest that an optimum two-pulse-tube arrangement requires well matched individual pulse tubes.

## ACKNOWLEDGMENTS

This study was funded in part by the U.S. Defense Nuclear Agency under DNA MIPR 92-719, work unit CD:00014, RCC: 7010.

## REFERENCES

1.    Radebaugh, R., Zimermann, J., Smith, D.R. and Louie, B.," A Comparison of Three Type of Pulse Tube Refrigerators: New Methods for Reaching 60K" *Advances in Cryogenic Engineering,* Vol.31, Plenum Press, New York, 1986, p.799

2.    Radebaugh, R., "A Review of Pulse Tube Refrigeration", *Advances in Cryogenic Engineering,* Vol.35, 3.Plenum Press, New York, 1990, p.1191

3.    Mikulin, E.I., Tarasov, A.A., and Shkrebyonock M.P., "Low Temperature Expansion Pulse Tubes," *Advances in Cryogenic Engineering,* Vol.29, Plenum Press, New York, 1984, p.629.

4.    Zhu, S., Wu, P. and Chen, Z. "Double Inlet Pulse Tube Refrigerator: An Important Improvement", *Cryogenics* Vol.30, 1990, p.514

5.    Pfotenhauer, J.M. and Yuan, J., "Multiple Cold Finger Cryocooler with Voltage Isolation," *Advances in Cryogenic Engineering vol. 41,* p. 1445 .

6.    Storch, P.J., Radebaugh, R. and Zimmerman, J., "Analytical Model for the Refrigerator Power of the Orifice Pulse Tube Refrigerator," *NIST Technical Note 1343* , 1991.

# UCLA Pulse Tube Investigations

**K. V. Ravikumar, S. Yoshida, N. Myung, Paul Karlmann, S. Sapida, B. Dransart, Tu Nguyen, and T. H. K. Frederking**
*In Collaboration with*
Nell Papavasiliou, Aung Win, Kuo-Wei Huang, Yolanda Henry, Francis Pan, Mohan Sankaran

Chemical Engineering Department.
University of California,
Los Angeles, CA 90095

## ABSTRACT

We have constructed and investigated a linear, low frequency pulse tube system in the basic [BPT] and orifice [OPT] configurations with a three-way electro-mechanical valve. Prior to manufacturing, regenerator characteristics have been studied using a cellular model with repetitive cells in the asymptotic range of laminar boundary layer-dominated flow at relatively large (subcritical) Reynolds numbers. The ratio of heat transfer rate to throughput rate of the model system agrees in the order of magnitude prediction with screen data noting quantitative superiority of the cell system. We have determined for the basic and the orifice pulse tube frequencies at the minimum temperatures: 1 Hz for the BPT, 2 Hz for the OPT, both operating with air.

## INTRODUCTION

In the pulse tube area, considerable progress has been reported recently[1]. The refrigeration load has approached the Gifford-McMahon cooler range[2]. A low temperature of 2.15 K has been achieved[3]. Aiming at basic phenomena, we have studied thermoacoustic oscillations[4] and thermodynamic cooler performance[5]. We have built a pulse tube for operation as Basic Pulse Tube [BPT] and Orifice Pulse Tube [OPT]. We report frequency data and related measurements for these two PT versions.

## REGENERATOR MODELLING

Repetitive sections of regenerator stacks have been modeled using the unit cell as base unit. The Darcy length is employed as reference length $L_c = \kappa^{1/2}$. The regime of operation is laminar boundary layer flow at relatively large velocities, i. e. the hydrodynamic asymptote of duct flow[6]. We make use of these flow results[6] to model the transport as modified ratio [$T_r$] of the dimensionless heat transfer to the dimensionless flow rate. $T_r$ is expressed as the Chilton-Colburn factor divided by the permeability ratio [$\kappa_{eff}/\kappa$] ; ($\kappa_{eff}$ effective permeability, $\kappa$ Darcy permeability). The assumptions are continuum flow of Newtonian fluid, wall material of high heat capacity, quasi-steady transport, laminar asymptote of flow. The Nusselt number [$Nu_{Lc}$] for aspect ratio unity is

Cryocoolers 9, Edited by R.G. Ross, Jr.
Plenum Press, New York, 1997

$$Nu_{Lc} / Pr^{1/3} = 0.28\ Re_{Lc}^{1/2} \qquad (1)$$

(Pr Prandtl number, aspect ratio = (cell length)/(cell size)). The related permeability ratio is

$$k_{eff}/k = 2.9/Re_{Lc}^{1/2} \qquad (2)$$

The transport ratio $T_r$ of our model is

$$(T_r)_m = Nu_{Lc}\ Pr^{-1/3}/[\kappa_{eff}/\kappa] \qquad (3)$$

The experimental transport ratio of screens is[7]

$$(T_r)_{exp} = 0.3\ Re_{Lc}^{1/2}/[Re_{Lc}^{1/2}/1.4] = 0.42 \qquad (4)$$

Results (3) and (4) for the boundary layer asymptote are in good order of magnitude agreement with each other. Figure 1 shows the transport ratio $T_r$ versus $Re_{Lc}$. Included in Fig. 1 is the duct function $T_r$ with the Dittus-Boelter transport regime, the laminar boundary layer range and the transition region. The function is transformed in comparison to the NBS plot reported by Nilles et al.[8]. The inset of Fig. 1 presents the permeability ratio of screen stacks[7].

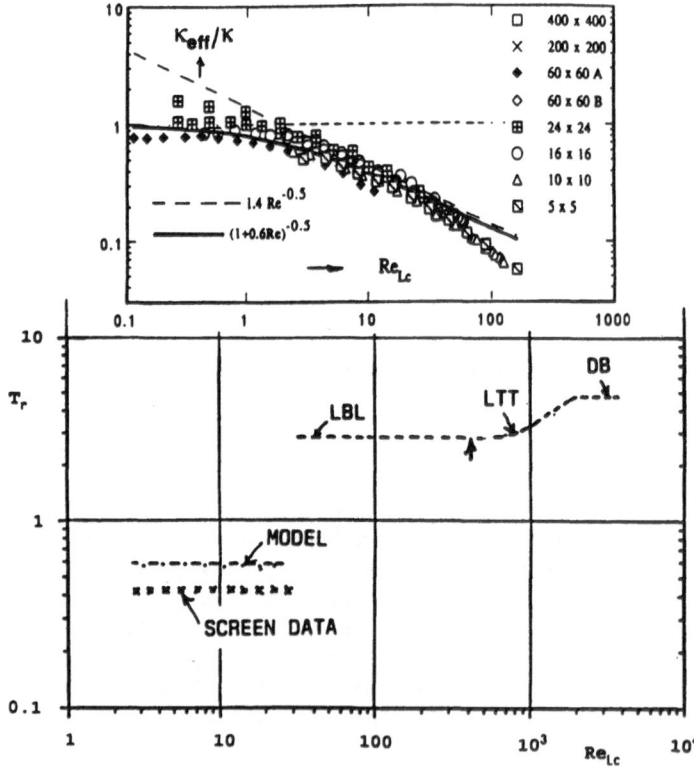

**Figure 1.** Transport ratio $[T_r]$ = as a function of the Darcy length-based Reynolds number
-.-.-.-. model prediction:                    xxxxxx data compilation of L. S. Wong et al.[7]
          duct function based on Nilles et al[8].    [DB] = turbulent Dittus-Boelter range,
[LTT] = laminar-to-turbulent flow transition,  [LBL] = laminar boundary layer-dominated flow regime.
                              Inset: effective permeability ratio.

## REGENERATOR SIZING

It has been shown by Miyabe et al.[9] that there is a relatively small influence of porosity on Darcy permeability for consistently packed screen stacks. We have calculated transport rates using this data set[9] for mesh numbers 100 to 400 in order to select the screens. The base lengths of the unit cells extend from 254 to 64 micron. The temperature (T) distribution has been based on the postulate of minimum energy expenditure rates for the compensation of the axial regenerator heat leak. Fig. 2 displays pressure drops ($\Delta P$). Fig. 3 shows the average heat transfer rates. We have selected 180 mesh for the upper T-range stack and 250 mesh for the lower T-range stack.

## PULSE TUBE SYSTEMS

Three different designs have been considered prior to construction: 1. a linear setup, e.g.[10]; 2. a U-geometry cooler; 3. a linear pulse tube. Version 3 has been chosen for the present runs.

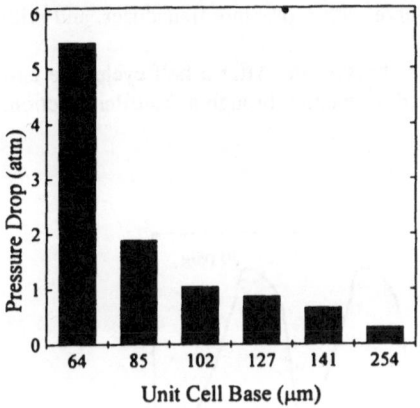

**Figure 2.** Pressure drop of screen stacks for different base length: mesh # 100-400.

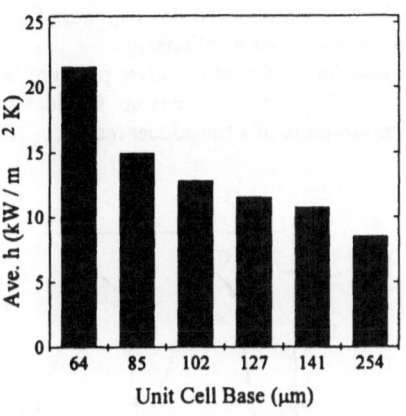

**Figure 3.** Average heat transfer of screen stacks for different base length: mesh # 100-400.

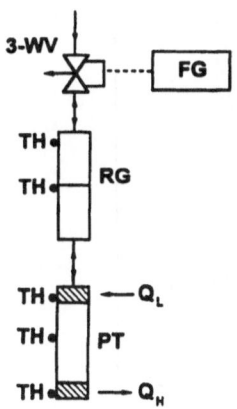

**Figure 4.** Basic Pulse Tube[BPT], schematically 3-WV three-way valve; F filter:
FG Frequency generator system;
PT pulse tube section; RG regenerator;
TH thermometers

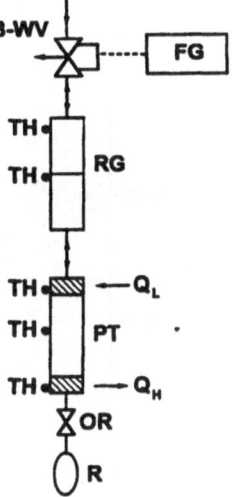

**Figure 5.** Orifice Pulse Tube, schematically, OR Orifice; R Reservoir;
other designations as in Figure 4.

The pulse tube proper and the regenerator section have the same diameter (I.D. 2.46 cm, wall 0.041 cm; stainless steel 304). Figures 4 and 5 show main components of the basic pulse tube (BPT) and the orifice pulse tube (OPT). In the initial run the compressed air line available in the building was used in the range from 30 to 50 psig [3.1 to 4.4 bar absolute].The gas enters a filter, a 3-way valve and the pulse tube sections. The dc-actuated solenoid of the valve is controlled by a frequency function generator (Wavetek Model 130 function generator).

The gas passes through the valve during its "open" position. It enters the regenerator section, a final distribution plate, and subsequently the cold section. The latter is equipped with electrical heater windings for refrigeration load simulation. Finally, a heat rejection section terminates the BPT. Either a blind flange has been used or a short tube section (volume 8 cc). leading to a needle valve [Nupro "L" series metering valve: maximum flow coefficient 0.15; 3.25 mm orifice (maximum); internal volume 570 mm$^3$].

In the OPT, the needle valve represents the orifice. Several reservoir volumes have been employed up to 34 Liters. Thermometers measure mean temperatures on the pulse tube wall, and pressure has been monitored downstream of the 3-way valve with a pressure transducer, and with Bourdon gages at other locations.

After opening of the inlet valve, pressure builds up in the system. After a half-cycle, the exit section of the 3-way valve opens up. Cooldown is initiated. Gas exits through a "muffler" section. Fig. 6 is an example of a transducer record.

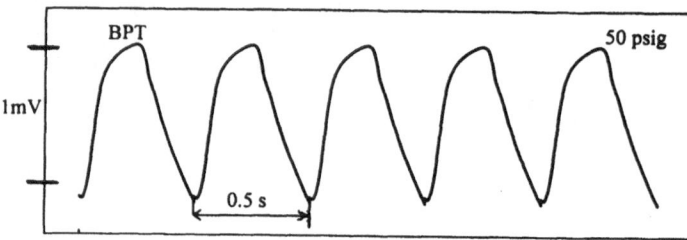

**Figure 6.** Pressure transducer record:

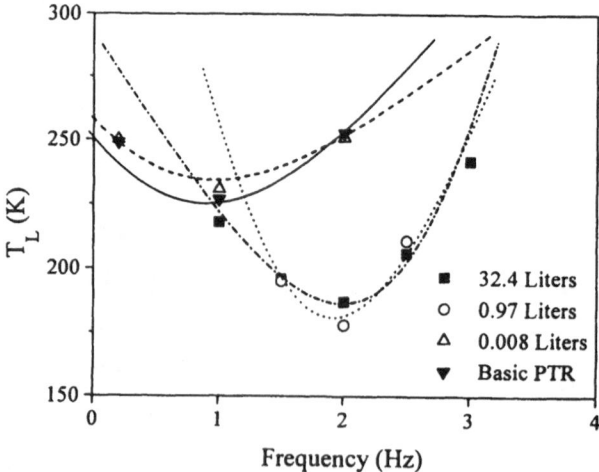

**Figure 7.** Temperature (voltage signal) of cold section after cool down as a function of frequency: Parameter: volume of reservoir for OPT

## FREQUENCY RESULTS

Steady temperatures are obtained after a cooldown on the order of magnitude of 30 minutes. During the runs, the valve position is kept constant for each frequency. Fig. 7 is a plot of the minimum temperatures $T_{min}$, achieved in the cold section, versus frequency. A clear T-minimum is seen at different locations for the two setups. The BFT has a low frequency optimum $T_{min}$ near 1 Hz. The OPT has an optimum frequency near 2 Hz with a minimum temperature near 200 K. It is distinctly lower than the $T_{min}$ value of the BPT.

In figure 8 the OPT frequency (f) is plotted versus the reciprocal square root of the reservoir volume ($V_r$). This function is rather flat indicating no f-change within data resolution. Figure 9 shows the BPT frequency versus the reciprocal square root of pulse tube volume (blind flange case and closed orifice valve respectively). Again there is little frequency variation within data uncertainty.

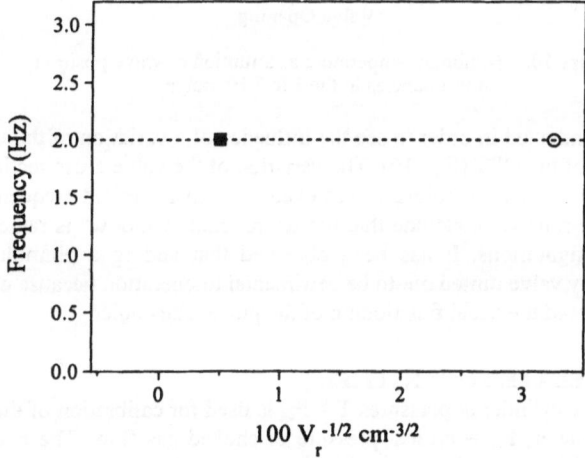

**Figure 8.** Frequency at minimum temperature: OPT data versus the

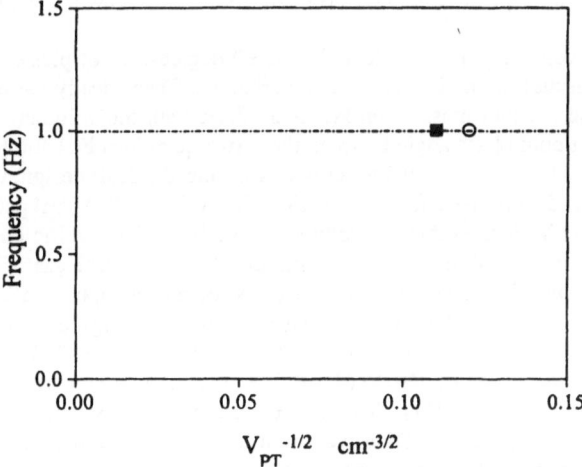

**Figure 9.** Frequency at minimum temperature: BPT results versus reciprocal square root of the pulse tube section volume ($V_{pt}$).

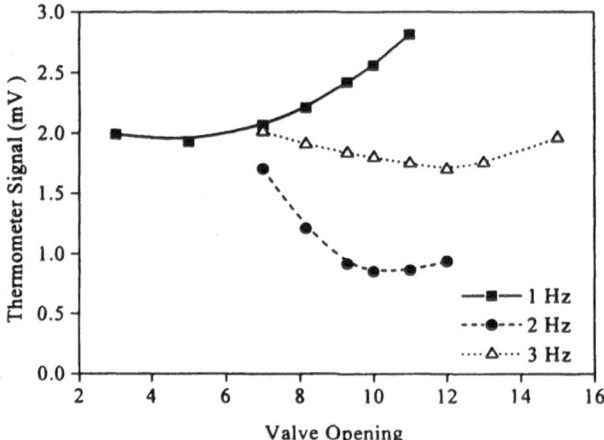

**Figure 10.**   Minimun temperature as a funtion of valve position
at frequencies in the 1 to 3 Hz range

Additional runs were conducted in order to see the influence of a variation of the valve position on minimum temperatures of the OPT (Fig. 10). The variation of the valve cross section is in a narrow range near the optimum of $T_{min}$. Cooling is achieved only in a limited frequency/cross section range. From all of these runs we conclude that the operational "window" is rather narrow for the present pulse tube configurations. It has been observed that adding a plumbing section for a transducer near the 3-way valve turned out to be detrimental to operation because of heating effects. Careful rearranging restored the usual functioning of the pulse tube cooler.

## MASS FLOW AND RELATED QUANTITIES

Outflow of gas from a cylinder at pressures $T > P_{cg}$ is used for calibration of the muffler section which is used as flow meter; $P_{cg}$ = critical pressure of choked gas flow. The mass flow rate is a linear function of P in this range of constant line impedance. Exponential decay as a function of time is observed for short times. Fig. 11 is an example. Upon noticeable cooling after a finite time (t), we see departures from the function $P = P_o$ -const t. The departures indicate an enhanced mass flow rate compared to the initial rate of P-drop $[(dP/dt)_{t = 0}]$. The order of magnitude [OOM] of the input power is assessed: we separate the mass flow rate [dm/dt] = OOM 0.2 g/s from the specific input power [dW/dt].

First an ideal volume [V] is considered. It is 90 degrees out of phase with the pressure wave. We restrict the discussion to sinusoidal time variations. Then ideally we have a maximum work input per cycle and per unit mass when P(t) is in phase with the volumetric rate [dV/dt]. A finite (experimental) optimum phase angle between these two quantities P(t) and [dV/dt] leads to a reduction factor $[f_r < 1]$. The factor is equal to unity in a balanced (ideal) design. Our result for the ideal case is written as product of the effective quantities $[P_e, V_e]$, i.e. $(P_e\ V_e/m)$ per unit mass. $P_e$ is the effective pressure and $V_e$ is the effective volume. Multiplying the specific power by the mass flow rate (dm/dt) we arrive at $dW/dt = P_e\ V_e = (dm/dt)\ R_i\ T_e$ for ideal gas; ($R_i$ individual gas constant 8.314/$M_i$; $M_i$ molecular mass). The related OOM of power input is dW/dt = [OOM 0.2 g/s] (p.287) x $300/(2)^{1/2}$ = OOM 12 Watts. For the data shown in figures 6 and 10 the cold section Carnot coefficient of performance is $COP_c$ = OOM 2. The COP value of heater simulation runs reached the OOM of 1W/12W = 0.083. Thus the Carnot fraction [CF] attains the OOM of 0.04. We scale up to a "first design" pressure and density which are significantly above the present range of operation. Then we attain a Carnot fraction of 0.1. This value is consistent with conventiona single-stage refrigerators.

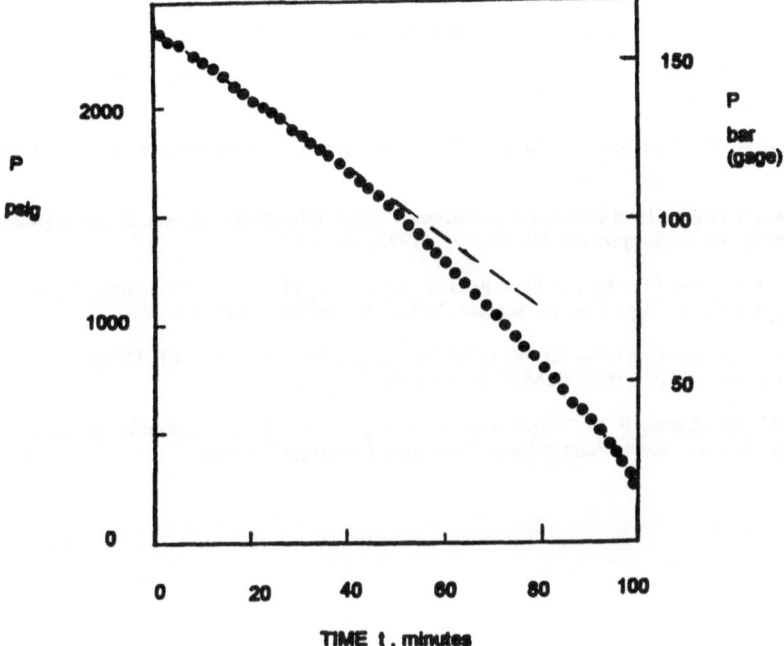

**Figure 11.** Outflow for mass flow rate determination: Pressure as a funtion of time.

## CONCLUDING REMARKS

The present pulse tube coolers tests are encouraging for applications with moderate cost specifications. It is necessary to operate at the optimum frequency and with matching components, subject to the specifications imposed. The optimum frequencies of 2 Hz [OPT] and 1 Hz [BPT] indicate significant sensitivity to non-linear thermo-acoustic conditions influenced by valve details, regenerator stack selection and component matching.

## ACKNOWLEDGMENTS

This work has been supported in part by Hughes Aircraft Co. We are indebted to Dr. Robert Schaefer and his collaborator team for interest and encouragement. N. S. Myung, Steve Sapida and Bryan Dransart have received Hughes Scholarship support in the course of the work. We are thankful for this support. Visiting Scholar Dr. Frithjof Mastrup has been instrumental in the pulse tube startup through his enthusiastic presentations and discussions. His inputs are gratefully acknowledged. Richard Van Affelen helped with screen SEM pictures. S. Grubwieser and collaborating workshop staff were very supportive in the course of the pulse tube startup. We thank our industrial supporters and contributing SRP students.

## REFERENCES

1. Radebaugh, R., "Advances in Cryocoolers", ICEC16, Kitakyushu, May 1996, paper PL6.

2. Zhu, S. W., Kakimi, Y., Fujioka, K. and Matsubara Y., "Active Buffer Pulse-Tube Refrigerator", ICEC16, Kitakyushu, May 1996, paper OB6-5.

3. Thummes G., Bender, S. and Heiden C.,"Nitrogen-Pecooled Multi-Stage Pulse Tube Refrigerator", ICEC16, Kitakyushu, May 1996, paper OB6-4.

4. Yoshida S., Ravikumar, K. V. and Frederking, T. H. K., "Thermoacoustic Taconis Oscillations in Liquid Helium-4 Level Sensors", ICEC16, Kitakyushu, May 1996, paper OC2-4.

5. Pinsky C. et al., "Evaluation of the Thermodynamic Performance of Pulse Tubes", *Adv. Cryog. Eng.* 41 (1996) pp 1365-1372.

6. Yoshida, S. et al., "Friction Factor of Stacks of Perforated Regenerator Plates", *Cryocoolers 8*, Plenum Press, New York (1995) pp 259-268.

7. Luna, J. et al., "Compact Heat Exchanger Phenomena", Proc. 8th Intersoc. Cryog. Sympos. [ICS-8, Houston 1991], *AIChE Sympos. Ser* 89, No. 294, (1993) pp 80-88.

8. Nilles M. J. et al.,"Heat Transfer and Flow Friction in Perforated Plate Heat Exchangers", *Aerospace Heat Exchanger Technology*, Elsevier Science Public., Amsterdam (1993) pp. 293-313.

9. Miyabe et al.,"An Approach to the Design of Stirling Engine Regenerator Matrix Using Packs of Wire Gauzes", Proc. 17th IECEC, *IEEE Pub82*CH1789-7 (1982) pp 1839-

10. Soloski , S.C. and Mastrup F. N., "Experimental Investigation of a Linear Orifice Pulse Tube Refrigerator", *Cryocoolers 8*, Plenum Press, New York (1995) pp 321-328.

# Reversible Cycle Piston Pulse Tube Cryocooler

Alfred L. Johnson

Electro Thermo Associates
Deer Harbor, Washington 98243

## ABSTRACT

The replacement of the orifice and reservoir elements of a fixed orifice pulse tube cryogenic refrigerator with an expander piston results in a reversible cycle pulse tube machine. At the 7th International Cryocooler Conference, Ishizaki and Ishizaki presented the experimental performance of a machine where the phase relationship between the compressor and expander pistons was maintained by a kinematic drive. An alternate implementation can be based on flexure bearing supported reciprocating linear drive engines wherein the phase shift between the compressor and the expander pistons is maintained by either appropriate plumbing of several dual piston engines across a like number of regenerator/pulse tube elements, or by separate compressor and expander engines with appropriate phase control electronics.

A generalized lumped parameter model of thermally regenerated engines has been developed and programmed to predict the comparative thermodynamic performance of Alpha type Stirling cycle, fixed and modulating orifice pulse tube, and piston-pulse tube machines. Subject to the modeling approximations, the analysis predicted the following: Alpha Stirling Cycle - 35%; Piston-Pulse Tube - 29%; Modulating Orifice Pulse Tube - 19.5%; Fixed Orifice Pulse Tube - 16.5% of Carnot for a 5 to 10 watt at 60K cryocooler with a 300K heat rejection temperature.

The basis of the analytical model and computer program, the results of the computer analysis, a preliminary design of the flexure supported dual piston integrated compressor-displacer engine, and a schematic of the fluid circuit required for phase shift control will be presented. The prime cryocooler application of this technology is in multistage applications.

A major attraction of piston-pulse tube machines is that all moving parts are operating at effectively the heat rejection temperature, which simplifies the design and fabrication of cryogenic refrigerators, and which allows increased temperature ratios (with resulting higher Carnot limits) for power producing integrated alternator-heat engines.

## INTRODUCTION

The pulse tube refrigerator was invented by Professor W.E.Gifford in 1961[1]. The research from which the pulse tube refrigerator concept evolved was reportedly initiated in an effort to determine the cause of an unanticipated temperature difference which occurred on some blanked-off gas distribution tubing connected to a reciprocating compressor.

By 1962, Gifford and R.C.Longsworth had experimentally demonstrated a pulse tube with a 205K cold temperature rejecting to a 297K sink[2]. In 1964, Gifford and Longsworth reported obtaining 183K with a single stage pulse tube refrigerator and 85K with a three stage unit[3], and by 1965 Longsworth had obtained 123K with an optimized single stage pulse tube refrigerator[4]

The designation "pulse tube" was originally coined to denote the mode of pneumatically exciting the refrigerator in a cyclic manner by periodically pressurizing the hot end of the regenerator from a high pressure gas source followed one half cycle latter by exhausting the hot end of the regenerator to a low pressure gas sink refrigeration The theory proposed to explain the phenomena demonstrated in the Gifford/Longsworth type pulse tube refrigerator introduced the concept of "surface heat pumping". Surface heat pumping is attributed to the interaction between viscous fluid interaction along the wall of the pulse tube, energy exchange within the fluid in the pulse tube, and heat exchange between the surface of the pulse tube and the fluid contained therein due to periodic pressure changes introduced at the open end of the pulse tube segment. Further research on variations to the Gifford/Longsworth pulse tube configuration were reported by Ackerman in 1970[5] and Lechner in 1971[6]; however due to the low thermodynamic efficiency of the device, there was little further interest until the 1984 report on the research by Mikulin, Tarasov, and Shkrebyonock of the Moscow High Technical School which attained 103K with a single stage unit[7]. This variation incorporated an orifice and receiver at the ambient temperature end of the pulse tube, which enhanced the enthalpy flow from the cold to the hot end thereby increasing the thermodynamic efficiency of the refrigerator. In 1986, Radebaugh, et al, reported the attainment of 60K with a similar single stage pulse tube refrigerator configuration[8].

The results of a performance comparison between a basic configuration (Gifford-Longsworth), an orifice/receiver end configuration (Mikulin-Tarasov-Shkrebyonock), and a moving plug end configuration, based on using a single stage common test rig, was published by Matsubara and Miyake in 1988[9]. They reported cold end temperatures of 163K, 78K, and 73K for these three variations respectively. Zhou, et al, reported in 1988 on a two stage, dual compressor pulse tube experimental unit which attained a temperature of 31K[10]. In 1989, Burt, et al, reported on obtaining 53K with a single stage and 26K with a two stage, single compressor orifice/receiver pulse tube cryogenic refrigerator[11]. Subsequent work incorporating multiple by-pass orifices (Zhou, et al [12]), a modulating orifice Pulse Tube (Peckham, et al[13]), and a kinematically driven phase shift controlled ambient temperature expander piston (Ishizaki and Ishizaki[14]) continues to demonstrate the improvements that can be made in this technology. Figure 1 graphically summarizes the evolution of this technology.

## PISTON PULSE TUBE CRYOGENIC REFRIGERATOR CONCEPT

In considering the evolution of the Pulse Tube cryogenic refrigerator, it is interesting to note that the work of Ishizaki and Ishizaki has taken the concept almost full circle back to the basic Stirling cycle. The warm displacer piston and pulse tube are, in effect, a "soft" displacer-piston replacing the "hard" displacer piston of the Alpha class Stirling cycle. This implementation of the Pulse Tube thermodynamic cycle is reversible, the power produced at the expansion end being recoverable. Thus it should also be possible to build a heat-to-power engine using the "soft" displacer-piston for the expansion engine power output. Viewed from this perspective, there are some immediate advantages which can be derived. First, the existing Stirling power cycle analysis programs are applicable. Second, the configurations which proved to be superior for Stirling power units should be examined as potential configurations for the Piston Pulse Tube Cryogenic Refrigerator.

A major problem with the configuration used by Ishizaki and Ishizaki was the need to use a kinematic mechanism to coordinate the motion of the power and displacer pistons and to recover the power produced by the displacer piston. Experience with spacecraft class cryogenic refrigerators has shown the problems inherent with kinematic drives. A design which maintains the long life, high reliability mechanical advantages of the flexure bearing supported linear drive system while incorporating the thermal and thermodynamic advantages of the warm

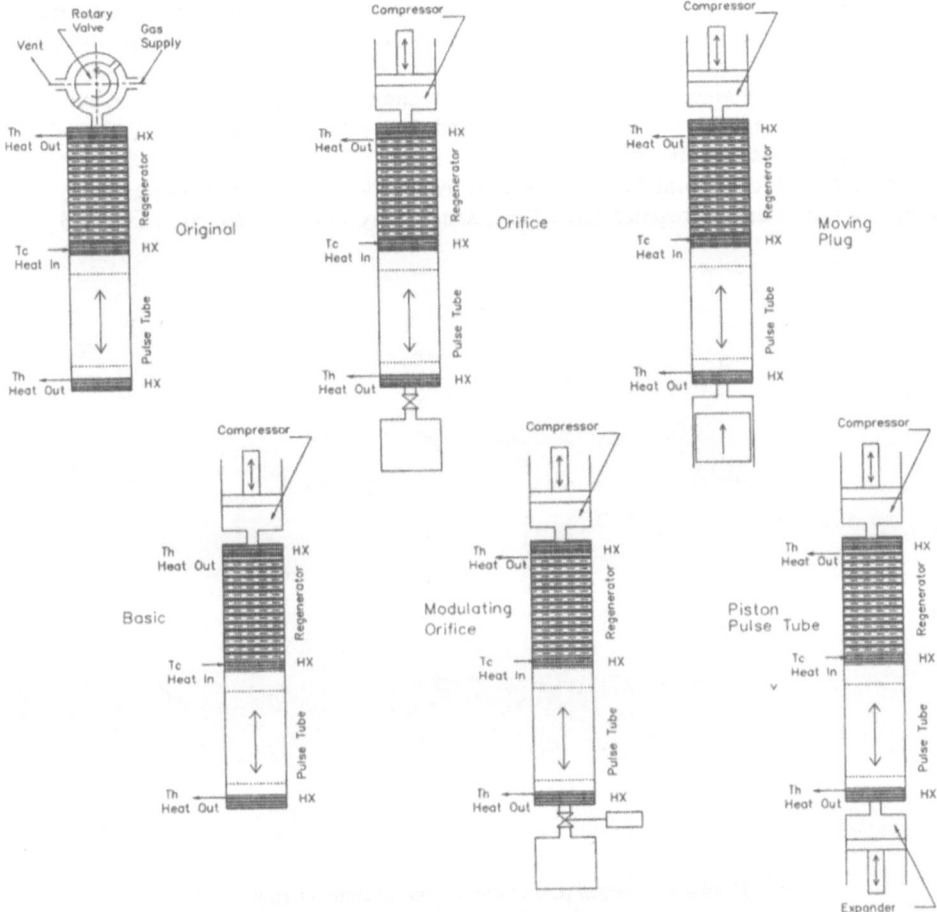

**Figure 1** - Evolution of the Pulse Tube Cryogenic Refrigerator

displacer piston/pulse tube configuration is needed. There are two configurations which can provide these desirable characteristics:

I. A flexure bearing supported, linear drive, clearance sealed pressure wave generator (compressor) coupled via a compressor/hot end heat exchanger/regenerator/cold end heat exchanger/pulse tube/expander/hot end heat exchanger assembly (sans orifice), which is henceforth called a "Pulse Tube Assembly", with a flexure bearing supported, linear drive, clearance sealed expander. The compressor and expander are driven at the same frequency, with the expander electronically controlled to phase lead the compressor by 30° to 120° depending on the swept volume ratio of the compressor/expander pistons, the system dead volume, the pulse tube volume, and the regenerator/heat exchanger thermal and fluid dynamic characteristics.

II. A set of flexure bearing supported, linear drive, dual piston/cylinder clearance sealed compressor/expanders appropriately plumbed to an equal number of pulse tube assemblies. The phase relationship between the compression and expansion cycles across any given pulse tube assembly is fixed by the specific fluid circuit plumbing.

Configuration I is the appropriate choice for fundamental experimental work, however for multi-stage applications, configuration II is more appropriate due to the combined advantages of intrinsic vibration balance and the reduction in the number of moving parts.

To obtain the capability of fluid coupled thermodynamic phasing required to realize multi-cycle Piston Pulse Tube Cryocooler configuration, it is necessary to develop a particular integrated compressor/expander thermodynamic engine. Figure 2 shows such a piston pulse tube thermodynamic engine. This particular layout shows a moving sleeve version. It is apparent that a moving piston version and a mixed (one end a moving piston, the other a moving sleeve) are possible alternatives.

The simplest configuration which can provide adequate phase shifting using fluid circuiting is a triplex engine/regenerator/pulse tube assembly such as depicted in Figure 3 .

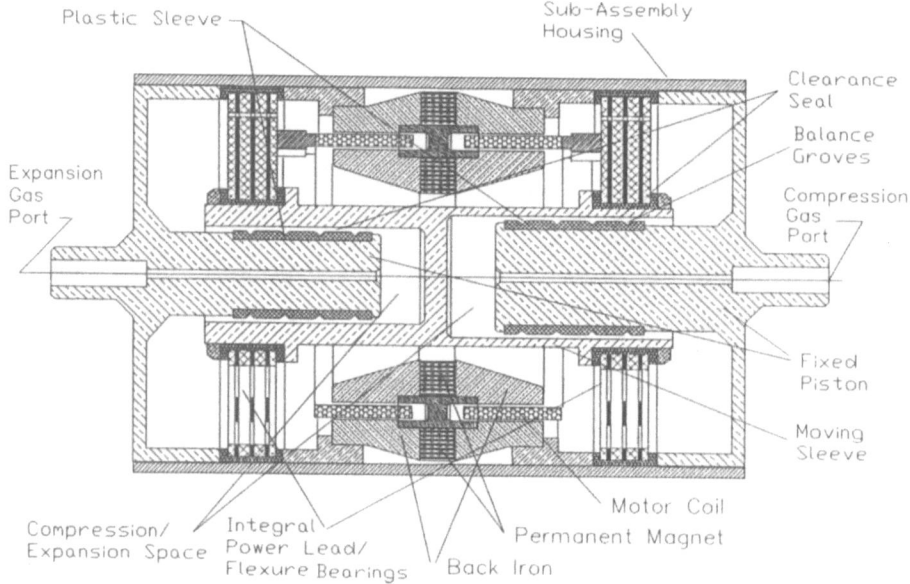

**Figure 2** - Piston pulse tube thermodynamic engine

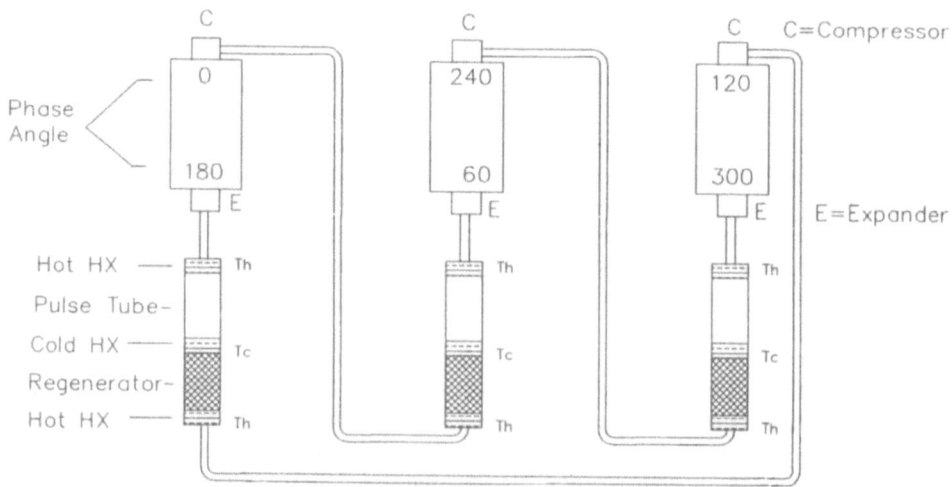

**Figure 3** - Triplex piston pulse tube cryogenic refrigerator

The triplex system offers some potential intrinsic advantages:

1. Three separate heat sink temperatures can be accommodated

2. A single stage and a two stage capability can be jointly provided, which could be useful in applications such as sensor cooling wherein it is necessary to cool both the optics at a moderate cryogenic temperature and the detector at a much colder temperature.

3. A three stage capability can be provided, with sink temperatures as low as 10K being a feasible goal.

These alternatives are graphically summarized in Figure 4.

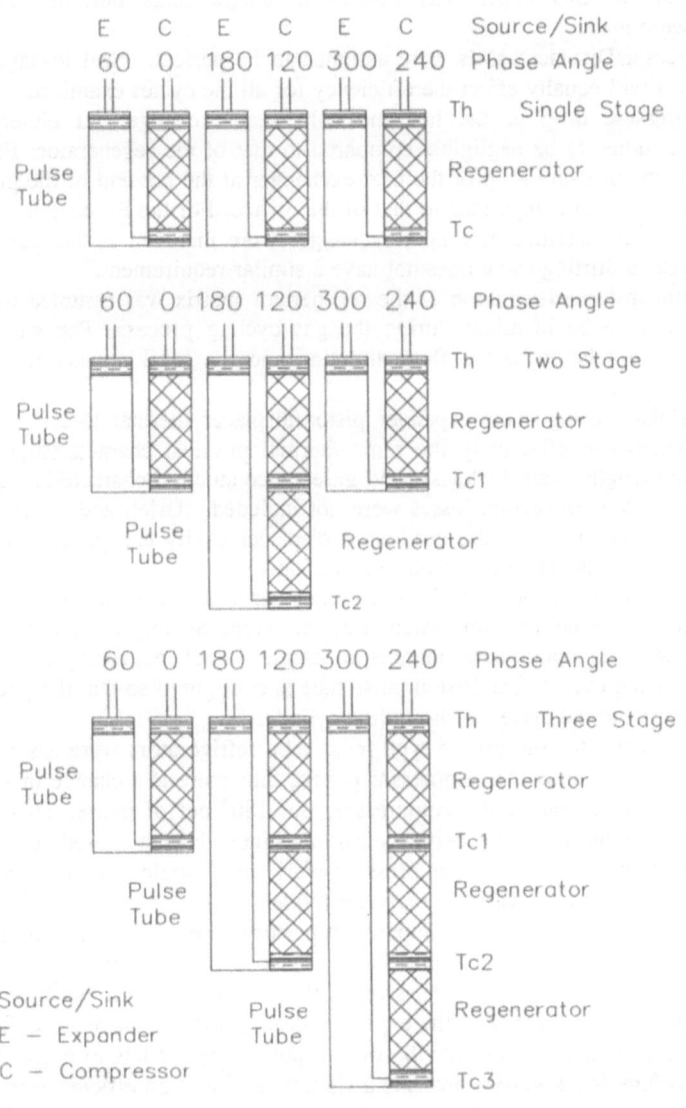

**Figure 4** - Triplex regenerator/piston pulse tube alternatives

By adding another Piston Pulse Tube engine and Pulse Tube Assembly, a four cycle version is obtained. It is apparent that if the engines are arranged it a cruciform pattern, it is possible to dynamically cancel the unbalanced forces. The fourth stage can fabricated to use a recuperator rather than a regenerator, which is advantageous when operating at temperatures below 20K. Figure 5 depicts this variation of the Reversible Piston Pulse Tube Cryocooler.

## COMPARATIVE CYCLE ANALYSIS

The ETA Regenerative Cycle Thermodynamic Analysis Program is a lumped parameter thermodynamic circuit performance prediction program which can analyze fixed and modulating orifice pulse tube designs, piston pulse tube designs, and Alpha configuration Stirling cycle cryogenic refrigerator designs. The code runs under a Microsoft Windows NT environment. The model is based on the teachings of Urieli and Berchowitz[15]. The version of the program used for this paper was limited to single stage designs. The following approximations were made:

1. Compressor/Displacer seals were assumed to be perfect. Seal leakage will reduce performance, but it will equally effect the efficiency for all the cycles examined

2. The pressure drop of the hot and cold heat exchangers at either end of the regenerator was assumed to be negligible compared to that of the regenerator. For the orifice pulse tube models the pressure drop of the heat exchanger at the hot end of the pulse tube was assumed to be negligible in comparison to that of the orifice. For the Piston Pulse Tube model, the pressure drop characteristics of a heat exchanger were included at the warm end of the pulse tube. The Alpha Stirling cycle does not have a similar requirement.

3. The temperature distribution of the regenerator matrix was assumed to run linearly from Th to Tc, and to be invariant during the gas cycling process. The gas temperatures cyclically changed in relation to the flow direction and the local regenerator heat transfer coefficients (screen by screen).

4. The Alpha Stirling cycle expander piston/displacer thermal losses were calculated using a model which had effectively the same thermal physical characteristics as the pulse tube, i.e. the same length, wall thickness, and gaseous conduction characteristics. Shuttle and displacer internal forced convection losses were not included. (Urieli and Berchowitz suggest that the forced convection within the oscillating displacer cavity can product a heat leak of approximately 10 times that of pure gaseous conduction.)

5. The work output of the cycle from the expander piston was subtracted from the work input to the cycle for the compressor piston. For the Alpha Stirling cycle configuration, this may be an unrealistic assumption since there is no simple way of recovering the expander work for linear drive Stirling cycles. The Piston pulse tube is configured so that the pneumatic work output directly returns to the cycle via the shuttle cylinder.

6. The reservoirs for the orifice type pulse tube refrigerators were assumed to be the engine cavity which surrounds the compressor piston, thus there is a change in volume of the reservoir that is equal to that of the compressor, but 180° out of phase. This results in an additional work input term for the orifice class machines. Had a closed, constant volume reservoir been used, the compressor compression work term would have been greater by the amount of work required by the variable reservoir model.

The model was used to conduct a comparative performance analysis for the four different cycles. For the Piston pulse tube and Alpha Stirling class cryocoolers, the phase shift term was varied between 30° and 120°, the expander swept volume was allowed to be different from that of the compressor swept volume, and the expander dead volume term reduced from about 20 times the compressor stroke volume for the orifice pulse tube models to a about 10% of the expander swept volume(s). Figure 6 presents a Carnot refrigeration efficiency mapping of the four cycles. Table 1 summarizes the physical characteristics assumed in this analysis, and presents the performance prediction at maximum Carnot efficiency for each of the cycles.

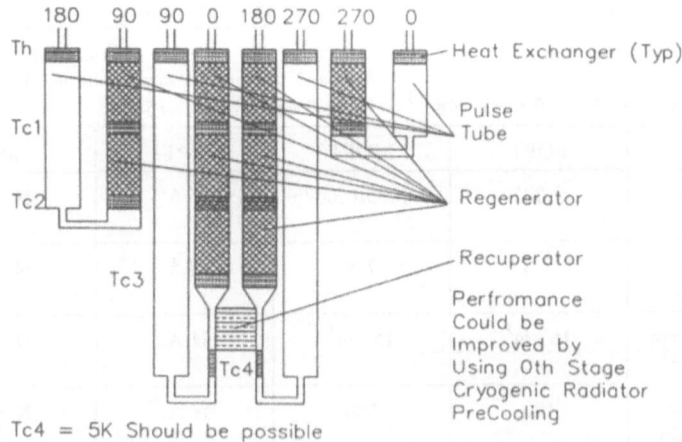

**Figure 5** - Four cycle/four stage piston pulse tube cryocooler

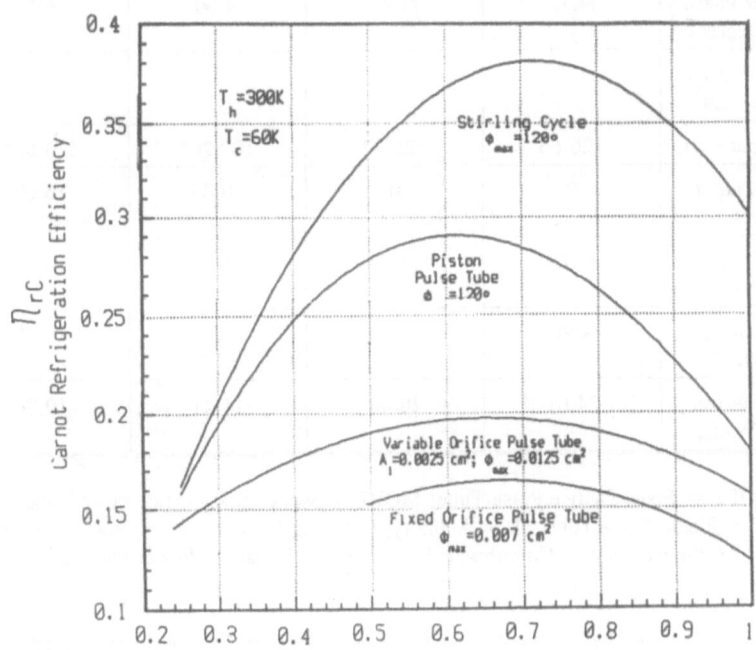

**Figure 6** - Carnot refrigeration efficiency versus phase shifting parameter

**Table 1** - Regenerative cycle thermodynamic analysis performance prediction

| Cryogenic Refrigerator Performance Prediction - $T_h$=300K; $T_c$=60K | | | | |
|---|---|---|---|---|
| **Regenerator** | | | **Compressor** | |
| Area $4 cm^2$   Length 20 cm   Spacer 25%   Matrix 250 mesh #48 | | | Dead Volume $1.5 cm^3$   Swept Volume $15 cm^3$   Frequency 30 cps | |
| Item | FOPT | MOPT | PPT | aS |
| Orifice Area - $cm^2$ | 0.005 | 0.0025/0.0075 | N.A. | N.A. |
| Pulse Tube Volume - $cm^3$ | 7.5 | 7.5 | 7.5 | N.A. |
| Reservoir Swept Volume - $cm^3$ | $15 cm^3$ | $15 cm^3$ | N.A. | N.A. |
| Reservoir Phase Shift - Degrees | 180 | 180 | N.A. | N.A. |
| Reservoir Dead Volume - $cm^3$ | 300 | 300 | N.A. | N.A. |
| Expander Swept Volume - $cm^3$ | N.A. | N.A. | 2.25 | 2.25 |
| Expander Phase Shift - Degrees | N.A. | N.A. | +60 | +90 |
| Expander Dead Volume - $cm^3$ | N.A. | N.A. | 0.225 | 0.225 |
| Power Input watts | 120.66 | 125.17 | 100.26* | 115.83* |
| Power Output Watts | 0 | 0 | 10.35 | 16.09 |
| Gross Cooling Watts | 11.68 | 12.77 | 11.95 | 16.09 |
| Net Cooling Watts | 4.90 | 6.13 | 6.96 | 10.79 |
| Specific Power $Watts_p/Watts_c$ | 24.61 | 20.40 | 14.41 | 10.74 |

FOPT = Fixed Orifice Pulse Tube; MOPT = Modulating Orifice Pulse Tube;
PPT = Piston Pulse Tube; aS = Alpha type Stirling Cycle; N.A. = Not Applicable
* Power Input = Compressor Power Input - Expander Power Output

Arguably, the losses neglected in the Piston Pulse Tube modeling are smaller than the losses neglected in the Alpha type Stirling Cycle modeling, thus the Carnot Refrigeration efficiencies for real hardware should not exhibit as great a difference as is predicted by this model. If it is not feasible to recover the expansion energy in the Alpha Stirling configuration, the relative performance of the Piston Pulse Tube approach will be further enhanced.

The ETA Regenerative Cycle Thermodynamic Analysis program is being revised to include some of the elements which have been ignored or grossly approximated. The extension to include multi-stage analysis will follow, subject to the availability of additional funding.

## CLOSURE

The modular, fluid circuit defined, Reversible Cycle Piston-Pulse Tube Cryocooler concept can be developed to meet a great number of different cryogenic cooling needs with a relatively small number of compressor/expander thermodynamic engine designs. Assuming that these generic engines are developed as stock items, this modular feature can introduce significant cost savings for applications with low procurement quantities. The development of individually sized pulse tube assemblies is a low cost, low risk technology development issue.

For multi-stage, low temperature (> 15K), the 4 cycle/4 stage configuration may prove to be the most reliable, cost effective approach for small quantity, low heat load applications.

The thermodynamic efficiency of this technology is high enough to justify its selection over the simpler orifice pulse tube designs for moderate to high heat load applications. This technology also appears to offer significant advantages over Stirling cycle alternatives when multi-stage cooling is required.

The application of this technology to the field of heat-to-electrical energy conversion appears promising. Of particular interest is the ability to operate over high temperature ratios since the moving parts are all at effectively ambient temperature, thus producing a high thermal efficiency.

It would appear that this technology could be adapted to function as a simulated Vuilleumier cycle cryocooler.

## ACKNOWLEDGMENT

This research was funded by NASA/JPL, Pasadena, California 91109 as an SBIR program.[16]. The program technical monitor was Dr. Jose Rodriguez.

## REFERENCES

1. Gifford,W.E.;"Pulse Tube Method of Refrigeration and Apparatus Therefore"; U.S.Patent 3,237,421

2. Gifford,W.E.,Longsworth,R.C.; "Pulse Tube Refrigeration"; Journal of Engineering and Industry, Transactions of the ASME, Series B; August 1964

3. Gifford,W.E.,Longsworth,R.C.; "Pulse Tube Refrigeration Progress"; Advances in Cryogenic Engineering, Volume 10; Plenum Press, New York; 1965

4. Longsworth,R.C.; "An Analytical and Experimental Investigation of Pulse Tube Refrigeration"; M.E. PhD. Thesis; Syracuse University; 1966

5. Ackerman,R.A.; "Investigation of Gifford-McMahon Cycle and Pulse Tube Cryogenic Refrigerators"; Technical Report ECOM-3245; US Army Electronics Command; Fort Monmouth, New Jersey; March 1970.

6. Lechner,R.A.; "Investigation of Regenerators and Pulse Tube Cryogenic Coolers"; Technical Report ECOM-3409; US Army Electronics Command; Fort Monmouth, New Jersey; May 1971

7. Mikulin,E.I.,Tarasov,A.A.,Shkrebyonock,M.P.; "Low-Temperature Expansion Pulse Tubes";Advances in Cryogenic Engineering; Volume 29; 1984

8. Radebaugh,R.,Zimmerman,J.,Smith,D.R.,Louie,B.; "A Comparison of Three Types of Pulse Tube Refrigerators: New Methods for Reaching 60K"; Advances in Cryogenic Engineering, Volume 31; Plenum Press, New York, 1986

9.      Matsubara,Y.,Miyake,A.; "Alternate Methods of the Orifice Pulse Tube Refrigerator"; Proceedings of the 5th International Cryocooler Conference; Monterey California, August 1988

10.     Zhou,Y,Wenxiu,Z.,Jingtao,L.; "Two-Stage Pulse Tube Refrigerator"; Proceedings of the 5th International Cryocooler Conference; Monterey California, August 1988

11.     Burt,W.,Chan,C.,Tward,E.; "Pulse Tube Refrigerator Performance Investigations"; 1989 Space Cryogenics Workshop; CIT; Pasadena, California, August 1989

12.     Zhou,Y.,Han,Y.H.; "Pulse Tube Refrigerator Research", Proceedings of the 7th International Cryocooler Conference; Sante Fe, New Mexico, November 1992

13.     Peckham,V.A., et al; "Novel Pulse Tube  Cryogenic Refrigerator" Final Report, Phase I SDIO SBIR  90-003; November 1990

14.     Ishizaki,Y.,Ishizaki,E.;. "Experimental Performance of Modified Pulse Tube Refrigerator Below 80K Down to 23K"; Proceedings of the 7th International Cryocooler Conference; Sante Fe, New Mexico, November 1992

15.     Urieli,I., Berchowitz,D.M.; "Stirling Cycle Engine Analysis" Adam Hilger Ltd., Bristol, 1984.

16.     Johnson, A.L.; "Reversible Cycle Pulse Tube Cryogenic Refrigerator", NASA/JPL Phase I SBIR Final Report, 6 June 1995

# Isothermal Model of a Warm Expander Pulse Tube

**M. M. Peters, G. D. Peskett and M. C. Brito**

University of Oxford, Department of Physics
Clarendon Laboratory, Oxford OX1 3PU, UK

## ABSTRACT

This paper describes a numerical model of a novel type of cryocooler for long life applications, the Warm Expander Pulse Tube (WEPT). The model was used as a design tool for the initial sizing of a breadboard miniature cooler.

The model is of the second order, based on an isothermal analysis with losses considered separately. Although it is more complex than the isothermal model of a Stirling refrigerator, it is much simpler than a full nodal analysis.

The model shows that the power dissipated at the warm end of the pulse tube via the warm expander is equal to the gross heat lift. In principle this power may be partially recovered as electrical energy. The action of the warm expander also produces a high pressure ratio, which enables a miniature machine to maintain a relatively high work rate.

An expression is developed for the cyclic pressure drop in the regenerator. This incorporates empirically determined parameters of the regenerator matrix to predict both pressure drop and phase shift. Thermal conduction measurements of the regenerator matrix under different loading conditions are also incorporated into the model.

The model demonstrates that the WEPT can achieve an overall efficiency comparable to or exceeding that of the Stirling cycle refrigerator, with the advantages of no moving components at cryogenic temperatures and the elimination of shuttle heat transfer.

## INTRODUCTION

The warm expander pulse tube (WEPT) is a novel type of cryocooler for long life applications. It is a development of the 'moving plug' or 'hot piston' pulse tube first described by Matsubara[1]. In the standard orifice pulse tube, the function of the orifice and reservoir is to optimise the phase shift between the pressure and mass flow oscillations in order to maximise the heat flow down the tube. The same effect may be achieved by replacing the orifice and reservoir with a small secondary piston, originally referred to as the moving plug or hot piston. This device closely resembles the two piston Stirling machine in layout, with the addition of a pulse tube between the cold end of the fixed regenerator and the secondary piston, the 'warm expander' (see **Figure 1**).

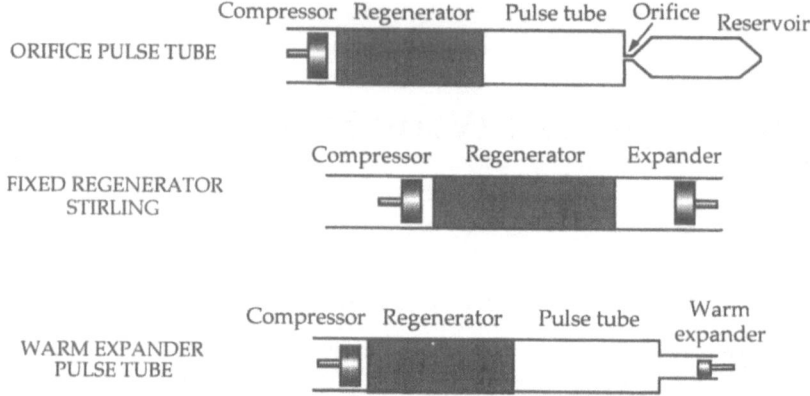

**Figure 1.** Schematic diagram comparing the layout of the orifice pule tube, the fixed regenerator Stirling machine and the WEPT.

The hot piston concept has to date not been developed into a machine suitable for long life cryogenic applications, despite the design's many attractive features. In the WEPT described here, a clearance seal is provided by a spiral arm flexure bearing to give the device an Oxford heritage for long life capability.

## ISOTHERMAL MODEL

A numerical model of the WEPT was developed with the following objectives:

- To verify the WEPT concept.
- To determine the performance of the WEPT compared to the Stirling cycle cooler and the orifice pulse tube.
- To establish the optimum geometry and operating conditions for a miniature device.

A novel modelling approach was developed to produce an efficient closed form solution for the thermodynamic operation of the WEPT. The core of the model is based on an ideal isothermal analysis, and loss processes are considered separately. The machine is paramaterized as six connected gas volumes, as shown in **Figure 2.**

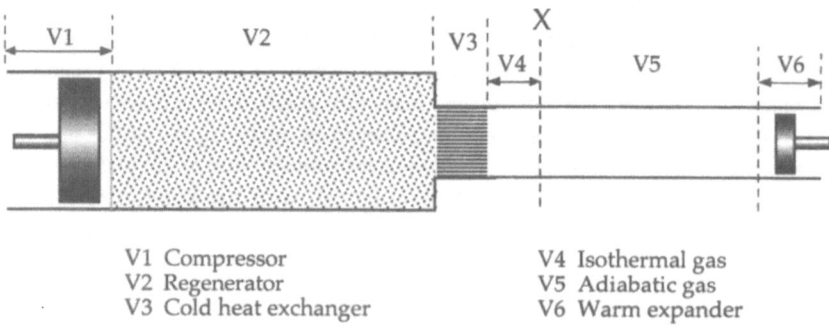

V1 Compressor                V4 Isothermal gas
V2 Regenerator                V5 Adiabatic gas
V3 Cold heat exchanger      V6 Warm expander

**Figure 2.** Model paramaterization of the WEPT

All volumes are variable, with the exception of the regenerator, $V_2$, and the cold heat exchanger, $V_3$. The pulse tube itself is divided into two regions, an isothermal region, V4, and an adiabatic region, $V_5$, separated by a boundary, X, across which there is no mass flow. The swept volume of the warm expander, $V_6$, may be regarded as an extension to the adiabatic region of the pulse tube.

The compressor volume, $V_1$, is assumed to be at a constant ambient temperature. The regenerator is initially assumed to be perfect, with no pressure drop or axial conduction. The operation of the model is as follows: The temperature of the cold heat exchanger, $V_3$, is set to the required refrigeration temperature. As the compressor generates the pressure cycle, the boundary, X, between the isothermal region, $V_4$, and the adiabatic region, $V_5$, oscillates back and forth along the pulse tube. The isothermal region, $V_4$, is defined to contain any gas which has passed into the cold heat exchanger at any point during the cycle. The minimum value of $V_4$ is therefore zero by definition. The governing equations for the model are given below (subscripts refer to the volume elements shown in **Figure 2**):

$$V_1 = V_{10} + V_{11} \sin(wt + \phi_1) \qquad \text{(compressor volume)} \qquad (1)$$

$$V_6 = V_{61} (1 + \sin(wt + \phi_1 + \phi_6)) \qquad \text{(warm expander volume)} \qquad (2)$$

$$V_{4(min)} = 0 \qquad \text{(isothermal tube volume)} \qquad (3)$$

$$V_5 = V_{tube} - V_4 + V_6 \qquad \text{(adiabatic tube volume)} \qquad (4)$$

$$T_1 = T_2 = T_0 \qquad \text{(ambient temperature)} \qquad (5)$$

$$T_3 = T_4 = T_c \qquad \text{(refrigeration temperature)} \qquad (6)$$

$$T_2 = (T_0 - T_c) / \ln(T_0/T_c) \qquad \text{(mean regenerator temperature)} \qquad (7)$$

$$M_1 = P_0 \Sigma_i V_i/T_i \ (i=1 \text{ to } 4) \qquad \text{(isothermal mass)} \qquad (8)$$

$$P = M_1 / \Sigma_i V_i/T_i \ (i=1 \text{ to } 4) \qquad \text{(system pressure)} \qquad (9)$$

$$PV_5^\gamma = \text{constant} \qquad \text{(adiabatic constant)} \qquad (10)$$

$$W_{in} = \int P \, dV_1 \qquad \text{(gross input power)} \qquad (11)$$

$$W_c = \int P \, dV_4 \qquad \text{(gross refrigeration power)} \qquad (12)$$

where $V_{10}$ = compressor rest volume $\qquad$ $V_{61}$ = expander peak-to-peak amplitude
$\qquad\qquad$ $V_{11}$ = compressor amplitude $\qquad\quad$ $\phi_6$ = phase angle between $V_1$ & $V_6$
$\qquad\qquad$ $\phi_1$ = phase angle between $V_1$ & $V_4$ $\quad$ $P_0$ = mean pressure

The numerical method involves solving for the motion of the boundary, X. This is achieved as follows: The starting point for the cycle, $t_0$, is chosen where $V_4=0$ (its minimum value by definition). It is assumed that the system pressure $P_s$ is known at this point (this can be adjusted once the solution has been obtained, as described below). If the warm expander were stationary, then $V_1$ and $V_4$ would be in antiphase, and $t_0$ would correspond to the compressor bottom-dead-centre position (i.e. $V_1(max)$). With the warm expander moving, however, a phase shift, $\phi_1$, is introduced between the motion of the compressor and the boundary X, so that $V_1$ is not known at $t_0$. To find $\phi_1$, an initial guess is made of $V_1$ at $t_0$, and a complete cycle is computed by evaluating equations (1) to (12). If the isothermal volume $V_4$ becomes negative at any point during the cycle (which by definition it cannot, as the boundary X would then be inside the heat exchanger $V_3$), or if $V_4$ does not return to zero at the end of the cycle, then a new starting value for $V_1$ is chosen. This procedure is repeated until a value of $V_1$ is found such that $V_4$ begins and ends the cycle exactly at zero, having at no point passed into the cold heat exchanger. At this stage the mean cycle pressure $P_0$ can be determined, which may be different from the initial starting value $P_s$. The starting pressure, $P_s$, may now be adjusted if required, and the whole procedure repeated until the desired mean pressure, $P_0$, is obtained.

Next we include a number of loss processes. The regenerator is the principal source of losses in a miniature cryocooler, and we consider three components: pressure drop; thermal conduction and regenerator inefficiency. The regenerator efficiency is modelled after an analysis by Radebaugh[2]. For the pressure drop and thermal conduction losses, none of the analyses in the literature were found to be adequate, and new expressions and measurements are presented.

## REGENERATOR PRESSURE DROP

We assume a relation between the pressure drop and the mass flow rate as follows:

$$\frac{\partial p}{\partial x} = -\frac{k \mu \dot{m}}{\rho A_f} \tag{13}$$

$p$ = pressure;                    $x$ = flow direction;             $\mu$ = gas viscosity;
$\dot{m}$ = mass flow rate;        $\rho$ = gas density;             $A_f$ = cross sectional flow area.
$k$ = constant determined by regenerator geometry with (dimensions: $area^{-1}$).

For a perfect gas the density is given by       $\rho = \frac{pM}{RT}$ $\qquad$ (14)

where  $M$ = molar mass;   $R$ = gas constant;   $T$ = temperature.

Using the continuity equation $\qquad$ $\frac{\partial p}{\partial t} + \frac{1}{A_f}\frac{\partial \dot{m}}{\partial x} = 0$ $\qquad$ (15)

we find $\qquad$ $\frac{\partial p}{\partial t} = \frac{1}{k} T \frac{\partial}{\partial x}\left(\frac{p}{T\mu}\frac{\partial p}{\partial x}\right)$ $\qquad$ (16)

If we consider short regenerator elements in which the temperature may be assumed constant, equation (16) becomes

$$\frac{\partial p}{\partial t} = \frac{1}{k\mu}\frac{\partial}{\partial x}\left(p\frac{\partial p}{\partial x}\right) \tag{17}$$

This non-linear differential equation has no analytic solutions. However, by approximating the cyclic pressure $p$ by its mean value $\bar{p}$, we have

$$\frac{\partial p}{\partial t} = -\frac{\bar{p}}{k\mu}\frac{\partial^2 p}{\partial x^2} \tag{18}$$

Applying the boundary conditions: $\qquad$ $p = p_0 + p_1 \cos(\omega t)$ $\qquad$ at $x = 0$ $\qquad$ (19)

$\qquad\qquad\qquad\qquad\qquad\qquad\qquad$ $\frac{\partial p}{\partial x} = 0$ $\qquad\qquad$ at $x = \infty$ $\qquad$ (20)

gives the solution $\qquad$ $p = p_0 + p_1 e^{-\lambda x}\cos(\omega t - \lambda x)$ $\qquad$ (21)

where $\lambda$ is given by $\qquad$ $\lambda = \sqrt{\frac{\omega k\mu}{2 p_0}}$ $\qquad$ (22)

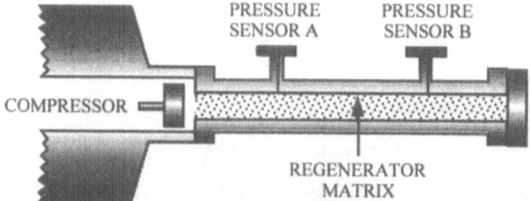

**Figure 3.**  Schematic of experiment to measure regenerator pressure drop.

Equation (21) only suggests a general form for the regenerator pressure drop. The constant $k$ in equation (22) was determined empirically by measuring the pressure drop at ambient temperature over a wide range of frequencies and fill pressures using the experimental arrangement shown schematically in **Figure 3**. The regenerator tested comprised a 400-mesh stainless steel screen matrix of length 60mm and diameter 7.2mm.

There was found to be some variability in $k$ over the range of operating conditions measured. Furthermore, it was found that at any given operating condition, the value of $k$ derived from the exponential term in equation (21) was different from that derived from the phase shift term. Equation (21) was therefore modified to include two independent parameters $\lambda_1$ and $\lambda_2$, thus:

$$p = p_0 + p_1 \, exp(-\lambda_1 x) \, cos(\omega t - \lambda_2 x) \tag{23}$$

$$\text{where} \quad \lambda_1 = \sqrt{\frac{\omega k_1 \mu}{2 p_0}} \quad \text{and} \quad \lambda_2 = \sqrt{\frac{\omega k_2 \mu}{2 p_0}} \tag{24}$$

The 'constants' $k_1$ (in the exponential term) and $k_2$ (in the phase shift term) were determined by experiment over the required operating range of frequency and gas pressure. The effect of regenerator temperature on the pressure drop is included by using a temperature dependent value for the viscosity $\mu$ in equations (24). The validity of this approach will be determined when pressure drop measurements are repeated with the test regenerator cooled at one end. If these measurements show the approach to be inadequate, the constants $k_1$ and $k_2$ may be determined as a function of the regenerator cold-end temperature, as well as frequency and pressure.

**Figure 4** shows the modelled and measured pressure cycles at either end of the 60mm test regenerator filled with 6 bar of helium at ambient temperature, at a frequency of 60 Hz. Both the amplitudes and phase shift are well represented by the model. The mismatch in the shape of the pressure waves may be due to non-harmonic motion of the compressor, which is not accounted for in the model.

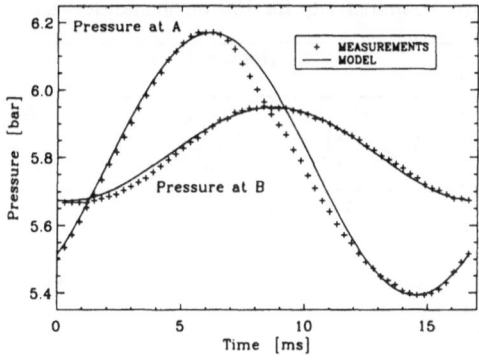

**Figure 4.**  Modelled and measured pressure cycles for a test regenerator at ambient temperature.

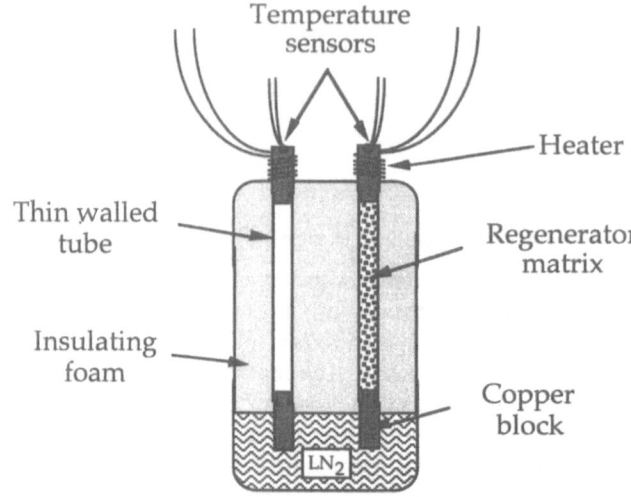

**Figure 5.**  Schematic of the regenerator thermal conductivity experiment.


## REGENERATOR THERMAL CONDUCTIVITY

To enable an accurate calculation of the conductive heat load on the cold end, the thermal conductivity of the regenerator matrix was measured using the experimental arrangement shown schematically in **Figure 5**.

Two thin walled titanium alloy (6Al-4V) tubes (diameter 6mm; length 60mm; wall thickness 0.05mm) were suspended in insulating foam inside a dewar with their lower ends immersed in $LN_2$, while their upper ends were maintained at room temperature using electrical coil heaters. One of the tubes was lightly packed with the regenerator matrix (400-mesh stainless steel screens), while the other remained empty. The conductivity of the matrix was determined from the difference in power, $\Delta Q$, required to maintain the upper ends of the two tubes at ambient temperature. The measurements were made in air at ambient pressure.

The effect of the packing density on the matrix conductivity was investigated by increasing the number of screens by approximately 8% and packing down firmly using a press. The results are shown in **Table 1**.


**Table 1.**  Thermal conductivity for 400-mesh stainless steel screens

| Packing density | Heater Power | | | Conductivity k |
|---|---|---|---|---|
| | Q1 (empty tube) | Q2 (packed tube) | $\Delta Q$ | W/mK |
| 250 discs/cm  (lightly packed) | 150 mW | 180 mW | 30 mW | 0.24 |
| 270 discs/cm  (firmly packed) | 150 mW | 210 mW | 60 mW | 0.48 |

The thermal conductivity of the matrix is only 2% of the bulk material for a lightly packed regenerator, and 4% for a firmly packed one.

## MODEL RESULTS

The results of a model run are shown in **Table 2**, and **Figure 6** illustrates the use of the model to optimise a number of parameters in designing a breadboard WEPT cooler.

In order for heat to flow in the required direction, energy must be dissipated at the warm expander. It can be seen from **Table 2** that this energy is equal to the gross heat lift. This is also true at the orifice of a conventional pulse tube. The ideal coefficient of performance (COP) for the WEPT can be shown to be $T_c/T_w$ ($T_c$ and $T_w$ being the heat absorption and rejection temperatures respectively), which is less than the Carnot COP by a factor $(T_w - T_c)/T_w$. Thus, for a cooler operating between 80K and 300K, the ideal COP is 73% of Carnot. However, one advantage of the WEPT over the OPT is that the energy lost at the warm end of the pulse tube may be partially recovered as electrical energy if the warm expander is coupled to a small linear moving coil generator.

**Table 2.** Sample model output.

| | | |
|---|---|---|
| Specific power | 22.8 | Watts/Watt |
| Total input power | 8.53 | W |
| Power dissipated at warm expander | 1.01 | W |
| Gross heat lift | 1.01 | W |
| Total conduction loss | 0.33 | W |
| Regenerator loss | 0.15 | W |
| Net heat lift | 0.53 | W |
| Pulse tube length | 40.0 | mm |
| Pulse tube radius | 5.00 | mm |
| Regenerator length | 80.0 | mm |
| Regenerator radius | 5.00 | mm |
| Regenerator efficiency | 99.91 | % |
| Compressor motor efficiency | 80.0 | % |
| Irreversible cycling losses | 15.0 | % |
| Compressor swept volume | 1.08 | cc |
| Warm expander swept volume | 0.57 | cc |
| Heat rejection temperature | 300.0 | K |
| Cold block temperature | 80.0 | K |
| Frequency | 60.0 | Hz |
| Mean pressure | 0.68 | MPa |

## CONCLUSION

A numerical model has been developed to demonstrate the feasibility of a novel type of cryocooler for long life applications, the Warm Expander Pulse Tube (WEPT). The model includes a semi-empirical expression for the cyclic pressure drop in the regenerator matrix, and thermal conductivity measurements for a wire screen matrix.

The advantages of the WEPT over the orifice pulse tube are: (i) a high degree of control over the phase between pressure and mass flow oscillations to maximise heat flow; (ii) the energy dissipated at the warm end of the tube may be partially recovered as electrical energy, and (iii) the action of the warm expander produces a high pressure ratio, giving the WEPT a greater cooling capacity for a given compressor swept volume.

The advantages of the WEPT over the Stirling cycle cooler are: (i) the elimination of moving parts at cryogenic temperatures, and (ii) the elimination of shuttle heat transfer.

These features could provide the WEPT with an overall efficiency comparable to or exceeding that of a Stirling cycle cooler, making it an attractive option where efficiency and size are the principal constraints.

A breadboard WEPT has been built and is currently being tested and optimised. This work will be reported at a later date.

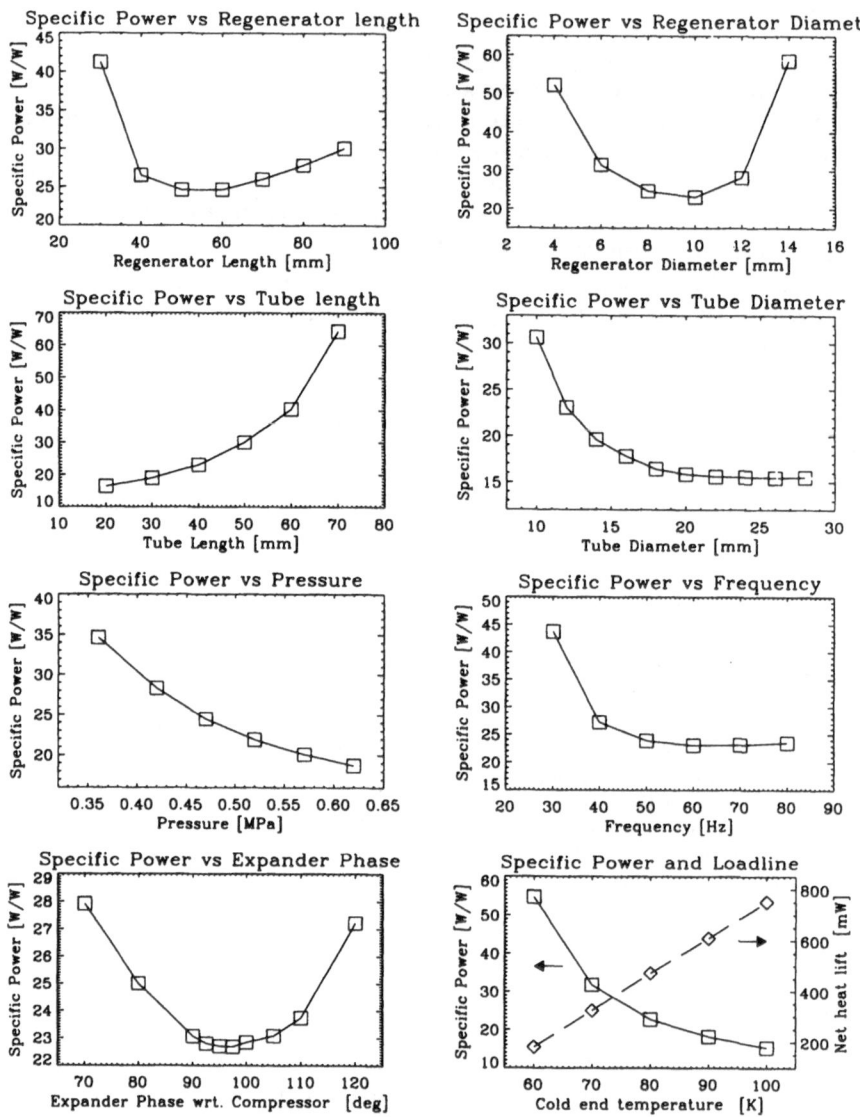

**Figure 6.** Results showing use of the model as a design tool.

## ACKNOWLEDGEMENT

M.C. Brito is funded by JNICT/PRAXIS XXI.

## REFERENCES

1. Matsubara, Y., "Alternative methods of the orifice pulse tube refrigerator", *Proceedings of the 5th International Cryocooler Conference*, (1988).

2. Radebaugh., R., "A simple first step to the optimisation of regenerator geometry", *Proceedings of the 3rd International Cryocooler Conference*, (1984).

# A Simple Modeling Program for Orifice Pulse Tube Coolers

**Pat R. Roach**

NASA Ames Research Center
Moffett Field, CA 94035

**Ali Kashani**

Atlas Scientific
Sunnyvale, CA 94086

## ABSTRACT

We have developed a calculational model that treats all the components of an orifice pulse tube cooler. We base our analysis on 1-dimensional thermodynamic equations for the regenerator[1] and we assume that all mass flows, pressure oscillations and temperature oscillations are small and sinusoidal. The resulting mass flows and pressures are matched at the boundaries with the other components of the cooler: compressor, aftercooler, cold heat exchanger, pulse tube, hot heat exchanger, orifice and reservoir. The results of the calculation are oscillating pressures, mass flows and enthalpy flows in the main components of the cooler.

By comparing with the calculations of other available models, we show that our model is very similar to REGEN 3 from NIST and DeltaE from Los Alamos National Lab for low amplitudes where there is no turbulence.

Our model is much easier to use than other available models because of its simple graphical interface and the fact that no guesses are required for the operating pressures or mass flows. In addition, the model only requires a minute or so of running time, allowing many parameters to be optimized in a reasonable time.

## INTRODUCTION

Pulse Tube coolers are complex systems that require careful optimization of their many components to achieve the best cooling performance. In particular, the regenerator, where a large surface area and high heat capacity are needed to damp out temperature oscillations, is a component that must be designed to maximize heat exchange with the gas passing through it while minimizing the pressure drop across it. The fact that the gas flow is oscillating back and forth while the gas pressure is also oscillating with a different phase makes the analysis of this part of the system too complex for a simple analysis.

We wanted to understand this interaction in the regenerator and, at the same time, we wanted a tool to help us design pulse tube coolers. Toward that end we developed a computer model of

Cryocoolers 9, Edited by R.G. Ross, Jr.
Plenum Press, New York, 1997

the regenerator that would capture the fundamental behavior of the gas-matrix interaction while remaining simple enough to be quick and easy to use. We realized that in optimizing the performance of the regenerator we would be affecting the performance of the other parts of the cooler. Since it is only the net cooling power of the system that is ultimately important, we extended the model to include the other major components of a typical cooler. Now the model (called ARCOPTR for Ames Research Center Orifice Pulse Tube Refrigerator) treats the entire system starting with the compressor and includes the heat exchangers, the regenerator, the pulse tube section itself and also the orifice and reservoir that provide the phase shift of the mass flow that is necessary for cooling to occur.

To simplify the calculations, we limited the model to the consideration of only the fundamental frequency component of the oscillating parameters and we take the limiting case of infinitesimal amplitude oscillations. This allows us to study all the fundamental processes that affect the performance of the pulse tube cooler, but we do not expect highly accurate estimates of performance of actual systems having large amplitudes of pressure oscillation or mass flow. Nevertheless, we feel that the model provides a very useful guide to the optimum trade-off between the many conflicting requirements, especially since higher accuracy is hard to justify without a better understanding of some of the loss mechanisms that occur.

## FEATURES OF ARCOPTR

Figure 1 shows the components of an orifice pulse tube cooler that the model treats. Details for the various parts are as follows:

**Compressor:** The compressor is taken to be adiabatic. All temperature oscillations are assumed to be completely damped in the aftercooler. The equation describing the compressor is:

$$\frac{1}{P_c}\frac{\partial P_c}{\partial t} = -\frac{\gamma}{\zeta_c}\frac{\partial \zeta_c}{\partial t} - \frac{\gamma}{m_c}\frac{\partial m_c}{\partial t} \tag{1}$$

where $P_c$ is the pressure in the compressor, $\zeta_c$ is the piston position and $m_c$ is the mass of gas in the compressor.

**Regenerator:** Here, the 1-D equations developed in a previous paper[1] are used. The basic differential equation in P that resulted is:

$$0 = \frac{\partial^2 P_d}{\partial z^2} + \left[\frac{(1+\alpha\, i)\,\Gamma\, T_0 + \alpha^2 + 1}{([\Gamma\, T_0 + 1]^2 + \alpha^2)\, T_0} - \frac{\lambda\,(\lambda\, T_0 - i)}{\lambda^2\, T_0^2 + 1}\right]\frac{\partial T_0}{\partial z}\frac{\partial P_d}{\partial z}$$

$$+ \frac{4\,\pi^2\, M^2\,(\lambda\, i\, T_0 - 1)\{(\gamma - 1)[(1+\alpha\, i)\,\Gamma\, T_0 + \alpha^2 + 1] - \gamma[(\Gamma\, T_0 + 1)^2 + \alpha^2]\}P_d}{\varepsilon^2\,[(\Gamma\, T_0 + 1)^2 + \alpha^2]\, T_0} \tag{2}$$

where $P_d$ is the dynamic (oscillating) pressure amplitude (scaled by the average pressure),

$\alpha = \dfrac{2\,\pi\,(1-f)\,\rho_m^*\, C_m\, r_h}{f\,\tau^*\, h}$, $\quad \Gamma = \dfrac{(1-f)\,\rho_m^*\, C_m}{f\,\rho_0^*\, C_p}$, $\quad \lambda = f\,\mu\,\tau^*/2\,\pi\, K_p\,\rho_0^*$, $T_0$ is the

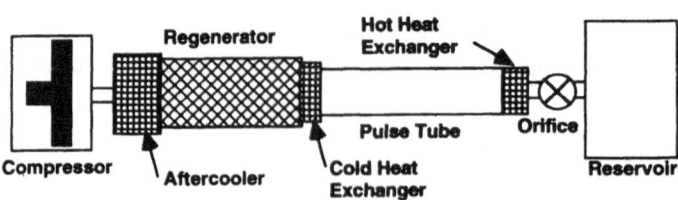

**Figure 1.** Main components of an orifice pulse tube cooler.

steady component of the local temperature, $M$ = Mach number = $q_0^*/\sqrt{\gamma R T_0^*}$, $q_0^*$ is the effective velocity of the gas at the regenerator inlet, $\gamma = C_p/C_v$, $R$ is the gas constant, $f$ is the void fraction, $\varepsilon = \tau^* q_0^*/L_0^*$, $\mu$ is the viscosity, $r_h$ is the hydraulic radius of the matrix, $\tau^*$ is the period of oscillation, $L_0^*$ is the length of the regenerator, $h$ is the heat transfer coefficient, $\rho_m^*$ is the matrix density, $C_m$ is the matrix heat capacity, $\rho_0^*$ is the gas density at the warm end, and $K_p$ is the Darcy permeability of the matrix (from the friction factor—see below).

Equation (2) is solved numerically to arrive at a fundamental-frequency amplitude for the pressure; temperature and velocity are then derived from this pressure solution. An initial linear temperature gradient from the hot end to the cold end is used to estimate temperature-dependent values of $\mu$, $h$ and $\alpha$. The heat flow to the solid matrix is taken to be in phase with the temperature difference between the gas and matrix. This assumes that the dimensions of the pores containing the gas are small enough that the gas is in fairly good contact with the matrix; in other words, the hydraulic radius of the pores is less than the diffusion length in the gas.

The parameters for heat transfer to the matrix and for friction factor come from Kays and London[2]. Since these data show considerable non-linearities at higher velocities, it was felt important to include some high-velocity effects. An initial guess to the velocity in the regenerator is used to estimate the heat transfer coefficient and the friction factor; the regenerator equations are then solved and much more accurate values of the velocity are found. From these velocities the final values for the heat transfer coefficient and the friction factor are found and the calculation is repeated. This is the only instance where some allowance for finite-velocity effects is included. In calculating the effect of thermal conduction axially through the metal matrix, a correction factor of 0.3 is used to reflect the poor contact between adjacent screens.

The boundary conditions for the regenerator solution are that the pressures and mass flows at the ends of the regenerator match those of the compressor and pulse tube (as modified by the aftercooler and cold heat exchanger, respectively).

**Heat Exchangers:** The aftercooler and the cold and hot heat exchangers are assumed to be isothermal. The primary effect they have on the modeling is to introduce a pressure drop due to their impedance and a phase shift in the mass flow due to their void volume. No attempt is made to assess their adequacy for heat transfer or to calculate the amount of heat that flows through them. It is assumed that all temperature oscillations are completely damped in passing through them. The equation describing the flow in the heat exchangers is just Eq. (2) with the second term on the right missing, since $\partial T/\partial z = 0$:

$$0 = \frac{\partial^2 P_d}{\partial z^2} + \frac{4\pi^2 M^2 (\lambda i T_0 - 1)\{(\gamma - 1)[(1 + \alpha i)\Gamma T_0 + \alpha^2 + 1] - \gamma[(\Gamma T_0 + 1)^2 + \alpha^2]\}P_d}{\varepsilon^2 [(\Gamma T_0 + 1)^2 + \alpha^2] T_0} \quad (3)$$

where $\lambda$, $\Gamma$ and $\alpha$ take on appropriate values for the heat exchanger being considered. Since $T_0$ is independent of $z$, the coefficient of $P_d$ is just a constant; this equation is solved analytically.

**Pulse Tube:** In the pulse tube there is no pressure gradient. The mass conservation and the energy conservation equations can be used to arrive at an equation for the effective velocity, $q_d$:

$$\frac{\partial}{\partial z} q_d - \frac{\Gamma_{pt} q_d}{\Gamma_{pt} T_a + 1 + \alpha_{pt} i}\left(\frac{T_{hot} - T_{cold}}{L_{pt}}\right) = \frac{2\pi f_{pt} (\gamma \Gamma_{pt} T_a + 1 + \alpha_{pt} i) P_d}{\gamma i (\Gamma_{pt} T_a + 1 + \alpha_{pt} i)} \quad (4)$$

where $T_a$ is the z-dependent steady temperature and $L_{pt}$ is the length of the pulse tube (scaled by the length of the regenerator). $\Gamma_{pt}$ and $\alpha_{pt}$ in this equation are defined in the same way as $\Gamma$ and $\alpha$ for the regenerator, above, where heat exchange with the pulse tube wall has replaced the heat exchange with the regenerator matrix. This equation can be solved analytically if it is assumed that $T_a$ is linear in z and that $\alpha_{pt}$ and $\Gamma_{pt}$ are independent of z.

However, the assumption, made for the regenerator, that the heat transfer to the wall is in phase with the temperature difference between the gas and the wall is no longer adequate since

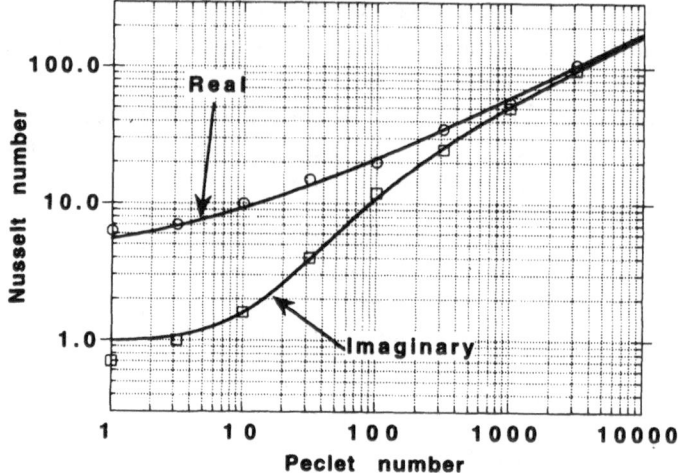

**Figure 2.** The real and imaginary Nusselt numbers from the data of Kornhauser.

the gas in the center of the pulse tube can be many diffusion lengths from the wall. When that is the case it is possible to have the heat transfer occurring at a phase different from that of the phase of $\delta T$ between the wall and the average temperature of the gas. This leads to the concept of a complex Nusselt number as discussed by Kornhauser[3]. The effect on Eq. (4) is to make $\alpha_{pt}$ complex since $h = k\,Nu/r_h$ where $k$ is the thermal conductivity of the gas, $Nu$ is the complex Nusselt number and $r_h$ is the hydraulic radius of the pulse tube ($r_h$ = cylinder volume/cylinder surface = $0.5\,r_{pt}$). A typical result from Kornhauser's data is shown in Fig. 2. The Peclet number used here is $Pe = \rho\,C_p\,\omega\,r_h^2 / 4\,k$, which is just $(r_h / 2L_d)^2$ where $L_d$ is the diffusion length in the gas. The curves we fit to the above data are:

$$\mathrm{Re}\{Nu\} = 5.603 + 1.170\,\sqrt{(Pe)} \quad \text{and} \quad \mathrm{Im}\{Nu\} = \frac{(16.43 + 1.140\,Pe^{1.5})}{25.02 + Pe} \qquad (5)$$

**Orifice and Reservoir:** In a pulse tube section with no wall interaction and no flow out the orifice, all the mass flow into the cold end goes into compressing the gas in the pulse tube. This mass flow will be 90° out of phase with both the pressure and the temperature and there will be no work flow or enthalpy flow at the cold end. If there is flow out the orifice, this will add a component of flow throughout the pulse tube that is in phase with the pressure and leads to an enthalpy flow that is the basis for the cooling in an orifice pulse tube cooler. If the orifice flow is too large, however, the pressure drop in the regenerator will yield a very small pressure oscillation in the pulse tube, and result in very little cooling. Clearly, there is an optimum orifice setting that produces the best cooling.

Our model treats the orifice as an impedance with flow proportional to pressure; the flow is symmetrical with flow direction and has no dependence on velocity. The reservoir is assumed to be an infinite volume as far as the interaction with the rest of the system is concerned. For convenience, the mass flow into the reservoir is expressed in terms of a pressure oscillation in a specified finite volume; for correct results, the reservoir volume must be big enough to make these pressure oscillations negligible compared to those in the rest of the system.

**Solution of the model.** Equation (2) for the regenerator is solved numerically. We have two versions of the model calculation; they each solve the equation by a different technique. One version uses a 'shooting' method where the slope of the pressure at the warm end of the regenerator is adjusted until the pressure and mass flow at the cold end, as determined by integrating the differential equation, agree with the boundary conditions. The other method is a two-point-boundary-condition technique that guesses a parabolic pressure profile for the regenerator that satisfies the boundary conditions but doesn't satisfy eq. (2). This pressure profile is then iteratively adjusted until it satisfies eq. (2). These two very different methods give identical results.

The analytic solutions for the compressor and for the pulse tube provide relationships between pressure and mass flow at each end of the regenerator; these relationships (rather than fixed values of pressure or flow) make up the boundary conditions at each end of the regenerator for the solution of eq. (2). The analytic solutions for the heat exchangers modify these boundary conditions in a way that accounts for the pressure drop and the phase shift in the mass flow that occurs in the heat exchangers. The general form of the relationship between pressure and mass flow is not changed by the heat exchangers.

## COMPARISON WITH OTHER MODELS

### The Modeled Cooler

For the comparisons, a pulse tube cooler of the following dimensions was modeled:
Aftercooler: L = 1.25 cm, I. D. = 3.62 cm, mesh = # 87, wire diam. = 0.0115 cm.
Regenerator: L = 5.0 cm, I. D. = 1.22 cm, mesh = (see tables), wall thickness = 0.05 cm.
Cold heat exchanger: L = 0.2 cm, I. D. = 1.128 cm, mesh = # 87, wire diam. = 0.0115 cm.
Pulse tube: L = 3.0 cm, I. D. = 1.128 cm, wall thickness = 0.0085 cm.
Hot heat exchanger: L = 0.5 cm, I. D. = 1.128 cm, mesh = # 87, wire diam. = 0.0115 cm.
Orifice setting: 0.247 g/s bar.
Reservoir volume: 500 cm$^3$.
System pressure: Helium at 20 bar.
Hot temperature: 300 K.
Cold Temperature: 80K.
Operating frequency: 55 Hz.

The aftercooler, cold heat exchanger and hot heat exchanger are copper, the regenerator (including screens) and the pulse tube are stainless steel. The screens in the regenerator have a void fraction of 0.69.

### The Comparison Models:

**REGEN3**[4] from NIST is a model of the regenerator only. It treats the full time-dependence of the oscillating parameters, so it can capture distortions to the waveforms. It handles large amplitude oscillations. It assumes a small pressure drop in the regenerator so it might not be suitable for very restrictive regenerators. The other parts of the cooler such as the compressor, the pulse tube and the orifice must be modeled by other means.

**DeltaE**[5] from Los Alamos National Laboratory models an entire cooler (except the compressor). Like ARCOPTR it is a linearized model that treats only the lowest order sinusoidal component of the oscillations. It is based on thermo-acoustic wave equations so it is especially suitable for higher-frequency systems where the dimensions are comparable to an acoustic wavelength. It can treat regenerators with large pressure drops and it uses correlations for heat flow and friction factor that include nonlinear effects that occur at high velocities.

### Results of the Comparison

For the ARCOPTR calculation, the compressor stroke was adjusted to give the desired inlet

**Table 1.** Model Comparison for $P_{in}$ = 0.600 bar and # 400 Mesh Screens.

| | REGEN3 (NIST) | DeltaE (LANL) | ARCOPTR (NASA Ames) |
|---|---|---|---|
| **Input** | | | |
| Regenerator mesh = # 400 wire = 0.0025 cm | $\dot{m}_{in}$= 0.497 @ 30.2° $\dot{m}_{out}$= 0.411 @ 13.2° | $P_{in}$ = 0.600 bar @ 0° | $P_{in}$ = 0.600 bar @ 0° |
| **Output** | | | |
| Compressor PV work (W) | 4.32 | 4.36 | 4.03 |
| $\dot{m}$ into regenerator (g/s) | (input) | 0.492 @ 29.7° | 0.497 @ 30.2° |
| $\dot{m}$ out of regenerator (g/s) | (input) | 0.415 @ 13.3° | 0.411 @ 13.2° |
| $\Delta P$ across regenerator (bar) | 0.215 | 0.242 @ 25.1° | 0.227 @ 25.6° |
| Regenerator energy flow (W) (gas enthalpy + matrix conduct.) | 0.95 | 1.00 | 1.26 |
| Pulse tube pressure (bar) | (not available) | 0.394 @ −15.2° | 0.408 @ −13.9° |
| Pulse tube enthalpy flow (W) | (not available) | 0.45 | 0.49 |

pressure into the regenerator. For DeltaE, the inlet pressure is an input parameter. For REGEN3, mass flows in and out of the regenerator are the starting parameters. The energy flows quoted for REGEN3 are not exact because the calculation did not completely converge to give a consistent value of total energy flow (enthalpy flow in the gas plus conduction in the matrix) at the two ends of the regenerator in runs lasting overnight. Note: all of the pressure and mass flow values in the table are the amplitudes of oscillating variables; peak-to-peak values would be twice the quoted values. Enthalpy flows and PV work are time-averaged steady values.

Table 1 shows the comparison for the regenerator filled with # 400 mesh screens when the warm-end pressure oscillations are ±0.6 bar for a 20 bar ambient pressure. The ARCOPTR results agree within 7% of both REGEN3 and DeltaE for $\Delta P$ and within 1% of DeltaE on the mass flows in the regenerator. For the energy flow (loss) in the regenerator, ARCOPTR is about 30% higher than the other models, while the compressor PV work is 8% lower than the others. In the pulse tube, ARCOPTR has 10% more enthalpy flow (cooling power) and 4% more pressure oscillation than DeltaE.

Table 2 shows the comparison for the regenerator filled with # 400 mesh screens when the

**Table 2.** Model Comparison for $P_{in}$ = 2.400 bar and # 400 Mesh Screens.

| | REGEN3 (NIST) | DeltaE (LANL) | ARCOPTR (NASA Ames) |
|---|---|---|---|
| **Input** | | | |
| Regenerator mesh = # 400 wire = 0.0025 cm | $\dot{m}_{in}$= 1.75 @ 28.4° $\dot{m}_{out}$= 1.43 @ 9.0° | $P_{in}$ = 2.400 bar @ 0° | $P_{in}$ = 2.400 bar @ 0° |
| **Output** | | | |
| Compressor PV work (W) | 64.5 | 60.1 | 57.5 |
| $\dot{m}$ into regenerator (g/s) | (input) | 1.71 @ 27.9° | 1.75 @ 28.4° |
| $\dot{m}$ out of regenerator (g/s) | (input) | 1.42 @ 8.9° | 1.43 @ 9.0° |
| $\Delta P$ across regenerator (bar) | 1.159 | 1.208 @ 21.5° | 1.136 @ 23.0° |
| Regenerator energy flow (W) (gas enthalpy + matrix conduct.) | 3.50 | 3.20 | 3.92 |
| Pulse tube pressure (bar) | (not available) | 1.352 @ −19.9° | 1.424 @ −18.1° |
| Pulse tube enthalpy flow (W) | (not available) | 6.15 | 6.03 |

**Table 3.** Model Comparison for $P_{in}$ = 0.600 bar and # 200 Mesh Screens.

| | REGEN3 (NIST) | DeltaE (LANL) | ARCOPTR (NASA Ames) |
|---|---|---|---|
| **Input** | | | |
| Regenerator mesh = # 200 wire = 0.0025 cm | $\dot{m}_{in}$ = 0.626 @ 33.9° $\dot{m}_{out}$= 0.527 @ 22.0° | $P_{in}$ = 0.600 bar @ 0° | $P_{in}$ = 0.600 bar @ 0° |
| **Output** | | | |
| Compressor PV work (W) | 4.82 | 5.24 | 4.87 |
| $\dot{m}$ into regenerator (g/s) | (input) | 0.638 @ 34.7° | 0.626 @ 33.9° |
| $\dot{m}$ out of regenerator (g/s) | (input) | 0.543 @ 22.6° | 0.527 @ 22.0° |
| ΔP across regenerator (bar) | 0.096 | 0.101 @ 31.8° | 0.092 @ 31.0° |
| Regenerator energy flow (W) (gas enthalpy + matrix conduct.) | 1.95 | 2.10 | 2.52 |
| Pulse tube pressure (bar) | (not available) | 0.516 @ –5.9° | 0.523 @ –5.2° |
| Pulse tube enthalpy flow (W) | (not available) | 0.83 | 0.81 |

warm-end pressure oscillations are ±2.4 bar for a 20 bar ambient pressure. The ARCOPTR results agree within 6% of both REGEN3 and DeltaE for ΔP and within 2% of DeltaE on the mass flows in the regenerator. For the energy flow (loss) in the regenerator, ARCOPTR is 23% higher than DeltaE and 12% higher than REGEN3, while the compressor PV work is 5% lower than DeltaE and 11% lower than REGEN3. In the pulse tube, ARCOPTR has 2% less enthalpy flow (cooling power) and 5% more pressure oscillation than DeltaE.

Table 3 shows the comparison for the regenerator filled with # 200 mesh screens when the warm-end pressure oscillations are ±0.6 bar for a 20 bar ambient pressure. The ARCOPTR results agree within 9% of both REGEN3 and DeltaE for ΔP and within 3% of DeltaE on the mass flows in the regenerator. For the energy flow (loss) in the regenerator, ARCOPTR is 20% higher than DeltaE and 29% higher than REGEN3, while the compressor PV work is 7% lower than DeltaE and 1% higher than REGEN3. In the pulse tube, ARCOPTR has 3% less enthalpy flow (cooling power) and 1% more pressure oscillation than DeltaE.

**Summary.** ARCOPTR results agree well with REGEN3 and DeltaE for all three cases studied. Pressure drops, mass flows and compressor PV work of ARCOPTR are within about 10% of the other models. The energy flow (loss) term for the regenerator is 12-30% higher than the other models, while the pulse tube enthalpy flow (cooling power) is within 10% of that of DeltaE. It is not clear why ARCOPTR is consistently higher than the other models for the regenerator energy flow.

## CONCLUSIONS

Our model treats all the components of an orifice pulse tube cooler using rigorous 1-D thermodynamic equations. It linearizes the equations, looking only at the fundamental frequency terms in the limit of small amplitudes. Since most of the important phenomena in a pulse tube cooler should be evident in such a treatment, the model should have great utility in suggesting ways to optimize performance of an actual cooler. It isn't clear that the use of higher harmonics to describe the waveforms is a great advantage when there are important loss mechanisms due to 2-D flow effects in the pulse tube section which are not being treated yet.

Comparing our model with REGEN3 from NIST and DeltaE from LANL, we find quite good agreement for the three cases studied. In addition, we have found that our model is much faster and easier to use than the other two.

We are currently looking at the effects of convection and other 2-D phenomena[6] in the pulse tube section and we plan to incorporate these results into the model when they become available. We feel that this should greatly improve the ability of the model to describe real-world coolers.

## REFERENCES

1. Roach, Pat R., Kashani, A. and Lee, J. M., "Theoretical Analysis of a Pulse Tube Regenerator", in *Advances in Cryogenic Engineering*, vol. 41, p. 1357.

2. Kays, W. M. and London, A. L., *Compact Heat Exchangers*, McGraw-Hill, New York (1955).

3. Kornhauser, A. A., "Gas-Wall Heat Transfer During Compression and Expansion", thesis for Doctor of Science, MIT (1989); also, Kornhauser, A. A. and Smith, J. L. Jr., "Application of a Complex Nusselt Number to Heat Transfer During Compression and Expansion", *J. of Heat Transfer*, vol. 116, (1994), pp. 536-542, and Kornhauser, Alan A. and Smith, Joseph L. Jr., "Heat Transfer with Oscillating Pressure and Oscillating Flow", *Proc. 24th Intersociety Energy Conversion Engineering Conf.*, Aug. 6-11, 1989.

4. Gary, J., O'Gallagher, A. and Radebaugh, R., *A Numerical Model for Regenerator Performance*, Technical Report, NIST-Boulder, (1994).

5. Ward, W. C. and Swift, G. W., "Design Environment for Low-Amplitude Thermoacoustic Engines", *J. Acoust. Soc. Am.*, vol. 95, (1994), p. 3671.

6. Lee, J. M., Kittel, P., Timmerhaus, K. D. and Radebaugh, R. " Useful Scaling Parameters for the Pulse Tube", in *Advances in Cryogenic Engineering*, vol. 41, p. 1347.

# A One-Dimensional Model of High-Frequency Pulse Tube Heat and Mass Flows

**C. S. Kirkconnell**

Hughes Aircraft Co.
El Segundo, California, U.S.A., 90245

**G. T. Colwell**

Georgia Institute of Technology
Atlanta, Georgia, U.S.A., 30332

## ABSTRACT

Pulse tube cryocoolers have received considerable attention in recent years due to their advantages in cost and reliability and reduced cold-end vibrations compared to other regenerative cryocoolers. In particular, high-frequency (> 60 Hz) pulse tube cryocoolers are of interest for many applications since the same amount of gross refrigeration can be obtained using more compact, lighter weight compressors. Accurate modeling of these high-frequency cryocoolers is greatly facilitated by a thorough understanding of the oscillatory heat and mass flows inside the pulse tube component. Towards this end, a numerical model was developed to solve the one-dimensional, non-linear governing equations for these types of flows. The model utilizes the method of lines. Dimensional analysis was performed to identify the important dimensionless groups and to facilitate the reduction of the governing system of equations to a dimensionless system. Given a different scaling of the governing equations, the methods described herein are applicable to other pulse tube flow regimes.

## INTRODUCTION

The high-reliability and low-cost advantages of pulse tube refrigeration have been recognized since the development of the basic pulse tube refrigerator (BPTR) in the 1960's[1],[2], but the poor thermodynamic efficiency of the BPTR precluded most practical applications. The addition of an orifice at the warm end by Mikulin[3] in 1984 resulted in much more efficient devices, called orifice pulse tube refrigerators (OPTRs), and significant research in the field has followed. Pulse tube refrigerator technology has today reached the point that it is a viable, practical alternative for many applications. For example, Hughes is presently manufacturing an orifice pulse tube refrigerator for the Low Cost Cryocooler (LCC) program which, based upon the performance of the prototype, should deliver 4 watts of net refrigeration at 70 K for less than 150 watts of compressor input power[4].

The cost of developing efficient pulse tube refrigerators can be greatly reduced if numerical and/or analytical modeling tools are available to guide the design process. Several such tools have been developed and discussed over the years. The earliest, popularly called the SRZ model in its most detailed formulation, is due to Storch, Radebaugh, and Zimmerman[5]. The SRZ model provides a technique to calculate the enthalpy flow through the pulse tube component given certain operating conditions of the pulse tube refrigerator (phase angle between the hot and cold-end mass flows, operating frequency, etc.). This technique has also been discussed by Radebaugh[6] and Storch and Radebaugh[7]. Dropping the assumption of small pressure fluctuations used in the SRZ model, a more generally applicable model was developed by David, et al.[8] in 1993. Distinct from these pulse tube component models, numerical tools have been developed to guide pulse tube refrigerator design from a system-level approach. For example, Wu and Zhu[9] used control volume energy and mass conservation equations for each system component to predict refrigerator performance.

The focus of this paper is the development and discussion of a component-level, one-dimensional pulse tube model, but also of present interest is the implementation of this model in conjunction with a system-level model. The need for a system-level model is obvious since the goal of the engineer is to develop an efficient refrigeration system, not just an efficient pulse tube. Recent discoveries regarding the complex flow patterns inside the pulse tube[10] and the role of the pulse tube design on the overall system performance[11] make it apparent that the accurate modeling of the pulse tube cryocooler cannot be achieved without a similarly accurate representation of the pulse tube itself and the refrigeration losses occurring therein. The eventual goal of this research is to dynamically link a system-level and a detailed pulse tube component model to obtain a single pulse tube refrigerator design tool. At present the system-level model is used to help obtain the boundary conditions for the pulse tube model.

The one-dimensional model of the heat and mass flows inside the pulse tube considers the variation in the solution during the operating cycle, a distinguishing characteristic from the time-averaged component models discussed above. The most distinctive feature of this model, however, is the inclusion of inertial forces in the mathematic formulation. The motivation for including the inertial terms in the formulation arises from the desirability of higher-frequency, smaller pulse tubes for applications where size and weight are at a premium. As frequencies increase and pulse tube sizes decrease, velocities inside the pulse tube increase and inertial and compressibility effects can no longer be ignored. The inclusion of the inertial forces results in a numerically-challenging, non-linear set of equations. A numerical solution to those equations is discussed.

## List of Symbols

| | **English** | |  | |
|---|---|---|---|---|
| | | | $\overline{T_h}$ | temperature of rejection HX |
| | | | $u$ | velocity |
| $A$ | offset function in Dirichlet boundary function | | $\overline{U}$ | solution vector $(\rho, u, T)$ |
| $B$ | amplitude function in Dirichlet boundary function | | $z$ | axial coordinate |
| $C$ | phase shift function in Dirichlet boundary function | | | **Greek** |
| $C$ | amplitude of cold-end temperature wave | | | |
| $c_p$ | constant-pressure specific heat | | $A$ | amplitude of cold-end mass flow rate |
| $D$ | amplitude of hot-end temperature wave | | $B$ | amplitude of hot-end mass flow rate |
| $H$ | general form of equation for boundary functions | | $\varepsilon$ | coefficient of numerical dissipation terms |
| $k_{th}$ | thermal conductivity | | $E$ | amplitude of cold-end pressure wave |
| $L$ | length of pulse tube | | $\Phi$ | phase angle: hot-end to cold-end mass flow |
| $m$ | mass flow rate per unit area | | $\Gamma$ | aspect ratio of pulse tube |
| Ma | Mach number | | $\gamma$ | ratio of specific heats |
| $P$ | thermodynamic pressure | | $\Lambda$ | compressibility parameter |
| $\overline{P}$ | average pressure; surge tank pressure | | $\mu$ | dynamic viscosity |
| $R$ | gas constant | | $\mu$ | phase angle: cold-end temperature to mass |
| $t$ | time | | $\rho$ | density |
| $T$ | temperature | | $\tau$ | period (= 1/frequency) |
| $\overline{T_c}$ | temperature of cold HX | | $\omega$ | circular frequency |

## MATHEMATICAL FORMULATION

### Governing Equations

The physical model is shown in Figure 1. A finite cylinder of length $L$ is exposed to periodic temperature, density, and velocity waves at each end. The radial velocity and pressure gradients are zero, and radial variations in the temperature and velocity profiles are neglected. The 1-D formulation requires no boundary data along the walls, only at the ends of the cylinder.

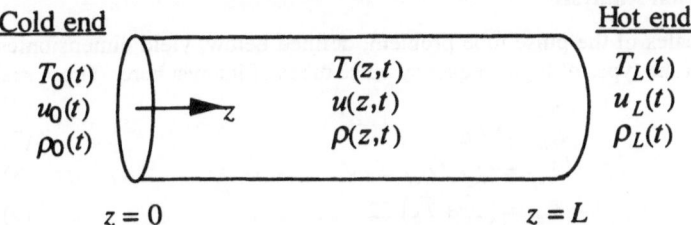

Cold end

$T_0(t)$
$u_0(t)$
$\rho_0(t)$

$T(z,t)$
$u(z,t)$
$\rho(z,t)$

Hot end

$T_L(t)$
$u_L(t)$
$\rho_L(t)$

$z = 0$         $z = L$

Figure 1. One-dimensional, time-dependent pulse tube model.

The thermal system shown in Figure 1 is governed by the full Navier-Stokes equations (NSE) for transient, viscous, compressible flows. For most applications of practical interest, the ideal gas approximation is appropriate and is used here. The thermophysical properties are taken as constants. As shown by the dimensional analysis which follows this section, the inclusion of the second-order terms containing these properties is important only from the standpoint of obtaining a stable numerical system, thus maintaining precise accounting of their numerical values is unnecessary. Given these assumptions, the system is described by the following conservation equations:

mass
$$\frac{\partial \rho}{\partial t} + u\frac{\partial \rho}{\partial z} + \rho\frac{\partial u}{\partial z} = 0, \tag{1}$$

momentum
$$\frac{\partial u}{\partial t} + \frac{RT}{\rho}\frac{\partial \rho}{\partial z} + u\frac{\partial u}{\partial z} + R\frac{\partial T}{\partial z} - \frac{\mu}{\rho}\frac{\partial^2 u}{\partial z^2} = 0, \tag{2}$$

energy
$$\frac{\partial T}{\partial t} + (\gamma-1)T\frac{\partial u}{\partial z} + u\frac{\partial T}{\partial z} - \frac{k_{th}}{\rho c_p}\frac{\partial^2 T}{\partial z^2} = 0, \tag{3}$$

where $\gamma$ is the ratio of specific heats, a constant for ideal monatomic gasses such as helium, a common pulse tube refrigerant. $\gamma$ is therefore assumed to be constant in the model.

Periodic Dirichlet boundary conditions are prescribed at the ends of the cylinder for the temperature, pressure, and "specific mass flow rate" (mass flow rate per unit of cross-sectional flow area) waves. The density and velocity functions required by the model are derived from the prescribed boundary conditions, while the temperature functions are provided directly. The above constitute a set of six independent boundary conditions in $z$ for a second-order system of three equations, hence the system is spatially well-defined. The general form for each boundary function $H(t)$ is

$$H(t) = A(t) + B(t)\sin(\omega t - C(t)). \tag{4}$$

The boundary condition in time is given by the periodicity of the solution, i.e., only steady-state solutions are considered. The transient problem of cool down is not. Time-averaged solutions to the steady-state problem are readily obtainable[4,7], but in an effort to gain a more thorough understanding of the effects of the changing heat and mass flow profiles during the operational cycle on performance, the periodic solution to the governing equations was sought. For a solution with period $\tau$, the periodicity condition simply requires

$$\overline{U}(t = 0) = \overline{U}(t = \tau) \tag{5}$$

where $\overline{U}$ the solution vector defined by the expression $\overline{U} = (u, T, \rho)$.

## Scaling/Dimensional Analysis

The natural scales of the pulse tube problem, defined below, yield dimensionless variables which are O(1) for the types of high-frequency pulse tubes of interest here. Those scales are:

$$z_{ref} = L \tag{6}$$
$$t_{ref} = 1/\omega \tag{7}$$
$$u_{ref} = z_{ref} / t_{ref} = \omega L \tag{8}$$
$$T_{ref} = \left(\overline{T_c} + \overline{T_h}\right)/2 \tag{9}$$
$$\rho_{ref} = \frac{\overline{P}}{RT_{ref}} \tag{10}$$

$L$ and $\omega$ are the length of the pulse tube and the characteristic frequency of the oscillatory flow. $\overline{T_c}$ and $\overline{T_h}$ are the temperatures of the heat exchangers at each end of the pulse tube, and $\overline{P}$ is the average pressure in the pulse tube over the cycle. The dimensionless system is obtained and the relevant dimensional parameters derived by substituting for the dimensional variables in equations (1) - (3) in terms of the reference values and the dimensionless variables. These substitutions yield the following set of dimensionless equations:

$$\frac{\partial \rho *}{\partial t *} + u * \frac{\partial \rho *}{\partial z *} + \rho * \frac{\partial u *}{\partial z *} = 0, \tag{11}$$

$$\frac{\partial u *}{\partial t *} + \Lambda \frac{T *}{\rho *} \frac{\partial \rho *}{\partial z *} + u * \frac{\partial u *}{\partial z *} + \Lambda \frac{\partial T *}{\partial z *} - \frac{\varepsilon}{\rho *} \frac{\partial^2 u *}{\partial z *^2} = 0, \tag{12}$$

$$\frac{\partial T *}{\partial t *} + (\gamma - 1) T * \frac{\partial u *}{\partial z *} + u * \frac{\partial T *}{\partial z *} - \frac{\varepsilon}{\rho *} \frac{\partial^2 T *}{\partial z *^2} = 0. \tag{13}$$

(The *'s are dropped from hereon for clarity.) The "compressibility parameter" $\Lambda$ appearing in the momentum equation is defined by the expression

$$\Lambda = \frac{RT_{ref}}{u_{ref}^2} = \frac{1}{\gamma Ma^2} , \tag{14}$$

and $\varepsilon$ is a numerical dissipation parameter included for numerical stability. Dimensional analysis reveals the axial thermal conduction and viscous diffusion terms to be several orders of magnitude smaller than the other terms for practical pulse tube flows, but a stable numerical scheme for the first order system, obtained by simply eliminating these terms, has proven elusive. It was determined that stable solutions could be obtained if small but finite coefficients were used for the second order terms in the immediate vicinity of the ends of the tube. A typical $\varepsilon$ vs. $z$ function, showing a maximum value of $\varepsilon$ two orders of magnitude smaller than the magnitude of the dimensionless solution vector, is provided in Figure 2. Numerical experiments have revealed the solution to be relatively insensitive to further reductions in the magnitude of the $\varepsilon$ function[12].

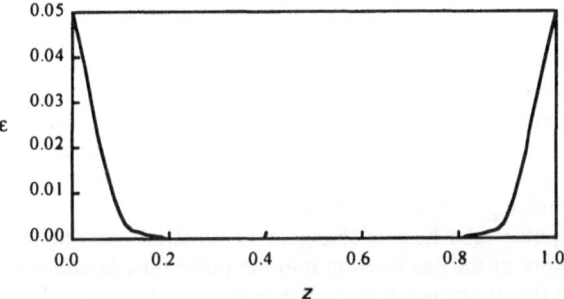

Figure 2. Numerical dissipation parameter. $\varepsilon = O(E\text{-}02)$ at ends. $\varepsilon = O(E\text{-}05)$ in the interior.

Boundary conditions for the one-dimensional model can be selected in a number of ways. For example, the model can be exercised with various mathematical expressions imposed at the ends of the tube which do not necessarily reflect realistic pulse tube flows if only the response of the model to various driving functions is of interest. Alternatively, realistic pulse tube boundary function equations can be implemented with parametrically varying coefficients to investigate the sensitivity of the solution to the tested parameters. A more physically relevant implementation is to use the model to describe a specific pulse tube flow, coupling the solution with the overall cryocooler operation through the use of a system-level model, correlation with experimental observation, etc.. Such an implementation is described below.

The functional forms of the mass flow and pressure boundary equations and the characteristic numerical parameters (amplitudes, phase angles, etc.) can be determined largely from a system-level pulse tube cryocooler model. The system-level model used in this effort, described in detail elsewhere[12], is essentially a control volume mass conservation model which determines flow rates and pressure waves given component geometries and operating conditions. Assuming sinusoidal compressor piston motion, the system-level model yields roughly sinusoidal pressure and mass flow rate waves at the ends of the pulse tube, the deviations from sinusoidal behavior being slightly more significant at the warm end. Even so, these deviations are insignificant for the purposes of this investigation. Therefore, sinusoidal mass flow rate and pressure waves are assumed at the ends of the pulse tube, and the mean pressure, amplitudes, and mass flow phase angles are provided by the system-level model. The mean mass flow rate at any given location is, by the conservation of mass, identically equal to zero.

Based upon experimental observation and analysis, additional assumptions regarding the boundary functions are made. Most significantly, the phase angle and amplitude differences between the boundary pressure waves are ignored. Even for pulse tubes of moderately high frequency (60 to 150 Hz) in which compressibility effects should not be neglected, the acoustic wave is typically much longer than the physical pulse tube length. For these cases, which are of present interest, the differences between the pressure waves at the ends of the tube, though finite, are so small as to be within the range of numerical error. The introduction of these small deltas into the boundary conditions offers no additional accuracy. However, the validity of this assumption breaks down for very high frequency pulse tubes in which the pulse tube length is no longer insignificant in relationship to the acoustic wave. (These flows are not considered but can be with no modification to the model other than the omission of the above assumption in the specification of the boundary conditions.) Another important assumption is that the temperature and mass flow rate are assumed to be in phase with the pressure wave at the warm end of the pulse tube. The mass flow rate is essentially in phase with the pressure wave due to the proximity of the warm end to the orifice. Experimental measurements on lower frequency pulse tube coolers have revealed small (< 15 degree) phase angles between the pressure and temperature at the warm end, so the approximation of a zero phase angle has been deemed adequate. However, non-zero phase angles can be accommodated.

The functional form of equation for the temperature boundary conditions was derived primarily by adapting waveforms obtained from experimental thermocouple measurements, and the parameters in the boundary functions are obtained from the system-level model. In-situ temperature measurements taken on the valve side of the warm heat exchanger for a lower frequency pulse tube cooler, which was described by Soloski and Mastrup[13], have shown that the out flow of the heat exchanger is essentially isothermal. This is consistent with the typical practice of designing pulse tube heat exchangers to function essentially as isothermalizers. Therefore, the temperature of the gas flowing into the pulse tube at either end is assumed to be isothermal at the operational temperature of the respective boundary heat exchanger. With respect to the gas flowing out of the pulse tube at either end, sinusoidal behavior is assumed, but different waveforms can be accommodated with relatively little additional effort. The warm end temperature wave is assumed to be in phase with the pressure wave as stated above, but the determination of the remaining temperature boundary parameters is somewhat more involved. Since enthalpy flux is the dominant heat transfer path in the pulse tube, a steady-state operating point can be approximated by forcing the enthalpy fluxes at the ends of the pulse tube to be equal. This yields an expression for the cold-end phase angle in terms of the mass flow and temperature wave amplitudes. The temperature wave amplitudes are defined such that the boundary enthalpy flux is equal to the gross refrigeration calculated using the system-level model. Thus the determination of the cold-end phase angle and temperature wave amplitudes must be iterative to insure both of the above criteria are satisfied.

Given the above assumptions, the dimensionless boundary functions can now be expressed in their functional form:

$$m_c(t) = A\cos t \tag{15}$$
$$P_c(t) = \overline{P} + E\cos(t - \Phi) \tag{16}$$
$$T_c(t) = \begin{cases} \overline{T_c} + C\cos(t - \mu) & m_c < 0 \\ \overline{T_c} & m_c \geq 0 \end{cases} \quad \text{for} \tag{17}$$
$$m_h(t) = B\cos(t - \Phi) \tag{18}$$
$$P_h(t) = \overline{P} + E\cos(t - \Phi) \tag{19}$$
$$T_h(t) = \begin{cases} \overline{T_h} & m_h \leq 0 \\ \overline{T_h} + D\cos(t - \Phi) & m_h > 0 \end{cases} \quad \text{for} \tag{20}$$

where

$$\mu = \arccos\left(\frac{BD}{AC}\right). \tag{21}$$

## NUMERICAL SOLUTION

The governing system of equations was solved using a method of lines (MOL) technique. The "method of lines" is a general descriptor assigned to numerical techniques which, for a system with two independent variables, discretize one and allow the other to vary continuously[14]. This technique translates the system of partial differential equations (PDEs) into a system of coupled ordinary differential equations (ODEs) which can then be evaluated using any one of a number of well-established methods, such as the Runge-Kutta method or Gear's method. For the pulse tube equations, the spatial coordinate is discretized into a linear combination of cubic Hermite polynomials, and the time coordinate is the continuous variable.

The particular numerical solver used, called MOLCH, is commercially available from IMSL. The algorithms upon which MOLCH is based are thoroughly discussed by Madsen and Sincovec[15] and Sincovec and Madsen[16]. In short, the following steps are executed each time the routine is called: the spatial coordinate is approximated in terms of a series function composed of the basis functions for the cubic Hermite polynomials; the system of PDEs is reduced to a system of ODEs by substitution of the spatial approximations into the PDEs; the ODE system is solved using Gear's predictor-corrector method. Gear's method is used because the pulse tube equations are often stiff.

Due to the tendency of the solution to diverge, a marching routine utilizing a four-step time relaxation technique is used to advance the solution from time $t'$ to time $t'+dt$. The first step is to advance the solution one full time interval using MOLCH to obtain the preliminary solution $\overline{U}_{guess}(t'+dt)$. A three-point polynomial curve fit, utilizing the solutions for each component at the times $t'-dt/2$, $t'$, and $t'+dt$, is then used to obtain a value for $\overline{U}(t'+dt/2)$. MOLCH is again used to advance the solution one full time step to $\overline{U}(t'+3dt/2)$. Finally, $\overline{U}(t'+dt)$ is obtained using the same curve-fitting technique with the coefficients this time determined from the solution at $t'$, $t'+dt/2$, and $t'+3dt/2$. The polynomial curve fit approximation described above is based upon the assumption that the solution for each component of the solution vector at a given axial location can be described over a short period of time ($2dt$) by a polynomial equation of the form

$$U^k = at^2 + bt + c . \tag{22}$$

The solution is known at three time steps, so equation (22) yields three equations in the three unknown coefficients $a$, $b$, and $c$ for each component of the solution vector. The calculated coefficients are then substituted into (22) and used to calculate the solution for $U^k$ for the desired value of $t$.

Apart from the four-step relaxation technique, the logic flow is relatively simple. To start the algorithm, a solution $\overline{U}(t = 0)$ is assumed. The solution is then advanced through two complete cycles using the methods described above, and the cycle-to-cycle differences between each component of the solution vector are compared to evaluate whether the solution has converged to within the user-prescribed tolerance (typical value is 5%). Since the initial guess of $\overline{U}(t = 0)$ is arbitrary, the solution never converges on the first two-cycle pass. For the second and all subsequent passes, the solution $\overline{U}(t = 2\tau)$ is used as the initial guess $\overline{U}(t = 0)$ for the next iteration. Typical cases take three to five iterations to converge.

## MODEL CORRELATION

The numerical model was validated against a test solution with boundary conditions characteristic of those for a real pulse tube. The test solution and its derivatives are substituted into the left-hand-side of equations (11) - (13) to determine the non-zero source/sink terms required to force the solution. The numerical solution in the interior of the tube was then calculated using the one-dimensional model. The exact and numerically-predicted solutions agree very well, as demonstrated in Figure 3.

## SAMPLE CALCULATIONS

Kirkconnell and Colwell[17] discuss the sensitivity of the model to several input parameters, such as the compressibility parameter $\Lambda$ and the phase angle $\Phi$, in a paper being written at present. As a first step if the sensitivity analysis, they establish a "baseline" set of boundary parameters based upon the system-level modeling predictions for a 200 Hz, 50 K OPTR. Those boundary parameters are provided in Table 1. A few plots are shown here to illustrate the types of solutions available from the 1-D model. The hypothetical pulse tube cryocooler upon which the baseline parameters are based is physically characteristic of a real PTR, so the physical features of the plots shown are also of interest.

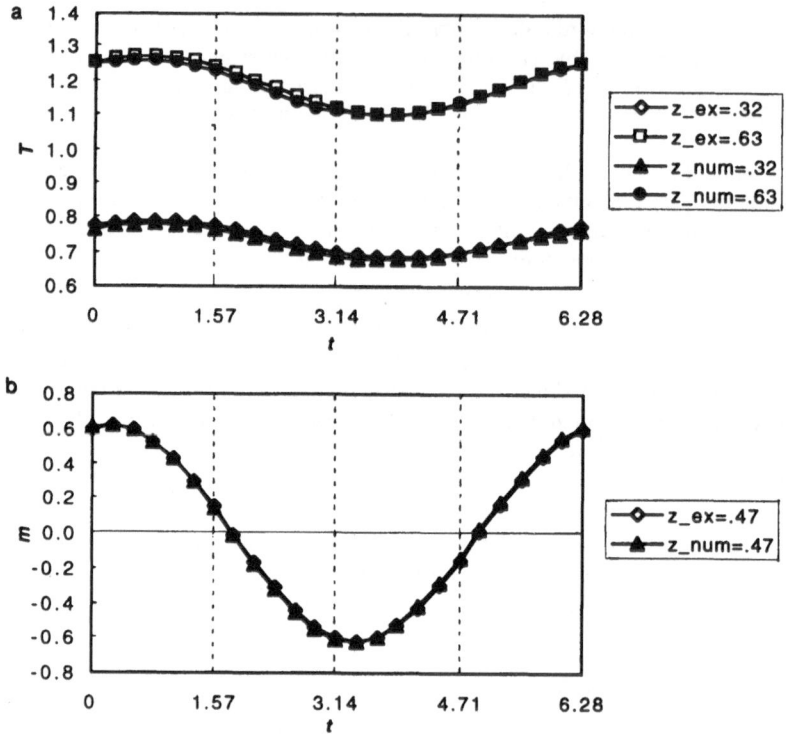

Figure 3. Model validation. (a) temperature, (b) specific mass flow rate. Filled symbols are results of numerical calculations.

Table 1. Baseline Parameters

| Parameter | Value | Reference |
|-----------|-------|-----------|
| A | 1.075 | eqn. (15) |
| B | 0.110 | eqn. (18) |
| C | 0.021 | eqn. (17) |
| D | 0.125 | eqn. (20) |
| E | 0.200 | eqn. (19) |
| $\overline{T_c}$ | 0.286 | eqn. (17) |
| $\overline{T_h}$ | 1.714 | eqn. (20) |
| $\overline{P}$ | 1.000 | eqn. (19) |
| F | $\pi/6$ | eqns. (16), (18), (19), (20) |
| L | 92 | eqn. (14) |

The model can be used to solve for the periodic temperature function, $T(z,t)$, which is plotted parametrically in Figure 4. As expected, the plot shows the maximum and minimum temperatures occurring within the interior of the pulse tube, not at the boundaries. Also of note are the large temperature fluctuations in the middle of the tube compared to the relatively small fluctuations at the ends.

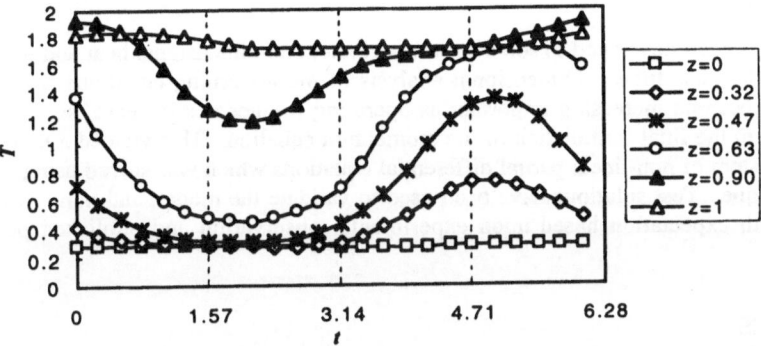

Figure 4. Temperature versus time.

The velocity solution is shown in Figure 5a. Note that the amplitudes of the curves are fairly consistent along the length of the tube and that the curves are essentially sinusoidal, even away from the ends. This is interesting because the mass flow rate waves, shown in Figure 5b, are clearly not sinusoidal in the core, and the amplitudes vary considerably along the length of the tube. The differences between the two are due to the influence of the density solution $(m = \rho u)$. As discussed by Kirkconnell and Colwell[15], the density solution is very sensitive to the value of the compressibility parameter $\Lambda$.

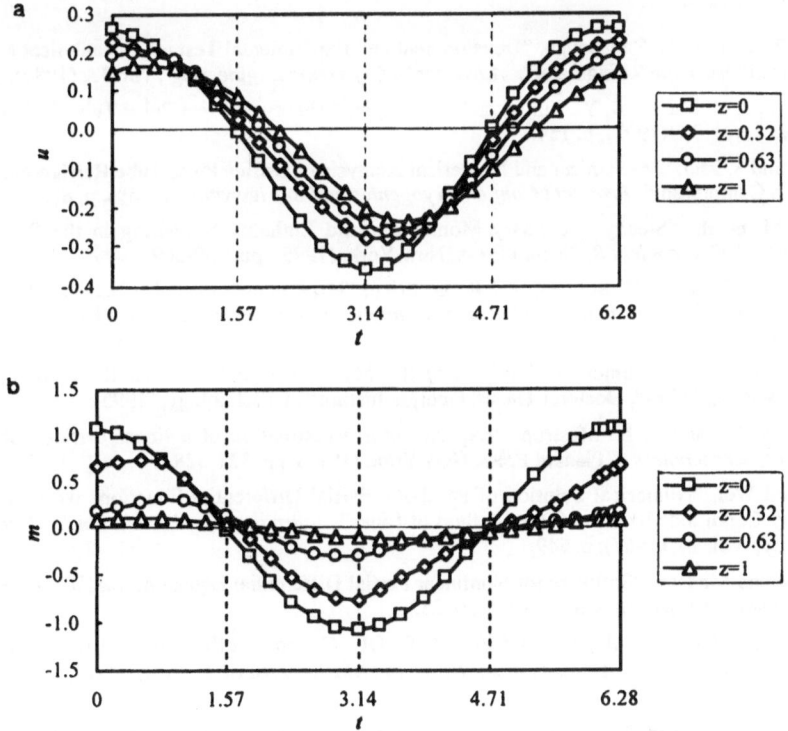

Figure 5. Velocity and mass flow rate versus time. (a) velocity, (b) mass flow rate.

## CONCLUSIONS

A model has been developed to solve for the periodic, one-dimensional heat and mass flows occurring inside pulse tubes. Dimensional analysis of the governing equations revealed that inertial forces become increasing important as operating frequencies increase, so these forces were included in the final formulation of the momentum equation. This yielded a numerically-challenging system of non-linear partial differential equations which was solved using a method of lines technique. Test solutions have been used to validate the model, and numerical results agree well with expectation based upon experimental observation and published pulse tube theory.

## REFERENCES

1.  Gifford, W. E. and R. C. Longsworth, "Pulse Tube Refrigeration Progress", *Advances in Cryogenic Engineering* vol. 10, (1965), p. 69.

2.  Longsworth, R. C., "An Experimental Investigation of Pulse Tube Refrigeration Heat Pumping Rates", *Advances in Cryogenic Engineering* vol. 12, (1967), p. 608.

3.  Mikulin, E. I., A. A. Tarasov, and M. P. Shkrebyonock, "Low-Temperature Expansion Pulse Tubes", *Advances in Cryogenic Engineering*, vol. 29, (1984), p.629.

4.  Russo, S. C., "Hughes Cryocooler Technology", M-CALC Meeting. Oral Presentation, (1996).

5.  Storch, P., Radebaugh, R., and Zimmerman, "Analytical Model for the Refrigeration Power of the Orifice Pulse Tube Refrigerator", *NIST Technical Note 1343*, Boulder, CO, (1991).

6.  Radebaugh , R. C., "Pulse Tube Refrigeration - A New Type of Cryocooler", *Proc 18th Int'l Conf on Low Temperature Physics - Japanese Journal of Applied Physic*, vol. 26 (supplement), (1987), p. 2076.

7.  Storch, P. J. and R. C. Radebaugh, "Development and Experimental Test of an Analytical Model of the Orifice Pulse Tube Refrigerator", *Advances in Cryogenic Engineering* , vol. 33, (1987), p. 851.

8.  David, M., J. C. Marechal, Y. Simon, and C. Guilpin, "Theory of Ideal Pulse Tube Refrigerator", *Cryogenics* vol. 33, (1993), p. 154.

9.  Wu, P. and S. Zhu, "Mechanism and Numerical Analysis of Orifice Pulse Tube Refrigerator with a Valveless Compressor", *Proc Int'l Conf on Cryogenics and Refrigeration*, (1989), p. 85.

10. Lee, J. M. et al., "Steady Secondary Momentum and Enthalpy Streaming in the Pulse Tube Refrigerator", *Cryocoolers 8*, Plenum Press, New York, (1995), pp. 359-369.

11. Kirkconnell, C. S., S. C. Soloski, and K. D. Price, "Experiments on the Effects of Pulse Tube Geometry on PTR Performance", submitted for publication in *Cryocoolers 9*, Plenum Press, New York, (1996).

12. Kirkconnell, C. S., "Numerical Analysis of the Mass Flow and Thermal Behavior in High-Frequency Pulse Tubes", Doctoral Thesis, Georgia Institute of Technology, (1995).

13. Soloski, S. C., and F. N. Mastrup, "Experimental investigation of a linear orifice pulse tube expander", *Cryocoolers 8*, Plenum Press, New York, (1995), pp. 321-328.

14. Hicks and Wei, "Numerical Solution of Parabolic Partial Differential Equations with Two-Point Boundary Conditions by Use of the Method of Lines", *Journal of the Association for Computing Machinery*, vol. 14, (1967), p. 549.

15. Sincovec and Madsen, "Software for Nonlinear Partial Differential Equations", *ACM Transactions on Mathematical Software*, vol. 1, (1975), p. 232.

16. Madsen and Sincovec, "Algorithm 540 - PDECOL, General Collocation Software for Partial Differential Equations", *ACM Transactions on Mathematical Software*, vol. 5, (1979), p. 326.

17. Kirkconnell, C. S. and G. T. Colwell, "A Numerical Analysis of Non-Linear Pulse Tube Flows", submitted for publication in the ASME annual meeting proceeding, (1996).

# Higher Order Pulse Tube Modeling

**J. M. Lee and P. Kittel**

NASA Ames Research Center
Moffett Field, CA 94035

**K. D. Timmerhaus**

University of Colorado
Boulder, CO

**R. Radebaugh**

National Institute for Standards and Technology
Boulder, CO

## ABSTRACT

A linearized model of the pulse tube is computed to second order to study heat transfer and steady mass streaming. A two-dimensional anelastic approximation of the fluid equations is used as the basis for this analysis. Anelastic theory applies because pressure drops in the open tube of the pulse tube are negligible; it allows the equations to describe compression and expansion of the gas without mathematical formation of shocks. The calculated results are given as functions of the dimensionless numbers appropriate for oscillating compressible anelastic flows. These dimensionless numbers were previously described at the 1995 Cryogenic Engineering Conference[1].

The model shows how transverse oscillating heat transfer influences enthalpy flow in the Orifice Pulse Tube. The model also quantifies the steady mass recirculation in the open tube that results from the higher order Reynolds stresses. An interesting result of the linearized approach is that steady mass streaming does not affect the steady refrigeration enthalpy flow at lower order; its effect on enthalpy flow is two orders higher.

## INTRODUCTION

The present two-dimensional pulse tube model is based on an asymptotic expansion (infinite series) solution of the differential fluid equations for mass and energy conservation, and momentum balance[2]. The postulated series solution is valid for very small magnitudes of the nonlinear advection terms and converges to an exact solution of the nonlinear equations for an infinite number of terms. The advantages of such an approach is the ability to approximate the nonlinear terms of the momentum equation as a higher order effect, thereby only having to solve a quasi-linear set of partial differential equations. A numerical solution to the linear set of equations can generally be obtained, and for a number of cases in which sinusoidal time dependence is assumed, an analytic solution is available. An asymptotic expansion reduces the effort and computation power necessary when solving the fluid equations compared to a fully discretized numerical approach.

Cryocoolers 9, Edited by R.G. Ross, Jr.
Plenum Press, New York, 1997

After obtaining the leading order solution, nonlinear quantities of higher order are usually directly calculated from the time-averaged products of the leading order solution. For the pulse tube, the relevant quantities are the time-averaged steady enthalpy flow, higher order steady secondary streaming due to the Reynolds stresses, and the steady temperature. Steady enthalpy flow is the refrigeration effect, steady secondary streaming is large scale mass circulation within the tube, and the steady temperature produces steady heat transfer between the tube wall and the gas. Steady secondary streaming and steady heat transfer represent loss mechanisms and so are important when optimizing pulse tubes.

In this paper, we will focus on how these effects interrelate. Our approach to modeling is a 2D linear anelastic approximation which can directly calculate: 1. enthalpy flow with thermal and viscous diffusion (important near the tube wall), 2. steady secondary streaming losses, and 3. steady heat transfer to the wall.

## MODELING APPROACHES

A qualitative comparison of different modeling approaches is useful for understanding advantages and disadvantages. This is shown in Table 1. Table 1 shows that 2D modeling can yield the highest return on modeling effort. With 2D modeling, transverse diffusion effects (which represent heat transfer between the gas the tube wall, and viscous effects) can be calculated to obtain transverse temperatures, steady mass streaming and steady higher order heat transfer. The information obtained from a 2D model is significantly more than 1D models, and anelastic approximations are computationally less intensive than fully compressible CFD (computational fluid dynamic) codes.

**Table 1.** Qualitative comparison of different modeling approaches.

| Model Comparison Chart | Integral | Differential Analytic | | CFD Compressible | | | Anelastic |
|---|---|---|---|---|---|---|---|
| | Phasor | 1D | 2D | 1D | 2D | 3D | 2D |
| **Primary Measures** | | | | | | | |
| Refrigeration (enthalpy flow) | √ | √ | √ | √ | √ | √ | √ |
| Temperature, axial profile | | √ | √ | √ | √ | √ | √ |
| Temperature, transverse profile | | | √ | | √ | √ | √ |
| Steady mass streaming losses | | | √ | | √ | √ | √ |
| Heat transfer between tube and gas | | | √ | | √ | √ | √ |
| **Secondary Measures** | | | | | | | |
| Temperature dependent properties | | † | † | √ | √ | √ | † |
| Oscillating temperature at tube ends | √ | √ | | √ | √ | √ | √ |
| Nonlinearities | | | | √ | √ | √ | √ |
| Buoyancy effects (free convection) | | | √ | | √ | √ | √ |
| Flow end effects | | | | | √ | √. | √ |
| **Qualitative Measures** | | | | | | | |
| System optimization | | √ | √ | √ | √ | √ | √ |
| Transient simulation | | | | √ | √ | √ | √ |
| Easy to understand physics | √ | √ | √ | | | | |
| Computer system required/speed (PC=Personal Computer, WS=WorkStation) | pencil/fast PC/faster | PC/fast | **PC/med WS/fast** | PC/med WS/fast | WS/med Cray/fast | Cray | **PC/med WS/fast** |
| Status on progress Scale 0-5 | developed | moderate | **moderate** | moderate | minimal | none | **none** |
| (0=no progress, 5=fully developed) | 5 | 5 | 4 | 4 | 1 | 0 | 1 |

($\dagger$ = correction at $O(\varepsilon)$)

Our present 2D model uses an anelastic approximation[3] of the equations. This allows decoupling between the momentum and mass conservation equations. The result is significantly decreased computation times since there is no numerical formation of shocks. Anelastic approximations are valid in the limit of small pressure drops relative to bulk pressure oscillations (due to bulk compression and expansion of the fluid). For a typical pulse tube, pressure drops are $O(10^{-5})$ that of bulk pressure oscillations, where "$O$" means of order. The results we present here are for an analytical solution to the anelastic equations for zero leading order temperature gradients. Future work will concentrate on directly solving for the leading order temperature through numerical solution of the 2D anelastic equations.

## SYSTEM

The system under consideration is shown in Figure 1. Gas contained in a cylindrical tube of thin but finite wall thickness extends over the domain $r = 0$ to $r = 1$, and $z = 0$ to $z = 1$. The tube wall domain, extends from $y = 0$ to $y = 1$, and $z = 0$ to $z = 1$. Planar geometry is used for the tube wall domain. This is valid in the limit of a very thin-walled tube relative to the tube radius. Solutions of the two domains are coupled through the boundary conditions between the gas and the tube wall interface such that the temperature and heat flux must be continuous across the interface.

The normalized velocity field for the gas domain is composed of axial velocity, $u$, and radial velocity, $v$. The periodic boundary conditions for the *amplitude* of $u$ are: at $r = 0$ and $z = 0$, $u = U_o = 1$; and at $r = 0$ and $z = 1$, $u = U_L$, where $U_L$ is normalized relative to $U_o$. The boundary conditions for temperature are: at the centerline ($r = 0$) the radial heat flux is zero; at the inner surface of the tube wall ($r = 1$, $y = 0$), the temperature and heat flux are continuous between the gas and the wall; and at the outer surface of the tube ($y = 1$) the radial heat flux is zero (adiabatic).

For the gas domain, the fact that the tube radius is much smaller than the tube length allows the $r$ - momentum equation to be decoupled from the $z$ - momentum equation by approximating $p_{,z} \gg p_{,r} \approx 0$, where $p_{,z}$ means partial derivative of $p$ with respect to $z$. Consequently, the pressure, $p$, depends only on $z$ and $t$. The thermodynamic variables of temperature, $T$, and density, $\rho$, are functions of spatial coordinates $r$ and $z$, and of time, $t$. For the tube wall domain, the tube wall temperature, $\theta$, is a function of $y$, $z$ and $t$.

The dimensionless form of the fluid equations for the 2-D axisymmetric system for an ideal gas are given in a previous paper[1]. Mass, and energy conservation, momentum balance, and the equation of state define the system, along with the boundary conditions. These equations are linearized for the leading order problem, which is valid for the small displacement length ratios, $\varepsilon$. This dimensionless number, along with the other relevant dimensionless numbers that fully describe the system are given in Table 2. Typical order of magnitudes for each dimensionless number applicable to pulse tubes are also given.

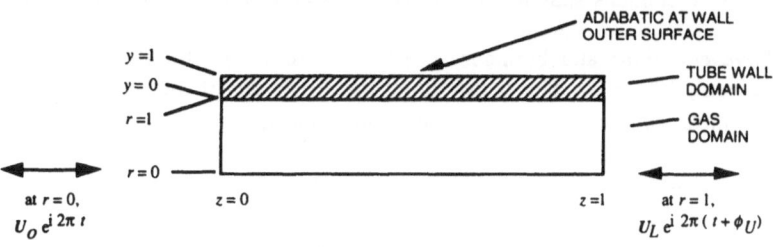

**Figure 1.** System

**Table 2.** Dimensionless Numbers

| | Order | Name | Definition | Comments |
|---|---|---|---|---|
| $\varepsilon$ | $10^{-1}$ | expansion parameter | $U_0^*/(\omega^* L^*) = d^*/L^*$ | ratio of displacement length, $d^*$, to tube length $L^*$ |
| $\lambda$ | $10^{-5}$ | "shock number" | $M^2/(\varepsilon\gamma) = \dfrac{1}{\gamma}\dfrac{U_0}{a}\dfrac{\omega L}{a}$ | ratio of constant pressure to constant volume heat capacities, $\gamma = C_p^*/C_v^*$ |
| $¥^2$ | 1 - 500 | Valensi Number | $r_w^{*2}\omega^*/v^*$ | ratio of tube inner radius to viscous diffusion length |
| $P^2$ | 0.7 | Prandtl Number | $v^*/\alpha^*$ | ratio of viscous to thermal diffusion length scales |
| M | $O(10^{-3})$ | Mach Number | $U_0^*/\sqrt{\gamma R T_0^*}$ | ratio of velocity at $z = 0$ to speed of sound |
| $F^2$ | $\approx 1$ | reciprocal Fourier Number | $\alpha_t^*/(\omega^* l_t^{*2})$ | ratio of thermal diffusion length to tube wall thickness |
| $r_w^*/L^*$ | $10^{-1}$ | gas domain length ratio | | ratio of tube radius to tube length |
| $l^*/L^*$ | $10^{-2}$ | tube wall length ratio | | ratio of tube wall thickness to tube length |
| $\bar{H}$ | | normalized enthalpy flow | $\bar{H}^*/\bar{H}_{ref}^*$ | ratio of enthalpy flux to reference enthalpy flux $\bar{H}_{ref}^* = \rho_c^* T_c^* U_0^* C_p^* \pi r_w^{*2}$ |
| $U_L$ | $\approx 1$ | velocity ratio | $U_L^*/U_0^*$ | ratio of velocity amplitude at $z = 1$ to amplitude at and $z = 0$ |
| $\phi_U$ | $\approx -0.05$ to -.15 | velocity phase angle | | velocity phase angle at $z = 1$ relative to $z = 0$ |

To illustrate the values of the dimensionless numbers, consider a small-sized, moderately-high frequency pulse tube operating at 50 Hz, with a thin-walled titanium steel tube of wall thickness $l^* = 0.02$cm, inner radius $r_w^* = 0.3$cm, length $L^* = 7$cm and thermal diffusivity $\alpha_w^* = O(10^{-5} m^2 / \text{sec})$; helium gas at 1MPa mean pressure; Prandtl Number $P^2=0.7$, and the speed of sound $\sqrt{\gamma R T_0} = 10^3 m/\text{sec}$. These values for $\varepsilon = 10^{-1}$ give $¥^2 = 225$, $M = 2.2\times10^{-3}$, $\lambda = 2.9\times10^{-5}$, and $F^2 = 1.67$, with $M^2/¥^2 = O(10^{-8}) \ll \varepsilon^2$. The last relation, which represents viscous dissipation, shows that it is not important.

The series expansion in the anelastic limit is

$$p(z,t) = 1 + \varepsilon\, p_1(t) + \lambda\, p_2(z,t) + \varepsilon\lambda\, p_3(z,t) + O(\varepsilon^2\lambda) \qquad (1)$$
$$\rho(r,z,t) = 1 + \varepsilon\rho_1(r,z,t) + \varepsilon^2\rho_2(r,z,t) + O(\varepsilon^3)$$
$$T(r,z,t) = 1 + \varepsilon T_1(r,z,t) + \varepsilon^2 T_2(r,z,t) + O(\varepsilon^3)$$
$$u(r,z,t) = u_0(r,z,t) + \varepsilon u_1(r,z,t) + O(\varepsilon^2)$$
$$v(r,z,t) = v_0(r,z,t) + \varepsilon v_1(r,z,t) + O(\varepsilon^2)$$

which shows that the first correction to pressure depends only on time while the second correction to pressure contains spatial dependence. This allows decoupling of the momentum equation.

The above equations and boundary conditions are solved for the case of no axial temperature gradient, $\nabla T_0 = 0$. The higher order quantities are of interest, and they follow directly from the linear solution. The steady momentum equation is

$$\frac{(\rho_0 \overline{v_0 u_0})_{,r}}{r} + (\rho_0 \overline{u_0 u_0})_{,z} = -\bar{p}_{3,z} + \frac{1}{¥^2}\frac{(r\bar{u}_{1,r})_{,r}}{r} \qquad (2)$$

where the overbars designate time-averaged quantities. The left-hand-side of Eq. 2 are the Reynolds stresses. The steady velocity $u_1$, and the resulting pressure field $p_3$ are both unknown.

The steady axial mass flux, when integrated over the area normal to the axial flow, must be zero for a closed system

$$0 = \int_0^1 \left( \rho_0 \overline{u_1} + \overline{\rho_1 u_0} \right) r \, dr \cdot \tag{3}$$

The above two equations are used to solve for $\overline{p_3}$ and $\overline{u_1}$. The mass conservation equation is then used to solve for $\overline{v_1}$

$$0 = \frac{\left( \rho_0 \overline{v_1} r + \overline{\rho_1 v_0} r \right)_{,r}}{r} + \left( \rho_0 \overline{u_1} + \overline{\rho_1 u_0} \right)_{,z} \tag{4}$$

The steady second order energy equation is used to find $\overline{T_2}$

$$\frac{1}{P^2 \Psi^2} \frac{\left( r \overline{T}_{2,r} \right)_{,r}}{r} = \nabla \cdot \left( \rho_0 \overline{u_1} + \overline{\rho_1 u_0} \right) \tag{5}$$

which shows how work flow is converted to heat conduction between the gas and the tube wall. In the more general case of $\nabla T_0 \neq 0$, axial heat conduction is also present. The secondary mass flux components are

$$\overline{j}_1 = \rho_0 \overline{u}_1 + \overline{\rho_1 u_0} \quad \text{and} \quad \overline{k}_1 = \rho_0 \overline{v}_1 + \overline{\rho_1 v_0} \tag{6}$$

which describes the steady velocity field for mass streaming. The steady enthalpy flux is

$$\overline{h}_1 = T_0 \left( \rho_0 \overline{u}_1 + \overline{\rho_1 u_0} \right) + \rho_0 \overline{T_1 u_0} \cdot \tag{7}$$

The enthalpy flow is of primary importance

$$\overline{H}_1 = \int_0^1 \left[ T_0 \left( \rho_0 \overline{u}_1 + \overline{\rho_1 u_0} \right) + \rho_0 \overline{T_1 u_0} \right] r \, dr \cdot \tag{8}$$

From the mass conservation equation of Eq. 3, and from the fact that $T_0$ is independent of $r$, the first term in parenthesis of Eq. 8 is zero, hence the enthalpy flow is

$$\overline{H}_1 = \rho_0 \int_0^1 \overline{T_1 u_0} \, dr \tag{9}$$

Eq. 9 is the enthalpy flow the characterizes refrigeration. Note that the losses due to steady recirculating enthalpy streaming, $\rho_0 \overline{u}_1 T_0$, does not affect the enthalpy flow for refrigeration at this order. The enthalpy streaming loss due to the steady mass streaming is, $\overline{u_1 T_2}$; it is an $O(\varepsilon^3)$ effect which is two orders higher than the refrigeration enthalpy flow.

## CALCULATIONS

Calculations for varying $\Psi^2$ and $\phi_U$ are given below. Parameters for the test case are $\varepsilon = 0.1$, $U_L = 1$, $r_w^*/L^* = 0.1$, $l^*/L^* = 8.5 \times 10^{-4}$, helium gas with $\gamma = 5/3$ and $P^2 = 0.7$. Changes for $\Psi^2$ where done by changing the frequency (as opposed to changing the tube radius). Changing the frequency changes the $\Psi^2$ and M. However, since $M = O(10^{-3})$, M never approaches shock conditions, hence any pressure wave energy (energy due to pressure gradients)

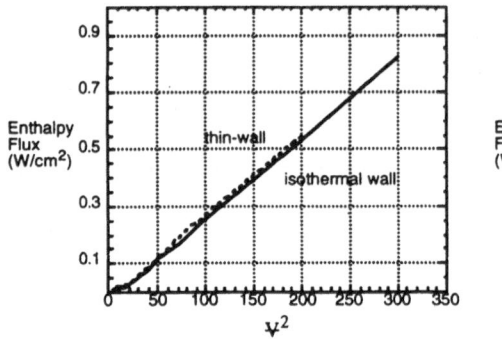

**Figure 2.** Enthalpy Flux vs. $\Psi^2$

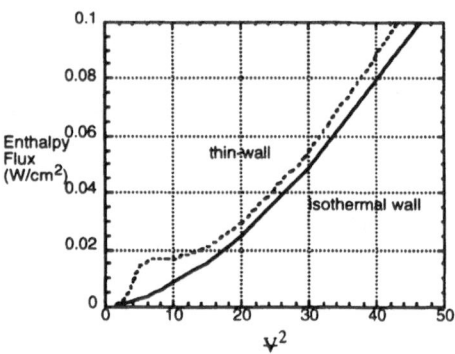

**Figure 3.** Enthalpy Flux vs. $\Psi^2$ for $\Psi^2 < 50$

is negligible throughout the range of $\Psi^2$ investigated, and therefore is not significant at the order of calculation.

Figures 2 and 3 show calculated heat flux vs. $\Psi^2$ for both isothermal tube walls and thin-wall systems. From Figure 2, for $\Psi^2 > 40$ it is linear, and from Figure 3 for $\Psi^2 < 40$ it is nonlinear. In both cases, heat transfer between the gas and the tube wall is present. However, for $\Psi^2 < 40$, the effect of heat transfer is more important because the thermal diffusion layer penetrates over most of the gas domain, whereas for $\Psi^2 > 40$, the diffusion layer is confined to a thin layer near the tube wall. For small pulse tubes with $\Psi^2 < 40$, heat transfer is significant.

Thin-walled tubes in which the tube wall thickness is less than the thermal diffusion length in the tube wall ($F^2 \ll 1$) can increase enthalpy flux. This is shown in Figure 3 where for $\Psi^2 \approx 5$ there is a large increase in enthalpy flux relative to the isothermal case. Figure 4 examines this more closely by plotting the ratio of enthalpy flow to work flow. This is a measure of efficiency. Large efficiencies may be obtained for small $\Psi^2$. This is due to the amplitude and phase shift changes of the temperature at the tube wall for a thin-walled system. Appropriate sizing the tube wall by sizing $F^2$ can increase efficiencies, particularly for small $\Psi^2$. This would be applicable to regenerators, thermoacoustic stacks, and very small, low frequency pulse tubes.

Figure 5 shows how $\phi_U$ and $T_w$ change for increasing $\Psi^2$, with $\Psi^2/F^2 = 5 \times 10^3$. At low $\Psi^2$ (high $F^2$), $T_w$ is small and has a phase angle that leads the velocity at $z = 0$. The reason why $T_w$ is small is because for small $\Psi^2$, the heat capacity of the gas is very small relative to the heat capacity of the tube wall, and so any temperature increase due to pressure increase is significantly damped by the tube wall throughout the gas domain. As $\Psi^2$ increases, $T_w$ steeply increases to a maximum and $\phi_U \approx 0$. Gas velocity and temperature are now in phase thereby maximizing enthalpy flow and efficiency. This is shown in Figure 4, with a maximum at about $\Psi^2 \approx 5$. Further increasing $\Psi^2$ results in decreasing temperature amplitude and $\phi_U$ turns negative which means that temperature at the wall now lags the velocity at $z = 0$. Still further increases in $\Psi^2$ results in a second increase in $T_w$ and $\phi_U$, which again increases efficiency. Further increases in $\Psi^2$ results in $T_w$ going to zero where now the thermal diffusion in the tube wall is less than the wall thickness. At this limit, the thin-walled case approaches isothermal conditions.

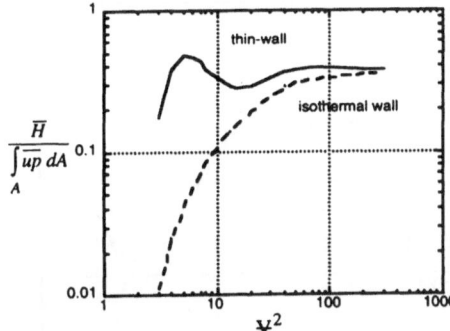

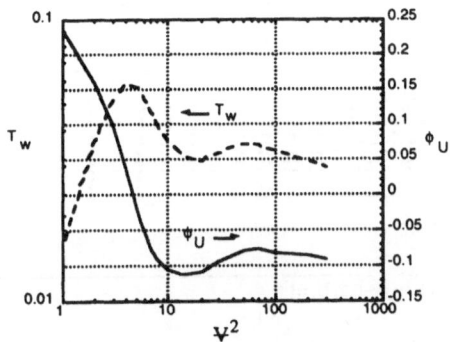

**Figure 4.** Efficiency vs. $\Psi^2$ for a thin-wall tube and an isothermal tube wall.

**Figure 5.** Temperature amplitude and phase angle at the interface between the gas and the tube wall.

Figure 6 shows the ratio of $\overline{H}$ to an ideal 1D enthalpy flow calculation[4], $\overline{H}_{id}$. $\overline{H}_{id}$ is calculated based on both adiabatic and isothermal pressure ratio limits. Figure 6 shows that for small $\Psi^2 < 50$, $\overline{H}/\overline{H}_{id}$ decreases significantly. This is because temperature and velocity, the product which is enthalpy flux, decrease to zero near the tube wall, in order to satisfy temperature continuity and no-slip. Hence, enthalpy flux decreases. For small $\Psi^2$ (small diameter tubes and low frequency systems) this effect is significant. For large $\Psi^2 > 100$ (large diameter tubes, high frequency systems), $\overline{H}/\overline{H}_{id}$ reaches a plateau and is a factor 0.7 (for isothermal pressure ratio calculations) and 0.4 (for adiabatic pressure calculations ) less than ideal flow, respectively. $\overline{H}/\overline{H}_{id}$ is significantly less than 1 for large $\Psi^2$ which is again due in part to decreased enthalpy flow in the diffusion layer. However this does not fully account for the difference, and may be due to the ability of the present model to directly calculate phase angles between oscillating velocity and oscillating temperature.

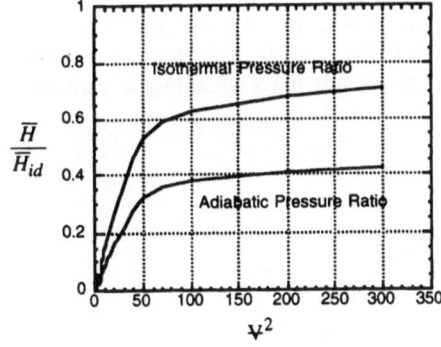

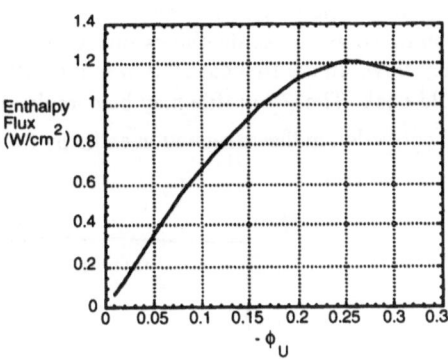

**Figure 6.** Enthalpy Flow ratio vs. $\Psi^2$. $\overline{H}_{id}$ calculations base on adiabatic and isothermal pressure ratios.

**Figure 7.** Enthalpy flux vs. $\phi_U$.

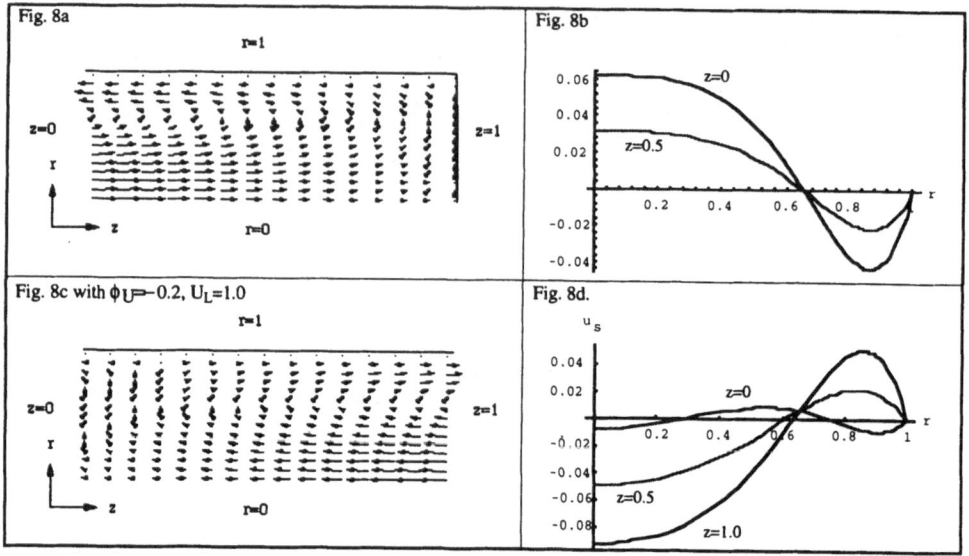

**Figure 8.** Steady secondary velocity fields and plots for a BPT and OPT. $\varepsilon = 0.1$, $\Psi^2 = 50$, $P^2 = 0.703$, $U_L = 1.0$, $F^2 = 0.10$, $\gamma = 5/3$.

Figure 7 plots enthalpy flux vs. $\phi_U$ for the conditions of $\varepsilon = 0.1$, $\Psi^2 = 250$, $P^2 = 0.703$, $U_L = 1.0$, $F^2 = 0.05$, $\gamma = 5/3$. The figure shows that a peak in enthalpy flux at $\phi_U \approx -0.25$. Radebaugh has reported that for a typical PT, $\phi_U \approx -0.1$, thus it would appear there is significant room for improving performance.

The capabilities of our 2D model can be extended to directly calculate higher order steady secondary mass flows. Figure 8 shows results for the case of $\varepsilon = 0.1$, $\Psi^2 = 50$, $P^2 = 0.703$, $U_L = 1.0$, $F^2 = 0.10$, $\gamma = 5/3$. Figure 8a and 8c show the velocity fields for basic and orifice pulse tube configurations, respectively. Figures 8b and 8d show the local axial velocity values for the corresponding fields. Large scale recirculation of $O(\varepsilon)$ is seen to be present.

The steady radial temperature for the above case is shown in Figure 9 for an isothermal wall and a thin-walled tube. For the isothermal condition, enthalpy flow is converted to heat conduction to the wall. For the thin-walled case, heat is transferred from the wall and converted to enthalpy flow. In the first case this would result in wall heating, while in the second case the wall would cool. The $\bar{T}_2$ profile may be combined with the steady mass streaming to estimate the $O(\varepsilon^3)$ correction in the refrigeration enthalpy flow.

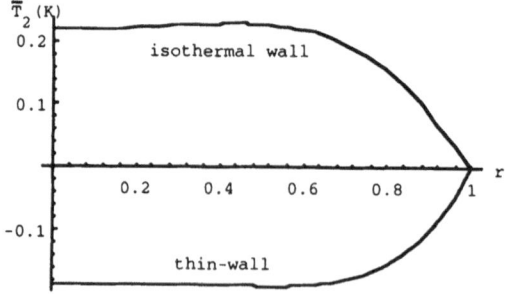

**Figure 9.** Steady $O(\varepsilon^2)$ radial temperature.

## SUMMARY

The core computation engine for a two-dimensional model has been developed. It is now being validated and refined for calculating enthalpy flow in the open tube of the pulse tube. It is capable of calculating diffusion effects within the gas, heat transfer between the gas and a tube wall of finite thickness, steady mass streaming, and steady radial temperatures. Optimizing the pulse tube using direct calculations of these higher order effects is now possible. Future work will incorporate these losses into our present 1D model[5]

## REFERENCES

1. Lee, J.M., Kittel, P., Timmerhaus, K. D. , Radebaugh, R., "Useful Scaling Parameters for the Pulse Tube", *Advances in Cryogenic Engineering*, Plenum Press, New, York (1996) p. 1347.

2  Lee, J.M., Kittel, P., Timmerhaus, K. D. , Radebaugh, R., "Steady Secondary Momentum and Enthalpy Streaming in the Pulse Tube Refrigerator", *Cryocoolers 8,* Plenum Press, New, York (1996) p. 359.

3. Sherman, F. S., *Viscous Flow,* McGraw-Hill, New York (1990) p. 82.

4. Radebaugh, R., "A Review of Pulse Tube Refrigeration", *Advances in Cryogenic Engineering,* Plenum Press, New, York (1996) p. 1196, Eq. 8

5. Roach, P. R. and Kashani, A., "A Simple Modeling Program for Orifice Pulse Tube Coolers", paper presented at the 9th International Cryocooler Conf., Waterville Valley, New Hampshire, June 25-27, 1996.

# Visualization Study of Velocity Profiles and Displacements of Working Gas Inside a Pulse Tube Refrigerator

**Masao Shiraishi,\* Nobuhiko Nakamura\*\***
**Kazuya Seo\*\* and Masahide Murakami\*\***

\*Mechanical Engineering Laboratory, Tsukuba, Ibaraki, 305 Japan
\*\*University of Tsukuba, Tsukuba, Ibaraki, 305 Japan

## ABSTRACT

The flow behavior inside a basic and an orifice pulse tube refrigerator has been observed during one compression-expansion cycle by means of a smoke-wire flow visualization technique. The pulse tube was operated with a comparatively large pressure amplitude, and the velocity profiles and displacements of the working gas were observed at the cold end, at the middle of the pulse tube, and at the hot end. The suction and blowout of the working gas between the hot end of the pulse tube and the reservoir was also observed.

The orifice pulse tube has larger velocities and displacements as compared to the basic pulse tube. The velocity oscillation in the basic pulse tube is mainly governed by the rate of increase (decrease) of pressure in the pressure oscillation, i.e. dP/dt. On the other hand, in the case of the orifice pulse tube, in addition to dP/dt, the pressure difference between the pulse tube and reservoir also has a large effect on the velocity oscillation. Typical velocity profiles of viscous oscillating flow are observed in the beginning of both processes. Moreover, the existence of a secondary flow is observed and its velocity is estimated.

## INTRODUCTION

The pulse tube refrigerator discovered by Gifford and Longsworth[1] is a new type of refrigerator. Many researchers have made strenuous efforts to improve its performance[2] and the lowest temperature that has been reached is in the range of under 4 K.[3] Recently, the pulse tube has attracted the attention of many researchers as a new cooling system applicable not only for space and cryogenics, but also for air-conditioning and commercial refrigeration; one advantage is that it has no CFC (chlorofluorocarbon) and HCFC (hydro-chlorofluorocarbon) refrigerants. While many efforts have been made to improve the performance of the pulse tube refrigerator and to develop a practical system, few systems are commercially available now. The main reason is that the basic refrigeration mechanism (and its potentiality) are not well understood.

Although several analytical approaches[4,5,6,7] for understanding pulse tube operation have been presented, only a few experimental approaches have been demonstrated for analyzing the detailed flow behavior. In one such effort, Lee, et al.[8] conducted flow visualization experiments

Cryocoolers 9, Edited by R.G. Ross, Jr.
Plenum Press, New York, 1997

that provided visualization of the internal flow in the pulse tube and observed a secondary flow phenomena; this secondary flow behavior during one compression-expansion cycle is not well understood, but is thought to be a loss mechanism. Because the basic refrigeration mechanism is strongly determined by the characteristics of the flow during one cycle, it is thought to be very important and essential to achieve a better understanding of these fundamentals.

The objective of this study is to further explore the visualization of flow in the pulse tube refrigerator, and to obtain information concerning the flow field throughout the pulse tube from the cold to the hot end during one cycle.

## EXPERIMENT

A schematic of the experimental apparatus is shown in Fig. 1; this is almost the same as a previous one[9] except for a small change in size of the pulse tube. The pulse tube is made of a transparent acrylate tube having a length of 315 mm, an inner diameter of 16 mm, and a wall thickness of 4 mm. Three pairs of tube fittings are fitted on the tube wall opposite to each other at the middle and near both ends of the tube; these are used to thread a smoke-wire across the tube diameter or to insert a thermocouple. The hot end of the pulse tube is closed by a shiftable end stop (or orifice) to observe the flow behavior near the inlet to the orifice valve by changing the distance between the smoke-wire and the orifice inlet. In this experiment the end stop is placed 5 mm from the smoke-wire at the hot end so that the effective inside length of the pulse tube from the cold side of the regenerator to the closed end is 280 mm. The inner diameter of the tubing between the hot end and the orifice valve is about 4 mm. The regenerator is composed of a stainless steel housing and a Bakelite tube—18 mm in diameter by 168 mm long—packed with 770 discs of 100-mesh stainless steel screen. The reservoir has a volume of about 0.3 m³, which is about 5 times of the pulse tube volume. The hot end of the pulse tube is connected to the warm side of the regenerator through a double inlet valve for future observations. The pulse tube is installed in a vertical orientation.

The smoke-wire is made of 0.1-mm diameter tungsten wire. Both ends of the wire are connected to copper supports that serve as the electrical connections for the wire as well as attachments to the tube. The length of tungsten wire is adjusted to have the same length as the inside diameter of the pulse tube. Air is used as the working gas and pressure oscillation is

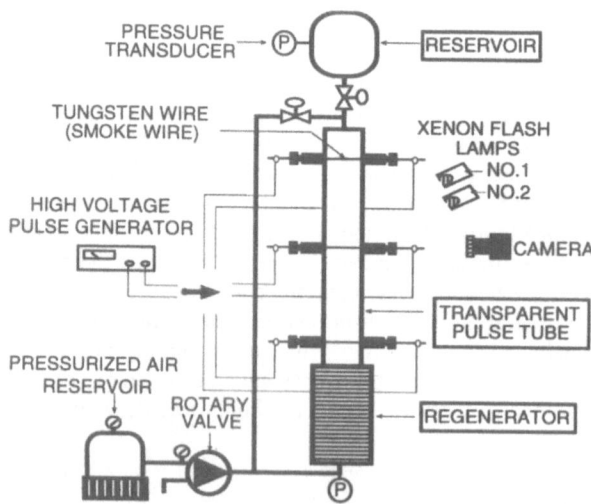

**Figure 1.** Experimental apparatus.

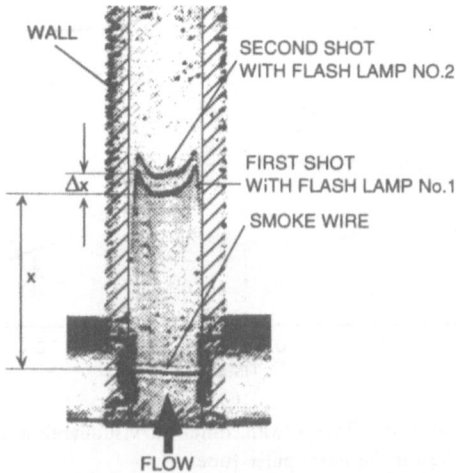

**Figure 2.** Smoke-wire flow visualization in the basic pulse tube. The dispalcement and the velocity are obtained by measuring x and Δx, respectively.

generated by introducing pressurized air of about 0.2 MPa into a rotary valve (Sumitomo Heavy Industries) and then venting the air to the atmosphere. The oscillation of pressure and temperature are monitored using a strain-gauge type pressure transducer (Kyouwa) and a Chromel/Alumel sheathed thermocouple of 0.15-mm diameter with exposed cold junction (Sukegawa Electric). The opening of the orifice valve is optimized by monitoring the difference in gas temperature between the cold and hot ends.

After applying a thin paraffin coating to the wire, the wire is suspended in the pulse tube and connected to a high voltage pulse generator (KI-Tech.). The pulse tube is operated at a frequency of 2 Hz and a compression ratio of about 1.3. The ratio is lower than in the previous experiments[9]; in the previous experiments the displacement was larger than the distance between the smoke-wire and the hot end, thus preventing observation of the full gas displacement.

The visualization is done separately for each half cycle . After a half cycle, the line of smoke is very distorted when the flow direction reverses. As a result, the smoke-lines become very indistinguishable in the next half cycle. From the results of preliminary experiments, the smoke is emitted at the moment when the pressure oscillation reaches its positive and negative peaks—for the basic pulse tube—and when it equals the pressure in the reservoir—for the orifice pulse tube. Because the direction of flow changes at these moments, the velocity of flow is expected to be almost zero. After the desired time delay, the smoke-line is illuminated twice—with a 15-msec time interval between flashes—using two xenon flash lamps, and is recorded by a still video camera (Minolta). A photodiode sensor is used to detect the exact time of illumination, and the correspondence with the pressure oscillation is obtained.

Figure 2 shows a typical result for the basic pulse tube. It must be noted that the shape of the smoke-line does not demonstrate directly the velocity profile at the particular observed position because the shape of the smoke-line is determined by its total history of encountered local velocities. To overcome this, the local velocity is obtained from the difference of displacement between two smoke-lines (Δx) corresponding to the 15-msec time interval between the first and second illuminations. The velocity profiles are also obtained from the radial distribution of Δx. The displacement of gas is determined directly by the distance (x) from the smoke-wire.

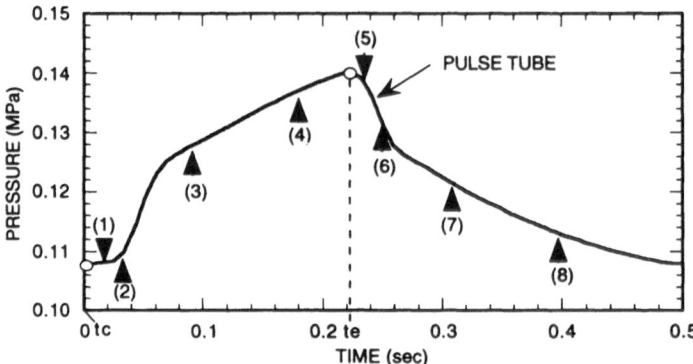

**Figure 3.** Pressure oscillation with time of visualization and emitting smoke for the basic pulse tube.

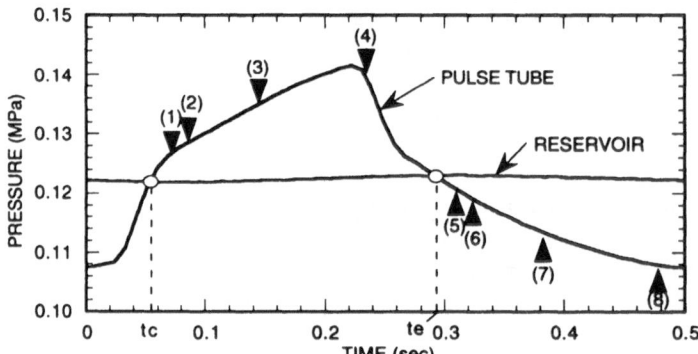

**Figure 4.** Pressure oscillation with time of visualization and emitting smoke for the orifice pulse tube.

## EXPERIMENTAL RESULTS

The basic pulse tube is operated at an optimized frequency of 2 Hz, and the difference in gas temperature between at the cold and hot ends is about 3 K. Figures 3 and 4 show the pressure oscillation during one cycle for the basic and the orifice pulse tube, respectively. The numbers from (1) to (8) correspond to numbers in Figs. 5&7 and 6&8 for each case. Times $tc$ and $te$ in Figs. 3 and 4 correspond to the time smoke is emitted to visualize the compression (points (1) to (4)) and expansion (points (5) to (8)) half-cycles, respectively. The flow is visualized separately at the cold end, at the middle of the tube, and at the hot end. As part of the data processing, the data are subsequently arranged geometrically in the same position as they appear in the apparatus. The results are shown in Figs. 5 and 6.

### Observation for Basic Pulse Tube

Figure 5 shows the typical flow behavior throughout one cycle in the basic pulse tube. One should note that the visualized fluid particles are different at different points in the process. At the onset of each process, the visualized fluid particles exist at the position of the smoke-wire so that the particles under the compression and expansion processes oscillate in the upper and the lower region near the smoke-wire, respectively. The reason why the smoke-line is not found in the expansion process at the cold end (see (6),(7),(8)) is that the smoke-line has flowed into the regenerator. The same is true in Fig. 6.

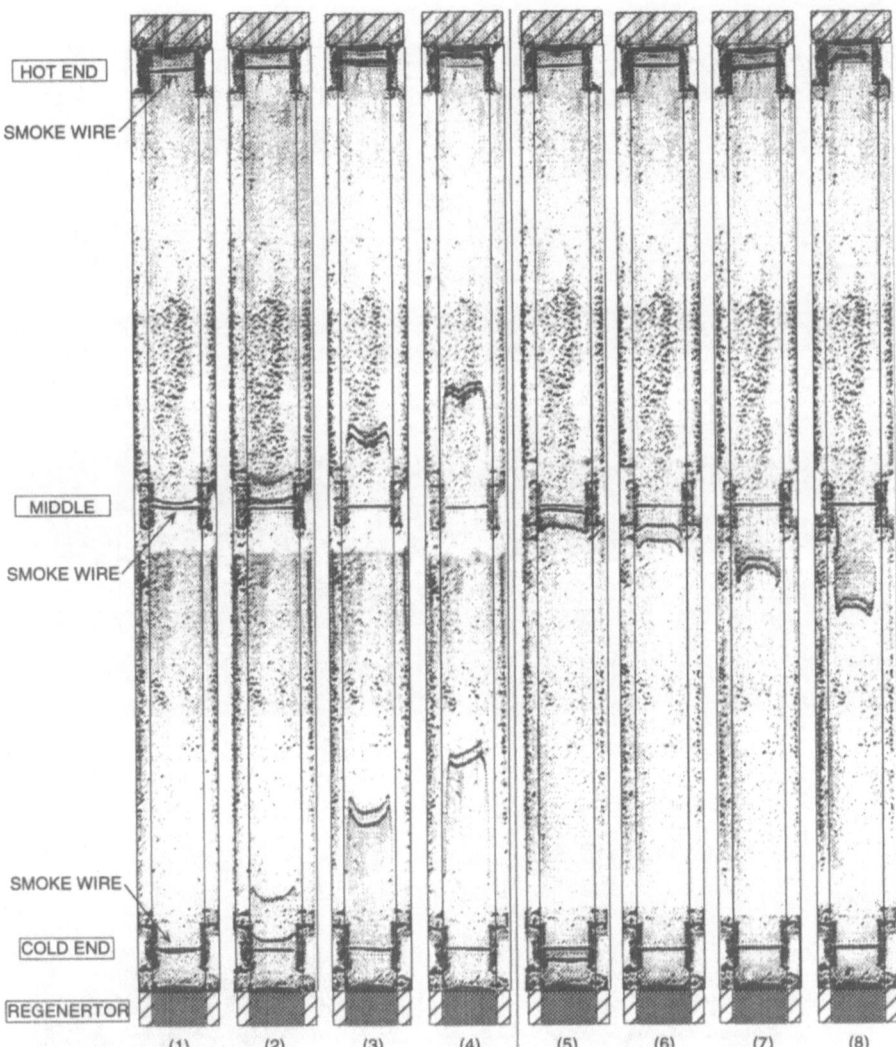

**Figure 5.** Smoke-wire flow visualization photographs in the basic pulse tube during one cycle.

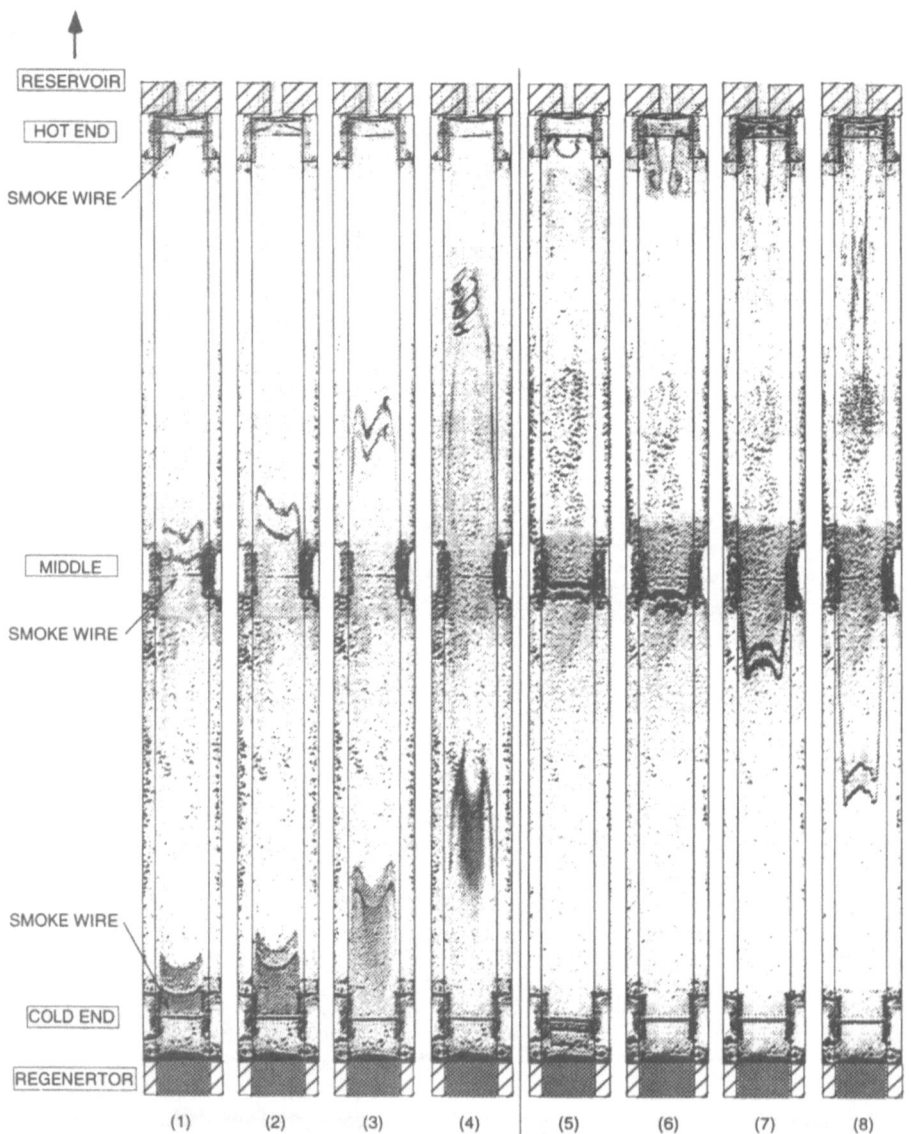

**Figure 6.** Smoke-wire flow visualization photographs in the orifice pulse tube during one cycle.

For the compression process, the displacements at the positions—except the hot end—increase as the process proceeds, but the distance between the smoke-lines decreases. Both the displacement and the distance between the smoke-lines at the cold end are larger than at the middle of the tube. The smoke-line at the hot end does not move except in photographs (4) and (8), in which case the smoke-line shifts slightly upward due to the effects of buoyancy, not due to the oscillating flow. Figure 5 indicates that the displacement and local velocity are strongly dependent on the position along the pulse tube. In the beginning of the process (see (1),(2) and (5),(6)), the smoke-line is concave. This means that the velocity near the wall is larger than at the center line. This behavior is typical for oscillating viscous flows.

## Observation for Orifice Pulse Tube

Figure 6 shows typical flow behavior for the orifice pulse tube under the same conditions as the basic pulse tube. The difference in gas temperature between at the cold and hot ends is about 25 K. There is a significant difference between Figs. 5 and 6. In Fig. 6, the displacement at the middle of the tube is almost the same as at the cold end, not decreasing like the basic pulse tube. The smoke-line at the middle of the tube in Fig. 6, photograph (4) is different from the other typical shapes, which are characterized by a concave front. The Reynolds number of oscillating flow, which is estimated to be about 80 from the velocity amplitude[10], is lower than the critical Reynolds number of 700. This unique pattern is not observed in the expansion process (see (8)). Therefore, the cause of the disturbance is thought to be associated with the sudden contraction of the cross-section at the hot end.

The suction and blowout of gas between the pulse tube and the reservoir is also visible from the flow photographs at the hot end. The direction of flow at the inlet is reversed between photographs (1) and (2). From the photographs of the expansion process (see (5)-(8)) a jet issuing into the pulse tube can be seen, and the formation of vortex rings is recorded (see (6)). Due to the rapid spreading of the smoke and the instability of the jet, the top of the jet is not visible; thus the penetration length of the jet is not clear in this experiment. From photograph (8) it is estimated roughly as over 5 times the pulse tube diameter. This length corresponds to a third of the effective pulse tube length. The typical velocity profiles of the viscous oscillating flow are also observed—in this case in the beginning of the processes.

## Velocity Profiles and Displacements

Based on the visualization results at the middle of the tube, Figs. 7 and 8 note the change in velocity profiles and displacements through one cycle for the basic and orifice pulse tubes, respectively. The numbers above the upper frame correspond to those in former figures. For the basic pulse tube, although the velocity near the wall is greater than at the centerline in the beginning of both processes, near the end of the processes the velocity is smaller than at the centerline. This means that there is a phase difference in the radial direction due to the influence of viscosity. For the orifice pulse tube, the velocity profiles in the beginning of the processes are similar to those of the basic pulse tube, namely, the velocity near the wall leads the velocity at the centerline. Compared to the results in Fig. 7, the change of the velocity profiles is not so large through a cycle, but the displacement is larger.

## Oscillation of Velocity

From the observed velocity profiles, the velocity oscillation at the centerline is obtained and compared with the calculation by assuming a thermodynamic mass balance in the pulse tube under isothermal conditions. Figures 9 and 10 show the results for the basic and the orifice pulse tube, respectively. Figure 9 shows a good agreement between the experiment and the calculation. These results suggest that the velocity oscillation in the basic pulse tube is governed mainly by the rate of increase (decrease) of pressure in the pressure oscillation (dP/dt). On the other hand, in the case of the orifice pulse tube, the agreement in Fig. 10 is not as good as in Fig. 9. In this calculation, an uncertainty factor associated with the orifice size is assumed and a most

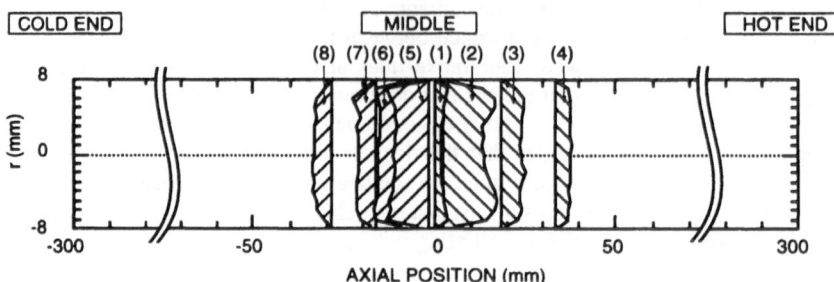

**Figure 7.** Changes of velocity profiles and displacements in the basic pulse tube.

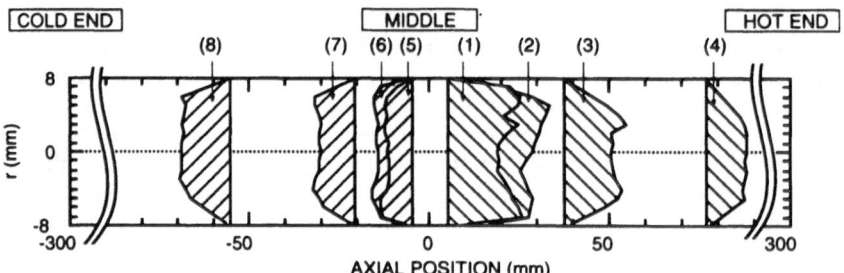

**Figure 8.** Changes of velocity profiles and displacements in the orifice pulse tube.

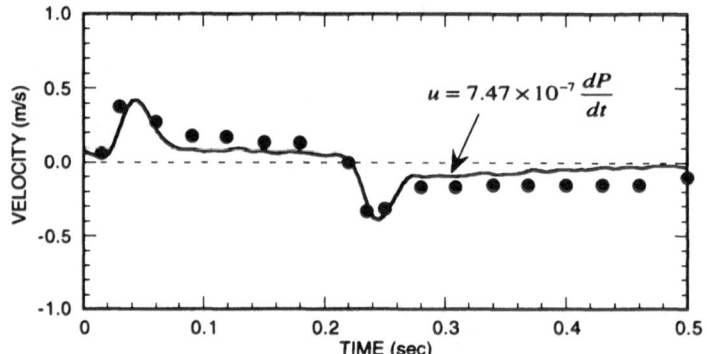

**Figure 9.** Observed velocity oscillation at the middle of the basic pulse tube with calculations.

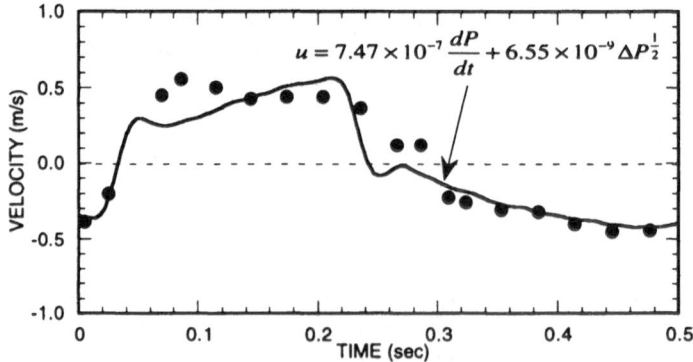

**Figure 10.** Observed velocity oscillation at the middle of the orifice pulse tube with calculations.

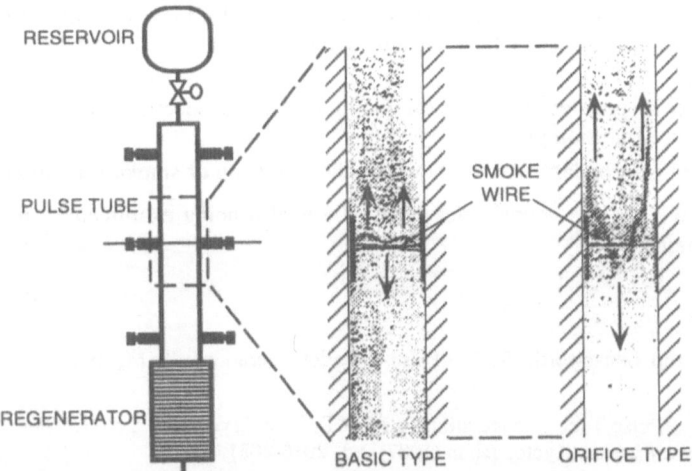

**Figure 11.** Distortions of the smoke-line after one cycle due to a
secondary flow in the basic and orifice pulse tube.

suitable value is selected through comparison with the experiment. Although some differences in the results arise from this, the results in Fig. 10 suggest that, in the orifice pulse tube, not only the rate (dP/dt), but also the pressure difference between the pulse tube and reservoir ($\Delta$P), has a large effect on the velocity oscillation.

### Secondary Flow in the Pulse Tube

A secondary flow, which is thought to be one of the loss mechanisms in a pulse tube, was observed[8] and recently analyzed[11,12] under conditions of small amplitude oscillation. Practical pulse tube refrigerators have a larger amplitude of oscillation so that it is also important to understand the secondary flow behavior under those conditions. Figure 11 shows the distortion of smoke-lines after one cycle at the middle of the pulse tube in which the smoke-line around a center line shifts towards the cold end, and near the wall it shifts towards the hot end. These distortions increase with every cycle and it can be observed as a large scale streaming in the pulse tube. That is, a secondary flow appears in which the fluid particles around a center line flow from the hot end to the regenerator, and the particles near the wall flow from the cold end to the hot end.

The relative velocities of the particles at the centerline and near the wall are estimated from the maximum difference of the displacements in the figures. Those are about 4 mm/sec for the basic, and about 50 mm/sec for the orifice pulse tube. This is only 1 % of the velocity amplitude of the main flow for the basic, and 9 % for the orifice pulse tube.

### SUMMARY

The smoke-wire flow visualization technique has been used to observe the flow behavior inside a basic and an orifice pulse tube refrigerator operated with a comparatively large amplitude of pressure oscillation. The flow field throughout the pulse tube has been observed during one cycle. In the case of the orifice pulse tube, the suction and blowout of the working gas at the hot end has also been visualized. Some results show that:

1. The orifice pulse tube has a larger gas velocity and displacement compared to the basic pulse tube.

2. The velocity oscillation in the basic pulse tube is mainly governed by the dP/dt of the pressure

oscillation. On the other hand, in the case of the orifice pulse tube, not only dP/dt, but also the pressure difference between the pulse tube and reservoir have a large effect on the velocity oscillation.

3. The typical velocity profiles of viscous oscillating flow are observed in the beginning of both compression and expansion processes.

4. The existence of a secondary flow is shown by the distortion of smoke-lines after one cycle.

The flow behavior in the double inlet pulse tube is also being examined by means of this visualization technique.

## REFERENCES

1. Gifford, W.E. and Longsworth, R.C., "Pulse-Tube Refrigeration", ASME paper, No. 63-WWA-290 (1963).

2. Radebaugh, R., "Pulse Tube Refrigeration—a New Type of Cryocooler", *Proc. 18th Int. Conf. on Low Temperature Physics*, Kyoto, Japan (1987), pp. 2076-2081.

3. Matsubara, Y. and Gao, J.L., "Multi-Stage Pulse Tube Refrigerator for Temperatures below 4K", *Cryocoolers 8*, Plenum Press, New York (1995), pp. 345-352.

4. Swift, G.W., "Thermoacoustic Engines", *J. Acoust. Soc. Am.*, vol. 84, no. 4, (1988), pp. 1145-1180.

5. Radebaugh, R., "A Review of Pulse Tube Refrigeration", *Adv. Cryog. Eng.*, vol. 35, (1990), pp. 1191-1205.

6. Kittel, P., "Ideal Orifice Pulse Tube Refrigerator Performance", *Cryogenics*, vol. 32, no. 9, (1992), pp. 843-844.

7. Tominaga, A., "Thermodynamic Aspects of Thermoacoustic Theory", *Cryogenics*, vol. 35, (1995), pp. 427-440.

8. Lee, J.M., Kittel, P. Timmerhaus, K.D. and Radebaugh, R., "Flow Patterns Intrinsic to the Pulse Tube Refrigerator", *Proceedings of the 7th International Cryocooler Conference*, PL-CP-93-1001, Santa Fe (1993), pp. 125-139.

9. Shiraishi, M., Nakamua, N., Seo, K. and Murakami, M., "Investigation of Velocity Profiles in Oscillating Flows Inside a Pulse Tube Refrigerator", Presented at the 16th ICEC/ICMC, May 20-24, 1996, Kitakyushu, Japan.

10. Merkli, P. and Thomann, H., "Transition to Turbulence in Oscillating Pipe Flow", *J. Fluid Mech.*, vol. 68, (1975), pp. 567-575.

11. Lee, J.M., Kittel, P., Timmerhaus, K.D. and Radebaugh, R. "Steady Secondary Momentum and Enthalpy Streaming in the Pulse Tube Refrigerator", *Cryocoolers 8*, Plenum Press, New York, (1995), pp. 359-369.

12. Jeong, E.S., Secondary Flow in Basic Pulse Tube Refrigerators", *Cryogenics*, vol. 36, no. 5, (1996), pp. 317-323.

# Investigation of Radial Temperature and Velocity Profiles in Oscillating Flows Inside a Pulse Tube Refrigerator

K. Seo*, M. Shiraishi**, N. Nakamura* and M. Murakami*

*Institute of Engineering Mechanics, University of Tsukuba, Japan
**Mechanical Engineering Laboratory, MITI, Tsukuba, Japan

## ABSTRACT

Measurements of velocity oscillations have been made in order to understand the basic refrigeration mechanism in typical pulse tube refrigerators. A hot wire probe, equipped with a fine thermocouple, has been used. The radial profiles of the velocity and the temperature in typical pulse tube refrigerators of the basic, orifice, and double-inlet types, have been measured at the cold end, the middle, and the hot end of the pulse tube, by traversing the probe in the radial direction. On the basis of the profiles, velocity distributions along the pulse tube have also been determined, and the work flux of fluid elements has been estimated. It is found that for the basic pulse tube, the radial distribution of the work flux has a maximum value near the wall, and for the orifice and double-inlet type pulse tubes, the radial profiles have a maximum value at the centerline of the pulse tube.

## INTRODUCTION

It is known that there are three typical types of pulse tube refrigerators, the basic[1], the orifice[2,3], and the double inlet[4]. Increasingly lower temperatures have been achieved as the technology has progressed from the basic type, to the orifice type, and on to the double inlet type. Recently, it has been demonstrated that pulse tube refrigerators have achieved comparable refrigeration performance to that of Stirling and G-M refrigerators; the lowest temperature has reached a level of 4 K[5]. Compared to the excellent progress in achieving improved refrigeration performance, the improvement in understanding of the basic refrigeration mechanism has not progressed at an equal rate for several reasons. One of the reasons is that there is little experimental data concerning the flow behavior in the pulse tube, though several analytical models have been presented[6,7]. Although many researchers have made measurements of pressure and temperature of the working gas, only a few reports concerning velocity measurements have been published[8,9]. Velocity measurements of the oscillating helium gas in the practical pulse tube are not easy to make compared to pressure and temperature measurements. The velocity of a gas element relates to its P-V work through its displacement, so velocity is one of the most basic parameters to analyze refrigeration performance. In a previous study[10], we paid much attention to the pressure and the temperature oscillations of the working gas and discussed the behavior of temperature

Cryocoolers 9, Edited by R.G. Ross, Jr.
Plenum Press, New York, 1997

oscillations. This time we try to measure the velocity and estimate the work of the gas element from the pressure oscillation in the practical pulse tube.

The objective of this study is to make velocity measurements and to estimate the work of the gas element on the basis of the velocity distribution in the pulse tube.

## EXPERIMENT

A schematic of the experimental apparatus is shown in Figure 1. It is designed for measurement of the temperature and velocity of the working gas at three points along the pulse tube — at the cold end, at the middle of the pulse tube, and at the hot end. A commercially available helium compressor from a G-M refrigerator (SRDD-208, Sumitomo Heavy Industries) and a rotary valve were employed to generate the pressure oscillation. The dimensions of the pulse tube are 15.6 mm in inner diameter and 288 mm long. The regenerator is composed of a stainless steel housing and a Bakelite tube, 35 mm in diameter by 160 mm long, packed with 865 discs of 100 mesh stainless steel screen. The volume of the reservoir is about $6.0 \times 10^{-4}$ m$^3$, which is about 11 times larger than that of the pulse tube.

The probe, which is composed of a platinum hot wire of 25 $\mu$m diameter, and an E-type sheathed thermocouple 0.15 mm in diameter, is attached to one end of a 4-mm diameter stainless steel tube (see Figure 2). By this probe, it is possible to measure the velocity and temperature variation simultaneously at the same position. The probe and the thermocouple are tightened by Swagelock fittings with Teflon rings for the sake of traversing the probe. The pressure is measured with a strain gauge type pressure transducer.

## CALIBRATION OF HOT WIRE ANEMOMETER

In the present study, it is necessary to calibrate the hot wire anemometer for pressure and temperature because the flow in a pulse tube is accompanied by pressure and temperature oscillations. Moreover, the working gas used is helium, not air. Figure 2 shows a schematic arrangement for the calibration. A constant temperature anemometer is adopted to measure the velocity. The pressure is controlled by a pressure regulator, and the temperature is controlled by heating or cooling the gas line. The relation between the output of the hot wire anemometer V and the velocity U is expressed on the basis of King's law, that is

$$(V^2 - V_0^2) / V_0^2 = a \, U^{0.5}$$

where $V_0$ is the output of the hot wire anemometer when $U = 0$, and $a$ is the coefficient. Both $V_0$ and $a$ depend on temperature and pressure so they are calibrated for helium gas in the tem-

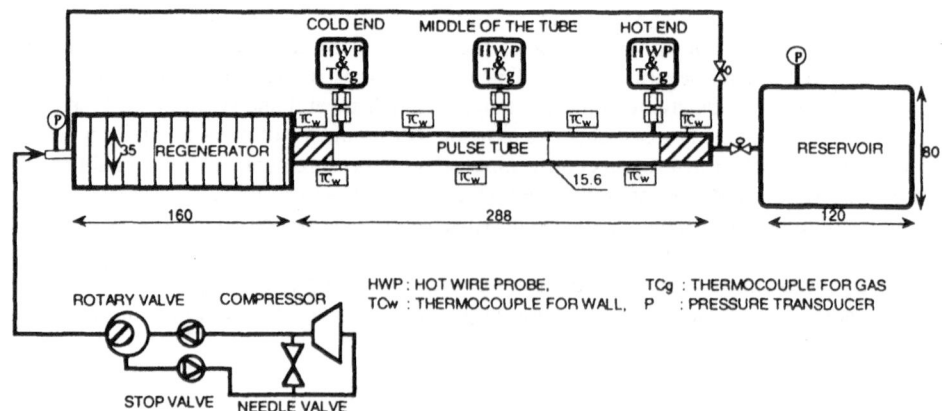

**Figure 1.** Experimental apparatus.

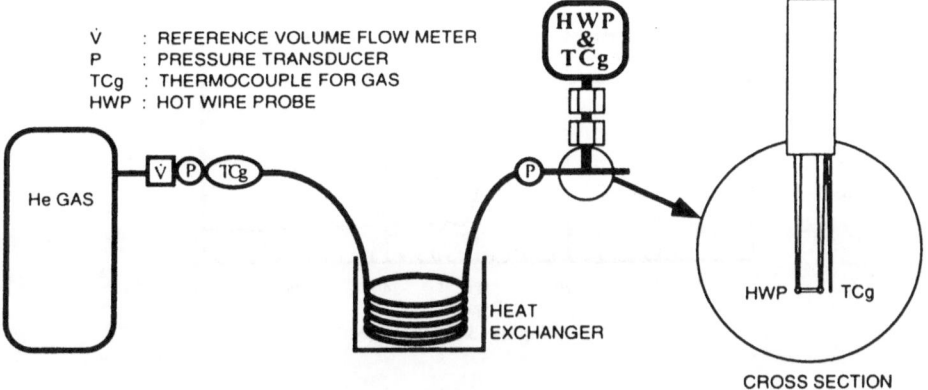

**Figure 2.** Calibration system.

perature range from 220 K to 320 K, and for pressures from 0.1 to 1.5 MPa. The accuracy of the velocity measurement is estimated to be within about ±25%.

The experimental procedure is as follows. After evacuation of the pulse tube, helium gas is charged, and cooling water is circulated through the heat exchanger at the hot end. Then, the compressor unit is started, and the driving frequency and the pressure ratio are adjusted to the prescribed values. After the wall temperature at the cold end and the heat out from the hot end reach steady values, data are acquired. The average pressure is about 1.3 MPa, and the compression ratio is 1.4. The temperature and the velocity of the working gas are measured at the position of the cold end, the middle, and at the hot end of the pulse tube.

## RESULTS AND DISCUSSIONS

Figures 3, 4 and 5 are typical examples of oscillations for each type of pulse tube. Figure 3-a is the pressure oscillation in the basic type pulse tube at 2 Hz, and Figures 3-b through 3-g are corresponding plots for the velocity and the temperature at the cold end, middle, and hot end of the pulse tube. The frequency of 2 Hz is the optimum condition — where the lowest wall temperature is obtained. The x-axis corresponds to the rotation angle of the rotary valve, and the y-axis denotes the radial position. The origin of the x-axis is defined as the onset of the compression process. The wall of the pulse tube is $r = 0$ mm, and the centerline of the pulse tube is $r = 7.8$ mm. It is found that the velocity in the basic type pulse tube rapidly increases and decreases corresponding to the time variation of the pressure, and the amplitude of the velocity oscillation decreases from Figure 3-b to 3-d, that is, with the distance from the cold end. From Figure 3-b the peak of the velocity oscillation at the compression process is in phase in the radial direction. There is a little disturbance near the wall just after the velocity peak at the expansion process. This disturbance causes a larger phase difference between the pressure and the velocity. On the other hand, it is seen from Figures 3-e, 3-f and 3-g that the amplitude of the temperature oscillation increases with distance from the cold end. The ratio of oscillation amplitudes between the velocity and the temperature in the longitudinal direction is consistent with the conservation of enthalpy flow in the pulse tube. From Figure 3-g, it can be seen that the profile of the temperature oscillation near the wall is similar to that of the pressure oscillation, but the data on the centerline differ from a square wave. It is considered that the gas element near the wall can not move owing to the no-slip condition; thus its temperature oscillation directly corresponds to the pressure oscillation. However, the gas element on the centerline is affected by the flow, so that its temperature oscillation is different from that near the wall.

Figure 4 presents the same plots of the velocity and temperature for the orifice type pulse tube at 10 Hz. The flow in the orifice type pulse tube is caused not only by the pressure oscillation, but also by the pressure difference between the pulse tube and the reservoir. The

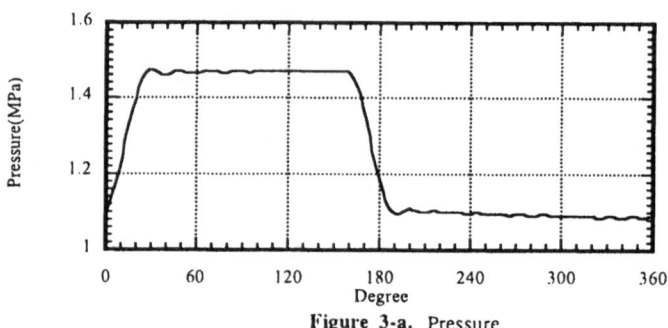

**Figure 3-a.** Pressure.

**VELOCITY**                                          **TEMPERATURE**

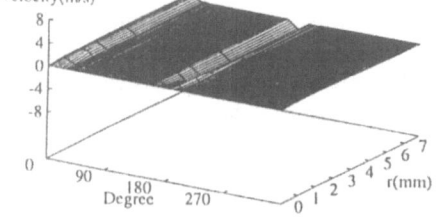 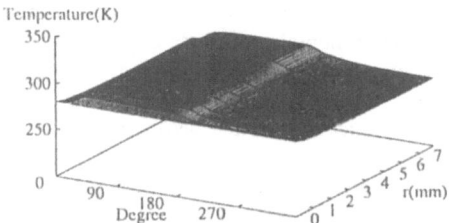

**Figure 3-b.** Velocity at the cold end.              **Figure 3-e.** Temperature at the cold end.

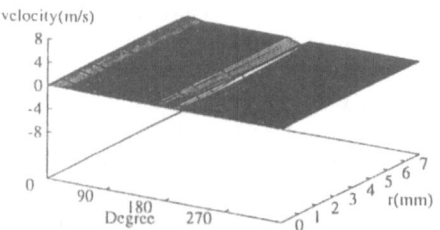 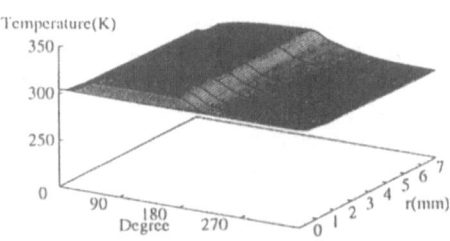

**Figure 3-c.** Velocity at the middle point of the tube.     **Figure 3-f.** Temperature at the middle point of the tube.

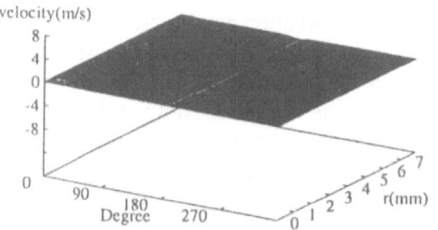 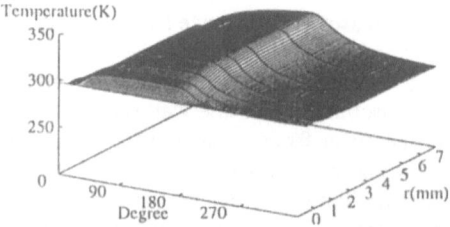

**Figure 3-d.** Velocity at the hot end.              **Figure 3-g.** Temperature at the hot end.

**Figures 3.** Basic type pulse tube at 2 Hz.

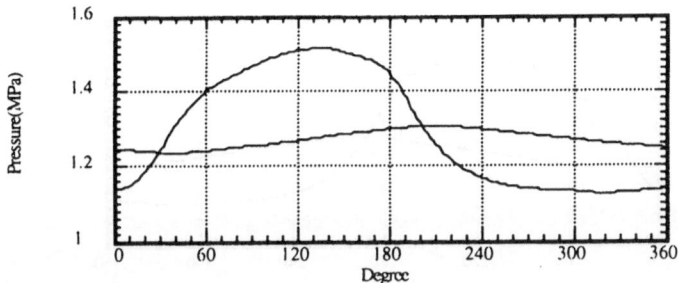

**Figure 4-a.** Pressure.

**VELOCITY**  **TEMPERATURE**

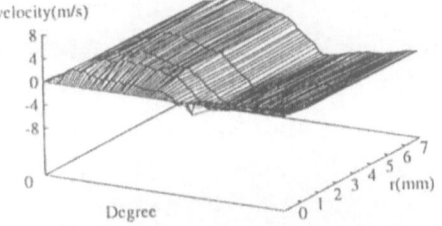

**Figure 4-b.** Velocity at the cold end.

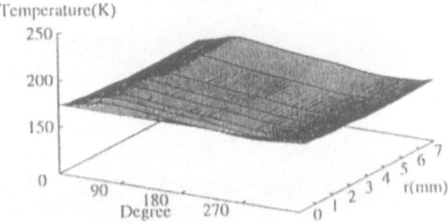

**Figure 4-e.** Temperature at the cold end.

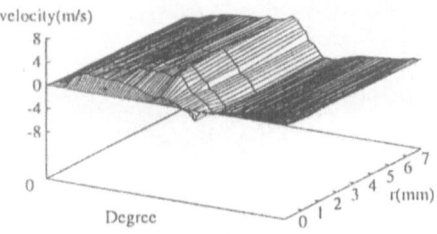

**Figure 4-c.** Velocity at the middle point of the tube.

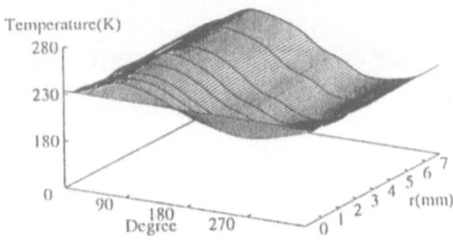

**Figure 4-f.** Temperature at the middle point of the tube.

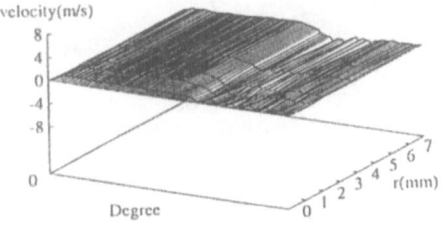

**Figure 4-d.** Velocity at the hot end.

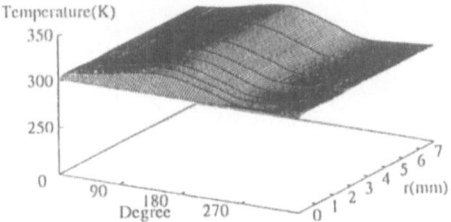

**Figure 4-g.** Temperature at the hot end.

**Figures 4.** Orifice type pulse tube at 10 Hz.

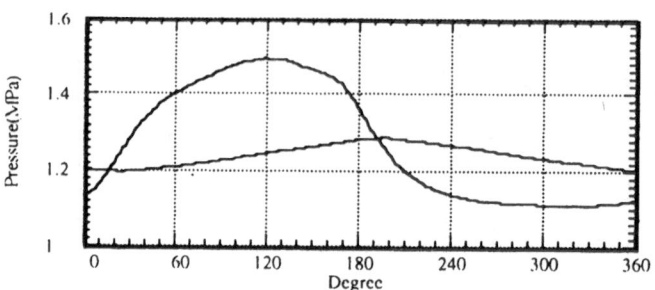

**Figure 5-a.** Pressure.

**VELOCITY**                                    **TEMPERATURE**

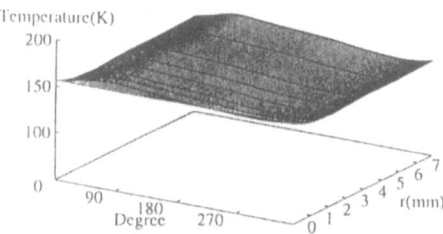

**Figure 5-b.** Velocity at the cold end.              **Figure 5-e.** Temperature at the cold end.

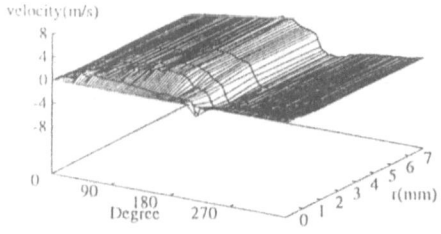

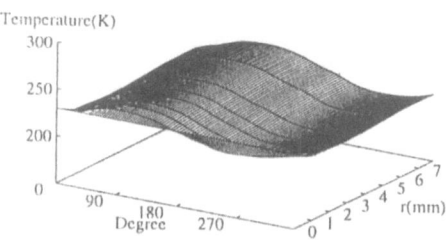

**Figure 5-c.** Velocity at the middle point of the tube.   **Figure 5-f.** Temperature at the middle point of the tube.

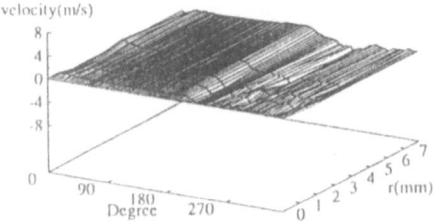

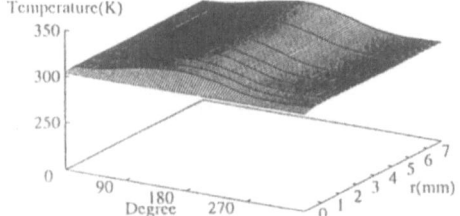

**Figure 5-d.** Velocity at the hot end.              **Figure 5-g.** Temperature at the hot end.

**Figures 5.** Double-inlet type pulse tube at 10 Hz.

amplitude of the velocity oscillation decreases from Figure 4-b to 4-d, that is, with the distance from the cold end; in particular, it decreases dramatically between the middle point and the hot end. The peak of the velocity oscillation on the centerline at the middle point shifts backwards about 70 degrees compared to that at the cold end. At the hot end, there are many disturbances in the expansion process compared to the compression process. There is some phase difference of the velocity in the radial direction in Figures 4-b and 4-c, and the velocity near the wall becomes the maximum value in the radial direction at the beginning of the compression process. This is a typical feature of viscous oscillating flow. As for the temperature, the amplitude at the middle point (Figure 4-f) is the largest of the three measuring positions. The peak of the temperature oscillation on the centerline at the cold end and the middle point shifts forwards compared to the hot end.

Figures 5 presents the same plots of velocity and temperature for the double-inlet type pulse tube at 10 Hz. The amplitude of the velocity oscillation decreases from Figure 5-b to 5-d, that is, with the distance from the cold end. The amplitude of the temperature oscillation at the middle point is the largest among the three measuring positions as in the orifice type pulse tube. The amplitudes of the velocity and the temperature oscillations at the middle point in the double inlet type are smaller than those in the orifice type, though the amplitudes of both types are almost the same at the cold end. The difference in the velocity and the temperature is greatest at the middle point.

Figures 6 and 7 show the radial profiles of the work flux at the cold end, the phase difference between the pressure and the displacement, and the amplitude of the displacement for the basic pulse tube. The flux is estimated from the area of the P-V diagram, while the displacement is derived by assuming a linear velocity distribution between the cold end and the middle point.

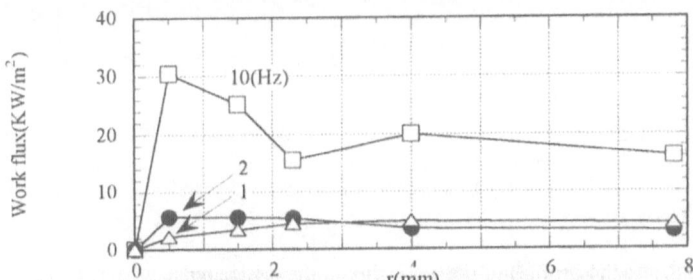

**Figure 6.** Radial profiles of the work flux in the basic type pulse tube.

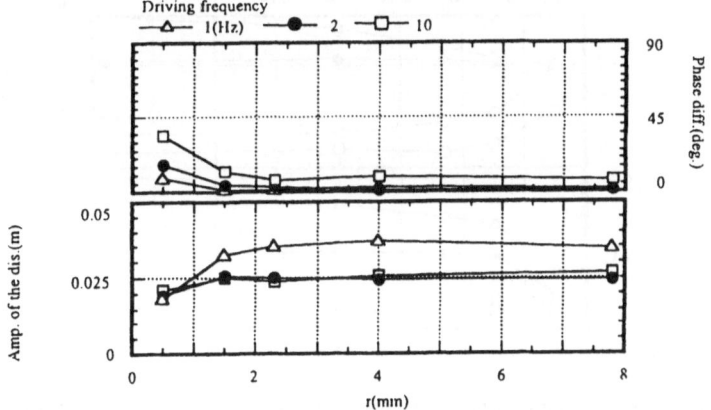

**Figure 7.** Radial profiles of the phase difference between the pressure and the displacement and the amplitude of the displacement in a basic pulse tube.

The driving frequency is used as the variable parameter for the basic pulse tube. The black circular data points in Figure 6 denote data at the optimum condition, where the lowest wall temperature is obtained. It is found that the radial profile depends on the driving frequency, and it has the maximum value near the wall except at 1 Hz. Therefore, the work flux near the wall plays the most important role. Moreover, it should be noted that the total work, which is the integrated value of the work flux in the radial direction, is not directly related to the refrigeration performance between 2 Hz and 10 Hz. It is considered that the total work increases proportionally with the driving frequency since the compression ratio is kept constant, so that the displacement is also constant for all driving frequencies. However, the velocity increases with the driving frequency owing to the constant displacement for all frequencies, so that the frictional loss due to viscosity increases rapidly, and thus the net refrigeration power decreases at higher frequencies.

Figure 7 shows the radial profiles of the amplitude of the displacement and the phase difference between the pressure and the displacement in the basic type pulse tube. The phase of each oscillation is obtained by comparing the fundamental mode. The amplitude of the displacement is almost the same in the radial direction except near the wall, and the phase difference is also the same for all driving frequencies, except near the wall. The phase difference near the wall is large due to the influences of viscosity, and it becomes larger as the driving frequency increases. It is found from Figures 6 and 7 that the phase difference affects the work flux significantly in the range of frequency between 2 Hz and 10 Hz.

Figures 8 and 9 show the radial profiles of the work flux at the cold end, the phase difference

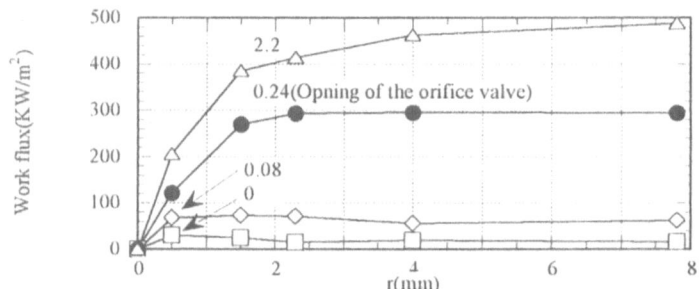

**Figure 8.** Radial profiles of the work flux in the orifice pulse tube at 10 Hz.

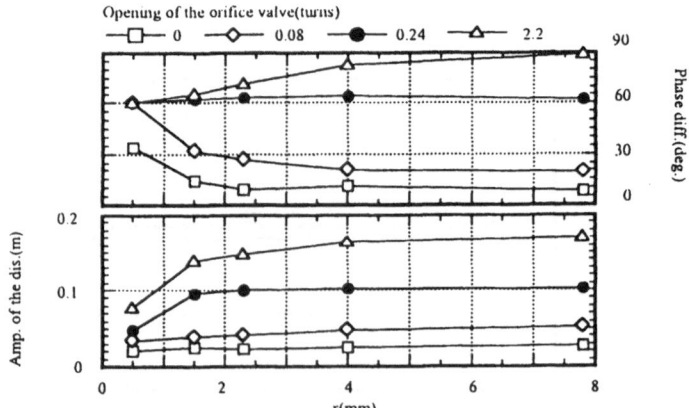

**Figure 9.** Radial profiles of the phase difference between the pressure and the displacement and the amplitude of the displacement in the orifice type pulse tube at 10 Hz.

between the pressure and the displacement, and the amplitude of the displacement for the orifice type pulse tube. The opening of the orifice valve is taken as the variable parameter because it is the most influential parameter on the wall temperature at the cold end. It is found that the radial profiles of the work flux are quite different from that of the basic type at 10 Hz. As shown in Figure 10, the work flux increases with distance from r = 0 mm to r = 2.3 mm at the optimum condition (valve opening = 0.24 turns) and the work flux on the centerline is one order larger than that of the basic type, so the total work is also one order larger than the basic type. However, the total work at the optimum opening in the orifice type is not the largest.

Figure 9 shows the radial profiles of the amplitude of the displacement and the phase difference for the orifice pulse tube. As the opening of the valve increases, the amplitude and the phase difference increase. In particular, the radial profile of the phase difference changes significantly between 0.08 and 0.24 turns. At the optimum, the phase difference and the amplitude are the same in the radial direction except near the wall.

Figures 10 and 11 show the radial profiles of the work flux at the cold end, and the phase and amplitude plots for the double-inlet type pulse tube. The opening of the double inlet valve is taken as the variable parameter in this case. As shown in Figure 10, the wall temperature at the cold end is almost the same over 2 turns. It is found that the work flux of the double inlet type pulse tube becomes smaller with the increased opening of the double inlet valve except near the wall, although it is almost constant over 2 turns. Accordingly, the total work also becomes smaller with increased opening of the double inlet valve. The decrease of the total work is caused by the decrease of the amplitude as shown in Figure 11.

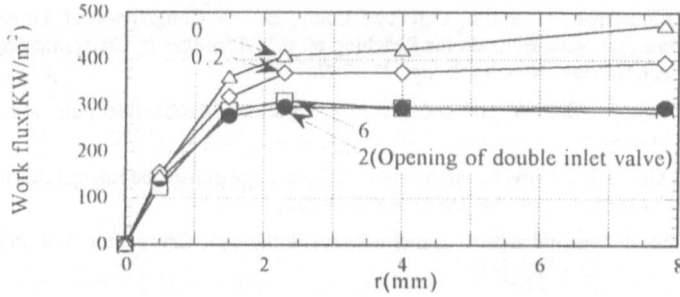

**Figure 10.** Radial profiles of the work flux in the double inlet pulse tube at 10 Hz.

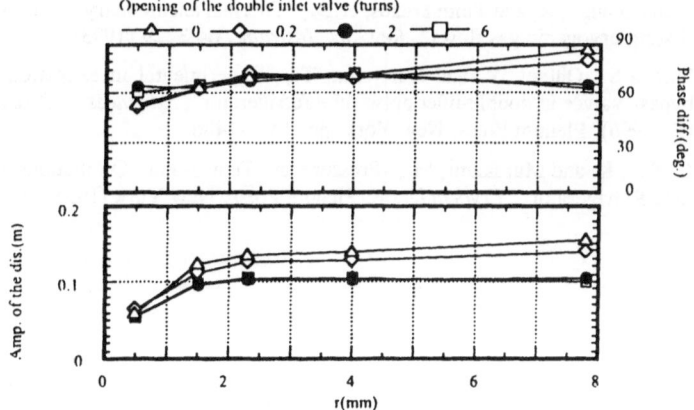

**Figure 11.** Radial profiles of the phase difference between the pressure and the displacement and the amplitude of the displacement in the double inlet pulse tube at 10 Hz.

## CONCLUSION

The velocity oscillations, in addition to the pressure and the temperature oscillations, are measured for three types of typical pulse tube refrigerators. The radial profiles of the work flux are estimated from the area of the P-V diagram by using the velocity distributions in the longitudinal direction. It is found that the radial profile of the work flux at the optimum condition of the basic type pulse tube has the maximum value near the wall, and it is significantly affected by the phase difference between pressure and displacement. On the other hand, the radial profile of the work flux in the orifice and double-inlet type pulse tubes differs from that of the basic type, and it increases with distance from the wall. The total work of the orifice and double-inlet type pulse tubes at the optimum condition is one order larger than that of the basic pulse tube.

## ACKNOWLEDGMENT

This research was partly supported by the Japanese Society for the Promotion of Science.

## REFERENCES

1.  Gifford, W.E. and Longsworth, R.C., "Pulse-tube Refrigerator", *ASME paper*, No. 63-WA-290 (1963).

2.  Mikulin, E.I., Tarasov, A.A. and Shkrebyonock, M.P., "Low Temperature Expansion Pulse Tubes", *Advances in Cryogenic Engineering*, Vol. 29 (1984), Plenum Press, New York, pp. 629-637.

3.  Radebaugh, R., Zimmerman, J., Smith, D.R. and Louie, B., "A Comparison of Three Types of Pulse Tube Refrigerators: New Methods for Reaching 60 K", *Advances in Cryogenic Engineering*, Vol. 31 (1986), Plenum Press, New York, pp. 779-789.

4.  Zhu, S., Wu, P., Chen, Z., Zhu, W. and Zhou, Y., "A single stage double inlet pulse tube refrigerator capable of reaching 42 K", *Cryogenics*, Vol. 30 September Supple. (1990), pp. 257-261.

5.  Matsubara, Y. and Gao, J.L., "Novel configuration of three-stage pulse tube refrigerator to temperatures below 4K", *Cryogenics*, Vol. 34 (1994), pp. 259-262.

6.  Tominaga, A., "Thermodynamic aspects of thermoacoustic theory", *Cryogenics*, Vol 35 (1995) pp. 427-440.

7.  Luo, E.C., Xiao, J.H. and Zhou, Y., "A Simplified Thermoacoustic Modeling For Pulse Tube Refrigerator", *Proc. of 4th Joint Sino-Japanese Seminar on Cryocoolers and Concerned Topics*, (1993), pp. 94-100.

8.  Rawlins, W., Radebaugh, R. and Timmerhaus, K.D., "Thermal anemometry for mass flow measurement in oscillating cryogenic gas flows", *Rev. Sci. Instrum.*, 64:3229 (1993).

9.  Inada, T., Nishio S., Ohtani, Y. and Kuriyama, T., "Experimental investigation on the role of orifice and bypass valves in double-inlet pulse tube refrigerator", *Advances in Cryogenic Engineering*, Vol. 41B (1996), Plenum Press, New York, pp. 1479-1486.

10. Shiraishi, M., Seo, K. and Murakami, M., "Pressure and Temperature Oscillations of Working Gas in a Pulse Tube Refrigerator", *Cryocoolers 8*, Plenum Press, New York (1995), pp. 403-410.

# An Experimental Investigation of How the Heat Pumping Mechanism in a Pulse Tube Changes with Frequency

**B.E. Evans and R.N. Richardson**

Institute of Cryogenics, University of Southampton,
Southampton, SO17 1BJ, UK.

## ABSTRACT

The mechanism of surface heat pumping that is responsible for thermal transport in a basic pulse tube is widely accepted and may be reconciled with the ideal representation of a refrigerator as a second law device. At the higher frequencies at which orifice pulse tubes operate the time available for heat exchange between the gas and tube wall is very short. This suggests that heat pumping is due to some other mechanism primarily within the pulsating gas column rather than by surface heat pumping. An apparatus has been constructed which allows a single pulse tube to be operated over a frequency range from 1 Hz to more than 50 Hz in an attempt to detect the transition between the modes of heat pumping.

## INTRODUCTION

Originally described by Gifford and Longsworth in 1964[1], the pulse tube has enjoyed renewed research interest in the past decade. This is undoubtedly due to the fact that it contains no moving parts at low temperature and thus has a potentially greater lifetime than any other cooler presently available. Many improvements have been made to the physical design of the pulse tube, and double inlet pulse tubes have now been constructed[2] that can cool to temperatures below 4 K. However, despite the advances in practical techniques, many aspects of pulse tube operation remain poorly understood.

In common with Stirling and G-M coolers, operation of a pulse tube refrigerator depends on a regenerative cycle. However the nature of this cycle and the mechanism of heat pumping would seem to be somewhat different, and certainly more difficult to explain. This is not helped by the fact that the mechanism appears to change as the frequency of operation increases. At low frequency it is now widely agreed that cooling results from the process know as surface heat pumping, first described by Gifford and Longsworth[3] and subsequently investigated in more detail by a number of other workers[4,5]. However, at higher frequencies there is insufficient time for appreciable heat transfer between the gas and adjacent tube wall suggesting that a different mechanism, operating wholly within the pulsating gas column, must be responsible for heat pumping under these conditions.

Cryocoolers 9, Edited by R.G. Ross, Jr.
Plenum Press, New York, 1997

It is quite marked that hitherto most research on pulse tubes has either been at the relatively low frequencies, typically up to about 5 Hz, that can be achieved using valves to produce the pressures pulses, or at around 50-60 Hz. Even where performance has been investigated as a function of frequency the range has been limited. This is undoubtedly a consequence of the hardware available to the researchers. Those with access to the compressors used in Stirling coolers have tended to concentrate on optimising pulse tubes at higher frequencies whilst when no such compressors are available research has tended to be at lower frequencies using simpler and cheaper valve systems. What would seem to be lacking is a systematic investigation of a single device over the full range of frequencies which will facilitate a better understanding of the changes which occur in the heat pumping mechanism as the pulse rate increases. The apparatus and experiments described here are intended to provide more information about the fundamental aspect of pulse tube operation.

## DEVELOPMENTS IN PULSE TUBE DESIGN AND THEORETICAL ANALYSIS

The three main pulse tube designs, the basic (BPTR), orifice (OPTR) and double inlet (DIPTR) are summarised in Figure 1. Gas, usually helium, is supplied to the tube in pulses causing alternate compression and expansion of the gas already within the system. This

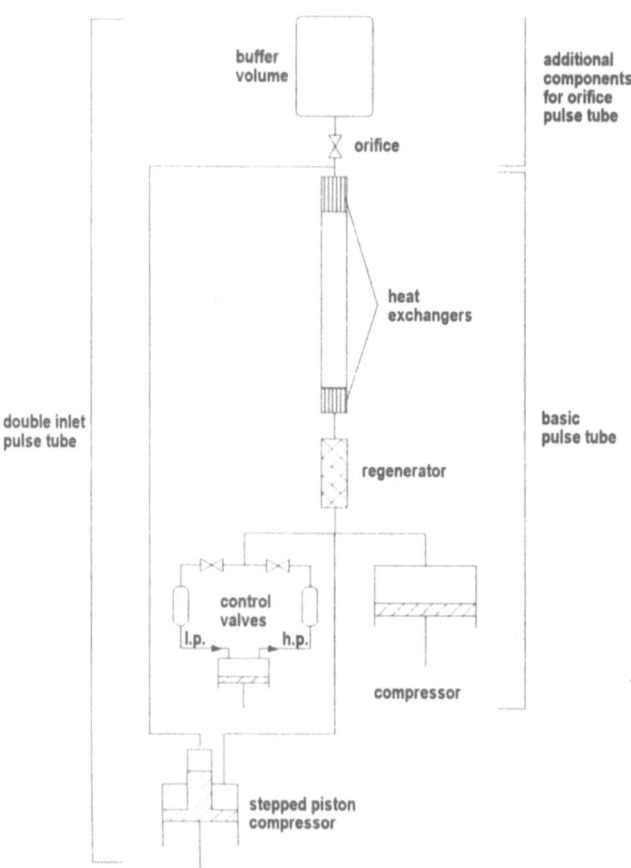

**Figure 1** The characteristics of basic, orifice and double inlet pulse tubes.

produces cooling in the "cold end" heat exchanger between the tube and the regenerator, while heat is rejected at the "hot end" heat exchanger at the other end of the tube.

The unconventional nature of the pulse tube is such that it is not as easy to reconcile with the ideal representation of a refrigerator as a second law device as, say, a Stirling cooler. The second law of thermodynamics states that for a system to function as a refrigerator, heat must be removed from a source at low temperature and rejected at a higher temperature. For this to be achieved there must be net work input in that region of the system in which heat pumping occurs. Thus any theory of pulse tube refrigeration must explain how heat is transported along the pulse tube, from "cold end" to "hot end", against a temperature gradient.

Despite the difficult nature of the subject, many advances have been made in the theoretical research of pulse tube behaviour. Some of the main ideas to have emerged during the past thirty years of investigation are summarised very briefly below...

**Surface Heat Pumping** In 1966 Gifford and Longsworth proposed the idea of surface heat pumping as an explanation of the cooling mechanism in the basic pulse tube[3]. This theory suggests that the gas can be divided into infinitesimal elements which carry heat along the pulse tube in a series of steps. During compression, these elements are heated and pushed along the tube by the incoming gas and come to rest further up the tube. Here the adjacent tube walls are cooler than the gas itself and thus the gas loses heat to the walls. During expansion, the gas is returned to it's original temperature and position and absorbs heat back from the tube wall. Thus the heat deposited by each element of gas after compression will be passed onto another element further along the tube after the following expansion stroke. In this way heat is transported along the pulse tube in a number of discrete steps and without the need for any elements of gas to travel between the cold and hot ends.

The mechanism of surface heat pumping is widely accepted as an explanation of heat pumping in a BPTR. Further work undertaken by Richardson[4,5] and more recently by de Boer[6] explains how the surface heat pumping process may be represented by a series of heat pumps (Figure 2). The Second Law does not specify how heat pumping is achieved, only that to "raise" heat against a temperature gradient, in this case from the cold end heat exchanger to the warm end heat exchanger, requires that work be done by a system operating between these

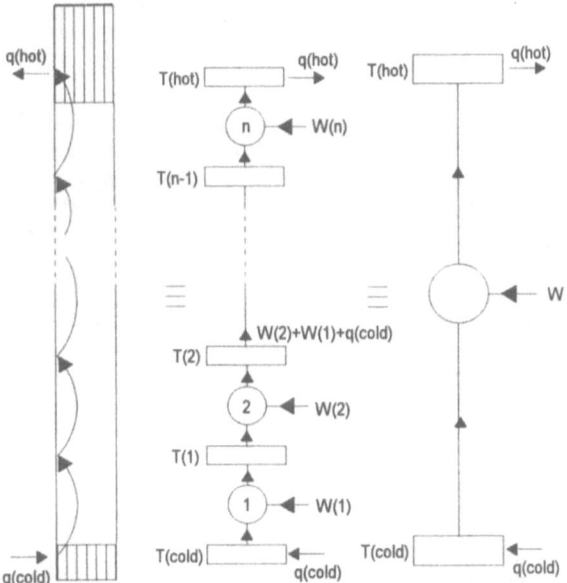

**Figure 2** Surface Heat Pumping and the Second Law

temperature limits.  Although it is difficult to isolate the thermodynamic work input to a pulse tube (there is no difficulty of course in quantifying the electrical/mechanical input to the compressor but this is not the same thing) there can be no doubt that there must be a work input to that part of the system in which heat pumping takes place if the Second Law is not to be violated.

**Orifice Pulse Tube**  With the introduction of the orifice pulse tube in 1984[7] came the need for a theory which could encompass this new design.  Two of the main groups of OPTR investigators have been those led by Radebaugh and Ravex who are both agreed that there are different mechanisms operating in the OPTR and BPTR.

Unlike the surface heat pumping mechanism which relies entirely on <u>gas/wall</u> heat interactions, it appears that the OPTR is predominantly dependent on <u>gas/gas</u> heat transfer to produce refrigeration.  Radebaugh[8] describes the mechanism using an "enthalpy flow analysis", whilst a different approach is taken by Liang[9] (Ravex group) in his "non-symmetry effect" explanation.

## THE ROLE OF FREQUENCY IN PULSE TUBE OPERATION

The idea of a different heat transport mechanism operating in the OPTR and the BPTR has been supported by experiments which clearly show how the optimum frequencies of these two refrigerators differ[10].  This in itself highlights the potential importance of frequency as an indicator of changes taking place in the mechanism at work in the pulse tube.

Liang[9], in a recent analysis of pulse tube function, formulated an expression for gross refrigeration power directly dependent on operating frequency.  Other investigations have produced useful, if sometimes conflicting, information about the role of frequency.  Figure 3 gives a brief summary of the results of some of the main studies in this area.

To summarise, Rawlins et al.[11] found the enthalpy flow along an orifice pulse tube increased with frequency.  This is as predicted by enthalpy flow analysis - an increase in frequency should produce an increase in mass flow rate which should in turn produce an increase in the enthalpy flow rate.  Wu[12] too found an increase in OPT performance with increased frequency.  Others have tested and compared several designs of tubes.  Shiraishi[13] tested a BPTR and OPTR and

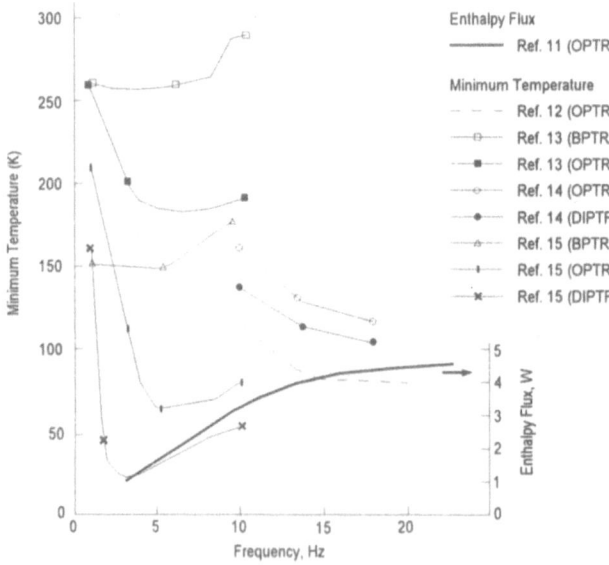

**Figure 3**   Sketch graph showing results of previous frequency investigations.

found that minimum temperature was obtained at a higher frequency in the OPTR than in the BPTR. Wang[14] found that the minimum temperatures reached by an OPTR and DIPTR decreased as frequency was increased, and Ravex[15], in tests involving all three main designs, noted that the OPTR obtained a minimum temperature at a higher frequency then either the BPTR or the DIPTR.

Thus different pulse tubes have been run and tested at many different frequencies. However, the pulse tubes themselves have varied greatly in terms of geometry and operating conditions and clearly this will have a significant effect on pulse tube performance. The size of the pulse tube significantly alters the dependency of performance on frequency and this has been shown by the difference in optimum frequencies recorded for the variety of BPTRs and OPTRs used in past experiments.

Pressure ratio (the ratio of high to low pressure over one cycle) and the average system pressure are two important factors which may also be influenced by frequency. In practical terms, the pressure ratio is likely to diminish as frequency increases as the time available for compression and expansion will decrease. The effect of the frequency on the average pressure will depend on the equipment used, for example, the pressure obtainable in a compressor-driven system may increase with frequency as the leak rate past the piston decreases. Thus, in order to clearly see the effect of a change in frequency on pulse tube operation it is imperative that concurrent changes in other factors are controlled as far as possible.

Most experimental results concerning the effect of frequency in pulse tube operation have so far been limited in range by the equipment and conditions available. Thus there does not appear to have been a continuous assessment of the performance of a pulse tube beginning at very low frequencies and extending into typical compressor frequencies of 50-60 Hz. This is our aim at Southampton, to plot the variation in performance of both a BPTR and OPTR over as wide a range of frequencies as possible and in particular to be able to show clearly the transition that occurs between the different heat transport mechanisms.

## EXPERIMENTAL SET-UP

Our experiments have been performed using the equipment and set-up shown in Figures 4,5 and 6.

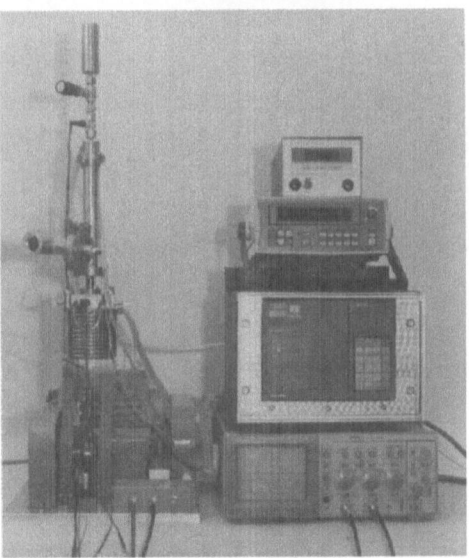

**Figure 4**   Photograph of the pulse tube, compressor and instrumentation.

**Pulse Tube Assembly**  The thin-walled stainless steel pulse tube measures 110 mm by 9 mm (inner diameter), while the regenerator has dimensions of 60 mm by 7 mm (i.d.) and is packed with 250-mesh phosphor bronze gauze disks.  Between the regenerator and one end of the tube is soldered a copper heat exchanger (the cold end) and at the other end of the pulse tube is the other copper heat exchanger (the hot end) over which fits a water-cooled jacket to maintain ambient temperature.  Everything except the hot end heat exchanger is enclosed in a cylindrical vacuum shell, the pressure in which is maintained below $1 \times 10^{-6}$ torr.  A micrometer head needle valve situated between the cylindrical buffer volume of dimensions 75 mm by 25 mm (i.d.) provides a variable impedance orifice when the system is run as an OPTR.

**Compressor**  The assembly described above is mounted directly above the compressor which has been designed and built in our laboratories specifically for these experiments.  The compressor is driven by an inverter controlled electric motor via pulleys and a timing (toothed) belt.  The basic components of the compressor are shown clearly in Figure 5 and 6.  A water-cooled copper coil, labelled (1) in Figure 5, is wrapped around the cylinder (2), while a network of channels in the cylinder head (3) provides additional water cooling.  Side loads on the piston are minimised through the use of a crosshead assembly (4) and linear guide bearings (5) for the piston rod (6).  In addition, the piston skirt has a guide ring (7), plus two standard piston rings (8).  Sealing between the piston rod and cylinder base is achieved effectively with a spring energised lip seal (9).  A crank disc (10) has been chosen in preference to a more conventional crank pin to enable stroke variation without the need to change the crankshaft, and to give a large bearing surface.  Several crank discs and piston crowns have already been constructed to allow stroke variation, and in combination with volume variations using different sized pistons and cylinders, provide a means of altering the overall pressure and pressure ratio obtainable within the system.

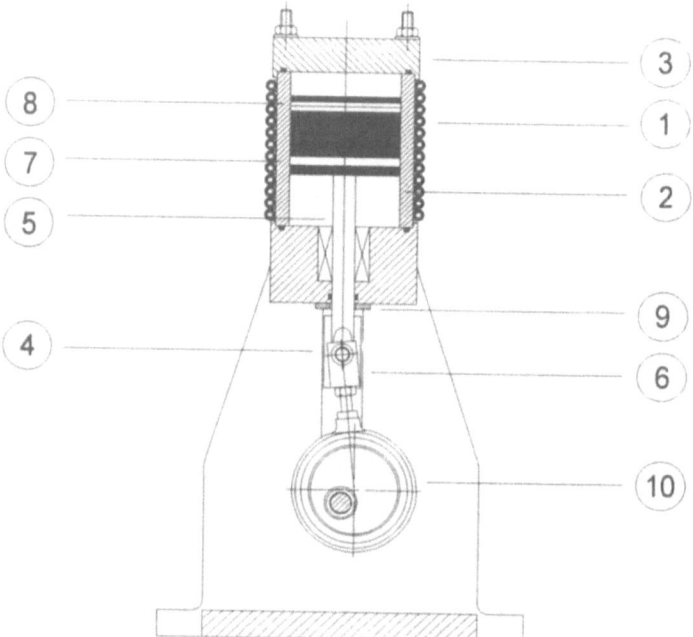

**Figure 5**  Cross-sectional view of compressor

The main measurements made on the pulse tube system are the variations in temperature and pressure.

**Temperature**  Copper/constantan thermocouples and a silicon diode are used to measure the temperature at various points in the system.  The temperature of the cold end heat exchanger is measured by both a thermocouple and a diode, while several thermocouples are placed at various points along the outside wall of the pulse tube and regenerator.  The voltages produced are read and stored by a data acquisition system linked to a personal computer.

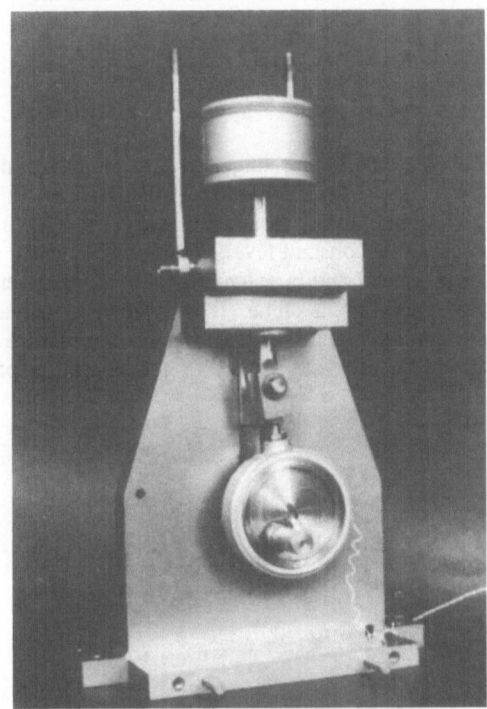

**Figure 6**  Photograph of the cross-sectional view of the disassembled compressor.

**Pressure**  Measurements are made at various points in the pulse tube system using Kulite diaphragm pressure transducers.  These are positioned at the compressor and before and after the orifice valve.  The pressure traces produced are viewed on a cathode ray oscilloscope and used to monitor the pressure ratio (maximum pressure divided by minimum pressure recorded) at the different points and also the average overall pressure and its variation along the tube.  It is also possible to see how the relative shape of the pressure pulses changes with frequency as well as with position.

## EXPERIMENTAL PROCEDURE AND RESULTS

Experiments have so far been performed on both the BPTR and OPTR operating at frequencies from 1 Hz up to and including 15 Hz. It is hoped to run the compressor and pulse tube assembly through a wide range of frequencies from a few Hz up to more usual compressor speeds of around 50-60 Hz. Results so far have indicated this to be possible. Some of the measurements made to date are shown in Figure 7. This graph clearly shows the optimum frequency of the BPTR as 7 Hz. Operating in OPTR mode performance increases with frequency, and as yet the optimum frequency has not been reached.

As discussed earlier, the influence of frequency on the pressure ratio and average pressure in the pulse tube is an important consideration and thus has been monitored. As expected, the average pressure produced in the compressor increased slightly with frequency as the leak rate past the piston decreased, but the pressure ratio itself did not show any dependency on frequency.

## CONCLUSIONS

These results support the idea that there is change of heat transport mechanism occurring between the BPTR and the OPTR. This is clearly seen by the difference in behaviour of the two tubes as frequency is increased while other parameters such as pressure ratio and average pressure are maintained at a reasonably constant level. The variable speed compressor designed for these experiments is performing well and is potentially able to run over the largest range of frequencies so far possible with a single pulse tube. The availability of variable stroke as well as variable speed for future measurements implies that changes in pressure parameters may also be accounted for, thus providing an accurate survey of the operation of pulse tubes over a complete range of frequencies.

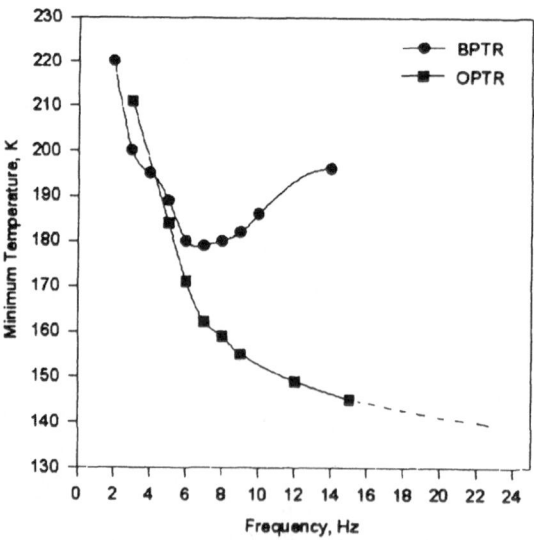

**Figure 7**   Relation between BPTR and OPTR minimum temperatures and operating frequencies

## REFERENCES

1. Gifford, W.E. and Longsworth, R.C., "Pulse Tube Refrigeration", *Trans. ASME J Eng Ind*, 63 (1964), pp. 264-268

2. Gao, J.L. and Matsubara, Y., "Experimental Investigation of 4 K Pulse Tube Refrigerator", *Cryogenics*, vol. 34, no. 1 (1994), pp. 25-30

3. Gifford, W.E. and Longsworth, R.C., "Surface Heat Pumping", *Adv Cryo Eng*, vol. 11, (1966), pp. 171-179

4. Richardson, R.N., "Pulse Tube Refrigerator - an Alternative Cryocooler?", *Cryogenics*, vol. 26, (1986), pp. 331-340

5. Richardson, R.N., "Pulse Tube Refrigeration - An Investigation pertinent to Cryocooler Development, D.Phil Thesis, University of Oxford, (1982)

6. de Boer, P.C.T., "Thermodynamic Analysis of the Basic Pulse Tube Refrigerator", *Cryogenics*, vol. 34, no.9 (1994), pp. 699-711

7. Mikulin, E.L., Tarasov, A.A. and Shkrebyonock, M.P., "Low Temperature Expansion Pulse Tubes", *Adv Cryo Eng*, vol. 29, (1984), pp. 620-

8. Radebaugh, R., "A Review of Pulse Tube Refrigeration", *Adv Cryo Eng*, vol. 35, (1990), pp. 1191-1205

9. Liang, J., Ravex A, and Rolland, P., "Study on Pulse Tube Refrigeration Parts 1-3", *Cryogenics*, vol. 36, no.2 (1996), pp. 87-106

10. Richardson, R.N., "Valved Pulse Tube Refrigerator Development", *Cryogenics*, vol. 29, (1989), pp. 850-853

11. Rawlins, W., Radebaugh, R., Bradley, P.E. and Timmerhaus, K.D., "Energy Flows in an Orifice Pulse Tube", *Adv Cryo Eng*, vol.39, (1994), pp. 1449-1456

12. Wu, P.Y., Zhang, Li., Qian, L.L. and Zhang, L., "Numerical Modelling of Orifice Pulse Tube Refrigerator by using the Method of Characteristics", *Adv Cryo Eng*, vol.39, (1994), pp. 1417-1423

13. Shiraishi, M., Seo, K. and Murakami, M., "Pressure and Temperature Oscillations of Working Gas in a Pulse Tube Refrigerator, *Cryocoolers 8* , Plenum Press, New York (1995), pp. 403-410

14. Wang, C., Wu, P. and Chen, Z., "Experimental Study of Three Types of Pulse Tube Refrigerator", *Proc. 4th JSJS*, (1993), pp. 64-68

15. Ravex, A., Bleuze, P., Duband, L., Poncet, J.M. and Rolland, P., "Pulse Tube Cooler Development at CEA/SBT", *Proc. 19th ICR*, vol. IIIb, (1995), pp. 1209

# DC Gas Flows in Stirling and Pulse Tube Cryocoolers

David Gedeon

Gedeon Associates
Athens, Ohio 45701

## ABSTRACT

Whenever a closed-loop flow path exists in a stirling or pulse-tube cryocooler, there is a potential for a DC gas flow — where the positive and negative flows over each half cycle do not quite cancel completely. A closed loop may be formed by a displacer seal in a stirling cooler, which connects the compression and expansion spaces in parallel with the regenerator, or by a bypass orifice in a double-inlet pulse-tube cooler, which connects the warm end of the compliance tube (pulse tube) back to the compression space.

DC gas flows were discovered in simulation models of double-inlet pulse-tube cryocoolers run with the Sage computer program. At the root of the problem are in-phase density and velocity fluctuations in the flow-resistive elements of the closed loop which tend to produce DC pressure gradients. The issue is whether or not these DC pressure gradients completely cancel each other. If they do not, then DC flow is the consequence and designers must contend with new thermal losses associated with components like regenerators and compliance tubes. A DC flow of less than one percent of the AC flow amplitude may be enough to cause significant heat lift degradation

## INTRODUCTION

For flow resistances dominated by viscosity, pressure drop is independent of fluid density, or only weakly dependent. So, it is clear that we can get a mass imbalance on either side of such a flow resistance simply by arranging for density to be higher in one flow direction than the other, even though pressure drop is the same in both directions. Either that or we can get a pressure drop imbalance as a result of mass balance. In general, we cannot expect the time-average mass flow rate and pressure drop to be both zero at the same time. The remainder of this paper explores this issue in more detail under conditions of sinusoidally-varying flow velocity and density.

Useful will be the trigonometric identity

$$2\sin(\omega t)\,\sin(\omega t + \alpha) \equiv \cos(\alpha) - \cos(2\omega t + \alpha) \tag{1}$$

which you can verify by expanding the second term on the right using the standard addition formula for cosines. In English, the product of two in-phase sine waves is a twice-frequency cosine wave — plus a constant. Constant $\cos\alpha$ happens to be one if phase angle $\alpha$ is zero and vanishes only for $\alpha$ an odd multiple of $\pi/2$. This constant will turn out to be the source of DC flow.

Cryocoolers 9, Edited by R.G. Ross, Jr.
Plenum Press, New York, 1997

## DARCY FLOW

For creeping flow (low Reynolds number) through porous materials, such as a regenerator matrix, pressure drop $\Delta P$ is a linear function of velocity $u$. This is known as Darcy flow and while not entirely accurate it serves to illustrate the point of this paper. Darcy flow also characterizes laminar developed flow, often found in clearance-type seals. If we introduce a factor of density $\rho$ on both sides of the equation, we can formulate Darcy flow as

$$\rho u \propto \rho \Delta P \tag{2}$$

The advantage of this form is that the quantity on the left is the mass flow rate per unit cross-sectional area — mass being a conserved quantity.

Now, in the typical stirling-like cycle, both pressure drop across any given flow resistance and the density within are time varying. Say, for the sake of argument, that the pressure drop varies sinusoidally as

$$\Delta P = D_0 + D_1 \sin \omega t \tag{3}$$

and that density likewise fluctuates sinusoidally as

$$\rho = \rho_0 + \rho_1 \sin(\omega t + \alpha) \tag{4}$$

where $\alpha$ is some phase angle.

If we substitute the right-hand sides of the preceding two equations for $\rho$ and $\Delta P$ into Darcy's equation, expand terms, then apply our original trigonometric identity, we wind up with

$$\rho u \propto \rho_0 D_0 + \rho_0 D_1 \sin \omega t + \rho_1 D_0 \sin(\omega t + \alpha) + \frac{\rho_1 D_1}{2} \cos \alpha - \frac{\rho_1 D_1}{2} \cos(2\omega t + \alpha) \tag{5}$$

Of all the terms on the right-hand side, only the first and fourth contribute to DC flow. The remaining terms time-average to zero. So if we time average the whole equation, denoted by { } brackets, the DC flow rate is just.

$$\{\rho u\} \propto \rho_0 D_0 + \frac{\rho_1 D_1}{2} \cos \alpha \tag{6}$$

So, to the extent that density fluctuation is in phase with pressure drop, we can always expect either DC flow or DC pressure drop within, say, the regenerator in a stirling or pulse-tube cooler. In the case where DC mass flow rate is physically impossible, perhaps because there is no return path parallel to the regenerator, this DC pressure drop will be exactly the amount required to cancel any DC flow. By setting $\{\rho u\}$ to zero in the previous equation and solving, we see that the normalized pressure drop that cancels DC flow in a single Darcy flow resistance is

$$\frac{D_0}{D_1} = -\frac{1}{2} \frac{\rho_1}{\rho_0} \cos \alpha \tag{7}$$

But what if there is a parallel return path?

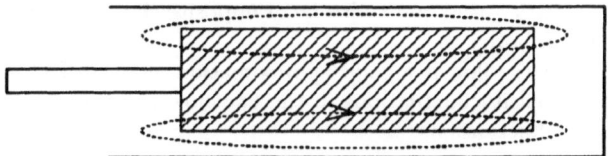

Figure 1: Closed flow loop in a stirling cooler.

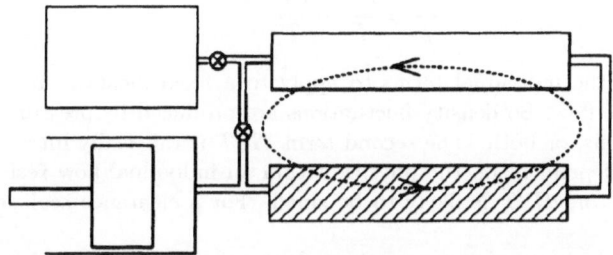

Figure 2: Closed flow loop in a double-inlet pulse-tube cooler.

## CLOSED LOOPS

For purposes of this paper, a closed loop comprises two distinct flow resistances, typified by a regenerator plus any flow pathway in parallel with it, such as the displacer seal in a stirling cooler or the bypass orifice in a double-inlet pulse-tube cooler. Schematically, a stirling closed-loop might look like figure 1 and a pulse-tube closed-loop like figure 2. For understanding the essential physics, both closed loops are special cases of the idealized closed-loop in figure 3, where the flow resistances are governed by Darcy's law.

With no further assumptions, we can already conclude one fact: If the density fluctuations in both flow resistances are identical, then there will be no DC flow. This follows by applying Eq. (6). Pressure drop mean and amplitude $D_0$ and $D_1$ are referenced to common spaces. So they are necessarily the same for both. With $\rho_0$, $\rho_1$ and $\alpha$, presumed the same, Eq. (6) implies that the DC flows in both paths must be proportional — with the same sign.

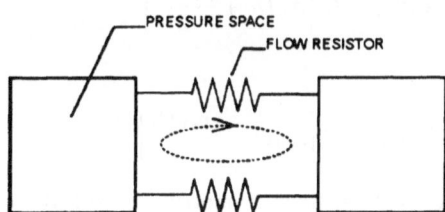

Figure 3: Idealized closed flow loop.

But to satisfy mass conservation they must also be of opposite sign. The only way this can be so is if DC flow is zero in both.

So to the extent that Darcy flow is a good approximation, nonzero DC flow in a closed flow loop implies unequal density fluctuations in the two flow resistances. So, what could cause differences in density fluctuations?

## DENSITY FLUCTUATIONS

According to the ideal-gas equation of state in differential form,

$$\frac{d\rho}{\rho} = \frac{dP}{P} - \frac{dT}{T} \tag{8}$$

where we understand the individual terms to apply to a fixed location, rather than to a frame moving with the flow. So density fluctuations are produced by pressure fluctuations, temperature fluctuations, or both. The second term $dT/T$ is especially interesting because it likely to be highly dependent on the physics within an individual flow resistance. For a very effective regenerator, $dT/T$ is likely to be small. For a clearance seal or orifice, it is likely to be higher.

The two limiting extremes for temperature fluctuation will be found in cases of an isothermal flow resistance and an adiabatic flow resistance. In the first case $dT/T = 0$, by definition. In the second case, $dT/T$ depends on upstream conditions, since temperature of a fluid particle (moving with the flow) within a flow resistance remains constant as a consequence of enthalpy conservation (throttling process). If we assume adiabatic upstream conditions, then the temperature fluctuation at the inlet of an adiabatic flow resistance will be $dT/T = (1 - 1/\gamma)dP/P$, where $\gamma$ is the ratio of specific heats and $P$ is the upstream pressure.

## AN IDEALIZED SOLUTION

We now consider in detail two Darcy flow resistors — one isothermal, one adiabatic. Looking back at Eq. (5), we can simplify matters by assuming pressure-drop and density fluctuation are in phase ($\cos \alpha = 1$), which will be the case if one of the spaces in the idealized geometry of figure 3 is producing all of the pressure variation and the other is essentially isobaric. This is a reasonable approximation if the left space is a compressor volume and the right space a large reservoir. It is also convenient to presume both flow resistors have identical geometry (same flow area, hydraulic diameter, etc.). If we use subscripts $i$ and $a$ to distinguish quantities in the isothermal resistor from those in the adiabatic resistor, we can from Eq. (6) write the time-mean flow for the isothermal path as

$$\{\rho u\}_i = C \left[ \rho_{i0} D_0 + \frac{\rho_{i1} D_1}{2} \right] \tag{9}$$

and for the adiabatic path as

$$\{\rho u\}_a = C \left[ \rho_{a0} D_0 + \frac{\rho_{a1} D_1}{2} \right] \tag{10}$$

where $C$ is some constant dependent on flow geometry. If we further assume time-mean temperature is the same everywhere, we can write $\rho_0$ for both $\rho_{i0}$ and $\rho_{a0}$. If we limit our attention to the half of the cycle where flow is from the compression space to the reservoir and the volume of the resistors is small compared to the compression-space swept volume, the differential equation of state (8), allows us to substitute $\rho_0 P_1/P_0$ for $\rho_{i1}$ and $\rho_0(P_1/P_0 - T_1/T_0)$ for $\rho_{a1}$, where $P_0$ and $T_0$ are the time-mean pressure and temperature, $P_1$ is the spatial-mean resistor pressure amplitude and $T_1$ is the compression-space temperature amplitude. Since the spatial average pressure amplitude in either resistor is roughly just half the pressure drop, we may substitute $D_1/2$ for $P_1$. And presuming the compression space to be nearly adiabatic, we may substitute $(1 - 1/\gamma)D_1/P_0$ for $T_1/T_0$. With all these substitutions, and introducing a factor of $1/2$ to account for the fact that we are considering only a half cycle, the two equations become

$$\{\rho u\}_i = C/2 \left[ \rho_0 D_0 + \frac{\rho_0 D_1^2}{4P_0} \right] \tag{11}$$

and

$$\{\rho u\}_a = C/2 \left[ \rho_0 D_0 + \frac{\rho_0 D_1^2}{4P_0}(2/\gamma - 1) \right] \tag{12}$$

Adding these two equations, the left-hand sides sum to zero, in order to conserve mass, leaving after some simplification

$$D_0 = -\frac{D_1^2}{4\gamma P_0} \tag{13}$$

If we substitute this for $D_0$ in Eqs. (11) and (12), we get for the time-mean flows in the two resistors

$$\{\rho u\}_i = C\frac{\rho_0 D_1^2}{8P_0}(1 - 1/\gamma) \tag{14}$$

and

$$\{\rho u\}_a = C\frac{\rho_0 D_1^2}{8P_0}(1/\gamma - 1) \tag{15}$$

which sum to zero, as they must. For the half cycle where flow is directed into the compression space, the flows in the two resistors are identical because of the isothermal entrance conditions from the reservoir. So the previous equations give the DC flow for the whole cycle as well. Interestingly, the DC flow in the isothermal resistance is away from the compression space while for the adiabatic resistance it is toward the compression space.

To eliminate the geometrical constant $C$ we can divide the DC flow rate by the AC amplitude of the first harmonic, which according to Eq. (5) is approximately $C\rho_0 D_1$ ($C\rho_1 D_0$ being negligible in comparison), to obtain

$$\frac{\{\rho u\}}{|\rho u|} \approx \pm\frac{(1/\gamma - 1)D_1}{8P_0} \tag{16}$$

With $1 - 1/\gamma = 0.40$ for helium and pressure drop amplitude $D_1$ on the order of $1/10$ the mean pressure $P_0$, the DC flow magnitude would be on the order of 0.5% of the AC flow amplitude.

It was relatively straight forward to setup this idealized problem using the Sage stirling simulation software. (See the previous ICC proceedings[1] for a discussion of Sage.) And, indeed, Sage came within about 5% of relation (16), suggesting that both the physical model within sage as well as the numerical solution to that model are adequate for resolving the DC flow phenomena. This was not achieve without a bit of fussing though. Darcy flow presumes constant viscosity so it was necessary to temporarily re-defining the properties of helium to have constant viscosity, so that it would not fluctuate along with temperature. And to satisfy the presumption that time-mean temperature be the same everywhere, it was necessary to adjust the mean temperature of the compression space by lowering its wall temperature.

## A REALISTIC SOLUTION

In reality closed loops are a lot more complicated than the previous idealized solution.

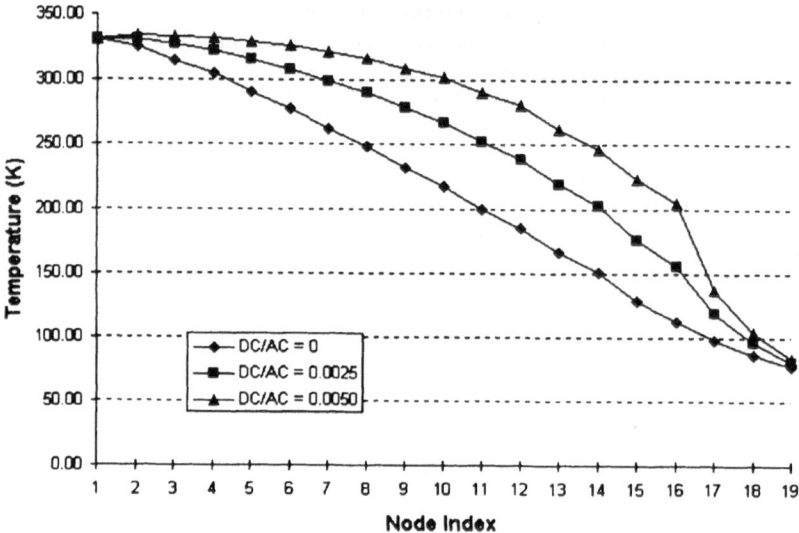

Figure 4: Temperature profiles in a regenerator for various values of DC flow as a fraction of the AC flow amplitude, showing an advancing warm temperature front in the direction of DC flow.

Flow resistances are neither exactly isothermal nor adiabatic. Nor are they exactly governed by Darcy's law. The dependence of viscosity on temperature cannot be neglected. And upstream temperature conditions are important, as are asymmetric local pressure drops, turbulence, nonsteady transients, and so forth.

A good example is DC flow in a double-inlet pulse tube of the type illustrated schematically in figure 2. It was in just such a device that DC flow first reared its ugly head during simulation with Sage. The symptoms were not immediately recognized as DC flow though. Only after detailed examination of the solution was it clear that the cause of the wildly-varying enthalpy flow through the regenerator was due to a small amount of DC flow within it. The amount of DC flow was dependent on the setting for the bypass orifice which was a tuning variable.

You can get an idea of the importance of DC flow in a regenerator by looking at figure 4, produced by introducing calibrated amounts of DC flow into the regenerator of an actual pulse-tube refrigerator — as simulated by Sage. As DC flow increases, a warm temperature front advances through the regenerator changing the temperature distribution from nearly linear to mostly warm, with a short transition region near the end. The worst-case DC/AC flow ratio of 0.005 is the order Sage predicts for the combination of a metal-felt regenerator and a sharp-edged bypass orifice.

The same sort of thing happens in the compliance tube (pulse tube), shown in figure 5, except this time a cold temperature front advances down the tube.

The following table shows the effect of DC flow on compliance-tube and regenerator enthalpy flows:

| DC flow / AC amplitude | 0 | 0.0025 | 0.0050 |
|---|---|---|---|
| Compliance-tube enthalpy flow (W) | 11.42 | 15.32 | 18.04 |
| Regenerator enthalpy flow (W) | 6.29 | 10.22 | 15.01 |
| Difference (W) | 5.13 | 5.10 | 3.03 |

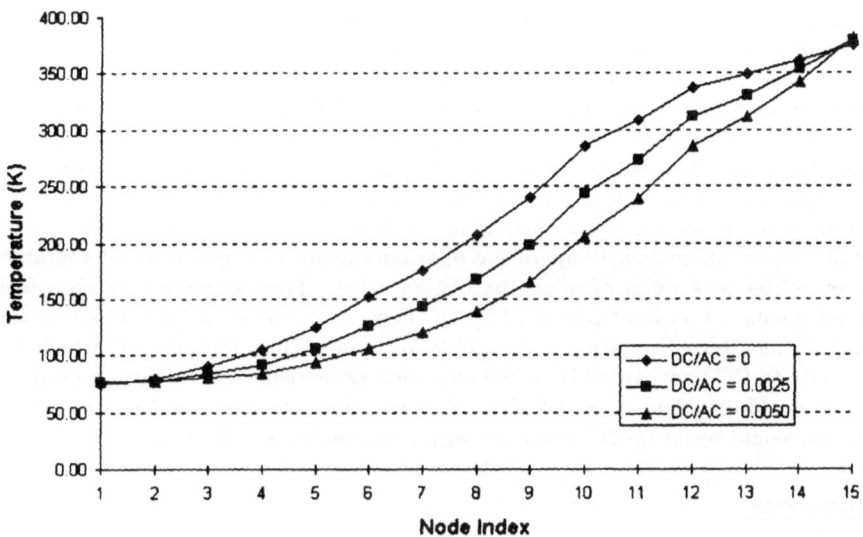

Figure 5: Temperature profiles in a compliance tube (pulse tube) at piston TDC for various values of DC flow as a fraction of the AC flow amplitude, showing an advancing cold temperature front in the direction of DC flow.

The compliance-tube enthalpy flow provides the gross refrigeration power while the regenerator enthalpy is a parasitic thermal loss. The difference between the two is roughly the net refrigeration power. So while the regenerator loss increases dramatically with DC flow, so does the compliance tube enthalpy flow. The result is that there is hardly any net refrigeration decrease until DC flow exceeds 0.0025 of the AC flow. But after that the decrease becomes quite significant. Curiously, for the particular pulse tube of this example, the maximum net heat lift occurred at a DC flow rate equal to 0.001 of the AC amplitude! Although there is no reason to expect this end result of a complex model applies to all cases.

There are two distinct mechanisms contributing to these observed enthalpy changes: The first is the enthalpy carried by the DC flow itself, which would be the case even under isothermal (zero temperature gradient) conditions. Since the positive gas mass displacement exceeds the negative displacement, the net enthalpy transport is necessarily shifted in the positive direction. As an empirical observation, the change in enthalpy for both the regenerator and compliance tube seems to be of the order $C_p \{\dot{m}\} T_m$ where $\{\dot{m}\}$ is the DC mass flow rate and $T_m$ is some intermediate effective temperature — 200 to 250 K for the present example. The second mechanism has to do with the losses associated with the regenerator and compliance tube — including axial conduction, gas-to-matrix or gas-to-wall heat transfer, flow streaming, etc. All of these losses tend to scale directly with temperature gradient. And temperature gradients increase with DC flow, as figures 4 and 5 show. In other words, DC flow effectively shortens the length of the regenerator and compliance tube.

## SHOULD WE WORRY ABOUT IT?

Is DC flow real or is it just an artifact of the Sage simulation model? The idealized solution previously worked out and the agreement of Sage with that solution are strong arguments that DC flow is real. However, Sage predictions for the more complicated case of the regenerator in an actual pulse-tube cooler cannot be compared against a theoretical solution — because there is none. The best we can hope for is experimental verification in a real stirling or pulse-tube cooler. This is not as straight-forward as it might first sound because measuring a DC mass flow on the order of 0.005 of the AC mass flow rate is not easy. In fact, measuring DC flow of any magnitude is not easy. Perhaps the best means of measurement will turn out to be indirect, through the equilibrium temperature profiles generated in the regenerator or compliance tube, which are so very sensitive to the DC flow.

Measurement difficulties not withstanding, evidence for DC flow is beginning to come in. Shigi et al.[2] report anomalous temperature distributions in the regenerator of a double-inlet pulse tube refrigerator upon opening the bypass valve. Their observations are consistent with an advancing warm front produced by DC flow. And Seki et al.[3] report a temperature instability in long-term operation of a double-inlet pulse-tube refrigerator, which they in fact attribute to DC flow caused by a flow-direction dependence in the bypass needle valve. While this would certainly cause DC flow, it is not the same mechanism proposed in this paper which would result in DC flow even with a symmetric needle valve.

## REFERENCES

1. Gedeon, D., "Sage: Object-Oriented Software for Cryocooler Design", *Cryocoolers 8*, Plenum Press, New York (1995), pp. 281–292.

2. Shigi, T., et al., "Anomaly of One-Stage Double-inlet Pulse Tube Refrigerator", *16th International Cryogenic Engineering Conference*, Japan (1996).

3. Seki N., et al., "Temperature Stability of Pulse Tube Refrigerators", *16th International Cryogenic Engineering Conference*, Japan (1996), paper PS3-E1-29.

# Convective Heat Losses in Pulse Tube Coolers: Effect of Pulse Tube Inclination

G. Thummes, M. Schreiber, R. Landgraf, and C. Heiden

Institute of Applied Physics, University of Giessen, Heinrich-Buff-Ring 16, D-35392 Giessen, Germany

## ABSTRACT

In order to investigate the effect of pulse tube inclination on the performance of a pulse tube refrigerator (PTR), we have built a test rig in which the angle $\theta$ between the pulse tube axis and the direction of gravity can be varied between 0 and $\pm 180°$. $\theta = 0°$ corresponds to the vertical orientation with the hot end up. The PTR was operated with orifice, reservoir and second inlet at the warm end using helium as working fluid. The pulse tube has a length of 250 mm and an inner diameter of 13.4 mm. Operating parameters are: average pressure 18 bar, peak to peak pressure variation 5.4 bar and frequency $f = 1.6 - 4$ Hz. Optimum cooler performance is obtained for $\theta = 0$ and $f = 2$ Hz with a minimum no-load temperature of $T(0°) = 52.5$ K and a net cooling power of $\dot{Q}(0°) \approx 2$ W at 80 K. Upon tilting the pulse tube, $T(\theta)$ initially increases moderately up to $T(70°)/T(0°) \approx 1.2$. Further increase of $\theta$ leads to a steep rise of $T(\theta)/T(0°)$ attaining a maximum of $\approx 3$ for $\theta \approx \pm 120°$ and finally a value of $T(\pm 180°)/T(0°) \approx 2$. The measured variation of $T(\theta)$ and $\dot{Q}(\theta)$ indicates that tilting results in excess heat loads of up to 6 W. These losses are ascribed to an enhanced heat transfer by natural convection of He-gas occurring for $\theta \neq 0°$, which is superimposed on the oscillatory gas displacement in the empty pulse tube. This interpretation is supported by the calculated Nusselt number $Nu(\theta)$ which can semi-quantitatively account for the observed inclination effect. At a frequency of 4 Hz the magnitude of $T(\theta)/T(0°)$ is reduced with a most pronounced effect at $\theta = \pm 90°$. The $\theta$-dependence from convection is considerably weakened by filling the pulse tube with a porous material, but this also leads to a degradation of the cooler performance at $\theta = 0°$.

## INTRODUCTION

The pulse tube refrigerator (PTR) is attracting an increasing degree of attention[1] due to inherent features like absence of moving mechanical parts in the cold stage leading to increased reliability, less costly manufacturing, and greatly reduced mechanical vibrations. This and the absence of magnetic interference signals, which would result e.g. from a moving mechanical displacer, appear to make the PTR an ideal choice for cooling of highly sensitive sensors like SQUIDs. This is corroborated by first results obtained with a coaxial PTR designed for SQUID operation[2-4]. The refrigerator in this system is connected to the pressure-wave generating assembly by means of a

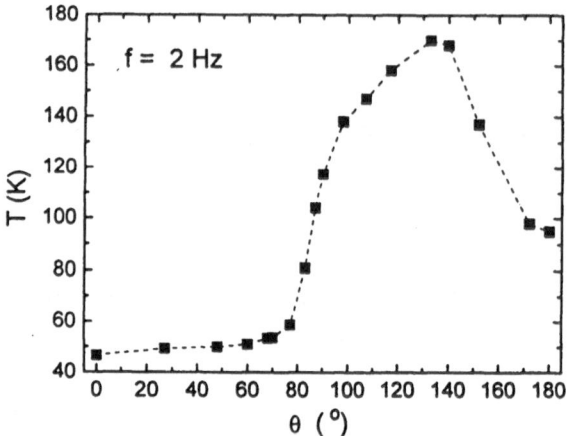

**Figure 1.** Minimum no-load temperature of the coaxial pulse tube refrigerator for SQUID operation (described in Refs. 2-4) as function of the inclination angle θ. θ = 0° corresponds to vertical orientation of the pulse tube axis with the cold end facing downwards. Length and inner diameter of the pulse tube are 250 mm and 13.4 mm, respectively.

flexible tube with lengths of up to 10 m, thereby reducing interference noise from the compressor and rotary valve to a negligible level, as sensed by a high-$T_c$ SQUID gradiometer attached to the cold stage[4].

The flexible connection between compressor and cold head allows one to change the orientation of the latter, which should be another attractive feature for applications. However, first observations made with different orientations of the cold stage indicated a dramatic change in net cooling power as function of the inclination angle θ against the vertical. Fig. 1 shows the dependence of the minimum no-load temperature of this coaxial PTR on the inclination angle θ for otherwise constant operating conditions: average pressure <p> = 18 bar, frequency f = 2 Hz, and peak to peak pressure difference Δp = 7.5 bar at the cooler inlet.

Related observations were made recently in the quiescent mode of a miniature PTR showing a considerable enhancement of the off-state heat load, when the pulse tube was oriented horizontally or with the cold end facing upwards, which was attributed to natural convection[5].

Since the frequency of the pressure wave in our PTR is much smaller than in the miniature PTR of Ref. 5, it appeared likely that even during normal operation, e.g. during the high and low pressure dwell times, heat transfer by gravity driven convection in the pulse tube could take place, which then would lead to the observed effect of θ on the minimum temperature. This conjecture led to the investigations reported in this paper.

## HEAT TRANSFER BY NATURAL CONVECTION

In this paragraph we give some basic notations and relations for estimating the heat transfer by natural convection. Natural convection in a fluid layer results from a temperature-induced density gradient and correlated buoyant forces due to gravity. The enhanced heat transfer from natural convection is characterized by the Nusselt number Nu which can be defined as

$$Nu = (\dot{Q}_{conv} + \dot{Q}_m)/\dot{Q}_m \quad , \tag{1}$$

where $\dot{Q}_{conv}$ is the heat transferred by convection and $\dot{Q}_m$ that by molecular conduction of the

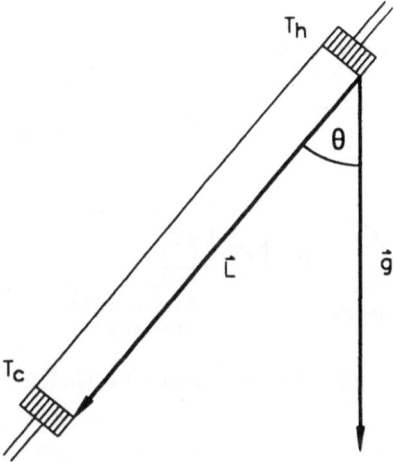

**Figure 2.** Geometry of the tilted pulse tube

fluid. In general Nu is a function of the Rayleigh number Ra, the Prandtl number Pr, and the geometry of the system. For an ideal gas[6,7]:

$$Ra(X) = g\,(T_h - T_c)\,X^3\,Pr/(v^2 <T>)\,. \tag{2}$$

Here X is a relevant length dimension, g the acceleration due to gravity, $T_h$ and $T_c$ the temperatures at the hot and the cold boundary, $<T>$ the average temperature, and $Pr = v/\alpha$, where v is the kinematic viscosity and $\alpha$ the thermal diffusivity of the gas.

The present case of a pulse tube, having a length L and inner diameter d, which is tilted with respect to the gravity vector $g$ by an angle $\theta$, is sketched in Fig. 2. For $\theta = 0°$ there is no buoyant force and therefore no natural convection, i.e. $Nu(0°) = 1$. For cylinders with high aspect ratios L/d, which for pulse tubes are typically $> 10$, a relation for the Nusselt number is only available for $\theta = 180°$ , i.e. for the orientation with the cold end up[6,7]. A concise summary of the appropriate relations for Nu at $\theta = 180°$ has been given in Ref. 8. In this case, $X = L$ is the relevant length and at high aspect ratios[6] the critical Rayleigh number for the onset of convection is $Ra_c(180°) \approx (5.75\,L/d)^4$, when heat exchange with the pulse tube walls is neglected. In the limit $Ra \gg Ra_c$, which corresponds to our experimental situation, Nu can then be written as[7]

$$Nu(180°) \approx 0.069\,Ra(L)^{1/3}\,Pr^{0.074} \tag{3}$$

At intermediate angles of inclination the Nusselt numbers may be estimated from relations for convection in enclosed spaces[9] . For the interval $100° < \theta < 180°$ we take from Ref. 9

$$Nu(100° < \theta < 180°) \approx 1.44 + \{Ra(L)\cos(180°-\theta)/5830\}^{1/3} \tag{4}$$

which holds for high Ra and L/d < 1. A justification for using Eq. (4) also for cylinders with high aspect ratios is given in Ref. 7: for high Ra the heat transfer is carried by convection cells with dimensions much smaller than the tube diameter and therefore the presence of the tube walls becomes less important.

For small angles of inclination the convection is driven by the gravity component $g\times\sin\theta$ along the hot and cold ends of the tube. In the range $2° < \theta \le 90°$ the heat transfer is estimated from[9]

$$Nu(2° < \theta \le 90°) \approx 0.58\,\{Ra(d)\sin\theta\}^{1/5} \quad , \tag{5}$$

where now the Rayleigh number has to be evaluated for $X = d$.

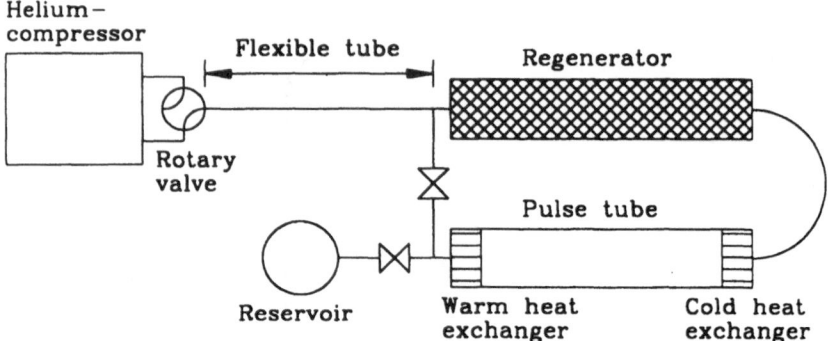

**Figure 3**. Schematic of the double-inlet pulse tube refrigerator

## EXPERIMENTAL DETAILS

A schematic of the cooler set-up is shown in Fig. 3. A standard compressor for Gifford-McMahon coolers (Leybold, RW 2000) provides the high pressure helium gas as working fluid. The pressure oscillation in the cooler is generated by means of a motor-driven rotary valve which periodically connects the high and low pressure sides of the compressor to the refrigerator inlets, as illustrated in Fig. 3 and described in more detail in Ref. 10. The regenerator and pulse tube are made of stainless steel tubes with outer diameter, wall thickness and length of $19\times0.3\times210$ mm³ and $14\times0.3\times250$ mm³, respectively. A standard U-shaped configuration of regenerator and pulse tube is used. The regenerator matrix consists of a stack of 940 pieces of no. 170 mesh stainless steel screens in the warm part and 560 pieces no. 280 mesh bronze screens in the cold part. Both heat exchangers, connected to the hot and cold ends of the pulse tube, consist of copper cylinders with thin axial flow channels providing heat exchange and flow straightening.

The PTR is operated in orifice and double-inlet configuration with the two needle valves and the reservoir (volume: 500 cm³) located outside the vacuum vessel. The temperature profile along the pulse tube wall is measured by means of Pt100 resistance thermometers. Piezoelectric pressure sensors are used for monitoring the dynamic pressures at the hot ends of pulse tube and regenerator. Net cooling powers are measured using a calibrated resistive heater attached to the cold end heat exchanger.

In order to facilitate the measurements of the cooler performance at different pulse tube inclinations, the refrigerator is connected to the rotary valve via a flexible polyimide tube with inner diameter of 4 mm and with a length of 1.2 m. Pulse tube and regenerator are contained in a cylindrical vacuum vessel with the cylinder axis parallel to the tube axis. The vacuum vessel is mechanically fixed to a rotatable horizontal axle. In this way the angle between pulse tube axis and vector of gravitation (see Fig. 2) can be adjusted between $\theta = 0°$ and $\pm180°$ within an uncertainty of $1°$.

## EXPERIMENTAL RESULTS AND DISCUSSION

Best performance of the PTR is achieved for vertical orientation of the pulse tube with the cold end facing downwards ($\theta = 0°$) and at a frequency of 2 Hz for the pressure oscillation. The  peak

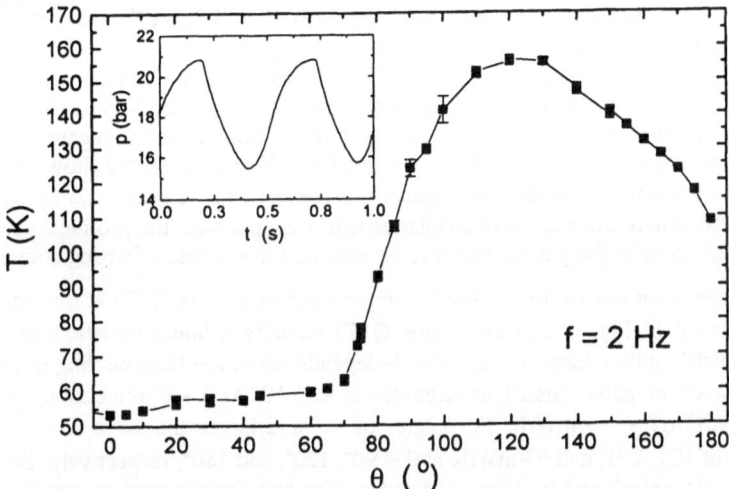

**Figure 4.** Variation of the minimum no-load temperature with angle θ of inclination of the pulse tube. Inset: Pressure wave at the cooler inlet.

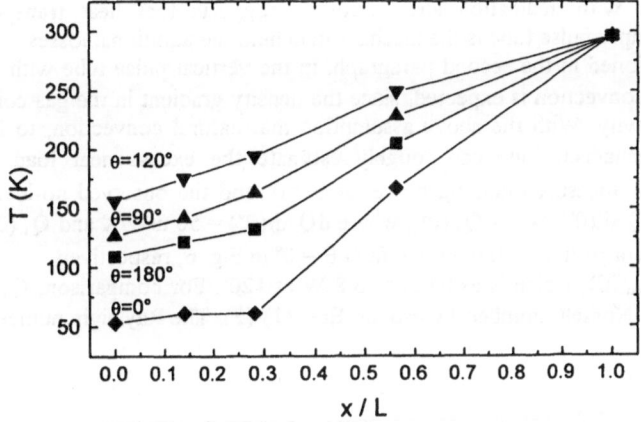

**Figure 5.** Axial temperature profiles along the pulse tube wall (L = 250 mm)

to peak pressure difference at the cooler inlet is then Δp = 5.4 bar and that in the reservoir at the optimum setting of the orifice valve is Δp₀ = 0.5 bar. The optimum flow coefficient of the second-inlet valve is 0.019. In all experiments the average pressure in the system is 18 bar.

**Inclination effect on minimum temperature and temperature profile.** For the above operating parameters the minimum no-load temperature at the cold end is T(0°) = 52.5 K with a reproducibility of ±1 K between different runs. Fig. 4 displays the variation of the minimum no-load temperature as function of the inclination angle θ. The observed variation of T(θ) is found to be symmetric around θ = 0° with a maximum deviation of 2 K for θ = ±90° , where the steepest change of T(θ) occurs. As seen from Fig. 4, at small angles T(θ) increases only moderately up to T(70°)/T(0°) ≈ 1.2. Further increase of θ leads to a steep rise of T(θ)/T(0°) attaining a value of 2.4

for the horizontal position ($\theta = \pm90°$) and a maximum of $T(\pm120°)/T(0°) \approx 3$. Finally, with the cold end up a value of $T(\pm180°)/T(0°) \approx 2$ is measured.

Comparison to $T(\theta)$ for the coaxial PTR in Fig. 1 shows that the variation of the minimum temperature with inclination angle is very similar to that of the present U-shaped PTR, indicating that the effect is not related to the particular configuration of pulse tube and regenerator.

The axial temperature profiles along the pulse tube for different angles are shown in Fig. 5. For $\theta = 0°$ the temperature profile in the part adjacent to the cold end is rather flat up to $x/L \approx 0.3$, while for $\theta = 120°$, where the maximum inclination effect is observed, the profile is approximately linear. A linear T-profile in the gas column is to be expected in the case of strong convection[6,7,9]

**Inclination effect on net cooling power.** The net cooling powers $\dot{Q}(T)$ at distinct inclination angles are displayed in Fig. 6. For all angles $\dot{Q}(T)$ exhibits a linear variation with cold end temperature, which can be related to the linear T-dependence of the ideal cooling power and to a linear T-variation of the cooler losses, as discussed in Ref. 10. At $\theta = 0°$ the cooling power varies with a slope of $d\dot{Q}/dT = 56$ mW/K. For angles of 90° and larger the slope strongly increases, attaining values of 121, 191, and 83 mW/K at $\theta = 90°$, 120°, and 180°, respectively. To first order, the loss mechanism introduced by tilting the pulse tube may be assumed to simply add to the losses at $\theta = 0°$, and that the ideal cooling power is independent of $\theta$. The strong rise of $d\dot{Q}/dT$ then reflects a strong temperature dependence of the additional losses.

**Discussion in terms of natural convection.** The observed variation of minimum temperature and cooling power with inclination angle strongly suggests that heat transport by natural convection in the empty pulse tube is the mechanism behind the additional losses.

As already mentioned in the second paragraph, in the vertical pulse tube with the hot end up ($\theta = 0°$) no natural convection is expected, since the density gradient in the gas column is parallel to the vector of gravity. With the above assumption that natural convection, to first order, is a decoupled loss mechanism, one can roughly estimate the excess heat load $\dot{Q}_{conv}(\theta)$ from convection using the measured cooling power at $\theta = 0°$ and the observed no-load temperatures $T(\theta)$: $\dot{Q}_{conv}(\theta) \approx d\dot{Q}/dt(0°)\, T(\theta) - \dot{Q}_1(0°)$, where $d\dot{Q}/dt(0°) = 56$ mW/K and $\dot{Q}_1(0°) = 2.9$ W are the slope and intercept from the straight line fit at $\theta = 0°$ in Fig. 6, respectively.

Fig. 7 shows $\dot{Q}_{conv}(\theta)$, which is as large as 5.8 W at 120°. For comparison, $\dot{Q}_{conv}(\theta)$ has been estimated from the Nusselt number by use of Eqs. (1)-(5). The Rayleigh numbers have been

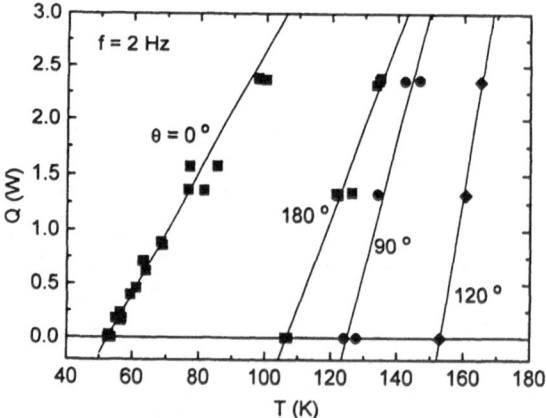

**Figure 6.** Net cooling powers vs temperature of the cold end heat exchanger at various angles of inclination. Straight lines represent linear fits. Operating frequency: 2 Hz.

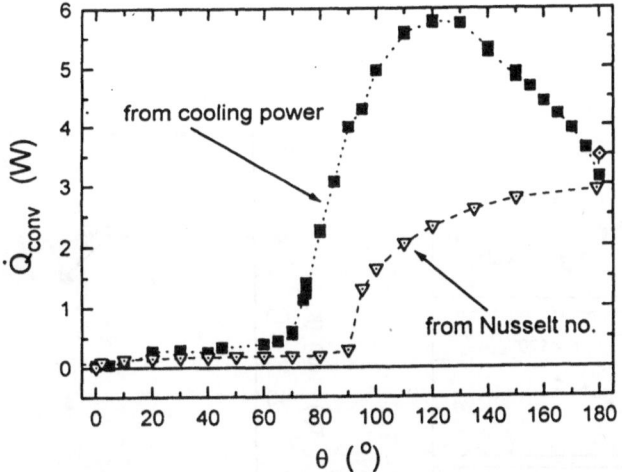

**Figure 7.** Estimated excess heat load as function of the inclination angle derived from cooling power measurements and from the Nusselt number evaluated at 52.5 K.

evaluated from Eq. (2) for $T_h = 300$ K, $T_c = T(0°) = 52.5$ K, $<p> = 18$ bar, and $Pr = 0.67$ for ${}^4$He-gas in the range 50 - 300 K. For the kinematic viscosity $v$ of ${}^4$He-gas the value at the (logarithmic) average temperature $<T> = 142$ K has been used in Eq. (2). The resulting Rayleigh numbers are $Ra(L) = 4.2×10^{10}$ and $Ra(d) = 6.4×10^6$. The critical Rayleigh number for $\theta = 180°$ is $Ra_c(L) = 1.3×10^8 << Ra(L)$.

The heat load $\dot{Q}_m$ from molecular conduction of the ${}^4$He-gas between the hot (300 K) and cold (52.5 K) end of the pulse tube is calculated to be $\dot{Q}_m = 15$ mW. For the excess heat load with the cold end at 52.5 K it then follows that $\dot{Q}_{conv}(\theta) \approx Nu(\theta, Ra)×15$ mW, which is plotted in Fig. 7.

As seen from Fig. 7, the heat loads $\dot{Q}_{conv}(\theta)$ estimated from $Nu(\theta)$ and from the experimental net cooling power agree nearly quantitatively for $\theta < 70°$ and at $\theta = 180°$. For $70° < \theta < 180°$ the Nusselt number only roughly describes the observed variation. A reason for this might be that Eqs. (4) and (5) are not quantitatively applicable to cylinders with a high aspect ratio ($L/d = 18.6$ for the present pulse tube). In particular, the higher losses at 120° compared to that at 180° indicate a different structure of the convective streaming in the two cases.

**Frequency dependence of the inclination effect.** The convective streaming induced by buoyancy is superimposed on the oscillatory gas displacement in the pulse tube, and therefore it is expected to be attenuated at higher operating frequencies.

Fig. 8 shows the variation of the minimum no-load temperature for frequencies between 1.6 and 4 Hz at distinct angles. All other operating parameters were held constant. At $\theta = 0$ the optimum frequency is 2 Hz, as already noted above. For other angles the minimum temperature is obtained at a higher frequency with the most pronounced effect at 90°, where the optimum frequency is shifted to 3.3 Hz. Fig. 9 displays the temperature rise $T(\theta)-T(0°)$, which is a measure of the heat loss from convection, at $f = 2$ and 4 Hz. The strong reduction of $T(90°)-T(0°)$ at 4 Hz, indicates that in the horizontally positioned pulse tube the convective streaming takes place on a larger time scale, and therefore is more affected by increasing the frequency, than in the 120°- and 180°-positions.

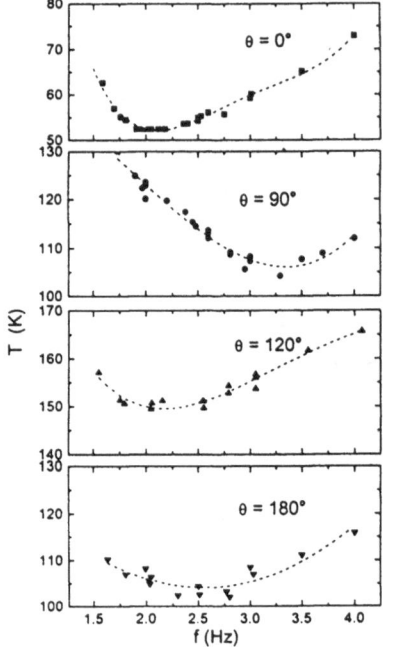

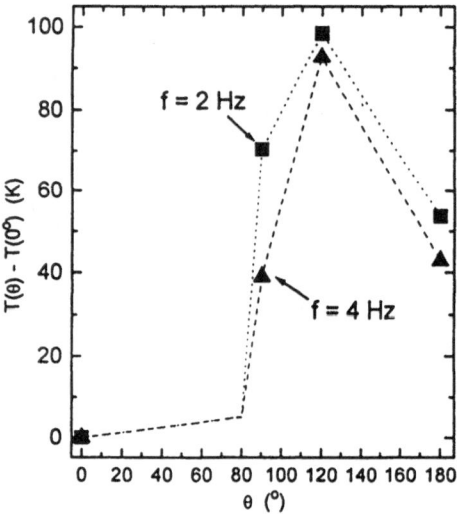

**Figure 8.** Frequency dependence of the minimum temperature at fixed angles of inclination.

**Figure 9.** $T(\theta)-T(0)$ at $f = 2$ Hz and 4 Hz. Lines serve as guides to the eyes.

**Experiments towards reduction of convection.** Natural convection in the pulse tube should be weakened or even suppressed by inserting an appropriate structure into the tube. In order to test this idea, two experiments with different fillings of the pulse tube were performed.

In the first test, the tube was completely filled with felt mats having a porosity of $\approx 0.98$. This led to a strong rise of the minimum no-load temperature to 200 K, but which now turned out to be completely independent of inclination angle.

In the second test, four porous plugs from plastic foam, each with a height of 1 cm, were mounted with regular spacing in the pulse tube. As seen from Fig. 10, by this structure the minimum temperature at $\theta = 0°$ is increased to 86 K as compared to $T(0°) \approx 53$ K for the empty tube (see Fig. 4). For $\theta \geq 90°$, however, the cold end temperatures are now lower than in the empty tube (see Figs. 4 and 10), which indicates that the natural convection is weakened by the porous plugs. The most striking effect is observed at $\theta = 120°$, where the cold end temperature is lowered by 50 K. From this it follows that the convective flow pattern which gives rise to the maximum at 120° in the $T(\theta)$-curve of the empty tube (Fig. 4) is strongly weakened by the presence of the plugs.

One problem with the internal structure in the pulse tube is the degradation of cooler performance at $\theta = 0°$. Among other possible reasons, such as reduction of the oscillating mass flow and distortion of the velocity profile, we believe that the regenerative effect of the plug material will reduce the dynamic temperature oscillation in the tube and therefore will reduce the enthalpy flow. Clearly, more work is needed to understand the effects of an internal structure on the cooler performance.

Another way of decreasing convective heat losses in tilted pulse tubes is to increase the operating frequency, as seen from Fig. 9 above. In this respect, miniature pulse tube coolers[5] operating at frequencies of 40 - 60 Hz should be most favorable. We note that recently[11] no inclination effect could be detected in a miniature pulse tube cooler operating at 50 Hz and reaching a minimum temperature of 78 K.

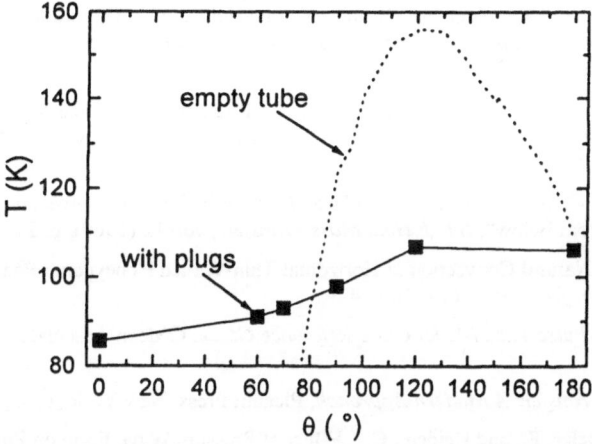

**Figure 10.** Minimum no-load temperature vs angle of inclination for the pulse tube filled with porous plugs (see text). Dashed line represents $T(\theta)$ for the empty tube (Fig. 4). Operating frequency: 2 Hz.

## CONCLUSIONS

The results in the present paper show that tilting of the pulse tube of a PTR with respect to the vertical direction ($\theta = 0°$) can result in a strong degradation of the PTR performance for angles $\theta > 70°$. The effect can be related to heat losses originating from gravity-induced natural convection in the helium working fluid, which is superimposed on the oscillatory mass flow in the pulse tube. From the experiments it follows that convective losses can be reduced in two ways. Either by increasing the frequency of the pressure oscillation or by introducing an internal structure in the pulse tube that hinders the development of the convective flow pattern. First results obtained with an internal pulse tube structure consisting of porous plugs show that for large inclination angles ($\theta \geq 90°$) the cooler performance is improved in comparison to an empty pulse tube.

In several potential applications of PTRs, such as cooling SQUIDs for biomedical investigations and non-destructive evaluation of defects in materials, the possibility of varying the orientation of the cold head is requested. Further investigations towards reduction of convective heat losses in PTRs therefore are required.

## ACKNOWLEDGMENTS

We thank D. Schwabe (Giessen) for useful discussions on convective heat transfer. Financial support by a BMBF grant (No. 13N6176 0) is gratefully acknowledged.

## REFERENCES

1.   Radebaugh, R. „Recent Developments in Cryocoolers", *Proc. 19th International Congress of Refrigeration*, The Hague, vol. IIIb (1995), p. 973

2.   Thummes, G., Landgraf, R., Giebeler, F., Mück, M. and Heiden, C., „Pulse Tube Refrigerator for High-T$_c$ SQUID Operation", *Advances in Cryogenic Engineering*, vol. 41 (1996), to appear

3.   Heiden, C., „Pulse Tube Refrigerators: a Cooling Option for High-Tc SQUIDs", *NATO Advanced Studies Institute*, Maratrea (1995), to be published

4.  Thummes, G., Landgraf, R., Mück, M., Klundt, K., and Heiden, C., „Operation of a High-T$_c$ SQUID Gradiometer by Use of a Pulse Tube Refrigerator", *Proc. 16th International Cryogenic Engineering Conference*, Kitakyushu (1996), to be published

5.  Collins, S.A., Johnson, D.L., Smedley, G.T., and Ross, R.G., Jr., „Performance Characterization of the TRW 35K Pulse Tube Cooler", *Advances in Cryogenic Engineering*, vol. 41 (1996), to be published

6.  Edwards, D.K. and Catton, I., „Prediction of Heat Transfer by Natural Convection in Closed Cylinders Heated from Below", *Int. J. Heat Mass Transfer*, vol. 12 (1969), p. 23

7.  Hollands, K.G.T, „Natural Convection in Horizontal Thin-Walled Honeycomb Panels", *Trans. ASME: J. Heat Transfer*, vol. 95 (1973), p. 439

8.  Gedeon, D., *Sage: Pulse Tube Model Class Reference Guide*, Gedeon Associates, Athens, Ohio, (1995), p. 5

9.  Becker, M., *Heat Transfer: A Modern Approach*, Plenum Press, New York (1986), pp. 220-227

10. Thummes, G., Giebeler, F., and Heiden, C., „Effect of Pressure Wave Form on Pulse Tube Refrigerator Performance", *Cryocoolers 8*, Plenum Press, New York (1995), p. 383

11. Klundt, K, „Untersuchungen zum Betriebsverhalten eines Miniatur-Pulsröhrenkühlers", Diploma Thesis, University of Giessen (1995)

# Advanced Compressor for Long-Life Cryocoolers

P. W. Curwen, Consultant

221 Charlton Road
Ballston Spa, NY

W. D. Waldron

Mechanical Technology Incorporated
Latham, NY

## ABSTRACT

This paper describes work performed to demonstrate the potential advantages (high reliability, low weight, long life) of a hermetically sealed diaphragm-type compressor that separates the gas cycle from the drive system and uses a moving-magnetic-type motor. To accomplish this objective, a "proof of concept" compressor was designed, built, and tested. The original intent was to use the compressor to drive a 30 K, 0.5-W orifice-pulse-tube cryocooler requiring 250 W of pressure-volume (PV) power. The compressor was developed to address these requirements and optimize the overall cryocooler performance by minimizing electrical input power and compressor weight. The proof-of-concept compressor differed from a flight-type design in that it used O-ring-type seals instead of welded hermetic closures and was made of bar stock construction. All other significant components were consistent with a space-flight hardware design. A pulse-tube simulator was used to load the compressor, allowing it to produce the desired pressure amplitude, phase angle, and PV power when operated at design stroke and frequency. With the compressor operating between 0.5 and 1.2% of design frequency, stroke, mean pressure, pressure amplitude, and phase angle, the PV power and estimated efficiency exceeded design values by approximately 4%, thus demonstrating the performance advantages of a diaphragm-type compressor.

## INTRODUCTION

The use of Oxford-style compressors (compressors with flexure-supported linear motors and noncontacting close-clearance piston seals) predominates in current space-based regenerative cryocooler applications. However, there is continuing interest in evaluating advanced types of compressors that might achieve significant improvements in reliability, weight, size, efficiency, and/or procurement cost. This paper describes one such compressor developed by Mechanical Technology Incorporated (MTI) under U.S. Air Force (Phillips Laboratory) Contract F29601-91-C-0112, entitled "Advanced Compressor for Long-Life Cryocoolers." The objective of the contract was to design and demonstrate a proof-of-concept cryocooler compressor with the potential for higher reliability, reduced specific weight, very long life, and reduced cost. This potential arose from two principal features of the compressor concept: 1) the use of diaphragms (instead of pistons) to achieve gas compression in a manner that hermetically isolates the cryocooler gas cycle from the compressor drive mechanism, and 2) the use of oil-flooded linear-motor drive modules to hydraulically actuate (drive) the diaphragms. This combination of features results in the following attributes:

1. The compressor drive system is separated from the cryocooler gas cycle by static seals (which can be hermetically welded), thus eliminating the need for close-clearance or rubbing-contact compressor seals within the gas-cycle environment. Consequently, the drive system cannot contaminate the gas cycle with particulate matter or outgassing vapors.

2. Each oil-flooded linear-motor drive module requires only two simple, low-cost, oil-lubricated journal bearings. Flexure bearings and mechanical springs are not used, thus eliminating motor stroke limitations imposed by such elements. The flooded drive modules can operate in any orientation and are not affected by a 0-g environment. As a consequence of the low bearing loads imposed by the linear motor, bearing life (in terms of allowable wear) is predicted to be in excess of 100,000 hr.

3. Hydraulic actuation of the diaphragms (as contrasted to direct drive) results in reduced weight motors (compressors) since the linear stroke of the motor does not have to match that of the diaphragms. Typically, minimum compressor weight is attained at a motor stroke significantly larger than diaphragm stroke. Additionally, hydraulic actuation results in minimum diaphragm stresses since the instantaneous gas-cycle pressure is balanced by the instantaneous hydraulic actuating pressure. The only differential-pressure loading across the diaphragms is the small loading required to elastically deflect the diaphragms.

4. The compressor concept can be applied to either recuperative cryocooler cycles (e.g., Gifford-McMahon, Joule-Thomson) or to regenerative cycles (e.g., Stirling, pulse-tube). For recuperative cycles, the compressor must be equipped with pressure-actuated valves or displacement-actuated ports to provide a non-zero average cycle flow. For regenerative cycles where time-averaged cycle flow is zero, flow control valves or ports are not required, resulting in less complex compressor designs.

For the proof-of-concept program, a conceptual two-stage pulse-tube application was used as the basis for the compressor design. Table 1 lists the design requirements for the compressor. Representative mission requirements were defined by Phillips Laboratory, with the pulse-tube design point and interface requirements defined by the National Institute of Standards and Technology (NIST). The following sections describe the compressor's mechanical and thermal design, its control and power electronics, and the proof-of-concept test results.

## COMPRESSOR MECHANICAL DESIGN

Figure 1 is a cross section of MTI's advanced 250-W (PV power) compressor as it would be configured for driving a flight-rated split-Stirling or pulse-tube cryocooler. Two identical 125-W (PV power) compression modules are assembled in back-to-back arrangement for cancellation of first-order unbalanced vibration forces. Welded static seal joints are used throughout to obtain hermetic containment of both the cryocooler helium gas and the hydraulic fluid within the compressor drive modules. However, as built and tested, the proof-of-concept compressor used static O-ring seals for ease of assembly, development modification, and post-test inspection. Additionally, to minimize development time and cost, the compressor housings were made from aluminum bar stock rather than drawn steel shells. Figure 2 is a cross section of one of the proof-of-concept compression modules.

The significant design-point parameters for the two-module, back-to-back compressor are as follows: 0.10 clearance volume ratio, 15.45-cc swept volume, 12-mm (0.47-in.) plunger stroke, 85.4% motor electrical efficiency, 97.7% compressor mechanical efficiency, 20.2-W compression space hysteresis loss, and 324-W motor input power. The 12-mm (0.47-in.) motor (plunger) design stroke was selected based on preliminary design optimization surveys to minimize compressor weight. Since compressor PV power and pressures were fixed, only three parameters were available for optimization: compressor (pulse tube) frequency, motor stroke, and linear

**Table 1.** Compressor Design Requirements.

| Phillips Laboratory Requirements | NIST Requirements |
|---|---|
| Operational Life: >100,000 hr | Gas: Helium |
| Reliability: 95% at 60% confidence level | Mean Pressure: 2.0 MPa |
| Power source: 28 V dc ±20% | Pressure Ratio: 1.5 nominal |
| Compressor Residual Vibration: < 0.1 N rms | Compressor Pressure Angle: 40° ±5° |
| Envelope: compatible with pulse tube installation | Compressor PV Power: 250 W nominal |
| Minimum nonoperating: -40°C (-40°F) ambient<br>Minimum operating: 0°C (-32°F) ambient<br>Nominal operating: 27°C (81°F) ambient<br>Maximum operating: 50°C (122°F) ambient | |
| Compressor Frequency: 40 Hz | |

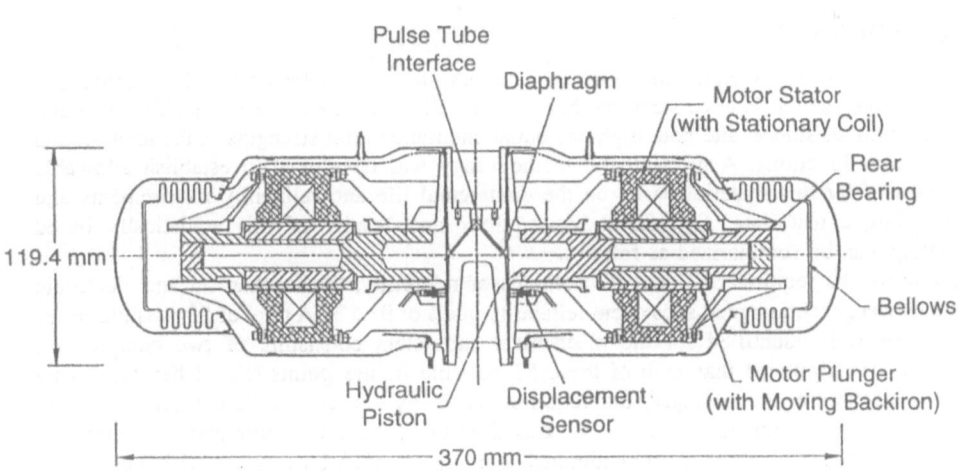

**Figure 1.** Cross section of advanced 250-W (PV power) diaphragm compressor.

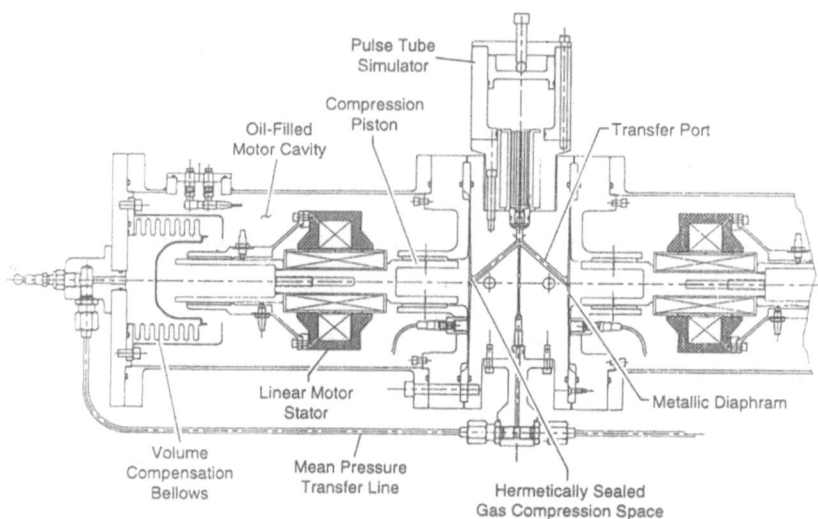

**Figure 2.** Cross section of 125-W (PV power) proof-of-concept compression module.

motor backiron arrangement (moving versus stationary backiron). The results of the optimization surveys showed that a 60-Hz compressor would yield a feasible, minimum mass compressor design weighing approximately 5 kg (11.02 lb). However, in view of the still-emerging state of pulse-tube technology, a 40-Hz compressor frequency was selected with an attendant 40% increase in compressor weight to approximately 7 kg (15.43 lb).

Major engineering tasks for each compression module were design of the metallic diaphragm, the linear motor, the metallic bellows, and the two boundary-lubricated journal bearings, plus selection of hydraulic fluid and method of module temperature control. Each of these tasks is described briefly below.

**Diaphragm Design**

Armco PH13-8Mo precipitation-hardened stainless steel was selected for the diaphragms. This is a double-vacuum-melted steel that has high metallurgical cleanliness, high chemical and stress corrosion resistance, and both high and equal endurance-limit strengths in the longitudinal and transverse directions. A probabilistic methodology was developed to establish allowable design stresses for this material based on the contractual life and reliability requirements and available fatigue test data. Significant assumptions and results of this statistically based methodology can be summarized as follows.

To achieve the required 95% overall compressor reliability, the mechanical and electronic subsystems were each assigned subsystem reliability goals of 97.5%. A total of 52 possible series failure points were identified for the mechanical subsystem consisting of two compression modules. It was stipulated that each of these 52 possible failure points should have the same probability of failure. Accordingly, the required reliability for each failure point, based on 100,000-hr life, was computed to be 0.9995. Based on this reliability requirement and applying a 60% confidence level to the fatigue endurance strength of PH13-8Mo in the H1000 condition (several additional derating factors were also applied), the maximum allowable design stress for the diaphragms was established to be 273.5 MPa (39.6 ksi).

To eliminate fretting as an unpredictable source of fatigue failure, the diaphragms were designed with integrally machined OD rims. The nominal diaphragm thickness of 0.368-mm (0.0145-in.) is blended into the integrally machined rim via a 6-mm (0.24-in.) transition radius that effectively eliminates any stress concentration in the transition region.

## Bellows

A bellows component is used in each compression module to provide three functions. First, it maintains the average hydraulic pressure in the flooded motor drive cavity slightly above the average pressure of the module's compression cycle (i.e., slightly above average pulse-tube pressure). The gas side of the bellows is connected through a flow restrictor to the helium compression chamber. Since the bellows has very low axial stiffness, the average gas pressure is, in effect, transmitted through the bellows to the hydraulic fluid within the motor drive cavity.

Second, the bellows also compensates for volume changes in the hydraulic fluid due to compressibility and thermal expansion effects under all compressor operating and non-operating conditions. Thermal expansion effects are significant because of the -40°C to +65°C (-40 to 149°F) range in hydraulic fluid temperature that the compressor was designed to accommodate.

Third, the bellows provide dynamic accommodation of the cyclic changes in motor cavity volume due to the swept volume displacement of the inboard motor bearing (this bearing acts as the hydraulic piston which drives the diaphragm). These volume changes occur at compressor frequency, thus imposing a high-cycle fatigue stress on the bellows.

Due to time and cost constraints, a commercially available single-ply hydroformed Inconel 718 bellows was selected. The same probabilistic methodology used for the diaphragms was used to establish maximum allowable design stresses for the bellows. Since MTI had no control over the bellows manufacture, maximum bellows stress was further reduced by increasing the number of bellows convolutions to yield a bellows reliability greater than that of the diaphragms.

## Linear Motor

The electromagnetic principles of MTI's moving-magnet linear motor are basically the same as those of the more common moving-coil-type motor. However, in MTI's linear motor, the ac winding is mounted in the stator assembly of the motor. This eliminates the need for any mechanical or electrical connections to the reciprocating motor plunger. The motor air gap is small, and the amount of iron and copper (or permanent magnet material) required to drive dc flux across the gap is minimized, resulting in reduced motor weight for a given power and efficiency level.

Figures 1 and 2 show the cross-section arrangement of the MTI linear motor. Three circumferential rows of radially magnetized samarium cobalt ($Sm_2Co_{17}$) permanent magnets are mounted on the OD of a laminated cylinder formed from Hyperco-50 lamination steel. The laminations, which function as "backiron" for the magnets, are radially stacked around the axial centerline of the motor plunger and assembled as an integral part of the reciprocating assembly. The magnets and backiron thus move with the plunger in the motor's air gap, giving rise to the designation of this arrangement as a "moving backiron" motor.

The motor stator assembly consists of a single prewound and potted ac coil. Square cross-section wire is used to maximize coil packing factor and thus motor efficiency. Axially aligned Hyperco-50 laminations are radially stacked around the coil to form a rigid, cylindrical stator assembly. Hyperco-50 lamination material was selected for both the plunger and stator assemblies because of its higher flux saturation level, thus resulting in a minimum weight motor.

Nominal design-point operating parameters for the linear motor are: 145-W plunger output power, 12-mm (0.47-in.) plunger stroke; 40-Hz frequency; 85% electrical efficiency, 14.4 $V_{rms}$ ac line voltage, and 16.65 $A_{rms}$ ac line current. In addition to providing 145 W of output power, the motor is also designed to provide ±31 N (±7 lbf) of dc force. The compressor controller provides a dc component of current to the motor as required to maintain centered operation (zero offset) of the diaphragm. Figure 3 shows the linear motor hardware.

The compressor motors are driven by a pulse-width-modulated (PWM) inverter that provides both the ac and dc components of current as commanded by the compressor's controller. The PWM inverter is powered from a 28 V dc supply.

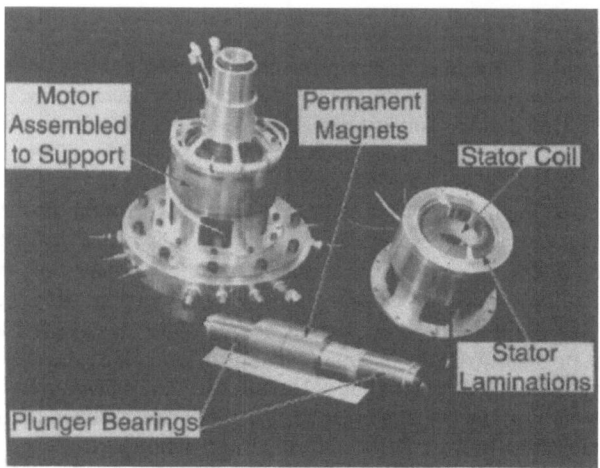

**Figure 3.** Linear motor hardware.

## Motor Bearings

Each compressor drive module contains two journal bearings that are an integral part of the motor's reciprocating assembly. The end face of the inboard bearing also serves as the hydraulic piston that actuates the compressor diaphragm. Since the bearings are located within the motor drive compartment, they are completely immersed in hydraulic fluid. The stationary part of each bearing consists of a bronze-filled carbon-graphite sleeve captured in the bearing housing. The reciprocating journal surface is coated with flame-sprayed tungsten carbide. Average bearing loading is 0.042 MPa (6.06 psi). Because of this very light loading, approximately half of the load is carried hydrodynamically and half by boundary-layer asperity contacts. For a wear factor of 3.83E-12 mm$^2$/N (2.64E-14 in.$^2$/lbf) and a permissible wear of 0.038 mm (0.0015 in.), bearing life is conservatively computed to be in excess of 116,000 hr. Using a conservative friction factor of 0.08, bearing friction loss due to asperity contacts is computed to be 1.5 W per bearing, while hydrodynamic fluid film shear loss is 0.5 W per bearing. Total loss for two bearings is 4.0 W. All of the compressor bearings performed flawlessly throughout the compressor test program.

## Hydraulic Fluid

Key requirements for the hydraulic fluid used in the flooded motor compartment are long-term stability and chemical inertness with the internal parts of the compressor at temperatures up to 65°C (149°F). Additional important characteristics are low viscosity, freeze or pour point well below -40°C (-40°F), low density, low coefficient of thermal expansion, high electrical resistivity, and lubrication properties at least equivalent to water.

Both silicone and synthetic polyalphaolefin (PAO) oils appear to be excellent hydraulic fluids for this compressor. Because of its lower density and lower thermal expansion coefficient, a PAO oil (Royco 602) was selected. This Mil-certified oil is used to cool aircraft avionics by direct flooding of the electronic boards. Although manufacturers express high confidence in the long-term stability of these oils, MTI has not yet found actual 10 to 20-year experience data to substantiate these opinions.

## Module Temperature Control

Each compression module is cooled by conductive heat transfer of module losses to the compressor center body. In the proof-of-concept compressor, liquid coolant channels were machined into the center body. Assuming water flow of 12.6 cc/sec (0.2 gpm) at 18°C (64°F) supply temperature, maximum predicted module temperature (which occurs within the motor ac winding) was a satisfactory 58°C (136°F). It appears that conductive heat straps could also be used to transfer module losses from the center-body region to a cold-plate heat sink.

## COMPRESSOR CONTROL AND POWER ELECTRONICS

The compressor control and power electronics were designed using standard laboratory and commercially available electronic and instrumentation components. The functions of the control and power electronics are to: 1) provide nominal 40-Hz electrical power to each compression module motor; 2) provide master/slave control functions for maintaining setpoint stroke amplitude and phase relationships between the two compression modules; 3) control dc voltage to each module to maintain centered diaphragm (zero offset) operation; 4) provide a means for manually trimming phase and amplitude between each module to minimize residual first-harmonic force transmitted to the mounting structure; 5) provide control algorithms for automatic start-up and shutdown of the compressor, as well as functions to detect abnormal operation and to initiate an emergency shutdown if necessary.

The control system consists of two subsystems: one controlling each of the two modules (master and slave). Each subsystem uses a capacitance probe to monitor diaphragm stroke amplitude and dc offset. The capacitance probes are mounted on the hydraulic fluid side of the diaphragms to eliminate possible contamination and leakage sites in the cryocooler gas cycle. Additional sensors are used to monitor motor plunger and bellows displacements, but these sensors were not part of the control system.

The compressor controller is designed as a state machine having the following five control states: inactive, start-up and ramp, stabilize, run, and calibrate. The control system is always operating in one (and only one) of these states. The control system software is written in Forth.

## Compressor Test Program

The compressor test program started with testing of the linear motors to compare predicted versus measured air-gap force, followed by dynamic testing of the individual compression modules to optimize module performance and to check operation of the PWM amplifiers. The complete back-to-back compressor configuration was then tested at its design-point conditions using a pulse-tube simulator to produce the required pneumatic loading of the compressor (i.e., pressure amplitude, pressure angle, and PV power). Finally, the compressor was used to drive an MTI-developed 50-W pulse-tube simulator.

## Motor Air-Gap Force Tests

Static tests were performed on each motor to measure force production capability. In these tests, a force transducer was used to measure air-gap force as a function of dc current in the ac coil and plunger displacement position. Predicted motor force gradient was 8.17 N/A (1.84 lbf/A). Measured force gradient was 8.45 N/A (1.90 lbf/A) and -8.27 N/A (-1.85 lbf/A) for currents of +25 and -25 A, respectively. This excellent agreement between predicted and measured coil forces indicated that the motors could produce a design output power of 145 W.

The predicted axial spring-force gradient due to the plunger magnets was 67 N/cm (38.3 lbf/in.) (restoring) as compared to measured values ranging from 44.5 to 58.4 N/cm (25.4 to 33.3 lbf/in.) depending on plunger position. The lower value of measured gradient was not considered a problem since compressor dynamics are dominated by the gas-spring stiffnesses of the compression cycle and the bellows dynamic displacement cycle.

**Individual Module Test Results**

During testing of the individual modules, only one significant design modification was made. The diameter of the helium flow passage connecting each compression cylinder to the compressor discharge port was increased from 2.94 to 3.96 mm (0.116 to 0.156 in.) (an 80% increase in flow area). Flow passage pressure drop was thereby reduced from about 13 kPa (1.9 psi) to about 3.4 kPa (0.5 psi). PV power available at the compressor discharge port was increased by about 11 W (10.4%).

**Dual-Module Compressor Test Results**

Each compression module was instrumented with sensors to measure dynamic motor and diaphragm strokes, compression cylinder pressure, bellows pressure, compressor discharge pressure, motor voltage and current, and various internal compressor temperatures, including motor coil temperature. A PC-based data acquisition system with high-speed analog-to-digital converters was used to record and compute compressor performance.

The dual-module, back-to-back compressor assembly was operated successfully in both manual (open loop) and automatic (with closed-loop controller) modes. Some difficulty was initially encountered in maintaining stable operation during automatic compressor start-up, but modifications to the controller software algorithms corrected this problem. Stable start-up and ramp up to steady-state design-point conditions under fully automatic control was achieved. The ramp-up interval currently takes about 10 min to reach design-point conditions. Further development of the control algorithms could probably reduce this time.

Figure 4 is a photograph of the fully assembled proof-of-concept compressor with the pulse-tube simulator installed. Table 2 compares overall measured performance of the compressor to design-point predictions. Because the test values of swept volume, pressure amplitude, and frequency slightly exceeded the design-point values, compressor output PV power exceeded the 250-W design value by 4.2%. The estimated accuracy of the measured data is ±2.0%.

The test value of overall compressor efficiency listed in Table 2 had to be estimated since accurate measurements of PWM input power to the motors were not obtained. The estimated value of test efficiency is based on measured values of cylinder PV power for each module and calculated values of module drive losses. The drive losses consisted of bearing and hydraulic pumping losses, bellows dynamic heat transfer loss, and motor losses ($i^2R$ coil loss plus eddy-current and hysteresis losses in the lamination iron and plunger magnets). The fact that estimated compressor efficiency is three points higher than the predicted efficiency may be due to lower-than-predicted dynamic heat transfer losses in the compression cylinders.

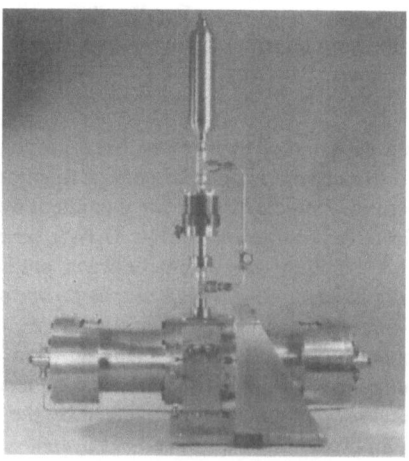

**Figure 4.** Proof-of-concept compressor assembly with pulse-tube simulator installed.

Table 2. Compressor Performance: Design Point Versus Test.

| Compressor Performance | Design | Test |
|---|---|---|
| Frequency, Hz | 40 | 40.2 |
| Piston Swept Volume, cc | 15.5 | 15.7 |
| Diaphragm Swept Volume, cc | 15.5 | 15.9 |
| Mean Pressure, psig[a] | 275 | 276.8 |
| Pressure Ratio | 1.5 | 1.5 |
| Pressure Amplitude, psi[a] | 58 | 58.3 |
| Pressure Angle (relative diaphragm), deg[a] | 40 | 40.2 |
| PV Power to Cryocooler, W[a] | 250 | 260.6 |
| Compressor Efficiency | 0.71 | 0.74 |
| **Module Performance[b]** | | |
| Piston Amplitude, mm | 6.0 | 6.07 |
| Diaphragm Amplitude, mm | 0.96 | 0.99 |
| Bellows Amplitude, mm | 1.04 | N/A |
| Cylinder Pressure Amplitude, psi | 58.0 | 59.0 |
| Cylinder Pressure Angle, deg | 45.5 | 42.9 |
| Cylinder PV Power, W | 138.5 | 135.2 |
| Motor Shaft Power, W | 150.8 | N/A |
| ac Current Amplitude, A | 23.5 | 23.6 |
| ac Voltage Amplitude, V | 20.5 | 12.0[c] |
| Motor Alpha Angle, deg | -3.1 | -11.7 |
| Motor Efficiency, % | 85.1 | N/A |
| Motor Input Power, W | 177.2 | 112.0[c] |

P476

[a]At compressor discharge port.
[b]Average of two modules.
[c]Inaccurate measurement of PWM drive voltage.

Table 2 also compares single module design versus test performance for the validation test point data shown in the upper portion of the table. Since the measured data for each module was almost identical, the average of the two modules is tabulated. Comparison (discharge pressure amplitude minus cylinder pressure amplitude) shows the compressor discharge pressure drop to be about 4.8 kPa (0.7 psi), consistent with the individual module test data. The fact that measured cylinder pressure angle is 2.6 degrees smaller than the design pressure angle further suggests that the actual dynamic heat transfer loss in the cylinders was less than allowed for in the design-point calculations. Table 2 also shows the inaccuracy of the PWM voltage measurement, resulting in measured motor input power being less than cylinder PV power (which, of course, is not possible).

A second proof-of-concept compressor was built under an MTI IR&D program, along with an MTI-developed orifice pulse tube to provide 80 W of cooling at 200 K. During these tests, the compressor was operated at pressure levels up to twice design-point levels. To maintain near-resonant operation at these higher pressures, additional mass was added to each motor plunger assembly. No other modifications were required.

## CONCLUSIONS

The compressor design and test program described herein achieved its design-point performance objectives without encountering any major development problems. Given these results, the program successfully demonstrated that hydraulically actuated, diaphragm-type compressors can provide the performance capabilities and ruggedness desired for space cryocooler applications while maintaining hermetic separation of the compressor drive system from the cryocooler gas cycle. The reduced specific weight characteristic of this type of compressor was also demonstrated via several design studies during the course of the program.

The potential for long life and high reliability was demonstrated by means of appropriate industry-accepted design analyses and reliability prediction techniques. Actual endurance and other proof-type testing will be required to confirm the compressor's life and reliability potential.

As would be expected from a proof-of-concept development program, several possible design improvements were identified. These include techniques for reducing hydraulic fluid inventory, use of a second diaphragm to perform the dynamic volume compensation function of the bellows (thereby eliminating a reliability concern arising from bellows inspection difficulties), and improvements to the compressor controller.

# Flexure Bearing Analysis Procedures and Design Charts

C. C. Lee and R. B. Pan

The Aerospace Corporation
Los Angles, California, USA 90009

## ABSTRACT

Flexure bearings are used in cryocooler compressors and expanders to maintain the ingegrity of the clearance seal between the piston and cylinder. At present, three-finger spiral flexure designs are widely used, but the tangential linear flexure design patented by The Aerospace Corporation is gradually being adopted by the industry.

In this study, analytical procedures were developed to predict flexure performance for any geometric configuration, considering complex flexure design requirements for operating amplitude, flexure radial and axial stiffnesses, fatigue stress allowables, and dynamic characteristics. Rotating a single flexure finger about the center of the drive shaft, a repeated multi-finger geometric model is generated. This model is then analyzed by a nonlinear finite-element analysis code, varying various geometric and material design parameters, to determine the impact of these parameters on flexure performance. The procedures to perform all analysis tasks have been automated. Results have also been verified with available test data.

A set of three-finger spiral flexure design charts was developed using these analysis procedures. The design charts provide basic design characteristics of the spiral flexures. The procedures and charts can be used for verification of design margins and effectiveness, for optimization of new designs with specified constraints, or for the design tangential flexures.

## INTRODUCTION

Long-life, high-reliability, space qualified cryogenic cryocoolers are a major enabling technology for longwave sensor systems on surveillance satellites. It has been shown that a linear drive, non-contacting bearing, Stirling cycle cryocooler concept is one of the best approaches to satisfy these needs.[1,2] The key technology developments that enable spacecraft applications of Stirling cycle cryocoolers are flexure bearings and gas seals. The flexure bearings provide the frictionless and non-wearing support for the reciprocating components of the cryocooler. The high radial stiffness of these flexure bearings allows for the use of non-contacting gas gap clearance seals. These bearings were first applied to cryocoolers by Oxford University,[3] although, they had previously been used in an artificial heart application and in an ultraviolet sensor shutter system.[4,5]

Figure 1 shows some of the typical flexure bearing configurations used in various appications. At present, three finger spiral flexure designs are widely used, but a new linear flexure bearing design patented by The Aerospace Corporation,[6,7] see Fig. 2, is being gradually adopted by the industry. In general, the flexure bearing assembly consists of a stack of axially flexible, cut-diaphragms with inner and outer rim/spacers. Figure 3 shows a single diaphragm assembly. The inner rim/spacer provides support to the piston assembly. The outer rim/spacer is fastened to the compressor housing, which provides support to the flexure diaphragm. An ideal flexure bearing should have large radial or in-plane stiffness, minimal axial or out-of-plane stiffness, and low operating stresses. The large radial stiffness is needed to maintain the extremely tight clearance between the piston and cylinder to form the gas gap clearance seal. The axial stiffness needs to be kept low to minimize the effect upon the natural frequency of the spring-mass system composed of the piston and the compressible gas in the compressor chamber. Low operating stresses in the flexure are required to assure that the flexure bearing will not fail due to fatigue. Therefore, the flexure bearing operating response characteristics have to be clearly defined before a preferred design can be properly chosen.

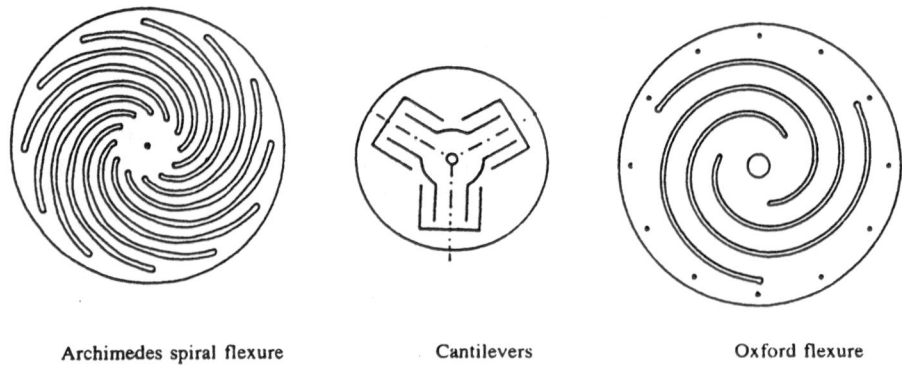

Archimedes spiral flexure                    Cantilevers                    Oxford flexure

**Figure 1.** Typical flexure bearings

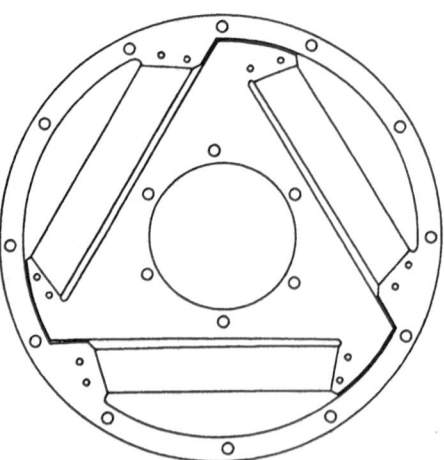

**Figure 2.** Tangential linear flexure bearing

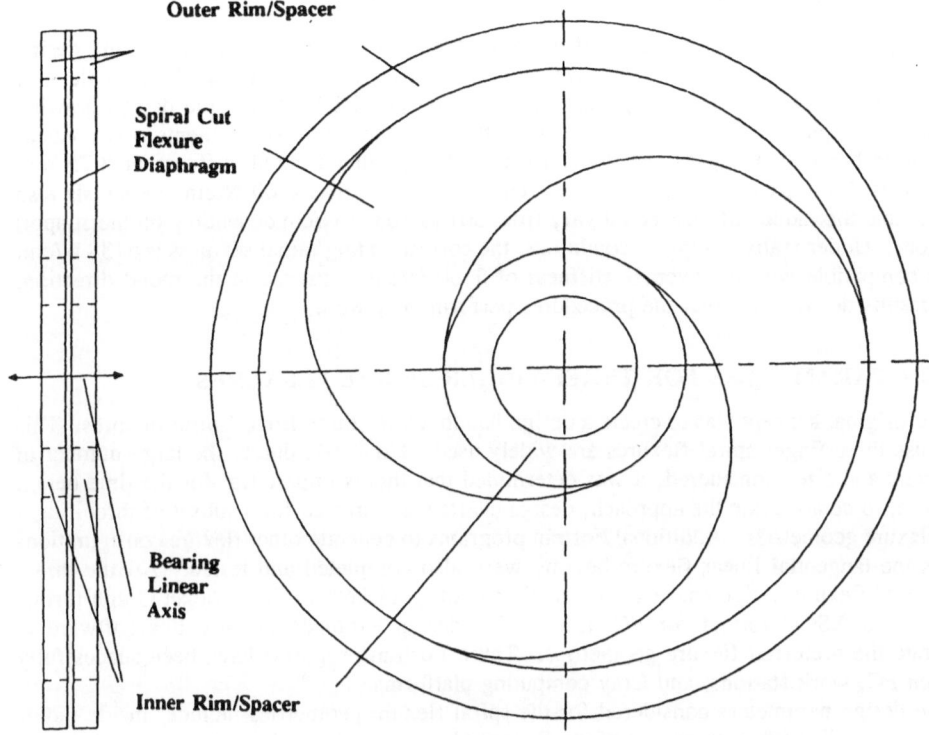

**Figure 3.** General topology - spiral flexure bearing

In this paper, our intention is to develop automated analysis procedures for predicting the flexure performance of any given flexure configuration and for a variety of design requirements. These requirements include the flexure operating amplitude, flexure radial and axial stiffnesses, and material fatigue stress allowable. The automated analysis procedures also have to be easy to use, and able to operate under any computing platforms.

## ANALYSIS PROCEDURES

The analysis procedures consist of special application Fortran programs developed for this study and commercially available stress analysis software. One of the Fortran programs generates the basic geometry of a single flexure finger. It can be a spiral, trapezoid, rectangle or any other shape as long as the multiple fingers model can fit into the specified working space without an interference problem. After finalizing the geometry of a single finger, another Fortran program, by repeated rotation of the single finger about the center of the drive shaft, generates a multi-finger flexure model. With the appropriate force and displacement boundary conditions applied to the multi-finger flexure model, the flexure response characteristics can be determined using commercial nonlinear finite element stress analysis software. By varying the finger geometry, the sensitivity of design parameters on flexure performance can be analyzed. Thus, a preferred design configuration can be determined. This procedure has been automated for flexure design applications.

## VERIFICATION BY TEST

To verify the numerical accuracy of the procedures, the Hughes tangential flexure design was analyzed. The Hughes design is based on the Aerospace patented tangential linear flexure. Due to its proprietary nature, design details cannot be provided. Using the design drawing provided by Hughes, a finite element model with 601 nodes and 504 elements was created. Each finger has essentially a trapezoidal shape. The predicted axial stiffness at 3.75 mm deflection is 3.93 N/cm. In comparison, Hughes's test result is 4.00 N/cm. Analysis also indicates that the radial stiffness could vary from 501 to 10378 N/cm depending on the support conditions. Under realistic support conditions, the corresponding radial stiffness is 5138 N/cm. This is compatible with the average stiffness of 5100 N/cm measured in the radial direction. These results demonstrate that the procedures perform very well.

## DESIGN PARAMETERS FOR THREE-FINGER SPIRAL FLEXURES

Our original concept was to create a design handbook for three-finger spiral flexures. This is because three-finger spiral flexures are widely used. However, due to the large number of parameters and sizes considered, it was determined that this is impractical for the time being. Therefore, to demonstrate the approach, design charts were created for a subset of three-finger spiral flexure geometries. Additional Fortran programs to generate other flexure configurations such as the tangential linear flexure bearing were also completed and tested. All the three-finger spiral flexure design charts are currently stored on diskettes using Cricket Graph format for Mac and ASCII format for PC users. The design engineer can use this software to regenerate the preferred flexure geometries. These Fortran programs have been successfully tested on PC, work stations, and Cray computing platforms.

The design parameters considered for the spiral flexure geometries include; inside radius (RI), outside radius (RO), thickness of the flexure (T), approaching slope at inside radius (SI), approaching slope at outside radius (SO), inflection point (FM), and total sweep angle (SA).

For the spiral flexure, the spiral configuration can be determined by the equation

$$\mathbf{R} = RI + (RO-RI)*(A1*\Theta^1 + A2*\Theta^3 + A3*\Theta^5 + A4*\Theta^7) \qquad \ldots\ldots\ldots(1)$$

where $\mathbf{R}$ is the polar distance from the center of the flexure, in terms of the relative sweep angle ratio $\Theta$, i.e. the angle divided by the sweep angle. $\Theta$ is 0 at the inner radius and 1 at the outer radius. The corresponding slope can be expressed as

$$\mathbf{S} = (dR/d\Theta)/(RO-RI) = A1 + 3.0*A2*\Theta^2 + 5.0*A3*\Theta^4 + 7.0*A4*\Theta^6 \qquad \ldots\ldots\ldots(2)$$

while the inflection point is defined by

$$FM = (R(FM)-RI)/(RO-RI) = A1*FM + A2*FM^3 + A3*FM^5 + A4*FM^7 \qquad \ldots\ldots\ldots(3)$$

Using equations (1) through (3), in conjunction with appropriate boundary conditions at $\Theta$ = 0. and $\Theta$ = 1., the spiral geometry can be determined for selected values of RI, RO, SI, SO, and FM. The coefficients A1, A2, A3 and A4 are defined by the expressions

$$A1 = SI \qquad \ldots\ldots\ldots(4)$$

$$A2 = - [(2*FM^2 + 1)/FM^2]*SI - [FM^2/2/(FM^2 - 1)]*SO$$
$$+ FM^2*(5*FM^2 - 7)/2/(FM^2 - 1)^2 + 1/M^2/(M^2 - 1)^2 \qquad \ldots\ldots\ldots(5)$$

$$A3 = [(2 + FM^2)/FM^2]*SI + [(FM^2 + 1)/2/(FM^2 - 1)]*SO + (7 - 3*FM^4)/2/(FM^2 - 1)^2$$
$$- 2/FM^2/(FM^2 - 1)^2 \qquad \ldots\ldots\ldots(6)$$

and

$$A4 = -SI/FM^2 -SO/2/(FM^2 -1) +(3*FM^2 -5)/2/(FM^2 -5)/2/(FM^2 -1)^2$$
$$+1/FM^2/(FM^2 -1)^2 \qquad\qquad\qquad\qquad\qquad\qquad .........(7)$$

These notations are also used in the Fortran programs.

In the present study, spiral flexure inside radius ( 0.59"), outside radius ( 1.54" ), flexure material ( stainless steel ) and thickness ( 0.005"and 0.01"), and the target maximum axial deflection ( 0.2" ) are fixed. The notation Sijkl is used to represent the remaining four parameters. The range of these parameters considered in the present study is given in Table 1. A list of cases analyzed is shown in Table 2.

## ANALYSIS RESULTS AND DESIGN CHART

Analysis results for representative case, S5556, are shown in Figure 4 and Figure 5. These results illustrate the performance characteristics of this flexure in terms of the radial stiffnesses, axial stiffnesses, and von Mises stresses at various magnitudes of axial deflection. A summary of other cases and results at the maximum axial deflection of 0.2" are shown in Table 3 and plotted in Figure 6. According to the design chart in Figure 6, at the lower stiffnesses, it can be seen that there are many flexure configurations possible for a given stiffness. For higher stiffnesses, options become limited. In either case, it can be seen that for a given design material fatigue stress level and flexure stiffness requirement, it is possible to identify the best flexure configuration from the design chart.

**Table 1.** Three-finger Spiral Flexures Sijkl

| Index | SI (i) | SO (j) | FM (k) | SA (l) |
|-------|--------|--------|--------|--------|
| 1 | 0.8 | 0.2 | 0.4 | 60.0 |
| 2 | 0.9 | 0.4 | | 120.0 |
| 3 | 1.0 | 0.6 | | 180.0 |
| 4 | 1.1 | 0.8 | | 240.0 |
| 5 | 1.2 | 1.0 | 0.5 | 300.0 |
| 6 | 1.3 | 1.5 | | 360.0 |
| 7 | 1.4 | 2.0 | | 420.0 |
| 8 | 1.5 | 2.5 | | 480.0 |
| 9 | 1.6 | 3.0 | 0.6 | 540.0 |

**Table 2** CASES RUN

S1154 S1155 S1156 S1157 S1158 S1159 S1554 S1555 S1556 S1557 S1558 S1559
S1754 S1755 S1756 S1757 S1758 S1759 S1954 S1955 S1956 S1957
S1958 S1959 S5154 S5155 S5156 S5157 S5158 S5159 S5515 S5554 S5555 S5556
S5557 S5558 S5559 S5595 S5754 S5755 S5756 S5757 S5758 S5759
S5954 S5955 S5956 S5957 S5958 S5959 S9154 S9155 S9156 S9157 S9158 S9159
S9554 S9555 S9556 S9557 S9558 S9559 S9754 S9755 S9756 S9757 S9758 S9759
S9954 S9955 S9956 S9957 S9958 S9959

**Table 3.** Axial & Radial Stiffness of Design Charts

| Design Chart | Plate Thickness mils | Axial Stiffness lbs/in | Radial Stiffness x100.lbs/in | Von Mises ksi |
|---|---|---|---|---|
| S5959 | 5 | 0.0260 | 0.0302 | 7.021 |
|  | 10 | 0.1900 | 0.2375 | 7.179 |
| S9159 | 5 | 0.0260 | 0.0394 | 6.780 |
|  | 10 | 0.1880 | 0.2637 | 7.824 |
| S9758 | 5 | 0.0300 | 0.0666 | 6.051 |
|  | 10 | 0.2020 | 0.3085 | 8.925 |
| S5159 | 5 | 0.0330 | 0.0566 | 5.622 |
|  | 10 | 0.2540 | 0.3790 | 10.096 |
| S9558 | 5 | 0.0360 | 0.0837 | 8.382 |
|  | 10 | 0.2400 | 0.3962 | 8.970 |
| S1958 | 5 | 0.046 | 0.0842 | 8.834 |
|  | 10 | 0.340 | 0.4844 | 12.924 |
| S9557 | 5 | 0.072 | 0.1468 | 7.813 |
|  | 10 | 0.417 | 0.8445 | 10.727 |
| S9756 | 5 | 0.122 | 0.2713 | 13.169 |
|  | 10 | 0.623 | 1.5693 | 14.838 |
| S9556 | 5 | 0.158 | 0.3181 | 14.279 |
|  | 10 | 0.798 | 1.9318 | 16.934 |
| S9156 | 5 | 0.205 | 0.3914 | 17.539 |
|  | 10 | 1.030 | 2.4497 | 20.449 |
| S5756 | 5 | 0.210 | 0.3706 | 22.522 |
|  | 10 | 1.323 | 2.3375 | 26.613 |
| S5556 | 5 | 0.210 | 0.4024 | 23.885 |
|  | 10 | 1.810 | 2.4826 | 31.308 |
| S1956 | 5 | 0.212 | 0.3116 | 27.797 |
|  | 10 | 1.480 | 2.1061 | 34.126 |
| S1756 | 5 | 0.236 | 0.3429 | 29.818 |
|  | 10 | 1.722 | 2.1951 | 41.454 |
| S9555 | 5 | 0.352 | 0.6263 | 18.791 |
|  | 10 | 1.539 | 4.3013 | 25.028 |
| S5755 | 5 | 0.385 | 0.7344 | 26.102 |
|  | 10 | 2.408 | 4.4014 | 35.060 |
| S1955 | 5 | 0.442 | 0.6806 | 30.492 |
|  | 10 | 3.116 | 3.9957 | 42.301 |
| S1954 | 5 | 1.223 | 6.3480 | 33.928 |
|  | 10 | 8.535 | 37.627 | 60.980 |
| S1554 | 5 | 1.601 | 11.077 | 41.453 |
|  | 10 | 10.960 | 60.111 | 82.130 |

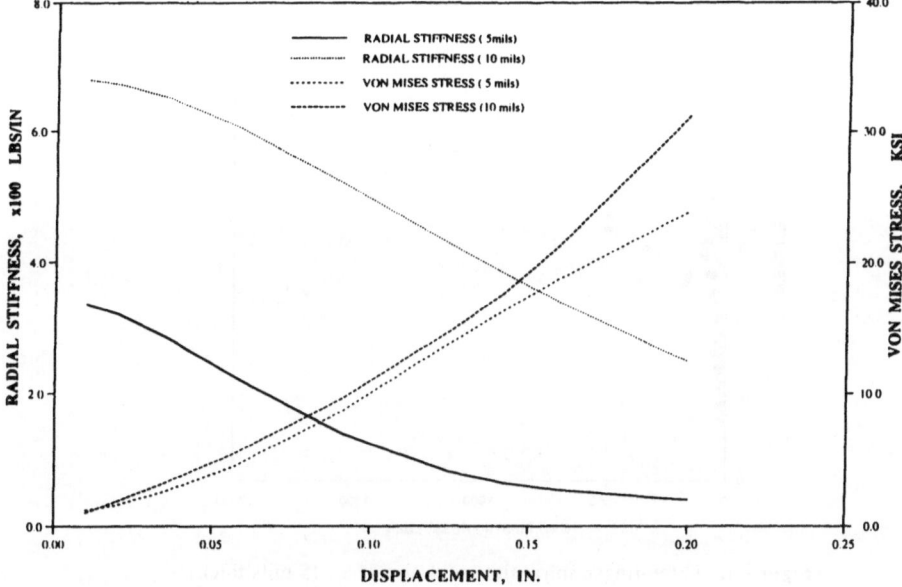

**Figure 4.** Design charts S5556 radial

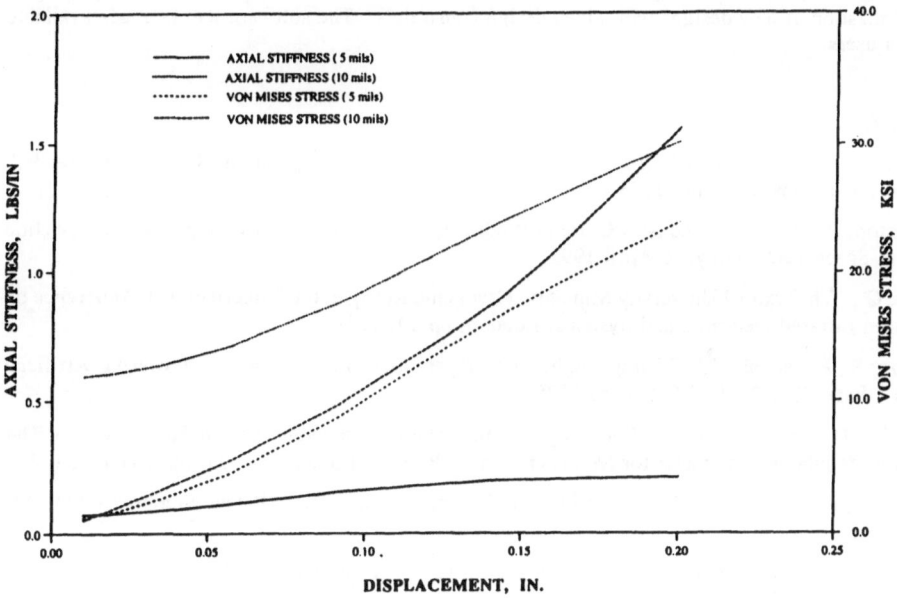

**Figure 5.** Design chart for S5556 axial

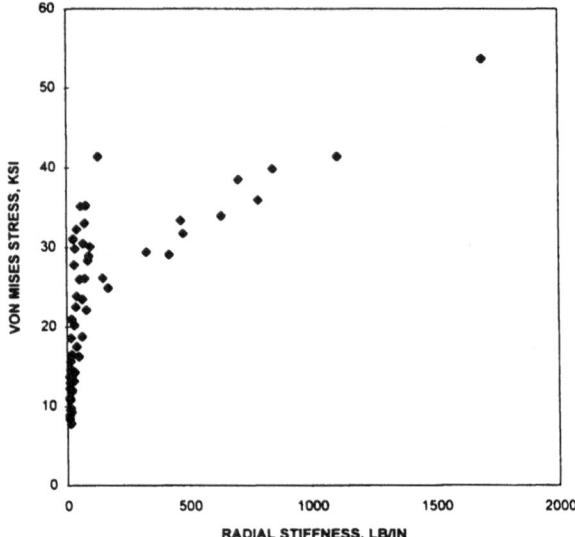

**Figure 6.** Three-finger spiral flexure design chart (5 mils thick)
Flexures at max. axial deflection of 0.20"

## CONCLUSIONS

Analytical procedures were developed to predict flexure bearing performance characteristics as a funtion of geometric configuration and design requirements. The procedures to perform all analysis tasks have been applied to three-finger spiral flexures. A set of design charts was developed using the analysis procedures. According to the design charts developed for the spiral flexure, a preferred design configuration can be chosen according to the material fatigue stress allowable and the flexure stiffness requirements. The procedures are generic in nature and can be used for verification of existing designs, or for optimization of new designs with given design constraints. The software will be made available to potential users.

## REFERENCES

1. Johnson, A. L., "Spacecraft Borne Long Life Cryogenic Refrigeration Status and Trends," Cryogenics, pp.339-347, July 1983.

2. Henderson, B. W., "U. S. Industry Close to Producing Long-Life Space Cooling System," Aviation Week & Space Technology, 6 April 1992.

3. Davey, C., "The Oxford University Miniature Cryogenic Refrigerator," International Conference on Advanced Infrared Detectors and Systems, London, pp.39, 1981.

4. Johnson, R. P., et. al., "A Stirling Engine with Hydraulic Power Output for Powering Artificial Hearts," Paper 75212, IECEC Record 1975.

5. Curtis, P. D., et. al., "Remote Sounding of Atmospheric Temperature from Satellites. V - The Pressure Modulator Radiometer for Nimbus F," Proc. R. Soc. (London), A 337, pp. 135-150, 1974.

6. Pan, R. B., et. al., "Tangential Linear Flexure Bearing," United States Patent Number: 5,494,313, 20 February 1996.

7. Wong, T. E., et. al., "Novel Linear Flexure Bearing," 7th International Cryocooler Conference, November 17-19, 1992.

# Investigation of Gas Effects on Cryocooler Resonance Characteristics

M. K. Heun, S. A. Collins, D. L. Johnson, and R. G. Ross Jr.

Jet Propulsion Laboratory
California Institute of Technology
Pasadena, CA    91109

## ABSTRACT

Cryocooler thermal and vibrational performance is determined, fundamentally, by the dynamic interactions between the mechanical system and the working fluid. This paper explores the effect of working-fluid characteristics on the mechanical response of the cooler. Experimental data collected from two coolers characterized under the Jet Propulsion Laboratory's extensive program of cryocooler testing and characterization show that a classical single-degree-of-freedom spring-mass-damper model does not capture the full frequency dependence of the mechanical response. The data from two modes of cooler operation (slosh and head-to-head) are used to motivate the explanation that working-fluid characteristics dominate at high frequencies, and mechanical system characteristics dominate at lower frequencies. Operating temperature is shown to be a significant factor in determining resonance behavior. Finally, the discussion provides a framework within which resonant parameters and cooler characteristics can extracted from the experimental data.

## INTRODUCTION

The growing demand for long-wavelength infrared imaging instruments for space observational applications has led to the ongoing development of cryocoolers, mechanical refrigerators which typically provide cooling at 30–100 K with refrigeration loads of 1–10 W.[1] Because these units are used to cool imaging system detectors, both the vibration of the coldhead assembly and the force transmitted to the structure are important for image integrity and overall instrument performance. And, characterization of cryocooler vibration and force signatures is an essential element of instrument modeling and design. Thus, an understanding of the parameters that affect coldhead vibration and transmitted force is required for optimized instrument design.

A previous publication[2] highlighted the theory behind cryocooler resonance characterization for single-piston Stirling coolers, focusing, in particular, on launch vibration response. The present paper extends the previous results by examining the resonance characteristics of a "back-to-back" cooler and identifying the helium working fluid as an important element affecting cooler dynamics.

## Cooler Designs

Fig. 1a shows a schematic diagram of the compressor motor of a typical single-piston Stirling cycle cryocooler, the subject of the earlier study.[2] Fig. 1b shows the back-to-back unit evaluated in the present paper. Back-to-back systems are said to be operating in "head-to-head" mode when the two pistons move toward the center simultaneously. Refrigeration is achieved only during head-to-head mode operation. For testing purposes, a back-to-back system may be driven in "slosh" mode: as one piston moves toward the center, the other piston moves toward the end. The slosh and head-to-head modes are illustrated in Fig. 2.

In head-to-head mode (normal operation), the compression work of the pistons on the helium is necessary to obtain refrigeration. Mechanically, the gas acts as a spring which compresses during volume reduction, and it provides damping as the gas is forced through the transfer tube, the coldhead, and the regenerator matrix. In slosh mode, no compression or expansion of the gas occurs. In effect, the gas spring and damper are eliminated from the system, leaving only the mechanical spring and its dynamic characteristics.

## Dynamic Models

The motion of the moving masses (pistons) may be described by the classical spring-mass-damper equation:[3]

$$m\ddot{x} + c\dot{x} + kx = F_d \sin(2\pi f t) \tag{1}$$

where $F_d = K_{mf} i$ is the amplitude of the drive force, $K_{mf}$ is the motor force constant, and $i$ is the zero-to-peak current supplied to the motor. This model is hereafter termed the "classical model," and the system which it describes is said to be the "classical system."

The zero-to-peak stroke response x of the classical system to the sinusoidal excitation at frequency $f$ is given by

$$\frac{x}{x_o} = \frac{1}{\sqrt{\left[1 - \left(\frac{f}{f_o}\right)^2\right]^2 + \left[2\zeta\left(\frac{f}{f_o}\right)\right]^2}}, \tag{2}$$

where $x_o = F_d/k$ is the static displacement, $\zeta = c/c_c$ is the damping ratio, $c_c = 2(km)^{1/2}$ is the critical damping coefficient, and $f_o = (k/m)^{1/2}/(2\pi)$ is the undamped natural frequency. The phase $\phi$ of the stroke relative to the applied force is given by

$$\tan\phi = \frac{2\zeta\left(\frac{f}{f_o}\right)}{1 - \left(\frac{f}{f_o}\right)^2}. \tag{3}$$

The force transmitted to the base $F_t$ is given by

$$\frac{F_t}{F_d} = \frac{\left(\frac{f}{f_o}\right)^2}{\sqrt{\left[1 - \left(\frac{f}{f_o}\right)^2\right]^2 + \left[2\zeta\left(\frac{f}{f_o}\right)\right]^2}}, \tag{4}$$

and the phase of the transmitted force relative to the applied force is identical to Eq.(3).

## Experimental Procedures

During laboratory testing, a sinusoidal drive current is applied to the cooler motors. The piston stroke, transmitted force, and the phases of both stroke and transmitted force relative to the drive current are measured. The above data is collected as a function of drive frequency $f$, drive current amplitude $i$, and coldblock temperature. The effective spring stiffness $k_s$, the

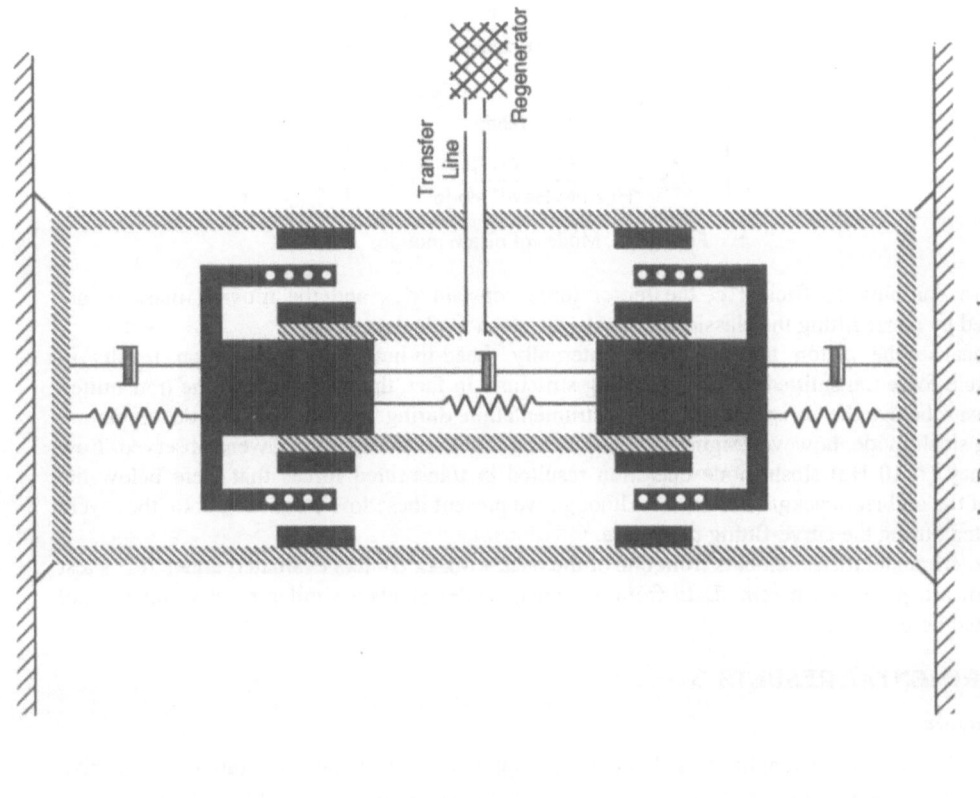

**Figure 1.** Cooler motor schematic.

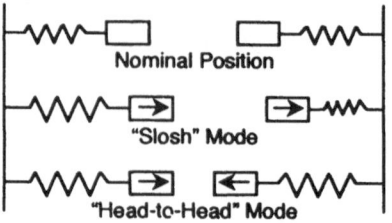

**Figure 2.**  Modes of piston motion.

effective damping coefficient c, the motor force constant $K_{mf}$, and the moving mass m are obtained by curve fitting the classical model to the slosh mode data.

Because the piston inertia cancels internally, head-to-head mode operation results in negligible force transmitted to the supporting structure. In fact, the magnitude of the transmitted force was below the sensitivity of our instrumentation during head-to-head mode operation. During slosh mode, however, significant and measurable transmitted forces were observed. Low frequency (< 10 Hz) slosh mode operation resulted in transmitted forces that were below the level of the ambient background noise. Although we present these low-frequency data, they were neglected during the curve-fitting procedure.

Experimental measurements from one of the back-to-back coolers evaluated under JPL's test program are presented herein. Data from a second cooler showed similar results and are not presented here.

## EXPERIMENTAL RESULTS

### Slosh mode

Fig. 3 shows stroke amplitude and phase vs. frequency for slosh mode operation at two drive current levels. The solid lines show the classical model curve fit, and the data points are experimental measurements. The cooler response conforms to the classic model, and the natural frequency of the system is about 24 Hz. The natural frequency exhibits a minor sensitivity to the drive current, and the effective spring stiffness increases slightly with amplitude.

In slosh mode, no gas compression occurs between the pistons, and the parameters derived from the slosh mode curve fit reflect the characteristics of the mechanical springs themselves. Thus, the mechanical springs are found to have a stiffness $k_s$ = 8 N/mm, a damping ratio $c/c_c$ = 0.010 and a motor force constant $K_{mf}$ = 22 N/A.

For completeness, Fig. 4 shows the transmitted force amplitude and phase vs. frequency. Again, the data follow the classical model presented above.

### Head-to-head mode

**Cooler Response.** Fig. 5 shows both the amplitude and the phase of the stroke relative to the drive current. Two classical models are presented on the graphs. The solid line is a classical model response generated with the parameters ($k_s$, $c/c_c$, and $K_{mf}$) derived from the slosh mode testing. Or, put another way, the solid line shows the predicted response of the system assuming no gas participation in the overall system stiffness or damping. We see that at low frequencies (< 10 Hz) the classical model nearly matches the experimental data, indicating that the parameters of the mechanical spring are sufficient to describe the system response in this frequency regime.

We see, however, that the resonant frequency (the frequency of maximum forced amplitude) is higher for head-to-head mode operation than slosh mode operation. This indicates a stiffer system in head-to-head mode operation. Furthermore, the response curve is flatter near resonance, indicating increased damping. The dashed line shows a curve fit to the data

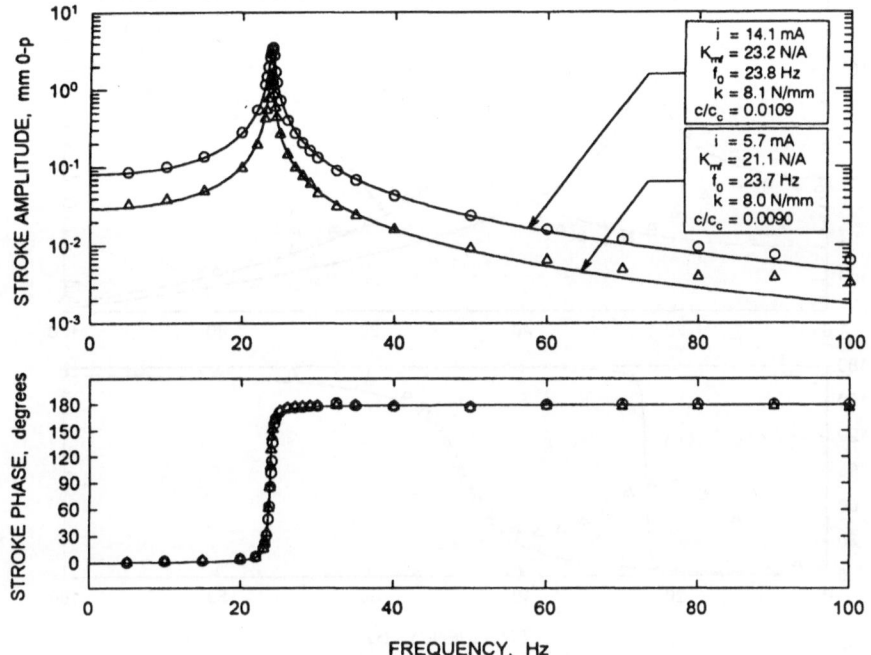

**Figure 3.** Slosh mode stroke.

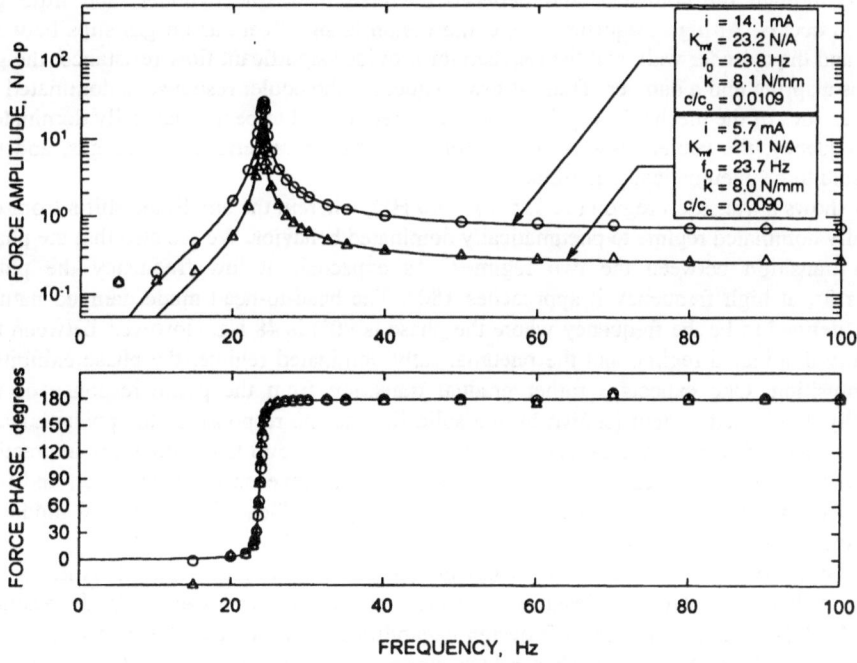

**Figure 4.** Slosh mode force.

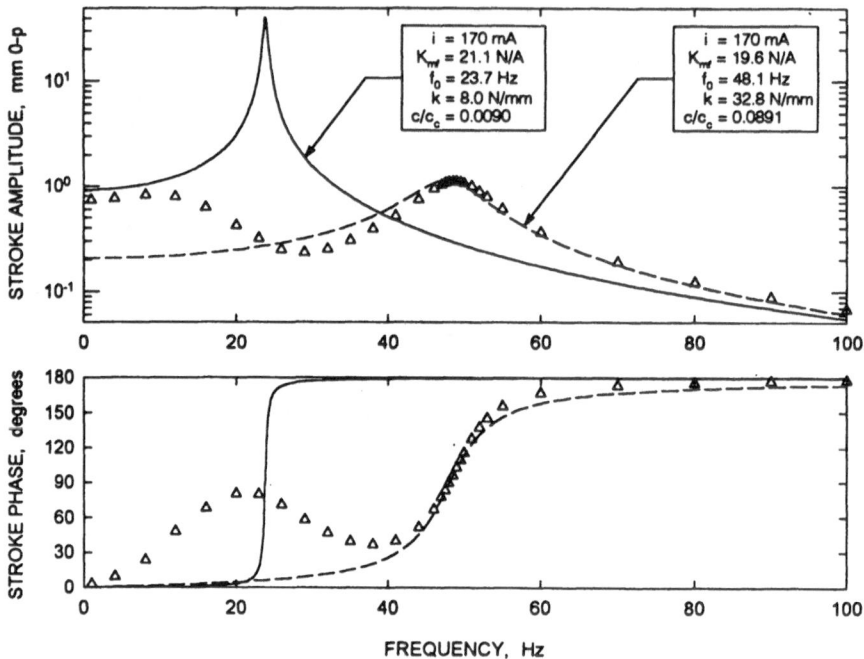

**Figure 5.** Head-to-head mode stroke, ambient temperature.

above 40 Hz. The high frequency data can be represented by a classical system with a stiffer spring and more damping compared to the low frequency data.

We postulate that at low frequency, gas effects are minimal. Gas flows readily through the regenerator, through the coldhead, and between the piston and the cylinder, and little gas compression occurs. At high frequency, gas compression is significant as no gas slips between the piston and the cylinder wall, and the regenerator provides significant flow resistance: the gas acts both as a spring and a damper. Thus, at low frequency, the cooler response is dominated by the characteristics of the mechanical spring, and the system is said to be mechanically dominated. At high frequency, the cooler response is controlled by the characteristics of the gas, and the system is said to be pneumatically dominated.

Fig. 5 shows a transition region (10 Hz $< f <$ 40 Hz), wherein the amplitude shifts from the mechanically dominated regime to pneumatically dominated behavior. We see also that the phase exhibits a transition between the two regimes. As expected, at low frequency the phase approaches 0°, at high frequency it approaches 180°. The head-to-head mode damped natural frequency (defined to be the frequency where the phase is 90°) is 48 Hz. However, between the mechanically dominated regime and the pneumatically dominated regime, the phase exhibits a peculiar transition. One expects a rather gradual transition from the phase response of the mechanically dominated system (shown by the solid line) to the response of the pneumatically dominated system (shown by the dashed line). Instead, Fig. 5 shows that a complex interaction exists between the two regimes. This effect is manifest as a wave on the phase plot, and it is termed the "transition phase lead" because the measured phase "leads" the classical model as frequency increases.

The implications of the observed behavior on cooler vibration control systems are significant. Although coolers are designed to work only at the head-to-head mode resonant frequency (in this case, $f$ = 48 Hz), off-design capabilities are an essential aspect of a robust control algorithm. It is clear that such a robust algorithm should not expect classical system

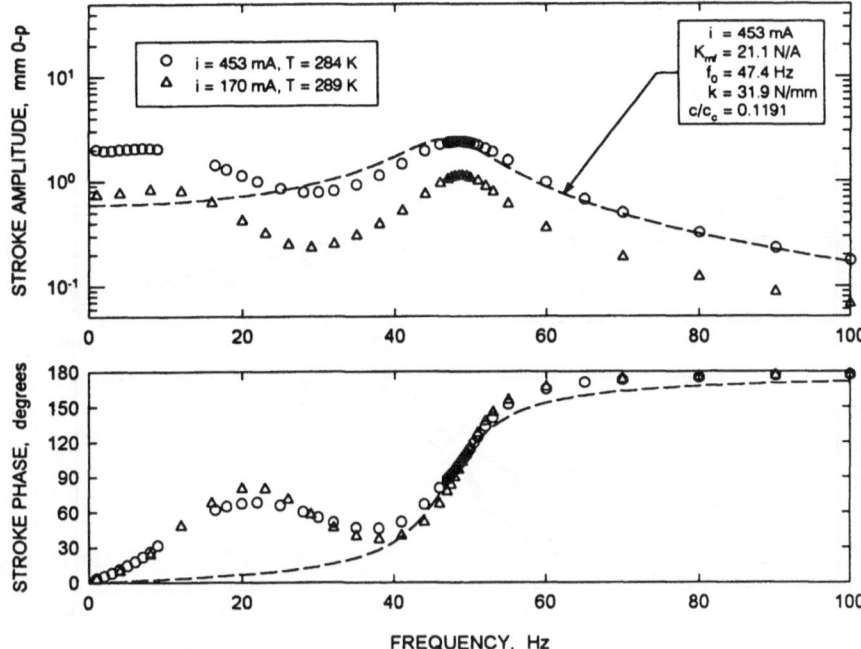

**Figure 6.** Effect of drive current on head-to-head mode stroke, ambient temperature.

behavior at frequencies below the operating frequency, particularly in terms of the phase response. Near-resonance phase control and dynamic modeling are significantly impacted by gas effects.

**Effect of Drive Current, Ambient Conditions.** Fig. 6 shows the sensitivity of cooler response to drive current. As expected, the higher drive current generates larger stroke. The resonant frequency is not significantly altered by the increased stroke, although the phase and the values of $c/c_c$ indicate a positive correlation between damping and amplitude. The larger stroke results in a more highly damped response, likely due to the difference in the dynamic behavior of the gas. The amplitude of the wave on the phase plot decreases as amplitude increases.

**Effect of Operating Temperature.** Fig. 7 shows the effect of operating temperature on the amplitude and phase of the system. The data represented by the triangles was collected with the coldhead at about 290 K (ambient conditions) whereas the data represented by the diamonds was collected with the coldhead at cryogenic temperatures (about 45 K). In each case, the motor drive current was selected to provide identical peak stroke (about 1 $mm_{o-p}$) at the resonant frequency (48 Hz). Clearly, operating temperature significantly affects the dynamic behavior of the cooler, and there are many differences between cryogenic and ambient operation.

Looking first at the amplitude data, we see that cryogenic operation requires substantially higher motor current (283 mA) compared to ambient conditions (170 mA) to achieve the same stroke at resonance. Relative to ambient operation, the resonant peak in the cryogenic data is broader than the corresponding peak in the ambient data, indicating a more highly damped system under cryogenic conditions. The higher damping is verified by the classical model fit to the high frequency data: damping increases significantly while the overall stiffness and the motor force constant are relatively unchanged when moving from ambient to cryogenic operation. A likely explanation for the increased damping is the increased density of the helium gas in the cold-block at cryogenic temperatures.[2] Once again, the gas characteristics are seen to impact significantly the resonant behavior of the coolers.

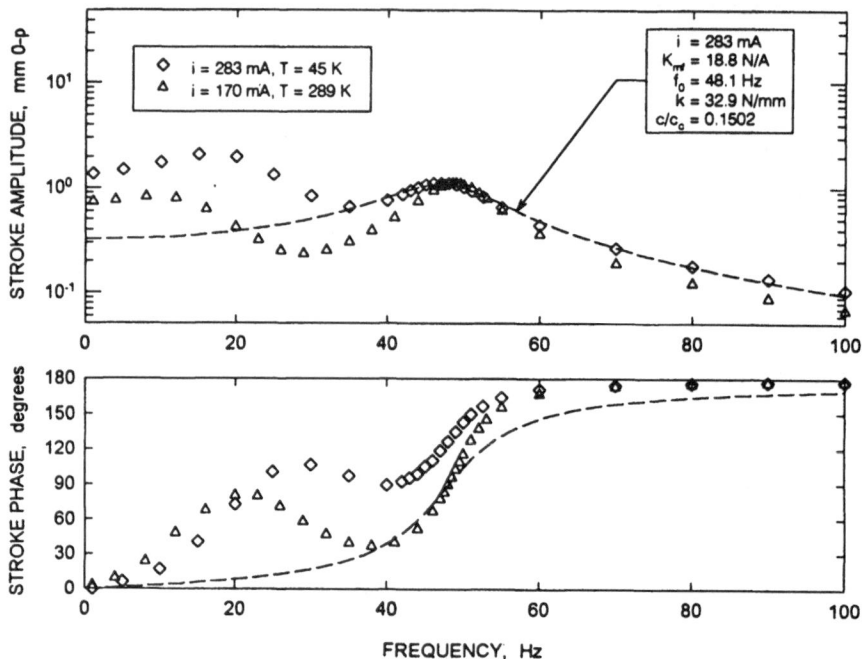

**Figure 7.** Effect of coldhead temperature on head-to-head mode stroke.

The dashed line in Fig. 7 is a curve fit of the classical model to the amplitude data above 40 Hz. Similar to the ambient data in Fig. 5, the classical model fits the amplitude data nicely. However, unlike the ambient data in Fig. 5, the classical model poorly represents the cryogenic phase data. The transition phase lead for the cryogenic data is higher in frequency and smaller in amplitude compared to the ambient data. Surprisingly, the phase crosses 90° three times. Near the resonant frequency, we observe a 90° crossing at 39.8 Hz and again at 40.2 Hz. Finally, a 90° crossing occurs at 23.0 Hz. Figure 7 shows that the phase behavior is a function of both the operating temperature and stroke amplitude. Clearly, the classical model does not capture the behavior of the system.

The cryogenic data shown in Fig. 7 amplify the conclusions drawn from the ambient data shown in Fig. 5, namely that active vibration control strategies must include gas effects if a complete model of the cooler dynamic behavior is required. This is more important with phase than with amplitude. The cryogenic data shows that the amplitude can be represented by a classical model, but that the phase behavior cannot be captured by parameters derived from the stroke data.

**Effect of Drive Current, Cryogenic Conditions.** Fig. 8 shows the impact of motor drive current at cryogenic conditions. Similar to Fig. 6, we see that higher drive current generates larger stroke. The lower drive current generates a larger amplitude transition phase lead. At both cryogenic and ambient operating conditions, higher drive current results in a more highly damped system. The dashed line represents a classical model curve fit to the high drive current data at frequencies higher than 40 Hz. As shown in the previous section, the classical model fails to capture both the stoke and phase response simultaneously.

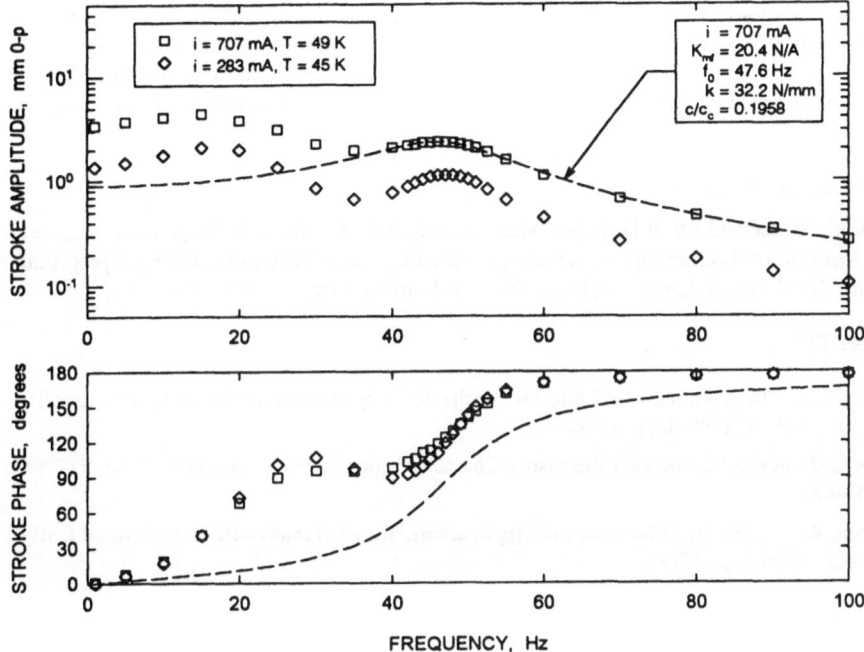

**Figure 8.** Effect of drive current on head-to-head mode stroke, cryogenic temperature.

## CONCLUSION

Previous papers on cryocooler resonance characteristics have neglected phase information when characterizing cooler parameters, in part because the previous studies focused on cooler response to launch-induced vibration. In this paper, we have examined both the amplitude and phase response of back-to-back cryocoolers to find that pneumatic effects have a significant impact on cooler resonance characteristics.

Three regimes were observed during head-to-head mode operation. (1) The mechanically dominated regime occurs at low frequencies where gas effects are minimized and the characteristics of the mechanical spring control the resonant behavior of the cooler. (2) The pneumatically dominated regime occurs at high frequencies where the gas characteristics predominate. (3) Between the mechanically dominated regime and the pneumatically dominated regime, a transition regime exists that is nearly 30 Hz in extent.

We showed that the classical model is insufficient to describe head-to-head piston motion across a broad range of frequency. At ambient conditions, two classical models may be used to describe both the amplitude and phase response of the coolers. A different set of parameters is required for each regime. At cryogenic conditions, a classical model may be employed to describe the amplitude but not the phase in the pneumatically-dominated regime. In the transition regime, the amplitude makes a smooth shift from the mechanically-dominated regime to the pneumatically-dominated regime. The phase exhibits peculiar behavior that is termed the "transition phase lead." Further work should be undertaken to describe completely the cooler behavior in the transition regime. The phase characteristics merit particular attention because of the importance of phase information for active vibration control algorithms.

Ultimately, a model that completely captures gas effects on cryocooler resonance may be used in conjunction with diagnostic testing. The results may allow determination of parameters that influence gas behavior such as the clearance between the piston and the cylinder, the piston length and diameter, and the effective flow resistance of the transfer tube, coldhead, and regenerator.

## ACKNOWLEDGEMENTS

The work described in this paper was carried out by the Jet Propulsion Laboratory, California Institute of Technology, and was sponsored by the NASA EOS AIRS project, under a contract with the National Aeronautics and Space Administration.

## REFERENCES

1.  Ross, R.G. Jr., "Requirements for Long-life Mechanical Cryocoolers for Space Applications", *Cryogenics*, vol. 30, (1990), pp. 233-238.

2.  Ross, R.G. Jr, et al., "Cryocooler Resonance Characterization", *Cryogenics*, vol. 34, no. 5 (1994), pp. 435-442.

3.  Thomson, W.T., *Theory of Vibration with Applications*, 3rd ed., Prentice Hall, Englewood Cliffs, New Jersey, (1988), pp.50-81.

# Thermodynamic Comparison between the Orifice Pulse Tube and the Stirling Refrigerator

**P.C.T. de Boer**

Sibley School of Mechanical and Aerospace Engineering
Cornell University, Ithaca, NY 14853, USA

## ABSTRACT

Using methods developed in previous papers, expressions are derived for the thermodynamic performance of a Stirling Refrigerator (SR). The cycle considered corresponds to the case where the pistons are driven by two-position linear motors, actuated alternately. Detailed results are derived for the temperature profiles of the gas expelled from both sides of the regenerator assembly. Knowledge of these profiles allows calculation of the temperature of the gas entering the cold heat exchanger, and hence the amount of heat $Q_c$ removed per cycle. The formulation leads to two non-linear equations with two unknowns. These equations can easily be solved numerically. The work done by the pistons is found from calculating the pressure as function of piston position. Results are presented for $Q_c$, for the coefficient of performance COP, and for the refrigeration efficiency $\eta$ as function of the ratio of temperatures $T_c/T_h$ at the two sides of the regenerator. These results are compared with corresponding ones for the Orifice Pulse Tube (OPT). For the case considered, the SR is found to remove about 10% more heat per cycle than the OPT. The efficiency of both the OPT and the SR are found to about 90% at very small $T_c/T_h$. While the efficiency of the SR remains close to this value over a wide range of $T_c/T_h$, the efficiency of the OPT decreases nearly linearly with $T_c/T_h$ till it equals zero at $T_c/T_h = 0$.

## NOMENCLATURE

| | |
|---|---|
| A | cross-sectional area of tube |
| COP | coefficient of performance |
| k | ratio of specific heats |
| L | physical length of initial expansion space ($0 \le x \le 1$ in Fig. 3) |
| $L_p$ | non-dimensionalized stroke of compression piston |
| $L_h$ | non-dimensionalized stroke of expansion piston |
| p | pressure |
| Q | amount of heat flowing per cycle, non-dimensionalized by $A\,L\,p_1$ |
| T | temperature |
| W | net work per cycle, non-dimensionalized by $A\,L\,p_1$ |
| x | axial coordinate, non-dimensionalized by L |
| $z_h$ | $1 + L_h$ |

### Greek symbols

| | |
|---|---|
| $\pi_1$ | $p_2/p_1$, compression ratio |
| $\pi_2$ | $p_2/p_3$, expansion ratio |
| $\pi_4$ | $p_4/p_1$ |
| $\rho$ | density |

### Subscripts

| | |
|---|---|
| 1-4 | conditions before steps 1-4, resp. |
| 3a | part a of step 3 |
| 3b | part b of step 3 |
| a | after compression or expansion |
| b | before compression or expansion |
| c | refers to (cold) heat exchanger |
| cp | compression piston |
| ep | expansion piston |
| h | refers to (hot) heat exchanger |
| he | heat exchanger |
| p | refers to piston |
| t | compression and expansion tubes |

Cryocoolers 9, Edited by R.G. Ross, Jr.
Plenum Press, New York, 1997

## INTRODUCTION

Since its invention in 1984 by Mikulin et al.[1], the Orifice Pulse Tube (OPT) has received a great deal of attention as a possible replacement for the Stirling Refrigerator (SR). Potential applications include the cooling of infrared instruments in outer space, the cooling of high temperature superconductors, the cooling of electronic components, and the cooling of superconducting magnets. A crucial aspect of the operation of both the OPT and the SR is the phase difference that must exist between pressure and velocity variations in the gas that serves as the working substance.[2] It was found by Zhu et al.[3] that a significant improvement in performance can be obtained by adding a second inlet to the pulse tube; this addition provides an increase in the necessary phase difference. Many experimental results with pulse tubes have been reported in recent issues of Cryogenics, and in Proceedings of Cryogenics Engineering Conferences such as the present one.

A number of analytical studies of the OPT have been presented. Radebaugh[2] developed an enthalpy flow model. Storch et al.[4,5] described the operation in terms of phasor diagrams. Work and heat flows were considered by Kasuya et al.[6] Kittel[7] proved that the coefficient of performance of an ideal OPT equals the ratio of low to high temperatures. David et al.[8] explained the operation of the OPT by means of the heat flow mechanism at the tube ends. Mirels[9,10] presented a theory for the performance of the OPT based on linearizing the pressure variations. Rawlins et al.[11] analyzed energy flows in the OPT. Zhu and Chen[12] presented an isothermal model of the OPT, analogous to the well-known isothermal model for the SR. De Boer[13] analyzed the performance of the OPT in terms of the mechanism of "pressure heat pumping," as opposed to the mechanism of "surface heat pumping" that is operative in the Basic Pulse Tube Refrigerator.[14] In the model of Ref. [13], the piston was assumed to move within the heat exchanger at the warm end of the pulse tube. The model was recently extended to the more realistic case that the piston moves in a tube outside this heat exchanger.[15]

In the present paper, the methods developed in Refs. [13] and [15] are applied to the Stirling Refrigerator. The principal difference between the previous treatments and the present one is that account is taken of the work delivered to the expansion piston of the SR. As a consequence, the COP and refrigeration efficiency of the SR are found to be larger than that of the OPT. A minor difference is that in the Stirling case there is no equivalent to the heat exchanger which is present adjacent to the orifice of an OPT. It is found that this difference has only a minor impact on COP and efficiency. For identical values of the various independent parameters of importance, the heat removed per cycle is found to be slightly higher in the case of the SR.

## THERMODYNAMIC ANALYSIS

### Formulation of the Problem

The model used for the OPT is sketched in Fig. 1. The heat exchangers at the left and right are maintained at temperature $T_h$ by appropriate cooling methods. Heat $Q_c$ is removed from the cold environment by the heat exchanger at $T_c$. In Refs. [13] and [15], the action of the orifice-reservoir assembly was accounted for by the motion of a virtual "orifice piston" moving within the heat exchanger at the left. Fig. 2 shows the corresponding model for the Stirling refrigerator. The virtual orifice piston here is replaced by an actual expansion piston. The cycle considered is identical for the two cases. Differences arise only from the presence of the heat exchanger in front of the orifice of the OPT. The present paper describes the theory for the Stirling refrigerator, following the corresponding theory for the OPT.[15]

Various simplifying assumptions are made. All frictional effects are neglected, as are all conduction effects within the gas and within the walls, the heat exchangers and the regenerator. The pressure is taken to be uniform during the entire cycle. The temperature of a gas element in a heat exchanger equals the temperature of the exchanger. The temperature of a gas element in the regenerator equals the local temperature of the regenerator. The assembly is filled with an ideal gas having constant specific heats.

The thermodynamic cycle is taken to consist of the following four steps:

1. The compression piston moves towards the left, while the expansion piston remains stationary. The pressure increases from $p_1$ to $p_2$.
2. The expansion piston moves towards the left, while the compression piston remains stationary at the entrance of the heat exchanger on the right. The pressure decreases from $p_2$ to $p_3$.

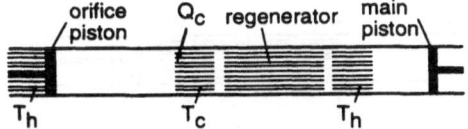

**Figure 1.** Layout of the OPT

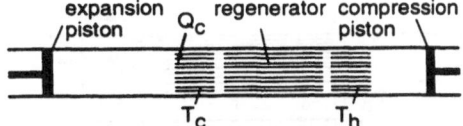

**Figure 2.** Layout of the Stirling refrigerator.

3. The compression piston returns to its original position by moving to the right, while the expansion piston remains stationary. The pressure decreases from $p_3$ till $p_4$.
4. The expansion piston returns to its original position while the compression piston remains stationary. The pressure returns from $p_4$ till its original value $p_1$.

The cyclic heat $Q_c$ removed from the cool area equals the heat taken up by gas entering the cold heat exchanger from the pulse tube.[16] Calculating $Q_c$ therefore requires knowledge of the temperature of the gas just before it enters the heat exchanger. This temperature can be found by following the cyclic temperature history of the column of gas that is expelled from the cold heat exchanger during step 1. After the temperature of the entering gas is found as a function of pressure, the determination of $Q_c$ is straightforward. Similarly, once the piston position $x_p$ is found as function of pressure, the determination of the cyclic work $W_{cp}$ and $W_{ep}$ of the two pistons is straightforward. The basic independent parameters are taken to be the compression ratio $\pi_1 = p_2/p_1$ ($>1$), the expansion ratio $\pi_2 = p_2/p_3$ ($>1$), and the temperature ratio $T_c/T_h$ ($<1$); all of these ratios are assumed given.

## Basic Thermodynamic Relations

There are only a few basic thermodynamic relations that are needed for the determination of the temperature profiles. These are summarized below; the derivation can be found in Ref. [17]. The first one pertains to the isentropic compression or expansion of a gas column of length $x$ in the pulse tube. As the pressure is changed from $p$ to $p + dp$, the length of the column is changed by an amount $dx = - x \, dp/(k \, p)$. Integration yields

$$x_a / x_b = (p_b / p_a)^{1/k} \quad \text{(column in pulse tube)} \tag{1}$$

where subscripts $a$ and $b$ indicate after and before the change, respectively. This result holds for any temperature distribution within the column. An element originally at temperature $T_b$ ends up at temperature $T_a$ given by

$$T_a / T_b = (p_a / p_b)^{(k-1)/k} \tag{2}$$

Analogously, a column of length $x$ in a heat exchanger has its length changed by $dx = - dp/p$ as the pressure changes by $dp$. Integration yields

$$x_a / x_b = p_b / p_a \quad \text{(column in heat exchanger)} \tag{3}$$

Another basic relation that is needed is the change in length that a gas element suffers when it enters a heat exchanger from either the expansion tube or the compression tube. The ratio of the length $dx_t$ of an element just before it enters to its length $dx_{he}$ just after is has entered equals the ratio of temperature $T_t$ of the element just before entering to the temperature $T_{he}$ of the heat exchanger:

$$dx_t / dx_{he} = T_t / T_{he} \tag{4}$$

Gas elements leaving the heat exchanger retain their length while exiting.

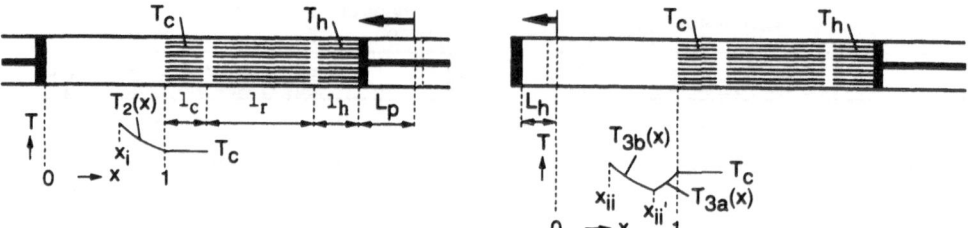

**Figure 3.** Piston motion during step 1, and resulting temperature profile of column expelled.

**Figure 4.** Piston motion during step 2, and resulting temperature profile of column expelled.

## Temperature Profiles

The motion of the piston during step 1 is indicated schematically in Fig. 3, as is the resulting temperature profile of the gas expelled from the cold heat exchanger during this step. Use of the results given in the previous subsection yields the temperature profile as:

$$T_2(x) = T_c / x^{k-1} \text{ (the adiabatic profile)}, \quad x_i = 1 / \pi_1^{1/k} \tag{5}$$

The non-dimensional axial coordinate $x$ here varies from $0$ at the initial position of the expansion piston till 1 at the entrance of the cold heat exchanger; the corresponding tube is referred to as the expansion tube. Equating the mass initially in the tube at the compression side (referred to as the compression tube) to the increase of mass in the regenerator assembly and in the expansion tube leads to

$$\int_0^{L_p} \frac{T_c}{T_1(x)} dx = \frac{\pi_1 - 1}{k}(1 + kL_r) \tag{6}$$

The integral here is over the gas in the compression tube the start of the step. $L_r$ is the non-dimensionalized length of the regenerator assembly, defined as

$$L_r = 1_c + \int_{1_r} \frac{T_c}{T} d1_r + \frac{T_c}{T_h} 1_h \tag{7}$$

The motion of the expansion piston during step 2 is indicated in Fig. 4, as is the resulting temperature profile of the gas expelled from the cold heat exchanger during steps 1 and 2. The temperature profile is given by

$$\frac{T_{3a}}{T_c} = \left(\frac{kL_r}{1 + kL_r - x}\right)^{k-1}, \quad \frac{T_{3b}}{T_c} = \frac{1}{\left[x - 1 - kL_r + \pi_2^{1/k}(1 + kL_r)\right]^{k-1}} \tag{8}$$

$$x_{ii} = 1 + kL_r - \pi_2^{1/k}\left(1 + kL_r - \pi_1^{-1/k}\right), \quad x_{ii}' = 1 - kL_r\left(\pi_2^{1/k} - 1\right) \tag{9}$$

All of the quantities appearing in Eqs. (6) - (9) are known, with the exception of $T_1(x)$ and hence $L_p$.

In considering step 3, two different possibilities must be distinguished. The first one is that all of the gas expelled during step 2 (i.e., segment **a** in Fig. 4) re-enters the heat exchanger at $T_c$ during the step. This case is designated as case A, the other one as case B. The pressure at which all the gas expelled during step 2 has re-entered is denoted by $p_{3a}$. It follows that $p_{3a} > p_4$ in case A, while $p_{3a} < p_4$ in case B.

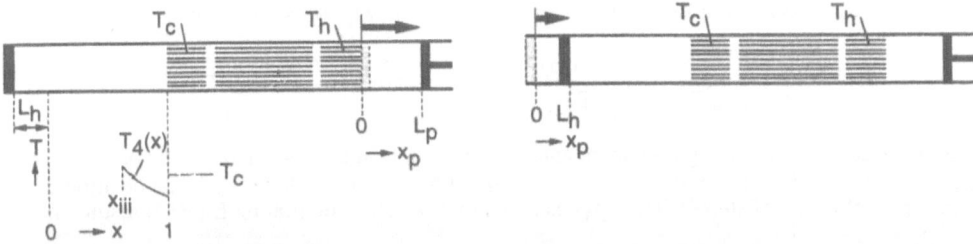

**Figure 5.** Piston motion during step 3, and resulting temperature profile of column expelled.

**Figure 6.** Piston motion during step 4, and resulting temperature profile of column expelled.

**Results for Case A.** The motion of the compression piston during step 3 and the resulting temperature profile of the remaining part of the gas expelled from the cold heat exchanger during step 1 are indicated in Fig. 5. The latter temperature profile is given by

$$\frac{T_4}{T_c} = \left\{ x + L_h + \left(\frac{p_3}{p_4}\right)^{1/k} \left[ -1 - L_h - kL_r + \pi_2^{1/k}(1 + kL_r) \right] \right\}^{-k+1} \tag{10}$$

$$x_{iii} = -L_h + (p_3 / p_4)^{1/k}(x_{ii} + L_h) \tag{11}$$

The temperature $T_3(1,p)$ of a gas element entering the cold heat exchanger during step 3 when the pressure is $p$ follows from the adiabatic relation $T_3(1,p) = (p/p_3)^{1-1/k} T_3(x_3)$, where $x_3$ is the position of the element at the start of step 3; $T_3(x_3)$ is given by Eqs. (8). After applying the principle of conservation of mass it is found that:

$$\frac{T_{3a}(1,p)}{T_c} = \left(\frac{kL_r}{z_h}\right)^{k-1} \frac{1}{(-1 + 1/s_1)^{k-1}}, \quad s_1 \equiv C_1 \left(\frac{p}{p_3}\right)^{1/k}, \quad C_1 \equiv \frac{z_h}{z_h + kL_r} \tag{12}$$

$$\frac{T_{3b}(1,p)}{T_c} = \frac{1}{z_h^{k-1}} \frac{1}{(1 - 1/s_2)^{k-1}}, \quad s_2 \equiv C_2 \left(\frac{p}{p_3}\right)^{1/k}, \quad C_2 \equiv \frac{z_h}{z_h + kL_r - \pi_2^{1/k}(1 + kL_r)} \tag{13}$$

Conservation of mass during step 3 while the pressure increases from p to p+dp (with dp < 0) yields

$$d\left(p^{1/k}x_p\right) = -\frac{T_h}{T_c} L_r \frac{dp}{p^{1-1/k}} - \frac{T_h}{T_3(1,p)} z_h \frac{dp}{k \, p^{1-1/k}} \tag{14}$$

Integrating this equation from $p = p_3$ till $p = p_4$ results in

$$L_p = \frac{T_h}{T_c} kL_r \left[ \left(\frac{p_3}{p_4}\right)^{1/k} - 1 \right] + \frac{T_h}{T_c}\left(\frac{p_3}{p_4}\right)^{1/k} \frac{z_h}{C_1(kL_r)^{k-1}} \int_{C_1C_3}^{C_1} (-1 + 1/s)^{k-1} ds$$

$$+ \frac{T_h}{T_c}\left(\frac{p_3}{p_4}\right)^{1/k} \frac{z_h^k}{C_2} \frac{C_2C_3}{C_2C_4} \int (1 - 1/s)^{k-1} ds \tag{15}$$

$$C_3 \equiv \left(\frac{p_{3a}}{p_3}\right)^{1/k} = \frac{z_h - (\pi_2^{1/k} - 1)kL_r}{z_h}, \quad C_4 \equiv \left(\frac{p_4}{p_3}\right)^{1/k} = \left(\frac{\pi_2\pi_4}{\pi_1}\right)^{1/k} \tag{16}$$

The equation corresponding to Eq. (14) for step 4 (sketched in Fig. 6) is

$$d\left(p^{1/k}x_p\right) = \frac{z_h}{k}\frac{dp}{p^{1-1/k}} + D\frac{T_4(1,p)}{T_c}\frac{dp}{kp^{1-1/k}}, \quad \text{where } D \equiv kL_r + \frac{T_c}{T_h}L_p \tag{17}$$

The temperature $T_4(1,p)$ of a gas element entering the cold heat exchanger when the pressure is p follows from the adiabatic relation $T_4(1,p) = (p/p_4)^{1-1/k}T(x_4, p_4)$, where $x_4$ is the position of the element at the start of step 4; $T(x_4, p_4)$ is given by Eq. (10). Integrating Eq. (17) results in

$$0 = 1 - \pi_4^{1/k}z_h - (D\pi_4)^{1/k}\left[(C_5 - 1/\pi_4)^{1/k} - (C_5 - 1)^{1/k}\right], \quad \text{where } C_5 \equiv 1 + \frac{z_h^k}{D}\left(1 - \frac{1}{C_2C_4}\right)^k \tag{18}$$

Eq. (18) is the first equation between the two unknowns $\pi_4$ and $z_h$. In order to find a second equation, the temperature profile in the compression tube needs to be determined. Integrating Eq. (14) with $x_p$ replaced by x, and solving for x yields the temperature profile in this tube at the end of step 3 in parametric form (with p being the parameter):

$$T_4(x) = T_h\left[p_4 / p(x)\right]^{1-1/k} \quad \text{(compression tube)} \tag{19}$$

$$x = \frac{T_h}{T_c}kL_r\left[\left(\frac{p}{p_4}\right)^{1/k} - 1\right] + \frac{T_h}{T_c}\frac{z_h^k}{C_2C_4}\frac{C_2C_3}{C_2C_4}\int\left(1 - \frac{1}{s}\right)^{k-1}ds$$

$$+ \frac{T_h}{T_c}\frac{z_h^k}{(kL_r)^{k-1}}\frac{1}{C_1C_4}\int_{C_1C_3}^{C_1(p/p_3)^{1/k}}\left(-1 + \frac{1}{s}\right)^{k-1}ds \quad \text{(part a)} \tag{20a}$$

$$x = \frac{T_h}{T_c}kL_r\left[\left(\frac{p}{p_4}\right)^{1/k} - 1\right] + \frac{T_h}{T_c}\frac{z_h^k}{C_2C_4}\int_{C_2C_4}^{C_2(p/p_3)^{1/k}}\left(1 - \frac{1}{s}\right)^{k-1}ds \quad \text{(part b)} \tag{20b}$$

These results can be used to find the mass entering the compression tube during step 3. The additional mass entering during step 4 is easily found to be $(T_h/T_c)(L_p/k)(1-\pi_4)$. Combining these results to evaluate the left hand side of Eq. (6) yields the second equation between the two unknowns $\pi_4$ and $z_h$:

$$0 = 1 + (1 - \pi_4)D - \frac{\pi_1}{\pi_2}z_h^k\left(C_4 - \frac{1}{C_2}\right)^k \tag{21}$$

Eqs. (18) and (21) can easily be solved numerically using a program such as MATLAB.

**Results for case B.** The treatment of case B is similar to that of case A. It results in Eqs. (18) and (21) being replaced by, respectively,

$$0 = 1 - \pi_4^{1/k}z_h + (D/\pi_{4a})^{1/k}(kL_r)^{1-1/k}\left[(C_6 + \pi_{4a})^{1/k} - (C_6 + 1)^{1/k}\right]$$

$$+ (\pi_4 D)^{1/k}\pi_{4a}^{-1+1/k}\left[(C_7 - \pi_{4a})^{1/k} - (C_7 - 1/\pi_4)^{1/k}\right] \tag{22}$$

$$0 = 1 + D(1 - \pi_4) - \pi_1(1 + kL_r) + kL_r\frac{\pi_1}{\pi_2} + \frac{z_h^k}{(kL_r)^{k-1}}\frac{\pi_1}{\pi_2}\left[\left(\frac{1}{C_1} - C_4\right)^k - \left(\frac{1}{C_1} - 1\right)^k\right] \tag{23}$$

$$\text{where } C_6 = -1 + \frac{z_h{}^k}{D(kL_r)^{k-1}}\left(-1 + \frac{1}{C_1 C_4}\right)^k, \quad C_7 = \pi_{4a} + \frac{\pi_1}{D\pi_4}$$

$$\tag{24}$$

$$D = \frac{kL_r}{C_4} + \frac{z_h}{(kL_r)^{k-1}} \frac{1}{C_1 C_4} \int\limits_{C_1 C_4}^{C_1} (-1 + 1/s)^{k-1} ds$$

$$\tag{25}$$

$$\pi_{4a} = 1 + \frac{kL_r}{D}\left[\frac{\pi_1}{\pi_4} - \left(\frac{z_h}{kL_r}\right)^k \left(-1 + \frac{1}{C_1 C_4}\right)^k\right]$$

$$\tag{26}$$

**Results for work and heat removed.** The nondimensional work done by a piston follows from

$$W_p = \int \frac{p}{p_1} dx_p$$

$$\tag{27}$$

while the heat removed is determined by

$$Q_c = \int \frac{c_p \rho}{p_1}[T_c - T(1,p)]dx = \frac{k}{k-1}\int \frac{p}{p_1}\left[\frac{T_c}{T(1,p)} - 1\right]dx$$

$$\tag{28}$$

Most expressions for these quantities agree with the corresponding ones for the OPT given in Ref. [15]. Lack of space prevents duplicating those here. The only results that differ are the ones for $Q_{c4}$, the heat removed during step 4, in both cases A and B.. These are given in the Appendix. Also given in the Appendix are the results for the work $W_{ep}$ done by the expansion piston $(W_{ep} < 0)$.

Finally, the coefficient of performance COP and the refrigeration efficiency $\eta$ are given by

$$COP = \frac{Q_{c3} + Q_{c4}}{W_{cp} + W_{ep}}, \quad \eta = \frac{COP}{COP_{Carnot}} = COP\left(\frac{T_h}{T_c} - 1\right)$$

$$\tag{29}$$

## RESULTS

Using the analytical expressions derived, numerical results have been obtained for the main performance parameters of the SR and the OPT. In computing these results, the following choice was made for the regenerator assembly: $l_c = 0.1$, $l_r = 0.2$, $l_h = 0.1$ (Cf. Fig. 3). The temperature profile in the regenerator was taken to be linear. The resulting expression for the nondimensionalized length of the regenerator assembly is $L_r = 0.1 + 0.2 \ln(T_h/T_c)/(-1+T_h/T_c) + 0.1T_c/T_h$. The ratio of specific heats was taken as $k = 5/3$. The values of the compression and expansion ratios were taken to be $\pi_1 = 4$ and $\pi_2 = 1.1$, respectively. Fig. 7 shows the cyclic heat removed $Q_c$ for the two devices. The SR result is about 10% higher than that for the OPT over the entire range of temperatures. Corresponding results for the refrigeration efficiency and the coefficient of performance are shown in Figs. 8 and 9. At very low values of the temperature ratio $T_c/T_h$, the difference between SR and OPT results is negligible. As $T_c/T_h$ increases, the difference becomes larger. For $T_c/T_h = 0.2$ it still amounts to only about 25%. At $T_c/T_h = 0.8$, the efficiency of the SR remains as high as 80%, while the efficiency of the OPT has dropped to about 20%. This leads to the conclusion that the OPT is primarily suited for cooling to very low temperatures, i.e., to cryocooling, while the SR in principle is suited for refrigeration over a wide range of temperature ratios. The maximum theoretical efficiency of the OPT for the parameter values used occurs at $T_c/T_h \approx 0.05$, and is on the order of 90%. Also indicated in Figs. 8 and 9 are the results for the Stirling refrigerator in case the work delivered to the expansion piston is not recovered; these are marked "Stirling (net)." It is seen that they are very close to the corresponding OPT curves. The "Stirling (net)" result for $Q_c$ coincides with the Stirling result in Fig. 7, and hence was not labeled separately. Finally, the nondimensionalized piston travel $L_p$ is plotted in Fig. 10. Again, the SR and OPT results for this case coincide. It is seen that at

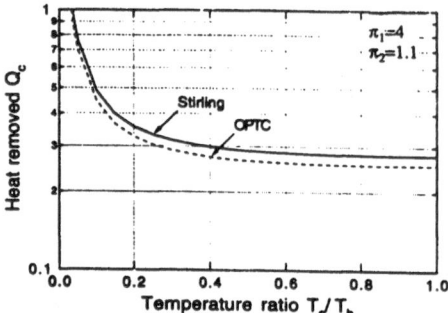

**Figure 7.** Cyclic heat removed as function of temperature ratio, for SR and OPT.

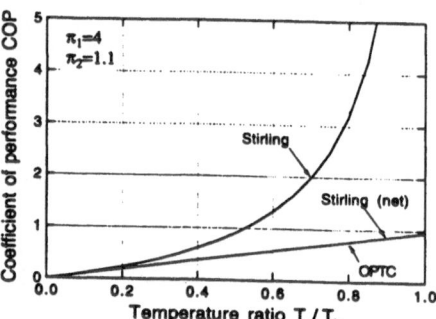

**Figure 8.** Coefficient of performance as function of temperature ratio, for SR and OPT.

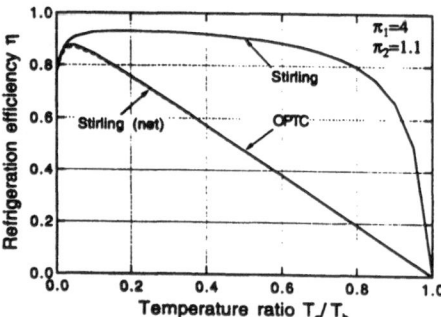

**Figure 9.** Refrigeration efficiency as function of, temperature ratio, for SR and OPT.

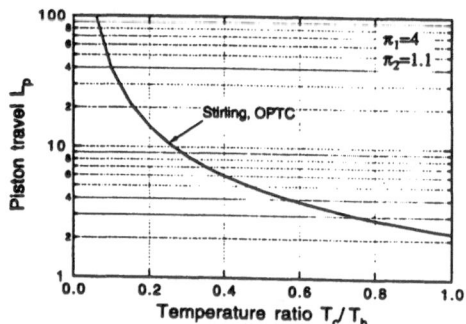

**Figure 10.** Piston travel as function of temperature ratio, for SR and OPT.

small temperature ratios, the piston travel required to maintain the compression ratio of 4 becomes very large.

The results obtained for the COP (Fig. 8) of the OPT and the SR without recovery of work from the expansion system are quite close to the result $COP = T_c/T_h$ derived by Kittel[7] for the "ideal" OPT, in which irreversibilities are neglected. This also holds for the COP results derived in Ref. [13], pertaining to the case that the compression piston moves within the heat exchanger on the compression side. It can be concluded that the value of the COP is determined primarily by the mechanism by which the OPT operates, rather than by irreversibilities neglected in the ideal case. The ideal case does not provide information on the amount of heat $Q_c$ removed per cycle; this amount is directly related to the irreversibilities that occur. The results for $Q_c$ obtained here are quite different from those obtained in Ref. [13]. This is due to the fact that the virtual piston invoked in the model of Ref. [13] remains at rest during step 2, so that no gas is leaving during this step. In the present model, the gas leaving the cold heat exchanger during step 2 plays an important role the cooling process, and leads to a considerable increase in $Q_c$ over the values found in Ref. [13].

## CONCLUDING REMARKS

In this paper the operation of the Stirling Refrigerator (SR) and the Orifice Pulse Tube (OPT) is analyzed using basic thermodynamic principles. The theory developed is non-linear, and is exact within the standard thermodynamic context. Results are found for the nondimensional heat removed per cycle as a function of the ratio of temperatures $T_c/T_h$ across the regenerator. For a compression ratio $\pi_1 = 4$ and an expansion ratio $\pi_2 = 1.1$, the SR removes about 10% more heat per cycle than the OPT. The coefficient of performance of the OPT and of the SR without work recovery from the expansion piston is close to the ideal value $COP = T_c/T_h$. As a consequence, the refrigeration efficiency of these devices is close to its ideal value $\eta = 1 - T_c/T_h$. Except at very low values of $T_c/T_h$, the efficiency of the SR with work recovery is significantly higher; it remains above 90% up to $T_c/T_h \approx 0.6$, and is about 80% at

$T_c/T_h \approx 0.8$. Correspondingly, the COP of the SR with work recovery from the expansion piston is significantly higher than that of the OPT when $T_c/T_h$ is not small. It can be concluded that the orifice pulse tube is primarily suited for cooling to very low temperatures, i.e. for cryocooling, while the Stirling device in principle is useful over a wide range of temperatures.

The present results establish ultimate limits on the performance of actual devices. The methods developed can provide guidelines for the optimization of actual devices by predicting the effect of changes in parameter values. Of course, in drawing conclusions with respect to actual devices it must be kept in mind that friction and other loss mechanisms will reduce the work that can be recovered from the expansion piston to well below its ideal value. Other losses will also lead to significant reductions in performance. Most of these losses can be expected to be of the same order of magnitude for the OPT as for the SR.

## APPENDIX

This Appendix lists results for the heat $Q_c$ removed during step 4 of the cycle considered for the SR. These results are different from the corresponding results for the OPT, given in Ref. [15]. Also listed here are the results for the work $W_{ep}$ done by the expansion piston. The net value of this work is $W_{ep} = W_{ep,4} - W_{ep,2}$ (< 0).

**Case A.**

$$Q_{c4} = \frac{1}{k-1}\left[D(1-\pi_4) - D^{1/k}\pi_4 C_5 \int_{1/C_5}^{1/(\pi_4 C_5)} \frac{1}{(-1+1/s)^{1-1/k}}ds\right] \tag{A1}$$

$$W_{ep,2} = \frac{1+kL_r}{k-1}\pi_1\left(1-\pi_2^{-1+1/k}\right) \tag{A2}$$

$$W_{ep,4} = \frac{\pi_4}{k-1}$$

$$\times\left\{\left[z_h - D^{1/k}(C_5-1)^{1/k}\right]\left(\frac{1}{\pi_4^{1-1/k}}-1\right) + \frac{k-1}{k}D^{1/k}C_5 \int_{1/C_5}^{1/(\pi_4 C_5)}\frac{1}{s(-1+1/s)^{1-1/k}}ds\right\} \tag{A3}$$

**Case B.**

$$Q_{c4} = \frac{1}{k-1}\left\{D(1-\pi_4) - D^{1/k}\pi_4 C_6(kL_r)^{1-1/k}\int_{1/C_6}^{\pi_{4a}/C_6}\frac{1}{(1+1/s)^{1-1/k}}ds\right.$$

$$\left. -D^{1/k}\pi_4 C_7 \int_{\pi_{4a}/C_7}^{1/(\pi_4 C_7)}\frac{1}{(-1+1/s)^{1-1/k}}ds\right\} \tag{A4}$$

$$W_{ep,2} = \frac{1+kL_r}{k-1}\pi_1\left(1-\frac{1}{\pi_2^{1-1/k}}\right) \tag{A5}$$

$$W_{ep,4} = \frac{\pi_4 z_h}{k-1}\left(\frac{1}{\pi_4^{1-1/k}}-1\right) - \frac{(kL_r)^{1-1/k}}{k}D^{1/k}\pi_4 C_6 \int_{1/C_6}^{\pi_{4a}/C_6}\frac{1}{s(1+1/s)^{1-1/k}}ds$$

$$-\frac{\pi_4^{1/k}}{k-1}\left[1-(\pi_4\pi_{4a})^{1-1/k}\right](kL_r)^{1-1/k}D^{1/k}(C_6+\pi_{4a})^{1/k}$$

$$+\frac{\pi_4^{1/k}}{k-1}\pi_{4a}^{1-1/k}\left[1-\pi_4^{1-1/k}\right](kL_r)^{1-1/k}D^{1/k}(C_6+1)^{1/k} \tag{A6}$$

$$-\frac{\pi_4^{1/k}D^{1/k}}{k-1}\left[1-(\pi_4\pi_{4a})^{1-1/k}\right](C_7-\pi_{4a})^{1/k} - \frac{\pi_4}{k}D^{1/k}C_7 \int_{\pi_{4a}/C_7}^{1/(\pi_4 C_7)}\frac{1}{s(-1+1/s)^{1-1/k}}ds$$

## REFERENCES

1.  Mikulin, E. I., Tarasov, A. A., Shkrebyonock, M. P., in *Advances in Cryogenic Engineering*, Plenum Press, New York (1984), Vol. 29, pp. 629-637.

2.  Radebaugh, R., "Pulse Tube Refrigeration - A New Type of Cryocooler," *Japanese J. Appl. Phys.*, Vol. 26, Suppl. 26-3 (1987), pp. 2076-2081.

3.  Zhu, S. W., Wu, P. Y., Chen, Z. Q., "Double Inlet Pulse Tube Refrigerators: An Important Improvement," *Cryogenics*, Vol. 30 (1990), pp. 514-520.

4.  Storch, P. J., Radebaugh, R. *Development and Experimental Test of an Analytical Model of the Orifice Pulse Tube Refrigerator*, Plenum Press, NY, Proceedings of the 1987 Cryogenic Engineering Conference, St. Charles, IL (1988), Vol. 33, pp. 851-859.

5.  Storch, P. J., Radebaugh, R., Zimmerman, J. E. "Analytical Model for the Refrigeration Power of the Orifice Pulse Tube Refrigerator," National Institute of Standards and Technology, Technical Note 1343, December 1990.

6.  Kasuya, M., Nakatsu, M., Geng, Q., Yuyama, J., Goto, E., "Work and Heat Flows in a Pulse-Tube Refrigerator," *Cryogenics*, Vol. 31 (1991), pp. 786-790.

7.  Kittel, P., "Ideal Orifice Pulse Tube Refrigerator Performance," *Cryogenics*, Vol. 32, No. 9 (1992), pp. 843-844.

8.  David, M., Maréchal, J.-C., Simon, Y., Gulpin, C., "Theory of ideal orifice pulse tube refrigerator," *Cryogenics*, Vol. 33 (1993), pp. 154-161.

9.  Mirels, H., "Linearized theory for pulse tube cryocooler performance," *AIAA J.*, Vol. 32, No. 8 (1994), pp. 1662-1669.

10. Mirels, H., in *Advances in Cryogenic Engineering*, Plenum Press, New York (1994), pp. 1425-1431.

11. Rawlins, W., Radebaugh, R., Bradley, P. E., in *Advances in Cryogenic Engineering*, Plenum Press, New York (1994), pp. 1449-1456.

12. Zhu, S. W., Chen, Z. Q., "Isothermal Model of Pulse Tube Refrigerator," *Cryogenics*, Vol. 34, No. 7 (1994), pp. 591-595.

13. de Boer, P. C. T., *Cryogenic Engineering Conference*, Columbus, Ohio, Plenum Press, New York (1995), Paper TU-B1-7.

14. Gifford, W. E., Longsworth, R. C., in *Advances in Cryogenic Engineering*, Plenum Press, New York (1966), pp. 171-179.

15. de Boer, P. C. T., "Thermodynamic Analysis of the Orifice Pulse Tube Cryocooler," to be published.

16. de Boer, P. C. T., "Analysis of basic pulse-tube refrigerator with regenerator," *Cryogenics*, Vol. 35, No. 9 (1995), pp. 547-553.

17. de Boer, P. C. T., "Analysis of the basic pulse-tube cryocooler," *Cryogenics*, Vol. 34, No. 9 (1994), pp. 699-711.

# Application of the Periodic Temperature Variation Technique to the Measurement of Heat Transfer Coefficients in High-NTU Matrices

J.A. Ramirez[1], K.D. Timmerhaus[1] and R. Radebaugh[2]

[1]Department of Chemical Engineering,
University of Colorado at Boulder,
Boulder, Colorado 80309-0424, USA
[2]National Institute of Standards and Technology,
Boulder, Colorado 80303, USA

## ABSTRACT

The existing periodic temperature variation technique for determining heat transfer characteristics of heat exchanger surfaces is extended to the high NTU (over 50) ranges. The initial formulation of the problem considers the effects of axial conduction in both fluid and matrix and the effects of the regenerator casing. The model is solved analytically by using the complex temperature technique. This solution is used in conjunction with experimental measurements to obtain the regenerator NTU at the given conditions. Results from experiments conducted on regenerators consisting of packed lead spheres and stainless steel screen under varying Reynolds numbers are presented. Very good agreement is observed between the results obtained and those from established correlations for the heat transfer coefficient.

## INTRODUCTION

A variety of experimental techniques are currently available for evaluating the heat transfer characteristics of regenerative heat exchangers. A common approach used in the cryocooler field is that based on the method of Gifford et al.[1,2] where the NTU is estimated through the measurement of the regenerator ineffectiveness. Although the method is suitable for high NTU matrices it involves a certain degree of sophistication in the experimental apparatus and procedure. The other approach is based on the transient techniques developed in the power generation[3,4] and iron and steel industries[5,6], where the test fluid maintained at a constant flow rate is passed through the heat exchanger matrix while a temperature disturbance is applied at the inlet. The outlet fluid temperature response is recorded and a mathematical model of the system is used to evaluate the heat transfer coefficient. Different disturbance functions have been used, including step functions, exponential rises or decays[7,8], pulsed perturbations[9] and periodic excitation[10,11,12].

---

Cryocoolers 9, Edited by R.G. Ross, Jr.
Plenum Press, New York, 1997

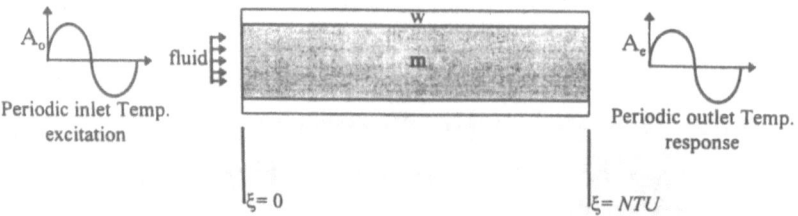

**Figure 1.** Model system schematic

In all of the studies mentioned above, it has been observed that the transient methods are unsuitable for higher NTU ranges, thus virtually all of the research work has focused on test matrices with NTUs of less than ten. Recently, the applicability of the transient techniques for higher NTU matrices with geometries typical of those used in cryocooler regenerators was reviewed by Ramirez[13]. According to his investigations the potentially most useful method for higher NTU matrices is the periodic temperature variation technique. In this paper, the application of the above mentioned method to high NTU matrices (50 and higher) is addressed. The mathematical model for the process is formulated and an analytical solution is presented. Tests are performed on two well characterized regenerator geometries from which the heat transfer coefficients are obtained. It is observed that the results obtained compare favorably with established heat transfer correlations.

## MATHEMATICAL MODEL

A fluid flowing steadily through a heat storage matrix subject to a sinusoidal inlet temperature perturbation can be modeled by applying the energy conservation equation separately to the fluid, solid matrix casing and the solid matrix, depicted in Figure 1.

In terms of dimensionless variables one obtains

$$\kappa_f \frac{\partial \Theta_f}{\partial \tau} + \frac{\partial \Theta_f}{\partial \xi} = \Lambda_f NTU \frac{\partial^2 \Theta_f}{\partial \xi^2} + (\Theta_m - \Theta_f) + \frac{NTU_i}{NTU}(\Theta_w - \Theta_f) \tag{1}$$

$$\kappa_w \frac{\partial \Theta_w}{\partial \tau} = \Lambda_w NTU \frac{\partial^2 \Theta_w}{\partial \xi^2} + \frac{NTU_i}{NTU}(\Theta_f - \Theta_w) + \frac{NTU_o}{NTU}\Theta_w \tag{2}$$

$$\frac{\partial \Theta_m}{\partial \tau} = \Lambda_m NTU \frac{\partial^2 \Theta_m}{\partial \xi^2} + (\Theta_f - \Theta_m) \tag{3}$$

subject to

$$\Theta_f(0,\xi) = 0 \qquad\qquad \Theta_m(0,\xi) = 0 \qquad\qquad \Theta_w(0,\xi) = 0 \quad (4)$$

$$\Theta_f(\tau,0) - \Lambda_f NTU \left[\frac{\partial \Theta_f}{\partial \xi}\right]_{(\tau,0)} = \cos \Omega\tau \quad \left[\frac{\partial \Theta_m}{\partial \xi}\right]_{(\tau,0)} = 0 \qquad \left[\frac{\partial \Theta_w}{\partial \xi}\right]_{(\tau,0)} = 0 \quad (5)$$

$$\Lambda_f NTU \left[\frac{\partial \Theta_f}{\partial \xi}\right]_{(\tau,NTU)} = 0 \qquad \left[\frac{\partial \Theta_m}{\partial \xi}\right]_{(\tau,NTU)} = 0 \qquad \left[\frac{\partial \Theta_w}{\partial \xi}\right]_{(\tau,NTU)} = 0 \quad (6)$$

Certain assumptions are implicit in this model, namely:
- Constant heat transfer coefficients and heat capacities
- Constant transport properties
- No radial dependency of variables, that is, infinite thermal conductivies in radial direction

- Axial symmetry
- Axial dispersion model is applicable

For very high NTU matrices, the ratio of the number of units transferred to (or from) the casing wall to the number of units transferred to (or from) the matrix ($NTU_t/NTU$) is a very small number; thus the effect of the last term on Eq. (1) may be neglected. The disappearance of the dependent variable $\Theta_w$ from the fluid energy balance, permits the solution of the system of equations without using the casing energy balance. The problem now becomes

$$\kappa_f \frac{\partial \Theta_f}{\partial \tau} + \frac{\partial \Theta_f}{\partial \xi} = \Lambda_f NTU \frac{\partial^2 \Theta_f}{\partial \xi^2} + (\Theta_m - \Theta_f) \tag{7}$$

$$\frac{\partial \Theta_m}{\partial \tau} = \Lambda_m NTU \frac{\partial^2 \Theta_m}{\partial \xi^2} + (\Theta_f - \Theta_m) \tag{8}$$

subject to the initial and boundary conditions

$$\Theta_f(0,\xi) = 0 \qquad\qquad \Theta_m(0,\xi) = 0 \tag{9}$$

$$\Theta_f(\tau,0) - \Lambda_f NTU \left[ \frac{\partial \Theta_f}{\partial \xi} \right]_{(\tau,0)} = \cos \Omega \tau \qquad\qquad \left[ \frac{\partial \Theta_m}{\partial \xi} \right]_{(\tau,0)} = 0 \tag{10}$$

$$\Lambda_f NTU \left[ \frac{\partial \Theta_f}{\partial \xi} \right]_{(\tau,NTU)} = 0 \qquad\qquad \left[ \frac{\partial \Theta_m}{\partial \xi} \right]_{(\tau,NTU)} = 0 \tag{11}$$

These equations may be solved by the complex temperature technique, which permits the transformation of the system of partial differential equations on space and time into a system of partial differential equations in space only, with the added advantage that one of the dependent variables (the fluid or the matrix temperature) may be eliminated by substitution. The methodology followed here is based on the treatment of Ramirez[13] and a more detailed explanation of the technique can be found in Arpaci[15]. One begins by defining a supplementary problem governed by the same differential equations and initial and homogeneous boundary conditions. The periodic nonhomogeneous boundary condition is modified so as to lag by an angle of $\pi/2$ with respect to the original nonhomogeneous boundary condition. The supplementary problem is then

$$\kappa_f \frac{\partial \Theta^*_f}{\partial \tau} + \frac{\partial \Theta^*_f}{\partial \xi} = \Lambda_f NTU \frac{\partial^2 \Theta^*_f}{\partial \xi^2} + (\Theta^*_m - \Theta^*_f) \tag{12}$$

$$\frac{\partial \Theta^*_m}{\partial \tau} = \Lambda_m NTU \frac{\partial^2 \Theta^*_m}{\partial \xi^2} + (\Theta^*_f - \Theta^*_m) \tag{13}$$

subject to

$$\Theta^*_f(0,\xi) = 0 \qquad\qquad \Theta^*_m(0,\xi) = 0 \tag{14}$$

$$\Theta^*_f(\tau,0) - \Lambda_f NTU \left[ \frac{\partial \Theta^*_f}{\partial \xi} \right]_{(\tau,0)} = \sin \Omega \tau \qquad\qquad \left[ \frac{\partial \Theta^*_m}{\partial \xi} \right]_{(\tau,0)} = 0 \tag{15}$$

$$\Lambda_f NTU \left[ \frac{\partial \Theta^*_f}{\partial \xi} \right]_{(\tau,NTU)} = 0 \qquad\qquad \left[ \frac{\partial \Theta^*_m}{\partial \xi} \right]_{(\tau,NTU)} = 0 \tag{16}$$

The complex temperatures are defined in terms of the dependent variables of the original problem and the supplementary problem as

$$\psi_f(\tau,\xi) = \theta_f(\tau,\xi) + \iota \Theta^*_f(\tau,\xi) \tag{17}$$

$$\psi_m(\tau,\xi) = \theta_m(\tau,\xi) + \iota \Theta^*_m(\tau,\xi) \tag{18}$$

where $\iota$ is the conventional imaginary unit $\iota = \sqrt{-1}$. The differential equations and the initial and boundary conditions satisfied by both $\psi_f(\tau,\xi)$ and $\psi_m(\tau,\xi)$ are obtained by multiplying the supplementary problem of Eqs. (12) and (13) by $\iota$ and adding this result to the original problem of Eq. (7) and Eq. (8). It follows then that

$$\kappa_f \frac{\partial \psi_f}{\partial \tau} + \frac{\partial \psi_f}{\partial \xi} = \Lambda_f NTU \frac{\partial^2 \psi_f}{\partial \xi^2} + (\psi_m - \psi_f) \tag{19}$$

$$\frac{\partial \psi_m}{\partial \tau} = \Lambda_m NTU \frac{\partial^2 \psi_m}{\partial \xi^2} + (\psi_f - \psi_m) \tag{20}$$

subject to the initial and boundary conditions

$$\psi_f(0, \xi) = 0 \qquad\qquad\qquad \psi_m(0, \xi) = 0 \tag{21}$$

$$\psi_f(\tau, 0) - \Lambda_f NTU \left[ \frac{\partial \psi_f}{\partial \xi} \right]_{(\tau, 0)} = e^{\Omega \tau} \qquad \left[ \frac{\partial \psi_m}{\partial \xi} \right]_{(\tau, 0)} = 0 \tag{22}$$

$$\Lambda_f NTU \left[ \frac{\partial \psi_f}{\partial \xi} \right]_{(\tau, NTU)} = 0 \qquad \left[ \frac{\partial \psi_m}{\partial \xi} \right]_{(\tau, NTU)} = 0 \tag{23}$$

Solutions of the type $\psi_f(\tau, \xi) = \phi_f(\xi) e^{\Omega \tau}$ and $\psi_m(\tau, \xi) = \phi_m(\xi) e^{\Omega \tau}$ valid for large times can be proposed, and after substitution in Eqs. (19) to (23) one obtains

$$\Omega \kappa_f \phi_f(\xi) + \frac{d \phi_f(\xi)}{d \xi} = \Lambda_f NTU \frac{d^2 \phi_f(\xi)}{d \xi^2} + \left[ \phi_m(\xi) - \phi_f(\xi) \right] \tag{24}$$

$$\Omega \phi_m(\xi) = \Lambda_m NTU \frac{d^2 \phi_m(\xi)}{d \xi^2} + \left[ \phi_f(\xi) - \phi_m(\xi) \right] \tag{25}$$

subject to

$$\phi_f(0) - \Lambda_f NTU \left[ \frac{d \phi_f}{d \xi} \right]_{\xi=0} = 1 \qquad \left[ \frac{d \phi_f}{d \xi} \right]_{\xi=NTU} = 0 \tag{26}$$

$$\left[ \frac{d \phi_m}{d \xi} \right]_{\xi=0} = 0 \qquad \left[ \frac{d \phi_m}{d \xi} \right]_{\xi=NTU} = 0 \tag{27}$$

Solving for $\phi_m(\xi)$ in Eq. (24) and substitution in Eq. (25) and in the boundary conditions of Eqs. (26) and (27) yields the following ordinary differential equation

$$\sum_{j=0}^{4} \alpha_j \frac{d^{(j)} \phi_f}{d \xi^{(j)}} = 0 \tag{28}$$

with the coefficients in Eq. (28) given by

$$\alpha_0 = -\left[ \Omega^2 \kappa_f - \Omega(1 + \kappa_f) \right]$$
$$\alpha_1 = [1 + \Omega]$$
$$\alpha_2 = -\left[ NTU(\Lambda_f + \Lambda_m) - \iota \Omega NTU(\Lambda_f + \kappa_f \Lambda_m) \right]$$
$$\alpha_3 = -\Lambda_m NTU$$
$$\alpha_4 = \Lambda_m \Lambda_f NTU^2 \tag{29}$$

The solution to this problem is straightforward and is given by an expression of the form,

$$\phi_f(\xi) = \sum_{j=1}^{4} K_j e^{r_j \xi} \tag{30}$$

where the $r_j$ are the roots of the characteristic polynomial

$$\sum_{j=0}^{4} \alpha_j r^j = 0 \tag{31}$$

and the constants $K_j$ are given by the four boundary conditions

$$\sum_{j=1}^{4} r_j K_j e^{r_j} = 0 \qquad\qquad \sum_{j=1}^{4} \left[ \Lambda_f NTU\, r_j^{3} - r_j^{2} - \left(1 + \iota \Omega \kappa_f \right) r_j \right] K_j = 1$$

$$\sum_{j=1}^{4} (1 - \Lambda_f\, NTU\, r_j) K_j = 1 \qquad \sum_{j=1}^{4} \left[ \Lambda_f NTU\, r_j^{3} - r_j^{2} - \left(1 + \iota \Omega \kappa_f \right) r_j \right] K_j e^{r_j} = 1 \quad (32)$$

The solution for the fluid temperature ($\Theta_f(\xi)$) sought is given by the real part of the complex temperature defined in Eq. (17), which may be written in polar form as $\psi_f(\tau,\xi) = \phi_f(\xi) e^{\Omega \tau}$. Since the solution $\phi_f(\xi)$ is given by the complex number of Eq. (30), it can be seen by combining all of the above arguments, that the solution to the original problem is then given by an expression of the form

$$\Theta_f(\xi) = \rho(\xi)\left[\cos\left(\alpha(\xi) + \Omega\tau\right)\right] \tag{33}$$

where $\rho(\xi)$ and $\alpha(\xi)$ are the real and imaginary parts of the complex function $\phi_f(\xi)$ and represent the inlet to outlet temperature amplitude ratio and phase shift as a function of position, respectively. Thus in theory, if one measures experimentally the temperature amplitude ratio and phase shift between points $\xi = 0$ and $\xi = \xi_o$, one could obtain two unknown parameters from solving the system of equations

$$\rho(\kappa_f, \Lambda_f, \Lambda_m, \Omega, NTU, \xi = NTU) - A_{exp} = 0 \tag{34}$$

$$\alpha(\kappa_f, \Lambda_f, \Lambda_m, \Omega, NTU, \xi = NTU) - \Delta_{exp} = 0 \tag{35}$$

as suggested by Roetzel, et al.[12] For example, for known excitation frequencies and regenerator dimensions and a fixed fluid flow rate (this sets $\Omega$ and $\kappa_f$) one could solve for the NTU and $\Lambda_f$ or $\Lambda_m$ depending on which is selected as a known parameter. For the type of regenerators evaluated in this study however, it was observed that the flow regime deviated only slightly from plug flow (thus $\Lambda_f \to 0$). An important feature observed at higher NTUs and discussed later in this work is that the NTU vs. phase shift curves become very sensitive to uncertainty in the phase shift measurement. In this case Eq. (35) becomes unreliable and one is left with just the amplitude ratio as the only useful measurement. In the experiments that were performed, the matrix conduction parameter $\Lambda_m$ was estimated and Eq. (34) was used to solve for the NTU.

## EXPERIMENTAL METHOD

Each experiment involves the solution of Eq. (34), where the parameters $\Lambda_f$, $\Lambda_m$, $\kappa_f$ and $\Omega$ are known or can be set. Once the desired range of Reynolds number over which a regenerator will be tested is decided upon, the optimal attainable frequency to use in the test is sought. The optimal frequency value is obtained with the aid of a computer code, but is based on the reasoning discussed in the previous section.

The solutions of Eqs. (34) and (35) for the case of a regenerator constructed of stacked stainless steel screens of mesh no. 150 are shown in Figures 2 and 3, showing NTU as a function of the outlet to inlet temperature amplitude ratio and as a function of the phase lag between said waves respectively. In order to be able to present the solution in graphical form for a given temperature excitation frequency, it is convenient to introduce a modified dimensionless frequeny $\Omega^{*}$ given by $\Omega^{*} = \Omega NTU$, thus eliminating the dependence on the NTU. Three representative dimensionless frequencies $\Omega^{*}$ are shown for comparative purposes. The actual dimensionless frequency used in the experiments was $\Omega^{*} = 10.05$, which corresponded to a temperature excitation frequency of 0.4 Hz. It can be inferred from studying the slopes of the curves from the solution, that for this particular case, (where an NTU of about 100 was expected at the given flow conditions), higher dimensionless frequencies are desirable. This comes from observing the nature of the slope of the solution curves of Figure 2 at the different frequencies.

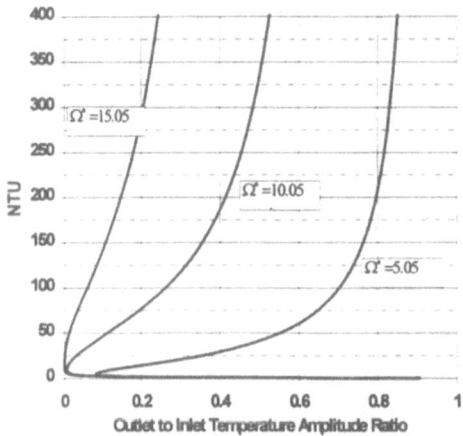

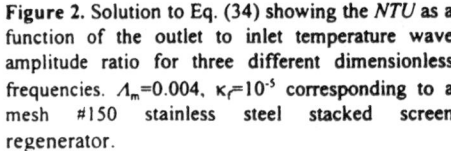

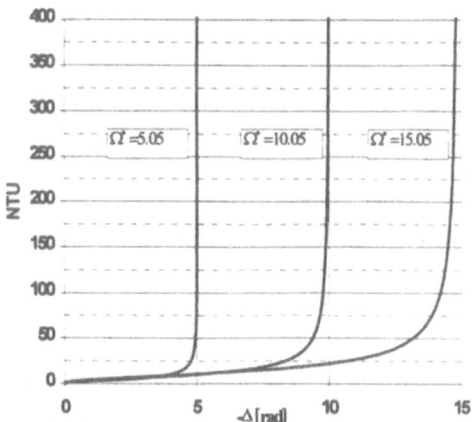

**Figure 2.** Solution to Eq. (34) showing the *NTU* as a function of the outlet to inlet temperature wave amplitude ratio for three different dimensionless frequencies. $\Lambda_m$=0.004, $\kappa_f$=10$^{-5}$ corresponding to a mesh #150 stainless steel stacked screen regenerator.

**Figure 3.** Solution to Eq. (35) showing the *NTU* as a function of the phase lag between the outlet and inlet temperature waves for three different dimensionless frequencies. $\Lambda_m$=0.004, $\kappa_f$=10$^{-5}$ corresponding to a mesh #150 stainless steel stacked screen regenerator.

The normalized values of these slopes for varying *NTU* give the sensitivity of the value of *NTU* to an uncertainty in the measurement of the amplitude ratio. The sensitivity index is given by

$$Sensitivity = \frac{A_R}{NTU} \frac{\partial NTU}{\partial A_R} \tag{36}$$

and is plotted in Figure 4 as a function of varying *NTU* for a fixed value of $\Lambda_m$. It is observed that at the expected *NTU* value for this particular run (~100), higher frequencies than $\Omega^*$=10.05 will yield less uncertain results. However, with the current experimental setup, physical limitations such as the heater and upstream pipe time constants prohibit operation at these higher frequencies.

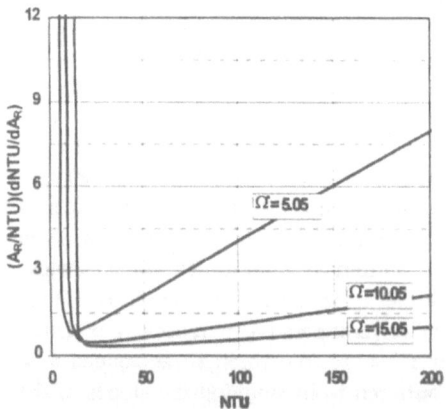

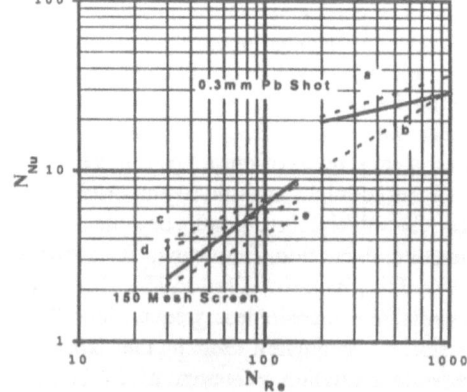

**Figure 4.** Sensitivity parameter as a function of the NTU. $\Lambda_m$=0.004, $\kappa_f$=10$^{-5}$ corresponding to a mesh # 150 stainless steel stacked screen regenerator.

**Figure 5.** Heat transfer correlations expressed in the form of the Nusselt vs. Reynolds number obtained for the regenerators tested. a[16], b[17], c[18], d[19], e[20].

On the other hand, it is seen from Figure 3 that in the higher $NTU$ ranges the solution for the phase lag has no practical use, since the curves are extremely sensitive to an uncertainty in the phase lag measurement. One possible use for these curves, however, is in estimating the value of the matrix heat capacity. It is possible to run the experiment at a given temperature excitation frequency and a fluid flowrate which will result in a very high $NTU$ value. The fact that at very high $NTU$s the value of the phase lag approaches asymptotically to that of the dimensionless frequency, can be used to experimentally determine the dimensionless frequency $\Omega^{*}$, from which the total solid matrix heat capacity can be calculated from its definition

$$\left(mc_{p}\right)_{m} = \frac{\left(wc_{p}\right)_{f} \Omega^{*}}{\omega} \tag{37}$$

Experiments were carried out on regenerators with matrices of both stacked woven stainless steel screens and packed lead spheres. These two geometries have been widely studied in the past and reliable correlations exist for predicting the heat transfer coefficients[14]. The fluid used was industrial grade nitrogen at a pressure of 200 kPa. An approximately sinusoidal temperature wave was achieved by applying a sinusoidal voltage wave to a resistance heater placed within the flow stream and upstream to the test regenerator, closely simulating the model of Figure 1. It is noted that it is not imperative to provide a perfect sine temperature excitation, since the process is described by linear equations and thus only the fundamental harmonics need to be considered for any type of periodic disturbance. Temperatures were measured with type E thermocouples located upstream and downstream from the test regenerator, that is at $\xi=0$ and $\xi=NTU$ in Figure 1. The thermocouples were wired in parallel and positioned across the flow area in sets of five junctions so as to obtain the average temperature of the flowing stream. Mesh screen flow straighteners were also used for ensuring plug flow both upstream and downstream from the measuring points. Data acquisition was performed automatically with a personal computer. Experiments were performed with different flow rates in order to obtain data over a range of Reynolds numbers.

## DISCUSSION AND RESULTS

A Fourier analysis was performed on the measured inlet and outlet temperature data with the aid of a FFT algorithm in the data acquisition software. The amplitude ratio and phase shift between inlet and outlet waves were obtained from a comparison of their fundamental harmonics. Once Eq. (34) is solved with the appropriate values of the known parameters particular to each run, an $NTU$ is obtained for the conditions of the experiment. Experimental runs on each of the two test regenerators were performed at various Reynolds numbers. The $NTU$ estimates were correlated and are presented in terms of the $N_{Nu}$ and $N_{Re}$ as shown in Figure 5.

$N_{Nu} = 0.134 \, N_{Re}^{0.840}$      for the # 150 mesh screen regenerator

$N_{Nu} = 5.310 \, N_{Re}^{0.246}$      for the 0.3 mm diameter lead sphere regenerator

These correlations yield heat transfer coefficients in good agreement with values cited by Kays and London[14,18], amongst others.

It was observed that the uncertainty of the calculated heat transfer coefficient will depend chiefly on two dimensionless quantities, namely the $NTU$ and the product $\Lambda_{m}NTU$. Large values of the latter dimensionless parameter can yield highly uncertain results, even if the $NTU$ value is relatively low. In such cases, it will be necessary to modify the dimensions of the test regenerator such that it will have a low value of $\Lambda_{m}NTU$.

## CONCLUSION

The periodic temperature technique was found to be applicable at higher $NTU$ ranges (up to 100) as long as the above mentioned limitations are taking into account. In general, it was observed that for high values of $NTU$ (over 50), it is desirable to conduct the experiments at as high a frequency as possible. For regenerators exhibiting a high value of the matrix thermal conductivity parameter ($\Lambda_m NTU$), the periodic method could yield unacceptable results. This may be solved by altering the dimensions of the test core so as to decrease the value of $\Lambda_m NTU$.

## ACKNOWLEDGMENT

The authors express their thanks to the National Institute of Standards and Technology Boulder Laboratories for providing the facilities and equipment for performing the experiments. The first author acknowledges FUNDAYACUCHO for providing him with financial support during the duration of this work.

## NOTATION

### Greek Symbols

| | | | |
|---|---|---|---|
| $\kappa_f$ | ratio of heat storage capacity | $(\kappa_f = (mc_p)_f/(mc_p)_m)$ | |
| $\Delta_{exp}$ | measured phase lag | rad | |
| $\sigma$ | volume fraction of regenerator not occupied by matrix | dimensionless | |
| $\Lambda_f$ | dimensionless dispersion parameter | $(\Lambda_f = k_f \sigma S/(wc_p)_f L)$ | |
| $\Lambda_m$ | dimensionless matrix conductive parameter | $(\Lambda_m = k_m(1-\sigma)S/(wc_p)_f L)$ | |
| $\xi$ | dimensionless axial coordinate | $(\xi = x\,NTU/L)$ | |
| $\Theta$ | dimensionless temperature | $(\Theta = (T-T_0)/A_o)$ | |
| $\tau$ | dimensionless time | $(\tau = t\,(hA)_m/(mc_p)_m)$ | |
| $\phi_f$ | complex fluid temperature function | dimensionless | |
| $\Omega$ | dimensionless excitation frequency | $(\Omega = \omega(mc_p)_m/(wc_{pf}NTU)$ | |
| $\Omega^*$ | dimensionless excitation frequency | $(\Omega = \Omega^* NTU)$ | |
| $\iota$ | conventional imaginary unit | $(\iota = \sqrt{-1})$ | |
| $\Delta$ | phase angle shift | rad | |
| $\omega$ | angular excitation frequency | $(\omega = 2\pi/P)$ | |

### Roman Symbols

| | | |
|---|---|---|
| A | amplitude ratio or surface area | dimensionless or m$^2$ |
| $A_e$ | outlet temperature amplitude | K |
| $A_{exp}$ | measured amplitude ratio | dimensionless |
| $A_o$ | inlet temperature amplitude | K |
| $A_R$ | amplitude ratio | dimensionless |
| $c_p$ | heat capacity at constant pressure | J/(kg-K) |
| h | heat transfer coefficient | W/(m$^2$-K) |
| k | thermal conductivity | W/(m-K) |
| $K_j$ | constants given by Eq. (32) | dimensionless |
| L | length of regenerator | m |
| m | mass of matrix or instant fluid mass | kg |
| NTU | number of heat transfer units | $(NTU = (hA)_m/(wc_p)_f)$ |
| $NTU_i$ | number of heat transfer units due to heat transfer to ambient from the casing wall | $(NTU_i = (hA)_i/(wc_p)_f)$ |

| | | |
|---|---|---|
| $NTU_o$ | number of heat transfer units due to heat transfer to ambient from the casing wall | $(NTU_o = (hA)_o/(wc_p)_f)$ |
| P | period of excitation wave | s |
| S | regenerator cross sectional area | $m^2$ |
| $T_o$ | excitation temperature dc component | K |
| w | mass flowrate of test fluid | kg/s |
| x | axial distance along regenerator | m |

**Subscripts**

| | |
|---|---|
| exp | experimentally measured |
| f | referring to the fluid |
| i,o | inward/outward |
| m | referring to the matrix |
| w | referring to the casing wall |

## REFERENCES

1. Gifford,W.E.,Acharya,A. and Ackermann,R.A., "Compact Cryogenic Thermal Regenerator Performance", *Adv. Cryog. Eng.,*34, 353 (1969).

2. Gifford,W.E. and Acharya, A., "Low Temperature Regenerator Test Apparatus", *Adv. Cryog. Eng.,*35, 436 (1969).

3. Locke,G.L.,"Heat Transfer and Flow Friction Characteristics of Porous Solids",*Tech. Rept. No. 10,* Dept. of Mech. Eng., Stanford University , 1950.

4. Pucci, P.F., Howard, C.P. and Piersall Jr., C.H., "The Single Blow Transient Technique for Compact Heat Exchanger Surfaces", *J. Eng. Pow.,* **89**, 1, 29 (1967).

5. Furnas, C.C., "Heat Transfer to a Broken Bed of Solids", *Ind. Eng. Chem.,* **22**, 721 (1930).

6. Saunders,O.A. and Ford,H., *J. Iron & Steel Inst.,* **141**, 291 (1940).

7. Liang, C.J. and Yang, W.J., "Modified Single Blow technique for Performance Evaluation of Heat Transfer Surfaces", *J. Heat Transfer.,* **97**, 16 (1975).

8. Cai, Z.H., Li, M.L., Wu, Y.W. and Ren, H.S., "Modified Selected Point Matching Technique for Testing Compact Heat Exchanger Surfaces", *Int. J. Heat and Mass Transfer.,* **27**, 7, 971 (1984).

9. Oldson, J.C., Knowles, T.R. and Rauch, J., "Pulsed Single Blow Method with Visual Curve Matching", *Proceedings of the 27th IECEC,* **5**, paper No. 929386 (1982).

10. Bell, C.J. and Katz, E.L., "A Method for Measuring Surface Heat Transfer Coefficients Using Cyclic Temperature Variations", Heat Transfer and Fluid Mech. Inst., Berkeley, 1949.

11. Stang, J.H. and Bush, J.E., "The Periodic Method for Testing Compact Heat Exchanger Surfaces", *J. Eng. Pow.,* **96A**, 2, 87 (1974).

12. Roetzel,W., Xing,L. and Yimin,X., "Measurement of Heat Transfer Coefficient and Axial Dispersion Coefficient Using Temperature Oscillations", *Exp. Therm. Fluid Sci.,* 7, 345 (1993).

13. Ramirez, J.A., " Application of the Periodic Temperature Variation Technique to the Measurement of Heat Transfer Coefficients in High NTU Regenerative Heat Exchangers", M.S. Thesis, University of Colorado, Boulder, 1995.

14. Kays, W., M. and London, A., L., "Heat Transfer and Flow Friction Characteristics of Some Compact Heat Exchangers Surfaces", *Trans. ASME,* **72**, 1075-1097 (1950).

15. Arpaci, V., Conduction Heat Transfer, Addison Wesley, Reading, MA, 1996.

16. Jeshar, J., "Wärmeübergang in Mehrkornschüttungen aus Kugeln", *Arch. Eisenhüttenw.*, 35, (1964).

17. Timmerhaus, K., D. and Flynn, T., *Cryogenic Process Engineering*, Plenum Press, New York, 1989.

18. Kays, W., M. and London, A.,L. , *Compact Heat Exchangers*, 3rd ed., McGraw-Hill, New York, 1984.

19. Tanaka, M., Yamashita, I. and Chisaka, F., "Flow and Heat Transfer Characteristics of the Stirling Engine Regenerator in an Oscillating Flow", *JSME, Int. Jour. Series II*, 33, (2), 283-289 (1990).

20. Rice, G., Thonger, M. and Dadd, W., "Regenerator Effectiveness Measurements", *Proc. 20th IECEC*, paper No. 859144 (1985).

# 3-D Flow Model for Cryocooler Regenerators

J. S. Nigen[2], R. Yaron[1], K. C. Karki[2], S. V. Patankar[2], and R. Radebaugh[3]

[1] Yaron Consulting
1935 Golden Way
Los Altos, CA 94024

[2] Innovative Research, Inc.
2520 Broadway Street N.E., Suite 200
Minneapolis, MN 55413

[3] National Institute of Standards and Technology
Boulder, CO 80303

## ABSTRACT

A 3-D computational fluid dynamics (CFD) model for cryocooler regenerators has been developed. The model was used to analyze an etched-foil regenerator. In the etched-foil regenerator, slots and channels are photoetched into the foil in precise patterns. The etched foil is then rolled upon itself to form a well defined three-dimensional, non-isotropic, porous structure having clean aerodynamic profiles. The regenerator flow resistance and heat transfer characteristics are analyzed by the three-dimensional, steady-state computational fluid dynamics model. Comparison between the model predictions and measured properties of an etched-foil regenerator is presented.

## NOMENCLATURE

G-modulus

$$G = \rho \bar{v}_x$$

Hydraulic Diameter

$$D_h \equiv 4r_h = \frac{4pA_{fr}L}{A} = 4\frac{(V_{total} - V_{solid})}{A_{wetted}}$$

Heat Transfer Coefficient

$$h \equiv \frac{\left(\dfrac{q}{A}\right)}{(\bar{T}_w - \bar{T}_b)} = \frac{q''}{(\bar{T}_w - \bar{T}_b)}$$

Nusselt Number

$$Nu \equiv \frac{h4r_h}{k} = \frac{hD_h}{k}$$

Reynolds Number

$$Re \equiv \frac{4r_h G}{\mu} = \frac{\rho \bar{v}_x D_h}{\mu}$$

| | |
|---|---|
| Prandtl Number | $\mathrm{Pr} \equiv \dfrac{\mu c_p}{k}$ |
| Stanton Number | $\mathrm{St} \equiv \dfrac{h}{G c_p} = \dfrac{\mathrm{Nu}}{\mathrm{Re}\,\mathrm{Pr}}$ |
| Friction Factor | $f \equiv \dfrac{\rho \tau_o}{G^2 / 2 g_c}$ |

## INTRODUCTION

This study is complementary to the work presented in the publication "Etched Foil Regenerator.[1]" In the previous work, a novel approach to fabricating a regenerator using photoetching was described. Using this procedure, a regenerator can be fabricated from a single piece of foil material, into which holes are etched in precise patterns. The performance of the regenerator is determined by the specific geometric configuration produced by the photoetching procedure.

Numerous geometric configurations have been demonstrated to augment heat transfer. Many of these configurations, such as grooved channels and louvered plate-fins, do so by presenting a repetitive sequence of entrance regions, in which boundary layers initiate, grow, and separate in a periodic fashion. The resulting thinner boundary layers reduce transport resistance and improve performance without the inducement of turbulence. Such geometries are particularly well suited for compact heat transfer applications, where the achievement of turbulent flows would require prohibitively large pressure drops. In the case of regenerators, surface segmentation provides an additional benefit by interrupting the conduction path, thereby reducing heat flow in the axial direction.

A Computational Fluid Dynamics (CFD) model provides a convenient and cost effective means of analyzing various geometric configurations of regenerators. In addition, visualization of local heat and fluid flow patterns can provide guidance for design modification. This paper will demonstrate the capability of CFD analysis in the evaluation of a specific regenerator composed of an etched foil. Values of Stanton number and friction factor are presented, as well as the effective conductivity matrix.

## THE ETCHED-FOIL REGENERATOR

As shown in Fig. 1, the regenerator is fabricated from a single piece of foil material. Holes are etched into the foil in precise patterns, opening highly defined flow passages.

The horizontal strips in the background (lines) are about half the thickness of the original material. The vertical connection straps (connectors) in the foreground are also approximately half the original thickness. Only at the places where the vertical and horizontal elements overlap does the material still retain its original thickness.

Foils from 25 μm to 127 μm thick have been successfully etched. Materials used to date include stainless steel, copper and lead alloys. Preliminary analysis suggests that neodymium foil can also be successfully etched.

## NUMERICAL FORMULATION

Equations governing the conservation of mass, momentum, and energy are solved using the commercial software package, COMPACT-3D. This software package is a finite-volume based simulation package that solves a generic convection-diffusion equation (Eq. 1). In this equation, $\phi$ is the dependent variable, $\bar{\upsilon}$ is the velocity vector, $\Gamma$ is the effective diffusion coefficient, and S is the source or sink term.

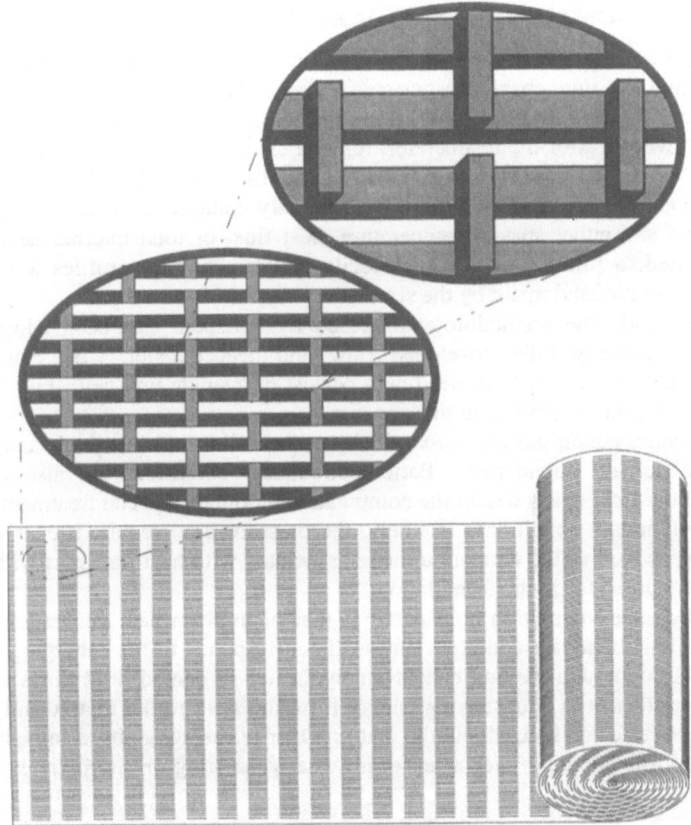

**Figure 1**. The etched-foil regenerator.

$$\frac{\partial(\rho\phi)}{\partial t} + \overline{\nabla}\cdot(\rho\overline{\upsilon}\phi) = \overline{\nabla}\cdot(\Gamma\overline{\nabla}\phi) + S \tag{1}$$

The computational domain is divided into discrete control volumes. A main grid point is located at the center of each control volume. A staggered grid arrangement is used for the storage of the dependent variables: the scalar quantities are located at the main grid points and the location of the velocity components are displaced to lie at the control volume faces.

The discretization equations are obtained by integrating the partial differential equations over the relevant control volumes. The resulting numerical solution satisfies the conservation principles over each of the control volumes and, hence, over the entire computational domain.

In COMPACT 3D, the power-law differencing scheme is used to represent the variation of a dependent variable between any two grid points. The coupling of the continuity and momentum equations is handled using the SIMPLER[2] algorithm. The linearized set of equations are solved using a modified version of the Stone's algorithm, supplemented by a block-correction procedure. Complete details of the numerical method employed in COMPACT-3D and its implementation are available in the Reference manuals[3].

## COMPUTATIONAL DOMAIN AND BOUNDARY CONDITIONS

The regenerator consists of a large number of unit cells. Due to the repetitive nature of the geometry, the thermal and fluid characteristics repeat after a certain entrance length. Such a flow is commonly referred to as periodically fully developed. Since the periodically fully developed conditions prevail over most of the regenerator, reliable estimates of the pressure drop and heat transfer can be obtained by solving the relevant equations only within this region. This is accomplished through the application of periodic boundary conditions in each direction. Only the overall flowrate and either surface temperature, heat flux, or total internal heat generation needs to be specified, a priori. Then, the velocity and temperature profiles are determined throughout the computational domain by the simulation.

In the present work, the methodology proposed by Patankar[4] has been adopted for the computation of periodically fully developed flow and heat transfer. By employing this formulation, the computational domain need only consist of a single unit cell. Fig. 2 shows the unit cells for the configuration studied in this paper.

The fluid boundary conditions are zero velocity at the solid-fluid interfaces and periodicity of velocity at the domain boundaries. Periodic boundary conditions are also imposed on pressure. Since a bulk flow exists within the computational domain, special treatment is required to apply periodic boundary conditions. Namely, the pressure drop required to drive the flow through the matrix, as well as the temperature rise associated with heat transport within the unit cell, needs to be accounted for in the formulation.

The thermal boundary condition considered is that of uniform heat generation within the etched-foil mesh. This boundary condition physically corresponds to the "charging" of the foil by the oscillatory flow through the unit cell. Namely, heated or cooled fluid flows alternatively through the regenerator matrix, transferring energy from the fluid to the matrix and vice versa. This periodic process effectively results in the matrix either releasing or absorbing energy, which can be idealized as fluid flowing over a uniformly heat generating or alternatively, absorbing, matrix.

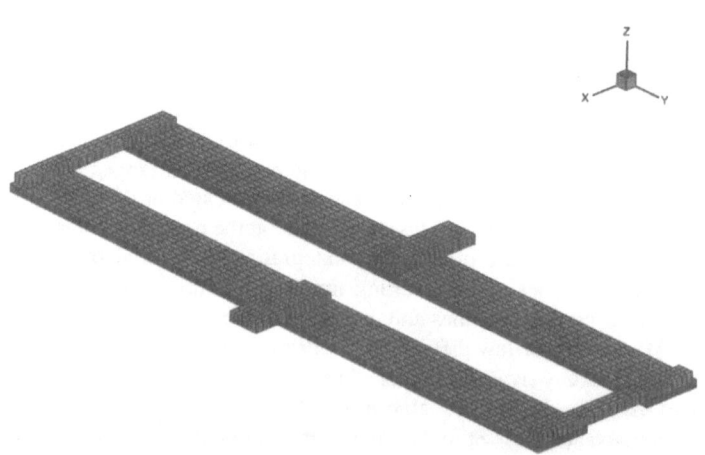

**Figure 2.** Computational representation of one unit cell of the etched foil.

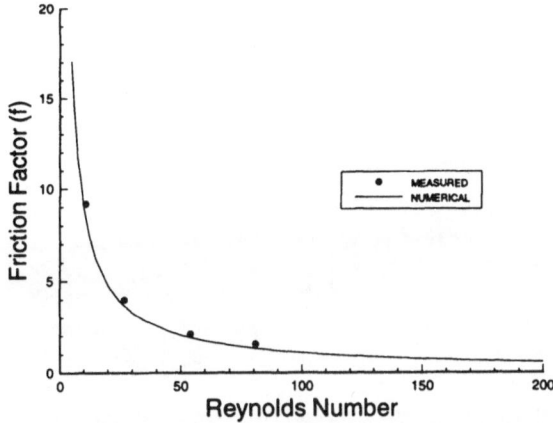

**Figure 3.** Comparison between the measured and computed friction factor versus Reynolds number.

As with pressure, periodicity is imposed at the computational boundaries and special treatment is required to account for the temperature increase in the flow direction. Following the approach of Kelkar and Patankar[5], the temperature at the exit plane is adjusted so as to account for the change in bulk temperature associated with internal heat generation.

## COMPUTATIONAL PREDICTIONS AND MEASURED PROPERTIES

A series of steady, three-dimensional computations were conducted with operating conditions of 30K and 1MPa over a range of typical Reynolds numbers. The computed values of friction factor were within 10% of the measured values[1], as shown in Fig. 3.

Comparisons of computed Nusselt or Stanton numbers with measured values is unfortunately not possible. This is because the extremely high efficiency of the etched-foil material exceeds current measurement limitations. However, research is currently being conducted to extend the range for which measurements can be taken[5].

## FLOW PATTERNS

Flows in cryocooler regenerators exhibit typical Reynolds numbers of less than 200 and the resistance offered by the etched foil will tend to make the flow uni-directional. However, instabilities can occur during that cold-blow phase that induce regions of multi-dimensional flow. It is, therefore, of interest to characterize the pressure drop and heat transfer performance in all three directions.

The photoetching process provides significant control over geometric features. Design modifications can be conceived by visualizing the flow patterns associated with a specific geometric configuration. In this manner, CFD can foster design innovation through increased understanding.

As shown in Fig. 4a, cool fluid flows through the unit cell in the x-direction. Energy is transferred from the foil surface to the fluid, which increases in temperature and exits the computational domain. This pattern is contrasted by the recirculating fluid trapped between the lines and connectors. Fluid within these regions is partitioned from the bulk flow by a shear layer, which significantly diminishes convective transport.

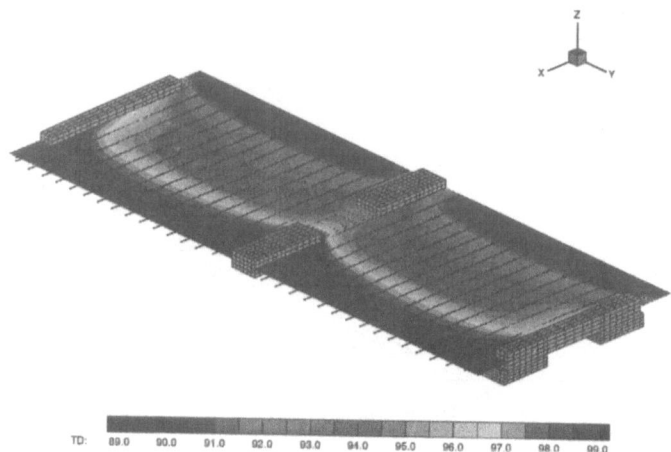

**Figure 4a.** Temperature contours and velocity vectors at a representative xy-plane. The
bulk flow is in the x-direction and the Reynolds number is 50.

In Fig. 4b, cool fluid flows through the unit cell in the y-direction. In this orientation, the
connectors present much greater frontal area to the bulk flow, inducing very large recirculating
zones. These recirculating zones reduce convective communication between the bulk flow and
the back of the connectors. In should be noted that if the Reynolds number become large enough,
vortex shedding would result, as with flow over bluff bodies. In this flow regime, convective
communication would increase, as would overall heat transfer performance. To properly model
flows at these Reynolds numbers, a time dependent formulation would need to be employed.

In Fig. 4c, the cool fluid flows through the unit cell in the z-direction. Since the gaps
between the lines and connectors are relatively small (approximately half the foil thickness), the
separation zones are not as prominent as in the other flow directions. In essence, the
recirculating zones are confined to the regions between adjacent lines or connectors.

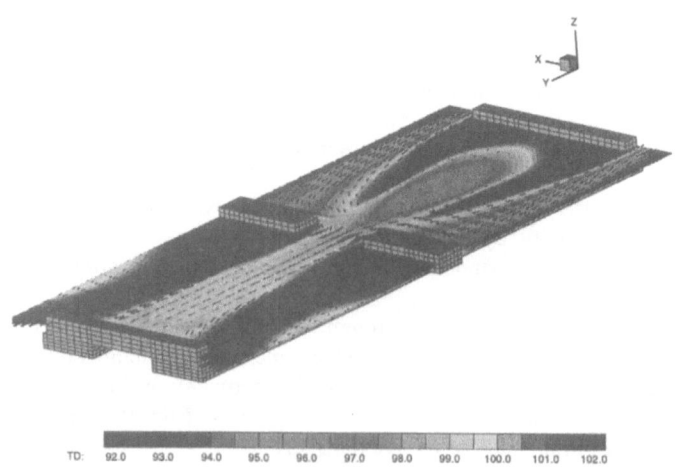

**Figure 4b.** Temperature contours and velocity vectors at a representative xy-plane. The
bulk flow is in the y-direction and the Reynolds number is 50.

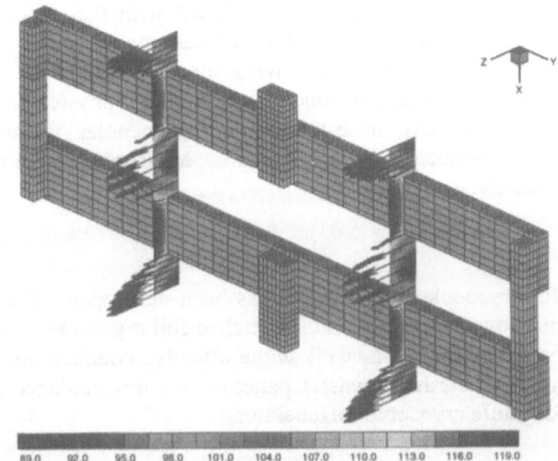

TD:  89.0  92.0  95.0  98.0  101.0  104.0  107.0  110.0  113.0  116.0  119.0

**Figure 4c.** Temperature contours and velocity vectors at representative xz-planes. The bulk flow is in the x-direction and the Reynolds number is 50.

The top and bottom surfaces of the lines, which contributes significantly to the overall surface area of the unit cell, are now surrounded by recirculating zones. This flow pattern effectively removes convective communication between the surfaces of the lines and the bulk flow.

Table 1 summarizes the computed friction factor and Stanton number for the three flow directions at a Reynolds number of 50. In accordance with the previous flow pattern descriptions, the y-direction case exhibits the largest pressure drop and the z-direction the lowest heat transfer.

It should be noted that these directions, being non-principal, are not of primary concern. However, the characteristics exhibited by flows in the non-principal directions provide insight into the effect specific geometric configurations have on transport performance.

**Effective Conductivity Predictions**

In addition to the above flow related characteristics, the effective conductivity of the etched-foil mesh was determined. The calculation of effective conductivity involves the solution of the conduction equation with periodic boundary conditions and an imposed temperature gradient in one direction. The effective conductivity matrix is given below:

$$
\frac{k_{eff,xy}}{k_{xy}}\Bigg|_{solid} =
\begin{bmatrix}
2.5878 \times 10^{-2} & 1.5394 \times 10^{-9} & -9.8855 \times 10^{-8} \\
-1.1485 \times 10^{-7} & 2.1020 \times 10^{-1} & -1.3580 \times 10^{-12} \\
-4.5365 \times 10^{-6} & -5.1149 \times 10^{-8} & 4.7473 \times 10^{-2}
\end{bmatrix}
$$

Since the etched foil is aligned with the principal axes, the off-diagonal elements are zero (at least within the numerical tolerance).

**Table 1.** Summary of computed performance characteristics by flow direction at Reynolds number 50.

| Flow Direction | f | $StPr^{2/3}$ | $StPr^{2/3}/f$ |
|---|---|---|---|
| x | 2.0620 | 0.32289 | 0.15659 |
| y | 6.2236 | 0.07398 | 0.01189 |
| z | 0.11320 | 0.01952 | 0.17244 |

Should a etched-foil pattern be analyzed that is not aligned with the principal axis, the off-diagonal elements would be non-zero. It should be noted that application of periodic boundary conditions prohibits net flux from entering or leaving the unit cell in all but the principal direction. This may result in a lower effective conductivity than that physically realized.

Longitudinal thermal conductivity measurements of cryocooler regenerators made of ordinary wire screens were conducted by Kuriyama[6]. Measurements for the etched-foil regenerator are similarly possible.

## CONCLUSIONS

A 3-D CFD model for cryocooler regenerators has been developed. The model has been used to analyze thermal and flow characteristics of an etched-foil regenerator. Values of Stanton number and friction factor are presented, as well as the effective conductivity matrix. Further, the ability to visualize the flow and heat transfer patterns provides guidance in improving the geometric configuration for future cryocooler regenerators.

## ACKNOWLEDGMENTS

This research was supported in part by the Ballistic Missile Defense Organization, managed by the United States Air Force, Air Force Materiel Command, Phillips Laboratory, Kirtland AFB, New Mexico.

## REFERENCES

1. Yaron, R., Shokralla, S., Yuan, J., Bradley, P.E., and Radebaugh, R., 1995, "Etched Foil Regenerator", *Proc. Cryogenic Engineering Conf.*, (to be published).

2. Patankar, S.V., 1980, *Numerical Heat Transfer and Fluid Flow*, Hemisphere Publishing Corp., Washington, D.C.

3. *Reference Manual for COMPACT-3D Version 4.0*, 1996, Innovative Research, Inc.

4. Patankar, S.V., Liu, C.H. and Sparrow, E.M., 1977, "Fully Developed Flow and Heat Transfer in Ducts having Streamwise-Periodic Variations in Cross-Sectional Area", *Journal of Heat Transfer*, Vol. 99, pp. 180-186.

5. Kelkar, K.M. and Patankar, S.V., 1992, "Numerical Prediction of Three-Dimensional Periodically Fully Developed Laminar Flow and Heat Transfer in a Channel with Heated Blocks", EEP-Vol. 3, *Computer Aided Design in Electronic Packaging*, Agonafer, D. and Fulon, R.E. (eds.), ASME Winter Annual Meeting, Anaheim, CA.

6. Ramirez, J.A., Timmerhaus, K.D., and Radebaugh, R., 1996, "Application of the Periodic Temperature Variation Technique to the Measurement of Heat Transfer Coefficients in High NTU Materials", *Proc. 9th International Cryocooler Conf.*, (to be published).

7. Kuriyama, T, Kuriyama, F., Lewis, M., and Radebaugh, R., 1996, "Measurement of Heat Conduction Through Stacked Screens", *Proc. 9th International Cryocooler Conference*, (to be published).

# Measurement of Heat Conduction through Stacked Screens

T.Kuriyama[1], F.Kuriyama[2], M.Lewis and R.Radebaugh

National Institute of Standards and Technology
Boulder, Colorado, USA 80303

## ABSTRACT

This paper describes the experimental apparatus for the measurement of heat conduction through stacked screens as well as some experimental results taken with this apparatus. Screens are stacked in a fiberglass-epoxy cylinder, which is 24.4 mm in diameter and 55 mm in length. The cold end of the stacked screens is cooled by a Gifford-McMahon (GM) cryocooler at cryogenic temperature and the hot end is kept at room temperature. Heat conduction flow is estimated from the temperature gradient in a calibrated heat flow sensor mounted between the cold end of the stacked screens and the GM cryocooler cooling stage. The sample used for these measurements consisted of 400 mesh stainless steel screens with an overall porosity of 66.7 %. The experimental results showed that the helium gas between each screen enhanced the heat conduction through the stacked screens by several orders of magnitude compared to that in vacuum. The conduction degradation factor, which is the ratio of actual heat conduction and heat conduction where the regenerator material is assumed to be a bulk, was about 0.1. This factor was almost constant for the temperature range between 40 K and 80 K at the cold end.

## INTRODUCTION

Stacked screens are commonly used as a cryocooler regenerator for in the approximate temperature range between 300 K and 80 K[1]. Because of its large temperature gradient, heat conduction through the stacked screens is a significant regenerator loss. Some experimental studies were reported in which an apparent thermal conductivity for stacked screens was estimated from electrical conductivity[2,3]. But in an actual regenerator, a working fluid of helium gas can transport a large fraction of the heat between screens. As far as we know, only Schumann and Voss investigated experimentally the thermal conductivity of packed beds without fluid flow[4]. Both experimental and analytical works for heat conduction loss, however, are not available for practical

---

1)Guest researcher from Toshiba Co., Japan
2)Guest researcher from EBARA Research Co., Japan

Cryocoolers 9, Edited by R.G. Ross, Jr.
Plenum Press, New York, 1997

use in the design of a cryocooler regenerator. The purpose of this study is to directly measure the heat conduction loss from room temperature to cryogenic temperature through stacked screens.

## EXPERIMENTAL APPARATUS AND PROCEDURE

Figure 1 shows the experimental apparatus used for this study. The apparatus consists mainly of a test section, a two-stage Gifford-McMahon (GM) cryocooler, a heat flow sensor and a vacuum vessel (not shown in Fig. 1).

The details of the test section are shown in Fig. 2. Two identical regenerators are used in this apparatus. Regenerator cylinders are made of fiberglass-epoxy, with an inner diameter of 24.4 mm and a length of 55 mm. The wall thickness of each cylinder is 1 mm. Heat conduction flow through the cylinder wall of a single regenerator from room temperature to 80 K is estimated to be 0.21 W from published thermal conductivity data[5]. Stainless steel screen of # 400 mesh and 0.025 mm wire diameter, was used in this study. The screens are stacked in the regenerator cylinder. The height of the stacked screens is approximately 45 mm. Helium gas lines are connected to the regenerators to change the filling pressure in the regenerators. The pressure can be varied from vacuum condition to 1.0 MPa. Multi-layer-insulation was wrapped around the test section to reduce radiation heat loss.

The cold ends of both regenerators are connected to a cold plate, which is cooled by the GM cryocooler via the heat flow sensor. The hot ends of both regenerators are capped by piston-shaped water jackets. Flowing water maintains the hot end temperature at room temperature. A bellows is attached to the lower water jacket. By changing the filling pressure of the helium in the bellows, the lower water-cooled piston can be moved. The cold plate and two regenerators are free to move with respect to the water jackets, so the force exerted by the bellows is applied equally to the two regenerator screen stacks. The stroke of the water jacket is about 5 mm. Thus, the porosity can be varied about 3 % by this mechanism.

The heat flow sensor is mounted between the cold plate and the first stage of the GM cryocooler. A flexible thermal link between the cold plate and the heat flow sensor allows for movement of the cold plate when the bellows pressure is changed. The heat flow sensor consists of a copper bar and two silicon diode thermometers. The copper bar is made of OFHC with a 72 mm$^2$ cross-sectional area and 135 mm length. The distance between two thermometers is 91.3 mm. The relationship between heat flow through copper bar and temperature difference of those two thermometers is calibrated before experiments.

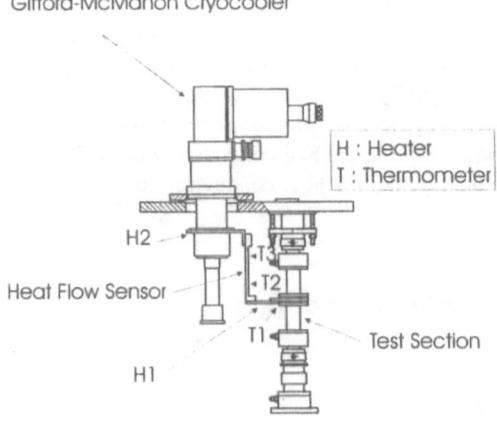

**Figure 1.**   Schematic view of experimental apparatus.

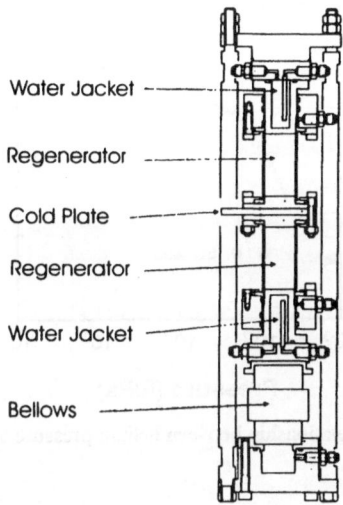

**Figure 2.** Schematic view of test section.

The experimental procedure is as follows. After pumping the vacuum vessel, the two-stage GM cryocooler is turned on. Both the cold plate and the heat flow sensor are cooled by the first stage of the GM cryocooler. The cold plate temperature is kept at a constant temperature by a temperature controller using a silicon diode thermometer at the cold plate and a electric heater mounted on the GM cryocooler first stage. The cold plate temperature can be varied from 40 K to 80 K and the temperature stability is within ±0.05 K. Once the cold plate temperature is set and the temperature difference at the heat flow sensor is measured, an additional heat load is supplied to a heater mounted on the cold plate and also the temperature difference at the copper bar is measured. The heat flow through the heat flow sensor is calculated by the heat loads, the temperature differences and a calibration curve obtained previously. The calculated heat flow here includes the heat conduction through both stacked screens, the two fiberglass-epoxy cylinders, and other heat losses, such as radiation loss and heat conduction loss through instrument wires. In a separate experimental run the heat flow through the regenerators without stacked screens is measured to determine the heat flow through the stacked screens only.

In this study, we have not changed the screen size, screen material and the porosity. The porosity used in this study is 66.7 %. The helium gas pressure inside the regenerator and the cold end temperature of the regenerator were varied.

## EXPERIMENTAL RESULTS AND DISCUSSION

First, the influence of the helium pressure in the regenerator was investigated. The cold plate temperature was 80 K and the hot end temperature was 285 K in this experiment. Before measuring the heat conduction through stacked screens, the heat leak through the two regenerator cylinders without screens was measured. The heat leak without stacked screens was 0.57 W, where the calculated value of the heat conduction through the cylinder wall for two regenerators is 0.42 W.

Figure 3 shows the relationship between helium pressure in the regenerator and heat leak from the hot end to the cold plate. The pressure was varied from $4.7 \times 10^{-3}$ MPa to 1.0 MPa. As shown in Fig.3, the heat leak through the regenerator in vacuum is very small, but the existence of helium gas strongly influences the heat leak. The heat leak is almost constant above about 0.5 MPa. The heat leak, however, decrease rapidly with decreasing pressure at pressures lower than 0.5 MPa.

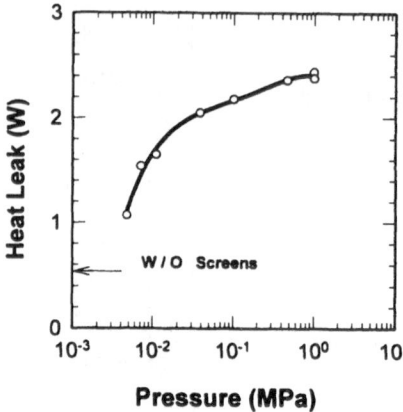

**Figure 3.** Relationship between helium pressure and heat leak.

Figure 4 shows the pressure dependence of mean free path and thermal conductivity of helium gas at 80 K, 200 K and 300 K. Though the thermal conductivity of bulk helium in this pressure range is almost constant, the heat leak does not show a constant value below about 0.5 MPa. In order to understand the test results, we discuss the heat transfer mechanism in the regions of molecular flow, viscous flow and intermediate flow. Generally, gas can be treated as molecular flow where the mean free path (ℓ) is larger than 10 times the distance (d) between the two solid plates. On the other hand, where d is larger than 100 ℓ, the gas can be treated as viscous flow. In the viscous flow region, heat transfer between two solid plates is proportional to the thermal conductivity and is constant for an ideal gas. In the molecular flow region, the heat transfer is proportional to the gas pressure. Figure 3 shows that the deviation from viscous conduction occurs at pressure below about 0.5 MPa. The mean free path for helium in the regenerator varies from 0.010 to 0.039 μm at 0.5 MPa. Therefore, a representative distance between each screen is estimated to be about 1 to 4 μm, or 100 times the helium mean free path. Since the wire diameter of the screen is 25 μm, helium which exists very close to the wire is considered to play the dominant role in transporting heat between the screen layers.

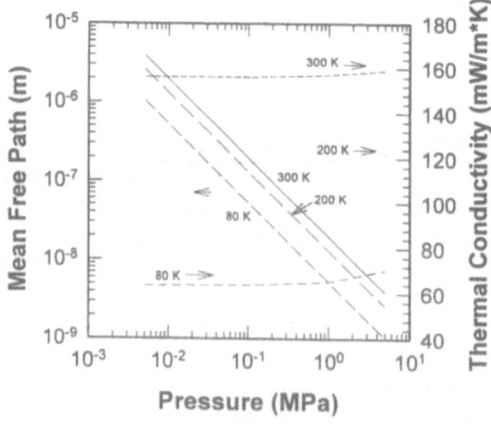

**Figure 4.** Mean free path and thermal conductivity of helium.

To estimate the heat leak through stacked screens generally, a conduction degradation factor f is applied. The factor f is determined as follows:

$$Q = f \times A/L \times (1-ng) \times \int_{TL}^{TH} k\, dT \tag{1}$$

where
Q    : Heat leak through stacked screens
f    :Conductivity degradation factor
A    :Total cross-sectional area of screens
L    : Length of staked screens
ng   :Porosity of stacked screens
TH   :Hot end temperature of regenerator
TL   :Cold end temperature of regenerator
k    :Thermal conductivity of regenerator material

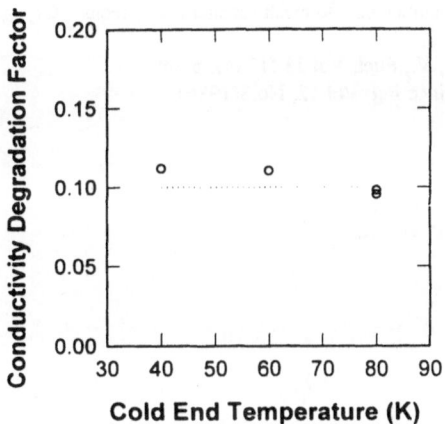

**Figure 5.**   Conductivity degradation factor.

The right hand term except f shows the heat leak if we assume the regenerator material is a bulk solid with no contacts. Then, the factor f shows the ratio of the actual heat conduction to the heat conduction through bulk material.

Figure 5 shows the relationship between the cold end temperature of the regenerator and the conductivity degradation factor. The factor is about 0.1 and almost constant in the cold end temperature range between 40 K to 80K. A previous study[2] showed that the factor f in vacuum condition is less than $1 \times 10^{-4}$. But our results show that helium gas between each screen increases the heat conduction loss by several orders of magnitude. On the other hand, our data agree with results from reference 4 within about 20 % of scattering.

## CONCLUSIONS

The heat conduction loss through stacked screens from room temperature to cryogenic temperature is measured experimentally. The experimental apparatus allows for a change in the screens, the porosity, the cold end temperature and the pressure in the regenerator. These measurements were performed using #400 stainless steel screens with a porosity of 66.7 %.

The experimental results showed that the helium gas between each screen play an important role in transporting the heat. The heat conduction through the stacked screens was enhanced by helium gas by several orders of magnitude compared to that in vacuum. The conduction degradation factor, which is the ratio of the actual heat conduction to the heat conduction where the regenerator material is assumed to be a bulk, was about 0.1. This factor was almost constant for the temperature range between 40 K and 80 K at the cold end.

The experiment will be continued for various porosity, screen material and screen size.

## ACKNOWLEDGMENT

This study is supported in part by EBARA Research Co.,Ltd.

## REFERENCE

1. Walker, G., Cryocoolers, Plenum Press, New York(1983).
2. Organ, A.J. et al., "Connectivity for quantifying regenerator thermal shorting", to be published.
3. Lee, A.C. et al., "Contact resistance of 500 mesh regenerator screens", Cryogenics, Vol. 34, No. 5 (1994), pp.451-456.
4. Schumann, T.E.W. and Voss, V., Fuel, Vol.13 (1934), p.249.
5. Takeno et al., Cryogenic Engineering, Vol.12, No.3(1986).

# A Single Stage Reverse Brayton Cryocooler: Performance and Endurance Tests on the Engineering Model

Francis X. Dolan
Walter L. Swift

Creare Incorporated
Hanover, NH, USA 03755

Lt. B. J. Tomlinson
Alvin Gilbert
Jeffrey Bruning

Phillips Laboratory
Kirtland AFB, New Mexico 87117

## ABSTRACT

A single stage reverse Brayton (SSRB) cycle cryocooler is being developed for applications in space that require high performance, low vibration and reliability. The cooler is designed for a refrigeration load of 5 W at 65 K. It is capable of refrigeration capacities from 0.1 W at 35 K to 10 W at 70 K. The cooler uses miniature turbomachines with self-acting gas bearings to achieve long life without wear or vibration.

An Engineering Model version of the cryocooler has been fabricated and is being tested at the U.S. Air Force Phillips Laboratory. Initial tests involved thermodynamic system performance measurements over a range of loads, load temperatures, power levels and rejection temperatures. Following the performance tests, the system has been set to operate at a constant load and load temperature for an extended period of time. A five year endurance characterization test was initiated in April 1995. As of the end of May, 1996, the cooler has accumulated a total of approximately 9900 hours of operation. Approximately 6900 hours of this total have been in endurance testing at steady conditions.

This paper presents a description of the cooler, a summary of the test program and test results.

## INTRODUCTION

Creare has been developing technology for components in single-stage, reverse-Brayton cycle cryocoolers for several years. These cryocoolers have potential applications for sensor cooling over a wide range of loads and temperatures. Among the key design goals for such a cryocooler are low power and high efficiency since these factors significantly affect the size and weight of the spacecraft power supply and the heat rejection system. Just as important are the reliability of the cryocooler and the requirement that the cooler not introduce vibrations to the sensor or instrument being cooled. Recent successful component developments aimed at these goals include: an efficient inverter and induction motor to drive a high speed compressor operating on self-acting gas bearings; an all-metal high effectiveness heat exchanger; and a miniature turboexpander also operating on gas bearings for long-life and no vibration [1]. The design basis for these components is for a cryocooler with a cooling load of 5 W at 65 K and an operating life time of at least 50,000 hours. Additional developments are underway as well that are aimed at smaller cooling loads and even lower input power [2,3].

The design of an Engineering Model (EM) cryocooler based on a single-stage reverse-Brayton cycle was described at the 8th International Cryocooler Conference two years ago [1]. Performance tests conducted up to that time confirmed that the cryocooler efficiency was very close to the design predictions at an input power of 200 W and a heat rejection temperature of 280 K. For example, the measured specific power of the EM at the design load was 43 W of input power per W of load at 65 K as compared with the design goal of 40 W/W.

The EM cooler was delivered to Phillips Laboratory at Kirtland Air Force Base, New Mexico for endurance testing in June 1994. An initial test of the EM two days after delivery duplicated the cooldown rate and steady-state load results of the Creare tests, and confirmed that the cryocooler was operating normally. For the next several months, until early 1995, a comprehensive test program was conducted that demonstrated the flexibility and durability of the EM over a wide range of loads, load temperatures and input power. Since the completion of these performance "mapping" tests, the EM has been in an endurance test at Phillips Laboratory with a goal of 5 years of total operating time [4].

## DESIGN OF THE ENGINEERING MODEL

The EM cryocooler in its original testing configuration is shown in Figure 1. On top of the bench is a vertical panel that supports the inverter and the motor/compressor on a cold plate. DC power for the inverter is obtained from a power supply in the rack to the left of the bench. The vacuum bell jar at the right end of the bench contains the recuperator, turboexpander, a load heater, particulate filters and connecting tubing. A vacuum pumping system is located under the bench directly beneath the bell jar. To the left of the bench is an instrumentation rack that has: (1)

**Figure 1.** The engineering model test facility.

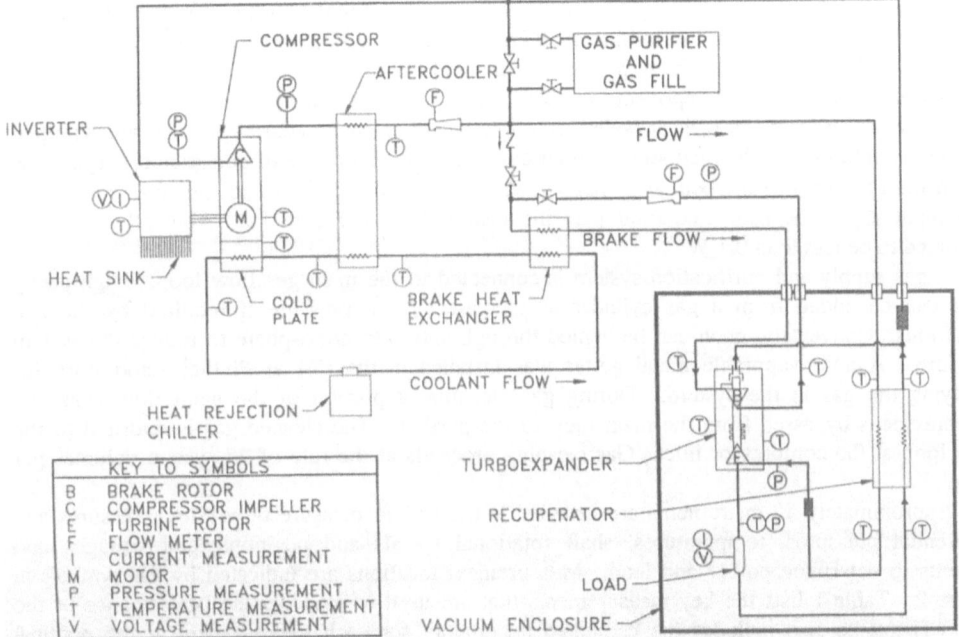

**Figure 2.** Flow and instrumentation schematic for the engineering model cooler test facility.

controls to start and stop the EM and to regulate the input power and the load, (2) a safety system to automatically shut down the EM in the event of accidents that could possibly damage any of the cryocooler components, and (3) a measurement system consisting of signal conditioning and a computerized data acquisition system to measure, display and record data from the EM. Other components which are not visible in the photograph include flow meters, temperature and pressure sensors, heat exchangers, valves, a neon gas supply and gas purification system, and a heat rejection system including an air-cooled chiller and coolant circulating pump.

Figure 2 is a schematic of the EM that shows the major components in the neon flow loop and in the heat rejection system. Neon gas is circulated through the flow loop by means of the compressor driven by a high speed induction motor. The motor/compressor housing conducts heat to a cold plate where it is rejected to the circulating liquid coolant. Remaining heat from the compression process is carried by the gas to an aftercooler where it is also rejected to the circulating coolant. The inverter is located inside an aluminum housing that is mounted adjacent to the compressor on the cold plate panel. Heat from the inverter is conducted by the housing to a finned heat sink where it is rejected at the ambient air temperature.

Following the aftercooler, the neon gas flows through the recuperator where it is cooled by the returning cold gas stream before entering the turboexpander. A particulate filter is placed in the flow line immediately ahead of the turbine inlet. A similar filter, also located inside the bell jar for convenience, is used in the return gas stream ahead of the compressor inlet. These filters are made from a single layer of wire cloth with a maximum opening of approximately 8 microns. Pressure drop through each filter is less than about 0.2% of the system pressure at the design flow rate. After being cooled by expansion through the turbine, the low pressure gas flows through the load which consists of an electric heater that is attached to the outside of the 1.27 cm diameter tubing. The load power is set by manually adjusting the DC voltage supplied to the heater. The maximum heater power is 10 W and regulation is within about ±10 mW. From the load, the gas returns through the recuperator where it pre-cools the incoming high pressure stream and flows back to the compressor inlet.

A separate flow path is established in the brake circuit for the turboexpander. The brake impeller extracts work from the turbine shaft by circulating neon gas through a flow restriction and then to a heat exchanger where the gas is cooled before it is returned to the brake.

All tubing and other components inside the bell jar are shielded to reduce heat input by thermal radiation from the bell jar walls. Multiple layers of aluminized Mylar film are wrapped around the cold tubing, the recuperator and around the cold end of the turboexpander. Typically, the vacuum level inside the bell jar is maintained at or below $1 \times 10^{-6}$ torr to minimize convective heat input [5]. The estimated total parasitic heat leak to the cold load from all sources is estimated to be less than 0.1 W.

A gas supply and purification system is connected to the main gas flow loop. High purity neon can be added from a gas cylinder to increase system pressure if required by the test conditions. Conversely, neon can be vented through valves to atmosphere to reduce the system pressure. A non-evaporable metal getter was installed in the EM at Phillips Laboratory for purifying the gas in the system. During gas cleaning, a portion of the neon flow from the compressor is bypassed from the main loop to the purifier. The cleaned gas is returned to the flow loop at the compressor inlet. Gas cleaning proceeds at the rate of 25 system volumes per hour.

Approximately 50 instruments are installed in the EM to measure flow rates, pressures and differential pressures, temperatures, shaft rotational speeds and positions, and voltages and currents to determine power and load. Measurement locations are indicated by the symbols in Figure 2. Table 1 lists the key measurements that are used to evaluate the performance of the EM. The table also includes the estimated uncertainty for each measurement at the nominal design point condition. Temperature probes (both platinum resistance sensors and thermocouples) are mounted on the outer surfaces of the flow tubes rather than inside the tubes to reduce the risk of neon gas leaks from the system. Sensing elements and lead wires are in close contact with the tube walls to minimize the thermal resistance between the sensor and the gas stream. The design goal for the sensors that measure the turbine inlet, turbine exit and load temperatures is a total installation error of less than 10 mK between the gas temperature and the sensor.

**Table 1.** Key Measurements in the Engineering Model

| MEASUREMENT | TYPE OF INSTRUMENT | RANGE OR NOMINAL VALUE | ESTIMATED UNCERTAINTY |
|---|---|---|---|
| System Pressures | pressure transducer | 0 to 207 kPa<br>0 to 345 kPa | ±0.80 kPa<br>±1.34 kPa |
| Load Temperature | platinum resistance sensor | 35 to 325 K | ±0.05 K |
| Heat Rejection Temperature | thermocouple at compressor inlet | 260 to 350 K | ±1.5 K |
| Compressor Flow | venturi meter and pressure transducer | 1.5 g/s | ±0.75 % |
| Brake Flow | venturi meter and pressure transducer | 0.15 g/s | ±2.2 % |
| Rotational Speed | capacitance probe | 3000 to 10,000 rps | ±1 % |
| Inverter Power | current shunt and meters | 18 amperes<br>12 volts | ±0.30 % |
| Load Power | voltage divider and current shunt | 0.28 amperes<br>18 volts | ±0.32 % |

One of the key indicators of EM performance is the calculated overall cycle efficiency which is defined as the load-to-power ratio normalized by the Carnot efficiency (also see [1]):

$$\eta_{cycle} = \frac{Q_L/P_E}{T_L/(T_R - T_L)} \times 100\% \qquad (1)$$

where:  $T_L$ is the cold end temperature (K)
$T_R$ is the heat rejection temperature (K)
$Q_L$ is the electric power supplied to the cold end heater (W)
$P_E$ is the rms electric power supplied the inverter (W)

Based on the measurement uncertainties listed in Table 1, the uncertainty in the calculated cycle efficiency is estimated to be ±0.8% of the efficiency at the nominal 5 W/65 K operating conditions.

## PERFORMANCE CHARACTERIZATION RESULTS

Performance characterization tests on the EM were carried out in two time periods. The first series of developmental performance tests was conducted at Creare during March and April, 1994. The tests consisted of eight steady-state conditions with loads ranging from 0 to 7.5 W, cold end temperatures between 43 K and 65 K, and input power from 167 W to 273 W [1]. The results demonstrated that the EM met most design point goals, with the exception of specific power which was slightly higher than the target 40 W/W. These initial performance results served as a baseline for comparisons in future testing both at Creare and Phillips Laboratory. Prior to shipment to Phillips Laboratory, Creare engineers operated the EM approximately 110 hours in various shakedown and burn-in tests, including nearly 100 start/stop cycles.

After installing the EM at Phillips Laboratory, several checkout tests were conducted to verify the system's normal operation after delivery. Tests included duplicating operational conditions conducted previously at Creare, such as compressor speed and load settings. However, the heat rejection temperature in the Phillips Laboratory test was 13°C higher and the system pressure was 5% greater than Creare tests (following adjustments after gas cleaning). At steady-state, the overall cycle efficiency ($\eta_{cycle}$) at a cold end temperature of 65 K was nearly identical in the Creare and Phillips Laboratory initial tests: 7.7% at Creare while 7.6% in the Phillips Laboratory tests.

The second major test period began in November, 1994 when Phillips Laboratory initiated its Performance Characterization Program. This program was comprised of an extensive battery of both steady-state and transient performance tests having the following primary objectives:
1. identify the best cooler operating conditions (nominal);
2. establish an in-depth database on the cooler's performance under various nominal and off-nominal operating conditions for reference and comparisons during future testing;
3. provide system developers with information on the cooler's inherent strengths and limitations under simulated environmental conditions.

Approximately 70 steady-state and transient performance tests were conducted over a five month period from November, 1994 to April, 1995. This array of tests included numerous repeat performance demonstrations of the design point nominal operating conditions established for the cooler. Tests were also conducted numerous times to demonstrate repeatability of the results. Table 2 lists the parameters varied and their variation ranges for Phillips Laboratory's performance characterization program. Although not all possible parameter variations were investigated, enough were to establish performance trends and inherent operating capabilities. Figure 3 displays the load-lines that were developed from these performance characterization tests. The plot shows the range of cold end temperatures that can be achieved in the EM as a function of the cold load, inverter power and heat rejection temperature. The load-lines can be reduced to the cycle efficiency as a figure-of-merit according to the definition of Eq. 1. Figures 4 and 5 illustrate some of the performance results in the form of cycle efficiency comparisons.

**Table 2.** Parameter Variations for the EM Performance Tests

| TEST PARAMETER | PARAMETER RANGES | NUMBER OF VALUES |
|---|---|---|
| System Pressure | 1.1 to 1.4 atm | 3 |
| Turbine Speed | 8000 to 9000 rev/s | 3 |
| Cold End Heater Power | 0 to 10 W | 4 |
| Inverter Power | 190 to 270 W | 3 |
| Rejection Temperature | 280 to 310 K | 4 |

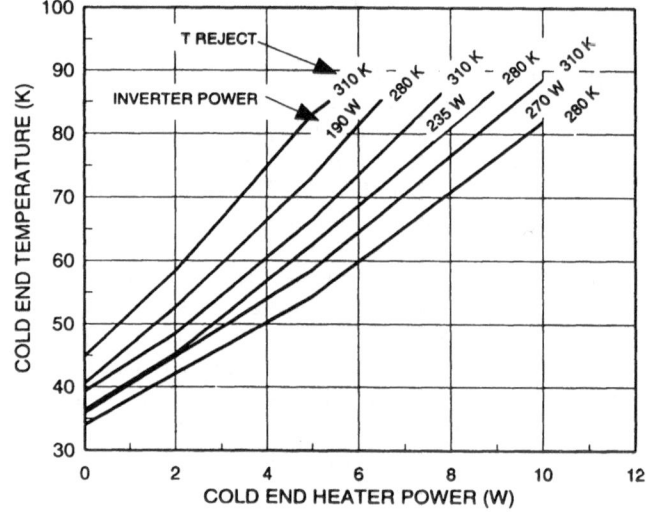

**Figure 3.** EM cooler load-lines.

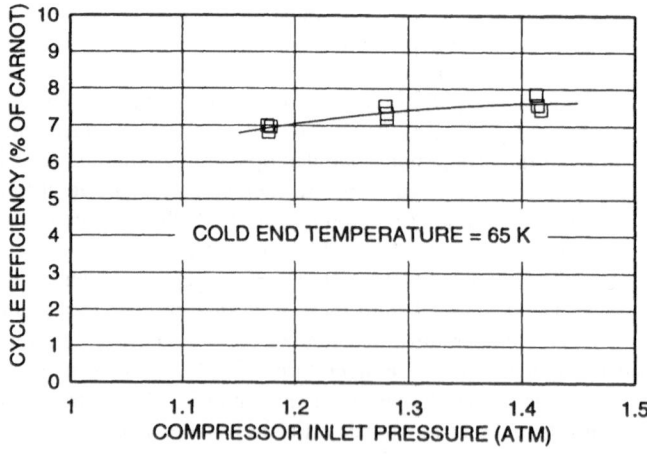

**Figure 4.** Cycle efficiency improves with an increase in the operating pressure level.

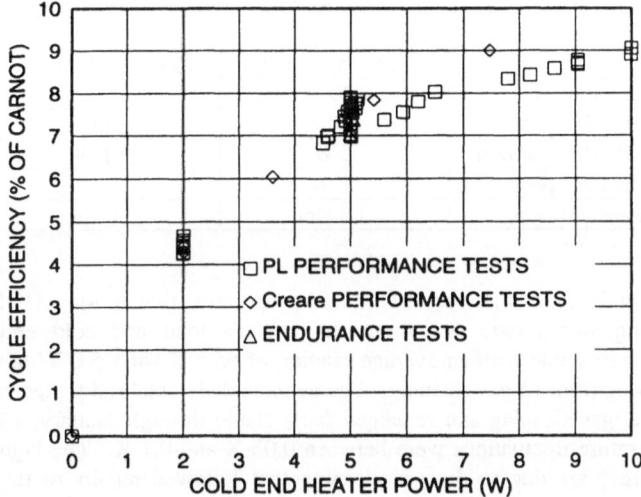

**Figure 5.** Performance comparisons during various periods of EM testing cycle.

In calculating cycle efficiency, the heat rejection temperature, $T_R$ is taken as the temperature of the gas stream at the inlet to the compressor. Because of the amount of tubing at the warm end of the system that is exposed to ambient air, the compressor inlet temperature more accurately reflects the cycle rejection temperature than does the temperature of the liquid coolant. The difference between cycle performance referenced to the coolant and the compressor inlet temperatures is about ±3% of the cycle efficiency.

As shown by the data plotted in Figure 4, EM efficiency is affected by the system pressure. In one series of tests, the pressure was systematically varied by addition or removal of purified neon gas while maintaining constant load temperature and inverter power. In the EM, which has a fixed gas volume, increasing the mass of gas raises the system pressure as exemplified here by the compressor inlet pressure. The increased efficiency results from improvements in turbomachinery efficiency at higher density because of better flow matching characteristics at these conditions.

Figure 5 summarizes the results from all the performance tests (including the tests at Creare prior to delivery and a few points from the endurance testing at Phillips Laboratory in 1995) as a single plot of cycle efficiency as a function of cold end heater power. This figure incorporates the multiple effects of variable heat rejection temperature, inverter input power and system pressure. The combined effect of these parameters on cycle efficiency is illustrated by the range of results for the tests at 2 W and 5 W of load.

## ENDURANCE TESTING

The cryocooler is currently undergoing endurance testing as part of its final phase of testing at Phillips Laboratory. The endurance test was started on April 7, 1995 and is scheduled to run for at least five years to demonstrate the cooler's long-life potential. The objective of the endurance test is to surface any wearout, drift, or fatigue-type failure mechanisms inherent in cryocooler's design. Since the start of the test, the EM accumulated over 6900 hours of steady-state operation through the end of May, 1996. The EM is being operated in a near-continuous, steady-state mode at its nominal test conditions as listed in Table 3. After reaching steady-state, operation is essentially unattended except for routine data recording and database archival activities.

**Table 3.** EM Operating Conditions for Endurance Testing

| PARAMETER | NOMINAL VALUE | TYPICAL RANGE |
|---|---|---|
| Inverter Power | 240 W | ±5 W |
| Rejection Temperature | 300 K | ±2 K |
| Cold End Heater Power | 5 W | ±0.01 W |
| Cold End Temperature | 65 K | ±2 K |
| System Pressure | 1.3 atm | ±0.15 atm |

The plot of cold end temperature versus cumulative operating time to May 31, 1996 is shown in Figure 6. During steady-state operation, the cooler's load and cold end temperature performance remain very stable with an average reading of 65.5 K for a 5.0 W heat load during the first phase of testing prior to gas cleaning. The average daily cold end temperature was 66.5 K following the initial gas cleaning and remained fairly stable through January, 1996. Average daily cold end temperature fluctuations were between 0.05 K and 0.1 K. The high spikes in the chart shown in Figure 6 are due to the gas cleaning that followed repairs to the test station's power system. Additional temperature fluctuations shown were due to intermittent unexpected facility power interruptions. Although the cooler continued to run normally during power interruptions, the shop air supply depleted, affecting the cooler's vacuum system operation with impacts on cold end temperature.

Figure 7 shows the EM cycle efficiency over the full duration of the operating period from April 7, 1995 through May, 1996. The efficiency remains essentially constant in the period from April until July and again in the period from August through December, although at a slightly lower value than earlier. In January and April through May, the cycle efficiency was essentially the same as in the initial period. These shifts in cycle efficiency may be due to differences in the extent to which the gas was cleaned in the period before steady-state testing was resumed following each shutdown.

Other periods of anomalous operation are marked by the data points of significantly reduced cycle efficiency in Figure 7. These brief excursions were due to the intermittent facility power interruptions mentioned earlier.

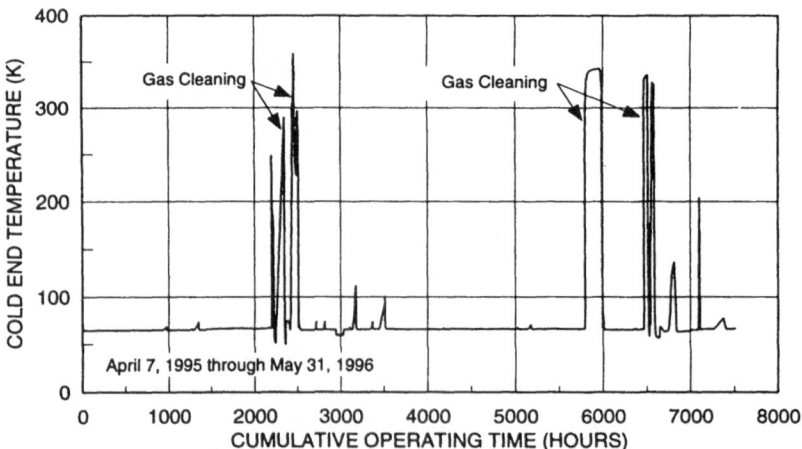

**Figure 6.** EM cooler endurance test time/temperature performance.

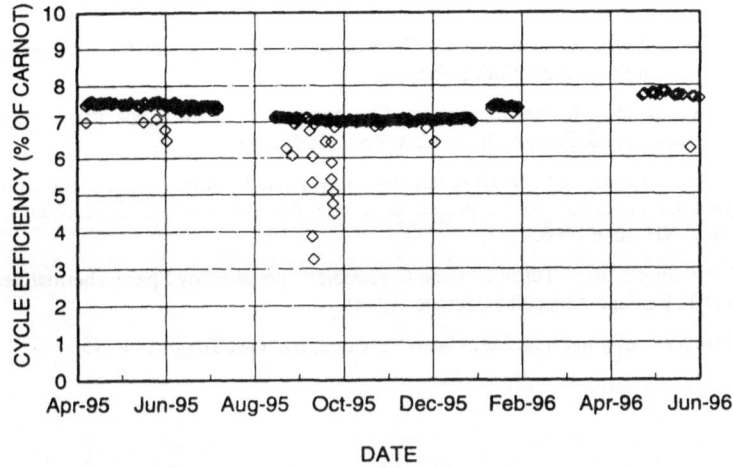

**Figure 7.** Cycle efficiency during the EM endurance test.

## CONCLUSIONS

The EM has proven to be rugged and reliable. It survived a cross-country trip without problems and within a few hours of its arrival at Phillips Laboratory it had been assembled and all systems were checked out. Within two days the EM had been cooled down to the design temperature and the previously measured performance was duplicated. The EM has been operated over a wide range of test conditions that have exceeded the design load by a factor of 2 and the input power by nearly 50%, without incident. Numerous planned and unplanned shutdowns have occurred during endurance testing, and in each case the system successfully restarted.

The performance data acquired in the EM closely matches the predicted cycle performance and trends. Within the scatter caused by variations in test conditions, there is no detectable difference in the EM performance between the Creare tests and the Phillips Laboratory tests.

For more than 6900 hours thus far of steady-state endurance testing and more than 9000 total hours of operation, the EM performance has not changed within measurement uncertainty. Overall cycle efficiency is affected by system pressure and by the actual heat rejection temperature which are both somewhat variable over the long term. System pressure is adjusted after each gas cleaning cycle, but not always to the same value. Gas cleaning prior to cooldown can also affect the cycle performance. However, when a standard cleaning procedure is followed the measured efficiency is repeatable over the long-term when the other variables are taken into account.

## ACKNOWLEDGMENTS

The work reported in this paper has been supported by BMDO, the Air Force Phillips Laboratory and NASA/GSFC. We also wish to acknowledge the significant management and engineering contributions made by Captain Christina L. Cain (U.S.A.F., retired) and the technical support provided by Mike Martin (Orion International Technologies) to the testing program for the SSRB Engineering Model cryocooler at Phillips Laboratory.

## REFERENCES

1. Swift, W.L., "Single Stage Reverse Brayton Cryocooler: Performance of the Engineering Model", *Cryocoolers 8*, Plenum Press, New York (1995), pp.

2. McCormick, J.A., et al., "Miniaturization of Components for Low Capacity Reverse Brayton Cryocoolers", *Cryocoolers 8*, Plenum Press, New York (1995), pp.

3. McCormick, J.A., "Progress on the Development of Miniature Turbomachines for Low Capacity Reverse Brayton Cryocoolers", to be presented at the 9th International Cryocooler Conference, Waterville Valley, NH, June 1996.

4. "Test Plan, Creare Single-Stage Turbo-Brayton Cryocooler", prepared by Space Thermal Technologies Branch, PL/VTPT, Kirtland AFB, NM (October 1994).

5. Scott, R.B., *Cryogenic Engineering*, Met-Chem Research, Inc. (1963), pp. 144-145.

# Progress on the Development of Miniature Turbomachines for Low-Capacity Reverse-Brayton Cryocoolers

**J.A. McCormick, W.L. Swift and H. Sixsmith**

Creare Inc.
Hanover, NH, USA 03755

## ABSTRACT

Developments are being carried out to extend the practical use of turbomachine-based reverse Brayton cycle cryocooler technology to lower capacity applications. These developments focus primarily on cooling loads in the range of 0.35 W to 2 W at temperatures from about 35 K to 70 K in a single stage cycle with neon. The technology is readily adaptable to multistage loads with either neon or helium as the working fluid. A multistage helium cycle could provide cooling at loads of approximately 100 mW at temperatures as low as 10 K. The goal is to produce systems incorporating vibration-free turbomachines with performance characteristics comparable to those that have been demonstrated for larger cycles - i.e., the 5 W 65 K single stage reverse Brayton cycle cryocooler. This paper reports recent progress on this effort, which should result in highly reliable turbomachine based cryocoolers with input powers less than 150 W.

The critical components for a low capacity reverse Brayton cycle cryocooler include the expansion turbine, the compressor and the recuperator. Advanced designs presently under development for the two turbomachines use high speed gas bearings and high energy permanent magnets for electromechanical energy conversion. An alternator operating at cryogenic temperatures converts the turbine power from the expansion of the gas to electric power that is dissipated through a resistive load at room temperature. The miniature centrifugal compressor is driven by a high efficiency three-phase AC motor using features that are similar to those of the turboalternator. Recent developments on an advanced parallel plate recuperator concept promise a substantial weight and size reduction in comparison to present slotted plate heat exchangers. The paper describes test results for these components and presents performance projections for several low capacity cryocooler cycles.

## INTRODUCTION

There are a number of critical requirements which drive the design of mechanical coolers for long term space applications. The more important ones are:
- high reliability,
- long life,
- high thermodynamic efficiency,

- very low vibration,
- flexibility in packaging and integration,
- low weight, and
- simplicity and robustness.

Among the candidate systems currently being pursued, the turbomachine-based reverse Brayton cryocooler (TRBC) offers the promise of meeting each of these requirements over a broad range of temperatures and loads. For cooling loads in excess of 5 - 10 W at temperatures above 4 K, TRBCs that achieve overall performance levels of 10% of Carnot efficiency or more can be built with today's technology. Existing gas bearing technology provides for no-wear, vibration-free operation without the need for auxiliary counter balancing. Simplicity is inherent in these systems in that there is only one moving part in each machine, the rotating shaft, which rotates continuously at constant speed. And because of the high degree of inertial balancing that is required for successful high speed operation, there is no detectable mass vibration. Depending on the particular hardware and performance requirements, the required control system can be very simple, increasing the reliability of the primary life limiting element in these systems, the electronics.

The high power density of turbomachines makes them very attractive in terms of system weight and packaging, and an important attribute of this system is that it is component based. Each of the components in the system is connected within a fluid loop. This modularization of the system allows for development and optimization at the component level, flexibility in packaging, and integration of individual components or subsystems with other reverse Brayton components or with cooling stages consisting of totally different systems.

The problem to date has been that for low cooling loads (below about 5 W), the state of current technology has limited the thermodynamic cycle performance levels which can be achieved with TRBCs. This is a consequence of one general factor: *The relative effect of parasitic losses in turbomachines increases as cycle capacity and size are reduced.* In order to improve the performance of TRBCs so input power meets acceptable levels for very low cooling capacities, parasitic losses at the component level must be reduced.

The single stage TRBC consists of four major components:

- an electronic controller and converter drawing unregulated dc power from the spacecraft bus and converting it to a controlled power output to drive a compressor,
- a compressor that consists of a drive motor and centrifugal compressor to pressurize and circulate the flow through the cycle,
- a counterflow heat exchanger, or recuperator that precools the high pressure gas flowing to the cold end of the cooler, and
- a turbine that provides the net refrigeration for the cycle.

The performance of each component can be characterized by an efficiency or effectiveness that is a function of the component's size, overall design configuration, materials used and the operating parameters such as pressure, temperature and flow rate. The performance of the compressor and its electrical drive system is characterized by the power train efficiency, which is the ratio of the isentropic compression power to the electrical power drawn from the spacecraft bus. This is the product of individual efficiencies for the compressor, motor and electrical drive. The performance of the recuperator is characterized by its thermal effectiveness, which is the ratio of the actual to the maximum possible heat transfer between the high pressure and low pressure streams. The performance of the turbine is determined by its net efficiency, which is the ratio of its actual cooling power to the power available in an isentropic expansion. The power train efficiency, recuperator effectiveness and turbine efficiency, along with the temperature and pressure ratios of the cycle and gas properties, determine the input power requirement for a given cooling load.

The most useful benchmark for the desired low power system is the single stage TRBC presently under life test at the Air Force Phillips Laboratory.[i][ii] This cooler provides 5 W of cooling at 65 K with approximately 200 W of input power at a 280 K heat rejection temperature.

The recuperator is a slotted plate heat exchanger; the compressor uses a solid rotor induction motor, and the turbine consists of a turboexpander, in which the power transferred to the shaft by the expanding gas is absorbed by an aerodynamic brake at the heat rejection temperature. Because of the nature of the losses in the induction motor and the turboexpander, it is not possible to scale the designs down to, for example, a cooling load of 2 W at 65 K with an input power of 100 W. The electrical efficiency of the induction motor decreases as it is scaled down in size and power, resulting in unacceptable decrease in power train efficiency. As the turboexpander is scaled down in size and power, the conductive heat leak from the brake to the turbine, which is approximately 2.5 W for the 5 W cooler, could decrease only slightly, actually becoming larger as a fraction of the turbine power. This would unacceptably decrease the net turbine efficiency.

To address the reduced efficiency in the turboexpander in low power coolers, a turboalternator is being developed in which the turbine power is absorbed by a miniature electrical generator. Current from the generator flows through leads of low thermal conductance to a resistive load at the heat rejection temperature. The entire turboalternator assembly operates at cryogenic temperature. This eliminates the conductive heat leak of the turboexpander, leaving a smaller parasitic loss comprised of bearing and rotor friction and electrical loss. The parasitic loss is on the order of 1 W, versus 2.5 W for the turboexpander.

The rotating assembly in the turboalternator is comprised of a radial inflow turbine rotor and a permanent magnet generator rotor. The permanent magnet is a solid cylinder of high energy material, magnetized across a diameter. It is contained in a hollow titanium shaft which is machined in one piece with the turbine rotor. A toothless stator provides high power density and avoids radial magnetic forces on the gas bearings. Figure 1 shows a conceptual layout of this machine configuration. The radial turbine runs at less than one-half the rotational speed of the turboexpander, which improves the stability of the gas bearings at low temperature and reduces their drag.

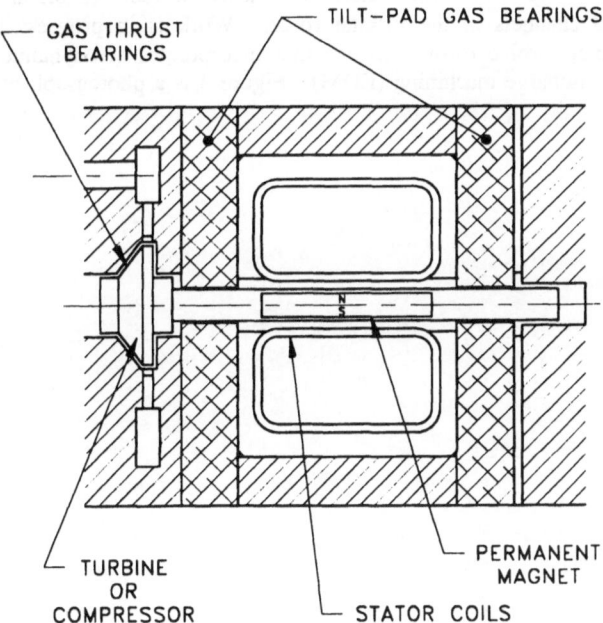

**Figure 1.** Mechanical arrangement of turboalternator or PM Motor compressor.

If the stator, in a machine mechanically identical to the turboalternator but somewhat larger, is driven as a synchronous motor, the machine becomes a centrifugal compressor well matched to the requirements of the desired low capacity TRBC. Importantly, the permanent magnet synchronous motor has a dramatically higher electrical efficiency than the induction motor at the low power levels of interest. It achieves higher efficiency by avoiding the induced rotor currents and iron teeth that are responsible for major losses in the induction motor. It also provides higher power density than the induction motor, allowing the shaft diameter to decrease, and, consequently, bearing drag loss to decrease, further increasing the power train efficiency.

Although the slotted plate heat exchanger used in the 5 W system provides equally high effectiveness at lower power levels (approximately 0.99), its large size and weight are a disadvantage. Work is underway on a novel parallel plate configuration that will be substantially smaller and lighter for the same effectiveness. Preliminary studies and subscale tests on the parallel plate concept indicate that the weight may be reduced by up to a factor of four in comparison to the slotted plate heat exchanger. An added attraction of the parallel plate concept is its potential scaleability to systems of even lower capacity.

The following sections of this paper describe current progress on a low capacity TRBC system that uses a turboalternator and permanent magnet motor compressor (PMMC). Design descriptions and preliminary test results are given for brassboard models of the two turbomachines. Projections are then given for overall cycle performance. The system can use either the present slotted plate heat exchanger, or, when it becomes available, the advanced parallel plate heat exchanger.

## TURBOALTERNATOR DEVELOPMENT

A brassboard turboalternator assembly was recently tested in neon in a true reverse Brayton cycle at temperatures down to 52 K. The rotor assembly has a turbine disk diameter of 7.62 mm (0.300 in.) and a shaft diameter of 3.56 mm (0.140 in.), and operates at speeds from 2000 to 5000 rev/s. Figure 2 is a photograph of the rotor assembly that was tested. Visible in the photograph are rudimentary drilled channels in the turbine rotor. Work is in progress to replace this preliminary low efficiency turbine rotor with one that incorporates aerodynamic flow channels produced by electrical discharge machining (EDM). Figure 3 is a photograph of the stator and the cylindrical magnet.

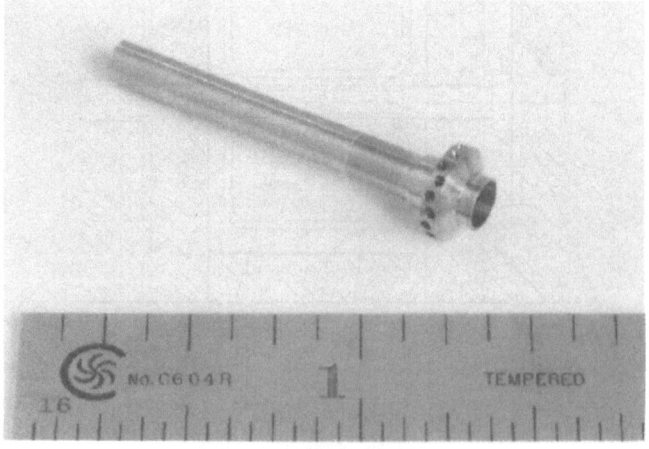

**Figure 2.** Shaft with drilled rotor.

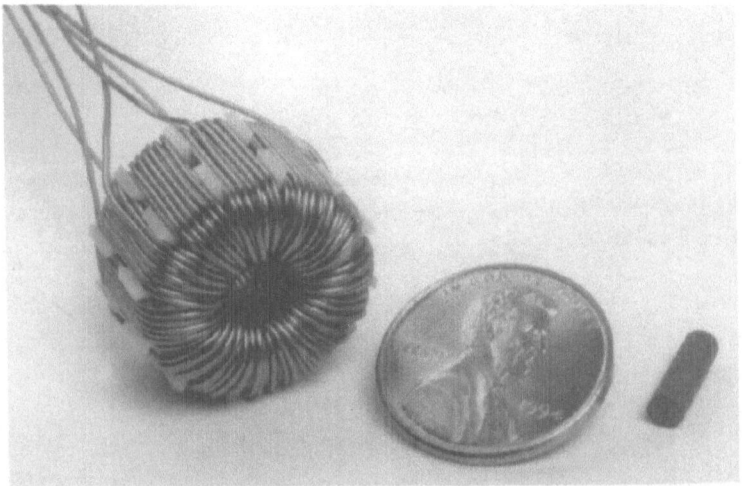

**Figure 3.** Stator and magnet.

Figure 4 is a photograph of the turboalternator installation at the cold end of the cryogenic facility  This was taken with the vacuum vessel removed and prior to wrapping the cold end with MLI.  The large vertical cylinder is the slotted plate recuperator.  The turboalternator is oriented horizontally next to the recuperator.  Three instrument flanges containing platinum resistance thermometers (PRTs) are visible in the tubing between the turboalternator and the recuperator. These measure the turbine inlet, turbine exit and load temperatures.  Electrical leads penetrating the turboalternator housing are three generator power leads near the aft end and two shaft sensors on the top.  The power leads run to a three phase load rheostat located outside the vacuum vessel.

Performance tests consisted of measurements of temperatures, flows and pressures over a range of speeds and  flow rates.  Speed was fixed by the setting of the load rheostat, and mass flow rate was fixed by the system pressure.  Both quantities were easily adjustable in the facility. Net turbine power $W_T$ was calculated from the voltage $V_L$ and current $I_L$ measured at the load rheostat according to:

$$W_T = 3/8 \ V_L \ I_L$$

where $V_L$ and $I_L$ are peak to peak readings.  Net efficiency of the turboalternator is the ratio of $W_T$ to the isentropic power $W_S$ which is calculated from measurements of the inlet temperature $T_{IN}$, mass flow rate m, and pressure ratio Pr, according to:

$$W_S = m \ C_P \ T_{IN} \ (1-Pr^{-(\gamma-1)/\gamma})$$

where $C_P$ is the heat capacity at constant pressure and $\gamma$ is the ratio of constant pressure to constant volume heat capacities.  Test results expressed as power and efficiency over a range of speed and turbine inlet pressure are shown in Figure 5 at a turbine inlet temperature of 67 K for two separate stator designs.  The abscissa in the efficiency plot is the velocity ratio $U/C_0$ where U is the turbine tip speed and $C_0$ is the isentropic spouting velocity given by

$$C_0 = (W_S / m)^{1/2}$$

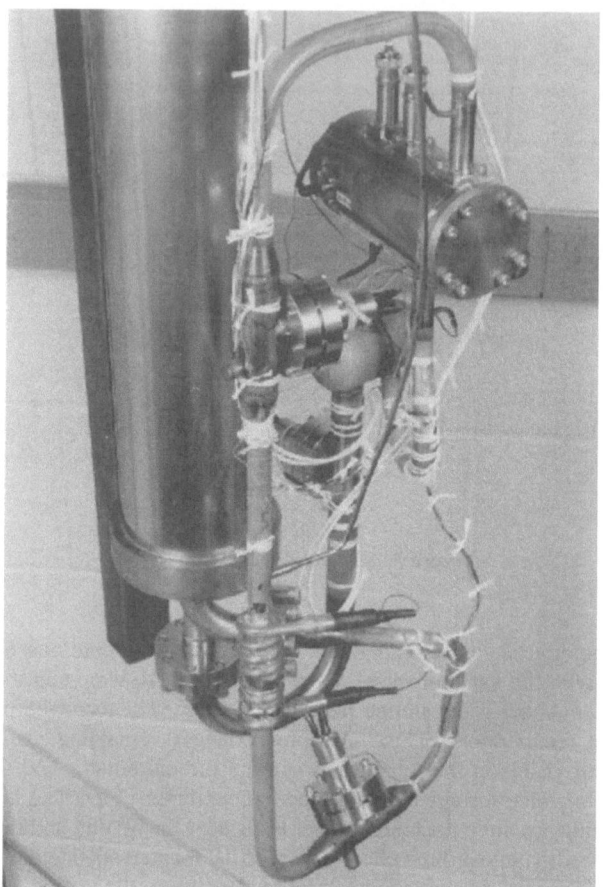

**Figure 4.** Turboalternator in cryogenic test installation.

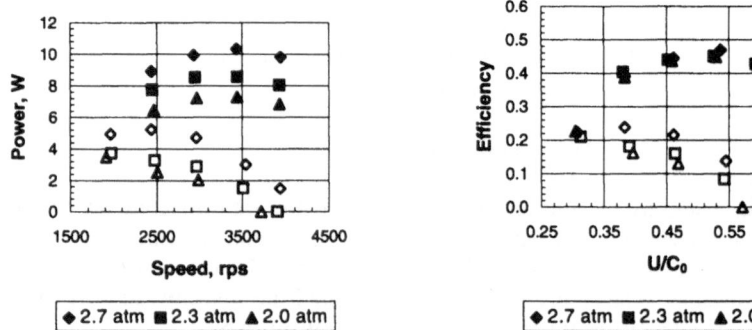

**Figure 5.** Turboalternator performance at 67 K. Open symbols are first generation stator and closed symbols are second generation stator. Power is measured electrical output of generator.

The relatively low power and efficiency shown by the open symbols in the figure reflects an early stator design with high eddy current losses. The 48% efficiency obtained with the second generation stator is comparable to the efficiency of the 5 W turboexpander. In the turboexpander, the turbine has a high specific speed and a relatively high aerodynamic efficiency of 70-75%. A large parasitic loss due to conductive heat leak lowers the net efficiency to around 50%. The turboalternator has a much lower aerodynamic efficiency due its primitive drilled hole rotor, but is showing reasonable net efficiency because the parasitic loss is low ($\approx 1$ W at 67 K). When the aerodynamic rotor becomes available in the near future, the aerodynamic efficiency is expected to increase to about 70-75% with a corresponding increase in net turboalternator efficiency to 55-65%.

## PM MOTOR COMPRESSOR DEVELOPMENT

A brassboard PMMC that is nearly identical mechanically to the turboalternator is presently being developed. The compressor impeller diameter is 12.7 mm (0.5 in.); the shaft diameter is 4.32 mm (0.17 in.) and the design speed is 9000 rev/s. The electrical drive will be the same high frequency inverter that drives the induction motor in the 5 W TRBC.

A benchtop test article for the turboalternator was used to demonstrate the motor principle and evaluate its performance. The test article used pressurized air bearings and contained the smaller rotating assembly of the turboalternator. In its use as a motor driven compressor, the rotor of the test article circulated ambient air. Inlet and exit flow paths to the rotor were made as wide as possible to maximize the air flow and thus maximize the shaft power drawn from the motor. Shaft power at 9000 rev/s was 54 W.

Detailed performance measurements were made at fixed speed and shaft power under varying electrical input. Figure 6 shows a plot of electrical efficiency, given by the ratio of shaft power to the DC power supplied to the inverter, against the DC voltage input. The optimum efficiency of 0.9, occurring at 9 V, corresponds to the condition of minimum AC current, occurring when the AC current is exactly in phase with the magnet's generated voltage. The 0.9 efficiency can be compared with an expected value of less than 0.7 for an induction motor of comparable size. At higher voltage the motor remains in synchronism but higher current decreases the efficiency. The same thing happens at lower voltage, down to approximately 8 V, below which the motor will pull out of synchronism and stall.

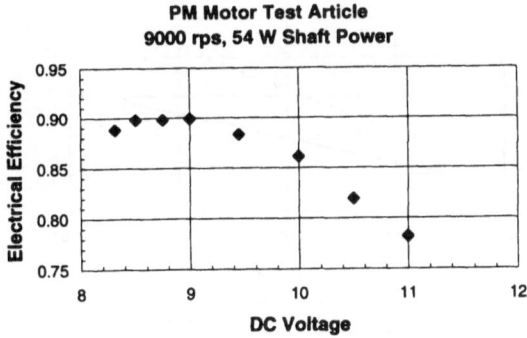

**Figure 6.** Electrical efficiency vs. inverter input voltage for PMMC test article.

The 0.9 electrical efficiency shown by the subscale test article (3.56 mm shaft diameter) accounts for electrical loss in both the motor and the inverter. Most of the loss in the motor is the electrical resistance loss in the stator coil. The more powerful brassboard unit (4.32 mm shaft diameter) can be expected to show approximately the same electrical efficiency, since a small reduction in stator resistance loss relative to the shaft power will be nearly offset by a small increase in the inverter loss relative to the shaft power.

The 4.32 mm brassboard machine will provide a pressure ratio of approximately 1.8 with neon at 9000 rev/s and heat rejection temperature of 280 K. At this speed the loss due to drag in the journal bearings and motor gap is approximately 20 W. A reasonable estimate for compressor efficiency, accounting for losses due to flow friction, leakage and disk friction is 0.6. Typical mass flow rates for neon cycles presently under consideration range from 0.5 to 0.75 g/s. For this range of mass flow rate, with 0.6 compressor efficiency, 20 W bearing drag and 0.9 electrical efficiency, the range of input power is 90 to 130 W. The next section presents a projection of overall TRBC cycle performance over this power range.

## PERFORMANCE OF LOW CAPACITY CYCLES

Advanced TRBC's based on the turboalternator and the PM Motor compressor can operate in single and multistage cycles with neon or helium. Calculations of net refrigeration (load power) have been carried out for various load temperatures and input power levels for a single stage neon cycle, based on the turbine and compressor performance projections discussed above. Built into these calculations is a recuperator effectiveness of 0.99, which is easily attainable with either the present slotted plate heat exchanger, or the advanced parallel plate heat exchanger presently under development. Results in the form of plots of load power vs. load temperature at fixed input power are shown in Figure 7. The condition of 1 W at 35 K for 95 W input corresponds to a net cycle efficiency of 7.4% of Carnot.

This technology is also applicable to very low temperature applications where helium is the working fluid. Calculations and system studies have been performed to define the requirements for several reverse–Brayton configurations that would be capable of providing about 100 mW of refrigeration at about 8 K. For one case, heat rejection is available at 80 K. In this system, the electrical input power to the cycle would be 15 W. The cycle would use a pressure ratio of 1.3 and a mass flow rate of 0.1 g/s. Compression at 80 K would be provided by two compressors of the same size as the present turboalternator. An advanced parallel plate heat exchanger would

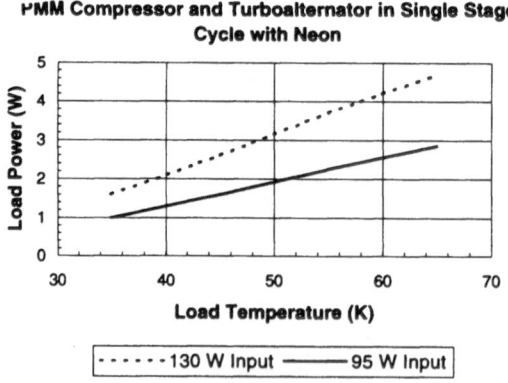

**Figure 7.** Performance of single stage cycles at various load temperatures and input power levels.

have a thermal effectiveness of 0.996, operating between 80 K and 8 K. The turboalternator would be scaled down in size from the present turboalternator by approximately a factor of two. In another version of the system for the same cooling load at 8 K, the heat rejection temperature would be somewhere between 220 K and 260 K. There would also be a parasitic heat load of about 100 mW at 60 K. This would require a second turboalternator at the 60 K load temperature. The pressure ratio, flow rate, 8 K turboalternator and 80-8 K heat exchanger would be the same as in the 80 K heat rejection cycle. The warm turboalternator would be the same size as the present turboalternator. There would be two compressors in series at 220 to 260 K, of the same size as the present PMM compressor. Input power to the cycle would be 100 to 150 W.

## CONCLUSIONS

The advanced miniature turbomachines described in this paper enable the efficient use of low vibration reverse-Brayton cryocoolers for low capacity spaceborne instrument cooling. Experimental results reported for brassboard and benchtop models of a cryogenic turboalternator and a permanent magnet motor compressor demonstrate levels of performance that can be projected to meet the requirements of several advanced spaceborne systems, including single stage cycles with neon for 1 to 5 W of refrigeration at 35 to 65 K at input powers from 75 to 150 W and heat rejection at 260 to 300 K, and one and two stage cycles with helium for 100 mW of refrigeration at 8 K and either 15 W of input power for heat rejection at 80 K or 100 to 150 W of input power for heat rejection at 220 to 260 K.

## ACKNOWLEDGMENTS

We would like to acknowledge support from NASA/GSFC in the development of the cryogenic turboalternator. The development of the parallel plate recuperator is also funded by NASA/GSFC. The development of the permanent magnet motor compressor and the single stage system is supported by BMDO and AF Phillips Laboratory.

---

[i] Swift, W.L.; *Single Stage Reverse Brayton Cryocooler: Performance of the Engineering Model*; Proc. 8th Intl. Cryocooler Conf., Vail, CO, June 28-30, 1994, Plenum Press, New York, NY, 1995.

[ii] Dolan, F.X., et al; *A Single Stage Reverse Brayton Cryocooler: Performance and Endurance Tests on the Engineering Model*; Presented at the 9th Intl. Cryocooler Conference, Waterville Valley, NH, June, 1996.

# Experimental Study of the Isentropic Efficiency of the Pulse Tube Expansion Process

J. Liang, C.Q. Zhang and Y. Zhou

Cryogenic Laboratory, Chinese Academy of Sciences
Beijing 100080, China

## ABSTRACT

The pulse tube is a simple and reliable new refrigeration device. But so far it is used only in regenerative cryocoolers with quite small refrigeration powers. For applications that require larger refrigeration powers, such as gas separation and liquefaction, the flow diagram of the regenerative pulse tube refrigerator may be inadequate. This investigation examines the use of a pulse tube as a Brayton-cycle expander whereby the regenerator is replaced with a recuperative heat exchanger, and a switching valve is installed at the cold end of the pulse tube.

In such a system, the pulse tube—coupled with the switching valve—works as a new kind of expander, and removes heat rather than work. Except for the switching valve, it has no moving parts at low temperature. An experimental apparatus has been set up to study the isentropic efficiency of this pulse tube expansion process. The influence on isentropic efficiency of pressure ratio, frequency, and orifice opening have been investigated. An isentropic efficiency of 42% has been obtained with air as the working fluid. Analysis of the experimental results indicates that higher isotropic efficiencies can be achieved.

## INTRODUCTION

The pulse tube refrigerator is an innovative method for producing cooling. Compared with other mechanical refrigerators, it eliminates the reciprocating piston at low temperature, which is an important source of unreliability; therefore it is simple and reliable. With the significant evolution in recent years, pulse tubes are now capable of replacing the cold piston/cylinders of Stirling and G-M refrigerators without much degradation of performance.

The success of applying pulse tubes in the field of regenerative cryocoolers encourages us to consider the possibility of using pulse tubes for applications that require larger refrigeration powers, such as gas separation and liquefaction. In principle, the pulse tube component can be enlarged, but the flow diagram of the regenerative pulse tube refrigerator may be inadequate for these applications for two reasons. One, some liquid may appear in the matrix of the regenerator and increase flow resistance, and two, the large void volume in the regenerator is alternatively pressurized and depressurized, which reduces the pressure oscillation amplitude in the pulse tube, and consequently reduces the refrigeration power. In order to avoid these problems, it is proposed to replace the regenerator with a recuperative heat exchanger, and to install a switching valve at the cold end of the pulse tube. Such a system is a Brayton refrigerator with a pulse tube expander based on the low temperature switching valve[1].

To understand the feasibility of this new pulse-tube-based Brayton expander, it is necessary to experimentally investigate the isentropic efficiency of the pulse tube expansion process. Although the refrigeration efficiency of a pulse tube refrigerator, which is defined as the ratio of cooling power to power input, has been studied[2], so far no experimental results on the isentropic efficiency of the pulse tube expansion process have been reported. In this paper, the results of measurements of the isentropic efficiency are presented and analyzed.

## PULSE TUBE EXPANDER

In the Brayton pulse tube refrigerator with low temperature switching valve, the pulse tube (including its hot end heat exchanger, orifice, gas reservoir, and switching valve) actually works as a new type of expander. The enthalpy, and hence the temperature of the gas, decreases after expansion in the pulse tube, since heat is removed via the hot end heat exchanger as shown in Figure 1. The physical behavior of the gas in the pulse tube is described in detail in previous papers[3, 4]. This kind of expander is called a pulse tube expander. Unlike piston expanders and turbo expanders commonly used in refrigeration and gas liquefaction cycles, the pulse tube expander removes heat instead of work from the working substance. Pulse tube expansion is also different from Simon expansion, because the pulse tube is a thermodynamically uneven system.

The isentropic efficiency of the pulse tube expansion process is defined as:

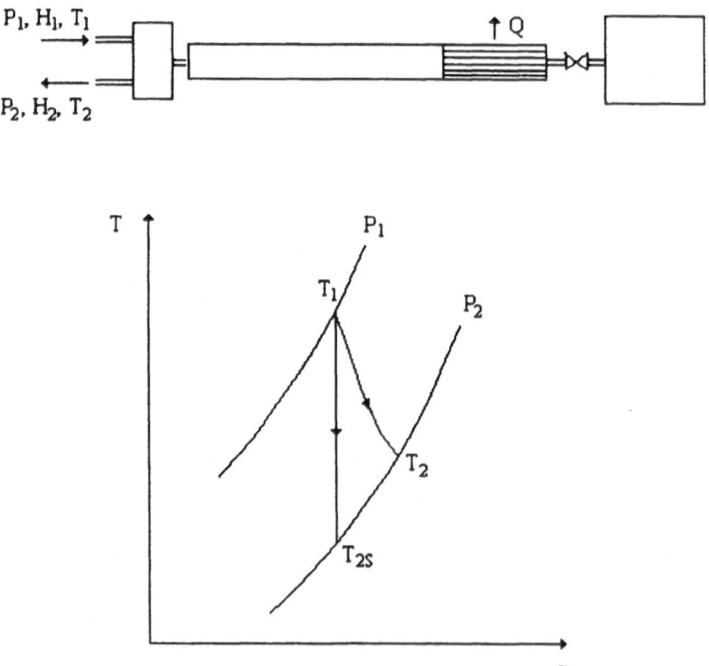

**Figure 1.** Pulse tube expander and pulse tube expansion process on T-S diagram

$$\eta_s = (H_1 - H_2) / (H_1 - H_{2s}) = (T_1 - T_2) / (T_1 - T_{2s}) = \Delta T' / \Delta T_s \qquad (1)$$

where $H_1$ and $T_1$ are the enthalpy and temperature of the gas flowing into the pulse tube at pressure $P_1$, $H_2$ and $T_2$ are the enthalpy and temperature of the gas flowing out of the pulse tube at pressure $P_2$, and $H_{2s}$ and $T_{2s}$ are the enthalpy and temperature of the gas after isentropic expansion from $(P_1, T_1)$ to $P_2$. In short, the isentropic efficiency is the ratio of the practical temperature drop $\Delta T$ to the ideal temperature drop $\Delta T_s$.

Compared with piston and turbo expanders, the pulse tube expander has the following advantages:

(1) It has no reciprocating piston and no high speed rotating components at low temperature. Thus, it is simple, reliable, easy and cheap to fabricate, and flexible for spacial arrangement.

(2) It can work under two-phase conditions, and it is not very sensitive to impurities.

(3) It can work at quite large pressure ratios.

However, in order to find practical applications, it must have sufficiently high isentropic expansion efficiencies.

## EXPERIMENTS

### Experimental Apparatus

The experimental apparatus for the measurement of the isentropic efficiency of the pulse tube expansion process is shown schematically in Figure 2. All the experiments have been

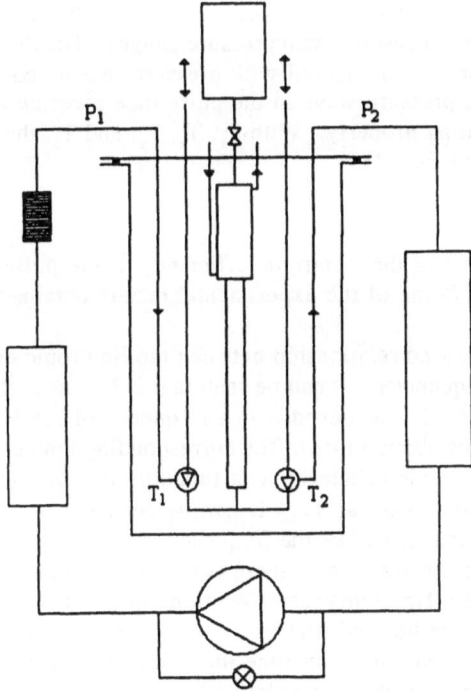

**Figure 2.** Experimental apparatus for the measurement of isentropic efficiency.

carried out with air as the working fluid. Air enters the pulse tube at ambient temperature; no precooling is used for these preliminary experiments. A membrane compressor with a maximum capacity of 22 $Nm^3/h$ is used to pressurize the air from $P_2$ (1 bar) to the higher pressure $P_1$. A bypass valve between the inlet and outlet tubes is used to adjust $P_1$. Two large buffer volumes are installed to stabilize $P_1$ and $P_2$. To eliminate water and other impurities from the air, a filter is installed next to the high pressure buffer. The low temperature switching valve, driven by cams, was salvaged from an old piston-type helium liquefier, and was adapted and used for the present experiments. Most of the experiments have been carried out with a stainless steel pulse tube of 20 mm OD, 19 mm ID and 400 mm length. Two other pulse tubes of 14 mm ID and 24 mm ID, respectively, with the same length, have also been tested. An optional flow straightener at the cold end of the pulse tube has also been investigated.

In the hot end heat exchanger there are about 60 stainless steel tubes of 2 mm ID, 3 mm OD, and 200 mm length. These tubes are mounted in parallel inside a water jacket, and cooled by water. Copper tubes would surely have been better, but would have been much more difficult to weld, given our present fabrication conditions. Since the enthalpy drop of the gas equals the amount of heat removed at the heat exchanger, special care has to be taken in the design of the hot end heat exchanger. The principle is to have the maximum specific heat transfer surface, i.e. the maximum heat transfer surface per unit void volume.

The orifice is made with a continuously adjustable needle valve with an aperture of 3 mm. The gas reservoir has a volume of 2 $dm^3$. The connecting tubes for the working gas circuit are copper tubes of 10 mm ID. Vacuum, provided with a vacuum jacket and a mechanical pump, was at first used to provide thermal insulation for the cold parts. Subsequently, as the temperature is not very low, the cold parts were simply wrapped with perforated plastic for thermal insulation, and no vacuum was used.

The inlet and outlet temperatures of the pulse tube, as well as the temperature gradient along pulse tube, were measured with calibrated copper-constantan thermocouples. The inlet and outlet pressures were measured with pressure gauges. The dynamic pressures in the pulse tube and gas reservoir were measured with pressure sensors connected to an oscilloscope and a computer. The pressure wave in the pulse tube gives an indication of whether the switching valve is operating properly. With $P_1$, $T_1$, $P_2$, and $T_2$, the isentropic expansion efficiency is determined with Eq. (1).

## Results and Discussion

Systematic measurements of the isentropic efficiency of the pulse tube expansion process have been conducted. Some of the experimental results obtained with the 19 mm ID pulse tube are given below.

Figure 3 shows the measured relationship between the isentropic efficiency and the orifice opening at different frequencies. It can be seen that 2-3 turns is the optimum opening. An isentropic efficiency of 42% is achieved with a frequency of 2.6 Hz, a pressure ratio of 11, and an orifice opening of about 3 turns. The corresponding temperature drop is 63.6 K.

Figure 4 shows the isentropic efficiency as a function of frequency at different orifice openings; the pressure ratio is kept at 11. It can be seen that the isentropic efficiency increases with increasing frequency when the frequency is lower than 2.6 Hz, and decreases when the frequency exceeds 2.6 Hz. Thus, there exists an optimum frequency for the isentropic efficiency, which is 2.6 Hz in this case. We think that this effect is due to the limited heat dissipation capacity of the hot end heat exchanger. Other parameters being the same, the amount of heat to be removed per unit time increases linearly with the increase in frequency. When the heat flux exceeds the maximum heat flux of the hot end heat exchanger, heat will be accumulated at the hot end, and consequently the outlet temperature $T_2$ will increase. Thus, the isentropic efficiency is mainly limited by the heat transfer capacity of

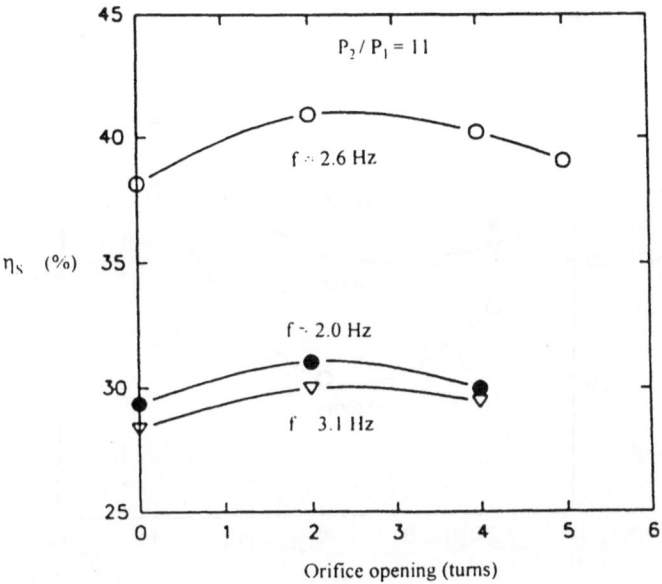

**Figure 3.** Isentropic efficiency versus orifice opening.

the hot end heat exchanger. If the hot end heat exchanger is improved, e.g. with greater specific heat transfer surface, better material, better configuration, etc., higher isentropic efficiencies can be achieved.

Figure 5 demonstrates the dependence of the isentropic efficiency on the pressure ratio at different orifice openings at 2.6 Hz. It can be seen that with optimum orifice opening, the isentropic efficiency generally increases with increasing pressure ratio.

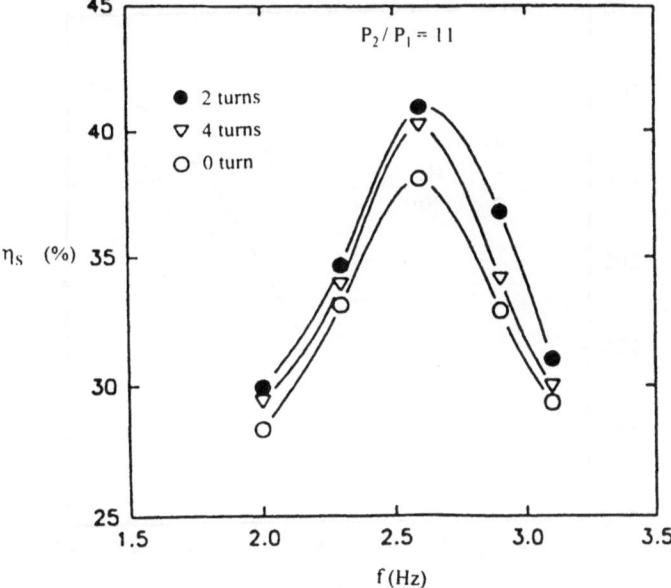

**Figure 4.** Isentropic efficiency versus frequency.

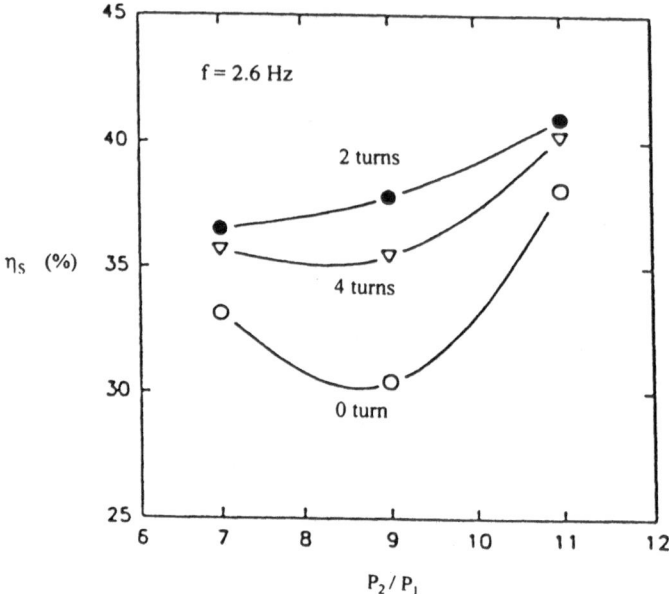

**Figure 5.** Isentropic efficiency versus pressure ratio.

Figure 6 illustrates the temperature gradient along the pulse tube with the orifice closed. The data indicate that the decrease in temperature mainly occurs in the upper half of the pulse tube. It should be noted that thetemperature measured at the cold end of pulse tube is more than 10 K higher than the temperature $T_2$ measured at the outlet tube. This means that there is a large temperature gradient in the cold block.

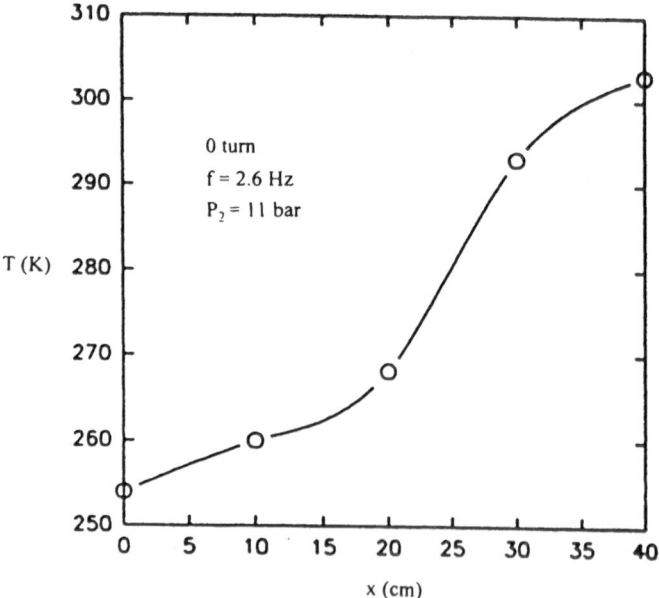

**Figure 6.** Temperature distribution along the pulse tube.

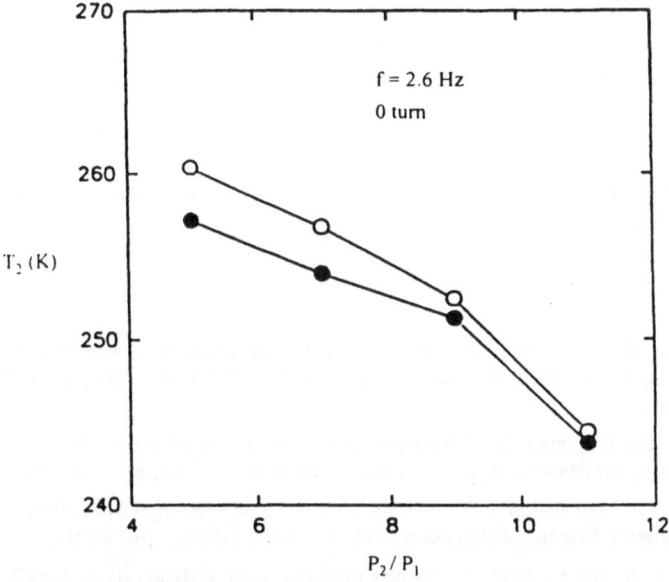

**Figure 7.** The influence of the flow straightener on outlet gas temperature.

The effect of a flow straightener on the outlet gas temperature $T_2$ was also investigated. Figure 7 shows the measured relationship between $T_2$ and $P_1$ in the cases with and without a cold end flow straightener. $T_2$ is slightly lower when a flow straightener is installed at the cold end of pulse tube. For the given pulse tube diameter, the effect of the flow straightener is not significant, especially at higher frequencies and higher pressure ratios.

For these preliminary experiments, the test condition were not optimized, which also has some influence on the experimental results. As the mass flow rate is small, the thermal losses are important with respect to the gross cooling power. There is conduction loss from the room temperature inlet tube to the low temperature outlet tube via the large cold block on which the switching valve and the pulse tube were fixed and supported. There is also heat conduction from room temperature along the outlet copper tube. The operation of the switching valve is not very stable, and some leakage through the switching valve persists; this allows part of high pressure gas to shortcut the pulse tube and increase the outlet gas temperature.

Although the isentropic efficiency of the pulse tube expander is not yet as high as that of a piston or turbo expander, it has demonstrated the potential for further improvements. With its appealing advantages such as simplicity and reliability, the pulse tube expander offers another choice besides the conventional piston and turbo expanders for Brayton-cycle refrigerators.

## CONCLUSIONS

The concept of using a pulse tube as an expander for non-regenerative refrigeration and gas liquefaction cycles, such as the reverse-Brayton cycle, is proposed. The isentropic efficiency of the pulse tube expansion process has been experimentally investigated. An isentropic efficiency of 42% has been achieved in the preliminary experiments with air as the working fluid and with a room temperature inlet. Under each working condition, there always exists an optimum frequency and an optimum orifice opening. The isentropic efficiency is generally greater at high pressure ratios. The heat transfer capacity of the hot end heat exchanger is the main factor that limits the isentropic efficiency of the pulse tube expander. Higher efficiencies can be achieved with better hot end heat exchangers and improved working conditions.

## ACKNOWLEDGMENT

This work is supported by the national natural Science Foundation of China under contract number 59506003.

## REFERENCES

1.  Liang, J., Zhang, C.Q., Xu, L., Cai, J.H., Luo, E.C. and Zhou, Y., "Pulse Tube Refrigerator with Low Temperature Switching Valve", presented at *ICEC16/ICMC*, May 20-24, 1996, Kitakyushu, Japan.

2.  Radebaugh, R. and Herrman, S., "Refrigeration Efficiency of Pulse Tube Refrigerators," *Proceedings of the 4th International Cryocooler Conference*, 1987, pp. 119-132.

3.  Liang, J., Ravex, A. and Rolland, P., "Study on Pulse Tube Refrigeration, Part 1: Thermodynamic Non-symmetry Effect," *Cryogenics*, vol. 36, no. 2 (1996), pp. 87-93.

4.  Liang, J., Ravex, A. and Rolland, P., "Study on Pulse Tube Refrigeration, Part 2: Theoretical Modelling," *Cryogenics*, vol. 36, no. 2 (1996), pp. 95-99.

# Joule-Thomson Cryocooler Development at Ball Aerospace

**R. Levenduski and R. Scarlotti**

Ball Aerospace Systems Division
Boulder, Colorado 80306

## ABSTRACT

Ball Aerospace is developing a two-stage Joule-Thomson (J-T) cryocooler for space applications. The cryocooler provides uniform and constant temperature cooling at two temperatures for varying heat loads and relatively large interface surfaces. The on-going Cryogenic On-Orbit Long Life Active Refrigerator (COOLLAR) program began in 1992 and includes producing two J-T cryocoolers, a scalability study, and a test effort to demonstrate the technology's ability to uniformly cool large interface surfaces. The first cryocooler produced, called the Engineering Development Model (EDM), has undergone ground-based testing and is being configured for a shuttle flight experiment to verify zero-g performance. The second cryocooler, called the EDM-DI (for distributed interface), will incorporate the large sensor interface technology. It will be tested for requirements compliance and then undergo reliability testing. This paper presents the performance test results of the EDM and the distributed interface technology development. It also presents an overview of the COOLLAR Flight Experiment (CFE) and EDM-DI projects.

## INTRODUCTION

Through a series of contracts, Ball Aerospace is developing a cryocooler for space applications based on the Joule-Thomson refrigeration cycle. The current program, COOLLAR, began in March of 1992. The primary objective of COOLLAR is to develop and demonstrate a flight-capable cryocooler. The first unit built, the EDM, recently completed its ground-based performance testing. The EDM is currently being configured as a shuttle flight experiment. It is on schedule to be delivered to NASA in January 1997 for a July 1997 launch. A second EDM class cryocooler, the EDM-DI, is currently being built. It is similar to the EDM but includes our recently demonstrated large interface cooling technology. Both cryocoolers will be available for reliability testing in the fall of 1997.

## JOULE-THOMSON TECHNOLOGY OVERVIEW

The Joule-Thomson cycle is a commonly used and well-understood refrigeration cycle. Variations of this cycle are used in many applications such as home refrigerators and

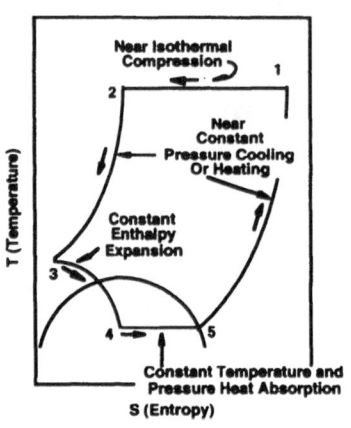 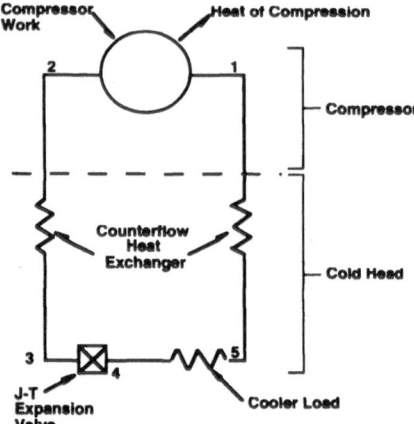

Figure 1. **Basic Joule-Thomson cycle.**

air conditioning because of its simplicity, reliability, and efficiency. Different working fluids can be used to achieve a wide range of cooling temperatures. (We have tested cryocoolers from 4.1K to 120K.)   The basic Joule-Thomson cycle is described in **Figure 1**. A compressor is used to compress the working fluid (the EDM uses nitrogen) to a high pressure. The high pressure gas passes through a counterflow heat exchanger and is precooled by the returning low pressure gas stream. The precooled gas is expanded isenthalpically through a J-T valve which for our applications produces a two-phase fluid. The liquid is used to absorb the heat load at constant temperature. The low pressure gas returns to the compressor to complete the cycle.

## ENGINEERING DEVELOPMENT MODEL CRYOCOOLER OVERVIEW

### Requirements

The EDM cryocooler is designed to meet an extensive set of technical requirements. The requirements address performance, reliability, and integration issues. The major requirements are listed in **Table 1**.

### System Operation

The EDM cryocooler consists of four subsystems: compressor, gas purifier, cold head, and electronic control/power conditioning unit (EC/PCU). The gas purifier is mounted to the compressor and the assembly is called the compressor unit. **Figure 2** is a photograph of the EDM cryocooler on its test stand.

The oil-lubricated compressor produces high pressure gas (1100 psia, 75 bar) and sends it to the gas purifier. Within the gas purifier, the passive oil scavenger removes all but a trace amount of oil from the gas stream and returns it to the compressor via the oil-pressure reducer. The gas stream then enters the coalescer which permanently removes the trace oil (less than 5 oz. over 5 years). The oil-free gas then enters the getter and particle filter. The getter permanently removes the gaseous contaminants, and the particle filter prevents particulate contamination from entering the cold head.

The ambient temperature clean gas enters the cold head and is cooled by counterflow heat exchange and the TEC in the hybrid heat exchanger. Upon exiting the hybrid heat exchanger, the flow splits into two streams and is expanded to produce cooling at 120K and 65K. Each stream returns to the compressor to complete the cycle.

**Table 1.** COOLLAR Program Major Requirements

| | |
|---|---|
| Cooling Temperature | $T_1$: 65 +1, -3 K<br>$T_2$: 120 K |
| Cooling Capacity | $T_1$: 0.75 W (low load), 3.5 W (high load)<br>$T_2$: 3.0 - 5.0 W |
| Temperature Stability | $T_1$: ±0.25 K, within 2 sec of load change<br>$T_2$: ±4.0 K |
| Input Power (average and peak) | < 325 W |
| Operating Environment<br>    Thermal<br>    Gravity | <br>267 - 322 K<br>0-g, any orientation in 1-g |
| Vibration<br>    Compressor<br><br>    Cold Head Load Interface | <br>< 0.035 lb. (0.133 N) linear force<br>< 0.35 lb-in (0.035 N-m) rotational force<br>< 0.002 lb. (0.009 N) linear force (1 Hz - 1 kHz)<br>< 0.002 lb-in (0.002 N-m) rotational (1 Hz - 1kHz) |
| Weight | < 125 lb. |
| Reliability | > 0.99 p for 5-year life, 0.5 confidence |
| Interchangeability | Components designated as "line replaceable units" shall be interchangeable with spares |

The EC/PCU contains a microprocessor that controls various aspects of the cryocooler. It maintains constant temperatures at the load interfaces through two closed-loop temperature control servos. The 120K interface is maintained at a constant temperature by adjusting the average compressor speed to keep the saturation pressure in the load tank constant at 360 psia. The 65 K interface is maintained at a constant temperature by using a motor driven needle valve (flow control valve) to keep the saturation pressure in the load tank constant at 2.5 psia. The EC/PCU also adjusts power to the thermoelectric cooler to maintain a constant cold side temperature regardless of environmental temperature and does the same for the getter to keep it at proper operating temperature.

## VERIFICATION TESTING OVERVIEW

The verification test program focused on demonstrating the cooler's performance in critical areas. Verification testing was conducted at the subsystem and system levels. Environmental vibration testing and weight measurements were performed at the subsystem level, since there were no system level implications to these tests and conducting them at the subsystem level was convenient and had schedule benefits. All performance requirements were verified at the system level.

**Figure 2.** EDM cryocooler

## SUBSYSTEM VERIFICATION TESTS

### Size and Weight

The compressor unit, comprised of the compressor and gas purifier is 66 cm (26 in) long, 40 cm (16 in) in diameter, and weighs 44 kg (96 lb). Removal of engineering instrumentation (pressure and temperature sensors and associated brackets and cables) would bring it to a flight weight of 42 kg (92 lb). The cold head fits within a 20 cm x 12 cm (8 in x 5 in) footprint and is 14 cm (5 in) high. It weighs 3.5 kg (8 lb) with each load tank having a dry weight of approximately 0.7 kg (1.5 lb).

### Vibration Environment Testing

The compressor unit and the cold head were each mounted to a vibration table and tested in all three axes to the levels shown in **Figure 3**. **Figures 4** and **5** show the compressor unit and cold head on the test fixtures. The test began with a 1/4g sine sweep to identify any high risk areas. None were found on the cold head but the testing revealed the need for an additional bracket to support the flow control valve on the compressor unit. Random vibration testing was followed by a 15g sine burst to verify structural integrity. The test ended with a repeat of the 1/4g sine sweep to verify the structural response had not changed.

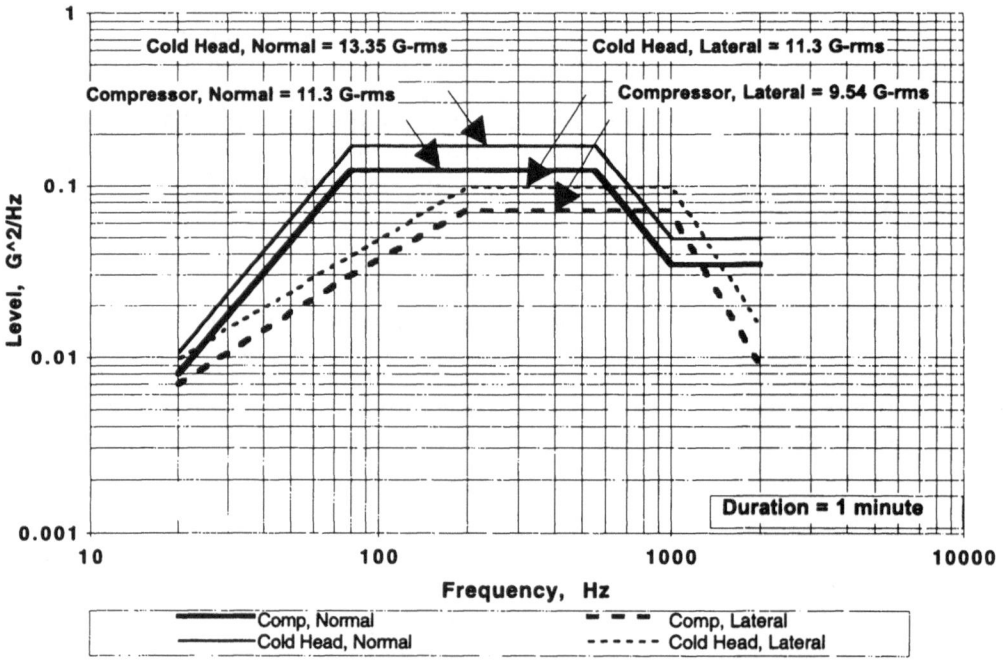

**Figure 3.** Vibration levels of compressor unit and cold head

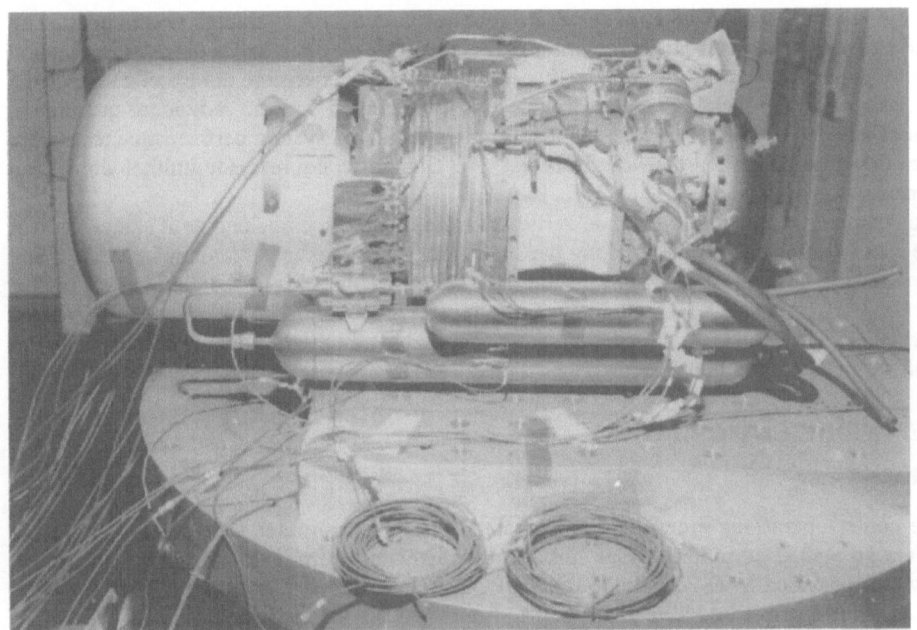

**Figure 4.** Compressor unit on vibration test fixture.

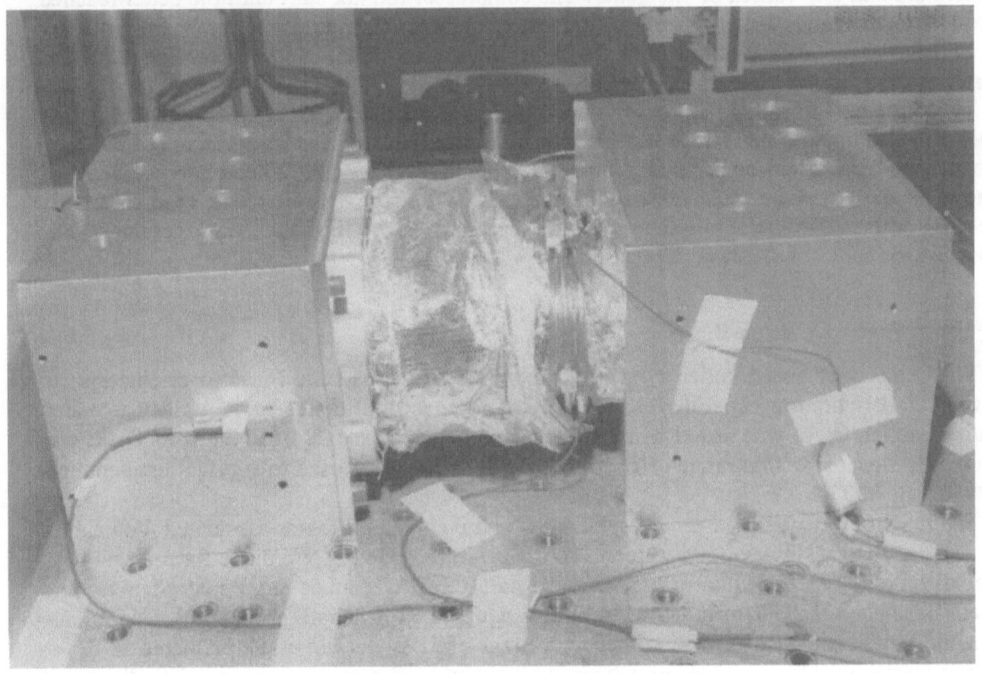

**Figure 5.** Cold head on vibration test fixture.

Instrumentation checks were made before each test and only one change was noticed. The resistance of the tachometer phase A windings changed from 7 ohms to 25 ohms after the first test and remained unchanged thereafter. Inspection could not determine the exact cause of the change, and since the tachometer performance was unchanged and the redundant tachometer was not affected, we proceeded with the test program as is. Subsystem performance tests were conducted before and after the vibration test. No change in compressor unit or cold head performance was seen.

The compressor unit x axis extends vertically from the compressor unit as it sits on its feet and the z axis extends horizontally along the shaft axis. The y axis is perpendicular to the x and z axes. The sine-sweep test data showed the first natural frequencies of the compressor unit to be 352 Hz, 105 Hz, and 118 Hz in the x, y, and z axes, respectively. The first mode was dictated by the mounting feet and the x axis fundamental occurred at a higher frequency because they are much stiffer in the x direction than the y and z directions.

## SYSTEM VERIFICATION TESTS

### Operations

The cooler operations proceeded as expected. The test computer sent commands to the cooler via an RS232 communications link. The "power on" command closed the bus power relays which provided 24 to 32 volts to the EC/PCU and the onboard computer. When powered up, the computer performed a health check and returned the information to the test computer upon command. When the health data was reviewed and everything deemed nominal, the "start" command was sent.

The start command set the initial position of the flow control valve used to control the $T_1$ temperature and then turned on the getter. The compressor did not start until the getter reached operating temperature. This was done to prevent contamination in the compressor from passing through the getter and reaching the cold head. Within an hour of compressor start, design pressures were achieved and the cooler began to cool down.

The software controlled the cooler's operation as it proceeded through cooldown and the tanks collected liquid. The software adjusted power to the TEC to keep it operating at 220 K and activated the trim heaters as necessary to maintain liquid level control. When the cooler reached steady operation, it was tested under a load profile, and the software maintained the $T_1$ and $T_2$ interfaces well within the requirements. At the completion of testing, the "stop" command was given and the cooler shut down without incident.

### Thermodynamic Performance

The cooler was tested under a variety of conditions. It was tested in four orientations to prove multi-orientation capability in 1g. This data adds confidence in its ability to meet requirements in 0g. It was tested with power bus voltages of 24, 28 , and 32 volts dc. It was tested over a heat rejection range of 267 K to 322 K and was exposed to survival temperatures of 255 K and 333 K.

The input power was measured "at the wall" and included all losses associated with cables and the electronics. However, adjustments to the measured power were made for a meaningful comparison to the flight requirements. The first adjustment was made to the getter power. The getter is designed to operate in a vacuum but was operated in laboratory ambient conditions. Since the getter is not sealed, it consumed an average of 51 W instead of the predicted 18 W for vacuum operation. Thus, a 33-W adjustment to the measured power was made. An adjustment also was made to resolve the difference between the as-built non-flight electronics and flight electronics. The adjustment was based on a board-by-board analysis and by assuming an efficiency improvement for the TEC power supply from 71% to 80%. This adjustment ranged from 20 W to 27 W depending on operating conditions. The last adjustment was 4.5 W and accounts for the difference between the tested and flight vent valve power.

The cooler successfully demonstrated key performance requirements under the above conditions. With the exception of power, the performance remained unchanged in all orientations and over the full range of bus voltage and heat rejection temperature. The cooling capacities and tank temperatures were insensitive to heat rejection temperature because the TEC stabilized cold head operation. The software adjusted the power to the TEC to keep it operating at 220K regardless of heat rejection temperature and in doing so, kept the cold head performance the same. However, the TEC required more power when rejecting heat to a higher temperature, as did the compressor and electronics. The input power increased by 15% when the heat rejection temperature increased from 295 K to 322 K.

The cooler provided simultaneous cooling of 5W at 120 K and from 0.75 to 3.5 W at 65 K. The $T_2$ tank cooled to 120 K in 4.5 hours and the $T_1$ tank reached 65 K 16 hours later. **Figure 6** shows key cold head temperatures during cooldown and **Figure 7** shows corresponding compressor speed and cooler input power. **Figure 8** shows the applied heat loads and the power of the $JT_1$ and $JT_2$ trim heaters. Cooldown was followed by 12 hours of continuous operation with 5W at 120 K and 0.75 W at 65 K. The steady state data appears at the end of the same figures.

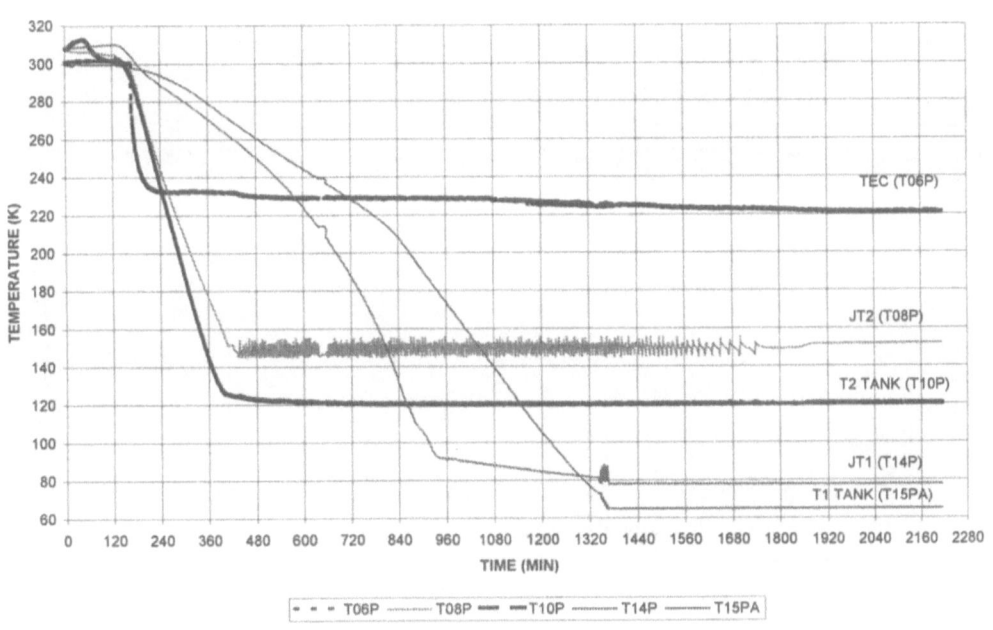

**Figure 6.** Cold head temperatures during cooldown.

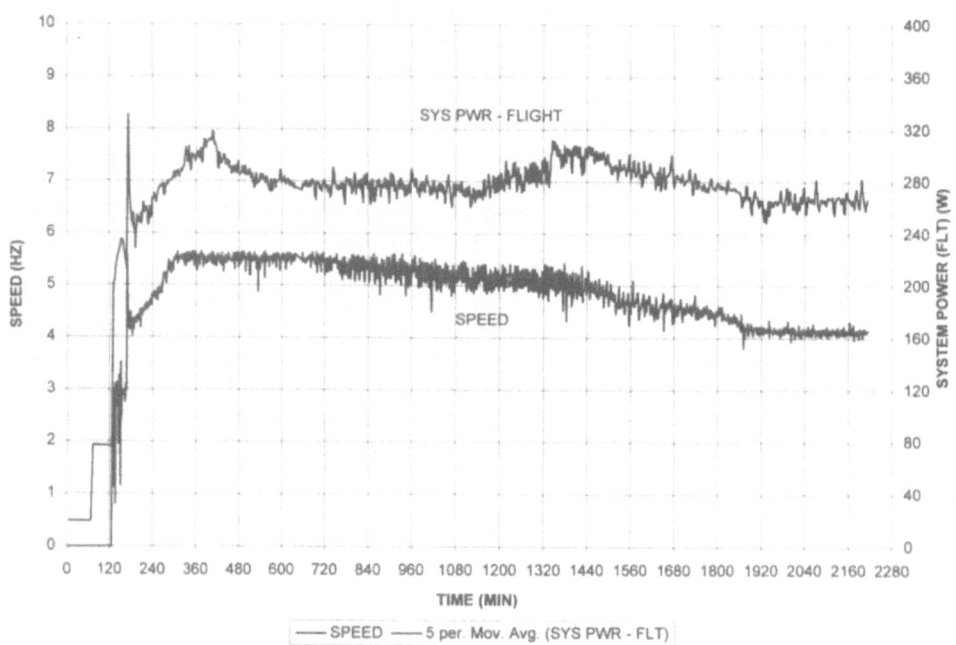

**Figure 7.** Compressor speed and cooler input power during cooldown.

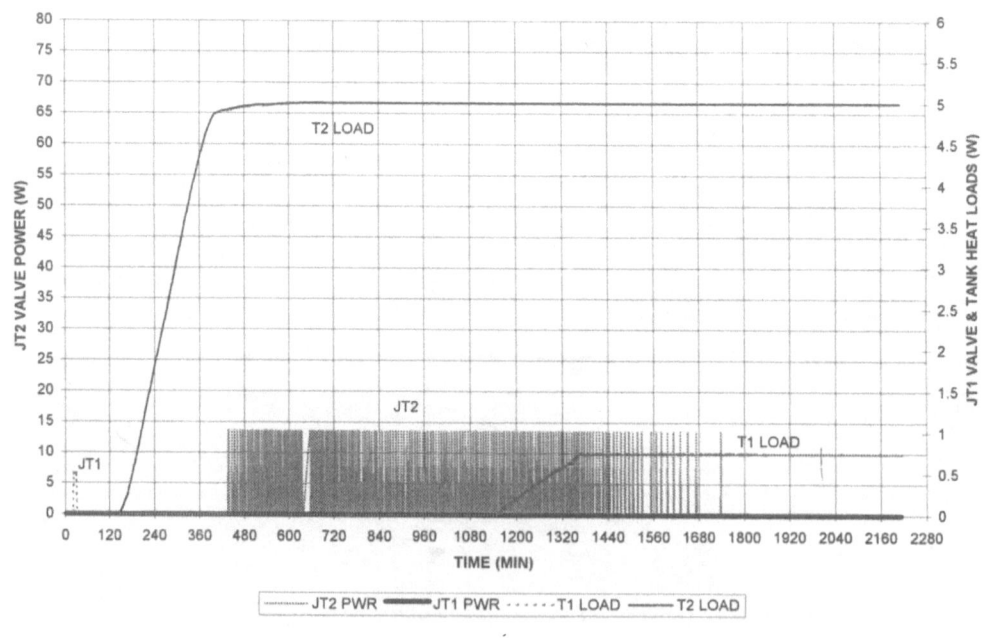

**Figure 8.** Heat loads during cooldown.

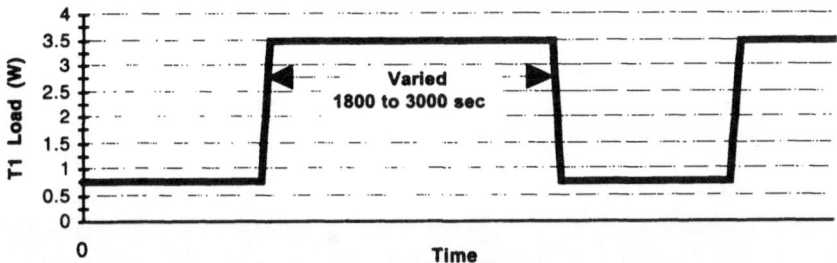

**Figure 9.** $T_1$ load profile.

The ultimate test of the cryocooler's performance was to demonstrate its capability to maintain constant temperature at two cold interfaces under conditions of varying load at the T1 interface. A test operation T1 load profile was used to test the cryocooler's capability to provide the required precision temperature control of these interfaces. This load profile consisted of a series of step load changes from a low load of 0.75 W to a high load of 3.5 W for varying periods. The duration of high load operation was varied from 1800 seconds to 3000 seconds as shown in **Figure 9**.

**Figure 10** shows the applied heat loads for one test on the $T_1$ and $T_2$ interfaces along with the cooler trim loads. Repeated steps of high load operation, at 3.5 W, were demonstrated for a total accumulated time of 260.5 minutes in a 12-hour period. Minor changes to the cooler would allow it to operate for longer periods of high load operation if desired. These changes could be increase in initial charge pressure, JT valve flow rate(s), or reduction in the TEC operating temperature. All would result in more liquid storage in the $T_1$ tank, thus extending the 3.5 W high load operating time.

The cooler met the critical temperature stability requirements of both interfaces. **Figure 11** shows the application of the 3.5W load and the corresponding effects on the $T_1$ and $T_2$ interfaces. The $T_1$ interface was consistently stable within 0.08 K throughout testing compared to the requirement of 0.25 K. The $T_2$ interface was stable within 1K compared to the 4K requirement.

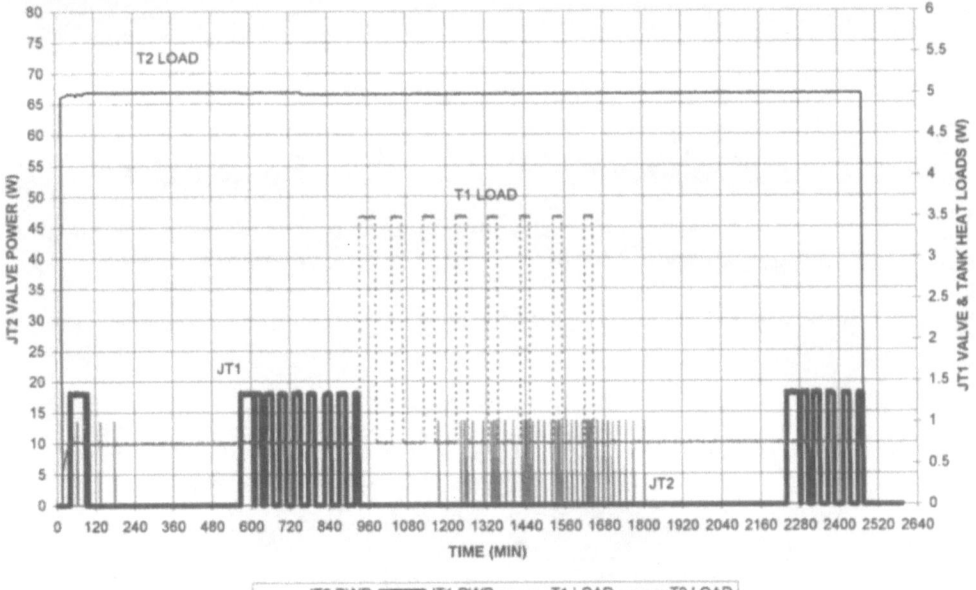

**Figure 10.** Heat loads during load profile.

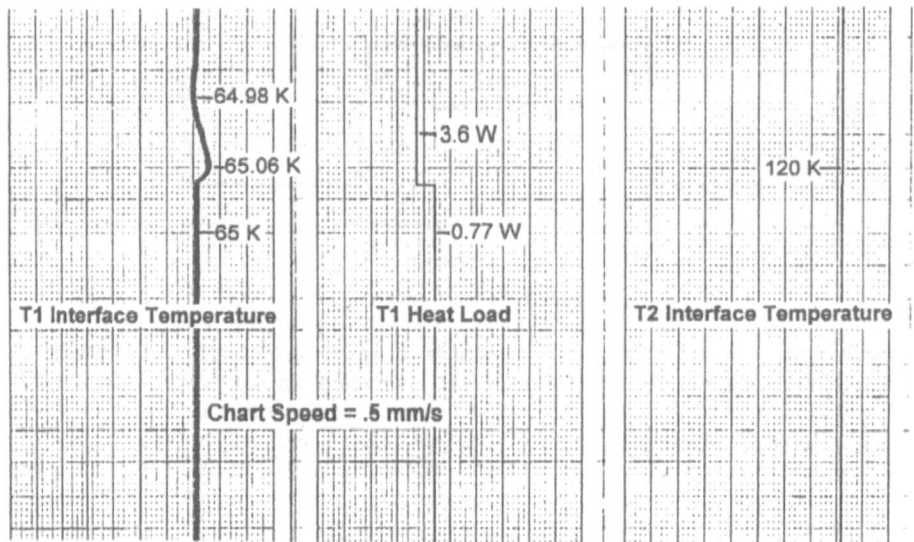

**Figure 11.** $T_1$ and $T_2$ interface temperature.

The cooler met the input power requirements. **Figure 12** shows the cooler input power during load profile operation. The average input power projected for flight was 280 W as compared to the requirement of 325 W. This data was taken with a 295-K heat rejection temperature.

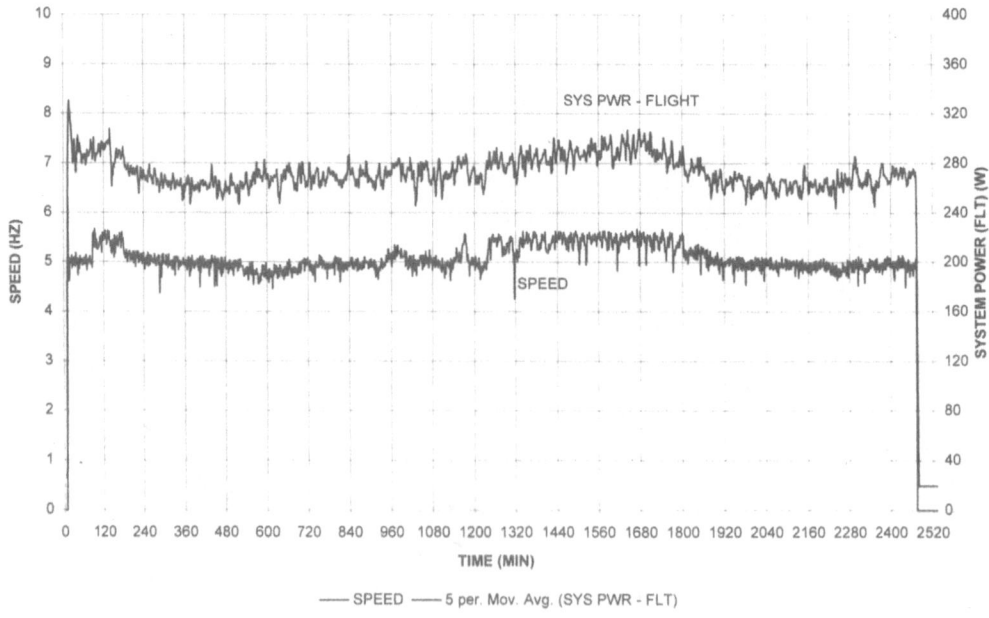

**Figure 12.** Compressor speed and cooler input power during load profile.

The cooler was tested to verify it could operate at the minimum heat rejection temperature of 267 K. Since the cooler is oil lubricated, oil collection and distribution at low temperature needed to be verified along with adequate motor torque at 24 volts. The cooler was put into a thermal chamber and cold soaked to 255 K to verify low temperature survival, then warmed to 267 K for the cold start test. The compressor unit started normally and the oil pressure quickly reached design conditions. The test continued until the compressor passed through peak torque operation and pressures stabilized.

### Induced Vibration

The vibration output of the compressor unit was measured with a custom designed, calibrated dynamometer. Forces and moments for all three axes were measured at the mounting feet. The cold head vibration output was not measured because it was measured during the preceding Advanced Breadboard (ABB) program and was proven not to be an issue. The peak vibration output from the ABB cold head was 2.75 EE-4 N (62 EE-6 lb.).

**Figure 13** is a sketch of the dynamometer. Measurements were taken during various modes of operation: startup, cooldown, the peak torque period and during load profile operation. Data taken with 3.5 W on $T_1$ is shown on the left of **Figure 14**. The linear vibration met requirements at low frequencies where forces caused by static and dynamic unbalance appear. This proves the machine is well balanced. A z-axis moment (not shown) is predominantly caused by compressor speed changes resulting from variations in shaft torque. The EC/PCU contains a speed control circuit that alters the current command to the motor in response to changes in speed. Thus, the z-axis moment is an indication of the effectiveness of the speed control servo at maintaining a constant speed. The rotational vibration met requirements except for small peaks at 50 Hz and 100 Hz.

The vibration output did not meet requirements at higher frequencies. The largest forces were seen at 180 Hz in the x axis. Those forces were most likely created by the ripple torque of the motor exciting the natural frequency of the getter. The getter weighs approximately 1.8 kg (4 lb) and is cantilevered off the gas purifier mounting plate. Since the mounting feet are very stiff in the x axis, the forces were transmitted into the dynamometer. However, the mounting

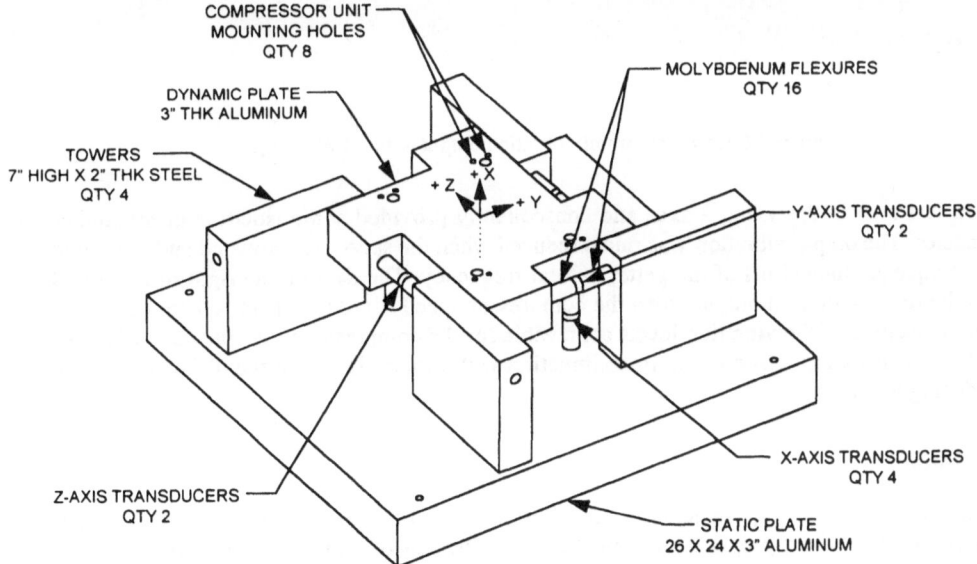

**Figure 13.** Sketch of dynamometer.

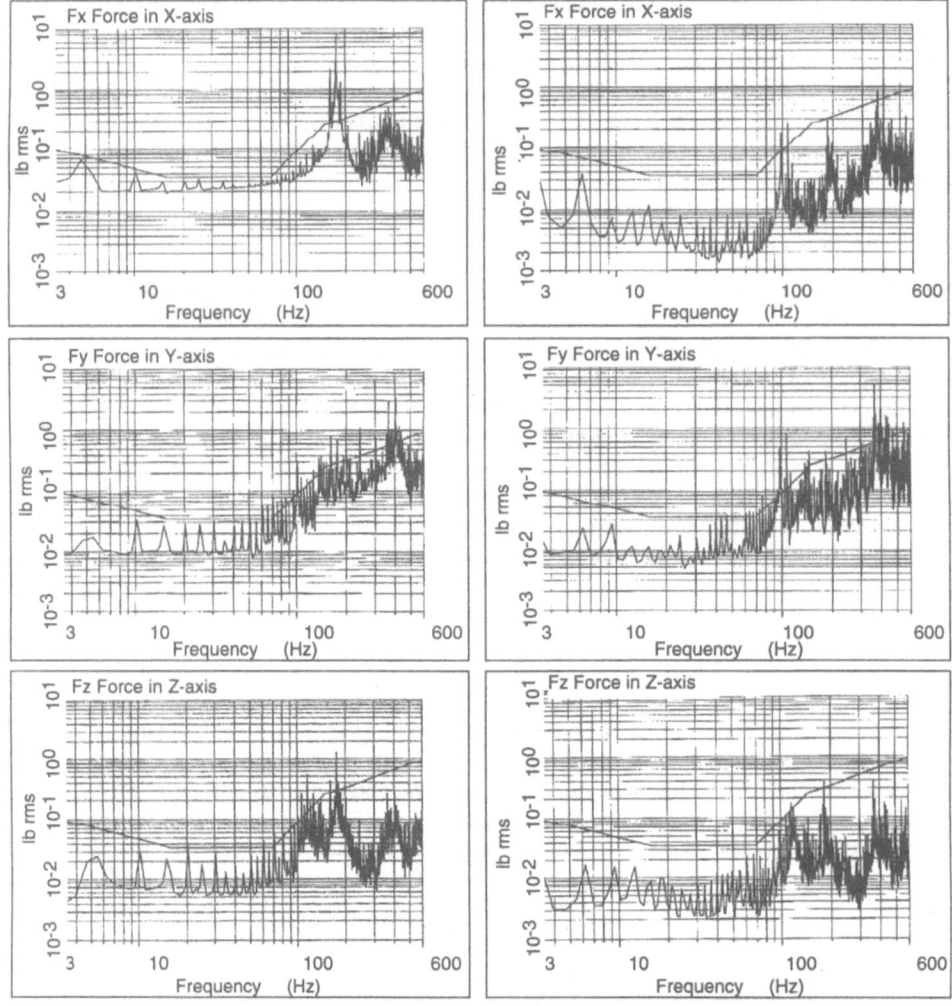

**Figure 14.** Compressor unit vibration output with 3.5W on $T_1$.

feet are less stiff in the y and z axes and consequently provided some isolation at the higher frequencies. The output vibration was much reduced when the speed was lowered and the ripple torque frequency moved off of the getter natural frequency. The data on the right of Figure 14 shows the linear vibration output when the compressor speed is reduced from 4.7 to 3 Hz. This is more indicative of the vibration levels achievable for the compressor unit. Even so, it is most probable that isolation will be used in conjunction with proper design in the ultimate solution for reducing vibration.

## EMI Testing

There were no EMI requirements for the EDM cooler so the electronics were not designed to a flight level. Without proper EMI design, EMI testing was of limited use but was conducted for investigative purposes. The primary objective of the tests was to establish the EMI levels of the cold head since it could potentially be integrated with a sensor. The secondary objective was to measure what was produced by the compressor and electronics to identify areas that required attention.

The EMI tests were limited to radiated emissions; no conducted emissions were measured. The tests were conducted in accordance with MIL-STD 461, Part 3, Class A2a and included RE01, RE02 and RE04.

**Figure 15** shows the cold head set up in the EMI chamber. The cold head was placed inside a Lexan cover with a nitrogen purge. The vacuum housing was removed to eliminate any attenuation it might provide. The tests were performed with the cryocooler operating at full power but the cold head warm. However, the TEC and trim heaters were active during the cold head tests so a meaningful measurement could be made. The radiated magnetic field measurements taken 7 cm from the $T_1$ and $T_2$ interfaces (RE01) showed the levels to be well below MIL-STD 461C requirements. The same is true for RE04 results, which made measurements at a distance of 1 meter.

A foil over-shield was added to the cables for RE02 testing. The radiated electric fields measured at a distance of 1 meter also met requirements. An over-shield would be used for a flight cold head.

The compressor and electronics were placed inside the EMI chamber for additional EMI measurements. The test setup is shown in **Figure 16**. The RE01 and RE04 test results exceeded MIL-STD 461C levels, as expected. The predominate EMI sources were the PWM motor driver and on-board computer. Steps that could be taken to reduce the compressor and electronics' magnetic emissions are: use shorter cables and less unshielded wire lengths, 2) add shields or conduit to absorb the magnetic fields, and 3) reduce the current ripple on the motor phase

**Figure 15.** Cold head EMI test setup.

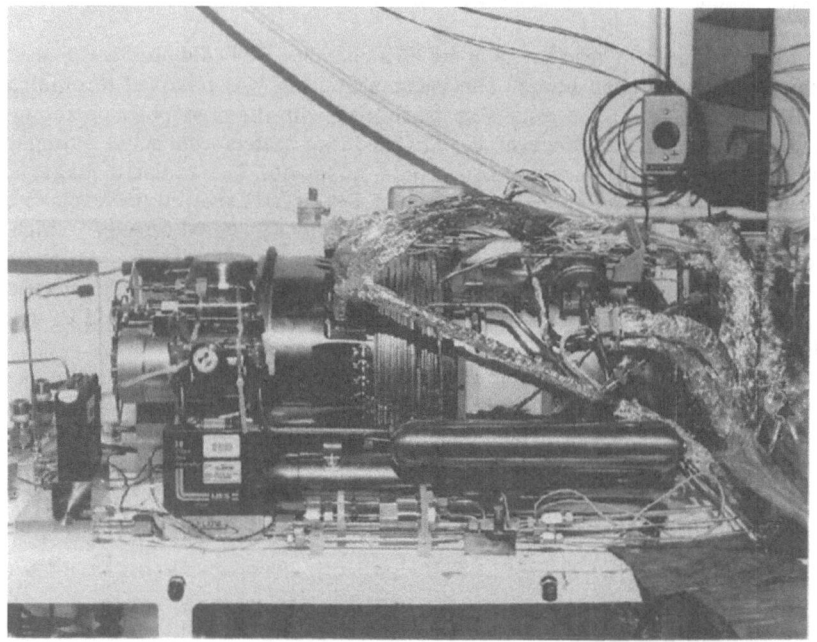

**Figure 16.** Compressor unit EMI test setup.

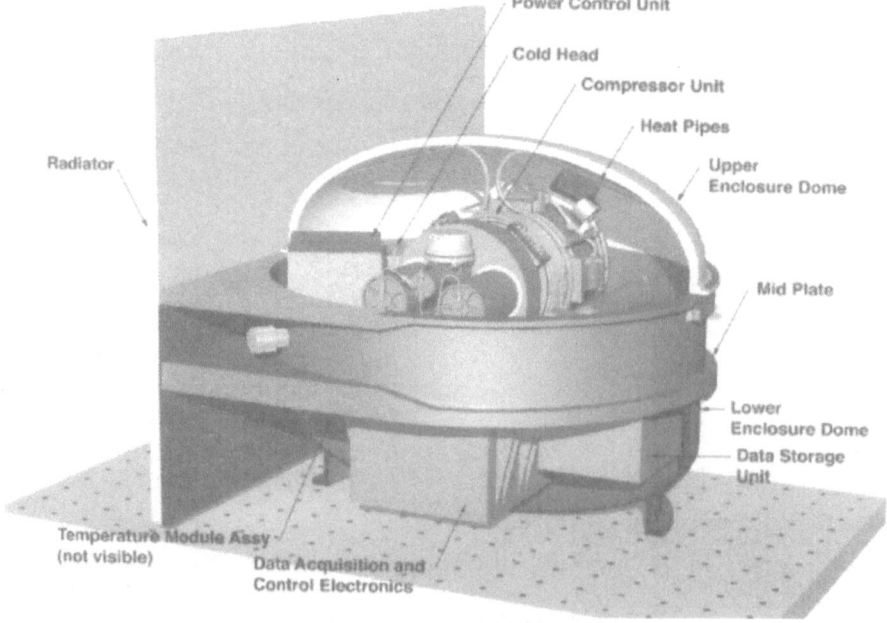

**Figure 17.** CFE test configuration

windings. To reduce electric field emissions, the common mode impedance between driver return and chassis should be reduced and the single board computer and other digital circuitry should be placed in its own EMI tight compartment.

## FLIGHT EXPERIMENT OVERVIEW

A program called the COOLLAR Flight Experiment (CFE) began in early June 1995 to flight test the EDM cryocooler aboard the Space Shuttle (STS). The experiment has been manifested aboard STS-85, scheduled to launch July 17, 1997. The primary objective of the experiment is to characterize the EDM cryocooler's steady state and transient operation in microgravity after being subjected to a launch environment. Evaluation of oil control and liquid nitrogen management is of particular interest. Other objectives are to correlate flight performance data with ground data and identify any flight related issues or constraints with this technology.

The cryocooler will be thoroughly characterized in a thermal vacuum chamber before delivery. Particular tests will be repeated on orbit and performance data compared. Temperature stability at the load interfaces under changing heat loads is a key requirement to be demonstrated.

**Figure 17** shows our design for the flight experiment. The EDM is completely enclosed to meet Shuttle safety requirements. Heat is rejected from the enclosure using a large integral heat pipe radiator and the external surface of the enclosure. Heat is transported from the cryocooler and electronics to the radiator and enclosure through the use of conductive thermal links and heat pipes. The experiment is fully instrumented (over 90 sensors) to completely characterize its performance on-orbit.

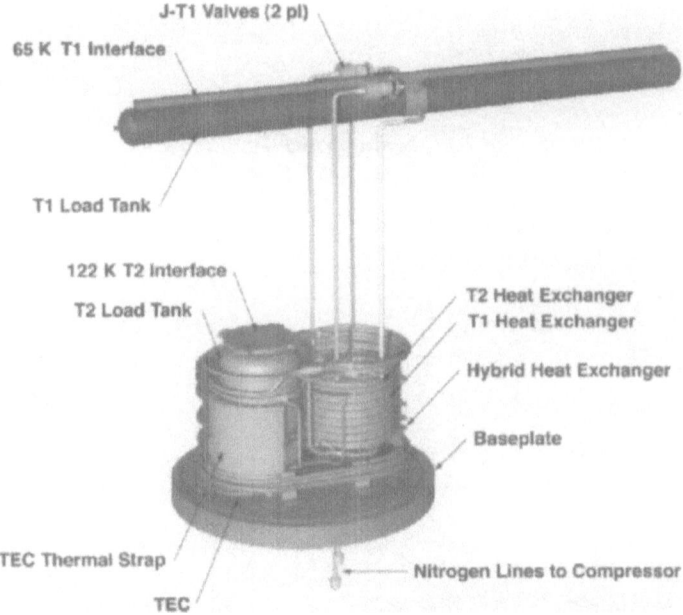

**Figure 18.** EDM-DI cold head.

## ENGINEERING DEVELOPMENT MODEL DISTRIBUTED INTERFACE (EDM-DI)

We began the build of a second EDM-class cryocooler in October of 1995. A few changes were made to the design to improve it based on our experience with the EDM. The most significant difference between the two cryocoolers is the distributed cooling at the 65K interface. We wanted to demonstrate the technology developed as part of scalability work into a complete cryocooler. Our cold head design for the EDM-DI is shown in **Figure 18**. The fabrication of the EDM-DI is expected to be completed in the summer of 1997 and life testing begins in the fall.

## SUMMARY

A J-T cryocooler is being developed for space applications where a benign sensor interface, precise temperature control and integrateability are key requirements. The EDM cryocooler has demonstrated compliance to a demanding set of requirements and sets the industry standard for temperature stability, vibration, and EMI at the cold interfaces. Future testing is underway to demonstrate its on-orbit performance and system reliability.

# Low Cost Cryocoolers for Cryoelectronics

**W. A. Little and I. Sapozhnikov**

MMR Technologies, Inc.
1400 N. Shoreline Blvd, #A5
Mountain View, CA 94043

## ABSTRACT

The development of cryoelectronics has been hampered by the high cost of refrigeration systems required for cooling to 80 K to 150 K with reasonable efficiency and high reliability. This has now changed. We report on the development of a simple cryogenic cooling system based on early work in the former Soviet Union, that uses components from the commercial refrigeration industry to provide cooling at a cost an order of magnitude lower than any other. This it does with good efficiency and high reliability. It can be expected to open up the market in superconducting electronics, low noise amplifiers and cooled CMOS processors. We describe the system, the reasons the costs are low, and performance figures to date. This work was sponsored by the Advanced Research Projects Agency through the Naval Research Laboratory, Washington.

## INTRODUCTION

A major barrier to the development of cryoelectronics in the past has been the lack of a low cost, low noise, closed-cycle refrigeration system capable of providing cooling between 80K and 150 K with acceptable efficiency and high reliability. Interest in this temperature regime is particularly strong today because of the development of superconducting microwave devices using the recently discovered cuprate, high temperature superconductors and the possibility of extending such developments to superconducting digital devices. And there are other areas where there is also a need for a low-cost cooler of high reliability. These include cooled CMOS, HEMT, and GaAs devices; Low Noise Amplifiers and refrigerators for the small scale storage of biological materials.

We report on the development of a simple cryogenic cooling system that goes a long way to addressing these needs. Our work is based on early work in the former Soviet Union by Kleemenko[1] and Alfeev et al[2]. Kleemenko pointed out in a paper in 1959 that the use of a properly chosen multi-component mixture of refrigerants in a single stream, cascade, throttle-expansion refrigeration cycle could reduce the degree of irreversibility of the heat exchangers. And, in addition, showed that by separating the liquid from the vapor phase and throttling the expansion of the liquid, efficient refrigeration could be obtained at temperatures below 150 K. Alfeev and co-workers in 1971 showed that certain mixtures of the lighter hydrocarbons, methane, ethane and propane, with nitrogen when used in a Joule-Thomson refrigerator gave a greatly enhanced refrigeration performance at temperatures down to below 80 K as compared with that of pure nitrogen.

Cryocoolers 9, Edited by R.G. Ross, Jr.
Plenum Press, New York, 1997

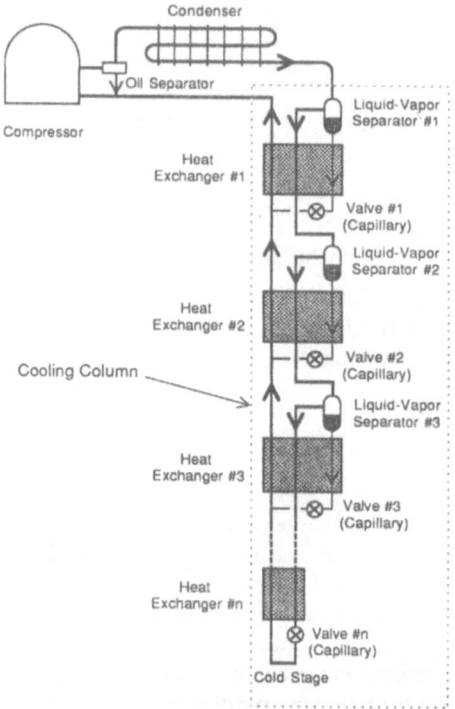

**Figure 1.** Kleemenko Cycle Cooler

A Kleemenko cycle cooler is illustrated in Fig.1 . A compressor compresses the multi-component refrigerant mixture and passes it through the air-cooled condenser where some of the higher boiling fraction condenses. This is separated from the vapor phase and both pass down a counter-current heat exchanger, cooled by the cold return stream. The liquid fraction then expands through a throttling valve or capillary restrictor into the return stream. Expansion causes a drop in temperature, cooling the vapor fraction, and causing more liquid to condense. The process is repeated in one or more stages as illustrated. For operation down to temperatures ≈ 115K, which is of interest for cooling of semiconductor devices we have found that it is possible to simplify considerably the above cycle and still get efficient and reliable performance. This simplified cycle is shown in Fig. 2.

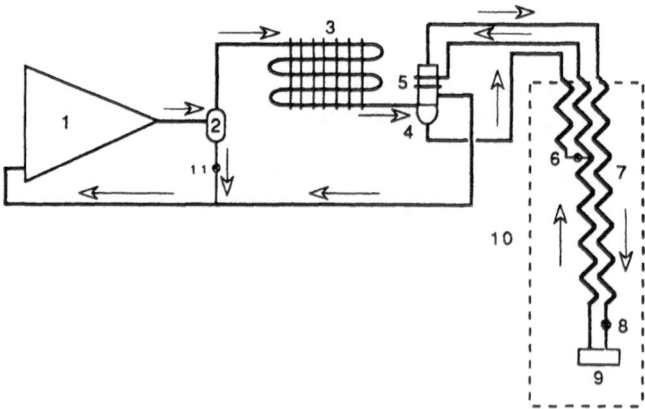

**Figure 2.** Simplified Kleemenko Cycle Cooler with a single liquid-vapor separator[4, 5]

## KLEEMENKO CYCLE COOLERS FOR 150K OPERATION

In this embodiment, warm vapor from the compressor [1] passes through an oil separator [2], where most of the oil entrained in the vapor is removed and is returned to the compressor, the cleaned vapor is then cooled in the air-cooled condenser [3], where some of the high boiling point fraction condenses. This liquid fraction is separated from the vapor stream in the second separator [4] and is fed, via a restricting capillary [6] to a point about a third of the way down the main heat exchanger. Here the liquid vaporizes, cools, and cools the incoming vapor from the top of the vapor/liquid separator [5]. This cooled vapor condenses as it moves down the exchanger [7] and eventually expands through the capillary restrictor [8] cooling the cold stage [9] before passing back up the exchanger. The vapor, after leaving the cooling column [10], cools the top of the separator [5] causing more of the high pressure vapor to condense and be fed via [4] to the heat exchanger at [6].

The refrigerant we use is an eight component mixture of hydrocarbons, fluorinated hydrocarbons, plus argon and nitrogen. This is designed to minimize the irreversibility losses of the heat exchanger over the operating temperature of the exchanger and, in addition, to be compatible with the lubricating oil of the compressor and with the materials of construction of the compressor and refrigeration system.

Our first prototypes were built eighteen months ago and were designed for operation at 120 - 150 K. These were run over a period of six months during which some were run continuously, some were used for refinement of the refrigerant mixture, and some for testing different designs of the heat exchangers. Based on the experience gained, a second set of prototypes of various forms were built and several of these have been running since that time. The general performance of these "150K" coolers is given in Table 1.

A typical cooler is shown in Fig. 3. This unit has been designed for cooling a Low Noise Microwave Amplifier for operation at 140 K. Electronic control and readout of the temperature is provided. The Temperature vs Load for this cooler is shown in Fig 4. and the operating performance is given in Table I. This design has gone through several iterations and has reached a reasonably mature status. The cost of the refrigeration unit itself is low, determined largely by the compressor ($60), condenser ($100), heat exchanger ($50), and dewar ($50) and labor, which is a strong function of production volume. This cost is lower, by at least one order of magnitude, than that for any other refrigeration system for this temperature range. The cost of a complete system can be appreciably higher than that of the basic refrigeration unit, of course, depending upon the details of the engineering interface to the cold stage, the instrumentation provided, and the enclosure.

## KLEEMENKO CYCLE COOLERS FOR 80K OPERATION

To achieve efficient operation at these lower temperatures, a different multi-component refrigerant mixture had to be designed to that used for 150 K operation. The mixture had to be prepared so that it contained only components that did not freeze at the lowest temperature or if they did, would remain in solution in the remaining mixture. In addition, improvements had to be made

**Table 1.** "150 K" Kleemenko Cycle Coolers

| | |
|---|---|
| Minimum Temperature | ≈ 113 K |
| Capacity at 120 K | ≈ 10 W |
| Cooldown Time to 130 K (No Load) | ≈ 20 min |
| Power Input (Compressor and Fan) | ≈ 190 W |
| Number of units built to-date | ≈ 20 |
| Operating time to-date | > 7,500 hrs |
| Noise Level at 1 m | < 56 dB |

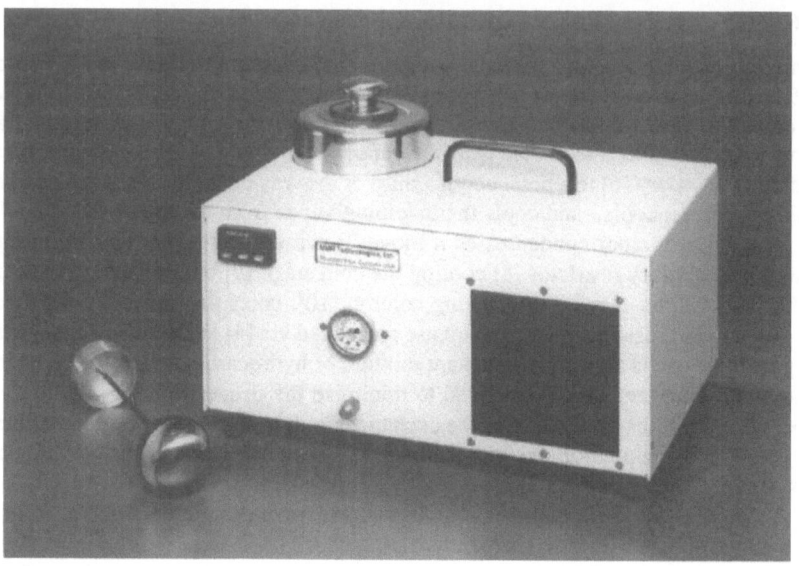

**Figure 3.** "150 K" Cooler for Cooling of Low Noise Amplifiers and BioMedical Materials.

in the separation efficiency of the oil, and vapor/liquid separators to keep the residual oil within the solubility limit of the mixture. This work has been under way for only a few months but promising results have been obtained thus far. A minimum temperature of 76 K has been achieved with a cooling capacity of 5 W at 90 K, and continuous operation at temperatures below 80 K for more than 1,250 hrs to-date. See Table 2. Oil contamination resulting in partial clogging of the capillary restrictors has been the most critical problem. This usually is a consequence of a surge of oil passing through the separators during startup, contaminating the upper heat exchanger, this contaminant then finds its way to the low temperature end over time. Improvements in the efficiency of separation has extended the time of continuous operation from <100 hours on the first prototypes, to 400 and now

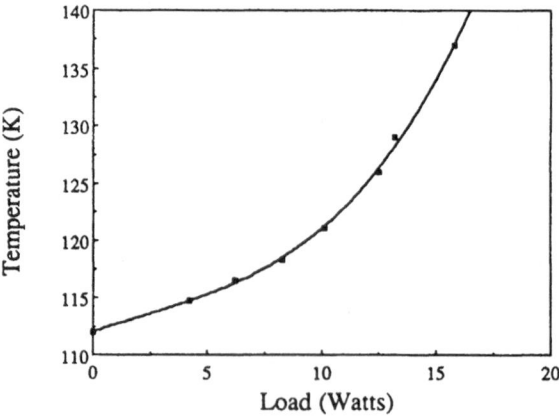

**Figure 4.** Plot of Temperature vs Load of "150 K" Cooler

**Table 2.** "80 K" Kleemenko Cycle Coolers

| Minimum Temperature | ≈  76 K |
|---|---|
| Capacity at 90 K | ≈   5 W |
| Cooldown Time to 90 K | ≈  90 min |
| Power Input at 76 K (No Load) | ≈ 160 W |
| Number of Units built to-date | 3 |
| Operating Time to-date | >1,250 hrs |
| Noise Level at 1 m | <    54 dB |

to more than 1,250 hrs. We believe this problem will be solved shortly with some engineering changes in the vapor/liquid separators. In addition, refinements in the heat exchangers and the composition of the refrigerant mixture are expected to improve substantially the efficiency of refrigeration at these temperatures without significant increase in cost.

## CONCLUSION

The first phase of this program has shown conclusively that efficient, low noise, closed cycle refrigerators using the Kleemenko cycle can be built, at low cost, even in reasonably small quantities, to provide cooling down to about 115 K with excellent efficiency and high reliability. The second phase has shown the feasibility of extending the operating range of these coolers to below 80 K. Work is needed here to improve further the efficiency of the oil, and liquid/vapor separators to allow continuous operation at these temperatures.

## ACKNOWLEDGEMENTS

We are indebted to the Naval Research Laboratories for support of this work under Contract N00014-94-C-2164.

## REFERENCES

1.   A. P. Klecmenko, Proccedings Xth Int. Congress of Refrigeration, Copenhagen, 1, 34 - 39 (1959) Pergamon Press, London

2.   V. Alfeev, V. Brodyansky, V. Yagodin, V Nikolsky and A. Ivantsov, U. K. Patent # 1,336,892 November, 1973.

# A Throttle Cycle Cryocooler Operating with Mixed Gas Refrigerants in 70 K to 120 K Temperature Range

Ajay Khatri and M. Boiarski

APD Cryogenics, Inc.
Allentown, PA 18103 USA

## ABSTRACT

A low-cost, throttle-cycle cryocooler based on a single-stage, oil-lubricated compressor has been designed to meet requirements for cooling electronic devices, water traps, medical and many other applications. Mixed refrigerants of different types are used to operate in the appropriate temperature range within 70 to 120 K.

The adaptability of the basic system configuration to diverse applications is due to the ability of the system to use different refrigerant mixtures and the possibility to customize the system configuration. Requirements such as no periodic maintenance, long life, low noise level and no sensitivity to orientation can be satisfied by such a cooler.

The basic configuration of the cooler was tested at different conditions to define the cooling capacity at different refrigeration temperatures. Computer calculated characteristics of the cooler and the experimental data are in good agreement. It allows us to modify blend compositions and the system modules to satisfy customer needs.

## INTRODUCTION

A cryocooler based on a single-stage, oil-lubricated compressor has been designed to meet different application requirements. The basic design and the initial test results showing the heat load vs. temperature map were described earlier.[1] A commercial cryocooler of this type consists of a compressor assembly, a cryoblock assembly and two connecting gas lines. The cryoblock assembly consists of a counter-flow heat exchanger and a coldplate evaporator which provides the interface with the object to be cooled.

Several mixed-gas refrigerants have been designed to provide operation in appropriate temperature ranges selected from 70 K to 120 K according to the customer needs. The ability to design a mixed-gas refrigerant ( MR ) to fit a desired temperature range of operation without any change in the hardware is due to a unique computer software analytical capability.

The cooler is maintenance free, has high reliability and very low vibration levels.[2] This cryocooler provides a customer with flexibility and adaptability for variety of applications.

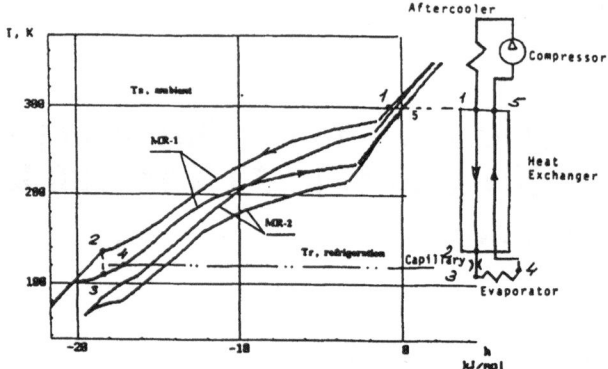

**Figure 1.** Cycle schematic and temperature - enthalpy diagram
MR-1 : He+ Ar + ΣCH;  MR-2 : He + N2 + ΣCH

## BACKGROUND

The flow schematic of the cooler and the throttle-cycle parameters in the temperature-enthalpydiagram is presented in Fig. 1. The simplicity of this cycle makes it attractive to build a low-cost cryocooler. This goal is unattainable when using pure refrigerants. The reason is that the Carnot efficiency of such a cycle is extremely low. In addition, a multi-stage compressor is required to provide high pressure (Ph) in the cycle during the operation. The use of multi-component refrigerants makes it possible to improve the cycle performance and reduce the Ph value. That is why an idea to use MR consisting of both low-boiling and high-boiling components was attractive in the cryocooler design. This brings up the question of how to carry the high-boiling components through the evaporator at the temperatures which may be below the freezing point of the pure components. Many different solutions of this problem were proposed. All of them were based on the cycle modifications which assumed either two compressor cycles[3,4] or the use of phase separators.[5]

An essential step for using MR in a single-circuit cryocoolers based on one compressor was made due to research which began in the 70's under the supervision of Dr. V . Brodianski.[6] Typically the freezing temperatures of the HCs are above 90 K. Blends of nitrogen and hydrocarbons (HC) were found which did not form a solid phase at temperatures below the triple-point temperature of the hydrocarbons included in the blends while improving the Carnot efficiency of the coolers.

Development of these ideas lead to commercially practical cryocooler based on single-stage, rotary piston, oil-lubricated compressor.[7] Complex optimization of the MRs were made providing not only a high refrigeration performance[8] but a low noise level and a relatively short cooldown time.

## CRYOCOOLER DESIGN AND PERFORMANCE

This cryocooler is a closed-cycle system. The system design emphasizes simplicity in fabrication and operation. The flow schematic of the cooler is close to the one presented in Fig. 1. The detailed description of the system setup has been published elsewhere.[1] The oil-lubricated compressor takes in MR at about 1 atm. to 3 atm. and discharges it at about 18 to 20 atm. Gas passes through an after-cooler and an oil separator. Oil is returned to compressor via an oil return line. The high-pressure gas then goes through the counter flow heat exchanger in the cold end. The cold end is designed such that it can tolerate a low level of oil carryover in the gas flow. The high-pressure MR undergoes expansion through a throttling device which is a capillary and produces a cooling effect in the evaporator. It then flows through the heat exchanger and cools the incoming high-pressure gas. Depending upon the MR composition at the cryogenic temperatures (120 K or less) the high-pressure gas can be mostly liquid as it enters the throttling device. The throttling action does not produce any significant change in temperature as the liquid refrigerant passes through the throttle. The refrigerant remains mostly liquid before and after the throttling process.

The cryoblock is designed to allow to run different MR compositions without any change in the hardware. The evaporator design allows possibilities to customize the interface with any object to be cooled. A customer can interface with a standard coldplate or the cold refrigerant can be supplied to the point of use where the cooling is required by a special interface setup. The MR composition depends on customer requirements such as refrigeration temperature, cooldown time, heat-load capacity and the noise/ vibration level required. In most cases the basic ingredients are helium, nitrogen, high-boiling components such as hydrocarbons (HC) and freons (R). There is quite a complex interaction between the gases in the charging blend and the oil in the system. This results in substantial change in the composition from the charge blend to the steady-state blend during the system cooldown.

The system was tested with different gas blends to cover the temperature range from 70 K to 120 K. The hardware remained the same during the tests. The performance is affected by the heat exchanger design, gas-blend design, pressure ratio of the compressor, flow through the throttling device etc. Once the hardware is finalized, then the gas blend composition is the only major variable that can be changed to get the desired performance. All the tested MRs were designed to provide a gaseous state MR composition to the cold end during normal operation within the ambient temperature range of 285 K to 320 K. This makes the cryocooler insensitive to the cryoblock orientation.

The performance data presented here were obtained by testing a few cryocoolers taken from production and prototype units. The present manufacturing technology provides an excellent reproducibility of the data. For the same applied heat load, deviation in the refrigeration temperature between different units is less than ±1 K.

Fig. 2 shows the refrigeration temperature vs. applied heat load at the cold plate. The plot shows the performance of the basic configuration. The system was charged with three different refrigerants MR-1 (He+Ar+$\Sigma$CH), MR-2 (He+N$_2$+$\Sigma$CH) and MR-3 (He+N$_2$+$\Sigma$CH with different molar proportion than MR-2). Testing was done at normal ambient temperature of 290 K to 300 K. The refrigerant was at the static pressure of 19 to 20 atm (270 to 280 psi.) before start.

For the operating range from 70 K to 85 K MR-3 was tested. This range is suitable for high-temperature, superconducting-type devices that need to operate close to liquid nitrogen temperature of 77 K. It carried between 3 to 4 watts at 77 K and had the minimum temperature of about 70 K. It carried a maximum of 8 to 9 watts at about 86 K. To operate in the temperature range of 80 K to 95 K MR-2 was tested. This range is suitable for operating devices like IR detectors, gamma-ray detectors, electron microscopes etc. The minimum temperature reached was 76 K and carried a maximum load of 13 watts at about 96 K. MR-1 was developed to operate in the range of 90 K to 120 K. This range is suitable for operation of water-trap type devices. The minimum temperature of 88 K was reached at no-load condition and carried about 25 watts at about 120 K. The compressor has a power consumption below 500 watts normally and uses maximum power (about 500 W) when operating at about 120 K and is producing 25 watts. The system efficiency, relative to Carnot, at this point is about 7.5%. This is acceptable for a cryocooler with small capacity. If necessary, it is possible to optimize the refrigerants and find a higher-efficiency compressor to further improve the system efficiency.

Fig. 3 shows the performance of the system using MR-2 at different ambient temperatures. The complete system was placed inside an environmental chamber and ambient was varied. The performance curve experienced a small shift with rising ambient temperature as shown.

The cryocooler not only provides acceptable refrigeration capacity but also has a relatively short cooldown time of 40 to 50 minutes, as shown in Fig. 4. The cooldown time of the customer interface will depend on the mass attached to the cold plate. The figure shows the changing parameters during the cooldown for the cooler running the MR-2 blend. The cooler must also be sized to handle active heat load as well as parasitic heat load. The cryocooler can provide from 50% to 80 % of the steady state cooling capacity during the cooldown at any intermediate temperature depending upon the MR selected as shown in Fig. 5.

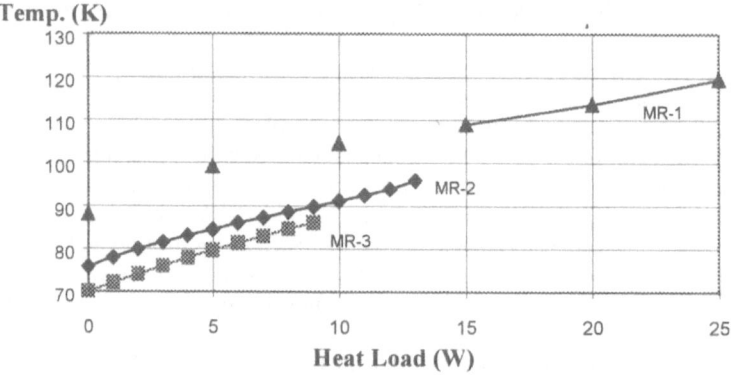

**Figure 2.** Refrigeration map of the cooler based on different refrigerants.

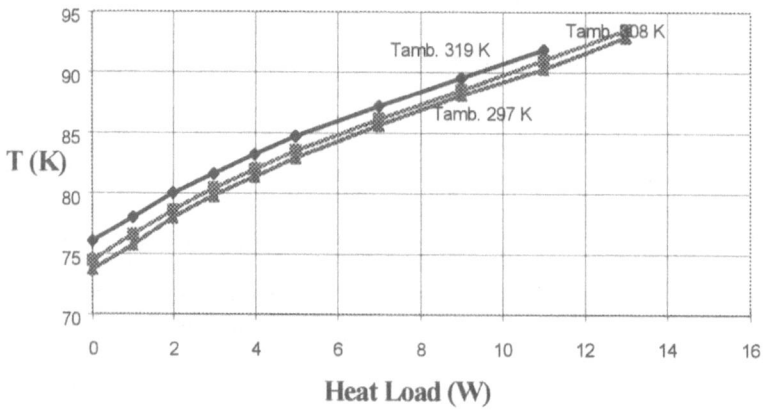

**Figure 3.** Refrigeration map at different ambient temperatures using MR-2.

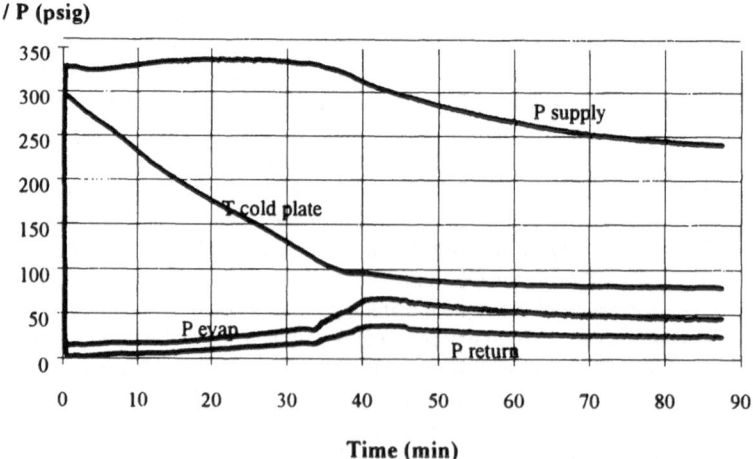

**Figure 4.** Cooldown performance using MR-2.

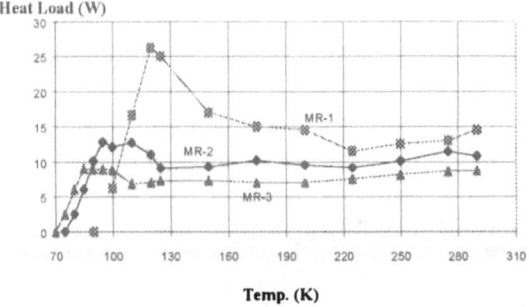

**Figure 5.** Refrigeration map during cooldown.

## CRYOCOOLER ANALYSIS

The characteristics described above correspond to the particular system configuration which has been designed to reach a reasonable compromise with regards to different requirements of customers such as operating temperature, cooldown time, refrigeration capacity, physical size, low vibration, vacuum compatibility, ease of operation, no maintenance, high reliability, low cost, etc.

To satisfy this sometimes conflicting requirements a unique computer software has been designed to predict the cryocooler characteristics and effect of possible variations of parameters. It calculates multi-phase equilibrium and thermodynamic properties of the MRs within wide range of temperatures and pressures using equation of state.[9] It also calculates transport properties and oil-refrigerant solubility. On this base it is possible to predict the cryocooler performance if the configuration of the system, volumetric efficiency of the compressor and heat exchanger parameters are known.

To calculate the refrigeration capacity (Qr), we need to define the refrigeration temperature (Tr), compressor discharge pressure (Ph), evaporator pressure (Pl), the heat exchanger dimensions and the compressor characteristics. The heat exchanger warm end temperature difference (dT) between the high and low pressure gas stream was measured which helped us to correlate the heat transfer coefficients for the calculations. Table also shows the pressure drop between compressor suction point and evaporator as (dPlow) which is the result of the heat exchanger pressure drop and the gas connecting gas lines.

Comparison of experimental and calculated data are presented in table 1 below for the basic cryocooler described above using MR-2. The calculated and the measured values of Qr provide reasonably good match for system performance.

**Table 1.** Comparison of Experimental and Computer Calculated Data for Basic Cryocooler Using MR-2.

| # | \multicolumn{5}{c}{Experimental Data} | | | | | \multicolumn{5}{c}{Calculated Data} | | | |
|---|---|---|---|---|---|---|---|---|---|---|
| | Ph atm | Pevap. atm | dPlow atm | dT warm end K | Pcomp W | Heat Load W | Tr K | Qr W | Pcomp. W | dP-HX atm | Tamb. K |
| 1 | 19 | 3.4 | 1.2 | 8.0 | 430 | 0.0 | 74 | 0.8 | 380 | 0.55 | 298 |
| 2 | 19.6 | 3.7 | 1.4 | 8.0 | 440 | 3.0 | 80 | 2.6 | 395 | 0.65 | |
| 3 | 20.1 | 3.8 | 1.4 | 6.6 | 450 | 5.0 | 83 | 5.2 | 400 | 0.70 | |
| 4 | 21.6 | 4.2 | 1.5 | 5.0 | 470 | 11.0 | 90 | 9.6 | 430 | 0.80 | |
| 5 | 22.0 | 4.0 | 1.3 | 9.0 | 470 | 0.0 | 76 | 1.3 | 415 | 0.64 | 316 |
| 6 | 22.6 | 4.2 | 1.4 | 9.0 | 480 | 2.0 | 80 | 2.8 | 425 | 0.75 | |
| 7 | 23.5 | 4.4 | 1.6 | 8.0 | 490 | 5.0 | 85 | 5.5 | 440 | 0.8 | |
| 8 | 25 | 4.9 | 1.8 | 6.3 | 520 | 11.0 | 92 | 11.1 | 470 | 0.9 | |

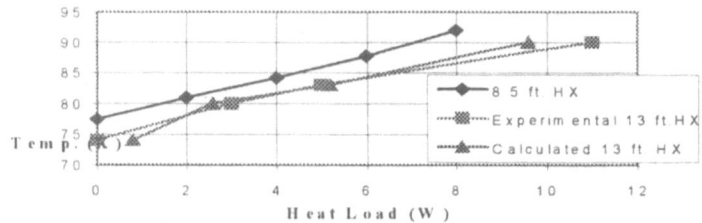

**Figure 6.** Refrigeration map with the different size heat exchangers using MR-2.

It is possible to improve some particular parameter at the cost of some other parameter. For example it is possible to reduce cryocooler dimensions using a smaller heat exchanger at the cost of some refrigeration capacity. Fig. 6 shows the performances of the basic cryocooler (13 ft. long heat exchanger) and shorter heat exchanger (8.5 ft. long) using same gas blend MR-2. If we need higher Carnot efficiency then it is possible to have a refrigerant blend with more high boiling components but the system becomes more sensitive to the ambient temperature variations. In a controlled environment this may be acceptable.

## SUMMARY

Cryocooler of this type offers unique flexibility in system design. The refrigerant mix can be designed for a specific application and desired performance can be achieved without making any change to hardware. With the proper mixed-gas refrigerant design it is also possible to limit the minimum temperature the system could reach, thus eliminating any need for a temperature controller. The cryocooler is producable in high volume at lower cost compared to other type of cryocoolers. It is maintenance free and has high reliability, long life, no moving parts in the cold end, low noise and vibration levels and separation of the coldend and the compressor. It is also possible to customize the coldend configuration and deliver the refrigeration at the point of use. The cold end could also be designed as an integrated part of the customer device. It is possible to use such a cryocooler where the use of liquid nitrogen is not convenient or desired. Possible applications include infra-red detectors, gamma ray detectors, HTS filters, medical applications, electron microscope, material science studies, non-destructive testing and many more.

## REFERENCES

1. Longsworth R., Boiarski M., Klusmier L., "80 K Closed-Cycle Throttle Cycle Refrigerator", Cryocoolers 8, Plenum Press NY, 1995 p. 537.

2. Dennis Hill, "Throttle Cycle Cooler Vibration Characterization", 9th International Cryocooler Conference, June 1996.

3. Gaumer.L. "LNG Processes," Advance in Cryogenic Eng'g, Vol. 31, p. 1095, Plenum Press, NY 1986.

4. Missmer, D. "Self-Balancing, Low-Temperature Refrigeration System", US Patent 3,768,273; 1973.

5. Klimenko, A., "One Flow Cascade System", Progress in Refrigeration Science and Technology, Pergamon press, 1960.

6. Brodianski, V. and all, "High Efficient Cryocoolers Working with Multi-Components Refrigerants", Chimicheskoe and Neftianoe Mashinostroenie. No. 12, p.16, 1971, Moscow, Russia.

7. Longsworth R.C., "Cryogenic Refrigerator with Single Stage Compressor", US Patent 5,337,572; 08/94.

8. Boiarski M., Yudin B., Mogorichny V., and all, "Cryogenic Mixed Gas Refrigerant for Operation within Temperature Ranges of 80 K - 100 K", US Patent 5,441,658; 1995.

9. Boiarski M., Podchevniaev O., "Methods of Calculations of Phase Equilibria and Thermodynamic Properties for Analyzing Throttle Refrigeration Cycles on Multi-Component Mixtures", Viscotemperaturnaia Sverkhprovodimost, V. 3-4, 1990, Moscow, Russia.

# 80 K Throttle-Cycle Refrigerator Cost Reduction

**R. C. Longsworth**

ADP Cryogenics
Allentown, PA, USA

## ABSTRACT

Throttle cycle refrigerators have recently been introduced to the market which use a single oil lubricated air conditioning type compressor with an air cooled aftercooler and oil separator to supply a mixed refrigerant to a counter flow heat exchanger and throttle valve in a vacuum housing. The basic refrigerator presently sells for about $6,000 and is being used to replace LN2 dewars in applications such as X Ray and Gamma Ray detectors because no maintenance is required.

This paper describes the work being done to reduce the manufacturing cost in larger production volumes in order to facilitate applications of new high temperature superconductor, HTS, technology. One of the refrigerant mixtures that was tested with the experimental heat exchanger produced 7 W of cooling at 80 K with 450 W input power. An interface was built to cool an HTS-Sapphire Resonator which added several different options for reducing vibration to the basic cooler. Tests with the resonator showed good performance with the basic cooler.

## INTRODUCTION

New HTS electronic devices have a lot of promise to find important applications in filters for cellular communication and resonators for improved radar systems. HTS SQUIDS are also being tested for non destructive testing and biomedical applications. Typical cooling requirements range from a fraction of a watt for SQUIDS to about 5 watts for multiple filters.

One of the most important drivers in reducing the cost of a suitable refrigerator is to be able to build it in large quantities. The throttle cycle refrigerator that we have designed has a lot of flexibility in using different mixtures in the same hardware to be used in many different applications including SQUIDS, HTS filters and resonators, X ray and Gamma ray detectors, IR detectors, laboratory instrumentation, and vacuum system cold traps. The system has been designed to use a standard cooler with adapters for these different applications.

Cryocoolers 9, Edited by R.G. Ross, Jr.
Plenum Press, New York, 1997

Funding for this program was provided by the Navy and included the integration of an HTS electronic device. We have been working with Du Pont to cool a high power 5.5 Ghz HTS-Sapphire Resonator which is made with thallium based material[1]. They have been testing it in a LHe and LN2 cooled dewars. The need for a low level of vibration and stable temperature were recognized but the values had not been quantified. The experimental cryostat that was built was designed to enable vibration limits and temperature stability requirements to be quantified.

## STANDARD REFRIGERATOR

The standard Cryotiger® refrigerator which was just being introduced to the market when this program was started was previously described[2]. Figure 1 is taken from that paper for reference, with dashed line boxes around the components that were changed. The compressor system has an oil lubricated air conditioning type compressor,[3] an air cooled after cooler, an oil separator with capillary oil return, an adsorber, and electrical controls for 115 V 60 Hz or 230 V 50 Hz operation. Two flexible gas lines with self sealing couplings connect the compressor to the cold end which has a multi-tube heat exchanger and an adjustable throttle valve. The gas mixture provided a minimum temperature of about 78 K with the valve nearly closed and carried 10 W at 90 K with the valve further open. Cool down time was minimized by starting with the valve full open and reducing it during cool down to < 1 turn open for operation below 90 K. A fixed setting at a mid position resulted in a slightly longer cool down time and a temperature of about 100 K until the valve was closed further.

In addition to reducing the system cost we wanted to reduce the minimum temperature, increase the cooling capacity, and design an automatic throttle valve that would minimize cool down time and reach the desired minimum temperature.

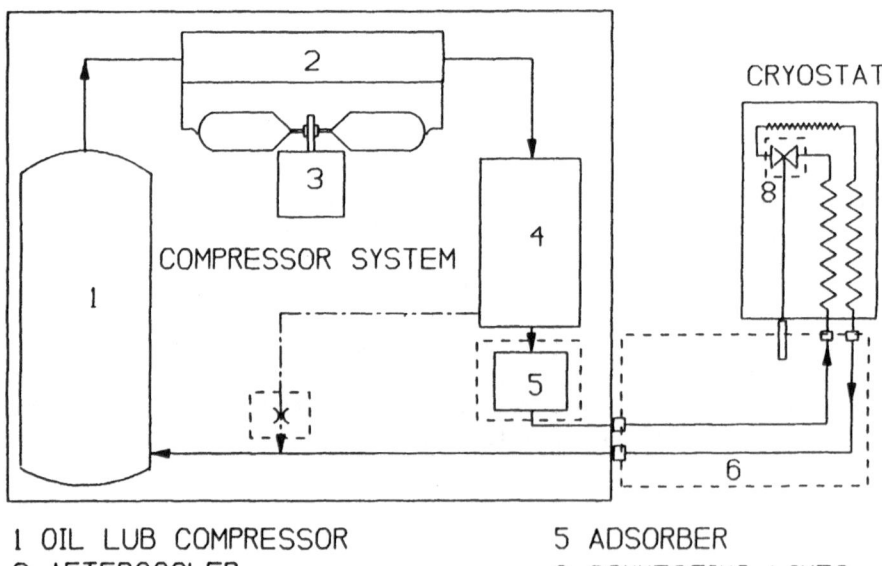

| | |
|---|---|
| 1 OIL LUB COMPRESSOR | 5 ADSORBER |
| 2 AFTERCOOLER | 6 CONNECTING LINES |
| 3 FAN/MOTOR | 7 HEAT EXCHANGER |
| 4 OIL SEPARATOR | 8 THROTTLE VALVE |

**Figure 1.** Piping schematic of a standard system. Components with dashed boxes were changed.

## EXPERIMENTAL UNIT DESIGN

A lot of work was done under APD funding to develop an automatic throttle valve which would have a large orifice at room temperature and would reduce the flow area as the unit cools down until it reaches a preset minimum area corresponding to the desired minimum temperature. Several designs were tested with good results in minimizing cool down time. In parallel we studied the cool down characteristics with capillary throttle tubes and discovered how to have reasonably fast cool down and a low temperature after cool down*. This has enabled the use of a simple capillary tube which has enhanced the reliability of the present standard cooler and simplified the construction.

A capillary throttle tube was incorporated in the experimental cryostat that was built to study the cooling requirements of the DuPont resonator. Figure 2 shows a cross section of the experimental cryostat that was built. The heat exchanger is a coiled multi-tube type which has more surface area than the standard heat exchanger. Since sensitivity to vibration was one of the major unknowns to be evaluated this unit incorporated a number of vibration isolation mechanisms. The cryostat was set on a soft pad to isolate vibration from its support. It was known from previous work that we had done that the flow of fluid through the throttle device generates a low level of vibration which can be a problem in some applications[4]. This vibration can be attenuated by flexible braided copper straps between the refrigerator cold plate and the plate to which the resonator is attached*. The mass of the arrangement helps to isolate vibration but results in a long cool down time. A split G10 tube is clamped outside the heat exchanger coil to support the cold mass. Gas was supplied from a standard compressor.

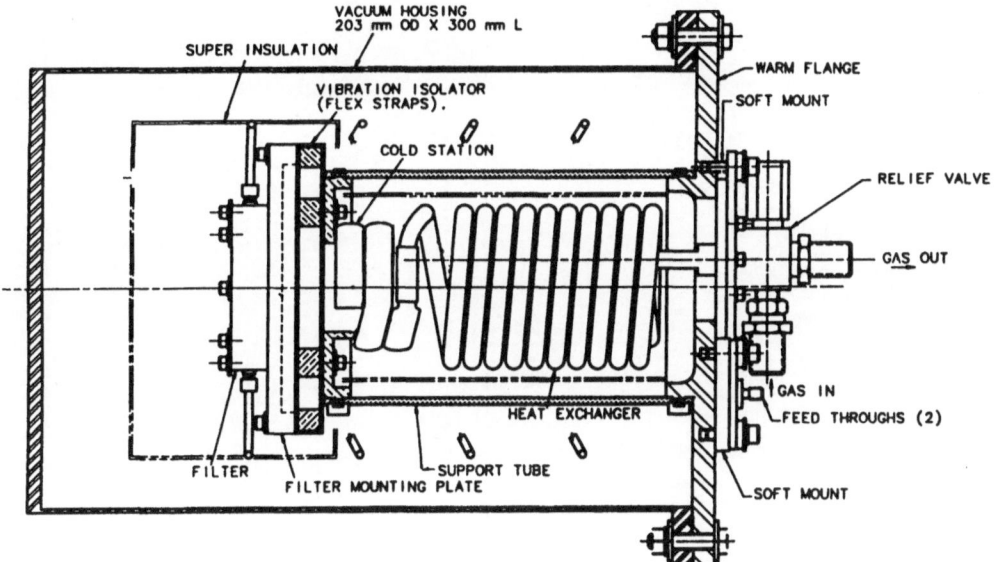

**Figure 2.** Cross section of experimental cryostat.

---

*    Patents Pending

## EXPERIMENTAL UNIT TEST RESULTS

Refrigeration capacity tests were run with three different mixtures, designated 1, 2, and 3, that were designed to operate over different temperature ranges. The vibration isolator and dummy resonator were not mounted during these tests and the heat exchanger was wrapped with aluminized mylar. Results are plotted in Figure 3. Parasitic losses are estimated to be about 2 W based on warm up data after the unit is turned off. Mixture #1 had a minimum temperature of 69 K and carried 7 W at 80 K. Mixture #2, which was selected for cooling the resonator, had a minimum temperature of 72 K and carried 13 W at 88 K. The third mixture carried 25 W at 115 K. When the vibration isolator and dummy resonator were mounted the minimum temperature was 75 K and a maximum heat load of 12 W was carried at 93 K.

Selected cool down plots are shown in Figure 4 for mixture #2 and the basic cold end with out the vibration isolator and dummy resonator. The equivalent weight of Cu in the heat exchanger and cold plate that is being cooled is approximately 1,100 g. The top plot in Figure 4 shows temperature of the high pressure fluid entering the throttle tube and temperature at the cold plate. Temperatures are nearly the same while gas is entering the throttle tube but spread apart when liquid starts to enter. When it is cold and the flow in the throttle tube is mostly liquid there is very little temperature drop during the throttle process. The process involves a very small change in entropy and contributes to the high thermodynamic efficiency of this cycle relative to a conventional JT process. For this reason we refer to this as a throttle cycle.

The high and low pressures in the system change as the flow rate through the throttle tube changes and as liquid collects in the heat exchanger and cold end. The bottom plot in Figure 4 shows a calculation of the rate at which heat is being removed from the cold mass assuming it to be 1,100 g of Cu. This has some inaccuracy because the material is not all Cu and because the heat exchanger temperature profile is changing at a different rate than the cold mass. The refrigerator is producing more cooling than this to cover parasitic losses.

Power input readings that were taken during cool down are shown on the plot with low pressure. Under no load conditions power input was 360 W and when maximum heat load was applied the power input was 460 W, the same as the peak during cool down. A lot of liquid

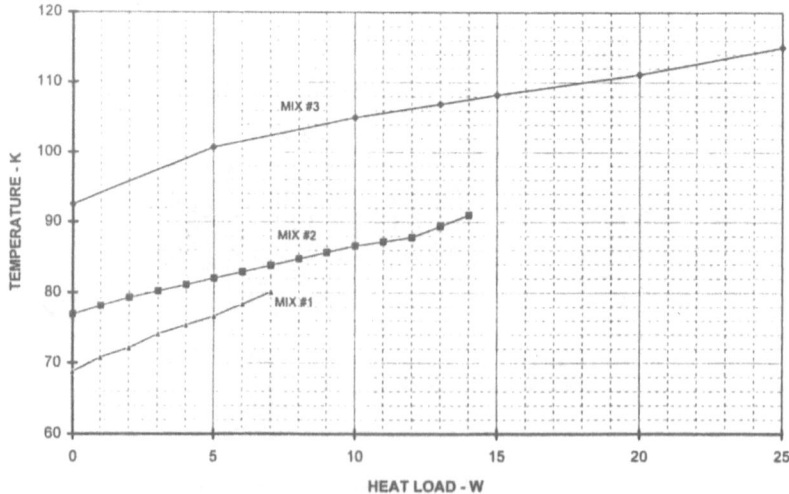

**Figure 3**        Temperature versus cooling capacity for three different mixtures in experimental cryostat

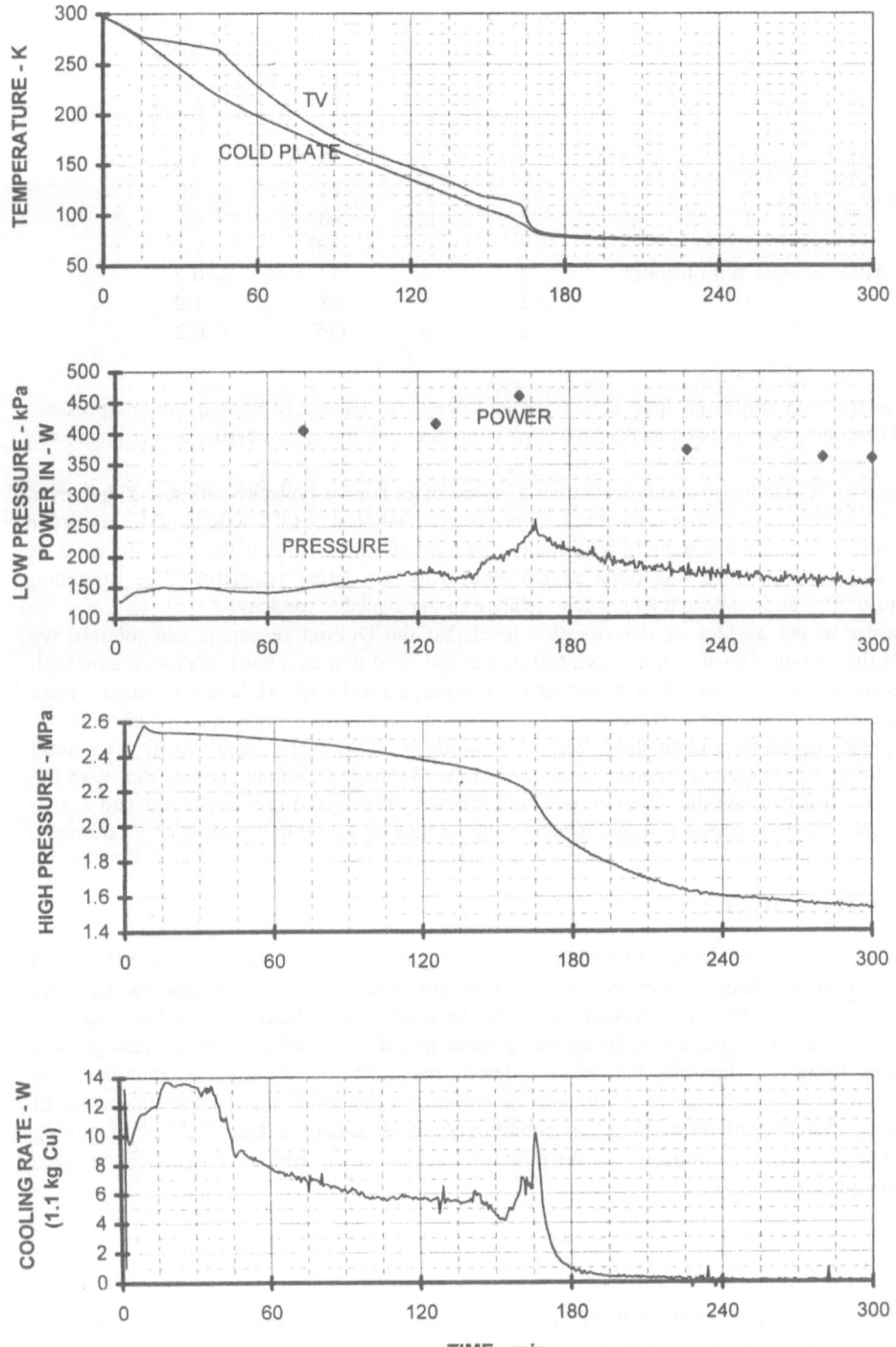

**Figure 4** Cool down plots for mixture #2 and the basic cold end.

**Table 1**   Summary of Vibration Measurements

| Configuration | Mixture | Load - W | Temp. - K | Vibration re LN2 |
|---|---|---|---|---|
| Du Pont LN2 Dewar | | 0 | 77 | 1.1 |
| | | 2 | 77 | 1.0 |
| | | 0 | 291 | 0.3 |
| APD Cryostat Without Isolator | 2 | 0 | 73 | 2.8 |
| | 2 | 5 | 81 | 3.7 |
| | 2 | 13 | 88 | 4.7 |
| | 2 | 0 | Off | 1.0 |
| APD Cryostat With Isolator | 2 | 0 | 77 | 0.7 |
| | 2 | 5 | 88 | 1.0 |
| | 2 | 0 | Off | 0.2 |

collects in the cold end at no load conditions and pressures change to unload the compressor. Some of this liquid is vaporized as the heat load is increased so the high pressure and input power increase.

Temperature stability in steady state over a period of an hour is typically within .3 K however changes in ambient temperature cause the temperature to change about 0.1 K/K at 80 K. In order to maintain the temperature stability required by the DuPont resonator it is necessary to have an active controller. Testing was done at Du Pont with an actual resonator. The operating temperature of < 80 K reflected a light load relative to the available capacity.

In order to get an idea of the vibration level that the DuPont resonator can tolerate we measured the vibration level of their LN2 test dewar and used that as a basis of comparison with the experimental refrigerator. Both cryostats were sitting on soft pads while measurements were made. Results of these tests are listed in Table 1. We also made up a voice coil oscillator driven by a frequency generator and amplifier which we mounted on the warm flange so that we could input vibration with a specific frequency and amplitude. Testing at DuPont showed that with the cryostat on a soft pad and the vibration isolation mechanism defeated their resonator had a 3 % degradation. The basic cooler with out isolators should thus be acceptable in most applications.

## FINAL SYSTEM DESIGN

The final system was designed to minimize cost in quantities from about 500 to 5,000 units per year with minor changes. For example in lower quantities all of the compressors will have transformers so that a single compressor model can be used in any country. At higher quantities the transformer is eliminated and different compressor models are used for specific voltages and frequencies. There is an interplay between the design, manufacturing tooling, and manufacturing processes, all of which change as production quantities are increased. Given the similarities of this cooling system with household air conditioners it is expected that the manufacturing processes and costs, and probably the distribution / selling costs, will be comparable in large production quantities.

## Cryostat Design

In designing the final cryostat it was decided to have the basic heat exchanger, cold plate, warm flange, and cold plate support, be a standard design that can be used for different applications. Adaptations for specific applications can then be added to it. Figure 5 is a layout of the final cryostat with the adapter flange for the resonator feedthroughs and a vacuum housing.

The edge of the warm flange can either be welded to a vacuum housing or bolted with an "O" ring seal as shown. The most significant change from a cost standpoint is the replacement of multi-tubes in the heat exchanger with a single shaped piece. For a given heat transfer efficiency this weighs more than the multi-tube design so cool down time is increased slightly. The use of a capillary throttle tube that is integral to a flattened section of tubing at the cold end of the heat exchanger further simplifies the construction at the cold end relative to previous designs. This cold end heat exchanger is coiled around the outside of the cold station and soldered as shown in Figure 5. In addition to cost savings associated with fewer pieces in the cold end there are important cost savings in the manufacturing processes associated with cleaning, leak checking, and pressure testing. A support tube is located inside the heat exchanger coil and the center area is left open for possible applications that would require access along the axis. It is important to note that there are no critical tolerances in the construction of the cryostat assembly.

## Concentric Gas Line

A self-sealing concentric coupling has been designed that enables the use of a single gas line between the cold end and the compressor. Having the high pressure gas line inside the low pressure line extends the heat exchanger back to the compressor, but it is uninsulated so it will not contribute much to the cooling. The cost of the concentric gas line and couplings does not drop below the present cost of separate lines and standard couplings until production quantities exceed about 1,000 units per year. The single line is more convenient and flexible so it is attractive even at low production quantities.

## Compressor Assembly

With reference to the standard unit shown in Figure 1 the significant changes in the compressor are the elimination of the adsorber and the use of an oil float return mechanism in the oil separator in place of a restrictor tube. Removal of the adsorber is possible if the procedures for purifying the oil and cleaning the system are maintained at high levels. The processes are known from the fabrication of our 4 K GM/JT systems and require an additional investment in equipment to implement in the throttle cycle systems. The oil float return costs about the same as the present capillary tube but increases the capacity slightly because it avoids having any gas return with the oil. Other design changes were aimed at reducing the number of components, reducing the number of assembly steps, and enabling the compressor assembly to be continuous.

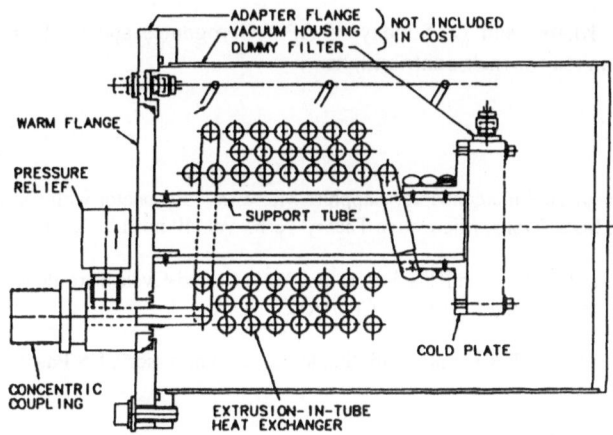

**Figure 5.** Cross section of final cryostat.

## COST

System cost, taken as cost to the customer, includes the compressor, concentric gas line, the basic cold end assembly, and a shipping container. It does not include the warm flange adapter with instrumentation, vacuum housing, or shipping. Annual operating cost for the user would be the cost of power, about 400 W, but no maintenance cost. Life expectancy would be the same as a standard refrigerator.

APD Cryogenics has set up a separate assembly area for Cryotiger products but there is still interaction with the rest of the shop in processing materials. Some of the cost savings of the new design are based on new material handling procedures and processes that reduce the time required to assemble and test the system. Pricing of the basic system in small quantities is presently about $6,000. When the volumes approach those of the commercial refrigeration industry the resulting system pricing is anticipated to be in the range of $700. This is about 50 % more than the cost of a window air conditioner with the same size compressor.

The continued reduction in prices from the current levels assumes that costs are reduced in larger production volumes for many reasons including changes in design, assembly tooling, assembly processes, labor efficiencies, overhead, sales, etc.. It requires an active program to implement the changes which are encouraged by market demand.

## SUMMARY

The present Cryotiger throttle cycle refrigerator is gaining good market acceptance in applications that require cooling of several watts above 75 K because of its simplicity, lack of moving parts in the cold end, and no need for service. In the present application to cooling a HTS resonator it has been found that the vibration level of the basic cooler is low enough that it should be acceptable to mount the resonator directly on the cold plate. Manufacturing costs are being reduced by the use of a shaped insert in the heat exchanger, an integral throttle tube in a formed cold heat exchanger, and the elimination of an adsorber in the compressor as a result of improved cleaning processes. A single concentric gas line with a new self-sealing coupling will reduce the cost in larger quantities but provides more convenience now.

It is hoped that the present efforts in developing a new refrigerator which is simpler, requires no maintenance, should be more reliable, and is less expensive than previous cryogenic refrigerators will enable new developments in cryo-electronics to be commercialized.

## ACKNOWLEDGEMENTS

L. A. Klusmier, A. Khatri, and D. Sunday of APD Cryogenics and C. F. Collier and C. Wilker of DuPont are thanked for their contributions.

## REFERENCES

1. Shen, Z. Y., et. al., "High Tc Superconductor-Sapphire Microwave Resonator with Extremely High Q-Values up to 90 K", IEEE Trans. Microwave Theory Tech., vol. 40 no. 12, Dec 1992.

2. Longsworth, R. C., Boiarski, M. J. and Klusmier, L. A. "80 K Closed-Cycle Throttle Refrigerator" Proceedings of the 8th International Cryocooler Conference 1994.

3. Longsworth, R. C. "Cryogenic Refrigerator with Single Stage Compressor", US Patent 5,337,572, (Aug. 1994).

4. Hill, D. " Throttle Cycle Cooler Vibration Characterization" Proceedings of the 9th International Cryocooler Conference 1996.

# Experimental Investigation of an Efficient Closed-Cycle Mixed-Refrigerant J-T Cooler

**E.C. Luo, J. Liang and Y. Zhou**

Cryogenic Laboratory, Chinese Academy of Sciences

P. O. Box 2711

Beijing 100080, China

**V. V. Yakuba and M. P. Lobko**

V. Berkin Institute for Low Temperature Physics and Engineering

Ukraine Academy of Sciences

Kharkov 310204, Ukraine

## ABSTRACT

The thermodynamic performance of a closed-cycle Joule-Thomson cryocooler with gas mixtures is experimentally investigated. Good refrigeration performances have been achieved with rather low pressure ratios. The high pressure of the two-stage oil-free compressor ranges from 2.5MPa to 4.5MPa, but the low pressure of the compressor is fixed as 0.1MPa. And the temperature of the inlet of high pressure flow for heat exchanger is fixed as 300K. In the experiment, two kinds of gas mixtures are used. The first one is $N_e + CH_4 + C_2H_6 + C_3H_8 + i - C_4H_{10}$ and the second one is $N_e + CH_4 + C_2H_6 + C_3H_8$. Under the above-mentioned operating conditions, the mixed-refrigerant J-T cooler achieved the lowest temperatures of around 69K-70K and 25W-35W of cooling capacity at liquid nitrogen temperature. The excergy efficiency of the whole cooling system is about 5-9%. The experimental results show that using gas mixtures as working substance is indeed a very potential way to improve the efficiency of Joule-Thomson cryocooler.

## INTRODUCTION

It is well known that Joule-Thomson cooler has very low efficiency compared with Stirling cooler at the level of liquid nitrogen temperature[1]. The excergy efficiency of a closed-cycle Joule-Thomson cooler with pure nitrogen gas approximately ranges from 0.5% to 1.5%. However, a Stirling cooler has rather high excergy efficiency of around 10% or sometime more. The reason of Joule-Thomson cooler still being widely used is because it has simpler structure, higher reliability and flexibility, lower vibration and noise, much more easier miniaturization, etc.. In terms of thermodynamics the low efficiency of Joule-Thomson cooler is determined and limited by its intrinsically thermodynamic processes and thermal properties of working substance. About from 1930's, some scientists tried to use gas mixtures as the working substances of Joule-Thomson cooling system, but the most successful experiment for liquefying natural gas with gas mixtures was made by Professor Klimenko A. P.[2]. In 1970's to 1980's, some other Russian and American scientists focused on liquid-nitrogen temperature J-T cooler with gas mixtures[3,4,5]. Recently, some scientists began to be interested in mixed-refrigerant throttling cooler with commercial oil-lubricated air-conditioning compressor[6,7].

The efficiency of Joule-Thomson cooler using gas mixtures as working substances is increased dramatically, but its internal working mechanism ( hydraulic process, heat transfer process, phase equilibrium transition process, etc. ) also becomes complicated extremely. Due to lack of adequate experimental information about reliable flowing and heat transfer coefficients and reliable properties of multiphase multicomponent mixtures, it is difficult, up to now, to theoretically and quantitatively optimize components, compositions and working conditions of gas mixtures used in Joule-Thomson cooler. At present, the optimization work is still ongoing. Experimental investigation is still a direct and effective way to reveal the working process of mixed-refrigerant Joule-Thomson cooler.

Experimental investigation of a closed-cycle Joule-Thomson cooler with gas mixture when it operating at relatively low working pressure from 2.5MPa to 4.5MPa, is presented in this paper. Some important parameters for revealing the internal working process of Joule-Thomson cooler are measured. And two parameters focusing only on gas mixture, specific cooling capacity and excergy efficiency are discussed. Taken into consideration of practical performance of compressor used in the experiment, the total excergy efficiency of the closed-cycle Joule-Thomson cooler with gas mixture is also discussed.

## EXPERIMENTAL SET-UP

The scheme of experimental set-up is shown in fig. 1. The experimental set-up is made up of four subsystems: the cooler system, the gas-mixture-prepared system, the measuring system and the auxiliary system. The cooler system mainly consists of compressor, recuperative heat exchanger, throttling valve. In the experiment a two-stage oil-free compressor is used. The compressor can supply discharging pressure up to 6.0MPa and sucking pressure from 0.07MPa to 0.4MPa. The maximum volumetric rate can be up to 3.0Nm$^3$/h. The gas-mixture-prepared system comprises couples of balloons and controlling valves. The measuring system includes several copper-constant thermocouples, pressure gauges, flowing rate meter, power meters and vacuum meter. The auxiliary system is made up of water cooling system, cold nitrogen gas cooling system , heater system and vacuum system.

The main components of the whole refrigeration system in figure 1 are: 1-Two-stage Oil-free Compressor ; 2-After-cooler No.1 (with water); 3-Mechanical Filter; 4-After-cooler No.2 (with cold nitrogen gas); 5- Electrical Heater; 6- Absorption Filter; 7-Counter-flow Heat Exchanger; 8-Adjustable Throttling Valve; 9-Evaporator; 10-Simulating Load Heater; 11-Mass-flow Meter; 12-Balloon for Gas Mixture; 13-Vacuum Pump No.1 for Refrigeration System; 14-Vacuum Pump No.2 for Vacuum Chamber.

In the experiment the given working condition is as follows.
The output pressure of compressor:          2.5MPa - 4.5MPa
The inlet pressure of compressor :           0. 1MPa

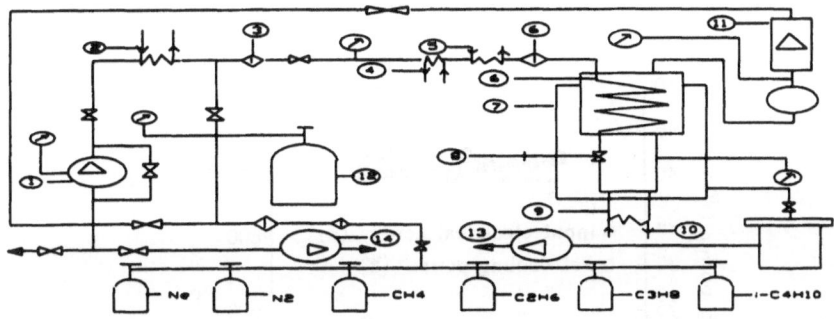

**Figure 1.** Experimental set-up of mixed-refrigerant Joule-Thomson cryocooler

The inlet temperature of heat exchanger : 300. 0K

The two samples of gas mixtures are prepared to be investigated in the experiment. They are gas mixtures No.1 $N_e + CH_4 + C_2H_6 + C_3H_8 + i - C_4H_{10}$ and No.2 $N_e + CH_4 + C_2H_6 + C_3H_8$. To develop a long-life and efficient mixed-refrigerant Joule-Thomson cooler at liquid-nitrogen temperature, the mixture No.1 is optimally selected based on Peng-Robinson equation of state at the following given working conditions: 2.5MPa for high pressure, 0.1MPa for low pressure, 300K for the inlet temperature of recuperative heat exchanger, 77K for cold-end temperature.

## DISCUSSIONS AND CONCLUSIONS

Generally speaking, two important parameters, specific refrigeration capacity and excergy efficiency, are employed to appreciate thermodynamic performance of mixed refrigerant used for Joule-Thomson cooler. The total thermodynamic excergy efficiency of the whole refrigeration system depends not only the used gas mixtures but also the practical performances of compressor, counter-flow exchanger, heat leakage, etc.

Based on the originally measured data, we can calculate such thermodynamic parameters as specific refrigeration capacity, excergy efficiency, COP and so on. Fig.2 and Fig.3 are the experimental curves of net cooling capacity and input electrical power of two mixtures respectively. From this two curves, one can find that net cooling capacity of two mixtures depend upon both the cold end temperature and working pressure but the input electrical power nearly most keep constant values. For No.1 gas mixture, the net cooling capacity of 15W to 25W at liquid nitrogen temperature is achieved when working pressure from 2.5MPa to 3.5MPa, but the electrical power keeps around 1.0kW during the experimental process. From lowest temperature (~69K-70K) to liquid nitrogen temperature the net cooling capacity increase continuously, and up to liquid nitrogen temperature the net cooling capacity doesn't increase and then keeps constant valve of about 25W. For No.2 mixture it has similar behavior with No.1 mixture. Fig.4 and Fig.5 are the experimental curves of specific cooling capacity and the total excergy efficiency of two gas mixtures working at different cold end temperatures and pressures. Fig.4 and Fig.5 show that the specific cooling capacity increase with the growth of cold end temperatures and working pressures. The total excergy efficiency firstly increases with the growth of cold end temperature up to around liquid nitrogen temperature, then decrease with the growth of cold end temperature. In addition to the total excergy efficiency increases with the growth of working pressure. For No.1 mixture the total excergy efficiency of 5% to 7% and the specific cooling capacity of 7.0W.h/Nm³ to 9.5W.h/Nm³ at liquid nitrogen temperature when working pressure from 2.5MPa to 3.5MPa are achieved. For No.2 gas mixture the total excergy efficiency of 4.5% to 8% and the specific cooling capacity of 6W.h/Nm³ to 14W.h/Nm³ at liquid nitrogen temperature when working pressure from 2.5MPa to

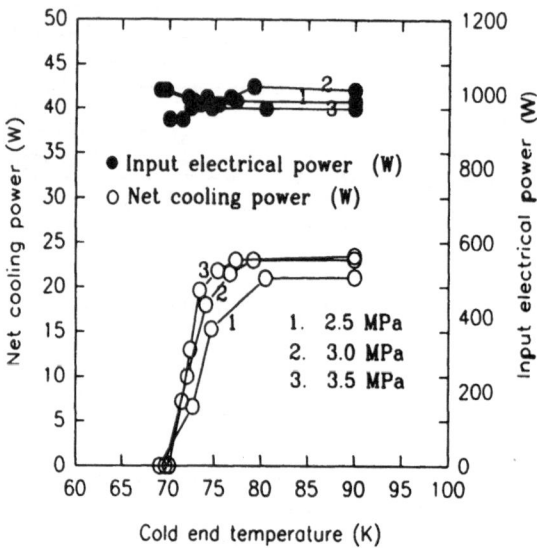

**Figure 2.** Net cooling power and input electrical power vs. cold-end temperature

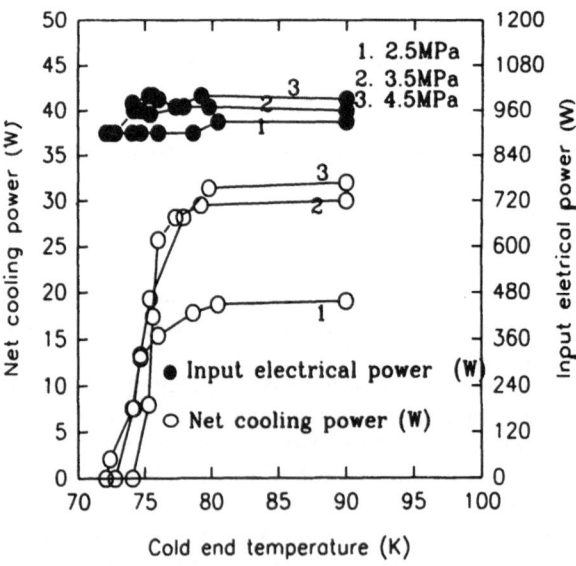

**Figure 3.** Net cooling power and input electrical power vs. cold-end temperature

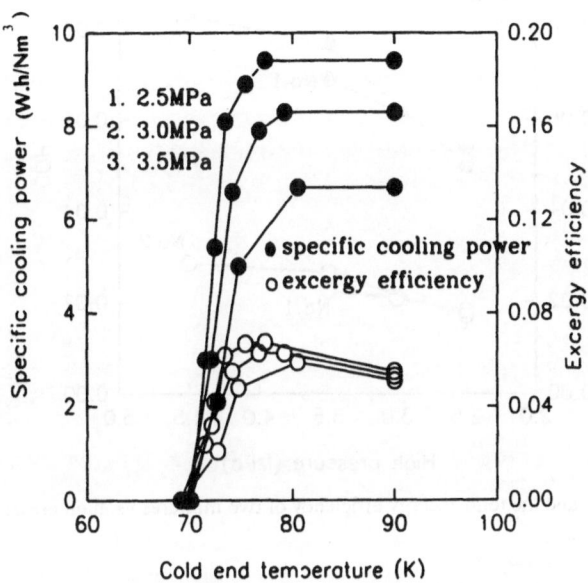

**Figure 4.** Specific cooling capacity and excergy efficiency vs. cold-end temperature

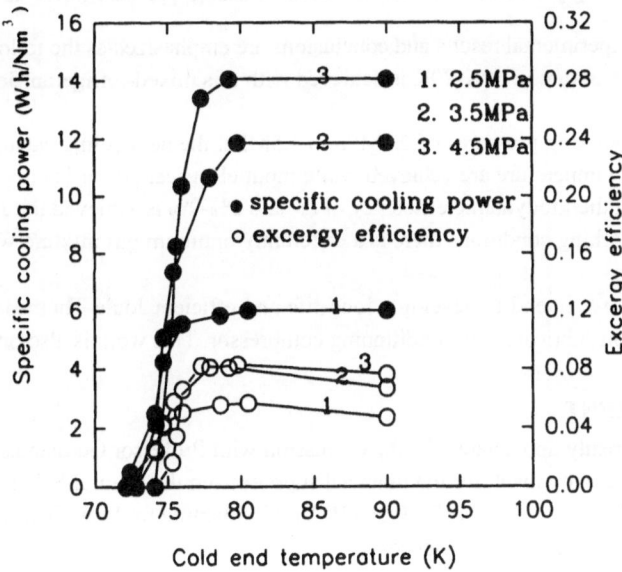

**Figure 5.** Specific cooling capacity and excergy efficiency vs. cold-end temperature

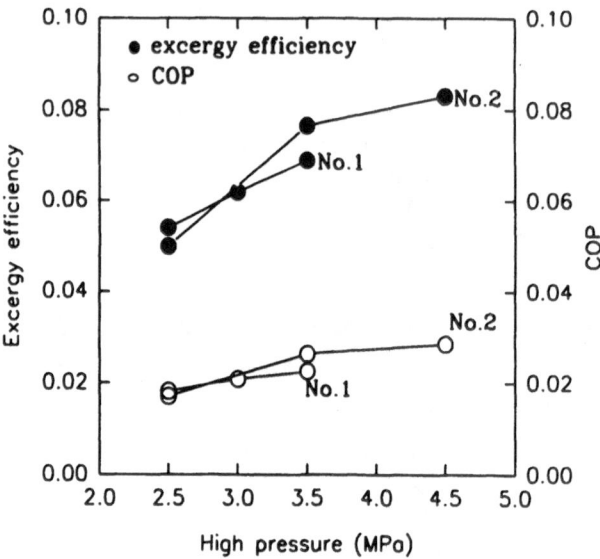

**Figure 6.**   COP and the total excergy efficiency of two mixtures vs. high pressure at 77K

4.5MPa are achieved. Fig.6 is the experimental curve of excergy efficiency and COP of two gas mixtures working at different pressure but at constant cold end temperature of liquid nitrogen temperature. From this figure we find that at 2.5Mpa of working pressure the excergy efficiency and COP of No.1 gas mixture is slightly higher than that of No.2 gas mixture but at other working pressure the excergy efficiency and COP of No.1 gas mixture is lower than that of No.2 gas mixture. It possibly and interestingly means that for different working pressure there is a differently optimum gas mixture.

Some important experimental results and conclusions are emphasized as the following.

1. The lowest temperature of 69K-70K is achieved with this mixed-refrigerant Joule-Thomson cooler.

2. At rather low working pressures of 2.5MPa to 4.5MPa, the net cooling capacity of 15W to 30W at liquid nitrogen temperature are achieved while input electrical power is only about 1.0kW. Therefore, a rather high thermodynamic efficiency of around 5%-9% is achieved the J-T cooler.

3. For different working conditions there is a differently optimum gas mixture which can have highest efficiency.

In the near future we intend to develop a long-life and efficient Joule-Thomson cooler using gas mixture driven by oil-lubricated air-conditioning compressor. This work is also ongoing.

## ACKNOWLEDGEMENT

The authors are greatly appreciated for the discussion with Professor Getmanets V. F. and Dr. Arhipov V. T. , and especially thank for the financial support from the Natural Sciences Foundation of China under granted number No.59476009 and B. Verkin Institute for Low Temperature Physics & Engineering, Ukaraine Academy of Sciences.

## REFERENCES

1.    Gresin, A. K. , et al, Miniature Cooler , Mechanical Press, Moskva (1980) (in Russian)

2.    Klimenko A. P. , "One Flow Cascade Cycle ", International Institute of Refrigeration (1959),
      Paper 1-a-6

3.    Brodynsky, V. M. et al. , "The Use of   Mixture as the Working Gas in Throttling   "Joule-
      Thomson" Cryogenic Refrigerator ", Proceedings of the 13th International   Refrigeration
      Conference , New   York (1973) , Vol. 1 , pp43

4.    Little, W. A. , "Recent Developments in J-T Cooler:Cooling Gases, Cooler and Compressor",
      Proceeding of the 5th Cryocooler Conference, Phymouth, U. S. A (1988), pp3

5.    Arhipov, V. T. , et al, "Effect of Properties of Nitrogen-hydrocarbon Mixture on Performance
      of Compressed-throttling Cooling Cycle", B. Verkin Institute for Low Temperature Physics
      and Engineering, Ukraine Academy of   Sciences, Kharkov (1989), Preprint 36-89 ( in
      Russian)

6.    Lavrenchenko, G. K. , Doctor Dissertation, Odessa   (1985), pp459   (in Russian)

7.    Longsworth G., et al, "80K closed-cycle mixed-refrigerant throttling refrigerator", Cryocooler
      8, Plnum press, New York (1995), p537

# 65 K Two-Stage MR/O$_2$ Throttle Cycle Refrigerator

**R. C. Longsworth and D. Hill**

APD Cryogenics Inc.
Allentown, PA, 18103, USA

## ABSTRACT

Cascade refrigerators using sorption compressors have been proposed for use in space because they have no active components and should thus have long life. A two circuit cryostat was built which assumed thermal cycling of a chemisorption adsorber system to compress oxygen, O$_2$, from 2 to 300 kPa in one circuit and used a mixed refrigerant, MR, in the other circuit to condense the O$_2$ at 100 K prior to expanding it to 65 K.

The refrigerator that was built removes about 2.5 W of heat from the O$_2$ at 100 K to provide 2 W of cooling at 65 K. O$_2$ is pumped using a pair of single stage Displex® GM refrigerators operating out of phase to liquefy the O$_2$ return at about 55 K in a small pressure vessel, then heat it to vaporize it into a storage bottle at the supply pressure. A standard Throttle cycle compressor with a special mixture was used for the precooling circuit.

Pneumatically actuated valves were designed with different objectives for the two circuits. The O$_2$ valve was designed to control the flow rate to match the load and rapidly clear clogs. The MR valve was designed to control the flow to maintain a constant temperature.

## INTRODUCTION

Design studies of multi-stage regenerative sorption cooler systems operating in the 65 K range have shown that these systems are more efficient than Stirling coolers for cooling loads from 0.25 W to 1 W and are competitive with Stirling coolers for cooling loads up to 5 W. Studies have shown that a three-stage cooler that would provide 0.5 to 1.0 W of third stage cooling at 65 K can be designed which has about 60 W of input power[1]. Except for the valves, which do not contain precision, wearing surfaces, the cooler has no moving parts. Consequently, the cooler would produce no measurable vibration and would be expected to last 10 years or more in operation. Merits of the regenerative sorption cooler systems relative to Stirling coolers are greater simplicity, high potential reliability, and minimal vibration. One of the development issues with regard to operation of the system is the susceptibility of the Joule-Thomson (JT) valves to clogging by minute concentrations of impurities produced by the sorption compressors due to the low flow rates required for efficient operation.

Cryocoolers 9, Edited by R.G. Ross, Jr.
Plenum Press, New York, 1997

The present work had as its objectives the development of JT valves, one suitable for minimizing the flow rate of $O_2$ at 65 K while being able to clear clogs, and a second to control the flow of a mixed refrigerant to maintain a constant temperature of 100 K. These valves were incorporated in a two stage cryostat having multi-tube-in-tube heat exchangers with mixed refrigerant gas supplied from an air conditioning type compressor[2] and $O_2$ supplied by using two closed cycle refrigerators to alternately liquefy the $O_2$ then heat it to repressurize it.

A mixed refrigerant operating at relatively low pressures and a low pressure ratio was selected for the first stage to demonstrate that such a scheme might be considered for future use in a sorption cycle system to replace the two upper stages of the present concept. A sorption compressor would have to be developed for this to be implemented.

## PROCESS

Figure 1 is a process schematic of the system that was built. The JT valves for both circuits are actuated by the high to low pressure difference of the gases being circulated and controlled by pairs of room temperature solenoid valves. The solenoid valves are actuated by electrical signals from controllers which sense key temperatures. Oxygen is pumped at low pressure, < 1 kPa, using a special dual cryopump that liquefies $O_2$ at < 60 K in one pump while the other pump is warmed to deliver gas at pressures up to 2 MPa to a storage tank which has a pressure regulator on the outlet. The role of the pumps is then switched. The JT heat exchanger thus sees a nearly constant supply and return pressure. Studies of the relations between pressure drop in the return side of the heat exchanger and heat exchanger efficiency showed that the penalty in efficiency for a poor heat exchanger with low pressure drop is small.

Process studies were carried out to define the temperatures, pressures and mass flow rates for a two stage JT cooler using $O_2$ to produce 2 W of cooling at 65 K. The $O_2$ is precooled by mixed refrigerant (MR) cooling a cold stage at a nominal value of 100 K. The design of the MR heat exchanger is based on high efficiency and low pressure drop in order to minimize the compressor power input.

Precooling is supplied from a throttle-cycle refrigerator that uses an oil-lubricated compressor having a displacement of 28 L/ m to compress a MR from a low pressure in the range of 0.1 to 0.3 MPa to a high pressure of 1.0 to 2.0 MPa. A blend was selected that has a minimum temperature of about 90 K and can produce about 15 W at 110 K.

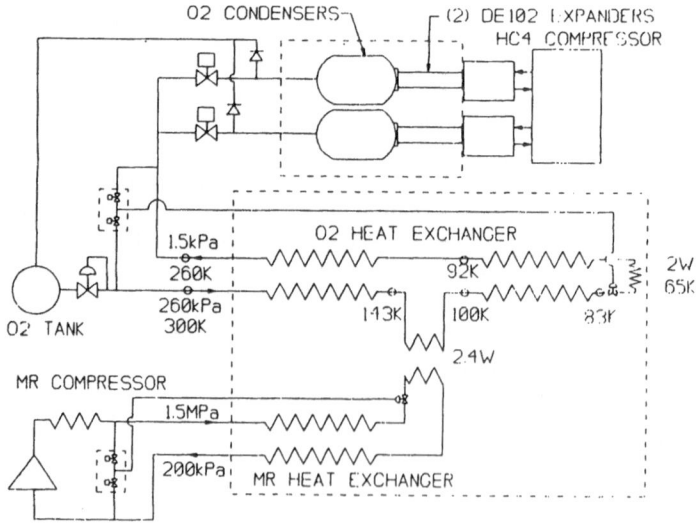

**Figure 1.**    Two stage 65 K cooler process schematic

## CONTROL CONCEPTS

The second stage is controlled to maintain cooling capacity within 10 percent of the heat load. This minimizes gas use and consequently power consumption. The first stage is controlled to maintain a constant temperature within ±0.5 K.

Analysis of the second stage heat exchanger showed that when the flow rate is higher than necessary to support the applied heat load, part of the low pressure return fluid will be liquid. This causes a downward shift in the temperature of the low pressure return in the lower part of the heat exchanger from that of the ideal temperature profile. The analysis predicted that the temperature change would be on the order of 19 K, easily measured with conventional diodes. Two temperature differential inputs are used in the second stage control. The primary control loop uses the temperature differential between a point on the low pressure return line where the fluid first exits the load heat exchanger to a point 10 percent of the way up the heat exchanger. The control program is set up to increase the flow rate if the temperature differential rises or decrease the flow rate if the temperature differential falls. The clog sensing part of the controller uses the temperature differential across the heat station to measure the heat load into the cryostat. The clog sensor compares the heat load as measured from the temperature differential across the heat station with the cooling capacity as measured by the pressure in the second stage control line (which gave the needle position and theoretical flow rate). If the two measurements differ by more than 50 percent, a clog is assumed to have formed and the clog sensor activates the unclog actuator.

The first stage control program makes use of the temperature-pressure relationship displayed by the mixed gas and the compressor. The first stage compressor is a positive displacement type for which the suction pressure varies inversely with flow rate. The flow rate in turn is set by the JT valve. Closing the JT valve reduces the flow rate and the temperature and conversely opening the JT valve increases the flow rate and the temperature. No attempt is made to match first stage heat load with first stage flow rate as the first stage has more than adequate capacity to cool the second stage.

## CONTROL VALVE CONSTRUCTION

The flow control hardware isolates the power input from the cold end by separating the system into a warm end actuator, in which electrical energy is converted to pneumatic pressure, and a cold end actuator, in which the pneumatic pressure is utilized for the displacement of a needle by a bellows. The primary warm end actuator for both the first and second stages, depicted in Figure 2, is simply a pair of solenoid valves tied to the high and low pressure gas streams. The two solenoid valves are commercially available oxygen compatible valves with a 15 millisecond response time. The valves were modified such that a capillary tube restricts the gas flow into or out of the pneumatic control line so that a minimum amount of gas is expended by the control system. The second stage warm end hardware includes a small stepper motor which drives a bellows through a screw. Since one of the goals of this program was to clear clogs rapidly, the stepper motor is used to affect rapid pressure changes in the closed system made up of the large bellows at the warm end, a connecting capillary tube, and a small bellows in the second stage control valve. As the stepper motor expands the large bellows, the pressure in the closed system drops; this compresses the small bellows in the cold end, opening the valve and clearing the clog. As soon as the clog is cleared, the stepper motor is rotated back to its original position, which in turn returns the cold end valve to its original position.

The second stage cold end actuator is depicted in Figure 3. The top bellows forms the pressure boundary for the mechanical adjustment linkage. The mechanical adjustment allows fine tuning of the zero position after the valve is assembled. The needle assembly is guided by an archimedes-cut spiral flexure. The flexure is in the undeformed condition when the needle is in the fully closed position. This configuration overcame the sticking phenomenon encountered with an early prototype cryostat. The bellows closest to the needle is the flow control bellows.

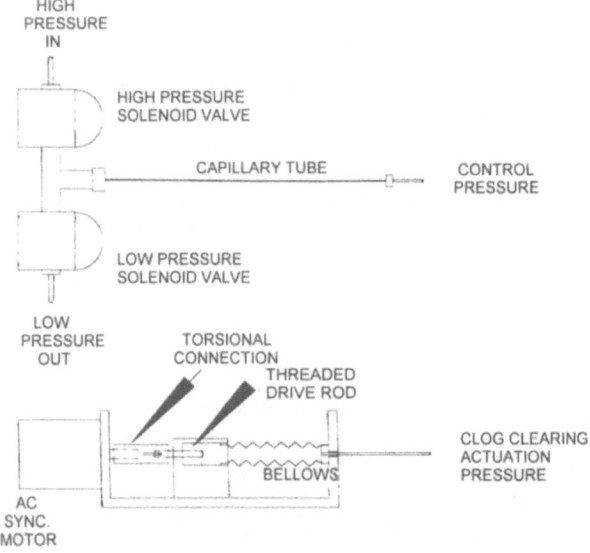

**Figure 2.** Warm end actuators

In the second stage, the control bellows is surrounded by low pressure oxygen and the needle moves into the orifice if the pressure in the control line is increased and out of the orifice if the pressure in the control line is reduced. For stability reasons, the first stage bellows is surrounded by the high pressure gas, so the needle moves out of the orifice if the control pressure is decreased. The center bellows in the second stage JT valve is the clog clearing bellows. The first stage cold end actuator is identical to the second stage actuator except no center clog clearing bellows is necessary because of the relatively large first stage orifice size and large first stage flow rate.

Design of the first stage needle/ orifice proved challenging because the refrigerant density increases over an order of magnitude from gas at room temperature to liquid at the operating temperature. Thus the needle was designed with a double angle such that the first part of the valve's range of motion (from closed) throttled the cold liquid refrigerant and the last part of the valve's range of motion controlled the refrigerant in the gaseous state. The second-stage needle / orifice was designed for the steady-state condition since the first stage cools down fairly rapidly, in turn cooling the second-stage. The JT needle and orifice were sized for 100 percent liquid oxygen entering the valve.

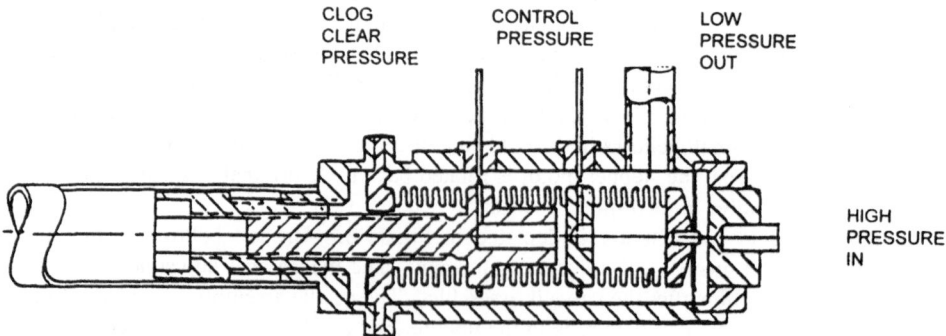

**Figure 3.** Second stage cold end actuator

## CRYOSTAT CONSTRUCTION

Figure 4 is the layout of the cryostat assembly. The first stage heat exchanger is on the outside and is made up of multiple small tubes carrying high pressure mixed gas inside a larger tube. The low pressure gas returns through the gaps between the small tubes. The second stage heat exchanger fits on the inside of the first stage and is made up of four low pressure return lines around a high pressure line in the center. This type of heat exchanger provides easy access to the low pressure tubes for the temperature measurements used in controlling the cryostat. The first and second stages are thermally and mechanically tied together at the precooling station, where the high pressure line of the second stage is broken out and wrapped around the precooling station.

The cryostat is designed to be modular for ease of assembly, ease of testing as a subassembly, and for ready access to make adjustments. Thus the first stage can be removed from the second stage. The JT actuators are both accessible. Pneumatic and electrical interfaces for both stages are through the warm flanges. Each stage is mounted on its own warm flange, which have independent instrumentation feedthroughs so that they can be separated after being instrumented. A blanking plate was be made to replace the second stage warm flange so that the first stage could be tested independently.

## OXYGEN PUMP DESIGN

A schematic of the second stage oxygen pump and associated tank and valve hardware is included in Figure 1. The bulk of oxygen in the system is held in pressure vessels mounted on two APD Cryogenics DE102 expanders. One expander head is heated to drive off oxygen at high pressure through the check valve and into the reservoir while the other expander is held at about 55 K to condense oxygen which returns from the cryostat through a solenoid valve at < 1 kPa. Pressure out of the reservoir is regulated to provide a constant supply pressure to the cryostat.

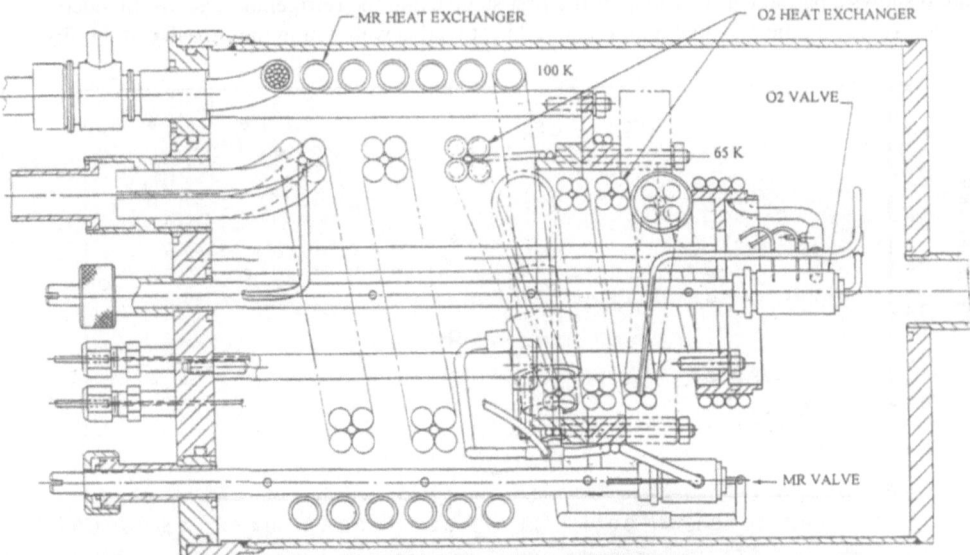

**Figure 4.** Cryostat cross section showing $O_2$ and MR heat exchangers.

The oxygen pumping cycle starts with one station warming up and the other station condensing oxygen at a constant 55 K. When the warming station reaches 122 K, the pressure in the station exceeds 1 MPa, and oxygen starts to flow through the check valve into the reservoir. After a few minutes, the temperature in the station reaches a maximum of 133 K and the pressure in the station and in the reservoir is 2 MPa. Oxygen continues to flow until the cold tank is empty at which time the heater on the station is turned off and the temperature drops to 55 K. Since neither station is supplying oxygen to the reservoir at this point, the pressure inside the reservoir drops as oxygen continues to flow to the cryostat. When the temperature of the cooling station stabilizes at 55 K, the solenoid valves are switched. The heater on the other station is then switched to full power to raise the temperature of the station to 133 K. The cycle is repeated, thus providing a constant supply of oxygen to the cryostat and a constant suction pressure exiting the cryostat. The entire cycle, which is designed to repeat every 65 minutes at maximum flow rate, is very similar in function to the pumping cycle used with sorbent pumps.

## TEST RESULTS

Component level tests were performed to verify the performance of each part of the system before final integration. The first stage was characterized on its own under both open and closed loop control. The second stage oxygen pumping station and test panel and the two stage cryostat were also run as components before final integration and test.

### First Stage Test Results

The first stage was first run open loop to verify smooth operation of the control hardware. Flow rate (measured by the return pressure) responded in a well defined, repeatable manner to control bellows pressure as shown by Figure 5, in which a number of operating data points are plotted. Pressure P2 was measured just after the JT flow control valve and pressure P3 was measured just before the return flow entered the compressor. The valve functioned as designed with the needle moving out of the orifice (thus increasing the flow rate and return pressure) as the control pressure was decreased (pressure differential was increased).

The cryostat first stage was designed to provide at least 5 W cooling from 100 to 120 K. Figure 6 shows a partial capacity map of the first stage using the refrigerant that was blended for this test. As shown, the 5 W maximum required heat load is well within the capacity of the first

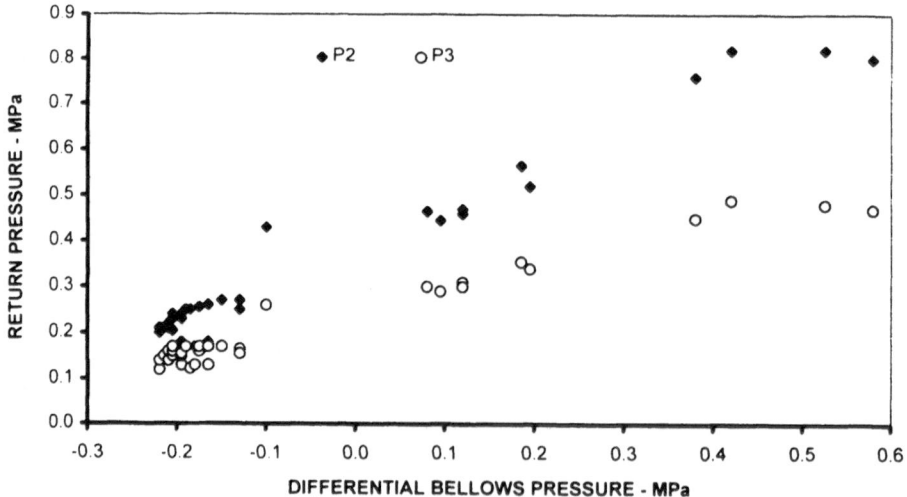

**Figure 5.** Smooth function of first stage JT control valve demonstrated by smooth relationship between control pressure and return pressure.

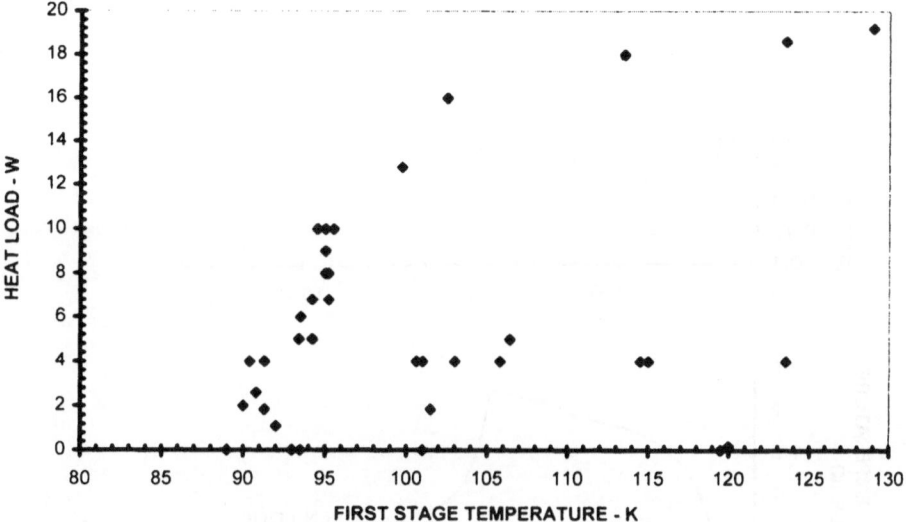

**Figure 6.** First stage capacity map with mixed refrigerant.

stage. The figure shows the highest capacity achieved over the range of operating temperatures as well as some intermediate points at lower heat loads. It was found that a constant temperature could be maintained at less than the full heat load by simply adjusting the bellows control pressure to maintain a constant suction pressure, P2.

Computer control of the first stage employed a simple PID algorithm on temperature. The closed loop response of the cryostat to a step function from 2 W to 6 W and return to 2 W is compared with the open loop response in Figure 7. After the heat load was applied, the control system closed the valve slightly, which in turn reduced the return pressure. The lower return pressure resulted in a lower temperature mixed refrigerant exiting the JT valve and maintained the temperature error in the heat station to less than ± 0.3 K, meeting the ± 0.5 K design goal. In comparison, the open loop temperature error was still increasing to 2.5 K when the heat load was switched off.

## System Test Results

The system was assembled after successful first stage, oxygen pump, and two stage cryostat component tests. The system was first run with nitrogen in the second stage in order to verify safe operation before running with oxygen. Initially, the system checked out well and both the clog clearing actuator and JT control valve functioned as designed. Figure 8 shows second stage flow rate as a function of control pressure, demonstrating smooth control to very low flow rates. It was planned to establish initial PID control coefficients for the second stage while still running with nitrogen, and then switch to oxygen for final testing and a long term endurance test. However, while the control coefficients were being established, disaster struck in the form of a cracked capillary tube which provided control pressure to the second stage JT control valve. Upon investigation, it was found that the tube was brittle and that the crack had formed inside the JT control valve.

The JT valve was removed from the cryostat and was rebuilt with a capillary tube from a different mill run of tubing. Unfortunately, the valve was compromised from rebuilding. A series of failures including a scratched orifice seat, leaking bellows, and leaking bellows solder joint hampered further testing.

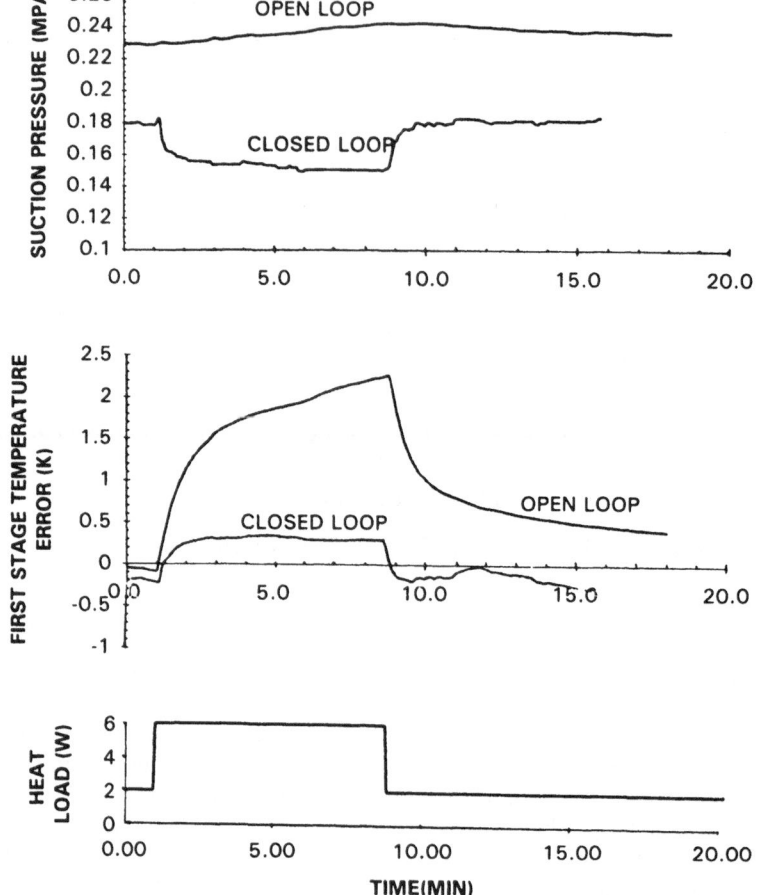

**Figure 7.** First stage open and closed loop response to a 4 W step change.

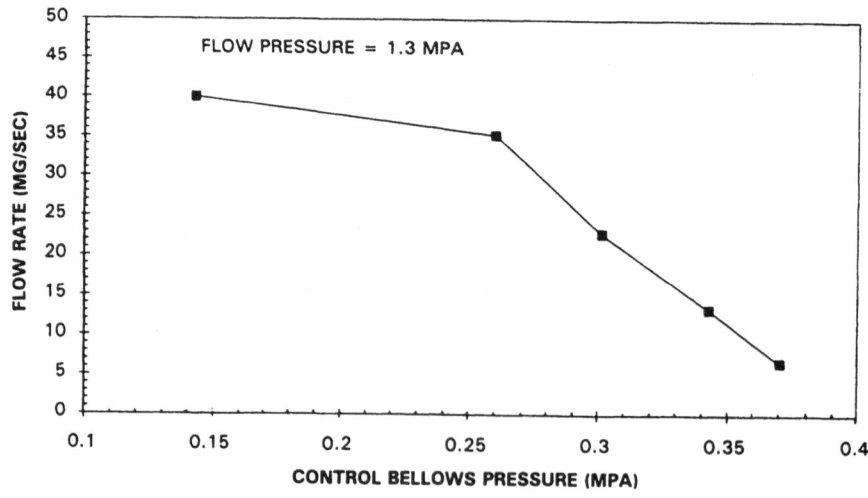

**Figure 8.** Smooth function of JT control valve demonstrated by observed relationship between control pressure and flow rate

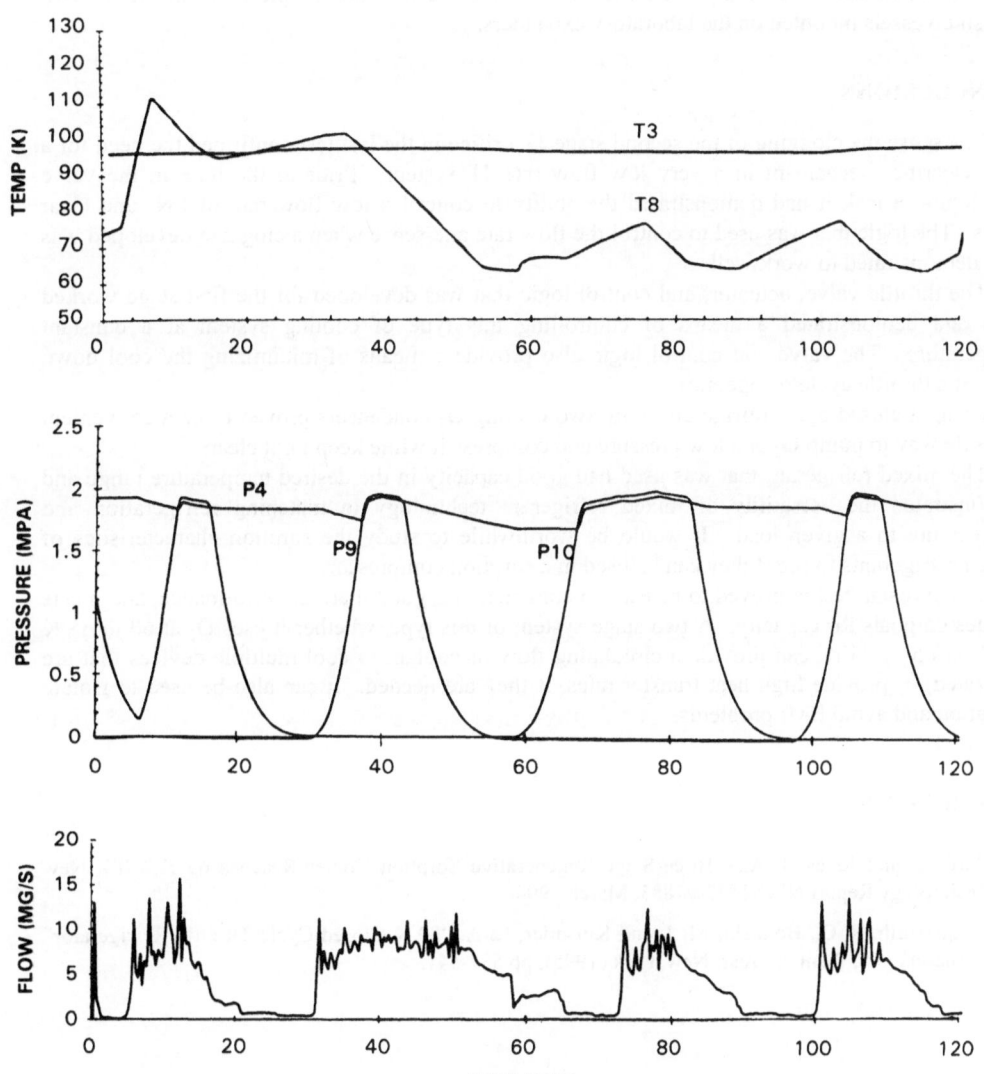

**Figure 9.** Temperature Cycling Caused by Freezing and Thawing of Contaminants

Because of time limitation, the nitrogen in the second stage was replaced with oxygen. The cryostat reached 65 K as designed, but the valve did not respond to changes in the control pressure. Contaminants in the oxygen caused the cold station temperature, T8, to cycle between 65 K and 80 to 95 K as shown in Figure 9. The thawing of the contaminants at 80 to 90 K suggest that $CO_2$ was causing the blockage. The rest of the system functioned properly. The first stage temperature, T3, was maintained at between 96 and 100 K. The oxygen pump cycled to maintain an adequate supply pressure, P4. P9 and P10 indicate the pressures inside the two pressure vessels mounted on the laboratory expanders.

## CONCLUSIONS

In a sense the clogging of the second stage JT orifice in the last test confirmed the need for a clog clearing mechanism in a very low flow rate JT system. Prior to the tube in the valve developing a leak it had demonstrated the ability to control a low flow rate of $LN_2$ and clear clogs. The logic that was used to control the flow rate and sense when a clog had developed was also demonstrated to work well.

The throttle valve, actuator, and control logic that was developed for the first stage worked well and demonstrated a means of controlling this type of cooling system at a constant temperature. The valve and control logic also provide a means of minimizing the cool down time of a throttle cycle refrigerator.

Using a closed cycle refrigerator with two cycling $O_2$ condensers proved to be a convenient and safe way to pump $O_2$ at a low pressure and compress it while keeping it clean.

The mixed refrigerant that was used had good capacity in the desired temperature range and demonstrated the versatility of mixed refrigerant technology in matching refrigeration and temperature to a given load. It would be worthwhile to study the sorption characteristics of mixed refrigerants to see if they can be used in a sorption compressor.

The cryostat design proved to be easy to construct, has good thermal performance, and meets the design goals for capacity. A two stage system of this type, whether it uses $O_2$ at 60 to 65 K or $N_2$ at 65 to 75 K, can provide a circulating flow of coolant to cool multiple devices that are separated, or provide high heat transfer rates, if they are needed. It can also be used to isolate vibration and avoid EMI problems.

## REFERENCES

1. Bard, S. and Jones, J. A., "Three-Stage Regenerative Sorption Cooler Reaches 65 K," JPL New Technology Report NPO-18336/7883, March 1994.

2. Longsworth, R. C., Boiarski, M. J. and Klusmier, L. A. "80 K Closed-Cycle Throttle Refrigerator" *Cryocoolers 8*, Plenum Press, New York (1995), pp 537-541.

# Closed-Cycle Neon Refrigerator for High-Temperature Superconducting Magnets

**P.E. Blumenfeld and J.M. Pfotenhauer**

Applied Superconductivity Center
University of Wisconsin-Madison
Madison, WI, USA 53706

## ABSTRACT

A closed-cycle refrigerator using neon as a working fluid has been designed and is under construction at the Applied Superconductivity Center of the University of Wisconsin-Madison. The refrigerator is intended for use in testing small superconducting magnets formed from BSCCO and other high temperature superconducting materials. A hybrid design is used, with a helium cryocooler operating on the Gifford-McMahon cycle precooling a separate neon stream. The neon stream is driven by a rotary vane compressor that provides a flow rate of 0.4 g/s between high and low side pressures of 450 kPa and 50 kPa, respectively. The high and low sides are separated by a Joule-Thomson metering valve, which maintains low side pressure, and thus cold end temperature via a closed-loop PID controller. The refrigerator is designed to provide a cooling capacity of 10 Watts at 25 K. At the cold end, the neon will be partially condensed into a two-phase fluid which is transferred out of the refrigerator vessel to a separate magnet test vessel. The liquid fraction will be available for an isothermal convective interface with the magnet. This paper details the considerations and constraints that went into the design, and discusses the hardware chosen to perform the required processes. Special emphasis is placed on design and testing of the tube-in-tube and cable-in-conduit heat exchangers that have been built, or are presently being built for this project.

## INTRODUCTION

Developments in superconducting magnet technology point toward production of magnets operating in the 20 to 30 K temperature range. The Applied Superconductivity Center (ASC) at the University of Wisconsin-Madison wishes to establish laboratory test facilities to accommodate these magnets and materials. There is thus a need for a new refrigeration system to be used in testing small magnets formed from BSCCO and other high Tc superconductors. Design considerations and constraints are similar to those given in Richardson, et. al.[1]. These constraints and others are discussed briefly here.

Cryocoolers 9, Edited by R.G. Ross, Jr.
Plenum Press, New York, 1997

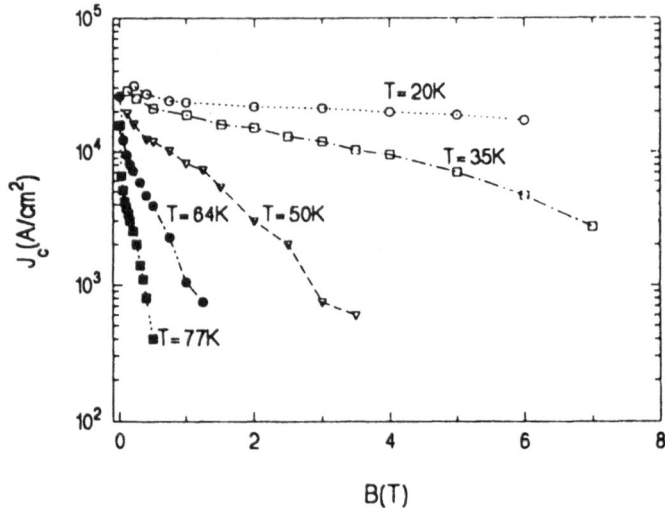

**Figure 1.** Interaction Between Operating Parameters for BSCCO Material.

## Design Constraints

1) Operating temperatures for BSCCO magnets range up to 40 K before magnet performance is critically degraded by loss of flux pinning centers and flux creep. As shown in Figure 1[2], the result is a marked decrease in critical current density, $J_c$, with increasing applied magnetic field, B, for temperatures, T, above 35 K.

2) An isothermal heat transfer interface between the magnet and the cryogen is desirable to allow testing under controlled conditions over a range of heat transfer rates.

3) A high heat transfer rate with small temperature gradients is desirable to expand the range of quench testing conditions. Unintentional quenches are not likely in high Tc magnet testing[3], but intentional, controlled quench testing is an important area of study for these materials and magnets.

Two constraints motivated more by practice than by physics are:

4) Mechanical isolation of the magnet from refrigerator vibration, and

5) Closed-cycle operation to avoid user handling of cryogens.

## Potential Working Fluids

As per constraint (1), the possible choices for a working fluid include helium, hydrogen, and neon. Isothermal heat transfer, constraint (2), is most easily achieved via evaporation of a liquid cryogen, ruling out helium as an option for this temperature range. Hydrogen exists as a liquid over a range of temperatures that is well suited to this application, but there are safety concerns associated with this choice that make it an unlikely candidate. Neon is an inert substance, liquid in the desired range of temperatures, and also has a very high volumetric cooling capacity (the product of density and latent heat of evaporation). Neon has been chosen as the appropriate working fluid for this application.

## Heat Transfer Interface

Constraints (2), (3), and (4) all mitigate toward a convective, rather than a conductive interface. Pool boiling can provide isothermal heat transfer at rates significantly greater than can be achieved through a comparable conductive path. Cryogenic jet spray cooling is another interesting convective cooling option. Either of these interfaces, or liquid cooling through flow channels in the magnet, offers a low thermal impedance over a wide surface area. Vibration isolation can be achieved via liquid or two-phase transfer of the working fluid from the refrigerator to a separate magnet test chamber, where the convective interface of choice can be realized.

## GM-JT Hybrid

A hybrid design, with a Joule-Thomson neon cycle precooled by helium in a Gifford-McMahon cycle has been selected for the ASC's neon refrigerator. Fundamental features of this system include a two-stage GM cryocooler and its associated helium compressor, a separate compressor to circulate neon in a closed loop, two recuperative counterflow heat exchangers, two gas-to-surface heat exchangers to transfer thermal energy from the neon stream to the cryocooler cold stages, and a metering valve to provide a controlled, isenthalpic pressure drop with an consequent drop in temperature. Figure 2 is a schematic of the flow path for the neon loop. Thermodynamic state points are labeled in Figure 2, and are used to identify heat transfer and flow processes. Table 1 presents the operating conditions anticipated for the neon refrigerator.

### Thermodynamic Model

A portion of the thermodynamic cycle for the neon loop is plotted on the P-h plane in Figure 3. Processes are defined between each of the state points marked on the diagram; when reference is made to "design" or "nominal" operating conditions, it is these state points and processes that are referenced.

## REFRIGERATOR COMPONENTS

### GM Coldhead

A two-stage G-M coldhead (Cryomech Model GB-04) has been selected as appropriate for use in the refrigerator design. The coldhead performance data[4] (cold stage temperature as a function of heat load) provided by the manufacturer is the starting point in mapping the thermodynamic cycle shown in Figure 3. This data helps to fix state points in the flow stream by specifying the lowest temperature attainable at the outlets of the cold stage heat exchangers, given the heat transfer rates for those heat exchangers. The coldhead is powered by a compressor (Cryomech model CP640) drawing 5.0 kW.

### Neon compressor

A modified helium compressor (APD model HC-2) has been selected as the driver for the neon loop. This compressor draws 1.7 kW of electrical power. The manufacturer has determined that when charged with neon to its design equilibrium pressure, it will circulate the gas at a flow rate of 0.4 g/s between a high side pressure of 435 kPa and a low side pressure of 50 kPa[5]. The sub-atmospheric pressure on the low side is an attractive feature for this design, as it allows the liquefied neon to boil at a pressure very close to its triple point. Thus the cold end temperature approaches the lowest liquid temperature attainable with this working fluid.

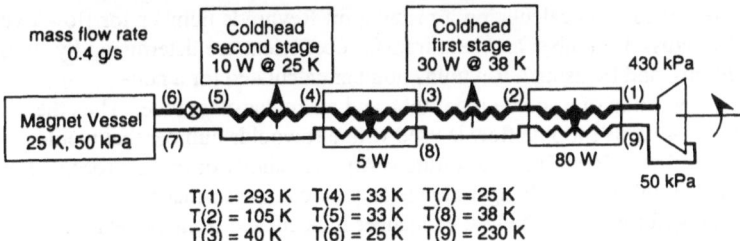

Figure 2. Neon Flow Loop, with Design Temperatures.

**Table 1.** Refrigerator Operating Conditions

| Neon Mass Flow Rate (g/s) | Electric Power Consumption (W) | Cooling Capacity (W) | Coefficient of Performance | Figure of Merit |
|---|---|---|---|---|
| 0.4 | 6700 | 10 | 0.0015 | 0.016 |

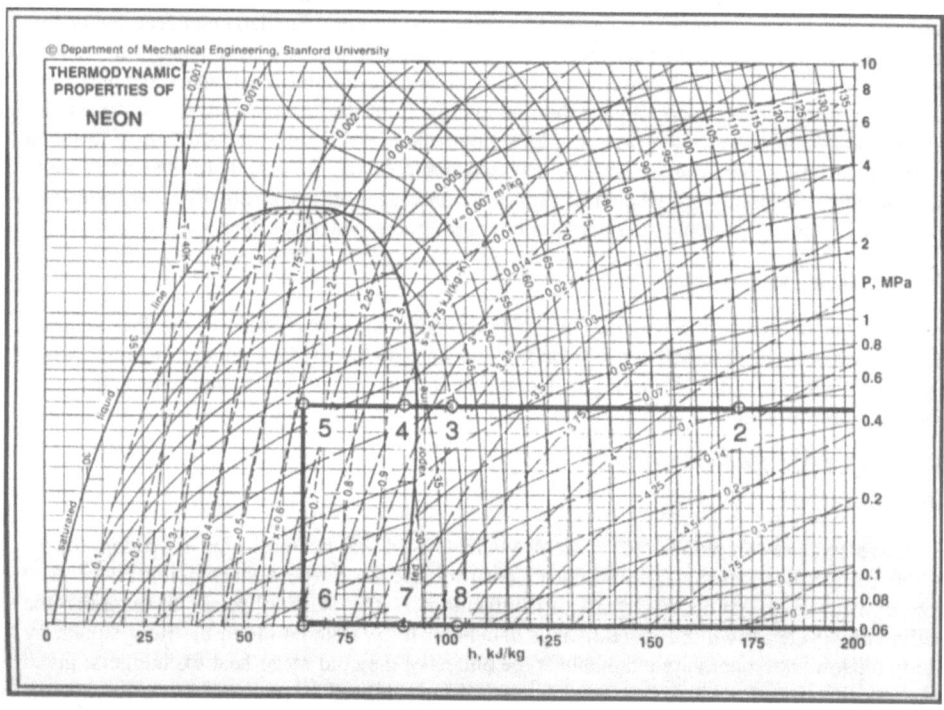

**Figure 3.** P-h Diagram for Neon[6], with State Points Indicated

### Gas-to-Gas Heat Exchangers

Two tube-in-tube counterflow heat exchangers have been constructed to perform the recuperative heat exchange function. Each heat exchanger is one meter long, and is wound into a helix with a diameter of 7.6 cm. The outer tube has a round cross-section with an outside diameter of 1.3 cm. The inner tube has a square cross-section with an outer side length of 6.35 mm. The square inner tube is axially twisted with a pitch of 6 turns per meter. This axial twist generates turbulence for flow over the outside surface, even under the low flow rate conditions in the neon stream.

Because of the difficulty in calculating a meaningful Reynolds number for flow over this structure, a theoretical Nusselt number and heat transfer coefficient are determined by assuming turbulent flow conditions, and by using a Reynolds number calculated for a non-circular cross section consisting of a square inner tube and a circular outer tube. Thus the Dittus-Boelter relation for turbulent pipe flow is used with a Reynolds number that would usually indicate laminar flow. Experimental results verify the validity of this approach. Figure 4 plots data from one test of this heat exchanger alongside predicted performance.

The theoretical performance of the heat exchanger is calculated from the relation

$$h = 0.023 \cdot C_p \cdot G \cdot Re^{-0.2} \cdot Pr^{-0.66} \cdot \left[ 1 + 3.5 \cdot \frac{D_e}{D_h} \right] \qquad (1)$$

This is a modified form of the Dittus-Boelter relation, which accounts for the effects of a helical flow path[7] (reference Timmerhaus), where h is the heat transfer coefficient, Cp is the specific heat of the fluid, G is the mass flux, Re is the Reynolds number based on hydraulic diameter (neglecting the effect of the axial twist), Pr is the Prandtl number of the fluid, $D_e$ is the inside tube diameter, and $D_h$ is the diameter of the helical tube winding. Eq. (1) is used to model heat transfer behavior for both

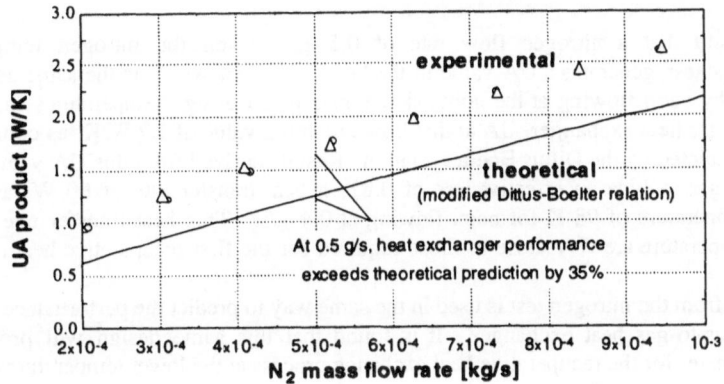

**Figure 4.** UA Product, Theoretical and Experimental, Varying with Flow Rate

the hot and the cold streams in the heat exchanger. Thus predictions are made for the hot side and cold side heat transfer coefficients; these combined with known geometry lead to a prediction for UA (the product of the overall heat transfer coefficient and the heat transfer surface area).

As is evident in Figure 4, the heat exchanger performs much better than theory predicts, presumably because of the underestimation of the Reynolds number used in eq. (1). The increase in UA with increasing flow rate is similar for the theoretical and the experimental plots, indicating that the flow is, indeed, made turbulent by the axial twist in the inner tube. Laminar flow would generate a constant Nusselt number, and thus UA would be constant with increasing flow rate.

This heat exchanger design has been tested with nitrogen gas flowing at rates ranging from 0.1 g/s to 1.0 g/s (identical flow rates were used for both hot and cold streams). The nitrogen enters the hot end at room temperature, and the cold end at temperatures ranging from 158 K to 231 K, depending on the mass flow rate. These flow conditions differ markedly from nominal operating conditions. An analysis of the parameters influencing heat transfer leads to predictions for heat exchanger performance with neon flow at low temperatures, based on the data collected with warmer nitrogen gas. It is found that there is a particular nitrogen flow rate at which the UA value developed in the heat exchanger can be expected to be the same for the warm nitrogen as for the colder neon flowing at the nominal 0.4 g/s. The value of UA at that flow rate is then used to calculate the theoretical effectiveness and heat transfer rate of the heat exchanger at design conditions. The UA product is used as a measure of heat exchanger performance, because it is a function of heat transfer conditions on both the hot and the cold sides of the heat exchanger.

Stated algorithmically, the analysis proceeds as follows:
1) evaluate neon properties at design operating conditions (from coldhead performance
   data)

2) determine design neon flow rate (from compressor performance specifications)
3) using eq. (1), solve for the hot and cold side heat transfer coefficients expected
   under design conditions
4) solve for UA from hot and cold side heat transfer coefficients and geometry
5) equate UA under design conditions to UA under nitrogen test conditions (in other
        words, assume that UA is now known)
6) evaluate nitrogen properties at test conditions
7) solve for nitrogen flow rate required to provide known UA

It is found that a nitrogen flow rate of 0.5 g/s, given the nitrogen temperatures encountered in the test, generates a UA value in the heat exchanger which is the same as the UA value generated by neon flowing at the nominal 0.4 g/s, given design temperatures in the first (warmer) gas-to-gas heat exchanger. UA at this flow rate has a value of 1.7 W/K, as compared to a value of 1.2 predicted by the Dittus Boelter relation. Results derived from this UA value for the first heat exchanger include an effectiveness of 0.81, a heat transfer rate of 80 W, and a hot stream outlet temperature of 98 K for neon flowing at 0.4 g/s. This heat transfer rate and hot stream outlet temperature are very close to those required for the first recuperative heat exchange process.

The data from the nitrogen test is used in the same way to predict the performance of the second gas-to-gas heat exchanger. It is found that this same design will provide the required heat transfer for the recuperative heat exchange process at the lower temperatures as well.

## Gas-to-Surface Heat Exchanger

Thermal energy is rejected from the system only through the cold stages on the GM cryocooler. However, the surface area available for heat exchange on these stages is not large, especially on the second stage. Thus a gas-to-solid surface heat exchanger has been designed that increases the surface area of the cold stages without introducing a large thermal resistance. A cable-in-conduit design is used, in which a bundle of many fine wires is thermally anchored to the cold stage at one end. The wires are contained in an insulating conduit which guides the gas flow axially through the wire bundle, from the free end of the wires toward the end that is anchored to the cold stage.

This design has been analyzed using a finite difference model. Figure 5 illustrates the heat transfer modes that are considered in writing an energy balance. In this sketch, $q_k$ is conductive heat transfer, $q_c$ is convective heat transfer, $m_{dot}$ is a mass flow rate, and N is the number of nodes. It is assumed that convection occurs uniformly throughout the bundle, with a constant heat transfer coefficient along the length of the wires. A constant thermal conductivity is also assumed for the copper wires. A reasonable value for the average convection heat transfer coefficient is sought through experiment.

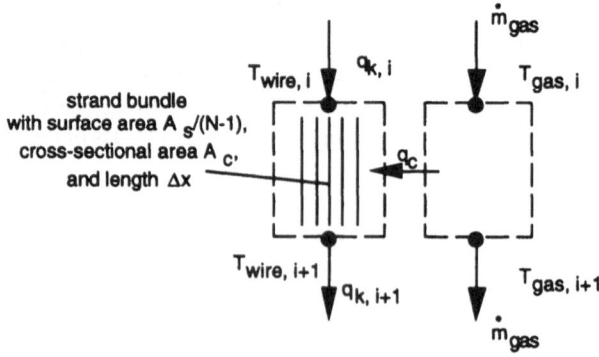

**Figure 5.** Finite Difference Model for Cable-in-Conduit Heat Exchanger

**Table 2.** Cable-in-Conduit Heat Exchanger Performance for Nitrogen and Neon

| Fluid | T_inlet (K) | T_outlet (K) | T_base (K) | Average Heat Transfer Coefficient (W/m$^2$-K) | Heat Transfer Rate (W) |
|---|---|---|---|---|---|
| N2 (data) | 293 | 159 | 125 | 16 | 66 |
| Ne (predicted) | 33 (sat. vap.) | 33 (2-phase) | 25 | 97 | 6.5 |

A prototype of this design has been built and tested with nitrogen gas. As with the recuperative heat exchanger design, test results are used to predict heat exchanger performance with neon flow at design conditions. The prototype was mounted on the first stage of the coldhead and cooled, with nitrogen entering the conduit at room temperature, and exiting at a temperature somewhat higher than the heat exchanger base temperature. Table 2 presents some data from this test, along with performance predictions for neon at design conditions.

The conversion of test results from warm nitrogen flow to cold neon flow is simplified for this design by the low Reynolds number, and thus the laminar flow prevailing in the wire bundle. In laminar flow, the Nusselt number, and thus the heat transfer coefficient is expected to be constant with changing flow rate. It is thus expected that the Nusselt number calculated from the nitrogen test data provides a reasonable estimate of the Nusselt number for cold neon flow. The average heat transfer coefficients in the two cases, then, are proportional to each other by the ratio of the thermal conductivities of the two fluids at their respective temperatures and pressures. The heat transfer coefficient for design conditions can be calculated from data collected with nitrogen flowing through the heat exchanger at any desired rate of laminar flow.

Using the finite difference model to predict heat exchanger performance in terms of heat transfer rate and fluid outlet temperature, it is found that a single cable-in-conduit assembly is not effective enough to achieve the heat transfer rate (28 W) and outlet temperature (40 K) required by process 2,3. However, using the same finite difference code to model multiple assemblies connected in series reveals that three such heat exchangers will provide the required heat exchange to accomplish process 2,3. The increase in high side pressure drop resulting from this series connection is expected to be small enough so as not to degrade the performance of the refrigerator.

A single cable-in-conduit assembly is predicted to provide a heat transfer rate of 6.5 W on the second cryocooler stage, to accomplish process 4,5. Results from the finite difference analysis are less reliable for this process, as the heat transfer coefficient for condensing flow is expected to vary from that predicted for single-phase flow. As an approximation, the heat exchange process is modeled using the predicted 97 W/m$^2$-K average heat transfer coefficient, with neon properties evaluated at saturated vapor conditions, and a high side pressure of 435 kPa. The refrigerator will be assembled with a single assembly on the second stage, but it is recognized that additional assemblies will increase the effectiveness of this process. Increased effectiveness at the very low temperatures in this heat exchanger will go a long way to increase the cooling capacity of the refrigerator.

**Joule-Thomson Metering Valve**

The isenthalpic expansion called for in process 5,6 is to be achieved and precisely controlled by means of a cryogenic metering valve. The pressure drop across this valve will be controlled by a tapered stainless steel needle 2 cm in length positioned within a bored valve body made of Ultem brand Epoxy resin. The bore diameter and the needle's maximum diameter are both 1 mm. The valve body is located about 40 cm below the top flange of the refrigerator vessel, to provide a long thermal path between room temperature connections and the coldest location in the neon flow loop. The needle is actuated by a thin-walled stainless steel push rod which travels vertically within the assembly's main tube. Sealing is accomplished with a double o-ring at the warm end of the rod. The main tube is a piece of G-10 tubing 2.54 cm in outside diameter, with thick walls for strength. Estimated heat leak through the main tube and push rod is on the order of tens of milliwatts.

Design constraints call for operation of this refrigerator at just above the triple point of the working fluid. Thus the pressure drop in process 5,6 must be precisely controlled to avoid solidification of the neon, and plugging of the needle valve. This control will be provided by a closed-loop PID controller and its associated stepper motor assembly. These components have been modified from their original service as a vacuum pressure control system by detaching the stepper motor assembly from the original butterfly valve mechanism, and installing a pinion on the axis of the stepper motor. This pinion is made to engage a rack set into the top end of the push rod. The stepper has a resolution of 1500 steps over its 1/4 turn rotation; the rack and pinion convert this angular motion into linear steps with a resolution of 10 microns per step.

## Diagnostics

Nine platinum resistance temperature sensors will be used to measure temperatures at state points 1 through 9. Six of these sensors have been calibrated for use at temperatures below 77 K. These small cylindrical sensors will be suspended in the flow stream, perpendicular to the flow direction, at the entrances and exits of each of the heat exchangers. Static pressure taps will be installed at the same locations. The pressure measurement at the outlet of the metering valve is especially important as this serves as the input to the feedback control loop for the valve. A static pressure line will be run from this point out of the dewar to the pressure sensor on the PID controller.

## WORK TO BE DONE

### Testing with Helium

After the refrigerator has been assembled, preliminary testing will be done using helium instead of neon. There are a number of good reasons for testing with helium. First, the APD HC-2 compressor was shipped with a full helium charge, so it will not need any preparation for this test. Second, leak testing should be the first order of business for a closed-cycle system that uses an expensive fluid such as neon. The leak testing will be simplified by use of a helium leak detector and a charge of helium gas in the neon loop. It will also be worthwhile to note the performance of the refrigerator with helium as a working fluid, as this may be an appropriate operating mode for some applications.

### Further Construction

After preliminary testing, three additional components will need to be installed to make the refrigerator into a useful laboratory tool. First, the refrigerator vessel must be reworked to allow for liquid or two-phase neon transfer out of the lower end of the vessel. Vacuum insulated transfer lines must be installed, and finally, the magnet test vessel must be designed and built.

## CONCLUSION

This refrigerator should prove to be a useful tool for laboratory testing of small superconducting magnets made from BSCCO and related materials. The liquid interface will provide a range of cooling options for the experimenter, including pool boiling, jet spray, and flow through cooling channels. For operation at the lowest temperatures, however, it will be important to minimize pressure drop within the magnet test chamber.

A second-law analysis of the neon loop is planned, to locate those components that would yield the largest benefit from design improvements. Improvements in cooling capacity and figure of merit are, however, limited by the operating characteristics of the coldhead used to cool the neon stream. Even a very large, very effective cable-in-conduit heat exchanger could not transfer more than 30 Watts into the coldhead's first stage at 40 K, for example. Another consideration is low side pressure drop: very effective recuperative heat exchangers with relatively high pressure drop will increase low side pressure at the cold end with a consequent increase in cold end temperature. Optimization of these components will prove worthwhile, but the present design should be sufficient to enable versatile magnet testing with this most interesting cryogen.

## REFERENCES

[1] Richardson, R.N., Scurlock, R.G., Tavner, A.C.R., "Cryogenic Engineering of High Temperature Superconductors Below 77 K", *Cryogenics*, vol. 35, no. 6 (1995), pp. 387-391.

[2] Maley, M.P., Kung, P.J., Coulter, J.Y., Carter, W.L., Riley, G. N., McHenry, M. E., "Behavior of Critical Currents in Bi-Pb-Sr-Ca-Cu-O / Ag Tapes from Transport and Magnetization Measurements: Dependence on Temperature, Magnetic Field, and Field Orientation", *Physical Review*,        vol. 45, no. 13 (1992), pp. 7566-7569.

[3] Ries, G., "Magnet Technology and Conductor Design with High Temperature Superconductors", *Cryogenics*, vol. 33, no. 6 (1993), pp. 609-614.

[4] CryoMech, Inc, *Model GB04 Operation and Service Manual*, CryoMech, Inc., Syracuse (1995), p. 54.

[5] Rerig, R., Personal Communication, APD Cryogenics, Inc., Allentown (Sept. 1995).

[6] Reynolds, W.C., *Thermodynamic Properties in SI*, Department of Mechanical Engineering, Stanford University, Stanford (1979), p. 53.

[7] Timmerhaus, K.D., Flynn, T.M., *Cryogenic Process Engineering*, Plenum Press, New York (1989), p. 194.

# Design and Testing of a Combined Stirling-Cycle Joule-Thomson Cryocooler System

L. Barry Penswick and Donald C. Lewis, Jr.

STC Stirling Technology Company
4208 West Clearwater Avenue
Kennewick WA 99336-2626 USA

## ABSTRACT

This paper describes the design, development, and testing of an integral Stirling cycle Joule Thomson Cryocooler (SJTC) carried out under a Phase II SBIR contract with NASA/Goddard Space Flight Center. The project goal is development of a long life cryocooler system capable of providing cooling at 4.5 K.

The SJTC system consists of a two-stage free-piston Stirling cycle cooler which is employed as a pre-cooler for the closed cycle Joule Thomson (JT) gas circuit and a pair of linear drive gas compressors to provide high pressure gas for the JT expander. Power for the combined JT gas compressors and the Stirling cycle pre-cooler is derived from common linear drive motors. The SJTC system design details are presented and test results to date discussed.

## INTRODUCTION

The SJTC evolved in response to the need for a liquid helium temperature (approximately 4 K) refrigeration system which would incorporate technology applicable to long life maintenance-free operation. A unique aspect of the SJTC system in comparison to other combined Stirling cycle/JT expander configurations [1,2] is the integration of the gas compressors with the drive system for the Stirling cycle pre-cooler. As shown schematically in Figure 1, the gas compressors for the JT gas circuit are directly coupled to the same linear motor that drives the piston of the Stirling cycle pre-cooler. In this configuration, the compressors increase the pressure for the JT circuit gas from the approximately 0.1 MPa pressure present downstream of the JT expander to the mean pressure of the Stirling cycle pre-cooler, approximately 1.2 Mpa. While this configuration eliminates the need for separate linear drive motors for the JT gas compressors, improving overall system reliability, it does place a number of interesting constraints on the design of the gas compressors and integration with the pre-cooler/compressor combination.

Cryocoolers 9, Edited by R.G. Ross, Jr.
Plenum Press, New York, 1997

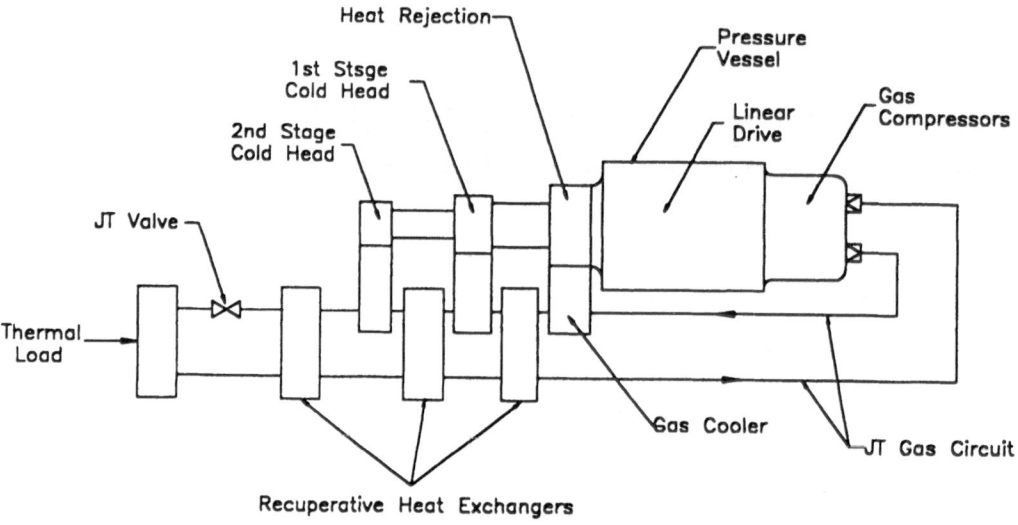

Figure 1.     SJTC System Configuration

The fundamental operating characteristics of the SJTC system and its key components are defined in Table 1 with the prototype hardware shown in Figure 2.  It is important to note that the input power noted in Table 1 is for the pre-cooler and gas compressors combined and includes the linear drive motor losses.

Table 1
SJTC System Characteristics

| PARAMETER | VALUE |
|---|---|
| JT Circuit Cooling Capacity (W) | 0.2 @ 4.5 K |
| Second Stage Capacity (W) | 0.5 @ 20 K |
| First Stage Capacity (W) | 5 @ 65 K |
| Mean Operating Pressure (MPa) | 1.2 |
| Operating Frequency (Hz) | 45 |
| JT Expander Pressure Ratio | 12 |
| Drive Voltage (VAC) | 120 |
| Rejection Temperature (K) | 300 |
| SJTC System Input Power (W) | 576 |

Figure 2.     SJTC Prototype Hardware

The following subsections provide an overview of the development of the SJTC system, a detailed description of the critical components making up the system, and a description of test results.

## System Design

Due to the highly integrated nature of the SJTC concept, it is not possible to optimize the Stirling cycle pre-cooler and JT gas compressor systems independently as in [2]. By definition the mean pressure of the Stirling cycle pre-cooler is the source pressure for the JT gas circuit and the gas pressure downstream of the JT expander the inlet pressure for the gas compressors. This latter pressure is essentially defined by the desired final refrigeration temperature. For the 4.5 K target temperature, this gas pressure is essentially atmospheric (0.1 Mpa) and the required compressor ratio approximately equal to the mean pre-cooler pressure. For reasonable JT circuit performance, minimum pressure ratios on the order of 8 to 10 are required clearly indicating the need for multi-stage compression of the JT working fluid.

Because the gas compressors are mounted within the buffer space of the Stirling cycle pre-cooler, gas leakage past the clearance seals into the compressors plays a critical role in the optimization process. This closely couples the system design to the flexural bearing's ability to accurately align the compressor piston. Additionally, the free-piston linear drive motor, pre-cooler power piston, and gas compressor assembly must be properly sized to ensure that the near resonate operating point for the drive is maintained under all operating conditions. This factor played a significant role in defining the basic gas compressor mechanical configuration.

Throughout the pre-cooler design process, the GLIMPS Stirling cycle simulation model, developed by Gedeon Associates, was employed to evaluate the thermodynamic characteristics and performance of the various two-stage configurations. A significant number of simulation model modifications were incorporated by STC to reflect actual SJTC hardware characteristics and better define thermal parasitic losses based on STC test results. During the preliminary optimization process, the GLIMPS based optimization model, GLOP, was employed. Because of the large number of potential variables in the optimization process, the initial phase of the design effort focused on defining reasonable ranges for the various optimization variables. This process involved the development of submodels for the gas compressor, JT heat exchangers [3,4], and a number of the critical components such as the flexural bearings. These models were employed for initial screening purposes and as the system definition process proceeded, were actually incorporated into the optimization process. From this initial screening process the second stage cold head temperature was fixed at 20 K. Table 2 defines the basic parameters evaluated in the optimization process.

Table 2
SJTC System Optimization Variables

| PARAMETER | OPTIMIZATION RANGE |
|---|---|
| Pre-Cooler Mean Pressure (MPa) | 0.8 to 2.0 (1.2) |
| Operating Frequency (Hz) | 30 to 60 (45) |
| First Stage Operating Temperature (K) | 100 to 50 (65) |
| Piston Amplitude (mm) | 4 to 7 (6.0) |
| Displacer Amplitude (mm) | 2 to 6 (3.0) |
| Piston-Displacer Phase Angle (degrees) | 40 to 90 (70) |
| JT Circuit Mass Flow Rate (kg/sec) | 6 to 10E-5 (9.4E-5) |
| First Stage Compressor Ratio | 2 to 4  (3.4) |
| Second Stage Compressor Ratio | 2 to 4 (3.55) |
| Recuperative Heat Exchanger Effectiveness | 0.95 to 0.995 (0.99) |
| Interstage Gas Cooler Temperature (K) | 300 to 350 (330) |

Parameters in parentheses ( ) indicate selected values

## Pre-Cooler

The free-piston Stirling cycle pre-cooler, shown in Figure 3, employs dual opposed power pistons and a single two-stage displacer. Both pistons and the displacer are supported on flexural bearings which ensure non-contacting operation of the critical gas clearance seals. Two modular, moving iron, permanent magnet linear motors provide the input power for the pre-cooler (and gas compressors). Motors are wired in a series manner and are driven by a common switching power supply. The moving portion of the motor and the power piston are directly coupled to the piston rod. An adapter is provided on the rear (outboard) end of the piston rod for attachment of the compressor. The flexure stack axial spring, the magnetic spring of the linear drive motor, the spring effect of the Stirling cycle, and compressor gas spring effect provide the necessary spring to resonate the power piston assembly. External capacitors are employed for electrical tuning of the drive motors. Power piston flexures are optimized to provide sufficient radial spring stiffness to ensure proper alignment of the piston during normal system operation while also capable of overcoming the potential magnet-induced side loads that can occur with an eccentric motor mover. The individual motors are contained within a power module housing made up of the pressure vessel and associated support structure for the gas compressor. Each power module is assembled, all alignment checked, and tested prior to attachment to the expander module. With the exception of the gas compressors, each power module is identical. This approach to the pre-cooler drive system design has resulted in considerable flexibility in the test program and in efforts to employ the basic SJTC pre-cooler to alternative commercial cooler applications.

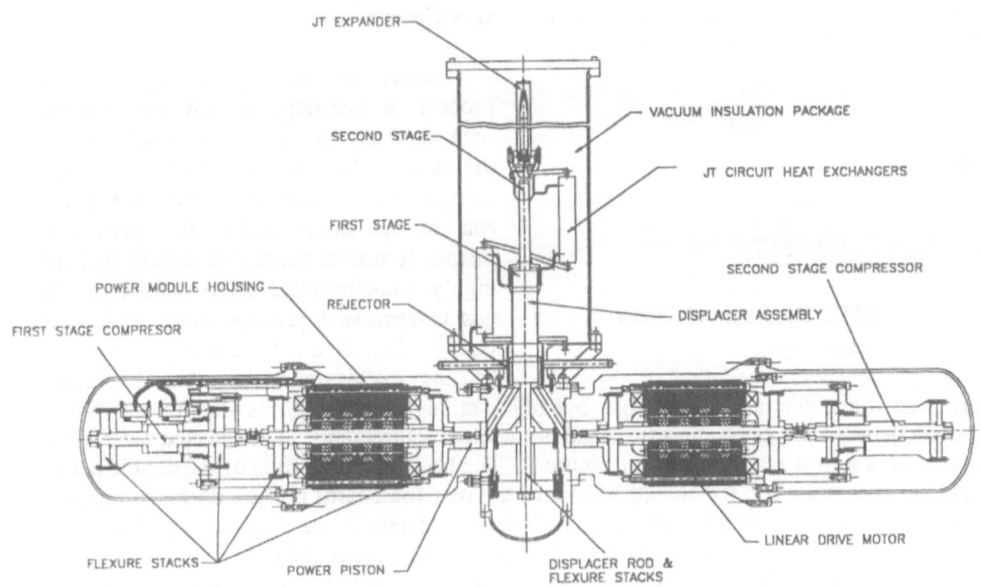

Figure 3.    SJTC Pre-Cooler

The two-stage displacer is mounted perpendicular to the axis of the linear drive system. As noted in Figure 3 the moving displacer contains the first and second stage regenerator stacks and incorporates a total of three gas clearance seals. A sintered, random fiber (22 micron diameter) material is employed in the first stage and a sintered woven screen (100 micron diameter) in the second stage. The flexure stack is mounted in the lower portion of the expander housing module. To simplify the hardware development process, the same flexures employed on the JT gas compressors were utilized on the displacer. Annular gap heat exchangers are employed on both the first and second cold stages. A slotted heat exchanger configuration is employed for the system heat rejection and is an integral part of the displacer cylinder assembly. In the prototype system, a laboratory water supply is employed for final heat rejection. All components of the prototype displacer assembly were fabricated from stainless steel and joined via vacuum brazing. Final machining of the critical clearance seals was carried out after the brazing process. The displacer cylinder, shown in Figure 4, incorporates the first and second stage external heat exchangers which provide pre-cooling of the JT working gas. Tubing for the JT gas is attached to the copper heat exchanger exterior after vacuum brazing of the cylinder assembly.

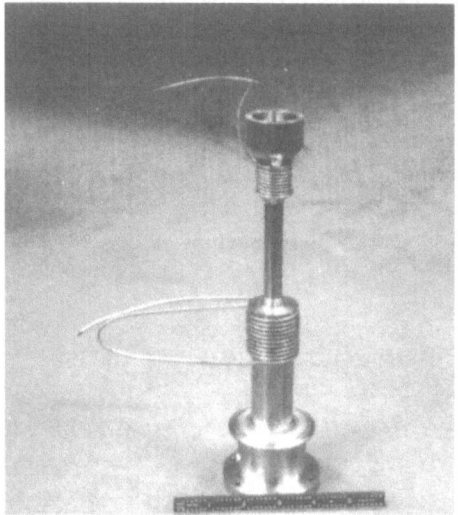

Figure 4.    SJTC Displacer Cylinder
             Assembly

As in the case of the linear drive power modules, the expander module is assembled and aligned as an independent component prior to final system assembly. During pre-cooler testing, an additional gas spring piston was attached to the rear (outboard) end of the displacer drive rod. This very weak gas spring served as a variable damper on the displacer simplifying the mechanical tuning of the displacer relative to the piston.

**Gas Compressor**

Based on the system optimization process, a two-stage compression process with interstage gas cooling was selected for the baseline JT gas compressor. Since the compressors are attached to the same piston rod as the linear motor and pre-cooler piston, it was necessary to ensure that the highly non-sinusoidal loads caused by the gas compression process would not cause poor performance of the linear motor and/or Stirling cycle pre-cooler. An additional concern was the potential for unstable operation at high power levels due to the power distribution between the gas compressors and pre-cooler requirements. The selected mechanical configuration is shown schematically in Figure 5. The basic compressor is of the double-acting type and incorporates a rod which extends out of the rear (outboard) portion of the compressor housing. This eliminates the strong bias force which would exist if one single acting piston were employed [5]. With the selected configuration a symmetric, nearly sinusoidal motor current is required, greatly simplifying the power supply, insuring centered piston operation under all conditions, and significantly improving system dynamic stability.

The first stage compressor housing and piston hardware are shown in Figure 6; the second stage compressor differs only in the diameter of the piston. The circular pockets contain the exhaust reed valves for each compression space. A fundamentally similar set of intake valves are located on the lower portion of the compressor housing. The piston, also shown in Figure 6, is supported on flexures which are mounted to the compressor housing. This allows the compressor to be fully assembled and aligned as an independent module prior to coupling to the linear drive. Nominal radial clearances between the piston and housing range between 15 and 20 microns on the various gas clearance seals employed.

**TEST RESULTS**

The overall test program of the SJTC is based on a series of sequential component test efforts which will lead to testing of the complete system. Due to the modular nature of the SJTC components, the linear drive motors could be employed in independent testing of the

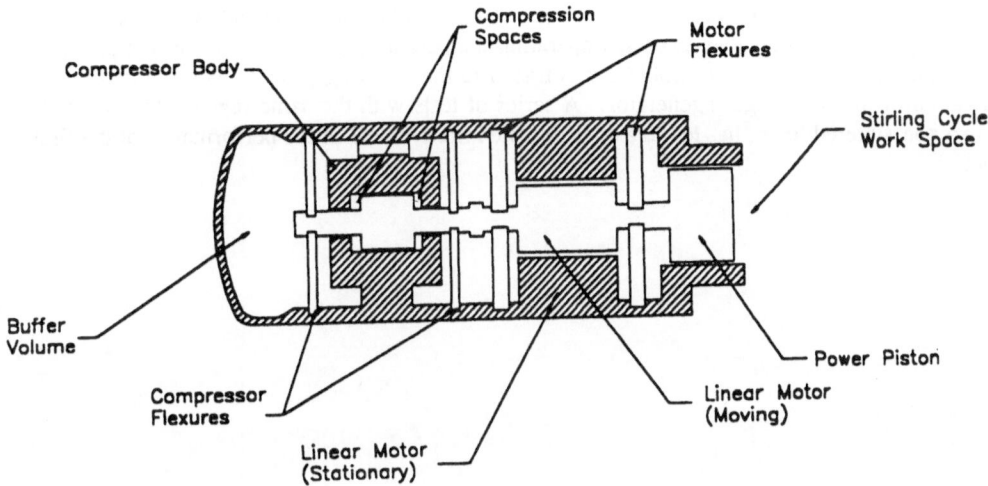

**Figure 5.     SJTC Gas Compressor Configuration**

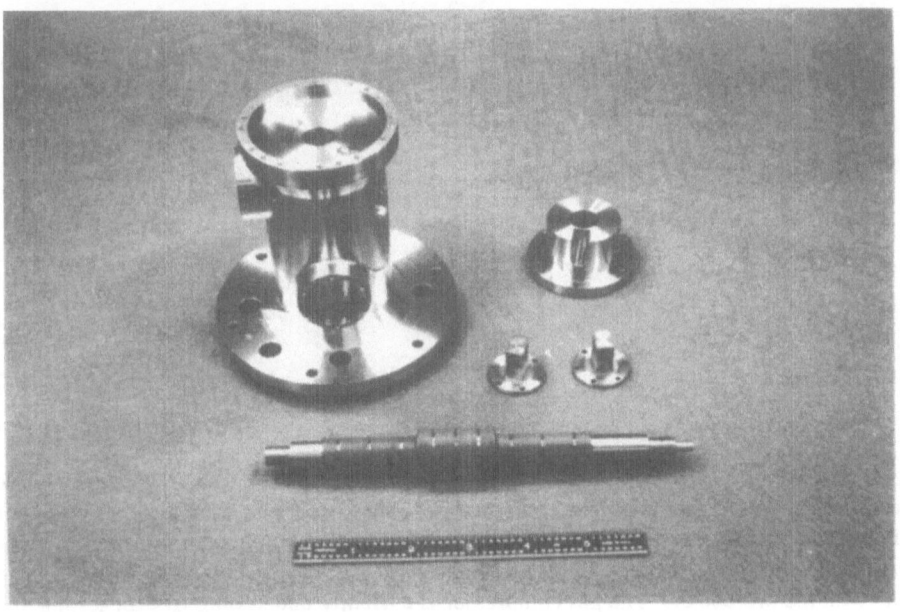

**Figure 6.     First Stage Compressor Housing and Piston**

pre-cooler and gas compressors. The pre-cooler was tested initially without the JT gas compressors to simplify the testing procedure. Figure 7 is an example of the cooldown characteristics of the pre-cooler when operating at approximately design input power. A discrepancy between predicted and actual first stage Stirling cycle operating characteristics was noted during this phase of pre-cooler testing. This resulted in somewhat lower first stage cooling capacity, a higher than expected operating temperature, and a greater thermal parasitic load on the second stage. The problem was traced to the sintering process incorporated in the fabrication of the first stage regenerator. A series of tests with the same regenerator material, but in a non-sintered form, has led to a considerable improvement in the performance of the first stage.

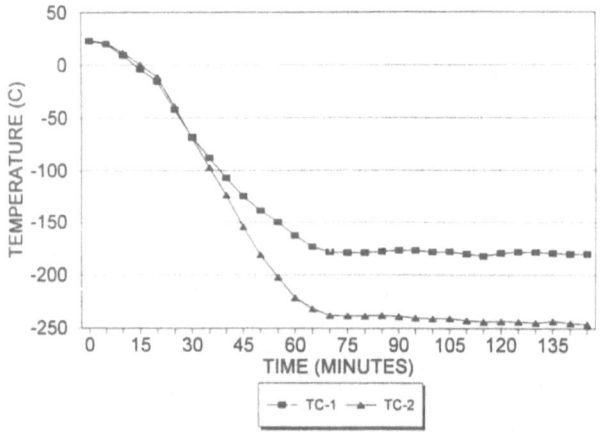

Figure 7.        Pre-Cooler Cooldown Test Results

The JT gas compressors have been evaluated during tests in which a single linear drive motor is utilized to drive an individual compressor. The compressor and linear drive motor were fully instrumented during these tests to investigate fundamental compressor operation, valve characteristics, and to fully define the dynamic characteristics of the gas compressor for effective integration with the pre-cooler. The gas compressor is currently undergoing free-air testing and pressurized testing with helium prior to integration with the pre-cooler and JT gas circuit.

## COMMERCIALIZATION

An integral part of the SBIR program is the commercialization of the technology or hardware developed during the Phase II effort. STC's efforts in this area have focused on commercialization of various pre-cooler variants and the linear drive gas compressor. Because of the modularity of the SJTC pre-cooler hardware, both single and dual drive Stirling cycle and pulse tube units are well along in the development process. For example, Figure 8 depicts the beta unit of a single linear drive/single stage Stirling cycle unit currently in low volume production. These units target cooling capacity requirements in excess of 15 watts at 80 K.

Figure 8.    SJTC Derived Cryocooler

## CONCLUSIONS

The SJTC system represents an innovative approach to long life, low temperature refrigeration systems. The ability of the JT expander to be mounted remotely from the pre-cooler/gas compressor combination will greatly simplify the integration of the final cooling stage with high sensitivity sensors for both space-based and ground-based commercial applications. The prototype hardware has demonstrated the that the combined JT gas compressor/Stirling cycle power piston can be integrated with a single linear drive. The flexural bearing system is capable of maintaining the necessary gas clearance seal gaps for good compressor performance. Continued testing will evaluate the performance of the integrated Stirling cycle pre-cooler, JT gas compressors, and JT gas circuit.

## ACKNOWLEDGMENTS

We wish to acknowledge and express our gratitude for the support provided by NASA Goddard Space Flight Center, the technical assistance provided by Dr. Max Gasser, and the strong support by NASA Goddard for commercialization of hardware developed during this program.

## REFERENCES

1.   Orlowska, A. H., Bradshaw, T. W., and Hieatt J., "Development Status of a 2.5 K - 4 K Closed-Cycle Cooler Suitable for Space Use", *Proceedings of the 8th International Cryocooler Conference*, Vail, Colorado, 1995.

2.   Jones, B. G., and Ramsay, D. W., "Qualification of a 4 K Mechanical Cooler for Space Applications", *Proceedings of the 8th International Cryocooler Conference*, Vail, Colorado, 1995.

3.   Shige, T., Baba, H., Kuraoka, Y., et al., "4.2 K Refrigerator for SQUID Magnetometer", *Proceedings of the Fourth International Cryocooler Conference*, Easton MD, 1986.

4.   Koizumi, T., Kuroki, K., Tomita, Y., et al., "Recondensing Refrigerator for Superconducting NMR-CT", *Proceedings of the Fourth International Cryocooler Conference*, Easton MD, 1986.

5.   Soedel, W., Mechanics, Simulation and Design of Compressor Valves, Gas Passages, and Pulsation Dampers, Purdue University, 1992.

# Flight Demonstration of a 10 K Sorption Cryocooler

**S. Bard, P. Karlmann, J. Rodriguez,**
**J. Wu, L. Wade, P. Cowgill, and K. M. Russ**

Jet Propulsion Laboratory
California Institute of Technology
Pasadena, CA, 91109

## ABSTRACT

The Brilliant Eyes Ten-Kelvin Sorption Cryocooler Experiment (BETSCE), flown on STS-77 in May 1996, was the first-ever space flight of chemisorption cryocooler technology. BETSCE measured and validated critical microgravity performance characteristics of a hydride sorption cryocooler designed to cool long-wavelength infrared and submillimeter-wavelength detectors to 10 K and below. The technology flight validation data provided by BETSCE will enable early insertion of periodic and continuous-operation long-life, low-vibration, low-power consumption, sorption refrigeration technology into future earth-observation, surveillance, and astrophysics space missions.

BETSCE produced solid hydrogen at 10 K in its first attempt on-orbit, cooling down from 70 K to 10 K in under 2 minutes and sustaining a 100 mW $I^2R$ heat load for 10 minutes, thus meeting the primary system performance objectives. In addition, a total of eight quick-cooldown liquid hydrogen cycles were completed, achieving a minimum temperature of 18.4 K and a maximum cooling duration of 32 minutes. Total cycle times ranged from 8 to 11 hours, depending on Shuttle orbiter attitude.

BETSCE successfully validated sorption cooler operation in a microgravity environment. Flight data obtained for a total of eighteen compressor cycles demonstrated their ability to consistently re-compress the hydrogen refrigerant fluid in a repeatable manner, and to the same high pressures as achieved in ground testing. No microgravity supercooling was observed of the n-hexadecane phase change material in the Fast Absorber Sorbent Bed, as it changed phase at its expected melting temperature of 291 K. Also, no adverse microgravity effects were observed in the cryostat cold head, as it demonstrated the ability to effectively retain liquid and solid hydrogen.

## INTRODUCTION

Many future space instruments depend on the successful development of long-life, low-vibration space cryocoolers. Sorption cryocoolers using hydrogen as the refrigerant and metal hydrides as the compressor chemisorbent material can meet the needs of future space instruments containing long wavelength infrared and submillimeter imaging detectors requiring operating temperatures in the 10 to 30 K temperature range. Examples include astrophysics space telescopes, earth and planetary atmospheric, geologic and resource mapping satellites, space superconducting electronics, and military surveillance satellites.

## Background

The concept for a continuous 10 K sorption cryocooler was originally developed by Jones in 1984.[1] In 1991, Johnson and Jones recognized that a periodic 10 K sorption cryocooler offers repeated quick cooldowns and low average power consumption for applications where intermittent operation is sufficient.[2] The proof-of-principle of a periodic 10 K sorption cooler stage was experimentally demonstrated and analytical concept design studies were conducted in 1992.[3] In 1994, ground testing of the Brilliant Eyes Ten-Kelvin Sorption Cryocooler Experiment (BETSCE) provided the first-ever ground-based performance demonstration of a complete closed-cycle 10 K sorption cryocooler.[4] The BETSCE flight demonstration described here provided the space performance validation that will enable the early insertion of this new technology into future space missions, and established the microgravity database required for development and production of periodic and continuous 10 to 30 K sorption cryocooler systems. Although the BETSCE cooler is configured for periodic operation, its space performance demonstration validates all of the important technology elements of a continuously operating cooler as well.

## Concept

Description of the concept and operation of a 10 K periodic sorption cryocooler have been presented in detail elsewhere.[2-5] The concept is based on sequentially heating beds containing metal hydride powders to circulate hydrogen as the refrigerant fluid in a closed cycle Joule-Thomson (J-T) refrigeration system. On command, it periodically cools the hydrogen to 10 K from an upper stage temperature of $\geq$ 65 K that can be provided by one of the emerging long-life mechanical cooler systems.[6]

## BETSCE Objective

This paper describes how BETSCE successfully achieved its primary objectives of: (1) providing a thorough end-to-end characterization and performance validation of a hydride sorption cryocooler in the 10 to 30 K temperature range, (2) acquiring the microgravity database needed to provide confident engineering design, scaling and optimization, (3) identifying and resolving interface and integration issues, and (4) providing hardware qualification and safety verification heritage.

## Microgravity Issues

BETSCE is designed to acquire needed microgravity data in each of these key areas: (1) phase separation and capture of liquid hydrogen during the formation of two-phase gaseous and liquid hydrogen at 18-30 K, (2) phase separation and capture of solid hydrogen during the formation of three-phase gaseous, liquid and solid hydrogen at 13.8-9.0 K, (3) liquid and solid hydrogen retention in the wicked cold head reservoir assembly when subjected to zero-gravity effects (e.g. Marangoni surface tension variations with temperature), (4) heat and mass transfer mechanics within the hydride powder beds and phase change materials in the absence of gravity-dependent natural convection and one-g contact forces, (5) supercooling of the phase change materials in microgravity, and (6) Joule-Thomson (J-T) efficiency and heat exchanger effectiveness in the absence of gravity-dependent mixed natural and forced convection.

A detailed understanding of these microgravity issues will enable elimination of excessive design margins, which is expected to result in significant weight savings for future periodic and continuous 10 K sorption cryocooler systems.

**Figure 1.** BETSCE mounted on test stand.

## EXPERIMENT HARDWARE OVERVIEW

Figure 1 shows the BETSCE instrument on its test stand. BETSCE mounts on the Shuttle orbiter payload bay side-wall to a Small Payload Accommodation (SPA) Get Away Special (GAS) adapter beam carrier, as shown in Figure 2. Power from the Shuttle 28 Volt DC power bus and crew inputs from the Standard Switch Panel (SSP) reach BETSCE via Shuttle cabling. BETSCE data/ commands are downlinked/ uplinked via standard Shuttle Payload Data Interleaver (PDI)/ Payload Signal Processor (PSP) interfaces. Waste heat from BETSCE is radiated to space by flat plate passive radiators oriented out (+Z) of the Shuttle bay. The BETSCE instrument contains four integrated subsystems: Sorbent Bed Assembly, Cryostat Assembly, Tank and Valve Assembly, and Control Electronics Assembly. The design of this hardware was described in detail in Reference 4.

BETSCE was developed as a collaborative team effort between industry, university and government. The Sorbent Bed Assembly was developed by Aerojet Electronic Systems Division

**Figure 2.** BETSCE mounted in Space Shuttle Endeavour payload bay.

(AESD), the Cryostat Assembly was developed by APD Cryogenics, Inc., the upper stage Stirling coolers for the Cryostat Assembly were provided by Hughes Aircraft Corp., the n-hexadecane phase change material was furnished by ESLI, Inc., and the sorbent materials were characterized by the University of Vermont. JPL is responsible for overall project management, principal investigator team, system design, design and fabrication of the Control Electronics Assembly, Tank and Valve Assembly, Structure and Cabling subsystems, structural and thermal systems analyses, system integration and test, flight and ground software, Shuttle integration support, mission operations, and postflight data analysis.

## FLIGHT EXPERIMENT RESULTS

The BETSCE cycle operation can be visualized with the aid of the fluid schematic shown in Figure 3. At launch, more than 93% of the ~35 g total fill of hydrogen in the BETSCE system is in hydride form. In order to begin cycling operations, BETSCE must first complete pre-conditioning sequences that first cool the thermal storage device to 65 K, and then desorb the sorbent beds to pressurize the hydrogen tank.

### Conditioning Sequences

BETSCE was powered-up on Flight Day 1 on May 19, 1996, and assumed a low power configuration that enabled thermostatically controlled heaters to keep hardware from cold-soaking at cold attitudes. On Flight Day 3, the BETSCE flight control team powered-on the BETSCE mechanical cryocoolers to precool the BETSCE upper stage to 65 K. Figure 4 shows the initial cooldown of the thermal storage device (TSD).

After the TSD was cooled to below 70 K, the BETSCE flight control team uplinked commands to begin heating and desorption of the sorption compressor beds. First, the High Pressure Sorbent

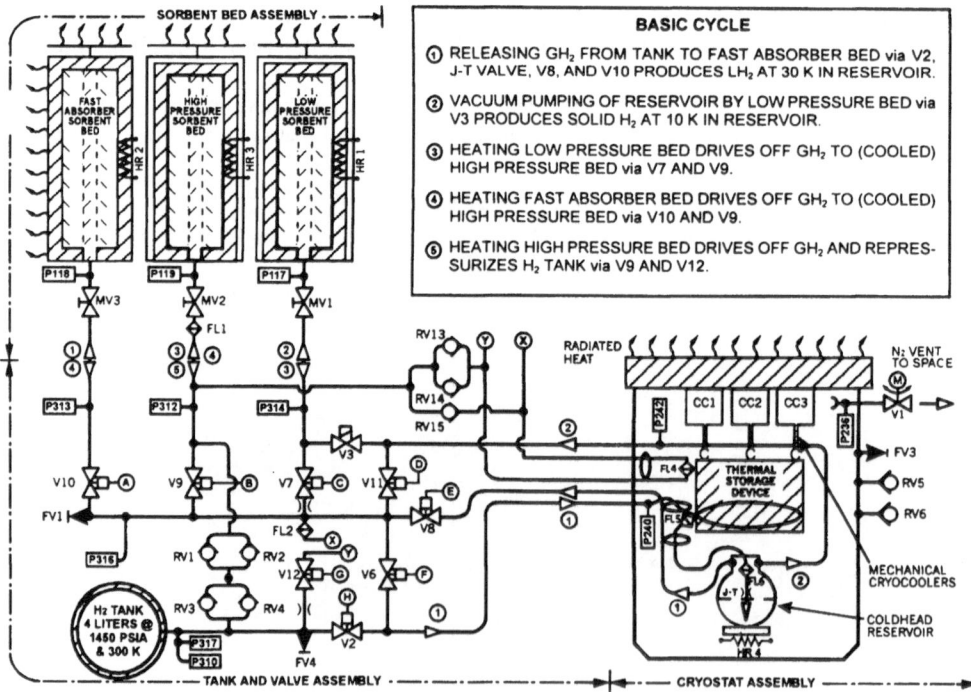

**Figure 3.** BETSCE fluid schematic.

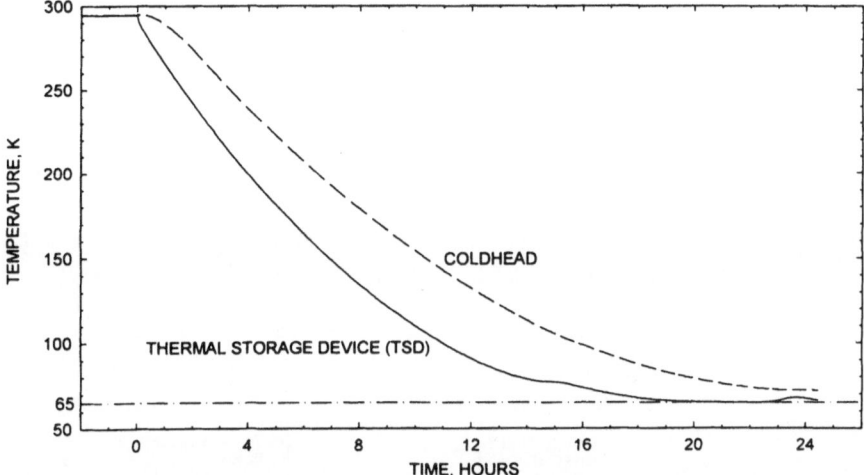

**Figure 4.** Initial on-orbit cooldown of cryostat thermal storage device.

Bed was heated to transfer its hydrogen to the 4 liter storage tank, and then allowed to cooldown. Next, the Low Pressure and Fast Absorber Sorbent Beds were heated in turn to transfer their hydrogen to the High Pressure Sorbent Bed, and then allowed to cool down. Finally, the High Pressure Sorbent Bed was heated again to bring the $H_2$ storage tank to operating pressure of 9.8 MPa (1425 psia) at 300 K, and again allowed to cool down. Figure 5 shows the flight desorption temperature and pressure histories. These results demonstrate that the sorbent beds were able to effectively compress hydrogen to identical pressures as achieved in ground tests.

## 10 K Cooldown

Cooldown was initiated by releasing pressurized $H_2$ gas from the storage tank to the Fast Absorber Sorbent Bed by way of the cryostat, whose pre-chilled TSD heat sink, heat exchangers, and J-T valve produced liquid $H_2$ at $\approx$ 27 K in the cold head reservoir. Next, solid $H_2$ at $\leq$ 11 K was produced by vacuum pumping the cold head reservoir with the Low Pressure Sorbent Bed. Under simulated sensor heat load provided by an electrical resistance heater, and continued vacuum pumping, the solid $H_2$ sublimated and was absorbed by the Low Pressure Sorbent Bed until it was consumed. Next, hydrogen was returned to the tank and pressurized by sequentially heating the sorbent beds as before, to complete the cycle. The total cycle time is between 8 and 12 hours, depending on the Shuttle orbiter attitude.[7]

Figure 6 shows the initial 10 K cooldown. The cooldown from 65 K to 10 K was completed in under 2 minutes, and a 100 mW $I^2R$ heat load was sustained at below 11 K for 10 minutes, thus meeting the primary system performance objectives.

After the cooldown was completed valve V2, which isolates the tank from the cryostat, did not re-seal properly. This allowed hydrogen to leak from the tank to the cryostat. The valve never completely re-sealed, although the leak rate was lowered by periodically cycling the valve and flowing through in forward and reverse directions.

## 20 K Cooldown

Although the problem with valve V2 was to prevent more 10 K cooldown cycles, pressure-cycling of the compressors continued, and valuable data on their space performance was obtained. The compressors demonstrated outstanding performance, repeatedly achieving high compression ratios by thermal compression despite the leak through V2. Figure 7 shows some of the sorbent bed desorption cycles on orbit.

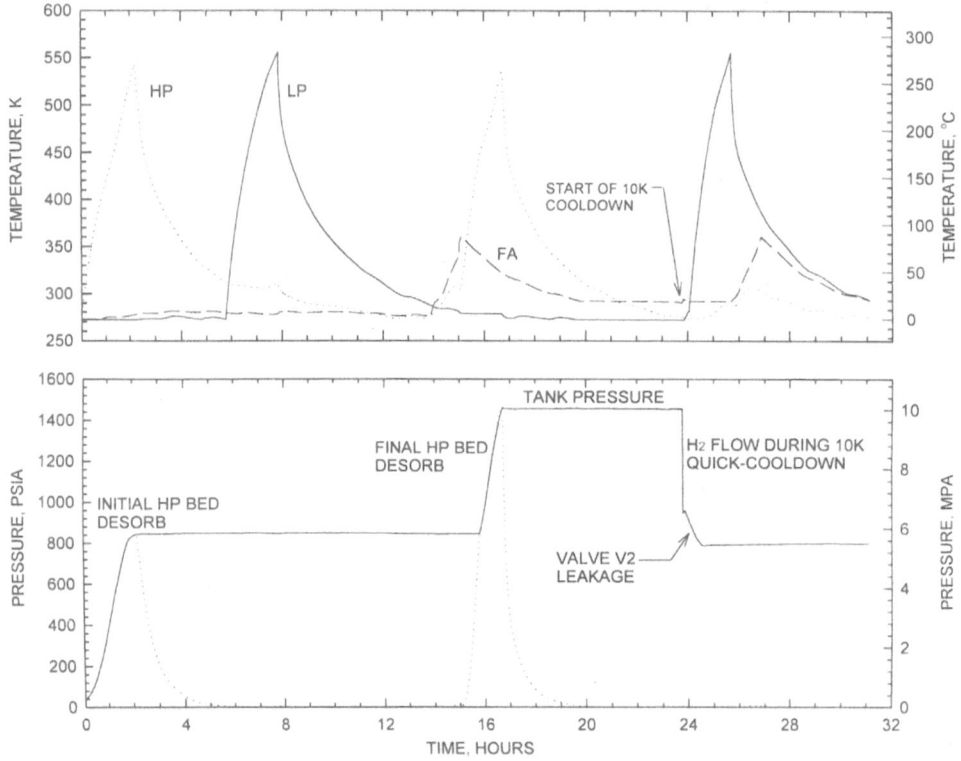

**Figure 5.** Initial on-orbit sorbent bed desorption cycles.

By Flight Day 6, the valve configuration and timing were tuned to where BETSCE was able to produce liquid hydrogen at 21 K. Further refinement of valve timing and sequence resulted in even lower temperatures and longer cooling duration. The coldest temperature achieved in these cycles was 18.4 K and the longest cooling duration was over 32 minutes. A total of eight liquid hydrogen

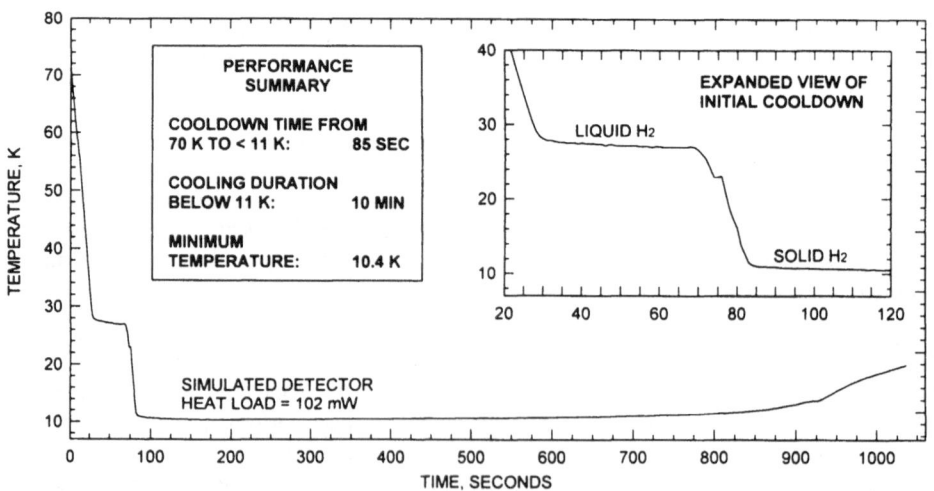

**Figure 6.** On-orbit 10 K cooldown.

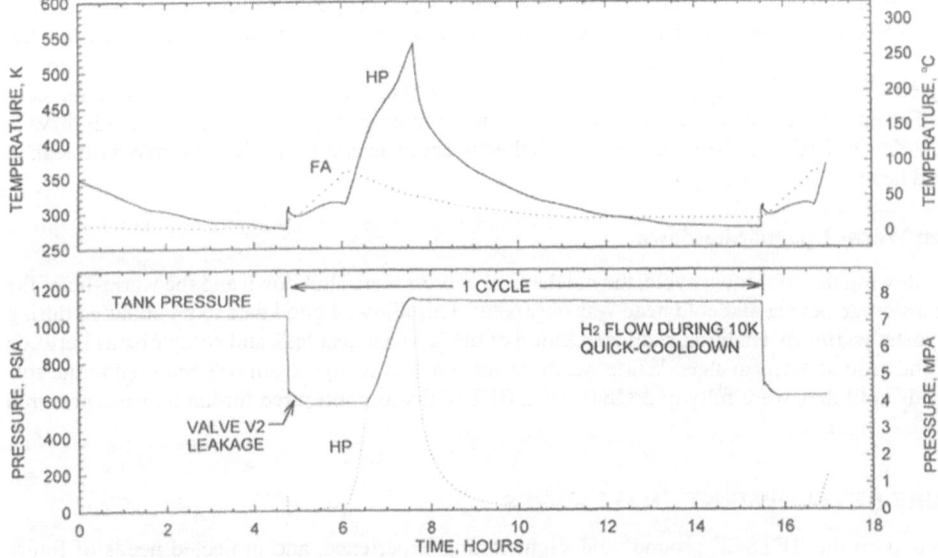

**Figure 7.** Sorbent bed desorption cycles on orbit (with valve V2 leakage).

cycles were achieved by the end of the mission. Results from some of these cooldowns are shown in Figure 8.

### Parasitic Heat Leak

The cooling duration at liquid or solid hydrogen temperatures with different $I^2R$ heat loads enables the effective parasitic heat leak to the cold head to be estimated. In ground testing, this effective heat leak was about 45 mW with the cold head at 10 K.[4] The parasitic heat leak due to conduction and radiation is reduced by the cooling effect of the vapor from the sublimating or evaporating cold hydrogen that flows away from the cold head. After valve V2 didn't seal properly, the warm hydrogen leaking from the tank to the cryostat created an additional heat leak. That

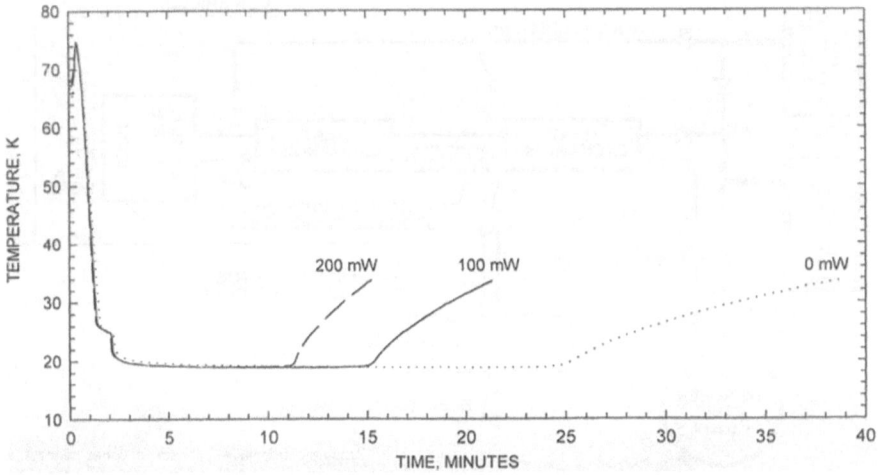

**Figure 8.** On-orbit liquid hydrogen cooldown cycles.

is why the minimum temperature during the initial solid hydrogen cooldown cycle was 10.4 K, while 9.5 K was repeatedly achieved in ground testing. Although warm hydrogen leaked into the cryostat, the heat load on the cold head was small because the TSD and cold heat exchangers are highly efficient at low mass flow rates resulting in low hydrogen temperatures reaching the cold head. The on-orbit effective heat leak during the liquid hydrogen hold time was about 130 mW at 18.4-20 K. The hydrogen leak through V2 resulted in about an additional 80 to 90 mW heat leak to the cold head.

## System Warm-Up and Shutdown

Following the cooldown cycle, the mechanical coolers were shut down and the warm-up of the thermal storage device and cold head was observed. This allowed good data to be obtained during the cryostat warm-up, enabling characterization of the cryostat heat leak and comparisons between flight and ground performance. While warm-up was underway, hydrogen was returned to the sorbent beds until they were fully hydrided. Then BETSCE was configured for landing and powered off.

## FUTURE DEVELOPMENT CHALLENGES

Based on the BETSCE ground[4] and flight testing experience, and projected needs of future missions,[9] a number of valuable lessons-learned can be applied to improve the design of future 10 K sorption cryocooler systems.

First, considerable reductions in size and weight are possible for future 10 K sorption coolers. As described in Reference 4, the weight of an operational 10 K sorption cryocooler can be reduced from BETSCE's 79 kg to 25-30 kg, which includes 15 kg allocated for an upper stage 60 K mechanical cooler. Figure 9 is a schematic of a periodic 10 K sorption cooler cycle without all of the extra test valves used to characterize BETSCE. Note that it was only because BETSCE contained these extra valves that achievement of 20 K liquid hydrogen cycles was made possible after the leak developed in valve V2.

Figure 9 shows a periodic 10 K sorption cooler without all of the test valves, without a separate Fast Absorber Sorbent Bed, and containing a single-stream regenerative TSD instead of the nearly isothermal BETSCE TSD. In addition to reducing the size and mass of the TSD, this regenerative design also reduces the heat load that needs to be removed by the upper stage mechanical cooler.

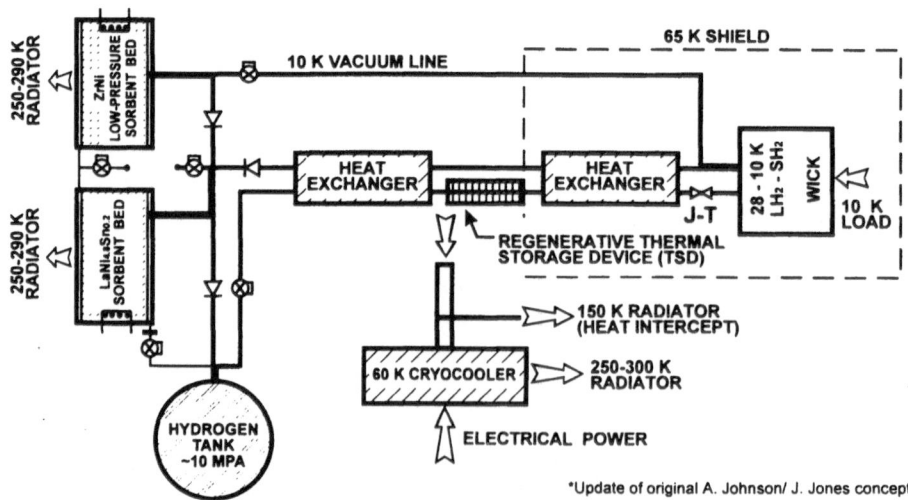

**Figure 9.** Periodic 10 K sorption cooler schematic.

Before sorption coolers can be confidently implemented in operational systems, reliable, long-life operation must be confirmed. Long-term heater cycling tests have confirmed over 10-year life capability.[9] Long-term hydride cycling tests were initiated and gave results consistent with long-life capability, but were stopped due to funding constraints.[9] Additional sorbent testing is certainly required.

Although removal, disassembly and detailed examination of BETSCE valve V2 has not yet been performed, it is likely that the leakage was caused by a particulate that was able to flow from the tank, fittings or plumbing in microgravity conditions. This particulate may have been trapped and unable to flow to the valve during numerous 1-g ground tests. But, in microgravity it was able to float and flow to the valve seat. Another factor may be that although it was constructed of electropolished 316L stainless steel tubing, and carefully cleaned and verified to Aerospace standards, the BETSCE tank was dead-ended. For future systems, it is recommended that tanks contain both an inlet and an outlet to aid in effective cleaning. Finally, placement of filters at each valve inlet and/ or outlet is expected to prevent the type of contaminant-caused leakage observed in BETSCE valve V2.

The likely direction of hydride sorption cooler technology is towards continuous cooling applications instead of periodic. A continuous sorption cooler is expected to be very efficient at 10 K, requiring about 700 W input power per W of cooling, which includes power for the upper stage mechanical cooler. References 1, 8, and 10 describe continuous 10 K sorption cooler designs. Reference 8 also describes a continuous 25 K hydride sorption cooler that will be integrated with an infrared detector and flight tested in a long-duration balloon flight over Antarctica in 1997. The successful BETSCE space flight test results described here validated all of the critical technologies for both periodic and continuous configurations of hydride sorption coolers, for 10 K to 30 K operation.

## SUMMARY AND CONCLUSIONS

BETSCE achieved all of its flight objectives. It demonstrated the ability to cooldown and produce solid hydrogen at 10 K in under 2 minutes on its first try on orbit while sustaining a simulated $I^2R$ detector heat load of 100 mW for 10 minutes, the ability to repeatedly produce liquid hydrogen below 19 K, and the ability to repeatedly recycle the sorption compressors. By the end of mission, valuable flight data was obtained for a total of eighteen compressor cycles.

Except for a valve internal leak, which would be prevented by filter placement in future systems, the performance of the BETSCE flight hardware was excellent. Microgravity operation has no adverse effect on sorption cooler performance, as the cold head demonstrated the ability to retain liquid and solid hydrogen, and the sorption compressors demonstrated similar heat and mass transfer characteristics as in ground testing.

Finally, the successful flight testing of BETSCE has provided the end-to-end characterization and validation, and the quantitative microgravity database that will enable early insertion of 10 to 30 K sorption cryocooler technology into future long-life, low-vibration, space remote sensing missions.

## ACKNOWLEDGMENT

The work described in this paper was carried out by the Jet Propulsion Laboratory, California Institute of Technology, and was sponsored by the Ballistic Missile Defense Organization (BMDO), the Air Force Space and Missiles Systems Center, and Air Force Phillips Laboratory through an agreement with the National Aeronautics and Space Administration. This work was made possible by the financial support provided by Erwin Myrick, Dr. Dwight Dustin, and Col. Ralph Gajewski. The dedication, skill, and hard work of the many engineers, technicians, and support personnel who were members of the BETSCE team at JPL, Aerojet, and APD Cryogenics are greatly appreciated. The BETSCE team would also like to thank the crew of the Endeavour and the entire Space Shuttle Program team at the Johnson Space Center in Houston and the Kennedy Space Center in Florida for all of their outstanding support during the STS-77 mission.

## REFERENCES

1.  Jones, J. A., "Hydride Absorption Refrigerator System for Ten Kelvin and Below," *Proceedings of the Third Cryocooler Conference,* NBS Special Publication 698, NBS, Boulder, CO, (1984).

2.  Johnson, A. L. and Jones, J. A., "Evolution of the 10 K Periodic Sorption Refrigerator Concept," *7th International Cryocooler Conference Proceedings,* Air Force Phillips Laboratory Report PL-CP-93-1001, Kirtland AFB, NM, (1993), pp. 831-853.

3.  Wu, J. J., Bard, S., Boulter, W., Rodriguez, J., and Longsworth R., "Experimental Demonstration of a 10 K Sorption Cryocooler Stage," *Advances in Cryogenic Engineering,* Vol. 39, Plenum Press, New York, (1994), pp. 1507-1514.

4.  Bard, S., Wu, J. J., Karlmann, P., Cowgill, P., Mirate, C. and Rodriguez, J., "Ground Testing of a 10 K Sorption Cryocooler Flight Experiment (BETSCE)," *Cryocoolers 8,* Plenum Press, New York, (1995), pp. 609-621.

5.  Bard, S., Cowgill, P., Rodriguez, J., Wade, L., Wu, J. J., Gehrlein, M., and Von Der Ohe, W., "10 K Sorption Cryocooler Flight Experiment (BETSCE)," *7th International Cryocooler Conference Proceedings,* Air Force Phillips Laboratory Report PL-CP-93-1001, Kirtland AFB, NM, (1993), pp. 1107-1119.

6.  Ross, R. G. Jr., "JPL Cryocooler Development and Test Program Overview," *Cryocoolers 8,* Plenum Press, New York, (1995), pp. 173-184.

7.  Bhandari, P. and Bard, S., "Thermal Systems Design and Analysis for a 10 K Sorption Cryocooler Flight Experiment," AIAA 28th Thermophysics Conference, AIAA 93-2825, Orlando, FL (1993).

8.  Wade, L., Levy, A., Bard, S., "Continuous and Periodic Sorption Cryocoolers for 10 K and Below," *9th International Cryocoolers Conference,* Waterville Valley, NH, (1996).

9.  Bard, Wu, J. J., Karlmann, Mirate, C., and Wade, L., "Component Reliability Testing of Long-Life Sorption Cryocoolers," *Cryocoolers 8,* Plenum Press, New York, (1995), pp. 623-636.

10. Longsworth, R. C. and Khatri, A., "Continuous Flow Cryogen Sublimation Cooler," *Advances in Cryogenic Engineering,* Vol. 42, Plenum Press, New York, (1996), in press.

# Continuous and Periodic Sorption Cryocoolers for 10 K and Below

Lawrence A. Wade,[1] Alan R. Levy,[2] and Steve Bard[1]

[1]Jet Propulsion Laboratory
California Institute of Technology
Pasadena, CA 91109

[2]University of California at Santa Barbara
Department of Physics
Santa Barbara, CA 93106

## ABSTRACT

This paper presents the current status of both continuous and periodic operation sorption cryocooler development for astrophysics missions requiring refrigeration to 10 K and below. These coolers are uniquely suited for cooling detectors in planned astrophysics missions such as the Exploration of Neighboring Planetary Systems, the Next Generation Space Telescope, and Darwin. The cooler requirements imposed by these missions include ten year life and the ability to scale designs to provide only a few milliwatts of refrigeration while consuming only a few watts of input power. In addition, the ExNPS and Darwin missions add stringent requirements for zero-vibration and zero EMI/EMC operation.

Spaceflight test results are summarized for the Brilliant Eyes Ten-Kelvin Sorption Cryocooler Experiment. This periodic operation sorption cooler is ideal for applications that require only intermittent operation at 10 K with quick cooldown capability (under 2 minutes). The experiment successfully provided flight characterization of all sorption cooler design parameters which might have shown sensitivity to microgravity effects. Full ground test performance was achieved with no indications of microgravity induced changes.

Ground test results from a continuous 25 K cooler planned for use in a long duration airborne balloon experiment are also presented. This 25 K cooler, which is in final integration and test, can be used as an upper stage for a continuous 10 K sorption cooler. The potential benefits of using a 10 K sorption cooler as an upper stage for a 4 K cooler are additionally described. Finally, a NASA program to develop 30 K, 10 K and 4 K vibration-free coolers for astrophysics missions, which is planned to start in FY 1997, is outlined.

## INTRODUCTION

The heritage provided by the many successful dewar cooled missions (e.g. ISO, IRAS, COBE, and the now underway WIRE and SIRTF) has enabled the serious consideration and development of a new generation of actively cooled space instrument design concepts. The interest in cryocoolers being shown by the designers of these missions is a result of the substantial maturation of cryocooling technologies, which has occurred over the past ten years, and of an increasing awareness within the scientific community of the potential benefits offered by these technologies. The utilization of long-life cryocoolers allows mission designers to refrigerate large format detector arrays during ten year missions. The volume and mass saved through the use of active coolers in combination with passive radiators enable mission designers to pack much larger telescope apertures into a given launch vehicle than would be possible in a dewar cooled mission. Thus, many of the missions that launch after 2005 will incorporate cryocoolers.

Astrophysics missions, now in the early design phase of development, which incorporate long-life, vibration-free cryocoolers include the Exploration of Neighboring Planetary Systems (ExNPS), the Next Generation Space Telescope (NGST) and Darwin. In addition to these precision pointing missions, moderate resolution missions such as FIRST and COBRAS/SAMBA are incorporating low-vibration cryocoolers. This paper gives a discussion of the state-of-the-art in sorption cooler technology and how recent work in the field is being directed toward the goal of producing sorption coolers for future space based astrophysics missions.

## FUTURE MISSION CRYOCOOLER REQUIREMENTS

Most of the mission concepts now under development will operate in thermally advantageous orbits for scientific and engineering reasons first pointed out by the EDISON team[1]. Observing strategies and telescope/spacecraft configurations are being developed to fully exploit these orbits, which place all of the ~300 K devices and structure together on a warm spacecraft bus oriented towards the sun and earth. The cold telescope and science instruments are remotely located from the warm spacecraft bus and thermally isolated by several radiative surfaces. This enables optical structures to be radiatively cooled to as low as 20 K without the use of coolers or dewars.[2]

While passive radiative cooling is very effective when providing environmental shielding of extended structures and optics, it is often not very effective for absorbing actively generated loads (e.g. electronics, high bandwidth actuators, and detectors) at temperatures below 50 K. Typical requirements for astronomical telescope applications which require active refrigeration include cooling at one or more of the following temperatures: approximately 25 K for high bandwidth actuators, InSb, and QWIP detectors; between 4 and 8 K for Si:As BIB arrays; at 4 K for Si:Sb BIB arrays, SIS heterodyne receivers, and for thermal sinking of magnetic, dilution, and Helium-3 coolers used to cool bolometers to 0.1 K.

Several of the more challenging requirements for active coolers are well illustrated by the ExNPS mission. The ExNPS program is tasked to detect, image, and characterize planets around other solar systems. The ExNPS Planet Finder Array (PFA) consists of four 1.5 m telescopes, passively cooled to 30 K, on a 75 m baseline which are operated as a nulling interferometer, a beam combiner that is also cooled to 30 K, and a detector that is cooled to 4 K. The PFA, along with its spacecraft bus and connecting structure, will be launched out to 5 AU using a Venus-Earth-Earth gravity assist trajectory on its nine year mission. The PFA must fit within an Atlas IIAS shroud (3.65 m diameter by 9.4 m long with 5.3 m of the length tapering to a 0.81 m diameter tip) and within a lift capability of only 1824 kg. Due to the long duration of this mission, initially unfavorable thermal environment (especially during Venus flyby), and limited launch vehicle shroud volume and lift capability it is not feasible to use a dewar to support this mission.

The ExNPS PFA is designed to operate at 10 microns with a 20% bandwidth using destructive interference to 'remove' the light from the central star, which is 1,000,000 times brighter than an earth-like planet would be. Combining the high resolution of this array and the need to null the target solar system central star leads to a pointing requirement of approximately

$10^{-6}$ arcseconds. Stirling and Pulse Tube coolers, or coolers using similar compressors, can not be used to actively cool the ExNPS PFA focal plane as the residual vibration perpendicular to the compressor axis is typically 0.25 N despite $9^{th}$ harmonic vibration nulling electronics.[3] To do better than this, three axis stabilization actuators would have to be incorporated into the compressor and expander, along with redundant actuators. Due to the high dimensional stability requirements ($\sim 10^{-10}$ m), stringent pointing requirements, and the difficulty of integrating this assembly into the beam combiner, use of these coolers is not deemed feasible.

The solar array for this mission must be sized to enable observations at 5 AU when power is 25 times tougher to come by than at 1 AU. Therefore any cooler incorporated into the ExNPS design must have very low power requirements. To achieve this requirement, the cooler must be capable of taking full advantage of the favorable thermal environment enjoyed during observations and to scale down to a size commensurate with the ExNPS PFA mission detector cooling requirement of approximately 5 mW at 4 K. An additional stringent requirement is imposed by the desire to do spectroscopy on the detected neighboring planetary systems. In this operational mode, the final signal is measured in electrons per hour. As a result, essentially no cooler induced EMI/EMC is acceptable.

Only sorption coolers can meet the stringent combination of life, vibration, mass, volume, power, and EMI/EMC requirements posed by missions such as the ExNPS PFA.

## STATUS OF SORPTION CRYOCOOLING TECHNOLOGY

### Sorption Technology Summary

Several review papers have been published which describe the history and basic concepts behind the various kinds of sorption coolers.[4,5] Only a brief summary will be provided in this paper.

Sorption coolers are comprised of a sorption compressor, containing a sorbent material, and a Joule-Thomson (J-T) expander. The refrigerant is selected to correspond with the required cooling temperature and the sorbent material is selected based on the choice of refrigerant and the available thermal environment. Cooling between 30 K and 8 K is achieved by use of metal hydrides as the sorbent material and hydrogen as the refrigerant. Cooling at 7 K and below is achieved using activated charcoal as the sorbent and helium as the refrigerant.

During operation of the cooler, compressed refrigerant, desorbed by a heated sorption bed, is expanded through a J-T orifice to create a gas/liquid refrigerant mixture. The liquid evaporates as it absorbs heat from the detectors and is then absorbed and repressurized in a cool bed, thus creating a fully reversible closed-cycle system. Due to the physics of the sorbent materials, the compressors work most efficiently when operated over a large pressure ratio at low mass flow rates. The combination of a sorption compressor with a Joule-Thomson expander provides a cooler which operates without cold moving parts and has a capacity that can be scaled linearly to below 1 mW. All of the sorption cryocooler designs being considered for future astrophysics missions utilize passive radiative cooling at between 50 and 65 K to precool the refrigerant gas.

The characteristics of sorption coolers which are important to mission designers include:
1) The ability to locate all warm components directly on the preferred heat rejection surfaces to both minimize system mass, simplify the mechanical design, and to prevent thermal parasitics into the passively cooled regions of the telescope;
2) Minimized cryostat size to simplify integration into the complicated final beam combiner or focal plane assembly regions;
3) Dimensional stability on the order of the amplitude of lattice vibrations in a simple block of stainless steel (i.e. no vibration imposed beyond that normal to a passive piece of stainless);
4) Zero EMI/EMC effects on the science instruments;
5) Extremely low power usage. This can be achieved through taking full advantage of the thermal environment to minimize environmental loads, intercept parasitics, and to precool the

refrigerant. Combining the aggressive use of the thermal environment with the ability to linearly scale the size and thereby the input power to the cooler results in extremely small system power requirements. Predicted rule-of-thumb performance ranges for coolers that provide less than 100 mW of cooling, reject their input power at approximately 300 K, and are designed for flight are:[a]

a) 300-400 W/W at 20 K
b) 700-900 W/W at 9 K
c) 3,000-5,000 W/W at 4K

Recent advances have substantially improved the flight readiness level of sorption technology. The Brilliant Eyes Ten-Kelvin Sorption Cooler Experiment (BETSCE)[6,7], which operated in orbit in May, 1996, examined all of the design characteristics which could be affected by the microgravity environment. The resulting flight dataset provides flight validation for the design of future periodic and continuous sorption coolers.

A continuous operation 25 K cooler is being developed for the University of California at Santa Barbara (UCSB) Long Duration Balloon (LDB) experiment[8]. The 25 K LDB cooler is the first hydride sorption cooler to help 'do' science instead of 'being' the science. This single-stage cooler was designed to robustly achieve stable performance while dramatically improving contamination tolerance.

## Summary of BETSCE Flight Results and Accomplishments

BETSCE,[3,4] shown in Figure 1, is a periodic operation cooler developed to achieve a cold end temperature of less than 11 K in under 2 minutes from a starting temperature of 65 K. This experiment was flown on the space shuttle Endeavour during the May, 1996 STS-77 mission. As the first hydride sorption cooler flight experiment, it offered a unique opportunity to measure microgravity effects on a wide variety of performance characteristics.

The in-flight performance of BETSCE has completely validated the use of hydride sorption coolers in space as no on-orbit degradation was found. As shown in Figure 2, the cooler successfully achieved a cold tip temperature of 10.4 K in less than two minutes from an initial temperature of 70 K. The cooler provided 100 mW of cooling for 10 minutes. This exceeded the BETSCE performance goals. In addition, a total of 8 quick cooldown cycles to liquid hydrogen temperatures were accomplished, achieving a minimum temperature of 18.4 K. A total of 18 compressor cycles were completed and the ability to repeatably achieve the 10.1 MPa high pressures achieved in ground testing was successfully demonstrated.

The measured microgravity effects on characteristics of interest to all sorption cooler designers were:

1) $LaNi_{4.8}Sn_{0.2}$ and ZrNi hydride powder thermal conductivities were perhaps the most important properties to characterize well. The in-flight conductivities were determined, from the rate of absorption and the absorption pressure, to have been identical to those measured in a one-g environment.

2) Supercooling of the n-hexadecane phase change material in the Fast Absorber Sorbent Bed is important to most periodic operation cooler designs and to some continuous cooler designs. No change in its expected 291 K melting temperature was observed.

3) The ability of the cryostat liquid reservoir to separate and retain both liquid and solid hydrogen substantially affects cooler capacity and temperature stability. Again, no adverse microgravity effects were observed in the cryostat cold head.

In summary, the BETSCE flight data shows that no additional design margin is required to design a hydride sorption cooler for space missions. In addition, BETSCE clearly demonstrated the feasibility of successfully developing and flying a sorption cryocooler in space.

---

[a] The estimates are based on a 60 K precooling temperature and designs incorporating full flight and ground test safety margins. Therefore, a 5 mW, 4 K cooler can be built for flight which requires less than 25 W of input power. Similarly, a 20 mW, 9 K requirement can be met with less than 18 W of input power.

**Figure 1.** BETSCE mounted on ground support equipment cart at the Kennedy Space Flight Center just prior to integration with the space shuttle Endeavour.

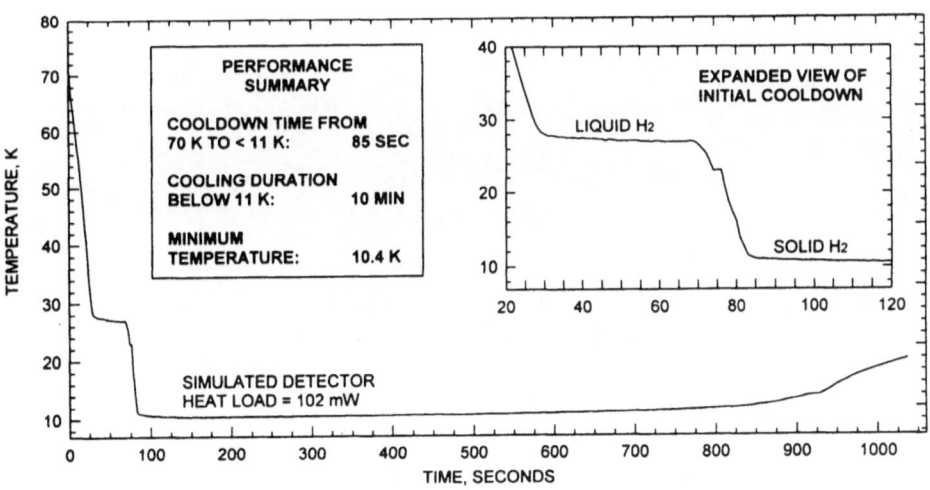

**Figure 2.** BETSCE 10 K flight cooldown data.

## Status and Design of 25 K UCSB Long Duration Balloon Cooler

A continuous operation 25 K single-stage cryocooler, shown in Figure 3, is currently in final integration and test at JPL in support of a long duration balloon experiment to measure anisotropy in the Cosmic Microwave Background radiation. The 25 K LDB cooler is designed to provide 480 mW of refrigeration using a measured 220 W of input power. Precooling of the hydrogen and thermal shielding of the focal plane is provided by two Sunpower Stirling cryocoolers. The final integration and performance testing of this cooler will be completed in fall 1996. Delivery and integration of this cooler into the UCSB dewar package will occur late in 1996. This UCSB LDB experiment is scheduled to fly over Antarctica for two weeks in December, 1997.

Since this cooler is the first hydride sorption cooler to be used to help gather science data, other than on the performance of the hydride cooler itself, it is also the first to be designed to support science instrument requirements. The use of this cooler in the LDB experiment has enabled the team at UCSB to realize substantial mission benefits by replacing their baselined 500 liter helium dewar. Flights of one to three months in duration, planned for future experiments, would be impossible without active cooling.

For a balloon flight experiment such as LDB, the primary requirement is to achieve reliable and safe operation. This has to be achieved in tests conducted in a open lab environment by people familiar with only the basics of cooler operation, after transportation and test flights in New Mexico or Texas, and later transport to, and flight in, Antarctica. At the conclusion of the test flights, the balloon is separated from the payload and the science instrument parachutes to the earth from an altitude of approximately 40 km. The landing is not always smooth and substantial repairs are often required after these flights. In recognition of the rigors this cooler will be subjected to, its design permits all of the major components to be isolated by hand valves; thereby permitting removal and repair as required. Additionally, all refrigerant sealing joints are

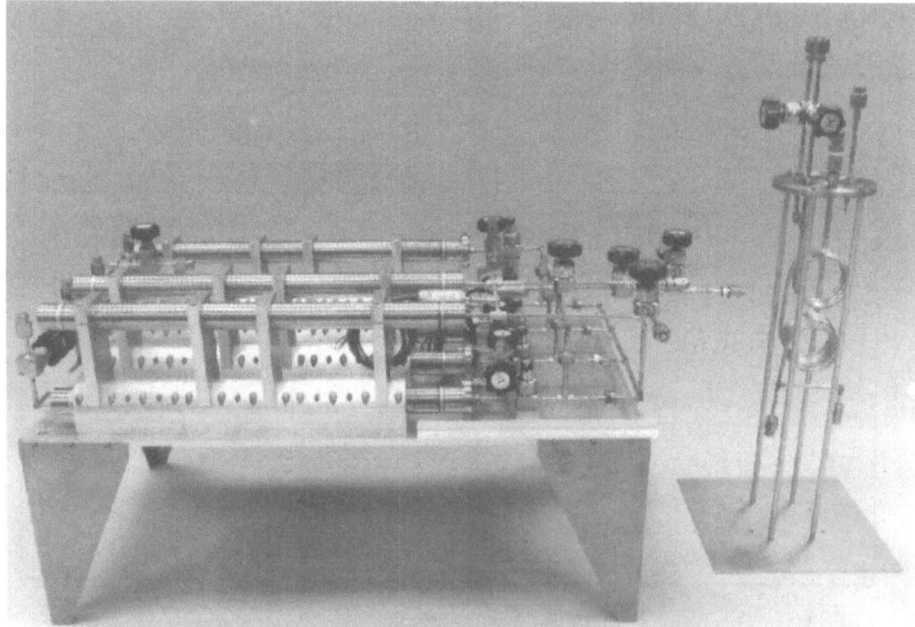

**Figure 3.** The 25 K UCSB Long Duration Balloon cooler is shown when assembly was nearly complete. The cryostat is to the right and is shown without the flight liquid reservoir and J-T attached.

welded to support the high vibration levels and abusive handling expected during transportation and test flights.  Because of the overriding concern to make the UCSB LDB cooler safe and rugged, it has been built at a level equivalent to flight engineering model hardware.  Hence, all of the materials selected, fabrication and assembly techniques, and design and safety margins are consistent with flight hardware requirements.

The most significant innovations in this effort, when compared to previous sorption coolers, are in the materials selection and fabrication processes used to minimize contamination levels. This is especially important since the primary reliability concern for any J-T cooler is contamination.  To achieve high reliability and to provide a better foundation for future flight missions, the cooler structure was entirely made of 316L VIM/VAR stainless steel.  The Department of Energy's Ames Laboratory at Iowa State University provided $LaNi_{4.8}Sn_{0.2}$ material with a purity level over 10,000 times better than that used for fabrication of any sorption cooler before this.  Assembly of the cooler was conducted entirely in a purified and monitored Argon glovebox.  Vacuum pump out ports were provided to each volume within the refrigerator to simplify contamination removal.

The cooler cryostat also has features incorporated which should enormously increase its tolerance to contamination.  Porous plugs with a diameter of approximately 0.2 cm are used for the actual expansion rather than the <0.002 cm diameter orifices more commonly chosen and are expected to be much more contamination tolerant.  Since the refrigeration capacity of a sorption cooler depends linearly on the mass flow rate, future long-life sorption coolers with refrigeration requirements less than 100 mW will have very low refrigerant mass flow rates and will thus need very restrictive Joule-Thomson expanders.  The J-T expanders for long-life coolers must also be extremely contamination tolerant to ensure reliability.  These two requirements preclude the use of any J-T expanders other than porous metal flow restrictors.

To further increase the reliability of the cryostat, a 0.01 micron filter is placed at the inlet of the J-T expansion element.  The temperature of the refrigerant at this point will be approximately 35 K.  At 35 K even 1 ppm of air constituents, such as oxygen and nitrogen, are frozen and can be filtered out of the refrigerant stream.

To reduce the cooler mass and power, novel, miniature ZrNi hydride sorbent beds (0.6 cm diameter by 2.5 cm long) are used to activate the compressor element gas gap thermal switches. The other novel features are the inclusion of a high pressure refrigerant reservoir and a low pressure sorbent bed to ensure the 1 mK/s stability of the cold temperature required by the UCSB radiometer.

This demonstration that a hydride sorption refrigerator can be fabricated in a lightweight, integrable package, and operate reliably despite a challenging environment will substantially advance the state-of-the art.  The proof of detector compatibility, as demonstrated by the quality of the science data gathered, and verification of cooler reliability and ruggedness will significantly add to the heritage of sorption cooler development for future astrophysics missions.

## Cooling to Ten Kelvin and Below

Continuous operation expanders for use below 10 K with hydrogen, originally proposed by Jones,[9] are currently under active development.  As hydrogen is a solid at this temperature with a vapor pressure of only 1.9 torr, a novel expander is used.  Longsworth and Khatri[10,11] recently described a successful laboratory test of such a device.  Operation of this expander is initiated by using a standard J-T expansion technique to collect a small amount of liquid hydrogen in a reservoir.  If a pump (or sorbent bed) is then used to evacuate the liquid reservoir which has a porous filter at its exit, a solid is formed.  Stable continuous operation is then achieved with a temperature gradient in the refrigerant reservoir.  The result is that both liquid and solid hydrogen are in the cold end.  As the solid sublimates, the heat of fusion conducts back to the liquid reservoir to freeze replacement refrigerant and the heat of vaporization serves to provide useful refrigeration. Tests demonstrated that this 'glacier cooler' operated in a stable and repeatable manner at 9.7 K.

A minor variation of this device has been proposed by L. A. Wade which permits two-stage operation of this cooler. This can be achieved through either of the following two methods. The conceptually simpler of these uses a common high pressure gas supply which is manifolded into two J-T expanders. One of these expands to a standard two-phase gas/liquid hydrogen mixture. The liquid hydrogen is separated from the two-phase flow and retained by a wick to provide thermal shielding for the cold stage and perhaps to provide active cooling for devices as needed. The vaporized hydrogen from this reservoir is returned to a compressor containing a medium pressure hydride sorbent, such as $LaNi_{4.8} Sn_{0.2}$, at a nominal 0.1 MPa. The second J-T expander is identical to the Longsworth and Khatri design. This reservoir vents into a low pressure hydride sorbent such as ZrNi. The alternate configuration would combine these two devices by using a single J-T to form a single, and therefore common, liquid reservoir. Two vents are provided: one to the medium pressure sorbent bed and the other to the low pressure bed. A heat exchanger at the medium pressure vent location is used to connect to a thermal shield about the low temperature part of the cryostat.

4 K sorption coolers have been proposed for many years.[12] In the past however the cooling requirements envisioned were usually between 0.1 and 1 W. The resulting power requirement quickly halted further development efforts. Reduced cooling requirements, coupled with the recent availability of hydride sorption coolers, have made use of these helium/charcoal coolers possible.

A typical three-stage cooler concept is to use an activated charcoal, such as Saran carbon, cooled to 16 K by the first hydride stage as the sorbent material. The high pressure helium refrigerant is then precooled to 9 K using the second hydride stage of the cooler. To further improve efficiency during desorption, the charcoal compressors are heated by the high pressure hydrogen refrigerant to approximately 40 K. This improves the overall cooler efficiency by precooling the hydrogen refrigerant from the nominal 60 K to 40 K thereby increasing the amount of refrigeration per unit mass flow in the two hydrogen refrigerant stages. The hydrogen gas manifolding for accomplishing this can be done either by separately plumbed lines which directly connect the appropriate compressors or through use of valves.

## FUTURE SORPTION COOLER DEVELOPMENT PLANS

The primary thrust for the continued development of sorption cryocooler technology will be provided by a NASA Code X research and development funded program planned to start in FY 1997. This effort will be focused on developing a series of vibration-free cryocoolers at 30 K, 6 to 10 K and 4 K in support of precision pointing NASA astrophysics missions such as ExNPS and NGST. These coolers will be developed at an engineering model level and integrated into a series of challenging science experiments in a manner similar to that followed in the 25 K LDB cooler effort.

The quality of the science data derived from these experiments will prove the most stringent characterization of the ability of these coolers to compatibly integrate with the science instruments. Following this path is considerably more useful and cost effective than developing separately the laboratory facilities required. As an example, integrating such a cooler with a ground based infrared interferometer provides far more useful information as compared with developing a 30 picometer cryogenic metrology station for dimensional stability testing in the laboratory.

The planned FY'97 effort will start with component development to determine the two major open issues remaining in sorption cooler development:

1) Can a sorption compressor operate with stable performance for ten years of continuous operation?

2) Can the continuous operation, sub-10 K, hydrogen sublimation cryostat developed by Longsworth and Khatri provide stable long term cooling? And if so, at what minimum temperature?

It appears that the last remaining challenge to operating a hydride sorption compressor for ten years is caused by the slow disproportionation of La from $LaNi_{4.8}Sn_{0.2}$ into $LaH_2$ and similarly of Zr from ZrNi into $ZrH_{1.5\ to\ 2}$. Measuring the rate of reaction of these processes will permit a maximum temperature to be selected at which the disproportionation reaction rate will be too slow to significantly affect compressor performance over a ten year period. Calculations by Wade indicate that this is so for $LaNi_{4.8}Sn_{0.2}$ when operating at a maximum temperature of 480 K. This lowered maximum operating temperature reduces the maximum operating pressure to 8 MPa from the typically baselined 10 MPa. Once an accurate measurement of these rates of reaction with flight grade hydrides has been made and the heats of reaction calculated from the data, compressors will then undergo life testing to verify the stability predictions.

The second major issue, cryostat minimum temperature and temperature stability, will be addressed though development and testing of a series of two-stage cryostats at 10 K and below. In ground tests, the BETSCE cooler and the Proof of Principle cooler[13] both demonstrated operation to 9 K. Neither of these cooler designs were optimized for operation below the 1.9 torr vapor pressure of 10 K hydrogen. The 0.0018 torr, 290 K equilibrium pressure of the ZrNi sorbent used in these coolers is 70 times lower than the 0.129 torr vapor pressure of normal hydrogen at 8.11 K. It is therefore not unreasonable to think that operation to below 8 K is possible for a hydrogen sorption cooler at low refrigeration load.

The results of these efforts will then be fed into the development of a continuous sub-10 K cooler which is planned to start in FY 1998. This effort in turn will support the future development of a continuous operation 4 K cooler.

## SUMMARY

Most of the sorption cryocooler development being actively pursued is focused on refrigerators which provide continuous cooling at temperatures below 30 K and at loads of well under 0.1 W. The successful flight of the BETSCE cooler has clearly demonstrated the suitability of sorption technology for spaceflight applications. The transition of these coolers from a development level primarily concerned with technology demonstration to a level primarily concerned with supporting aggressive science missions has been initiated with the development of the 25 K LDB cooler.

It seems reasonable that, with the planned development efforts, sorption coolers will reach maturation and, in doing so, enable several of the most ambitious and exciting scientific missions yet conceived.

## REFERENCES

1.  Thronson, Jr., H. A., et al, "*EDISON: The Next Generation Infrared Space Observatory,*" Space Science Reviews, Kluwer Academic Publishers (1992), vol. 61, pp. 145-169.

2.  Wade, L. A., Kadogawa, H., Lilienthal, G. W, Terebey, S., Rourke, K., Harwarden, T., "Mid-InfraRed Optimized Resolution Spacecraft (MIRORS)," To be published in the SPIE proceedings of the Space Telescopes and Instruments II meeting in Denver, CO, August (1996).

3.  Smedley, G. T., Johnson, D. L., Ross Jr., R. G., *TRW 1W-35 K Pulse Tube Cryocooler Performance Characterization*, JPL Internal Document D-13236, Jet Propulsion Laboratory, Pasadena, CA (1995), Fig. 7.7.

4.  Wade, L. A., "Advances In Cryogenic Sorption Cooling," *Recent Advances In Cryogenic Engineering-1993*, ed. By J.P. Kelley and J. Goodman, American Society of Mechanical Engineers, New York, NY, HTD-Vol. 267 (1993), pp. 57-63.

5.   Wade, L. A., "An Overview of the Development of Sorption Refrigeration," *Advances in Cryogenic Engineering*, Plenum Press, New York(1991), vol. 37 , pp. 1095-1105.

6.   Bard, S., et al, "10 K Sorption Cryocooler Flight Experiment (BETSCE)," *7th International Cryocooler Conference Proceedings*, Phillips Laboratory, Kirtland Air Force Base, NM (1993), PL-CP—93-1001, pp. 1107-1119.

7.   Bard, S., et al, "Flight Demonstration of a 10 K Sorption Cryocooler," to be published in *Cryocoolers-9*, Plenum Press, New York (1997).

8.   Wade, L. A., and Levy, A. R., "Preliminary Test Results For A 25 K Sorption Cryocooler Designed For The UCSB Long Duration Balloon Cosmic Microwave Background Radiation Experiment," to be published in *Cryocoolers-9*, Plenum Press, New York (1997).

9.   Jones, J. A., "Hydride Absorption Refrigerator System for Ten Kelvin and Below," *Proceedings of the Third Cryocooler Conference*, NBS special Publication 698, NBS, Boulder, CO (1984).

10.  Longsworth, R. C. and Khatri, A., "Continuous Flow Cryogen Sublimation Cooler," to be published in *Advances In Cryogenic Engineering*, Plenum Press, New York (1996), vol. 41.

11.  Longsworth R. C. and Khatri, A. N., "Continuous Flow Cryogen Sublimation Cooler," *U.S. Patent 5,385,027*, Jan. 31, 1995.

12.  Hartwig, W. H., "Requirements for and Status of a 4.2 K Adsorption Refrigerator Using Zeolites," *Proceedings of the Conference on Refrigeration for Cryogenic Sensors and Electronic Systems*, Boulder, Colorado, NBS SP 607 (1980).

13.  Wu, J.J., et al, "Experimental Demonstration of a 10 K Sorption Cryocooler Stage," *Advances in Cryogenic Engineering*, Plenum Press, New York (1991), vol. 39 , pp. 1507-1514.

# Preliminary Test Results for a 25 K Sorption Cryocooler Designed for the UCSB Long Duration Balloon Cosmic Microwave Background Radiation Experiment

**Lawrence A. Wade[1] and Alan R. Levy[2]**

[1]Jet Propulsion Laboratory
California Institute of Technology
Pasadena, CA 91109

[2]University of California at Santa Barbara
Department of Physics
Santa Barbara, CA 93106

## ABSTRACT

A continuous operation, vibration-free, long-life 25 K sorption cryocooler has been built and is now in final integration and performance testing. This cooler will be flown on the University of California at Santa Barbara Long Duration Balloon Cosmic Microwave Background Radiation experiment in Antarctica in December 1997. The cooler will refrigerate a focal plane composed of eight microwave feed horns, two working at 30 GHz and six at 42 GHz, with InP High Electron Mobility Transistor amplifiers. This will be the first hydride sorption cooler used to support an astrophysics experiment. As such, it is an important milestone in the development of vibration-free coolers for astrophysics applications.

The cooler uses hydrogen as the refrigerant and $LaNi_{4.8}Sn_{0.2}$ as the hydride sorbent. The materials, components, design margins, and assembly procedures are entirely consistent with space flight qualification requirements. Several features have been incorporated into the cooler design for long term reliability and temperature stability. A high pressure tank and low pressure sorbent bed are used to stabilize the cold end temperature to better than 1 mK/sec. Small ZrNi compressors are utilized to activate the compressor element gas-gap thermal switches without valves. To greatly enhance contamination tolerance, commercially available porous metal flow restrictors are used as the Joule-Thomson plug. Passive check valves direct the refrigerant flow, simplifying cooler operation enormously. A design description and preliminary test results are presented. Also presented are the results of flow tests conducted to determine the relationship between pressure drop and hydrogen mass flow rate as a function of temperature for a range of commercially available flow restrictors.

Cryocoolers 9, Edited by R.G. Ross, Jr.
Plenum Press, New York, 1997

## INTRODUCTION

Since the detection of anisotropy in the Cosmic Microwave Background (CMB) by the COBE DMR instrument[1], there have been a number of experiments to measure anisotropy in the CMB at higher spatial resolution. A map of CMB anisotropies with much higher spatial resolution (<0.5°) than the 7° achieved by the COBE DMR can be used to determine the Hubble constant (and thus the age of the universe), the density of the universe, and the fraction of that density that is made up of baryons. Finally, this information provides a crucial test for theories predicting how structure originated in the universe.

A team at the University of California at Santa Barbara (UCSB) is now constructing a payload to map the anisotropy of the CMB at the 0.3 degree angular scale. One of the requirements for this experiment to succeed is the use of active cryogenic cooling. The advantages of hydrogen sorption cooler technology make it the most desirable cryogenic cooler for this experiment, and thus a sorption cooler has been constructed and is now being tested in preparation for delivery to UCSB in late 1996. The cooler was designed and assembled to offer reliable, stable, and efficient long term cooling. Presented in this paper is a summary of the design and assembly processes for the UCSB hydrogen sorption cooler and preliminary test results. Included are results from the hydrogen flow tests that were required to design the Joule-Thomson expander used in the cooler cryostat.

## COSMIC MICROWAVE BACKGROUND EXPERIMENT

The Cosmic Microwave Background is understood to be the remnant of a hot, dense, early phase of the universe when matter and energy were in thermal equilibrium, and its existence and characteristics remain the most convincing evidence for the Big Bang. After the Big Bang, the universe expanded and the temperature of matter and radiation decreased. When the mean energy of the photons fell below the ionization energy for hydrogen, free electrons combined with protons to form hydrogen. This caused matter and radiation to decouple because neutral hydrogen only weakly scatters photons as compared to free electrons. As the universe has continued to expand, the CMB photons have continued to cool. Since the CMB photons have interacted primarily through gravitation since recombination, measurements of the CMB can be used to probe density fluctuations during recombination as well as gravitational processes since then.

On large angular scales, the CMB points to physical conditions right after the Big Bang. On smaller angular scales, anisotropy in the CMB is influenced by factors that control the expansion rate of the universe and formation of large scale structure such as the cosmological constant, the matter density, and the existence and nature of dark matter. Measurements of the anisotropy of the CMB will hopefully be able to discriminate between competing theories that predict the primordial mass distribution and help advance understanding of the gravitational collapse that led to galaxy formation.

Unlike normal 8 to 48 hour balloon flights, Long Duration Balloon (LDB) flights will be able to make two dimensional maps of the sky with substantially reduced atmospheric noise as compared to ground based experiments. Since liquid cryogens are impractical in a balloon borne experiment longer than two weeks, a cryogenic cooler will be required to refrigerate High Electron Mobility Transistor (HEMT) receivers. Hydrogen sorption cooler technology uniquely fulfills the refrigeration requirements for the UCSB Long Duration Balloon CMB experiment. Not only will the LDB flight give interesting scientific results but, since this is the first hydride sorption cooler to be used to help gather science data other than on the performance of the refrigerator itself, the LDB flight will also provide valuable experience in the use of sorption cooler technology that will be useful for any mission, balloon borne or space based, that requires active cooling for cryogenic detectors.

In June 1997 the UCSB LDB payload will test fly in the U.S., and it will fly in December, 1997 on a zero-pressure balloon in a two week long Antarctic circumpolar trajectory. The experiment will measure the CMB anisotropy with a beam full width at half maximum of about 0.3 degrees using an off-axis Gregorian telescope. Two receivers will measure a band centered at 30 GHz and another six will measure a band centered at 42 GHz. Having eight antennae will not only allow for observation at more than one frequency, but will also give better sky coverage and increase the signal to noise ratio. The signal from the antennae will be fed to HEMT amplifiers which will be cooled to 25 K by the hydrogen sorption cooler to take advantage of their low noise properties. The sky coverage for the LDB experiment will be $10^3$ to $10^4$ square degrees.

## SORPTION COOLER REQUIREMENTS

The primary mission requirement for the UCSB LDB cooler is to achieve reliable and safe operation for at least two years with little maintenance and for up to three months at a time, while in flight, with no maintenance. This has to be achieved after repeated transport and test flights in New Mexico or Texas and transport to and flight in Antarctica. At the conclusion of each flight, the balloon is separated and the science instrument parachutes to the earth from an altitude of approximately 40 km. Substantial repairs between flights are often required due to the generally rough nature of the landing. All of the major components of the cooler can be isolated by hand valves to permit removal and repair as required.

The UCSB LDB experiment requires 480 mW of refrigeration at 25 K to cool the radiometer. The radiometer focal plane temperature must remain constant to within 1 mK per second to avoid introducing noise into the experiment. This requirement is derived from the measured sensitivity of the InP HEMT amplifiers which varies linearly with temperature. As the cryostat and compressor may be separated by as much as two meters, the pressure drop through the intermediate lines must be low. In addition, the cryostat must be able to operate in any orientation and the compressor must be able to rotate on two axes. The compressor should be relatively light and compact due to LDB payload size and weight limitations.

## 25 K LDB COOLER OPERATION AND DESIGN

The 25 K LDB cooler follows the classic continuous operation sorption cooler concept design. Figure 1 shows a schematic of this cooler integrated with the LDB radiometer. The cooler is comprised of a compressor assembly, made up of four $LaNi_{4.8}Sn_{0.2}$ hydride sorption beds, connected to a J-T cryostat assembly. During operation of the cooler, hydrogen at a pressure of 10 MPa is desorbed by a heated sorption bed at 530 K. The hydrogen, at a mass flow rate of 0.0025 g/s, is expanded through a J-T orifice to create a gas/liquid refrigerant mixture. The liquid evaporates as it absorbs heat from the detectors and is then absorbed at pressure of 0.12 MPa in a cool, 300 K, bed. When the cool bed is reheated, the hydrogen is repressurized thus creating a fully reversible closed-cycle system. During each 800 second quarter cycle, one bed is hot (desorption), one is cool (absorption), one is heating up (pressurization), and one is cooling down (depressurization). The cooler requires a measured 220 W of input power which will be rejected in flight at 275 K in a 3 torr environment and at approximately 300 K in one atmosphere in the laboratory. A cold end temperature of 23.0 K is predicted from the sorbent composition, operating temperature, and measured cryostat performance.

To achieve high reliability and to provide a better foundation for future flight missions the cooler structure is entirely made of 316L VIM/VAR stainless steel. All refrigerant sealing joints are welded to support the high vibration levels and abusive handling expected during transportation and test flights. The 25 K LDB cooler was built at a level equivalent to flight engineering model hardware. All of the materials selected, fabrication and assembly techniques and design and safety margins are consistent with flight hardware requirements.

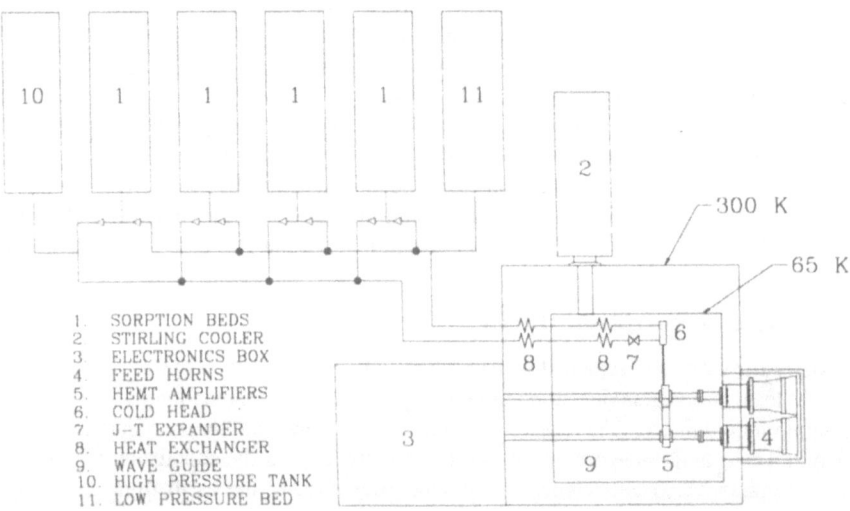

1.  SORPTION BEDS
2.  STIRLING COOLER
3.  ELECTRONICS BOX
4.  FEED HORNS
5.  HEMT AMPLIFIERS
6.  COLD HEAD
7.  J-T EXPANDER
8.  HEAT EXCHANGER
9.  WAVE GUIDE
10. HIGH PRESSURE TANK
11. LOW PRESSURE BED

**Figure 1.** Schematic of the sorption cooler and focal plane assembly.

The primary reliability concern for any J-T cooler is contamination. Hence, particular attention was paid to minimizing contamination levels through proper materials selection and fabrication process design. The approach selected was to make the entire system as clean as possible. This was accomplished by starting with clean materials and hardware, building the compressor in a monitored pure argon atmosphere glovebox, filling the beds with ultra-high purity refrigerant, filtering the cold refrigerant during operation, and using a porous plug for the J-T. The Department of Energy Ames Laboratory at Iowa State University provided the $LaNi_{4.8}Sn_{0.2}$ hydride with a purity level over 10,000 times better than that used for fabrication of any sorption cooler before this. The hydrogen refrigerant is supplied by a 20 ppb total contamination semiconductor grade facility.

Cold tip temperature stability is primarily set by the liquid reservoir pressure as the heat load into the focal plane is nearly constant. Therefore, to meet the instrument temperature stability requirement, high pressure tanks are incorporated into the compressor assembly. These tanks limit pressure variations at the inlet to the J-T flow restrictor to less than 0.5 MPa over the 800 second desorption. A constant temperature, low pressure sorbent bed, identical in mechanical design to the four active compressor elements, is used to regulate the pressure over the liquid reservoir by acting as the equivalent of a 120 liter plenum. Approximately 2 W of power is required to stabilize the low pressure sorbent bed temperature such that the average low pressure seen during an 800 second absorption phase corresponds to the equilibrium pressure at the middle of the absorption plateau. Acting in concert with an absorbing compressor, the low pressure sorbent bed will maintain the low pressure stability such that the calculated liquid reservoir temperature stability is 0.36 mK/s. As the focal plane assembly mass attached to the liquid reservoir is greater than 2.0 kg, the final temperature stability at the HEMT amplifiers will be far better than the 1 mK/s requirement. Finally, several 200-mesh copper screen cylinders are incorporated into the liquid reservoir to ensure, through wicking, even distribution of liquid hydrogen over the reservoir surface and to prevent pool boiling with its associated temperature fluctuations.

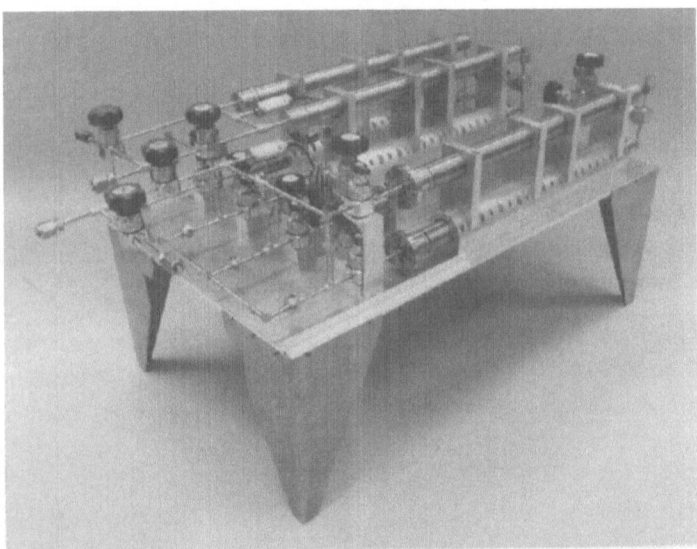

**Figure 2.** The compressor assembly is shown complete just prior to the installation of the cabling.

The compressor assembly can be remotely located from the cryostat without performance penalty because of the low 0.0025 g/s mass flow rate. The compressor has a mass of only 21.5 kg and dimensions of 87.5 cm by 50 cm by 23 cm. The cryostat masses 3 kg and easily fits within a volume 15 cm in diameter and 46 cm long.

**Sorption Bed Design**

The compressor assembly, pictured in Figure 2, has four active $LaNi_{4.8}Sn_{0.2}$ hydride compressor elements, each surrounded by a gas-gap thermal switch. The compressor elements are connected to the high and low pressure manifolds through passive check valves, which direct gas flows along appropriate paths. The gas-gap thermal switch in each compressor element assembly is activated when it is pressurized to 14 torr by heating a small (2.5 cm long x 0.6 cm dia.) ZrNi compressor to 450 K via a Kapton tape heater. This requires approximately 2.5 W of input power. The gas-gap switch is turned 'off' by allowing the ZrNi compressor to cool to 300 K which causes it to evacuate the gas-gap volume to 0.0044 torr.

Figure 3 shows a cross sectional view of an individual compressor element. The outer gas-gap shell is 3.81 cm in diameter. The 1.91 cm o.d. by 0.165 cm wall inner compressor contains 284 g of $LaNi_{4.8}Sn_{0.2}$ hydride. In the center of the compressor is a hollow 0.635 cm o.d. sintered 316L filter. The filter allows the hydrogen refrigerant to flow longitudinally through the compressor and then flow a short radial distance into the hydride powder contained between the filter and the 1.91 cm tube. Surrounding the hydride containing compressor element are five band heaters which are not shown in Figure 3. The back end of the compressor element is loosely supported by three pins, thereby accommodating longitudinal differential thermal expansion between it and the outer gas-gap shell as the temperature cycles during operation. Radiation shielding of each compressor element is provided by two concentric 0.005 cm thick stainless steel tubes. These shields are separated from each other by a helically wound 0.008 cm diameter NbTi wire which is periodically spot welded to the exterior surface of the inner cylindrical shield. The four gas-gap compressors, each filled with 0.8 g of ZrNi hydride, are pictured in Figure 4.

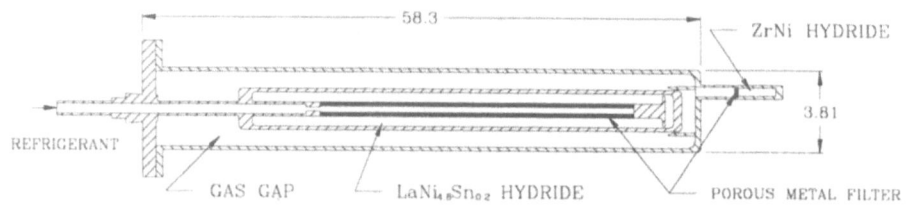

**Figure 3.** Cross sectional view of a compressor element assembly with dimensions in centimeters. The ZrNi bed is shown schematically.

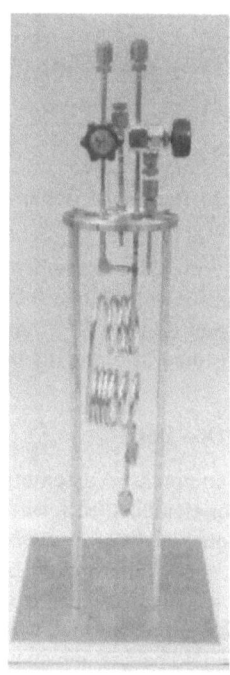

**Figure 4.** A view of the rear portion of the compressor assembly shows the ZrNi compressors used to activate the gas-gap thermal switch.

**Figure 5.** The cryostat tube-in-tube counterflow and precooling heat exchangers are shown prior to attachment of the J-T and liquid reservoir assembly.

## Cryostat Design

The cryostat assembly, as shown in Figure 5, consists of a coiled 1.0 m long tube-in-tube warm heat exchanger, a 15 cm long precooling heat exchanger, which is connected to a Sunpower Stirling cooler, a coiled 1.6 m long cold tube-in-tube heat exchanger, a 0.01 micron contamination trap, the J-T assembly and the cryogen liquid reservoir. The tube-in-tube heat exchangers operate at >0.98 effectivity despite their simplicity (just a 0.3175 cm o.d. tube inside a 0.635 cm o.d. tube) due to the low refrigerant mass flow rate. The high pressure gas stream flows in the outer annulus, eliminating the need to weld components at the precooling heat exchanger. The penalty for this unusual configuration was found to be only a 3 Pa pressure drop per meter of heat exchanger in the low pressure gas stream. The temperature of the refrigerant at the large surface area 0.01 micron filter, located at the inlet of the J-T expansion assembly, will be approximately 35 K. At that temperature, even one ppm of air constituents, such as oxygen and nitrogen, are frozen solid and can be filtered out of the refrigerant stream before reaching the J-T assembly.

The commercially available porous metal flow restrictors[2] used in the cryostat have a diameter of approximately 0.2 cm for J-T expansion, rather than the <0.002 cm diameter orifices more commonly chosen. Flow testing of a wide range of flow restrictors has been initiated to enable the selection of the appropriate combination of flow restrictors to meet the specific pressure, mass flow rate, and temperature specifications of this cooler. The flow test results will also be extremely useful for the design of future sorption coolers as these porous metal flow restrictors are the best choice for the J-T expander for any sorption cooler designed to refrigerate loads of less than 100 mW at cryogenic temperatures.

**Control and Data Acquisition Hardware**

The electronics system to control the sorption cooler and to collect data was designed to be simple and reliable. The cooler is controlled from a commercial data acquisition and control system manufactured by Elexor Associates. It includes a 16 bit digital I/O card, a 15 channel, 12 bit A/D card, and a CPU. Currently, commands are sent to the CPU via RS-232 using the LabVIEW software package from an IBM compatible PC. The rest of the cooler control hardware is contained in an electronics box that was designed and constructed at UCSB. It contains 16 latching relays, 12 temperature sensor channels, and two pressure sensor channels.

During the operation of the cooler, the 14 sensor values are measured every two seconds. The temperature is measured at each of the four sorption beds and the low pressure bed using platinum resistance thermometers. The temperature is also measured using silicon diode sensors at the compressor support/heat sink plate, two points on the refrigerant precooler in the cryostat, the J-T inlet, two points on the cold end, and the outlet from the cold end. Using pressure transducers, the pressure at the high pressure tanks and low pressure bed is also measured. The relays are used to turn on and off power to the heaters on the sorption beds as the cooler switches to the next quarter cycle as dictated by a simple timing sequence.

**PROGRAM STATUS AND RESULTS**

The cooler is in final integration and test. The compressor assembly input power, gas-gap parasitics and gas-gap thermal conductivity have been measured. The input power was measured as 220 W. The parasitics are approximately 25 W. The gas-gap thermal performance matched the required compressor element cool down time of 800 second with 13 torr of hydrogen in the gas gap. The $LaNi_{4.8}Sn_{0.2}$ hydride absorption plateaus were within 1 K of those predicted by the isotherms. One check valve was found to leak during these tests and will be replaced shortly. Final bake out of the assembly is currently underway with cryostat integration scheduled for the end of Summer 1996.

The cryostat heat exchanger performance has been measured to be ideal within the error of the instrumentation. The cold tip temperature stability was measured as being within 2 mK over many minutes under applied heat, indicating that the liquid reservoir performs very well. The conductance of the liquid reservoir from the focal plane interface to the liquid refrigerant was found to be 0.379 W/K. Figure 6 shows the cold tip temperature as a function of applied load.

The selection of the correct combination of porous plug flow restrictors has necessitated conducting a flow measurement experiment as the refrigerant flow rate at high pressure ratio (10 MPa to 0.12 MPa), coupled with real gas and phase change effects, is not easily predicted with any accuracy.  The flow tests conducted used hydrogen over a very wide range of pressures, flow rates, and temperatures.  A flow test cryostat was constructed that was identical to the flight cryostat except for the addition of VCR terminations at the cold heat exchanger high pressure outlet and low pressure inlet.  This permitted interchange of different J-T assemblies.  The porous metal flow restrictors were rated by the manufacturer for nitrogen gas flows over 30 psid between 250 sccm and 1 sccm.  These J-T plugs were pressed into holders, with VCR fittings attached to both ends, to permit them to be combined as needed.  For the 1 and 10 sccm plugs, samples were prepared that were pressed into holders and either welded or not welded.

High pressure hydrogen gas used in the flow tests was supplied from commercial gas cylinders; some of which had been stored outdoors for four years without cover.  No filtering of the hydrogen was done beyond the 0.01 micron filter at the J-T inlet.  In none of the tests, some lasting 3 days, did the J-T show any signs of plugging.  These tests were used to determine mass flow as a function of temperature (including the effects of real gas properties and liquefaction) and pressure difference/ratio (e.g. sonic exit effects).

While the flow tests are not yet complete, as the flow restriction was somewhat lower than was expected, several interesting observations can be made from the preliminary results.  A substantial increase in mass flow rate, for a given pressure drop, upon initiation of hydrogen liquefaction is seen in the data presented in Figure 7.  Figure 8 shows that the mass flow rate across a porous metal orifice is a linearly dependent on pressure drop for room temperature hydrogen in the ranges measured.  No low pressure ratio measurements have been made to date.  Therefore it is not yet possible to determine, through comparison with the high pressure ratio data presented in this paper, whether there is an additional sonic exit effect on the flow rate.

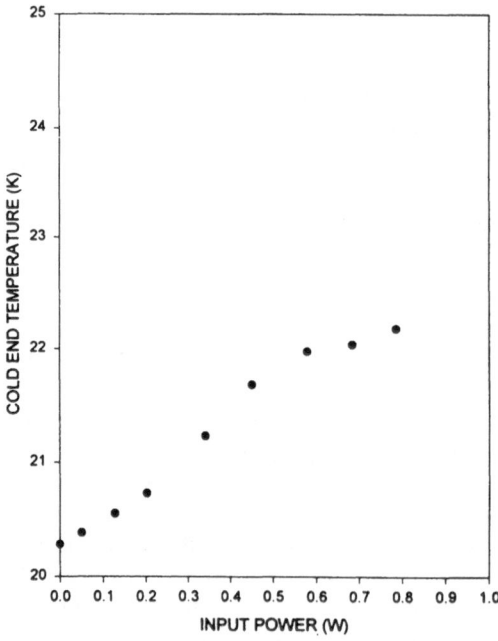

**Figure 6.**  Cold tip temperature at the focal plane as a function of heat input.

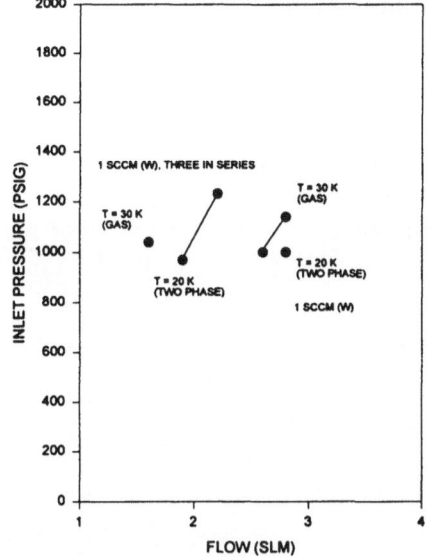

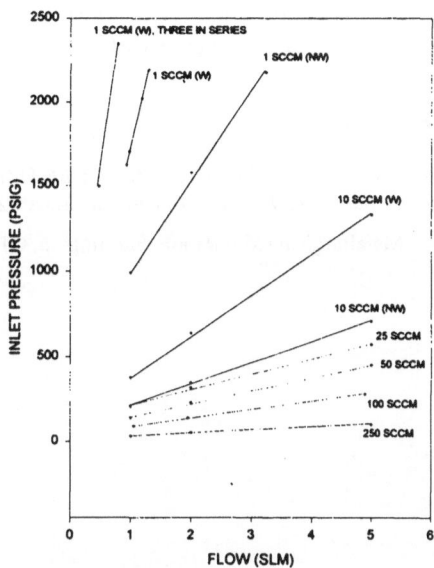

**Figure 7.** Cryogenic flow test results, with liquefaction, for 1 SCCM flow restrictors.

**Figure 8.** Room temperature flow test results for a series of different porous plug flow restrictors.

## SUMMARY

A continuous operation 25 K single-stage cryocooler is currently in final integration and test at JPL in support of a long duration balloon experiment to measure anisotropy in the Cosmic Microwave Background Radiation. Integration of this cooler with the flight radiometer is scheduled for late 1996. The UCSB LDB experiment is scheduled to fly in December of 1997 in Antarctica.

This is the first hydride sorption cooler to be designed to support science instrument requirements. The demonstration that a hydride sorption refrigerator can be fabricated in a lightweight, integrable package, and operate reliably despite a challenging environment will substantially advance the state-of-the art. The materials selected, and the operational and assembly procedures used are compliant with flight quality standards. The proof of detector compatibility, as demonstrated by the quality of the science data gathered, and verification of cooler reliability will provide substantial heritage as sorption coolers are developed for future astrophysics missions.

## ACKNOWLEDGMENTS

The work reported was done at the Jet Propulsion Laboratory and at the University of California at Santa Barbara. This effort was supported by the National Aeronautics and Space Administration through the JPL Director's Discretionary Fund, NASA grant NAGW-1062, the LICA NASA NRA grant, and JPL Primordial Structure Investigation (PSI) Bid and Proposal funding. The authors wish to express their appreciation to UCSB Physics Machine Shop for manufacturing excellent hardware quickly and efficiently. The authors would also like to thank Mike Schmelzel, Steve Elliot, Bill Boulter, Tim Conners, Matt Schalit, Irma Schneider, and Bob Losey for their hard work in helping to assemble the cooler. A.L. would also like to thank Phil

Lubin, Peter Meinhold, Todd Gaier, Mike Seiffert, John Staren, Jeff Childers, and Geoff Cook for help with work carried out at UCSB.

## REFERENCES

1.  Smoot, G., et al, "Structure in the *COBE* Differential Microwave Radiometer First-Year Maps," *The Astrophysical Journal*, vol. 396, no. 1, part 2 (1992), pp. L1-L5.

2.  Mott Metallurgical Corporation, Farmington, CT 06032-3159

# Experiments on a Charcoal/Nitrogen Sorption Compressor and Model Considerations

S.A.J. Huinink, J.F. Burger, H.J. Holland, E.G. van der Sar,
J.G.E. Gardeniers, H.J.M. ter Brake and H. Rogalla

University of Twente, Faculty of Applied Physics,
P.O.Box 217, 7500 AE Enschede, The Netherlands

## ABSTRACT

Sorption compressors are based on the cyclic adsorption and desorption of a working gas on a sorption material, and can be used to supply a gas flow or pressure wave to a cryocooler. An advantage of sorption compressors is the absence of moving parts, except for some check valves. This minimises electromagnetic and mechanical interference and also contributes to a long life time.

We constructed a test set-up for characterising sorbents and gases to be applied in a sorption compressor. This set-up is also used to develop thermodynamic models of sorption compressors. The central element of the test set-up is a sorption compressor with an Inconel sorber vessel (height 20.1 cm and inner diameter 3.9 cm) equipped with a central heater. The vessel is heat-switched to the environment via a 0.7 mm gas-gap heat switch. Experiments were performed with standard activated charcoal and nitrogen gas, the temperatures of the charcoal ranging from 300 K to 550 K and pressures from 1 to 30 bar.

In the paper the test set-up and the characterisation of the materials used in this set-up are described. Three different models describing the thermodynamic behaviour of a sorption compressor are presented and discussed. In the simplest model the centre parts of a sorption compressor are considered to be at a uniform temperature. The next model considers a specific temperature for every element in the compressor. The most complex model calculates a temperature profile in the sorbent material. The simplest model considered gives reasonable results, but due to the approximations made, the thermodynamic behaviour differs from experimentally determined behaviour. The two other models both show thermodynamic behaviour similar to that in experiments.

With the temperature distribution in the sorbent and the sorption behaviour of the sorbent the pressure can be evaluated. The pressures obtained with the simplest model are not completely correct, as can be expected. The two other models give accurate pressure values. The most complex model is expected to give the most accurate predictions of the thermodynamic behaviour and the pressure behaviour of a cylindrical sorption compressor. Therefore, this model can be used as a tool in the development of new sorption compressors.

Cryocoolers 9, Edited by R.G. Ross, Jr.
Plenum Press, New York, 1997

## INTRODUCTION

Sorption compressors are very suitable for use in cryogenic coolers [1]. Since they have no moving parts they can be operated continuously without wear and tear resulting in a long lifetime. The absence of moving parts and coils also brings about the advantage of low electromagnetic and mechanical noise.

At the University of Twente research is done to develop a cryogenic cooling system of small dimensions for cooling superconducting devices or low temperature electronics [2]. In this research sorption compressors are considered to be used as a compression stage. The research presented in this paper concentrates on characterisation experiments in a test set-up and on simulations of the thermodynamic behaviour of sorption compressors. With the test set-up the operation parameters that affect the dynamic behaviour of the temperatures and pressure in a sorption compressor can be measured, such as the thermal conductivity of the elements and the adsorption factor of the gas-sorbent combination. The combination used in the tests is nitrogen with activated charcoal.

A sorption compressor operates in a cycle as illustrated in figure 1. This figure shows the amount of gas adsorbed by the sorbent versus the pressure (C is the adsorption factor: the amount of gas adsorbed per amount of sorbent). The dotted lines are isotherms going from low temperature on the upper left to high temperature on the lower right. In the cycle two regeneration phases are present indicated with A and C. In these phases the compressor is heated up to increase the pressure (A) or cooled down to decrease the pressure (C). In the cycle two pumping phases can be seen, indicated with B and D. In these phases high pressure gas is flowing out of the compressor (B), or low pressure gas is flowing in (D).

For predicting and optimising the performance of sorption compressors a complete numerical model of their operation is made. In this model the time dependence of the temperature is calculated. With the adsorption factor and this dynamic temperature, a pressure is evaluated. The validity of the model is verified by comparing simulations and experiments.

In this paper the constructed test set-up is described and the characterisation experiments carried out with nitrogen gas and activated charcoal are considered. The sorption properties as well as some thermal properties are discussed. This is followed by a presentation of models that predict the thermodynamic behaviour of these sorption compressors. The combination of these models with the sorption properties results in a complete model predicting the operation of a sorption compressor. The validity of these models is checked by comparing simulation results with the actual behaviour of the sorption compressor in the test set-up.

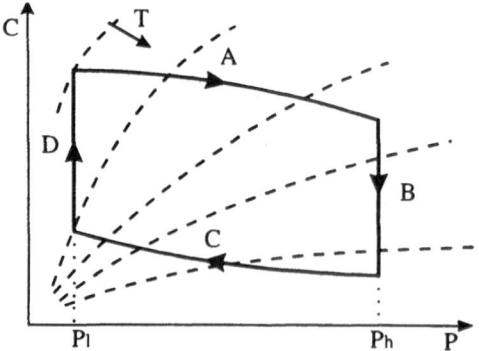

**Figure 1.** The operating cycle of a sorption compressor. (see text)

## SORPTION COMPRESSOR TEST SET-UP

In order to determine thermal properties and sorption properties of the materials used in a sorption compressor, a test set-up was built. It is also used for model validation purposes, yielding experimental data that can be compared with simulation results.

A schematic view of the compressor in the test set-up is given in figure 2. The dimensions of the inside container are 201 mm inner height and 39 mm inner diameter. A heater with a diameter of 10 mm and a length of 175 mm, is placed in the centre of the inside container. This container can be thermally connected to the environment by a gas gap heat switch of 0.7 mm. The operating region for the compressor is limited to a maximum temperature of 850 K and a maximum pressure of 50 bar. With this operating range it is also possible to use the compressor for other sorbents such as PCO [3].

The temperatures in the compressor are measured with thermocouples at four positions: at the outside container, at the inside container, inside the sorbent, and at the heater surface, as indicated in figure 2. The thermocouples are accurate within 2 Kelvin. The heater is fed by a PID-controlled power supply. This control sets the maximum heating rate and the maximum temperature at the heater. The maximum heating power allowed in the heater is 450 W. Before the measurements are started, the system is pumped vacuum to remove all residual gas in the system. Next, the compressor is filled with gas stored in a high pressure container (200 bar). In the connecting tubes a mass-flow controlling and measuring device is installed in order to measure the exact amount of gas flowing into the system (accuracy 5%). The pressure in the complete system is measured in the connecting tubes between mass-flow controller and sorption compressor. To obtain a large dynamic range in combination with a high accuracy for the pressure measurements, three pressure transducers are used, all with a different measuring range. This results in a pressure measuring range from 40 mbar up to 50 bar with an accuracy of 0.5% at full scale to 5% at the pressure scale minimum. All measuring data is recorded and stored by computer. The data file can be processed further in order to obtain material properties. For obtaining correct results the volumes of all tubes and the total dead volume inside the sorbent were measured very accurately and were corrected for in the data processing. A photograph of the complete measuring system is shown in figure 3.

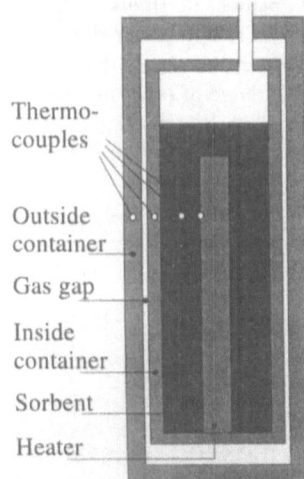

**Figure 2.** Schematic cross-section of the test compressor.

**Figure 3.** Photograph of the test set-up. (From left to right: test compressor, connecting tubes and measuring panel, heater power supply and measuring computer)

## CHARACTERISATION

With the test set-up the sorption behaviour of an activated charcoal-nitrogen combination was analysed. For modelling purposes also the thermal conductivity of the activated coal and the gas gap was examined.

### Sorption Characterisation

For characterising the sorption behaviour the theory of Polanyi is used [4]. This theory states that the adsorbed amount of gas is a unique function of the adsorption potential. For a sorbent of the so-called second structural type [5] this functional form is:

$$C = \frac{\alpha \cdot M}{b} \cdot e^{\gamma \cdot \varepsilon / \beta} \tag{1}$$

In this equation C is the adsorption factor, $\alpha$ en $\gamma$ are fitting parameters, b is the vanderWaals gas constant, M the molar mass of the gas, $\varepsilon$ the adsorption potential, and $\beta$ the affinity coefficient. The adsorption potential is given by:

$$\varepsilon = R \cdot T \cdot \ln\left( \frac{P_c}{P_g} \cdot \left( \frac{T}{T_c} \right)^2 \right) \tag{2}$$

In this equation the estimates for the pressure in the adsorbed phase given by Dubinin [5] are used. Further $P_c$ is the critical pressure, $P_g$ the pressure of the free gas, R the universal gas constant, $T_c$ the critical temperature, and T the absolute temperature. The adsorption factors have been determined for a system of activated charcoal and nitrogen. This is done in six series with in each series a different but fixed temperature of the heater and varying pressures from 1 bar up to 30 bar. The total amount of adsorbed gas is calculated as the difference between the total amount of gas in the system and the amount of free gas. The latter is calculated by means of the ideal gas law.

Taking the temperature profile in the sorbent into account, the relation between the local temperature in the sorbent, the system pressure and the local adsorption factor is calculated. This results in an adsorption factor for each pressure and temperature combination. Because these pressure-temperature combinations determine the adsorption potential according to Eq. (2), a relation between the adsorption factor and the adsorption potential results. In figure 4 the experimental results are shown, in combination with a theoretical fit. The fitting parameters $\alpha$ and $\gamma$ used in this fit are $\alpha = 3,59 \ 10^{-4} \ \text{m}^3\text{mol/g}$, $\gamma = 2.17 \ 10^{-4} \ \text{mol/J}$. It can be seen that the adsorption factor is a unique function of the adsorption potential in a large range of pressures and temperatures, thereby confirming the sorption theory of Polanyi. Only at low pressures in combination with low temperatures, below 4 bar and 350 K, there is a deviation between theory and experiments.

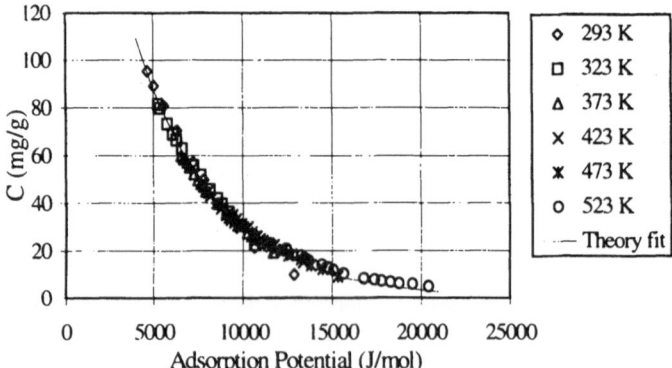

**Figure 4.** Adsorption factor versus the adsorption potential for an activated charcoal-nitrogen system.

## Thermal Conductivity

Most of the materials used are well known and their material properties are well tabulated, except for the properties of the sorption materials and the gas gap. With the test set-up it is possible to estimate the effective thermal conductivities of these two parts. The heating power necessary to maintain a stable temperature profile in the compressor is a measure for the thermal conductivity of the compressor. Since the geometry of the compressor is well known, the heat transport can be evaluated.

The heat transports taken into account in the gas gap evaluation are those due to radiation, conduction by the gas in the gap, and conduction via the mechanical connections between inside and outside container. The radiative term is calculated with Planck's radiation law, the two conductive terms are taken together in one effective thermal conductivity, resulting in $\lambda$=0.001 W/Km in the 'off' state and $\lambda$=0.03 W/Km in the 'on' state.

For the determination of the conductivity of the activated coal, two parallel transports are taken into account, one through the activated coal, and one through the bottom lid of the inside container. The resulting thermal conductivity of the activated coal is $\lambda = 0.25$ W/Km, this is comparable to values found in literature for activated charcoal, $\lambda = 0.15$ - $0.30$ W/Km [6] or $\lambda = 0.13$ W/Km [7].

## COMPRESSOR MODELS

The total gas contents of the compressor can be derived from the starting conditions of temperatures and pressure, incorporating the sorption behaviour. Using this gas contents, any temperature distribution can be used to evaluate the corresponding pressure in the compressor, provided the sorption behaviour of the sorbent-gas combination is sufficiently well determined. This pressure calculation, in which the heat of adsorption/desorption is taken into account, is an iterative process since the sorption affects the pressure and the pressure affects the sorption.

We have developed three numerical models that describe the thermodynamics of a cylindrical sorption compressor. Based on a temperature profile and a pressure in the compressor at a given moment in time, all models evaluate a new temperature profile a small time step later by means of an explicit method. From this new temperature distribution the corresponding new pressure is derived. By considering time step after time step the thermodynamic behaviour of the compressor is explicitly determined. The three temperature models differ in complexity. The results are expected to be more accurate with the more complex models. From simple to complex the three models are: the lumped capacitance model, the node model and the radial model. Schematic presentations of these models are given in figure 5 and are explained briefly in the following sections.

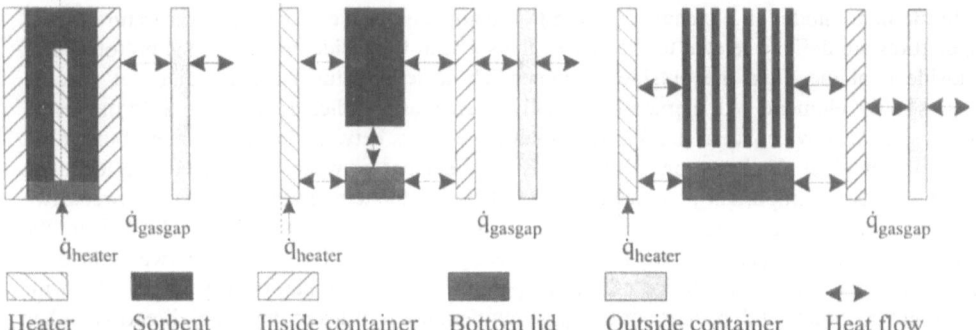

**Figure 5.** Schematic presentations of the three temperature models, from left to right: the lumped capacitance model, the node model and the radial model. (see text)

The basic equations used in the models are those of the heat transfers due to conduction, radiation and convection:

$$\dot{q}_{conduction} = -\lambda \cdot A \cdot \frac{\partial T}{\partial r} \tag{4}$$

$$\dot{q}_{convection} = h \cdot A \cdot (T_1 - T_2) \tag{5}$$

$$\dot{q}_{radiation} = \varepsilon \cdot \sigma \cdot A \cdot (T_1^4 - T_2^4) \tag{6}$$

In these equations $\lambda$ is the thermal conductivity, h the convection constant, $\varepsilon$ the effective emissivity, A the area through which the heat flows, $\sigma$ Planck's radiation constant, $T_1$ is the absolute temperature of the element, and $T_2$ is the temperature of the element to which the heat flows. Further, heat generation and heat storage are taken into account. The new temperatures are calculated by solving an energy balance: the heat added to a volume element minus the heat extracted from it gives the change in stored heat. This is directly related to a change in temperature by the heat capacity of a volume element:

$$\rho \cdot V \cdot c_p \frac{\Delta T}{\Delta t} = \sum \dot{q} \tag{7}$$

In this equation $\rho$ is the density of the material, V the volume of the element, $c_p$ the specific heat, $\Delta T$ the temperature change, $\Delta t$ the time step, and $\dot{q}$ are the different heat flows. With the calculated temperatures resulting from the three models a dynamic pressure can be evaluated for each model. For these pressure calculations the previously described sorption model is used.

## Lumped Capacitance Model

The lumped capacitance model is a small extension of previously presented thermodynamic models [8-12]. In this model the complete centre part, comprising the heater, the sorbent and the inside container, is considered to be at a uniform temperature. The outside container is the second element in this model. There are three heat flows taken into account, firstly the heat generation in the heater, secondly the heat transfer due to radiation and conduction through the gas gap and finally the radiation and convection from the outside container to the environment.

The total heat capacity of the centre part is the sum of the heat capacities of the three elements in it. Due to the assumption of a uniform temperature the modelled heat flows will deviate from experimental situations, resulting in errors in the thermodynamic behaviour. Another disadvantage of this model is the absence of the heater as a separate part. Due to the averaging of the temperature in the complete centre of the compressor it is impossible to simulate the temperature regulation at the heater that would be applied in an experimental set-up. The clear advantage of this model, however, is its simplicity resulting in a high calculation speed.

## Node Model

In the node model all elements of a sorption compressor have a specific temperature. Thus, temperatures are defined related to the heater, the sorbent, the inside container, the bottom lid of the inside container, and the outside container. These temperatures are evaluated at specific locations in the elements. All elements thermally interact as is indicated in figure 5. In the centre parts these heat flows concern the thermal conductive flows between heater, sorbent, bottom lid and inside container. In these conductive terms the distances between the temperature locations in the elements are important parameters. This is because the temperature gradient between two locations is considered to be constant. Thus, with a given heat generation in the heater, the temperature difference between two locations directly scales with the distance between them, as can be seen in Eq. (4). The choice of the temperature location is particularly important in the sorbent and the bottom lid, because of the large temperature profiles in the radial direction. When this position is taken close to the heater, the corresponding temperature will be much higher than the average temperature in the element during experiments. When this position is taken close to the inside container wall, the temperature will be much too low. This strongly affects the calculated pressure resulting from these temperatures. The radial positions chosen in the temperature evaluation of the sorbent and the bottom lid are the same as where the sorbent temperature is measured in the test set-up (see figure 2).

Other heat flows are a heat generation term in the heater, the radiative and conductive flows through the gas gap that connects the inside container and the outside container, and finally, the radiative and convective flows by which the outside container interacts with the environment. The total heat capacity of the elements in the compressor is directly related to the total volume of the separate elements and the specific heat of the materials used.

Although this model is more extensive than the lumped capacitance model, it remains very simple. Since all elements are present as separate parts, the dynamic temperature response is expected to be more realistic. The calculation speed is not noticeably slower than that of the lumped capacitance model.

## Radial Model

As in the node model, all elements of a sorption compressor are present as separate parts. In addition, a radial temperature profile in the sorbent is calculated. For this purpose the sorbent is divided into several shells as shown in figure 5. A temperature and a volume are related to each separate sorbent shell and the conductive heat flows between these shells are calculated. The radial length of the shells and the time step used in the explicit method are determined in such a way that a stable and correct solution of the differential equations is found.

The other heat flows considered in the radial model are the conductive flows between: the heater and the first sorbent shell, the heater and the bottom lid, the last sorbent shell and the inside container, and the bottom lid and the inside container. Further the heat generation term in the heater, the radiative and conductive flows through the gas gap and the convective and radiative flows from the outside container to the environment are taken into account.

As the other models do, this model also ignores the temperature profile in the axial-direction. On the other hand, all other processes related to heat transfer are taken into account. This makes the radial model a very accurate one at the expense of a longer calculation time.

## RESULTS

The three temperature models are compared with experimental results obtained with the test set-up. Several experiments were done varying the heating power, the starting pressure and the set point for the maximum heater temperature. In each experiment the compressor was cycled from room temperature to the maximum heater temperature and, after a short period at maximum heater temperature, back to room temperature. The differences in the temperatures of the elements between experiments and simulations are comparable in all sets of experiments and simulations. The conclusions related to the thermodynamics do not depend on the starting or cycle conditions. For that reason a single set of representative examples is shown in figure 6. In these figures the temperatures of the heater, the sorbent, the inside container and the outside container are plotted versus time. The sorbent temperature is located in the middle between the heater and the inside container.

The main conclusion for the lumped capacitance model is that its temperature is a reasonable average of the three temperatures of the centre elements. The temperature is always in-between the inside container temperature and the heater temperature. In the model the heat leak to the outside container and the environment is larger than in the experiments, which is caused by the higher inside container temperature.

The node model and the radial model show a similar behaviour. The main difference between the two models is the faster rise in temperature of the heater in the node model. The heat flow out of the heater is smaller in the node model since the average temperature profile in the first part of the sorbent, obtained from the temperature difference and the distance between the

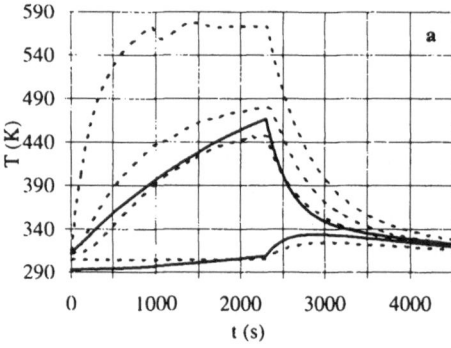

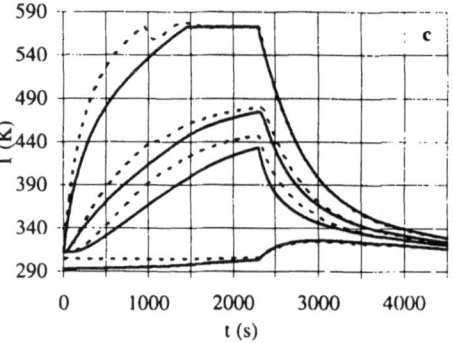

**Figure 6.** Comparison between experiments (dashed lines) and the three different models (solid lines). Lines from top to bottom: heater temperature, sorbent temperature, inside container temperature, and outside container temperature.
With a. the lumped capacitance model, b. the node model and c. the radial model.

temperature locations, is smaller than the local temperature profile at the heater surface calculated in the radial model. This results in a smaller conductive heat flow from heater to sorbent. In turn this should also yield a lower sorbent temperature. However, there is hardly any difference in the sorbent temperatures between the node model and the radial model. This effect can be related to the assumption of a uniform temperature of the sorbent in the node model.

The correspondence between experiment and simulations with the node model and the radial models is very promising. In the heating phase the sorbent temperature differences between experiment and simulations are smaller than 20 Kelvin (i.e. < 5%). We anticipate that the difference is mainly caused by the absence of the temperature profile in the axial direction of the compressor.

The differences that occur in the first part of the heating phase decrease in the second part when the heater has reached its set point temperature, 573 Kelvin in this case. The thermal conductivities in the model are therefore correct, since they determine the equilibrium temperature profile. In the cooling phase the simulations with the node model, the radial model, and the experiments almost overlap. This indicates that the gas gap and the heat flow to the environment are modelled correctly.

The pressures calculated out of these dynamic temperatures are depicted in figure 7. In this figure the pressures in the three models and the experiment are plotted versus time. Two extreme cases are shown: starting pressures of 1 bar, and 10 bar. In the first case the starting conditions are room temperature and 1 bar pressure, the conditions for which the sorption model is known to be inaccurate. In this region the sorption model gives values of the adsorption factor that are 40% too high, this results in a higher calculated gas contents in the simulated compressor. When more gas is present in the simulated compressor, the pressure change due to a temperature change is also higher than in the experimental case. The error in the pressure response is comparable to the error in the adsorption factor, meaning the pressure response is about 40% too high. In the second case the starting conditions are room temperature and 10 bar. In this situation the sorption model is accurate. In both cases the lumped capacitance model shows deviant behaviour, which is due to the averaging of the temperature over the complete centre of the

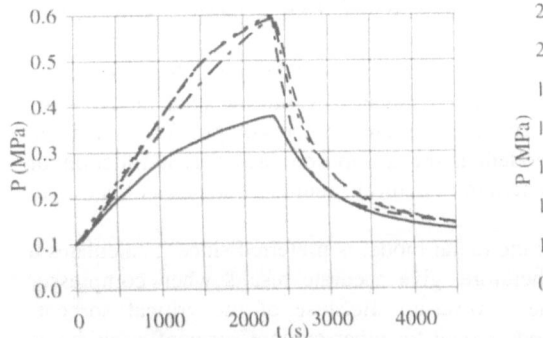 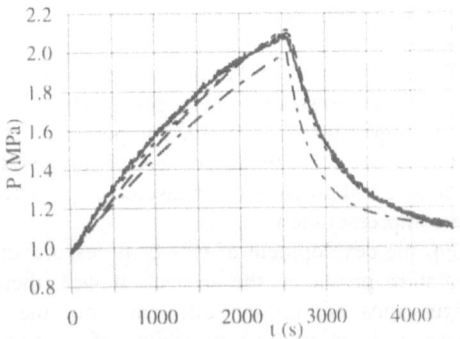

**Figure 7.** Simulated and experimental pressures. Starting conditions room temperature and 1 bar (left) and room temperature and 10 bar (right).Experiments (solid lines), Lumped capacitance model (dash-dot lines), Node model (dashed lines), and Radial model (dotted lines)

compressor. Between the node and the radial model no pressure differences can be seen. The experiment and the node and radial models overlap in the complete time interval. This makes the node model and the radial model useful tools in designing new sorption compressors.

As mentioned above, in the node model the position to which the sorbent temperature (as an average) is related is very important for the pressure calculations. In addition to this temperature dependence, the amount of adsorbed gas is also dependent on the volume of the sorbent. Therefore, in the pressure calculations, the sorbent temperature location in the node model was chosen such that the volume of sorbent between the heater and the location is the same as the volume of sorbent between this location and the inside container. Note that this position differs from the position in the temperature simulations depicted in figure 6. As can be seen in figure 7, the results of the node model coincide with those of the radial model for the test set-up configuration. However, because of the critical sorbent-temperature location, it is questionable whether the node model yields equally good results for other compressor configurations as well.

## CONCLUSIONS

We have constructed a test set-up in which we can characterise the operation of a sorption compressor. The sorption behaviour of an activated charcoal-nitrogen system and the thermal conductivities of the sorbent and the gas gap were experimentally determined. Three models were developed to predict the functioning of a sorption compressor. Experimental results obtained with the test set-up were used for validating these models

The description of the sorption behaviour of an activated charcoal-nitrogen system with the theoretical Polanyi model gives good results in very large temperature and pressure ranges. Only at lower temperatures in combination with a low pressure (<350 K and <4 bar) deviations between theory and experiments can occur. The Polanyi model can be used in the sorption compressor model, but the deviations in the sorption behaviour should be noted in the resulting calculated pressures. The thermal conductivity of the activated charcoal is $\lambda = 0.25$ W/Km. The gas gap in 'off' and 'on' position has thermal conductivities of 0.001 W/Km and 0.03 W/Km respectively.

The lumped capacitance model gives rough estimates of the average temperature in the centre parts of the compressor. Although not completely correct, this model is still reasonably accurate. The node model and the radial model both give very accurate results. In the heating phase inaccuracies occur due to the absence of a temperature profile in the axial direction of the compressor. The maximum sorbent temperature deviation between experiments and simulations is 20 Kelvin (i.e. < 5%). When the heater has reached its temperature set point, the differences between the models and experiments decrease. The cooling phase of the cycle is described very accurately in both models: the temperatures in the simulations and the experiments overlap in this phase.

The combination of the temperature model and the sorption model describes the operation of a sorption compressor. In the region where the sorption model is valid, the total model gives accurate approximations of the temperatures and the pressure in the compressor. When the sorption behaviour can not be adequately modelled, the adsorption factor as a function of temperature and pressure (as measured or presented in literature) should be used as a direct input in the compressor model.

For the development of new compressors, the radial model is preferred since it calculates a temperature profile in the sorbent. It will, therefore, give accurate results when compressor configurations are chosen different from the test set-up. Because of the critical sorbent-temperature location the applicability of the node model for other compressor configurations is questionable.

## ACKNOWLEDGEMENTS

This research is supported by the Dutch Technology Foundation (STW). The authors thank L.A. Wade (Jet Propulsion Laboratories, Pasadena), M. Feidt and R. Boussehain (both LEMTA, Nancy) for their contributions in the discussions on sorption experiments and compressor modelling.

## REFERENCES

1.  Wade, L.A., 'An overview of the development of sorption refrigeration', Advances in Cryogenic Engineering, Vol. 37 B, Plenum Press, New York (1992) , pp. 1095-1106.

2.  Burger, J.F., et al, 'Microcooling: Developments at University of Twente', to be published. Cryocoolers 9 (This conference), Plenum Press, New York (1996).

3.  Jones, J.A., Blue, G.D., 'Oxygen Chemisorption Compressor Study for Cryogenic Joule-Thomson Refrigeration', Journal of Spacecraft and Rockets, Vol. 25, no. 3 (1988), pp. 202-208

4.  Polanyi, M., Verhandl. Deutsch. Physik., Ges. 16 (1914), pp. 1012

5.  Dubinin, M.M., 'The potential theory of adsorption of gases and vapors for adsorbents with energetically nonuniform surfaces', Chemical reviews, Vol. 60, no. 1 (1960), pp. 235-241.

6.  Onyébuéké, L.C., 'Contribution a l'etudes des transferts de chaleur au générateur de machines trithermes a adsorption', Thesis, L'Institut National Polytechnique de Lorraine (1989).

7.  Chan, C.K., 'Improved heat switch for gas sorption compressor', Cryocoolers 3 (1984), Proceedings NBS special publication 698, US Government Printing (1985), pp. 42-46.

8.  Chan, C.K., 'Design and performance analysis of gas sorption compressors', Cryogenic Processes and Equipment-1984, ASME, New York (1984), pp. 65-70.

9.  Sigurdson, K.B., 'A general computer model for predicting the performance of gas sorption refrigerators', Proceedings 2nd Biennial Conf. on Refrigeration for Cryogenic Sensors and Electronic Systems, Nasa Conf. Publ. 2287 (1982), pp. 343-355.

10. Suhanan, et al, 'Etude Experimentale d'un refrigerateur Azote a cycle Joule Thomson a Adsorption', 19th Intern. Congres on Refrigeration, Proceedings Vol. IIIb (1995), pp. 1247-1256.

11. Bard, S., 'Development of an 80-120 K charcoal-nitrogen adsorption cryocooler', Proceedings Fourth International Cryocoolers Conference, Easton (1986), pp. 43-56.

12. Wade, L.A., 'Operating characteristics of a hydrogen sorption refrigerator part ii: A comparison between a second order analysis and experimental data', Proceedings Fourth International Cryocoolers Conference, Easton (1986), pp. 17-29.

# Low Cost Gifford-McMahon Cryocooler Development Program

John S. Kurtak

CTI-Cryogenics
Helix Technology Corporation
Mansfield, MA 02048

## ABSTRACT

CTI-Cryogenics is under contract with ARPA/NRL to develop a low cost, long life cryocooler to meet the cryogenic cooling needs of emerging applications for high temperature superconductive (HTS) microwave components. The design for this application was targeted to produce a minimum of 15 watts of refrigeration at 50K. Early prototype testing of this low cost design has shown performance to be > 25 watts at 50K.

The selection of a cryogenic cooling system for this HTS application started with a comprehensive study to evaluate both Stirling and Gifford-McMahon refrigerator design approaches. After completing a trade-off analysis that assessed and ranked refrigeration capacity, reliability, size, efficiency and price, the Gifford-McMahon (G-M) cycle was selected as the preferred design approach.

The design and manufacture of the cold head component represented the greatest potential for reducing the cost of the cryocooling system. Early in the design process, the development team members focused their efforts primarily on the cost drivers and design approaches that would reduce the total number of refrigerator parts. This effort led to the elimination of the traditional drive motor and a complete redesign of the associated mechanical linkages and valving. The subsequent redesign reduced our standard refrigerator part count by approximately 40%.

Material selection and construction methods, such as the use of "near net shape" fabrication techniques, were the keys to reducing the cost of the refrigerator parts. To substantiate that our cost goals were achieved, we solicited and received quotations on over 90% of the component parts. Analysis shows that, for large volume procurements, the price of a new cryogenic system can be reduced significantly from what is currently available.

## INTRODUCTION

It is important that low cost, long life cryogenic cooling support systems are developed and in place to meet the needs and requirements of what is anticipated to be a large market for HTS microwave applications. The cryogenic system for this program was designed to meet the demanding requirements for world-wide installations of HTS filters for use in the telecommunication market. In this regard, we have worked closely with a telecommunication service provider that is considering the use of HTS microwave components in their filter design. As part of this development effort, environmental and operational requirements as well as the servicing and packaging constraints of this application were defined early in the design selection process.

This program was structured into two phases. Phase I (Analysis and Design) included all the trade-off studies, cost analyses, concept evaluations, and the design of the cryogenic system. During Phase II (Fabrication and Test), three prototype systems were fabricated. These units were used as the test beds for evaluating the low cost concept and to validate the performance of the system.

## CYCLE/TRADE-OFF ANALYSIS

The initial effort of this program was focused on determining the best thermodynamic cycle for the identified application. Therefore, the purpose of the trade-off analysis was to compare the Stirling and Gifford-McMahon (G-M) cycles and to determine which cycle was best suited for this HTS application. A number of dependent variables, including capacity, size, efficiency, reliability, price and quantity were evaluated.

A baseline refrigerator model was selected for each of the two cycles being considered. Reliability for each baseline refrigerator was derived from actual test and field data using a 95% success rate of survival. For this analysis, the life of the G-M refrigerator model was baselined at two years and the Stirling cooler model was baselined at six (6) months. In extrapolating from this baseline, we examined a number of independent variables such as speed (RPM), charge pressure, internal volume, regenerator volume, length of stroke, compressor and expander diameters, pressure differentials, ambient temperature, length of seals (leak paths), etc. Relationships were then developed using these independent variables and tested against the baseline cases. The resulting relationships were then modeled to predict how the dependent variables (capacity, reliability, price, etc.) behave for a given set of performance requirements. To check the results of the analysis, data from production refrigerators was compared with the predictions to confirm the performance trends. The production data was consistent with the trends predicted by our model.

The information generated from this study was not intended to be used as accurate quantitative data for predicting performance, but only to establish trends for the criteria evaluated. The results of this study are summarized in Figures 1 through 4 which interrelate price, life, size, and efficiency of both Stirling and G-M cycle refrigerators for various capacities and quantities.

For the proposed telecommunication study, we assumed that the refrigerator would be mounted on a remote tower. Figure 1 is a plot of refrigerator size (cubic inches of volume on the tower) against life in years (R = 0.95) for a range of refrigeration capacities from 10 to 120 watts at 50K for both Stirling and G-M cryocoolers. To meet a reliability of 0.95, the operating speed of the baseline Stirling cycle model had to be lowered. This action resulted in a larger size unit. Thus, the Stirling volume on the tower is larger than the G-M Refrigerator for a given capacity. The G-M compressor was assumed to be remotely located on the ground and is, therefore, not included in the volume shown on Figure 1. Based on tower volume and reliability, the G-M cycle is the better choice.

Figure 2 is a plot of 'relative prices' versus life for a range of refrigeration capacities from 10 to 120 watts at 50 K for both Stirling and G-M cryocoolers. To maintain a reliability requirement of 0.95, the size (volume) of the Stirling cycle baseline model must be increased which, in turn, increases the price of the Stirling cycle configuration. This plot shows that the G-M refrigerator is less expensive to produce than the Stirling version for the same capacity and reliability.

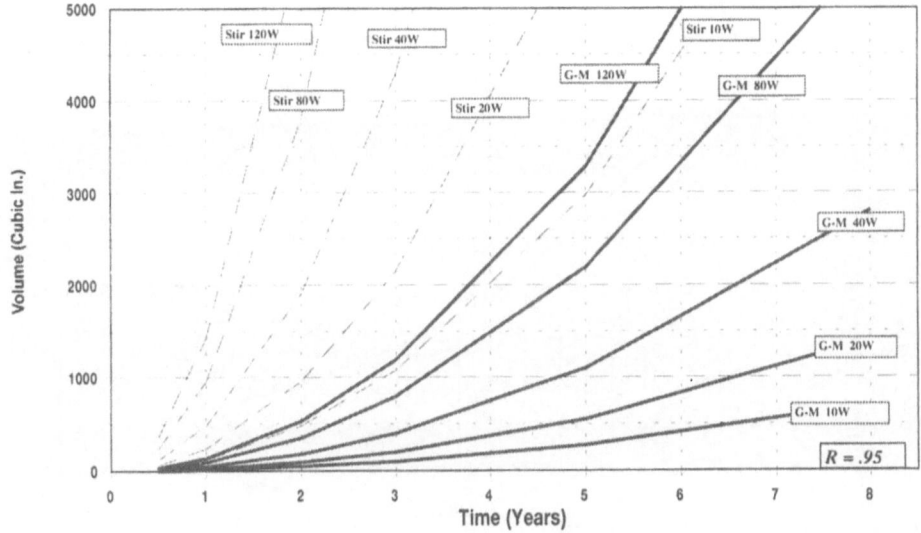

**Figure 1.** On-Tower Volume vs. Life

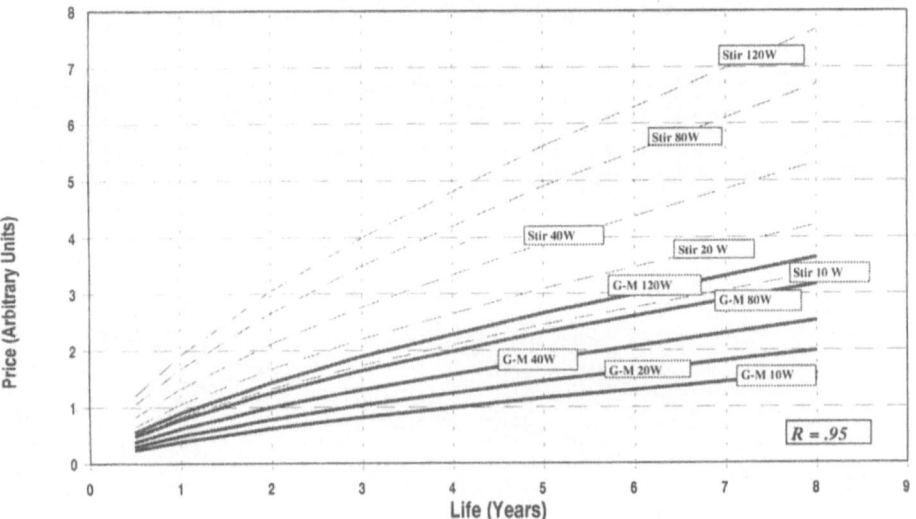

**Figure 2.** Price vs. Life

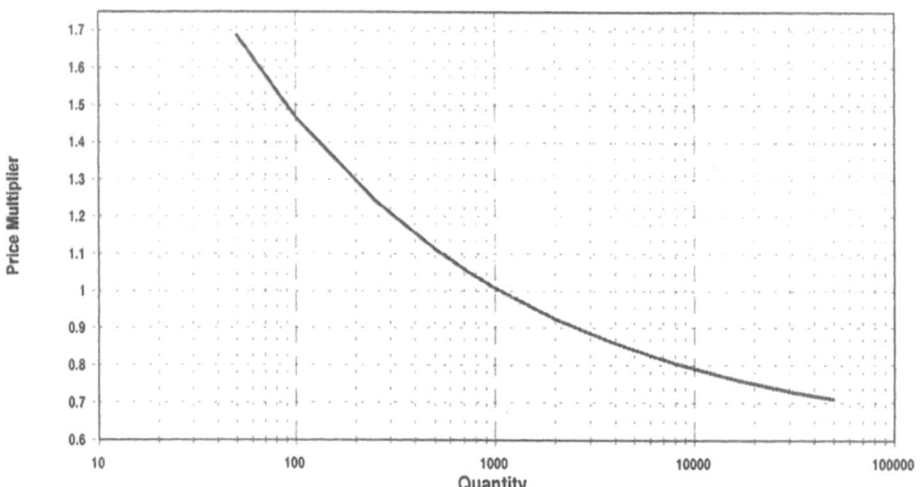

**Figure 3.** Price vs. Quantity

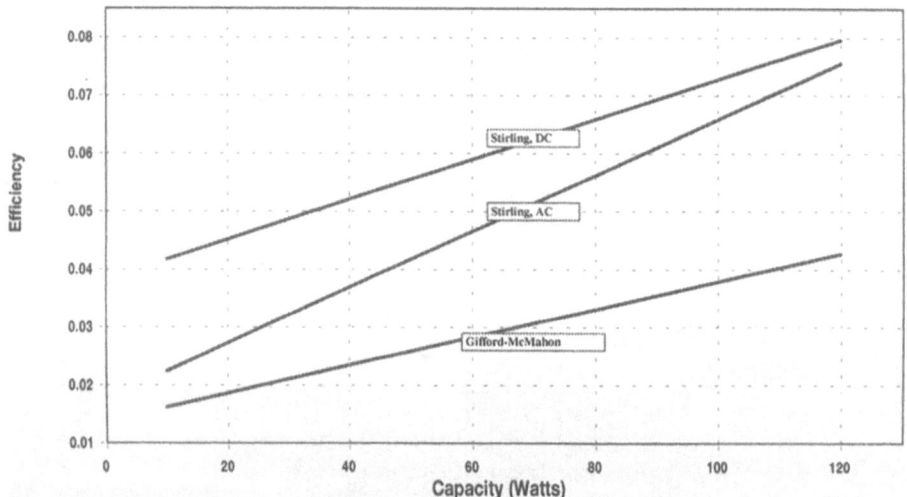

**Figure 4.** Efficiency vs. Capacity

Figure 3 shows the price multiplier based on the quantity purchased. This curve can be applied to either a Stirling or G-M cycle cryocooler. The fourth plot, Figure 4, shows the efficiency (capacity/input power) of both a G-M and an AC powered or DC powered Stirling cryocooler as a function of capacity. As expected, the Stirling cycle is inherently more efficient.

A review of the information shown on these four (4) plots indicates that the G-M cycle, other than the inherently higher efficiency of the Stirling cycle, is the preferred choice to meet the requirements for minimum price and size on tower for this HTS application.

## MECHANICAL DESIGN

During the design trade-off, we carefully evaluated the G-M cryocooler and concluded that the refrigerator, or cold end, portion of the design, represented the greatest opportunity for cost reduction. Our compressor design was already leveraged for high volume and highly engineered parts, and therefore, had the least opportunity for cost reduction.

The goal on this program was to simplify and reduce manufacturing costs on the new refrigerator design. This was accomplished by combining refrigerator functions, reducing the number of parts and leveraging high volume commercial parts. In this regard, our major focus was directed to the drive motor and the associated linkage and valves required for operating the cold end displacer.

A number of concepts were carefully assessed but discarded due to either complexity or high manufacturing costs. In evaluating how to control the displacer motion within the refrigerator, the design team agreed that the least expensive approach should take advantage of the differential pressure from the compressor as the driving force. Technical solutions were then focused on how to meter high and low pressure gas between the compressor and the cold end in a regulated and controlled sequence. The technical approach that was selected replaced the electrical drive motor and mechanical linkages with a pneumatic valving scheme that regulates the flow of gas into the displacer and uses the pressure differential from the compressor to control the phasing of the displacer.

Identifying the valves that meter gas into the displacer and control the motion of the displacer was critical to the design of a low cost and long life G-M cryocooler. In this selection process, a variety of valve designs and configurations (e.g., poppet, needle, spool) were considered. The objective was to select valves that are currently being produced in large enough quantities so that we could leverage the high volume pricing advantage. Two valve types were successfully identified and procured to support this program.

For the balance of the mechanical parts for the new design (displacer, cold finger housing, valve plate, dome, shroud), emphasis was placed on selecting a method of fabrication that would produce the lowest possible manufacturing cost. More information on this approach is covered later in the discussion of Cryocooler Price Analysis.

The control electronics was designed and fabricated at CTI-Cryogenics. The electronics provides the power and the control logic for the actuation and timing of the valves. The circuit can be either housed integrally with the refrigerator or mounted separately.

The mechanical arrangement of the refrigerator is illustrated in the cross sectional view shown in Figure 5, The refrigerator assembly with the shroud is approximately 13.5 inches long.

## TEST RESULTS

Three prototype coolers of the new design were fabricated and tested. Figure 6 is a load curve of the units indicating that the refrigeration capacities of all the units are consistent and that each unit has excess margin for the operating requirement of 15 watts at 50K. Figure 7 is a plot of cooldown time using a 2,500 gram copper mass on the cold end to simulate the HTS microwave package for this application.

PAPER #111

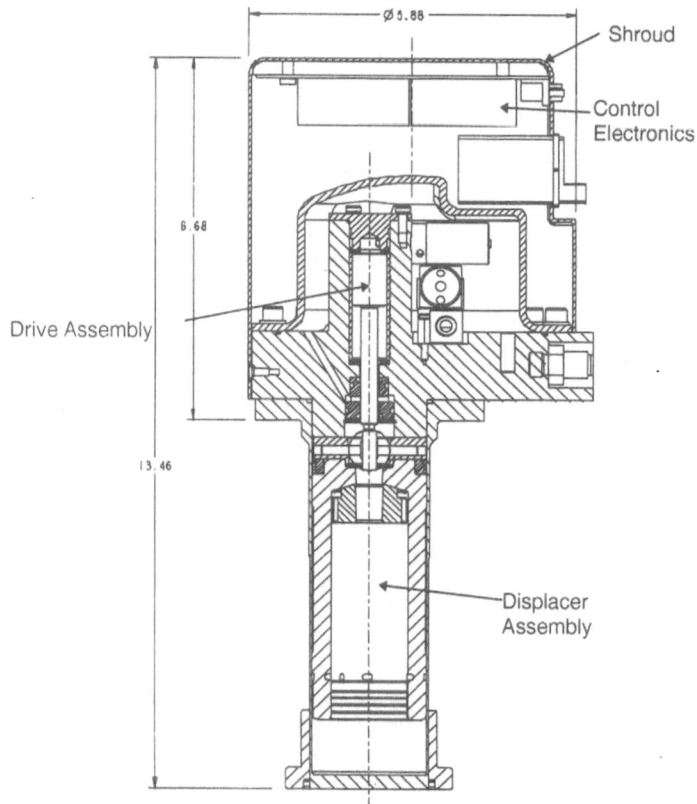

**Figure 5.** Cross Sectional View

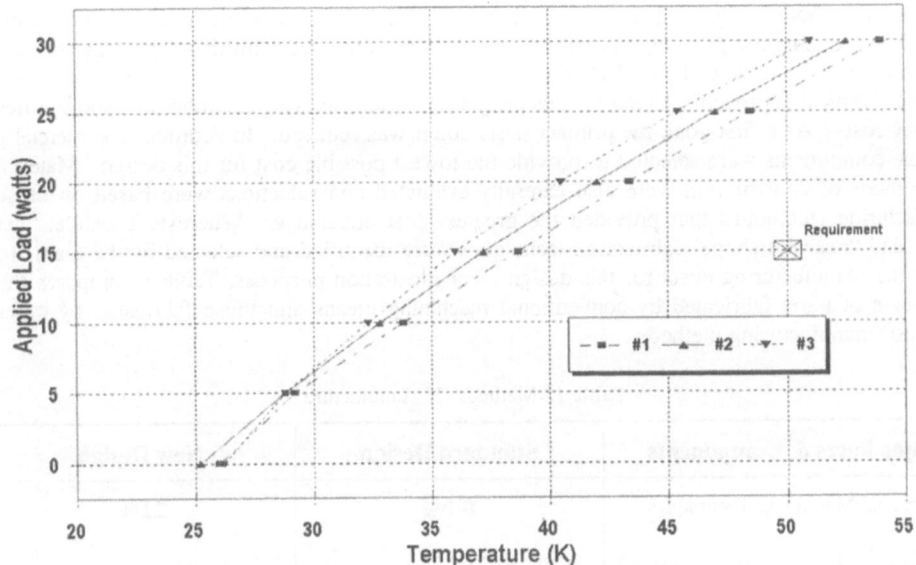

**Figure 6.** Load Curve-Prototype Units

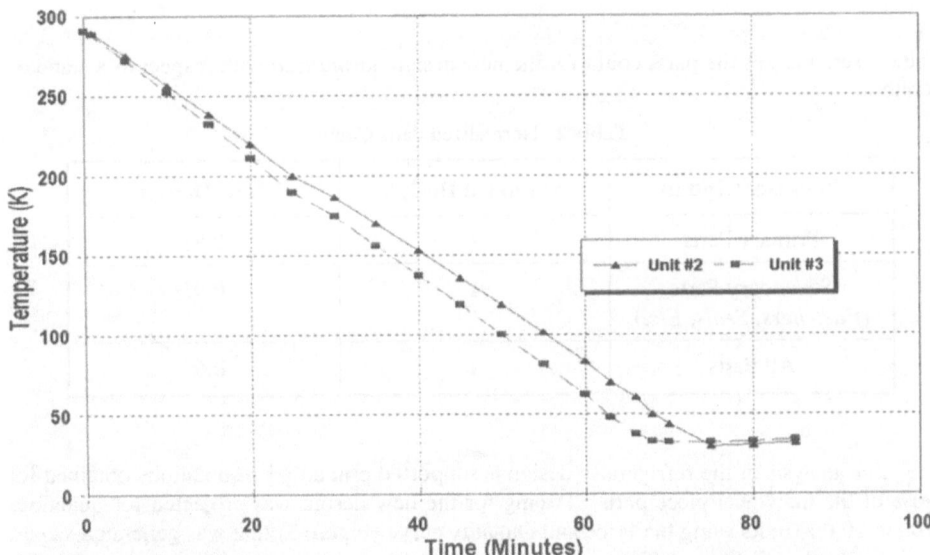

**Figure 7.** Cooldown Time with 2500 gram Copper mass

## CRYOCOOLER PRICE ANALYSIS

An aggressive cost reduction effort was implemented early in the design process by team members from both design and manufacturing engineering. This concurrent engineering effort focused on reducing the parts count of the new design as compared with the current mechanical refrigerator design produced by CTI-Cryogenics.

The technical approach selected for this program provides us with a number of opportunities to lower costs. As a first goal, the primary parts count was reduced. In addition, commercially available components were selected to provide the lowest possible cost for this design. Material and methods of construction were also carefully evaluated and selections were based on those manufacturing techniques that provided the greatest cost advantage. Wherever possible, "net shape" and "near net shape" fabrication techniques were identified and selected by the team to reduce the manufacturing costs for this design. For illustration purposes, Table 1 compares the percentage of parts fabricated by conventional machining means and those fabricated by other 'low cost' manufacturing methods.

**Table 1.** Methods Of Manufacture

| Major Parts & Components | Standard Design | New Design |
|---|---|---|
| Conventional Machining Techniques | 100% | 22% |
| Net or Near Net Shape Techniques* | 0% | 78% |

*Techniques such as molding, forming, casting, and forging were assessed for all candidate parts.

Table 2 summarizes the parts count for the new design *normalized* with respect to a standard refrigerator.

**Table 2.** Normalized Parts Count

| Item Description | Standard Design | New Design |
|---|---|---|
| Primary Parts | 1 | 0.55 |
| Secondary Parts *(Fasteners, Seals, Etc.)* | 1 | 0.93 |
| All Parts | 1 | 0.60 |

The price analysis of the refrigerator design is supported primarily by quotations obtained for over 90% of the individual piece parts. Pricing for the new design was projected for quantities from 100 to 10,000 units using the Price and Quantity curve (Figure 3) that was generated as part of the Cycle/Trade-off Study. Figure 8 represents the *normalized* price for the new design cryocooler (includes both the refrigerator & compressor) and is based on the price of a standard G-M design that is manufactured at a rate of 10,000 units per year. For example, the projected price of the new design cryocooler at quantity 10,000 is 65% of a standard unit produced at the same quantity.

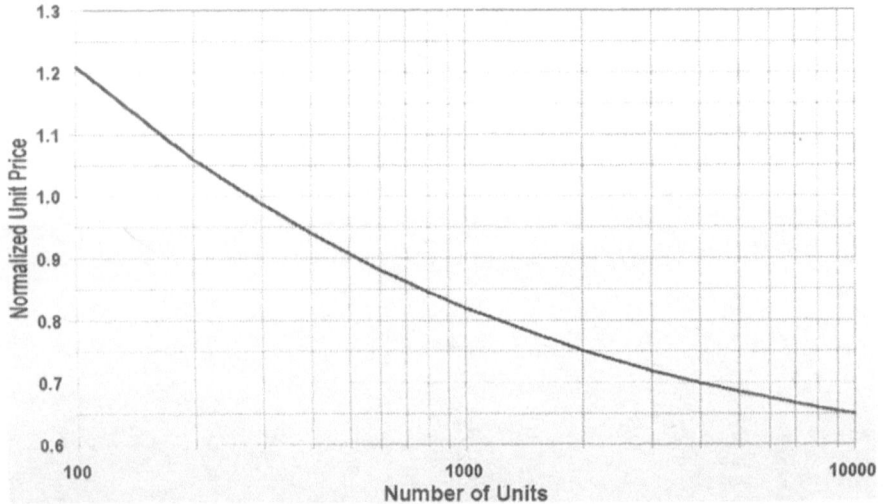

**Figure 8.** Projected System Pricing

## SUMMARY

The cryogenic system developed for this program was successful in meeting the technical goals for the proposed HTS application and provided the basis for an approach that substantially reduced the cost of the next generation G-M cryocooler. The approach used by our development team during this program leveraged the opportunity to take a 'fresh look' at a new refrigerator design, with the major objective to reduce manufacturing costs. After completing the trade-off study, the focus was placed on the refrigerator or cold head, which represented the greatest opportunity for reducing cost. Functions were combined wherever possible and emphasis was placed on reducing the number of parts and selecting low cost components that were currently being produced in large quantities. Selecting the right materials and identifying the methods of manufacture that would produce 'near net shape' parts were critical to reducing costs. This entire process resulted in a low cost G-M cryocooler that is designed to meet the HTS application described in this paper.

# Development of a 2W Class 4 K Gifford-McMahon Cycle Cryocooler

**Inaguchi Takashi, Nagao Masashi, Naka Kouki, and Yoshimura Hideto**

Advanced Technology R&D Center, Mitsubishi Electric Corporation, Tsukaguchi-Honmachi 8-chome, Amagasaki, Hyogo, 661 Japan

## ABSTRACT

This paper describes the principal design features and performance of the Gifford-McMahon cycle cryocooler by which we could obtain a cooling capacity of 2.2W at 4.2K. The main features of this machine are its large size expansion space, its use of rectifiable meshes which are packed in a regenerator at equal intervals, and its use of the combination of $Er_3Ni$ and $ErNi_{0.9}Co_{0.1}$ as regenerator materials.

## INTRODUCTION

Since the success of helium liquefaction by the Gifford-McMahon cycle cryocooler (hereinafter called 4K GM cryocooler)in 1989[1], the 4K GM cryocooler has been given attention as the cryocooler which will replace the JT cryocooler because the 4K GM cryocooler has higher reliability and much easier handling[2].

In particular, 4K GM cryocoolers actively have been actively applied to superconducting magnets such as helium free superconducting magnets[3,4,5] and MRI magnets[6] that don't have to supply liquid helium.

However, the use of the 4K GM cryocooler has been limited to small magnets or magnets whose heat load is small. In order to further extend the range of applications of the 4K GM cryocooler, the cooling capacity of the 4K GM cryocooler would have to be improved.

During this experimental period we obtained a cooling capacity of 2.2W at 4.2K. The main features allowing us to achieve this result are the machine's large size expansion space; the use of rectifiable meshes which were packed in a regenerator at equal intervals  in order to rectify helium gas flowing flows in the regenerator; and the use of the combination of $Er_3Ni$[7] and $ErNi_{0.9}Co_{0.1}$[8] as regenerator materials. In this paper we report on the experimental apparatus used and the results.

## EXPERIMENTAL APPARATUS

Fig. 1 shows a schematic of the expander of a two-stage 4K GM cryocooler, which is an experimental machine. The 4K GM cryocooler has two cooling stages through which heat loads are transferred into the cylinder. Displacers which contain regenerators reciprocate in the cylinder. As a result of the positioning of the displacers and the cylinder, two expansion spaces and a room temperature space are formed.  The first regenerator has a two-layer structure. We stacked

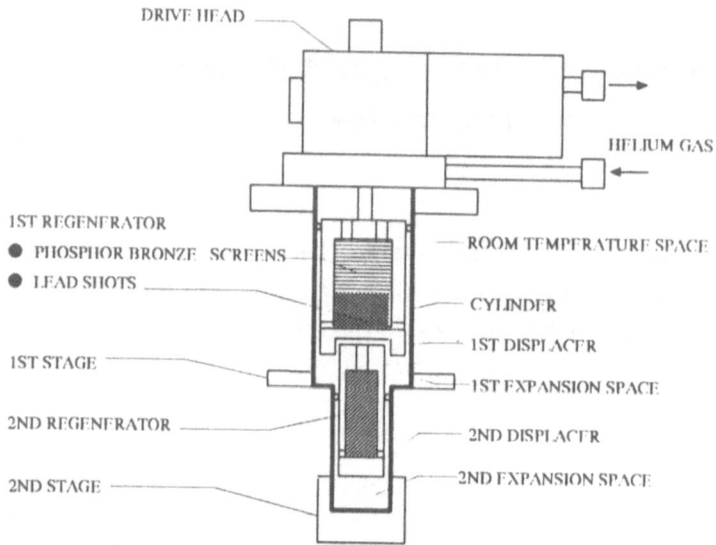

DRIVE HEAD

HELIUM GAS

1ST REGENERATOR
● PHOSPHOR BRONZE SCREENS
● LEAD SHOTS

1ST STAGE

2ND REGENERATOR

2ND STAGE

ROOM TEMPERATURE SPACE

CYLINDER

1ST DISPLACER

1ST EXPANSION SPACE

2ND DISPLACER

2ND EXPANSION SPACE

**Figure 1.** Schematic of expander of two-stage 4K GM cryocooler.

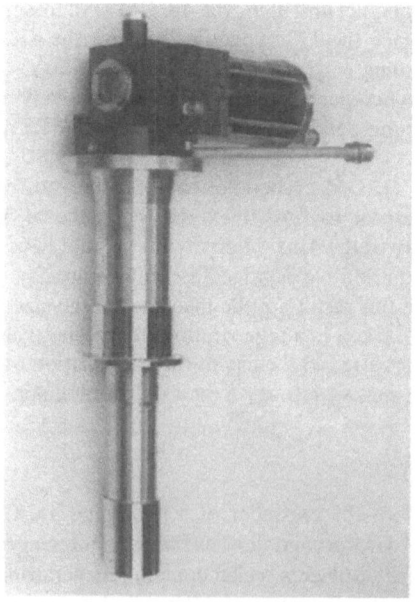

**Figure 2.** Photograph of expander of two-stage 4K GM cryocooler.

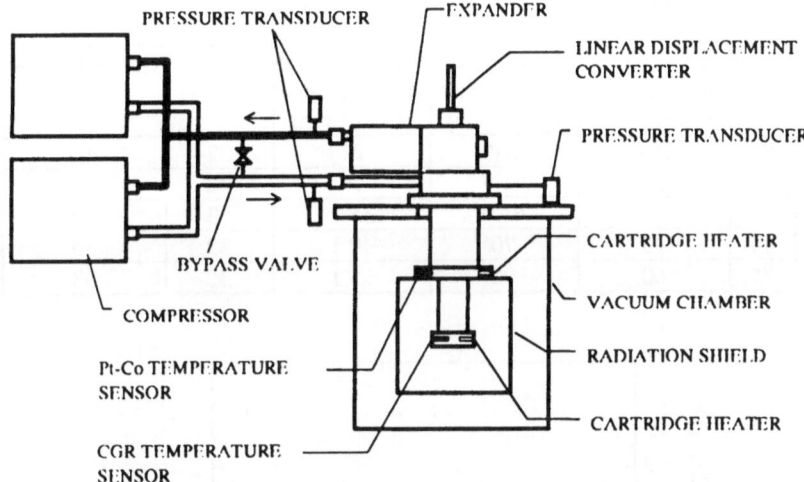

**Figure 3.** Schematic of experimental apparatus

phosphor bronze screens in the high temperature   part and lead shots in the low temperature part. In the second regenerator we packed Er₃Ni, or the combination of Er₃Ni and ErNi₀.₉Co₀.₁ as regenerator material, then investigated the effect on cooling capacity. Fig.2 shows a photograph of the expander of the two-stage 4K GM cryocooler.

Fig. 3 shows a schematic of the experimental apparatus. The cooling stages of the 4K GM cryocooler were installed in a vacuum chamber. A radiation shield was attached to the first stage, and it enclosed the second stage to prevent the radiant heat from room temperature from entering the second stage. The temperature of the first stage was measured using a Pt-Co resistance sensor and the temperature of the second stage was measured using a carbon glass resistance sensor. Cartridge heaters were installed at each stage to measure cooling capacity.

In order to secure enough flux to the expander, two compressors (each with a rated input power of 6kW) were arranged in a row. For this experiment we kept the differential pressure of the room temperature space at 1.2MPa~1.3MPa by adjusting a bypass valve which was installed between the high pressure pipe and the low pressure pipe.

A pressure transducer was installed into the room temperature space of the expander and a linear displacement converter was set up on the drive mechanism of the displacers in order to measure the indicated work. Pressure transducers were also installed at the gas exit and entrance of the expander, and inflowing or outflowing gas pressure was measured.

## EXPERIMENTAL RESULTS

### Effect of Volume of Expansion Space on Cooling Capacity

Table 1 shows the main parameters of the 4K GM cryocoolers that were used in this experiment. We changed the diameters of the second expansion space from 25.4mm to 60mm in order to investigate the effect of volume of the expansion space on cooling capacity. Inner diameters of the second regenerators were increased in accordance with the increase of the diameters of the second expansion space. The diameters of the first expansion spaces of #1 through #3 were 61.9mm; that of #4 was 70mm; and that of #5 was 90mm. The stroke of the cryocoolers were all 32mm. We employed Er₃Ni as the regenerator material.

**Table 1.** Main parameters of 4K GM cryocooler

|  | Diameter of 2nd expansion space (mm) | Diameter of 1st expansion space (mm) | Inner diameter of 2nd regenerator (mm) | Stroke (mm) | Optimum cycle frequency (rpm) |
|---|---|---|---|---|---|
| #1 | 25.4 | 61.9 | 21 | 32 | 45 |
| #2 | 30.2 | 61.9 | 26 | 32 | 45 |
| #3 | 40 | 61.9 | 35.1 | 32 | 45 |
| #4 | 45 | 70 | 40.1 | 32 | 33 |
| #5 | 60 | 90 | 55.1 | 32 | 33 |

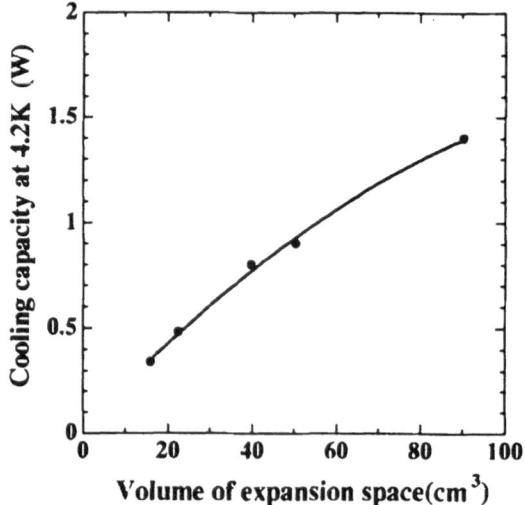

**Figure 4.** Effect of volume of expansion space on cooling capacity at 4.2K

The cycle frequency employed was the cycle frequency at which cooling capacity at 4.2K became optimum. The optimum cycle frequency was 45rpm in the cases of #1~#3 and the optimum cycle frequency was 33rpm in the cases of #4~#5.

Fig. 4 shows the effect of volume of the expansion space on cooling capacity at 4.2K. When the volume of the expansion space is 16.1cm² (diameter: 25.4mm), the cooling capacity is 0.34W.

The cooling capacities increase in accordance with the increase of volume of the expansion space. When the volume of the expansion space is 90.5cm² (diameter: 60mm), the cooling capacity becomes 1.4W. The optimum cycle frequency decreases from 45rpm to 33rpm in accordance with the increase of the volume of the expansion space. The rate of the increase of cooling capacity is therefore not proportional to the rate of the increase of volume.

## Effect of Diameter of Expansion Space on Cycle Frequency Dependence of No Load Temperatures

The optimum cycle frequency depends on the diameter of the expansion space and it is apt to decrease in accordance with the increase of diameter. We have investigated the effect of diameters of expansion spaces on cycle frequency dependence of no load temperatures.

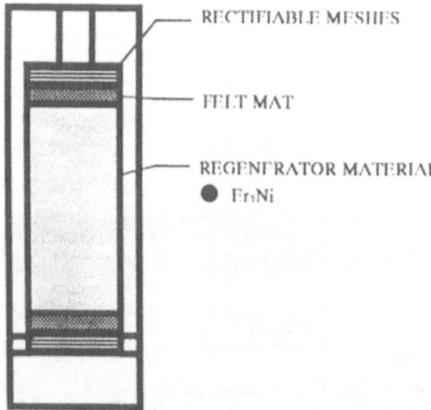

**Figure 5.** Schematic of the conventional second regenerator

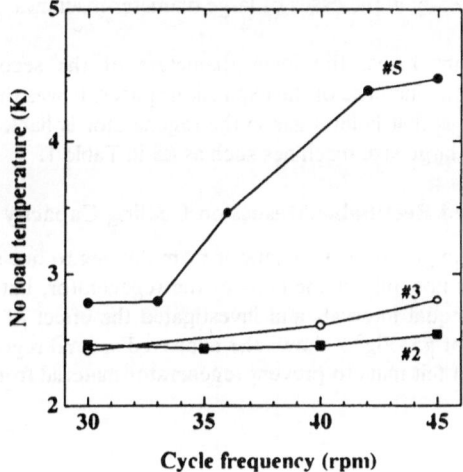

**Figure 6.** Effect of cycle frequency on no load temperature in the cases of
#2,#3,and #5 in Table 1.

Fig. 5 shows a schematic of the employed conventional second regenerator. Er₃Ni was employed as regenerator material and it was fixed by felt mats. Rectifiable meshes were packed at the ends of the regenerator to rectify helium gas.

Fig. 6 shows the effect of cycle frequency on no load temperatures in the cases of #2, #3,and #5 in Table 1. When the cycle frequency was 30rpm, the no load temperature of #2 was 2.46K and that of #3 was 2.42K. When the cycle frequency was 45rpm, the no load temperatures of #2 and #3 were respectively 2.54K and 2.80K. The temperature rise of #2 was 0.08K and the temperature rise of #3 was 0.38K. In the cases of #2 and #3, the no load temperatures do not depend so much on cycle frequency.

However, the no load temperatures in the case of #5 depend much on cycle frequency. The no load temperature was 2.78K at the cycle frequency of 30 rpm and it was 4.47 K at the cycle frequency of 45 rpm. The temperature rise was 1.68K. The no load temperature was liable to

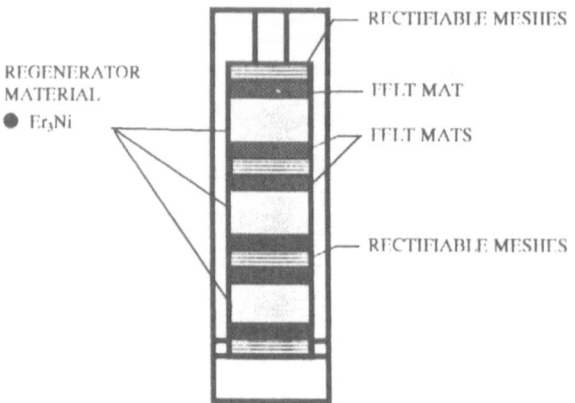

**Figure 7.** Schematic of the improved second regenerator

depend on cycle frequency greatly in the cases of large diameter machines, such as indicated by #5 in Table 1.

In these experimental machines, the inner diameters of the second regenerators were increased in accordance with the increase of the expansion spaces. It was therefore considered that the cause of temperature rise is that helium gas in the regenerator is liable to flow to one side of the regenerator in the case of large size machines such as #5 in Table 1.

### Effect of Regenerator Packed Rectifiable Meshes on Cooling Capacity

In order to prevent helium gas in the regenerator from flowing to one side of the regenerator, we packed rectifiable meshes not only at the ends of the regenerator, but also in two additional places of the regenerator at equal intervals and investigated the effect of cycle frequency on no load temperature in the case of #5. Fig. 7 shows the improved second regenerator. The rectifiable meshes were put between two felt mats to prevent regenerator material from blocking meshes.

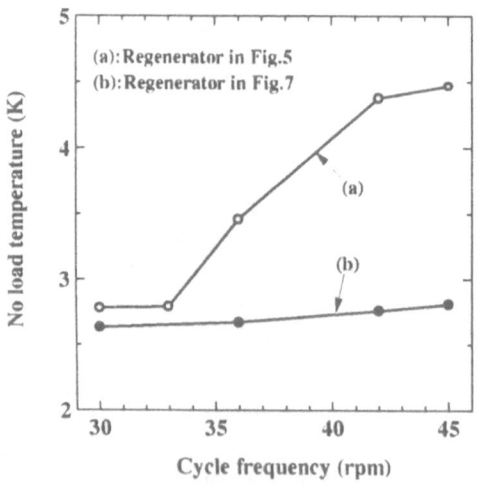

**Figure 8.** Effect of cycle frequency on no load temperature of #5 in Table 1 using regenerator in Fig.5 and Fig.7.

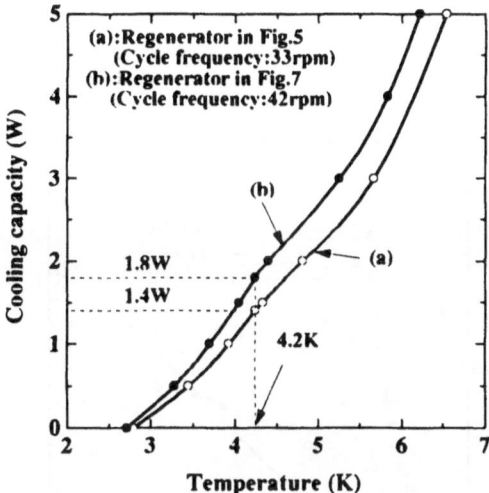

**Figure 9.** Cooling capacity of GM cryocooler using regenerator in Fig.5 and Fig.7

Fig. 8 shows the effect of cycle frequency on no load temperature of #5 in Table 1. When we employed the regenerator in Fig. 5, the no load temperature was 2.78K at the cycle frequency of 30 rpm and it was 4.47 K at the cycle frequency of 45 rpm. The temperature rise was 1.68K. When we employed the regenerator in Fig.7, the no load temperature was 2.63K at the cycle frequency of 30rpm and it was 2.81K at the cycle frequency of 45 rpm. The temperature rise was only 0.18K. The cycle frequency dependence of no load temperature was decreased and it is considered that the rectifiable meshes in two additional places of the regenerator could prevent helium gas in the regenerator from flowing on one side of the regenerator.

Fig. 9 shows the cooling capacity of the 4K GM cryocooler using the regenerators in Fig. 5 and Fig.6. When we employed the regenerator shown in Fig.4, the optimum frequency was 22rpm and the cooling capacity at 4.2K was 1.4W. When we employed the regenerator in Fig.7, the optimum frequency was 42rpm and the cooling capacity at 4.2K was 1.8W. The optimum cycle frequency of the 4K GM cryocooler using the regenerator in Fig.7 was 1.3 times higher than that of the 4K GM cryocooler using the regenerator in Fig.5, and the cooling capacity at 4.2K in the case of Fig.7 also improved to 1.3 times greater than that in the case of Fig.5. A regenerator packed with rectifiable meshes in two additional places of the regenerator at equal intervals can improve the optimum cycle frequency and the cooling capacity at 4.2K.

### Effect of Regenerator Material on Cooling Capacity

The effect of specific heat of regenerator material on cooling capacity was investigated, using $Er_3Ni$ and $ErNi_{0.9}Co_{0.1}$ as regenerator materials. Fig. 10 shows specific heat of the regenerator materials. Two cases were examined. In case one, experiments were carried out using a regenerator packed only with $Er_3Ni$ as shown in Fig.7. The other case used a regenerator packed with $Er_3Ni$ in two thirds of the high temperature side of the regenerator and $ErNi_{0.9}Co_{0.1}$ in one third of the low temperature side of the regenerator as shown in Fig.11. The experimental machine employed was the 4K GM cryocooler shown in #5 of Table 1.

Fig. 12 shows the cooling capacity of the 4K GM cryocooler using only $Er_3Ni$ and the cooling capacity of the 4K GM cryocooler using a combination of $Er_3Ni$ and $ErNi_{0.9}Co_{0.1}$. The cycle frequency was fixed at 42rpm.

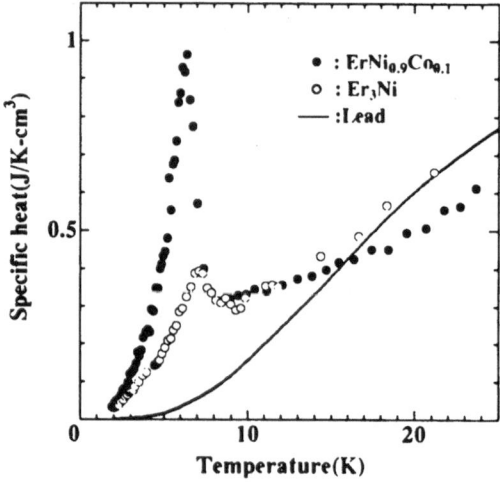

**Figure 10.** Specific heat of regenerator material

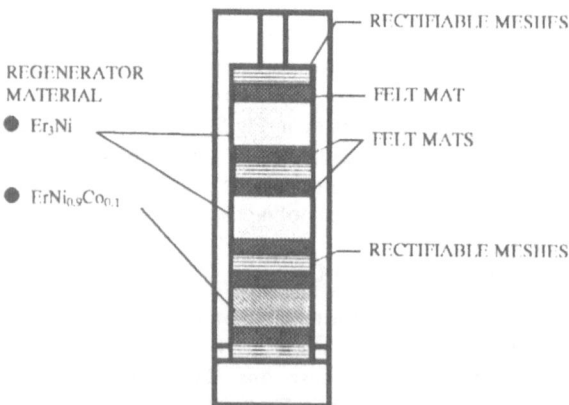

**Figure 11.** Schematic of the regenerator in which $Er_3Ni$ and $ErNi_{0.9}Co_{0.1}$ were packed.

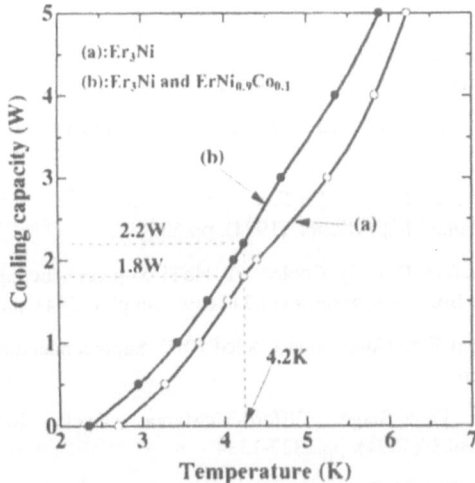

**Figure 12.** (a) Cooling capacity of 4K GM cryocooler using only $Er_3Ni$ and (b) cooling capacity of 4K GM cryocooler using combination of $Er_3Ni$ and $ErNi_{0.9}Co_{0.1}$

When only $Er_3Ni$ was employed, the cooling capacity was 1.8W, and when the combination of $Er_3Ni$ and $ErNi_{0.9}Co_{0.1}$ was employed, the cooling capacity was 2.2W. The cooling capacity of the 4K GM cryocooler using the combination of $Er_3Ni$ and $ErNi_{0.9}Co_{0.1}$ was 1.2 times greater than that of the GM cryocooler using only $Er_3Ni$.

When we could obtain the cooling capacity of 2.2W at 4.2K, the temperature of the first stage was 50K with the heat load of 25W and the input power of the compressors was about 12kW.

## CONCLUSIONS

(1) The cooling capacity of 2.2W at 4.2K was obtained by the 4K GM cryocooler . The main features of this machine are its large size expansion space, the use of rectifiable meshes which are packed in the regenerator at equal intervals, and the use of the combination of $Er_3Ni$ and $ErNi_{0.9}Co_{0.1}$ as regenerator materials.

(2) The optimum cycle frequency depends on the diameter of the expansion space and it is apt to decrease in accordance with the increase of diameter. A regenerator packed with rectifiable meshes in two additional places of the regenerator at equal intervals can improve the optimum cycle frequency and the cooling capacity at 4.2K.

(3) A regenerator packed with rectifiable meshes in two additional places of the regenerator at equal intervals can improve cooling capacity 1.3 times greater than a regenerator packed with rectifiable meshes only at the ends of the regenerator.

(4) The cooling capacity of the 4K GM cryocooler using a combination of $Er_3Ni$ and $ErNi_{0.9}Co_{0.1}$ can be improved 1.2 times greater than that of the 4K GM cryocooler using only $Er_3Ni$.

## REFERENCES

1. Yoshimura,H.,et al.," Helium Liquefaction by a Gifford-McMahon Cycle Cryogenic Refrigerator ",Rev.Sci.Instrum., vol.60(1989), pp.3533-3536

2. Inaguchi,T.,et al.," Two-Stage Gifford-McMahon Cycle Cryocooler Operating at about 4.2K ",Cryocooler 6 (1990), pp.25-36

3. Watanabe,K.,et al.," Liquid Helium-Free Superconducting Magnets and Their Applications ",Cryogenics vol.34 ICEC Suppl. (1994), pp.639-642

4. Kuriyama,T.,et al," Cryocooler Directly Cooled 6T NbTi Superconducting Magnet System with 180mm Room Temperature Bore ",Cryogenics vol.34 ICEC Suppl. (1994), pp.643-646

5. Yokoyama,S.,et al," Cryogen Free Conduction Cooled NbTi Superconducting Magnet for a X-band Klystron ",IEEE(1996) in press

6. Nagao,M.,et al.," 4K Three-Stage Gifford-McMahon Cycle Refrigerator for MRI Magnet ",Adv.Cryog.Eng. vol.39(1994), pp.1327-1334

7. Tokai.,Y.,et al.," Magnetic Field Influence on $Er_3Ni$ Specific Heat ",Jpn.J.Appl.Phys.Part1 vol.31 (1992), pp.3332-3335

8. Onishi,A.,et al.," Development of a 1.5W-Class 4K Gifford-McMahon Cryocooler ",Cryogenic Engineering vol.31 (1996), pp.162-167

# Geometric Scaling of a 4.2 K Gifford-McMahon Refrigerator

J. N. Chafe, G.F. Green, and J. W. Stevens

Naval Surface Warfare Center, Carderock Division
Annapolis, MD, USA 21402

## ABSTRACT

An approach is presented to increase the cooling power of small regenerative cycle cryogenic refrigerators (cryocoolers). While several researchers have demonstrated Gifford-McMahon refrigerators capable of producing 0.5W of cooling at 4.2K, the U.S. Navy is interested in several applications, including mine countermeasures and electric ship propulsion, that have cryogenic refrigeration requirements of several times that amount. These applications are the impetus for scaling the existing 0.5W, 4.2K Gifford-McMahon refrigerators to a size where they can produce more cooling power. The approach is to properly increase the size of the various components based on the basic thermodynamic processes involved and the known performance of the present systems. This scaling method allows the estimation of increased compressor requirements as well as other important parameters. A new larger refrigerator, 1.5W at 4.2K, has been designed and operated. Test results of that refrigerator are presented.

## INTRODUCTION

The work presented in this paper is sponsored by the Office of Naval Research (ONR). It is part of the Advanced Technology Demonstration program to develop a technologically advanced mine-sweeping system. This system is referred to as the Advanced Lightweight Influence Sweep System (ALISS).

ALISS is composed of two separate subsystems, one produces an acoustic energy pattern that closely resembles Naval landing craft, and the other produces a magnetic profile that mimics the profile of the same craft. The purpose of ALISS is to trigger mines typically used in shallow water for the defense of beaches and coastal areas. These mines are designed to sense Naval landing vessels by measuring their acoustic and/or magnetic characteristics. The ALISS is carried on board a Naval mine sweeping vessel and emits a simulated Naval craft profile far enough away so that when the mine detonates it causes no damage to the mine sweeper.

The subsystem that produces the magnetic profile uses a superconducting magnet to generate a large magnetic field. The cooling requirements of this magnet are the focus of this paper.

Cryocoolers 9, Edited by R.G. Ross, Jr.
Plenum Press, New York, 1997

Instead of cooling the magnet with liquid helium as has typically been done in the past for superconducting magnets, a cryocooler is used to conductively cool the magnet. In this method of cooling, the superconducting magnet is suspended directly in a vacuum. There is no fluid that bathes the magnet, but rather all the refrigeration passes directly from the refrigerator to the magnet through solid copper bus work, bolted or soldered joints and possibly heat pipes.

Operating a superconducting magnet in this way, an approach successfully used in the past[1], requires a cryogenic refrigerator that is small enough to be mounted directly on the magnet assembly. The refrigerator must provide low enough temperatures and enough cooling power to operate the superconductor, and it must be reliable. Due to these constraints, a Gifford McMahon (GM) refrigerator was chosen for the task. For the ALISS currently under development, the specific cooling requirements are estimated to be approximately 1.3 W of cooling power at 4.2 K and 50 W of cooling power at a temperature of 50 K. Commercially available coolers are not at this time capable of meeting these requirements. So this research was aimed at finding a way to increase the cooling capacity of low temperature GM refrigerators.

## DESCRIPTION

A GM cooler consists of two major modules; a compressor module and a cold head module. A compressor module is generally operated at room temperature, requires electrical input power and is either water or air cooled. The compressor takes in helium gas at pressures around 100 psig (0.79 MPa) and compress it to pressures around 300 psig (2.17 MPa). It must have enough capacity to supply the required helium flow rate. In addition, the gas must be free of contaminants. Therefore the compressor module is equipped with special filters to clean lubrication oil and other contaminants from the gas. The cold head module expands the compressed helium in a way that produces very low temperatures. This paper focuses on the cold head module of the GM cooler as opposed to the compressor module. It is assumed that scaling the cold head module to a larger size will increase gas flow requirements, and that no new technology development work is required to construct a compressor module capable of fulfilling those increased requirements.

## SCALE-UP METHOD

The simplest way to increase the available cooling power for a system is to increase the number of coolers being used. If commercially available coolers are capable of producing 0.5 W of refrigeration at 4.2 K and there is a cooling requirement of 1.5 W, then 3 coolers will satisfy that need. This increased cooling power is expensive. It requires both the purchase and the maintenance of multiple coolers. In addition, for some applications it might not be possible to fit the required number of coolers into the available space. Increasing cooling capacity of commercially available coolers is a much better solution.

One approach to increase the capacity of a Gifford-McMahon refrigerator is to make the refrigerator operate more efficiently. This is the approach we took when initially developing low temperature GM refrigerators[2]. In that effort, we modified the second stage regenerator of a refrigerator by replacing some of the lead (Pb) spheres in the regenerator with neodymium (Nd) spheres. We achieved increased refrigeration performance for both lower temperatures and higher cooling powers. In addition to that work, we made small changes to the refrigerator such as valve timing, regenerator construction methods and second stage seal configurations and were also able to achieve modest improvements in cooling power.

Another approach to increasing the cooling power of a refrigerator is to make it physically larger. This concept is similar to the use of multiple coolers. The goal is to "roll" multiple units up into one larger unit. The difficulty with this approach is accurately estimating the effect of scaling. As various parts of the cooler are increased in size, the thermodynamic processes that occur in those parts may change. This is the approach chosen for this effort.

## Sizing the Cooler

Carefully measuring the performance of the Balzer Kelcool 4.2™ refrigerator (the baseline cooler for this work) was the initial step for scaling up the GM cooler. Figure 1 is the result of the baseline measurements shown as a load map. This type of load map indicates the available cooling power and the corresponding operating temperature when both stages of the cooler are loaded simultaneously. One important feature of this figure is that the cooling power available from the second stage is virtually independent of the load applied to the first stage.

This data is the starting point for our scaling effort. It indicates that 0.5 W of cooling is available from the second stage at 4.2K and 50 W of cooling is available from the first stage at a temperature of 55K. Our efforts concentrated on keeping the same cooling capacity on the first stage while trying to increase the capacity to 1.3 W of cooling at 4.2K on the second stage. Thus increasing the capacity of the second stage by a factor 2.67.

A simplified sketch of a GM cold head is shown in Figure 2. This figure has been labeled to show the major components and regions of a two stage GM cooler. As shown by this figure the

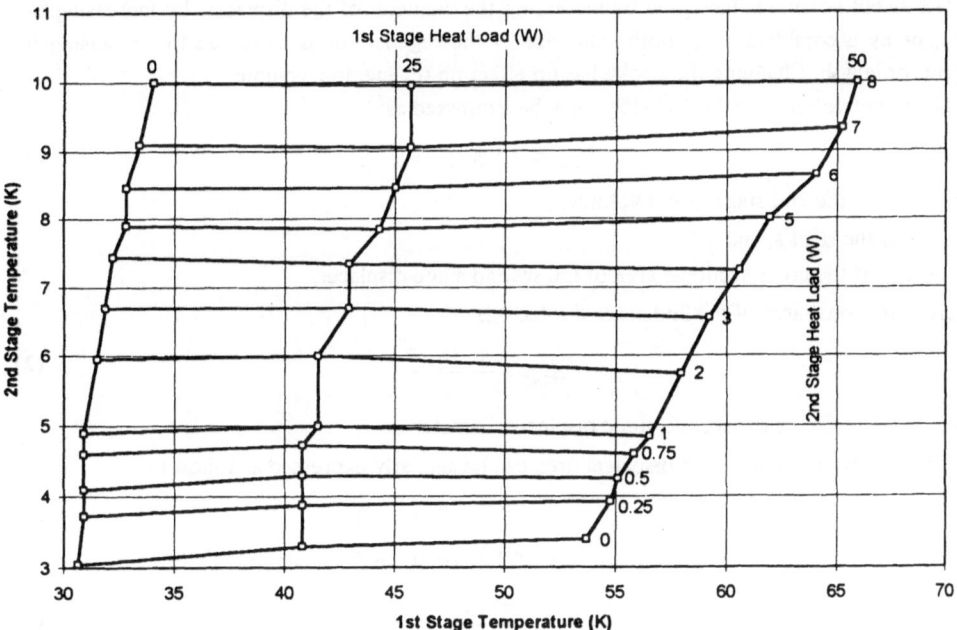

Figure 1 Baseline load map.

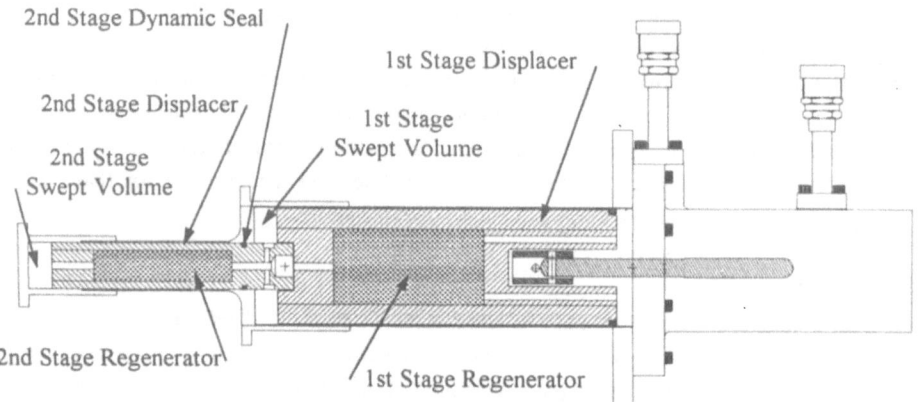

**Figure 2** Sketch of typical two stage GM cooler cold head module.

regenerators are located internally to their respective displacers. The swept volumes for each stage are also pointed out.

The expansion of the gas located in the swept volume of each stage of the refrigerator produces refrigeration (see Figure 2). The more gas available in each stage, the more cooling is produced. Consequently, to increase the performance of the second stage by a factor of 2.67, the swept volume should be increased by a factor of 2.67. If this is done, then the amount of gas that flows through the second stage regenerator will be 2.67 times greater. Therefore, the size of the regenerator must be increased by a factor of 2.67 to keep all the processes in the regenerator unaltered.

The swept volume is increased by increasing the diameter of the displacer, by increasing the stroke, or by a combination of both. The size of the regenerator is increased by increasing its diameter or length. Changing the stroke has no effect on regenerator volume.

The swept volume for the 2nd stage may be expressed as:

$$\vartheta_{2nd} = S * A_{2nd},$$                            (1)

where $\vartheta_{2nd}$ is the 2nd stage swept volume,

$S$ is the stroke, and

$A_{2nd}$ is the cross sectional area of the second stage displacer.

The cross sectional area of the 2nd stage displacer is:

$$A_{2nd} = \frac{\pi \cdot D_{2nd}^{2}}{4}$$                            (2)

where        $D_{2nd}$ s the second stage displacer diameter.

The 1st stage swept volume and displacer area can be similarly expressed as follows:

$$\vartheta_{1st} = S \cdot \left( A_{1st} - A_{2nd} \right)$$                            (3)

where        $A_{1st}$ is the cross sectional area of the first stage displacer and is given by:

$$A_{1st} = \frac{\pi \cdot D_{1st}^{2}}{4}$$                            (4)

where $D_{1st}$ is the first stage displacer diameter.

Similar expressions may be written for the first and second stage regenerators. However they are dependent on their respective lengths and diameters and are not affected by the stroke of the refrigerator. Thus the following two expressions:

$$V_{2nd} = L_{2nd} \cdot \frac{\pi \cdot d_{2nd}^2}{4} \tag{5}$$

$$V_{1st} = L_{1st} \cdot \frac{\pi \cdot d_{1st}^2}{4} \tag{6}$$

where $V_{2nd}$ and $V_{1st}$ are the second and first stage regenerator volumes,

$L_{2nd}$ and $L_{1st}$ are the second and first stage regenerator lengths, and

$d_{2nd}$ and $d_{1st}$ are the second and first stage regenerator diameters.

Equations 1 and 3 specify that if the diameter of the second stage displacer is increased (in order to alter the size of the swept volume), the swept volume of the 1st stage is inadvertently decreased. In addition, if the size of the second stage swept volume is increased by increasing the stroke, then the swept volume of the first stage is also increased. Thus no matter what is changed to increase the second stage swept volume, it will also change the first stage swept volume. To increase the swept volume of the second stage and leave the first stage unaltered, both the stroke and the diameter of the second stage displacer must be increased in the correct proportion. The above equations were used to accomplish this.

It is also necessary to ensure that the functioning of the regenerators and that the behavior of the gas that passes through them is not altered. Both regenerators were sized so that the following ratio was unaltered:

$$\frac{M_{he}}{M_{matrix}}, \tag{7}$$

where $M_{he}$ is the mass of helium that flows through the regenerator per cycle and $M_{matrix}$ is the mass of regenerator matrix material. Equations 5 and 6 are the expressions for volumes of the regenerators. These expressions, in part, determine the necessary regenerator sizes. The Apparatus section below presents a listing of the important sizing parameters of both the modified and unmodified coolers.

The scale-up presented in this paper is completely empirical because neither the thermodynamic process that produce refrigeration nor the losses that tend to limit it have been quantified. This scaling technique assumes that the losses and the refrigeration effects are scaled proportionally with the swept volume. One purpose for performing the experimentation phase of this work is to test this assumption.

**The Cooling Process and Losses**

Taking an empirical approach to scaling a GM cooler can be justified by examining both the process that produces refrigeration and the losses that tend to reduce the availability of that refrigeration. The cooling effect in a GM cooler is the result of the expansion of a gas from a rigid vessel[3]. This process is modeled as isentropic during the expansion part of the cycle for the gas that occupies the swept volume. Using this representation, the cooling effect should scale with the mass of gas contained in the swept volume as assumed earlier. The thermodynamic losses within a cooler are numerous and often interrelated to one another. Table 1 lists many of these losses along with a factor that indicates the way a particular loss scales with the physical size of the pertinent component of the cooler. These scaling factors assume that the length dimensions of the cooler are held constant. The isentropic cooling process scales as $D^2$ since it is dependent on the amount of gas contained in the swept volume, which is proportional to $D^2$ and given by equations 1 and 2. Table 1 shows that no losses scale at a rate greater than $D^2$. This means that the empirical approach to scaling that we used is reasonable and even conservative, but only if all the *major* loss mechanisms have been accounted for.

Two losses that are not listed in Table 1 are flow maldistribution in the regenerators and dynamic-seal leakage. Flow maldistribution (uneven flow through the regenerators) is caused by a regenerator that is too loosely packed allowing the gas to flow through open regions of the regenerator and preventing the gas from utilizing all the heat capacity of the regenerator. Flow maldistribution could also be caused by azimuthal and radial temperature gradients in the matrix material. This is common for multi-passage recuperative heat exchangers however for regenerators, which experience good thermal mixing because of reversing flows, this is not a likely cause of flow maldistribution. The second unaccounted for loss, dynamic-seal leakage, occurs at the second stage seal (see Figure 2) of two stage GM coolers. The performance of this seal, low leakage and low friction, is critical to achieving low temperatures. A small amount of leakage past this seal can severely limit cooling power.

**Table 1** Loss Mechanisms in a GM cooler

| Loss Mechanism | Scale Factor (D = diameter) |
|---|---|
| conduction | D and $D^2$ |
| radiation | D |
| pressure drop | $D^2$ |
| shuttle | D |
| gap flow | D |
| expansion space dead volume | D |
| regenerator dead volume | $D^2$ |
| regenerator heat exchange | $D^2$ |
| cold end heat exchange | $D^2$* |

* Losses may scale at a higher rate than indicated

## APPARATUS

A larger version of the Balzer Kelcool 4.2™ refrigerator was built based on the scaling method presented above. The apparatus is comprised of the cold head module and a compressor module. The compressor module is a Balzer Model UCC 110S which is water cooled and consumes approximately 6.5 kW of electrical power. This module is unaltered from its factory configuration with the exception of a modification to the cold head motor controller, located in the compressor module, to allow cycle speed changes in the cold head.

The cold head module, originally a Balzer model UCH-130 cold head, has been extensively modified. The dimensions of the parts of the cold head that have been modified are presented in Table 2. The first stage regenerator has an increased diameter and uses not only phosphor bronze screens but also lead (Pb) spheres for a matrix material. The second stage regenerator uses Pb spheres and neodymium (Nd) spheres for a matrix material. The second stage displacer/regenerator tube is made of stainless steel with a plastic sleeve covering it to prevent

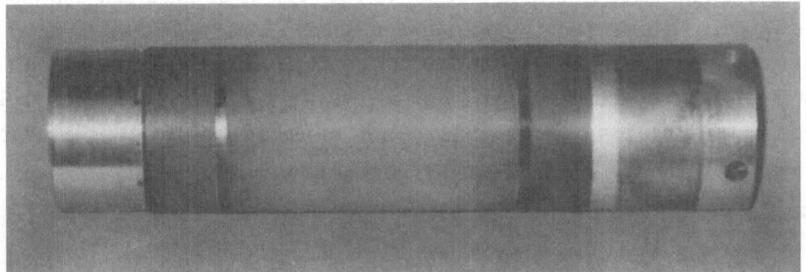

**Figure 3** Second Stage Displacer/Regenerator Tube.

galling and friction between it and its mating cylinder. A photograph of this component is shown in Figure 3.

In order to accommodate the larger second stage displacer, the second stage cylinder had to be enlarged. A photograph of the complete cold head module is shown in Figure 4. This photograph shows the enlarged second stage cylinder tube as well as the two regenerators

The cold head was outfitted with instrumentation to provide the ability to measure the refrigerator's performance. Two silicon diode temperature sensors were used. One was attached to each refrigeration stage along with a 25 ohm heater. This instrumentation, allowed the monitoring of temperatures while heat was being applied to each stage.

**Table 2.** Dimensions for modified GM cooler as compared to unmodified cooler.

|  | Modified | Unmodified |
|---|---|---|
| Stroke length, in (mm) | 0.625 (15.9) | 0.440 (11.2) |
| First stage displacer diameter, in (mm) | 3.00 (76.2) | 3.00 (76.2) |
| First stage regenerator diameter, in (mm) | 2.375 (60.3) | 1.75 (44.5) |
| First stage regenerator length, in (mm) | 4.375 (111) | 4.375 (111) |
| First stage mass of Pb (g) | 250 | 0 |
| Second stage displacer diameter, in (mm) | 1.75 (44.5) | 1.188 (30.2) |
| Second stage regenerator diameter, in (mm) | 1.633 (41.5) | 1.0 (25.4) |
| Second stage regenerator length, in (mm) | 4.75 (120.7) | 4.75 (120.7) |
| Second stage mass of Pb (g) | 500 | 200 |
| Second stage mass of Nd (g) | 390 | 150 |

## RESULTS

After making all the modifications indicated in Table 2, the performance of the cooler was quite poor. Initially the cooler would not even reach temperatures below 4.5K. However after some "fine tuning", the cooler performed as indicated by the load map presented in Figure 5. A comparison of this load map to the load map of the unmodified cooler (Figure 1) shows a significant increase in second stage cooling power. The cooler now produces 1.5 W of cooling at temperatures below 4.5 K. This performance increase is what was hoped for in scaling up the second stage. However we did find an unexpected result. The first stage is no longer capable of producing 55W at 50K. The modified cooler can only be loaded with about 25 W on the first stage before the second stage temperature begins to rise. This is an important and interesting effect of scaling.

The reduction of cooling power of the first stage can be explained by theorizing that the cooling power produced by the first stage is used in two ways. The first use is for cooling an external load, we represent this power as $P_a$. The second use of cooling power is for the pre-cooling of the gas which is used by the second stage, this power is $P_{2nd}$. Thus an expression for the total cooling power produced by the first stage, $P_{total}$, can be written as:

$$P_{total} = P_a + P_{2nd}.$$

(8)

$P_{2nd}$ is dependent on the amount of gas the second stage handles. Thus, the larger or more powerful the second stage, the more power would be required for pre-cooling from the first stage. The data from the modified cooler shows that no more than about 25 W of power is available from the first stage, while the data from the unmodified cooler shows that more than 50 W of power is available from the first stage. This suggests that for each 1.0 W of second stage cooling

**Figure 4** Modified cold head module.

power more than 25 W of first stage pre-cooling is required. More extensive testing of both the modified and unmodified coolers is required before it is possible to estimate the necessary incise in the size of the first stage to return its performance to 50 W at 50 K.

**Repeatability and Seal Performance**

Initially, after the cooler was modified, it failed to even reach the desired second stage operating temperature of 4.2K. There are many factors that restrict the cooler's performance, many of which are not well understood, but the proper functioning of certain components is

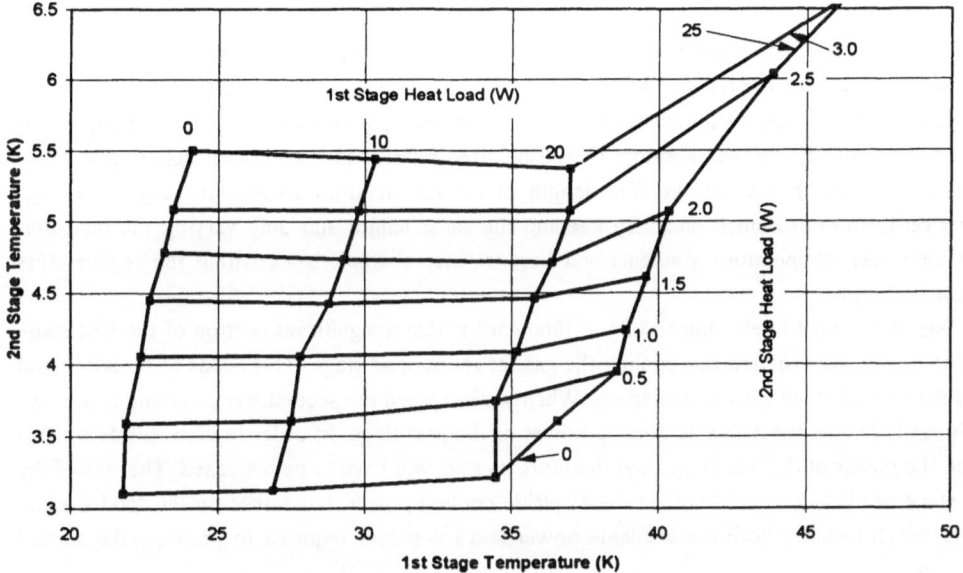

**Figure 5** Load map for modified cooler.

critical. First and foremost is the second stage seal. This seal is located at the top of the 2nd stage displacer and it forces the gas to flow through the second stage regenerator, not around it. As the second stage diameter is scaled up or down, the seal diameter also must be changed. Its ability to seal gas and not cause too much friction may be altered by the change in size, but it is difficult to predict in what way. This study has shown that the larger seal is more difficult to operate than the smaller one but the difficulty is not insurmountable. Through trial and error we were able to find materials and designs that function adequately.

Another component, critical to the performance and the repeatability of low temperature coolers, is the second stage regenerator. The construction details of this component are particularly important. The sphere beds must be packed tightly and the device used to separate the Nd from the Pb spheres must prevent mixing of the two types of spheres. If mixing of the spheres or channeling of gas through loosely packed spheres occurs, the regenerator will function poorly.

## Other Changes

Many small adjustments were made to the cooler to achieve the best performance. Among these changes were valve timing changes, cycle speed changes and stroke length changes. While any of these changes have the potential to seriously alter the performance of the cooler they did not. These parameters had been optimized for the standard cooler and only minor changes were made for the modified cooler.

## CONCLUSIONS

This work demonstrates that the cooling performance for a low temperature GM cooler can be increased by increasing the size of the cooler. It is possible to estimate the required increase in size by keeping the size of the swept volume of each stage proportional to the cooling power needed. Additionally, the regenerators should be sized relative to the mass of helium that flows through them during each stroke. The length of the regenerators should not vary from large power refrigerators to small ones. By keeping the same length and only varying the diameter, mass velocities, temperature gradients and heat transfer characteristics within the regenerators remain unchanged.

One of the most interesting results of this work is that a significant portion of the first stage cooling power goes toward pre-cooling the gas for the second stage. This needs to be accounted for if the second stage is to be scaled up. When we increased the second stage cooling power we significantly limited the available cooling power of the first stage. In order to compensate for this effect, the power of the first stage, and therefore its size, will have to be increased. The size of the first stage needs to be based not on the available cooling power, but rather on the total cooling power which includes both the available power and the power required to pre-cool the second stage gas.

## REFERENCES

1 M. Heiberger, A. Langhorn, T. Hurn, D. Morris, J. Walters, L. Apprigliano, D. Waltman, and G. Green, "A Lightweight Rugged Conduction-cooled NbTi Superconducting Magnet for U.S. Navy Mine Sweeping Applications", *Advances in Cryogenic Engineering*, Vol. 41, Plenunm Press, New York, (1966).

2 J. Chafe, G. Green, and R. Riedy, "Neodymium Regenerator Test Results in a Standard Gifford-McMahon Refrigerator", *Cryocoolers 7*, Report # PL-CP-93-1001, Phillips Laboratory, Kirtland Air Force Base, New Mexico, (1993), pg. 1157.

3 K. Wark, *Thermodynamics*, McGraw-Hill Book Co., New York, 1977, pp. 206-210.

# The Effect of Pressure Loss at Intake/Exhaust Valve on Performance of a 4 K G-M Cryocooler

Toshimi Satoh, Rui Li, Atsushi Onishi, Yoshiaki Kanazawa
and Yoichiro Ikeya*

Research & Development Center, Sumitomo Heavy Industries, Ltd.
63-30, Yuhigaoka, Hiratsuka, Kanagawa, 254, JAPAN

* Precision Products Division, Sumitomo Heavy Industries, Ltd.
2-1-1, Yato-cyo, Tanashi, Tokyo, 188, JAPAN

## ABSTRACT

The influence of pressure loss at intake/exhaust valve on performance of a 4K Gifford-McMahon(GM) cryocooler has been investigated. Two types of valves with different flow conductance were tested. The performance of 4K GM cryocooler at 4.2K was improved by using the new valve which generates lower pressure loss than the usual one. The maximum cooling capacity of 2.02W at 4.2K was obtained, where a hybrid second regenerator with lead and $ErNi_{0.9}Co_{0.1}$, a cylinder with first inner diameter 82mm and second inner diameter 40mm and two compressors were used. Power consumption was 11.3kW.

## INTRODUCTION

The performance of a 4K GM cryocooler has been much improved by development of a hybrid type second regenerator with magnetic regenerator material such as $ErNi_{0.9}Co_{0.1}$[1] or $Er_{0.9}Yb_{0.1}Ni$[2]. Now a 4K GM cryocooler has a cooling capacity more than 1W at 4.2K and is applied to many uses, such as SIS mixer cooling for radio astronomy[3], helium recondensation for MRI magnet[4], cryocooler cooled superconducting magnet[5] and so on. The application area is expected to expand by improving the cooling capacity.

The optimization of cryocooler component other than regenerator material is very important factor to improve the performance of a 4K GM cryocooler, and intake/exhaust valve is one of such factors. But it was optimized for a conventional GM cryocooler so far, not for a 4K GM cryocooler. Then the authors investigated the influence of intake/exhaust valve timing on the performance of a 4K GM cryocooler and found that the valve timing effects the cooling capacity not only of second stage but also of first stage[6]. The cooling capacity 1.5W at 4.2K was achieved by applying the optimum valve timing[7].

Not only intake/exhaust valve timing but also valve conductance is very important to improve the cooling performance. Pressure loss generated at intake/exhaust valve causes reduction of

Pressure-Volume(PV) work in the expasion volume and influences the cooling capacity. The authors have investigated the effect of pressure loss on the cooling capacity at 4.2K to improve the performance of a 4KGM cryocooler.

## EXPERIMENTAL  APPARATUS

Fig. 1 shows a schematic diagram of the experimental apparatus. A two stage type cold head is used in this experiment. The first displacer contains copper screen meshes in the high temperature side and lead spheres in the low temperature side as the regenerator. The second regenerator is a hybrid type and is composed of lead spheres and $ErNi_{0.9}Co_{0.1}$ spheres. Lead (500g) is stuffed in the higher temperature region and $ErNi_{0.9}Co_{0.1}$ (500g) in the lower temperature region of the second displacer[7]. The first and the second cylinder diameter were 82mm and 40mm, respectively.

A set of rotary intake/exhaust valve composed of crank, plate and body is installed in the cold head. Valve timing is represented by an angular measure, like -42deg./-46deg. for open timing and 124deg./173deg. for interval. The minus sign of open timing means that the intake/exhaust valve opens before the expansion volume becomes minimum/maximum, respectively. The valve interval shown above means that the intake valve interval is 124deg. and the exhaust one is 173deg. Two types of rotary valves having different size of flow path were prepared for this experiment. The new valve has about 35% larger cross sectional area of flow path than the usual one. The optimized intake/exhaust valve timing shown in Fig. 2 was employed for both valves.

Two types of compressor systems were prepared to examine the compressor capacity dependence of cooling capacity. One is single compressor CSW-71(Type S) of Sumitomo Heavy Industries, Ltd., which is shown in Fig. 1. The other is the two compressor system CSW-71+CSW-51 (Type D). Displacer stroke was 30mm.

Germanium resistance sensor was used to measure the second stage temperature, and plutinum-cobalt alloy resistance sensor to measure the first stage temperature. Electric heater was installed to measure cooling capacity on each stage. Three pressure sensors were installed to measure the intake pressure, the exhaust pressure and the pressure of the room temperature space of the first cylinder. Displacement of displacers was also measured to obtain the PV work.

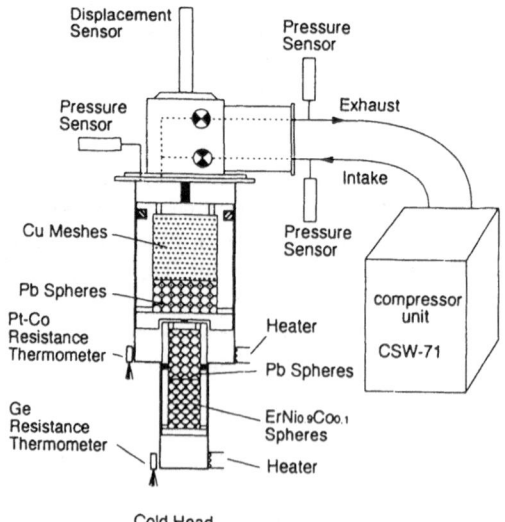

**Figure 1.**   Schematic diagram of experimental apparatus.

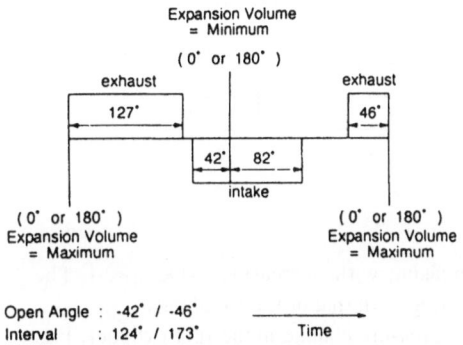

**Figure 2.** Definition of intake/exhaust valve open timing and interval by angular measure.

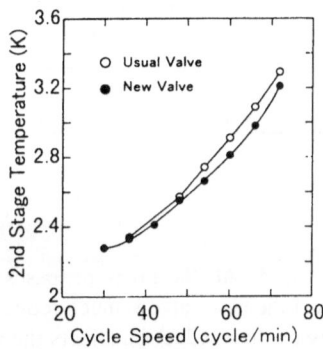

**Figure 3.** No load second stage temperature as a function of cycle speed.

## EXPERIMENTAL RESULTS AND DISCUSSIONS

### Improvement of Cooling Performance

The relationship between no load second stage temperature and cycle speed is shown in Fig. 3. This experiment was performed with Type S compressor. Input power was about 7kW. The second stage temperature obtained with the new valve is lower than that with the usual valve. The temperature difference is clearly the effect of flow path size of the valve. Fig. 4 shows an example of the measured data of the intake pressure, the exhaust pressure and the pressure of the room temperature space in the first cylinder. The minimum difference between the intake/exhaust pressure and the first cylinder pressure is considered as the pressure loss at the intake/exhaust valve, respectively. Cycle speed dependence of pressure loss with no heat load

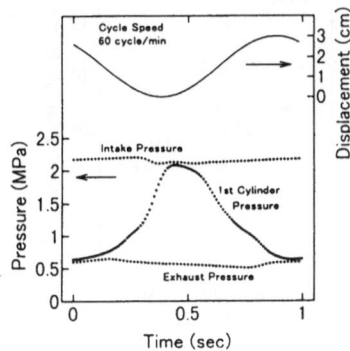

**Figure 4.** Intake/exhaust and 1st cylinder pressure variation and displacer movement at 60 cycle/min.

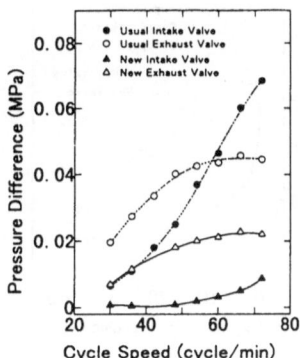

**Figure 5.** Pressure loss at usual and new intake exhaust valve as a function of cycle speed.

**Table 1.**   Comparison of cooling capacity with two types of valves with Type S compressor.

| 1st stage temperature (K) | | 23K | 30K | 40K |
|---|---|---|---|---|
| cooling capacity | new valve | 1.29 | 1.40 | 1.53 |
| at 4.2K (W) | usual valve | 1.25 | 1.36 | 1.49 |

is shown in Fig. 5. All the pressure losses are decreasing with decreasing cycle speed. The pressure loss at the new valve is much reduced comparing with that at the usual valve.

This pressure loss reduction affects the maximum pressure change in the first cylinder. Fig. 6 shows an example of PV diagram in the second stage with no heat load. As a valve timing is same for both valves, the shape of PV diagram is similar each other. Therefore the area of the PV diagram is related with the pressure change    at the expansion volume. The no load second stage temperature drop is understood by the PV work increase, which is caused by the increase of pressure change in the expansion space.

Table 1 shows the relation between cooling capacity at 4.2K and the first stage temperature. Cycle speed was 60 cycle/min. Cooling capacity at 4.2K was improved by using the new valve for all the first stage temperatures shown. The cooling capacity 1.53W at 4.2K was obtained at the first stage 40K with the new valve.

Cycle speed dependence of the first stage cooling capacity at 40K and the second stage cooling capacity at 4.2K is given in Fig. 7. The first and second stage cooling capacity vary with cycle speed. But the cycle speed dependence of the first cooling capacity and the second one of the same valve are different each other. Each valve has the optimum cycle speed for the second cooling capacity and the new valve has the higher optimum cycle speed.The optimum cycle speed for the first stage cooling capacity is not same with that for the second stage, and is shifted to higher value. When the valve open timing changes to earlier, according to Li et al., the optimum cycle speed shifts from lower value to higher[6]. This means that the reduction of pressure loss at the valve gives the same effect as making the valve open timing earlier.

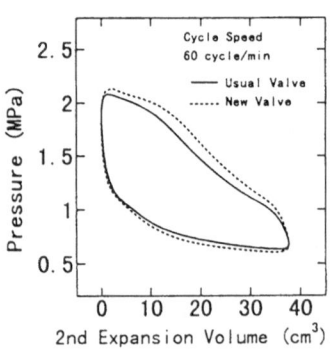

**Figure 6.** PV diagrams of 2nd expansion space for the usual and new valve with no heat load.

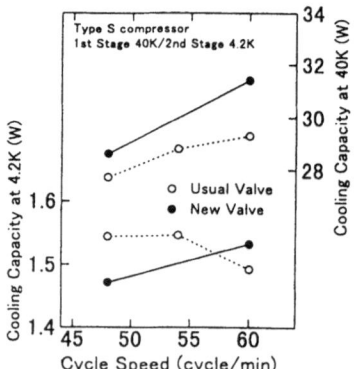

**Figure 7.**   First and second cooling capacities with two valves as a function of cycle speed.

**Table 2.**  Comparison of cooling capacity by two types of valves with Type D compressor.

| 1st stage temperature (K) | | 23K | 30K | 40K |
|---|---|---|---|---|
| cooling capacity | new valve | 1.51 | 1.63 | 1.74 |
| at 4.2K (W) | usual valve | 1.49 | 1.60 | 1.72 |

## Effect of Compressor Capacity on 4.2K Cooling Capacity

Valve timing shown in Fig. 2 is the optimized valve timing for the system with Type S compressor. The influence of valve timing on cooling performance of the 4K GM cryocooler with type D compressor was also investigated. The usual valve was used for this experiment. Fig. 8 shows the effect of valve open timing on the cooling capacity of the first and second stage.   Input power was 10kW. Valve interval was 124deg./173deg. Definition of valve open timing and interval were same as in Fig. 2.

The optimum open timing for the first stage seems same as that for the system with Type S compressor[7], though not so apparent in Fig. 8. On the other hand, the optimum open timing for the second stage cooling capacity was nearer to the point at which the expansion volume becomes maximum or minimum. The valve open timing -42deg./-46deg. was better for both the first and the second cooling capacity from Fig. 8. Then valve interval was examined for the timing -42deg./-46deg., and 124deg./173deg. was found optimum. From the above mentioned, the same open timing and interval shown in Fig. 2 was chosen for the experiment below.

As was reported already, the cooling capacity at 4.2K was improved up to 1.74W by using two compressor system and the same cylinder with the used one in this work[8]. The pressure loss at the intake/exhaust valve is expected to increase when the compressor capacity is increased, as the helium flow rate becomes larger. Since the pressure loss causes PV work reduction, the effect of valve conductance on the cooling capacity is more important. Cooling capacity at 4.2K with the new valve was measured for comparison with the cooling capacity by the usual valve. The measured data are shown in Table 2. Cooling capacity was improved.

To improve the performance much more, another valve of interval 133deg./179deg. and large flow path was tested. Cooling capacity 1.94W, which was larger than that with the valve

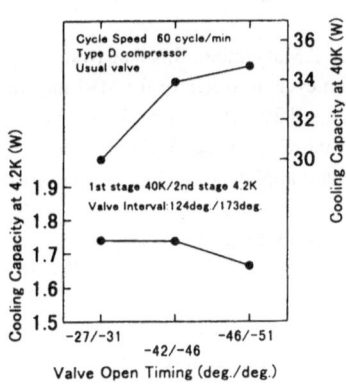

**Figure 8.** Cooling capacities variation with valve open timing for Type D compressor.

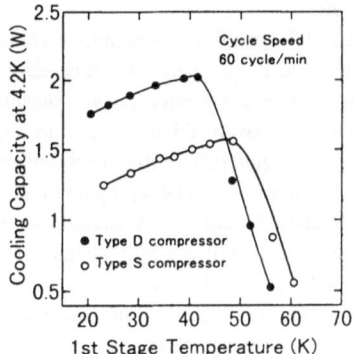

**Figure 9.** Second cooling capacities as a function of 1st temperature for two types of compressors.

of interval 124deg./173deg., was obtained at the first stage temperature 40K. This means the optimum interval varied. The obtained value 1.94W is about 10% larger than that with the usual one. The pressure losses at the intake/exhaust valve were reduced from 0.084MPa/0.072MPa for the usual valve to 0.044MPa/0.014MPa for the new valve of interval 133deg./179deg. at 60 cycle/min. Pressure loss reduction amount was larger than Type S compressor case, which is the reason why the cooling capacity improvement was larger for Type D compressor.

Finally input power was increased to obtain higher cooling capacity. Fig. 9 shows the relationship between the second stage cooling capacity and the first stage temperature. Cycle speed was 60 cycle/min. in this experiment. Solid circles were obtained with Type D compressor and with the new valve of interval 133deg./179deg. Open circles are the data with Type S compressor and the usual valve, which is shown for comparison. Input power for Type D compressor was 11.3 kW. Maximum cooling capacity 2.02W at 4.2K was obtained at the first stage temperature 41.6K.

## CONCLUSIONS

The influence of pressure loss at intake/exhaust valve on the performance of a 4KGM cryocooler has been investigated. Two types of valves were prepared for comparison. Not only second cooling capacity but also first stage cooling capacity were increased by using the new valve which has larger flow path than the usual one. The optimum cycle speed shifted to higher value comparing with that for the usual valve.

The maximum cooling capacity 2.02W at 4.2K was obtained. Two compressors and the new intake/exhaust valve were used in the experiment. Input power was 11.3kW.

## REFEFENCES

1. G.Ke, H.Makuuchi, T.Hashimoto, A.Onishi, R.Li, T.Satoh and Y.Kanazawa. "Improvement of two-stage GM refrigerator performance using a hybrid regenerator ". *Advances in Cryogenic Engineering*, vol. 40, Plenum Press, New York, (1994). pp.639-647.
2. T.Kuriyama, M.Takahashi, T.Nakagome, T.Hashimoto, H.Nitta and M.Yabuki. "Development of 1watt class 4 K GM refrigerator with magnetic regenerator materials". *Advances in Cryogenic Engineering*, Plenum Press, vol.39, (1994), pp.1335-1342.
3. M.Takahashi, H,Hatakeyama, T.Kuriyama, H.Nakagome, R.Kawabe, H.Iwashita. G.McCulloch. K.Shibata and S.Ukita, "A compact 150GHz SIS receiver cooled by a 4 K GM refrigerator". Cryocooler7, Air Force Phillips Laboratory Report PL-CP-93-1001. Kirtland AFB. NM. (1993). pp.495-507.
4. M.Nagao, T.Inaguchi, H.Yoshimura, S.Nakamura, T.Yamada, T.Matsumoto, S.Nakagawa. K.Moritsu and T.Watanabe, "4K three-stage Gifford-McMahon cycle refrigerator for MRI magnet". *Advances in Cryogenic Engineering*, Plenum Press, New York, vol.39, (1994). pp.1327-1334
5. K.Watanabe, S.Awaji, T.Fukase, Y.Yamada, J.Sakuraba, F.Hata, C.K.Chong. T.Hasebe and M.Ishihara, "Liquid helium-free superconducting magnet and their applications". *Cryogenics*. vol.34 15th ICEC supplement, (1994), pp.639-642.
6. R.Li, A.Onishi, T.Satoh and Y.Kanazawa, "Influence of valve open timing and interval on performance of 4K Gifford-McMahon cryocooler", *Advances in Cryogenic Engineering*. Plenum Press, vol.41, (1996), in press.
7. T.Satoh, A.Onishi, R.Li, H.Asami and Y.Kanazawa. "Development of 1.5W 4K G-M cryocooler with magnetic regenerator material", *Advances in Cryogenic Engineering*. Plenum Press. New York. vol.41, (1996), in press.
8. A.Onishi, R.Li, H.Asami, T.Satoh and Y.Kanazawa, "Improvement of 1.5W-class Gifford-McMahon cryocooler", Proceedings of 16th International Cryogenic Engineering Conference. in press.

# The Valve Timing Effect on Cooling Power of 4 K Gifford-McMahon Cryocooler

Masashi Nagao, Takashi Inaguchi, Naka Kouki and Hideto Yoshimura

Mitsubishi Electric Corp.
Advanced Technology R&D Center
Amagasaki, Hyogo, Japan

## ABSTRACT

The valve timing effect on cooling power of a Gifford-McMahon cryocooler was studied theoretically and experimentally. In this paper we describe first the theoretical refrigeration efficiency assuming an ideal gas and argue that the Carnot efficiency can be achieved if the valve timing is optimized.

Then we describe an experimental cryocooler fabricated to demonstrate an effect that can control the valve timing freely and report the results of the experiment using this equipment. The experiment demonstrated that the refrigeration efficiency can be enhanced by about 35%.

## INTRODUCTION

We have already succeeded in developing a 3-stage G-M cryocooler that can reduce the evaporation of helium from a superconducting MRI magnet to zero[1]. These cryocoolers must have a higher efficiency to be used more widely in the future. A higher efficiency will reduce both the initial and the running costs making it possible to make our cryocooler more competitive with a conventional shield type cryocooler.

For this purpose in this research we attempted to improve the refrigeration cycle of the G-M cryocooler. The efficiency of G-M cryocoolers is not very high, because an irreversible mix of high pressure and low pressure gases occurs[2, 3]. We got a hint for our development from the fact that the pressure can vary if the working fluid is moved by a displacer, through a regenerator, from the room temperature area to the low temperature one. This pressure variation, which is reversible, can enhance the refrigeration efficiency. Ideally this adjustment could realize a refrigeration cycle resembling the Ericsson cycle (referred to below as a "pseudo-Ericsson cycle") with a theoretical refrigeration efficiency equivalent to the Carnot efficiency. Although Li and other researchers have made some investigations of the optimization of valve timing, these studies all lay emphasis on experiments[4]. These has not yet been a report on the variation of the basic efficiency.

In this paper we describe first the theoretical refrigeration efficiency assuming an ideal gas and argue that the Carnot efficiency can be achieved if the valve timing is optimized and that this

Cryocoolers 9, Edited by R.G. Ross, Jr.
Plenum Press, New York, 1997

operation methodology is suited to a cryocooler whose operating temperature is of the order of 4 K.

Then we will describe an experimental cryocooler fabricated to demonstrate an effect that can control the valve timing freely and report the results of the experiment using this equipment. The experiment demonstrated that the refrigeration efficiency can be increased by about 35%.

## PSEUDO-ERICSSON CYCLE

Figure 1 shows the P-V diagram of a pseudo-Ericsson cycle with a reversible process. For comparison, the P-V diagram of a conventional G-M cycle is depicted by the dotted lines.
The pseudo-Ericsson cycle consists of processes 1-2-4-5-1.

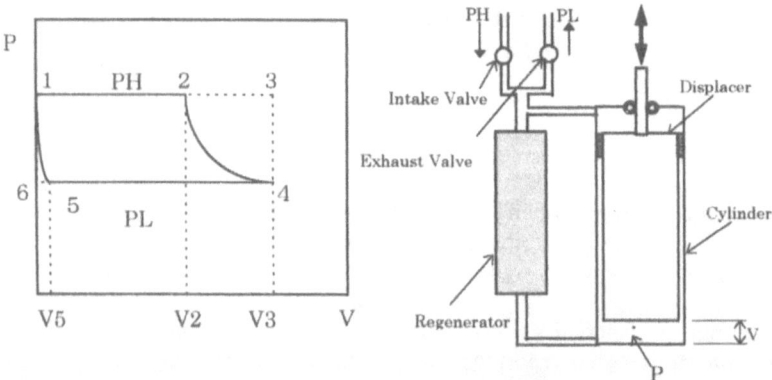

Fig. 1 The P-V diagram of a pseudo-Ericsson cycle

## PROCESS 1-2

Process 1-2 is a suction process where a Ph working fluid is introduced into the expansion space after cooling down by the regenerator with the exhaust valve closed and intake valve open.

## PROCESS 2-4

Process 2-4 is an expansion process. In status 2, the volume of the expansion space is V2, where the working fluid, cooled down by the regenerator, attains the operating temperature Tl of the cryocooler. The volume of the room temperature space is V3 minus V2 and the temperature of the working fluid is the environment temperature Ta. In process 2-4, the displacer is moved and the working fluid in the room temperature space is cooled down to the operating temperature of the cryocooler into the expansion space through the regenerator. Since the volume of the cylinder is constant, the pressure decreases by the displacement of the gas in the room temperature area. V2 is so adjusted that the pressure in the cylinder matches Pl when the displacer comes to the top dead point.

## PROCESS 4-5

Process 4-5 is a exhaust process where a Pl working fluid is exhausted after heating by the regenerator with the exhaust valve open and intake valve closed.

## PROCESS 5-1

The volume of the expansion space is V5, and the temperature of the working fluid is Tl. The volume of the room temperature space is V3 minus V5, while the temperature of the working fluid is the environment temperature Ta.    The displacer is moved in process 5-1, and the working fluid in the expansion space is heated, through the regenerator, up to the operating temperature of the cryocooler ready to be introduced into the room temperature space.    Since the volume in the cylinder is constant, the pressure rises as the gas of the expansion space is displaced.    V5 is adjusted so that the pressure in the cylinder coincides with the Ph when the displacer comes to the bottom dead point.    This allows the expansion process to be reversible when the gas is introduced into the room temperature space.

The processes above enable a reversible pseudo-Ericsson cycle to be realised.

## CALCULATION OF THE EFFICIENCY OF PSEUDO-ERICSSON CYCLE

We will now look at the efficiency of this refrigeration cycle.    Each process is presumed to be done isothermally.    The compression of the working gas is assumed to be isothermal too.
Calculation of Generated Refrigeration
In process 2-4, both the intake and exhaust valves are closed.    Therefore, the following equation holds from the conservation of mass in the cylinder:

$$V = \frac{P_L V_3 T_A}{P(T_A - T_L)} - \frac{V_3 T_L}{(T_A - T_L)}$$

area 1  2  4  6

$$= \frac{P_L V_3 T_A}{T_A - T_L}(\ln P_H - \ln P_L) - \frac{V_3 T_L}{T_A - T_L}(P_H - P_L)$$

$$V_2 = \frac{V_3(P_L T_A - P_H T_L)}{P_H T_A - P_H T_L}$$

Also, in process 5-1 both the intake and exhaust valves are closed as in the case of process 2-4. Similarly, the following equation holds from the conservation of mass in the cylinder:

$$V = \frac{P_H V_3 T_L}{P(T_A - T_L)} - \frac{V_3 T_I}{(T_A - T_L)}$$

area 1 5 6

$$= \frac{P_H V_3 T_L}{T_A - T_L}(\ln P_H - \ln P_L) - \frac{V_3 T_L}{T_A - T_L}(P_H - P_L)$$

$$V_5 = \frac{V_3(P_H T_L - P_L T_L)}{P_L T_A - P_L T_L}$$

The generated refrigeration Q can be obtained by solving the above equation.

$$Q = area\ 1\ 2\ 4\ 5 = area\ 1\ 2\ 4\ 6 - area\ 1\ 5\ 6$$

$$= \frac{V_3(P_L T_A - P_H T_L)}{T_A - T_L}(\ln P_H - \ln P_L)$$

In consideration of the isothermal compression, the power W required for the compressor can be obtained from the following equation:

$$W = \left( \frac{P_L V_3 T_A}{T_L} - P_H V_3 \right) \ln \frac{P_H}{P_L}$$

The COP of this cycle is obtained by Q/W.

$$\frac{Q}{W} = \frac{\dfrac{V_3 \left( P_L T_A - P_H T_L \right)}{T_A - T_L} \ln \dfrac{P_H}{P_L}}{\dfrac{V_3}{T_L} \left( P_L T_A - P_H T_L \right) \ln \dfrac{P_H}{P_L}}$$

$$= \frac{T_L}{T_A - T_L}$$

This agrees with the Carnot efficiency.

## COMPARISON OF A G-M CRYOCOOLER WITH PSEUDO-ERICSSON CYCLE

Gifford et al. reported the efficiency of a conventional G-M cryocooler[2]. Figure 2 shows the COP of the G-M cryocooler and that of the pseudo-Ericsson cycle as a function of the

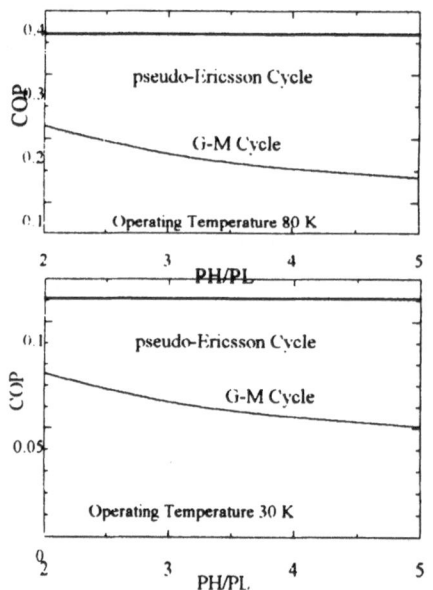

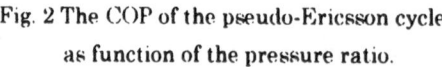

Fig. 2 The COP of the pseudo-Ericsson cycle
as function of the pressure ratio.

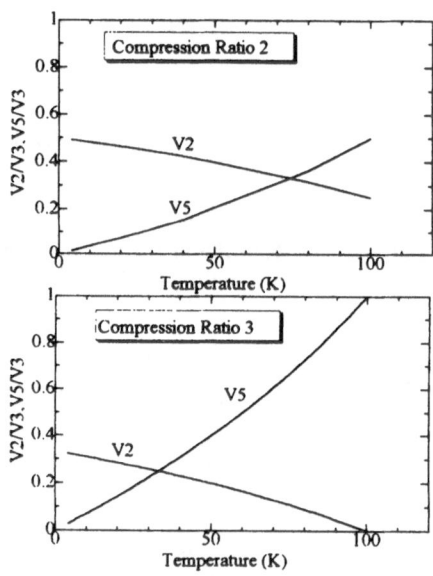

Fig. 3    V2 and V5 as functions of
the operating temperature.

compression ratio. The figure illustrates two cases where the operating temperatures are 80 K and 30 K. As is clear from this figure, the greater the compression ratio of the G-M cryocooler, the smaller the COP is. By contrast, the efficiency is constant irrespective of the compression ratio in the new cycle, which is higher than in the G-M cycle.

The most frequent operating pressure in a commercial G-M cryocooler is 21 bars on the high pressure side and 7 bars on the low pressure side, and the compression ratio is 3 under these conditions.    With a compression ratio of 3, the COP at an 80 K operating temperature is 0.177 in the G-M cycle, while the same is 0.364 in the new cycle.    It is thus understandable that the efficiency of the latter is twice that of the former.    The COP at an operating temperature of 30 K is 0.0627 in the G-M cycle, while the same is 0.111 (1.8 times) in the new cycle.

The volumes of V2 and V5 depend on the operating temperature and the pressure ratio. Figure 3 illustrates the relationship of V2 and V3 to the operating temperature. The environment temperature Ta was set at 300 K.    Though the machine is operable even at 100 K if the compression ratio is 2, V2 is almost zero near 100 K if it is 3. Therefore, the valve timing ought to be adjusted in terms of the operating temperature in this cycle. In the conventional G-M cryocooler where the helium is not liquefied, V2 is small and V5 is too large due to there being too high an operating temperature.    This makes the area under the P-V curve smaller and thus limits the generated refrigeration.    The probability is therefore high that the overall efficiency will be low.    By contrast, the operating temperature is low enough in, for instance, a 4 K G-M cryocooler, which is more suited to the cycle we now propose.

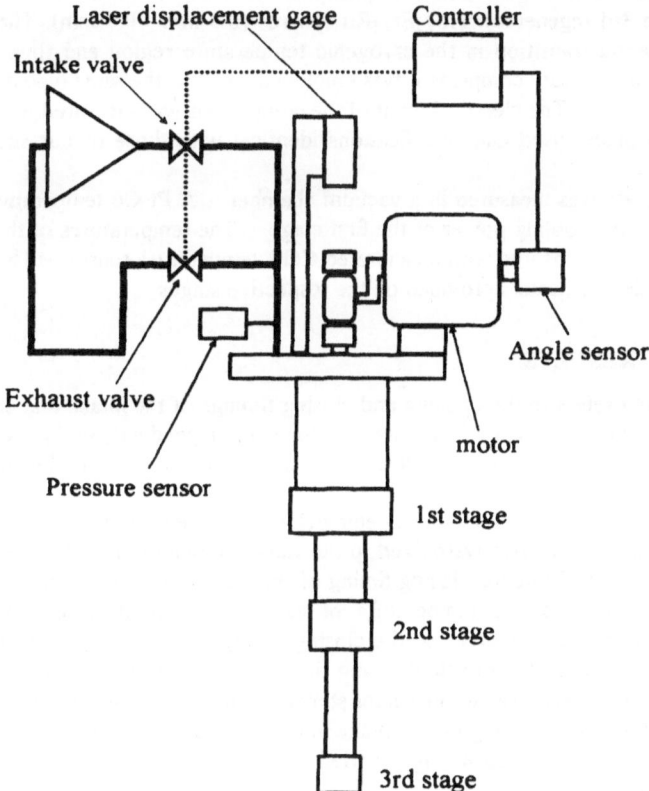

Fig. 4 Experimental Equipment

## EXPERIMENTAL EQUIPMENT

We fabricated experimental equipment that can freely control the valve timing in order to verify the effectiveness of our proposition in an actual cryocooler.    Figure 4 shows the experimental equipment schematically.    Two pilot type solenoid valves were used as intake valves. The orifice diameter of the solenoid valves was 6 mm.    Three identical solenoid valves were used as the exhaust valves.    The reason why we used plural solenoid valves is to enlarge the area of the channels.

The scotch yoke and motor were left to stand in the atmosphere.    Oil lubricated 0-rings were used to seal between the working fluid and the atmospheric space.    A laser displacement gage was provided on the upper portion of the scotch yoke, from the displacement of which the displacement of the displacer was given.    A stepping motor was used as the drive motor.    An angle sensor was mounted on the shaft of this motor and the sensor signal was forwarded to the controller so that the valve timing could be freely controlled.    A strain gage type pressure sensor was installed in the room temperature area to measure the pressure in the cylinder.

The cylinder was made of stainless steel and had three refrigeration stages brazed onto the outside. The displacer assembly, which contains three kinds of regenerator reciprocates in the cylinder. The displacer assembly consists of three displacers.    The 1st and 2nd displacers were made of phenolic resin, and the 3rd displacer was made of stainless steel tube.    The displacers were mutually connected using pins. Each displacer incorporates a regenerator.    As for regenerator materials, the 1st regenerator used phosphor bronze screen disks, the 2nd regenerator lead spheres and the 3rd regenerator $Ho_{1.5}Er_{1.5}Ru$ spheres (0.3 mm - 0.5 mm). The $Ho_{1.5}Er_{1.5}Ru$ has a magnetic ordering transition in the cryogenic temperature region and thus has a greater specific heat than lead. A rotary compressor was employed which is the same type of conventional 2-stage shield cryocooler.    The electrical input of the compressor was measured with a wattmeter. The cylinder and displacer used had specifications identical with those of our superconducting MRI magnet.

The cooling power was measured in a vacuum chamber.    A Pt-Co temperature sensor was employed to measure the cooling power of the first stage.    The temperatures of the 2nd and 3rd refrigeration stages were measured with a calibrated CGR temperature sensor.    The thermal load was given from the electric heaters provided on the respective stages.

## EXPERIMENTAL RESULTS

Adjustable timing refers to the opening and closing timings of the intake and exhaust valves. The effects of the intake and exhaust valves were examined independently in this experiment. The valve timing is represented with the lower dead point as zero degree.    A rated load was given to each stage.

Figure 5 shows the results. First the opening/closing timings of the intake valve and the opening timing of the exhaust valve were fixed to the values shown in Fig. 5 to change the timing to close the exhaust valve. While the closing timing of the intake valve was made to change from 230 degrees to 310 degrees, the temperature of each stage varied largely. The 3rd stage temperature was lowest at 290 degrees so the closing timing of exhaust valve was fixed at 290 degrees as an optimum. Then the opening/closing timings of the exhaust valve and the opening timing of the intake valve were fixed to the values shown in Fig. 6 to change the timing to close the intake valve. While the closing timing of the intake valve was made to change from 40 degrees to 120 degrees, the temperature of each stages varied largely. The optimum closing timing of the intake valve was 80 degrees.

These results are listed in Table 1 along with the design refrigeration capacities and the design heat loads of the superconducting MRI magnet. The experimental result show the cooling powers of each stage of this cryocooler performed at the design value. Electric power of 4.26 kW is required for the rated operation. This is the same electric power of the conventional 2-stage shield cryocooler The electric power of our conventional 4K 3-stage cryocooler is 6.5 kW So the experiment demonstrated that the refrigeration efficiency can be enhanced by about 35 %.

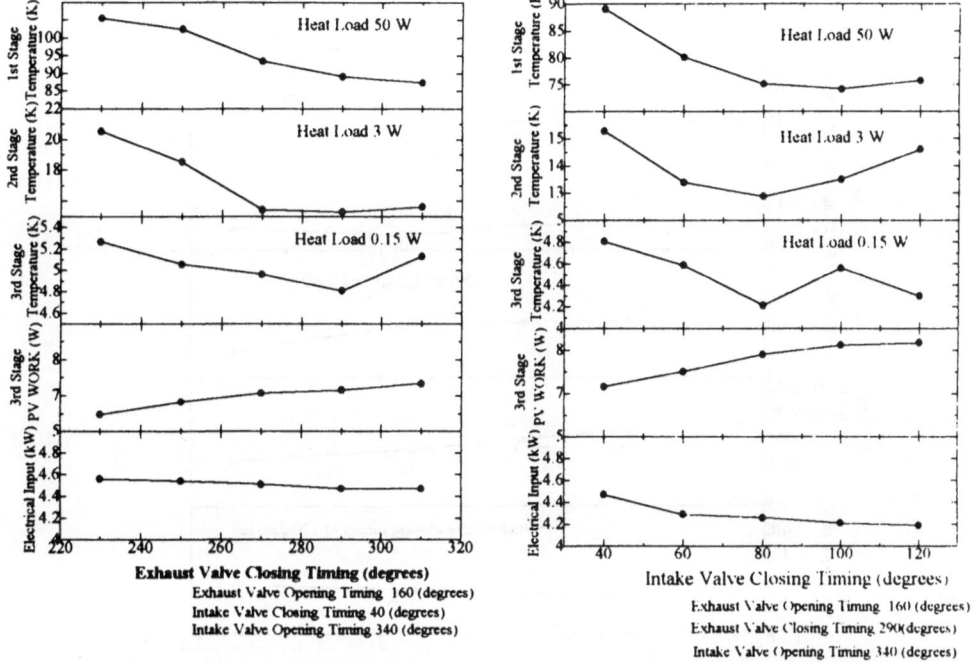

Fig. 5 Exhaust valve timing effects     Fig. 6 Intake valve timing effects

Table 1. Design refrigeration capacities and design heat loads of the superconducting MRI magnet

|  | Design | This work | Heat loads |
|---|---|---|---|
| 1st stage | 50 W at 75 K | 50 W at 75.1 K | 45 W at 70 K |
| 2nd stage | 3 W at 20 K | 3 W at 12.9 K | 2 W at 20 K |
| 3rd stage | 0.15 W at 4.2 K | 0.15 W at 4.2 K | 0.07 W at 4.2 K |

With these valve timings the cooling performance is good but efficiency(PV-Work/Electrical-Input) is not at a maximum. This tendency does not agree with our simple theory. This is because a real cryocooler cycle does not consist only of an isothermal process and the optimum condition is a mixture of a G-M cycle and a pseud-Ericsson cycle. Thus the efficiency decreases as the compression ratio increases.

Figure 7 show the compression ratio effect on the cooling powers. The compression ratio was controled by the operating speed. The efficiency increases as the compression ratio decreases, and the temperature of 1st stage decreases greatly. At the same compression ratio, the efficiency of intake valve closing timing at 80° is larger than that of the intake valve closing timing at 120

However the temperature of 3rd stage increases as the compression ratio decreases.
This is because the 3rd regenerator capacity is not sufficient in this condition.

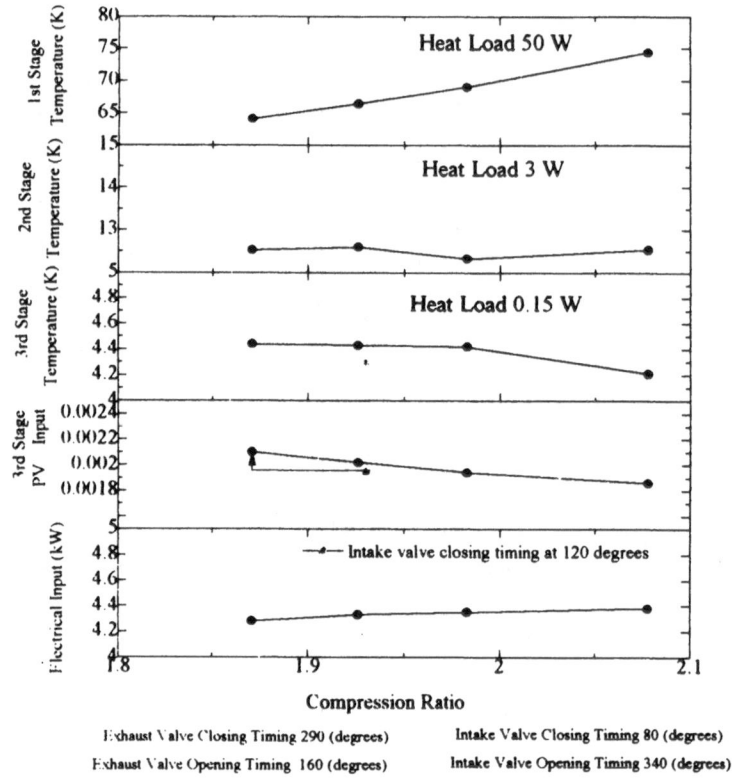

Fig. 7 Compression ratio effects

## CONCLUSION

Because of the fact that the pressure can be varied by displacing the working fluid using displacer from the room temperature area to the low temperature area through a regenerator with the valves left closed to improve the efficiency of G-M cryocooler, a pseudo-Ericsson cycle was proposed. In order to demonstrate the effectiveness of this cycle, the theoretical efficiency assuming an ideal isothermal variation was proved to be equivalent to the Carnot efficiency. Comparisons were made with regard to the G-M cycle and theoretical efficiency, which allowed us to make sure that the efficiency of the new cycle is higher by a factor of 2.

To examine this effect experimentally, we fabricated a trial cryocooler that can freely control the timing of the solenoid valves, and an experiment was performed with this machine, which demonstrated that the efficiency improved by about 35 %.

We will apply this result to our 4K 3-stage G-M cryocooler for superconducting MRI magnet in the future.

REFERENCES

1. M. Nagao, T. Inaguchi, H. Yoshimura, T. Matsumoto, S. Nakagawa and K. Moritsu, "4K Three-Stage Gifford-McMahon Cycle Refrigerator for MRI Magnet", Advances in Cryogenic Engineering, Vol. 39, Plenum Press, New York, 1994, pp.1327.

2. Gifford, W. E., and McMahon, H. O."A New Low Temperature Gas Expansion Cycle-Part II ." Advances in Cryogenic Engineering, Vol. 5, Plenum Press, New York,1960, pp.368

3. G. Walker, "Cryocoolers Part 2", Plenum Press, New York ,1983, pp.30-36.

4 Li, R., Satoh, T. and Kanazawa, Y., "Optimization of intake and exhaust valves for 4K-GM cryocooler", Cryogenic Engineering (Journal of the Cryogenic Society of Japan), 1996, 31 pp.172-181

# A Neodymium Plate Regenerator for Low-Temperature Gifford-McMahon Refrigerators

J. N. Chafe, G. F. Green and J.B. Hendricks*

Naval Surface Warfare Center, Carderock Division
Annapolis, MD, USA 21402
*Alabama Cryogenic Engineering, Inc.
Huntsville AL. USA 35804

## ABSTRACT

The present technology in low-temperature Gifford-McMahon cryogenic-refrigerators utilizes a regenerative type heat exchanger that is fabricated with rare-earth metal spheres having diameters of about 0.2 mm. A more effective regenerative heat exchanger could be constructed by precisely forming flow holes through neodymium (Nd) disks. Perforated disks were formed using a technique similar to that used to make superconducting wire. The disks were then stacked into a tube and this whole assembly was used as a regenerative heat exchanger. This type of regenerator should provide better heat transfer properties, lower pressure drop characteristics and reduced thermodynamic losses when compared to regenerators that utilize spherical particles. Pressure drop and performance test results of both types of regenerators are presented in this paper.

## INTRODUCTION

The work presented in this paper is sponsored by the Office of Naval Research (ONR). It is part of the Advanced Technology Demonstration program to develop a technologically advanced mine-sweeping system. This system is referred to as the Advanced Lightweight Influence Sweep System (ALISS).

ALISS is composed of two separate subsystems, one produces an acoustic energy pattern that closely resembles Naval landing craft, and the other produces a magnetic profile that mimics the profile of the same craft. The purpose of ALISS is to trigger mines typically used in shallow water for the defense of beaches and coastal areas. These mines are designed to sense Naval landing vessels by measuring their acoustic and/or magnetic characteristics. The ALISS is carried on board a Naval mine sweeping vessel and it emits a simulated Naval craft profile far enough away so that when the mine detonates it causes no damage to the mine sweeper.

Cryocoolers 9, Edited by R.G. Ross, Jr.
Plenum Press, New York, 1997

The subsystem which produces the magnetic profile uses a superconducting magnet to generate a large magnetic field. The technology needed to cool this magnet is the focus of this paper. Instead of cooling with liquid helium as has typically been done in the past for superconducting magnets, a cryocooler is used to conductively cool the magnet. In this method of cooling, the superconducting magnet is suspended directly in a vacuum. There is no fluid that bathes the magnet, but rather all the refrigeration passes directly from the refrigerator to the magnet through solid copper bus work, bolted or soldered joints and possibly heat pipes.

Operating a superconducting magnet in this way, an approach successfully used in the past[1], requires a cryogenic refrigerator that is small enough to be mounted directly on the magnet assembly. The refrigerator must provide low enough temperatures and enough cooling power to operate the superconductor, and it must be reliable. Due to these constraints a Gifford McMahon (GM) refrigerator was chosen for this task. For the ALISS currently under development, the specific cooling requirements are estimated to be approximately 1.3 W of cooling power at 4.2 K and 50 W of cooling power at a temperature of 50 K.

Over the past five years or so, several researchers have modified some GM coolers in such a way that they are capable of producing refrigeration at temperatures approaching 2 K. In fact GM coolers that produce 0.5 W of refrigeration power at 4.2 K are now commercially available. The research described in this paper is directed at increasing the cooling power of these refrigerators. In particular, by changing the way the second stage regenerator is fabricated we hope to increase the efficiency of the refrigerator.

## DESCRIPTION

The second stage regenerative heat exchangers in GM coolers have been made from lead (Pb) spheres for many years. By replacing some of these Pb spheres with materials that have higher heat capacity at temperatures below 10K, GM coolers have progressed from producing refrigeration at around 8 K to producing refrigeration at temperatures below 4 K. This improvement in performance has provided the ability to cool low temperature superconductors with small, low cost, reliable GM coolers.

The work presented here is another step in advancing the low temperature performance of GM refrigerators. By using regenerator materials that are formed into precise shapes, the heat transfer coefficient between the refrigerant gas and the regenerator will hopefully be increased while still maintaining a reasonably low pressure drop through the regenerator. In addition important performance gains are anticipated by reducing the dead volume of the regenerator. The regenerator dead volume is space in the regenerator that is occupied by the refrigerant gas. This dead volume is detrimental to the performance of the cooler, so if it can be reduced or eliminated, then there should be an increase in cooler performance.

When using spherical regenerator material of uniform size, the dead volume is fixed at about 35% of the total regenerator volume. This is independent on the particular size of the spheres that are used. The only way to decrease the amount of dead volume while using spheres is to use spheres of varying sizes within the same regenerator. When doing this, some of the spaces between larger spheres are filled by spheres of a smaller size. Fabricating a regenerator of this type is difficult because the spheres tend to separate into regions of large and small size spheres.

The dead volume, the amount of open flow area, and the amount of heat transfer surface area can all be tailored by changing from a packed bed of spheres to a more versatile geometry such as parallel plates or cylindrical holes through disks. This paper addresses the concept of producing a stack of disks that have many small holes passing through them. The disks are made from Nd which has relatively high heat capacity at temperatures below 10 K. Improvement in the performance of the regenerator and therefore the GM refrigerator should be possible by developing a process that can control the hole size and spacing, and by developing a technique to assemble them into a regenerator,.

## NEODYMIUM DISKS

Neodymium was chosen as the regenerator matrix material because it provides increased heat capacity at temperatures below 10 K. This makes it useful for low temperature regenerators.

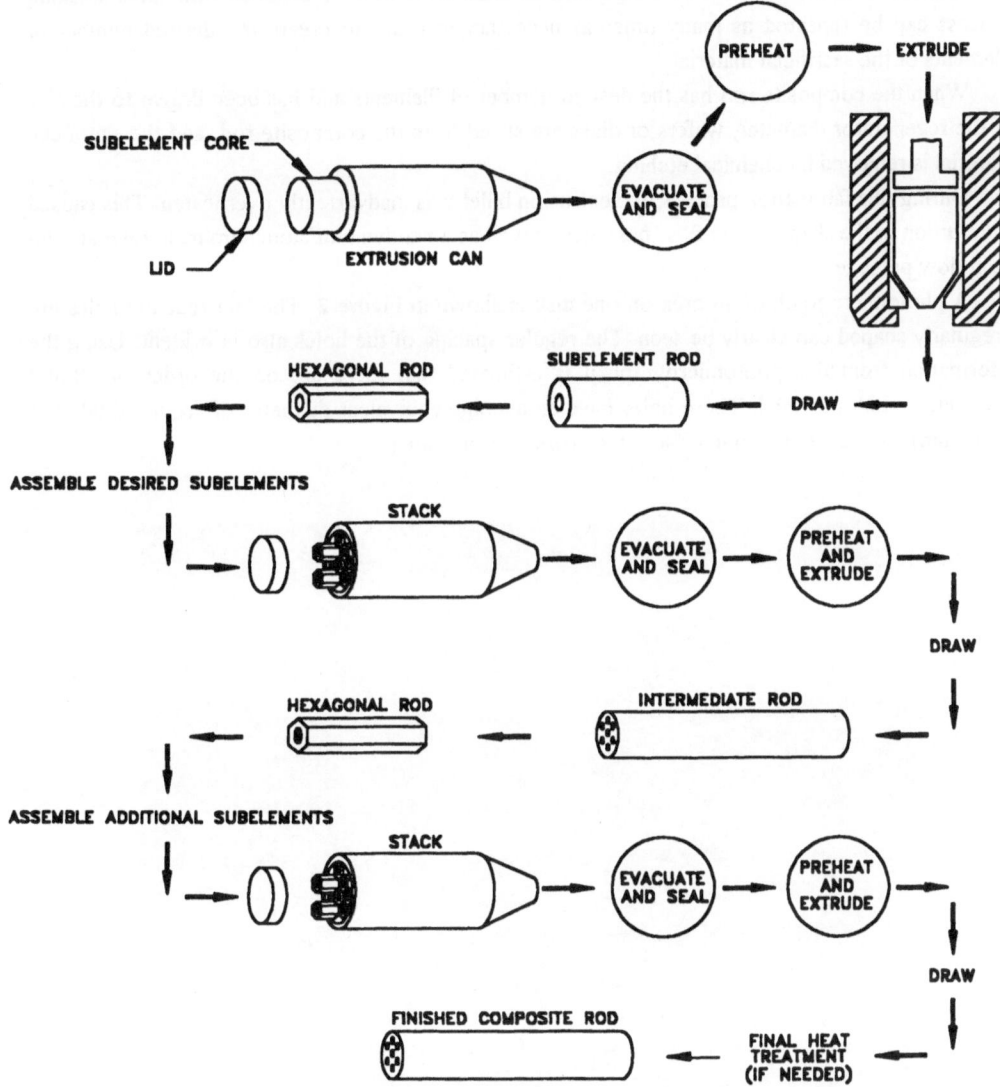

**Figure 1** Schematic of the stack/re-stack process that is used in the production of multi-filamentary superconducting wire.

While other materials also provide high heat capacity at these temperatures[2], Nd is beneficial because it is a pure metal that has good ductility. This means Nd can be formed with relative ease using processes such as extrusion, rolling and drawing.

The Nd perforated plates were produced using a process that is closely related to the one that is used to produce multi-filamentary superconducting wire. The process is illustrated in Figure 1. A cylinder of a sacrificial material (this material will eventually be removed to form the flow holes) is placed in a tube of Nd. The cylinder/tube combination is then reduced in cross-section by extrusion and/or drawing forming core rods. The next step is drawing through a hexagonal die, so individual core rods are produced with a hexagonal cross-section.

These individual core rods are then cut to length and stacked, in parallel, inside another can of Nd. This assembly is then extruded/drawn in order to reduce its cross-section. This stacking process can be repeated as many times as necessary in order to create the desired number of filaments of the sacrificial material.

When the composite rod has the desired number of filaments and has been drawn to the size of the regenerator diameter, wafers or disks are sliced from the composite rod, and the sacrificial material is removed by chemical etching.

During this fabrication process, the extrusion billet was inadvertently overheated. This caused a distortion in the shape of the flow passages as well as a neodymium/aluminum melt zone around each flow passage.

A photomicrograph of an area on one disk is shown in Figure 2. The fact that the holes are irregularly shaped can clearly be seen. The regular spacing of the holes also is evident. Using the information from this photomicrograph it is estimated that there are on the order of 28,000 filaments or holes formed. These holes have an average equivalent diameter of around 0.0019 in (0.05 mm) and make up about 10% of the cross-sectional area.

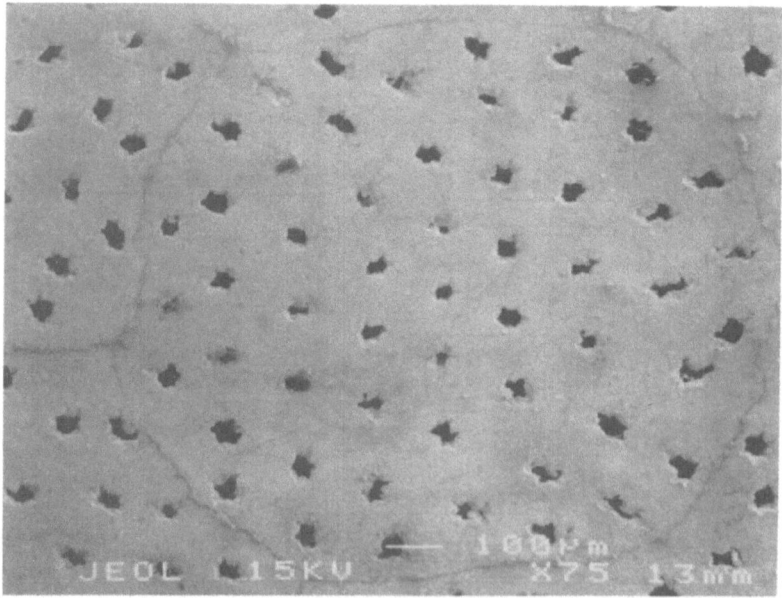

**Figure 2** Neodymium disk with flow holes.

After forming, the disks were machined to a diameter of 0.992 in ±0.001in (25.2 mm ±0.025 mm). They have a thickness of about 0.040 in (1.0 mm) which varies by about ±0.005 in (0.13 mm) and each disk weighs about 2.75 g.

## PRESSURE DROP TESTS

Measuring the pressure drop that occurs when gas flows through the disks provides some important data for the development of regenerators with new matrix geometry. Pressure drop measurements provide an indication of how good the Nd disk manufacturing process is because hole size, shape and distribution all effect the pressure drop across the disks and the heat transfer that occurs between each disk and the gas. By measuring each individual disk we get an idea of the consistency of the manufacturing process. Measuring each disk individually also allows the elimination of particular disks that have manufacturing flaws. A simple visual inspection of the disks shows that some disks have blotchy spots and rough surfaces. One way to determine if these visual flaws effect their performance in a regenerator is to make sure they allow an unrestricted flow of gas. An additional reason for performing flow tests is to establish a reference point for future efforts in building and testing regenerators that use matrix material formed into unique configurations. These measurements provide empirical data for the comparison of such configuration and manufacturing processes.

We constructed the test fixture shown in Figure 3 to measure the pressure drop of each disk at room temperature. We inserted each Nd disk and applied pressurized helium to one end of the test fixture. The other end of the test fixture was attached to a hot wire anemometer that measures flow rate. The pressure of helium was adjusted to a level that yielded a flow of 1.0 scfm (0.079 g/s). The outlet of the anemometer was open to atmospheric pressure so the pressure drop across the disk was simply the gage pressure at the inlet to the test fixture.

A total of 52 disks were tested. The resultant pressure drops ranged from a maximum of 100 psi/scfm ($8.77 \times 10^6$ Pa-s/g) to a minimum of 1.6 psi/scfm ($1.4 \times 10^5$ Pa-s/g). These results are shown in Figure 4. This figure clearly illustrates that five disks have a much higher pressure drop than the rest. These five disks had likely been damaged in the manufacturing process and many of the holes are either closed or too small to allow the free passage of helium. These five disks were

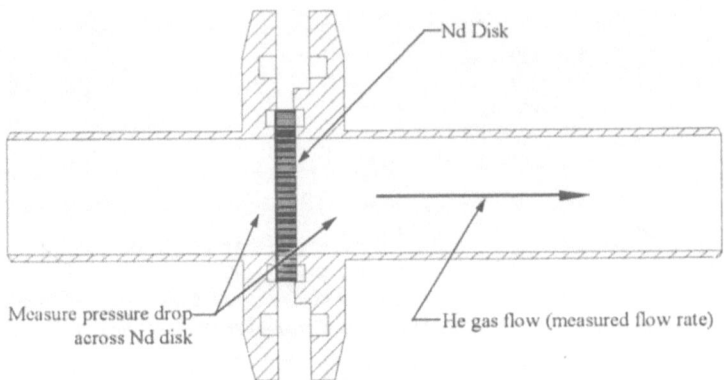

**Figure 3** Pressure Drop Test Fixture

discarded because of these flaws. A graph showing the pressure drop for the remaining 48 disks are shown in Figure 5. Of this group, no disks had a pressure drop of more that 5 psi/scfm ($4.39 \times 10^5$ Pa-s/g). Had these tests not been performed, then the disks that were eliminated would have blocked the flow of gas in the assembled regenerator.

## REGENERATOR CONSTRUCTION

Over the past few years it has been shown that careful construction of the regenerator/displacer assembly is critical to the performance of low temperature GM refrigerators. The displacer should be made of a stable material that can be machined to tolerances on the order of ±0.002 in (0.05 mm). This is important for reliable seal performance and the tolerance minimizes the wearing of the displacer against the cylinder wall. Additionally, the regenerator material must be packed tightly so that the matrix material cannot move. If the material is packed too loosely, then "preferred" paths for the gas form. If this happens then much of the regenerator matrix is bypassed and is therefore not useful for heat exchange.

Figure 6 shows a cross section of the regenerator/displacer that was packed with the Nd disks. The main tube sections (indicated by cross-hatching in this figure) are made of stainless steel. As shown, the majority of the displacer is covered with a plastic sleeve to prevent galling between the displacer and the cylinder. The 48 Nd disks occupy a region approximately 1.9 in (48.3 mm) long. Each disk weighs approximately 2.75 g and therefore a total of 132 g of Nd is being used. During the assembly process, no effort was made to align each disk so that the holes in one disk match up with the holes in the next disk. Instead each disk is separated from its neighboring disks by a circular piece of spun polyester cloth of trade name Cerex™, commonly used in super-insulation systems. These Cerex™ disks thermally isolate the Nd disks from each other and also provide space for the gas leaving one Nd disk to enter the next. If the Cerex™ disks were not present, one disk might block some or all of the holes in the next disk.

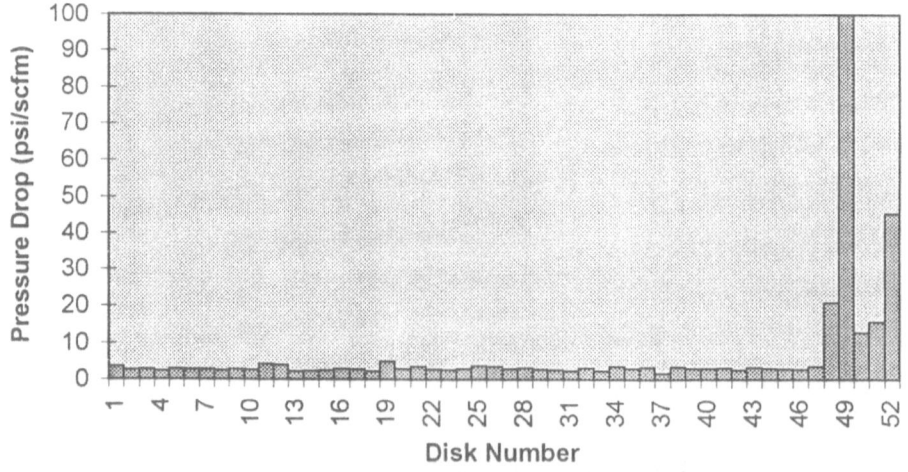

**Figure 4** Pressure Drop Measurements (all disks).

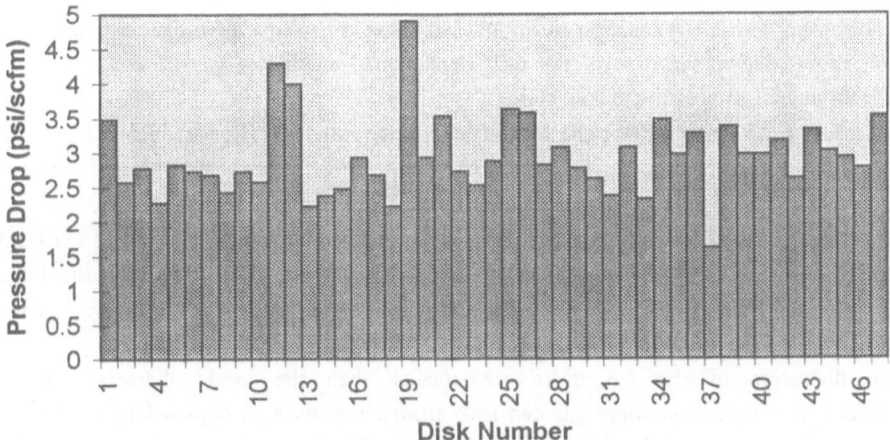

**Figure 5** Pressure Drop Measurements (selected disks).

Adjacent to the Nd disk stack is a 2.7 in (68.6 mm) length of packed Pb spheres. The total mass of these spheres is about 265 g. These spheres occupy the remaining space available for regenerator matrix material and are located at the end of the regenerator that operates at temperatures around 30K (the warm end). At this extreme end of the regenerator, Pb has a higher heat capacity, due to the high temperatures, than Nd and is, therefore, the preferable material to use. However, somewhere along the length of the regenerator the temperature reaches a level where it becomes advantageous to use Nd. Through experience we have found this to be around 1/2 the length and we normally do not assemble regenerators with less than 50% Nd. Less than 50% of Nd was used in this regenerator due to the limited number of disks available.

After the regenerator was constructed, pressure drop measurements were made on it in a similar way that each individual disk was tested. However, instead of taking a single datum point as done for each disk, the regenerator was tested at several flow rates. The results of these measurements as well as some comparative data for a regenerator made with Nd spheres is shown in Figure 7. In the test of the Nd disk regenerator, pressure differences across the regenerator of

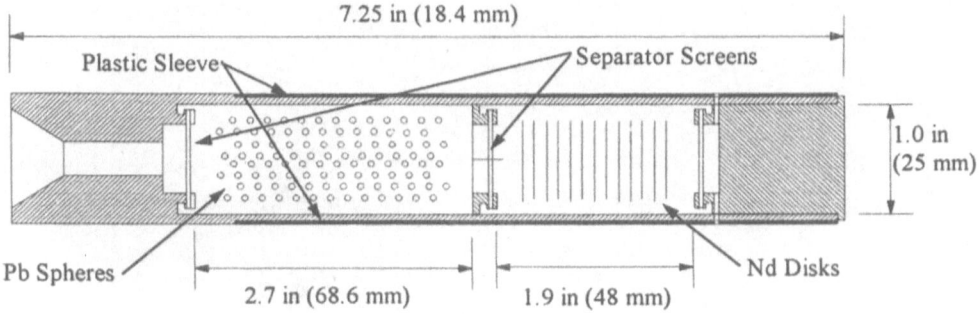

**Figure 6** Cross Section of Nd Disk Regenerator

up to 15 psi were measured at room temperature. As a comparison some data from a similar regenerator packed with a combination of Pb and Nd spheres is also shown in this figure. This comparison shows that in the case of the disk regenerator, much less gas flows through the regenerator for any given pressure drop.

The pressure drop measured for the assembled regenerator is 51.7 psi/scfm ($4.53 \times 10^6$ Pa-s/g). If we add the pressure drops measured for each disk we get an expected pressure drop of 136.3 psi/scfm ($1.195 \times 10^7$ Pa-s/g) for the stack of disks. The assembled regenerator incorporates Pb spheres, Cerex™ disks, which have pressure drops associated with them, as well as the stack of Nd disks. The pressure drop measure for the regenerator should be higher than the 136.3 psi/scfm ($1.195 \times 10^7$ Pa-s/g) of the stack of disks, but is not. This indicates that there is some significant leakage of gas around the Nd disks.

If the diameters of the Nd disks are smaller than the inside diameter of the regenerator/displacer tube then some gas can leak around the disks as opposed to all the gas flowing through holes in the disks. Not only does this tend to lower the overall pressure drop across the regenerator (this is good) but it also reduces the effectiveness with which the regenerator can exchange and store heat (this is bad). The Nd disks are intentionally made slightly smaller than the tube to facilitate assembly. If this were not done the disks would need to be forced into the tube possibly damaging them. How tightly the disks fit in the tube is also affected by operating temperature. As the regenerator cools to its operating temperature both the Nd disks and the regenerator/displacer-tube shrink. The tube which is made of stainless steel shrinks by approximately 0.003 in (0.076 mm) while the Nd disk shrinks by around 0.002 in (0.051 mm). This means that the leakage path around the disk diminishes as the regenerator reaches its operating temperature.

## REFRIGERATOR TESTS

The assembled regenerator, as described in the above section, was installed in a two stage GM refrigerator and the performance of that refrigerator was measured. The refrigerator used

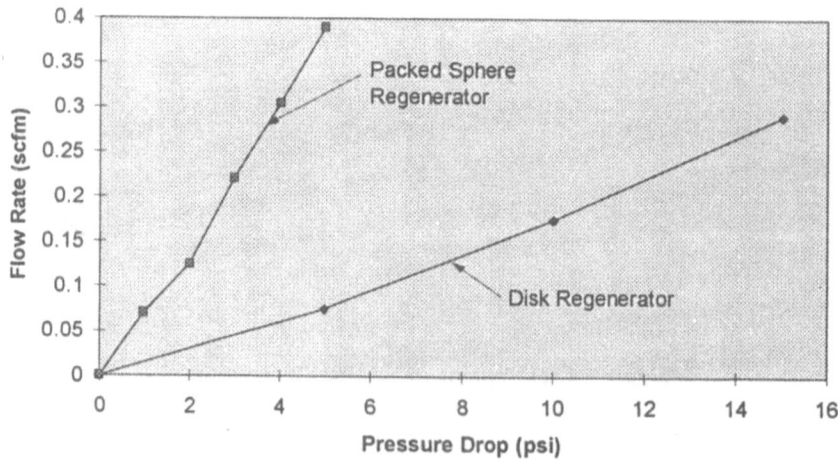

**Figure 7** Regenerator Pressure Drop Test

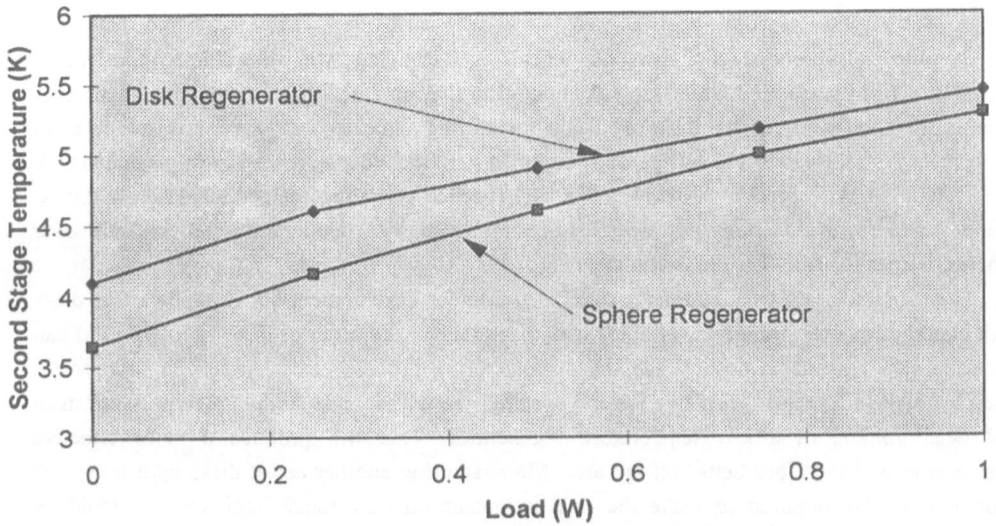

**Figure 8** Second Stage Temperatures with Heat Loads to 1 Watt.

consists of a modified Balzer UCH-130™ cold head and a Balzer UCC 110S™ compressor module. The modifications to the cold head consisted primarily of extending the length of the second stage cylinder to accommodate the regenerator described above. The length of the stroke was shortened from 0.625 in (15.9 mm) to 0.440 in (11.2 mm) and the first stage regenerator was modified to incorporate some Pb spheres in its lower portion. These modifications had been shown to improve the low temperature performance of this type of cold head[3].

The cooling performance of the second stage of the refrigerator is shown in Figure 8. This figure shows the second stage temperature as a function of heat applied to the second stage. The cold head was operated at a speed of 85 cycles per minute and no load was applied to the first stage for the data shown in this figure. The refrigerator reached a temperature of 4.1 K with no load applied to the second stage. With a maximum of 1.0 W applied, the second stage operated at 5.45 K. Although not shown in this graph, slightly lower temperatures were reached when a load of 20 W and 30 W was applied to the first stage.

In order to get an idea of how well this disk regenerator was performing relative to the more commonly used sphere regenerators that have much lower pressure drop characteristics, similar second stage load data is shown for a regenerator that was assembled from Nd and Pb spheres and packed into a similar regenerator/displacer tube. This regenerator used roughly equal volumes of Pb and Nd (200 g of Pb and 150 g of Nd). Recall that the disk regenerator used about 130 g of Nd and 265 g of Pb. Thus the disk regenerator had about 15% less Nd in it and 30% more Pb. As can be seen from Figure 8 the performance of the sphere regenerator was better than the disk regenerator. It produced lower temperatures and more cooling power. The differences, however, were rather small.

## CONCLUSIONS

By utilizing superconducting wire manufacturing techniques, we were able to construct a regenerator from perforated Nd disks, a configuration that potentially has many advantages over spheres. Results from the 48 disks available for testing indicate that good low temperature performance is achievable. The refrigerator, modified with the Nd disk regenerator, produced 1 W of refrigeration at temperatures below 5.5 K and reached operating temperatures below 4.0 K. While this performance does not demonstrate improved low temperature performance over spheres, it does indicate that the potential exists.

The manufacturing process for the disks is still under development but it appears that disks with hundreds or even thousands of holes with diameters on the order of 0.004 in (0.10 mm) can be made at a reasonable cost. When the manufacturing process has been refined to produce circular holes with controlled diameters and spacing, we will be able to tailor the pressure drop and dead volume of a low temperature regenerator. This will produce a more effective regenerator, and therefore better refrigerator. Manufacturing another set of disks with improved geometry is now required to make the final judgment on how much improvement could be expected.

## REFERENCES

1 M. Heiberger, A. Langhorn, T. Hurn, D. Morris, J. Walters, L. Apprigliano, D. Waltman, and G. Green, "A Lightweight Rugged Conduction-cooled NbTi Superconducting Magnet for U.S. Navy Mine Sweeping Applications", *Advances in Cryogenic Engineering*, Vol. 41, Plenunm Press, New York, (1996).

2 M. Sahashi, Y. Tokai, T. Kuriyama, H. Nakagome, R. Li, M. Ogawa, and T. Hashimoto, "New Magnetic Material $R_3T$ System with Extremely Large Heat Capacities Used as Heat Regenerators", *Advances in Cryogenic Engineering*, Vol. 35, Part B, Plenum Press, New York, (1989), Pg.1175.

3 J. Chafe, G. Green, and R. Riedy, "Neodymium Regenerator Test Results in a Standard Gifford-McMahon Refrigerator", *Cryocoolers* 7, Report # PL-CP--93-1001, Phillips Laboratory, Kirtland Air Force Base, New Mexico, (1993), pg. 1157.

# Influence of Alloying on the Behavior and Properties of Er$_3$Ni

V.K. Pecharsky, K.A. Gschneidner, Jr., R.W. McCallum, and K.W. Dennis

Ames Laboratory DOE and Department of Materials Science and Engineering
Iowa State University, Ames, IA 50011-3020, USA

## ABSTRACT

A study on processing and alloying of the passive magnetic regenerator material Er$_3$Ni revealed that the cooling rates of the order of $10^6$ K/sec are insufficient to quench it into a glassy state. Small additions of Ti (as low as 0.5 at.%) make it possible to quench the alloys into a partially amorphous state, which leads to an improved ductility and the possibility of being able to process this brittle low temperature regenerator material into ribbons and/or wire mesh. Even though the crystalline Er$_3$(Ni$_{1-x}$Ti$_x$) shows practically no improvement nor deterioration in the low temperature volumetric heat capacity, the partially amorphous alloys are inferior to that of the crystalline alloy.

## INTRODUCTION

The need for an improved and inexpensive low temperature regenerator material with high volumetric heat capacity below ~10 K is quite urgent for a number of commercial applications, such as MRI units. The magnitude of the heat capacity of the regenerator is one of the major limiting factors with respect to the attainability of low temperature and the refrigeration power. Following the discovery of Er$_3$Ni regenerator [1-3] the lowest temperature limit for a commonly used Gifford-McMahon-type (GM) cryocoolers was quickly extended down to 4.2K and below, with a substantial refrigeration power of close to 1W at 4.2K (for example see [4,5]). Almost immediately after the Er$_3$Ni had been found and tested, the search for new passive magnetic regenerative materials with large low-temperature volumetric heat capacity yielded to suggestions that Er$_3$Co, ErNi, and (Er$_{1-x}$,R$_x$)(Ni$_{1-y}$,Co$_y$)-based intermetallides (where R=Yb or Gd) may also be useful[5,6] for low temperature applications since they order magnetically between ~5K and ~12K and, thus, are characterized by the enhanced heat capacity in the vicinity of the magnetic phase transitions. Another prospective candidate material is the pure lanthanide metal, Nd, which has two magnetic phase transitions at ~8K and ~18K.

Studies on processing these materials into a usable form (typically spheres) [2,7-9] showed that at least two of them (Er$_3$Ni and Nd) can be readily processed into spheres using different atomization and rapid solidification techniques. Nevertheless, practical use of any of the above mentioned materials is somewhat limited, because the first (Er$_3$Ni) is extremely brittle (as are many other intermetallic compounds), and the second (Nd), although being ductile, requires special handling due to its high chemical activity and ease of oxidation in air. A recent report on evaluating the mechanical behavior of the Er$_3$Ni atomized spherical particulate[9] shows that the mechanical strength of this brittle material is critically dependent on the degree of sphericity and

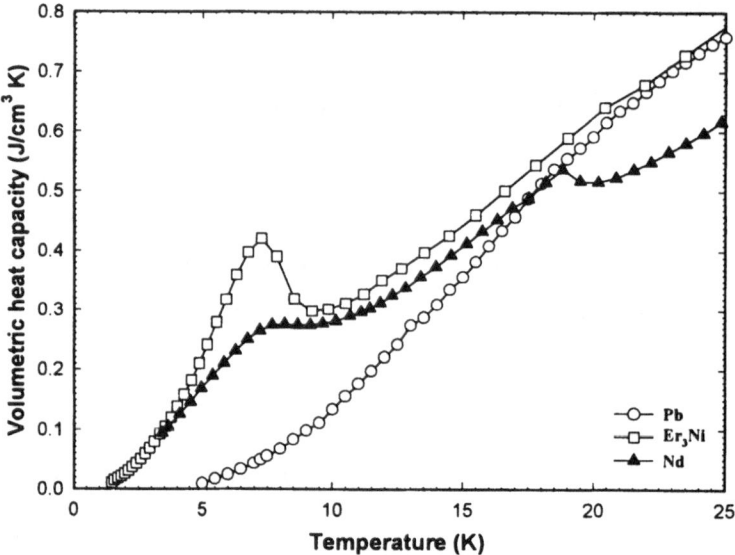

**Figure 1.** The low temperature volumetric heat capacity of Pb (upper stage regenerator material), Er₃Ni, and Nd (lower stage regenerator materials). The heat capacity of Nd was measured on gas atomized commercial Nd (reported to be 99+ wt.% pure).

also on porosity of the packed cartridge. These two factors (the necessity of selecting only the best spherically shaped particles, and also the achievement of the highest packing density for each regenerator cartridge) raised the cost of the Er₃Ni regenerator significantly.

Figure 1 shows volumetric heat capacities of these two regenerator materials together with that of pure Pb. The latter is commonly used as an upper stage regenerator material. It is obvious that both Er₃Ni and Nd are superior to Pb below about 15K with the heat capacity of Er₃Ni being consistently larger than that of Nd. Other known low temperature magnetic regenerator materials (see above) typically have large heat capacities at the magnetic phase transition temperature, but their lattice heat capacities and/or Schottky anomalies due to the crystalline electric field effects are not as favorable as in the case of Er₃Ni. Therefore, their heat capacities level-off rapidly above and below the transition temperature, making it impossible to reduce amount of Pb and, therefore, constrain the attainable refrigeration power at the lowest temperature.

Considering the excellent low temperature heat capacity of Er₃Ni we carried out a study on the alloying and the rapid solidification of this intermetallic compound to find out whether or not it is possible to improve mechanical properties of this brittle substance and to process it into sheets or wire mesh for further use as a low temperature magnetic regenerator material.

## EXPERIMENTAL

The alloys with nominal compositions $Er_3Ni$, $Er_3(Ni_{0.98}Ti_{0.02})$, and $Er_3(Ni_{0.90}Ti_{0.10})$ were prepared by arc melting the appropriate amounts of the metals in an argon atmosphere. The Er (99.9+ at.%) was obtained from the Materials Preparation Center at the Ames Laboratory, and the Ni and Ti (both 99.99 wt.% pure) were purchased from commercial vendors. The three alloys were remelted in quartz glass crucibles and rapidly solidified into ribbons using a melt spinning technique. All melt spin processing was done under an argon atmosphere. The crucible to wheel distance was 5mm. The molten alloy was ejected through a 0.8mm diameter orifice after the alloy had reached 1150°C (275°C superheat) onto a copper chill block rotating at 45 m/s. Cooling rates achieved were on the order of $10^6$ K/sec.[10] The product is ribbon pieces approximately 25 μm thick and 2 mm wide with lengths varying from 2 to 8 cm. X-ray diffraction patterns were obtained from powdered ribbons employing a Philips diffractometer equipped with a vertical θ-2θ goniometer and graphite monochromator using Cu $K_\alpha$ radiation. The x-ray examination of all as-cast alloys revealed that they were essentially single phase materials with the $Fe_3C$-type crystal structure of original $Er_3Ni$ intermetallide. The heat capacity of as-cast and rapidly solidified materials was measured using an automated adiabatic heat pulse calorimeter[11] in the temperature range from ~3.5 to 350K. The heat capacity of the as-cast alloys was measured on samples in the bulk form, while the heat capacity of rapidly solidified samples was measured on powdered samples compacted with an addition of 25-28 wt.% silver powder. The heat capa-city of silver then was subtracted off to determine the heat capacity of the rapidly solidified alloys.

## RESULTS AND DISCUSSION

According to the Er-Ni[12] binary phase diagram, the $Er_3Ni$ intermetallic compound melts peritectically at approximately 870°C with a liquidus temperature exceeding that of the solidus by about 30-40°C. Also, the $Er_3Ni$ compound exists close to a deep eutectic ($Er_3Ni+Er_5Ni_3$) and thus appeared to be a good candidate to quench-in as an amorphous alloy.

Nevertheless, the attempts to process the pure $Er_3Ni$ alloys into a metastable glassy or partially amorphous state by using the melt spinning technique failed even at cooling rates of the order of $10^6$ K/sec. The powder diffraction patterns of the two $Er_3Ni$ samples are shown in Fig. 2 and they prove that no detectable amount of glassy phase has been formed, eventhough the increased average width of the diffraction peaks is evidence that the effective grain size was smaller in the case of the rapidly solidified alloy. Therefore, we tried to frustrate the crystallization process by introducing an alloying element into the ordered $Er_3Ni$ crystal lattice. Figure 3 displays the x-ray powder diffraction patterns of as cast and rapidly solidified $Er_3(Ni_{0.90}Ti_{0.10})$ and rapidly solidified $Er_3(Ni_{0.98}Ti_{0.02})$ alloys. It is obvious that introducing even the smallest amount of Ti as an alloying element has a significant effect on the crystallization of $Er_3Ni$. Alloys containing Ti require a significantly smaller degree of undercooling than pure $Er_3Ni$, and, indeed, they can be quenched-in as an amorphous (i.e. nanocrystalline) materials. To achieve this, the cooling rates must be increased over ~$10^6$ K/sec, since the presence of a small amount of crystalline $Er_3(Ni_{0.98}Ti_{0.02})$ and $Er_3(Ni_{0.90}Ti_{0.10})$ is obvious in the x-ray diffraction

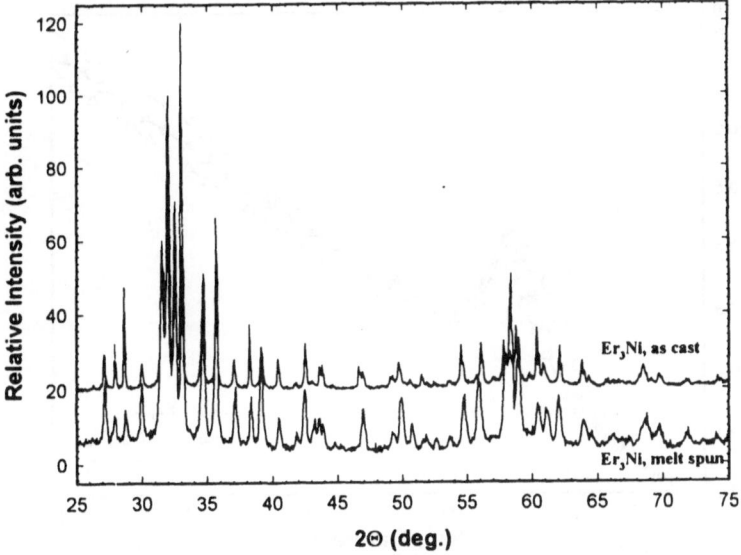

**Figure 2.** The x-ray powder diffraction patterns of as-cast and melt spun $Er_3Ni$. The cooling rate was ~$10^6$ K/sec for the rapidly solidified alloy.

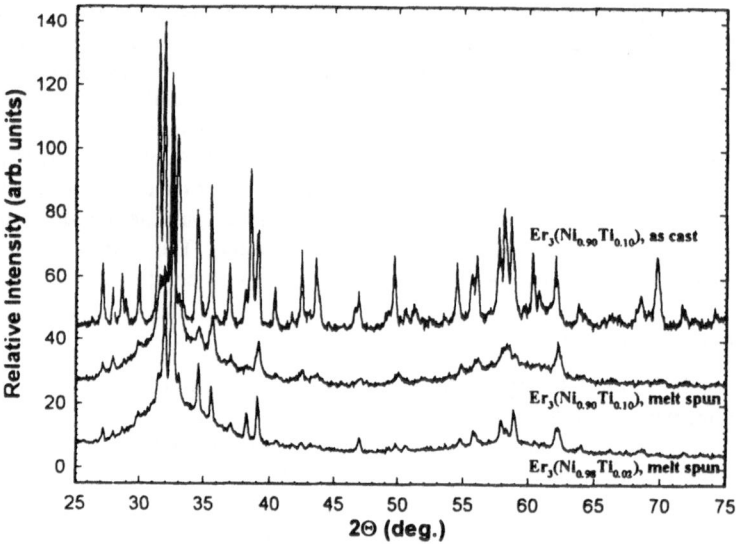

**Figure 3.** The x-ray powder diffraction patterns of as-cast $Er_3(Ni_{0.90}Ti_{0.10})$ and melt spun $Er_3(Ni_{0.90}Ti_{0.10})$ and $Er_3(Ni_{0.98}Ti_{0.02})$. The cooling rate was $\sim10^6$ K/sec in the case of rapidly solidified alloys.

patterns. The alloy $Er_3(Ni_{0.98}Ti_{0.02})$ appears to contain a larger portion of a crystalline phase, which is consistent with the amount of alloying element.

The low temperature volumetric heat capacity of crystalline (as-cast) $Er_3(Ni_{0.98}Ti_{0.02})$ and $Er_3(Ni_{0.90}Ti_{0.10})$ is shown in Fig. 4, and the heat capacity of the same rapidly solidified alloys is shown in Fig. 5 together with the heat capacity of pure $Er_3Ni$. Even though the effect of alloying with Ti on the low temperature heat capacity of the crystalline $Er_3Ni$ is minimal, the general tendency is that the magnetic transition temperature slightly increases, and that the volumetric heat capacity slightly decreases. The first is not quite understood, since introducing an alloying element with larger atomic size (Ti) compare to Ni and introducing of a certain degree of

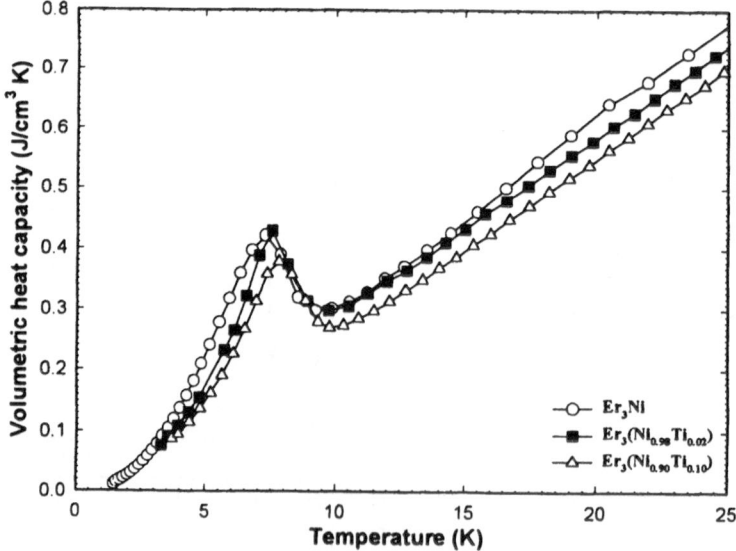

**Figure 4.** The low temperature volumetric heat capacity of as-cast (crystalline) $Er_3Ni$, $Er_3(Ni_{0.98}Ti_{0.02})$, and $Er_3(Ni_{0.90}Ti_{0.10})$ alloys.

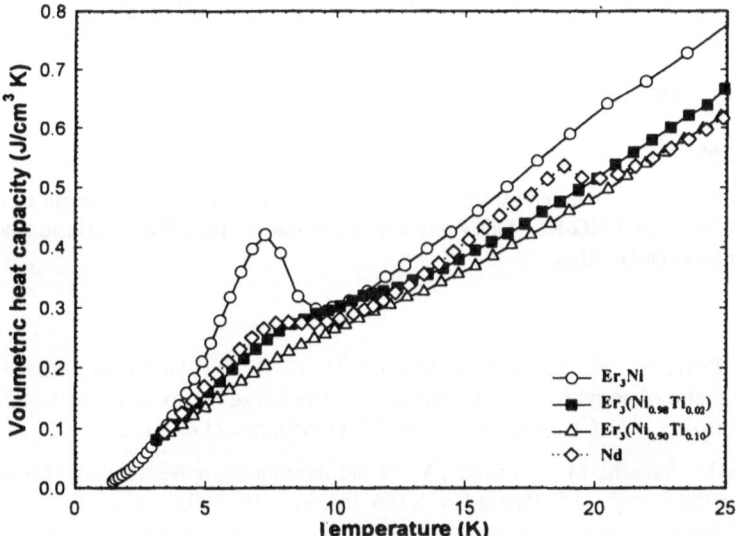

**Figure 5.** The low temperature volumetric heat capacity of as-cast Er₃Ni, gas atomized Nd, and rapidly solidified, $Er_3(Ni_{0.98}Ti_{0.02})$, and $Er_3(Ni_{0.90}Ti_{0.10})$ alloys.

randomness in the arrangement of magnetic (Er) atoms would be expected to lower the magnetic ordering temperature. The experimental observation (Fig. 4), however, is opposite to this expectation. The second, i.e. decreasing of the volumetric heat capacity of the crystalline $Er_3(Ni_{1-x}Ti_x)$ when the amount of Ti increases, is expected, since substituting Ti for Ni slightly expands the crystal lattice and since Ti is lighter than Ni, a slight decrease in the materials density is normal and this will automatically reduce the volumetric heat capacity.

The rapid solidification and partial amorphization of the $Er_3(Ni_{1-x}Ti_x)$ alloys has a much more significant effect on the volumetric heat capacity, as shown in Fig. 5. Unfortunately, this effect has just the opposite sign of the desired outcome: -- the heat capacity of partially amorphous alloys is significantly lower than that of the crystalline alloys, although the ordering temperature seems to remain practically unchanged. It is quite reasonable to assume, that the presence of a substantial amount of the amorphous phase in the rapidly solidified alloys spreads the magnetic entropy associated with the magnetic ordering of pure Er₃Ni over a much wider temperature range with the immediate effect of causing the lowering of the material's heat capacity.

The rapidly solidified alloys containing Ti shows significantly better mechanical behavior compared to that of Er₃Ni; the obtained ribbons were quite flexible and, in fact, we experienced difficulties grinding the rapidly solidified material into a powder for the x-ray diffraction studies and for the heat capacity measurements. This implies, that if they were used as the passive magnetic regenerator materials for the low temperature stage in a GM refrigerator, then their mechanical stability and ductility would not cause a dusting problem, as is often the case when using the pure Er₃Ni. The use of rapidly solidified $Er_3(Ni_{1-x}Ti_x)$ can be recommended as a substitute for pure Nd metal, because their low temperature heat capacities are practically the same (see Fig. 5). The chemical stability of rapidly solidified alloys is distinctly better than that of Nd, since they do not oxidize in air and, thus, do not require special precautions. Furthermore, the cost of rapidly solidified $Er_3(Ni_{1-x}Ti_x)$ may be significantly lower than that of gas atomized Nd powders, because rapid solidification offers close to 100% yields.

## CONCLUSION

Although the volumetric heat capacity of the partially amorphous alloys $Er_3(Ni_{1-x}Ti_x)$ is lower than that of crystalline Er₃Ni, and crystalline $Er_3(Ni_{1-x}Ti_x)$, their mechanical and chemical stability makes them attractive candidates as a substitute for another low temperature regenerator material

– Nd. The $Er_3Ni$ alloys with substituted Ti can be made partially (and probably completely) amorphous by using an inexpensive melt spinning rapid solidification technique, and they can be produced in form of ribbons or in form of a wire mesh for use as a lower temperature stage regenerator material in GM-type refrigerators.

## ACKNOWLEDGMENT

The Ames Laboratory is operated by Iowa State University by the US Department of Energy under contract No. W-7405-ENG-82. This study was supported by the Office of Basic Energy Science, Materials Science Division.

## REFERENCES

1.  Sahashi, M., Tokai, Y., Kuriyama, T., Nakagome, H., Li, R., Ogawa, M., and Hashimoto, T., "New Magnetic Material $R_3T$ System with Extremely Large Heat Capacities Used as Heat Regenerators", *Adv. Cryogenic Eng.*, vol. 35, (1990), pp. 1175-1182.

2.  Arai, T., Sori, N., Sahashi, M., and Tokai, Y., "Cold Accumulating Material and Method of Manufacturing the Same", U.S. Patent No. 5,186,765, Feb. 16, 1993.

3.  Kuriyama, T., Takahashi, M., Nakagome, H., Eda, T., Seshake, H., and Hashimoto, T., "Helium Liquefaction by a Two-Stage Gifford-McMahon-Cycle Refrigerator Using New Regenerator Material of $Er_3Ni$", *Jpn. J. Appl. Phys.*, vol. 31, (1992) pp. L1206-L1208.

4.  Kuriyama, T., Ohtani, Y., Takahashi, M., Nakagome, H., Nitta, H., Tsukagoshi, T., Yoshida, A., and Hashimoto, T. "Optimization of Operational Parameters for a 4K-GM Refrigerator", paper presented at Cryo. Eng. Conf., Columbus, Ohio, July 17-21, 1995.

5.  Satoh, T., Onishi, A., Li, R., Asami, H., and Kanazawa, Y., Development of 1.5W 4K-GM Cryocooler with Magnetic Regenerator Material", paper presented at Cryo. Eng. Conf., Columbus, Ohio, July 17-21, 1995.

6.  Tsukagoshi, T., Nitta, H., Yoshida, A., Matsumoto, K., Hashimoto, T., Kuriyama, T., Takahashi, M., Ohtani, Y., and Nakagome, H., "Refrigeration Capacity of a GM Refrigerator with Magnetic Regenerator Materials", paper presented at Cryo. Eng. Conf., Columbus, Ohio, July 17-21, 1995.

7.  Osborne, M.G., Anderson, I.E., and Gschneidner, K.A., Jr., "Centrifugal Atomization of Rare Earth Metal and Intermetallic Compounds", *Advances in Powder Metallurgy and Particulate Materials - 1993: Powders, Characterization, Testing, and Quality Control*, vol. 1, Lawley, A. and Swanson, A., eds., Metal Powder Industries Federation - American Powder Metallurgy Institute, Princeton, NJ (1993), pp. 65-73.

8.  Osborne, M.G., Anderson, I.E., Gschneidner, K.A., Jr., Gailloux, M.J., and Ellis, T.W., "Centrifugal Atomization of Neodymium and $Er_3Ni$ Regenerator Particulate", *Adv. Cryogenic Eng.*, vol. 40 (1994), pp.631-638.

9.  Okamura, M., Sori, N., Kuriyama, T., Saito, A., and Sahashi, M., "Evaluation of Mechanical Properties for Spherical Magnetic Regenerator Materials Fabricated by Rapid Solidification Process", paper presented at Cryo. Eng. Conf., Columbus, Ohio, July 17-21, 1995.

10. Froes, H., "Production of Rapidly Solidified Metals and Alloys", *J. Metals*, vol. 36, (1984), pp. 20-33.

11. Pecharsky, V.K., Moorman, J.O., and Gschneidner, K.A., Jr., "3.5 to 350 K, High Magnetic Field Automatic Small Sample Calorimeter", *to be published*.

12. Buschow, K.H.J., "Crystal Structure, Magnetic Properties and Phase Relations of Erbium-Nickel Intermetallic Compounds", *J. Less Common Metals*, vol. 16 (1968), pp. 45-53.

# Processing and Testing of the Low-Temperature Stage $Er_6Ni_2Sn$ Cryogenic Regenerator Alloy

**K. A. Gschneidner, Jr.[1], V. K. Pecharsky[1], M. G. Osborne[1], J. O. Moorman[1], I. E. Anderson[1], D. Pasker[1] and M. L. Eastwood[2]**

[1]Ames Laboratory, DOE and Department of Materials Science and Engineering, Iowa State University, Ames, Iowa 50011-3020, U.S.A.
[2]Eastwood Associates, 1954 Fletcher Ave., South Pasadena, CA 91030, U.S.A.

## ABSTRACT

Low temperature heat capacity measurements have shown that $Er_6Ni_2Sn$ might be a useful low temperature, <20 K, cryocooler regenerator material. Here we report on (1) our efforts to prepare small particles for use in Gifford-McMahon (GM) cryocoolers, and (2) the initial test results using $Er_6Ni_2Sn$ in the low temperature stage of a GM displacer. Replacement of some of the Nd spheres by $Er_6Ni_2Sn$ in the second stage achieved a modest improvement in the no load temperature of ~ 0.5 K over the all-Nd second stage.

## INTRODUCTION

Two years ago at the 8th Cryocooler Conference we reported on several new Er-based ternary alloys as potential cryogenic regenerators for operation as a low temperature stage for cooling below ~15 K. The best alloys, according to the volumetric heat capacity measurements, were $Er_6Ni_2Sn$ and $Er_6Ni_2Pb$ both of which had been annealed at 700°C for 2 weeks.[1] A comparison of the heat capacity of the $Er_6Ni_2Sn$ with that of Pb and $Er_3Ni$ are presented in Fig. 1. As one can see the heat capacity of $Er_6Ni_2Sn$ is significantly better than lead below 18 K, and is better than that of $Er_3Ni$ between 9 and 18 K. Thus, the $Er_6Ni_2Sn$ could be used as the low temperature stage along with Pb as the upper stage to reach ~10 K with significantly more cooling power than Pb alone or a combination of $Er_3Ni$ (low temperature stage) and Pb. Another possible application is to use a three stage device with $Er_3Ni$ as the lowest temperature stage, $Er_6Ni_2Sn$ as the middle stage, and Pb as the highest temperature stage. This should enable one to cool down to ~4 K with more cooling power than just a two stage device containing $Er_3Ni$ and Pb. Because of these potential applications, we have focused our efforts on the fabrication of $Er_6Ni_2Sn$ into useful form for packing it into a powder regenerator bed. Our progress to date is described below.

## PROCESSING TECHNIQUES

Several different processing techniques were explored because the phase relationships involving Er, Ni and Sn near the 66.7 at.% Er - 22.2 at.% Ni - 11.1% at.% Sn composition are

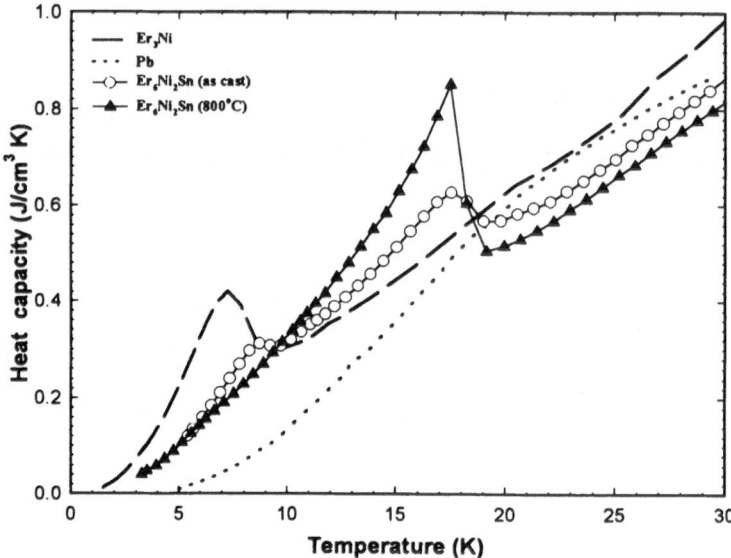

**Figure 1.** The volumetric heat capacities of as-cast and annealed $Er_6Ni_2Sn$. The volumetric heat capacities of Pb and $Er_3Ni$ are shown for comparison.

unfavorable for forming spheres using proven technologies. The compound $Er_6Ni_2Sn$ melts incongruently at ~925°C disproportionating into liquid and an unknown intermetallic phase with a high melting temperature of ~1400°C. The amount of liquid formed at the peritectic temperature is small, probably no more than 10%; the remaining portion of the sample is this high melting intermetallic compound, and possibly another solid phase.

**Centrifugal Atomization**

Centrifugal atomization has been successfully used to prepare $Er_3Ni$ and other rare earth intermetallic compounds[2,3] and thus was our first choice for attempting to prepare spheres of $Er_6Ni_2Sn$. Two different attempts were made to produce such spherical powders. In the first case the pour temperature ~1350°C was too low for a satisfactory yield of material. In the second trial a pour temperature of ~1475°C was used with a substantial increase in the yield, however, a significant portion of the melt remained in the $Al_2O_3$ crucible. Analysis of the residual alloy in the $Al_2O_3$ crucible suggested that the molten 66.7% Er - 22.2% Ni - 11.1% Sn alloy had reacted with the $Al_2O_3$. We could easily see the erosion of the crucible and the $Al_2O_3$ stopper rod. X-ray analysis of the resultant powder indicated that it consisted of two or three unknown phases. The low temperature heat capacity (Fig. 2) clearly indicates that the resultant powders do not contain the $Er_6Ni_2Sn$ phase (i.e., the absence of the peak at ~17K) which is in good agreement with the x-ray results. X-ray diffraction measurements also indicated that prolonged heat treatments, several weeks at 800°C, of the atomized powders did not result in the formation of the $Er_6Ni_2Sn$ phase. These results suggest that the starting 66.7% Er - 22.2% Ni - 11.1% Sn composition was shifted during the melting operation by the reaction of the Er in the alloy with the $Al_2O_3$ crucible and possibly by the loss of Sn due to evaporation.

Electron microscopy (SEM) revealed that for the smaller size particles, i.e. <125 μm, about 25% were spherical and 75% had irregular shapes, while the particles >125 μm were almost exclusively irregularly shaped.

Thus, it would appear that centrifugal atomization is not a viable method for producing $Er_6Ni_2Sn$ spheres, unless a costly and ambitious developmental program were undertaken to overcome the technical obstacles due to the unfavorable phase relationships in the Er-Ni-Sn

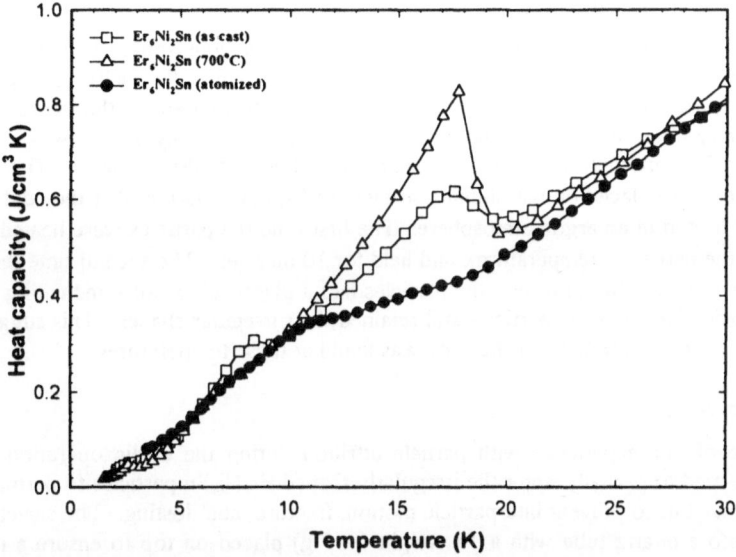

**Figure 2.** The volumetric heat capacity of the 66.7 at.% Er, 22.2 at.% Ni, 11.1 at.% Sn sample which had been centrifugally atomized. The volumetric heat capacities of as-cast and annealed $Er_6Ni_2Sn$ samples are shown for comparison.

system near the 66.7% Er - 22.2% Ni - 11.1% Sn composition.

## Crushed Particles

The $Er_6Ni_2Sn$ intermetallic phase is quite brittle, although not as brittle as $Er_3Ni$, and can be easily crushed into fine particles. A minimal amount of grinding leads to a high fraction of particles smaller than 150 μm, see Table 1. This is unfortunate, since particles between 150 and 300 μm are preferred for regenerator beds. Extensive grinding will yield nearly 100% of the particles with a size <150 μm. Thus, even under optimum conditions the best yield one can expect for particles within the desired range of 150 to 300 μm is about 50%, and even less if one needs a narrower range of particle sizes. There is also another drawback of the crushed $Er_6Ni_2Sn$ particles, they have a tendency to absorb moisture from the air while sitting around on a desk or laboratory bench.

The tendency for $Er_6Ni_2Sn$ to form fine particles was also evident in the first trials using a packed bed of 200-350 μm mesh size particles in a Gifford-McMahon (G-M) cryocooler. Within a short period of operation the $Er_6Ni_2Sn$ particles fractured and formed fines and dust which reduced the efficiency of the cryocooler (see below).

**Table 1.** Size Distribution of $Er_6Ni_2Sn$ Particles

| After Initial Crushing and Minimal Grinding | | After Sifting and Regrinding of Particles >300 μm | |
|---|---|---|---|
| Particle Size (μm) | Fraction (wt.%) | Particle Size (μm) | Fraction (wt.%) |
| <150 | 30 | <150 | 41 |
| 150-250 | 19 | 150-250 | 24 |
| 250-300 | 15 | 250-300 | 17 |
| 300-500 | 36 | 300-500 | 17 |

**Liquid-Phase Heat Treating**

In an attempt to spheroidize the $Er_6Ni_2Sn$ particles, a liquid-phase heat treatment was tried. Since the $Er_6Ni_2Sn$ phase melts incongruently and since metallic melts usually have high surface tensions, we thought that if the metal particles were heated above the peritectic melting temperature the resultant liquid plus solid might form spheres, especially if there was ~50% liquid phase present. Two attempts were made to spheroidize the $Er_6Ni_2Sn$ particles. The 106 - 150 μm size particles were placed on flat alumina surface, and spread apart so that they did not touch each other and heated in an argon atmosphere. The first time the particles were heated to 950°C (~25°C above the peritectic temperature), and held for 10 minutes. The second time the particles were heated to 1050°C. In both cases the particles had a glassy appearance indicating that some melting had taken place, but the particles still retained their irregular shapes. This suggested that probably no more than 10 to 20% of the alloy was liquid at these temperatures.

**Solid State Sintering**

Because of our experience with particle attrition during the cyclic operation of a GM cryocooler, we tried to partially sinter the irregularly shaped $Er_6Ni_2Sn$ particles to permit gas flow and heat exchange but to prevent interparticle motion, fracture, and dusting. The sieved powders were poured into a quartz tube with a Ta weight (~65 g) placed on top to ensure a reasonably compact, continuous, and homogeneous cylinder of the $Er_6Ni_2Sn$ powder. Because of the hygroscopic nature of the $Er_6Ni_2Sn$, the powders were heated in a vacuum to ~100°C to drive-off the absorbed water before sintering.

After several trials it was found that a minimum sintering temperature of ~850°C and a minimum time of 3 hours are required to give a fairly strong porous rod of $Er_6Ni_2Sn$. Generally we sinter the material between 850 and 900°C for about 20 hours, i.e. overnight. The heat capacity of the sintered $Er_6Ni_2Sn$ material is shown in Fig. 3. As seen, the thermal properties are essentially the same as that of the initial bulk heat treated material except just in the vicinity of the ordering peak, indicating that the solid state sintering process has no major adverse affect on its heat capacity below 30 K, as a cryocooler regenerator material.

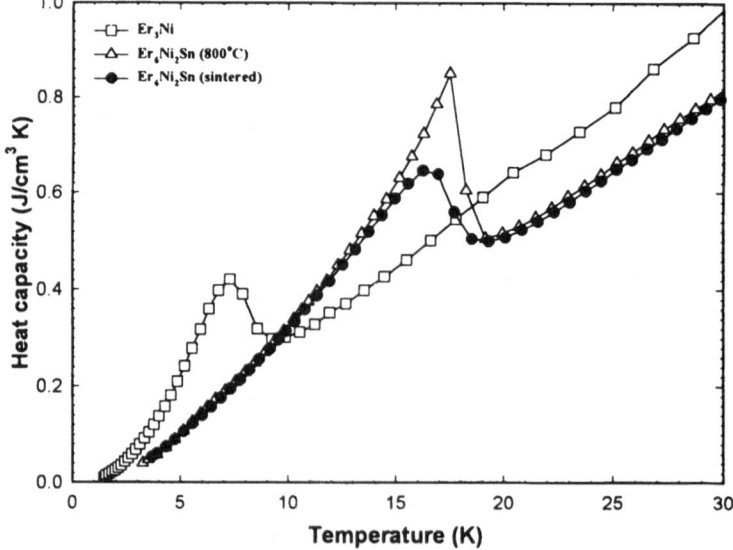

**Figure 3.** The volumetric heat capacity of the sintered $Er_6Ni_2Sn$ material. Also shown for comparison purposes is the heat capacities of $Er_6Ni_2Sn$ annealed at 800°C and $Er_3Ni$.

## CRYOCOOLER TEST RUNS

Testing the $Er_6Ni_2Sn$ irregular shaped powder consisted of replacing a portion of the Nd spheres in a second stage displacer of a modified CTI model 22 cold head, and running the unit under no-load conditions to determine bottom temperature achievable. The head had previously achieved no-load temperatures between 6.6 and 7.3 K in our test dewar with Nd spheres in the second stage displacer.

In the runs described below, 200-300 μm mesh $Er_6Ni_2Sn$ particles were used in two different packing schemes. In both cases there was a significant amount of fine dust produced over a short run time, far more than would be generated by spherical particles, such as $Er_3Ni$. The production of large amounts of dust indicated that during the alternate flow of the heat transfer gas through the particle bed the granular $Er_6Ni_2Sn$ worked loose and begun to pulverize. It is possible that the absorbed water may have also contributed to the attrition process due to freezing as the cryocooler is cooled below 273 K. The dusting likely interfered with seal seating and may have impeded the flow of gas through some portions of the matrix. The loosening of the pack would have certainly degraded the cooler's performance by allowing non-uniform gas flow through the second stage (i.e. the gas flow would be able to bypass the matrix in loosening regions).

The first packing scheme utilized the replacement of one-third of the Nd sphere layer at the high temperature end of the second stage by the $Er_6Ni_2Sn$ particles. This partitioning was chosen from a dead-reckoning of the temperature profile along the length of the second stage and temperature range of the $Er_6Ni_2Sn$ heat capacity characteristics. No improvement in cooling power was observed.

The second packing scheme utilized the Nd spheres for the lowest temperature layer (one-third), while the upper two-thirds consisted of a mixture of the Nd spheres plus the $Er_6Ni_2Sn$ particles. The same $Er_6Ni_2Sn$ material from the first packing arrangement was used for this second packing scheme, hence the Nd to $Er_6Ni_2Sn$ ratio was the same for both cases. The first cooldown run achieved a no-load temperature of 5.8 to 6.9 K, a modest improvement (~0.5 K) from the previous runs, which did not utilize the $Er_6Ni_2Sn$ powder. Significant dusting and pack settling was again observed upon disassembly. The pack was then topped-off with Nd spheres and a second run yielded temperatures of 6.4 to 7.4 K. A third run after inspection and cleaning of the seal yielded a temperature of 7 K.

Recent tests using $Er_3Ni$ and Pb spheres suggest a 1/3 $Er_3Ni$, 1/3 $Er_6Ni_2Sn$, 1/3 Pb layer scheme would better position the $Er_6Ni_2Sn$ in the temperature profile of the second stage matrix by placing it in the region where its heat capacity is higher than $Er_3Ni$ or Pb. Granular powder degradation can hopefully be mitigated by sintering the $Er_6Ni_2Sn$ particles (as noted above in the previous section) or by successfully fabricating $Er_6Ni_2Sn$ spheres. Once the dusting problem is solved a more comprehensive set of well-controlled experiments will be performed.

An electron microscopy examination of the $Er_6Ni_2Sn$ plus Nd mixture revealed that the $Er_6Ni_2Sn$ particles ranged in size from 50 μm to 300 μm with most of the particles (>75%) being smaller than 170 μm, and irregularly shaped. The Nd spheres had a uniform size of about 200 μm.

## ACKNOWLEDGMENT

The Ames Laboratory is operated for the U.S. Department of Energy by Iowa State University under contract no. W-7405-ENG-82. This work was supported in part by the Office of Basic Energy Sciences.

## REFERENCES

1. Gschneidner, K.A. Jr., Pecharsky, V.K., and Gailloux, M., "New Ternary Magnetic Lanthanide Regenerator Methods for the Low-Temperature Stage of a Gifford-McMahon

(GM) Cryocooler", *Cryocoolers 8*, Ross, R.G., Jr. ed., Plenum Press, New York (1995), pp. 685-694.

2. Osborne, M.G., Anderson, I.E. and Gschneidner, K.A., Jr., "Centrifugal Atomization of Rare Earth Metal and Intermetallic Compounds", *Advances in Powder Metallurgy and Particulate Materials - 1993: Powders, Characterization, Testing, and Quality Control*, vol. 1, Lawley, A. and Swanson, A., eds., Metal Powder-Industries Federation-American Powder Metallurgy Institute, Princeton, NJ (1993), pp. 65-73.

3. Osborne, M.G., Anderson, I.E., Gschneidner, K.A., Jr., Gallioux, M.J. and Ellis, T.W., "Centrifugal Atomization of Neodymium and $Er_3Ni$ Regenerator Particulate", *Adv. Cryogenic Engin.*, vol. 40, (1994), pp. 631-638.

# Promising Refrigerants for the 4.2 - 20 K Region

**L.P. Bozkova and A.M. Tishin**

Physics Department
M.V. Lomonosov Moscow State University
119899 Moscow, Russia

## ABSTRACT

Changes in the magnetic entropy of the rare-earth magnets $DyAlO_3$ and $ErAlO_3$ and $Dy_xEr_{1-x}AlO_3$ are calculated. It is shown that lanthanide orthoaluminates compounds $Dy_xEr_{1-x}AlO_3$ are promising working materials for magnetic refrigerators operating in the temperature region 4.2-20 K.

## INTRODUCTION

One of the major problems in magnetic refrigeration is the search for efficient working materials for refrigerators. Among the most important characteristics of highly efficient working materials in the low temperature region are the following: the extended temperature range where there is a large change in the entropy magnetic; a low temperature magnetic phase transition; sufficiently high thermal conductivity. It should be noted that the listed parameters are interconnected, although indirectly. At a sufficiently low Nèel temperature, $T_N$, the efficiency of using a refrigerant compound is to a great extent determined by the value of the magnetic moment of the lanthanide atom and the number of magnetic atoms per unit volume.

At the beginning of the century, Langevin showed for the first time that changing the magnetisation of a paramagnet will generally result in a reversible temperature change. Achievements in the past decade allowed the extension of both the temperature range and the spectrum of magnetic materials under study. Materials with more exotic structures have been considered [1]. Today investigations have been concentrated mainly on the following temperature subranges: 4.2-20 K [2,3], 20-77 K [4,5] and the room temperature region [6,7]. Whereas some suitable working materials have been found in the first and second subranges, the extensive research is currently under way for the third subrange and the work is still far from being complete in all temperature regimes.

## MATERIALS FOR USE BELOW 20 K

The two garnets $Gd_3Ga_5O_{12}$ (GGG) and $Dy_3Al_5O_{12}$ (DAG) have been investigated both experimentally and theoretically, and are considered as the most promising working materials for the temperature region below 20 K [2, 3]. Although considerable success has been achieved in developing magnetic refrigerants in the 4.2-20 K temperature region, the search for novel working materials is still an important task.    Certain progress in this direction was made by Kuz'min and Tishin [8, 9], who have presented detailed calculations for orthoaluminates with perovskite structure - $RAlO_3$, where R is a lanthanide metal. It was shown that the lanthanide orthoaluminates $DyAlO_3$ and $GdAlO_3$ are novel and effective refrigerants for the 4.2-20 K region. The calculations were made using the mean-field approximation (MFA) method. The results of the first experimental studies of this class of working materials were presented in Refs. [10, 11]. The data reported by Kimura et al [10] indicate that $ErAlO_3$ (ErOA) is an efficient working material for magnetic refrigerators. However, $DyAlO_3$ (DOA) does not seem to have been studied in detail by these authors.

The purpose of the present study is to search for working materials that would be more efficient than DOA and ErOA. We have utilised the MFA to calculate magnetic entropy changes in DOA, ErOA and the system $Dy_xEr_{1-x}AlO_3$. These compounds have the   perovskite type structure. Making use of the results reported in Ref. [8] the expression for change in the magnetic part of entropy can be written in the following form:

$$\Delta S_m = S(B,T) - S(0,T) = Nk \, (\ln \cosh x - x \tanh x) \quad (1),$$

where $x = \sigma \, \Theta /T + \mu \, B/kT$, and N is the concentration of Dy (Er or $Dy_xEr_{1-x}$) ions, k is the Boltzmann constant, $\sigma$ is the reduced magnetisation, $\Theta$ is the paramagnetic Curie temperature. $\mu$ is the magnetic moment, and B is the magnetic field. To calculate the temperature dependence of the magnetic entropy change at different values of the field we used expression (1). The values of the parameters $\Theta$, $\mu$, $T_N$ used in our calculations are given in table 1.

**TABLE 1.** Model parameters used in our calculations

| Compound | DOA | ErOA | $Dy_{0.9}Er_{0.1}AlO_3$ |
|---|---|---|---|
| Nèel point (K) | 3.52[a] | 0.6[a] | 2.86[b] |
| Paramagnetic Curie temperature (K) | -1.5[a] | -0.6[a] | -1.0[c] |
| Magnetic moment along the easy-axes field direction ($\mu_B$) | 6.88[a] | 3.44[a] | 6.62[d] |

[a] Taken from Ref. [9]

[b] Calculated as quadratically dependent on x, according to   assumptions reported in  Ref. [9].

[c] Calculated from the data Ref. [9].

[d] Calculated  by the formula : $\mu = (6.88^2 x + 3.44^2 (1-x))^{1/2}$

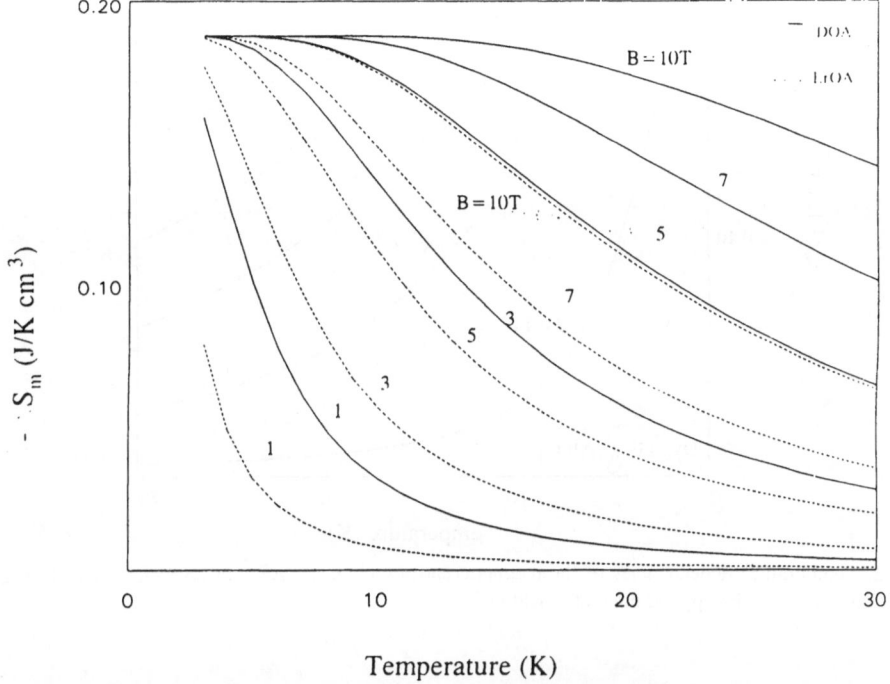

**Figure 1.** Temperature dependencies of the magnetic entropy change $\Delta S_m$ in DOA (solid lines). and ErOA (dashed lines). The numerical values indicate the applied magnetic field in T.

For comparison Fig. 1 represents the results of calculations of the temperature dependence of the magnetic part of entropy for DOA and ErOA in magnetic fields ranging from 1 to 10T. We find it convenient to calculate the entropy changes per $cm^3$. This allows comparison of the $\Delta S_m$ values obtained with those measured experimentally. To this end, the dimension of $\Delta S_m$ in Fig. 1 and 2 is J/K $cm^3$. It follows from Fig. 1 that the change in the magnetic part of entropy for DOA is considerably greater than that for ErOA for the same magnetic field. This is more evident at high fields. For example, at T = 20 K and the field B = 5 T : $\Delta S_m$ ~111.85 mJ/K $cm^3$ for DOA and 40.43 mJ/K $cm^3$ for ErOA. At the same temperature in the field B = 7 T : $\Delta S_m$ ~148.25 mJ/K $cm^3$ for DOA and 68.6 mJ/K $cm^3$ for ErOA.

Nèel point is given, in accordance with the data reported in Ref. [9], as quadratically dependent on x. When calculating the concentration dependence of the magnetic moment, we relied on the results reported in Ref. [12]. According to that study, one might expect the magnetic moment of Er atoms in the DOA matrix to have the values close to 3.44 $\mu_B$. It can be assumed that in the case under consideration the magnetic moment of Er atoms is even larger than this value [13]. Unfortunately, the authors of the present study were unable to find in the literature any direct experimental data concerning the concentration dependencies of the Nèel point and the magnetic moment in compounds of the $R_xR_{1-x}AlO_3$ type (see, for example, Ref. [14]). We used the reported dependencies to calculate the temperature dependence of magnetic entropy changes for the given system. As an illustration, Fig. 2 shows the calculated results for the compound $Dy_{0.9}Er_{0.1}AlO_3$.

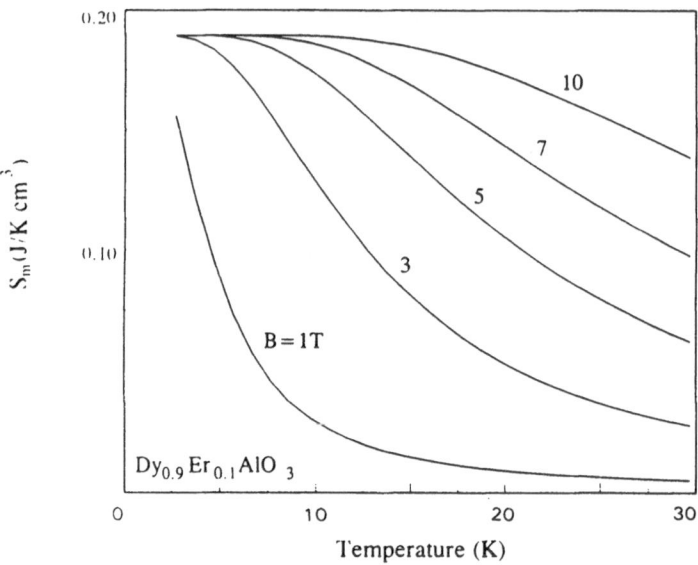

**Figure 2.** Temperature dependencies of the magnetic entropy change $\Delta S_m$ in $Dy_{0.9}Er_{0.1}AlO_3$. The numerical values indicate the applied magnetic field in T.

**Table 2.** The absorbed heat $Q$, from the load at T=4.2 K by DOA, ErOA and $Dy_{0.9}Er_{0.1}AlO_3$ in the ideal Carnot cycle from 4.2 to 20 K under B = 5,7,10 T

| Compound | B, T | DOA | ErOA | $Dy_{0.9}Er_{0.1}AlO_3$ |
|----------|------|------|------|------------------|
| $Q(J/cm^3)$ | 5 | 0.47 | 0.17 | 0.44 |
| $Q(J/cm^3)$ | 7 | 0.62 | 0.29 | 0.60 |
| $Q(J/cm^3)$ | 10 | 0.73 | 0.45 | 0.72 |

Analysis of the data represented in Figs. 1 - 2 makes it possible to conclude that the substitution of Er atoms for Dy atoms results in a noticeable lowering of the Nèel point without any significant reduction of the value of $\Delta S_m$. Thus the introduction of Er atoms in DOA produces a desirable effect. We have calculated the value of $\Delta S_m$ as dependent on the concentration of Dy at the temperature T = 4.2 K for various values of the field. These results show that as the concentration of Dy decreases from 100% to 85% $\Delta S_m$ changes from 120 to 114 mJ/K cm$^3$ in the field B = 1 T, and from 185.7 to 185.0 mJ/K cm$^3$ in the field B = 3 T. As the field increases and exceeds 5 T the value of $\Delta S_m$ only changes in the third decimal point.

We have calculated the absorbed heat, Q, from the load at T = 4.2 K by DOA, ErOA, and $Dy_{0.9}Er_{0.1}AlO_3$ in the ideal Carnot cycle ranging from 4.2 to 20 K (see Table 2). In accordance with the data reported in Ref.[8], we have assumed that the lattice specific heat can be neglected.

In conclusion it should be pointed out that our calculations show that the system

$Dy_xEr_{1-x}AlO_3$ is promising as a working material for magnetic refrigerators as compared with DOA in the 4.2-20 K temperature region. This conclusion is substantiated, first, by the fact that the proposed materials possess the magnetic moment per unit volume practically equal to that of DOA (and hence the change in the magnetic part of entropy of these materials is close to that of DOA). Second, these compounds have a lower Nèel point. It is suggested that $Dy_{0.9}Er_{0.1}AlO_3$, or a compound of similar composition, is a promising magnetic refrigerant.

## ACKNOWLEDGEMENTS

We would like to thank  Prof. Karl A. Gschneidner, Jr. and  Dr. Vitalij K. Pecharsky for valuable corrections and changes in the text of the manuscript. The authors wish also to thank Prof. N.P. Kolmakova for helpful discussion. This work was supported by the  NATO Linkage Grant.

## REFERENCES

1.  M.D. Kuz'min and A.M. Tishin, *Cryogenics*, vol. 32 (1992), p. 545.

2.  A. Barclay, O. Moze, and L. Paterson, *J. Appl. Phys.*, vol. 50 (1979), p. 5870.

3.  R. Li, T. Numazawa, T. Hashimoto, A. Tomokiyo, T. Goto, and S. Todo, *Advanced in Cryogenic Engineering*, Edited by R.P. Reed and A.F. Clark, vol. 32 (1986), Plenum, New York, p. 287.

4.  H. Takeya, V.K. Pecharsky, K.A. Gschneidner, Jr., and J.O. Moorman, *Appl. Phys. Lett.*, vol. 64 (1994), p. 2739.

5.  K.A. Gschneidner, Jr., H. Takeya, J.O. Moorman, and V.K. Pecharsky, *Appl. Phys. Lett.*, vol. 64 (1994), p. 253.

6.  G.V. Brown, *J. Appl. Phys.*, vol. 47 (1976), p. 3673.

7.  M.D. Kuz'min and A.M. Tishin, *Cryogenics*, vol. 33 (1993), p. 868.

8.  M.D. Kuz'min and A.M. Tishin, *J. Appl. Phys.*, vol. 73 (1993), p. 4083.

9.  M.D. Kuz'min and A.M. Tishin, *J. Phys. D: Appl. Phys.*, vol. 24 (1991), p. 2039.

10. H. Kimura., T. Numazawa and M. Sato, T. Ikeya and T. Fukuda, *J. Appl. Phys.*, vol. 77 (1995), p. 432.

11. H. Kimura, T. Numazawa and M. Sato, T. Ikeya and T. Fukuda, *Preceed. of Japan U.S. Workshop on Functional Fronts in Advanced Ceramics*, Tsukuba, Japan, (1994), p. 262.

12. P. Bonville, J.A. Hodges and P. Imbert, *J. Physique*, vol. 41 (1980), p. 1213.

13. Prof. N.P. Kolmakova, private communication.

14. Landolt-Bornstein, *Magnetic and Other Properties and Oxides and Related Compounds*, New Series, Springer, Berlin, Group III.

# The Los Alamos Solid-State Optical Refrigerator

R.I. Epstein, B.C. Edwards, C.E. Mungan and M.I. Buchwald

Los Alamos National Laboratory
Los Alamos, New Mexico 87545

## ABSTRACT

Optical refrigeration may provide the physical basis for solid-state cryocooling. Devices based on this physics would be vibrationless, compact, and free of electromagnetic interference. Having no moving parts and being pumped by a diode lasers, optical refrigerators would be rugged with operating lifetimes of years. Experiments at Los Alamos National Laboratory have demonstrated the basic physical principles of optical refrigeration.[1] Design studies suggest that optical refrigerators with efficiencies comparable to other small cryocoolers should be realizable with existing technologies.[2]

## INTRODUCTION

The Los Alamos Solid-State Optical Refrigerator (LASSOR) concept has a laser diode pumping a *cooling element* which then radiates its heat through anti-Stokes fluorescence. That is, the cooling element absorbs photons at one wavelength and then emits photons at shorter wavelengths. The energy difference between the absorbed and emitted radiation is supplied by annihilating thermal phonons and thereby cooling the solid.[3,4] The complete refrigerator also requires some way of disposing of the fluorescence and a means of attaching the load to the cooling element.

One LASSOR design is described below, but many others can be imagined for specific applications.[5] Since optical refrigeration occurs on the atomic scale, it offers great flexibility in geometry and size. It should be possible to construct 10-100 watt cooling capacity LASSORs for cooling large mirrors as well as a milliwatt LASSORs for small electronic components. The optical refrigerators may even be integrated directly into the device to be cooled. For example, the glass of a mirror may be the LASSOR cooling element[6] or electronic components may be deposited directly on a cooling element.[7]

All LASSORs would be vibration-free. Their lifetimes would be limited by that of the diode laser (which can be many years[8]), and additional reliability can be achieved by using redundant lasers. The cooling efficiency, operating temperature and power density could vary over wide ranges depending on the atomic properties of the cooling element.

The remainder of this paper is organized as follows. We first discuss the basics of optical refrigeration and then describe the laboratory confirmation of these concepts. Finally, we present a design for a one-watt LASSOR device . Additional information on the LASSOR program can be found at the web site <http://sst.lanl.gov/~edwards/cooling.html>.

Cryocoolers 9, Edited by R.G. Ross, Jr.
Plenum Press, New York, 1997

## BASIC PRINCIPLES

The essentials of optical refrigeration. can be illustrated by the example of a three-level atom imbedded in an otherwise transparent solid (see Fig. 1). Laser light of photon energy $E_L = E_3 - E_2$ pumps an atom from energy level 2 to level 3. A subsequent radiative deexcitation moves the atom either to level 2 or to the ground state, level 1. In the former case the emitted photon has the same energy $E_L$ as the absorbed laser photon and the system is unchanged. In the latter case the fluorescing photon carries away energy $E_F = E_3 - E_1 > E_L$ and there is a net shift of an excitation from level 2 to level 1. The relative populations of levels 1 and 2 are thus pushed out of thermal equilibrium. To restore thermal equilibrium the atom jumps from level 1 to 2 by absorbing a phonon of energy $Ep = E_2 - E_1$, thereby decreasing the solid's thermal energy.

The three-level optical refrigerator cycle highlights the necessary and desirable properties of cooling materials. The foremost requirement is that the fluorescent efficiency be nearly unity. That is, an atom in level 3 must nearly always deexcite to level 1 or 2 by emitting a photon . The alternative is that the atom *nonradiatively* decays producing phonons whose total energy is much greater than $E_3 - E_2$ and can overwhelm the optical refrigeration effect. High fluorescent efficiencies are possible if the energy gap, $E_3 - E_2$, is large compared to the maximum phonon energy $Ep,max$ the solid can support. When this condition is satisfied, a nonradiative decay requires that at least $(E_3 - E_2)/Ep,max$ phonons are emitted simultaneously; such high-order processes are strongly inhibited. Additionally, it is vital that the transparent host be exceedingly pure. Small levels of certain contaminants can quench the fluorescence and lower the quantum efficiency.

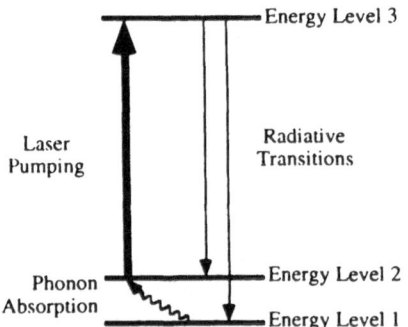

**Figure 1.** A three-level optical refrigerator

The next requirement is that the populations of the lower energy levels remain close to thermal equilibrium through the emission and absorption of phonons. This is usually possible if the energy spacing $E_2 - E_1$ is not large compared to $Ep,max$. However, at low temperatures the phonon density falls off and the thermalization rate of the low lying levels can slow, limiting the performance of the optical refrigerator.

If the decay rates from level 3 to levels 1 and 2 are equal, then the average heat lift per pump photon is $0.5(E_2 - E_1)$, and the coefficient of performance of the refrigerator is

$$COP = 0.5 \ (E_2 - E_1 )/(E_3 - E_2) . \tag{1}$$

The refrigerator's efficiency is thus improved if the energy gap, $E_3 - E_2$, is decreased while the gap, $E_2 - E_1$, is increased. We see that there may be a trade off where the criteria for a high COP is in conflict with the conditions for a high fluorescent efficiency and rapid thermalization rate of the lower levels.

The optimal operating temperature of the refrigerator is in large part set by the size of the energy gap, $E_2 - E_1$. As the temperature decreases, the fraction of atoms in level 2 decreases approximately as the Boltzmann factor, $\exp[(E_1 - E_2)/kT]$, where k is the Boltzmann constant. Eventually, there would be too few atoms in level 2 to pump with the laser, and the refrigerator cannot operate. The minimum operating temperature would depend on the fluorescent efficiency, the atomic cross sections as well as the design of the cooler. As a rough guide, we expect the useful operating temperature Top to be approximately

$$T_{op} \sim 0.1(E_2 - E_1)/k. \tag{2}$$

Comparing Eqs. (1) and (2) shows the tradeoff between efficiency and operating temperature, $COP \propto 1/T_{op}$.

## LABORATORY STUDIES

In our experiments to date we use a heavy-metal-fluoride glass (ZBLAN) doped with trivalent ytterbium ions. Each $Yb^{3+}$ ion possesses only two groups of energy levels below the UV absorption edge of the host glass.[9] These groups are separated by an energy of 1.3 eV corresponding to a wavelength of ~1000 nm. The ground-state group is split into four Stark levels and the excited-state group is split into three levels, as shown schematically in the inset of Fig. 2. The main part of this figure shows the measured absorption and fluorescence spectra of a sample of Yb doped ZBLAN. The mean energy of the fluorescent photons corresponds to a wavelength $\lambda_F = 995$ nm, as indicated by the vertical line. Pumping this glass in the long-wavelength tail of the absorption spectrum moves excitations from the top of the ground-state group to the bottom of the excited-state group. The relative populations of the Stark levels *within* each group are thereby shifted slightly out of thermal equilibrium. By absorbing phonons

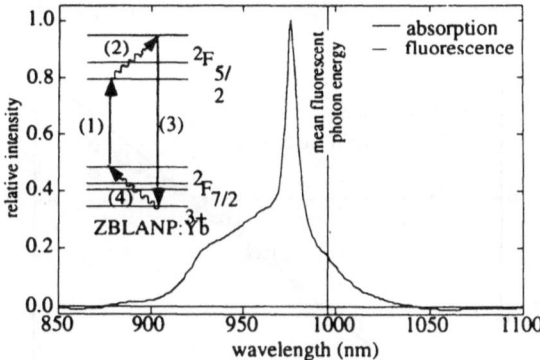

**Figure 2.** *Main plot*—Absorption coefficient (dashed curve) and fluorescence spectrum (solid curve) at room temperature for Yb doped ZBLAN glass  The wavelength corresponding to the mean fluorescent-photon energy, $\lambda_F$, is indicated by a vertical line at 995nm. *Inset plot*—The schematic energy-level structure of Yb doped ZBLAN; the splittings within each group have been exaggerated for clarity. The arrows denote a typical cooling cycle: (1) laser pumping, (2) phonon absorption, (3) radiative decay, and (4) additional phonon absorption.

from the host material, thermal equilibrium within each group is restored. Radiative decays from the excited-state to the ground-state groups produce photons that carry off both the absorbed radiative and thermal energies.

Each fluorescent photon carries off, on average, thermal energy equal to the difference between the pump-photon and the mean fluorescent-photon energies. In the ideal case where there are no nonradiative relaxations from the excited- to the ground-state groups, the cooling power, $P_{cool}$, is proportional to the absorbed pump power, $P_{abs}$, and to the average difference in the photon energies of the pump and fluorescence radiation. In terms of wavelength $\lambda$ of the pump radiation, the cooling power is

$$P_{cool}(\lambda) = P_{abs}(\lambda)(\lambda - \lambda_F)/\lambda_F.$$ (3)

Two separate experimental arrangements were used to test the validity of the optical refrigeration relation, Eq. (3). The specifics of these two experiments can be found in ref. 1. The main results of these experiments is shown in Fig. 3. which shows the cooling efficiency, defined as the ratio of the cooling power to the absorbed laser power, that was measured with the two experiments. In the first of these experiments, photothermal-deflection spectroscopy[10] was employed to measure the temperature changes induced by the pump laser in the interior of a sample. The pump beam from a ~ 1 micron cw laser was focused into the sample and heated or cooled a thin cylindrical column. The temperature change produced a small density change in the illuminated region thereby generating a weak lens. A helium-neon laser probe beam was directed through the sample so that it could be deflected by the thermally induced lens. Angular deflections of this probe beam were measured with a position-sensitive photodetector. An optical chopper placed in the path of the pump beam modulated the photothermal-deflection signal. The direction of the deflection indicated whether the sample was heating or cooling, and the amplitude of the deflection was proportional to the magnitude of the temperature change.

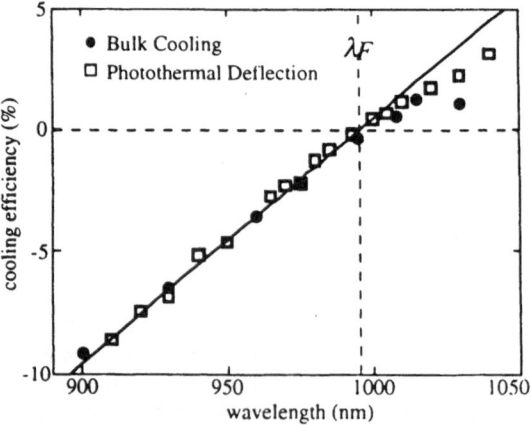

**Figure 3.** Cooling efficiencies measured in the photothermal-deflection (filled circles) and bulk-cooling (open squares) experiments. Negative efficiencies correspond to heating. The solid curve is a plot of $Pcool/Pabs$ from Eq. (3).

In the second experimental arrangement, the equilibrium temperatures of a bulk 2.5 × 2.5 × 6.9 mm³ sample were measured when the sample was continuously pumped by a laser beam. The sample was positioned at the center of a vacuum chamber and rested on two thin, vertical glass slides that limited the conductive heat transport from the chamber walls. The inner wall of the chamber was painted black to absorb stray pump radiation as well as the fluorescence emitted by the sample. The pump-laser beam is directed along the long axis of the sample. The temperature of the sample was monitored with an infrared camera. To account for the changes in the sample temperature arising from temperature drifts of the chamber, a reference sample was positioned ~1 cm away from the test sample, outside the pump-beam path but on the same glass supports in the chamber. Temperature differences of 0.02 K between the pumped and reference samples were resolvable.

The equilibrium temperature, $T_S$, of the sample is established by a balance between the optical refrigeration and the heat load from the environment. In our setup, the dominant heat load was from the radiative coupling between the walls of the vacuum chamber at ambient temperature $T_C$ and the sample. If the sample and walls radiate as blackbodies, then

$$P_{cool} = \sigma A (T_C^4 - T_S^4),$$ (4)

where $\sigma$ is the Stefan-Boltzmann constant and $A$ is the surface area of the sample.

Fig 3. shows that the results of the photothermal deflection and bulk cooling experiments were in good agreement with each other and with the predictions of Eq. (3). In agreement with the theory, most of the data lie on a straight line. If the quantum efficiency were unity, the zero crossing would occur at $\lambda_F = 995$ nm. Experimentally, the zero crossing agrees with this prediction to within 3 nm, indicating that the fluorescence quantum efficiency is at least 0.997. The deviations from linearity for $\lambda \geq 1020$ nm may be due to parasitic heating, perhaps from surface contamination.

Our most recent photothermal deflection experiments, which were carried out at cryogenic temperatures, have shown cooling efficiencies comparable to those of Fig. 3. Additionally, recent bulk cooling experiments using ZBLAN fibers have shown temperature drops more than a factor of 40 greater than those discussed in ref. 1. The specifics of these experiments will be reported elsewhere.

## LASSOR DESIGN

Since a solid-state refrigerator is particularly well suited for space applications, our initial designs have focused on these applications. Figure 4 illustrates a 1-watt version of the Los Alamos Solid-State Optical Refrigerator (LASSOR) that might be used for cooling a space-borne infrared detector or a germanium gamma-ray spectrometer to 77K.

This design is based on ytterbium doped ZBLAN glass. A diode laser coupled with an optical fiber delivers ~1020 nm pump radiation to the cooling element. The fiber carrying the pump radiation passes through the wall of the cryocooler and to a cylindrical block of Yb doped ZBLAN. Dielectric mirrors that reflect the pump radiation are deposited on both ends of the cooling element. The pump radiation enters the cooling element through a small hole in one of the mirrors. This radiation is directed parallel to the axis of the cylinder so that it is repeatedly reflected from the dielectric mirrors and from the sides of the glass cylinder (by total internal reflection). Ultimately, the pump radiation is absorbed by Yb ions, and the doped glass cools as it fluoresces. Most of the isotropic fluorescent radiation escapes from the cooling element and is absorbed by the warm walls of the cooling chamber. A metal mirror is attached on the back of the upper dielectric mirror creating a completely shadowed region where the cold finger can be mounted. The inner surfaces of the chamber walls contain a coating that readily absorbs 1 micron radiation while having a very

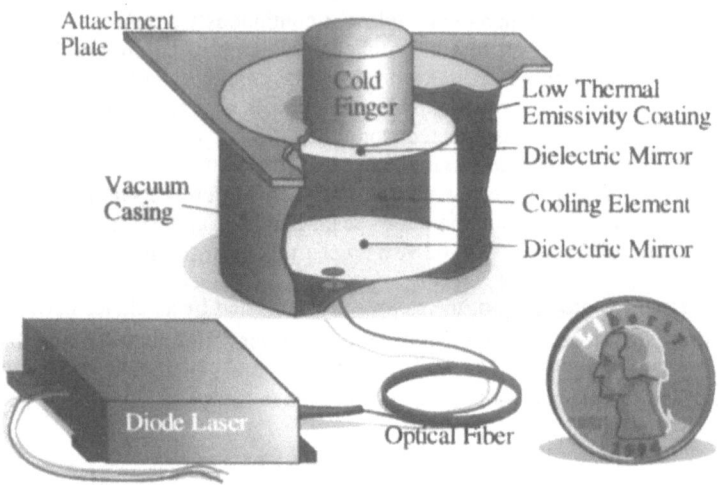

**Figure 4.** A design for a 1-watt LASSOR

high reflectivity for ~10 micron thermal emission. This coating would lessen the radiative heat load from the warm chamber walls.

We estimate that a practical, first-generation, optical refrigerator of the type described here could operate near 77 K with almost 0.5% efficiency (DC power to cooling power), weigh less than 2 kg/W and have a lifetime of 10 years of continuous operation.[2]


## ACKNOWLEDGMENT

This work was carried out under the auspices of the U.S. Department of Energy and was supported in part by IGPP at LANL.


## REFERENCES

1.    Epstein, R. I., Buchwald, M.I., Edwards, B. C., Gosnell, T.R., & Mungan, C.E. "Observation of Laser-Induced Fluorescent Cooling of a Solid", *Nature*, **377** (1995) 500.

2.    Edwards, B. C., Buchwald, M. I., Epstein, R. I., Gosnell, T. R. & Mungan, C. E. "Development of a Fluorescent Cryocooler" *Proceedings of the 9th Annual AIAA/Utah State University Conference on Small Satellites*, (ed. Redd, F.), (Utah State University, Logan, 1995).

3.    Pringsheim, P. "Zwei Bemerkungen über den Unterschied von Lumineszenz- und Temperaturstrahlung", *Z. Phys.* **57** (1929) 739.

4.    Kastler, A., "Quelques Suggestions Concernant la Production Optique et la Détection Optique D'une Inégalité de Population des Niveaux de Quantification Spatiale de Atomes", *J. Phys. Radium* **11** (1950) 255.

5.    Epstein, R. I., Edwards, B. C., Buchwald, M.I., Gosnell, T.R. "Fluorescent Refrigeration" (1995) U.S. Patent #5,447,032.

6.    Gustafson, E. private communication (1995).

7.    Razeghi, M. private communication (1995).

8.    Razeghi, M. "InGaAsP-based High Power Laser Diodes" *Optics & Photonics News* **6**, No. 8 (1995) 16.

9.    Dieke, G.H. *Spectra and Energy Levels of Rare Earth Ions in Crystals* (Interscience, NY, 1968).

10.   Boccara, A. C., Fournier, D., Jackson, W. & Amer, N. M. "Sensitive Photothermal Deflection Technique for Measuring Absorption in Optically Thin Media", *Opt. Lett.* **5** (1980) 377.

# Microcooling: Study on the Application of Micromechanical Techniques

**J.F. Burger, H.J.M. ter Brake, M. Elwenspoek and H. Rogalla**

University of Twente, Faculty of Applied Physics,
P.O. Box 217, 7500 AE Enschede, The Netherlands

## ABSTRACT

Small scale low temperature electronic applications would largely benefit from the development of a closed-cycle microcooler, with a cooling power in the milliwatt range. In this article first the requirements for microcooling of LT electronics are considered. A review is presented of micromechanical technologies and their possible application in a microcooler. Conventional cooling cycles are briefly reviewed on their suitability for miniaturisation, and as an illustration, some first ideas on a miniature regenerative and a miniature recuperative cooling cycle are presented.

## INTRODUCTION

### Low Temperature Electronics

Low temperatures provide an excellent operating environment for conventional and for superconducting electronics. It increases the speed of digital systems, it improves the signal to noise ratio and the bandwidth of analog devices and sensors, and it reduces aging. For superconducting electronics, it is an essential condition that the operating temperature is well below the critical temperature of the superconductor. There is a broad range of applications making use of cold electronics, such as computers, amplifiers, mixers, fast AD and DA converters, IR detectors and SQUID magnetometers[1]. Many of these systems need only a modest cooling power, as low as 10 mW or even lower. As illustrated by figure 1, a range of cooling techniques is available for cooling such devices, but these cooling techniques are often largely oversized. For that reason the development of a *microcooler* would be of great use for low temperature electronics.

### Small Scale Cryocoolers

Walker[1] has given an overview of small scale cryocoolers. The most striking results in this field were obtained by Little who scaled down the conventional Joule Thomson cold stage to miniature size by patterning gas channels in glass layers using photolithographic techniques[3]. This cooler, however, has the disadvantage that high pressure gas is needed from a relatively large storage bottle. To our knowledge, the smallest closed cycle cooler realised so far is manufactured by Inframetics[4]. This company makes integral Stirling coolers of 0.3 kg weight, 9

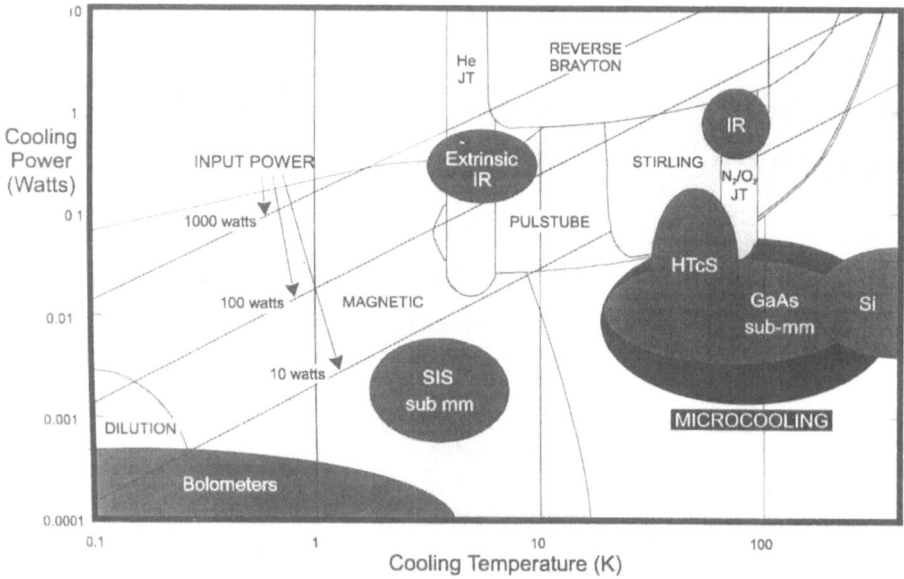

**Figure 1.** Overview of cooling capacity and cooling temperatures for different coolers (light gray), low temperature applications (dark gray) and intended microcooling regime (black). Input power lines are based on a 5 % Carnot efficiency and 300 K at the warm end. Based on data of Jet Propulsion Lab[2].

cm maximum size, and 0.15 W cooling power at 80 K with an input of 3 W. A further reduction in size will be limited by the manufacturing techniques, and in this respect micromechanical techniques have been suggested by Walker as an alternative to build micro-scale cryocoolers[5].

**Microcooling Project**

We recently started a project with the aim to realise a closed-cycle microcooler. In such a project a close collaboration is desired between different engineering disciplines: cryogenic engineering, fine-mechanics and micro-mechanics. The project is, therefore, carried out in a cooperation between the Low Temperature Division and the Research Institute for Micro Electronics, Materials Engineering, Sensors and Actuators (MESA) at the University of Twente. This institute is experienced in micromechanical technologies. These technologies use special deposition, etching and bonding techniques - originating from the IC-technology - with which very small mechanical structures can be made, in or on planar substrates. The project is not strictly limited to micromechanical technologies, however.

In this paper a study is presented on the application of micromechanical techniques, which will be used as a starting point in the microcooling project. Firstly, the main requirements for microcooling are given. Secondly, a short survey of micromechanical technologies is presented, as well as some typical applications that may be of importance for microcooling. Thirdly, some considerations about miniaturisation of commonly used cooling cycles are given. As an illustration, a concept for a small regenerative cold head is presented, as well as an idea for a recuperative sorption cooler.

**REQUIREMENTS**

The requirements for microcooling of low temperature electronics can be summarised as follows:

**1. Working temperature:** *60 - 80 K.* Our first aim is cooling of HTc superconducting devices, that operate around $LN_2$ temperature. The temperature of conventional electronics is not very critical. In general, it should not be lower than 40 K because of carrier freeze-out below that temperature. Therefore, $LN_2$ temperature is very suitable for many applications, also for hybrid electronics. For some applications, however, lower or higher temperatures may be required (see also figure 1).

**2. Cooling power (net):** *5 - 50 mW.* The net required cooling power is determined by three factors: (a) the power dissipated in the cold device, (b) the heat leakage from the environment and (c) the desired cool-down time. Each of these factors are discussed below.

(a) The power dissipated by superconducting electronics is very small, even a microprocessor with thousands of junctions on a substrate of 5 mm x 5 mm may dissipate less than a few milliwatts[6]. Dissipation of conventional electronics is more severe and, therefore, only small circuitry can be cooled. (b) The heat leakage is determined by conduction through the supporting structure and the wires, and by radiation. Conduction through the wires and radiation do not play a major role. As an example, one manganin wire between 80 K and 300 K with a length of 10 cm and a diameter of 100 μm gives a heat leakage of 0.2 mW. Further, radiation at 60 - 80 K can be reduced to roughly 1 mW via appropriate shielding. Note that the required *gross* cooling power may be much higher in order to deal with losses by conduction through the cooler itself, which puts severe restrictions on the design of the cooler. (c) The cool-down time is determined by the combination of the heat capacity at the cold side and the cooling power. The specified net cooling power is for the stable end temperature, at higher temperatures usually a higher cooling power is available which decreases the cool-down time.

**3. Closed cycle with long life-time.** We aim to realise a life-time of several years. This puts restrictions on the use of moving components, especially if friction is involved.

**4. Low electromagnetic and mechanical interference.**

**5. Other issues of interest are:** low heat rejection (i.e. high efficiency), long life, low price.

## MICRO SCALE DESIGN ASPECTS

In this section first a general review of micromechnical materials and processing techniques is presented. After that, a brief review is given of micromechanical components that may be of use for a microcooler.

### Micromechanical Materials and Processing Techniques

The basis of micromechanics is that silicon, besides its traditional role as an electronic material, can be exploited as a high-precision high-strength high-reliability mechanical material. Much advantage can be taken from the advanced microfabrication, originating from the IC technology. Furthermore, Single Crystal Silicon (SCS) is available in an extremely high quality, both electrical and mechanical. SCS is a brittle material that yields by fracturing, unlike most metals that deform inelastically. However, silicon is not as fragile as it may seem. The Young's Modulus ($7.0 \cdot 10^9$ Pa) is similar to that of stainless steel, and the yield strength is several times higher. Therefore, if surface and bulk defects are prevented[7], elastic structures made of SCS do not show mechanical fatigue, and mechanical structures can be obtained with strengths and lifetimes exceeding that of the highest strength alloy steels. With respect to the application of silicon in microcooling devices, it is important to note the high thermal conductivity of silicon, which may be a condition or a limitation for certain applications.

Other important micromechanical wafer materials are glass and quartz. In contrast to silicon, these materials show a low thermal (and electrical) conductivity. Glass wafers can be obtained with thermal expansion coefficients that match the expansion of silicon over a wide temperature range (T > 300 K), so that bonding between these materials is possible. Quartz and glass can be structured in similar ways as silicon.

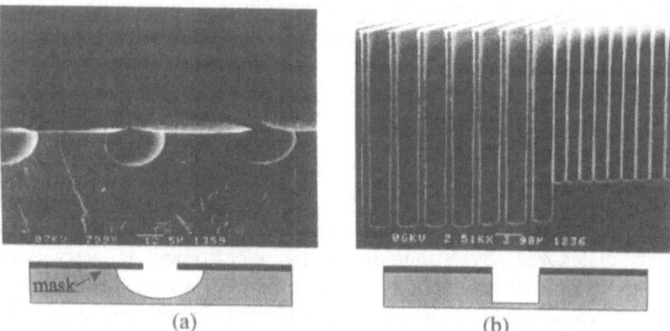

**Figure 2.** The difference between isotropic (a) and anisotropic (b) etching. The profiles are etched using chemical plasma etching (a) and synergetic reactive ion etching (b)[7].

Patterns are transferred to the substrate or a thin film by means of photolithographic techniques, after which it can be processed further, for instance by means of etching. For this, a patterned glass layer is used as a mask during the exposure of a photosensitive layer that is applied on top of the material. The wavelength of the light that is used determines the ultimate pattern resolution (typically about 1 μm).

**Etching techniques.** A key technology in micromechanics is the removal of (substrate) material at non-covered locations: etching. In general, one distinguishes isotropic and anisotropic etching (see figure 2) with wet and dry agents. Another important etch parameter is the etch *selectivity*, which is the ratio between the etch rate of the material to be etched and that of the mask material.

Numerous chemical etchants can be used for *wet etching* of silicon. KOH is a common etchant; it etches anisotropic along the crystal surfaces, making it possible to realise unique geometries. The etch rate is also strongly dopant dependent, so that the etch rate can be controlled as a function of the etched depth by varying the doping level.

*Dry etching* stands for the use of a gas plasma to dissociate and ionisate relatively stable gas molecules, forming chemically reactive and ionic particles[8]. Under the condition that a proper chemistry is chosen, these particles can be used to react with the solid substrate to form volatile products. The main advantage of dry etching over wet chemical etching is its capability to etch highly anisotropic patterns, *not* limited by crystal anisotropy - also in polycrystalline and amorphous materials. In addition, etch profiles and mask-underetching can be controlled by varying the plasma chemistry. Very high selectivities with respect to mask materials can be obtained (see figure 2b).

Another etching technique, *abrasive etching*, was developed by Little[3] for the realisation of gas channels in glass layers for a small Joule-Thomson cooler. In this method the substrate is coated with an extremely resilient photoresist and then abrasively etched with fine particles of $Al_2O_3$. With this method deep directive etching of any amorphous or crystalline material is possible, but the walls and surfaces of the etched structure become rather rough.

**Thin film technologies.** Besides removal of bulk silicon, the deposition and selective removal of thin films is also very important in micromechanics. Thin films can be deposited on a substrate, or realised in or grown from an existing material (e.g. oxidation), with a thickness of nanometers to several micrometers. Deposition can be done in a physical process - like sputtering - or in a chemical oriented process - like chemical vapour deposition. The physical processes are used to deposit many different metals, the chemical for materials such as silicon nitride, silicon oxide and polysilicon. The choice of an appropriate thin film technology depends on factors such as: desired material, deposition rate, maximum substrate temperature, adhesion between layer and substrate, layer morphology. In micromechanics, thin films are used as: electrical isolation, electrical signal tracks, membranes, strain sensors, capacitor plates, mask materials, passivation/protection layers, intermediate layers for bonding of two substrates, sacrificial layers, heater material, junction material, optical mirrors, valve-seat material, or actuator material.

**Substrate bonding.** Different bonding techniques can be used to fix two or more wafers to each other, a technology that is of great importance for micromachining. Bonding techniques make it possible to realise constructions in three dimensions, besides the conventional planar constructions. Both silicon and glass wafers can be used, as well as a combination of both. The wafers may be processed partly or completely prior to a bonding step, giving an enormous freedom in design.

Apart from these more or less conventional techniques, there is a wide range of alternative techniques currently being developed in the field of micromechanics[9]. We anticipate that these technologies will also become of interest for microcooling.

## Micromechanical components for microcooling

The application field of micromechanical technologies is very wide. One can think of membrane pressure sensors, microminiature microphones, thermal gas flow sensors, micro pumps, accelerometers, friction force sensors for nanotribology research, (linear) micromotors, xy-stages for nanopositioning, microgrippers, liquid handling devices for chemical analyse systems, micro-surgery components[9]. With respect to microcooling, it is of interest to focus on (sub)components which may be used in a microcooler.

**Membranes.** In many conventional small scale cryocoolers moving pistons are used as a compressing or expanding element, mainly in regenerative cycles. In micromechanics, the use of nonflexible constructions - such as the above-mentioned pistons, but also hinge points - is highly undesirable because of the presence of rubbing surfaces that cause wear. As an alternative, one may use miniaturised membranes on which much experience exists. Materials that are often used for flexible membranes are silicon and silicon-nitride. Thicknesses may vary from less than 1 $\mu$m to tens of micrometers, where diameters have been used up to 1 cm. As an example, a silicon nitride membrane with a diameter of 1 mm and a thickness of 1 $\mu$m deflects about 30 $\mu$m at a pressure difference of 1 atm without breaking. Corrugations have succesfully been added in the membrane to reduce the membrane stiffness and extend the linear deflection range of the membrane[10].

**Valves.** Basically, valves can be used in cryocoolers in two ways: to create an oscillating gas pressure from a static pressure generator by *actively* switching between the high and low pressure side (GM-coolers), or to create a static gas pressure from a varying pressure generator by *passively* switching to the high and low pressure sides (analogue to an electrical rectifier, JT-coolers). Many different types of microvalves have been fabricated, both passive and active types, and for fluid and gas systems[11]. In our opinion there are no principal limitations for a succesful design of an active low pressure valve, nor of a passive high pressure valve.

**Actuators.** Actuators are needed for the compression of gas (as used in most conventional cryocoolers) and for the actuation of active valves. Especially for the compression of gas an actuation principle is desired that is energy efficient, because it is directly related to the efficiency of the whole system and the heat rejection at the hot side. Different studies have been carried out to compare the most obvious actuation principles in micromechanics[12]. There are not so many principles that are energy efficient. For instance, electromagnetic and electrostatic actuation are not very suitable on micro-scale. Piezoelectric actuation is a fundamentally efficient actuator at micro scale dimensions, but only large forces and small displacements can be generated, and relatively large voltages are needed. We consider also thermal actuation by heating a gas or a gas/liquid in phase transition. This may be of special interest in relation to the Vuillemier cycle.

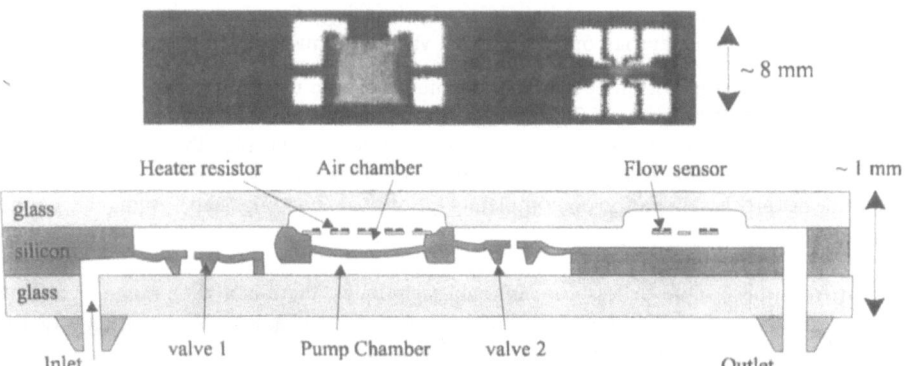

**Figure 3.** Photograph (top side) and cross section of a liquid dosage system with micropump and flow sensor[13].

**Glass/silicon constructions.** It is very likely that (components of) a microcooler will be constructed of a combination of glass and silicon substrate material, in such a way that the attractive thermal properties of these two materials are matched to the specific (thermal) requirements at each posistion in the microcooler. For instance, one may think of a regenerator constructed of silicon and glass in such a way that the desired low thermal conduction in the gas flow direction is combined with the desired high heat conduction in the other. However, a basic requirement for this is that the bond between silicon and glass survives thermal cycling to low temperatures. We performed tests on a commonly used silicon/glass combination with matching thermal expansion coefficients. After bonding at 400 °C thermal cycling to 77 K appeared to be no problem.

**Example of a micromechanical application.** To obtain an impression of the possibilities of micromechanical technologies, figure 3 gives a photograph and a cross section of a system in which most of the above-mentioned components can be found. The figure shows a liquid dosage system with a micropump and a thermal flow sensor, integrated on one chip (developed at MESA)[13]. The pump actuator is formed by a gas cavity in which the gas can be heated, thus deflecting the pump membrane. Membranes are also used as constructive elements for the check valves, heaters and flow sensor. They are all etched in one wafer by wet etching from both sides.

## COOLING CYCLES

Commonly used cryocoolers can be divided in recuperative (circulating gas flow: JT, Brayton) and regenerative (oscillating gas flow: Stirling, GM, pulse tube) cycles. Most of these cycles use the reversible expansion of an ideal gas, performing work on the environment and taking up heat at low temperature. Only Joule Thomson expansion is an irreversible process, performing *internal* work on a non-ideal gas. For microcooling one has to search for cooling principles which are particularly suitable if applied on micro-scale. Below some considerations with respect to the miniaturisation of conventional cooling cycles are given:

**1. Stirling.** Miniaturised membranes may be an attractive alternative for the use of pistons, but the relatively low pressure differences require highly efficient regenerators. In this respect, an etched foil regenerator as patented by Yaron and Mitchell may be useful[14]. Bowman et al.[15] recently patented an idea for a microminiature Stirling cooler with membranes running at a high frequency (500 Hz or more).

**2. Vuillemier.** This cycle may be of special interest because thermal energy is used as input energy. Thermal energy input is not limited by miniaturisation.

**3. Pulse tube.** This cycle is very interesting because no moving parts are used at the cold side. Membrane actuation may be possible. Scaling down of the pulse tube diameter will require a lower thermal penetration depth to prevent thermal losses to the wall. This requires an increase of the drive frequency. Higher drive frequencies put strict requirements on the regenerator: a high efficiency in combination with a low flow impedance is needed (similar to Stirling!). A patent of Cabanel et al. exists on a micro pulse tube[16].

**4. Brayton.** A very small turbo expander requires a very high rotation speed, and seems not feasible.

**5. Joule Thomson.** Little[3] has proved that miniaturisation is possible. However, a small reliable compressor being able to generate high pressure differences is a major requirement for closed cycle operation. A possible candidate for this is a sorption compressor, and the microcooling project therefore started with research in this field[17].

Interesting alternatives worth considering are: thermo-electric cooling and the desorption of a gas from a solution.

## REGENERATIVE CYCLE: MICRO COLD HEAD

In a regenerative cycle a working gas is compressed by a compressor and expanded in a cold head. The gas flow in the cold head is controlled by a displacer and the temperature gradient from the cold tip to the warm end is maintained via a regenerator[1]. In micro-dimensions the regenerative cooler will have to operate at relatively high frequencies since the pV-term per cycle will be small. A small high-frequency compressor may be realised with piezo-actuated membranes[12,15]. In combination with a small cold head a Stirling-type microcooler can be designed. As an alternative, a low-frequency compressor can be combined with high-frequency operated active valves in order to establish a pressure-wave generator. With such a pressure-wave generator a Gifford-McMahon cooler can be configured. In both designs, however, a micro cold head is needed.

A concept for such a micro cold head is depicted in figure 4. In our design, the displacer is realised via silicon diaphragms supported by glass tubes or rods, and it behaves like the free displacer of a split-Stirling cooler. Consider the cold head to be connected to a pressure-wave generator. The proper phase of the displacer movement with respect to the pressure wave is obtained by means of the displacer mass-spring dynamics. This dynamics is designed such that the displacer moves down when the high pressure valve is open: the gas is compressed at the hot side. The high pressure valve should be closed before the displacer moves up again (due to the dynamics) and forces the compressed gas to flow through the regenerator to the cold side.

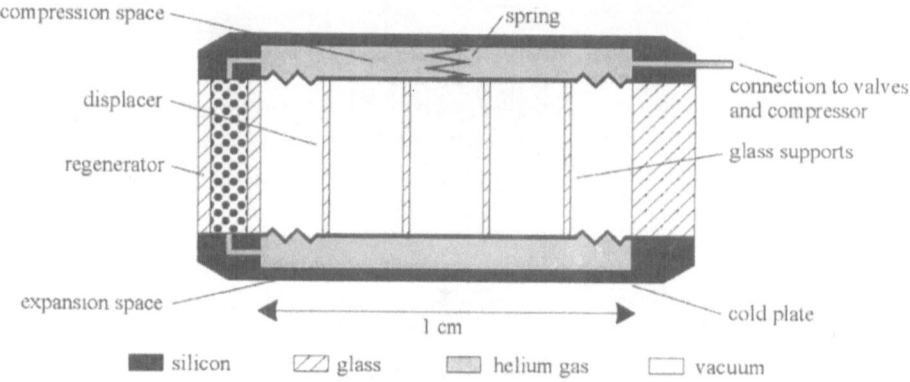

**Figure 4.** Design concept for a micro cold head.

Cooling occurs when the gas is expanded at the cold side by opening the low pressure valve.

The feasibility of flexible silicon diaphragms has been demonstrated by Y. Zhang et al., who constructed a 1 mm diameter diaphragm with a deflection of 30 µm at a pressure of 1 atm.[18]. With diaphragms of 10 mm in diameter we, therefore, expect to realise a displacer amplitude of roughly 0.3 mm. Assuming the full deflection of 0.3 mm to hold for a centre area of 6 mm diameter, an expansion volume of 16 mm³ results. With a pressure drop of, for example, 2 atm in the expansion, a cooling energy of 3.2 mJ per cycle results. Operation at 100 Hz would give a gross cooling power of 0.32 W. A major heat loss contribution will be due to the conductive heat flow via the glass parts that connect the cold and warm ends. For example, if a circular glass cylinder between the hot and cold side is considered of 5 mm length, 200 µm wall thickness and 1 cm diameter, the conduction through this material is already 0.22 W. However, the cross-sectional area of the glass can be largely reduced by means of holes and grooves, and also the length of the unit can be increased (both measures decrease the heat load in a linear way).

Special attention will have to be paid to the regenerator. This element should have a low flow impedance combined with a high efficiency in the heat exchange with the gas. These two requirements are somewhat contradictory. A further problem is the vacuum space in the displacer unit. If helium is used as the working gas, it will diffuse through the glass and the silicon into the vacuum space. The disadvantages of this are a reduction of the operating pressure and an increase in the conductive heat load to the cold plate. Depending on the diffusion rate and the required time of cooler operation, measures will have to be taken. Some possible approaches are: sputtering of a sealing layer on the silicon and glass surfaces, the use of other (larger-particle) working gases like neon, or the application of some sorption material on the cold side of the diplacer. Also a connection to an external sorption pump unit is possible.

## RECUPERATIVE CYCLE: SORPTION COOLER

Figure 5a gives the set-up of a sorption cooler[1,19]. It consists of a compressor unit, a counterflow heat exchanger, and a Joule Thomson expansion valve. The compressor unit contains four sorption compressors and several check valves to control the gas flows. Figure 5b gives a sketch of a sorption compressor. Low and high pressures are generated by the cyclic ad- and desorption of a working gas on a sorption material, which is accomplished by cooling and heating of the sorption material. Cooling is done with a gas gap heat-switch between the sorption material and a cold source on the outside.

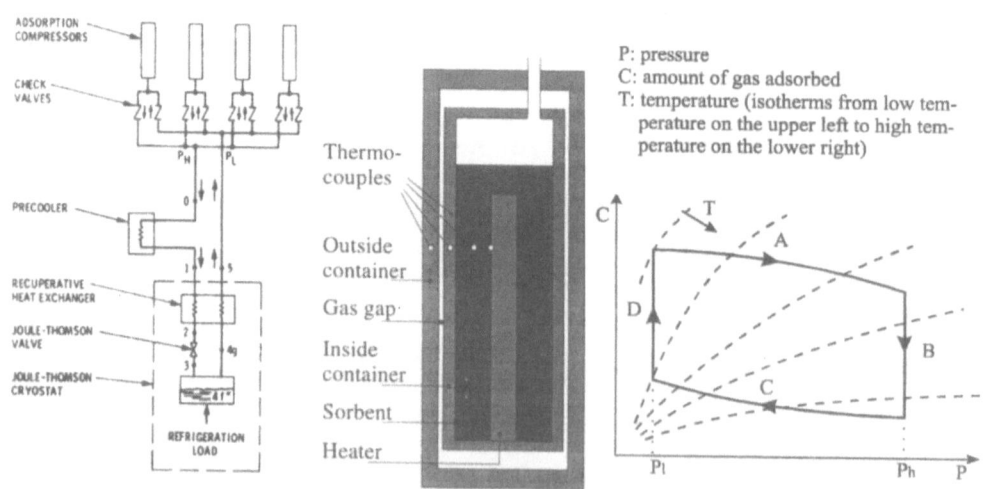

**Figure 5a.** Sorption cooler[1]     **Figure 5b.** Sorption compressor     **Figure 5c.** Compressor cycle

A compressor cycle is schematically drawn in figure 5c. The compressor is heated during sections A and B, and cooled during C and D. During sections À and C both valves of the compressor are closed, and the compressor is in a regenerating phase. During sections B and D one of the valves is opened; the compressor generates a high pressure gas flow during B, and a low pressure gas flow into the compressor during D. For predicting the behaviour of a miniaturised sorption compressor, a thermodynamic model for a cylindrical compressor was developed, as well as a test set-up for model validation and gas/sorber research. The test set-up, the model and the experiments are discussed in more detail in another contribution to this conference[16].

Except for some check valves, the cooler has no moving parts in it. This minimises electromagnetic and mechanical interference, and it contributes to a long life time. Moreover, distributed spot cooling can be realised very well with several small micromachined JT coolers.

Most important for the feasibility of a sorption cooler is an efficient and suitable gas/sorber combination. When such a combination is found, scaling down of the compressor seems feasible. The heat stored in the compressor should be as small as possible, since it is parasitic heat. The heat capacity is determined by that of the sorber and that of the container material, and it can be shown that the ratio of those two is independent of the compressor size. Therefore, the heat required for warming up the compressor scales linearly with its volume. *In first order*, when the cycle frequency of the compressor is kept constant, the input power and the resulting mass flow (with which the cooling power is proportional) scale linearly with the compressor volume (implying a constant efficiency).

Besides an efficient gas/sorber combination, other essential aspects for the miniaturisation of a sorption compressor are: the choice of a gas or a gas mixture with a high JT efficiency[20], the development of small compressors with an efficient heat management (miniaturisation of the heat switch), and the development of small micromachined high pressure check valves.

## CLOSING REMARKS

The combination of micromechanical techniques and small-scale cooling is an interesting field of research. In this field a 'microcooling' project was started at the University of Twente, aiming at a cooler that may be integrated with a high-Tc superconducting device and which offers a net cooling power of 5-50 mW at an operating temperature of 60-80 K. Introductory work has been done on the possibilities of micromechanical technologies related to the requirements of different miniaturised cooling cycles. There appear to be possibilities for the miniaturisation of regenerative and recuperative closed cooling cycles. For regenerative cycles, research is needed on the application of membranes in a pressure generator, and on the possibilities of an efficient microminiature regenerator. For a recuperative sorption cooler, the availability of a gas/sorber combination that is efficient at room temperature is a crucial requirement.

## ACKNOWLEDGEMENTS

This research is supported by the Dutch Technology Foundation (STW).

## REFERENCES

1.   Walker, G., *Low-capacity cryogenic refrigeration*, Oxford University Press (1994).

2.   Private communication with R.G. Ross, Jr.

3.   Little, W.A., Microminiature Refrigeration, *Rev. Sci. Instrum.*, vol. 55, no. 5 (1984), pp. 661-680.

4.   Inframetics, Inc., 16 Esquire Road, North Billerica, MA 01862-2598, U.S.

5.  Walker, G. and Bingham, R., Micro and nanno cryocoolers: speculation on future development, $6^{th}$ *Int. Cryocooler Conf.* (1990), pp. 363-375.

6.  Kirschman, R.K., Low-Temperature electronics, *IEEE Circuits and Devices*, vol. 6, no. 2 (1990), pp. 12-14.

7.  Petersen, K.E., Silicon as a mechanical material, *Proceedings of the IEEE*, vol. 70, no. 5 (1982), pp. 420-457.

8.  Jansen, H.V., Gardeniers, H., de Boer, M., Elwenspoek, M. and Fluitman, J., A survey on the reactive ion etching of silicon in microtechnology, *Proc. Micro Mechanics Europe*, Copenhagen, Denmark, (1995).

9.  See e.g. the Proceedings of the IEEE Micro Electro Mechanical Systems Workshop, San Diego, California, USA, 1996 (IEEE Catalog Number 96CH35856).

10. Spiering, V.L., Bouwstra, S., Burger, J.F. and Elwenspoek, M., Membranes fabricated with a deep single corrugation for package stress reduction and residual stress relief, *J. Micromech. Microeng.*, vol. 3 (1994), pp.243-246.

11. Shoji, S. and Esashi, M., Microflow devices and systems, *J. Micromech. Microeng.*, vol. 4 (1994), pp. 157-171.

12. van de Pol, F.C.M., *A pump based on micro-engineering techniques*, PhD thesis, University of Twente, the Netherlands (1989).

13. Lammerink, T.S.J., Elwenspoek, M., Fluitman, J., Integrated liquid dosing system, *Proc. of the Micro Electro Mech. Systems Workshop 1993*, Fort Lauderdale, Fl, USA (1993), pp. 254-259.

14. Yaron, R., Alto, P., Mitchell, M.P., Foil regenerator, U.S. Patent 5 429 177 (1995).

15. Bowman, L., Berchowitz, D.M., Urieli, I., Microminiature Stirling cycle cryocoolers and engines, U.S. Patent 5 457 956 (1995).

16. Cabanel, R., Friederich, A., Refroidiseur à gaz pulsé, demande de brevet european No. 0672 873 A1 (1995).

17. Huinink, S.A.J., Burger, J.F., Holland, H.J., van der Sar, E.G., Gardeniers, H., ter Brake, H.J.M., Rogalla, H., Experiments on a charcoal/nitrogen sorption compressor and model considerations, *This conference* (1996).

18. Zhang, Y. and Wise, K.D., Performance of non-planar silicon diaphragms under large deflections, *J. of Microelectromechanical Systems*, vol. 3, no. 2 (1994), pp. 59-68.

19. Wade, L.A., An overview of the development of sorption refrigeration, *Adv. in Cryogenic Eng. 37* (1992), pp. 1095-1106.

20. Alfeev, V.N., Brodyansky, V.M., Yagodin, V.M., Nikolski, V.A., Ivatsov, A.V., *Refrigerant for a cryogenic throttling unit*, UK Patent 1 336 892 (1973).

# Investigation into Vibration Issues of Sunpower M77 Cryocoolers

**Edward F. James and Stuart Banks**
McDonnell Douglas Aerospace
Seabrook, Md. USA

**Stephen Castles**
NASA, GSFC
Greenbelt, Md. USA

## ABSTRACT

NASA/Goddard Space Flight Center (GSFC) has been evaluating Stirling cycle coolers produced by Sunpower, Inc. The particular coolers being evaluated are two versions of the Model M77 with active counterbalancers. This paper reports on an investigation into the residual vibration characteristics of these machines using several different techniques for activating the counterbalancer.

The two versions of the M77 cryocooler are nearly identical except that the counterbalancer on the latest generation is adjustable to allow both axial and lateral vibration levels to be minimized. A comparison of the residual vibration levels of the two versions of the M77 is provided.

The relative change in the vibration signature (for the lowest 6 harmonics) is measured over time for both versions of the M77 cryocoolers, with the counterbalancer driven by a set, characteristic waveform chosen to initially minimize the residual vibration. These measurements are made as a function of heat sink temperature, power level, and load. Also, the effect that machine run time has on the vibration signature is determined. In addition, possible low cost and simplified methods for driving the counterbalancer will be discussed. These techniques result in various levels of vibration reduction. Typical results are presented.

## INTRODUCTION

Under contract to NASA/GSFC, Sunpower Inc. of Athens, Ohio, has developed inexpensive, single stage, integral Stirling cycle cryocoolers with counterbalancers. The cryocoolers are designed for operational lifetimes greater than 50,000 hours. The cryocoolers, use a Sunpower Inc. proprietary gas bearing/flexure support system to prevent contact between moving parts. An investigation of the flight worthiness of this cryocooler design for low cost space missions has been under way at GSFC since December 1994[1]. To date, thermodynamic, vibration, and electrical characterizations have been performed on this model cryocooler. In addition, two of the machines underwent launch load vibration tests. One unit was tested to a 14.1 g rms full qualification, random vibration level and the other was subjected to 10.0 g's rms. This test is specified in the NASA/GSFC General Environmental Verification Specification (GEVS). No apparent degradation resulted from these tests.

Cryocoolers 9, Edited by R.G. Ross, Jr.
Plenum Press, New York, 1997

**Table 1.**   Run Time Hours of 9 Sunpower Cryocoolers at GSFC as of June 12, 1996.

| Cryocooler # | Run time (hrs) | Testing |
|---|---|---|
| M77B - 4 | 4384 | Launch shake, thermal vac, life test |
| M77B - 5 | 4345 | Launch shake, thermal vac, stability, life test |
| M77B - 6 | 3864 | Life test, stability test, drive experiment |
| M77B - 7 | 1135 | Life test |
| M77B - 8 | 120 | PWM drive (Sun Power Inc. Electronics) |
| M77B - 11 | 1717 | Life test |
| M77B - 12 | 2239 | Life test, stability test, drive experiment |
| M77B -13 | 2986 | Capillary pump loop, intercept testing |
| M77C - 5 | 86 | Vibration characterization, drive experiment |

In this paper, the two versions of the M77 cryocoolers procured by Goddard are referred to as the M77B (with internal counterbalancer) and the M77C (with external counterbalancer). Both the M77B and the M77C have been designed to lift 4 watts at 77 kelvin with 90 watts of input power, with a reject temperature of 313 kelvin[2]. The run time hours of each of the 8 M77B and 1 M77C coolers at GSFC are specified in Table 1.

GSFC has operated the M77B cryocoolers using several different methods for counterbalance actuation. They have been operated with a fixed pure sinewave driven balancer with the phase shift relative to the compressor chosen to minimize the fundamental component of vibration. They have also been driven with a fixed characteristic waveform with the necessary harmonics required to cancel the first 6 harmonics, and they have been operated with a nonreal-time software algorithm that is designed to continually suppress the first 10 harmonics.

The graph in Figure 1 shows the comparison of drive methods on the first 4 harmonics of the axial reaction forces, with compressor power at approximately 60 watts. The levels shown in Figure 1 are typical values and depend on several parameters, such as heat sink temperature, power level, load, and cold-tip temperature. Typically, the vibration component forces roll off significantly after the fourth harmonic. With the balancer in a nonactive (passive) mode, harmonics 2, 3, and 4 will generally have levels between 1 and 6 newtons. As the fundamental is driven down with a sinusoidal balancer actuation, those harmonic forces will generally stay within the 1- 6 newton window.

When the cryocoolers are operated in life test mode, a reduction of the fundamental vibration component to less than 3 newton is desired to reduce coupling of vibration forces into adjacent coolers being tested. Use of a pure sinewave balancer drive with the correct phase has proven sufficient. Generally, at operating cryogenic temperature, the phase shift between the compressor and balancer required to suppress the fundamental is about 40 degrees.

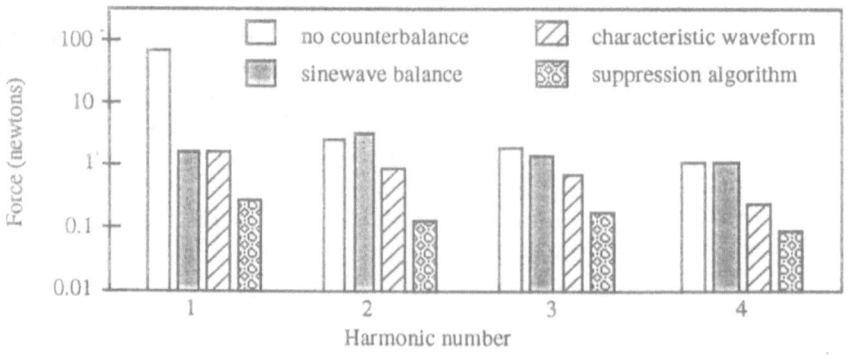

**Figure 1.** Typical vibration forces using different drive techniques.

The lateral forces generated by the M77B cryocooler are consistently in the 0.5 to 3.0 newton region. It is believed that these forces are to some extent caused by axial misalignment of the balancer, compressor, and displacer moving masses. The M77C cooler incorporates an external and adjustable balancer. Adjustment of this alignment and the effect on both axial and radial reaction forces will be investigated in the future. A picture of the M77C cooler in its dynamometer is shown in Figure 2.

## COMPARISON OF THE M77C WITH THE M77B COOLERS

The M77C cryocooler is almost identical to the M77B cryocooler with the exception of its external and adjustable counterbalancer. The location of the counterbalancer external to the pressure vessel allows for fine tuning of both the angular alignment of the balancer to the compressor as well as radial alignment. This fine tuning is accomplished by adjusting a series of set screws. The balancer on the new M77C cooler was initially aligned by Sunpower Inc. prior to delivery.

The M77C cryocooler underwent both thermal and vibration characterization. This unit was ordered without the standard external cooling fins in order to allow integration with a GSFC designed and manufactured heat sink which permits operation in the thermal vacuum chamber. Thermodynamically, the M77B and the M77C performed the same over the low input power region of interest to Goddard.

Electrical measurements were made on the M77C cooler. The compressor motor impedance during operation is slightly higher for the new cooler. An increase from typically 1.5 ohms to about 1.7 ohms was measured. Figure 3 shows the comparison of the typical axial forces with both machines running at about 60 watts using sinewave balancer. Figure 4 shows the comparison of lateral forces. The axial forces of the M77C were consistently lower than those of the M77B while operating at the same levels, particularly harmonics 2, 3, and 4. The lateral forces of the two machines measured are very similar.

GSFC has implemented vibration suppression control on over ten cryocoolers, resulting in force levels generally below 0.2 newtons (all harmonics). As shown in Figure 1, GSFC has operated the M77B coolers with suppression algorithm to vibration levels of less than 0.3 newton (all harmonics) at 60 W compressor power. Driving the M77B with a fixed characteristic waveform (counterbalance drive signal composed of harmonics required for vibration suppression at those frequencies), it is possible to operate the cooler at a slightly higher vibration level - fundamental at 1 to 3 newtons and all harmonics less than 2 newtons.

**Figure 2.** M77C cryocooler in dynamometer.

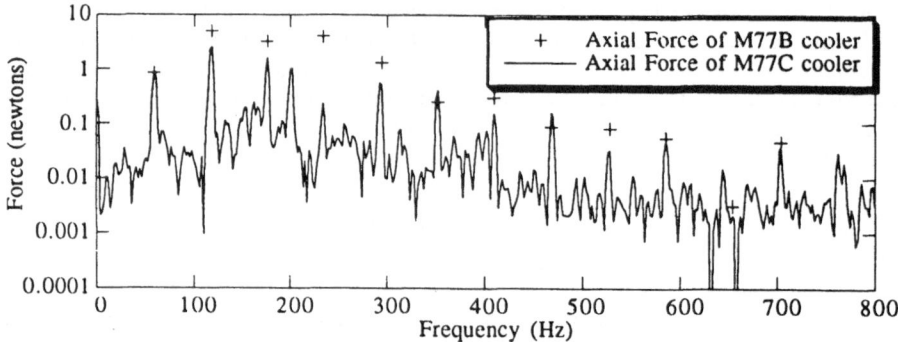

**Figure 3.** Comparison of typical axial forces of the M77B and M77C cryocoolers.

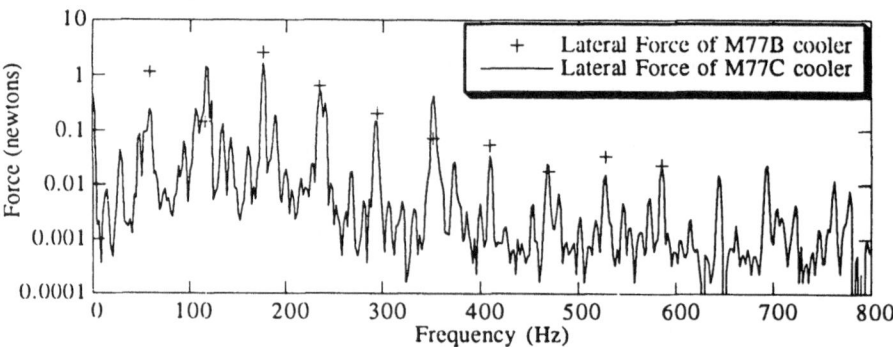

**Figure 4.** Comparison of typical lateral forces of the M77B and M77C cryocoolers.

Although this level of vibration is probably an order of magnitude greater than levels required for some spaceflight applications, this level is acceptable for other spaceflight applications. GSFC has integrated certain types of spectral imagers to coolers which have been tested with vibration levels (fundamental) of about 1 newton, using the same instrumentation as tests reported herein, and have found this to be acceptable. Such levels are easily attainable with the M77 version cryocoolers. In addition, axial force levels of back-to-back compressors of tactical coolers typically have similar vibration levels when driven with sinewave drive[3]. Also, some coolers which utilize sophisticated controllers to reduce axial vibration levels have off-axis vibration levels as high as 1 newton of force.

## VIBRATION STABILITY

During life testing of the M77B cryocoolers, it was observed that the vibration signature of the machine is very stable over time. Detailed measurements have been made to determine just how stable the M77B is and what parameters affect its stability.

The purpose of these tests is to determine if it is possible to implement a very simple cooler drive technique, which results in a considerable reduction in parts count and sophistication, and at the same time maintains vibration levels considered acceptable for certain applications.

Several separate experiments have been implemented. First, data was obtained at constant compressor power in a thermal vacuum chamber at 50 watts and 40 watts compressor power. The counterbalancer was operated with two fixed characteristic waveforms.

Figure 5 shows the result of approximately 1000 hours of operation with two different fixed counterbalancer waveforms, waveform 3 (wf #3) and waveform 3A (wf #3A). The heat sink temperature varies about a set point, and the cold tip temperature changes accordingly. At about 3400 hours the power level was reduced to 40 W and waveform 3A was established. Waveform 3 was utilized upon return of power to 50W. In Figures 5, 6, and 7, the rms force level shown is determined by digitizing the axial force signal and utilizing a software routine to calculate the root mean squared (rms) of the digitized time based force signal. In Figures 5 and 7, the force level includes background noise caused primarily by the chiller pumping fluid through the heat sink jacket which contributes about 1 newton rms force to the level. Generally, the fundamental component is the primary contributor to the rms force. The rms force is intended only to quantify, using a single number, the relative force level. Table 2 shows the typical rms force levels for each drive method.

Two additional tests were performed, the first test was to determine the effect of changing power levels required to maintain constant cold tip temperature with the load changing from 0 to 2.5 watts. This test (results in Figure 6) was performed on an air cooled unit, during the test the heat sink varied between 28 celsius and 37 celsius. The second test (results in Figure 7) was to determine the effect of power level changes on force when varying the heat sink +/- 10 celsius while maintaining constant cold-tip temperature. This second test was performed on a fluid cooled unit in the thermal vacuum chamber.

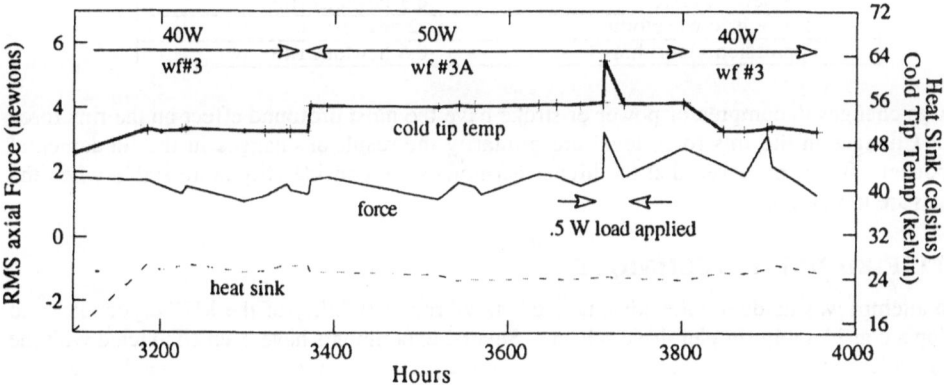

**Figure 5.** Cooler M77B-5 during 1000 hours of constant power operation

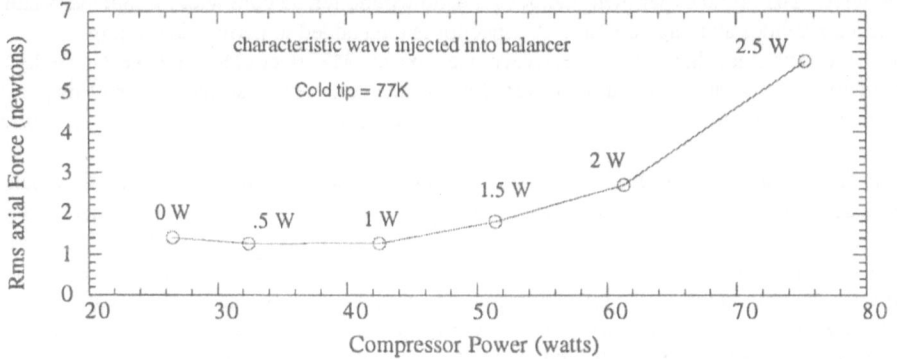

**Figure 6.** Change in rms force with compressor power to maintain constant cold tip temp for varying heat load from 0 to 2.5 watts.

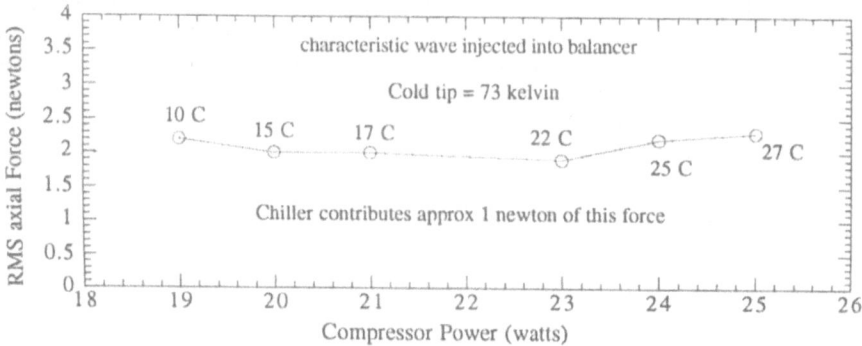

**Figure 7.** Change in rms force with compressor power to maintain constant cold-tip temp for varying heat sink from 10C to 30C.

**Table 2.** Typical rms force levels using 4 different drive techniques

| Drive technique | Typical rms force (newtons)@ 60 W |
|---|---|
| Passive balancer | 60 newtons rms |
| sinewave drive | 6.5 newtons rms |
| characteristic waveform | 2 newtons rms |
| suppression algorithm | 0.5 newtons rms |

Large changes in compressor power or stroke have the most profound effect on the rms force level. Changes in the rms force level are primarily the result of changes in the fundamental component. It was observed that the higher harmonics are considerably more stable when the changes are introduced.

## SIMPLIFIED DRIVE TECHNIQUE

An attempt was made to take advantage of the vibration stability of the M77 cryocoolers, to develop a considerably simpler drive scheme. Several experiments have been conducted with the M77B cooler in which a passive resistor / inductor / capacitor (RLC) network is used between the compressor motor and the balancer motor. The RLC network creates a phase shift such that the balancer is driven to suppress vibration. This arrangement results in the need for only 1 waveform source and motor power linear amplifier. The single power amplifier drives both motors simultaneously. The initial experiment included a hand wound ferrite core with multiple taps and a resistor which provides a voltage divider. A capacitor is then added to permit fine tuning.

The drive technique has been effectively used on an M77B cooler for several weeks of continuous operation with a characteristic waveform in the compressor signal. During this period no adjustment was made, and the rms of the time based force stayed within a 1 to 3 newton window.

It was found that by using this network it is possible to maintain the RMS of the time based force to less than 5 newtons (down from about 60 newtons with passive balance only). Furthermore, it is possible to utilize a characteristic compressor signal with the necessary harmonics to suppress the forces at the harmonics. When injecting harmonics into the compressor waveform, those components of the injected signal are attenuated across the RLC network.

Figure 8 shows the RLC network used and the frequency response of the network. Once the value of inductance is set for maximum reduction at the desired operating level, then only adjusting the capacitor and the potentiometer is required to keep the fundamental component between 1 and 3 newtons.

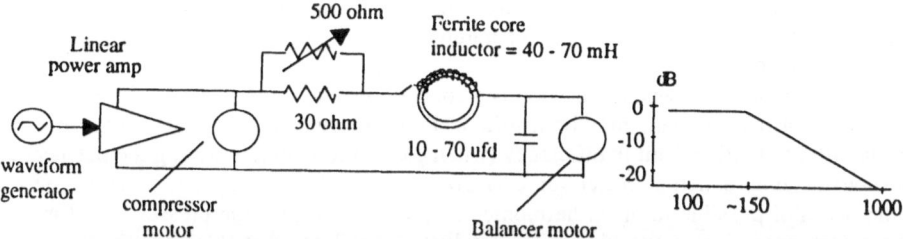

**Figure 8.** Diagram of RLC network with compressor motor and balancer motor.

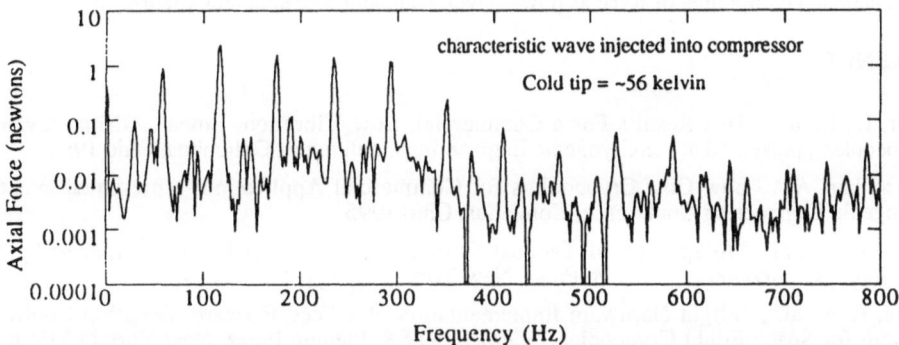

**Figure 9.** Axial force with compressor at 40 watts, while using RLC network.

The frequency response plot shown in Figure 8 was performed with resistors in place of the motors. The resistors were sized the same as the motor impedances. This network has also been demonstrated to successfully operate both motors when using the Sunpower Inc. supplied electronics which include a pulse width modulation (PWM) drive and cooler control electronics which incorporate cold tip temperature control. Figure 9 shows the axial force with the compressor input power at 40 watts after the network was tuned.

## CONCLUSION

The initial external balancer alignment performed by Sunpower Inc. on the M77C prior to delivery has proven effective at slightly reducing axial vibration levels. Measurements of the lateral vibration forces showed no appreciable change relative to the M77B. A second alignment is presently planned on the M77C to attempt to reduce the vibration levels further. The inherent stability of the M77 cooler further reinforces the fact that nonreal-time controllers are sufficient for axial vibration suppression[4]. The vibration signature of both the M77B and M77C cryocoolers is very stable over time as compared with some other cryocoolers tested at GSFC.

When operating the counterbalancer with a sinusoidal drive, the fundamental component of vibration does tend to drift with changes in heat sink and cold tip temperature, but correction is achieved simply by changing the phase angle of this sine wave relative to the compressor generally no more than +/- 3 degrees for constant power operation. Operation of the counterbalancer with a characteristic waveform is effective at reducing the harmonic force components to about 1 newton. These harmonics are considerably more stable than the fundamental. From the data presented in Figure 5, taken over a 1000 hour run, the maximum standard deviation of any harmonic was found to be +/- 0.47 newtons rms.

The constant cold tip temperature tests that were conducted show that even during large swings in power level, load, and heat sink temperature, the rms force can generally be maintained between 1 and 5 newtons.

For low cost cryocoolers where vibration is a concern, use of the RLC network to drive the balancer actuator of these coolers may be justified. Preliminary results indicate no sacrifice in vibration stability, a small 3% hit in efficiency (when compared with not driving the balancer at all), and a considerable reduction in electronic hardware relative to feed forward control systems. In addition, it is still possible to inject harmonic components into the compressor waveform to reduce those component forces, resulting in force reduction similar to that obtained with the use of a characteristic balancer waveform. Although the roll off response of those injected signals across the network is slow, sufficient attenuation does occur for this technique to be effective. An effort is currently underway to utilize the RLC network in a controlled fashion in order to minimize drift in the fundamental component of force as power level, load, and heat sink temperature change.

**REFERENCES**

1.  Sparr, L., et. al., "Test Results For a Commercial, Low Vibration, Linear, Stirling Cycle Cryocooler", presented at the Cryogenic Engineering Conference, Columbus, Ohio 1995.

2.  Karandikar, A. , "Low Cost Cryocoolers for Commercial Applications", presented at the Cryogenic Engineering Conference, Columbus, Ohio 1995.

3.  Sparr, L., et. al., "Adaptation of Tactical Cryocoolers for Short Duration Space-Flight Missions", *Cryocoolers 8*, Plenum Press, New York (1995) pp 163-172.

4.  Boyle, R. et. al. , "Flight Hardware Implementation of a Feed Forward Vibration Control System for Space Flight Cryocoolers" *Cryocoolers 8*, Plenum Press, New York (1995) pp 449-454.

# Summary and Results of Hughes Improved Standard Spacecraft Cryocooler Vibration Suppression Experiment

Michael Kieffer, Andy Wu and Shaun Champion

Hughes Aircraft Company
El Segundo, CA

## ABSTRACT

The application of split Stirling cycle cryogenic coolers to space-borne infrared (IR) surveillance sensors requires that the vibrations imparted to the optical sensor be kept to an absolute minimum, while still allowing the cooler to operate with power sufficient to cool a 1~2 W load to temperatures on the order of 60K. One method of accomplishing this is by applying an Adaptive Feedforward (AFF) algorithm to generate an auxiliary motor drive signal which, when summed with the sinusoidal drive signal in the motor drive, cancels out the residual cooler vibrations.

This paper reports test results achieved at Hughes with the Hughes Improved Standard Spacecraft Cryocooler (ISSC) #4. Using the AFF algorithm on the cooler's compressor, all of the controlled harmonics (first five, including the fundamental drive frequency) were reduced from levels ranging as high as 740 mN, to less than 10 mN. The operating point presented is a 2 Watt load at 60K, however, once optimized, this implementation of the algorithm exhibits robustness, yielding excellent cancellation results with a variety of cooler operating points.

## INTRODUCTION

This testing of the AFF algorithm on the Hughes ISSC #4 was performed during integration of the first implementation of AFF on an embedded-processor system at Hughes. Previous versions of the algorithm had been hosted on the SPROC ,which is strictly a non-productized development tool. This implementation of AFF was also novel in that it was used to control the vibrations of the fundamental drive frequency, as well as higher harmonics; previous implementations of the algorithm did not cancel the fundamental frequency. In addition to performing tests of the algorithm at a static cryocooler operating point, the robustness of the algorithm was also evaluated, by observing its behavior in the presence of large changes in the operating point of the cryocooler.

## TEST SETUP

For this test, ISSC#4 was mounted on the SMTS (Space Missile Tracking System) Heat Intercept Test Fixture, which was bonded to a granite slab (Figure 1). Residual vibration forces were measured with one of the three Kistler load cells used to mount the compressor to the test fixture. The electronics used for this implementation were developed for the PSC (Protoflight Spacecraft Cryocooler) program originally, which in turn used an Digital Processor Board that had been developed for another program. The PSC electronics were used to generate the motor drive signals for ISSC#4, as well as to provide the analog Piston Position Control (PPC) servos, and perform the data acquisition and processing for AFF. The PSC electronics chassis, together with all power supplies required for driving ISSC#4, were contained in an electronics rack (Figure 2).

Although a PID temperature control algorithm (using compressor stroke modulation) has been implemented on the same electronics, this feature was disabled to conserve processing power during AFF integration and testing, and temperature control was instead provided with a cold end heat load, controlled by a LakeShore 330 Temperature Controller. Cancellation was performed only on harmonics 1 through 5 because the residual vibration levels beyond the fifth harmonic were only on the order of 30 mN. The PSD of the residual vibration forces was observed on an HP 3562A dynamic signal analyzer.

**Figure 1.** ISSC#4 on SMTS heat intercept test fixture

**Figure 2.** PSC cryocooler electronics rack

## IMPLEMENTATION

The processing devices on the Digital Processor Board are an Analog Devices ADSP-2100, which is a fixed-point 16-bit digital signal processor running at 8.0 MIPS, and a Motorola 68000, a general-purpose 16-bit fixed-point processor running at approximately 0.3 MIPS. Because of the processing throughput limitations of the 68000, AFF vibration cancellation was performed only on the compressor. Communication between the two processors is accomplished by using shared RAM, which can be written to and read from by both processors. The memory resources required for this implementation are as follows:

CPU RAM: less than 30K words
DSP (Shared) RAM: less than 2K words
CPU EEPROM: less than 25K words (excluding C libraries)
DSP EEPROM: less than 2.5K words

The ADSP-2100 generates the sinusoidal motor drive signals from a lookup table, which serve as commands to the PPC loops. These loops use feedback from the piston LVDTs to drive the pistons in near-perfect sinusoidal motion, and the pistons are paired (two compressor pistons, one expander piston with a balancer) such that the net external force due to the piston motion should be zero at the fundamental drive frequency. However, some residual forces due to imperfect piston matching, and nonlinerar gas spring and flexure effects remain, and are sensed by the load cells that mount the cooler to the test fixture.

The load cell signals are sampled by the A/D converter at a rate equal to the motor drive frequency x 180 points/cycle (for this implementation, 46.0 Hz x 180 samples/cycle = 8.28 kHz). The load cell signals are sampled in synchronization with the motor drive signals, such that exactly two complete periods of the fundamental frequency are captured in a data "snapshot," and the phase relationship of the data snapshot with respect to the motor drive signals is invariant. A 12-bit A/D converter was used to capture the load cell data, and 12-bit DACs were used to output the AFF-generated vibration cancellation signals .

The software that controls the ADSP-2100 is written in the assembly language for that processor. The ADSP-2100 performs the sampling of the data when commanded to do so by the 68000. It is also responsible for outputting all motor drive signals as well as the AFF-generated correction signals. The software for the 68000 is written in C, and includes the AFF algorithm and various user interface functions.

## TEST CONDITIONS

- Expander Displacer stroke: 1.90 mm p-p

- Thermodynamic phase angle: 69 deg

- Motor drive frequency: 46.0 Hz

- Expander Dewar Vacuum: $5 \times 10^{-6}$ torr

- Reject Temperature: 280K

## TEST METHODOLOGY

Testing of this implementation of the AFF algorithm proceeded directly from integration of the software with the electronics, and determination of the proper gain and phase parameters for each harmonic, used to synthesize the vibration cancellation signal. After first establishing a performance baseline at 60K with a compressor stroke of 6.96 mm p-p, the algorithm was tested at various other operating points without changing the AFF gain and phase parameters. System robustness was evaluated by reducing the compressor stroke by 50% during operation, and also by varying the fundamental frequency by ±2 Hz.

## TEST RESULTS

Vibration data with and without AFF cancellation for two cryocooler operating points are shown in Tables 1 and 2. Table 1 shows vibration data for the cryocooler operating at 60.0 K with a 2.0W load, and Table 2 shows vibration data for 35.0 K with a 513 mW load. The robustness of the algorithm is demonstrated via the vibration data in Table 3, wherein the operating point was changed drastically during continuous operation of the algorithm (no changes were made to any gain or phase coefficients), to demonstrate the adaptive capabilities of the AFF algorithm.

Table 1 - Vibration Cancellation Results at 60.0 K, with 2.0 W load,
6.96 mm p-p Compressor Stroke, and 74.6 W Compressor Power

|  | Frequency (Hz) | | | | | | |
|---|---|---|---|---|---|---|---|
|  | 46 | 92 | 138 | 184 | 230 | 276 | 322 |
| Compressor Force along Z-axis (mN) without AFF | 439.0 | 734.9 | 275.4 | 51.94 | 164.0 | 28.64 | 34.33 |
| After 1 AFF iteration | 176.5 | 181.9 | 198.1 | 36.37 | 127.5 | 34.35 | 29.80 |
| After 11 AFF iterations | 2.71 | 4.60 | 7.73 | 4.37 | 2.40 | 23.33 | 27.84 |

Table 2 - Vibration Cancellation Results at 35.0 K, with 513 mW load,
6.45 mm p-p Compressor Stroke, and 71.2 W Compressor Power

|  | Frequency (Hz) | | | | | | |
|---|---|---|---|---|---|---|---|
|  | 46 | 92 | 138 | 184 | 230 | 276 | 322 |
| Compressor Force along Z-axis (mN) without AFF | 202.4 | 620.1 | 225.5 | 43.02 | 125.7 | 21.66 | 35.60 |
| After 1 AFF iteration | 79.11 | 192.2 | 159.7 | 33.94 | 95.99 | 22.46 | 35.11 |
| After 11 AFF iterations | 1.31 | 3.09 | 3.59 | 4.40 | 5.19 | 17.99 | 34.87 |

Table 3 - Vibration Cancellation Response to Change in Cryocooler Operating Point

|  | Frequency (Hz) | | | | | | |
|---|---|---|---|---|---|---|---|
|  | 46 | 92 | 138 | 184 | 230 | 276 | 322 |
| Compressor Force along Z-axis (mN); 6.9mm stroke (70.0W input; 1.75W load) | 6.52 | 9.48 | 8.72 | 2.72 | 7.06 | 21.66 | 36.77 |
| Stroke changed to 4.6mm (37.4W input; 468 mW load) | 111.2 | 263.3 | 283.4 | 58.74 | 78.07 | 12.29 | 11.84 |
| After 8 AFF iterations | 4.78 | 5.97 | 8.07 | 10.34 | 5.66 | 7.22 | 9.63 |

Because in this implementation, the number of points in one cycle of the motor drive waveform is fixed, and hence the drive frequency is determined by the rate at which the waveform data points are output on the DACs, the performance of the AFF algorithm is unaffected by changing the motor drive frequency during operation. Change of frequency did result in a transient increase in harmonic levels in the PSD, however this was due to the changing vibration profile of the cryocooler rather than the AFF algorithm itself.

## SUMMARY AND CONCLUSIONS

The residual vibration for the motor drive fundamental was maintained below 5 mN RMS, and the residual vibration for the first four higher harmonics was reduced to less than 10 mN RMS each using the AFF algorithm.

When subjected to large changes (~50%) in cryocooler operating point while running, the AFF algorithm was able to regain the previously achieved vibration cancellation levels within 10 iterations.

The SMTS Heat Intercept Test Fixture was clearly not the optimum test bed for vibration testing, however it was felt that the collected data for axial vibration would be sufficient to judge the performance of the algorithm, if not the cooler itself. Further testing of the Hughes ISSC#4 on the JPL dynamometer is planned for the near future.

Although the throughput of the 68000 processor used to implement the AFF algorithm was quite low, the performance of the algorithm was not significantly impaired. This is due to the inherently constant nature of the cryocooler vibration profile at a given operating point, once the system has equilibrated. The AFF algorithm performed satisfactorily with an iteration period of 80 seconds.

The Adaptive Feedforward algorithm will perform well with approximate knowledge of the plant for the frequencies of interest, and will adapt readily to variations in the plant over time.

## ACKNOWLEDGEMENTS

The funding for development and testing of this cryocooler vibration cancellation method was provided by the USAF SMC SMTS Program.

# MOPITT Stirling Cycle Cooler Vibration Performance Results

**E. L. Cook**

COM DEV Atlantic, 328 Urquhart Ave., Moncton, NB, Canada E1C 9N1

**J. Hackett**

COM DEV Limited, 155 Sheldon Drive, Cambridge, ON, Canada N1R 7H6

**Dr. James R. Drummond and G. S. Mand**

University of Toronto, 60 St. George St., Toronto, ON, Canada M5S 1A7

**L. Burriesci**

Lockheed-Martin, Bldg 250, 3251 Hanover St., Palo Alto CA 94304

## ABSTRACT

Both of the Measurements Of Pollution In The Troposphere's (MOPITT) instrument detector subassemblies require cooling to 90 K with 850 mW head load per subassembly. The stringent spacecraft level mechanical disturbance specification requires compressor forces to be $\leq 200$ $mN_{rms}$ and displacer forces to be $\leq 50$ $mN_{rms}$. For the first seven (7) harmonics, an acceleration feedback system using two Matra Marconi Space (MMS) 50-80 K Stirling Cycle Coolers (SCCs) and Lockheed-Martin Low Vibration Drive Electronics (LVDE) was used to meet these requirements.

Two 50-80 K SCCs were integrated with flight bracketry and the LVDE to form the MOPITT cooler subsystem. The final testing resulted in measured residual piston forces all less than 145 mN, the maximum transverse to the piston axis.

These encouraging results led to the integration of the cooler subsystem with the engineering qualification model (EQM) MOPITT instrument. Detector performance and field of view jitter were measured with the cooler subsystem operational and temporarily non-operational with the detectors at 100K. It was concluded from the tests that system vibration performance was adequate to meet the needs of the MOPITT instrument. Earth Observation System (EOS) satellite jitter analysis confirmed that the system met spacecraft level requirements.

## BACKGROUND

The (MMS) 50-80K SCC and Lockheed-Martin Low Vibration Drive Electronics were selected to meet the cooling requirements of the MOPITT instrument. The compatibility of this configuration and its ability to meet the MOPITT and (EOS) performance requirements was demonstrated with representative hardware during the fall of 1993 (Reference 1).

The requirements of the cooler system were to provide 850 mW of heat lift at 90 Kelvin while maintaining compressor forces as measured at the instrument baseplate below 200 mN and displacer forces below 50 mN in all axes.

In the fall of 1994, COM DEV took delivery of two MMS 50-80K coolers and one Lab Drive Electronics unit. The acceptance testing demonstrated full compliance with the MOPITT and MMS specifications.  The equipment was transported to Lockheed-Martin where a functionality check was successfully completed first with the MMS Lab Drive Electronics, then with the Lockheed-Martin LVDE.  The coolers were then transferred to the MOPITT EQM bracketry to commence the vibration performance testing  **(See Figure 1)**.

## TEST CONFIGURATION

The compressors were mounted head to head in aluminum brackets which include an assembly of copper braids to facilitate heat transfer to the baseplate. A block mounted at the mid point of the bracket was used to mount accelerometers in the 3 major axes.

The displacers were mounted with the cold fingers opposing as the MOPITT configuration includes redundant detectors. Three aluminum tye rods connect the two main displacer supports. An adapter is fitted on the detector interfacing side of the bracket so that the MMS vacuum canisters can be utilized for testing. The heater and temperature sensors supplied by MMS were left in place. Accelerometers were mounted on blocks to measure acceleration in the three major axes.

The entire assembly was mounted on a large aluminum base plate (which doubled as a shipping fixture) and the coolers were mated to the EQM LVDE. The accelerometers measured system acceleration which is input to the control system.  Independent accelerometers were utilized to monitor vibration reduction performance.

**Figure 1.  MOPITT Cooler System Test Configuration (Aluminum Baseplate)**

## OBSERVATION AND PROBLEM RESOLUTION

The transfer of the coolers from the shipping bracketry to the EQM bracketry is not a trivial process. Despite comprehensive planning and review of interfaces there were some unplanned setbacks. The MOPITT cooler interface was further complicated since it was modified from the standard configuration to include a displacer rotated $180^0$, utilized a custom transfer tube configuration and contained cabling modified to include EMC shielding. A handling fixture was built to ensure there was no stress applied to the transfer tube joints during integration.

The aluminum baseplate greatly affected the system performance. The plate, which was designed oversize for shipping purposes, produced accelerations which were picked up by the feedback acceleromters. An effort to resolve this problem through system software adjustments improved performance but not enough to meet specification requirements. Finally, the aluminum plate was replaced with a thicker magnesium plate. **(See Figure 2)**. This change resulted in the system producing the best results.

The MOPITT configuration requirements resulted in a small envelop for the displacers which were required to be at a height of 7.5 inches from the baseplate. The bracket cross section at the displacer interface was restricted to be relatively thin hence the design included tye rods to stiffen the structure. However, motion in the structure was still observable, and a cylindrical structure between displacers would be more effective. Such a structure will be utilized for the flight configuration.

## MEASURED RESULTS

Several separate tests are reported. The first from January 22, 1995 was with the standard configuration including the aluminum baseplate. During February, several adjustments were made to the control software in an effort to improve performance. Finally, it was concluded that

**Figure 2. MOPITT Vibration Test Configuration (Magnesium Baseplate)**

## MEASURED RESULTS

the baseplate needed to be replaced and the March 22 results were observed with the replacement magnesium plate in place.

Off axis forces were measured in the horizontal (parallel to the baseplate) and vertical (perpendicular to the baseplate) axes. Data was collected from monitoring accelerometers mounted near the displacers and compressors during the 27 Jan 95 and 22 Mar 95 tests and from a monitoring accelerometer mounted on the baseplate for the 27 Mar 95 test. The vibration performance requirement is defined at the baseplate, however, initial monitoring was done near the displacers to better characterize the mounting brackets.

The results, as presented in Table 1, show a significant amount of motion in both the compressors and displacers prior to the 27 Mar 95 test. This is attributed to three factors; 1) the response of the displacer brackets, 2) the response of the aluminium baseplate, 3) the response of the GSE vacuum cans cantilevered off the displacers.

### Table 1 System Off-Axes Vibration Performance

| Test Date | 1st | 2nd | 3rd | 4th | 5th | 6th | 7th | 8th |
|---|---|---|---|---|---|---|---|---|
| Compressor Off-Axis Horizontal Using A-DECS    mNrms | | | | | | | | |
| 27 Jan 95 Test | 379 | 100 | 200 | 450 | 1100 | 1200 | 200 | 100 |
| 22 Mar 95 Test | 2 | 2 | 2 | 2 | 2 | 2 | 125 | 15 |
| 27 Mar 95 Test | 9 | 7 | 10 | 16 | 14 | 6 | 36 | 4 |
| Displacer Off-Axis Horizontal Using A-DECS  mNrms | | | | | | | | |
| 27 Jan 95 Test | 75 | 260 | 130 | 500 | 240 | 150 | 120 | 50 |
| 22 Mar 95 Test | 50 | 750 | 150 | 150 | 75 | 150 | 200 | 10 |
| 27 Mar 95 Test | 9 | 7 | 10 | 16 | 14 | 6 | 36 | 4 |
| Compressor Off Axis (Vertical) Forces Using A-DECS  mNrms | | | | | | | | |
| 27 Jan 95 Test | 88 | 150 | 150 | 225 | 1500 | 1500 | 100 | 200 |
| 22 Mar 95 Test | 80 | 100 | 60 | 90 | 40 | 40 | 40 | 260 |
| 27 Mar 95 Test | 13 | 22 | 12 | 20 | 3 | 10 | 46 | 7 |
| Displacer Off Axis (Vertical) Using A-DECS  mNrms | | | | | | | | |
| 27 Jan 95 Test | 430 | 360 | 120 | 380 | 460 | 550 | 160 | 160 |
| 22 Mar 95 Test | 150 | 600 | 100 | 150 | 220 | 200 | 1000 | 200 |
| 27 Mar 95 Test | 13 | 22 | 12 | 20 | 3 | 10 | 46 | 7 |

Table 2.   Compressor Axial Vibration Performance

| COMPRESSOR AXIAL (Z) Using A-DECS   mNrms | | | | | | | | |
|---|---|---|---|---|---|---|---|---|
| Test Date | 1st | 2nd | 3rd | 4th | 5th | 6th | 7th | 8th |
| 20 Jan 95 Test | 260 | 80 | 60 | 40 | 20 | 20 | 60 | 50 |
| 27 Jan 95 Test | 208 | 30 | 30 | 60 | 40 | 150 | 25 | 10 |
| 20 Feb 95 Test | 170 | 75 | 40 | 35 | 20 | 40 | 20 | 20 |
| 22 Mar 95 Test | 247 | 60 | 40 | 30 | 30 | 20 | 40 | 20 |
| 27 Mar 95 Test | 57 | 145 | 25 | 18 | 15 | 15 | 36 | 30 |
| Notes:  Measured Acceleration Multiplied by Full System Mass 27 Mar Test Measured at Base Plate | | | | | | | | |

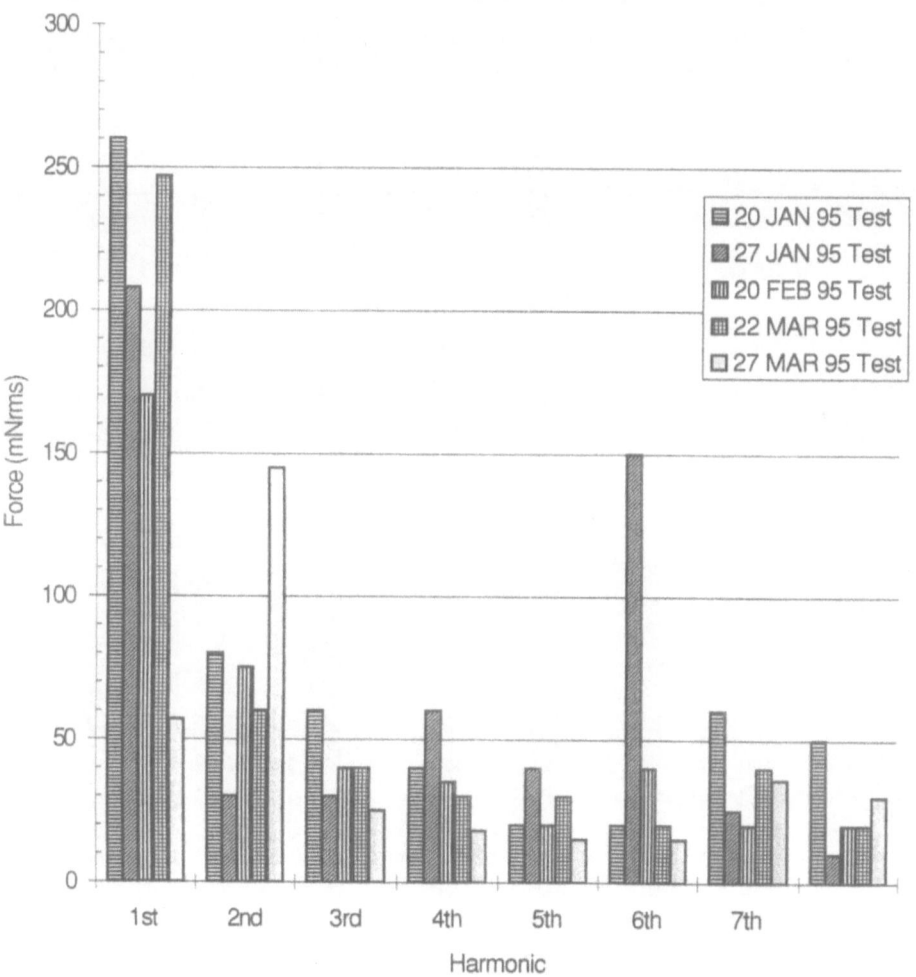

Figure 3.   Compressor Axial Vibration Performance

**Table 3  Displacer Axial Vibration Performance**

| Displacer Axial (Z) Using A-DECS    mNrms | | | | | | | | |
|---|---|---|---|---|---|---|---|---|
| Test Date | 1st | 2nd | 3rd | 4th | 5th | 6th | 7th | 8th |
| 20 Jan 95 Test | 240 | 160 | 40 | 160 | 480 | 320 | 240 | 170 |
| 27 Jan 95 Test | 200 | 350 | 400 | 700 | 100 | 500 | 600 | 180 |
| 20 Feb 95 Test | 400 | 140 | 60 | 120 | 90 | 150 | 30 | 50 |
| 22 Mar 95 Test | 119 | 60 | 70 | 60 | 60 | 30 | 280 | 30 |
| 27 Mar 95 Test | 57 | 145 | 25 | 18 | 15 | 15 | 36 | 30 |
| Notes:  Measured Acceleration Multiplied by Full System Mass  27 Mar Test Measured at Baseplate | | | | | | | | |

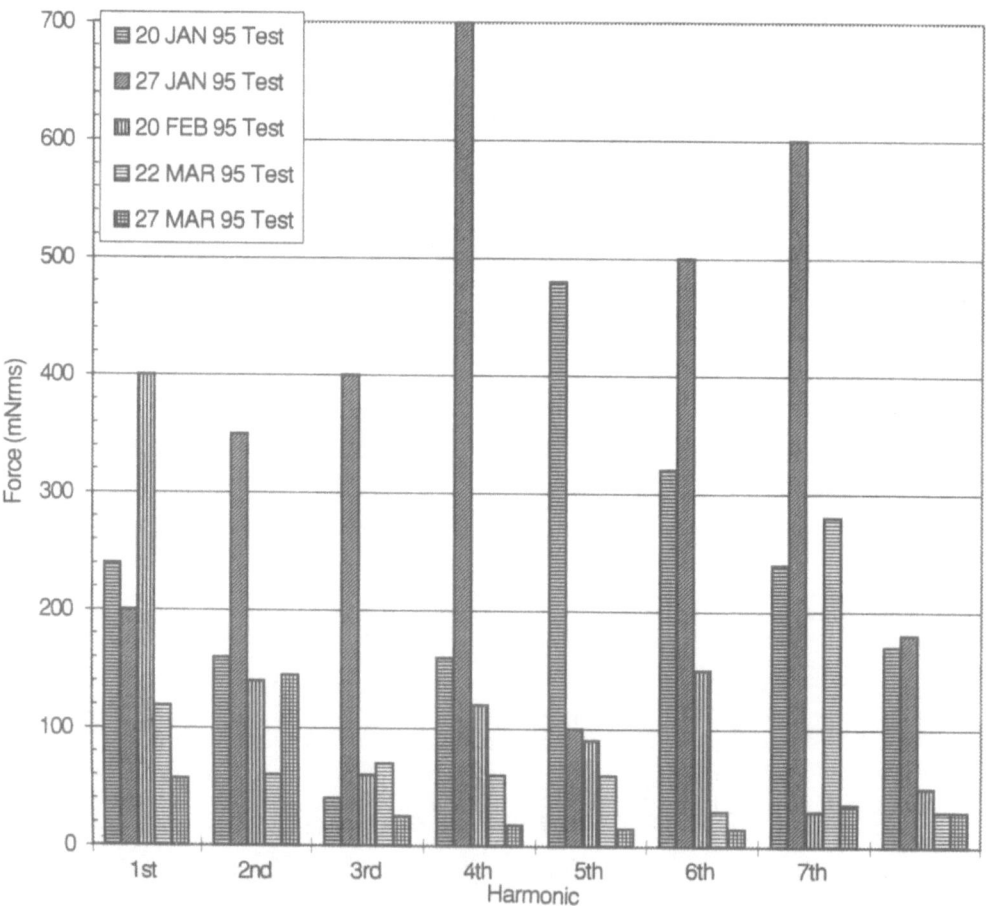

**Figure 4.  Displacer Axial Vibration Performance**

## MEASURED RESULTS

The replacement of the aluminum plate with the magnesium plate resolved the plate response issue. Schedule and budget limitation prevented the resolution of the displacer bracket issue, however, modifications have been implemented for the FM design. The vacuum canister issue was not addressed.

On 27 Mar 95, the test was done with off axes forces measured in accordance with the requirements at the baseplate interface and the results were well within requirements.

Axial vibration performance was measured on five (5) separate tests. The compressor axial vibration performance is presented in Table 2 and Figure 3. The displacer axial vibration performance is presented in Table 3 and Figure 4.

The displacer response presented a more challenging situation. As previously outlined, the displacer bracket design was not as stiff as the compressor brackets for several reasons; the required height above the baseplate, limited mass; opposing configuration and thin cross section due to envelop restrictions. Additionally, the vacuum canisters were cantilevered from the brackets which further affected performance. The vacuum canisters will not be in place for the FM and improved results are anticipated.

The results were very good, however, not optimal as indicated by the control parameters. Several software adjustments were made, and when performance was measured as specified at the baseplate, the performance was acceptable. However, improved performance is anticipated with the modified flight brackets.

## EQM INSTRUMENT INTEGRATION AND TEST

The acceptable performance during subsystem testing led to the integration of the cooler subsystem with the EQM instrument. Instrument system level performance testing was successfully completed.

Detector noise and jitter levels were tested with the coolers operational and while the coolers were briefly turned off. The detector noise level was not affected by the cooler operation. To assess jitter, a collimator was first aimed at MOPITT to position the image spot on an edge of a pixel. The noise level was then monitored with the coolers briefly turned off and while operational. An increase was detected which corresponds to a boresight jitter of 0.1 arcseconds. These results were measured utilizing the position control system which was sufficient to meet science requirements.

## EOS JITTER PERFORMANCE

The vibration measured at the base of the coolers was provided to the NASA-EOS team who maintain a spacecraft jitter model. The measured results were lower than the values input to the model which showed payload performance was acceptable and met all EOS requirements.

## SUMMARY AND DISCUSSION

Several observations were made during the test and integration program. Key issues are noted as below.

1) Test Configuration (bracketry, mounting plates, GSE) is critical to system performance;
2) Defining allowable disturbance levels requires careful consideration;
3) Cooler Configuration/Interfacing needs comprehensive planning.

In conclusion, the MOPITT cooler subsystem surpassed EOS spacecraft jitter requirements, science performance detector noise requirements and MOPITT jitter requirements.

**ACKNOWLEDGEMENTS**

The team effort of the University of Toronto (Physics Department), COM DEV, Lockheed-Martin, Matra Marconi Space and the Canadian Space Agency has resulted in a successful solution for the MOPITT instrument. This work was made possible through the support of the Canadian Space Agency whose contribution is gratefully acknowledged.

**REFERENCES**

1.  Cook, "MOPITT Stirling Cycle Cooler and Cooler Drive Electronics Evaluation," Cryocoolers 8, Plenum Press, New York N.Y., 1995.

# Summary and Results of a Space-Based Active Cryocooler Vibration Suppression Experiment

C.E. Byvik [1] and J. Stubstad [2]

[1] W.J. Schafer Associates, Inc.
1901 North Fort Myer Drive
Arlington, VA 22209

[2] Ballistic Missile Defense Organization
7100 Defense Pentagon
Washington, D.C. 20301

## ABSTRACT

Space-based MWIR and LWIR sensors for defense and civilian applications require cooling to temperatures below 100K. Stirling cycle coolers are typically used for small cooling loads based on power and mass considerations. In addition to demanding that the space-based cryocoolers be efficient, light-weight, long life, and reliable, there is the demanding requirement for low vibration. The approach to meeting the low vibration requirement, particularly for Stirling coolers, has been to null compressor vibrations by, e.g., balancing and phasing opposing piston motions. This approach dramatically effects the cryocooler cost. An alternative technique is to employ active vibration control using, e.g., piezo-ceramic materials with an electronic control system, to suppress compressor-induced vibrations.

The goal of this experiment was to demonstrate the design and qualification of a low-power, active vibration control system in a zero-g, limited mass, space flight experiment. Commercially available low voltage piezoelectric translators, eddy current transducers and digital controller were adapted to an off-the-shelf tactical cryocooler. On-orbit demonstration of vibration associated with the first eight harmonics was reduced by a factor of 75. The space-based cryocooler has been operated over 1,000 times to date for a total time in excess of 35 hours. The lowest temperature achieved was 77K when the cooler was operated at a maximum power for more than 30 minutes. This experiment demonstrates a low-cost approach to vibration control for space-based system.

Cryocoolers 9, Edited by R.G. Ross, Jr.
Plenum Press, New York, 1997

719

## INTRODUCTION AND BACKGROUND

Current MWIR and LWIR sensors require cooling to obtain practical signal to noise ratios. Systems relying on active cooling must be designed to isolate the vibration inherent in mechanical cryocoolers from the sensor. The degree of vibration isolation depends on a variety of system-level parameters. These include hardware and software components such as available power, mass limitations, focal plane array (FPA) wavelength response, signal processing algorithms, signal integration time, reliability, lifetime and cost. All of these parameters play a role in determining the allowable signal to noise ratio for the sensor system.

One approach to designing space-based systems is to specify requirements at the subcomponent level based on the mission. In the case of an actively cooled sensor, stringent vibration requirements have often been assigned to the cryocooler. Examples of vibration requirements for space-based sensors are given in Table 1. As a result, the cryocooler community has expended a considerable effort to meet this requirement. Recognizing that the principal vibration source in the coolers is the movement of the pistons in the compressor, the first order adjustment is to balance the compressor by, e.g., matching the pistons and controlling the compression phases. This approach dramatically reduced the vibration amplitudes but also dramatically effected the manufacturing cost and power consumption. Further vibration reduction strategies are under development focusing, in particular, on the mechanical operation of the cryocoolers.

**Table 1.** Typical Spacecraft Sensor Vibration Requirements

| System | Requirement |
|---|---|
| NASA, Atmospheric IR Sounder | 1 micron mechanical stability |
| NASA GSFC Cryocooler Program Goal | < 0.05 lb force max<br>< 0.02 lb force goal<br>(Soft mounting for vibration control unacceptable) |
| DoD for Stirling Cycle Systems | < 0.1 N (rms) compressor<br>< 0.01N (rms) displacer |

An alternative approach to cryocooler vibration effects on sensor systems is to employ active vibration control technology. Peizoelectric sensors and actuators as well as microprocessors and associated algorithms have matured sufficiently to be considered for cryocooler vibration control. Through Ballistic Missile Defense Organization support and program execution by the Airforce Phillips Laboratory, the Jet Propulsion Laboratory assembled a team to develop a space experiment to demonstrate the capabilities of on-orbit active cryocooler vibration suppression. The experiment was launched on June 17, 1994 and continues successful operation on a small British satellite; the Space Technology Research Vehicle 1b (STRV 1b). The results of this experiment reported here demonstrate the utility of active vibration suppression technology to meeting the requirements for space-based cryocoolers.

## EXPERIMENT DESCRIPTION

The details of the STRV 1b spacecraft and the cryocooler vibration suppression experiment have been described in previous reports.[1,2] Only a brief description is provided for completeness.

### The Spacecraft Platform

The STRV 1b spacecraft is a small British Royal Establishment satellite with limited, but adequate, capabilities for demonstrating emerging technologies in the space environment. Table 2 is a summary of the specifications and orbital parameters for the STRV 1b.

**Table 2.** STRV 1b Spacecraft and Orbital Parameters

| Parameter | Value |
|---|---|
| Spacecraft size | 48 cm X 48 cm X 40 cm |
| Spacecraft mass | 55 kg |
| Payload weight | 6.5 kg |
| Payload volume | 18.9 liters |
| Solar aspect angle | 90° ± 2° |
| Stabilization | Spin, 5 rpm |
| Payload power | 12 W, 28 V ± 2%, 40 W-hr/orbit |
| Data link | 1 kbps |
| On-board data handling | 8 bit parallel bus |
| Orbit characteristics | Geostationary transfer orbit Elliptical 200 km by 36,000 km Period of 10.5 hours |
| Thermal environment | - 50° C to + 50° C |
| Satellite design life | One year |

It should be noted that (1) the orbit takes the STRV 1b through the Van Allen radiation belts twice per orbit providing additional radiation stress (100 Krads/year) to the on-board electronics, (2) the only source of mechanical vibrations on board STRV 1b is the cryocooler, and (3) the spacecraft has operated one year beyond its design life.

### The Cryocooler Experiment

The experiment was allocated 1.75 kg for the mechanical elements, 4.75 kg for the electronics, and a volume measuring 40 cm X 15 cm X 7.5 cm. The cooler is a commercially available Texas Instruments 1/5-watt 80 K tactical Stirling cooler; a back-to-back, dual-piston, linear drive compressor with integral displacer mounted orthogonal to the compressor piston axis with a vibration level below 5 Newtons. The mechanical dimensions of the cooler are given in Table 3.

The principal components employed in the mechanical design of this experiment are given in Table 4.

**Table 4.** Component description for the STRV 1b Cryocooler Experiment[1,2]

| Component | Description |
|---|---|
| Stirling cryocooler | Texas Instruments, 1/5 watt tactical cooler |
| Piezoelectric translators | Low voltage (20 V) Physik P-842.10 |
| Ceramic appliqués | High voltage (700 V) Tubular Vernitron PZT5H |
| Displacement transducers | Eddy current Kaman SMU 9200-15N |
| Thermometers | Lake Shore Cryotronics PT-111 |
| Accelerometer | Triaxial Kistler 8692B5M1 |
| Microprocessor | UT69R000 |

The experiment provided two active motion control systems: (1) piezo-translators to move the entire cryocooler relative to the spacecraft and (2) ceramic appliqués to control the vibration of the cold finger. The vibration amplitude of the tip of the cold finger relative to the mounting case was monitored with three eddy current transducers having a measurement accuracy of 10 nm.

Initially, two control systems were considered; an analog control using a bandpass filter to track the drive signal obtained from the eddy current transducers and a narrow-band digital adaptive feed forward system that continually controlled the actuators to cancel the tip vibration. Digital control was adopted for the flight experiment due to flexibility compared to implementing analog control.

**RESULTS**

The uncontrolled vibration amplitudes of the cold tip as monitored by the three eddy current transducers in laboratory experiments is shown in Figure 2. The displacement amplitude for the tip in the z-direction (axial) is greater than 1.5 microns and exceeds the requirement for the NASA AIRS instrument.

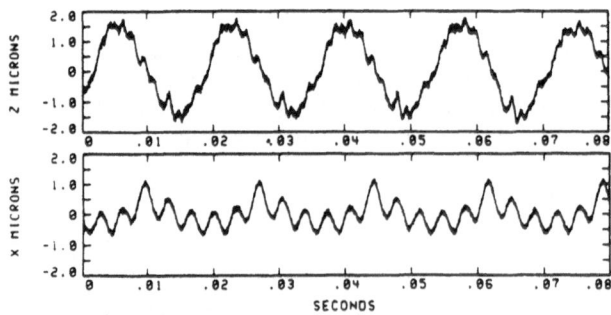

**Figure 2.** Measured cold tip motion.

Higher frequency components clearly accompany the fundamental drive frequency. This response is attributed to coupling higher order harmonics with the structural resonances of, e.g., the cold finger.

The digital control system was capable of managing up to eight harmonics of the fundamental drive frequency; limited by the control software developed in the resources (time and cost) allocated to this project. The vibration attenuation using the digital control on the piezoelectric translators is shown in Figure 3.

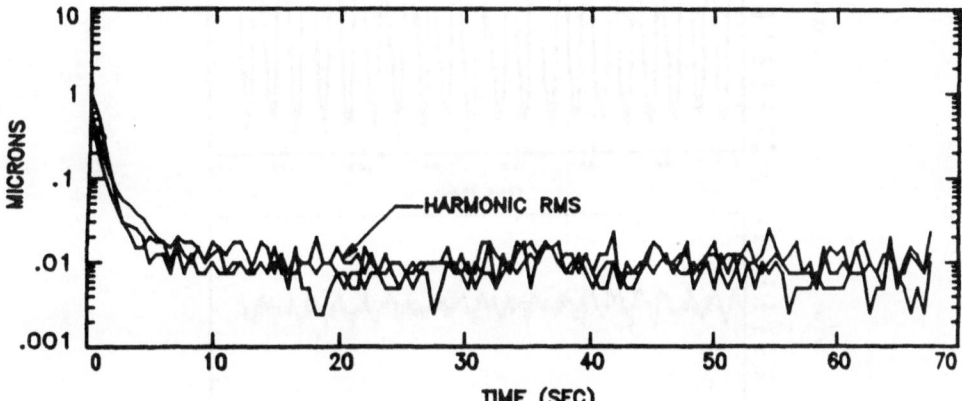

**Figure 3.** Harmonic RMS response of the cryocooler tip motion using the digital control of the piezoelectric translators.

The average attenuation by this approach is a factor between 75 and 80, reducing the vibration of the cryocooler tip by more than an order of magnitude below even the most aggressive of the requirements listed in Table 1. The cryocooler experiment has been exercised over 1,000 times for a total operating time in excess of 35 hours. The lowest temperature reported was 77K when the cooler was operated for more than 30 minutes.

The STRV 1b provided an opportunity to determine the impact of the operation of the cryocooler on the satellite itself. Since the satellite was relatively light weight (~55 kg) and did not have any other sources of mechanical vibrations, a triaxial accelerometer was attached to the base of the crycooler mounting base to monitor the motion of the STRV 1b satellite due to the operation of the cryocooler. The on-orbit accelerations measured parallel and perpendicular to the cold finger axis--the direction of maximum unbalanced cooler forces is shown in Figure 4.

The effect of the unbalanced cryocooler accelerations parallel to the cold finger axis on the motion of the STRV satellite can be estimated by noting that the relationship between the satellite vibration amplitude (y), the vibration force (F) and frequency ($\omega$) is

$$y \sim F/(m \cdot \omega^2) \sim 0.5 \ \mu m.$$

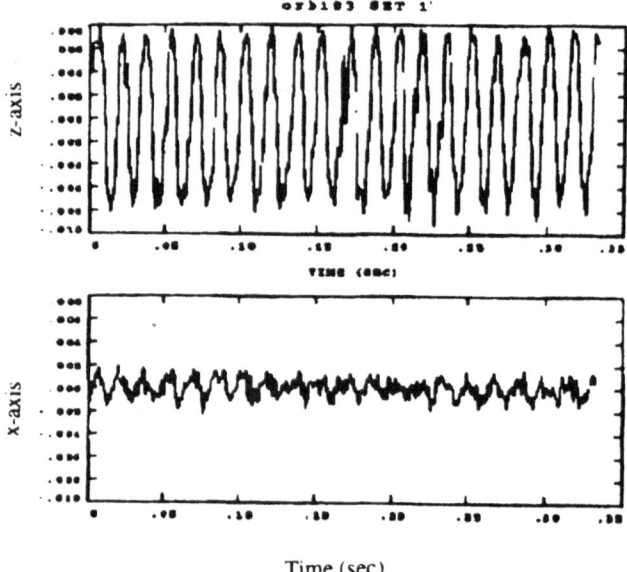

**Figure 4.** Satellite accelerations due to cryocooler operation.

Note that the resulting satellite vibration amplitude approaches the requirements listed in Table 1. The trend to smaller, lower mass surveillance satellite systems will, therefore, increase the demands on vibration control for moving mechanical assemblies.

Active vibration control using piezoelectric materials have been demonstrated in satellite structural[3] as well as commercial[4] systems. The application of this technology to cryocooler vibration control depends on the availability of useful piezoelectric materials and performance requirements. Table 5 presents the performance characteristics of available piezoelectric actuators. The piezoelectric materials are effective over frequency ranges required for most structural vibration regimes. The force, displacement and temperature limits will depend on the specific system application.

**Table 5.** Maximum performance limits of current piezoelectric actuators[5]

| Device | Force (N) | Displacement (mm) | Frequency (kHz) | Temperature (C) |
|---|---|---|---|---|
| Bimorph | 90 | 500 | 1-10 | 75 |
| Rainbow | 500 | 1000 | 1-10 | 100 |
| Patch | 100 | 10 | 20 | 175 |
| Tube | 1000 | 10 | 50 | 175 |
| Flextens | 800 | 250 | 1 | 50 |
| Lever | 500 | 750 | 0.5-1 | 50 |
| Mechanical | 2000 | 400 | 0.5 | 50 |
| Multilayer | 35,000 | 25 | 50 | 175 |

Performance comparisons of the available materials are listed as examples in Table 6. Pe and Pm are defined as the actuator reactive electrical input power and maximum actuator mechanical output power, respectively. Efficiency is defined as the ratio of the output mechanical power divided by the sum of the electrical input power and mechanical output power.

**Table 6.** Performance comparison of state-of-the art piezoelectric actuators[5]

| Device | Pe (watts) | Displacement ($\mu$m) | Block Force (N) | Pm (watts) | Efficiency (%) |
|---|---|---|---|---|---|
| Bimorph | 0.15v | 2130 | 0.19 | 0.0013v | 7.9 |
| Rainbow | 0.035v | 90 | 7.1 | 0.0020v | 5.4 |
| Patch | 0.005v | 8.25 | 15 | 0.00038v | 7.1 |
| Tube | 0.011v | 12.3 | 33.5 | 0.0013v | 10.5 |
| Multilayer | 4.1v | 28 | 12,200 | 1.1v | 21.1 |

The displacement vs force operational ranges for state-of-the art piezoelectric actuators is graphically illustrated in Figure 5. Multilayer devices are capable of forces in the tens of kilonewtons for displacements up to 10 micrometers. Mechanical amplifier devices can achieve displacements about ten times the multilayer at ten times lower maximum forces.

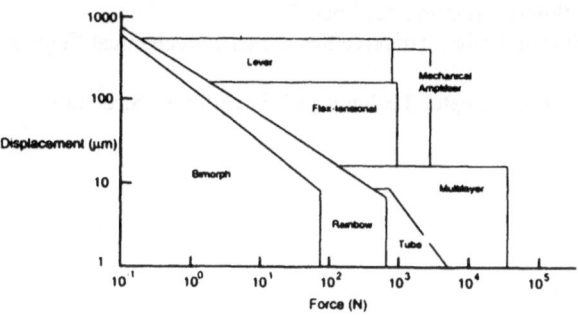

**Figure 5.** Operational ranges for state-of-the art piezoelectric actuators.

The rapid development of piezoelectric materials coupled with the availability of high speed, wide bandwidth digital processors provide an alternate approach to significantly reduce cryocooler vibrations. The STRV-1b cryocooler vibration suppression experiment demonstrates the efficacy of active vibration control.

## CONCLUSIONS

Cryocooler vibration control using piezoelectric actuators has been demonstrated in an on-orbit, low mass, power limited, satellite. The experiment continues to be operational after more than two years on orbit. The principal lessons-learned relevant to the cyrocooler applications are:

1. Active vibration suppression technology provides an alternative low cost approach to reduce cryocooler vibration effects for sensor applications.
2. Active vibration suppression technology is not off-the-shelf; commercially available components are readily available and will meet most structural frequency vibration regimes.
4. Active vibration suppression will have limits for satellite sensor applications; small, lightweight satellites will be sensitive to undamped moving mechanical assemblies.
5. The active vibration suppression system must be designed with a response accounting for structural resonances.

## ACKNOWLEDGMENT

The contributions of the technical staff of the Jet Propulsion Laboratory and the Airforce Phillips Laboratory to the success of this effort are greatly appreciated. This effort was supported in part by the Materials and Structures Program of the Ballistic Missile Defense Organization.

## REFERENCES

1.  Glaser, R., Ross, R.G., Jr. and Johnson, D.L., "Cryocooler Tip Motion Suppression Using Active Control Piezoelectric Actuators", *7th International Cryocooler Conference Proceedings*, Air Force Phillips Laboratory Report PL-CP-93-1001, Kirtland AFB, NM (1993), pp 1086-1097.

2.  Glaser, R.J., Ross, R.G., Jr. and Johnson, D.L., "STRV Cryocooler Tip Motion Suppression", *Cryocoolers 8*, Plenum Press, New York (1995), pp. 455-464.
3.  Nye, T., Casteel, S., Navarro, S., and Kraml, B., "Experiences with Integral Microelectronics on Smart Structures for Space".
4.  Ashley, S., "Smart Skis and other Adaptive Structures", Mechanical Engineering, November (1995).
5.  Near, C.D., "Piezoelectric Actuator Technology", SPIE Smart Structures and Materials Conference (1996).

# Vibration Reduction in a Set-Up of Two Split Type Stirling Cryocoolers

**A.P. Rijpma, J.F.C. Verberne, E.H.R. Witbreuk and H.J.M. ter Brake**

University of Twente, Department of Applied Physics
P.O. box 217, 7500 AE Enschede, The Netherlands

## ABSTRACT

At the University of Twente research is in progress on performing magnetocardiography in clinical conditions. For the magnetic measurements High-$T_c$ SQUIDs will be used, which will be cooled by commercial cryocoolers. These coolers consist of separate compressor and displacer modules. The compressors are of the dual opposed pistons type. Since the magnetic measurements are sensitive for movements of magnetic materials as well as for movements of the sensors, it is advantageous to minimise the vibrations caused by the cryocoolers. This is achieved by:
- Using coolers with dual opposed pistons in the compressor modules.
- Using two coolers and combining the two displacer modules into one rigid unit and operating them in a dual opposed manner.

The remaining vibrations in the axial direction are about 1 ms$^{-2}$ for the compressors and 0.2 ms$^{-2}$ for the displacer unit. In order to reduce these vibrations, the automated control system described in this paper was developed. It uses an adaptive feed-forward algorithm based on the work of Wu[1] and Boyle et al[2]. and is capable of controlling the vibrations of the two compressors and the displacer unit for multiple harmonics simultaneously. Thus the vibration levels are reduced down to 0.02 ms$^{-2}$ and 0.01 ms$^{-2}$ for the compressors and the displacer unit respectively.

## INTRODUCTION

The 'Low Temperature Division' and the 'System Dynamics Group' of the University of Twente are co-operating in a research project aiming at the application of magnetocardiography in clinical conditions, i.e. without the use of a magnetically shielded room. High-Tc SQUIDs will be used as the magnetometers and in order to realise a turn-key apparatus these sensors will be cooled by cryocoolers instead of liquid cryogen. The thermodynamical aspects of this development were reported elsewhere[3]. With respect to noise a number of contributions is relevant. Firstly, the noise contributions from the environment. For example, a bus passing at about 8 meters distance, was measured to produce a disturbance field of typically 200 nT, which is large compared to the field strength of 10-100 pT of an adult's heart. A second source of disturbances are the magnetic components of the magnetometer itself. Movement of these components, which are primarily found in the cryocoolers can lead to stray magnetic fields that are measured by the SQUID-magnetometers. The coils used to drive the compressors also give rise to a magnetic field. This field, however, can be reduced by properly positioning the cryocoolers[3]. Not only the movement of magnetic components, but also the movement of the sensors will lead to magnetic disturbances. A rotation of the sensors in the Earth's field with a mere $10^{-8}$ radians, will allready yield an equivalent field change of 0.5 pT in the magnetometers.

At the mains frequency of 50 Hz there will always be large environmental noise in the range of 10 nT. This will impede measurements at this frequency. Noise contributions resulting from vibrations with this frequency are therefore acceptable up to this limit.

The cryocoolers used are manufactured by Signaal Usfa[4] (type 7058). Their cooling power amounts to 1.5 W at 80 K and their drive frequency is 50 Hz. This is outside the frequency range of a healthy adult's heart: 0.1-40 Hz. Nonetheless, we would like to suppress the vibrations of the cryocoolers for this frequency and higher harmonics up to about 350 Hz. The reason for this is that due to bandwidening and non-linearities vibrations at 50 Hz may cause vibrations at other frequencies. Furthermore, in future applications such as the recording of heart disturbances or a foetal heart beat, the frequency band of interest will extend beyond 50 Hz.

Since the magnetocardiographic measurements are sensitive for vibrations, the aim is to minimise the cryocooler-vibrations. A set-up was already realised in which the two coolers were operated in an opposed arrangement[5]. This set-up, which was used to manually tune the compressor-coil currents to reduce vibrations, will now be used in an automated control system. The automated system is designed to suppress the vibrations of both compressors and the combined displacer unit simultaneously for multiple harmonics. This is achieved by using an adaptive feed-forward algorithm, that calculates the required control voltages from the measured vibrations and the previously determined transfer functions from voltage to vibration.

## PREVIOUS SET-UP

The starting-point for the automated control system was an earlier set-up[5], which is schematically shown in Fig. 1. Since the cryocoolers are equipped with one displacer each, it was decided to combine the displacer modules of the two cryocoolers into a single rigid structure. to which will be referred to as the displacer unit. The SQUID-magnetometers are not present in the current set-up. They will be situated about 30 cm from the displacer unit[3]. For experimental purposes, both the compressors and the displacer unit are mounted on thin plates to allow free

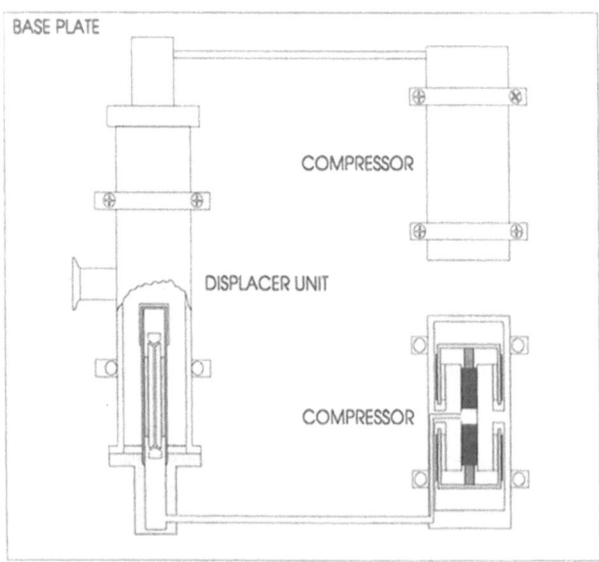

**Figure 1.** Experimental base plate with the two displacer modules in dual opposed arrangement on the left and the two compressor modules on the right.

motion along the axes. The vibrations are measured by acceleration transducers placed on the compressors and displacer unit. The combination of the two displacer modules to one unit reduced the vibrations at 50 Hz from about 4 ms$^{-2}$ per module to about 0.2 ms$^{-2}$ for the unit. The vibration of a compressor without any control is typically 1 ms$^{-2}$ at the 50 Hz drive frequency. With manual tuning of the coil currents it appeared to be possible to reduce the axial acceleration of a compressor to 0.05 ms$^{-2}$, whereas for the displacer unit a suppression down to 0.02 ms$^{-2}$ was achieved, both values for 50 Hz only[5].

## AUTOMATED CONTROL SYSTEM

### Basic Algorithm

A general representation of the control system is given in Fig. 2. In this figure $V_0$ stands for the voltage that drives the compressor and $V_c$ for the control voltage used to reduce the acceleration A. The transfer functions H and G, for the drive voltage and the control voltage respectively to the acceleration A, are not equal. This is due to the fact that the drive voltage is added to both coils of a compressor and the control voltage is not. The control voltage may e.g. be added to just one coil, or added to one and subtracted from the other. Assuming $V_0$, G and H are constant and G and A are known, it follows that the control voltage $V_c$ should be changed according to:

$$V_c(j\omega)(k+1) = V_c(j\omega)(k) - \sum_{n=1}^{N} G^{-1}(jn\omega_d) \cdot A(jn\omega_d)(k) \qquad (1)$$

Here, k is the control time step, $\omega_d$ is the drive frequency and n refers to the n-th harmonic. The algorithm used to approximate the control action as stated above, is based on an algorithm described by Wu[1]. The derivation of this algorithm starts with defining the control variable as shown in Eq. (2), (with respect to Wu's paper we changed the control variable from current to voltage):

$$V_c(t) = \sum_{n=1}^{N} V_n(k)\sin(n\omega_d t + \varphi_n(k)) \qquad for \quad k\Delta t < t \le (k+1)\Delta t \qquad (2)$$

By using goniometric relations Eq. (2) can also be expressed as Eq. (3), which results in changing the control parameters from V and $\varphi$ to $\beta$ and $\gamma$. The subsequently derived equations (4)..(7) are observed to have better physical and mathematical properties than Wu's equations. The physical improvement consists of Eqs. (4)..(7) being exact instead of a linear approximation; as for the mathematics: one no longer risks a division by zero ($V_n$).

$$V_c(t) = \sum_{n=1}^{N} V_{c,n}(t) = \sum_{n=1}^{N} [\beta_{c,n}(k)\sin(n\omega_d t) + \gamma_{c,n}(k)\cos(n\omega_d t)] \qquad for \quad k\Delta t < t \le (k+1)\Delta t \qquad (3)$$

Eq. (3) is only a valid approximation of the required control voltage under two conditions. Firstly, it is assumed that only the drive frequency and higher harmonics are relevant in the vibrations. Secondly, the cryocoolers respond as linear systems. Both conditions are met as will be shown in one of the following sections.

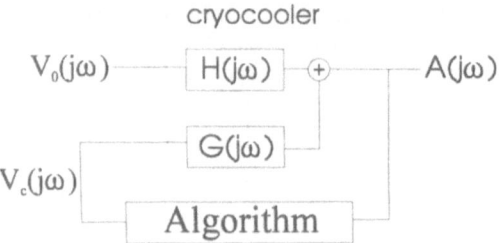

**Figure 2.** A representation of the control system. The drive voltage $V_0$ and the control voltage $V_c$ lead to a combined acceleration A of a cryocooler.

The calculations for the adjustment of the control voltage, which are to be carried out each control step, are:

$$\beta_{c,n}(k+1) = \beta_{c,n}(k) + \Delta\beta_{c,n} \qquad \text{and} \qquad \gamma_{c,n}(k+1) = \gamma_{c,n}(k) + \Delta\gamma_{c,n} \qquad (4,5)$$

with:

$$\Delta\beta_{c,n} = -\frac{B_n \cos(\psi_n) + C_n \sin(\psi_n)}{G_n} \qquad \text{and} \qquad \Delta\gamma_{c,n} = \frac{B_n \sin(\psi_n) - C_n \cos(\psi_n)}{G_n} \qquad (6,7)$$

Here, $G_n$ and $\psi_n$ represent the amplitude and phase of the transfer function $G(\omega)$ for the n-th harmonic frequency. The terms $B_n$ and $C_n$ in the equations are the Fourier sine- and cosine-coefficients of the acceleration A(t) as defined by equations 8 and 9, in which T is a multiple of the period of the drive frequency.

$$B_n = \frac{2}{T}\int_{-T/2}^{T/2} A(t)\sin(n\omega t)dt \qquad \text{and} \qquad C_n = \frac{2}{T}\int_{-T/2}^{T/2} A(t)\cos(n\omega t)dt \qquad (8,9)$$

To ensure stability of the control system, the errors $\Delta\psi_n$ and $\Delta G_n$ in determining the phase and amplitude of the transfer function G should satisfy the following inequality:

$$\left(\frac{\Delta G_n}{G_n}\right)^2 + (\Delta\psi)^2 \langle 1 \qquad (10)$$

**Implementation of the Algorithm**

Fig. 3 shows a schematic representation of the control system used. This is a simplified version of the general situation, in the sense that in principle each of the three sensors may lead to a unique control voltage for each of the four compressor coils. In that case twelve control variables would be required. The assumptions, which have led to the reduction, are:
1. The transfer functions from both coil voltages to the acceleration of a compressor are equal in magnitude, but differ $\pi$ in phase.
2. The transfer functions from both coil voltages of one compressor to the acceleration of the displacer unit are equal in size and phase.
3. The coil voltages of a compressor have no influence on the acceleration of the other compressor.

Because of assumption 1 the suppression of the displacer unit's vibrations is established by applying the same control voltage to both coils of a compressor. In this way the acceleration of

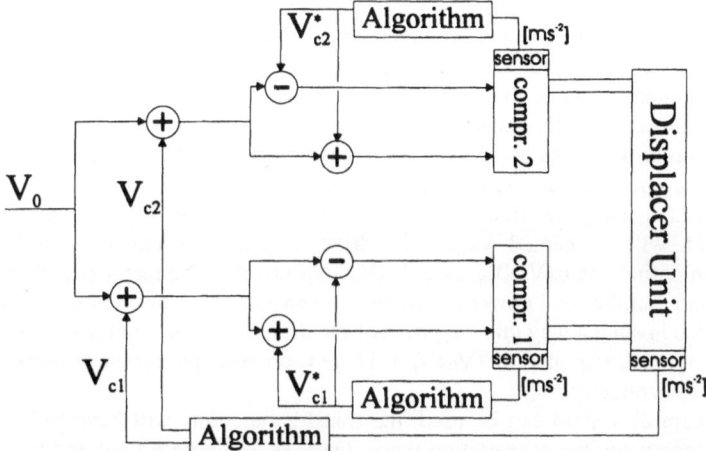

**Figure 3.** Schematic representation of the control system. Shown are the drive voltage $V_0$, the control voltages $V_{c1}^{*}$ and $V_{c2}^{*}$ for controlling the compressors and $V_{c1}$ and $V_{c2}$ for controlling the vibrations of the displacer unit.

the compressors is not disturbed, when controlling the acceleration of the displacer unit. Assumption 2 means that the compressor vibrations can be suppressed, without affecting the displacer unit, by adding control voltages to both compressor coils, which are equal in size but opposite in phase. Furthermore, assumption 3 implies that for controlling the vibrations of a compressor only the coil voltages of that compressor itself have to be controlled.

The algorithm used to calculate the voltages $V_{c1}^{*}$ and $V_{c2}^{*}$ is the basic algorithm given by Eq. (3)..(7) with $V_c$ replaced by $V_{c1}^{*}$ and $V_{c2}^{*}$ respectively. The transfer functions are determined experimentally, as will be discussed later.

In controlling the displacer unit there is still some freedom. The control can be done with just $V_{c1}$, with $V_{c2}$ or with some combination. For reasons of symmetry, there is a preference to use the following restriction for defining how $V_{c1}$ and $V_{c2}$ should be altered.

$$\left| V_{c1,n} \right|^2 = \left| V_{c2,n} \right|^2 \tag{11}$$

This equation states that the n-th harmonic components in $V_{c1}$ and $V_{c2}$ are equal in size. The changes in the control voltages $V_{c1}$ and $V_{c2}$ are given by Eq. (12)..(15). These can be derived from Eq. (3)..(7) by using both the restriction Eq. (11) and the condition that the control voltages $V_{c1}$ and $V_{c2}$ do not (partially) counteract.

$$\Delta\beta_{c1,n} = -\frac{G_{1n}}{G_{1n} + G_{2n}} \cdot \frac{B_n \cos(\psi_{1n}) + C_n \sin(\psi_{1n})}{G_{1n}} \tag{12}$$

$$\Delta\gamma_{c1,n} = \frac{G_{1n}}{G_{1n} + G_{2n}} \cdot \frac{B_n \sin(\psi_{1n}) - C_n \cos(\psi_{1n})}{G_{1n}} \tag{13}$$

$$\Delta\beta_{c2,n} = -\frac{G_{2n}}{G_{1n} + G_{2n}} \cdot \frac{B_n \cos(\psi_{2n}) + C_n \sin(\psi_{2n})}{G_{2n}} \tag{14}$$

$$\Delta\gamma_{c2,n} = \frac{G_{2n}}{G_{1n} + G_{2n}} \cdot \frac{B_n \sin(\psi_{2n}) - C_n \cos(\psi_{2n})}{G_{2n}} \tag{15}$$

Whereas restriction Eq. (11) is applied to the higher harmonics, it does not apply to the drive frequency. Manufacturing tolerances in the phase relation of the motion of the displacer modules with respect to the coil voltages, might necessitate significant shifts in phase of the drive component of these voltages. The total input power will change when this is done while satisfying Eq. (11). Therefore, the condition that is chosen for that component in $V_{c1}$ and $V_{c2}$ with the drive frequency, states that the total power input of the cryocoolers should remain unchanged. The procedure is illustrated for the first control step (i.e. $V_{c1}=V_{c2}=0$). The procedure for the following control steps is performed analogously. First, the drive frequency component of $V_{c1}$ necessary to suppress the acceleration is calculated. This calculation is done with the basic algorithm Eq. (3)..(7) using the transfer function from $V_{c1}$ to the displacer acceleration. The component found, will be named $V_{temp}$. The drive frequency components in $(V_0+V_{c1})$ and $(V_0+V_{c2})$ are then defined as $\alpha(V_0+V_{temp})$ and $\alpha V_0$ respectively. The factor $\alpha$ is then determined from the restriction that the total power has to remain constant. As a last action both components are shifted in phase in such a way that the phase of the drive frequency component in $(V_0+V_{c1})$ is minus the phase of that component in $(V_0+V_{c2})$. There is no principal reason to do so. It is merely done for reasons of symmetry.

Before the control system can be used, the transfer functions will have to be determined. This is done by measuring the acceleration twice. Once as a reference $(A_0)$ and the second time $(A)$ after having added a known test voltage $V_{test}$ to the control voltage. Then the transfer function can be determined from the change in acceleration:

$$G(j\omega) = \frac{A(j\omega) - A_0(j\omega)}{V_{test}(j\omega)} = \frac{\left[A_0(j\omega) + G(j\omega)V_{test}(j\omega)\right] - A_0(j\omega)}{V_{test}(j\omega)} = \frac{G(j\omega)V_{test}(j\omega)}{V_{test}(j\omega)} \qquad (16)$$

## EXPERIMENTAL SET-UP

Figure 4 shows a schematic representation of the experimental set-up. It contains the two cryocoolers with the displacer modules combined to a single unit. The same base plate as in the earlier experiments is used. The positioning of the compressors and displacer unit is also the same. The compressor vibrations are measured with acceleration transducers type Seika BDK3 (accuracy: $0.02$ ms$^{-2}$; sensitivity: $0.013$ V/ms$^{-2}$ and range: $30$ ms$^{-2}$). For the measurement of the displacer acceleration a Bruël&Kjær 4381 sensor with pre-amplifier was available: (accuracy and sensitivity $0.002$ ms$^{-2}$ and $0.32$ V/ms$^{-2}$ respectively). The output of the sensors is filtered by fourth-order Butterworth filters with cut-off frequencies of $500$ Hz. The filtered signals are then sampled by a DAP-board type 2400/e4 from Microstar Laboratories. For the BDK3 and B&K an on-board gain of $100$ and $10$ respectively is used, after which sampling takes place with a 12-bit DA-converter at $1024$ Hz. After performing a Discrete Fourier Transform the DAP-board sends

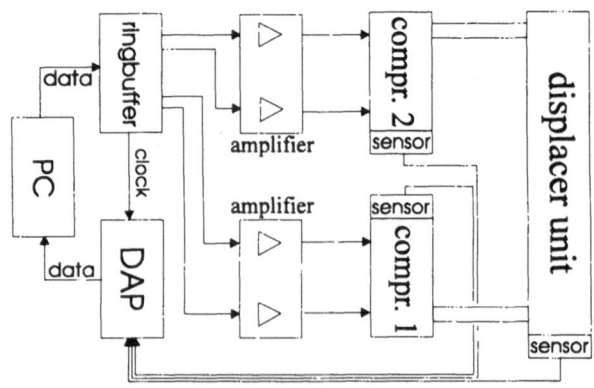

**Figure 4.** Schematic diagram of the experimental set-up.

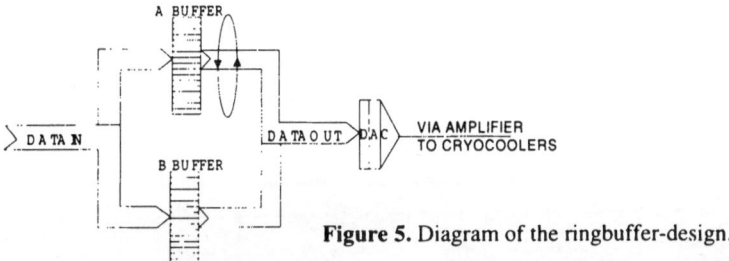

**Figure 5.** Diagram of the ringbuffer-design.

the data to the PC. The PC then determines the new values for the control variables and calculates the coil voltages for a full period. The next step is copying the calculated voltages to the 'ringbuffer'. This ringbuffer is an interface between the PC and two audio amplifiers, that was specially designed for this application. It consists of two sets of 4 buffers, 4 DA-converters (DAC) and some logic. While one set of buffers is circularly read to provide the coil voltages, the other set can be written by the computer (Fig. 5). When the new coil voltages are written, the computer can order the ringbuffer to switch sets and use the new data to define the coil voltages. The other set of buffers is then available for updating. Another feature of the ringbuffer is the clock-signal. The clock used by the DAP-board for sampling, is defined by the ringbuffer. This is done to ensure a lasting synchronisation between the coil voltages and the sampling. If this synchronisation does not exist the measured phase $\psi$ of the transfer functions is meaningless and the control system will fail.

## EXPERIMENTAL RESULTS

It should be noted that during the first 50 minutes of operation the transfer functions slowly change. This is probably due to changes in mechanical properties caused by the cooling down of the displacer and warming up of the compressor. Therefore, all measurements were performed after a period of about 50 minutes.

### Validation of Assumptions

Figures 6 and 7 show that the assumption of linearity is adequate for fast convergence. Figure 6 shows the change of the harmonic components in the acceleration of a compressor when

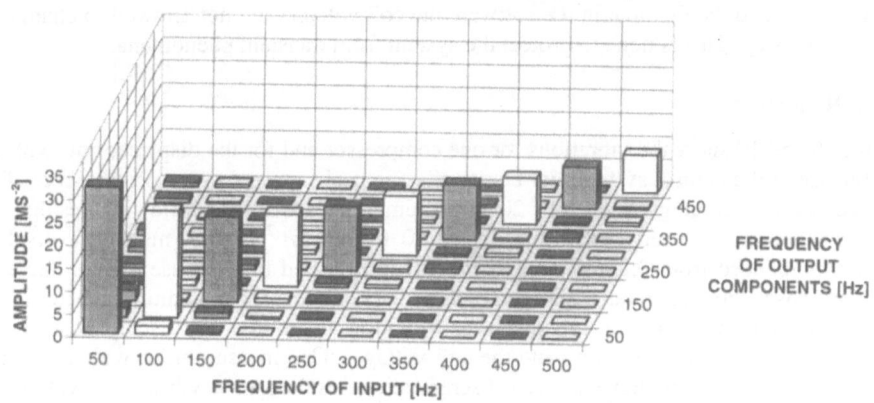

**Figure 6.** The change in the acceleration of a compressor split into the various frequency components for a number of input frequencies. The input variable is a coil voltage.

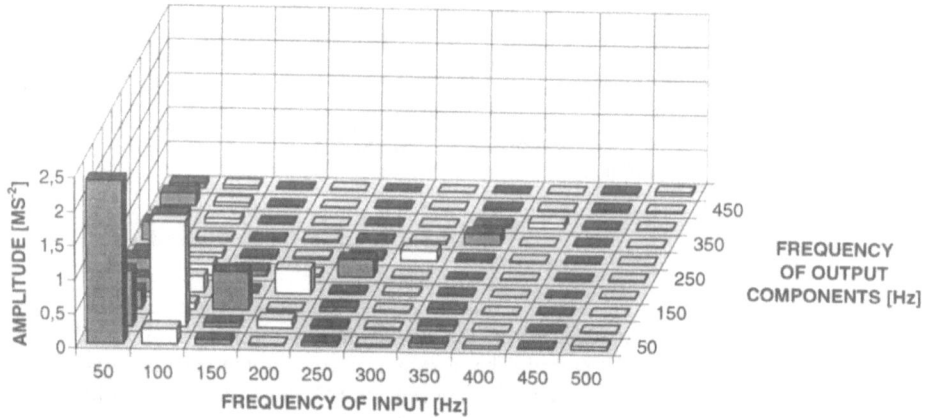

**Figure 7.** The change in the acceleration of the displacer unit split into the various frequency components for a number of input frequencies. The input variable is $V_{c1}$.

a certain frequency component is added to one of the coil voltages. The figure shows convincingly that a frequency component added to a coil voltage leads primarily to an additional acceleration with the same frequency. Figure 7 shows the same for the transfer function from the control voltage $V_{c1}$ to the acceleration of the displacer. The other assumptions are also verified and are found to be valid, although assumption 2 only appears to hold for the harmonics up to about 200 Hz.

## Convergence

The convergence of the control system is excellent. This can be seen in Fig. 8 for the 50 Hz component of the acceleration. For the higher harmonics similar graphs can be obtained. As the algorithm is a one-step control algorithm, it might be expected that the remaining acceleration is reduced to noise level in just one control step. According to the figure this is not the case. The observation that it takes multiple steps to reach the noise level however, is not due to the algorithm. It is due to the fact that in the software the coil voltages are not allowed to change too much in a single step. This is done to protect the system from transient phenomena.

## Vibration Reduction

Figures 9 and 10 show the vibrations for one compressor and for the displacer unit with and without the use of the control system. For clarity aliasing peaks present in the original recordings are omitted here. Averaging over about 20 measurements leads to the following results. The compressor vibrations are reduced with a factor 50 from 1 ms$^{-2}$ to 0.02 ms$^{-2}$. The displacer acceleration is reduced from 0.2 ms$^{-2}$ to 0.01 ms$^{-2}$. It is expected that the use of more accurate acceleration transducers may reduce the vibrations of the compressors an additional factor 5. It is anticipated that the reduction of the displacer vibration is limited by discrepancies in the synchronisation between the sampling and the coil voltages. One measurement, which was done with an ad hoc solution to prevent these discrepancies, resulted in a vibration level for the displacer unit of only 0.002 ms$^{-2}$, which corresponds to the noise level of the sensor used to measure the acceleration of that unit.

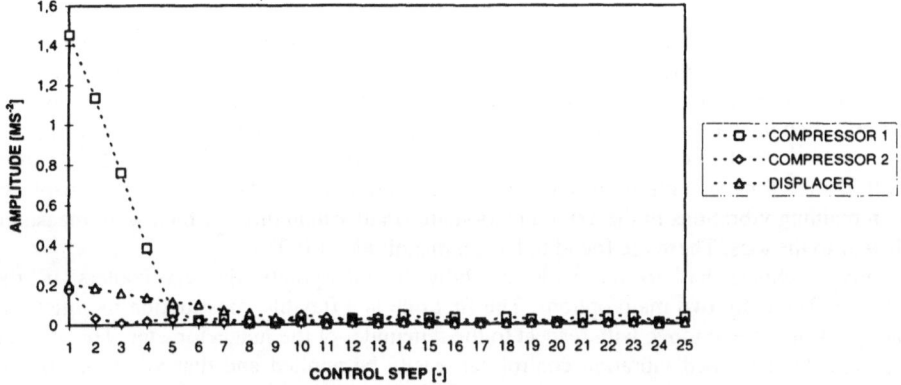

**Figure 8.** Measured acceleration for the first 25 control steps.

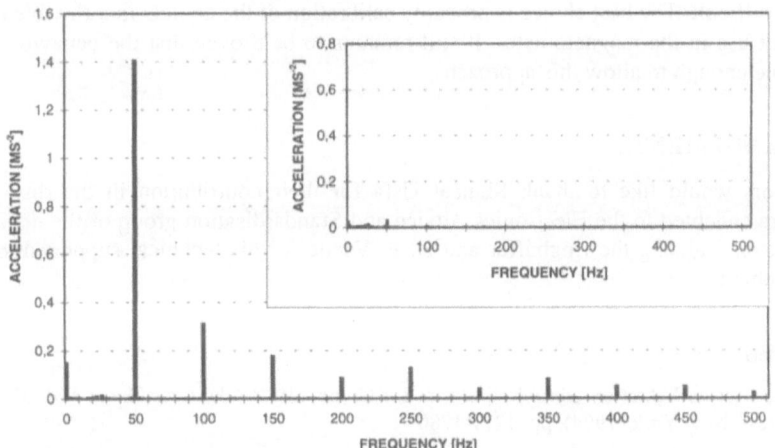

**Figure 9.** The acceleration measured on a compressor. The larger figure shows the vibration without control. The inset shows the remaining acceleration when the control system is used.

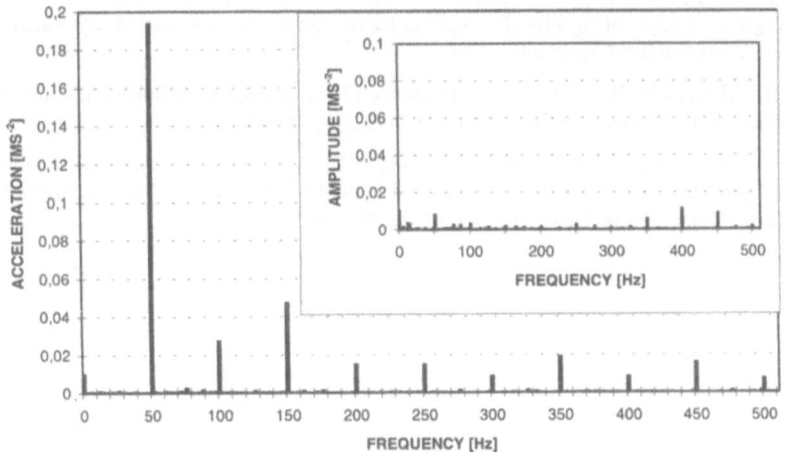

**Figure 10.** The acceleration measured on the displacer unit. The larger figure shows the vibration found without control. The inset shows the remaining acceleration when the control system is used.

## CLOSING REMARKS

Based on the algorithm described by Wu a control system was developed, which is capable of simultaneously reducing the vibration for all three components, i.e. the two compressors and the displacer unit. In principle, all vibrations can be reduced to the intrinsic noise level in just one control step, though the convergence may be limited by inaccuracies in measuring the acceleration or transfer functions. With this control system we reduced the vibration levels from 1 ms$^{-2}$ to 0.02 ms$^{-2}$ for the compressor modules and from 0.2 ms$^{-2}$ to 0.01 ms$^{-2}$ for the displacer unit. The remaining vibrations in the axial direction are smaller than those, which were measured perpendicular to the axis. These are found to have a magnitude of 0.05 ms$^{-2}$

Presently a choice has to be made on how to incorporate the cryocoolers in the magnetometer. There are two main options. The first one is a flexible construction in which the cryocooler parts are free to move with respect to the remainder of the magnetometer. This has the advantage that the described vibration control can easily be applied and that vibrations of the coolers do not act on the magnetometer as strongly as in a rigid integration of the coolers. This is the second option. A rigid integration has the benefit that vibrations will decrease as a result of the increased mass and is easier to construct. Measurement of the acceleration will however become more difficult. The best choice is probably calibration of the coolers in a flexible test bed and subsequent use in the magnetometer. It still remains to be proven that the behaviour of the coolers is stable enough to allow this approach.

## ACKNOWLEDGEMENTS

The authors would like to thank Signaal Usfa for their contribution in the discussions. Further, we are endebted to the Electronics Advice and Standardisation group of the department for their work on realising the ringbuffer and H. te Veene for his technical support during the experimental phase.

## REFERENCES

1.   Wu, Y.A., "Active Vibration control algorithm for Cryocooler", *Adv. Cryo. Eng., Vol. 39, part B*, Plenum Press, New York (1994), pp. 1271-1280

2.   Boyle, R., Conners, F., Marketon, J., Arillo,V., James, E., Fink, R., "Non-real time feed forward vibration control system development & test results", *Proc. of the 7-th Int. Cryocooler Conf.*, Santa Fe (1992), pp. 805-819

3.   Van den Bosch, P.J., Ter Brake, H.J.M., Holland, H.J., De Boer, H.A., Verberne, J.F.C., Rogalla, H., "Cryogenic design of a high-Tc SQUID-based heart scanner cooled by small Stirling cryocoolers", submitted to Cryogenics (1996)

4.   Verbeek, D., Helmonds, H., Roos, P., "Performance of the Signaal Usfa Stirling cooling engines", *Proc. of the 7-th Int. Cryocooler Conf.*, Santa Fe (1992), pp. 728-737

5.   Verberne, J.F.C., Bruins, P.C., Van den Bosch, P.J., Ter Brake, H.J.M., "Reduction of the Vibrations generated by Stirling Cryocoolers for cooling a high Tc SQUID magnetometer", *Cryocooler 8*, Plenum Press, New York (1995), pp. 465-474

# Throttle Cycle Cooler Vibration Characterization

**Dennis Hill**

APD Cryogenics
Allentown, Pennsylvania USA 18103

## ABSTRACT

Cryocoolers with low vibration are desirable in a number of applications including X-ray detectors, FTIR detectors, superconducting filters, and SQUID magnetometers. Vibration of a throttle cycle cooler is characterized and compared with vibration of other conventional forms of cooling, namely Stirling cycle, Gifford-McMahon and liquid nitrogen dewars.

Major components of the throttle cycle cooler include the compressor package, the cold end, and the interconnecting gas lines. All mechanical components are found to influence cold end vibration. The compressor package contributes forcing functions in the form of mechanical vibrations and pressure pulses which are transmitted by the interconnect lines. The major forcing function inside the cold end itself is the energy dissipated by the throttling process, creating vibration in the audible frequency range. The composition of the refrigerant mixture is also shown to affect the magnitude of vibration.

Vibration at the cold tip of the cooler is shown to be lower than that of Gifford-McMahon coolers but somewhat higher than that of boiling liquid nitrogen. Reaction forces at the cooler interface are more than an order of magnitude lower than those of Stirling coolers.

## INTRODUCTION

Cryogenic cooling is often employed to enhance the sensitivity of detectors, and of course superconducting devices must be cooled to function. By the nature of the data that they measure, cooled detectors must be held fixed in the earth's frame of reference; small perturbations in position add noise to the signal and reduce the resolution of the data being measured. The vibration imposed by the method of cooling must be considered in the system design so that resolution will not be compromised.

Liquid nitrogen has long been used to cool vibration-sensitive detectors. It is relatively inexpensive and the only source of vibration is the slow boiling of the fluid as heat is absorbed from the detector and the environment. Gifford-McMahon and Stirling coolers are more convenient because they operate on closed cycles and so they do not need constant replenishment, but they do produce higher vibrations because both employ moving masses to shuttle the refrigerant through its thermodynamic cycle.

The object of this investigation was to characterize the vibration of a throttle cycle cooler, manufactured under the product name CRYOTIGER® I. The system, depicted in Figure 1, consists of an air cooled compressor package, a pair of interconnecting lines, and a cold end.

Cryocoolers 9, Edited by R.G. Ross, Jr.
Plenum Press, New York, 1997

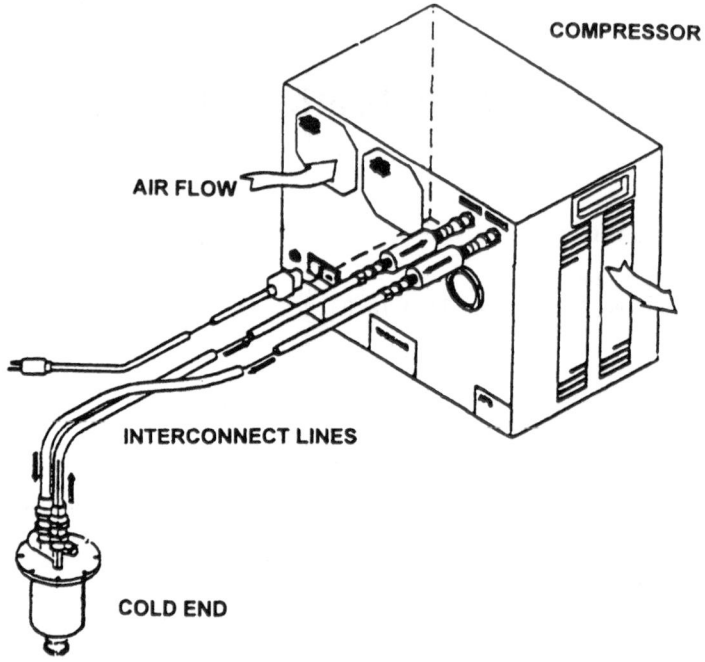

**Figure 1.** CRYOTIGER® I System: Compressor, Interconnect Lines, Cold End

The thermodynamic cycle has been discussed in detail in other works.[1,2] The compressor supplies high pressure mixed refrigerant via one of the interconnect lines to the cold end, where it flows through a counter flow heat exchanger, through the throttling device. The refrigerant expands through the throttling device, and flows through a cold tip where it absorbs heat from the device to be cooled. The low pressure refrigerant is then circulated back through the counter flow heat exchanger, the interconnect line, and back into the compressor. The excitation produced by all components was measured, and will be discussed in what follows.

## COMPRESSOR FORCING FUNCTIONS

The compressor contributes forcing functions to the system in the form of mechanical vibration and pressure pulses. Mechanical vibration of the compressor was measured perpendicular to the gas line at the compressor manifold with the system operating under 60 Hz, single phase power. The resulting acceleration profile, shown in Figure 2, shows a first harmonic at 57.5 Hz and clearly defined second and third harmonics. Some of this vibration is potentially transmitted to the cold end through the interconnect lines, the amount of vibration transmitted depending upon the length, routing, and fixity of the lines.

The compressor also creates pressure pulses in both gas streams. Figure 3 shows the dynamic pressure fluctuations of the discharge and suction-side gas streams measured near the cold end. Pressure fluctuation on the discharge side is relatively small, with only the first three harmonics evident in the measurements performed. Pressure fluctuation on the suction side is about one half of a percent of the total suction pressure, with harmonics clearly visible through 1000 Hz. The same 57.5 Hz fundamental harmonic is evident in both mechanical vibration measurements and dynamic pressure measurements.

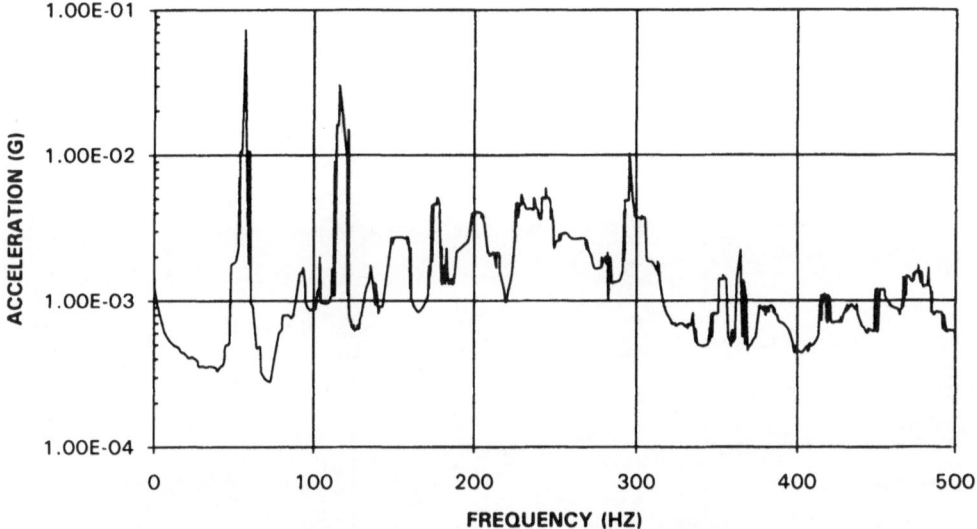

**Figure 2.** Compressor Mechanical Vibration

## COLD END VIBRATION MEASUREMENTS

Since the cooler interfaces with the sensor to be cooled, the magnitude of cold end vibration directly influences the sensor's resolution and thus cold end vibration was the primary interest of this study. Cold end vibration will be influenced by how the cold end is mounted in the system requiring cooling. Rigid mounting would minimize vibration, but maximize the interface force transmitted through the supports. Interface forces are presented in the next section.

For this investigation, the CRYOTIGER® I cold end was characterized in the "free-free" condition as depicted in Figure 4. The cold end weighed 1.7 Kg of the total 4.2 Kg supported

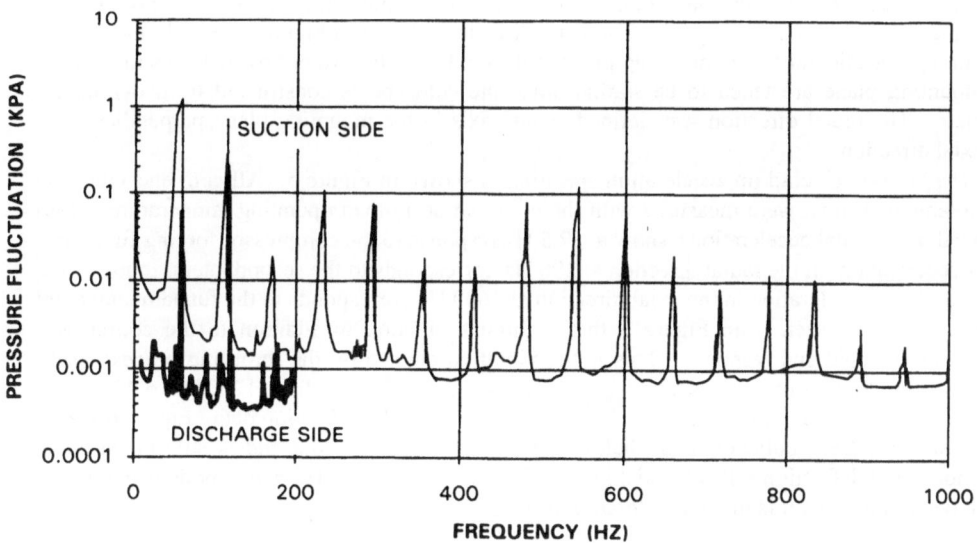

**Figure 3.** Compressor Discharge and Suction Dynamic Pressure Fluctuations

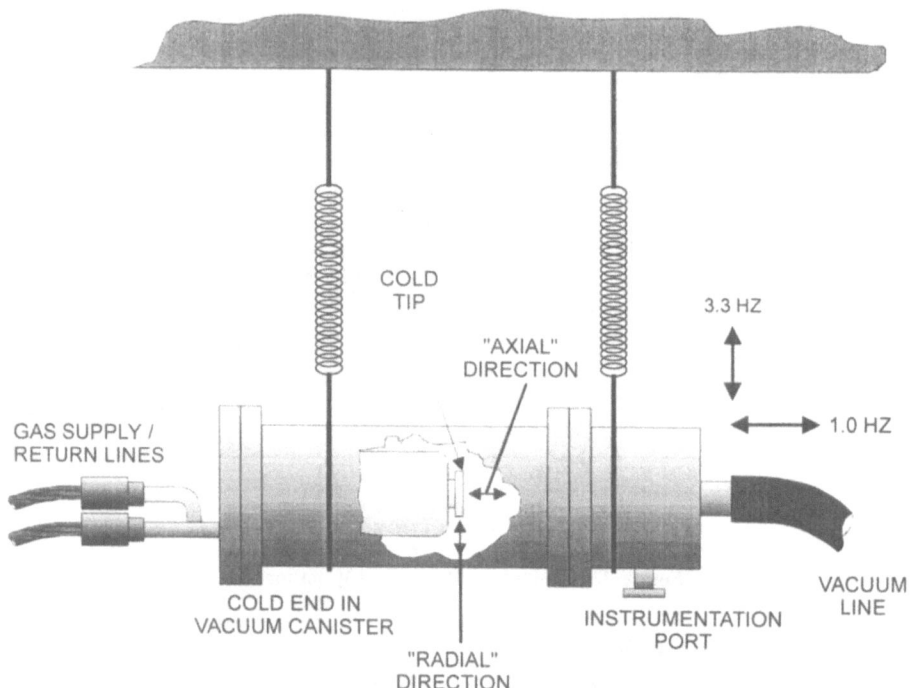

**Figure 4.** "Free-Free" Cold End Vibration Measurement Test Setup

weight which included the cold end, vacuum canister, and instrumentation portals. The canister was supported by two soft springs such that the system had a fundamental frequency of 1.0 Hz along the axis of the canister and 3.3 Hz along the axis of the springs. The "free-free" condition is a worst case because it yields the highest cold end vibrations. It is also the cleanest and most repeatable condition to perform testing because there is no question of interaction of the cold end with the support structure.

The cold end was instrumeted with a PCB Model 351B41 cryogenic accelerometer on the cold tip. Acceleration data was collected one axis at a time with HP 3562A and Spectral Dynamics SD345 spectral analyzers. The axial direction (shown in Figure 4) was defined as the direction perpendicular to the mounting plane at the cold tip. The two orthogonal axes in the cold tip mounting plane are taken to be similar since the cold end is constructed from cylindrical geometry. The radial direction was defined as any axis in the mounting plane perpendicular to the axial direction.

The "free-free" cold tip acceleration spectrum is shown in Figure 5. All cold end vibration measurements shown were measured with the cold end at no-load operating temperature. Both the axial and radial accelerations show a 57.5 Hz response to the compressor forcing functions. Peak acceleration in the radial direction at 200 Hz corresponds to the second lateral mode of the cold tip. Peak acceleration in the axial direction at 1600 Hz corresponds to the fundamental axial mode. It can be seen from Figure 5 that a forcing function in addition to the compressor influences the cold end response. This is the vibration created by the throttling process itself. The throttling forcing function is broad-band across the audible range from 1200 Hz to 4500 Hz.

The "free-free" axial response was integrated twice to yield deflection data. Figure 6 shows that peak "free-free" deflections are below 1 micron. Since the cold end is stiff in the axial direction, axial deflection rolls off quickly with frequency. The 200 Hz radial mode is evident in the deflection data, as it is in the acceleration response.

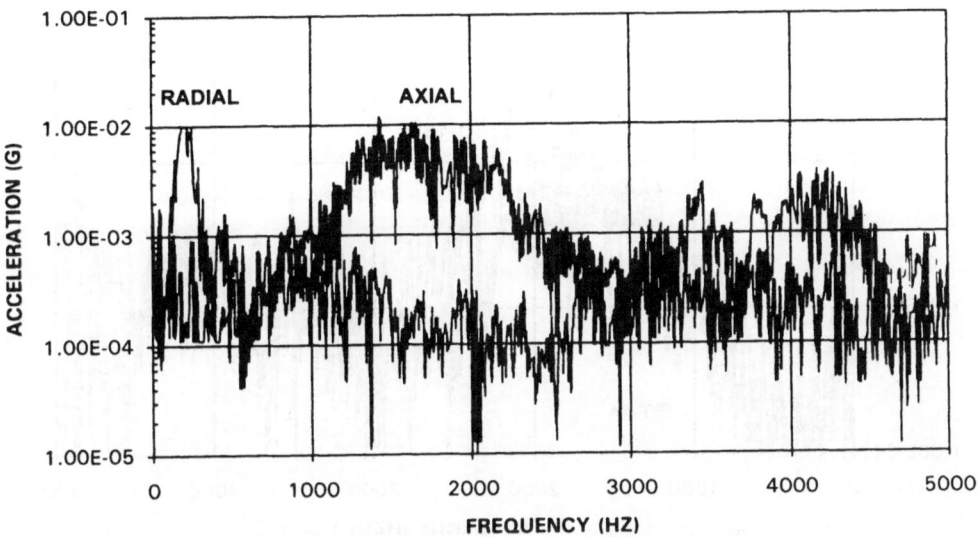

**Figure 5.** "Free-Free" Cold Tip Acceleration Response

Since the throttling process creates vibration energy, it may be expected that the refrigerant mixture itself has an influence on the cold end response. In Figure 7, the "free-free" axial acceleration response of two different mixtures is compared. Mixture PT-13 is optimized for low temperature, reaching 72 K with no load. Mixture PT-14 may be considered the general purpose mixture because it was designed to have higher heat lift and lower vibration than PT-13. The refrigerant mixture has the greatest influence on vibration in the high frequency part of the response curve, where the throttling effect is the dominant forcing function.

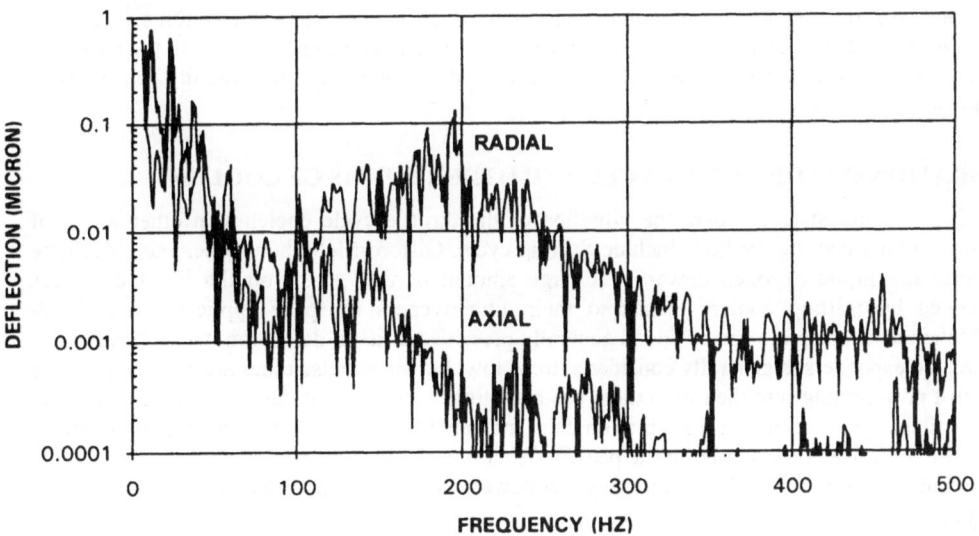

**Figure 6.** "Free-Free" Cold Tip Deflection

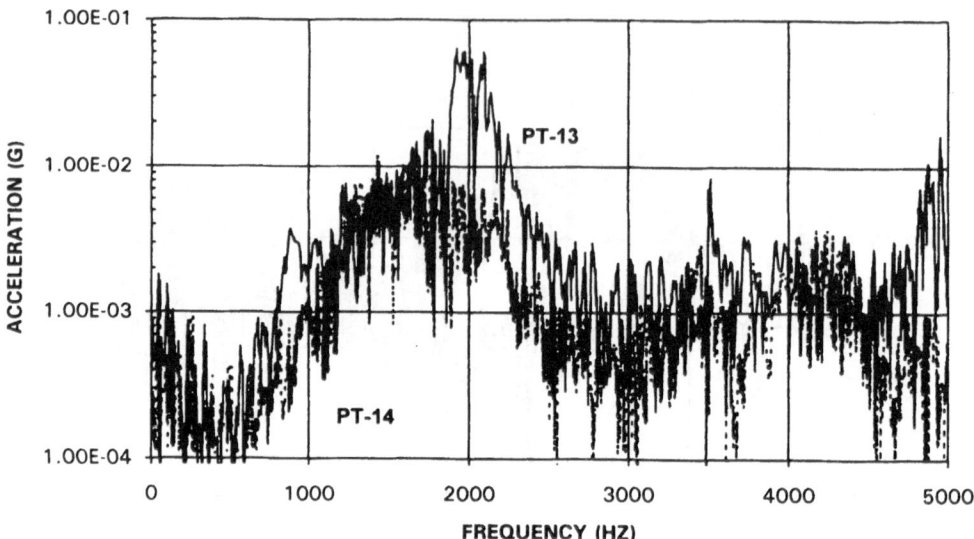

**Figure 7.** "Free-Free" Axial Acceleration Response of Two Different Refrigerant Mixtures

## COLD END INTERFACE FORCE MEASUREMENTS

The interface force transferred to the mounting structure by the cooler must be considered during the design of a detector system. If the cooler support structure cannot be isolated from the rest of the system, the interface force generated by the cooler will interact with the system, potentially degrading the detector resolution. CRYOTIGER® I interface forces were measured with a PCB 208A03 force transducer. The cold end was suspended through its center of gravity from a 40 Kg rigid mounting structure by a stinger with the force transducer in the load path. To verify the force measurement, a small DC motor with an eccentric mass was attached to the vacuum canister. The sinusoidal amplitude was calculated by measuring the motor speed with strobe light and measuring the eccentric mass and moment arm. The 50 mN calculated amplitude was in good agreement with the new peak which appeared in interface force spectrum.

CRYOTIGER® I axial and radial interface forces are shown in Figure 8. Both the axial and radial interface force traces show peaks at 57.5 Hz. Both interface forces rapidly roll off with frequency to 500 Hz.

## VIBRATION OF THROTTLE CYCLE AND OTHER FORMS OF COOLING

It is of interest to compare the vibration of the throttle cycle cooler with other forms of cooling. Other cooling methods include Stirling cycle, Gifford-McMahon cycle, and pulse tube coolers, and liquid nitrogen dewars. A large amount of data exists on the interface forces generated by Stirling coolers thanks to their attractiveness in space applications. Gifford-McMahon data was more scarce and is generally presented as axial deflection data. Vibration of pulse tube expanders is generally considered to be low, but since pulse tubes are rapidly evolving and just now coming into their own there has been little vibration test data for comparison. The one available piece of pulse tube vibration data[3] indicated that the peak deflection of two models of pulse tubes was very similar to the peak deflection of the throttle cycle cold end. Finally, the acceleration spectrum of a liquid nitrogen test dewar was measured for quantitative comparison with the throttle cycle cooler.

The interface force produced by several types of Stirling coolers is compared with that of the CRYOTIGER® I cold end in Figure 9. The data was condensed from axial and radial measurements by displaying the resultant force, defined as

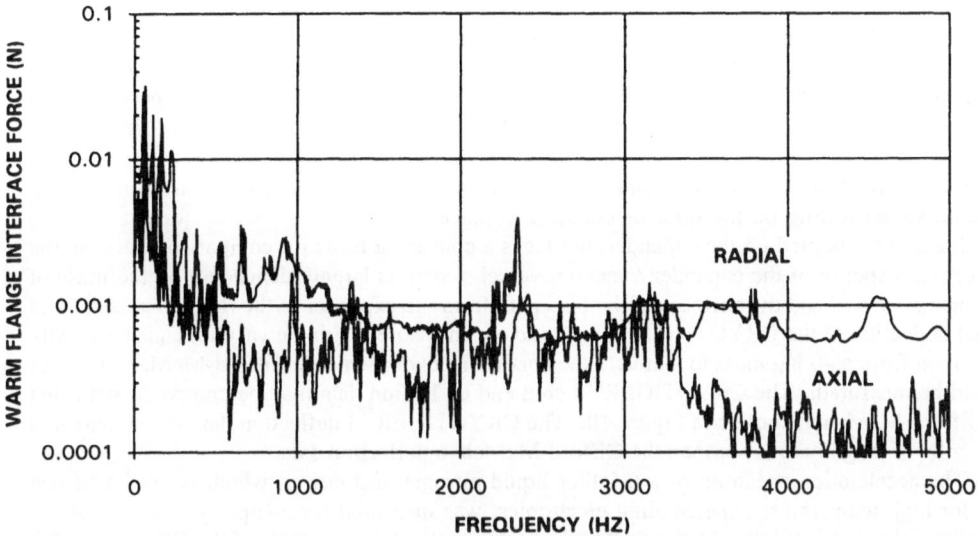

**Figure 8.** Axial and Lateral Cold End Interface Forces

$$F_R = \sqrt{F_a^2 + F_{r1}^2 + F_{r2}^2} \tag{1}$$

where $F_R$ is the resultant force, $F_a$ is the force in the direction along the axis of the cold tip, and $F_{r1}$ and $F_{r2}$ are forces in the two orthogonal directions perpendicular to the axial direction. Eq. 1 is accurate for forces which are well correlated as those of Stirling coolers, but generally yields conservative (higher) resultant forces for forces which are not as well correlated such as those of the CRYOTIGER® I.

The filled data points in Figure 9 are compressor vibration data while the unfilled data points are displacer data. Generally, compressor forces are higher than the displacer forces. The square data points are typical of a split Stirling cooler and the diamond data points represent a pair of split Stirling coolers with the compressors and displacers mounted back to back. The

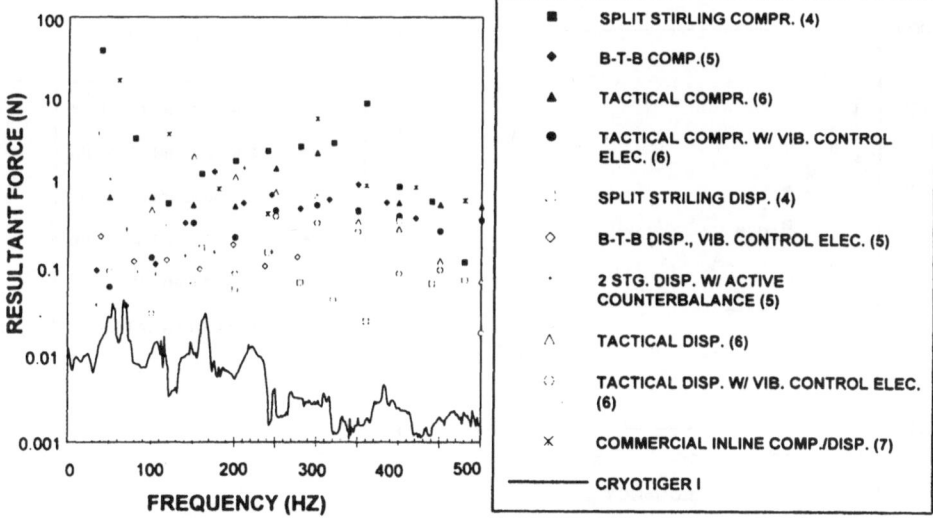

**Figure 9.** Stirling Cooler and CRYOTIGER® I Interface Forces

triangular data points represent a typical tactical Stirling cooler while the round displacer data shows the same displacer running under vibration control electronics. The star data is the resultant force of a commercial Stirling cooler for which the compressor and displacer are combined in one package. The continuous spectrum is the CRYOTIGER® I spectrum. In general, the CRYOTIGER® I spectrum is at least an order of magnitude below Stirling cooler data.

Gifford-McMahon deflection data is usually measured with the expander in an insulating vacuum vessel resting on the floor or clamped in place. This data is relevant for experiments which may be mounted on the expander, but lacks a consistent basis for comparison because the deflection response of the expander / vacuum vessel system is largely dependent on the mass of the vacuum vessel and the stiffness of the load path from the expander to the floor. The axial and radial deflection of the CRYOTIGER® I cold end was measured with the vacuum canister rigidly supported from a 40 Kg mass to simulate the conditions under which a Gifford-McMahon cooler would be measured. The CRYOTIGER® I cold end deflection data may be compared with that of Gifford-McMahon coolers in Figure 10. The CRYOTIGER® I deflection data is one half to a full order of magnitude lower than the Gifford-McMahon deflection data.

The acceleration spectrum of a 0.5 liter liquid nitrogen test dewar, which is used as a test bed for high temperature superconding electronics, was measured for comparison to that of the throttle cycle cooler. Figure 11 shows that the axial acceleration spectrum of the CRYOTIGER® I cold end is comparable to that of the liquid nitrogen dewar, but the lateral acceleration spectrum is higher than that of the liquid nitrogen dewar.

## CONCLUSIONS

Elements of a throttle cycle cooler have been characterized with regard to vibration. The compressor contributes forcing functions in the form of mechanical vibration and pressure pulses. The throttling process produces high frequency vibration energy, which is evident in the cold end response spectrum. Finally, the refrigerant mixture itself has been shown to influence cold end vibration.

Vibration of the throttle cycle cooler has been compared with that of other forms of cooling. Stirling cycle and Gifford-McMahon coolers produce higher levels of vibration than the CRYOTIGER® I cold end and boiling liquid nitrogen produces a lower level of vibration than the CRYOTIGER® I cold end.

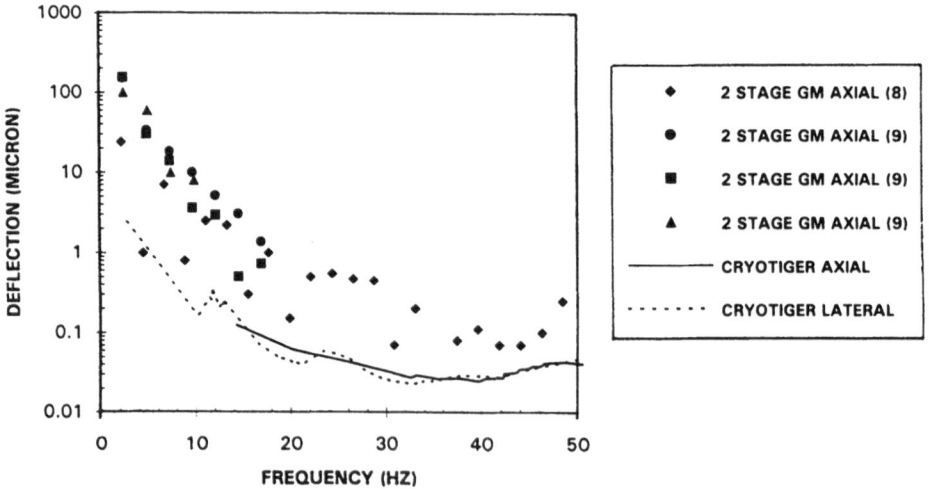

**Figure 10.** Gifford-McMahon and CRYOTIGER® I Deflection Data

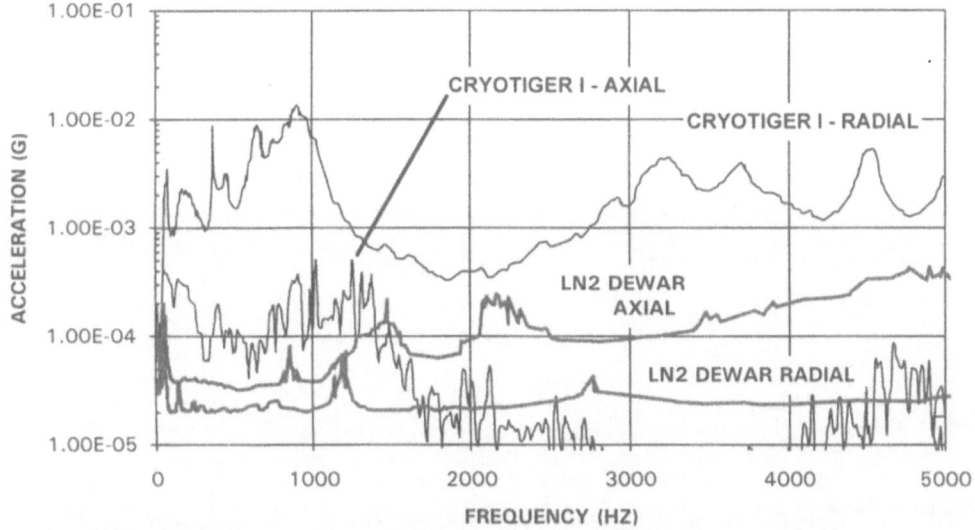

**Figure 11.** Liquid Nitrogen Dewar and CRYOTIGER® I Acceleration Data

## ACKNOWLEDGMENT

The author wishes to thank Ravi Bains for his help in tracking down vibration data. Professor Forbes Brown provided many insights to the vibration phenomenon. Ajay Khatri's help in designing low vibration refrigerant mixtures was invaluable. Thanks to Dale Layos for spending many hours taking data with the spectral analyzer and setting up experiments.

## REFERENCES

1. Khatri, Ajay, "A Throttle Cycle Refrigerator Operating Below 77 K," *Advances in Cryogenic Engineering*, vol. 41, Plenum Press, New York (1996).

2. Longsworth, R. C., Boiarski, M. J., Klusmier, L. A., "80 K Closed-Cycle Throttle Refrigerator," *Cryocoolers 8*, Plenum Press, New York (1995), pp. 537-542.

3. Iwatani International Corporation marketing specification for pulse tubes P201 and P301.

4. Mon, G. R., Smedley, G. T., Johnson, D. L., and Ross, R. G., "Vibration Characteristics of Stirling Cycle Cryocoolers for Space Applications," *Cryocoolers 8*, Plenum Press, New York (1995), pp. 199-200.

5. Sparr, Leroy, et. al., "NASA/GFSC Cryocooler Test Program Results for FY93/94," *Cryocoolers 8*, Plenum Press, New York (1995), pp. 223-225.

6. Sparr, Leroy, et. al., "Adaptation of Tactical Cryocoolers for Short Duration Space-Flight Missions," *Cryocoolers 8*, Plenum Press, New York (1995), pp. 168-170.

7. Smedley, G. T., Ross, R. G., Berchowithz, D. M., "Performance Characterization of the Sunpower Cryocooler," *Cryocoolers 8*, Plenum Press, New York (1995), pp. 155-157.

8. Kang, Yoon-Myung, et. al., "4 K GM/JT Cryocooler For Cryogenic Sensors," *Proceedings of the Sixth International Cryocoolers Conference*, David Taylor Research Center, Maryland (1991), pp. 19-21.

9. APD Cryogenics Inc. vibration data archives, Allentown, Pennsylvania.

# Development of a 60K Thermal Storage Unit

**David C. Bugby, Ronald G. Bettini, and Charles J. Stouffer**

Swales & Associates, Inc.
Beltsville, Maryland, USA 90705

**Marko Stoyanof**

Phillips Laboratory (PL/VTPT)
Kirtland AFB, New Mexico, USA 87117

**David S. Glaister**

The Aerospace Corporation
Kirtland AFB, New Mexico, USA 87117

## ABSTRACT

Thermal storage units (TSUs) with embedded phase change materials (PCMs) can improve the performance and reduce the weight of space-based IR sensor systems. This paper summarizes the findings of a Phillips Laboratory-sponsored SBIR program to develop TSU technology at an operational temperature of 60 K. During Phase I of this program, nitrogen trifluoride ($NF_3$) was identified as a promising PCM, despite a small degree of single-phase supercooling. So far in Phase II, an improved design methodology was developed which divides the TSU into two volumes: a heat exchanger volume and a storage volume. This approach increases TSU energy storage capacity and thermal stability while reducing both weight and volume. Testing in Phase II has shown that an aluminum foam core suppresses $NF_3$ supercooling and eliminates concerns over heat exchanger filling in microgravity. Current plans call for a 6000 J TSU to be tested on-orbit in a Hitchhiker GAS Canister experiment on the STS in mid-1998. In this flight system, the TSU heat exchanger will have a beryllium shell for CTE compatibility with beryllium focal planes and an aluminum foam core. The TSU storage tank will be made of thin-walled titanium. A low pressure (P < 100 psi) system will be utilized to minimize safety concerns.

## INTRODUCTION

Future space-based IR sensor systems will require reduced levels of vibration and tighter focal plane temperature control to meet performance goals. In concert with these increasing requirements, focal plane heat load variability for certain applications is expected to increase. To deal with these competing factors, cryogenic thermal storage units (TSUs) with embedded phase change materials (PCMs) have been proposed as a potential solution.[1-2] By using a PCM that has

Cryocoolers 9, Edited by R.G. Ross, Jr.
Plenum Press, New York, 1997

a phase transition at or near the focal plane operating temperature, the TSU can provide "thermal load leveling" so that the cryocooler can be sized for the average rather than the peak heating load. The cryocooler can then be made smaller and lighter, which presumably reduces launch costs and on-orbit power consumption. The TSU also provides a cryocooler "turn-off" capability which provides true "vibrationless" IR sensor operation and increased cryocooler lifetime due to reduced wear and periodic redistribution of regenerator contaminants. The benefits, therefore, of a TSU are improved IR sensor system performance, reduced cryocooler weight, lower system cost, and longer system lifetime.[3] Figure 1 illustrates these benefits.

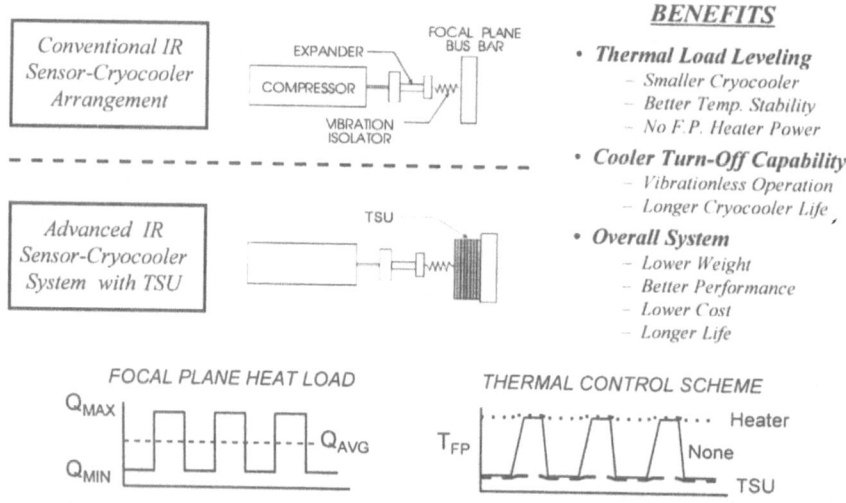

**Figure 1.** Benefits of Cryogenic TSU Technology to IR Sensor System

To provide the benefits indicated in Figure 1, the PCM embedded within the TSU must undergo a phase transition such as melting (solid-to-liquid), vaporization (liquid-to-gas), sublimation (solid-to-gas), or crystal structure realignment (solid-to-solid). Trade studies indicate that phase transitions that occur with small changes in density will require the least weight, volume, and complexity to implement. Thus, PCMs that undergo melt or solid-solid phase transitions (at or near the focal plane operating temperature) are the working fluids of choice for cryogenic TSUs.

Other favorable PCM characteristics include high transition energy, chemical stability, low toxicity, low flammability, negligible supercooling, and compatibility with potential container materials. Favorable characteristics of TSU container materials include CTE compatibility with IR sensor hardware, high strength to provide pressure containment at room temperature, and high thermal conductivity to maximize the transport of energy to (and away from) the PCM and to minimize the internal thermal gradients. This paper will describe a working fluid (nitrogen trifluoride), container design methodology, and overall testing procedure that enables nearly all of these criteria to be met.

With the foregoing in mind, the primary objective of this SBIR program is to develop a TSU that operates at or near 60 K and demonstrate its operation both on the ground and in space (via an STS flight experiment). A secondary objective, as required by all SBIR programs, is to commercialize cryogenic TSU technology. This paper details the progress made toward the former objective and covers the following topics in the order listed; requirements, development issues, design methodology, design calculations, test set-up, test results, flight experiment, and future plans. Ideas for commercial products with cryogenic TSUs will appear in a future paper.

## REQUIREMENTS

The requirements for this program were derived, in part, to fill the needs of a particular defense-based application involving the use of IR sensors in space. The key requirements are listed in Table 1. One additional implicit requirement is that any viable PCM-based TSU must weight significantly less than a similarly-capable sensible heat device. Figure 2 illustrates a comparison between a PCM-based TSU and a sensible heat device using the properties of nitrogen trifluoride ($NF_3$) and a Mg-Li alloy. The heat capacity of this Mg-Li alloy at 60 K is several times that of aluminum.

**Table 1.** Requirements for the 60 K Thermal Storage Unit

| TSU Operating Parameter | Performance Requirement |
|---|---|
| Operating Temperature | 55-65 K |
| Energy Storage Capacity | 6000 J |
| Temperature Stability | +/- 0.5 K |
| Weight | < 3.5 kg |
| Volume | < 1500 cc |
| Focal Plane Heat Load | 1-5 W, 3 W avg. |
| Focal Plane Interface Material | Beryllium |
| Gravity Environment Functionality | 1-G and 0-G |

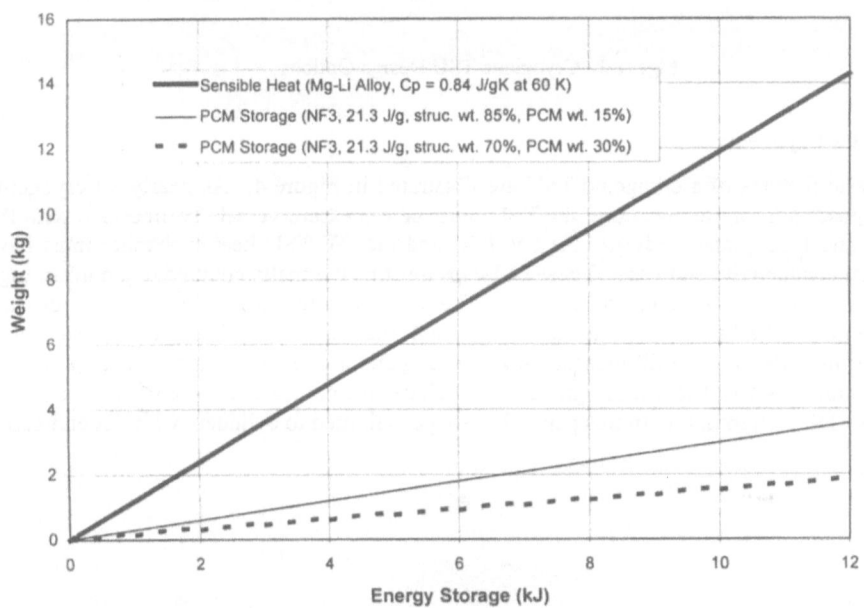

**Figure 2.** Weight Advantage of PCM-Based TSU over High $C_P$, Sensible Heat Device

## DEVELOPMENT ISSUES

The principal TSU development issues are discussed below. These include: (a) cryogenic TSU design options; (b) basic features of a cryogenic TSU; (c) identification of 60 K PCMs; (d) advantages of nitrogen trifluoride ($NF_3$) over nitrogen ($N_2$); (e) $NF_3$ supercooling from single-phase state; (f) heat exchanger core design options; (g) temperature stability; (h) core and containment materials; (i) interface considerations; and (j) drilled-hole TSU prototype.

## Cryogenic TSU Design Options

There are two basic types of cryogenic TSUs: single-volume (SV) and dual-volume (DV). Both types are pressurized with PCM gas at room temperature. An SV TSU uses the same volume for heat exchange and fluid storage, whereas a DV TSU has a separate heat exchanger (HE) and storage tank (ST). In an SV TSU, the PCM contained within the structure transitions from high pressure gas to a low pressure solid upon cooling. In a DV TSU, the HE is cooled to cryogenic temperature while the storage tank remains at room temperature. Figure 3 illustrates the two TSU design options. In general, DV TSUs are lighter, smaller, and have better temperature stability than SV TSUs. DV superiority stems from the fact that the density of the PCM as a condensed liquid or solid is typically much greater than its density as a room temperature gas. Thus, the portion of DV HE void volume unoccupied by PCM at cryogenic temperatures (denoted the "waste volume") can be minimized at all fill pressures, whereas the SV waste volume is much larger and a function of the fill pressure.

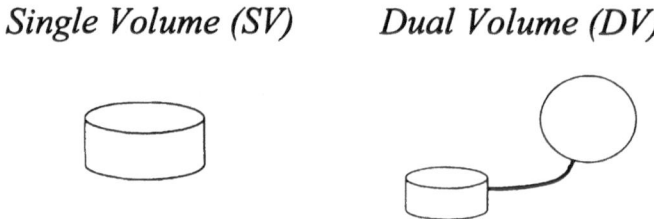

**Figure 3.** Cryogenic TSU Design Options

## Features of a Cryogenic TSU

The basic features of a cryogenic TSU are illustrated in Figure 4. As nearly all cryogenic PCMs are gases at room temperature, the TSU must be a pressure vessel. To interface with the cryocooler and focal plane hardware, the SV TSU and the DV TSU heat exchanger must have high thermal conductivity end-caps. These end-caps must be thermally coupled to a porous, high surface area, high thermal conductivity core. The void volume within the core provides the working space for the PCM. As indicated in Figure 4, there are four primary design variables: (1) working fluid (PCM); (2) fill pressure at room temperature; (3) heat exchanger core design; and (4) structure design. The last design variable includes the choice as to whether the TSU is to be an SV or DV configuration. In this paper, HE shape is limited to cylinders with flat end-caps.

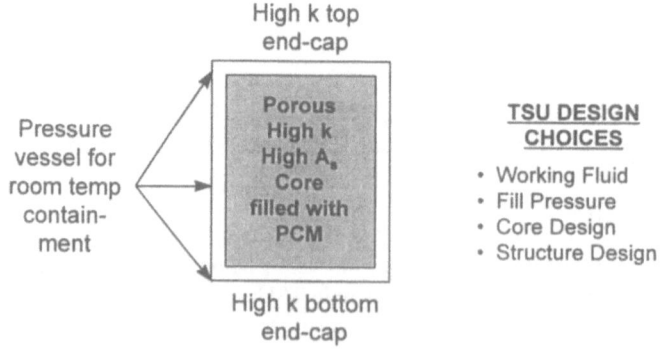

**Figure 4.** Principal Features of a Cryogenic TSU

## Identification of 60 K Phase Change Materials (PCMs)

One of the first tasks undertaken during Phase I of this SBIR program was to identify materials that have phase transitions in the range 55-65 K. Table 2 lists the potential PCMs and their thermophysical properties.[3-5] From this list, four PCMs – nitrogen, air, freon-22, and nitrogen trifluoride – were selected for testing in Phase I. The results of these initial tests indicated that NF$_3$ would be the most promising 60 K PCM. In particular, NF$_3$ has significant advantages over N$_2$, an issue which is discussed in the next subsection.

**Table 2.** Compilation of 55-65 K PCM Properties

| PC Temp (K) | Compound | PC Energy (kJ/kg) | PC Type | MP* (K) | BP* (K) | T$_c$ (K) | P$_c$ (MPa) | Liquid Density (kg/m³) | Safety |
|---|---|---|---|---|---|---|---|---|---|
| 55 | Palladium Hydride | | S-S* | | | | | | |
| 55.8 | Oxygen/Argon Peritectic | | Melt | 55.8 | | | | | |
| 56.7 | Nitrogen Trifluoride | 21.3 | S-S | 64.5 | 144 | 234 | 4.53 | 1530¹ | Toxic if Inhaled |
| 58 | Gold Zinc Compound | | S-S | | | | | | |
| 59 | Air | | Melt | 59 | 78.7 | | | | |
| 59 | Freon-22 | | S-S | 113 | 232 | 369 | 4.98 | 1060¹ | Toxic if Inhaled |
| 61.8 | Carbon Monoxide | 22.6 | S-S | 67.5 | 81.5 | 133 | 3.50 | 790¹ | Toxic, Flammable |
| 63.2 | Nitrogen | 25.7 | Melt | 63.2 | 77.5 | 122 | 3.40 | 870² | |
| 63.5 | Silane | 19.2 | S-S | 88.5 | 162 | 269 | 4.85 | 680² | Flammable in Air |
| 64.5 | Nitrogen Trifluoride | 5.6 | Melt | 64.5 | 144 | 234 | 4.53 | 1530¹ | Toxic if Inhaled |

\* MP = Melting Point, BP = Boiling Point, S-S = Solid-Solid; ¹ at BP; ² at MP; ³ at 327 K

## Advantages of NF$_3$ Over N$_2$ as a 60 K PCM

Nitrogen trifluoride undergoes a solid-solid transition at 56.7 K with a heat of transition of 21.3 J/gm whereas nitrogen undergoes a melt transition at its triple-point of 63.15 K with a heat of melting of 25.7 J/gm.[6] Because of its high molecular weight (71 gm/mol) and non-ideal behavior at high pressures, the energy density (J/cc) of NF$_3$ is 2-3 times that of N$_2$. So, for a given energy storage capacity, an NF$_3$-based TSU will be much smaller and lighter than an N$_2$-based TSU. Additionally, because of the large difference between its melt and boil temperatures, NF$_3$ is a much safer fluid to employ in an STS flight experiment. That is, the likelihood that any solid NF$_3$ will block the flow of gas or liquid from a DV TSU heat exchanger (when the system heats up) is orders of magnitude lower than it is for an N$_2$ system. Figure 5 illustrates these advantages.

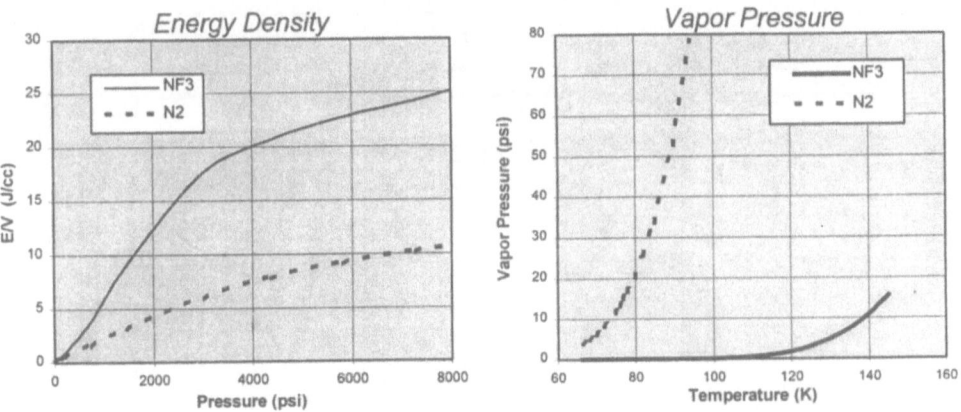

**Figure 5.** Energy Density and Vapor Pressure of Nitrogen Trifluoride and Nitrogen

## $NF_3$ Supercooling from Single-Phase State

One of the characteristics of $NF_3$ identified in Phase I of the SBIR program was its tendency to supercool when cooling from a single-phase state. In these tests, the TSU heat exchanger core was formed by drilling a large number of parallel, 1/8" holes in a block of 2219 aluminum.[7] Figure 6 illustrates the observed behavior. One of the objectives in Phase II of the program was to suppress this supercooling, and the first approach for doing so was to increase the core surface area. The next subsection details some of the core design options.

Test 20, SD 1
Test 22, SD 1
Test 24, SD 1
Test 26, SD 1
Temperature (K)
Time (min)
60
59
58
57
56
55
54
53
52
51
50
49
48
0
50
100
150
200
250
300

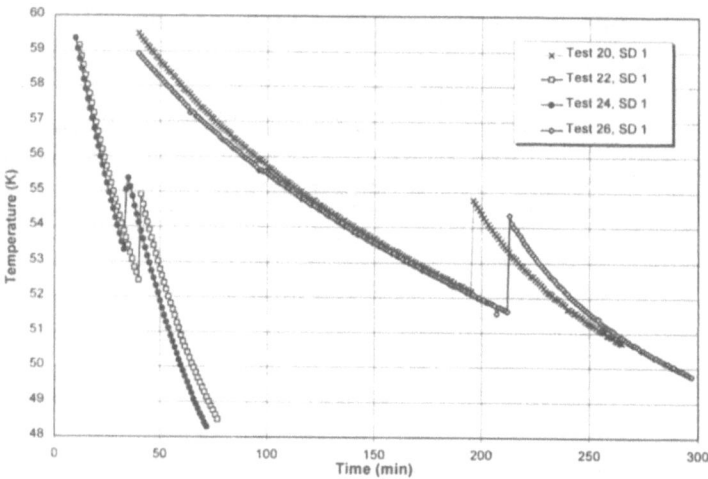

**Figure 6.** Supercooling Behavior of Nitrogen Trifluoride within 1/8" Drilled-Hole Core

## Core Design Options

To maximize the thermal coupling between the TSU core and the PCM, the specific surface area of the core (i.e., its surface area per unit volume) needs to be as high as possible. Some of the options that were investigated include the following: (1) drilled holes; (2) parallel plates; (3) metal foam; and (4) hybrid (combination of 1 or 2 and 3). Other options are possible, but this list covers the vast majority of potential core design situations. Figure 7 illustrates each of these options and the specific surface area of each. As indicated, from a specific surface area standpoint, metal foam is vastly preferable to any of the other core design options.[8]

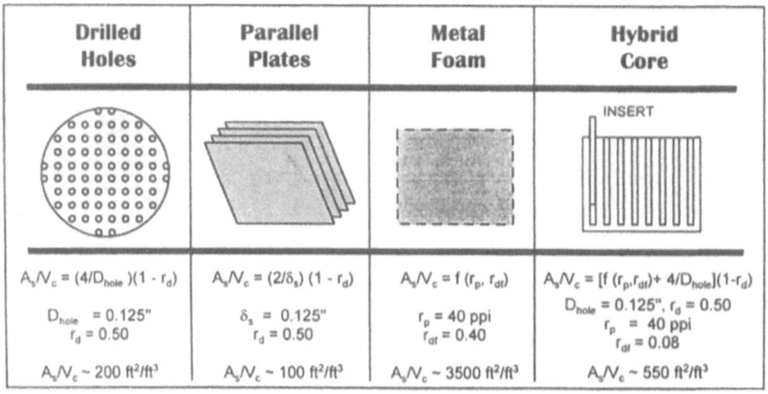

Drilled Holes
Parallel Plates
Metal Foam
Hybrid Core
INSERT
$A_s/V_c = (4/D_{hole})(1 - r_d)$
$D_{hole} = 0.125"$
$r_d = 0.50$
$A_s/V_c \sim 200\ ft^2/ft^3$
$A_s/V_c = (2/\delta_s)(1 - r_d)$
$\delta_s = 0.125"$
$r_d = 0.50$
$A_s/V_c \sim 100\ ft^2/ft^3$
$A_s/V_c = f(r_p, r_{df})$
$r_p = 40$ ppi
$r_{df} = 0.40$
$A_s/V_c \sim 3500\ ft^2/ft^3$
$A_s/V_c = [f(r_p, r_{df}) + 4/D_{hole}](1 - r_d)$
$D_{hole} = 0.125"$, $r_d = 0.50$
$r_p = 40$ ppi
$r_{df} = 0.08$
$A_s/V_c \sim 550\ ft^2/ft^3$

**Figure 7.** Cryogenic TSU Core Design Options and Specific Surface Areas

## TSU Temperature Stability

The primary function of the TSU is to maintain temperature stability of the focal plane hardware. For cylindrical TSU configurations with flat end-caps, the peak focal plane heat load will be applied at one end-cap (call it the "top") and the cryocooler cold head will be thermally coupled to the other end-cap. The resulting heat transfer within the TSU will be one-dimensional along the axis of the cylinder. Accordingly, the temperature rise of the focal plane hardware during peak heating will be the sum of the temperature drops across the top end-cap, core, and PCM. Figure 8 illustrates the situation. If the end-caps have a high thermal conductivity and the core has a large surface area, the temperature drop across the core is the dominant term and temperature stability of the TSU is +/- $\Delta T_{CORE}/2$. To attain the required end-cap and core properties at 60 K, a survey of TSU materials was performed, and this topic is addressed below.

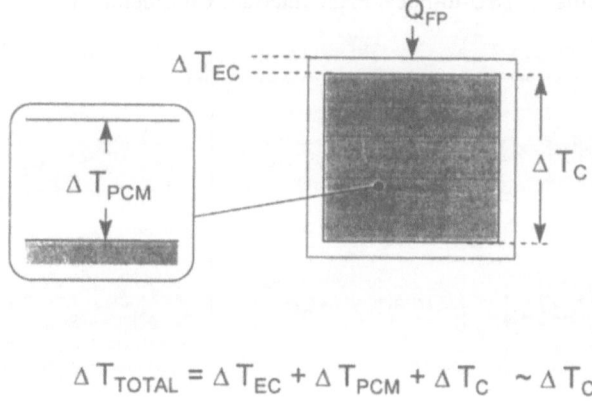

$$\Delta T_{TOTAL} = \Delta T_{EC} + \Delta T_{PCM} + \Delta T_C \sim \Delta T_C$$

**Figure 8.** Temperature Stability of a Cryogenic TSU

## TSU Structure and Core Materials

Several different materials were identified as possible candidates for TSU structure and/or core materials. The various materials and their key thermophysical and structural properties are provided in Table 3 below.[9] In addition to the properties listed in Table 3, another key TSU development issue is the CTE compatibility of the TSU structure and the focal plane hardware. This issue is addressed in the next subsection.

**Table 3.** Properties of Cryogenic TSU Structure and Core Materials

| Material | k at 60 K (W/mK) | $F_{ty}$ (psi) | $\rho$ (kg/m³) | Principal Use in TSU |
|---|---|---|---|---|
| 2219 Al | 39 | 30,000 | 2710 | Shell/Core |
| 1100 Al | 315 | 5,000 | 2710 | Core |
| 6063 Al | 270 | 7,000* | 2710 | Shell/Core |
| OFHC Cu | 580 | 11,000 | 8950 | Core |
| Be | 100-200 | 47,000 | 1800 | Shell/Core |
| Ti (6Al-4V) | 3.1 | 120,000 | 4500 | Shell |
| 15-5 S.S. | 7.2 | 145,000 | 7820 | Shell |
| Metal Foam** | $k_o\, r_d\, \tau$ | non-struc. | $\rho_o\, r_d$ | Core |

\* Increases to 31,000 psi in T-6 condition
\*\* $k_o$, $\rho_o$ = metal thermal conductivity, density; $r_d$ = relative density; $\tau$ = tortuosity

## TSU-to-Focal Plane Interface Considerations

There are two possible TSU-to-focal plane interface situations: (1) the TSU structure is CTE compatible with the focal plane hardware in which case the TSU can be mounted directly to the hardware (e.g., a beryllium bus bar); or (2) the TSU structure is not CTE compatible in which case a highly conductive, but only slightly flexible interface coupling device is required. Table 4 indicates the concerns of each situation. While the objective in this program was to strive for maximum TSU performance (which implies a directly-mounted TSU), a prototype coupling was developed with a conductance of 8 W/K and sufficient flexibility to ensure that an aluminum TSU could be mated to a beryllium bus bar with minimal deflection of the bar. If this coupling were necessary, a focal plane temperature rise during peak heating of 5 W would be at least 0.4 K higher than for a directly-mounted TSU.

**Table 4.** TSU-to-Focal Plane Interface Considerations

| I/F Configuration | TSU Issues | Comment |
|---|---|---|
| TSU Be<br>Direct Interface | • CTE Compatibility<br>• Fill Pressure Deflection<br>• Ti and Be Only CTE Compatible Materials for TSU Shell | Optimum from thermal standpoint |
| TSU Be<br>Compliant Joint | • Conductance of Joint<br>• Flexibility of Joint | • Only slight flexibility required<br>• Prototype with conductance of 8 W/K developed and tested |

## Phase I TSU Prototype

To evaluate potential PCMs, an SV TSU prototype was developed in Phase I of this program. While this prototype functioned as intended, it was not designed for optimum thermal performance or minimum weight. In proceeding to Phase II of the program, several areas of improvement were identified to arrive at an optimum TSU design. Figure 9 illustrates the Phase I TSU prototype and the areas which were identified for improvement. To achieve these goals, an overall TSU design methodology was needed. The next major section of the paper outlines this methodology.

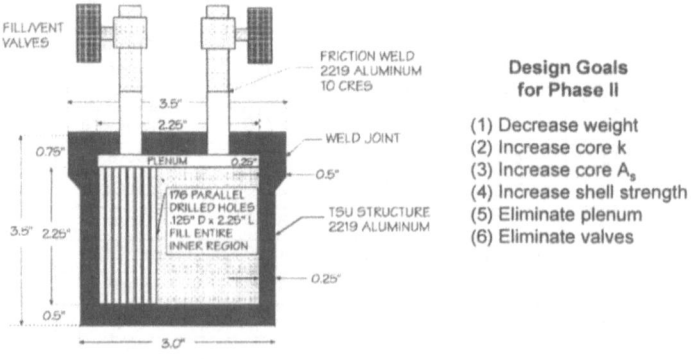

**Figure 9.** Phase I TSU Prototype and Design Goals for Phase II

## DESIGN METHODOLOGY

A generalized design methodology was developed for cryogenic TSUs. The methodology analytically combines the following three elements: (1) design constraints (i.e., the requirements); (2) independent design variables (i.e., the design options); and (3) design equations (i.e., the underlying physics). Figure 10 illustrates the analytical relationships. The equations indicated in Figure 10 assume a cylindrical, flat end-capped TSU heat exchanger and, in the case of DV TSUs, a spherical storage tank.

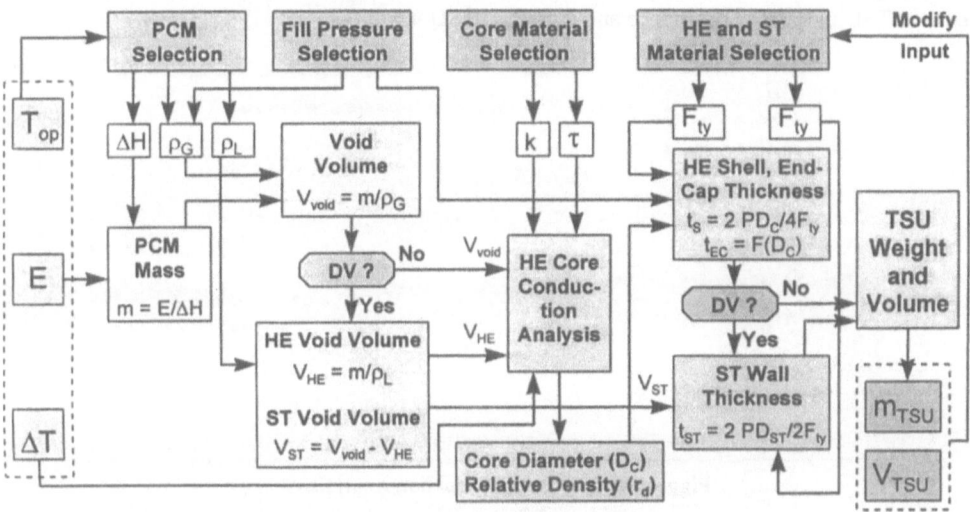

**Figure 10.** Generalized Cryogenic TSU Design Methodology

### Model Variables

The design constraints consist of the operating temperature ($T_{OP}$), energy storage capacity (E), temperature stability ($\Delta T$), weight ($m_{TSU}$), and volume ($V_{TSU}$). The first three constraints are model inputs and the latter two are model outputs. The independent design variables consist of the PCM selection, fill pressure ($P_{FILL}$), core design, and structure design. As stated, part of the structure design choice involves choosing whether the TSU is to be an SV or DV configuration. The design equations consist of calculations of PCM mass ($m_{PCM}$), total void volume ($V_{VOID}$), HE volume ($V_{HE}$), ST volume ($V_{ST}$), core diameter ($D_C$), core relative density ($r_D$), cylindrical HE shell thickness ($t_S$), HE flat end-cap thickness ($t_{EC}$), and spherical ST wall thickness ($t_{ST}$).

### Calculation Procedure

**Internal Void Volume.** The overall procedure begins with the calculation of the internal volume occupied by the PCM at room temperature and at cryogenic temperature. The PCM is limited to those materials with phase transitions near $T_{OP}$. Once identified, the choice of PCM defines the transition energy ($\Delta H$), liquid density ($\rho_L$), and (upon selection of $P_{FILL}$) the gas density ($\rho_G$). The first quantity to be computed is the PCM mass, which is simply the energy storage capacity divided by the transition energy ($m_{PCM} = E/\Delta H$). Once the required PCM mass is identified, the volume needed to contain the PCM gas at room temperature can be calculated ($V_{VOID} = m_{PCM}/\rho_G$). If the DV configuration is selected, the heat exchanger void volume is the next quantity to be calculated ($V_{HE,VOID} = m_{PCM}/\rho_L$), followed by the storage tank volume ($V_{ST} = V_{VOID} - V_{HE,VOID}$).

**Core Design.** Once the internal void volume within the core is defined, the solid part of the core must be specified. The two additional parameters that define the core for cylindrical, flat-ended cryogenic TSU heat exchangers are the core diameter ($D_C$) and the core relative density ($r_D$). In general, these two parameters must be selected to satisfy the TSU thermal stability constraint ($\Delta T$). Figure 11 illustrates an analysis that derives an analytical relationship between $\Delta T$, $D_C$, and $r_D$. In that analysis, G, A, and V respectively represent the conductance, cross-sectional area, and volume of the core. The parameters Q, $k_o$, and $\tau$ respectively represent the peak focal plane heat load, core material thermal conductivity, and core tortuosity. The analysis shows that the core relative density should never exceed 0.5. Increasing $r_D$ above 0.5 will result in reduced TSU thermal performance and increased TSU weight.

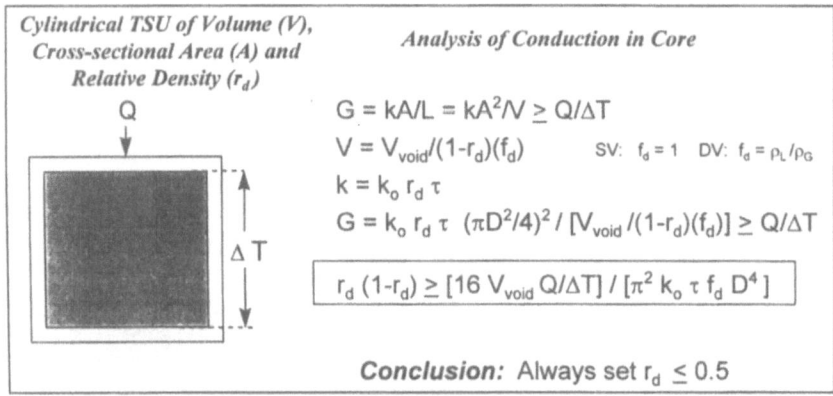

**Figure 11.** TSU Core Conduction Analysis

**Containment Structure.** Once the core shape is specified, the containment structure must be defined. This definition involves the HE wall thickness, the HE end-cap thickness, and (in the case of DV TSUs) the ST wall thickness. Figure 10 indicates three ideal equations for obtaining preliminary estimates of these thicknesses. Of course, detailed stress analysis is always required to verify TSU structural integrity, and such analysis usually requires adding slightly to these thicknesses, especially near weld joints, to account for degraded properties or non-ideal stresses.

## DESIGN CALCULATIONS

### Core Conduction Analysis Results

Using the analytical relationship between $\Delta T$, $D_C$, and $r_D$ given in Figure 11 (see the boxed-in expression), design curves can be generated for both the SV and DV TSU configurations. Figures 12 and 13 illustrate these results. What these curves provide are $D_C$-$r_D$ combinations that satisfy the $\Delta T$ constraint.

Some interesting results illustrated in Figures 12-13 include the following: (1) for a given $D_C$, there is a minimum achievable $\Delta T$ (i.e., when $r_D = 0.5$); (2) the slopes of the curves near this minimum $\Delta T$ value are very steep indicating that a reduction in $r_D$ below 0.5 will not appreciably decrease thermal performance but may save weight; and (3) the curves for the DV TSU configuration are invariant with respect to $P_{FILL}$ while those for the SV configuration are not. This last result stems from the fact that both the void volume ($V_{VOID}$) and the density ratio parameter ($f_D$) are inversely proportional to the gas density ($\rho_G$) for DV TSUs, but only the void volume is proportional to the gas density for SV TSUs ($f_D = 1$). Thus, the weight of a DV TSU is a much weaker function of $P_{FILL}$ than it is for an SV TSU.

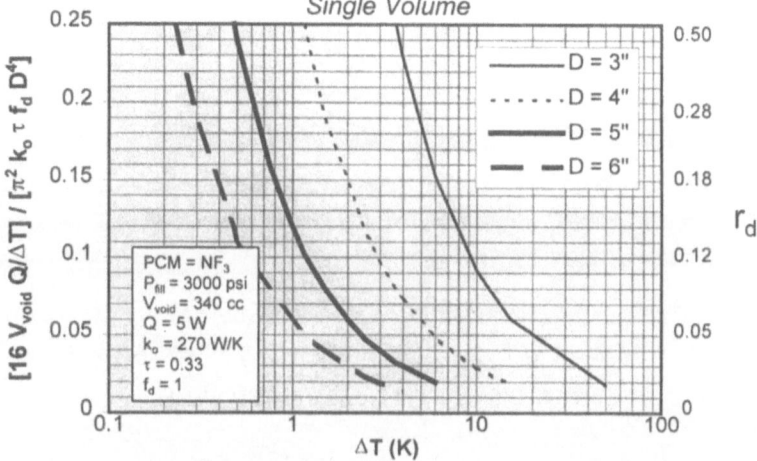

**Figure 12.** Core Conduction Analysis Results for SV TSU Configuration

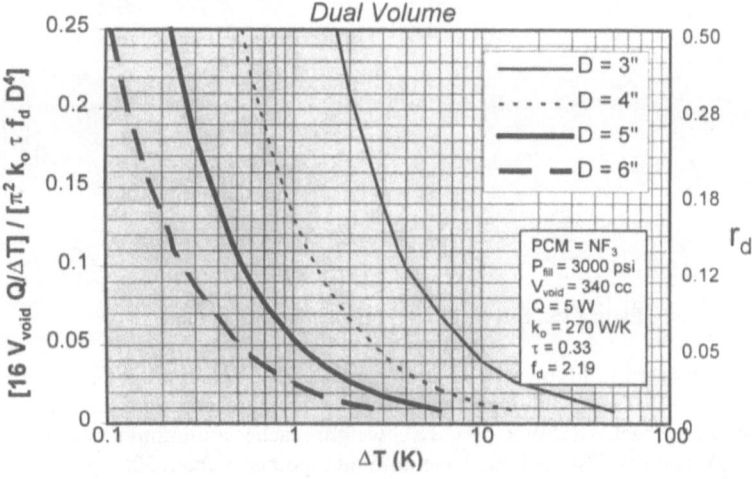

**Figure 13.** Core Conduction Analysis Results for DV TSU Configuration

## Trade-Off Calculations

Using the analytical relationships indicated in Figure 10, a set of trade-off calculations was carried out to determine the effects of fill pressure and core relative density on TSU weight and volume. The calculations were conducted for a 6000 J, $NF_3$-filled TSU in both SV and DV configurations. The fill pressure was varied from 14.7 to 7000 psi and the relative density was varied from 0.01 to 0.50. Other parameter values used include the following: (a) HE wall/lid material yield strength (30,000 psi); (b) ST wall material yield strength (100,000 psi); (c) core thermal conductivity (270 W/mK); (d) core tortuosity (0.33); and (e) core diameter (4.5"). Figures 14-15 illustrate the pressure-weight and pressure-volume results, respectively (note: the core tortuosity represents the deviation from ideal thermal conductivity that occurs in metal foam core materials).

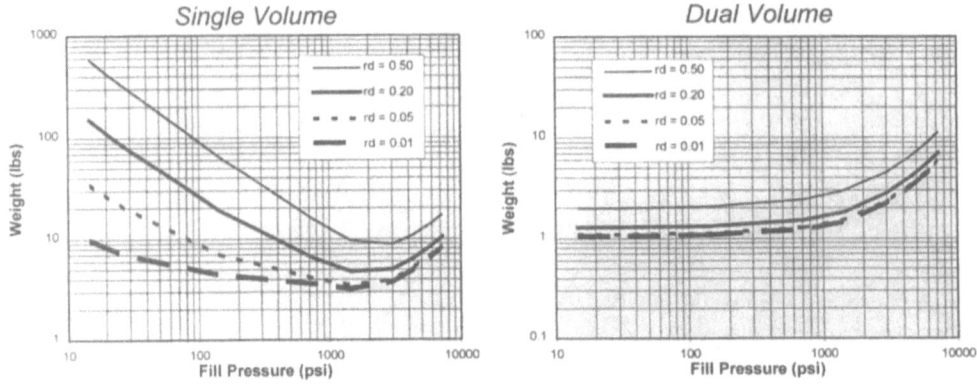

**Figure 14.** Effects of Fill Pressure on TSU Weight

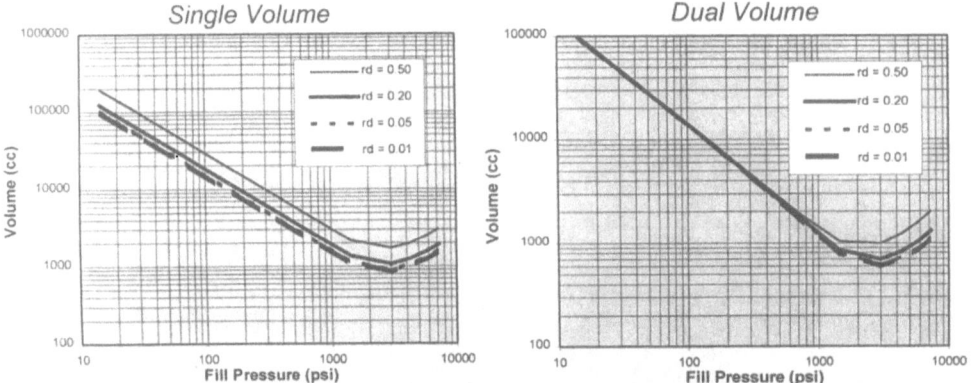

**Figure 15.** Effects of Fill Pressure on TSU Volume

The results illustrated in Figures 14-15 indicate that DV TSUs are lighter and smaller than equivalent capacity SV TSUs, although the effects of pressure and core relative density are somewhat different for each TSU type. SV TSU weight reaches a minimum at a fill pressure of about 2000 psi. SV and DV TSU volume have minimum points at about 3000 psi while DV TSU weight is a monotonically increasing function of fill pressure that is weak below about 1000 psi and stronger above that level. For each TSU type, weight and volume increase with core relative density, although the effect is stronger for SV TSUs, especially at low pressure. While many other parameter sensitivity calculations have been performed, those presented above are the most important. The focus of the paper will now shift to testing.

## TEST SET-UP

### Basic Cryogenic TSU Test Set-Up

Testing of $NF_3$ (and the other PCMs evaluated in this program) was carried out using the test set-up illustrated in Figure 16. The principal elements of this system are the cryocooler, vacuum chamber, liquid $N_2$-cooled shroud, temperature/pressure instrumentation, and personal computer. Using the LABVIEW computer program, a specialized software package was developed that dynamically displays system temperatures on a CRT and automatically records them to disk.[10] In this set-up, the TSU is connected to the cryocooler via a spool-shaped, aluminum cold finger.

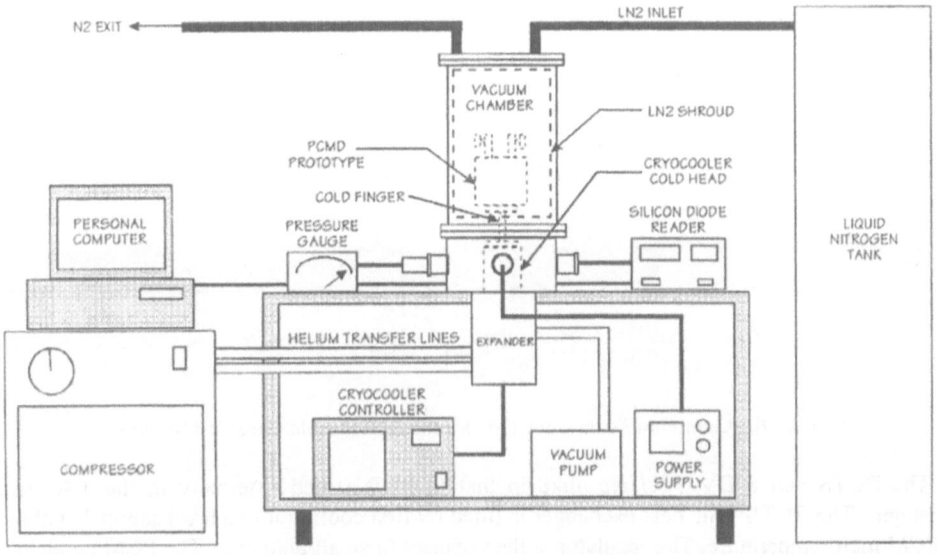

**Figure 16.** Cryogenic TSU Test Set-Up

## Replaceable Core Test System

To augment the basic test set-up, a system was developed (designated as the Replaceable Core Test System or RCTS) to allow TSU testing to be carried out without constructing a separate TSU pressure vessel each time a specific core design was designated for testing. The RCTS is depicted graphically in Figure 17 and consists of a 6061 aluminum canister with top flange and o-ring groove, 6063 aluminum canister lid with TSU core attached, flexible stainless steel bellows, vacuum feed-through, stainless steel tubing, PCM storage tank, pressure regulator, and various valves/welded fittings. An indium o-ring is used to maintain the TSU canister seal at cryogenic temperatures. Although not pictured, dual redundant relief-valves are incorporated in the RCTS due to $NF_3$ toxicity. Figure 18 illustrates the TSU canister lid with a 6101 aluminum foam core.

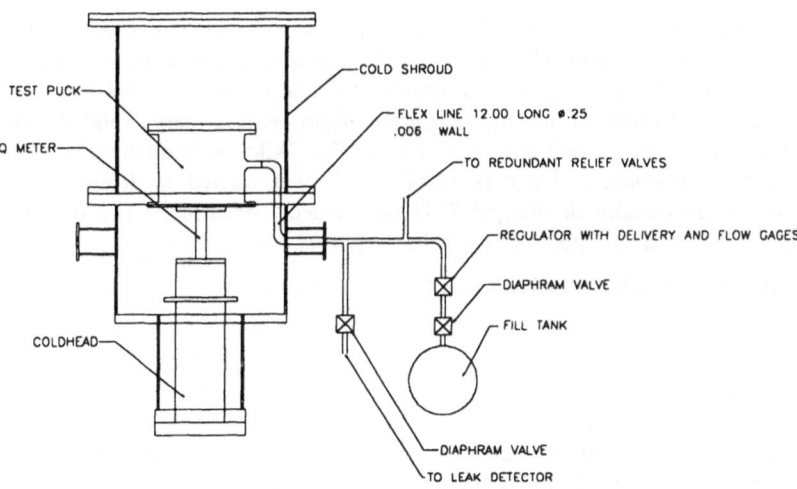

**Figure 17.** Replaceable Core Test System (RCTS)

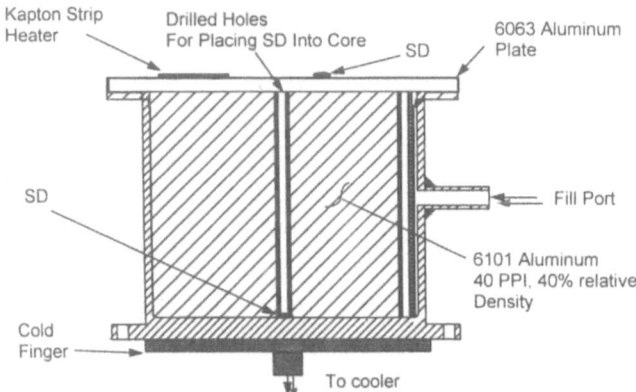

**Figure 18.** Aluminum Foam Core TSU for the Replaceable Core Test System

The RCTS and a DV TSU are alike in that PCM is stored externally to the TSU heat exchanger. The RCTS TSU heat exchanger is filled by first cooling it to a level somewhat above the PCM melt temperature. The regulator is then opened (a small amount). The PCM condenses within the heat exchanger and the heat exchanger temperature rises. The system filling continues until the (heat exchanger) temperature reaches a level somewhat below the PCM boiling temperature at which point the filling process is discontinued and the RCTS heat exchanger is allowed to cool. If additional PCM needs to be added, the regulator is reopened and the filling process commences until the TSU is filled to the desired level. For PCMs that have melting and boiling temperatures that are very close, this process might have to be carried out many times.

## TEST RESULTS

This section of the paper presents some of the more important test results that have been obtained over the course of the program. Included are descriptions of the following: (1) heating and cooling of $NF_3$ in an 1/8" drilled-hole core TSU; (2) heating and cooling of $NF_3$ in an aluminum foam core TSU; (3) simulation of a space-based IR sensor system; and (4) maximum filling of aluminum foam core TSU.

### Heating and Cooling of $NF_3$ in 1/8" Drilled-Hole Core TSU

As indicated earlier, $NF_3$ supercools slightly when it is cooled from a single-phase state. A test was conducted to determine whether cooling from a two-phase state would suppress or partly suppress this supercooling. Figure 19 illustrates the test results. In this test, the fractional discharge -- i.e., the fraction of the $NF_3$ transformed from the low-energy solid state to the high-energy solid state -- was varied between 20% and 95%. After each fractional discharge, a full discharge and single-phase recharge (FD-SPR) cycle was carried to determine whether the recharge of the fractionally discharged TSU had, indeed, occurred. The results show that supercooling of $NF_3$ can be suppressed by recharging from a two-phase state.

### Heating and Cooling of $NF_3$ in Aluminum Foam Core TSU

Despite the favorable results described above, operational solutions to suppress supercooling are not as desirable as design solutions. One obvious design change to suppress supercooling is to increase the core surface area. Based on the results indicated in Figure 7, the surface area of a metal foam core is at least an order of magnitude higher than that of an 1/8" drilled hole core. So, a test was conducted with an aluminum foam core TSU to examine this hypothesis. The results, illustrated in Figure 20, indicate that use of this particular variation of aluminum foam (40 ppi, 40% relative density) will almost completely suppress $NF_3$ single-phase supercooling.

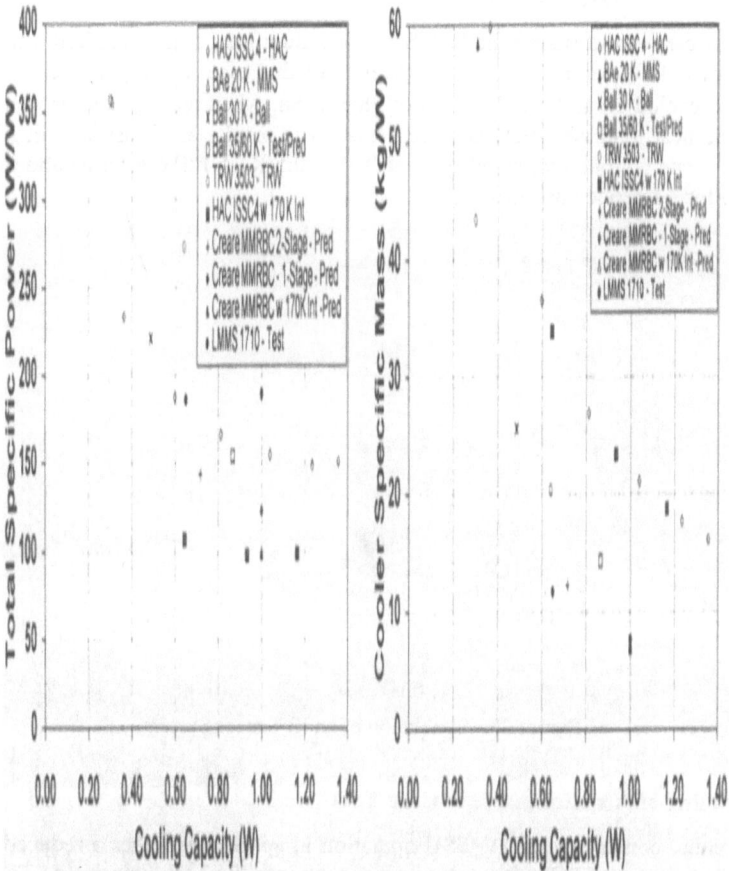

**Figure 19.** Heating and Cooling of $NF_3$ in 1/8" Drilled Hole Core

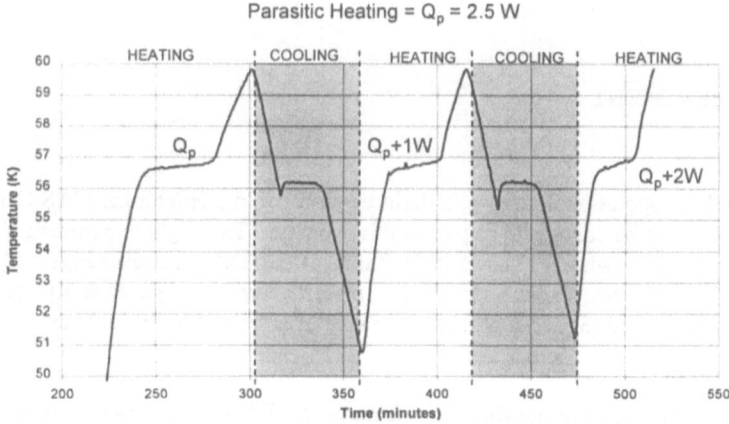

**Figure 20.** Heating and Cooling of $NF_3$ in Aluminum Foam Core

## Simulation of an IR Sensor System

A test was carried out (using the RCTS and the aluminum foam core TSU) to simulate the performance of a TSU-bearing IR sensor system. In this case, the cryocooler and focal plane heat load were cycled on and off, 180° out of phase. So, when the cryocooler was powered on, the focal plane heat load was off, and vice versa. The results are illustrated in Figure 21 and indicate focal plane temperature stability of about +/- 0.5 K. A full discharge and recharge cycle is also shown for reference purposes.

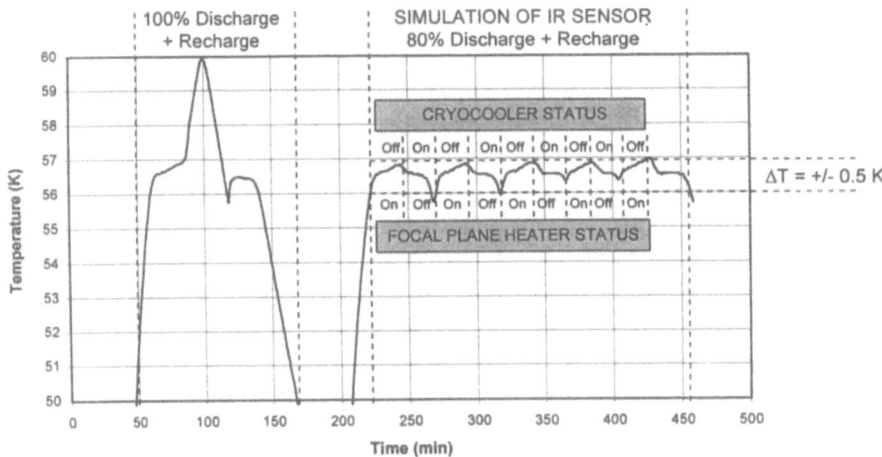

**Figure 21.** Simulation of an IR Sensor System

## Maximum Filling of Aluminum Foam Core TSU

One potential concern with DV TSU operation in space is whether a reduced gravity field will inhibit the transfer of PCM from the storage tank into the heat exchanger. To examine this concern, a test was carried out to determine the extent to which an aluminum foam core TSU can be filled with $NF_3$ with the fill port oriented downward (i.e., against gravity). The results indicate that the core could be filled to a level close to 95%. The reason for this filling behavior is the wicking action of the foam core. Thus, there should be no problem with filling the TSU heat exchanger in 0-g.

## FLIGHT EXPERIMENT

### Objectives

To verify TSU operation in zero-g, a flight experiment in a Hitchhiker GAS Canister on the STS in mid-1998 will be conducted as part of this program. The flight experiment objectives are to demonstrate the following: (1) acceptable filling of the TSU heat exchanger; (2) acceptable emptying of the TSU heat exchanger; (3) acceptable energy storage capacity; (4) acceptable temperature stability; and (5) acceptable performance over several operational cycles.

### Flight TSU Design

The flight TSU will have the following features: (a) DV configuration to minimize weight; (b) beryllium heat exchanger walls and end-caps for compatibility with beryllium focal plane structures; (c) aluminum foam core (40 ppi, 40% relative density) to minimize $NF_3$ supercooling; (d) spherical, thin-walled titanium storage tank to minimize weight; (e) 282 grams of $NF_3$ to meet the energy storage capacity requirement; and (f) low fill pressure ($P_{FILL}$ < 100 psi) to minimize safety concerns. The aluminum foam core will be brazed to the top and bottom end-caps and the shell will be autogenously electron-beam welded to the top and bottom end caps. Figure 23 illustrates the flight TSU configuration. The unit below will have an energy storage capacity of 6000 J, temperature stability of +/- 0.2 K, and weight of 2.1 kg.

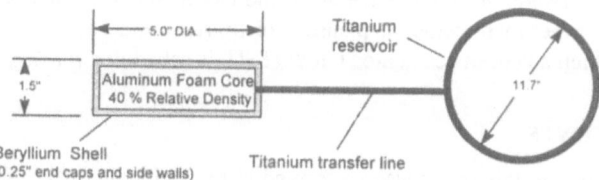

**Figure 23.** Flight TSU Design

## Layout of the Hitchhiker GAS Canister

The layout of the components to be utilized in the 60 K TSU flight experiment (designated as CRYOTSU) is shown schematically in Figure 24. The Cryogenic Test Bed (CTB) canister will be utilized for the fourth time in this flight.[11] As the figure illustrates, four Hughes tactical cryocoolers will be used to cool the TSU and a radiation shield. Flexible thermal coupling devices (used on previous CTB experiments) will be used to thermally link the cryocooler cold heads to the TSU and radiation shield. The purpose of the radiation shield is to reduce the parasitic heat load to the TSU and lessen the task of accounting for the total TSU heat load.

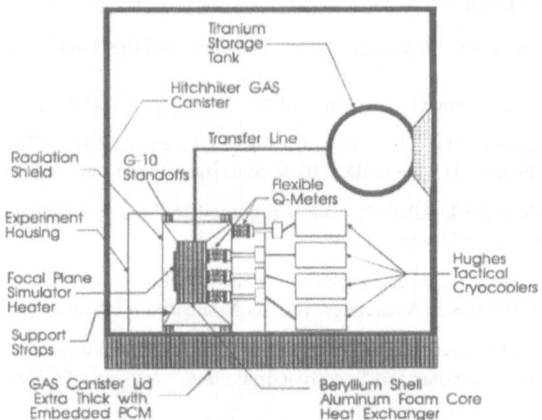

**Figure 24.** Layout of the 60K TSU Flight Experiment

## Flight Experiment Test Sequence

During the flight experiment tests will be conducted to meet the program objectives and compare with ground test results. A single on-orbit operational cycle, which may be repeated several times, consists of the following four step process: (1) beginning with 100 psi, ambient temperature, gaseous $NF_3$ contained within the TSU heat exchanger and storage tank, the $NF_3$ is liquefied by cooling the heat exchanger to 80 K; (2) the heat exchanger is then cooled to 60 K; (3) the heat exchanger is cycled between 60 and 52 K to demonstrate the TSU energy storage capacity and temperature stability; and (4) the heat exchanger is warmed back up to ambient temperature and the heat exchanger empties most of its $NF_3$ back into the storage tank.

## FUTURE PLANS

The plans for this program during the remainder of 1996 include the following: (1) complete TSU development testing; (2) complete the detailed design of the flight TSU and the hardware to incorporate it into the HH-GAS canister; (3) fabricate flight hardware; (4) fabricate and test a TSU engineering unit; and (5) fabricate and test a TSU flight unit. Integration of the flight TSU and accompanying hardware will commence when the Cryogenic Test Bed is returned from its use on the CRYOFD experiment, which is planned for flight in mid-1997. Again, the 60 K TSU flight experiment, which has been designated CRYOTSU, is scheduled for flight in mid-1998.

## ACKNOWLEDGMENTS

The work described in this paper was performed at the Beltsville, MD offices of Swales & Associates, Inc. under 1994 Phase I and 1995 Phase II SBIR contracts with Phillips Laboratory. The authors would like to thank Phillips Laboratory (PL/VTPT) and the SBIR Program Office for the opportunity to carry out this effort. The authors would also like to thank Mr. P. Brennan and Mr. F. Edelstein for providing valuable consulting and technical review services over the course of the program. Finally, the authors would like to acknowledge Mr. S. Schumacher, Mr. C. Lashley, Mr. D. Watson, Mr. N. Galassi, and Mr. T. Davis for their outstanding work and dedication over the course of the Phase I and Phase II programs.

## REFERENCES

1.  Glaister, D., K. Bell, M. Bello, and M. Stoyanof, "The Development and Verification of a Cryogenic Phase Change Thermal Storage Unit for Spacecraft Applications," *Cryocoolers 8*, Plenum Press, New York (1995), pp. 927-940.

2.  Bugby, D., R. Bettini, and M. Stoyanof, "60 K Thermal Storage Unit," Space Technology and Applications International Forum (STAIF-96), 2nd Spacecraft Thermal Control Symposium, Albuquerque, NM, AIP Conference Proceedings 361, Woodbury, NY (1996).

3.  Brennan, P. and F. Edelstein, "Development of a 60 K Cryogenic Phase Change Thermal Storage Device," Phase I SBIR Proposal to the DoD, PJB & Son, Inc., Timonium, MD (1994).

4.  Knowles, T., "Cryogenic PCM Database", Report prepared by Energy Science Laboratories, Inc. for Grumman Space & Electronics Division (1991).

5.  Grzyll, L., "Solid-Liquid Phase Change Materials Identification for Use Between 60 K and 65 K", Final report prepared for Swales & Associates, Inc. by Mainstream Engineering Corporation (1994).

6.  Pierce, L. and E. Pace, "Thermodynamic Properties of Nitrogen Trifluoride from 12K to Its Boiling Point. The Entropy from Molecular and Spectroscopic Data," *Journal of Chemical Physics*, v. 23, n. 3 (1965), pp. 551-555.

7.  Bugby, D., "60 K Phase Change Material Device Phase I SBIR Final Report," Technical Report No. SAI-RPT-051, Swales & Associates, Inc. (1995).

8.  Leyda, B., "Imagination Kit," Information Package Supplied by ERG (Materials Division) on Aluminum Foam Properties, Oakland, CA (1995).

9.  Lide, D. (Editor-in-Chief), "CRC Handbook of Chemistry and Physics, 73$^{rd}$ Edition," CRC Press, Inc., Boca Raton, FL (1992-1993).

10. LABVIEW Computer Program User's Manual (Version 3.01), National Instruments, Austin, TX.

11. Thienel, L. and C. Stouffer, "The Cryogenic Test Bed Experiments," 1995 Shuttle Small Payloads Symposium, Baltimore, MD (1995).

# Temperature Stabilization on Cold Stage of 4 K G-M Cryocooler

**Rui Li, Atsushi Onishi, Toshimi Satoh, and Yoshiaki Kanazawa**

R & D Center, Sumitomo Heavy Industries, Ltd.,
63-30, Yuhigaoka, Hiratsuka, Kanagawa, 254, Japan

## ABSTRACT

A simple method of temperature stabilization on cold stage of 4 K Gifford-McMahon (GM) cryocooler has been proposed.   A copper pot connected to compressor unit by a stainless steel capillary tube is mounted on the 4 K stage of a GM cryocooler.   Depended on the temperature of pot, pressurized helium can naturally go into or out of the pot.   The utilization of high volumetric specific heat of pressurized helium in the pot at 4 K region produces a good temperature stability on the 4 K stage.   The periodic temperature fluctuation of the 4 K stage is   ~ 0.5 K (peak-to-peak) typically, but the high pressure pot described in the present paper reduces the temperature fluctuation effectively down to ~ 0.05 K (peak-to-peak).   The method is safe to deal with, and the system is easy to operate.   This paper shows the experimental details, and discusses the temperature stabilization effect of the pot and other advantages of the method.

## INTRODUCTION

Because of the successful use of rare earth magnetic regenerator materials, Gifford-McMahon (GM) cycle cryocooler is developed remarkably in recent years[1, 2, 3].   For 4 K GM cryocoolers, the no load temperature has reached below 3 K, and the cooling capacity at 4.2 K has been over 1 W.   The progress in 4 K GM cryocooler provides a great possibility of operating superconducting devices, such as SQUIDs and SIS mixers, at 4 K region without liquid helium or helium liquefiers[4, 5, 6].

SQUIDs and SIS mixers, etc. have a potential of high sensitivity, and are widely applied in many science and technology fields.   The thermal requirement of such superconducting device operating systems is usually a small cooling capacity at 4 K region.   4 K GM cryocoolers are able to deliver an enough cooling capacity continuously for these operating systems, and can also sharply simplify the structure and the operation of such systems because the systems need no liquid helium or helium liquefiers.   Mechanical cryocoolers like GM cryocooler, however, produce temperature fluctuations, magnetic noises and mechanical vibrations in the operating systems.   For a highly sensitive measurement, it is necessary to reduce the disturbances mainly caused by these noise sources.

Kaiser, G., et al.[7, 8] have reported a noise reduction method for cryocooler cooled operating systems.   They used a conventional GM cryocooler with a latent heat collector of liquid nitrogen or liquid neon, and reduced the temperature fluctuations down to 0.04 K ~ 0.15 K.   In their system, however, the operating temperature was higher than 50 K, and the measurement had to be discontinued every few hours.

In the present work, we have developed a simple method of temperature stabilization on cold stage of 4 K GM cryocooler. The high volumetric specific heat of pressurized helium at 4 K region is applied to reduce the temperature fluctuation of the 4 K stage. The following sections describe the experimental details, and give discussions about the effect of temperature stabilization and other advantages of the method.

## EXPERIMENTAL DETAILS

The method developed in this work is to mount a small pot on the cold stage of GM cryocooler and to connect the pot to the supply line or the return line of compressor unit by a capillary tube. As is shown in figure 1, the experimental setup is very simple, and only a copper pot and a stainless steel capillary tube are needed besides a 4 K GM cryocooler.

Both the cold head and the compressor unit of the cryocooler are manufactured by Sumitomo Heavy Industries Limited. In the cold head, the second regenerator has lead spheres arranged at the hotter end, and spherical magnetic regenerator material, $ErNi_{0.9}Co_{0.1}$, placed at the colder end. The cooling capacity of the 4 K GM cryocooler is ~ 0.6 W at 4.2 K. The stainless steel capillary tube is 1.75 mm in inside diameter, and is arranged to have good heat contact with the first stage of the cryocooler. Figure 2 describes the details of the pot and the second stage. The copper pot is about 900 g in weight, and is 61 $cm^3$ in capacity. The temperature of the pot or the second stage is measured by a carbon glass resistor (Lake Shore Cryotronics, Inc.).

The temperature fluctuations are measured for four cases: (1) there is no pot mounted on the 4 K stage, (2) the pot is mounted on the stage but there is nothing in it, (3) the pot is mounted on the stage and is filled with low pressure helium (connected to the return line), (4) the pot is mounted on the stage and is filled with high pressure helium (connected to the supply line). We will simply call these cases at the following as: (1) no pot, (2) vacuum pot, (3) low pressure pot, and (4) high pressure pot. The carbon glass resistor and an electric heater of manganin wire are held on the second stage for the no pot case, and on the copper pot for other cases. For simplifying explanations in the following sections, the temperature measured at the pot will also be called as the second stage temperature.

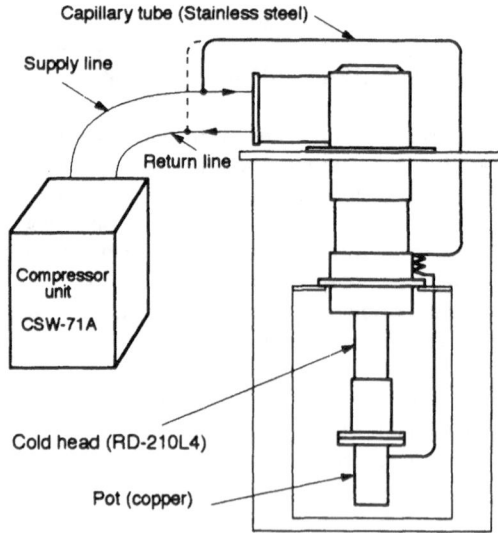

**Figure 1.** Schematic of experimental apparatus.

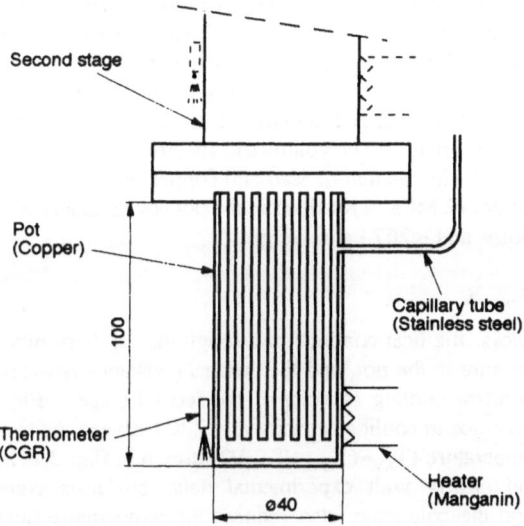

**Figure 2.** Details of copper pot and second stage.

## RESULTS AND DISCUSSIONS

### Effect of Temperature Stabilization

Figure 3 shows test results of temperature stabilization at 4.2 K. For the no pot case, the 4 K stage temperature fluctuates periodically, synchronized with the working gas cycle in the cold head. The peak-to-peak value of temperature fluctuation is 0.53 K, and the value is typical for the 4 K GM cryocoolers with a partial copper cylinder of the second stage. For the case of high

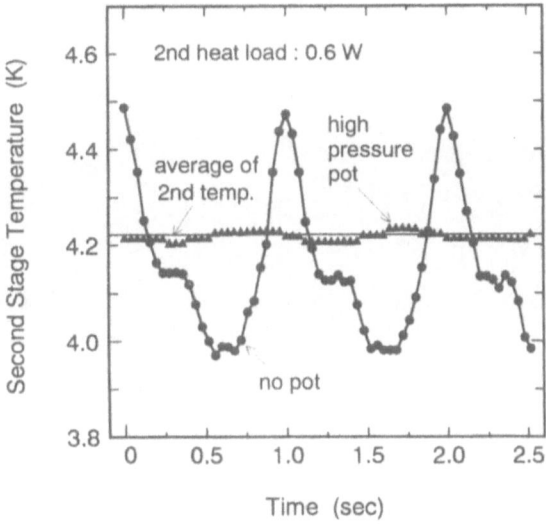

**Figure 3.** Periodic temperature fluctuations of second stage

pressure pot, however, the fluctuation is effectively reduced down to 0.054 K (peak-to-peak). The initial pressure of compressor unit is 1.76 MPa for both the cases, but the supply pressure is 2.35 MPa for no pot and 2.17 MPa for high pressure pot at 4.2 K.

The effect of temperature stabilization shown in Fig. 3 is brought about by the utilization of high volumetric specific heat of pressurized helium.    Figure 4 gives volumetric specific heats of helium and some cryogenic materials.    The volumetric specific heat of helium at 4 K for 2 MPa is ~ 49 times and ~ 490 times as high as that of lead and copper respectively.    In other words, the heat capacity of the helium gas (2 MPa, 4 K) filled in the pot equals that of a copper block which is ~ 0.03 (=0.31³) m³ in volume and ~ 267 kg in weight.

## Cooling Capacity at 4.2 K

There are several factors, the heat conduction through the capillary tube from the first stage, the small oscillation of pressure in the pot, and the thermal resistance between the pot and the 4 K stage, which may use up a few cooling capacity of the second stage.    Fig. 3 shows, however, that there is almost no difference in cooling capacity at 4.2 K between the two cases.    In fact, the variance of the average temperature $(T_{AV} = (T_{MAX} + T_{MIN})/2)$ drawn in Fig. 3 is less than 4 mK.    The reason is not understood clearly with experimental data, but it is considered that a good temperature stabilization on the cold stage also reduces the temperature fluctuation in expansion volume.    This means that the expansion process and the compression process of working gas may slightly shift to isothermal ones, and thus it is possible to increase a few cooling capacity for the high pressure pot case.    According to this consideration, more cooling capacity is expected if the pot is united with the cylinder in a body.

## Temperature Stabilization Below 20 K

Since pressurized helium shows a broad peak in specific heat below 20 K, Fig. 4 gives a hint that the pot can also be used to reduce the temperature fluctuations above 4 K.    For all the cases, the temperature fluctuations below 20 K are illustrated in figure 5 as a function of the second stage temperature.    The temperature fluctuation for the vacuum pot case is as same as that of the no pot case at 4 K region.    It means that the heat capacity of the copper pot is very poor and can be ignored.    At higher temperatures, the heat capacity of pot becomes larger, therefore the

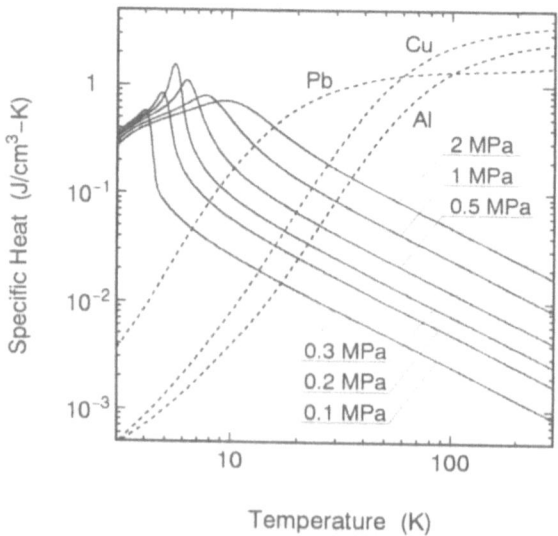

**Figure 4.**  Volumetric specific heats of helium and some cryogenic materials.

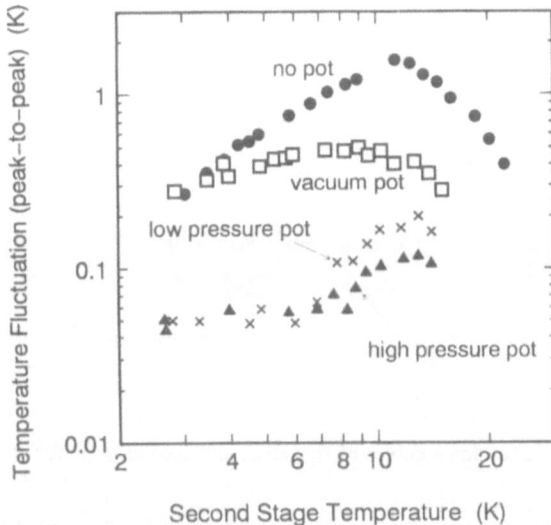

**Figure 5.** Temperature fluctuations as a function of second stage temperature.

temperature fluctuation of vacuum pot is smaller than that of no pot.

The effect of temperature stabilization is more superior for low pressure pot and high pressure pot. Because of the higher volumetric specific heat for higher pressure helium above 10 K, the high pressure pot reduces the temperature fluctuations more effectively than the low pressure pot case. Compared with the no pot case, the temperature fluctuation of high pressure pot is almost reduced by a factor of 10 below 20 K. The maximum temperature fluctuations shown in Fig. 5 are 1.57 K for the no pot case, 0.49 K for the vacuum pot case, 0.20 K for the low pressure pot case, and 0.12 K for the high pressure pot case.

## Cool Down Time

The difference in cool down time for the four cases is also an interesting matter. Figure 6 shows the cool down curves of the first stage and the second stage for all the cases. The second stage is cooled more rapidly than the first stage for the no pot case, but is changed more slowly when the pot is mounted on the 4 K stage. The total cool down time seems to be 140 minutes for the high pressure pot case, and 120 minutes for other cases. The fact indicates that there is no remarkable difference in cool down time whether the pot is mounted or not. Furthermore, because of the low specific heat of helium at higher temperatures, it is similar in cool down curves above 50 K whether the pot is filled with helium.

## Advantages of Pot

Since the working gas undergoes compression and expansion processes alternately in the cold head, the temperature of the cold stage certainly fluctuates periodically, synchronized with the working gas cycle. At low temperatures especially, the specific heat of solid materials is very poor, and then the temperature fluctuation becomes much larger.

In general, there are two ways to reduce temperature fluctuations on cold stage of cryocooler. One is to decrease thermal diffusibility between cold stage and cooled object (the object which is wanted to be cooled), for example, to insert a material with poor thermal diffusibility between the cold stage and the cooled object. Another way is to increase heat capacity on cold stage, for example, to mount a big copper block on the cold stage together with the cooled object. Both the

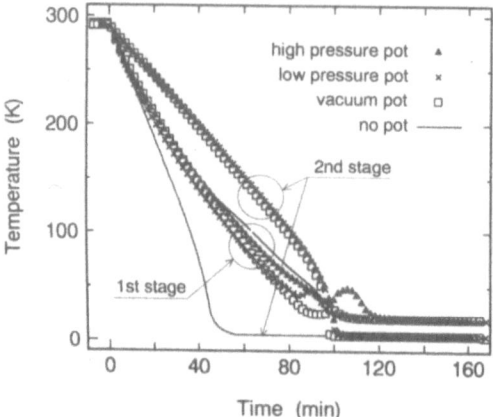

**Figure 6.** Cool down curves of first stage and second stage for all cases.

method are able to bring about a good temperature stabilization for the cooled object, but available cooling capacity may be decreased considerably owing to the large temperature difference between the cold stage and the cooled object for the former method, and it needs very long time to cool down the system from room temperature for the latter.    As is shown in Fig. 3, 5 and 6, the high pressure pot and the low pressure pot described above have both advantages of saving cooling capacity and saving cool down time, besides the great effect of temperature stabilization.    The high pressure pot and the low pressure pot belong to the method of increasing heat capacity on cold stage, but the important point is that the pot is not a closed one but connected to the compressor unit.    By connecting the pot to the supply line or the return line of compressor unit, the pressurized helium naturally goes into or out of the pot, depended on the temperature of pot. As a result, the pot has a great amount of heat capacity to stabilize the cold stage temperature because of the high volumetric specific heat of helium below 20 K, as well as the system is easy to cool down due to the low specific heat of helium at higher temperatures.

Furthermore, the pressure of pressurized helium in the pot is always equal to that of compressor unit (the supply pressure, the return pressure or the initial pressure) whether the system is under operation or not.    It means that the system is safe to deal with as same as a usual 4 K GM cryocooler.

## CONCLUSIONS

A simple method of temperature stabilization for 4 K GM cryocooler has been developed. With the copper pot connected to compressor unit, the second stage temperature fluctuation of 4 K GM cryocooler is reduced from 0.53 K down to 0.054 K at 4 K, and from 1.57 K down to 0.12 K at ~10 K.    The system is easy to cool down and is safe to deal with, as well as it also has advantages of saving cooling capacity and continuous operation.

Finally, the pot connected to compressor unit is effective for reducing temperature fluctuations of cryocooler, but for a highly sensitive measurement of superconducting devices and other applications, different techniques are needed for taking away the disturbances caused by mechanical vibrations and magnetic noises.

## REFERENCES

1. Hashimoto, T., Li, R., Kuriyama, T. and Nagao, M., " Recent progress in application of magnetic regenerator material" , *Cryogenic Engineering (Journal of the Cryogenic Society of Japan)*, Vol. 31 (1996), pp. 131-149.

2. Satoh, T., Onishi, A., Li, R., Asami, H. and Kanazawa, Y., " Development of 1.5 W 4 K G-M cryocooler with magnetic regenerator material" , *Adv. Cryog. Eng.*, Vol. 41, Plenum Press, New York (1996), in press.

3. Li, R., Onishi, A., Satoh, T. and Kanazawa, Y., " Influence of valve open timing and interval on performance of 4 K Gifford-McMahon cycle cryocooler" , *Adv. Cryog. Eng.*, Vol. 41, Plenum Press, New York (1996), in press.

4. Plambeck, R., Thatte, N. and Sykes, P., " A 4 K Gifford-McMahon refrigerator for astronomy" , *Proc. 7th International Cryocooler Conference*, Air Force Phillips Laboratory Report, PL-CP-93-1001, Kirtland, AFB, NM (1993), pp. 401-415.

5. Takahashi, M., Hatakeyama, H., Kuriyama, T., Nakagome, H., Kawabe, R., Iwashita, H., McCulloch, G., Shibata, K. and Ukita, S., " A compact 150 GHz SIS receiver cooled by 4 K GM refrigerator" , *Proc. 7th International Cryocooler Conference*, Air Force Phillips Laboratory Report, PL-CP-93-1001, Kirtland, AFB, NM (1993), pp. 495-507.

6. Fujimoto, S., Kazami, K., Takada, Y., Yoshida, T., Ogata, H. and Kado, H., " Cooling of SQUIDs using a Gifford-McMahon cryocooler containing magnetic regenerative material to measure biomagnetism" , *Cryogenics*, Vol. 35 (1995), pp. 143-148.

7. Kaiser, G., Seidel, P. and Thurk, M., " Noise reduction of cryo-refrigerators" , *Adv. Cryog. Eng.*, Vol. 39, Plenum Press, New York (1994), pp. 1281-1285.

8. Kaiser, G., Dorrer, L., Matthes, A., Seidel, P., Schmidl, F., Schneidewind, H. and Thurk, M., " Cooling of HTSC Josephson junctions and SQUIDs with cryo-refrigerators" , *Cryogenics*, Vol. 34, ICEC Supplement (1994), pp. 891-894.

# Reduction of Parasitic Heat Loads to Cryogenically Cooled Components

**Scott Jensen, J. Clair Batty, and David McLain**

Space Dynamics Laboratory
Logan, Utah USA 84341

## ABSTRACT

The Space Dynamics Laboratory at Utah State University, teamed with NASA Langley, are currently designing the SABER(Sounding of the Atmosphere using Broadband Emission Radiometry) instrument. The Focal Plane Assembly (FPA) must be cooled to 75 K for 2 years at a 100 percent duty cycle. Due to mass, size, and power constraints, expendable cryogen systems could not be used. The SABER team has selected the TRW miniature pulse tube refrigerator which provides roughly 250 mW of cooling at 72 K. Conventional methods for supporting the FPA were unacceptable due to the excessive conductive heat loads. Realizing the need for better thermal isolation while maintaining structural rigidity, work on tension support systems utilizing high performance fibers commenced. Utilizing Kevlar 49 fibers in an approach we refer to as Fiber Support Technology (FiST), we were able to reduce the conducted parasitic heat loads from 85 mW to less than 2 mW.

Various radiation suppression schemes coupled with wiring schemes were necessary to reduce the total parasitic heat loads on this system to less than 250 mW. This paper outlines the details of this development effort making the use of a low input power, small, mechanical cooler possible. This approach seems consistent with the "smaller, better, cheaper, faster" attitude of the nineties.

## INTRODUCTION

Certain components in optical systems, such as orbiting telescopes, must often be cooled to cryogenic temperatures to facilitate proper functionality of the optical sensors. These cooled components must be thermally isolated from their much warmer surroundings. Such thermal isolation is complicated by the necessity of supporting these components in a fixed and rigid position with respect to the warm structures to which they are attached.

Traditionally, cryogenically cooled components developed by Utah State University/Space Dynamics Laboratory (USU/SDL) have been supported by composite glass-epoxy (G-10) cylinders that provide a compromise between thermal isolation and the necessary rigidity to support typical launch loads (see Figure 1). This approach has two limitations that are becoming more serious with the "smaller, faster, cheaper, better" attitude of the 90s.

First, the parasitic heat loads due to conduction through this type of system, although small, have a significant thermal impact on systems with reduced cooling capacity. For systems using an

Cryocoolers 9, Edited by R.G. Ross, Jr.
Plenum Press, New York, 1997

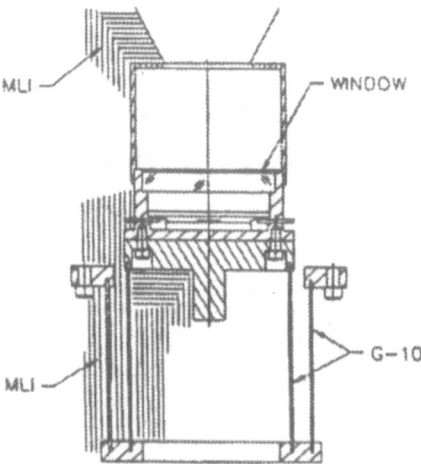

**Figure 1.** Conventional support method.

expendable cryogen as the heat sink, mission life is limited. In systems using mechanical coolers, refrigerator cooling capacity may be exceeded. Second, the first natural resonant frequency, for this application, is roughly 50-70 Hertz. Higher values are desirable to avoid the risk of resonant response during launch.

Increasing financial limitations, translating to ever more stringent mass and size constraints for satellites, provide motivation to develop alternative ways to support cooled telescope components in order to significantly reduce the parasitic heat load on the low temperature sink. This will make possible the use of smaller, lightweight mechanical coolers or perhaps less cryogen than would otherwise be necessary for systems using expendable cooling methods.

Seeing the need for improving cold component support, we began work on a tension support system which utilizes high performance fibers in tension to provide mechanical support and thermal isolation. We have called this effort FiST (Fiber Support Technology).

## FiST DEVELOPMENT

The research and design of FiST concepts began at USU in the spring of 1993 as a senior design project after learning of a similar approach utilized by Pat Roach at NASA Ames[1]. Initial results from analysis and testing on a prototype unit suggested that the conductive heat loads are significantly reduced while the mechanical stiffness of the system is dramatically enhanced[2].

Due to the overwhelming success of the student projects, engineers at the Space Dynamics Laboratory baselined the FiST design approach to support the focal plane assembly of a space-based infrared instrument called SABER (Sounding of the Atmosphere using Broadband Emission Radiometry). The Space Dynamics Laboratory, teaming with the National Aeronautics and Space Administration at Langley, Virginia (NASA LaRC), are currently in the process of designing the SABER instrument which will study the earth's limb. This instrument will fly on the TIMED (Thermoshpere-Ionosphere-Mesosphere Energetics and Dynamics ) mission sponsored by APL (Applied Physics Laboratory) at John Hopkins University.

Cooling of the infrared sensors on this instrument is provided by the TRW miniature pulse tube refrigerator, which provides roughly 250 mW of cooling at a cold block temperature of 72 K and a reject temperature of approximately 293 K[3]. If other parasitic heat loads calculated for the SABER Focal Plane Assembly (FPA) are added to the conduction through the G-10 tubing of the conventional support method (approximately 85 mW), it would preclude the use of this small, lightweight, low input power refrigerator. The SABER instrument would then not be possible under present mass and power constraints.

## MECHANICAL CONFIGURATION AND PERFORMANCE

The FiST system is composed of four basic sections: the strand, the strand fasteners, an inner support structure, and an outer support structure. The central component of the FiST system is the fiber. It provides the necessary thermal isolation for the cold components as well as the mechanical rigidity for the system. In determining the fiber to be used for the SABER FPA, many factors had to be considered. First, the thermal conductivity of the fiber selected must be low enough to limit the conducted heat loads incident upon the FPA to values less than approximately 10 mW. Second, the fiber selected must be sufficiently strong to withstand the rigorous loads incurred during launch. Third, molecular breakdown in certain environments and out-gassing of the material is critical due to the sensitive nature of the instrument. Optical systems must be virtually contamination free and high out-gassing materials are unacceptable. And Fourth, the fiber is unmanageable in its raw form, so a braiding scheme had to be determined.

Several different fibers were considered and a decision made based on overall properties of the fiber. Table 1 and Figure 2 are a summary of some of the critical properties of these fibers. Based on a compromise between mechanical and thermal performance, Kevlar 49 fiber was selected for use on the SABER instrument and proves to provide superior thermal isolation while maintaining sufficient mechanical strength.

Once the fiber selection was finished, an approach to taking hold of the strand, without inducing sufficient shear loading to weaken the strand, was needed. When designing a strand fastener, there were several things that had to be considered: First, shear loading must be minimized. Shear strength testing of the fiber showed that shear loads considerably weaken the strength of the fiber. Second, the SABER FPA sits in a very confined space so the strand fasteners must be very small. Room is not available for large radius fasteners such as eyebolts. Third, the strand fastener strength must be comparable to the strength of the strand. The current design of the strand fasteners shown in Figure 3 meets all of the requirements previously specified.

The outer support structure is the mechanism that provides an attachment point to the warm surroundings. The inner support structure usually provides a platform for the cooled components to be placed. However, this configuration need not be the case. Depending on the application, the inner support structure could serve as the warm support and the outer support structure is the cooled component.

**Table 1.** Material Properties of Selected Fibers

| Fiber Description | Thermal Conductivity (W/m-K) | Tensile Modulus (GPa) | Tensile Strength (MPa) | Tendency to Creep |
|---|---|---|---|---|
| T-300 Carbon | 13.0 | 231 | 3240 | Small |
| E-Glass (G-10) | 1.0 | 72 | 3100 | Small |
| Spectra 900 | 0.30 | 174 | 3128 | Large |
| Nomex | 0.13 | 13 | 614 | Large |
| Vectran HS | 0.20 | 65 | 2841 | Very Small |
| Kevlar 49 | 0.04 | 124 | 2800 | Small |

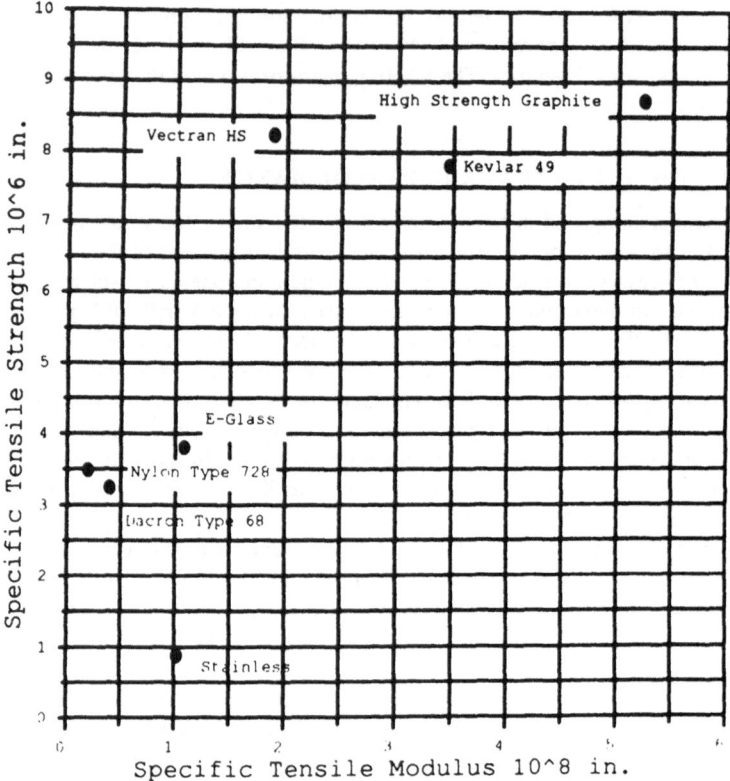

**Figure 2.** Specific tensile strength and specific tensile modulus of various materials.

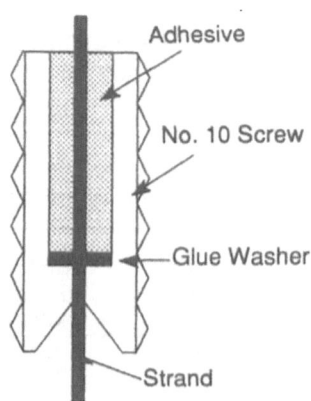

**Figure 3.** Strand fastener design.

With these thoughts in mind, a preliminary breadboard model of the SABER FPA was constructed using outer and inner support structures as shown in Figures 4 and 5.

We then subjected this prototype to a series of mechanical and thermal tests. The mechanical characteristics of the assembled FiST support system are very robust. The unit has undergone nearly 30 minutes of rigorous vibration testing. The first natural resonant frequency of the system was found to be nearly 700 Hz in all axes. Compared to the traditional G-10 tube approach this dramatically enhances the mechanical stiffness of the system while improving thermal isolation performance.

## THERMAL PERFORMANCE

The infrared optical sensors used on the SABER FPA must be cooled to approximately 75 K. Assuming a temperature difference of 3 K or less across the thermal path from the refrigerator cold block to the optical detectors can be achieved, the cold block on the cooler would need to run at 72 K. The TRW mechanical pulse tube refrigerator selected for the SABER instrument provides approximately 250 mW of cooling at this temperature.

A first-order estimate of the heat loads incident on the SABER FPA is shown in Table 2. If the conventional G-10 tube approach was selected, the SABER instrument would not be possible because the 250 mW cooling capacity of the TRW pulse tube refrigerator would be exceeded, and given the current power and temperature constraints, other alternatives are not feasible.

The modes of heat transfer specific to the SABER FPA are limited to radiation and conduction. The FiST system for the SABER FPA has a conductive thermal path from the warm outer support structure, through the Kevlar 49 support strands, and into the inner support structure. Due to the ultra-low thermal conductivity of Kevlar 49 fiber, the predicted parasitic heat loads conducted from the warm to cold support structures are extremely small. The modeling of this heat transfer mode is accomplished by utilizing Fourier's Law as shown in Figure 6.

Because the diameter of the strand is extremely small, radial conduction in the strand is neglected. The conduction problem then simplifies to a one-dimensional mode of heat transfer in which heat is conducted from the warm to the cold environment. As Fourier's Law dictates, the cross-sectional area (A) and length (L) of the strand are inversely related. The area must be kept to a minimum while the length should be maximized to achieve the maximum thermal isolation possible.

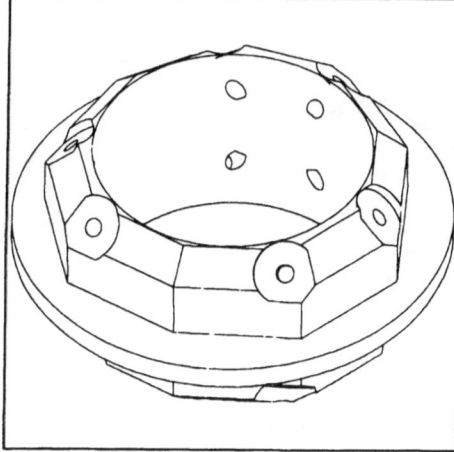

**Figure 4.** SABER FPA outer support structure.

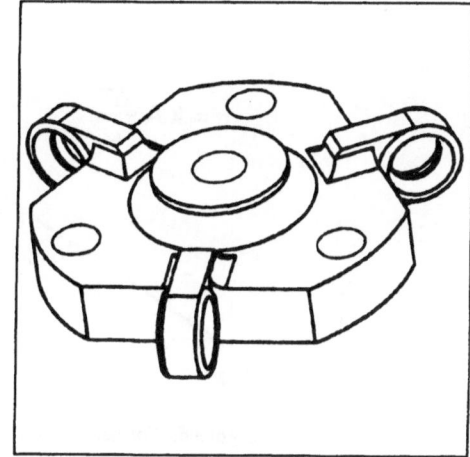

**Figure 5.** SABER FPA inner support structure.

**Table 2.** Predicted SABER FPA Heat Loads

| Description | Heat Load (mW) |
|---|---|
| Aperture | 50 |
| Wiring | 35 |
| Detectors | 50 |
| Radiation | 70 |
| Supports | |
|     Conventional G-10 | 85 |
|     Kevlar 49 FiST approach | 2 |
| TOTAL | |
|     Conventional G-10 | 290 |
|     Kevlar 49 FiST approach | 207 |

If another material, such as stainless steel, were selected for the strand material, this ratio would become critical due to the higher thermal conductivity of the material. However, because Kevlar 49 is an excellent insulator, diameter and length considerations of the supporting strands in the FiST system are of relatively little concern. Mechanical constraints on the SABER FPA will clearly drive the strand size.

By utilizing Fourier's Law, and strand length of .8" (.02032 m), which is the current design length for the SABER FPA FiST system, a plot of the conducted parasitic heat loads for various materials as a function of strand diameter is shown in Figure 7.

Conduction parasitic heat loads incident on the SABER FPA were not limited to the Kevlar 49 strands. The detectors used on the initial breadboard of the SABER FPA had 54-36 gauge stainless steel wires, which ran from the warm telescope environment to the cooled detectors. Each wire was made to approximately 12 inches long in an effort to reduce these parasitic heat loads. A detailed analysis of the wiring is underway at this time and preliminary results show thaι

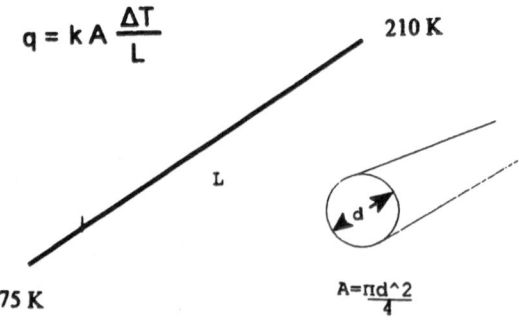

$$q = k A \frac{\Delta T}{L}$$

210 K

L

$$A = \frac{\pi d^2}{4}$$

75 K

**Figure 6.** Fourier's Law of conduction in one dimension.

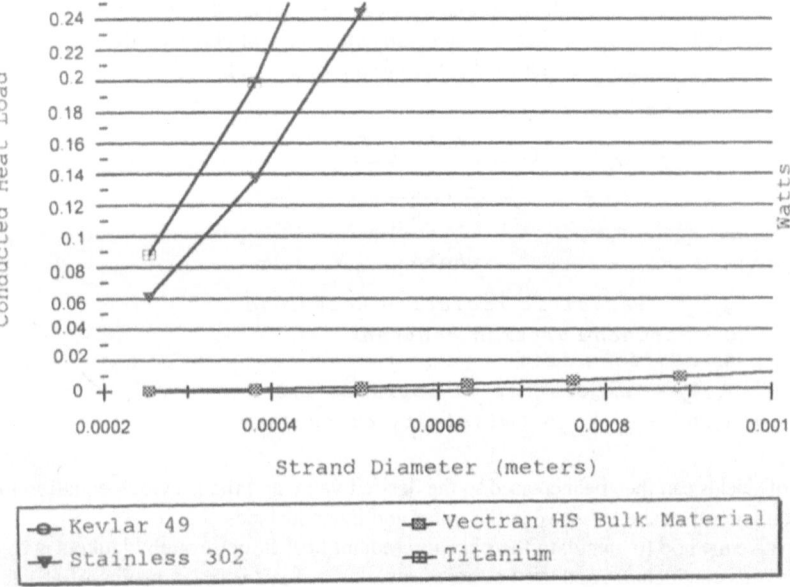

**Figure 7.** Conducted heat loads for various materials using SABER FPA parameters.

the conduction parasitics through the shielding could be significant. Efforts are currently being made to determine better ways of providing thermal isolation from the sensor wiring parasitics.

Typically, the radiation heat transfer for smaller FPA systems utilizing the conventional G-10 support approach is on the same order of magnitude as the loads conducted through the support. With the development of the FiST system, these radiation loads now appear to be monstrous compared to the small amounts conducted through the Kevlar 49 strands. The impact of these parasitic heat loads on the overall cooling capacity of the TRW mechanical refrigerator has motivated efforts to determine more effective ways to radiatively isolate warm components from cooler systems.

The cold components for the SABER FPA will be operating at a temperature of approximately 75 K. The surrounding warm components, for the first breadboard prototype, will run at approximately 210 K. If radiation suppression schemes were ignored, the total heat transfer between these temperature ranges to the cold components of the SABER FPA could be as much as 10 watts under certain circumstances where the heat was perfectly absorbed (blackbody). With an allowance of roughly 70 mW for parasitic radiation heat loads, measures had to be taken to reduce the effect of radiation on the SABER FPA.

Two methods where considered for suppressing the radiation effects on the breadboard model of the SABER FPA:

1. MLI (Multi-Layer Insulation) blanketing. MLI blanketing consists, in this case, of alternating layers of doubly aluminized Mylar and Dacron netting. The aluminized Mylar provides a radiation shield in which energy is directed away from the cooled components. The Dacron netting layers provide the thermal isolation necessary between alternating layers of Mylar to reduce conduction effects through the blanket.

2. Low emissivity surfaces. Low emissivity surfaces consist of gold, silver, aluminum, or any other material, which, when highly polished, becomes highly reflective to thermal energy. Typically, gold is the material of choice and provides the best surface with the lowest overall emissivity.

Because of limited funding for the breadboard FPA, and the abundance of MLI blanketing at SDL, blanketing was chosen to isolate the FPA from radiation effects.

The heat load allowed through these blankets can be calculated a few different ways. One method is to use radiation shield theory where an effective emittance of the entire blanket is calculated. The radiative parasitic heat load is then found by Eq. (1)[4].

$$q_{total} = \frac{A_1 \sigma (T_1^{\,4} - T_2^{\,4})}{\dfrac{1}{e_1} + \dfrac{1}{e_2} + \dfrac{1 - e_{3,1}}{e_{3,1}} + \dfrac{1 - e_{3,2}}{e_{3,2}}}$$

where

(1)

$q_{total}$ - Parasitic radiation heat load
$\sigma$ - Stephen-Boltzman constant
$A_1$ - Surface area
$T_1, T_2$ - Temperature of surfaces 1,2
$e_1, e_2, e_{3,1}, e_{3,2}$ - Emissivity of surface 1,2,3

The number of shields can then be increased to the desired value and the previous equation modified by simply adding another shield between the outer and inner surfaces.

An alternative method to calculate the parasitic radiant heat flow through blankets is to assume an effective thermal conductance and use Fourier's Law. Dr. J. Clair Batty, a professor at Utah State University, suggests that according to a study he did, "the values of the effective thermal conductivity of well designed and applied MLI blankets in cryogenic systems would usually fall in the range of 10 to 100 µW/m-K. Heroic measures are required to obtain values at the lower end of the range while blankets performing at the upper end of this range would be considered less than optimum under typical conditions".

The heat transfer can then be calculated by using Fourier's Law where the area is simply the cross-sectional heat transfer area of the blanket and the path length (L) is the thickness. If an effective conductivity of 55 µW/m-K is assumed, Figures 8 and 9 show the predicted parasitic radiation heat loads of each of these methods incident on the SABER FPA. Comparing the two methods yields similar values for the radiative parasitic heat loads. Using either method, the heat loads on the cooled components due to radiation are below the values allowed for in Table 2.

A complication in blanketing an extremely small focal plane assembly, such as the SABER FPA,

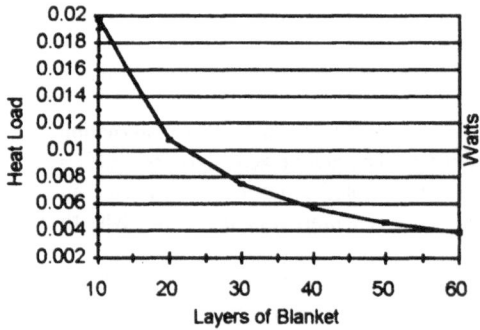

**Figure 8.** Calculated parasitic radiation heat load on SABER FPA using radiation shield theory.

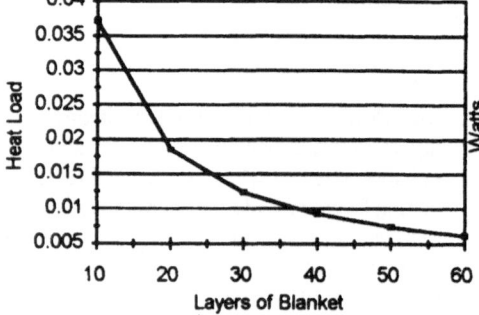

**Figure 9.** Calculated radiation parasitic heat loads on SABER FPA using effective conductivity method.

using the FiST approach, is that the blanketing may leave open gaps between the warm and the cold environments, along the strand, in which radiative heat loads could be excessive.
Blanketing around these strands while making sure that the blankets are not being thermally shorted was difficult.

To reduce these effects, we developed what we term "radiation rat stoppers." They are similar in concept to the large metal disks on the anchor chains of large ships which prevent rats from crawling up the chains and onto the ship. Small disks made of Teflon with low emissivity surfaces applied to either side are clipped over each strand to prevent direct radiation shots down the strand. Because a good portion of radiated energy is removed by a single shield, these radiation suppression disks will reduce added radiation loads from along the strand where blanketing could be insufficient.

### Other Parasitic and Operational Heat Loads

Two other heat loads contribute to the total heat load on the SABER FPA. The detectors during normal operation will continuously dissipate approximately 50 mW of heat. This is the only operational heat load incident on the SABER FPA.

The FPA aperture, which is a hole located in the cold stop assembly, is also a considerable parasitic heat source. The black aperture looking into a warm cavity at approximately 210 K will yield a heat load of approximately 50 mW. No methods of reducing this heat load are possible as the optical requirements of the SABER FPA prevent low emissivity coatings inside the detector cavity.

The total heat load on the SABER FPA can then be found by summing the various components. Table 2 is a representation of these summed heat loads. Of course, as the SABER instrument progresses in design, these numbers could change if operating temperatures or surrounding environment temperatures where to change.

## BREADBOARD THERMAL TEST RESULTS

The purpose of the thermal testing was to determine whether or not the thermal isolation schemes employed by the FiST concept were sufficient to enable the use of the TRW miniature pulse tube refrigerator. Figure 10 shows the FiST system assembled and blanketed before being integrated into the testing chamber.

The test ran for approximately 10 hours to achieve a steady-state condition. The thermal link used in the testing had been previously calibrated in a liquid nitrogen dewar and was found to have a thermal conductance of approximately .105 W/K. The total heat coming through the system can then be easily found by multiplying the conductance of the link by the measured temperature difference across the thermal link.

The temperature difference across the calibrated link was found to be 2.3 K. Multiplying this number by the calibrated link number yields a total heat load of approximately 240 mW. This is about 40 mW higher than predicted, however, with better care given to blanketing details and better isolating the sensor wiring from the cooled components should provide better correlation with predicted results on the next generation breadboard model.

## CONCLUSIONS

The design of the FiST system for the SABER FPA is well on its way to completion. A successful breadboard model has been built and tested with very promising results. The prototype breadboard unit is mechanically robust having a first natural resonant frequency of nearly 700 Hz in all axes. Thermally, the FiST approach is second to none. Conduction parasitic heat loads have been reduced to a few milliwatts. The successful development of this technology will revolutionize the cryogenic support industry. With the capability of reducing parasitic heat loads onto cooled components without compromising the structural integrity of the system, the systems are better able to the meet the approach of the 90s, which is "smaller, better, faster, cheaper" spacecraft.

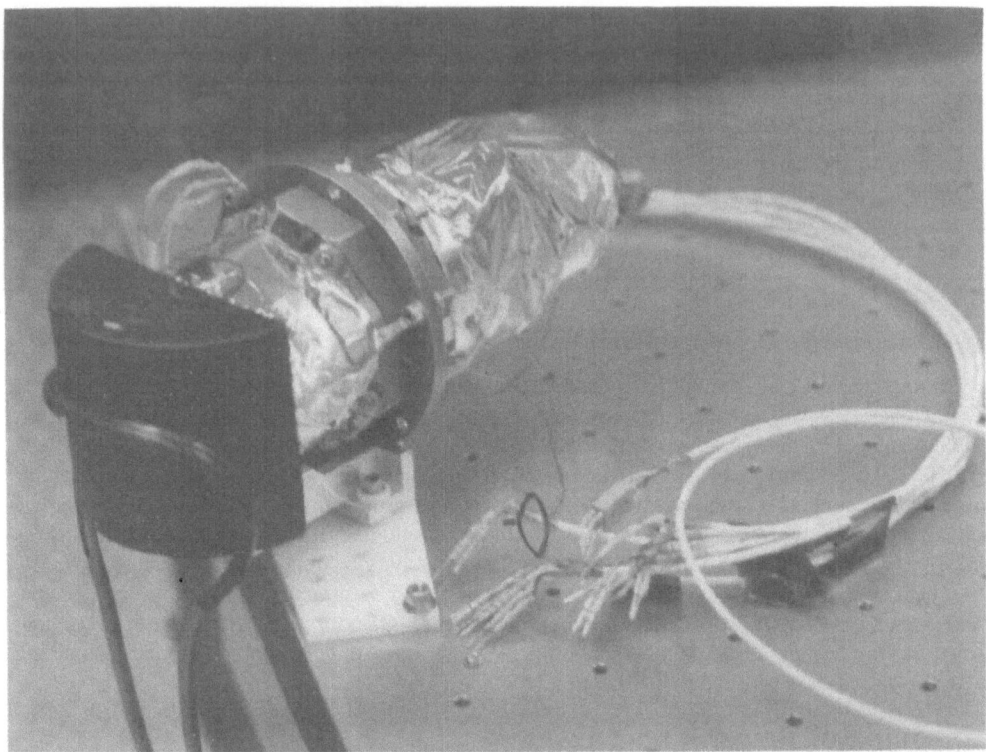

**Figure 10.** Assembled FiST system for the SABER FPA ready to be integrated into thermal vacuum chamber at SDL.

## REFERENCES

1.  Roach, Pat R., "Cryogenic attachment fixture with high strength and low thermal conduction", *Cryocoolers 7*, Plenum Press, New York (1993), pp. 349-354.

2.  Jensen, Scott, Ben Barrett, and Rick Hendrickson, *Cryogenic support fixtures.* Senior design review. Logan:Utah State University, (1993) Mechanical Engineering Department.

3.  Burt, William W., *SABER SRR, Volume I, 5 April 1995 to 6 April 1995.* Space Dynamics Laboratory Logan, (1995) Utah State University.

4.  Incropera F.P.and Witt D.P., *Fundamentals of Heat and Mass Transfer*: 3rd   edition. John Wiley and sons, New York, (1981).

# Cryocooler Heat Interceptor Test for the SMTS Program

**D. C. Gilman**

Hughes Aircraft Company
El Segundo, CA 90245

## ABSTRACT

The Space Missile Tracking System (SMTS) program has a sensor that is cooled by a cryocooler to 40K with cryocooler input power limited to 80 watts maximum. To enable the cryocooler to meet the input power requirement, a heat intercept strap was added to the cold finger of the cryocooler.

A heat intercept strap is a technique for improving the efficiency of a cryocooler by connecting a passive cryoradiator to the cryocooler cold finger.

This paper will discuss the testing that was done on a Hughes ISSC cryocooler with a heat intercept strap. Significant improvement in the thermal performance of the cryocooler with the heat intercept strap resulted in 40% reduction in cryocooler input power while maintaining the same refrigeration capacity or a 100% improvement in heat lift capacity while maintaining a constant cooler input power. This paper will also discuss the how to the cryocooler performance is affected by varying the heat intercept strap temperature from 200K to 140K and varying the location of the heat intercept attach point to the cold finger.

## INTRODUCTION

### Reason for the Test

The SMTS program's original design had a sensor that was cooled to 65K by a pair of Matra Marconi Space (MMS), formally British Aerospace (BAe), 80K cryocoolers. As part of risk reducing activities, the program installed two MMS 80K coolers into life test. To further reduce risk, the program also life tested two Hughes ISSC and two TRW mini stirling cryocoolers. As of April 1996, the MMS 80K coolers had accumulated 25 months each of life testing, Hughes ISSCs had 31 months each and TRW mini stirlings had 20 months each.

Cryocoolers 9, Edited by R.G. Ross, Jr.
Plenum Press, New York, 1997

Requirement changes dictated a change in sensor. The new sensor had to be cooled to 40K with a thermal load that is three times the original requirement. This increase in cooling load, and the programs desire to stay with cryocoolers that have been life tested on the program dictated that the Hughes ISSC cryocooler become the new baseline cooler. The ISSC had demonstrated the required capacity at 35K but with cooler power in excess of the requirements.

To reduce the input power to the cryocooler, the program decided to incorporate a heat intercept strap between an existing cryoradiator and the cold finger of the ISSC cryocooler. The goal of the heat intercept strap was to reduce the required input power by 20 to 30% while maintaining the refrigeration capacity[1]

## Test Plan

A test plan to demonstrate improvement to the ISSC cooler due to a heat intercept strap was developed. The goals of the test were: to determine the optimum location on the cold finger for the heat intercept strap, the optimum temperature to operate the heat intercept strap, the resultant heat load to the cryoradiator, and to measure the performance improvement to the cooler. To determine the optimum location for the heat intercept strap, the strap was attached to the thin wall section at three different positions. (See Figure 1 for the location of the strap on the thin wall section of the cold finger.) At each position, the temperature of the heat intercept strap was maintained at 140K, 170K and 200K ± 4K to evaluate the effect on the cooler performance. The cooler performance was evaluated at 60K, 35K and 30K for each strap location and temperature.

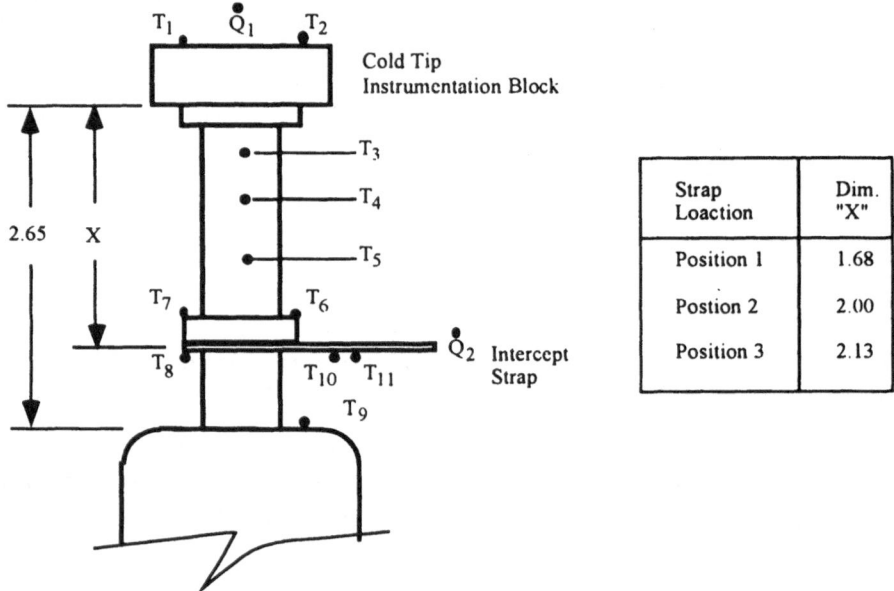

| Strap Loaction | Dim. "X" |
|---|---|
| Position 1 | 1.68 |
| Postion 2 | 2.00 |
| Position 3 | 2.13 |

**Figure 4.**    ISSC Cold Finger Instrumentation and Strap Locations

## Test Setup

The test setup consisted of a Hughes ISSC cooler, a 5 watt tactical rotary cooler to simulate a cryoradiator, a calibrated flexible copper strap for calorimetry measurements, an insulated vacuum dewar, cryocooler drive electronics, numerous temperature sensors and a data acquisition system. Figure 2 shows the ISSC cooler assembled into the heat intercept test setup.

**Figure 2.**     ISSC Heat Intercept Strap Setup

Inside the vacuum dewar were ten layers of MLI on the walls of the dewar, with five layers of MLI on the cold finger of the tactical cooler and on the flexible cooper strap, ten layer cup over the ISSC cold finger and thermal load. Figure 3 shows the major components of the heat intercept test, ISSC compressor, vacuum dewar with MLI and the ISSC expander, flexible thermal strap, tactical cooler subassembly.

Inside the vacuum dewar were fourteen temperature sensing diodes. Two temperature sensor were located on the ISSC cold tip, three temperature sensors were attached to the cold finger between the cold tip and the heat intercept strap to measure the thermal gradient along the cold finger, one temperature sensor mounted to the warm base of the cold finger, two temperature sensors located on the heat intercept collet with one of the sensors used as a feedback temperature sensor for the temperature control of the tactical cooler. Four temperature sensors were located on the flexible strap, two near the collet on the ISSC cold finger and two near the cold finger of the tactical cooler. These four temperature sensors were used to measure the delta temperature across the flexible strap to determine the heat flux through the strap. See Figure 1 for the approximate location of the temperature sensors on the ISSC cold finger. Finally, there was one temperature sensor at the cold tip of the tactical cooler.

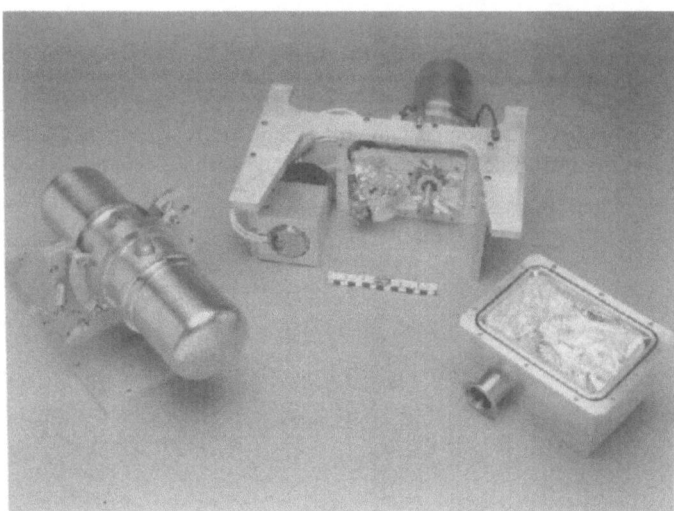

**Figure 3.**      Components of the ISSC Heat Intercept Test

Outside the vacuum dewar, the data acquisition system monitored the temperature of both the tactical and ISSC coolers expanders and compressor case temperature. The case temperature of the ISSC expander was maintained at 290 ± 4 K for all of the measurements by a recirculating coolant loop.

A multi-layer flexible copper strap was designed to attach to the thin wall section of the ISSC cold finger by means of a three jaw collet and the other end of the flexible copper strap was attached to the cold finger of the tactical cooler. At each attach point for the strap, Dow Corning thermal paste was inserted to improve the heat flow through the interface. A three jaw collet assembly was designed to attach to the ISSC thin wall section of the cold finger. The collet was designed to maintain a constant clamping force between the jaws and the thin wall section of the cold finger as the combined assemblies cycled from room temperature to cryogenic temperatures. The collet design also simplified repositioning the thermal strap on the cold finger during the test cycles. The strap was designed to have a 10K per watt thermal conductance. This thermal conductance was selected to provide adequate resolution of the heat flux from the ISSC cold finger through the strap to the tactical cooler. The thermal strap was calibrated prior to the start to testing.

**Thermal Strap Calibration**

The multi-layered flexible copper strap was calibrated by installing a copper slug with an internal heater into the collet jaws in place of the cryocooler cold finger. The thermal strap, collet assembly and calibration slug were then wrapped with multi-layer insulation to minimize the parasitic heat load to the strap and the tactical cooler was then turned on to maintain the copper slug at 170K. The delta T across the strap was measured while controlling the copper slug to 170K ± 2K and applying a known power to the internal heater. Thus, by measuring the delta T across the thermal strap for a known thermal load, we were able to calibrate the thermal strap for this experiment. Figure 4 shows the flexible copper strap attached to both the ISSC cold finger and the tactical cooler.

**Figure 4.**        Flexible Thermal Strap Attached to the ISSC and Tactical Coolers

**Performance of ISSC Cryocooler at Different Heat Intercept Strap Temperatures**

The performance of the ISSC cryocooler was measured at three different heat intercept strap temperatures, 200K, 170K and 140K. The temperature of the heat intercept strap was maintained at a constant temperature to within ±4K by a tactical cryocooler. The operating temperature of the strap was measured on one of the jaws in the three jaw collet that clamped to the thin wall section of the cold finger.

The performance of the ISSC cooler was evaluated by comparisons of constant cold tip temperature loadlines; i.e. varying the input power and measuring the thermal load at a constant cold tip temperature. The input power was varied from the minimum power required to more than 90 watts input power. The loadline data was then compared to a baseline constant temperature loadline.

The baseline loadlines were measured in the heat intercept test setup without the heat intercept strap attached to the ISSC cold finger. An attempt was made to perform a loadline with the intercept strap attached and the tactical cooler turned off. The effect was a direct thermal short of the cold finger to the ambient environment, thus increasing the parasitic load on the cold tip. This resulted in a cooler performance that was less than a cooler without a heat intercept.

The results showed that as the heat intercept strap operates at colder temperatures the performance of the cryocooler increases. The resulting cryoradiator heat load also increases for colder intercept strap temperatures. The most significant improvements in the cryocooler performance occur at lower cold tip thermal loads. The input power which can lift 0.25 watts at 35K without a heat intercept can cool a 0.5 watt load with a 200K intercept strap or a 0.85 watt load with a 140K intercept strap. For thermal loads of greater than 0.50 watts at 35K the thermal performance of the cryocooler improved by 30 to 100% based upon the temperature of the heat intercept strap. Significant improvements, 35 to 100%, in the thermal performance occurred for heat intercept strap operating between 200K and 170K as shown in Figure 5. Figure 6 shows the performance of the ISSC operating at 60K cold tip with the heat intercept strap at 200K, 170K and 140K.

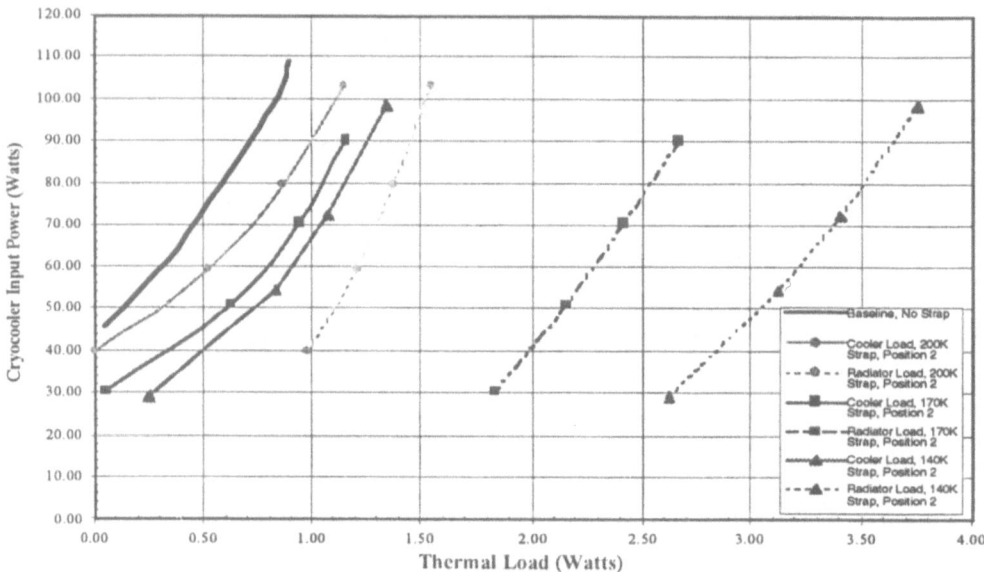

**Figure 5.**        35K Performance with Various Heat Intercept Strap Temperatures

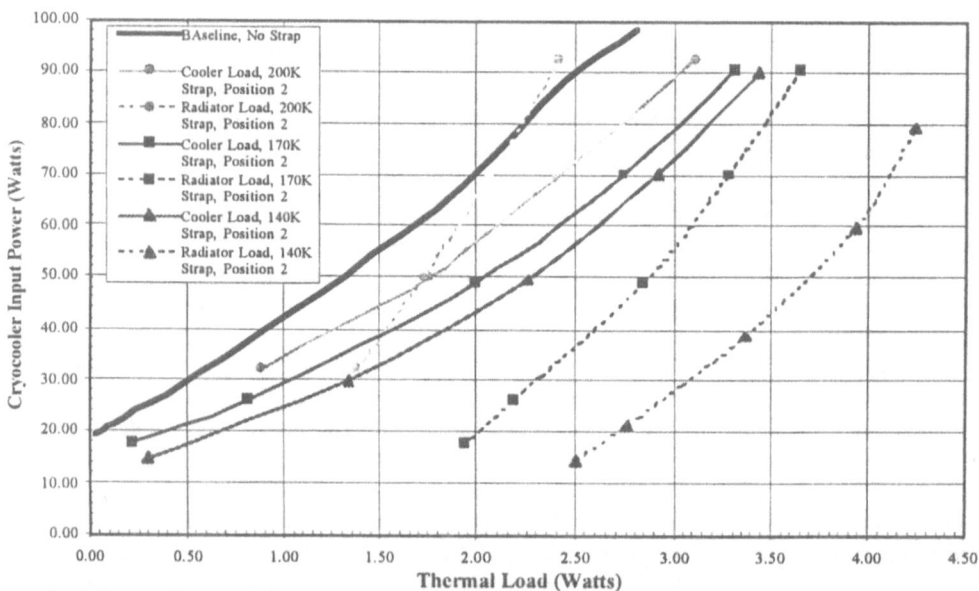

**Figure 6.**        60K Performance with Various Heat Intercept Strap Temperatures

The cryoradiator thermal load varied as a function of cryocooler thermal load or input power. The cryoradiator load varied from 1 watt to 1.5 watts for a 200K intercept strap, up to 2.5 to 3.5 watts for a 140K intercept. The cryoradiator load increased approximately 100% when the heat intercept strap was operated at 170K instead of 200K.

**Performance of ISSC Cryocooler at Different Heat Intercept Strap Locations**

The other major goal of this experiment was to evaluate the cryocooler performance as a function of the heat intercept strap location on the cold finger. The cryocooler was evaluated with the strap at three different positions on the cold finger as shown in Figure 1. At each location, constant temperature loadlines were performed for 30K, 35K and 60K cold tip temperatures. Also, at each location the operating temperature of the strap was held constant at 200K, 170K and 140K. The resulting loadlines were compared to baseline constant loadlines that did not have an intercept strap attached.

The performance of the cryocooler increased as the location of the heat strap moved toward the warm base of the cold finger. The improvement to the performance was about 100% in refrigeration capacity when compared to the baseline data. When comparing heat strap to heat strap performance data, the performance increased from strap position 1, closest to the cold tip, to position 3, closest to the warm end, the performance increased from 45% for thermal loads close to 0.25 watts to about 25% for thermal loads around 1.00 watts with a cold tip temperature of 35K. The corresponding cryoradiator loads had a significant increase. The radiator loads for the heat intercept strap attached at the warm end of the cold finger were 300 to 400% greater than the strap attached closer to the cold tip.

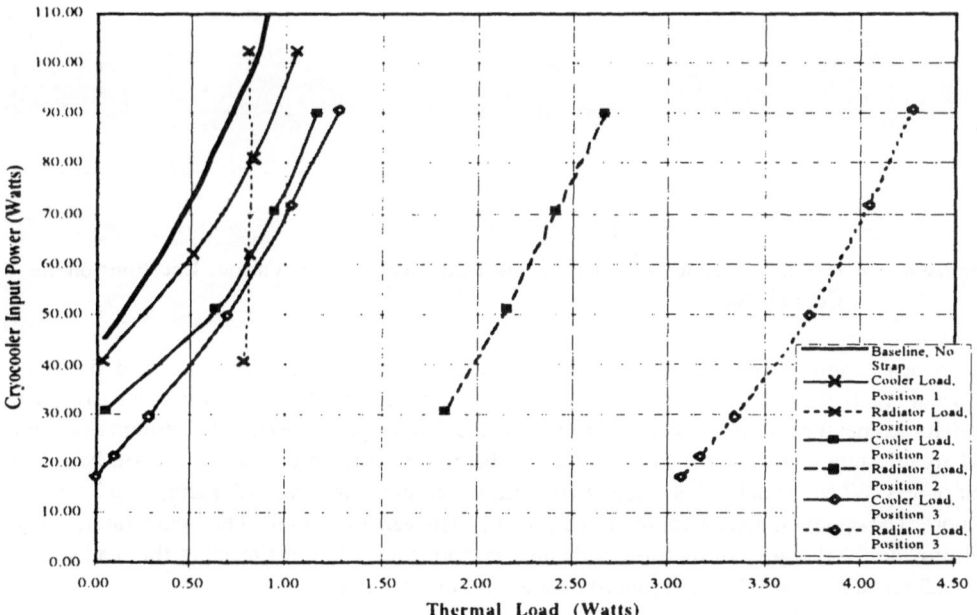

**Figure 7.**     35K Performance with a 170K Heat Intercept Strap at Various Locations on the Cold Finger

Results from the 200K intercept tests when the strap was attached at position 1, closest to the cold tip, showed that the heat flow in the strap was reversed. Under these test conditions, the heat intercept strap add approximately 0.25 watts parasitic load to the cold finger.

Figures 7 and 8 shows the changes in performance to the cryocooler and the change in the thermal load to the cryoradiator due to changing the attach point of the heat intercept strap to the cold finger. Figure 7 shows the results for the cooler operating at 35K and Figure 8 for a cooler at 60K.

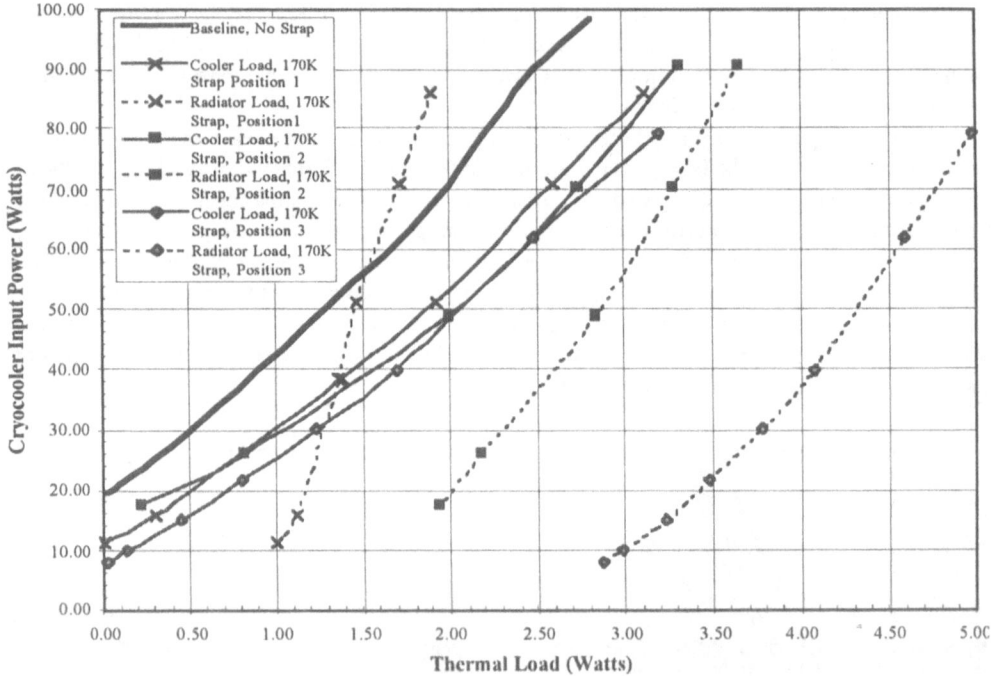

**Figure 8.**      60K Performance with a 170K Heat Intercept Strap at Various Locations on the Cold Finger

When reviewing the data as a function of temperature gradient along the cold finger as shown in Figure 9 and 10, it provides a clear view as to the effect of the heat intercept strap. As the delta temperature from a location on the baseline cold finger increases, the resulting radiator load significantly increases. When the delta T is below 25K the cryoradiator load will be at 1.0 watt or less. When the delta T is greater than 70K, the cryoradiator load will increase to 3.5 watts or more. The optimum delta T for the ISSC cooler is between 35 to 60K. This temperature range will provide for the greatest increase in thermal performance while maintaining the load to the cryoradiator at manageable levels, approximately 1.5 to 2.5 watts.

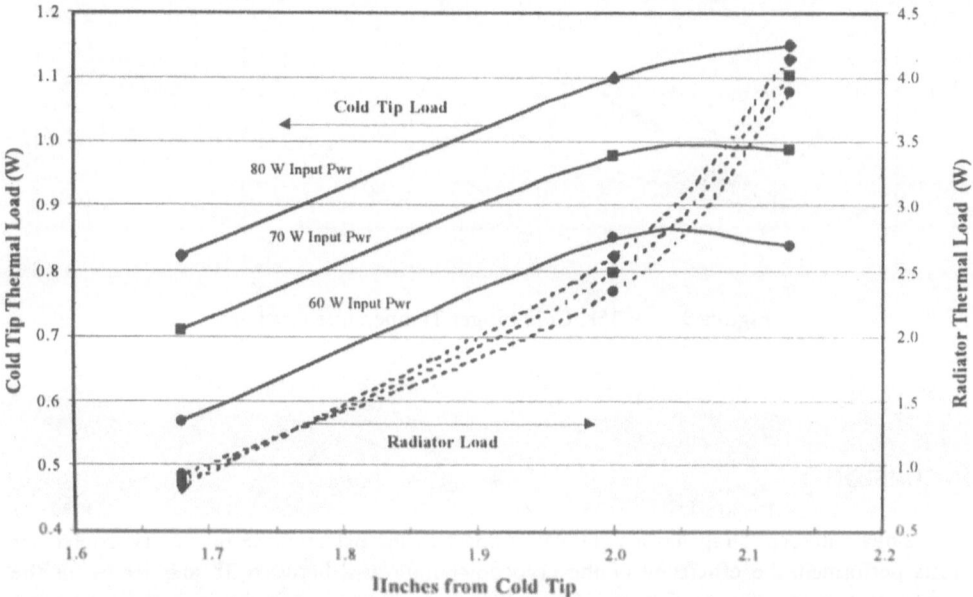

**Figure 10.**  35K Cyrocooler Performance with 170K Heat Intercept Strap and Radiator Loads as a Function of Distance from Cold Tip and Cryocooler Input Power

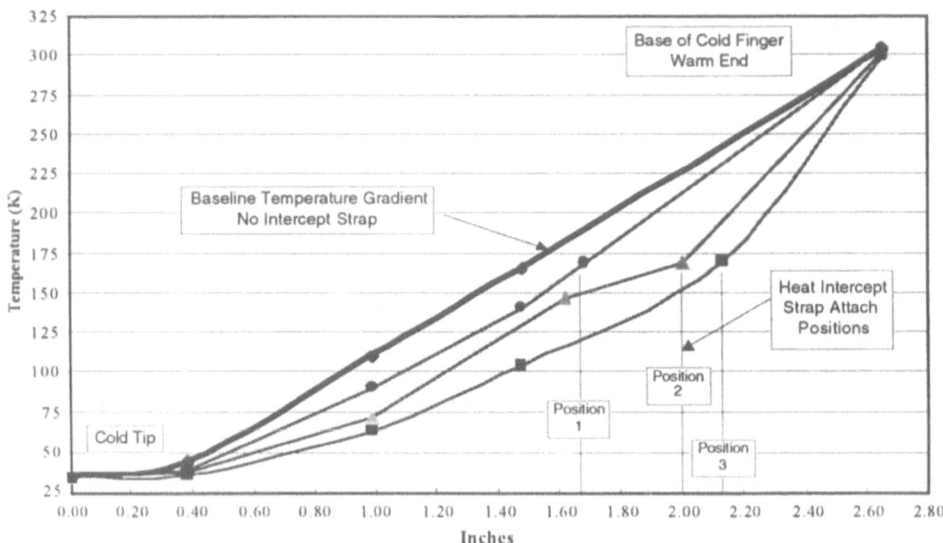

**Figure 9.**      35K Cold Finger Temperature Gradient

## CONCLUSION

The heat intercept strap has a significant impact to the performance of the cryocooler. Of the tests performed, the efficiency of the cryocooler improved between 38 and 100% for the same refrigeration capacity or a 30 to 35% reduction in power required to maintain the same capacity at 60K. At 30 and 35K cold tip temperatures, the cooler efficiency increased by 50 to grater than 150% for the same refrigeration capacity or a 35 to 50% reduction in power required to maintain the same capacity.

The data showed that if the strap was attached too close to the warm end of the cold finger, the resulting cooler performance improvement is approximately 15% in thermal capacity, but the cryoradiator heat load increases by 50 to 100%. Conversely, if the strap is attached too far from the warm end of the cold finger, the performance of the cryocooler will be start to approach the performance of the cooler without a heat strap and the resulting cryoradiator heat load will approach zero. If the operating temperature of the heat strap is too warm, the heat flow through the strap will be reversed and the heat strap will actually be adding to the parasitic heat load to the cold finger, thus reducing the efficiency of the cryocooler.

The addition of the heat intercept strap to the SMTS cryocooler design enabled the cryocooler to meet or exceed all the system requirements for the SMTS program in terms of thermal performance and input power. The heat intercept strap reduced the amount of input power to the cryocooler for a 1.0 watt thermal load at 35K from greater than 120 watts input power to less than 80 watts, 30% reduction in power. The heat intercept strap increased the thermal load to the cryoradiator by 2.5 watts. Maintaining the input power to the cryocooler at or below 80 watts, provides an additional benefit by having a significant reduction in the spacecraft mass compared to the mass required to support the 120 watts input power.

## ACKNOWLEDGMENTS

The work described in this paper was performed by Hughes Aircraft Company for the USAF SMC. The author would like to thank Lt. Charles Light, USAF SMC, Drs. Ron Ross and Dean Johnson, NASA JPL, and Drs. David Curran and Vini Mahajan, Aerospace Corporation for their support and overview. The author would like to acknowledge Mike Barr and Ken Price for the experiment design and Rich Kimpel and Tom Pollack for the many hours of data collection required.

## REFERENCES

1.  Johnson, D. L., Ross, R. G., Jr., "Cryocooler Coldfinger Heat Interceptor," *Cryocoolers 8*, Plenum Publishing Corp., New York (1995), pp 709-718.

# Feasibility Demonstration of a Thermal Switch for Dual Temperature IR Focal Plane Cooling

D. L. Johnson and J. J. Wu

Jet Propulsion Laboratory
California Institute of Technology
Pasadena, California 91009

## ABSTRACT

A conical-shaped hydrogen gas-gap thermal switch using a metal hydride sorption bed to control the hydrogen gas supply was designed to provide the thermal isolation link between two cooling systems operating at different temperatures. Test results of the thermal switch show a nominal 1 K/W thermal resistance in the on-state conduction mode for temperatures above 100 K and for heat flows to 8 W, and a temperature dependent 600-900 K/W thermal resistance in the off-state conduction mode. In particular, the off-state resistance of the switch for operation between 60 K and 140 K was 660 K/W, and with a 0.92 K/W on-state resistance at 140 K, the off-state/on-state switching ratio of 717 was achieved. The switching time to convert to the on-state or off-state mode was on the order of 5 minutes and 12 minutes, respectively.

The thermal switch was integrated with a Matra Marconi Space Systems cryocooler operating at 60 K, a G-M cryocooler simulating a 140-K cryoradiator, and a simulated focal plane heat source to demonstrate the thermal isolation capability of the switch. The focal plane heat source was directly coupled to the MMS cryocooler and indirectly coupled to the G-M cryocooler through the thermal switch. Heat flows to either cooling system were quantified in the two cases of the operating and non-operating cryocooler; as well, the cooling performance of the cryocooler in the integrated configuration was determined. The test results show the successful operation of the thermal switch as the isolation link between the cryocooler and the passive cryoradiator.

## INTRODUCTION

In applications where redundant cryocoolers are required to enhance reliability, the non-operating standby cooler presents an added parasitic load to the operating cooler. The non-operating cooler needs to be thermally isolated from the focal plane array, while the active cooler is thermally connected to the system. For spacecraft using intermediate temperature cryogenic radiators to provide thermal (radiative) shielding or cooling to the optical bench, the cryogenic radiator may also serve as a passive cooling option for a backup focal plane when the primary coolers are not operating. Dual IR focal plane arrays (FPAs) are considered in this payload configuration, with the primary FPA being cooled by 60-K mechanical cryocoolers. The co-located second FPA would be passively cooled to 145 K by a cryoradiator to ensure continuous, albeit different, infrared sensing capabilities when the mechanical cryocoolers are not operating. The capability to thermally isolate the two cooling systems and be able to rapidly switch between the two cooling systems can be accomplished with gas-gap thermal switches.

This paper describes the design and development of a hydrogen sorption gas-gap thermal switch needed to thermally isolate two cooling systems having different operating temperatures, and details the feasilbility demonstration of an integrated cryogenic system that utilizes a cryocooler to provide 60-K cooling to a primary focal plane arrary, while using the 140-K passive cooling of the cryogenic radiator for a backup FPA[1]. The 140-K cryogenic radiator on the satellite provides the opportunity to utilize the cryocooler coldfinger heat interceptor to enhance the 60-K cooling performance of the cooler. The success of this integrated cryogenic system depends on the on/off switching ratios of the thermal switch located between the radiator and the FPA to provide continuous detector/focal plane surveillence sensing when going to the backup focal plane cooled with the passive cryogenic radiator.

## THERMAL SWITCH DESIGN REQUIREMENTS

One application for the thermal switch is on a candidate Space Missile Tracking System (SMTS) payload configuration[2] for the Flight Demonstration System (FDS) as shown in Fig. 1. The configuration uses a pair of Matra Marconi Space (MMS) coolers to provide 65-K cooling to a primary focal plane. A co-located focal plane, designed to operate at 145 K, is cooled by a 140-K precooling cryoradiator. In the proposed configuration, heat pipes connect the cryoradiator to the cryocooler coldfinger heat interceptor[3], the thermal switch, and to the aft optics. When the primary coolers are in the non-operating mode, the thermal switch is turned on, cooling the focal plane to 145 K via the cryoradiator. In the off-state conduction mode, the thermal switch passes only a small parasitic load from the 140-K cryoradiator to the 60-K operating coolers.

The hydrogen gas-gap thermal switch design was selected for this application because it provided the most flexibility with respect to the range of operating temperatures and heat transfer capability required. The switch needed to be operated between the nominal 60-K and 140-K temperature extremes, but also be able to extend down to 35 K. The switch also needed to demonstrate heat transfer capabilities as high as 0.7 watts, and a heat transfer goal of 8 watts capability was desirous for other applications. Off-state/on-state switching ratios of 500 were required to assure adequate isolation between the two cooling systems.

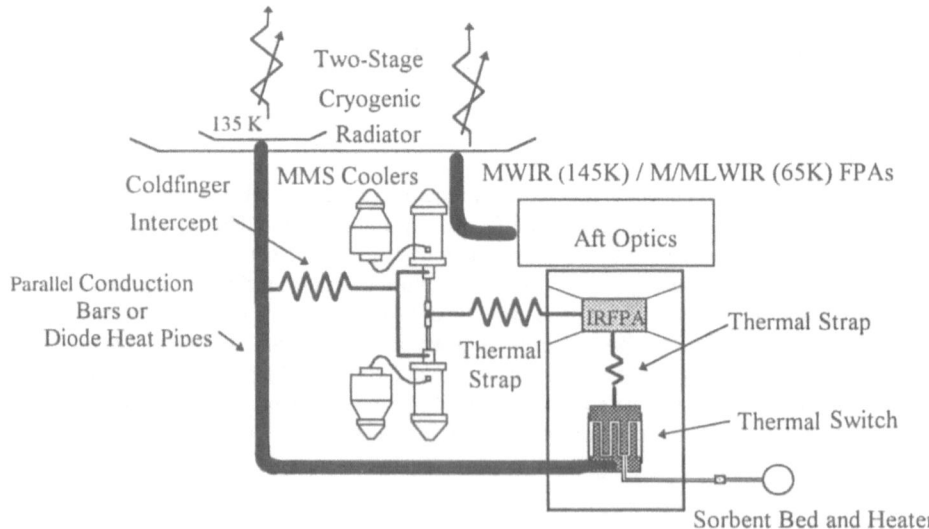

**Figure 1.** Candidate payload configuration schematic for the Flight Demonstration System (FDS).[1]

## THERMAL SWITCH DESIGN

### Thermal Switch Description

The thermal switch design, shown in Fig. 2, consists of two nested copper cones precisely separated by a narrow gap using six high-strength low-conduction G-10 pins. A retaining spring with a G-10 connecting rod is used to apply an axial force to assure the stability of the gap width. The conical surfaces provide a large (10 cm$^2$) interface area, and are gold plated to reduce the radiation heat transfer between these two surfaces to negligible levels. The angle of the mating surfaces helps reduce the sensitivity of the positioning of parts during assembly, since a slight axial misalignment results in a much smaller change in gap separation between the two pole pieces.

The outer shell that encloses the gas gap is a folded combination of thin-wall stainless tubing and flexible, thin-walled stainless bellows. For hydrogen applications, 300-series stainless steel is the preferred material even though there are materials with lower thermal conduction.

Physically, the thermal switch weighs 295 grams including the sorption pump, and has the dimensions: 3.8 cm diameter by 7.6 cm length with an external surface area of 91 cm$^2$. To facilitate the early thermal performance tests the switch length and weight are more than twice that needed in an optimized flight unit.

### Hydrogen Sorbent Bed

Hydrogen is a highly conductive gas. Its low triple point temperature, 13.9 K at 7.20 kPa, is ideal for the wide temperature range of operation required for this thermal switch application. Zirconium-nickel hydride (ZrNiHx) has absorption/desorption characteristics ideal for the thermal switch application. The hydride absorbs quite well at 20°C, a temperature typical of spacecraft radiators, and the hydride needs to be heated to about 200°C to sufficiently desorb the hydrogen. ZrNiHx sorbent beds have been successfully demonstrated elsewhere[4,5,6].

For this switch a low power hydride pump containing approximately 0.12 g of ZrNi, was designed, having a reversible hydrogen capacity of more than 10 standard cc. Most of the characterization tests were conducted using a hydrogen fill of 14.1 std. cc. The heating element of the pump is a thin-film platinum resistance thermometer (PRT), serving both as heater and thermometer. The PRT heating element is separated from the hydride pump housing by a 1-mm thick ceramic substrate and has a power consumption on the order of 300 mW. The gold-plated hydride pump weighs only 0.5 g and is mounted on a small-diameter thin-walled tube used to transport heat from the hydride pump to a heat sink around 280 K. Thus cooling of the hydride

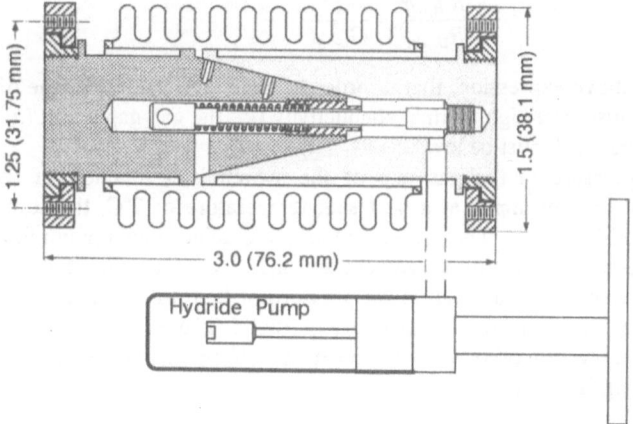

**Figure 2.** Cross sectional view of the thermal switch design.

pump to turn the thermal switch off is primarily conduction through this standoff. Typical times to cool from 230°C to <20°C were under 15 minutes. Application of the 300-mW heat load liberates the hydrogen from the hydride, filling the gas-gap of the switch. Since the switch becomes conductive at low pressures (<1 kPa), the time to turn the switch on is about 5 minutes.

**Overall Baseline Thermal Resistance**

**Off/on thermal resistance ratio of the gas gap.** Normally, the thermal switch is in an "on" conduction state when the gas pressure in the gap is sufficiently high so that the molecular mean free path is small compared to the gap width (the continuum flow regime). In the continuum flow regime, the conduction heat flux across two parallel surfaces of area A, separated by a small gap $d$ is given by

$$q = k\frac{A}{d}(T_2 - T_1) \tag{1}$$

In this case, the thermal resistance can be expressed as

$$q = \frac{T_2 - T_1}{R_{th,c}} \tag{2}$$

or

$$R_{th,c} = \frac{d}{kA} \tag{3}$$

When the gas filling the gap is evacuated to a pressure low enough that the mean free path of the gas is on the same order of magnitude as the gap width (the free molecular flow regime), the thermal switch is considered to be in the "off" conduction state. In this case, the heat flux across the gap can be expressed as

$$q = GpA(T_2 - T_1) \tag{4}$$

where p is the gas pressure measured with an ambient temperature gage, and G is a mass flow velocity term dependent on the specific gas-surface interaction and on the surface temperature[7]. In other words, the conduction thermal resistance for a gas gap when the gas is in free molecular flow regime is

$$R_{th,fm} = \frac{1}{GpA} \tag{5}$$

The off/on thermal resistance ratio can therefore be given as

$$\frac{R_{th,off}}{R_{th,on}} = \frac{k}{Gpd} \tag{6}$$

It is clear, from the above expression, that in order to have high values for the off/on resistance ratio, the filling gas must have high thermal conductivity ($k$); the gas-gap width ($d$) must be small; and the vacuum pressure ($p$) must be low.

The hydrogen pressure in the gas gap of the switch is a function of the hydride bed temperature (Fig. 3). For example, at a heat sink temperature of 7°C, the sorbent bed has an absorption pressure of 0.0685 Pa. The pressure required to achieve continuum flow for hydrogen is 13.2 kPa, which requires a desorption temperature of 233°C. However, full conduction within the switch has been achieved at a much lower pressure of 2.26 kPa. Therefore, the gas gap is probably fully conductive at a hydride temperature of 186°C. If the thermal switch is operating at 140 K with a gas-gap width of 0.038 mm, the ideal off/on resistance ratio of the gas gap, calculated from Eq. (6), is about 33,000.

**On-state thermal resistance.** There are two thermal conductions in series when the switch is in the on state. The first one is the gas gap, its thermal resistance can be calculated using the

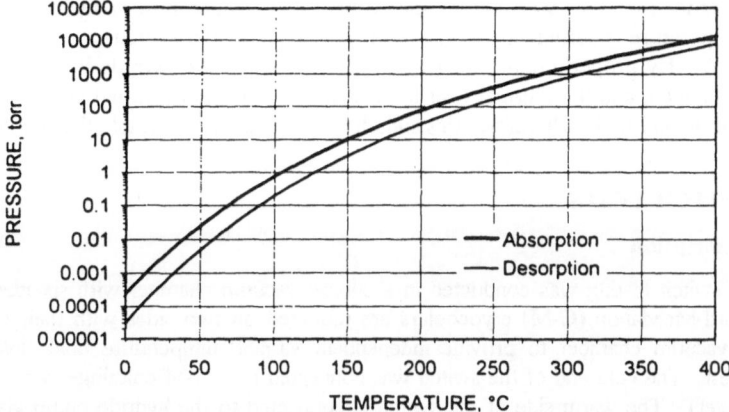

**Figure 3.** Hydrogen gas pressure in ZrNiHx sorbent bed as a function of bed temperature.

equation discussed earlier. For hydrogen gas at 140 K, $k = 0.09613$ W/m·K, the on-state resistance for the gas gap is

$$R_{th,gap} = \frac{d}{kA} = 0.376 \ K/W \tag{7}$$

The second thermal conductance is the two OFHC copper cones, their thermal resistance can be calculated using an approximate shape of 7.62 cm long tube with 2.03 cm OD and 0.635 cm ID. Since the thermal conductivity of OFHC copper is 417 W/m·K at 140K, the thermal resistance is

$$R_{th,Cu} = 0.624 \ K/W \tag{8}$$

Therefore, at 140 K, the thermal switch has an overall on-state resistance of

$$R_{th,on} = 1.0 \ K/W \tag{9}$$

The on-state resistance varies linearly with length for both the small gap separations and for the copper, therefore the desired on-state resistance can be selected accordingly.

**Off-state thermal resistance.** Neglecting the effects of the radiation heat transfer in the gap, there are four parallel conductive paths when the switch is in the "off" mode. Normally, the switch is "off" when the warm side is at 140 K and the cold side is at 60 K. The thermal conductivity of materials can be evaluated at the mean temperature of 100 K. Furthermore, the hydride bed is assumed to be at 15°C, so that the vacuum pressure in the gap is 0.156 Pa (1.2 milli Torr). The estimated thermal resistance of the four paths are[1]:

| | | | |
|---|---|---|---|
| (1) | $H_2$ Gap | 4331 | K/W |
| (2) | Formed Bellows (14 Convolutions) | 1996 | K/W |
| (3) | Six G-10 Gap Separating Columns | 4641 | K/W |
| | (0.76 mm Dia. x 3.18 mm long) | | |
| (4) | G-10 Rod - Retaining Mechanism | 28979 | K/W |
| | (2.36 mm Dia. x 31.75 mm long) | | |

Therefore, the nominal "off" resistance is

$$\frac{1}{R_{th,off}} = \frac{1}{R_{th,1}} + \frac{1}{R_{th,2}} + \frac{1}{R_{th,3}} + \frac{1}{R_{th,4}} \tag{10}$$

or

$$R_{th,off} = 1018 \ K/W \tag{11}$$

The gas gap resistance increases with decreasing sorbent bed temperature or hydrogen pressure. If the hydride temperature is sufficiently low, the off resistance of the switch can reach a limiting value of 1332 K/W. By properly controlling the heater power to the hydride bed, the hydrogen pressure can be changed, and the thermal resistance of the switch can be varied between "on" and "off" values as shown in Fig 4. Therefore, this switch can be used as a variable thermal resistor.

## THERMAL SWITCH TESTS

### Test Facility Description

The thermal switch testing was conducted in a cubical vacuum chamber with six identical side ports. Two Gifford-McMahon (G-M) cryocoolers are mounted on two sides with their coldfingers enclosed in the vacuum chamber to provide independent variable-temperature cold sinks for the device(s) under test. The cold end of the switch was connected to a G-M coldfinger via a heat flow transducer (Q-meter). The warm side of the switch is connected to the hydride pump via a 53-cm length of 3.175-mm diameter stainless tubing having a wall thickness of 0.305 mm. A heater block attached at the warm end of the switch was used to vary the switch temperature. The vacuum chamber wall provided the heat sink for the hydride pump. The chamber wall temperature was varied between 12°C and 30°C using a small circulating refrigerator to control the temperature. Wall temperatures were kept above 12°C to prevent condensation on the outside of the vacuum chamber. The experimental setup for the thermal switch characterization tests is shown in Fig. 5.

Measurements of the switch performance were made taking the temperature difference across the switch as a function of applied heat load to the switch. Multiple heat loads were applied for each cold-end switch temperature to produce a straight line fit through the data. The slope of the line indicates the thermal resistance of the switch. The Q-meter placed between the G-M cooler and the switch provided a check on the measured heat flows. At zero-applied heat load, the Q-meter gives a measure of the parasitic loads on the thermal switch.

### On-State Resistance Measurements

The on-state thermal resistance measurements were made at applied heat loads ranging from 0 W to 2 W for cold-end switch temperatures of 35 K, 60 K, 110 K, 140 K, and 160 K. An 8-watt heat load was applied at the 140 K temperature to show that no change in thermal resistance occurred for higher applied loads. The net thermal resistance results for the different switch temperatures are tabulated in Table 1.

The switch thermal resistance was also measured as a function of measured gas pressure for cold-end temperatures of 60 K and 140 K. This is done by varying the heat applied to the hydride pump (Fig. 6). It can be noted that the thermal resistance quickly falls to its on-state levels for pressures as

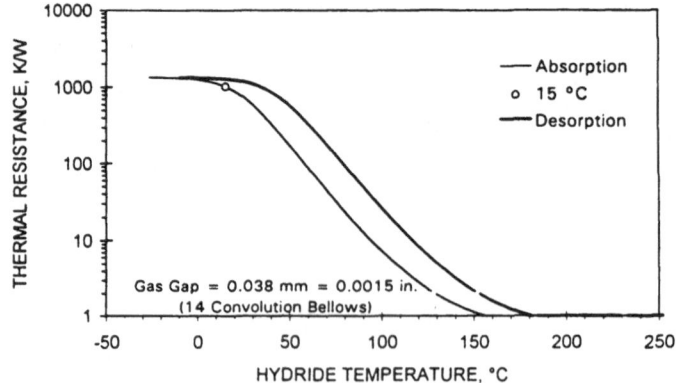

**Figure 4.** Thermal resistance of the thermal switch as a function of hydride pump temperature.

**Figure 5.** Experimental setup to characterize thermal switch.

low as 0.05 psia (0.34 kPa), and quickly reaches the minimum resistance for pressures around 0.3 psia (2.1 kPa). This implies that either the hydride need not be heated to as high a temperature as expected, or that the switch could operate with a smaller hydrogen charge.

## Off-State Resistance Measurements

The off-state resistance condition is attained by turning off the hydride pump heater and allowing the hydride bed temperature to cool to the ambient heat sink temperature. The low pressure attainable is dependent on the ultimate hydride bed temperature. See Fig. 3.

Measurement results for various combinations of warm-end and cold-end temperatures are shown in Table 1. The thermal resistances determined using the Q-meter values for the measured heat flows. Note that the thermal resistance for the switch is generally in the 600 to 900 K/W range. Note also that with the ambient sink temperature of 285 K there was sufficient residual hydrogen gas available in the gas gap to cause the thermal resistance to drop by 25 % to 30 % when compared to the evacuated condition. A comparison of the demonstrated thermal switch characteristics to the design requirements are shown in Table 2.

**Table 1.** Thermal Switch Characterization Results.

| Condition | Cold Side Switch Temperature | Warm Side Switch Temperature | Thermal Resistance |
|---|---|---|---|
| On-State | 35 K | | 1.49 K/W |
| | 60 K | | 1.17 K/W |
| | 110 K | | 1.04 K/W |
| | 141 K | | 0.92 K/W |
| | 159 K | | 0.92 K/W |
| Off-State (Charged) | 35 K | 140 K | 640 K/W |
| | 60 K | 140 K | 660 K/W |
| | 140 K | 250 K | 400 K/W |
| Off-State (Evacuated) | 60 K | 170 K | 850 K/W |
| | 140 K | 250 K | 600 K/W |

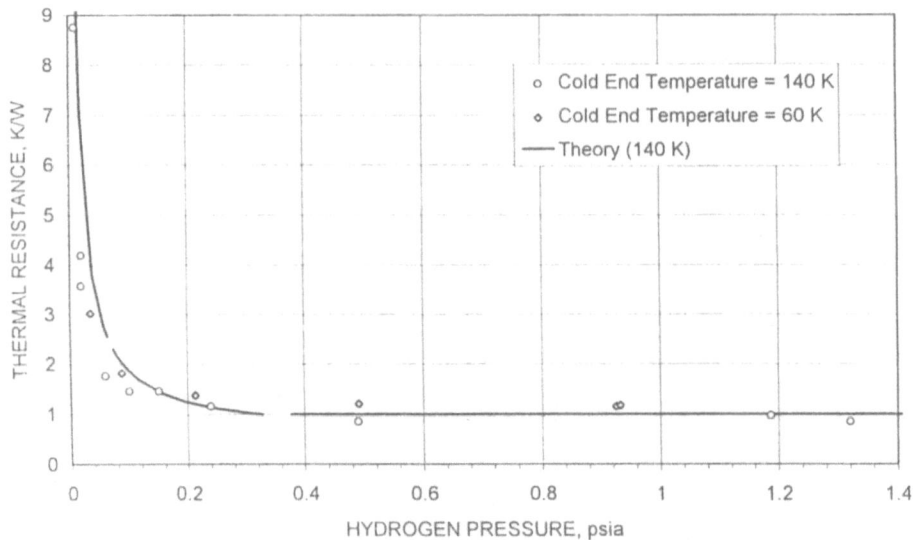

**Figure 6.** Thermal resistance as a function of hydrogen gas pressure with the gas gap.

**Table 2.** Thermal Switch Demonstrated Performance Compared To Design Requirements.

| Requirements | | Specification | Demonstrated |
|---|---|---|---|
| Temperatures | Nominal Hot Side | 145 K | 140K |
| | Nominal Cold Side | 60 K | 60K |
| | Cold Side Range | 35-60 K | 35-60K |
| | Alternate Hot/Cold | 250 K/170 K | 250K/170K |
| | Hydride Pump Heat Sink | >280 K | 288K-301K |
| Heat Loads | Nominal | 0.3 W | 0-8.0 W |
| | Range Requirement | 0.3-0.7 W | |
| | Range Goal | 0.3-8.0 W | |
| Thermal Resistance | Off Requirement | >1000 K/W | 650 K/W (60 K/140 K) |
| | Off Goal | >2000 K/W | |
| | On Goal | <2 K/W | 0.92 K/W (140 K) |
| Off/On Resistance Ratio | | > 500 | > 700 |
| Switching Time | Heating Requirement | <10 minutes | 5 minutes |
| | Heating Goal | Minimize | |
| | Cooling Requirement | <30 minutes | |
| | Cooling Goal | <10 minutes | <12 minutes |
| Switching cycles | | >100 | >124 |
| Life Goal by Analysis | | >10 years | N/A |
| External Surface Area | | <100 cm$^2$ | 91 cm$^2$ |
| Weight | | < 1 kg | 295 gm |

## Thermal Cycling Life Test

The thermal switch underwent a thermal cycling life test to check for metal fatigue and for detrimental levels of hydride material breakdown which may either limit performance or hydride pump life. The switch was thermally cycled 124 times between 60 K and 140 K to simulate operation in space. The switch was cycled every hour by alternately heating and cooling the hydride pump every 30 minutes. A constant heat load was applied to the warm end of the switch throughout the life test, which drove the temperature of the warm end of the switch to 140 K during the off-state condition. No degradation in performance was observed over the course of the life test.

## INTEGRATED THERMAL SWITCH/COOLER FEASIBILITY DEMONSTRATION

The thermal switch was integrated with the MMS cryocooler coldfinger (visible at the bottom of Fig. 5) to demonstrate the feasibility of isolating the two cooling systems to alternately provide 60-K and 140-K cooling to a simulated focal plane. The thermal switch is tied directly to the copper conduction bar coupled to the G-M cooler simulating the cryoradiator. Beneath the switch is mounted, in order: a Q-meter (Q-#1), the FPA heater block, a second Q-meter (Q-#2), and the flexible braid that attaches to the coldfinger of the MMS cooler. The Q-meters were placed on either side of the FPA heater block so that the distribution of the FPA heat load could easily be determined. The MMS cryocooler coldfinger incorporated the heat interceptor to provide additional performance to the cryocooler[3]. The heat interceptor was thermally tied to a second G-M cooler to provide an independent cold sink A third Q-meter (Q-#3) located on the heat interceptor measured the heat flowing from the MMS coldfinger to the G-M cooler.

## Off-State Thermal Switch Operation

The cooldown of the integrated system required several hours due to the mass of the Q-meters, flex braids, and thermal switch. The MMS cooler and the two G-M coolers were operated to provide the maximum cooling rate to the system. The MMS cooler was driven at near full stroke and with a 140-K heat intercept temperature at the MMS coldfinger. At equilibrium, with the warm side of the thermal switch at 140 K, the FPA reached a minimum temperature of 48 K. With the application of a 565-mW FPA heat load, the FPA temperature increased to 66 K.

Measured heat flows on either side of the focal plane heater were in the direction of the MMS coldfinger; these heat flow values are shown in Table 3. Q-#1 measured the heat flow out of the thermal switch; Q-#2 measured the sum of the thermal switch heat flow and the applied FPA heat load. The uncertainty in the Q-meter readings are approximately +/-20 mW. Using the Q-#1 values from Table 3, the thermal resistance of the switch measured 279 K/W and 259 K/W for the two data points. These thermal resistance values are about a factor of two low compared to the measured thermal resistances when testing the switch by itself. This change in thermal resistance was determined to be from a mechanical contact problem created within the switch during cooldown. There was sufficient differential thermal contractions within the experimental setup to create a side load force on one end of the thermal switch. This side load produced a bending moment about the G-10 standoff pins which, if the G-10 pins were not all the same length, could have caused the two switch halves to make contact.

**Table 3.** Off-state conduction mode for thermal switch.

| FPA heat load/temp | MMS Cooler input power | Thermal switch heat flow (Q-#1) | Coldfinger heat flow (Q-#2) | Interceptor heat flow (Q-#3) | Switch Thermal Resistance |
|---|---|---|---|---|---|
| 0W/48 K | 30 W | 0.330 W | 0.380 W | 2.50 W | 279 K/W |
| 565 mW/66 K | 26.5 W | 0.282 W | 0.905 W | 2.58 W | 259 K/W |

**Table 4.** On-State Thermal Switch Feasibility Test Results

| Heat Sink Temp. | Cryoradiator Temp. | FPA Load/Temp | Q-Meter Heat Flows | | | Switch Temperature Gradient | Switch Thermal Resistance |
|---|---|---|---|---|---|---|---|
| | | | Switch Q-#1 | Coldtip Q-#2 | Interceptor Q-#3 | | |
| 15°C | 110 K | 0W/111.3K | 0.119W | 0.110W | 0.97W | 0.31K | .967K/W |
| | | .318W/113.1K | 0.436W | 0.103W | 0.97W | 0.60K | |
| | | .765W/115.5K | 0.885W | 0.119W | 0.97W | 1.04K | |
| | | 1.56W/119.7K | 1.680W | 0.119W | 0.98W | 1.82K | |
| 15°C | 140 K | 0W/141.1K | 0.110W | 0.092W | 0.83W | 0.13K | .907K/W |
| | | .317W/142.8K | 0.430W | 0.101W | 0.82W | 0.39K | |
| | | .763W/145.0K | 0.884W | 0.099W | 0.82W | 0.78K | |
| | | 1.56W/149.0K | 1.675W | 0.108W | 0.82W | 1.55K | |
| 31°C | 140 K | 0W/141.2K | 0.120W | 0.102W | 0.92W | 0.13K | .887K/W |
| | | .316W/142.8K | 0.448W | 0.102W | 0.92W | 0.448K | |
| | | .763W/145.1K | 0.898W | 0.102W | 0.93W | 0.898K | |

## On-State Thermal Switch Operation

The on-state conduction mode represents the condition where the cryocooler is non-operational and thus the focal plane is cooled by the 140-K passive cryogenic radiator. The cryoradiator temperature applies to both the upper end of the thermal switch as well as the heat intercept temperature at the coldfinger flange of the non-operating MMS cooler, and are cooled separately by the two G-M coolers.

Data sets for various FPA heat loads at three different ambient heat sink and cryoradiator temperature combinations are shown in Table 4. The small thermal switch temperature gradient is the difference of the diode temperature readings across the switch. The thermal resistance of the switch is determined from the slope of the line formed when plotting the temperature difference across the switch as a function of the measured heat load from Q-meter #1 for each cryoradiator temperature. Heat flow is directed from the MMS coldfinger through the switch to the G-M cooler.

## SUMMARY

A hydrogen gas gap thermal switch was designed and demonstrated that would meet the Space Missile Tracking Systems requirement for size (both weight and surface area) and thermal performance. The thermal switch utilized a symmetrical, conical geometry for the gas gap to eliminate differential thermal contraction within the switch which could cause mechanical contact. Built for convenience of handling during laboratory testing, the present 295-g switch design is roughly 50% longer and more massive than anticipated for a flight design.

The switch design was capable of handling a large heat flux with minimal temperature gradient over a large temperature range. The on-state resistance of the switch measured a nominal 1K/W over the temperature range of 110 K to 170 K, and demonstrated the capability of conducting heat flows of at least 8 watts. The switch was also capable of operating in the on-state condition to temperatures as low as 35 K, where its thermal resistance had increased to only 1.49 K/W.

The off-state resistance of the switch was limited primarily by the thermal conductance through the stainless bellows and secondly by the conduction through the residual gas in the gap. The off-state thermal resistance measured 660 K/W over the temperature span of 60 K to 140 K, and measured 640 K/W over the extended temperature span of 35 K to 140 K. These resistances could be increased by 25 % if the hydride pump were cooled to lower temperatures, however the results would still be low compared to the design requirement of >1000 K/W. Even so, the switch did achieve an off-state/on-state resistance ratio of >700. The thermal switch also demonstrated the capability of switching to the on- state and off-state conduction modes in less than five minutes and 12 minutes, respectively.

The feasibility demonstration integrated the thermal switch with the two distinct cooling systems, the

MMS cryocooler and the 140-K simulated cryoradiator, to show that the cooling systems can be thermally isolated. In the on-state conductance mode, the thermal switch provided high thermal conductance to keep the FPA heater block near the cryoradiator temperature of 140 K. In the off-state conduction mode, the thermal switch provided sufficient thermal resistance to let the focal plane to be cooled by the MMS cryocooler to temperatures as low as 48 K, and be able to maintain a 66-K focal plane temperature when the heater block is producing 565 mW of heat.

## ACKNOWLEDGEMENTS

The work described in this paper was carried out by the Jet Propulsion Laboratory, California Institute of Technology, for the Air Force Phillips Laboratory, Albuquerque, New Mexico and the Los Angeles Air Force Base, Space and Missiles Systems Center. The work was sponsored by the Space Missiles Tracking Systems Organization under a contract with the National Aeronautics and Space Administration.

The successful demonstration of this feasibility demonstration of the thermal switch is due to the important contributions from M. Schmelzel, and T. Conners for fabrication, assembly, and vacuum bakeout/fill of the switch, R. Leland who provided assistance and support helped with the assembly and operation of the test facility, and M. Heun for curve fitting of Q-meter calibration data and heat flow measurement data. Special thanks are due M. Stoyanof of the Air Force Phillips Laboratory, and V. Mahajan, D. Curran, and D. Glaister of the Aerospace Corporation for their many comments and suggestions over the course of this development effort.

## REFERENCES

1. Johnson, D.L. and Wu, J.J, "Feasibility Demonstration of a Thermal Switch for Dual Temperature IR Focal Plane Cooling," Final Report JPL D-13423,, February 15, 1996.
2. Glaister, D.S., Curran, D.G.T., Mahajan, V.N., and Stoyanof, M., "Application of Cryogenic Thermal Switch Technology to Dual Focal Plane Concept for Brilliant Eyes Sensor Payload," presented at the IEEE Aerospace Applications Conference, Aspen, Colorado, February 3-10, 1996.
3. Johnson, D.L. and Ross, R.G., Jr., "Cryocooler Coldfinger Heat Interceptor," Cryocoolers 8, Ed. by R. Ross, Jr., Plenum Publishing Corp, (1995), pp 709-718.
4. Bard, S., et al. "10 K Solid Hydrogen Sorption Cooler." Final Report for Period January 1991 to December 1991, JPL Publication D-9326.
5. Wu, J. J., et al. "Experimental Demonstration of a 10 K Sorption Cryocooler Stage." Advances in Cryogenic Engineering, Vol. 39, Plenum Press, New York, 1994.
6. Wade, L., et al. "Performance, Reliability, and Life of Hydride Compressor Components for 10 K to 30 K Sorption Cryocoolers." Advances in Cryogenic Engineering, Vol. 39, Plenum Press, New York, 1994.
7. Barron, R. Cryogenic Systems, 2nd Ed. Oxford University Press, New York, Clarendon Press, Oxford, 1985.

# An Advanced Solderless Flexible Thermal Link

Brian Williams, Scott Jensen, and J. Clair Batty

Dept. of Mechanical & Aerospace Engineering
Utah State University
Logan, Utah, USA  84322-4130

## ABSTRACT

Thermal links play a vital role in the thermal management of space based cryogenically cooled instruments by connecting the cold heat sink with a cooled component.  When the heat sink is the cold block or cold tip of a cryocooler, usually mounted on a very fragile tube, the design becomes particularly challenging.  It is required that this link be highly flexible, have a low thermal resistance, and a low mass so as to not impose significant inertial loading during the harsh vibration environment of launch.  The authors have developed a simple technique for constructing solderless links to meet such stringent thermal, flexibility, durability, and mass constraints.

These links consist of flexible layers of thin metal foils "swaged" into end blocks. Swaging provides a fast, simple method for providing a low thermal impedance between the foils and blocks.  In addition, the swaging process does not present an out-gassing problem as is sometimes associated with solder

This paper describes the performance characteristics of an aluminum foil flexible thermal link in terms of length, mass, thermal resistance, flexibility, and survivability when subjected to vibration testing.

## INTRODUCTION

Flexible thermal links play an important role  in the thermal management of cryogenically cooled components.  These links are often used to thermally connect cooled sensors to cryogenic mechanical refrigerators (cryocoolers) and storable cryogen tanks in space applications.

Flexible thermal links are typically made of braided wire or foil strips.  A standard approach to making a link is to solder flexible foils or braids between two solid end blocks machined to the proper size and shape.  This approach has a few limitations, such as (1) the solder adds a thermal impedance to the link.  Solder is a poor conductor compared to typical link materials such as copper and aluminum; for systems with little margin for added temperature gradients, this can be prohibitive. (2) For highly sensitive optical systems, the outgasing of the solder could pose a potential problem by contaminating optical surfaces. (3) The solder could wick into the braid making it stiff unless special precautions are taken when soldering the link.  The soldering technique required to eliminate wicking, and to minimize the solders' effect on impedance, is very time consuming and requires the skill of a well practiced technician.

When a thermal link is attached to the cold end of a cryocooler, extreme care must be used to ensure that the delicate coolers is not damaged. For space applications, due to the fragile nature of cryocooler's cold finger, the thermal link must be extremely light weight to ensure that large inertial forces are not imposed on the cooler during launch. Some systems have the added requirement of vibration suppression between the sensor and cold sink, hence the need for a flexible thermal link to dampen out any vibrational effects which may effect proper operation of the sensor. Flexibility is also advantageous because it allows for relative motion between the sensor and the cold finger during assembly and operation. The link must also be capable of surviving the harsh vibrational launch environment.

Utah State University (USU) has developed a simple solderless process for securely attaching flexible foils or braids into solid end blocks. This process dramatically reduces the thermal impedance at the interface between the foil/braid and block while maintaining original flexibility. Previous research[1] at USU focused on the design and performance of a copper-braid link. However, recent work has concentrated on using aluminum foils instead of braids in an all-aluminum link.

This paper presents the design requirements specified and the performance characteristics achieved in the construction of an aluminum foil flexible thermal link. Also included in this paper are descriptions and results of testing used to validate the link design.

## DESIGN REQUIREMENTS AND PERFORMANCE CHARACTERISTICS

Utah State University, under contract from TRW in Redondo Beach, CA, has designed, fabricated, and tested an aluminum foil thermal link. Design requirements and achieved performance characteristics are listed in Table 1. Figure 1 shows a photograph of the completed link in its shipping support fixture.

**Table 1.** Design Requirements and Performance Characteristics of Aluminum Foil Link

| Specification | Design Required | Achieved |
|---|---|---|
| Thermal Resistance | 2.2 K/W @ 77 K | 1.9 K/W |
| Mass | < 100 g | 63 g |
| Dimensions: | | |
| diameter | < 50.8 mm | achieved |
| width of end block | < 31.8 mm | achieved |
| length from sensor to cold finger | 86 mm | achieved |
| Flexibility: | | |
| x axis | < 3 N force for | 1.85 N |
| y axis | 7mm deflection | 0.8 N |
| z axis | per axis | 0.75 N |
| Survivability | 32 $g_{rms}$ | 32 $g_{rms}$ demonstrated |

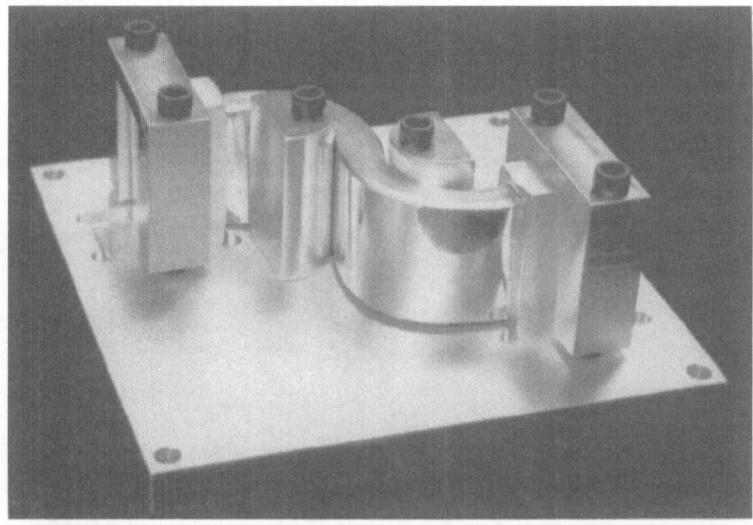

**Figure 1.** Photograph of the completed aluminum foil thermal link.

## LINK TESTING

Once a link has been fabricated, design qualification tests are performed. Typically there are three tests which are specified: thermal resistance, flexibility , and vibration qualification.

### Thermal Testing

The fabricated link is placed inside a cryogenic vacuum test dewar (see Figure 2) for thermal evaluation. One end of the link is then attached to the dewar's cold finger and the other is suspended from the top of the dewar by a piece of Kevlar 49 thread which acts as a thermal isolator so negligible amounts of conductive parasitic heat load make it into the link. A LakeShore DT-470-CY silicon diode temperature sensor is attached to each end of the link and connected via wires through the dewar to a Lakeshore 208 thermometer to display the temperatures. On the end opposite of the cold finger, a small electrical resistance heater is attached to simulate an instrument heat load. The entire thermal link is then insulated with multilayer insulation (MLI) to minimize undesired radiative loads or losses.

The test procedure is as follows: after the inside of the dewar is evacuated via a vacuum pump, the tank is filled with an appropriate cryogen. Usually liquid nitrogen is used, however other cryogens could be used depending on the desired temperature. The power source for the electrical heater is turned on and set to a predetermined value as measured by digital voltage and current meters (refer to Figure 2). Once steady-state is achieved, the temperatures at each end of the link are recorded. This procedure is then repeated for several different electrical power levels. If care has been taken to significantly reduce the parasitic heat leak into the link, then small powers levels can be applied to the system with a high degree of accuracy. The larger the heat load applied to the link, typically the more reliable the temperature readings will be due to the fact that the parasitic heat loads become a smaller fraction of the total heat through the link.

Since resistance is measured in K/W (which is independent of the power level applied), a uniform reading will be obtained for the resistance as dictated by the proportionality of Q

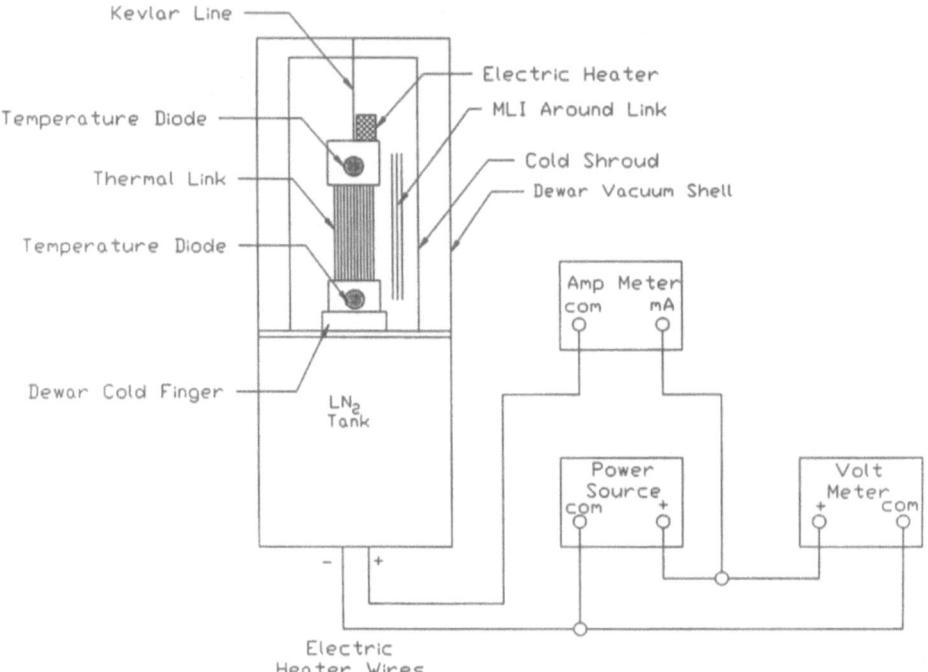

**Figure 2**. Schematic diagram of thermal conductivity test setup including wiring schematic for heater power supply and digital volt meters.

and $\Delta T$ in Fourier's law[2]. An effect of increasing the heat power that has to be closely watched is the fact that the thermal conductivity of the material typically changes with temperature. If the average temperature of the link has changed significantly from one power setting to the next, slightly different values of thermal resistance will be found and could be a direct result of the change in thermal conductivity of the material at the new temperature.

### Flexibility Test

A static flexibility test is performed on the link to determine the force required to displace the link a given amount in either the x, y, or z direction. Depending on the application, the range of required flexibility varies considerably.

The test procedure is as follows: the link is placed in a testing fixture, known as a "flex jig" (see Figure 3 for a schematic); one side of the link is bolted down to the jig and the other to a precision scale capable of milligram measurements. The plate of the flex jig in which the link is attached is displaced the desired amount, as measured by a dial indicator. Once the plate has been moved, the new reading on the scale is compared to the zero-displacement reading; this difference is the weight required to move the end of the link and can easily be converted to newtons of force. This process is repeated for a negative deflection.

Since the flexibility of the link is desired in all three axes, and the flex jig only moves in the one direction, the link must be rotated and the scale placement adjusted to measure the other two directions.

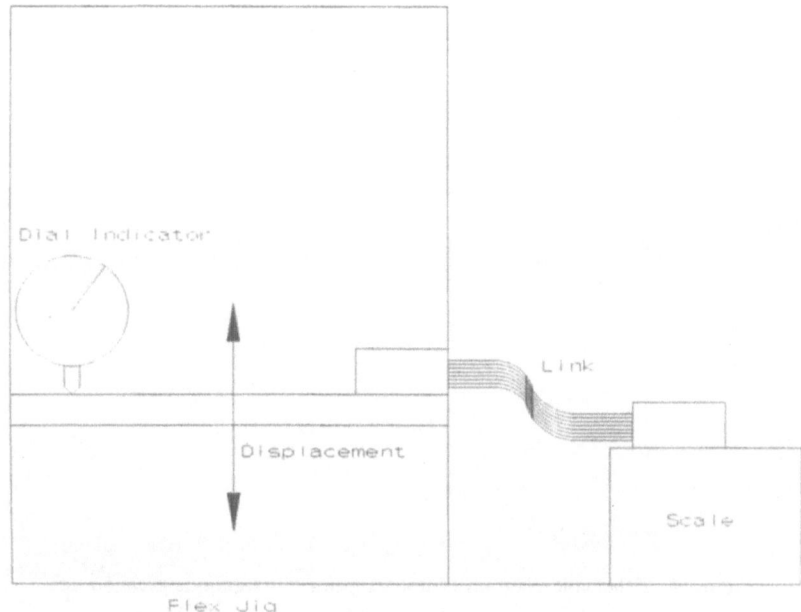

**Figure 3**. Schematic of test setup for flexibility measurement.

### Vibration Qualification Test

The next test is the vibration qualification. This test is carried out on a shaker table and will simulate the harsh environment of launch. The purpose of this test is to determine if the link will survive launch on a rocket.

The test is carried out in the following manner: one end of the flexible thermal link is connected to the shaker table and the other end mounted to an external rigid support. The shaker table is then run for one minute at a specified level. Similar to the flexibility test, this test must also be performed in all 3 axes. Since the shaker table only moves in the vertical direction, the link must be rotated in the other two axes and the test re-performed to obtain the desired results. The random vibration shake levels vary from link to link, but thus far accelerations up to 32 $g_{rms}$ have been tested successfully. Figure 4 is sketch of the basic test setup.

## CONCLUSIONS

A technology has been developed to fabricate reliable, efficient, solderless flexible thermal links using a swaging process. The swaging process essentially cold welds braid/foil and end blocks into one integral piece in which the contact resistance is dramatically reduced. Compared to the standard soldering approach, this process has proven to be relatively fast, simple, and low cost.

An aluminum-foil flexible link has been constructed and demonstrated a thermal resistance of 1.9 K/W at 77 K, flexibility in all three axes less than 3 N for a 7 mm deflection, and vibration survivability up to 32 $g_{rms}$. The link was also within the required dimensional envelope and mass limit.

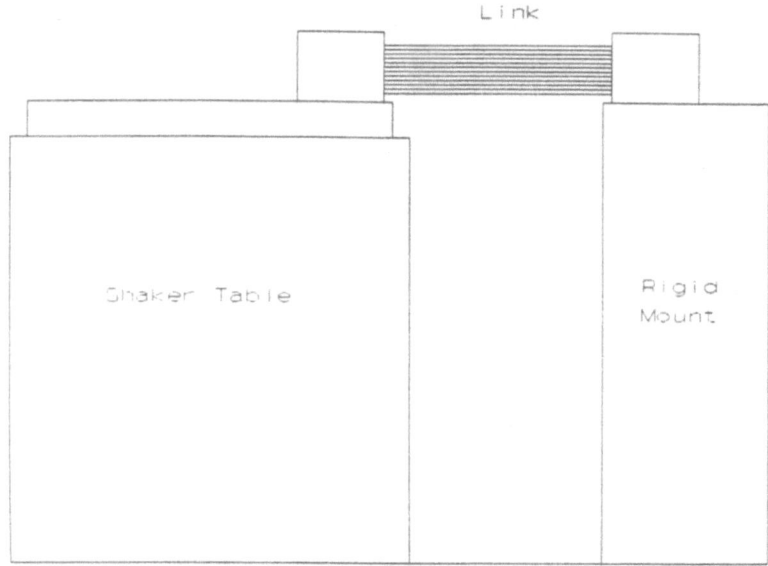

**Figure 4.** Schematic diagram of test setup for vibration qualification test.

## REFERENCES

1.  Williams, Brian, Scott Jensen, Mark Chadek, and J. Clair Batty, "Solderless Flexible Thermal Links", presented at the 1995 Space Cryogenics Workshop, 26 July, in Beltsville, Maryland, USA. Accepted for publication in *Cryogenics* (1995).

2.  Incropera, Frank P. and David P. Dewitt, *Fundamentals of Heat and Mass Transfer*, third edition, John Wiley & Sons, New York (1990), pp. 44-65.

# Development and Testing of a Demountable Cryocooler Thermal Interface

**A. R. Langhorn**

Startech Inc.
Solana Beach, CA 92075 USA

**J. D. Walters**

Annapolis Detachment, NSWCCD
Annapolis, MD 21402 USA

**M. Heiberger**

General Atomics
San Diego, CA 92121 USA

## ABSTRACT

A superconducting magnet system is being designed and built by General Atomics for the US. Navy's Advanced Lightweight Influence Sweep System (ALISS) program. The magnet system is intended to detonate magnetic influence sea mines in shallow water mine sweeping operations. Program goals include the requirement that the magnet be conduction cooled (no liquid cryogens allowed), and the coolers be replaceable without warming the magnet. Copper stabilized niobium-titanium has been selected for the coil, which is designed to operate in the 5 to 6.5 K temperature range. The cold components of the magnet weigh about 2800 lb. Two Gifford-McMahon (G-M) 4.2 K cryocoolers are incorporated into the magnet cryostat. Each G-M cooler is separated from the magnet insulating vacuum by a cold sleeve. Requirements for fast cool down and low steady-state coil operating temperatures put severe constraints on the design of the cryogenic interfaces of the cold sleeve. The G-M cooler second stage interface at the cold sleeve is of particular interest as it must be of high thermal conductance while providing adequate thermal isolation from higher temperature regions. This paper describes the development and testing of a unique thermal interface device which meets these needs. Details of the device including mechanical and thermal test results are presented.

## INTRODUCTION

The US Navy's ALISS program includes an investigation into the use of high field magnets to detonate magnetic influence mines in shallow water. Design and fabrication of a superconducting magnet system, shown in Figure 1 is under way at General Atomics. This system is designed to meet the arduous shock and vibration requirements and other strategic requirements of a unit deployed at sea.

The major design criteria are shown in Table 1, these include 100g shock with continuous vibration loads of up to 2g. Due to the difficult logistics involved with liquid helium, the magnet is cooled using closed loop refrigeration.

Cryocoolers 9, Edited by R.G. Ross, Jr.
Plenum Press, New York, 1997

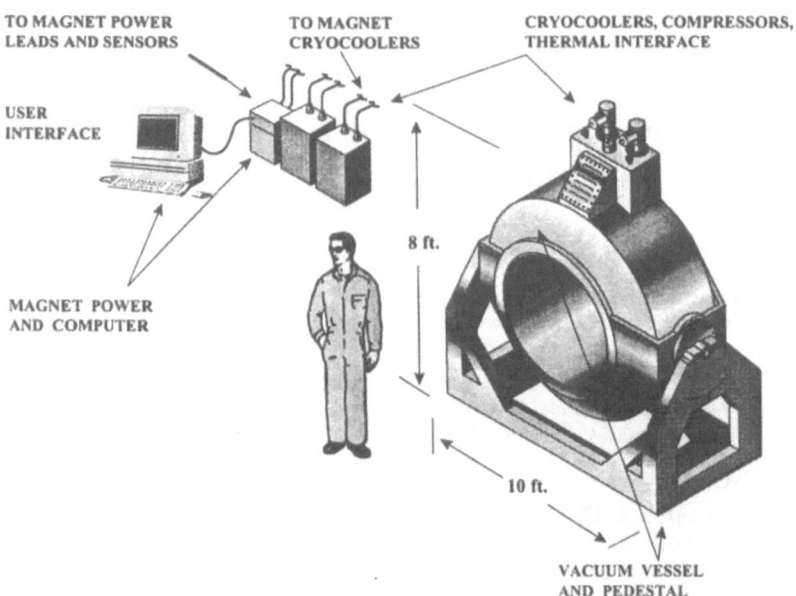

**Figure 1.** ALISS Magnet System

The system is designed to use redundant, replaceable G-M cryocoolers operating between 4.2 and 4.6 K. Niobium titanium is used in the superconducting coil which can operate at temperatures as high as 6.5 K while remaining energized. The steady state heat leak to the coil is about 1.5 W. Heat pipes are used to shunt between the 1st stage of the coolers and the magnet coil during cool down. Rapid cool down of the magnet from room temperature requires that full use is also made of the cryocooler's 2nd stage performance from 300 to 4 K.

The requirement for removable, redundant coolers presents several challenging design problems. In order to replace the cryocoolers without warming the coil, each cryocooler is housed in a sleeve to separate it from the main cryostat insulating vacuum space. This sleeve represents a thermal barrier in the cooling path to the coil. A design solution must incorporate an efficient thermal joint at the cryocooler interfaces which can be made and broken cold without degradation in thermal performance. A cold sleeve design, shown in Figure 2 has been developed to meet these requirements.

**Table 1.** Major design parameters of the ALISS magnet being built by General Atomics

| | |
|---|---|
| Magnetic Moment | $5 \times 10^5$ A-m$^2$ |
| Peak Field | 4.81 T |
| Cold Mass Weight | 2800 lb. |
| Total Weight | 3835 lb. |
| Static Shock Load | 100 g multi-directional |
| Vibrational Load | 2 g at 20 Hz |
| Coil operating Temperature | 5 - 6.5 K |
| Heat Load to 50 K | 56 W |
| Heat Load to 4.5 K | 1.5 W |

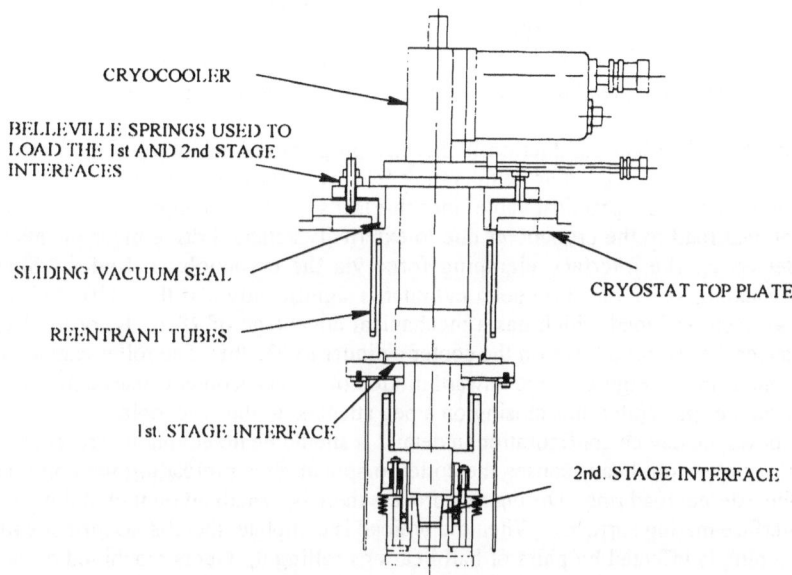

**Figure 2.** Cold Sleeve Assembly Mounted on Cryostat Top Plate

The cold sleeve is comprised of a series of reentrant tubes sized to withstand external vacuum loads and 100g side loads. The cryocooler is mounted on a flange which slides in an O ring sealed gland on the top flange. A total seating force of 1500 lbf. for the first and second stage cryocooler interfaces is applied externally by Belleville spring washers, located on the cryocooler mounting flange. A complimentary stack of Belleville springs on the second stage interface in the cold sleeve ensures correct load sharing between the first and second stage interfaces. These springs in the second stage are preloaded to 150 lbf. After the second stage is clamped, further motion seats and clamps the first stage which is a flat, indium plated copper surface. Adequate range is provided in the spring stacks to allow for any differential thermal contraction between cryocooler and cold sleeve.

The second stage uses a peg-in-collet interface configuration and force multiplying clamping mechanism. Although test data exists for contact conduction at ~ 4 K, there is little information in the literature regarding contact conductance of such joints over the range of 4 to 300K. In order to verify the mechanical and thermal performance of the second stage interface design, a test article has been made and tested.

## SECOND STAGE INTERFACE DESIGN

Interface conduction dictates the size of a thermal interface for a given heat flux and allowable temperature difference. A survey by Van Sciver et al.[1] of published data indicates that the conductivity of pressed contacts is enhanced by a soft, pure material at the interface. There is evidence that joint conduction is further enhanced by high interface pressures. In order to maximize the joint conduction and thus minimize device size, a copper to copper joint with an indium interface has been selected. Also, a mechanism has been designed to efficiently deliver a high clamping force to the interface.

Selection of the joint contact area was made based on preliminary calculations of the overall cryogenic buswork conductance between the cold sleeve and coil at a temperature of 4.5 K. A minimum value of ~4 W/K for the interface conductance was selected to achieve an overall buswork conductance of ~1.7 W/K. Salerno et al [2] measured the conductance of a clamped joint comprised of a thin indium gasket between two copper surfaces. They suggest a value of $h = 0.09$ W/cm$^2$ K at our design conditions. Tests performed by Nilles et al. [3] on a similar clamped joint suggest that 0.45 W/cm$^2$ K is achievable. Due to the scatter in the data, the design was set using the lower value, leading to the selection of a contact area of 45 cm$^2$ (approximately 7 in$^2$).

It is reported that contact conduction is directly proportional to interface pressure (P) at ~ 4 K [4] and varies by P$^{0.66}$ at room temperature [5]. An interface pressure of ~500 psi was chosen as being the maximum practical value since this leads to a total clamping force of 3500 lbf.

Additional heat load to the cryocooler due to externally actuated drive mechanisms can be avoided by delivering the interface clamping force via the cryocooler cylinder. Since the maximum allowable force on the cryocooler cylinder is significantly less than 3500 lbf., a roller cam system has been designed which has a mechanical advantage of 28:1. Ignoring frictional losses, this reduces the required force on the cooler cylinder to 125 lbf. The roller cam concept is illustrated schematically in Figure 3. Downward motion of the cryocooler causes ball bearings to ride up a taper on the split collet thus closing on a peg attached to the cryocooler.

Figure 4 shows the device configuration in detail. It should be noted that insertion of the peg (part of the cryocooler cold head) causes the collet to spread, thus preloading the collet against the balls and the external load ring. During this process there is a small amount of sliding between the thermal interface mating surfaces. When the preload is complete, there is no further sliding in the joint. Clamping is effected by pairs of ball bearings rolling up tapers machined on both the collet and load ring surfaces. Since the balls are grouped in pairs a rolling action occurs between these highly stressed surfaces as shown in Figure 5, thus avoiding galling due to sliding. Some sliding takes place at the interface between the ball carrier and the balls. However, the stresses are low in this area. All bearing surfaces are dry film lubricated with MoS$_2$ applied in a high pressure impingement process [6]. To account for frictional losses in the cam, the Belleville washers are set up to deliver 150 lbf. to the ball retainer. Motion of the whole assembly is permitted by metal bellows welded to the cold sleeve, this also permits self alignment during engagement of the peg.

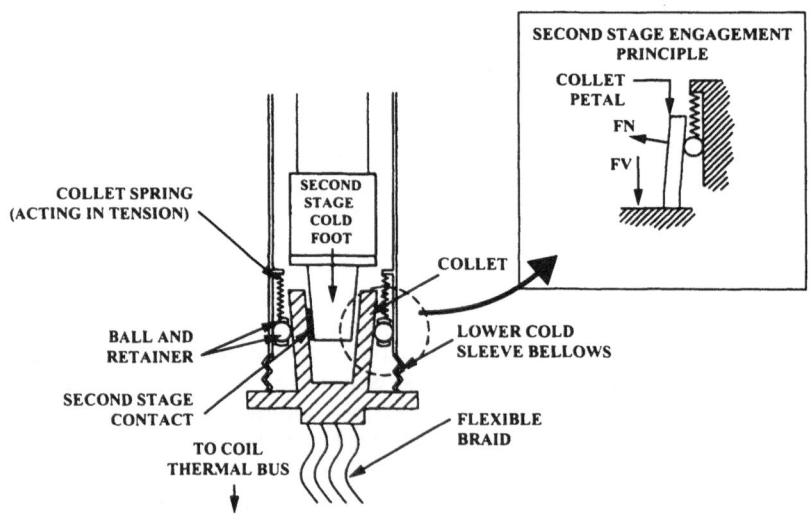

**Figure 3.** Schematic of the Peg-in-Collet Principle

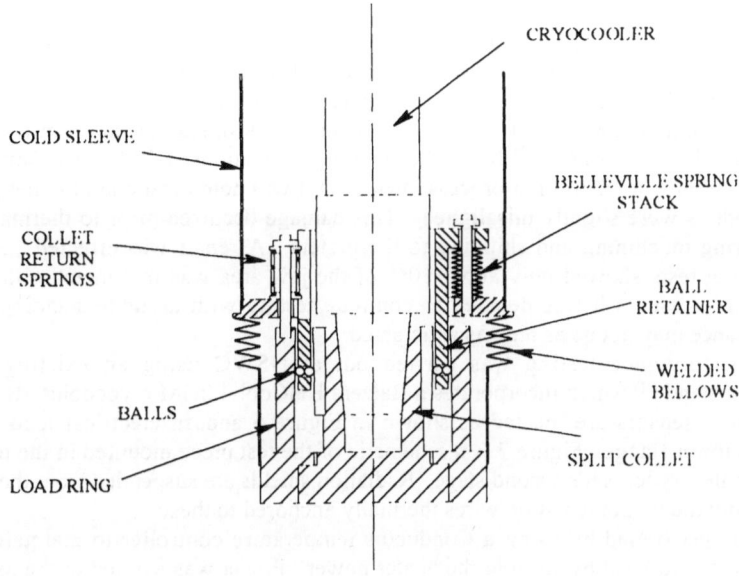

**Figure 4.** Details of the 2nd Stage Interface

Although both surfaces of the joint are precision machined copper, there is evidence that conduction is further enhanced by a thin layer of indium at the interface. Indium maintains some ductility at cryogenic temperatures and has an oxide layer which is softer than copper and more easily broken during joint clamp up. Due to the configuration of the peg-in-collet joint design an indium foil gasket is impractical. Therefore, a brush plating process [7] is used to apply 0.001/0.002 inches of indium to both mating surfaces.

This approach also avoids the problem of positioning the gasket during cooler insertion into the cold sleeve. An additional benefit results from the removal of the oxide layers from the copper surfaces during the brush plating process (the process includes an etch prior to indium plating). Nilles et al [3] report a factor of two increase in conductance between identical copper samples measured with and without an oxide layer.

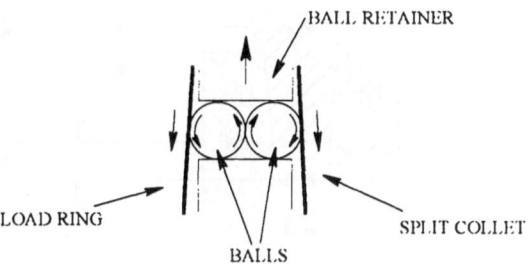

**Figure 5.** Detail of the collet cam showing the motion of the balls as they roll up tapers on the split collet and load ring.

**TEST ARTICLE**

A full scale test article was built in order to verify the mechanical and thermal performance of the design. The design details were slightly modified to allow operation without external actuation. However, the collet clamping load path remains the same.

Mechanical testing comprised of repeated make and break operations at room temperature followed by several cold shocks at 77 K. Make and break operations at 77 K were carried out in a helium gas environment. During in-process inspection it was noted that one of collet jaws was damaged and others were slightly misaligned. This damage occurred prior to thermal testing, most likely during machining and shipping to the platers. A repair was effected. However, subsequent bluing tests showed only about 50% of the jaw area was in contact with the peg. Due to schedule pressures it was decided to continue testing with an understanding that the contact conductance may not be as high as anticipated.

Thermal performance testing was carried out at NSWC using an existing thermal conductivity test stand [8] which incorporates a Balzers KelCool 4.2GM cryocooler. Ruthenium oxide temperature sensors are located as shown in Figure 6 and an electrical load heater is installed on the lower flange. Figure 7 is a schematic of the test piece mounted in the test stand, suspended from the cryocooler's second stage. Radiation shields are suspended from the first and second stage, with the instrumentation wires thermally anchored to these.

Testing was performed by using a Conductus temperature controller to maintain a fixed temperature at the cold head by varying the heater power. Power was applied to the lower load heater and $\Delta T$ measurements were made at selected temperatures between 4.5 K and 175 K. Measurements above 175 K were not made due to an instability in the temperature control loop. However, enough data is available to extrapolate to 300 K. Several series of test data were taken during two test periods, in the interim period the peg-in-collet joint was made and broken several times.

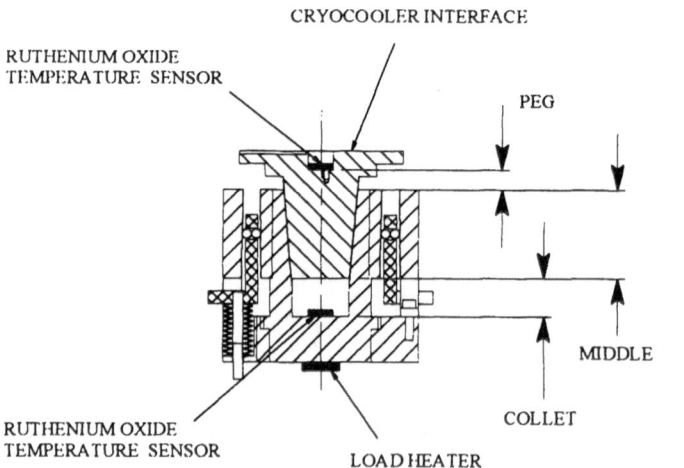

**Figure 6.** Sectional view of the Test Article showing the location of the temperature sensors and heater. The three sections used in calculating the conductance are also shown.

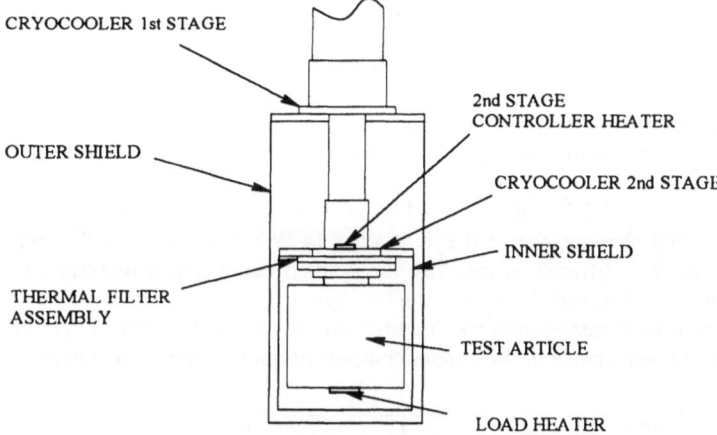

**Figure 7.** Schematic of the Thermal Test Stand with the outer vacuum jacket removed.

## RESULTS

The peg-in-collet mechanical concept worked as predicted. After repeated cold shocks and many make and break operations there was no permanent damage or unexpected wear.

The thermal performance exceeded our predictions. Figure 8 shows the test results for the conductance across the test article detailed in Figure 6. Conductances are calculated using:

$$C = \frac{Q}{\Delta T} \ , \tag{1}$$

where Q and $\Delta T$ are the heater load and temperature difference developed across the test article, respectively.

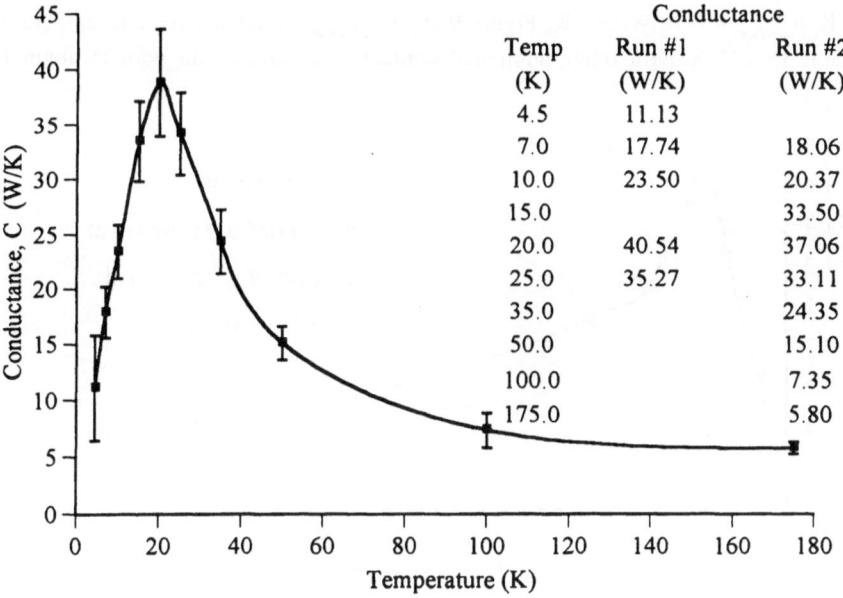

| | Conductance | |
| Temp | Run #1 | Run #2 |
| (K) | (W/K) | (W/K) |
| 4.5 | 11.13 | |
| 7.0 | 17.74 | 18.06 |
| 10.0 | 23.50 | 20.37 |
| 15.0 | | 33.50 |
| 20.0 | 40.54 | 37.06 |
| 25.0 | 35.27 | 33.11 |
| 35.0 | | 24.35 |
| 50.0 | | 15.10 |
| 100.0 | | 7.35 |
| 175.0 | | 5.80 |

**Figure 8.** Plot of conductance as a function of temperature of the Test Article

There is some experimental uncertainty, primarily due to scatter in the temperature data stemming from thermal noise at the sensors. Each individual temperature is constant and reproducible to within less than 1%. However, these small errors become significant when the temperature differences across the peg and collet interface are calculated. From these calculated differences, the cumulative uncertainty for a particular set-point temperature is obtained by combining all of the uncertainties using a root mean square average. These cumulative uncertainties are reflected in the error bars shown in Fig. 8.

The total conductance at 4.5 K was measured to be 10.9 W/K, which is considerably higher than the predicted contact conductance of the joint alone (4 W/K). Because of the test article's physical configuration, it is difficult to measure the contact conductance directly. However, noteworthy assessments can be made by examining the data.

Assuming the combined cross section and length of the peg and collet in the interface zone is fully active, the overall conductance of the copper components can be approximated by:

$$C_{copper} = \frac{1}{\dfrac{L_{peg}}{A_{peg} \cdot k_{copper}} + \dfrac{L_{middle}}{A_{middle} \cdot k_{copper}} + \dfrac{L_{collet}}{A_{collet} \cdot k_{copper}}}, \qquad (2)$$

where $L$ and $A$ are the effective length and cross sectional areas of the copper elements shown in Figure 6, and $k_{copper}$ is the thermal conductivity of copper. The thermal conductivity of OFHC copper as used in the test article varies with the degree of cold work and level of impurities contained in the material. Residual Resistance Ratio (RRR) values ranging between 50 to 520 have been reported [9]. Since the peg was machined from solid bar and the collet was in the fully annealed state, a value of RRR 200 was selected. Using this value to calculate $C_{copper}$, the contact conductivity at the interface can be established from the measured data using:

$$h_{int\,erface} = \frac{1}{A_{int\,erface}} \cdot \frac{1}{\dfrac{1}{C_{measured}} - \dfrac{1}{C_{copper}}} \qquad (3)$$

At 4.5 K $h_{int\,erface}$ = 0.38 W/cm² K, Figure 9 shows $h_{int\,erface}$ plotted over the temperature range of 4.5 K to 175 K with other published contact conduction data points shown for comparison.

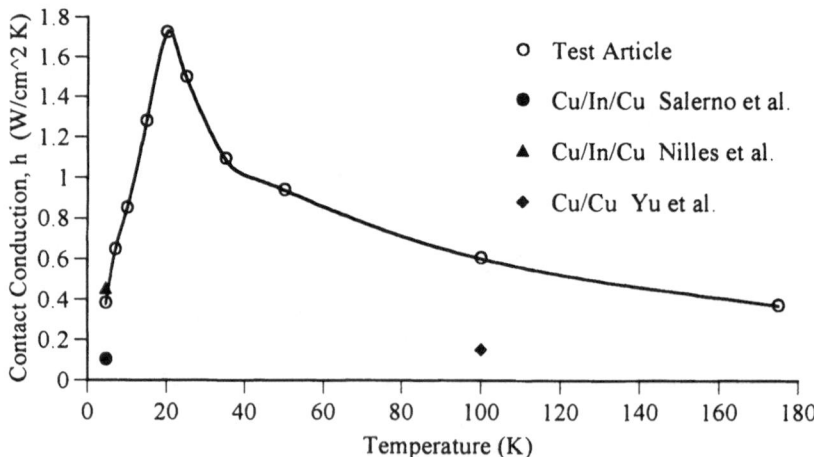

**Figure 9.** Plot of Test Article contact conductivity as a function of temperature, data points from other published reports are shown. References 2, 3, and 10.

## CONCLUSIONS

A thermal interface has been designed and tested, this will become part of the ALISS magnet cryocooler cold sleeve. The peg-in-collet mechanical design has demonstrated reliable and repeatable operation. Thermal tests indicate that the design is viable since the interface conductivity appears to exceed the design values. Additionally, the measured thermal conductances of the test article are consistent with previously reported copper/indium/copper interface data (as shown in Figure 9).

No detrimental effects were noticed due the apparent damage to the collet jaws.

An unexpected result is the large "knee" in the interface conduction in Figure 9 at 22 K, this appears to follow the conductivity curve of copper rather than indium. Although there is no published data on clamped copper/indium/copper joints in this temperature range, Radebaugh [11] reports a similar knee in a soldered copper/indium/copper joint.

The test article design will be incorporated into the cold sleeve without modification. A complete cold sleeve will be constructed and tested prior to installation into the ALISS cryostat. The cold sleeve test program will be an extension of the tests reported here and include measurements of cryocooler first stage interface performance.

## REFERENCES

1. S.W. Van Sciver, M.J. Nilles and J. Pfotenhauer, "Thermal and Electrical Conductance Between Metals at Low Temperatures". *Proc. Space Cryogenics Workshop.* 36 Berlin (1984)

2. L.J Salerno, P. Kittle and A.L. Spivak. "Thermal conductance of Pressed Metallic Contacts Augmented with Indium Foil or Apiezon Grease at Liquid Helium Temperatures". *Cryogenics* 1994 Volume 34. Number 8.

3. M.J. Nilles and S.W. Van Sciever. Effects of oxidation and roughness on contact resistance from 4K to 290K". *Advances in Cryogenic Engineering.* Vol. 34 (1988)

4. M.G. Cooper, B.B. Mikic and M.M. Yovanovich, Int. J. *Heat Mass Transfer 12,* 279 (1969).

5. *Handbook of Heat Transfer Fundamentals.* 2nd ed., Ed. W.M. Rohsenow, McGraw Hill Book Co., N.Y.

6. Lubelok 5306. E/M Corp. N. Hollywood, CA91605

7. Sifco Selective Plating. San Dimas, CA91773

8. J.D. Walters, T.H. Fikse, and T.L. Cooper. "Variable Temperature Thermal Conductivity and Conductance Measurements using a Gifford-McMahon Cryocooler". *Cryocoolers 8.* pp 923. Ed. R.G. Ross. Plenum Press NY 1995.

9. *"Materials at Low Temperatures, Electrical Properties".* pp 163 Ed. R.P. Reed, A.F. Clark American Society for Metals OH 44073

10. June Yu, A.L. Lee and R.E. Schwall. "Thermal conductance of Cu/Cu and Cu/Si interfaces from 85 K to 300 K". *Cryogenics* 1992. Volume 32, Number 7.

11. Ray Radebaugh. "Thermal conductance of indium solder joints at low temperatures". *Rev. Sci. Instrum.* Vol 48, No. 1, January 1977.

# Thermal Conductance of Multilam Make and Break Cryocooler Thermal Interface

J. D. Walters, T. H. Fikse, and Tracey L. Cooper

NSWCCD, Annapolis Detachment
Annapolis, Maryland, USA 21402

## ABSTRACT

The U.S. Navy is pursuing the design and fabrication of a superconducting magnet system cooled by closed cycle cryogenic refrigerators; no liquid helium will be used to cool the superconducting magnet. High thermal conductance interfaces between cryocoolers and the magnet, current leads, and radiation shields are critical because of the limited cooling capacity of cryocoolers being considered. A desirable attribute of a cryocooled magnet is that the coolers be easily removable for maintenance or replacement. Therefore, a first stage thermal intercept of a two stage cooler system has been designed using multicontact multilam($MC^R$-multilam) material. $MC^R$-multilam is a multi-louvered sliding contact material used in high current electrical applications. An interface using this material will facilitate easy extraction or insertion of the cooler into the superconducting magnet system. The thermal conductance of the $MC^R$-multilam has been measured over the temperature range of 8 - 154 K. The conductance was found to be significantly better than multilam material whose thermal conductance was previously reported. At 33 K, for example, the conductance of the as tested $MC^R$-multilam is approximately 0.64 W/K, whereas the previously measured conductance of multilam is 0.20 W/K. The implications of the conductance of the newly measured multilam on cooler interface designs will be discussed herein.

## INTRODUCTION

For the past three years the U.S. Navy has been engaged in an aggressive technology development program in support of the Advanced Lightweight Influence Sweep System(ALISS) technology demonstration. The goal of the demonstration is to sweep acoustically and magnetically influenced sea mines.[1] It has been the task of the Annapolis Detachment of the Naval Surface Warfare Center(NSWC) to develop a lightweight, extremely rugged superconducting magnet system to be used as the magnetic subsystem within ALISS.[2]

Magnetic subsystem requirements include logistical simplicity, shock and vibration tolerance, and craft independence. Of primary consideration are the logistical simplicity and ruggedness

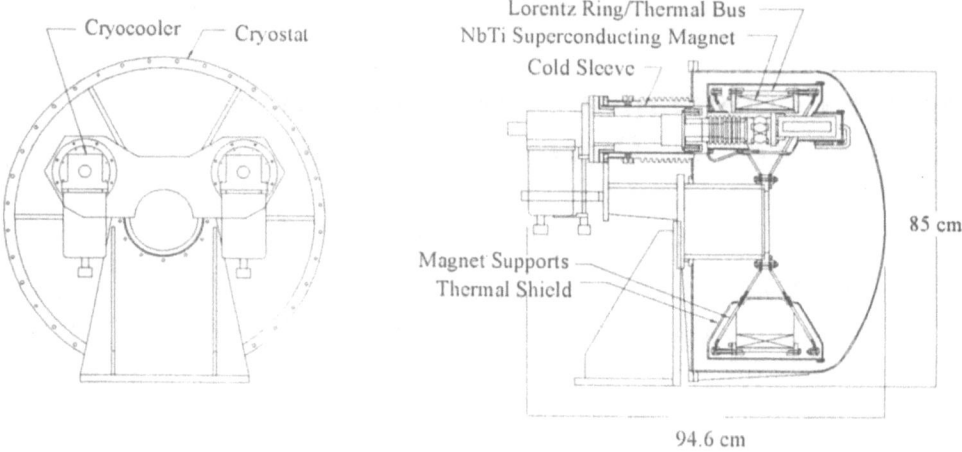

**Figure 1.** NSWC, Annapolis conductively cooled superconducting magnet component test system showing the cryocooler, vacuum isolation cold sleeve, superconducting magnet, and magnet mechanical supports.

requirements.  In order to minimize logistics of system handling, the superconducting magnet subsystem will be cooled by Gifford-McMahon cryocoolers eliminating the need to manage liquid cryogens.  The ALISS subsystem will be ruggedized in conformance with a stringent vibration spectra derived from transportation and operation standards MIL-STD-167 and MIL-STD-810. In the event that the system is exposed to an explosive shock, the magnet vacuum vessel and all encompassed components are designed to withstand, omnidirectionally, a 100g static equivalent shock load.

All of the above mentioned challenges are being addressed throughout the development phase of the ALISS program.  The superconducting magnet system represented in Fig. 1 has been designed and is being fabricated at NSWC, Annapolis as a test bed for ALISS component technology developments. This paper focuses on design aspects of the secondary vacuum

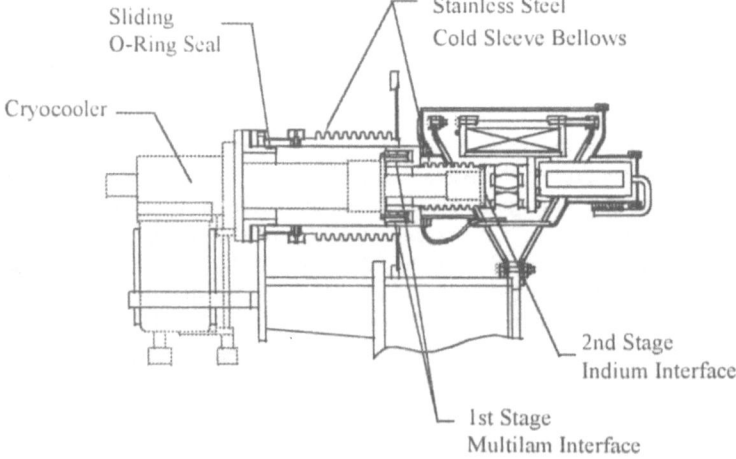

**Figure 2.** Vacuum isolation cold sleeve showing the stainless steel bellows, sliding multilam first stage thermal interface, flat press contact indium second stage thermal interface, cryocooler, and flexible cold sleeve to magnet thermomechanical interface.

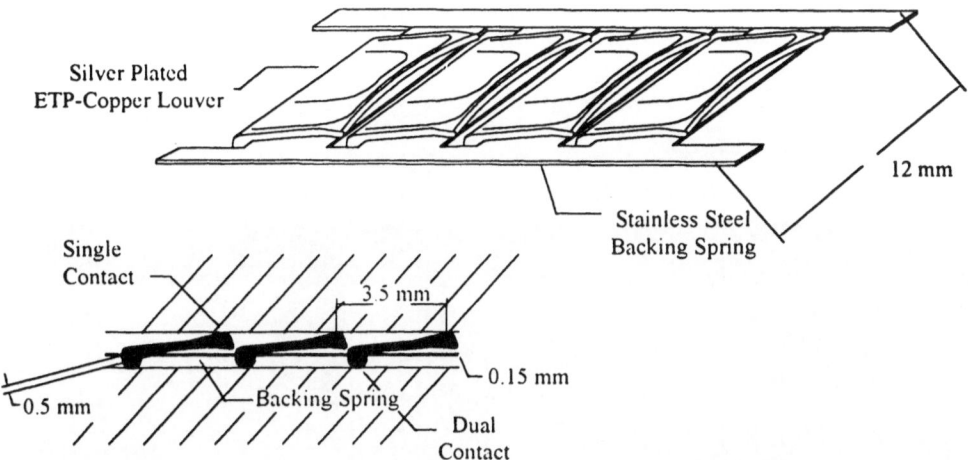

**Figure 3**. MC$^R$-multilam LA-Cu showing silver plated electrolytic tough pitch copper louvers and stainless steel backing spring.

isolation sleeve for the cryocoolers which is also shown in Fig. 1. The vacuum isolating cold sleeve allows cooler replacement when the magnet is cold, further reducing operational logistics. Details of the cold sleeve are shown in Fig. 2.

Components of the cold sleeve include stainless steel bellows down tubes providing vacuum isolation, first and second stage thermal interfaces, a sliding o-ring seal for cooler location adjustment, and a flexible cold sleeve to magnet thermal interface. In order to simplify the first and second stage heat intercepts, a sliding cryocooler interface has been designed for the first stage. This sliding interface simplifies the cooler alignment within the cold sleeve and engagement force adjustments at the second stage indium interface.

The first stage sliding heat intercept incorporates multilouvered multilam banded material which the manufacturer designates MC$^R$-multilam LA-Cu. We will refer to this specific material simply as multilam. The multilam is fabricated from silver plated electrolytic tough pitch(ETP) copper with louvers engaged by a stainless steel backing spring as shown in Fig. 3. As a cryocooler is inserted into a cold sleeve, the multilam actuates by having the louvers forced against the two adjacent sliding surfaces via the backing spring mechanism. Heat is then transferred from one body to the other through the multilam louvers. The insertion and spring contact forces are adjusted by lessening or increasing the gap between the two engaging members. The tighter the spacing between the two mated pieces, the higher the insertion force and a higher contact force results. Lessening the insertion force lowers the louver contact spring force and reduces contact pressure at the interface. This insertion force is applied by bolting the cooler to the upper plate of the cold sleeve and forcing the first and second stages to engage. The insertion force is 2.5 N(0.56 lb.-force) per louver resulting in a 7 N(1.6 lb.-force) contact force in the manufacture's recommended design configuration.[3]

## MEASUREMENT APPARATUS

The general details of the thermal conductance measurement apparatus used here have been reported elsewhere, and only those portions pertinent to the multilam conductance measurements will be discussed.[4,5] The test holder used in measuring the multilam conductance is shown schematically in Fig. 4a. The critical dimensions for this multilam holder are the peg diameter and multilam retaining groove within the receptacle which are 2.502 cm and 2.780 cm, respectively. These two dimensions define the spacing and contact force between the peg and multilam. The copper holder's surfaces which contact the multilam have been prepared by

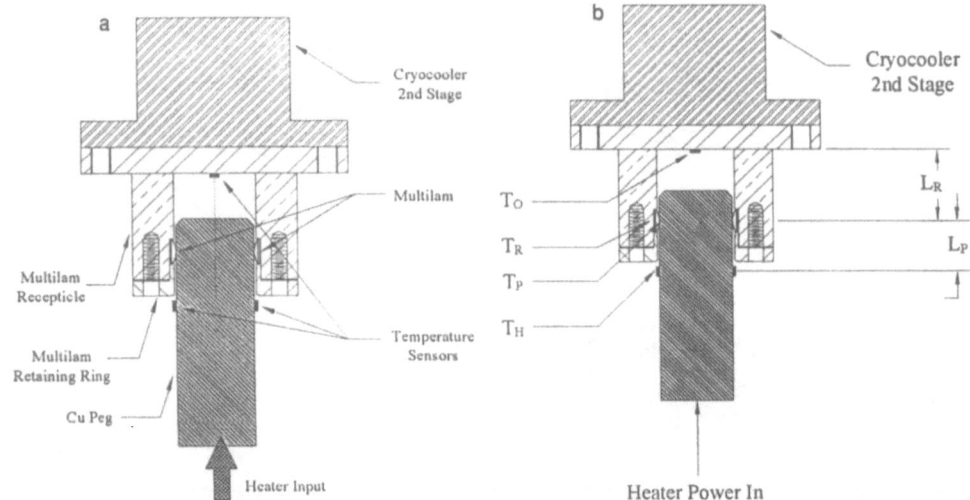

**Figure 4a.** Multilam test receptacle and peg showing a single band of multilam, cryocooler second stage, heater, and temperature sensors.

**Figure 4b.** Multilam test article showing positions at which temperatures are recorded, where $T_O$, $T_R$, $T_P$, and $T_H$ are the temperatures at the receptacle base, at the receptacle side of the interface, at the peg side of the interface, and at a known location along the peg, respectively. Heat transfer occurs through the lengths $L_P$ and $L_R$ of the copper peg and receptacle, respectively.

abrading the copper to break through oxides and then wiping the surfaces with acetone for degreasing and cleaning. No additional special surface preparations were applied in keeping with this projects general philosophy that components have to be developed as if they are being used in a harsh field environment.

Measurements were made by holding the base of the multilam receptacle at a constant temperature, $T_O$, using a Conductus LTC10-G temperature controller. Power is input through a 50 ohm cartridge heater located at the base of the peg, and the system is allowed to attain thermal equilibrium. The temperature difference developed across the multilam interface due to the known power input is then recorded for at least ten minutes in five second intervals. Ruthenium-oxide temperature sensors were positioned along the test piece as shown in Fig. 4b, and a channel has been machined into the peg in order for sensor wires to be routed out of the receptacle. The temperatures $T_O$, $T_R$, $T_P$, and $T_H$ are the temperatures at the receptacle base, at the receptacle side of the multilam interface, at the peg side of the multilam interface, and at a known location along the peg, respectively. The temperatures $T_O$ and $T_H$ are measured, and the temperatures $T_R$ and $T_P$ are calculated from the measured data as discussed below.

## DATA

Temperature data was acquired over a wide range of power inputs and resulting temperature differences. As described above, power is input to the base of the peg. The running length of the multilam test piece is 8.4 cm as measured along the backing spring, and excluding the groove for data acquisition wires, 21 louvers were fitted into the receptacle.

The multilam interface conductance is obtained from the following:

$$k_{ML} = \frac{\dot{Q}}{T_P - T_R},$$

(1)

**Table I.** Multilam Interface Thermal Conductances

| Temperature (K) | Power Input (W) | k (W/K) |
|---|---|---|
| 8.00 | 0.010 to 0.250 | 0.119 (0.001) |
| 10.0 | 0.049 to 0.993 | 0.172 (0.002) |
| 12.0 | 0.048 to 1.017 | 0.215 (0.003) |
| 30.0 | 0.045 to 5.11 | 0.599 (0.018) |
| 40.0 | 0.045 to 11.96 | 0.742 (0.031) |
| 50.0 | 0.045 to 15.16 | 0.701 (0.041) |
| 59.9 | 0.047 to 10.06 | 0.742 (0.049) |
| 99.9 | 0.265 to 15.34 | 0.692 (0.064) |
| 154 | 1.180 to 11.10 | 0.786 (0.115) |

where $k_{ML}$ and $\dot{Q}$ are the interface thermal conductance and heater power input, respectively. Small corrections to the measured thermal conductances are made to account for the temperature gradients in the OFHC copper peg and receptacle in order to obtain the thermal conductance of the multilam interface alone. Temperature gradients developed within the peg and receptacle are obtained once steady state has been achieved by considering the continuity of heat flow according to the following:

$$\dot{Q} = \frac{A_R}{L_R} \int_{T_O}^{T_R} \lambda_{cu}(T) \, dT = k_{ML}(T_O) \, (T_P - T_R) = \frac{A_P}{L_P} \int_{T_P}^{T_H} \lambda_{cu}(T) \, dT, \qquad (2)$$

where $A_R$, $A_P$, $L_R$ and $L_P$ are the effective cross sectional areas and lengths of the receptacle and peg, respectively, as shown in Fig. 4b. From left to right in Eq. (2) are the heat flows through the receptacle, multilam interface, and peg. $\lambda_{cu}(T)$ is the temperature dependent thermal conductivity of OFHC copper. Because the exact residual resistivity ratio(RRR) was not known, a RRR=100 was assumed for calculational purposes based on typical values used in the literature for OFHC copper.[6] Because the power input $\dot{Q}$ to the bottom of the peg and the temperatures $T_O$ and $T_H$ are measured, the temperatures $T_R$ and $T_P$ are obtained by solving the resulting integral equations. In the present case, solutions to Eq. (2) are obtained iteratively using MathCad.[7]

A summary of the multilam thermal conductances is given in Table 1 including uncertainty estimates which are shown in parenthesis. The experimental uncertainty is primarily due to variations in the temperature sensor data originating from sensor thermal noise and temperature controller instabilities. The multilam conductances are reported at the set point temperatures $T_O$ because the estimated temperature gradients within the receptacle are less than 1% different from the set point temperatures. At least one hundred temperature readings were taken at each heater power input. The standard deviations for each of these data sets is less than 1% over all set point temperatures and power inputs. However, when the differences between $T_R$ and $T_P$ are calculated for use in Eq. 1, the temperature variations become significant. The cumulative uncertainties for varied power inputs and set points are combined using a root mean square average. These uncertainties are reflected in the error bars shown with the conductances in Fig. 5. For comparison, also shown in Fig. 5, is the conductance of the previously measured beryllium copper based multilam which has significantly lower conductance primarily due to the beryllium copper rather than ETP copper used in the louvers.[5]

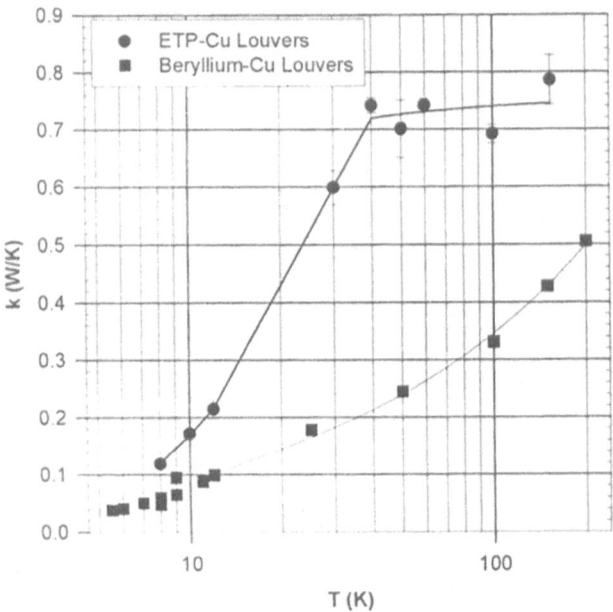

**Figure 5.** MC$^K$-multilam interface thermal conductance as a function of temperature. Also shown is the thermal conductance of the previously measured beryllium copper based multilam. The running length of the two samples along the backing spring is 8.4 cm.

## DISCUSSION

The first stage thermal interface shown in Fig. 2 is expected to pass 18W at 33 K. Using 350 multilam louvers this heat load will result in a 2 K temperature rise across the interface, and require an 875 N (197 lb-force) insertion force. The number of louvers was determined by balancing the desire for very low thermal resistance which requires a large number of louvers against added parasitic heat load, increased insertion force, and spacial constraints.

In an attempt to further reduce the interface thermal resistance, it is insightful to determine the fraction of the total interface resistance due to contacting surfaces and the fraction due to the copper and silver comprising the multilam louvers. The thermal resistance due to the louver material is given by

$$R_{Louvers} = \left( \frac{\alpha\, w_{Cu}\, t_{Cu}}{L_{Cu}} \int_{T_R}^{T_P} \lambda_{cu}(T)\, dT + \frac{\alpha\, w_{Ag}\, t_{Ag}}{L_{Ag}} \int_{T_R}^{T_P} \lambda_{Ag}(T)\, dT \right)^{-1}, \qquad (3)$$

where the first quantity on the right side of Eq. (3) is the louver conductance of the ETP copper and the second is the thermal conductance of the silver plating. Within Eq. (3), $w_{Cu}$, $t_{Cu}$, $L_{Cu}$, $w_{Ag}$, $t_{Ag}$, and $L_{Ag}$ are widths, thicknesses, and lengths of the copper and silver comprising the louvers. Additionally, $\lambda_{Cu}(T)$ and $\lambda_{Ag}(T)$ are the temperature dependent thermal conductivities of RRR = 50 copper and silver, respectively.[x]  For the case presently being examined, $w_{Cu} = w_{Ag} = 0.844\,cm$, $L_{Cu} = L_{Ag} = 0.438\,cm$, $t_{Cu} = 0.050\,cm$, and $t_{Ag} = 0.002\,cm$.

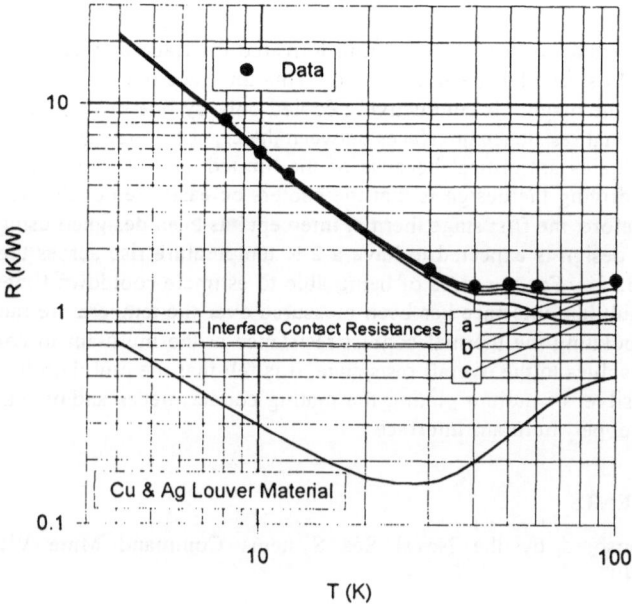

**Figure 6.** Thermal resistance of entire multilam interface, contacts only, and louver material alone as functions of temperature. Curves a, b, and c represent the contact resistance of a multilam interface effectively using only 100%, 50%, and 23% of the louver width for heat transfer. The contact resistance at the interface dominates heat transfer over the entire measured temperature range.

The contact resistance of the louvers is then obtained by using the results from Eq. 3 in the following:

$$R_{Contact} = R_{Data} - R_{Louvers} \tag{4}$$

where $R_{Contact}$ is the calculated contact resistance at the copper multilam interface, and $R_{Data}$ is the total interface thermal resistance as measured experimentally and taken from Table 1.

The actual cross sectional area of the louvers through which heat flows is not well defined due to ambiguity in knowing the contact areas at the interface. According to the manufacturer, the multilam louvers contact one of the two mating surfaces at "one point" and the other surface at only "two points". This uncertainty in the heat transfer path is addressed by varying $\alpha$ within Eq. (3) and adjusting the effective heat transfer width of the louvers. Curves a, b, and c in Fig. 6 represent the contact resistance at the multilam and metal interface for the cases of 100%, 50%, and 23%(i.e. $\alpha = (1.0, 0.5, 0.23)$) of the actual louver width being used for heat transfer. The undulations at the higher temperatures are a calculational artifact of the curve fitted to the data and are unimportant to the present discussion. The most important general feature to note is that over a wide range of temperatures and encompassing all reasonable heat transfer cross sections, the thermal contact resistance dominates heat transfer at the interface. This is a potentially important result because it has been shown by other researchers that carefully prepared surface finishes and increased intersurface contact force decreases interface contact thermal resistance.[9,10] Therefore, future interface improvements may be possible by, for example, plating the two copper pieces with indium or silver, and adjusting the dimensions of the interface so as to increase the louver contact force.

## CONCLUSION

The thermal conductance of a multicontacting louvered interface between a cryocooler and a vacuum isolation cold sleeve has been evaluated. This interface is used in a superconducting magnet system where the magnet is conductively cooled only by one or two Gifford-McMahon cryocoolers. Low thermal resistance interfaces between the cryocooler(s) and the magnet, current leads, and radiation shields are critical because of the limited cooling capacity of cryocoolers. The desired attribute driving the design is that the coolers be easily removable for maintenance or replacement. Therefore, the first stage thermal intercept has been designed using multicontact $MC^R$-multilam. This design is expected to have a 2 K temperature rise across the interface for a 18 W heat load at 33 K. For purposes of being able to estimate cooldown times, the thermal conductance of the multilam interface has been measured over the temperature range of 8 - 154 K. Finally, by extrapolating the thermal contact resistance at the multilam to copper interface and comparing those values to the overall resistance, it is felt that the multilam interface may be further improved by silver or indium plating the mating copper pieces and/or manipulating the contact force at the copper multilam interface.

## ACKNOWLEDGEMENTS

This work was sponsored by the Naval Sea Systems Command Mine Warfare Project Management Office 407.

## REFERENCES

1. E.M. Golda, J.D. Walters. and G.F. Green, "Applications of superconductivity to very shallow water mine sweeping", Naval Engineers Journal, May (1992).
2. M. Heiberger, A. Langhorn, T.W. Hurn, D.G. Morris, J.D. Walters, L.F. Aprigliano, D.J. Waltman, and G.F. Green, "A lightweight rugged conduction-cooled NbTi superconducting magnet for U.S. Navy mine sweeping applications", Advances in Cryogenic Engineering, Vol 41(Plenum Press, New York 1996).
3. Multi-Contact USA, Santa Rosa, California 95403, Company information package on $MC^R$-Multilam LA Cu.
4. J.D. Walters, T.H. Fikse, and T.L. Cooper, "Variable temperature thermal conductivity measurements using a Gifford-McMahon cryocooler", Cryocoolers 8, Proceedings of the International Cryocooler Conference, Ed. R.G. Ross, Jr., pp. 823-833 (Plenum Press, New York 1995).
5. T. L. Cooper, J. D. Walters, and T. H. Fikse, "Thermal conductance of heat transfer interfaces for conductively cooled superconducting magnets", Advances in Cryogenic Engineering, Vol 41(Plenum Press, New York 1996).
6. OFHC Copper and Silver Thermal Conductivities were obtained using Cryocomp Software V. 2.0, Cryodata, Inc. by Eckels Engineering.
7. MathCad V 5.0 Plus, Published by Mathsoft Inc. 1995.
8. The RRR for Electrolytic Tough Pitch copper was obtained from the Handbook on Materials for Superconducting Machinery, Sponsored by the Advanced Research Projects Agency, ARPA order no. 2569, Program code 4D10 (November 1975).
9. M.J. Nilles and S.W. Van Sciver, "Effects of oxidation and roughness on Cu contact resistance from 4.0 K to 290 K", Advances in Cryogenic Engineering, Vol. 39(Plenum Press, New York 1988).
10. Handbook of Heat Transfer Fundamentals, 2nd ed., Ed. W.M. Rohsenow, Mcgraw Hill Book Co., NY.

# Heat Pipes for Enhanced Cooldown of Cryogenic Systems

F.C. Prenger[1], D.D. Hill[1], D.E. Daney[1], M.A. Daugherty[1],
G.F. Green[2], J. Chafe[2], M. Heiberger[3], and A. Langhorn[4]

[1]Los Alamos National Laboratory
Los Alamos, NM, 87545, USA
[2]Naval Surface Warfare Center
Annapolis, MD, 21402, USA
[3]General Atomics
San Diego, CA, 92186, USA
[4]Startech
Solana Beach, CA, 92075, USA

## ABSTRACT

In many important cryogenic applications the use of liquid cryogens for system cooling are either not feasible or are unsuitable. In such cases a cryogenic refrigeration system or multi stage cryocooler must be employed to provide the necessary cooling. To shorten cooldown time for such a system, especially if the thermal mass is large, a thermal shunt directly connecting the first stage of the cryocooler to the load during cooldown is desirable. This thermal shunt allows effective utilization of the greater cooling power available from the first stage of the cryocooler early in the cooldown. Upon reaching operating temperature, the thermal shunt must exhibit a high resistance to thermally isolate the first stage of the cryocooler from the load. Heat pipes are well suited to achieve these objectives. The Advanced Lightweight Influence Sweep System (ALISS), under development by the U. S. Navy for shallow water magnetic mine countermeasures, employs a large, conductively cooled, superconducting magnet that must be cooled from 300 to 4.2 K. Cryogenic heat pipes acting as cryocooler thermal shunts are used to shorten the cooldown time. Ethane, nitrogen and oxygen were evaluated as possible working fluids. A thermal model of the ALISS was developed to evaluate the cooldown performance of various heat pipe combinations. In conjunction with heat pipe performance tests, this model was used to select a suitable design for the heat pipe thermal shunts.

## INTRODUCTION

Cooldown times for components or systems attached to multi stage Gifford-McMahon (GM) cryocoolers are significantly influenced by the thermal capacitance of the components, the thermal interface conductances and by parasitic heat loads on the system. For systems containing superconducting magnets with large thermal capacitance, the cooldown times can be considerable. These components are attached thermally to the second stage of the cryocooler because, ultimately, this is where the lowest temperatures are attained. However, because the cryocooler's second stage has a significantly lower refrigeration capacity compared with its first stage, cooldown times can be reduced by temporarily shifting part of the heat load during cooldown to the cryocooler's first stage where capacity is greater. Gravity-assisted, cryogenic heat pipes acting as thermal shunts are well suited for this task.

In the U. S. Navy's Advanced Lightweight Influence Sweep System[1] (ALISS), a superconducting magnet[2] will be used to sweep shallow water magnetic mines. The magnet is cooled by a pair of GM rare-earth cryocoolers. Figure 1 shows how the heat pipes connect the first stage of the cryocoolers directly to the load with the evaporators attached to the load and the condensers attached to the cooler's first stage. During cooldown the heat pipes act as thermal shunts connecting the load and the first stage of the cryocoolers. The heat removed from the load, which is mechanically attached to the second stage, is then shared thermally by the first and second stages of the cryocoolers. As the load temperature continues to decrease and falls below the triple point of the heat pipe working fluid, the heat pipe becomes inoperable as the working fluid freezes, thermally disconnecting the load from the first stage of the cryocoolers. By selecting different working fluids, the operating temperature range of the heat pipes can be tailored to the cooldown requirements of the application. There remains a parasitic heat leak along the heat pipe wall, when the working fluid is frozen, which can be made small by proper selection of the heat pipe material and geometry[3]. Because the optimum liquid fraction within the heat pipe is small, the freezing and thawing of the working fluid will not damage the heat pipe. At temperatures above the critical point of the heat pipe working fluid, the heat pipe is also inoperable because of the absence of a two phase interface within the heat pipe.

## SYSTEM PERFORMANCE

Ethane and either nitrogen or oxygen are being considered as heat pipe working fluids to shorten the system cooldown time between 300 and 4.2 K. Gravity-assisted heat pipes containing these working fluids operate in different temperature ranges as shown in Fig. 2. By combining an appropriate number of ethane and either nitrogen or oxygen heat pipes connecting the first and second stages of the cryocoolers, the relatively high cooling capacity of the first stage can be applied to the large second stage heat load that arises as a result of the cooldown process. Each type of heat pipe is subjected to a different operating environment during system cooldown. The ethane heat pipes turn on almost immediately upon cryocooler startup and operate nearly isothermally from startup at 295 K down to about 150 K where the vapor pressure of the ethane becomes too low to provide significant heat transport. The working fluid eventually freezes at 89.9 K preventing any further circulation and resulting in shutdown of the heat pipe as desired.

By contrast, the nitrogen or oxygen heat pipes turn on when their condenser temperature reaches approximately 126 and 155 K respectively. With their higher critical temperature compared with nitrogen, the oxygen heat pipes startup sooner overlapping with the operation of the ethane heat pipes. This overlap results in smaller heat pipe axial temperature gradients at startup. As a result of a lower critical temperature, the nitrogen heat pipe operation does not overlap with the ethane. Therefore, the nitrogen heat pipes operate initially at higher axial temperature gradients. Eventually, as the load cools, the evaporator and condenser approach the same temperature and nearly isothermal operation follows. Upon further cooldown of the load, the working fluid in the heat pipes freezes stopping circulation. For nitrogen this occurs at 63.1 K and for oxygen at a lower temperature of 54.3 K. Heat transfer is then limited to conduction along the heat pipe walls.

Regardless of the temperature gradient imposed along the heat pipe length, in all cases the internal operation of the heat pipes is the same. The large temperature differences only change the effective length of the heat pipe. In the non isothermal mode the condensate does not reach all the way to the evaporator before it is vaporized. The point at which the liquid film is depleted defines the effective length of the heat pipe, which for strongly non isothermal operation is less than the total length. Heat from the load flows by conduction along the wall of the heat pipe to the point above the evaporator where the liquid film disappears. This conduction length varies according to the amount of superheat present in the evaporator. Large superheats require large conduction lengths and result in shorter effective heat pipe lengths. Eventually, as the load cools the conduction length shrinks to zero and the heat pipe becomes nearly isothermal over its entire length. The condensation, circulation and evaporation inside the heat pipe is similar in all cases, only the point at which evaporation occurs differs and therefore, changes the effective heat pipe length. Non isothermal operation of the heat pipe should not be viewed as abnormal or detrimental as the heat pipe accommodates for this condition.

A thermal model for the ALISS system was developed to analyze the various cooldown scenarios and evaluate the optimum heat pipe configuration and selection of working fluids. The model is a lumped capacitance, finite-difference type using the SINDA[4] computer code. Temperature dependent heat capacities and interface conductances were used. Cryocooler performance was also temperature dependent and based on available test data.

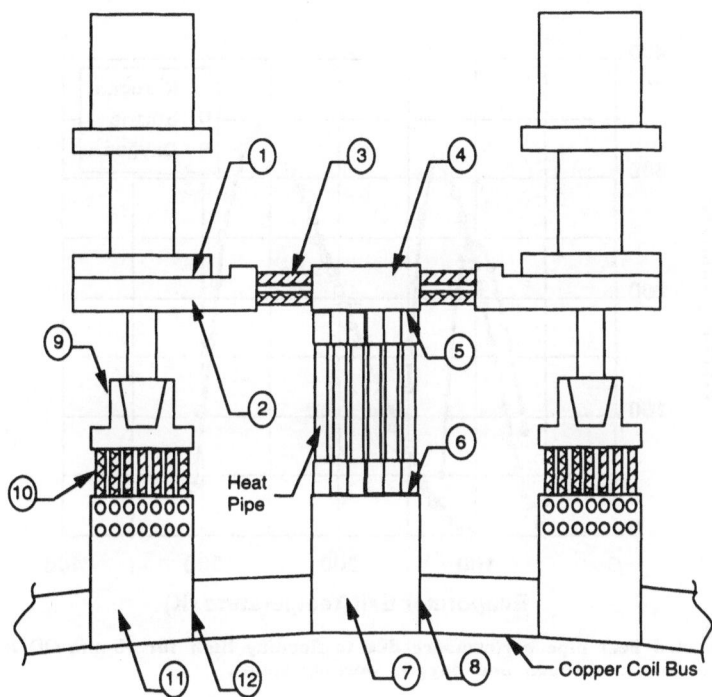

**Figure 1.** Schematic of ALISS thermal configuration showing shunt heat pipes connecting the first stages of the cryocoolers to the load with thermal paths identified as 1) 1st stage contact conductance; 2) 1st stage cryocooler bus bar; 3) 1st stage flexible braids; 4) 1st stage collector bus; 5) heat pipe condenser contact conductance; 6) heat pipe evaporator contact conductance; 7) heat pipe cold bus; 8) heat pipe cold bus to coil contact conductance; 9) 2nd stage interface conductance; 10) 2nd stage flexible braids; 11) cryocooler cold bus; and 12) cryocooler cold bus to coil contact conductance.

Use of the shunt heat pipes results in significant reduction in cooldown time for the ALISS system as shown in Fig. 3. The 961 kg load temperature is shown as a function of time for the case without any heat pipes and for the case with two ethane and four nitrogen heat pipes corresponding to the ALISS configuration. With heat pipes present the cooldown time is reduced to nearly half of the case without heat pipes. Figure 4 shows a heat load map for the two cooldown cases. Included in the figure is the predicted heat load on the first stage of the cryocoolers both with and without heat pipes present. These curves illustrate the effect of shunting a portion of the cooldown load to the first stages of the cryocoolers resulting in faster cooldown for the system. Also shown in Fig. 4 is the heat load carried by the ethane and nitrogen heat pipes. The ethane heat pipes operate for the first 65 h of the cooldown during which time the nitrogen heat pipes remain supercritical and do not operate. After a brief period of approximately 4 h when none of the heat pipes are operational because the ethane is frozen and the nitrogen is still supercritical, the nitrogen heat pipes begin operation in the non isothermal mode described earlier. The effect of the 4 h interruption in heat pipe duty can be seen in Fig. 4 as a drop in the first stage heat load. Finally, at 105 h the nitrogen also freezes and all heat pipes cease operation. Except for some heat conduction along the heat pipe wall, this terminates the shunting of heat between the load and the upper stages of the cryocoolers. From this point, only an additional 10 h are required for the magnet to reach 4 K. Control of the heat pipes is passive in the sense that no external action is required to initiate or terminate their operation. The heat pipes respond automatically to the imposed thermal conditions.

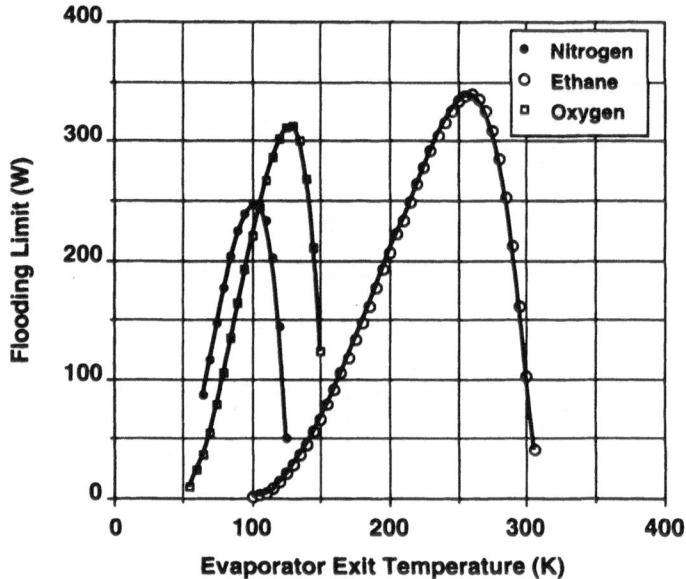

**Figure 2.   Predicted heat pipe performance due to flooding limit for 13-mm-OD heat pipe with ethane, nitrogen and oxygen working fluids.**

## HEAT PIPE DESIGN

The heat pipes are gravity-assisted and therefore require the evaporators to be below the condensers in a gravity field when operating. All of the heat pipes are identical and are approximately 300-mm-long with an outside diameter of 13 mm. The evaporators and condensers are copper while the adiabatic section is fabricated from low thermal conductivity, thin-walled tubing. The evaporators and condensers are sized to accommodate the radial heat load and the geometry of the thermal interfaces. A schematic of the heat pipes is shown in Fig. 5.

The heat pipe performance, as shown in Fig. 2, is expected to be limited by flooding of the condenser by the counter flowing vapor. At high vapor velocities the liquid return flow along the heat pipe wall is impeded by the liquid-vapor interaction. At the onset of flooding the liquid is pushed into the condenser by the upward flowing vapor, interrupting the flow of liquid to the evaporator. As a result, partial dryout of the evaporator occurs and the heat pipe capacity is reduced. At temperatures near the freezing point of the working fluid, the vapor pressure is low

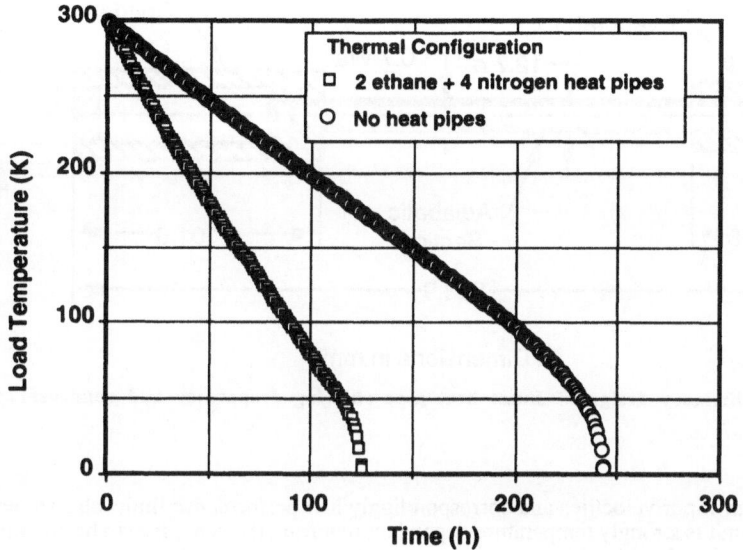

**Figure 3.** Cooldown performance of ALISS system with and without 2 ethane and 4 nitrogen 13-mm-OD heat pipes.

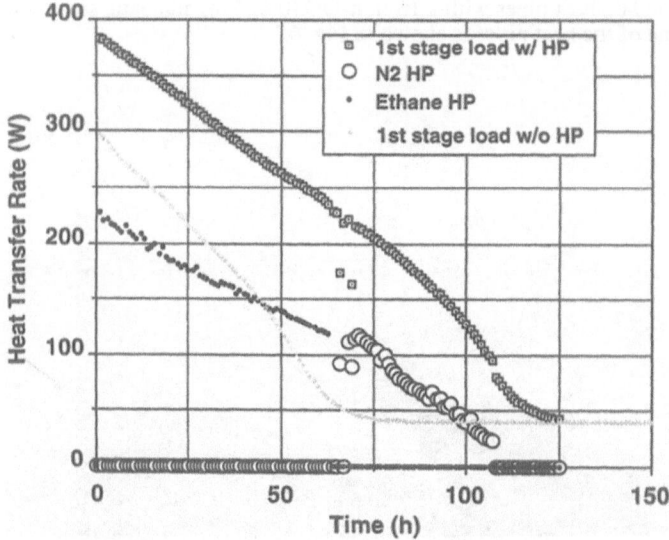

**Figure 4.** Load map of ALISS system cooldown with and without 2 ethane and 4 nitrogen 13-mm-OD heat pipes.

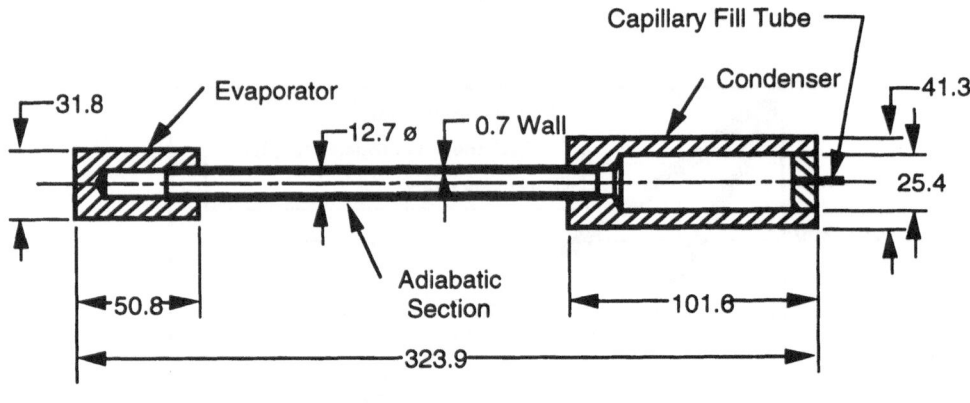

All Dimensions in mm

**Figure 5.   Preliminary design of shunt heat pipe showing evaporator and condenser geometry.**

resulting in high vapor velocities and correspondingly low performance limits also shown in Fig. 2. The flooding limit is strongly temperature dependent requiring the heat pipes to be slightly oversized to carry the required heat load over a wide temperature range.  The data in Fig. 2 were generated using the Los Alamos Heat Pipe Code HTPIPE[5].

## HEAT PIPE TESTS

Heat pipe performance testing was conducted at the Carderock Division of the Naval Surface Warfare Center, Annapolis Detachment (CDNSWC/A).  Based on the heat pipe design described previously, two scaled up heat pipes with a 16-mm-OD (0.625 in) adiabatic section were fabricated. A photograph of one of the heat pipes is shown in Fig. 6.

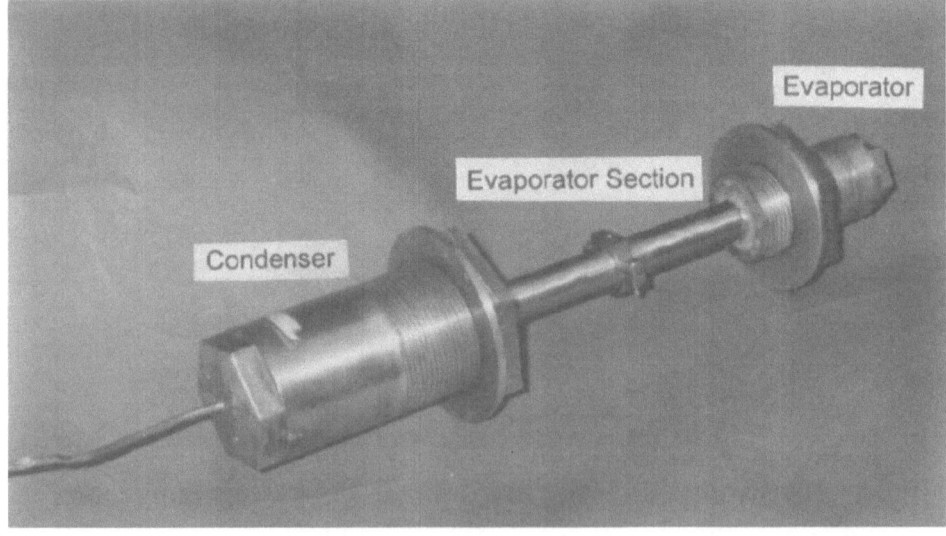

**Figure 6.   Heat pipe showing mounting provision for evaporator and condenser**

The isothermal performance of the 16-mm-OD heat pipe can be accurately predicted using the Los Alamos computer code HTPIPE.[5] Figure 7 shows the heat pipe performance predicted by HTPIPE for the various working fluids together with test data obtained during isothermal heat pipe operation for ethane and nitrogen. These measurements were limited by large temperature gradients between the cooler and the heat pipe interfaces and by the cooler capacity. Because of these limitations, only heat pipe performance limits near the working fluid critical point could be measured. The measured data are consistent with the performance predictions for isothermal operation.

However, the non-isothermal (evaporator and condenser at different temperatures) performance is not predicted by HTPIPE and; therefore, additional tests were performed to determine these characteristics. The heat pipe condenser was thermally connected to the cold plate of a single stage GM cooler from Cryomech (AL 200). The heat pipe was installed in a vacuum vessel with a 1-stage GM cooler, five Lake Shore diode temperature sensors, and two cartridge heaters. Three diode temperature sensors were located on the heat pipe: one at the condenser; one at the evaporator; and one at the middle of the heat pipe (adiabatic section). The remaining two temperature sensors were located on the cold plate and the hot plate respectively of the test apparatus. The first heater, which was computer controlled, was placed near the heat pipe condenser to provide a nearly constant condenser temperature. The second heater was located on the evaporator plate to provide the thermal load for the heat pipe. Figure 8 shows the arrangement of the heat pipe, cooler, and instrumentation.

The heat pipe was filled with nitrogen to a pressure of 136 atm (2000 psig) and the non-isothermal performance measured. In simulating non-isothermal operating conditions the heat pipe temperature was increased until partial dryout of the evaporator occurred. The condenser temperature was then reduced and maintained at 105 K while the evaporator temperature was varied from 160 K to 255 K by varying the power supplied to the evaporator.

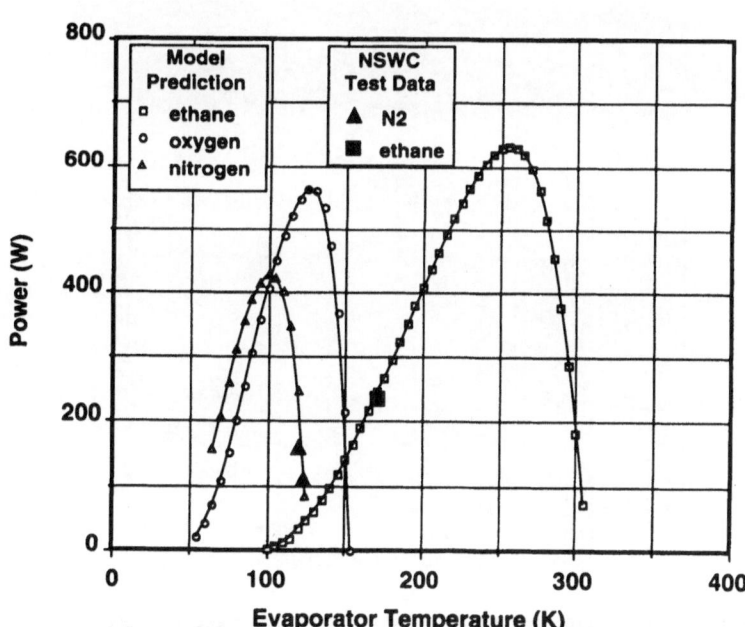

**Figure 7.**   Comparison of predicted isothermal heat pipe performance with test data for 16-mm-OD heat pipe.

The results of the non isothermal tests are shown in Fig. 9. The non-isothermal capacity of the nitrogen heat pipe is a strong function of the evaporator temperature. The length of the evaporator dryout zone increases non linearly with increasing evaporator temperature. Consequently, the heat transfer rate which is proportional to the temperature difference along the heat pipe and inversely proportional to the evaporator dryout length, goes through a minimum as the evaporator temperature increases. Initially, the increase in the evaporator dryout length increases faster than the heat pipe temperature difference causing the heat transfer rate to decrease. Eventually, the temperature difference increases faster than the dryout length and the heat transfer rate increases with increasing evaporator temperature. The non isothermal performance of the oxygen heat pipes is expected to be similar to the nitrogen heat pipes and testing of the oxygen heat pipes is planned. The heat pipe evaporator response determined from the non isothermal tests will be incorporated into the system cooldown model.

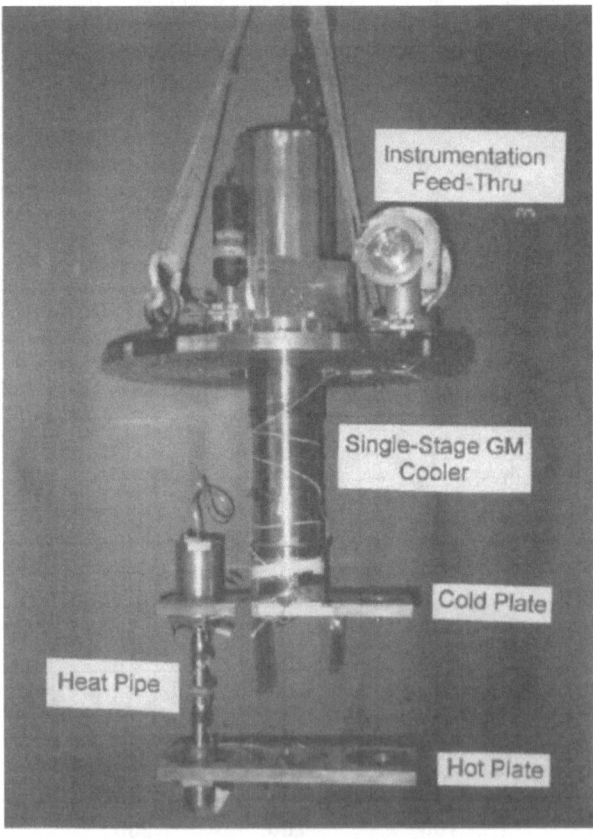

**Figure 8.   Heat pipe mounted on cryocooler first stage for testing.**

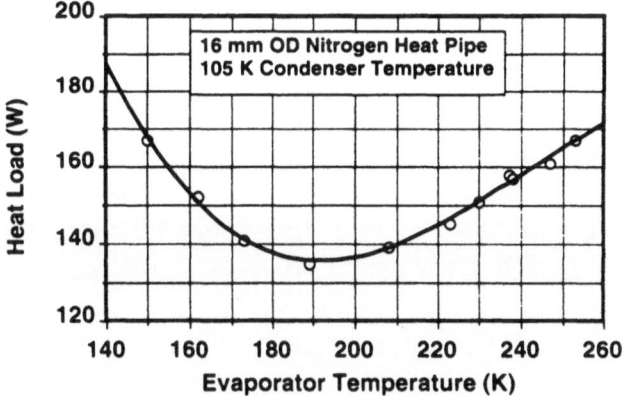

**Figure 9.** **Non isothermal heat pipe performance based on constant condenser temperature.**

## CONCLUSIONS

This study investigated the use of ethane, oxygen, and nitrogen heat pipes to enhance the cooldown of a large superconducting magnet system. High capacity ethane, oxygen, and nitrogen cryogenic heat pipes were modeled, designed, fabricated, and tested. In addition, the non-isothermal performance of the nitrogen and oxygen heat pipes, where the evaporator temperature exceeded the working fluid critical point, were experimentally determined.

The impact of the various types of heat pipes on the cooldown time for a large superconducting magnet system was determined through the use of a system cooldown thermal model. The thermal model predicted that the use of heat pipes as thermal shunts between the first and second stages of the cryocooler can reduce the cooldown time by 50 percent.

Testing and evaluation of the three candidate heat pipes centered around the determination of an effective heat pipe combination. The use of oxygen heat pipes in conjunction with ethane heat pipes provided a continuous, nearly isothermal shunt for the cryocooled system from room temperature down to about 60 K. In contrast, it was found that nitrogen heat pipes in combination with ethane heat pipes did not perform isothermally in the temperature range of 150 K - 125 K.

Because of oxygen's wider operating temperature range (60 - 155 K) and its greater transport properties compared with nitrogen, oxygen heat pipes were selected to operate with the ethane heat pipes as a thermal shunt between the first and second stages of the cryocooled system.

## REFERENCES

1. E. M. Golda, J. D. Walters and G. F. Green; "Applications of Superconductivity to Very Shallow Water Mine Sweeping," Naval Engineers Journal, May 1992.

2. M. Heiberger, et. al., "A Light-Weight Rugged Conduction-Cooled NbTi Superconducting Magnet for U.S. Navy Minesweeper Applications," Cryogenic Engineering Conference, Columbus, OH, July, 1995.

3. F. C. Prenger, et. al., "Nitrogen Heat Pipe for Cryocooler Thermal Shunt," Cryogenic Engineering Conference, Columbus, OH, July, 1995.

4. B. A. Cullimore, et. al., SINDA/FLUINT Systems Improved Numerical Differencing Analyzer and Fluid Integrator, Users Manual, Version 3.1, Cullimore and Ring Technologies, Inc. Sept 1995.

5. K. A. Woloshun, et. al., "HTPIPE: A Steady-State Heat Pipe Analysis Program", Los Alamos National Laboratory internal report no. LA-11324-M, Nov 1988.

# Control Electronics Development for Space-Based Cryocoolers

**P. M. Mayner**

Hughes Aircraft Company, Electro-Optical Systems
Santa Barbara Remote Sensing
75 Coromar Drive
Goleta, CA, USA 93117

## ABSTRACT

This paper presents the control electronics development program at Hughes to support space-based, long-life cryocoolers. The paper provides descriptions of electronics developed by Hughes to support cryocooler applications as well as sample test data obtained using these electronics.

A description is provided of the electronics used for the February 1995 flight aboard the Space Shuttle of the Hughes Improved Standard Spacecraft Cryocooler and Oxygen Diode Heatpipe. Various flight data obtained from the flight—including system power, cryocooler power, cooling capacity, and uncompensated vibration data—are provided. The system impact of starting and stopping the cryocooler during the flight experiment is examined.

A vibration and temperature control algorithm development and test system is described. The system uses an IBM-compatible personal computer and commercial circuit cards to provide the cryocooler drive signals and data sampling system. The control algorithms are developed in C++ and are designed so that the sample and update rates can be adjusted to measure the effect on vibration control scheme convergence.

## INTRODUCTION

This paper describes the Cryo System Experiment (CSE) electronics and some of the results obtained from the CSE flight in February 1995, and provides a description of a vibration and temperature control algorithm development system that is being used to study vibration control algorithms, provide possible electronics simplifications, and characterize cryocoolers. The CSE was developed under contract to JPL/NASA.

Hughes Aircraft has been involved in the design and manufacturing of long-life, low-vibration Stirling cycle cryocoolers and pulse tubes since the late 1980's. The cryocooler activities started with the Air Force sponsored 65K Standard Spacecraft Cryocooler (SSC) program, which developed a prototype Stirling cryocooler that was intrinsically balanced by having opposed pistons. This cryocooler was followed by the Improved Standard Spacecraft Cryocooler (ISSC) and the Prototype Space Cooler (PSC).

To support the cryocooler development and test activities, a laboratory set of electronics was developed and built. This laboratory electronics set formed the basis for future cryocooler electronics development activities including PSC brassboard electronics and the flight electronics for CSE. Each of these electronics sets had a different emphasis. The laboratory electronics is manually operated and designed with the flexibility to accommodate cryocooler variations and to support the needs of cryocooler testing and characterization. The PSC brassboard is processor controlled, uses flight-like circuits, and provides a vibration control and temperature control capability. The CSE electronics provided a flight-qualified design that emphasized the characterization of both cryocooler performance and the Oxygen Diode Heatpipe.

Using the lessons learned from its previous electronics, Hughes has embarked on a program to develop an improved cryocooler electronics that will be lighter, more power efficient, and more reliable. The electronics is intended to support the current ISSC-type coolers, and it is anticipated that it will support the new smaller coolers being developed. The electronics is being developed to be compatible with the Loral R6000 processor, although it is anticipated that the processor will be very lightly loaded and will be available to support other payload functions if needed. A working prototype is expected by the end of 1996.

## THE CRYO SYSTEM EXPERIMENT ELECTRONICS DESCRIPTION

**CSE Objectives.** The CSE system objectives were to flight-qualify the Hughes Aircraft Company Improved Standard Spacecraft Cryocooler and Oxygen Diode Heatpipe designs by demonstrating launch survival and on-orbit operation, to determine the micro-gravity behavior of these devices, and to provide cryo system integration experience. The CSE system flew successfully on Space Shuttle mission STS-63 in February 1995 as part of the NASA Hitchhiker program.

**CSE Subassembly Mounting.** The CSE consisted of two major subassemblies that were mounted on a cross-bay assembly with other Hitchhiker experiments as shown in Figure 1. One subassembly was a side plate for mounting the electronics units; the other was a "GAS" canister

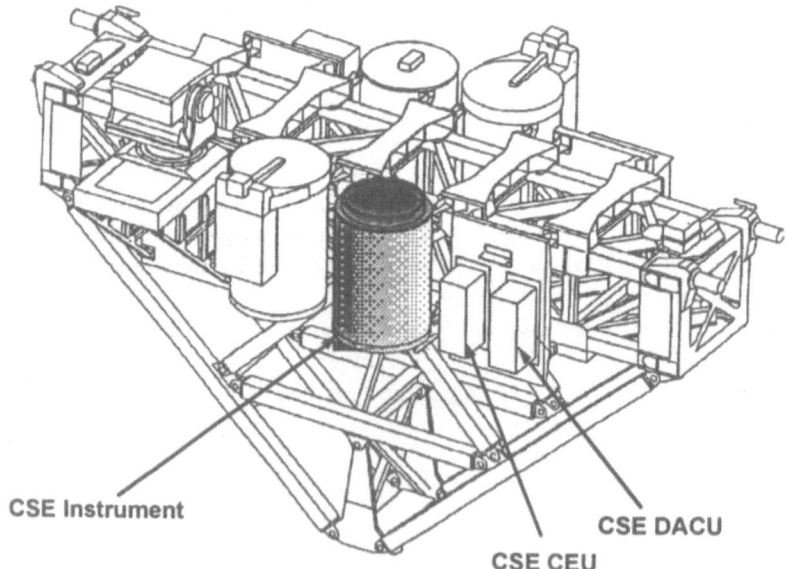

**Figure 1.** CSE location on cross bay assembly

which contained the experimental cryogenic system and a canister electronics module. The CSE system shared the cross-bay assembly with an experiment that measured the "Shuttle Glow," an ODERACS orbital debris experiment, and an IMAX camera.

**CSE Cryogenic System Instrumentation.** The CSE cryogenic system contained in the "GAS" canister had an ISSC, an Oxygen Diode Heatpipe, a thermal mass, a thermal radiation shield, two tactical cryocoolers, thermal straps, and instrumentation. Figure 2 is a functional diagram showing the heaters, coolers, and sensors in the experiment canister. The ISSC and Heatpipe were configured to cool the common thermal mass through thermal straps that were instrumented so that the heat flow could be measured. One tactical cooler minimized the parasitic heat load on the thermal mass by cooling the thermal radiation shield; the other provided cooling to the condenser end of the Heatpipe. The thermal mass heaters provided temperature stabilization and the Heatpipe heater transitioned the Heatpipe into reverse-conductance mode. The orthogonally mounted ISSC compressor and expander/balancer housings had accelerometers mounted with the axis of sensitivity aligned with the piston drive direction.

In order to create a quick-turnaround, cost-effective electronics subsystem package, the CSE electronics used proven, available designs. For the flight processor and telemetry subsystems, previously flow Military Standard circuit cards built by Titan Electronics in San Diego were

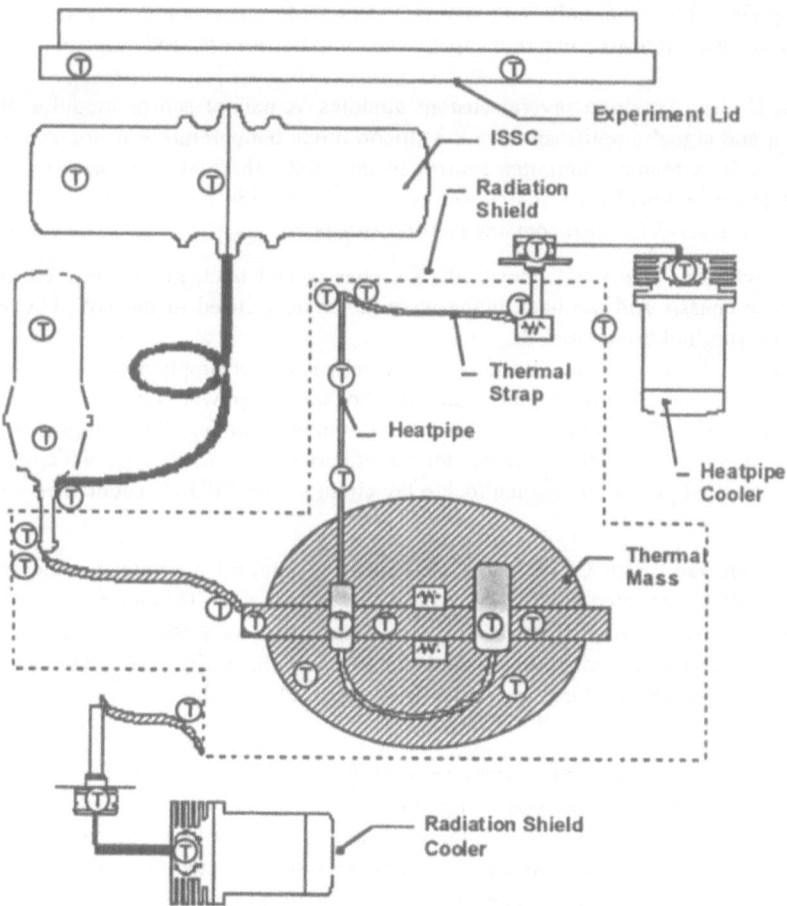

**Figure 2.** Telemetry block diagram

used. The CSE flight electronics enhanced available Hughes cryocooler electronics building blocks to add extra telemetry data points and provide the required interfacing with the flight processor.

**CSE Flight Electronics Partitioning.** The CSE flight electronics was partitioned into three units: the Data Acquisition and Control Unit (DACU), the Cryocooler Electronics Unit (CEU), and the Canister Electronics Module (CEM). The DACU was the command and control interface to the Hitchhiker system and provided the data acquisition and experiment control functions. The CEU provided cryocooler motor drive electronics and power control for the tactical coolers. The CEM provided signal within the experiment canister. Data from the experiment were transmitted by the Hitchhiker system to the Ground Support Equipment (GSE) using RS232 protocol at a 1200 baud rate. The GSE consisted of a standard IBM-compatible PC that logged experiment data to disk, provided numeric or graphic data displays, and provided off-line data retrieval.

**DACU Description.** The DACU consisted of a chassis, back plane, power supply, single-board computer, analog input/output, and bus termination electronics supplied by Titan Electronics. This hardware was selected because it had been used successfully on other shuttle flights. The back plane used a modified multibus format with wire wrap to allow easy customization. The chassis was a modified "air transport rack" allowing for side mounting on the Hitchhiker. The single-board computer used an Intel 80186 microprocessor with ultraviolet-erasable programmable read only memory (UV-EPROM), static random access memory (SRAM), serial data interface, discrete inputs and outputs, timers, and interrupt controller peripherals.

The DACU also contained several custom modules. A pair of sensor modules provided source current and signal amplification to the silicon diode temperature sensors. A waveform generation module extended computer control to the ISSC, thermal mass heaters, heatpipe heater, and tactical coolers. Two analog servo modules provide basic ISSC piston position control functions. The DACU also contained a survival heater.

**CEU Description.** The CEU consisted of a chassis and back plane supplied by Titan Electronics. The chassis and the back plane were the same as used in the DACU except the printed circuit layers had been removed.

The CEU modules were custom designs. A low-voltage power supply module use MIL-STD dc-dc power converters. A power control module housed the power relays, while a pair of switch-regulated motor drive modules provided the current drive to the ISSC compressor motors. A single expander/balancer module provided the current drive to the ISSC expander and balancer motors, and controlled power and launch lock relay circuits. The CEU also contained a survival heater.

**CEM Description.** The CEM was a chassis that contained a single module with two identical circuit cards mounted back-to-back on a single heat sink. This module provided the sensor drive and signal amplification for the differential temperature sensors and the accelerometers. The module also sensed and conditioned the ISSC piston contact signals, provided the thermal mass and heatpipe heater drives, and selected the heatpipe cooler set-point temperature.

**Power Distribution.** The CSE system was configured to use one Hitchhiker system power port for the main electronics power, consisting of two power lines. One power line was dedicated to the DACU; the other was split between the ISSC and tactical coolers. This arrangement was chosen so that the telemetry and control could not be adversely affected by cooler malfunctions. This resulted in a power distribution and fusing that allowed the radiation shield cooler to run continuously while either the heatpipe cooler or the ISSC could be operated. The ISSC was fused to allow 100W of power for the electronics and cryocooler.

**Cryocooler Interface.** The Hughes ISSC has two opposed compressor pistons, an expander piston, and a expander balance motor (or "balancer"). For each piston and the balancer, the cryocooler uses a Linear Variable Differential Transformer (LVDT) position sensor. Each piston has a piston contact sense interface. Each motor drive circuit was instrumented such the commanded piston position, the position servo error, the motor drive current, and the motor drive voltage could be sampled.

To control the cryocooler, the cryocooler drive electronics accepted a desired piston position input from the processor and closed a position servo loop to generate the motor control signals. This allowed the processor to set up the drive parameters once and eliminated the need for the flight software to actively monitor the piston position.

**Cryocooler Launch Survival.** One of the critical issues for the cryocooler system is the ability to survive the launch environment. The ISSC was designed to survive launch loads by "launch locking" the pistons with an electrical short on the motor windings. This eliminated the need for maintaining power to the electronics for the launch. The successful operation of the cryocooler on-orbit demonstrated that the cryocooler had sufficient damping and that relays used were adequate for protecting the cryocooler during launch.

During the design of the electronics, a trade-off was made between using latching and non-latching relays to provide the motor winding short. Non-latching relays were chosen because power and weight are not a major design factors for Hitchhiker payloads on the shuttle and, more importantly, the experiment had to survive the landing vibrations—so it was imperative that the cryocooler to be protected no matter how the experiment was shut down.

**Experiment Timeline.** The CSE experiment timeline consisted of five major segments: system checkout, thermal mass cooldown, main experiment, extended experiment, and shutdown/warm-up. Because the Hitchhiker payloads do not have guaranteed access to the command uplink and hence may have to operate without commands, the electronics was programmed to perform the experiment autonomously as a sequence of experiment. Each step controlled what coolers and heaters operated and what telemetry was gathered. Each step had a default duration and could be set to advance when certain telemetry conditions were met. Each timeline segment consisted of several programming steps.

The CSE experiment was powered up 5 hours after launch and was operated for 7 days and 19 hours. The system checkout phase lasted 1.5 hours but due to some problems with the shuttle engines the experiment was put on hold to minimize power consumption while the shuttle orbited with the bay facing the sun for extended periods. The thermal mass cooldown was resumed 11 hours after launch.

## CRYO SYSTEM EXPERIMENT FLIGHT DATA

Many of the results obtained from the CSE integration and flight are contained in the Final Report, Cryo System Experiment, Hughes Electro-Optical Systems. Presented here are some of the results that relate mainly to the electronics performance.

### Cryocooler Stiction Testing

One of the lessons learned during the CSE electronics integration was the need for an in-flight hysteresis or stiction test capability for the ISSC. The hysteresis test is an important diagnostic tool that detects piston to side-wall rubbing. This rubbing can occur without the contact-sensing interface detecting the contact because the interface relies on detecting an electrical short, and the rubbing can occur between an electrically insulated portion of the piston and the side-wall. Piston to side-wall rubbing is therefore detected by measuring and plotting the piston position versus the motor drive current.

During integration of the CSE, the cryocooler stiction test proved to be vital for detecting and correcting cooler cold-tip side loading problems. As a result of these experiences, the test was incorporated into the flight system and was conducted several times during the mission.

To perform this test, the CSE electronics configures the piston position servos into an open loop mode, then commands the motor drive signals to slowly ramp up and down. Piston to side-wall contact is detected as a difference in the piston position as a function of increasing or decreasing motor current. Any piston to side-wall contact would be visible on the GSE display as an opening up of the curve obtained by plotting current versus position.

During the CSE experiment flight, the cryocooler stiction test was run to completion a total of six times. The stiction test indicated before launch that compressor A had a slight amount of friction at the extremes of the compressor stroke, while the other pistons showed no sign of friction during testing. Figures 3, 4, and 5 show the Compressor A curves, the expander and balancer curves, and a close-up of the friction region of Compressor A. Each figure shows the data from all six hysterisis tests. Compressor B is not shown because it matches the shape of Compressor A but has none of the opening of the curve. The expander curve shows that the motor current was increased to the point that a piston stop was encountered (as shown by the vertical increase in current without a change in position).

Worst-case Compressor A friction was measured on mission day three while the cold-tip temperature was 66 Kelvin and after the cryocooler had been operated for about three hours. The compressor case and compressor rejection temperatures were 308.4K and 308K, the highest compressor case temperature at which a stiction measurement was made (the next to worst case was measured at the second highest compressor case temperature of 302K). Worst-case friction was severe enough that the motor current at the stroke limit was reduced by 0.9 amps before the piston motion resumed. The normal operation of the cryocooler, however, was commanding a stroke amplitude of less than 2.7 mm peak-to-peak which is not within the worst friction region.

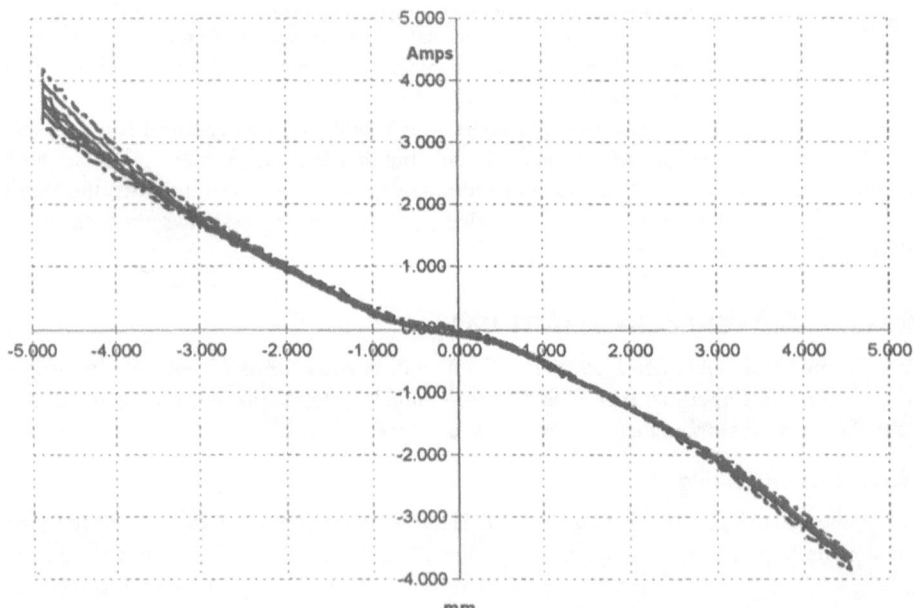

Figure 3 Compressor A Hysteresis Curve showing contact

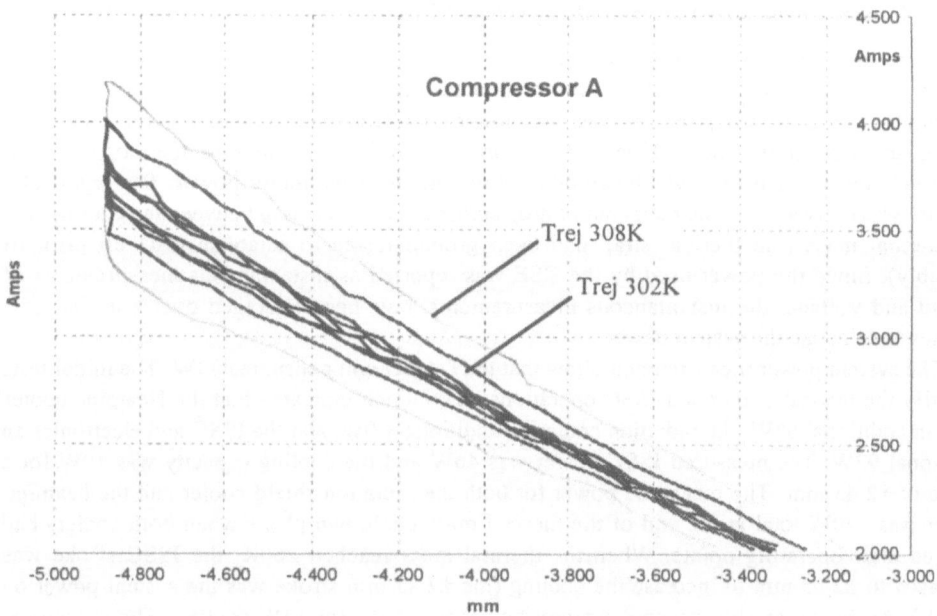

**Figure 4** Friction Region of Compressor A Hysteresis Curve

The hysterisis curves differ from those taken pre-launch in the starting position of the piston. These changes are due to the change to a micro-gravity environment. The center of the flat portion of the curve is the expected at-rest position for the piston, but the flight data indicated that the compressor piston started off center, possibly indicating that there may have been a slight pre-load on the compressor piston flexures.

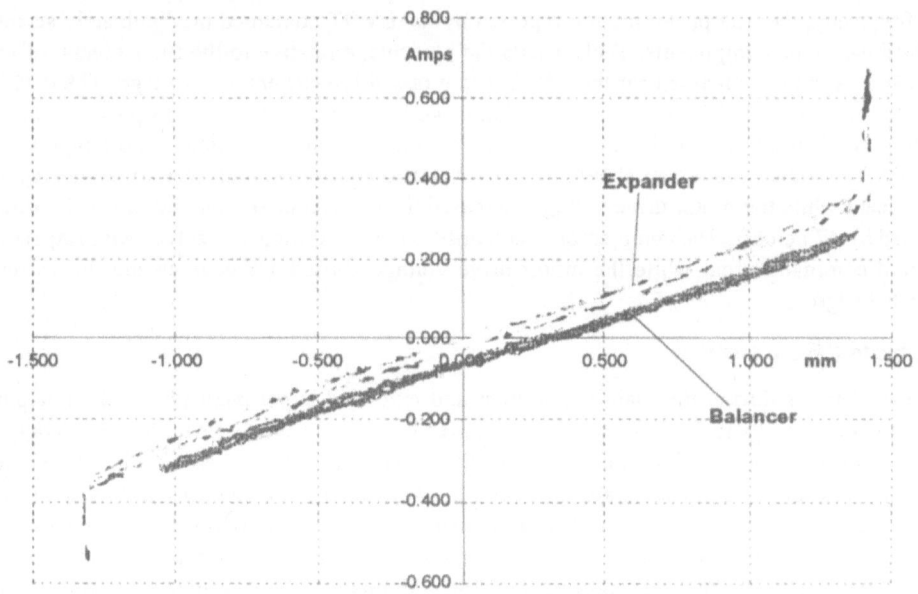

**Figure 5** Hysteresis Curves for Expander, and Balancer showing endstop contact

### System Power, Cryocooler Power, and Cryocooler Capacity

Three different power measurements could be made with the CSE experiment: the total system power, the power provided to the cryocooler motors, and the cooling capacity of the cryocooler. The CSE system power was measured by the Hitchhiker system and reported about once every 4 seconds. The CSE experiment provided individual measurements of the ISSC motor voltages and currents which can be used to determine the motor power. The cryocooler capacity was derived from the temperature drop across the thermal strap between the cold-tip and the thermal mass (the thermal strap had been ground tested to establish a dT/dP prior to assembly). Since the power used by the CSE was reported as instantaneous measurements of current and voltage, the instantaneous measurements have been averaged over a period of 5 minutes to calculate the system power.

The system power measurements show that the DACU unit consumed 83W. The initial tests to verify the tactical cooler and ISSC operations post-launch indicated that the Heatpipe cooler used an additional 65W, the radiation cooler an additional 60W, and the ISSC and electronics an additional 92W. The measured ISSC power was 46W and the cooling capacity was 10W for a stroke of ±2.43 mm. The measured power for both the radiation shield cooler and the heatpipe cooler was 120W total at the end of the thermal mass cooldown phase when both coolers had reached their operating points. When the thermal mass reached 260K, the ISSC stroke was increased to ±2.65 mm to increase the cooling (the ±2.43 mm stroke was the system power on default). At this point, the cryocooler motor power was 55W for 12W cooling. The cryocooler motor power increased to 64W for 6.5W cooling at 135K. At 65K the measured motor power was 76.5W for 1.4W cooling capacity.

### ISSC Electronic Drive Performance

The spectral content of the ISSC piston positions and the motor drive voltages and currents provide a measure of the effect of using a piston position servo loop in the control electronics and indicate the benefit of using current motor drivers versus voltage motor drivers.

The flight data indicate that with a servo position input that contained only the fundamental drive frequency, the true piston motion reported by the LVDT contained predominately second and third harmonic components. Table 1 lists the percentage relative to the fundamental of the first seven harmonics obtained for the LVDT and motor drive current and voltage. The data in the table were compiled from motor drive output voltage, motor current, and piston position sample sets taken during the thermal mass cooldown and when the thermal mass temperature was 65K. The same sample sets show that from warm to cold the compressor stroke decreased by 5 percent while the motor drive voltage increased by 12 percent and the motor drive current increased by 47 percent. The compressor piston position phase shifted 11 degrees with respect to the servo command input while the motor drive voltage shifted 1.5 degrees and the current shifted 8.3 degrees.

### Heatpipe to ISSC transition

Several times during the main experiment and extended experiment phase, the heatpipe cooler was turned off and the ISSC cooler was started to maintain the thermal mass temperature. In order to transition the diode heatpipe into the reverse mode, the heatpipe heater on the condenser was turned on. This heater provided 3.4W load to the condenser so that the large reverse-mode temperature drop could be established rapidly (as opposed to waiting for the parasitic loads). Figure 6 shows a detailed temperature vs time plot for the transition from heatpipe cooling of the thermal mass to the ISSC—indicating that the thermal mass rose in temperature by 3.4K in the next 35 minutes even though the ISSC cold-end temperature dropped within one minute by 5K. The test data show that the ISSC cooling capacity jumped from a 1W

Table 1 Relative Amplitude of Harmonics

| Harmonic | LVDT | | | | Motor Voltage | | | |
|---|---|---|---|---|---|---|---|---|
| | 65 K | Max | Min | Avg | 65 K | Max | Min | Avg |
| 2 | 1.67% | 2.20% | 1.67% | 2.04% | 1.63% | 1.63% | 1.16% | 1.44% |
| 3 | 0.46% | 0.46% | 0.09% | 0.21% | 0.20% | 0.50% | 0.20% | 0.42% |
| 4 | 0.02% | 0.05% | 0.01% | 0.03% | 0.08% | 0.08% | 0.01% | 0.04% |
| 5 | 0.05% | 0.07% | 0.04% | 0.06% | 0.09% | 0.13% | 0.09% | 0.11% |
| 6 | 0.01% | 0.02% | 0.00% | 0.01% | 0.00% | 0.03% | 0.00% | 0.02% |
| 7 | 0.02% | 0.02% | 0.00% | 0.01% | 0.06% | 0.06% | 0.03% | 0.04% |
| Harmonic | LVDT | | | | Motor Current | | | |
| | 65 K | Max | Min | Avg | 65 K | Max | Min | Avg |
| 2 | 1.67% | 2.20% | 1.67% | 2.04% | 4.87% | 10.13% | 4.87% | 8.86% |
| 3 | 0.46% | 0.46% | 0.09% | 0.21% | 2.19% | 2.19% | 0.44% | 1.12% |
| 4 | 0.02% | 0.05% | 0.01% | 0.03% | 0.25% | 0.60% | 0.18% | 0.37% |
| 5 | 0.05% | 0.07% | 0.04% | 0.06% | 0.35% | 0.92% | 0.35% | 0.62% |
| 6 | 0.01% | 0.02% | 0.00% | 0.01% | 0.08% | 0.48% | 0.07% | 0.25% |
| 7 | 0.02% | 0.02% | 0.00% | 0.01% | 0.05% | 0.39% | 0.03% | 0.22% |

load to a 1.3W cooler after the turn-on sequence. The measured piston power jumped to 77W. The thermal mass was constructed out of 8.7 lb of aluminum, so the rise in temperature is due to a sizable load. The heatpipe evaporator temperature did not "break free" for 32 minutes and when it did, the temperature of the heatpipe midpoint temperature dropped. It therefore appears that the temperature rise of the thermal mass was due to the heatpipe heater and not the ISSC startup.

This finding appears to be consistent with the heatpipe anomaly detected during the mission. After the final heatpipe forward mode test, it was observed that the parasitic loads were not adequate for reversing the heatpipe. The reason for this is unknown, but it was established that the heatpipe would switch into reverse mode with only parasitic heat loads if the tactical cooler for the radiation shield was turned off. It appears that the tactical cooler was providing enough vibration to keep the heatpipe active.

## Uncompensated Vibration Data

The CSE system had two accelerometers mounted on the ISSC to measure uncompensated piston on-axis vibration data. The tactical coolers used for the heatpipe and radiation shield produced enough vibration at the ISSC to make the ISSC vibration unreadable. To obtain just the ISSC vibration data, the system provided the ability to turn off the tactical coolers slightly before vibration data sampling and to resume the cooler operation shortly afterwards. Unfortunately, during integration and test of the experiment, while exercising the tactical cooler stop/measure/start, it was discovered that the tactical coolers would occasionally exceed the specified maximum surge current. Attempts to mitigate the surge currents by lengthening the cooler off-time (in case there was a pressure imbalance on restart) were not successful. It was determined that the magnitude of the surge was sufficient to place the experiment at risk if the stop/measure/start were used during the main mission. The net result is that there are only a few readings of the vibration of just the ISSC.

Because the vibration sampling was synchronized to the ISSC and the cooler fundamental frequencies were 28 Hz for the tactical coolers and 45 Hz for the ISSC, any calculation of the vibration spectral information while the tactical coolers were running is not meaningful because the tactical cooler data cannot be filtered out. (This is because the sampling interval was one ISSC stroke and hence the data set resulting from a Fourier transform of the data will be

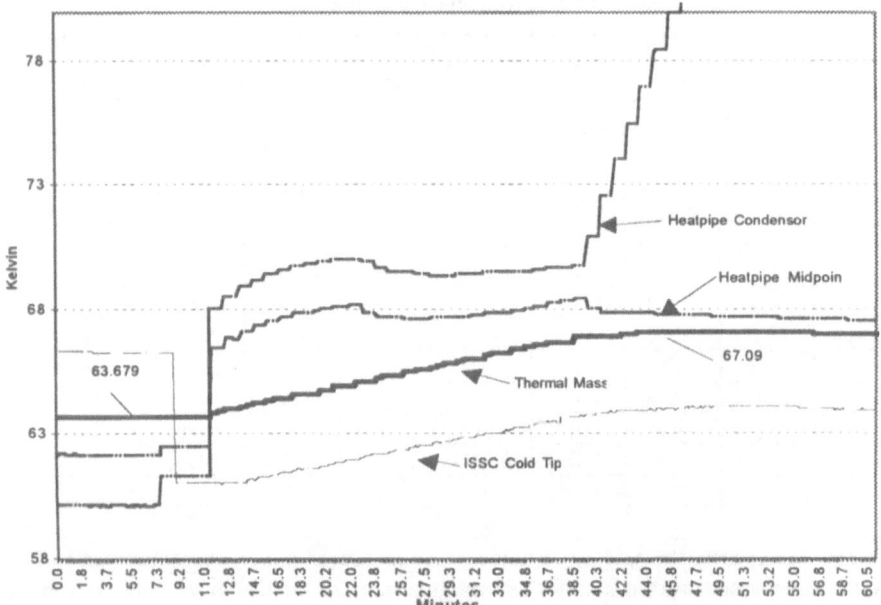

**Figure 6** Heatpipe to ISSC transition

multiples of the ISSC fundamental frequency. The tactical cooler vibration will be a convolution of the 28 Hz and harmonics with unknown phase into multiples of the ISSC fundamental.)

The accelerometers measured 3 mG peak-to-peak while no coolers were running in the CSE experiment. The heatpipe cooler produced a reading on the compressor accelerometer with a sigma of 47.6 mG and 52.36 mG at the expander. Similarly, the radiation shield cooler produced a sigma of 37.82 mG at the compressor and 56.2 mG at the expander. At the first startup of the ISSC—without any tactical cooler running, and without any fundamental or harmonic vibration control—the cooler produced a vibration sigma of 11.7 mG at the compressor and 3.81 mG at the expander.

**Piston Contact**

One of the features of the Hughes ISSC cryocoolers is that they contain a piston contact sense interface that can be used to detect when a piston is making electrical contact either with the piston end stops or the cylinder wall. The CSE electronics used this interface to detect contacts and would stop the cryocooler operation if such a contact occurred. Any contact event would be reported and after a short interval the electronics would try to resume operation of the cryocooler.

During the mission (7 days 6 hours), a total of 53 contact messages were recorded. These contacts occurred mainly when the cooler drive signals were being ramped up; however, others occurred when the shuttle was firing the main orbital maneuvering thrusters and when the shutter door was opened and closed on the IMAX camera.

**ALGORITHM DEVELOPMENT AND TEST SYSTEM**

The algorithm development and test system was created to solve a need for a relatively inexpensive and easily programmable way to drive cryocoolers while providing a friendly operator interface. The approach chosen was to use an IBM-compatible computer with some commercial data acquisition cards to generate the drive control signals. This setup could then be used to computerize existing laboratory electronics or drive brassboard electronics.

The setup's capabilities had to include outputting fundamental drive signals, vibration control signals, and timing signals at variable drive frequencies. It also had to be able to synchronously burst sample vibration data, motor position data, drive currents and voltages. The system needed to be able to sample temperature data and to provide a shutdown control input that could be tied to a piston contact sensing circuit.

The data acquisition cards chosen for the system were both made by National Instruments. The cards are provided with drivers for Lab-View and Lab-Windows as well as the NI-DAQ software. The AT-AO-10 was selected for output generation, while the Lab-PC was selected for timing control and data acquisition. Both cards support direct memory access (DMA) with the PC memory. The AT-AO-10 has a 16-bit wide interface while the Lab-PC is an 8-bit interface. DMA in the PC is controlled by two DMA controllers (one each for 16-bit and 8-bit transfers). Both the AT-AO-10 and Lab-PC have on-board FIFOs to provide data storage when the DMA channels are waiting for PC bus access. The PC DMA controllers can be set up to provide a circular buffer in the PC memory so that the DMA controller for the output channel need only be set up once by the processor. Once the output buffers are set up, any changes to the drive waveforms are made simply by updating a program array. The net result is that output waveform generation places no computational load on the PC.

The Lab-PC card has two programmable timers that have selectable clock inputs and provide external gating. One timer has a 2 MHz oscillator input that was used to generate the timing strobes used both for data sampling and for simultaneously updating the output waveforms. Reprogramming this timer effectively changes the fundamental drive frequency. The timer provides an external gate that is tied to the piston contact detection circuit. The intent is to freeze the drive waveforms at their current position in hardware. The contact signal is also routed to the digital input/output on the Lab-PC that is configured to interrupt the software when a contact occurs. This allows the software to remove the drive signal.

The Lab-PC card provides eight single-ended or four differential analog input channels. The card can be configured to sample a channel using DMA with external sample rate control and an external trigger. One of the Lab-PC timer channels is used to provide a sample count. For the algorithm development and test system, the sample rate was tied to the timing strobe used to update the AT-AO-10 output channels, and the external trigger was tied to an output channel that was programmed to provide a start of cycle trigger. This trigger allows all the sampling to be referenced to the drive signal timing and was made available externally for triggering scopes, etc. The trigger is essential if phase information is to be observed with external test equipment.

The AT-AO-10 card provides the capability to use either internal or external reference voltages for pairs of digital-to-analog converters. The Lab-PC provides two analog outputs that were tied to selected AT-AO-10 output channel converter reference inputs. In this configuration, drive signals can be ramped up by either changing the reference value or writing a new output waveform patterns.

The software for the system was written in C++ for the DOS environment. The drivers provided by National Instruments were not used because we were unable to use them to configure the AT-AO-10 correctly for the recirculating buffer mode. The card documentation provided sufficient detail so that it was relatively easy to program the card directly. The DOS environment was selected over Windows to avoid having to develop Windows device drivers and to avoid the virtual memory to physical memory translation that would be required for the DMA buffers.

The combination of the AT-AO-10 and Lab-PC has been successfully operated with seven analog outputs updating at a sustained 23 kHz each while sampling telemetry at the same rate. To date this rate has proved adequate for vibration-control needs.

The advantages of using a PC for the host for this implementation are many. Apart from being able to code up and test new vibration control/temperature control algorithms, the system can be used to synchronously sample test data and log the results, write sample data to Excel for off-line analysis, etc. As new cryocoolers are developed, these data bases could be vital for understanding system performance and could also help minimize the time spent tuning servos and debugging on flight electronics.

## SUMMARY

This paper has described the electronics developed for the NASA IN-STEP CSE program and presented some of the flight data obtained. It has also described an algorithm development and test system that is built with commercial circuit cards and a standard PC. Using the lessons learned from CSE and other cryocooler electronics development activities, Hughes has started development of a new set of drive electronics that will be smaller, lighter, more efficient, and more reliable. A working model of the new drive electronics is expected later this year.

## ACKNOWLEDGMENTS

Development of the CSE flight electronics subsystem was made possible through the active support of the Jet Propulsion Laboratory, California Institute of Technology under contract with the National Aeronautics and Space Administration. Thanks also to the NASA Hitchhiker team at GSFC for supporting the integration, testing, flight support, and post flight support. Both JPL and the GSFC Hitchhiker team made integrating and flying a payload on the shuttle a smooth, easy, and very enjoyable experience.

# Cooler Test Data Acquisition and Environmental Control Software

S. Arwood and T. Roberts

United States Air Force, Phillips Laboratory
Kirtland AFB, Albuquerque, New Mexico, USA 87117

## ABSTRACT

Cryocooler testing and performance mapping seeks to provide data concerning the ability of the cryocooler to meet program performance specifications as well as to provide input to the design of future coolers which will exceed current performance envelopes. An assumption of this testing process is that cooler performance can be measured within a controlled operating environment so that varying cooler performance can be directly attributed to the cooler itself, and separated from any variation in the testing environment. Similarly, control and instrumentation of the test stand must be implemented to minimize measurement error in comparison with operating variations due to the cooler and to not intrude in the operation of the cooler and test stand's environmental controls. Due to reduced costs for microcomputers and integrated circuits, these objectives can be met, provided that the testing stand and software is viewed as a complete design system.

The data acquisition and environmental control software designed for use in cryocooler testing at Phillips Laboratory is described with respect to its objectives, structure, and implementation. Notable aspects of these programs are: 1.) Emphasis on fully automated data acquisition and environmental control using microcomputers, 2.) Instrumentation error trapping for communications and operating errors, 3.) Integration of a wide range of test plan requirements into the environmental control program and instrumentation, 4.) Use of standardized subroutines for instrument control and data output and storage, and 5.) Active interaction between the data acquisition program and environmental control program to allow for the test stand to accommodate a range of test events or anomalies.

## INTRODUCTION

Many United States Air Force software development projects have pursued the concepts of modularity, reusability, and reliability. Lately these concepts have been mandated as a requirement in all software development. Modularity and reusability have been addressed on a very wide basis throughout the military by the use of Ada - the Department of Defense software development system. With Ada's restrictiveness, software development has matured and will continue to improve over time for mission critical or embedded real time systems which require computer certification. The cryocooler testing environment was not suitable for the use of Ada, however, as it would require the development of low level communications protocols which have already been provided in other languages by equipment vendors. The decision was made to use a high level (level 4 or

Cryocoolers 9, Edited by R.G. Ross, Jr.
Plenum Press, New York, 1997

higher) language that would make the task as easy as possible and produce a product that could be maintained without employing a dedicated team of people. The software chosen was LabVIEW® from National Instruments®.

LabVIEW® is a graphical design language that does not force a software developer to design to specific standards. After months of prototype software development the system became unmanageable due to its complexity and size. It also was not reliable, failing almost every test due to the non-predictability of the multiple 'black boxes' in the system. As system development began to slow down an alternative solution was sought. The concept of the Dataport was developed to meet the expectations of the test engineers and management.

## THE CRYOCOOLER TESTING PROBLEM

Phillips Laboratory supports a mission of managing satellite cryocooler technology development. A significant component of this mission is profiling the performance envelope of the coolers after development and delivery. this effort is divided into two interrelated components:

1. Direct cooler performance data acquisition (DAQ), and
2. Proactive control of the cooler testing environment.

To accomplish these tasks the laboratory uses computers to obtain digital property measurement data and then either:

1. Write this data to storage for DAQ purposes, or
2. Conducts real time processing using this data as part of a test stand or environmental control algorithm.

Implicit in this use of computers is the requirement that all measurement instruments be capable of digital communication with these computer systems. This is both a resource for exploitation and a latent system weakness. The largest source of system failures is due to communications network failures and the ensuing instrument or computer lockups. This is balanced by the ability to correlate the interrelationships between environmental variations, both on a systemic and post test analysis basis. A properly controlled environmental system which provides stable environmental conditions for the individual test stands allows proper conclusions to be made concerning the actual dependencies of the test data on the test's environment. By example, rejection temperature control is a primary test stand objective, yet any test stand has some variation in the rejection temperature of a cooler under test. This must be controlled within acceptable limits and then quantified so that the residual variations can then be correlated with the test data for the subject cryocooler.

## SOFTWARE DEVELOPMENT CONCEPT

The concept of a Dataport is based on the idea of a presentation table where the user may pick and choose the information that they wish to see. The Dataport is made up of five subsystems: Data Presentation, Instrument Drivers, Interface Software, Automated Controllers, and the Report & Control System. Four of the subsystems are part of the Dataport, and the fifth is a parallel package that depends upon the Dataport for information and command control capabilities.

A Dataport is a software structure that serves as a platform that encompasses all subfunctions except for control requirements. The idea is to have all information gathering routines come together to present their data via one software routine. This one software routine is like a table that displays all information needed by the user and by control routines. All information then may be viewed or retrieved via the Dataport.

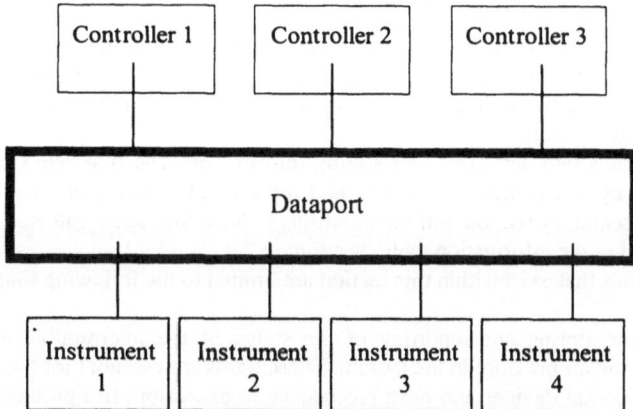

**Figure 1.** Example of the 'Port' for all information.

With all data being presented and controlled via one platform there is a distinct separation of the software functions of "data gathering" and of "system control". The only control functions within a Dataport are the control efforts to keep the data gathering subroutines operating correctly. The Dataport and its associated subroutines should be an independent and a stand alone routine. It can operate without any controlling software running.

The controlling software uses the Dataport as a source of information and as a means to get control requests executed. First, the controlling software will pick and choose the information needed to perform its function from the Dataport. And second, the controlling software will generate commands to be executed by the Dataport that will maintain a user specified test environment. As seen in Figure 1 above, the Dataport is a "DATA-PORT" through which all information flows.

## SOFTWARE DEVELOPMENT IMPLEMENTATION

### Interface Software

The Interface software is a small software routine that communicates with the external 'Black Boxes'. The software and protocol that the communication system used is a General Purpose Interface Bus (GPIB). LabVIEW® provides a GPIB driver that will meet most users needs, but our system required more capabilities. The standard LabVIEW® GPIB drivers were modified to be able to trap all write, read, or bus failure errors. These errors are trapped, translated to a readable text, logged, presented to the user, and in the event no user response is given the system passes the error to an evaluation subsystem. These lower level functions have never aborted or failed to execute, independent of what error was encountered due to GPIB, file I/O, or slave instrument failures.

### Instrument Drivers

An Instrument Driver is a complete software package that is capable of driving an identified 'Black Box'. The package should be capable of performing all associated functions that the external device can provide. Being able to perform all device functions gives the software routine the reusability desired in a true software package system. It detects any fault with the GPIB read or write and assists the Interface Software with all actions that are to take place when an error is encountered. The error message is passed through the routine to the display area. All display areas are character strings as opposed to real or integer values, and this allows either the returned error message or the character version of the returned value to be displayed in the same data window.

## Data Presentation and Process Controllers

The Data Presentation section of the Dataport is the hub for information control. This section displays all user information, performs system monitoring, provides data tables for the controllers, and writes data to files. For real time data gathering and project controlling, the Data Presentation section is the heart of the system.

This section maintains two data tables for storing information. The first data table contains all information provided by the Instrument Drivers and the second contains all commands passing between the Data Presentation section and the controllers. All information gathered by the Instrument Drivers is placed in the information table. See Figure 2.

The control functions that exist within this section are limited to the following four areas:

1. Constant monitoring and reporting of the status of the interruptible utility electrical power and the air pressure in the system. These items are essential for the functioning of all of the external devices and have precedence in execution. If a problem exists in one of these two areas the environmental system controllers are notified and the information is logged.

2. Evaluations are conducted every 30 seconds to determine if any of the Instrument Drivers, controllers, or monitoring systems have an error that requires attention. If a problem gets to the point where the evaluation decides that the experiment may be at risk, then the Data Presentation section will phone operation personnel at home. If user intervention does not take place before the evaluation system returns a 'System Critical' evaluation, the project will be shut down.

3. Whenever a GPIB error occurs the Data Presentation system initiates a System In Charge (SIC) command to inform all devices on the GPIB that they will reset and that the computer is in charge of the bus.

4. All data is converted into a series of data strings and transmitted to a graphical display computer. This second computer portrays the environmental system status graphic display, but, it also serves as the backup environmental control system. If the second computer does not receive a data string input from the primary control computer within a specified time interval (presumably due to an internal computer failure) it will initiate a SIC command and turn off the electrical supply to the other computer. This second computer will then run the environmental control system using a duplicate copy the software package that is on the primary computer.

Within the Data Presentation section the Instrument Drivers are executed according to precedence. This priority system may be adjusted at any time. Currently it gives the user the ability to have a specific instrument pulsed more frequently then any other instrument. Currently the control for all temperature control and vacuum valves, vacuum pumps, and fluid cooling loop chillers are driven with priority due to the fact that they are the items that the computer would need to manage in an emergency.

## Automated Controllers

The vacuum and fluid systems are the only systems that required automated controllers. Both work in the same manner but the vacuum controller is sequential and the fluid controller is dynamic. The controllers must function with the ability to interrupt any process and have the ability to recover, they must give the user the ability to change their options after the process has started, and they must be able to swap from the primary system to the back up hardware without any interruption in the system.

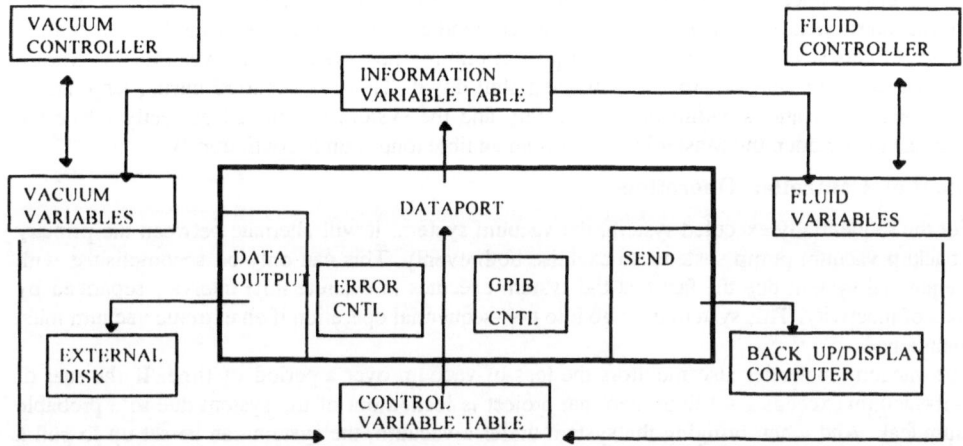

**Figure 2.** Data Presentation Conceptual View.

In order for a controller to be able to function as a parallel system and operate in a real time mode its software structure *must* be single point entry and exit. The data used by the controller and the controller commands *must not* be accessed only at these single point entries and exits. Single point entries and exits specify that a routine must always enter through a single " front door" and exit via the single "back door" with no exceptions. The data, in order to be real time, must be able to be retrieved at any time during the control routine when it is needed and only when it is required. An analog of such a process is ordering a pizza by telephone. The process is dialing, requesting prices and availability, ordering the pizza, and hanging up. There is a unique entrance and exit set. The intermediate steps may involve intricate looping (revising the order to use discounts, for instance) depending on the preferences of the "user" of this process. During the process, data is obtained on request. The ability to ask for decision data before dialing is very limited, and may well result in the user making the wrong decision based on obsolete pricing information. The value of hanging up the phone in this example is obvious, but it is analogously forgotten by many programmers. In complex parallel control systems, not terminating the subsidiary control process in a logically managed manner results in this subsidiary process using processing resources without benefit to the user.

Iteratively cycling through the controller routine, waiting for an event to occur which will need the controller's intervention is how the each controller spends most of its time. After a significant event, such as an environmental disturbance, has occurred the controller executes its process using the data it obtained in real time to make its decisions. The recovery of the project may take quite a bit of time over several control cycles, so the controller waits, watches, and makes minor adjustments based on updated data until a threshold of acceptable environmental conditions has been regained.

While the controller is waiting, watching, and making minor adjustments this controller needs to get to the Data Presentation section in real time to be able to control the system. The commands are constantly monitored to insure immediate action. The controllers also monitor execution of the commands to insure they got through. If a command does not function correctly the controller will make a determination as to whether to re-issue the command or to declare the primary system (e.g. fluid loop chiller or vacuum pump) malfunctioning and swap to its backup system.

## System Error Control

All adjustments made are determined by the user's set of inputs which define the operating limits for that particular project. The environmental control system manager will set group limits for the hardware within this particular configuration. If these limits stop the system from maintaining the proper environment for a particular project an error will register and a loud audible tone will be generated intermittently. If the system switch is set to indicate that no one is in the laboratory or if the problem gets severe enough to cause a project failure, then system will over-ride the Master User inputs to try to save the project from being shut down. This has occurred during large ambient temperature variations within our test facility and the system functioned correctly. Once the system has over-ridden the Master User inputs an audible tone sounds continuously.

## Sequential Controller Operations

For the sequentially executed system, the vacuum system, it will alternate between the primary and backup vacuum pump systems to exercise both evenly. This can only be accomplished with the sequential system due the fact that the system executes in discrete time intervals separated by periods of inactivity. This system does go into non-sequential operation if an extreme vacuum must be maintained.

The vacuum controller also monitors the loss-of-vacuum over a period of time. If the rate of loss-of-vacuum exceeds a set limit then that project is locked out of the system due to a probable vacuum leak. And when bringing the system under a vacuum, the system can be set up to pull a vacuum down to a steady state or to a specific desired vacuum.

The controllers are also notified when a non-UPS (Un-interruptable Power Supply) power failure has occurred. They will passively monitor the system for severe problems and will restart their active controlling when the power is restored. The Data Presentation section monitors and updates all software as to the disposition of non-UPS power.

## Report & Control System

This section manages and communicates with the subordinate systems which manage the cryocooler test stands. The GPIB is the primary environmental control computer's link to many external devices and the five subordinate computers. This has a master and slave relationship. The back up environmental control computer is connected to no projects directly. Its only function is to display graphics and to take over control of the environmental system if the master fails.

## Test Stand Data Acquisition

The test stand computers all have two GPIB cards due to the fact that they are on two different GPIB systems. Each one operates as the master computer for it's own small system consisting of a few devices. Refer to figure 3 for a simplified diagram of the two separate GPIB networks. The data acquisition (DAQ) computer's function is to run that particular cryocooler project. They have the same type of software functions that the Dataport has on a scaled down basis. They drive their projects and log all pertinent data. Their second GPIB card is a slave to the Dataport and responds when queried by the Dataport. The Dataport pulses all four of the test project computers for their status and for critical data. If one of the computers has a problem that it cannot handle, it signals the Dataport. The Dataport would then in turn try to resolve the problem on its own. If the Dataport could not resolve the problem within a specified amount of time then the Dataport would call someone at home for assistance. If the Dataport evaluated the problem to be so severe that the project could be damaged it would instruct the slave computer and test stand to shut down. Approximately three minutes after the shutdown command was issued the Dataport would shut off all power to that particular project. Shutting off power insures the cryocooler's projects survivability, as it has been determined that the primary danger to the test coolers is from protracted operation outside of a cooler's acceptable environmental or performance envelope.

## CONCLUSIONS AND PERFORMANCE SUMMARY

For an overall idea of the equipment that we use and how it is organized, refer to the system map contained in figure 4 below. The benefits which have accrued to this laboratory by organization of the test stand and environmental control along the comprehensive lines outlined above lie

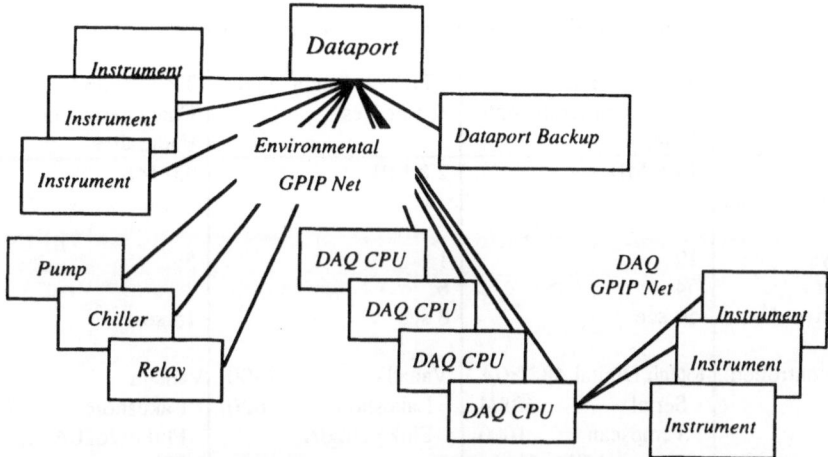

**Figure 3.** Schematic of dual GPIB net structure.

largely in the areas of software reliability and efficiency. Because the overall concept is based on the creation of a consolidated data or information pool on the environmental and DAQ levels, a failure of any particular instrument or equipment subsystem remains precisely a subsystem failure. It does not arbitrarily endanger the overall conduct of the test or of the other equipment items which coexist in the laboratory. Greater program stability leads to enhanced testing reliability.

While the computers used are not true multiprocessing computers, it is obvious that the entire environmental control and DAQ system is a multiprocessing system. Each instrument and computer has processing and communications abilities, to a greater or lesser degree. Each can be tasked to conduct its business in parallel so that the DAQ and environmental control systems can be operated on a multithreaded basis. This uses the processing resources and time more efficiently, and allows for both a denser test data set and effective real time control of the test stands and environment. Instrumentation can also be used more efficiently, as data from one instrument can be communicated across the communications networks as needed. The data derived from the instrumentation and parameters for hardware operation are part of this common pool of shared information, allowing for correlations such as: "the spike in cold end temperature and power draw at 1 AM was due to this vacuum turbo failing and #2 coming on line".

Documentation of reliability in software is notoriously subjective with the statistical samples available, as various organizations may define system failures differently. Using the conservative criteria of failure to control a critical environmental system, failure to monitor and record critical test data, or general program failure (e.g. lock-up of the CPU) , the following test programs shown in Table 1 provide representative program performance data. Notable are the failure mechanisms indicated in Table 1, which indicate that external or chaotic events provide the principal source of system down time. The two failures noted resulted in a general system shutdown without equipment damage in the case of UPS failure, and continuation of program execution and operator notification in the case of the Lakeshore 820 failure.

## FUTURE WORK

As this concept of system engineering is intrinsically hardware independent, and does not rely on the chosen software package, future needs and development missions can be addressed with relative ease. The problems inherent in the GPIB communications network operations have led to newer communications network structures such as VXI or placing all computers and instruments on a dedicated Ethernet network. In either of these options, the Dataport data structure would remain the same with the major transition being in removing the GPIB calls and replacing them with the new protocol.

**Table 1.** DAQ Equipment and Measured Properties

| | 24" Vacuum Chambers Environmental Control | STRV-2 Flight Qual. & Life Test | TRW 3585 Pulse Tube Characterization and Endurance |
|---|---|---|---|
| Prgm. Size | 14.6 MB | 2.8 MB | 3.0 MB |
| Number of Global Variables | | | |
| Arrays | 19 | 1 | 5 |
| Scalars | 54 | 6 | 7 |
| Min. Repetitive Cycle Time | 50 sec. | 8 sec. | 15 sec. |
| Instruments controlled | Iotech Digital 488/80A, Serial 488/4, Tempscan 1000; Athena AT16; Varian Multigage; Inficon IC3; FTS RC211C Chiller | Vahalla 2300, Lakeshore 820, Fluke 2620A | Vahalla 2300, Lakeshore 93C, Fluke 2620A, Lecroy 9304A |
| Execution time between failure | 7000 hours | 2500 hrs. | 1200 hrs |
| Cause for failure(s) | UPS power system circuit tripped | Lakeshore 820 GPIB card failure | no failures to date |

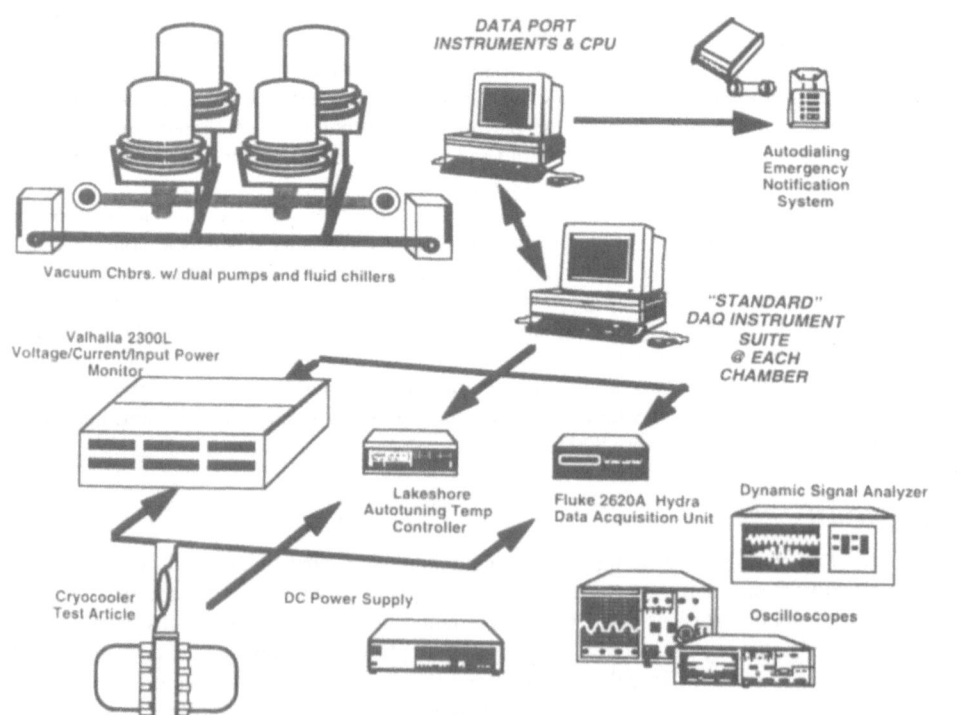

**Figure 4.** Dataport and External Devices.

# Cryogenic Systems Integration Model (CSIM)

**M. Donabedian and S.D. Miller**

The Aerospace Corporation
El Segundo, CA  90245

**D.S. Glaister**

The Aerospace Corporation
Albuquerque, NM  87119

## ABSTRACT

The Cryogenic Systems Integration Model (CSIM), is an interactive PC "Windows" based software tool for the simulation and analysis of spacecraft cryogenic mechanical refrigeration thermal control systems.  CSIM development was initiated in response to a need for an encompassing and efficient method for design and analysis of spacecraft cryogenic mechanical refrigeration systems.  Previous experience has shown that cryogenic systems exhibit large analytical uncertainties.  Historically, cryogenic thermal integration has been critical and often inadequately considered during the preliminary design phase resulting in significant revisions and system penalties.

CSIM includes design algorithms, subroutines, and databases to allow the user to conduct analyses and trade-offs necessary to complete the design integration and simulation of instruments, cryocoolers, thermal straps, thermal storage units, thermal switches, heat pipes, intermediate shields, radiators, and brackets. CSIM outputs provide a complete breakdown of temperatures, heat flows, dimensions, weight, power, and total system penalties.  Several databases are provided within the program including cryocoolers, thermal storage materials, and heat pipe fluids which can be enhanced or modified by the user.

A beta version 2.0 was completed in October 1995 and demonstrated to the Air Force Phillips Laboratory along with the distribution of a preliminary user's manual.  Several refinements including instrument cyclic load and compressor stroke and heat sink interpolation capability have been added since that time.  Also, the program was validated by testing it against a real hardware system. The current program, version 2.1, is described in this paper.

## INTRODUCTION

Integration, analysis, and evaluation of sensor cooling systems at cryogenic temperatures using mechanical refrigerators (cryocoolers) has become increasingly complex, difficult, and time consuming. The large number of variables and options available to the designer together with an increasing number of cryocoolers available or that are being developed begs for a more efficient method of integrating and analyzing these systems. Past experience has shown that systems integration is not adequately considered during the early design stages leading to significant system performance degradation or drastic design changes. In response to these needs, an interactive PC "Windows" based software tool has been developed at The Aerospace Corporation. It is intended that this program will be used as a preliminary design tool to aid in parametric and trade-off analysis and provide a means to simulate the performance of various systems options.

The preliminary beta version of CSIM was described in Ref. 1 and was presented at the 8th International Cryocooler Conference held at Vail, Colorado, June 26-29, 1994. A completed beta version was demonstrated to Phillips Laboratory in October 1995 and a preliminary user's manual distributed to selected individuals.

Since that time, a number of new features have been added including a cyclic load capability with more accurate thermal storage sizing algorithms, improved heat rejection system with a variable radiator aspect ratio, a cryocooler stroke interpolation capability, and improved output displays. This current program, version 2.1, was tested successfully against a real hardware system and an updated user's guide was developed with an expanded list of diagnostic messages to provide a more user friendly program. A further improved version 3.0 is expected to be completed near the end of fiscal year 96 and will be available for release to government agencies and industry users with selected government contracts.

This current paper focuses primarily on the changes and additions incorporated since the original concept was presented at the 8th Cryocooler Conference. [1]

## INTEGRATED THERMAL CONTROL SYSTEM

A schematic of the baseline integrated cryogenic thermal control system is shown in Fig. 1. The equivalent thermal network is shown in Fig. 2. This represents the normal screen display available to the user to select specific components, disable or enable certain options, and to activate input or output displays.

## CSIM IMPLEMENTATION

CSIM is a Microsoft Windows-based software tool for IBM PC-compatibles which provides a graphical interfaces for user input, design information, and system output. CSIM is programmed using the Borland C++ 4.5 compiler and the Borland Object Windows Library version 2.5. CSIM can be run on any PC with Windows 3.1 (which is pre-installed on most PCs). The existing CSIM program and its components require 3 Mbyte of disk space and can be

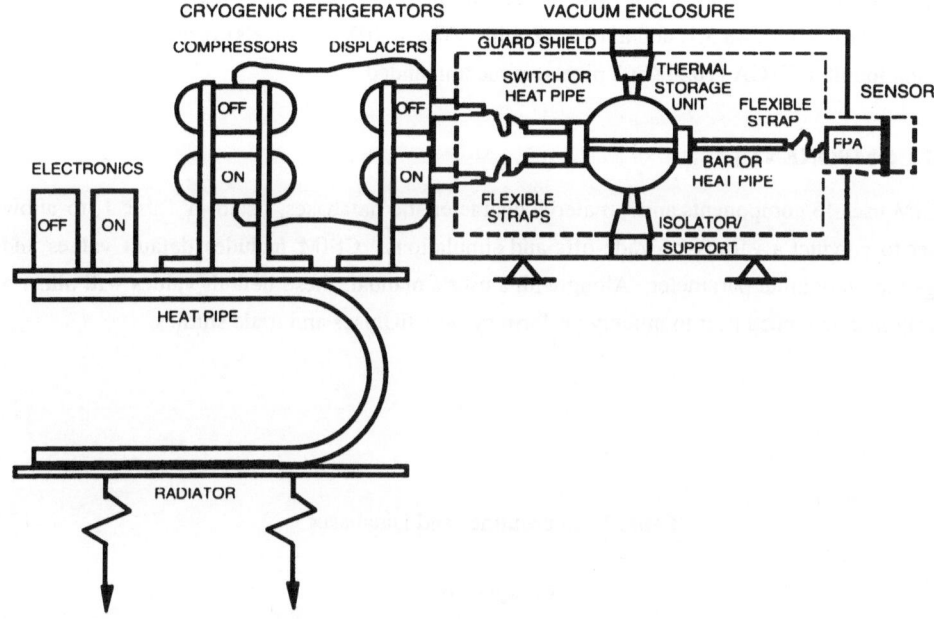

**Figure 1.** Integrated thermal control system

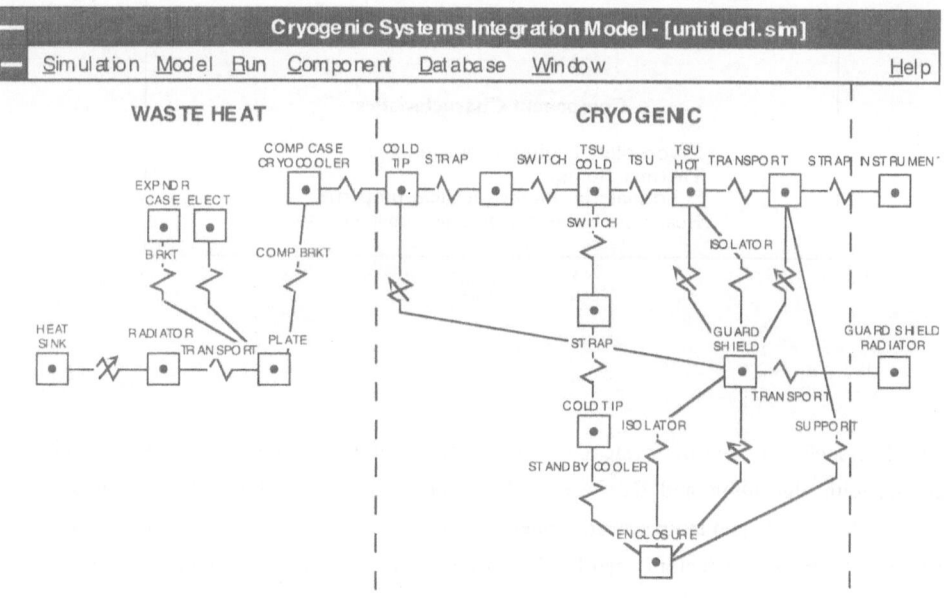

NOTE: ALL RESISTORS DEFINED AS TRANSPORT CAN BE EITHER HEAT PIPES OR CONDUCTION BARS

**Figure 2.** Systems thermal network

loaded from three high density 1.44 Mbyte 3.5 inch diskettes. The minimum computer system requirements are a 80486 processor with 8 Mbyte of RAM (random access memory), a hard drive, Microsoft Windows 3.1, and a monitor with VGA (640X480 pixels) resolution. A higher resolution monitor SVGA (1024X768 pixels) is recommended.

## CSIM COMPONENTS

CSIM uses 13 components and 4 material characteristic databases, listed in Table 1, to allow the user to conduct a variety of trade-offs and simulations. CSIM includes default values and settings for every input parameter. Along with a user's manual, these default values will allow a relatively inexperienced user to quickly perform system analyses and trade studies.

**Table 1.** Subroutines and Databases

**Components**

| | |
|---|---|
| Cryocoolers | Guard shields |
| Operating | Waste heat system |
| Stand-by | Brackets |
| Heat transport | Radiator |
| Flexible straps | Power penalty |
| Thermal storage units | Global outputs |
| Thermal isolators | Instrument load |
| Thermal switches | Constant |
| Cryogenic enclosure | Cyclic |

**Component Characteristics**

Cryocoolers cooling characteristics
Thermal storage
Solid material thermophysical properties
Heat pipe fluids-performance characteristics

CSIM models the thermophysical characteristics of the components under steady-state conditions with algorithms and databases. The algorithms are used for the computation of steady-state heat loads and temperatures under operational conditions, and the component sizing. Reference 1 describes the factors used by CSIM algorithms to model the characteristics of the components.

CSIM employs 4 databases; 1 database characterizes the cryocoolers being used, and 3 databases characterize the thermophysical properties of the materials used within the components.  The 3 material databases contain the characteristics of heat pipe fluids, Phase Change Thermal Storage Unit fluids, and the thermal conductivity, specific heat and density of solid materials as a function of temperature.  Presently, there are 20 different cryocoolers from 9 manufacturers, 9 heat pipe fluids for operation in the range of 16 K to 600 K, 14 PCM fluids for use from 14 K to 195 K and 160 materials which can be used for radiators brackets, conduction bars, thermal straps, etc.  Reference 1 also contains the details of these databases and the individual component subroutines.

## DESCRIPTION OF IMPROVEMENTS AND FEATURES ADDED

Due to limited space, a description of only the major improvements and features added to the original beta version are included herein.  The interested reader who desires more detail of the entire program is directed to Ref. 1.

### Instrument Cyclic Load Capability

The starting point of the network is the "instrument" which represents the interface of a focal plane assembly or sensor to be cooled.  Analysis of the parasitic load within the instrument are not included in CSIM.  The original program included only a constant load simulation wherein the user need input only the heat load and instrument temperature desired.  For the new cyclic load option added, shown in Fig. 3, the maximum and minimum heat load, the maximum instrument temperature, the period for the maximum load and the duty cycle (in percent) must be specified. When the cyclic load option is selected, the thermal storage unit (TSU) option must also be selected because CSIM assumes that the cryocooler operates continuously at the average cooling capacity required at the cold tip.  Although this value will be lower than the maximum instrument load, it will be somewhat higher than the average instrument load because of the parasitic loads along the network.

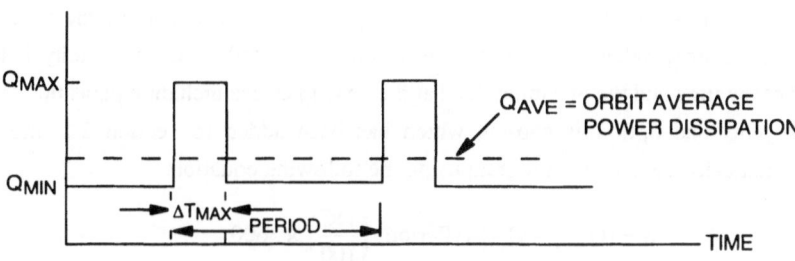

**Figure 3.**  Cyclic instrument load

Although the user must specify a maximum instrument temperature to initiate the simulation, CSIM will compute the heat flows, parasitics and temperatures, and reset the maximum instrument temperature so that it is compatible with TSU material selected and the network characteristics specified.  In addition, the average heat load and minimum instrument temperature is also computed and displayed.

The average load $Q_{ave}$ is calculated by

$$Q_{ave} = \left(Q_{max}\right)\left(\frac{DC}{100}\right) + \left(Q_{min}\right)\left(1 - \frac{DC}{100}\right) \qquad (1)$$

where

$$Q_{max} = \text{maximum instrument load, W}$$
$$Q_{min} = \text{minimum instrument load, W}$$

The duty cycle (DC) in percent is given by

$$DC\ (\%) = \Delta T_{max}\ (100)\ (\text{Period}) \qquad (2)$$

If the maximum instrument temperature is unacceptable, the operator has a number of options including reducing the load, decreasing the network resistances, or if phase change material (PCM) was selected for the TSU selecting one with a lower phase transition temperature.

## Thermal Storage Units (TSU) Sizing for Cyclic Loads

The inclusion of TSUs in thermal control systems can be used effectively to orbit average peak parasitic loads or instrument loads thereby decreasing required cryocooler capacity. Thermal storage can also be used to allow the sensor to operate even if cryocoolers must be turned off because of inadequate system power or operational constraints. CSIM allows the user the option to add or delete a TSU and also provides the choice of either a phase change material (PCM) or a sensible heat storage device (SHD). Subroutines and databases have been defined which allow the user several options depending on the specific system requirements. Material properties are provided from separate databases for each option. The PCM option accesses the phase change thermal storage unit (PCM) fluids database while the sensible heat option utilizes the Thermophysical Property database.

The use of a TSU is mandatory when the cyclic instrument load option is chosen but is optional when the constant load option is used. For the constant load option, the user must specify the operating time. When operating time is specified, the TSU storage capacity is based on the product of the time and the maximum load at the TSU interface including parasitic.

When the cyclic load option is chosen, which has been added to version 2.1, the TSU capacity is automatically sized by the program using the following equation

$$S = \left(Q_{max} - Q_{ave}\right)(\text{Period})\left(\frac{DC}{100}\right) \times 3600 \qquad (3)$$

where

| | | |
|---|---|---|
| S | = | TSU capacity, Joules |
| $Q_{max}$ | = | maximum load at the TSU interface, W |
| $Q_{ave}$ | = | average load at the TSU interface, W |
| T | = | period, seconds |
| DC | = | duty cycle, percent |

The average heat load at the TSU hot side is calculated by

$$Q_{ave} = (Q_{max})\left(\frac{DC}{100}\right) + (Q_{min})\left(1 - \frac{DC}{100}\right) \tag{4}$$

where

$Q_{min}$   =   The heat load at the TSU interface for the minimum instrument load, W

To provide for worst case scenarios, the temperature drops for the TSU and its interfaces are computed using the maximum heat flows. However, once the simulation is on the cold side of the TSU, the average heat flow is carried through to the cold tip of the cooler since it is assumed that the cooler is operating continuously at a capacity to meet the average heat load.

An iterative routine is used to compute the mass and volume of the TSU material and the containers accounting for the parasitic heat loads. For a PCM, the mass of material required is based on the capacity, heat of transformation, and a specified efficiency. For a sensible heat material, the mass required uses the specific heat, temperature rise (specified as part of the input) and an efficiency factor.

In practical designs, the weight and volume of the storage container for PCM materials can be very large compared to the actual PCM because of the need for metal fins due to the poor thermal conductivity properties of the PCM, metal matrices for fluid positioning, and in the case of specific gases such as $O_2$ or $N_2$, high storage pressure requirements at ambient or elevated temperatures. Therefore, to provide realistic designs, empirical constants are utilized based on flight, development, or test hardware when available.

## Cooler Stroke Interpolation

Compressor stroke (or power level) interpolation was added to version 2.1 to provide a better match of a specific cooler capability with cooling requirements. Although the beta version of CSIM did include a compressor case temperature interpolation and extrapolation feature, both the stroke and case temperature interpolation have been combined into one feature.

The existing cryocooler database provides load lines (i.e., capacity versus cold tip temperature) for a limited number of compressor strokes and heat rejection temperatures (case temperature). As a result, in many cases, especially for large capacity coolers, there could be a significant mismatch between the available and required cooling capacity preventing realistic simulation. The new feature allows the user to select virtually any stroke/or percent of maximum power) and compressor case temperature. CSIM will perform a linear interpolation and/or extrapolation necessary to carry out the simulation. Appropriate alert messages are given to insure that the user is aware of what is being done.

There are certain limitations and restrictions however. For example, CSIM will not extrapolate a stroke (or percent of maximum power) beyond the highest value given in the database. On the other hand, compressor case temperature extrapolation above the maximum or below the minimum value given is allowed in order that the waste heat rejection system simulation can be carried out independently if desired.

When these interpolations or extrapolations are performed, the display always notes the boundary values from the database and the user selected values during the simulation.

## VALIDATION OF CSIM

To verify the accuracy of CSIM, it was requested by Phillips Laboratory to test it against a real hardware system. After surveying several programs, it was decided to use the SPAS III IR radiometer experiment as the basis.

The Shuttle Pallet Satellite (SPAS III) is a joint space test effort between the Ballistic Missile Space Defense Organization (BMDO), Phillips Laboratory, Utah State University Space Dynamic Laboratory (SDL), and the SMTS (SPO). Its compliment of experiments includes an IR radiometer which uses a pair of Lockheed-Martin split cycle Stirling coolers (Model 2010) to provide approximately 1.2 W of cooling at 59 K or 2.0 W at 65 K at the IR instrument interface.

### Description of SPAS III

The entire SPAS III spacecraft is shown on Fig. 4 with the IR radiometer and cryocooler elements highlighted. The general arrangement of the IR imaging radiometer together with the experiment support platform (ESP) and the cryocooler is shown in Fig. 5. A more detailed view of the cryocooler compressor and displacer, support bracket, pedestal radiator and extended radiator which is part of the ESP is shown in Fig. 6 and 7. A plan view of the radiometer with details of the instrument interface, conductor bar, T-coupler, flexible straps, and the displacer cold tips are shown in Fig. 4. Note that the only portion of the SPAS that was simulated with CSIM is that point from the instrument interface to the cooler cold tips (which includes the conduction bar, coupler, flex straps, housing, MLI, and supports) together with the cryocooler compressors, supports, pedestal radiator, and ESP radiator.

The estimated weight of the IR radiometer cooling system weight based on Lockheed-Martin delivered hardware described in Ref. 4 and the SDL hardware described in Ref. 2 is 32.36 kg as defined on Table 2. The total input power is 100.6 W which includes 29.6 W for the cooler driver electronics (controller).

Due to some differences in the method of heat rejections, several adjustments had to be made in actual radiator weight. Depending on the exact method used, the results of the CSIM analysis produced a total system weight ranging from 31.6 kg to 34.1 kg. The total input power calculated by CSIM was 112 W (as compared to 100.6 for the actual systems) but provided a cooling capacity of margin of 14.77. When adjustment for the margin was made, the difference in power was less then 5 W.

Overall, the results of the comparative evaluation provides an excellent validation of the accuracy (within 5%) and usefulness of CSIM as a software tool for conducting preliminary analysis, integration, and simulation of sensor cooling systems.

**Table 2.** Estimated SPAS III Radiometer Cooling System Weight

| Item | Weight (kg) |
|---|---|
| **LMSC delivered hardware** | |
| Cryocooler assembly (includes controller) | 17.90 |
| All attachment hardware (includes brackets, compressor collar, flex links, housing MLI couplers, etc.) | 6.90 |
| Subtotal | 24.80 |
| **SDL hardware** | |
| Compressor pedestal support/radiator | 1.76 |
| Extended ESP radiator (compressor) | 4.70 |
| Drive electronics (controller) radiator (side panel) | 1.10 |
| Subtotal | 7.56 |
| Total | 32.36 kg |

**Table 3.** CSIM Output Summary for System Weight

| Component | Mass (kg) |
|---|---|
| Conduction bar | 0.529 |
| Thermal strap | 0.848 |
| Isolator support | 1.000 |
| LMSC 2010-D (SPAS III) | 17.900 |
| Compressor bracket | 1.678 |
| Expander bracket | 1.755 |
| Electronics bracket | 0.250 |
| Vacuum enclosure | 0.504 |
| Radiator | 9.591 |
| Total | 34.055 |

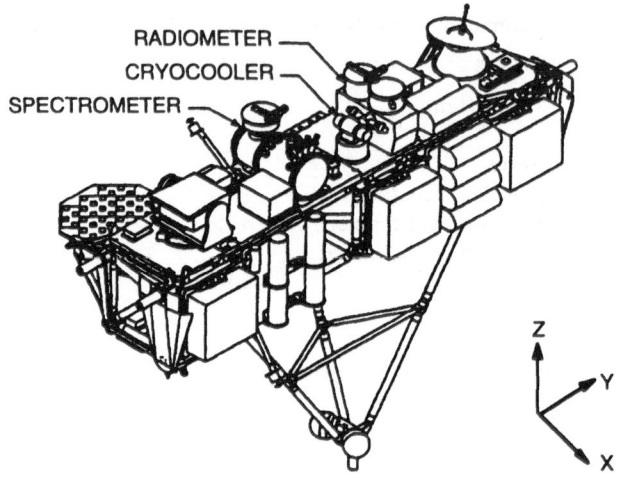

**Figure 4.** SPAS III spacecraft (courtesy of Space Dynamics Laboratory)

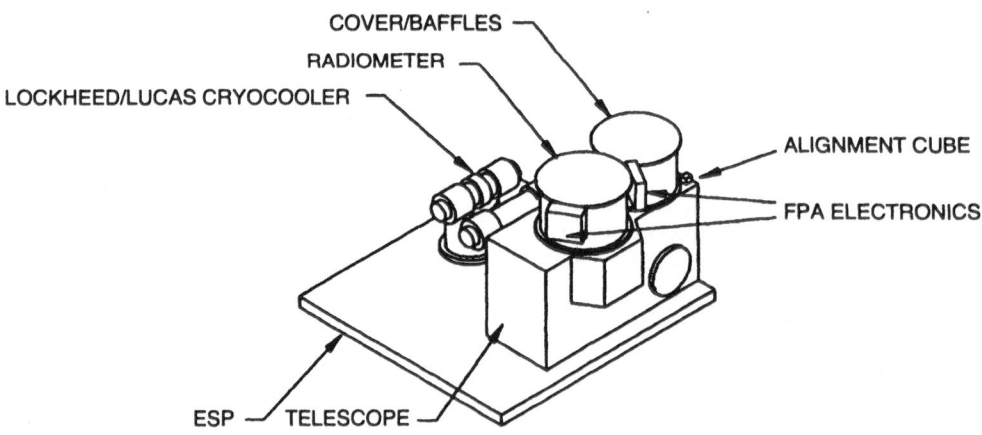

**Figure 5.** IR imaging radiometer with cryocooler (courtesy of Space Dynamics Laboratory/Utah State University)

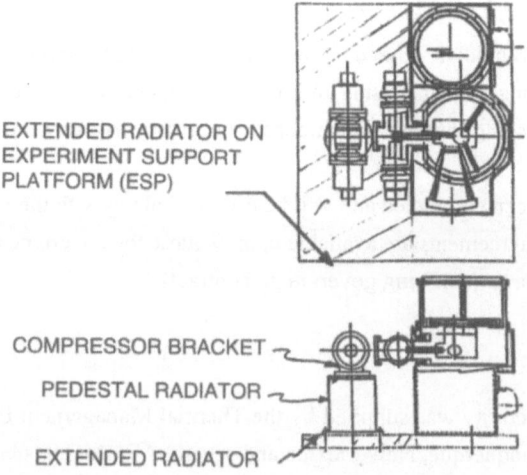

EXTENDED RADIATOR ON
EXPERIMENT SUPPORT
PLATFORM (ESP)

COMPRESSOR BRACKET

PEDESTAL RADIATOR

EXTENDED RADIATOR

**Figure 6.** Compressor supports, pedestal, and radiator

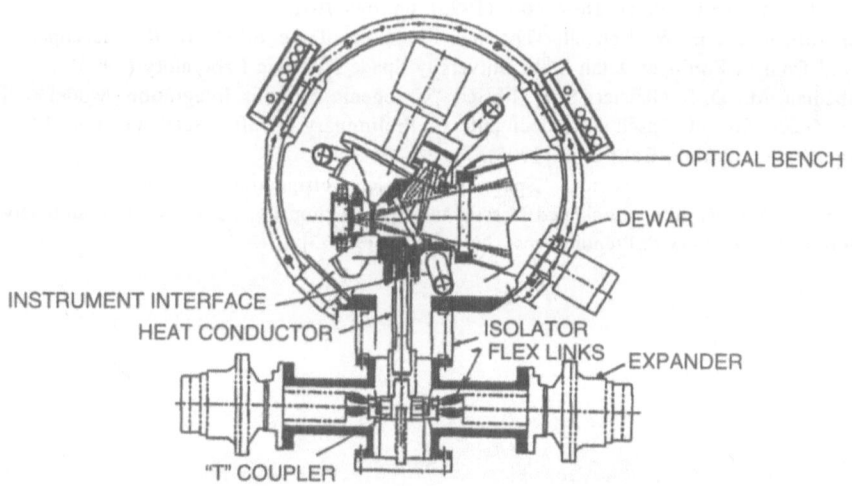

OPTICAL BENCH

DEWAR

INSTRUMENT INTERFACE
HEAT CONDUCTOR

ISOLATOR
FLEX LINKS

EXPANDER

"T" COUPLER

**Figure 7.** Radiometer/cryocooler assembly

## PREPARATION OF USER'S GUIDE

A user's guide for version 2.1 has been prepared and is available to qualified organizations who apply for and receive a licensing agreement. The user's manual describes the hardware and software requirements and the installation instructions in detail. Also provided are a description of each of the CSIM elements, program operation and network description, sensor system options, an overview of the four databases, a sample problem, and a section on troubleshooting.

## SUMMARY AND CONCLUSION

An interactive PC "Windows" based computer program for analysis and simulation of integrated cryogenic cooling systems using mechanical refrigeration has been developed. The program includes the logic, algorithms, and databases necessary to carry out a simulation of an entire system.

A user's manual has been completed and will be provided along with the program to qualified organizations. Licensing agreements are available upon request for government organizations and other organizations who have qualifying government contracts.

## ACKNOWLEDGMENT

The support for this activity was supplied by the Thermal Management Branch of the USAF Phillips Laboratory at Albuquerque, New Mexico and several AF/SMC System Program Offices.

## REFERENCES

1. Donabedian, M., D. S. Glaister, M. D. Bernstein, "Cryogenic Systems Integration Model (CSIM)," *Cryocoolers 8*, Plenum Press, New York (1995), pp. 695-707.
2. Bertagnolli, K. E., E. W. Vendell, "Thermal Modeling of the SPAS-III IR Telescopes," copy received from L. Zollinger, Utah State University Space Dynamic Laboratory (1995).
3. Donabedian, M., D. S. Glaister, S. D. Miller, "Cryogenic Systems Integration Model (CSIM)," User's Guide for PC-Based Microcomputers, Preliminary Draft, Beta Version 2.0," The Aerospace Corporation, California (1995).
4. Spradley, I. E., W. G. Foster, "Space Cryogenic Refrigerator System (SCRS) Thermal Performance Test Results," Lockheed Missile and Space Company, Palo Alto Research Division, California, *Cryocoolers 8*, Plenum Press, New York (1995), pp. 13-22.

# Spacecraft Cryocooler System Integration Trades and Optimization

**D. S. Glaister**

The Aerospace Corporation
Albuquerque, NM, USA 87119

**D. G. T. Curran**

The Aerospace Corporation
El Segundo, CA, USA 90245

## ABSTRACT

Recent progress in the development of cryocoolers for spaceborne operation provide the payload and spacecraft integrator with a number of options for cooling in the 20 K to 150 K cryogenic temperature range. These options are explored from the standpoint of both domestic and foreign cryocooler availability. The main integration issues concerning cryocooler performance including capacity at desired temperature, power, mass, life, reliability, maturity, vibration, temperature stability and mechanical integratability are discussed in context with the more general criteria involving system penalty. The total spacecraft penalties for the integration of a cryocooler are evaluated through historical and empirical data.

Performance data are presented for a number of cryocooler programs at 35 K and 60 K. The data are also evaluated using the total spacecraft penalty. Recent experimental results for reducing the cryocooler power requirements and increasing the cooling capacity through either reducing heat rejection temperatures or cryogenic precooling or heat intercepting of the cold end are also discussed. These integration methods for optimizing cryocooler performance are evaluated at the system benefit level using the system penalty to trade the additional radiator mass against the reduced system mass resulting from the decreased cryocooler input power.

## INTRODUCTION

There are many factors to be considered when selecting and integrating a cryogenic refrigerator (or cryocooler) for a space application. Some of these factors, such as cooling capacity at a specific load temperature, are relatively obvious, while others, such as the total spacecraft system penalty, are not as apparent or as easily evaluated. This paper will describe and discuss selection or integration criteria as well as provide data and algorithms for evaluating cryocoolers against the criteria. The paper will also evaluate integration methods for improving the performance of the cryocooler system.

Cryocoolers 9, Edited by R.G. Ross, Jr.
Plenum Press, New York, 1997

## INTEGRATION CRITERIA AND SYSTEM PENALTIES

The list of selection or integration criteria includes cryocooler cooling capacity at the desired interface temperature, input power, mass, reliability for the desired lifetime, hardware or technology maturity, system mass penalty, induced vibration, temperature stability, and general integratability. The relative prioritization and weighting of these criteria are dependent on the customer or user requirements and constraints. For example, some spacecraft applications (such as small payloads and the Air Force Space and Missiles Tracking or SMTS program) may be very limited in power or heat rejection capability, while others may have excess mass and power margins. Some applications (such as superconducting electronics) may be relatively insensitive to cryocooler induced vibration, while others may not. The level of allowable risk for an application is especially relevant to the life, reliability, and maturity criteria. There also may be application unique mechanical or load integration features which have a large impact on the selection or design.

### Cryocooler Capacity, Power, and Mass

Usually the first and the easiest criteria evaluated in the selection or integration of a cryocooler is the cooling capacity at the desired temperature, input power, and mass. With the assistance of the contractors and several government agencies (References 1 to 10), a database of long life spacecraft cryocooler performance has been compiled. The intent was to compile data (mostly from test, but some predicted or extrapolated) from numerous contractors and standardize the data to specific reference conditions to enable relative comparison. The data was extrapolated or normalized to a reference cooling load temperature and a 300 K rejection temperature using the following Carnot cycle ratio:

$$Q_{ref} = \frac{T_{ref}^{ct} * (T_{data}^{rej} - T_{data}^{ct}) * Q_{data}}{T_{data}^{ct} * (T_{ref}^{rej} - T_{ref}^{ct})} \tag{1}$$

where the Q is the cooling capacity and T is the temperature either at the cold tip (ct) or heat rejection (rej) interface. The use of this ratio assumes that the Carnot efficiency (or refrigerator coefficient of performance) is the same for both the data and reference conditions. The equation calculates the cooling capacity at the reference conditions for the same input power as the data. The Carnot ratio was also used to extrapolate the upper temperature cooling loads of multistage cryocoolers to allow a (very) rough comparison to single stage units.

There are many caveats to the database. Because of the numerous contractors and models, the database is not comprehensive and becomes quickly outdated. The error associated with the Carnot extrapolation increases as the data conditions deviate from the reference. For cooling load temperature differences greater than about 5 K or rejection differences greater than about 20 K, the Carnot extrapolation is suspect. Also, because the database for flight quality cryocooler electronics controllers is limited, the input power from the electrical bus for most of the data has been estimated from the measured motor power. The estimate of electronic power can easily result in inaccuracies of 5%. Larger errors can occur when the data are extrapolated or predicted to obtain the motor input power. Overall, the database is not capable of or intended to be used for exact, but only approximate (± 10-20% at best) comparisons between cryocoolers.

Also, the reference cooling load temperature is at the cryocooler cold block interface, which can be significantly (typically 5 K for 1 to 2 W loads) colder than the cooled instrument. For cryocoolers or applications where the cooler cold head can be directly (without a thermal strap) mounted to the instrument interface, this temperature gradient can be significantly reduced. Thus, the cryocooler can be operated at a significantly higher temperature when directly mounted. Because of their long flexible lines from their compressors to the cold head, the Joule-Thomson and Brayton cycle cryocoolers can more easily be mounted directly to the instrument interface. Because of difficulties with side loads on the cryocooler cold tip and forces on the

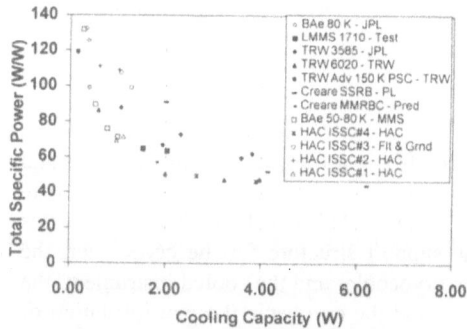

**Figure 1.** Cryocooler total specific power extrapolated to 60 K cooling and 300 K reject.

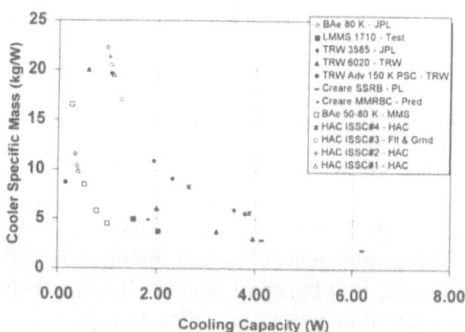

**Figure 2.** Cryocooler total specific mass extrapolated to 60 K cooling and 300 K reject.

instrument interface caused by differential material thermal contraction and launch vibration loads, other cryocoolers cannot usually be directly mounted. Thus, the Joule-Thomson and Brayton cryocoolers have the potential to be run at higher (about 3-5 K for 1-2 W loads) cold tip temperatures (and, thus, decreased input power and increased efficiency) while still maintaining the same instrument temperature. This potential advantage should be taken into account when evaluating the cryocooler input power.

Figure 1 is a plot of cryocooler specific (input divided by cooling capacity) total (motor and electronics) power (SP) for cooling at 60 K as a function of cooling capacity. The plot indicates the expected effect of efficiency increasing as the capacity increases. The plot also indicates the general trend at 60 K of the higher efficiency at low to medium loads of Stirling and Pulse Tube cryocoolers compared with the Reverse Brayton. Because of the limited existence of flight quality electronics, the motor specific power typically provides a more accurate basis for relative comparison of cryocoolers. However, due to space limitations, a motor specific plot is not presented in this paper. Also, the motor power may be misleading for units or cycles which have significantly different electronic power requirements. The relative efficiency of the Brayton cryocooler improves using total power compared to using only motor power because of less power consumption in the electronics.

Figure 2 is a plot of cryocooler specific mass (SM or mass divided by cooling capacity) for the total unit (including electronics) for cooling at 60 K as a function of cooling capacity. The trend of increased mass efficiency with increased capacity is indicated. The light weight nature of the Brayton (Creare SSRB) cryocoolers is also apparent.

Figures 3 and 4 are similar plots of the specific power and mass as a function of cooling capacity for cooling at 35 K. Of significance is the improved performance with the additional

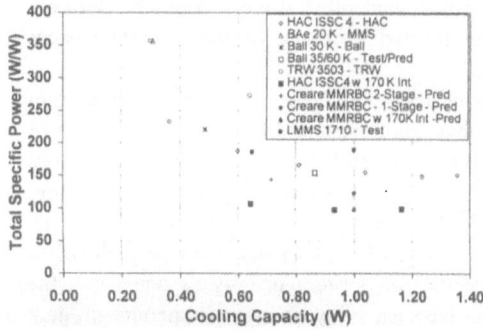

**Figure 3.** Cryocooler total specific power extrapolated to 35 K cooling and 300 K reject.

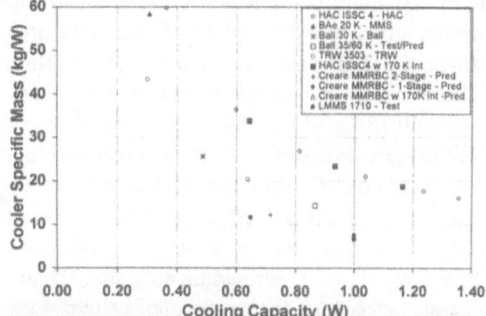

**Figure 4.** Cryocooler total specific mass extrapolated to 35 K cooling and 300 K reject.

use of a cryogenic heat intercepting strap (see data labeled 170 K Int.). This will be discussed in more detail later in the paper.

## Cryocooler System Penalties

The impact (or penalty) of a cryocooler on a spacecraft is not limited to its input power and mass. The cryocooler mass and power must be provided for by the spacecraft and translate through the entire vehicle to produce an overall system penalty. This penalty includes the masses of the power generation and distribution system, the support structure for the cryocooler, the cryogenic heat transport or cold plumbing between the cryocooler and the cooled instrument, the heat rejection system including transport and radiators, and the cryogenic thermal insulation or isolation. For some applications, there may also be additional cryogenic radiators to provide either heat intercept or reduced heat rejection.

Generic and parametric estimates for the cryocooler spacecraft penalties were developed using empirical and historical data. Previously, A. L. Johnson (see Reference 11) had developed a cryocooler system penalty based on parametric analytical evaluations of spacecraft thermal and power systems. The generic penalties are useful for early conceptual designs and trade studies. The generic penalties are also useful when a specific design or application is not available. However, for detailed designs, a more comprehensive analytical approach using tools such as CSIM (Cryogenic Systems Integration Model) (see Reference 12) and/or large thermal models would be recommended.

**Empirical System Penalties.** Data on system mass penalties was compiled for the following six cryocooler programs: MTI (Multispectral Thermal Imager), TCE (Third Color Experiment), CRYOTP (Cryogenic Two Phase), CSE (Cryogenic Systems Experiment), Advanced TRE (Teal Ruby Experiment), and a conceptual program. The MTI, TCE, Advanced TRE, and conceptual programs were and are for applications on free flying spacecraft, while CRYOTP and CSE were for Shuttle based Hitchhiker GAS Can experiments. The most accurate integration penalties are for the real hardware, free flyer MTI and TCE systems. The Shuttle experiment penalties are for real hardware, but are impacted by the GAS Can environments. For example, the heat rejection environment is significantly different than that of most spacecraft and impacts the radiator penalty.

The mass penalties were grouped into the following categories: structure and heat transport, radiator, cold plumbing and insulation, electrical cables and miscellaneous. After the integration components were grouped, the category masses were normalized by dividing by the total cryocooler input power. The input power is the leading driver of the mass penalties. The power directly sizes the heat rejection, heat transport, and cable systems. Because the power reflects on the cooler size and capacity, it also indirectly correlates to the structure and cold plumbing systems. A more complicated correlation to input power and other drivers could be derived. However, given the limited empirical data and the limited accuracy of the generic system penalties, the simple ratio to input power is sufficient.

Table 1 summarizes the empirical data mass penalties. The table shows a power system penalty of 0.15 kg/W which is an average from a database of current and future spacecraft end of life (EOL) power systems. The table also includes the average of the penalty categories across the six programs. Overall, though the individual category penalties may vary significantly from program to program (which is probably associated with the rough categorization method), the total system penalty was relatively consistent with a range of 0.31 to 0.42 and an average of 0.375 kg/W. It is important to note that the use of the full system penalty assumes that the spacecraft or host for the cryocooler system does not have existing mass or power margin and, thus, any cryocooler associated system mass is in addition to the baseline.

**Table 1.** Empirical cryocooler system mass penalties.

| Subsystem | Program System Mass Penalty (kg/W) | | | | | | |
|---|---|---|---|---|---|---|---|
| | MTI | TCE | CRYOTP | CSE | Adv TRE | Concept | Average |
| Structure & Heat Transport | 0.046 | 0.059 | 0.138 | 0.119 | 0.070 | 0.148 | 0.097 |
| Radiator | 0.087 | 0.029 | 0.103 | 0.078 | 0.058 | N/A | 0.071 |
| Cold Plumbing & Insulation | 0.003 | 0.005 | 0.009 | 0.026 | 0.082 | N/A | 0.025 |
| Cables & Misc | 0.025 | 0.074 | 0.014 | 0.008 | 0.064 | 0.007 | 0.032 |
| Local Subtotal | 0.161 | 0.167 | 0.264 | 0.231 | 0.274 | N/A | 0.225 |
| Power System | 0.150 | 0.150 | 0.150 | 0.150 | 0.150 | 0.150 | 0.150 |
| Total | 0.313 | 0.317 | 0.414 | 0.381 | 0.424 | N/A | 0.375 |

**Cryocooler System Specific Mass.** Using the 0.375 kg/W generic system penalty. relative comparisons can be made of the system impact of cryocoolers which incorporate both power and mass efficiency. The system specific mass is the total system mass associated with a cryocooler divided by the cooling capacity and can be calculated according to the following formula:

$$SM_{system} = SM_{total\ cooler} + SP_{total\ cooler} * SM_{system\ penalty} \tag{2}$$

Figure 5 contains a plot of the total system specific mass for cooling at 60 K as a function of cooling capacity. At 60 K, the power contribution to the system mass is dominant over the cooler mass portion. For a total cooler specific power of 60 W/W and specific mass of 8 kg/W, the system specific mass penalty associated with the power is 22.5 kg/W and is almost three times the specific mass of the cryocooler. Thus, in general at 60 K, the power efficient cryocoolers have lower system penalties than the mass efficient coolers.

Figure 6 contains a plot of the total system specific mass for cooling at 35 K as a function of cooling capacity. The cryocooler data utilizing cryogenic heat interceptors were not included as the system penalty does not account for the additional cryoradiator mass. The system penalties associated with cryogenic heat intercepting are discussed later in this paper. In contrast to 60 K, at 35 K, the power contribution to the system mass is approximately the same as the cooler mass portion. For a total cooler specific power of 150 W/W and specific mass of 50 kg/W, the system specific mass penalty associated with power is 56 kg/W.

## Cryocooler Lifetime And Reliability

The life expectancy and reliability of a cryocooler are two of the most important criteria for

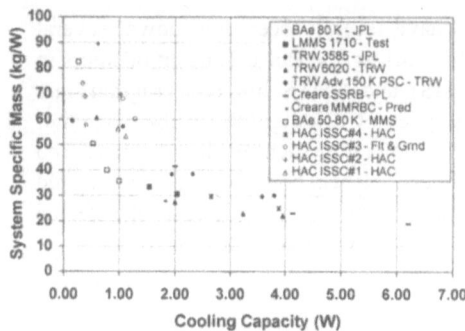

**Figure 5.** Cryocooler total system specific mass extrapolated to 60 K cooling and 300 K reject.

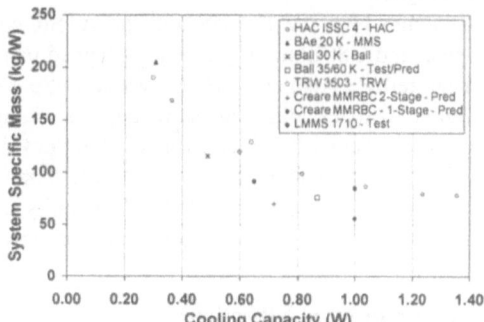

**Figure 6.** Cryocooler total system specific mass extrapolated to 35 K cooling and 300 K reject.

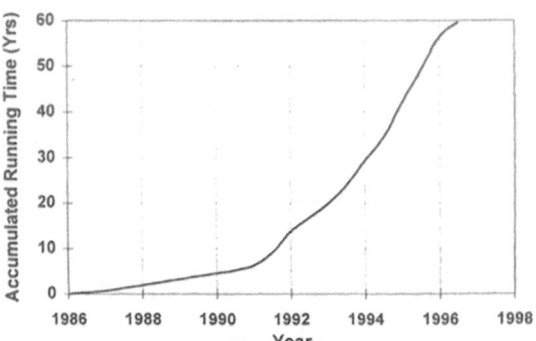

**Figure 7.** Total accumulated running time for Stirling
and pulse tube spacecraft cryocoolers.

most users.     However, due to the relatively small number of cryocoolers, a statistically significant number of operating hours on identical multiple units is not available.   However, confidence in the design of a cryocooler and data on long term failure mechanisms can be increased through the accumulation of life or endurance testing.   An important caveat on the accumulated running hours is that the test conditions (percentage stroke, heat rejection, etc.) vary significantly between tests.

The total running hours accumulated during ground testing and in flight over the past ten years for standard (Oxford heritage) Stirling and related pulse tube cryocoolers are shown in Figure 7.   This data began in January 1986 with the design and build of the Improved Stratospheric and Mesospheric Sounder (ISAMS) 80 K and Along Track Scanning Radiometer (ATSR) 80 K split-Stirling cryocoolers (see Reference 13).   The data include flight operation of the ATSR cryocoolers launched in July 1991 and two ISAMS launched in September 1991.

As shown in the figure a rise in running hours after 1991 occurs, in part, due to the successful flight operation and the establishment of test programs at MMS, NASA Goddard, JPL, and at TRW.   Also development programs at Phillips Laboratories (PL) for a 65 K Standard Spacecraft Cryocooler (SSC) and at NASA Goddard drew from the success of the Oxford/RAL design with similar linear motor driven, spiral flexure supports for compressor/displacer clearance seals. Starting in early 1993, a life test program was initiated by the Air Force Space and Missile Tracking System (SMTS), formerly known as the Brilliant Eyes Program.   The SMTS life test program involved life test at 60 K of two, third batch MMS 80 K units by Rockwell and two, first batch units by TRW, two mini-Stirling units at 60 K at TRW, three mini-pulse tubes at TRW at 60 K and 100 K, and two Hughes Improved Standard Spacecraft Cryocoolers (ISSC) in test at Hughes.

Although both the ISAMS and the ATSR coolers have recently been shut down, several European and US flight programs are planned using both Stirling and Pulse Tube cryocoolers in the near term.   Also, the Matra Marconi Space (MMS) and SMTS life test programs are continuing with PL plans to endurance test numerous types of Stirling/Pulse Tube and other types of cryocoolers delivered under their development programs such as the Creare Turbo Brayton 65 K unit which has achieved more than 9000 hours.   Hence, both the total accumulated hours on more than 40 Stirling/Pulse Tube units as well as total hours on single units is expected to rise significantly from the current levels of more than 500,000 and 50,000, respectively.

## Cryocooler Induced Vibration

Depending on the application, the induced vibration produced by a cryocooler can be of significant concern to the system.   Typically, most infrared sensors are extremely sensitive to jitter caused by vibration. In the Stirling and Pulse Tube cryocoolers, the reciprocating moving mass and support system for the displacer and/or compressor pistons generate force/moment disturbances that are transmitted through the cooler structural supports and the thermal hardware attachments. Because of its very high operating frequency, the Brayton cryocoolers typically do not create vibration disturbances which impact the cooled instrument. Both the Joule-Thomson and Pulse Tube cryocoolers have cold heads with no moving parts and minimal vibration sources. However, if a mechanical compressor is utilized, the associated vibrations can be transmitted either through the spacecraft structure or through connections (such as the transfer line or integral mounting) to the cold head and result in significant jitter at the instrument interface. These vibrations can be reduced significantly if either the compressor is mounted on an independent platform or a non-moving parts compressor (for example, a sorption unit) is utilized.

As a consequence of the above, it is often essential that an active vibration control be implemented in Stirling and Pulse Tube systems to reduce the drive axis residual vibrations to an acceptable level. Unfortunately, the cross-axis vibrations remain uncontrolled, but improved flexure tangential-arm bearings designed by The Aerospace Corporation and developed under a PL protoflight standard cooler (PSC) program at Hughes have shown that these forces can be substantially reduced. The induced vibration requirements entail a system penalty for electronics weight and power. This needs to be addressed when choosing a particular cooler.

### Mechanical And Load Interface Integration

An integration criteria which is often overlooked until the later stages of a cryogenic system design is the mechanical and/or load interface. Application specifics of the mechanical or load interface can have a large impact on the selection and integration of a cryocooler. For example, if the instrument load has a large variation (typically due to operating versus standby power consumption differences), a thermal storage unit may be optimum to significantly reduce the cooling load requirement. Two phase cryocooler cycles such as the Joule-Thomson have the ability to store liquid or solid working fluid and average the load, and, thus, may not require a separate thermal storage unit. This can facilitate integration, improve temperature stability, and reduce weight, parasitics, and power. Other mechanical or interface requirements which impact selection are interface area, deflection, temperature stability, and volume. There might also be volume and heat rejection constraints which can be alleviated through the remote mounting of the ambient temperature compressor. The Brayton and Joule-Thomson cryocoolers have this capability of mounting the compressor to the non-cryogenic hardware away from the cold head. Whereas the cooling capacity and efficiency of split-Pulse Tube and split-Stirling coolers can be significantly degraded if their transfer line length increases.

## CRYOCOOLER INTEGRATION PERFORMANCE ENHANCEMENT METHODS

Two integration methods for improving the performance of the cryocooler system were evaluated through parametric trade studies. The first method involves reducing the ambient heat rejection temperature to improve cryocooler power efficiency. The second method involves the utilization of a cryogenic radiator to intercept heat (or pre-cool the working fluid) in the cold head or regenerator. For both methods, the resulting savings in power and the associated system mass was evaluated against the increased radiator mass.

### Ambient Temperature Radiator Heat Rejection Trade Study

Extensive cooler data from characterization testing at both the 60 K and 35 K have been documented by R. Ross, D. Johnson, et al., at JPL on MMS 80 K and 50/80 K Stirlings and TRW 20 cc and 10 cc Pulse Tubes (see References 1, 6, 11, and 17). This testing has shown that the cooler power levels are substantially reduced as the ambient temperature level is decreased. However, the rejection capability of spacecraft radiators substantially decreases with the fourth power of radiator temperature. Thus, the net benefit needs to be determined between the decreased system penalty associated with the power reduction and the increased radiator penalty from larger area requirements.

In order to assess the radiator weight penalty as a function of heat rejection capability, the authors have performed ambient radiator and system analyses including a survey of past and current spacecraft designs (see Reference 16 for survey data). The upper or maximum weight penalty curve was based on an area density of 15 kg/m$^2$ and the LLC (long life cooler) at 195 K and several ambient radiators (including MTI) near 300 K. Based on this limited data, the following rough correlation was developed for maximum radiator specific weight (mass divided by heat rejection in kg/W):

$$\frac{W_{radiator}}{Q} = \frac{15.0}{2.83 * (T/100)^4 - 37.5} \tag{3}$$

The LLC is insulated from the structural parasitics, but does not include environmental shielding as required for lower temperature cryoradiators. The lower or minimum weight penalty curve was based on an area density of 6 kg/m$^2$ and a design study for high altitude, low sun angle (<23.5 degrees), and low solar absorptance/high emmittance (<0.3/0.8) unshielded radiators (see Reference 14). The following rough correlation was developed for the minimum radiator specific weight:

$$\frac{W_{radiator}}{Q} = \frac{6.0}{3.08 * (T/100)^4 - 8.93} \tag{4}$$

The maximum radiator penalty curve corresponds to conditions where the radiator has some significant solar views. This is in contrast to the minimum radiator penalty which corresponds more closely to continuous deep space radiator views. For most applications, the minimum curve can only be achieved without difficulty in a sun synchronous, low earth, polar orbit. It is expected that most ambient radiator designs will fall within the range of these estimates.

Table 2 presents the results of the net weight that might be achieved from reduced cryocooler heat rejection. The ambient radiator weight penalty is based on the correlation range for the

**Table 2.** Cryocooler heat rejection system mass trades.

| Cryocooler | Load | Nominal Heat Rejection | Decreased Heat Rejection | Radiator Net Gain* (kg) | System Net Loss** (kg) | System Specific Net Loss*** (kg/W) |
|---|---|---|---|---|---|---|
| MMS 50/80 K Stirling | 1.15 W@ 60 K | 56.0 W@ 293 K | 50.9 W@ 273 K | 0.3 to 1.5 | 0.4-1.4 | 0.007-0.025 |
| TRW 3585 K 20 cc P-T | 3.42 W@ 60 K | 184.0 W@ 303 K | 146.0 W@ 273 K | 1.0 to 4.6 | 9.7-13.3 | 0.053-0.072 |
| TRW 6020 K 10 cc P-T | 3.16 W@ 60 K | 127.8 W@ 293 K | 116.4 W@ 273 K | 0.8 to 3.4 | 0.9-3.5 | 0.007-0.027 |
| MMS 50/80 K Stirling | 0.16 W@ 35 K | 63.9 W@ 293 K | 45.3 W@ 273 K | 0.3 to 1.3 | 5.7-6.7 | 0.089-0.105 |
| TRW 3585 K 20 cc P-T | 0.25 W@ 35 K | 187.0 W@ 303 K | 92.0 W@ 273 K | -1.8 to - 2.4 | 33.2-33.8 | 0.178-0.181 |
| TRW 3503 K 10 cc P-T | 0.5 W@ 35 K | 133.7 W@ 293 K | 91.3 W@ 273 K | -0.3 | 15.6 | 0.117 |

\* Radiator range based on maximum and minimum radiator penalty curves
\*\* System mass evaluated using 0.375 kg/W penalty
\*\*\* System specific net loss = system net loss (kg)/cryocooler heat rejection (W)

appropriate temperature level. The net weight change is calculated based on the weight difference between the nominal and the reduced heat rejection temperature level. The results show that there is a net weight loss in each case. Of interest is that at 35 K, both the TRW 10 cc and 20 cc unit power reduction is sufficient, due to substantial inefficiencies at low cooling loads, that there is a net reduction in the radiator weight alone (as well as the overall system reduction).

The cryocooler test data provide estimates of the system benefits of reducing the rejection temperature to 273 K. To evaluate more extreme reductions in the ambient rejection temperature, a theoretical Carnot relation was combined with the radiator weight curves to estimate the system mass impact. The input power dependence on heat rejection temperature was calculated according to the following relationship:

$$\frac{P_{reject}}{P_{300K}} = \frac{T^{rej}_{reject} - T^{ct}_{reject}}{T^{rej}_{300K} - T^{ct}_{300K}} \tag{5}$$

where the P is the input power at either the nominal 300 K rejection temperature or the reduced "reject" temperature and T is the temperature either at the cold tip (ct) or heat rejection (rej) interface. Figures 8 and 9 (expanded scale) show the specific change in system mass (system mass change in kg divided by cryocooler heat rejection in watts) from the nominal 300 K rejection as a function of the reduced heat rejection temperature. The figures have curves for both a 60 and 35 K cold tip temperature. While the system mass continues to decrease with decreased rejection temperature for the minimum radiator weight penalty curve, the system mass increases sharply at low rejection temperatures for the maximum radiator penalty. As indicated in Figure 9, the point below which the maximum penalty curve results in system mass increases occurs at around 255 K for a 60 K cold tip and 265 K for a 35 K cold tip. For the maximum radiator penalty, the optimum system mass savings occurs at about 275 K rejection for 60 K cold tip and about 280 K for 35 K.

In summary, for 35 and 60 K cooling, both the cryocooler test data and the Carnot evaluation indicate that the system mass can be decreased by reducing the rejection temperature from 300 K to 273-280 K. If the radiator environment is sufficiently cold (typically in a deep space facing, sun synchronous, polar orbit), the system mass will continue to decrease as the rejection temperature is decreased below 275 K. However, if there is significant solar loads on the radiator, the Carnot evaluation indicates that the system mass will increase below about 275 K and exceed the 300 K rejection mass below about 260 K. It should be noted that some of this data are for off-design performance which might be improved by optimization of the cold head and/or compressor for the cooling load and temperature level. However, this heat rejection option illustrates what might be achieved without further cooler development.

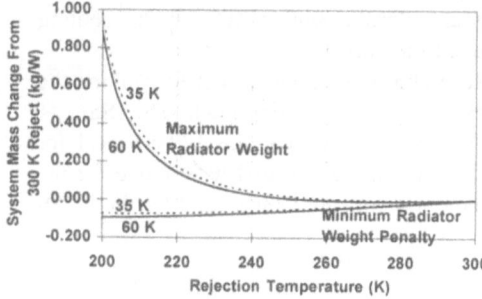

**Figure 8.** Cryocooler heat rejection system mass impact at 60 and 35 K cooling.

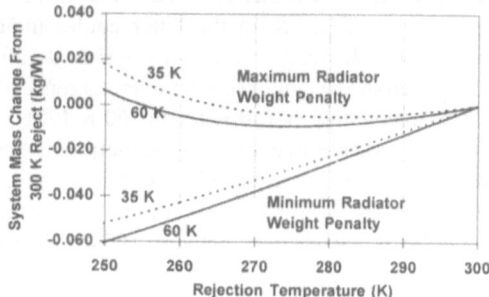

**Figure 9.** Cryocooler heat rejection system mass impact at 60 and 35 K cooling.

### Cryogenic Radiator Heat Rejection Trade Study

It was shown in the previous section that reducing the ambient radiator temperature for heat rejection from the warm-end of the cryocooler could produce a net system mass benefit. A similar benefit was evaluated using a cryogenic heat intercept technique previously reported by R. Ross and D. Johnson on their characterization testing of MMS 80 K and 50/80 K coolers (see Reference 15). This technique intercepts the parasitic load on the Stirling cold-finger, but has been applied to other types of coolers such as solid cryogen coolers, thermoelectric coolers, and multistaged radiators and can also be applied to coolers such as the Pulse Tube and Brayton. The heat intercept technique requires a cryogenic cooling source such as a cryogenic radiator (cryoradiator).

An assessment was made of cryoradiator technology that has been flown or tested between 100 K and 200 K and included the RM20B flight space test program, the CRTU (cryogenic radiator test unit), the MSR (multistaged radiator) study, the LLC (long life cooler) and the SMTS (Space and Missile Tracking System) programs (see Reference 16 for radiator data). Using this data, correlation equations were developed assuming a staged radiator system using an effective emmittance to relate the heat leak from the boundary structure (typically at 300 K) through the low conductance isolation supports and the multilayer insulation. The following correlation assumed an effective emmittance of 0.01 and an area density of 30 kg/m$^2$ for maximum radiator specific weight (mass or M divided by heat rejection or Q):

$$\frac{M_{radiator}}{Q} = \frac{30.0}{4.65*(T/100)^4 - 4.59} \tag{6}$$

The following minimum radiator specific weight correlation was based on an effective emmittance of 0.001 and an area density of 20 kg/m$^2$:

$$\frac{M_{radiator}}{Q} = \frac{20.0}{4.60*(T/100)^4 - 0.459} \tag{7}$$

While the 0.001 value is achievable in ground testing, the 0.01 is more typical of a flight qualified design. No incident environmental loads were assumed.

Table 3 presents the cryoradiator heat interceptor system mass trade results for several cryocoolers. The test data from Hughes was provided by D. Gilman and K. Price for their ISSC#4 cooler and published test data for the MMS 50/80 K cooler was provided by R. Ross and D. Johnson (see Reference 10). The Creare results are based on calculations provided by W. Swift and F. Dolan for the Phillips Lab (PL) 35 K Turbo Brayton cooler (see Reference 3). The heat intercept data was based on several temperature levels from 130 to 200 K. The heat intercept transport mass was based on two 1-meter aluminum heat pipes with two aluminum thermal straps. The cryoradiator rejection requirements vary between 0.8 and 4.1 W. Several other programs are currently investigating the use of this technique and include the Ball Aerospace 35/60 K cooler for PL and the MMS two-stage 20-50 K cooler. Information provided by B. Jones of MMS on the latter cooler indicates an improvement of 40% in the cooling capacity at 20 K using a heat intercept sink at 120 K (see Reference 13).

As seen in the tabulated data, there is a net benefit in almost all cases except for sinking the heat intercept strap at 140 K for the 60 K ISSC #4 tests with a low cooling load. The ISSC #4 data also indicate that there is an optimum heat sink temperature near the 170 K level for maximizing the weight benefit. A similar optimum appears in the 150 to 170 K range for the Turbo Brayton. More data are needed to determine the optimum sink temperature for the MMS cooler, but it appears to be between 150 and 190 K.

## SUMMARY AND CONCLUSIONS

This paper has described and discussed a number of selection or integration criteria as well as provided data and algorithms for evaluating cryocoolers against those criteria. The list of

**Table 3.** Cryocooler heat intercept system mass trades.

| Cryo-cooler | Load (W) | Nominal Heat Rejection (W) | Heat Intercept Load (W) | Heat Rejection Decrease (W) | Cryo-Radiator & Transport Increase* (kg) | System Net Loss** (kg) | System Specific Net Loss*** (kg/W) |
|---|---|---|---|---|---|---|---|
| HAC ISSC #4 Stirling | 3.0@ 60 K | 101.6 | 2.36@200 K 3.4@170 K 4.1@140 K | 11.9 22.6 27.9 | 1.53/1.90 2.79/3.99 5.76/10.27 | 2.6/2.9 4.5/5.7 0.2/4.7 | 0.026/0.029 0.044/0.056 0.002/0.006 |
| HAC ISSC #4 Stirling | 1.0@ 60 K | 42.0 | 1.42@200 K 2.3@170 K 2.88@140 K | 7.3 12.3 17.3 | 1.22/1.44 2.10/2.90 4.26/7.43 | 1.3/1.5 1.7/2.5 -0.9/2.2 | 0.031/0.036 0.040/0.060 -0.021/0.052 |
| MMS 50/80 K Stirling | 1.0@ 60 K | 42.5 | 1.0@190 K 1.5@150 K | 13.5 17.6 | 1.14/1.34 2.15/3.21 | 3.8/4.0 3.4/4.5 | 0.089/0.094 0.080/0.106 |
| MMS 50/80 K Stirling | 0.1@ 40 K | 47.6 | 0.82@190 K 1.35@150 K | 37.3 39.3 | 1.07/1.23 2.01/2.96 | 12.8/12.9 11.8/12.8 | 0.269/0.271 0.248/0.269 |
| HAC ISSC #4 Stirling | 1.0@ 35 K | 108.1 | 1.47@200 K 2.53@170 K 3.33@140 K | 18.1 36.4 40.6 | 1.23/1.46 2.23/3.12 4.81/8.48 | 5.4/5.6 10.5/11.4 6.7/10.4 | 0.050/0.052 0.097/0.105 0.062/0.096 |
| HAC ISSC #4 Stirling | 0.5@ 35 K | 66.9 | 1.2@200 K 2.15@170 K 2.8@140 K | 8.2 23.8 26.5 | 1.15/1.34 2.01/2.76 4.17/7.25 | 1.7/1.9 6.2/6.9 2.7/5.8 | 0.025/0.028 0.093/0.103 0.040/0.087 |
| Creare 35 K Brayton | 1.0@ 35 K | 112.9 | 2.0@170 K 2.0@150 K 2.0@130 K | 22.0 25.3 28.6 | 1.92/2.62 2.62/4.03 4.03/7.77 | 5.6/6.3 5.5/6.9 3.0/6.8 | 0.050/0.056 0.049/0.061 0.027/0.060 |

\* Radiator range based on maximum and minimum radiator penalty curves
\*\* System mass evaluated using 0.375 kg/W penalty
\*\*\* System specific net loss = system net loss (kg)/cryocooler heat rejection (W)

selection or integration criteria includes cryocooler cooling capacity at the desired interface temperature, input power, mass, reliability for the desired lifetime, hardware or technology maturity, system mass penalty, induced vibration, temperature stability, and general integratability. The relative prioritization and weighting of these criteria is highly dependent on the customer requirements and constraints.

Generic and parametric estimates for the cryocooler spacecraft penalties were developed using historical and empirical data. Overall, the total system penalty was relatively consistent with a range of 0.31 to 0.42 and an average of 0.375 kg/W (system mass increase divided by cryocooler input power). The generic penalties are useful for early conceptual designs and trade studies or when a specific design or application is not available. For detailed designs, a comprehensive analytical approach would be recommended. An integration criteria which can have a large impact on the system, but is often overlooked until the later stages of a design is the mechanical and/or load interface.

Two integration methods for improving the performance of the cryocooler system were evaluated through parametric trade studies. The first method involves reducing the ambient heat rejection temperature to improve cryocooler power efficiency. For both 35 and 60 K cooling, the trades indicate that the system mass can be decreased by reducing the rejection temperature from 300 K to 273-280 K. If the radiator environment is sufficiently cold (typically in a deep space facing, sun synchronous, polar orbit), the system mass will continue to decrease as the rejection temperature is decreased below 275 K. However, if there is significant solar loads on the radiator, the system mass will increase below about 275 K and exceed the 300 K rejection mass below about 260 K. The second method involves the utilization of a cryogenic radiator to intercept heat (or pre-cool the working fluid) in the cold head or regenerator. Based on a limited amount of test data, an optimum to the reduction in system mass appears to be heat intercepting at 160-170 K. The system mass savings from the cryogenic heat intercept method was similar in magnitude to those from the reduction of rejection temperature.

It is hoped that the material presented brings attention to the various cooling options and potential trades currently available to the user community.

## ACKNOWLEDGMENT

The authors wish to acknowledge the support of personnel of the Air Force Space Division SMTS SPO and the Kirkland AFB Phillips Labs including the Ballistic Missile Defense Organization and the respective program office personnel at The Aerospace Corporation. Special thanks are given to M. Donabedian of The Aerospace Corporation for empirical radiator data and R. Ross, D. Johnson, et al. of the Jet Propulsion Lab, W. Gully of Ball, W. Swift and F. Dolan of Creare, D. Gilman, K. Price, D. Makowski, et al. of Hughes, T. Nast of Lockheed Martin, B. G. Jones of Matra Marconi Space, and W. Burt, C. K. Chan, and M. Tward of TRW for providing the basis for much of the cryocooler data presented here.

## REFERENCES

1.  Johnson, D. L. and Wu, J. J., *Feasibility Demonstration of a Thermal Switch for Dual Temperature IR Focal Plane Cooling, Final Report*, JPL D-13423, 15 February 1996.
2.  Jones, B. G., *Review Meeting at MMS, Coolers*, 10 July 1995.
3.  Swift, W., *Creare Cryocooler Technical Interchange Meetings and Personal Correspondence*, 1995-1996.
4.  Russo, S., *SSC and ISSC Data Faxes and Correspondence*, January 1996.
5.  Nast, T. C., et al., "Design, Performance, and Testing of the Lockheed-Developed Mechanical Cryocooler," *Cryocoolers 8*, Plenum Press, New York 1995, pp. 55-67.
6.  Smedley, G. T., Johnson, D. L., and Ross, R. G., *TRW 1 W-35 K Pulse Tube Cryocooler, Performance Characterization*, JPL D-13236, December 1995.
7.  Ortiz, T., *Progress Status and Management Report, 35 K Cryocooler Program*, TRW S/N 60030, September 1995.
8.  Chan, C. K., *150 K Protoflight Spacecraft Cryocooler Program, Advanced Pulse Tube Cold Head Development Scientific and Technical Report*, CDRL A020, April 1995.
9.  Gully, W., *Ball Cryocooler Personal Correspondence and Faxes*, 1996.
10. Mon, G. R., Johnson, D. L., Ross, Jr., R. G., *BAe 50-80 K Cryocooler Performance Characterization*, JPL Cryogenics Technology Group, February 1994.
11. Johnson, A. L., *Cryocooler System Penalty Personal Correspondence*, May 1996.
12. Donabedian, M., Glaister, D. S., and Bernstein, M. D., "Cryogenic Systems Integration Model (CSIM)," *Cryocoolers 8*, Plenum Press, New York, 1995, pp. 695-709.
13. Jones, B. G., *Matra Marconi Space Telefax entitled "Cooler Life Test Hours,"* May 1996.
14. Curran, D. G. T. and Lam, T. T., "Optimal Sizing of Honeycomb Radiators with Embedded Heat Pipes," *30th AIAA Thermophysics Conference*, San Diego, Calif., AIAA-95-2136, 19-22 June 1995.
15. Johnson, D. L. and Ross, Jr., R. G. "Cryocooler Coldfinger Heat Interceptor," *Cryocoolers 8*, Plenum Press, New York, 1995.
16. Donabedian, M., *Satellite Thermal Control Handbook*, Gilmore, D. G., ed., *Chapter VIII, Cryogenic Systems*, 1994.
17. Johnson, D. L., *TRW 3503 Pulse Tube Cooler Characterization Data Personal Correspondence*, May 1996.

# AIRS Cryocooler System Design and Development

**R.G. Ross, Jr.**

Jet Propulsion Laboratory
California Institute of Technology
Pasadena, California 91109

**K.E. Green**

Lockheed Martin IR Imaging Systems
Lexington, MA 02173

## ABSTRACT

JPL's Atmospheric Infrared Sounder (AIRS) instrument is based on a cryogenically cooled infrared spectrometer that uses a pair of pulse tube cryocoolers operating at 55 K to cool the HgCdTe focal plane to 58 K; the instrument also includes cryoradiators at 150 K and 190 K to cool the overall optical bench to 150 K. The cryocooler system design is a key part of the instrument development and focuses heavily on integrating the cryocoolers so as to maximize the performance of the overall instrument.

The cryocooler system development activity is a highly collaborative effort involving development contracts with industry and extensive cryocooler characterization testing at JPL. In the first phases of the effort, the overall cryocooler integration approach was developed by Lockheed Martin, and TRW was selected to develop and produce the flight coolers. The selected state-of-the-art pulse tube cooler has excellent thermal performance, and has a number of attributes— particularly light weight—that greatly improve instrument integration.

This paper describes the AIRS instrument overall cryogenic system design and the results achieved to date with respect to integration of the TRW pulse tube cryocoolers into this demanding instrument. Results are presented detailing the cryogenic loads on the cooler, the overall cryocooler thermal performance margins achieved, and thermal heatsinking considerations. Mass properties of the cryocooler system, and thermal properties of the developed cold link assembly are also presented.

## INTRODUCTION

### Instrument Overview

The objective of the Atmospheric Infrared Sounder (AIRS) instrument is to make precision measurements of atmospheric air temperature over the surface of the Earth as a function of height above the Earth's surface. The technical foundation of the instrument is a cryogenically cooled infrared spectrometer that uses a pair of 55K cryocoolers to cool the HgCdTe focal plane to 58 K;

Cryocoolers 9, Edited by R.G. Ross, Jr.
Plenum Press, New York, 1997

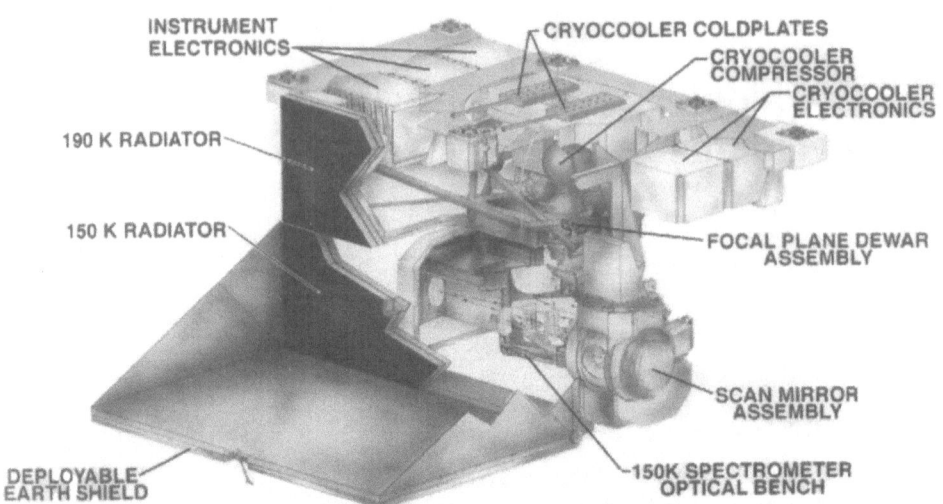

Figure 1. Overall AIRS instrument

the instrument also includes a 150K-190K two-stage cryogenic radiator to cool the optical bench assembly to 150 K. The spectrometer operates over a wavelength range from visible through 15.4 µm, and places particularly demanding requirements on the thermal and vibration performance of the cryocooler.

The AIRS instrument is scheduled to be flown on NASA's Earth Observing System PM platform in the year 2000, and is being designed and fabricated under JPL contract by Lockheed Martin IR Imaging Systems (formerly Loral LIRIS) of Lexington, MA; it is in the detail design and flight hardware buildup phase at this time.

Figure 1 illustrates the overall instrument and highlights the key assemblies. Physically, the instrument is approximately 1.4 m x 1.0 m x 0.8 m in size, with a mass of 150 kg and an input power of 220 watts. Configurationally, the 58K IR focal plane assembly is mounted integrally with the 150K optical bench, which is in-turn shielded from the ambient portion of the instrument by a 190K thermal radiation shield and MLI blankets. The ambient portion of the instrument contains the high power dissipation components including the instrument electronics and the cryocoolers. These power-dissipating components have their heat rejection interface to a set of coldplates that conduct the heat to spacecraft-mounted radiators via a system of heatpipes.

### Paper Organization

The remainder of the paper describes the details of the AIRS instrument cryogenic requirements and the approach used in the design of the cryosystem. This includes a detailed discussion of the cryocooler integration approach, performance predictions, and lessons learned to date with respect to the design and integration of the cryocoolers into the AIRS instrument.

## AIRS INSTRUMENT CRYOGENIC DESIGN AND REQUIREMENTS

### AIRS Cryosystem Conceptual Design

Early in the design of the AIRS instrument, key decisions of design philosophy were established that served as fundamental ground rules for the cryocooler system design. These included:

- Totally redundant cryocoolers—to avoid one cooler being a single-point failure
- No heat switches—to avoid increased complexity, cost and unreliability
- Ambient heat rejection to spacecraft-supplied cold plates operating between 10 and 25°C

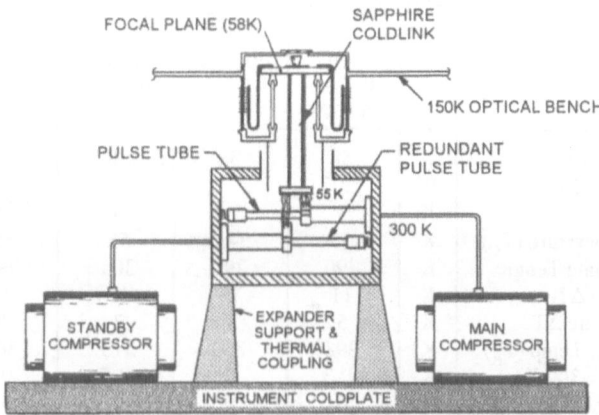

**Figure 2.** AIRS cryosystem conceptual design

- Cooler drive fixed at 44.625 Hz, synchronized to the instrument electronics—to minimize asynchronous vibration or EMI noise pickup from the cryocooler
- Cold-end load (focal plane) mechanically mounted and aligned to the 150 K optical bench with a maximum vibration jitter on the order of 0.2 μm
- Focal plane calibration (for temperature, motion, etc.) every 2.67 sec (every Earth scan)
- Cooler input power goal of 100 watts (22 to 35 volts dc), and mass goal of 35 kg
- Cooler drive electronics fully isolated (dc-dc) from input power bus; EMI consistent with MIL STD 461

Based on the above fundamental ground rules, the AIRS cryosystem conceptual design, shown in Fig. 2, was developed. This system incorporates two independent 55K cryocoolers, a primary and a non-operating backup, each connected to the 58K focal plane using a common high-conductance coldlink assembly. Ambient heat from the operating cooler is rejected to the coldplates located in the plane of the instrument/spacecraft interface. Table 1 provides a break-down of the approximate overall cryocooler refrigeration load for the AIRS instrument based on both the expected beginning-of-life (BOL) and possible end-of-life (EOL) properties of the cryosystem elements. A key determiner of these BOL/EOL loads is the temperature of the optical bench—assumed to be 145 K at BOL, and 160 K at EOL.

### Cooler Sizing Calculations

In order to provide an accurate understanding of both the beginning-of-life (BOL) and end-of-life (EOL) cryocooler system performance, a sensitivity analysis of the cryocooler/load sys-

**Table 1.** Breakdown of AIRS cryocooler loads at 58 K

| ITEM | Load (mW) | |
|---|---|---|
| | BOL | EOL |
| Focal plane radiation load | 70 | 100 |
| Focal plane electrical dissipation | 190 | 190 |
| Conduction down wires and cables | 100 | 120 |
| Focal plane structural support conduction | 130 | 160 |
| Pulse tube snubber conduction | 40 | 50 |
| Radiation to coldlink assembly | 90 | 120 |
| Off-state conduction of redundant cryocooler | 450 | 530 |
| Total cryocooler load | 1070 | 1270 |

**Table 2.** AIRS BOL/EOL performance margin analysis

| PARAMETER | Unit | BOL Perfor-mance | 200 mW Load Increase | 15°C Heatsink Increase | Cooler Wearout Degrad. | EOL Perfor-mance |
|---|---|---|---|---|---|---|
| Focal plane Temperature | K | 58 | 58 | 58 | 58 | 58 |
| Total Cooler Cold-End Load | W | 1.07 | 1.27 | 1.07 | 1.07 | 1.27 |
| Cooler Cold-tip $\Delta T$ to FP | K | 3 | 3.4 | 3 | 3 | 3.4 |
| Cooler Cold-tip Temperature ($T_C$) | K | 55 | 54.6 | 55 | 55 | 54.6 |
| Heat Rejection Coldplate Temp | K | 290 | 290 | 305 | 290 | 305 |
| Expander to Coldplate $\Delta T$ | K | 11 | 13 | 13 | 15 | 23 |
| Compressor to Coldplate $\Delta T$ | K | 5 | 7 | 7 | 8 | 11 |
| Avg. Cooler Rejection Temp ($T_R$) | K | 298 | 300 | 315 | 301 | 322 |
| $T_C$ Correction for $T_R \neq 300$ K | K | +0.3 | 0 | -2.5 | -0.2 | -3.7 |
| $T_C$ Correction for Cooler Wearout | K | 0 | 0 | 0 | -5.0 | -5.0 |
| Total Cold-tip Temp Correction | K | +0.3 | 0 | -2.5 | -5.2 | -8.7 |
| Effective 300K Cold-tip Temp ($T_{EC}$) | K | 55.3 | 54.6 | 52.5 | 49.8 | 45.9 |
| Cooler Specific Power at $T_{EC}$ | W/W | 57 | 55 | 62 | 72 | 83 |
| Cooler Compressor Power ($P$) | W | 61 | 70 | 66 | 75 | 105 |
| Total Input Power ($P/0.89 + 12$) | W | 81 | 90 | 86 | 96 | 130 |
| Compressor Stroke | % | 64 | 68 | 67 | 70 | 80 |

tem has been conducted and periodically updated using BOL and EOL estimates of the key governing parameters. This analysis is summarized in Table 2. In this table, the column labeled "BOL Performance" presents the predicted performance for the 1070 mW nominal BOL cryogenic load noted in Table 1, together with BOL estimates of the cryocooler heat rejection temperature, and the baseline BOL performance of the cryocooler (presented in Fig. 3 for 300 K heat rejection temperature). For heat rejection temperatures different from 300 K, the cold-tip temperature for a given load and input power rises approximately 1 K for each 6 K increase in heat rejection temperature. Note that this correction is included in the line "$T_C$ Correction for $T_R \neq 300$ K."

The middle three columns of Table 2 predict the performance of the AIRS cooler with the

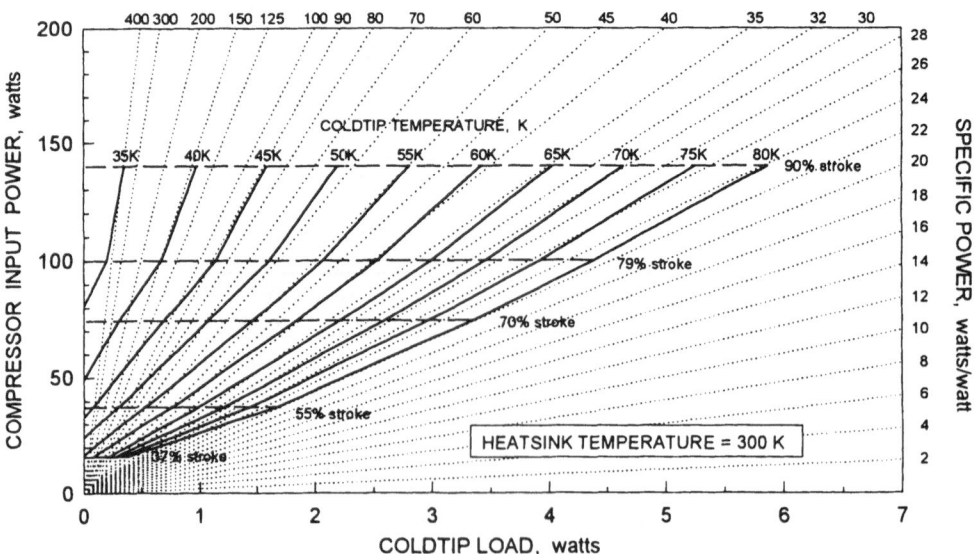

**Figure 3.** Baseline thermal performance of the AIRS pulse tube cooler with 300 K heat rejection temperature

individual EOL effects of a 200 mW increase in the cryocooler load (as noted in Table 1), a 15°C increase in the cryocooler heat rejection temperature, and a nominal value for EOL cryocooler degradation, respectively. End-of-life performance of the cryocooler is modeled as a 5K shift in the cryocooler load line, i.e. the EOL input power at 55 K is the same as the BOL input power at 50 K for the same cryogenic load. Based on lifetest experience to date, this 5 K degradation of performance at EOL appears to be a conservative, yet reasonable assumption.

The right most column of Table 2 represents the computed EOL performance of the AIRS cryocooler system. Note that the predicted difference between EOL and BOL performance is very significant — nearly a factor of two in compressor input power. Table 2 demonstrates the good match of the AIRS cooler to the requirements of the AIRS instrument over its total life cycle, including representative end-of-life degradation. With the assumed end-of-life degradation, the cooler performance satisfies the focal plane cooling requirement and remains within the nominal operating range of the compressor, i.e. less than 80% of maximum stroke.

## DEVELOPMENT WORK LEADING UP TO FLIGHT COOLER SELECTION

Because the required cryocooler performance, noted in Fig. 3, was greater that any existing cryocoolers at the beginning of the AIRS development effort, the AIRS Project established a collaborative in-house/contractor teaming approach to achieve the necessary cryocooler technology advances. This approach involved the establishment of an extensive cryocooler characterization program at JPL[1] to provide the foundation of cryocooler performance data needed, and a contractor-based effort lead by JPL's AIRS instrument systems contractor, Lockheed Martin IR Imaging Systems (formerly Loral LIRIS), to expand the performance of the first-generation coolers to meet the AIRS requirements. This contractual effort proved the feasibility of achieving the AIRS requirements, and fostered important design improvements associated with reduced off-state conduction down the cold finger, and high accuracy cold-tip temperature regulation via compressor piston stroke control.[2,3]

Following four years of extensive cryocooler characterization and development contracts, TRW was awarded the contract to develop and produce the flight coolers for the AIRS instrument. The selected state-of-the-art pulse tube cooler builds on the demonstrated performance of the successful TRW 1W-35K pulse tube cryocooler,[4,5] and promises excellent thermal performance, comparable to the best Stirling coolers; it also offers a number of features that greatly improve instrument integration, such as reduced mass, size and complexity, increased stiffness, and reduced vibration at the cold head.

## FLIGHT CRYOCOOLER SYSTEM DETAIL DESIGN

Upon selection of the TRW pulse tube cooler and the AIRS cryosystem concept illustrated in Fig. 2, work was focused on developing the flight cooler and resolving the details of a number of key design-integration trade-offs.

### Pulse tube Expander Integration Considerations

To minimize thermal conduction losses between the focal plane and the cryocooler, the pulse tube coldblock needs to be located close to the focal plane. Unfortunately, in addition to providing refrigeration, the expander of a modern high-efficiency Stirling or pulse tube refrigerator also dissipates a large amount of ambient heat — often 50% of the total compressor input power. Thus, the expander also needs to be mounted close to the instrument heat rejection system in order to minimize its operating temperature and maximize its efficiency. With the AIRS instrument, the distance between the focal plane and the instrument heat-dissipation cold plates is approximately 45 cm (18 inches). This distance has to be spanned by a combination of the cooler-focal plane coldlink assembly and the pulse tube expander heat-rejection mounts.

The AIRS pulse tube expander/coldlink integration design is illustrated in Fig. 4. This configuration uses high-cross-section aluminum structural members to heatsink the expanders to

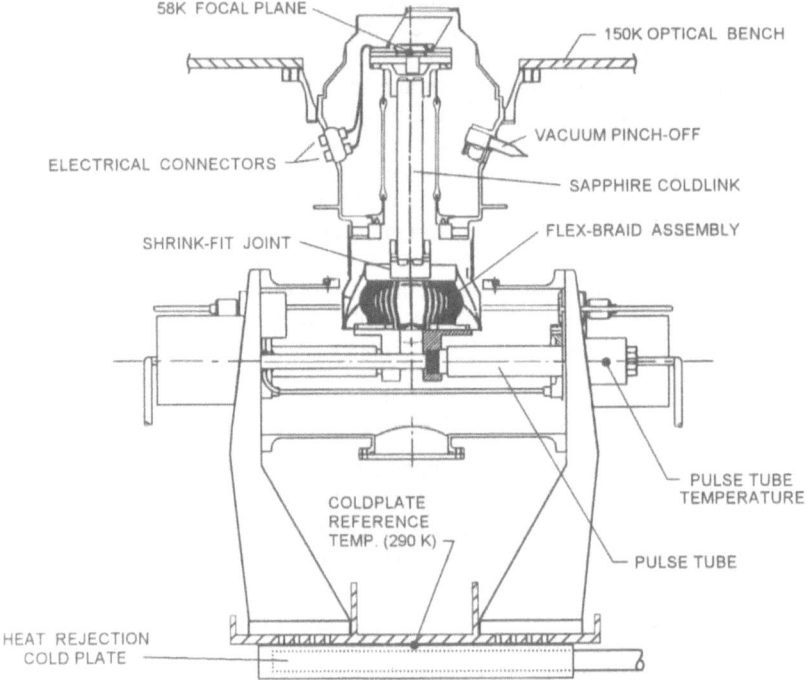

**Figure 4.** AIRS focal plane/cryocooler integration approach

the spacecraft coldplates, and a high-conductance Sapphire coldrod/flexlink assembly to connect the pulse tube coldblock to the instrument focal plane.

**Sapphire Coldlink Assembly.** As shown in Fig. 5, the sapphire coldlink assembly—designed and fabricated by Lockheed-Martin—contains a copper-braid flexlink section to accommodate the relative motion that occurs between the pulse tube and the focal plane dewar during launch and during cooldown of the instrument to cryogenic temperatures. The copper flexlink assembly bolts directly onto the two pulse tube coldblocks at one end, and at the other end attaches to the gold-plated sapphire coldrod using a molybdenum/aluminum shrink-fit interface.

**Figure 5.** AIRS focal plane/pulse tube coldlink assembly

**Table 3.** Breakdown of AIRS coldlink assembly thermal resistance

| ITEM | Resistance (K/W) |
|------|:---:|
| Focal plane to Sapphire rod | 1.57 |
| Conduction down Sapphire rod | 0.16 |
| Sapphire rod to moly coupling | 0.34 |
| Resistance across shrink-fit joint | 0.40 |
| Resistance across flex braid | 1.35 |
| Coldblock contact resistance | 0.30 |
| Total focal plane/pulse tube thermal resistance | 4.12 |

The total measured thermal resistance of the complete coldrod assembly from the pulse tube coldblock to the focal plane active elements is approximately 4K/W; the details of this resistance are shown in Table 3. In addition to the copper-braid section that connects the pulse tube coldblocks to the sapphire rod, the cold link assembly also contains copper braids that connect the coldblocks to one another so that the appreciable (~0.5 watt) off-state conduction of the redundant cryocooler pulse tube does not have to be conducted to the Sapphire rod and back to the operating cooler.

**Pulse Tube Expander Aluminum Heatsinks.** The pulse tube heatsink structural/thermal mount, also illustrated in Fig. 4, has been designed and fabricated by TRW as part of the structural support of the pulse tube/coldlink vacuum-housing assembly. This mount is required to conduct up to 40 watts from the operating expander to the cryocooler heat-rejection coldplate while simultaneously minimizing the rejection temperature of the pulse tube and the total required mass. The design achieves a thermal resistance of approximately 0.4 K/W from the pulse tube regenerator base to the 290 K coldplate interface.

## Compressor Thermal-Structural Mounting Considerations

As with the expanders, minimizing the temperature of the compressors is equally important to achieving high cryocooler efficiency. This has been accomplished by mounting the AIRS compressors as close to the instrument heat-rejection coldplates as possible, yet also as close to the expanders as possible so as to minimize the length of the interconnecting transfer line. A second important consideration has been to uniformly spread the thermal dissipation over the surface of the coldplates so as not to create local high-heat-flux areas within the heatpipe evaporators that might lead to heatpipe dryout and depriming. The AIRS compressor mounting ap-

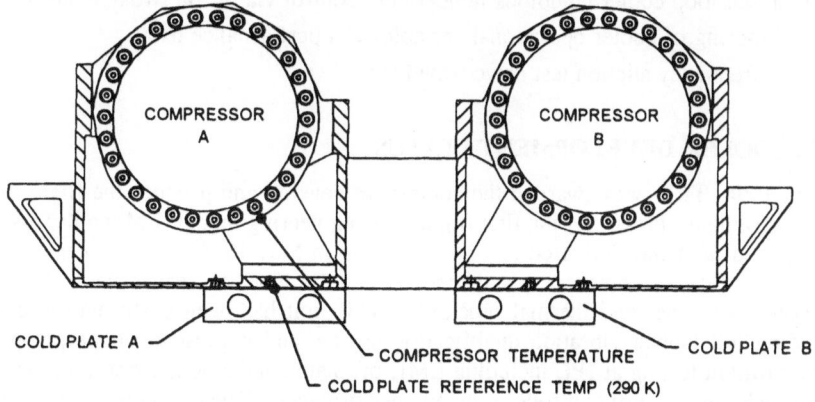

**Figure 6.** AIRS cryocooler compressor mounting approach and heat transfer interface

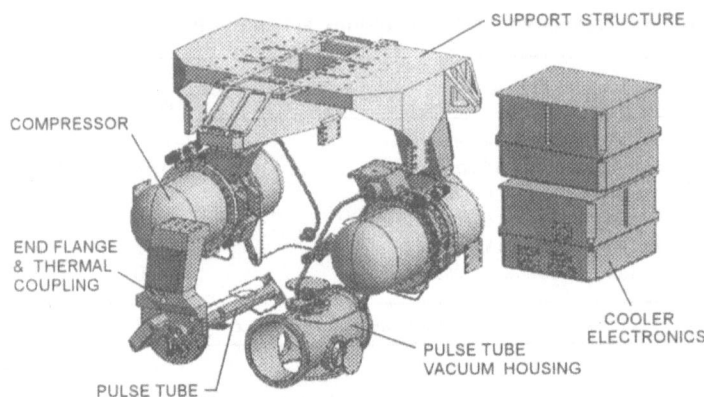

**Figure** 7. Exploded view of AIRS cryocooler assembly

proach, illustrated in Fig. 6, was a trade-off of mass against ΔT and resulted in a final thermal resistance of approximately 0.2 K/W between the compressor outer shell and the 290K coldplate reference temperature.

### Cryocooler Electronics

In addition to the cryocooler thermal/mechanical components, the cooler drive electronics is a key part of the overall AIRS cryocooler system and plays a critical role in the cooler integration. An exploded view of the total cryocooler assembly including the electronics is shown in Fig. 7. The cooler electronics not only drive the compressors with high electrical efficiency, but also perform a number of vital control, noise suppression, and data acquisition functions.[6] Key attributes of the AIRS cryocooler drive electronics include:

- Very high electrical efficiency (90% throughput to the compressors) including full (dc-dc) transformer isolation from input power bus (23 to 35 volts dc)
- Built-in shorting relays to suppress cooler piston motion during launch
- Cooler drive fixed at 44.625 Hz, synchronized to the instrument electronics—to minimize asynchronous vibration or EMI noise pickup from the cryocooler
- Very high degrees of EMI shielding, consistent with MIL STD 461
- Advanced feedforward vibration suppression system with accelerometer-based closed-loop nulling of the first 16 cooler vibration harmonics
- Precision closed-loop cooler coldblock temperature control via piston stroke control
- Built-in monitoring of cooler operational variables and performance data
- Built-in low-frequency stiction test drive waveform

### AIRS CRYOCOOLER DEVELOPMENT STATUS

In Spring 1994, TRW was awarded the contract to develop and produce the flight coolers for the AIRS instrument. Presently, the first flight-like engineering model (EM) cooler assembly has been completed, and was delivered to JPL for testing in May 1996. This cooler, shown in Fig. 8, has one flight-like compressor and associated pulse tube; to reduce cost, the second redundant cooler is a mass and thermal mock-up. This unit has been performance tested at TRW,[7] including full launch vibration qualification testing, and is presently undergoing additional characterization testing at JPL including EMI, off-state conduction, vibration modal testing, and coldblock temperature controller dynamic performance. The excellent mass properties of the AIRS cryocooler system are summarized in Table 4.

**Figure 8.** AIRS Engineering Model cryocooler assembly during testing at JPL

In addition to the AIRS EM cooler, TRW has completed a pair of similar pulse tube cryocoolers for the Air Force/BMDO.[8] These AIRS-like pulse tube coolers have been extensively characterized at JPL over the past year under joint AIRS/Air Force sponsorship and have led to a wealth of performance data generally applicable to the AIRS cooler design.[9]

Upon completion of the characterization testing at JPL, the AIRS EM cooler will be delivered to Lockheed Martin IR Imaging Systems in Lexington, MA, in summer 1996. There, it will start an extensive series of integration tests with the Lockheed-Martin coldlink/focal plane assembly and the complete Engineering Model AIRS instrument. The AIRS flight (PFM) coolers are scheduled for completion of full qualification testing and delivery to Lockheed Martin around the beginning of calendar 1997.

## SUMMARY AND CONCLUSIONS

The AIRS cryocooler system development activity is a key part of the AIRS instrument development and focuses on developing and integrating the cryocoolers so as to maximize the performance of the overall instrument; it is a highly collaborative effort involving development contracts with Lockheed Martin and TRW, and cryocooler characterization testing at JPL. To date, the overall cryocooler integration approach has been developed and refined, and the state-of-the-art TRW pulse tube cooler has demonstrated excellent thermal performance and light weight.

**Table 4.** Breakdown of mass of AIRS cryocooler assembly

| ITEM | | Mass (kg) |
|---|---|---|
| Total cryocooler A (primary) weight | | 12.5 |
|     Compressor A | 8.4 | |
|     Pulse tube expander A | 0.3 | |
|     Electronics A | 3.8 | |
| Total cryocooler B (backup) weight | | 12.5 |
| Pulse tube vacuum housing and heat sinks | | 3.8 |
| Integrating structure/coldplate support | | 5.2 |
| Compressor-to-electronics cables (2 sets) | | 1.0 |
| Total cryocooler assembly | | 35.0 |

Results have been presented detailing the cryogenic loads on the cooler, the overall cryo-cooler thermal performance margins achieved, and thermal heatsinking considerations. Mass properties of the cryocooler system, and thermal properties of the developed coldlink assembly have also been presented.

## ACKNOWLEDGMENT

The work described in this paper was carried out by the Jet Propulsion Laboratory, California Institute of Technology, Lockheed Martin IR Imaging Systems, and TRW, Inc; it was sponsored by the NASA EOS AIRS Project through an agreement with the National Aeronautics and Space Administration.

## REFERENCES

1. Ross, R.G., Jr., "JPL Cryocooler Development and Test Program Overview," *Cryocoolers 8*, Plenum Publishing Corp., New York, 1995, pp. 173-184.

2. Smedley, G.T., Mon, G.R., Johnson, D. L. and Ross, R.G., Jr., "Thermal Performance of Stirling-Cycle Cryocoolers: A Comparison of JPL-Tested Coolers," *Cryocoolers 8*, Plenum Publishing Corp., New York, 1995, pp. 185-195.

3. Clappier, R.R. and Kline-Schoder, R.J., "Precision Temperature Control of Stirling-cycle Cryocoolers," *SPIE Proceedings*, Vol. 2227 (1994).

4. Burt, W.W. and Chan, C.K., "Demonstration of a High Performance 35K Pulse Tube Cryocooler," *Cryocoolers 8*, Plenum Publishing Corp., New York, 1995, pp. 313-319.

5. Collins, S.A., Johnson, D.L., Smedley, G.T. and Ross, R.G., Jr., "Performance Characterization of the TRW 35K Pulse Tube Cooler," *Advances in Cryogenic Engineering*, Vol. 41, 1996.

6. Chan, C.K., et al., "AIRS Flight Pulse Tube Cooler System," *Cryocoolers 9*, Plenum Publishing Corp., New York, 1997.

7. Chan, C.K., et al., "Performance of the AIRS Pulse Tube Engineering Model Cryocooler," *Cryocoolers 9*, Plenum Publishing Corp., New York, 1997.

8. Burt, W.W. and Chan, C.K., "New Mid-Size High Efficiency Pulse Tube Coolers," *Cryocoolers 9*, Plenum Publishing Corp., New York, 1997.

9. Johnson, D.L., Collins, S.A., Heun, M.K. and Ross, R.G., Jr., "Performance Characterization of the TRW 3503 and 60K Pulse Tube Coolers," *Cryocoolers 9*, Plenum Publishing Corp., New York, 1997.

# AIRS Pulse Tube Cryocooler System

**C. K. Chan, J. Raab, A. Eskovitz, R. Carden III, M. Fletcher, R. Orsini**

TRW Space & Technology Division
Redondo Beach, CA 90278, United States of America

## ABSTRACT

The TRW AIRS cooler system features redundant pulse tube cryocoolers to control the temperature of the Atmospheric Infrared Imaging Spectrometer (AIRS) instrument focal plane array. The AIRS instrument mission is to acquire high resolution atmospheric temperature profile data to enhance atmospheric weather models; AIRS is scheduled to fly on the EOS-PM platform in 2000.

Each of the standby redundant cryocoolers consists of a compressor and split pulse tube cold head controlled by its own cooler electronics and integrated onto a structure that mates with the AIRS instrument. A vacuum housing, rigidly affixed to the mounting structure, surrounds the two cold heads. The cold interface couples the cold heads together through a flexible link to a sapphire rod that conducts heat from the focal plane assembly. The mounting structure couples heat from the compressors and cold head assembly to the spacecraft cold place interfaces.

The standby redundant coolers are designed conservatively to satisfy the 7-year lifetime requirement with a reliability exceeding 0.997. The system includes radiation-hardened control electronics.

## INTRODUCTION

The AIRS cooler is a standby redundant pulse tube cryocooler assembly that controls the temperature of the high performance HgCdTe IR focal plane arrays of the Atmospheric Infrared Imaging Spectrometer (AIRS) instrument (Figure 1). The AIRS instrument covers wavelengths of 3.74 to 15.5 microns in three bands and 0.4 to 1.0 microns in four bands; it will acquire high resolution atmospheric temperature profile data to enhance atmospheric weather models. AIRS is scheduled to fly on the EOS-PM platform in 2000.

Two coolers are integrated onto a structure that mates with the AIRS instrument. Each redundant cryocooler consists of a compressor and a separate, or split, pulse tube coldhead, and is regulated by its own electronics assembly (Figure 2). A housing, rigidly affixed to the mounting structure, surrounds the two cold heads. A flexible link couples the two cold heads together and to a sapphire rod that conducts heat from the focal plane assembly. The support structure conducts waste heat from the compressor and coldhead to the spacecraft cold plate interfaces. The system instrument-spacecraft interfaces are shown in Figure 3.

The AIRS pulse tube cooler builds on previously flight-qualified pulse tube cooler designs (references 1, 2). Pulse tubes offer significant advantages over Stirling coolers; the no-moving-part cold head produces lower vibration and is easier to interface because of its robust nature and its lower mass.

Cryocoolers 9, Edited by R.G. Ross, Jr.
Plenum Press, New York, 1997

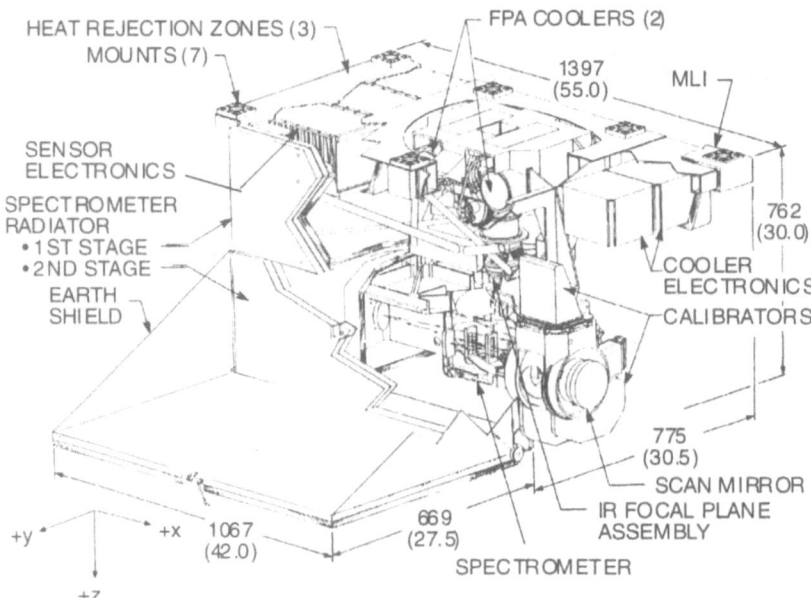

**Figure 1.** Pulse tube cooler system will fly on Atmospheric Infrared Sounder (AIRS) instrument in 2000

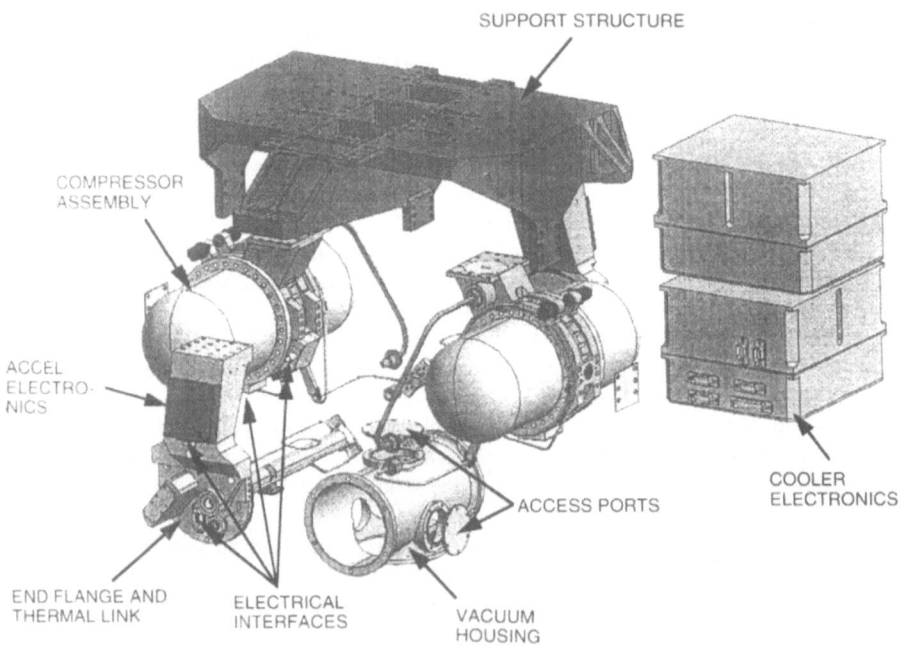

**Figure 2.** AIRS cryocooler system is simple with few components

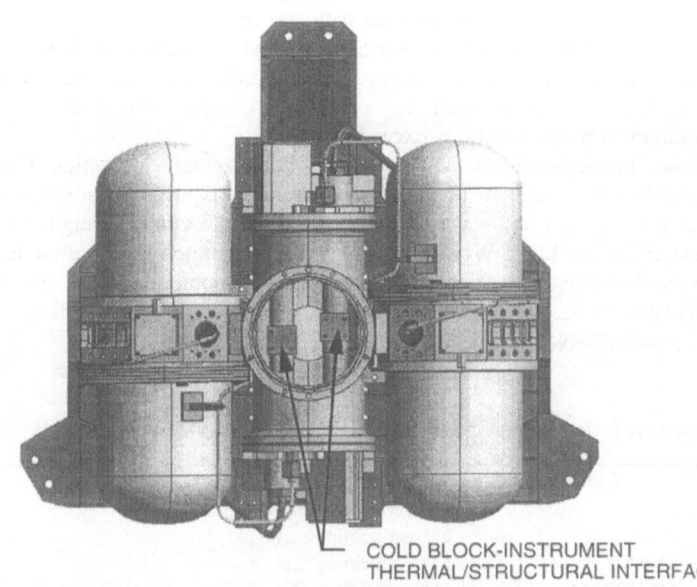

COLD BLOCK-INSTRUMENT
THERMAL/STRUCTURAL INTERFACE

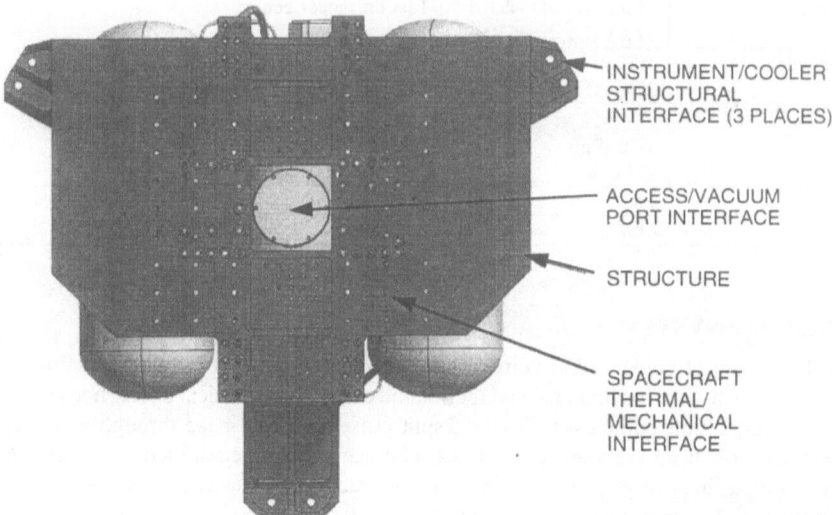

INSTRUMENT/COOLER
STRUCTURAL
INTERFACE (3 PLACES)

ACCESS/VACUUM
PORT INTERFACE

STRUCTURE

SPACECRAFT
THERMAL/
MECHANICAL
INTERFACE

**Figure 3.** Instrument and spacecraft interfaces feature flexibility

The cooler control electronics is based on a low-power, radiation-hardened microprocessor and highly efficient drive electronics. Compactness is achieved by implementing a large portion of the control logic in field programmable gate arrays. The power conversion and control electronics supplies compressor drive with an efficiency in excess of 90%, while providing bus isolation.

To minimize EMI-induced corruption of the spectrometer data, the operation of the cooler and its electronics is synchronized to a clock signal from the AIRS instrument. A flexible serial control and data interface permits the AIRS instrument to acquire cooler data, command the cooler operation, and modify the operating software in flight. The coolers are designed conservatively to satisfy the 7-year lifetime requirement with a reliability exceeding 0.997.

The redundant AIRS cooler system consists of two sets of coolers and electronics. Each compressor and the cold head subsystem weighs 8.65 kg, each electronic subsystem is estimated to weigh 4.33 kg, and the supporting structure and vacuum vessel that hold both coolers weighs 9.04 kg. Total bus power to the electronics is 112.2 W with 19.4 W being consumed by the drive and the control electronics. With the remaining 92.8 W of power input into the compressor, the system has a cooling capacity of 1.63 W at 55 K when the heat reject temperature at the vacuum flange is 320 K. Key capabilities are shown in Table 1.

**Table 1.** Key Capabilities are Low Power, Temperature Control, and Reduced Vibration and EMI

|  | Capability | Verification Method |
|---|---|---|
| Cooler temperature | 55 K | EM verification test |
| Temperature control | ±0.2 K | EM verification test |
| Temperature rate control | 0.5 mK/s | EM verification test |
| Net redundant cooler capacity | 1.25 W @ 55 K | EM verification test |
| Operating life | 50,000 hours | Analyses, life test, and burn-in test on similar cooler |
| Compressor vibration | <0.15 N (<150 Hz) | EM verification test |
|  | <0.44 N (≥150 Hz) |  |
| Cold head motion | ±4 μm (x, y, z) (≤45 Hz) | Test on similar cooler |
|  | ±2 μm (x, y, z) >45 Hz |  |
| Input power (28 Vdc) | 112.2 W | Prototype verification test |
| Mass | 35 kg | EM verification |
| EMI |  |  |
| -REO 1 ac | <129 dBpt | Analysis |
| -REO 4 ac | ≤73 dBpt |  |
| Redundancy | Yes | By design |
| Reliability | 0.997 | Analysis |

## MECHANICAL SUBSYSTEM

The AIRS mechanical cooler subsystem (Figure 4) consists of two mechanical coolers, the cooler integrating structure, a vacuum can, and thermal links. Each mechanical cooler houses a dual-opposed 10 cc swept volume compressor driving a split pulse tube cold head through a connecting pneumatic line. The cold head is mounted in the vacuum can which is connected to the integrating structure with thermal links (Figure 3). The integrating structure mechanically interfaces with the AIRS instrument and has a thermal/mechanical interface with the EOS spacecraft for heat rejection. The pulse tube cold block has a thermal/mechanical interface with the instrument for cooling the detector array. The mechanical cooler electrical interface is with the cooler drive and control electronics subsystem.

The cooler was designed to meet minimum power, volume and weight requirements with maximum life expectancy. The minimum power was accomplished with a high-efficiency cold head and compressor motor design and proper thermal management. Minimal weight and volume are

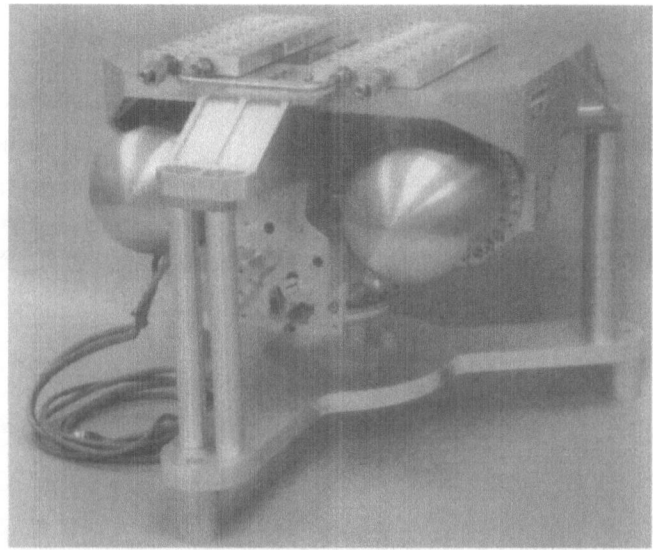

**Figure 4.** AIRS mechanical cooler system features standby redundancy

achieved with a compact design for the compressor that uses high-performance magnetic and structural materials and extensive lightweighting of cooler piece parts. The long life was accomplished by use of a passive cold head, flexure springs with infinite fatigue life, cooler processing to alleviate internal contamination, hermetic/metal sealing of the cooler, a non-wearing compressor design scaled from other proven TRW designs, and conservative derating of electronic components.

A schematic cross section of the dual-opposed compressor is shown in Figure 5. The compressor aluminum pressure vessel and centerplate assure good thermal conductance for removing the irreversible heat of compression while providing excellent thermal expansion, light weight, and ease of fabrication. To assure helium retention for the life of the cooler, hermetic metal-to-metal seals are located between the helium working fluid and ambient. The piston shaft is supported fore and aft on flexure bearings. Spring thickness arm geometry and arm termination were optimized for infinite fatigue life. Spring reliability has been quantified by test.

The piston, which is attached to a linear moving coil armature, is forced into oscillation against the helium in the compression space. This TRW-designed motor yields >89% conversion efficiency at the AIRS operating point (Figure 6).

High-strength NdBFe permanent magnets using a cobalt iron return flux path generate the stator motor field. An internal mu-metal shield further reduces the external field. These high-performance materials maximize the field in the motor gap for the least weight. All wiring for the position sensors and power exits the center plate through hermetic feedthroughs.

Differential capacitive position sensors determine the position of the compressor pistons. The sensors are insensitive to the ambient magnetic fields of the motor, have a high signal output, and are easy to calibrate during assembly. Sensor excitation and pickup electronics are contained in the compressor.

The AIRS cold head is comprised of an inlet copper mesh heat exchanger, a titanium regenerator with stainless steel screens, a copper cold block containing a copper mesh heat exchanger, a titanium pulse tube, an orifice block containing a warm copper mesh heat exchanger, a flexure designed into the orifice block to accommodate differential thermal contraction, orifice and bypass lines, and a mounting interface to the cryovac housing.

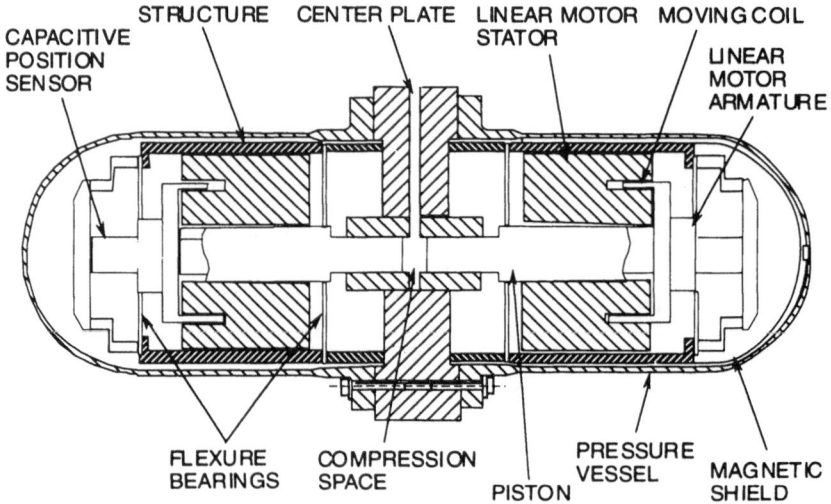

**Figure 5.** The dual-opposed compressor cross section

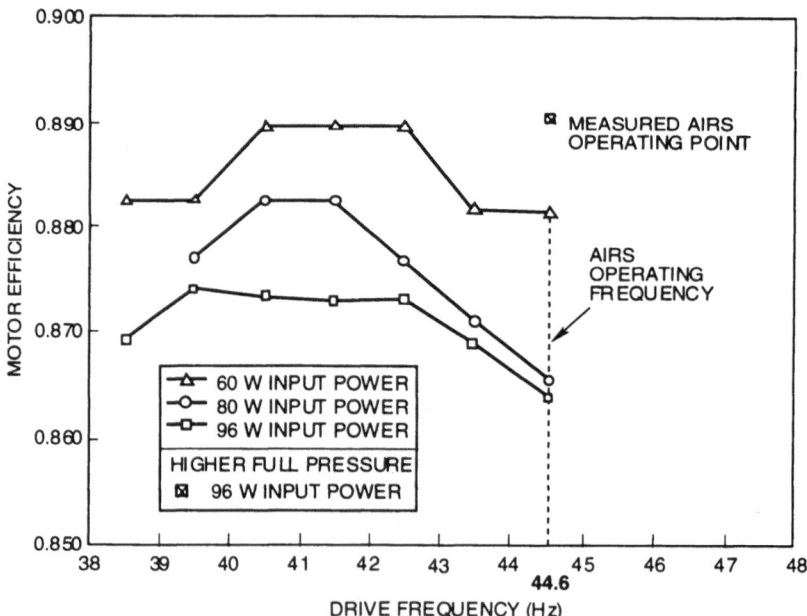

**Figure 6.** Increasing fill pressure increases AIRS motor efficiency to 89% at 44.6 Hz

## ELECTRICAL SUBSYSTEM

AIRS cooler electronics contain a fully integrated digital controller, power amplifiers and controllers, and a sensor processor, all of which meet space radiation environment requirements and minimize power loss. Cooler electronics functions are shown in the block diagram of Figure 7. The electrical interface to the electronics from the host (user) consists of an EIA RS-422A serial command and telemetry link, and a 23 to 35 V power bus. Commands control the launch caging relays; set the operating mode, maximum compressor stroke, and dc offset; and control temperature. Commands also request telemetry data and, if required, diagnostic data such as instantaneous vibration, current, voltages, and compressor piston stroke waveforms.

The AIRS electronics use TRW standard "slice" packaging with slice subassemblies stacked one above the other. Within the slices, low-power dissipation parts are surface-mounted on printed wiring boards (PWBs). The unit consists of two slice subassemblies, one for the central controller and one for the power converter and power amplifiers (power supply). Power dissipating components are mounted to heat-conducting assemblies within the power slice subassembly. The aluminum frames are designed for minimal weight, total ionization dose (TID) radiation shielding, and excellent thermal conductivity to the mounting surface. Feedthrough filters, "RF clean" cavities, polished mating surfaces, and double-shielded cabling meet EMC requirements above 2 GHz. Common-mode L-C baluns provided on every interface line eliminate digital noise, and 10 to 200 MHz radiation to the spacecraft. Use of field programmable gate arrays and surface-mounted components optimize packaging volume. Spacecraft flight-approved D subminiature connectors are employed for the user and cooler.

Measurements taken on the AIRS Engineering Model electronics show the excellent overall efficiency (>90%) of the dc/dc power converter and power amplifier

## COOLER AND GSE SOFTWARE

The major control routines are vibration reduction, temperature control, and dc piston position. The vibration program samples an acceleration waveform and determines, by Fourier analysis, transfer gain and error signals for 16 harmonic frequencies of the fundamental. These are used to compensate the compressor drive waveform. The temperature algorithm controls the cold head to within the AIRS requirement of 30 mK for 1 minute and 0.5 K for 24 hours. The processor features a 32K x 16 PROM for program storage and 64K x 16 RAM for program execution. An additional 2K x 12 RAM provides "ping-pong" buffering of output waveform data. Less than 60% of the available memory is used, allowing for future expansion or data storage.

AIRS commands are entered by 18 or 35-bit serial words and clocked by a nominal 1 MHz clock input. A command word can request any telemetry datum which is then output on the serial data output lines as an 18-bit word – the electronics uses the command clock input for output serial data clocking. Figure 7 shows the hardware implementation of the software.

The PC-based ground support equipment (GSE) (Figure 8), implements the AIRS instrument interfaces to support the cooler through integration and test. It commands the cooler electronics and aquires data. The key functional requirements of the AIRS GSE software are to: communicate with the AIRS cooler by transmitting commands to the cooler and receive serial data from cooler; communicate with test equipment by transmitting commands to the power supply and power meter and by receiving data from the power meter; display cooler data in tabular and graphical form; log cooler and power meter data to hard disk; display and log notification and error messages; and limit check specified cooler parameters.

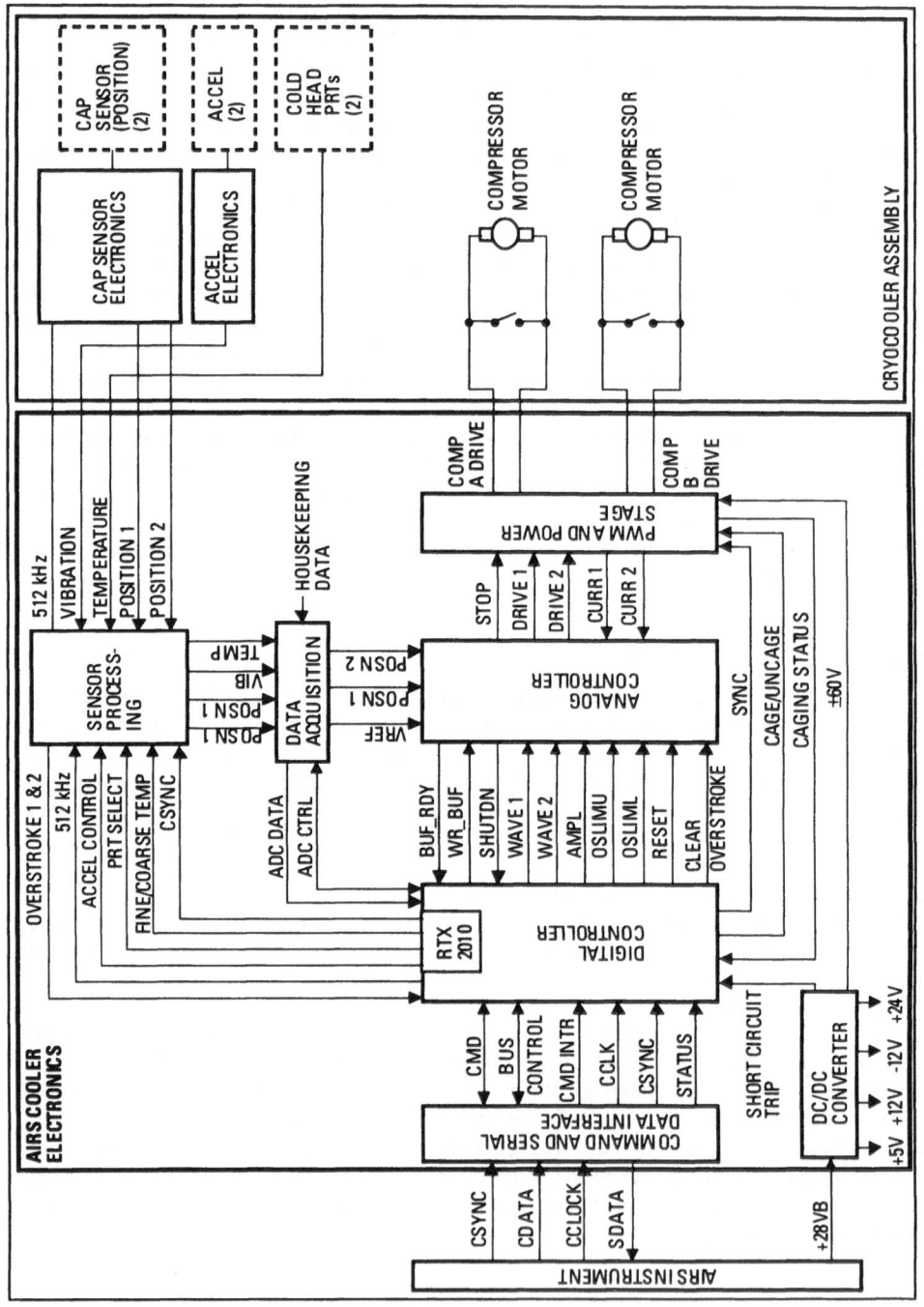

**Figure 7.** AIRS electronics hardware houses tailored software

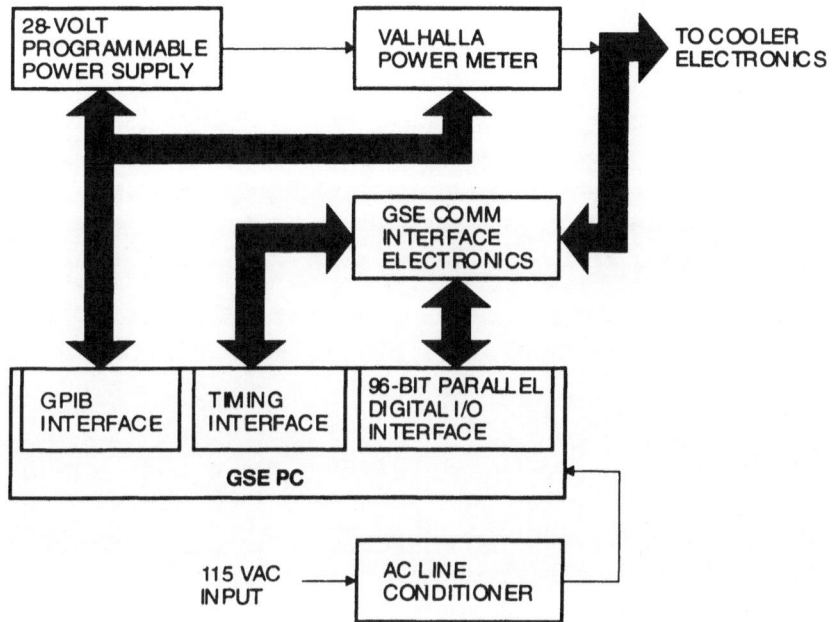

**Figure 8.** AIRS interfaces uses ground support equipment

## CONCLUSION

We have designed, fabricated and tested a reliable, compact, efficient, vibrationally balanced split pulse tube cooler and control electronics for the AIRS instrument. Our fully standby redundant pulse tube cooler produces a net cooling power of 1.25 W at 55 K for an input power of 112 W at the bus. A detailed description of the performance and the test data is presented in reference 3. Its compact size and high cooling capacity make this cooler system attractive for medium to large satellite applications.

## ACKNOWLEDGEMENT

This work was sponsored under contract by Lockheed Martin IR Imaging System, Inc., for the JPL AIRS Instrument Program.

## REFERENCES

1   W.W. Burt and C.K. Chan,, "Demonstration of a High Performance 35 K Pulse Tube, Cryocooler", 8th ICC, TRW, Redondo Beach, CA.

2   E. Tward, C.K. Chan, J. Raab, R. Orsini, C. Jaco and M. Petach, "Miniature Long-Life Space-Qualified Pulse Tube and Stirling Cryocoolers," TRW, Redondo Beach, CA.

3   C.K. Chan, C. Carlson, R. Colbert, T. Nguyen, J. Raab, and M. Waterman, "Performance Measurements of an AIRS Pulse Tube EM Cryocooler," 9th ICC, TRW, Redondo Beach, CA.

# The Qualification and Use of Miniature Tactical Cryocoolers for Space Applications

**K. S. Moser**

Nichols Research Corporation
Albuquerque, NM, USA, 87106

**A. Das**

United States Air Force, Phillips Laboratory
Kirtland AFB, NM, USA 87117

**M. W. Obal**

Obal Technologies Group, Inc.
Woodbridge, VA, USA 22192

## ABSTRACT

A novel cryogenic systems design approach for reducing the cost of strategic space based infrared (IR) surveillance and ballistic missile defense sensors experiments is presented. This approach uses space qualified tactical cryocoolers to replace extremely expensive long life ultra-quiet space cryocoolers. Tactical cryocoolers can be space qualified and used to provide adequate cryogenic environments for IR focal plane arrays (FPAs) for the duration of most space based IR system development experiments. Tactical cryocoolers have been flown on a number of Ballistic Missile Defense Organization (BMDO) space experiment platforms such as the Clementine, MSTI and STRV-1b. Their performance has been outstanding and far exceeded their design expectations. Tactical cryocoolers produce substantially more vibration that space cryocoolers but when combined with advanced adaptive "smart" structures isolation and innovative thermal designs, they have been shown to meet the stringent optical system jitter requirements.

In 1998, BMDO plans to field a 1.0 watt Texas Instrument (TI) tactical cryocooler to provide an 80 degree Kelvin cooling environment for a United Kingdom (UK) Mid-Wave Infrared (MWIR) target tracking sensor experiment on the Space Technology Research Vehicle number 2 (STRV-2) experiment module, The STRV-2 module will be flown on a Space Test Program flight mission designated as TSX-5. Life testing of the selected TI cryocooler has exceeded 4000 hours

Mean Time to Failure (MTTF). Prototype tests were conducted on both the mechanical components and the TI-provided drive electronics and included: thermal performance mapping, induced vibration characterization, EMI/EMC evaluation, random vibration testing and thermal vacuum cycling. The issues associated with using tactical coolers for limited duration space missions are presented. The selection process, test validation philosophy and methodology for the STRV-2 tactical cooler are also described.

## INTRODUCTION

Historically, a handful of simple and complex methods have been employed to bring orbiting extremely sensitive IR FPA's and optics to cryogenic operational temperature. The simplest, and least effective, of these is radiative passive cooling; and the most complex, and arguably the most effective, would be hybrid systems which employ mechanical coolers. Cooling needs for infrared sensors range from very low temperatures (4K) to relatively moderate temperatures (120K); and from very short durations (days) to years of maintenance free, nearly-continuous operation. The range of different needs has spawned the development of a range of solutions Table 1. is a summary of the first order trade space between cooling technologies. We recognize that the data in the table are qualitative, and somewhat coarse; but maintain that they provide a adequate description of the departure points for these various technologies.

The current trend is toward smaller, cheaper spacecraft, particularly in the realm of experimental platforms. An immediate consequence of this trend is that miniature mechanical cryocoolers provide an extremely attractive technology for 60K to 100K infrared sensors with moderate to long mission-life requirements and limited weight and power budgets.

## MSTI AND CLEMENTINE

A total of four RICOR K506-B coolers were flow between the MSTI-2 and Clementine spacecraft. MSTI-2 launched 8 May 1994 on a Scout-G1 and enjoyed a successful 5-month mission. Clementine had launched earlier in the year on 25 January, and spent 71 days in lunar orbit, mapping the lunar surface. Each of these coolers operated flawlessly until they were shut down at mission termination. Table 2. summarizes the cumulative hours of operation each of these coolers received prior to mission termination.

**Table 1.** Cooling Technology Trade Space

| Cooling Technology | Advantage | Disadvantage | Potential Orbit Life (Cumulative) |
|---|---|---|---|
| Passive Cooling | Reliability, Cost, Maturity | Limited Temperature | Years |
| Cryogenic Dewars | Reliability, Maturity, Very Low Temp. | Weight, Expendable | Weeks |
| Solid Cryogen | Very Low Temp. | Weight, Expendable | Months |
| Tactical Coolers | Cost, Maturity | Space Heritage, Complexity | Year |
| Long-Life Coolers | 10 Year Design-Life | Cost, Maturity, Complexity | Multiple Years |

**Table 2.** Operation Hours: MSTI-2 and Clementine Cryocoolers.

| Cooler ID | | Operational Hours (Cumulative) | | |
|---|---|---|---|---|
| | | Ground | Flight | Total |
| MSTI-2 | #1 | ? | >1200[i] | >1200 |
| MSTI-2 | #2 | ? | >1200[i] | >1200 |
| Clementine | #1 | 500-1000[ii] | 840[ii] | 1340-1840 |
| Clementine | #2 | 500-1000[ii] | 670[ii] | 1170-1670 |

Notes: (i) E. Grigsby (AFPL/SX), Telephone Interview by K. Moser, 8 Feb 1995.
(ii) R. Priest (LLNL), Telephone Interview by K. Moser, 9 Feb 1995.

In addition to these four flight coolers, Lawrence Livermore National Laboratory (LLNL) conducted life testing on three additional Ricor K506B units. Under simulated flight conditions, two of the three life-test articles had operated for more than 4000 hours when the testing was terminated for lack of funding. The third unit had over 3000 hours of life-test operations when it failed.

## SPACE TECHNOLOGY RESEARCH VEHICLE-1B (STRV-1B)

### Description

The STRV-1b spacecraft was developed by the UK's Defence Research Agency for the Ministry of Defence (MOD), and was one of a pair of experiment platforms which were launched as a piggyback payloads into a severe-radiation geosynchronous transfer orbit (200 km X 36000 km). These two 55 kg. satellites were launched on 17 June 1994. The U.S. contributed to the success of the STRV-1 program by populating the STRV-1b spacecraft with several small BMDO M&S (Materials and Structures)-sponsored experiments, the most notable of which was a cryocooler vibration suppression experiment. At the core of this experiment is a Texas Instrument 1/5 Watt (@ 80K) tactical cooler. As of June 1996, this experiment was still operational; had accumulated somewhat less than 100 hours of cumulative cryocooler on-orbit hours and over 1000 on-off cycles. A detailed description of the experiment which demonstrated a 2-order-of-magnitude vibration reduction using adaptive "smart" structures technology is given by Glaser, Garba and Obal.[1]

### Selection of the Texas Instrument 1/5 Watt Tactical Cooler

The extremely limited spacecraft power, weight and volume resources allocated to the STRV-1b cryocooler vibration suppression experiment drove the selection of the TI 1/5 Watt cooler which weighs 380 grams and draws 4 to 6 watts at a cold-tip temperature of 95K.[2] For comparison, the total weight allocation for the experiment (and three radiation exposure experiments) was only 6.5 kg. 4.75 kg was allocated to the electronics package. This gave the experiment a mere 1.75 kg allocation for its mechanical and structural components. Available power was comparably scarce with a total spacecraft allocation of 20 watts. Resource scarcity was the driving criterion behind the selection of the TI 1/5 watt tactical cryocooler for the cryocooler vibration suppression experiment.

## SPACE TECHNOLOGY RESEARCH VEHICLE -2 (STRV-2)

### Historical Overview

In 1993, the BMDO M&S program developed a flight experiment payload module concept known as STRV-2. STRV-2 was an offshoot from their earlier unsuccessful attempt to provide a low cost series of space experimental platforms for technology insertion demonstrations called TECHSAT program. The goal of the STRV-2 program is to design and develop a compact flight payload with numerous autonomous complex experiments that employ a simple structural, thermal, electrical and communication interface to a small satellite bus. The motivation to make the STRV-2 an international, multi-agency funded effort was derived form the successful collaboration between the U.S. and the UK on STRV-1b.

Early in 1993, the UK had identified the need for an experimental MWIR instrument whose primary objective was the detection of airborne targets from space. During this period, BMDO and the UK examined numerous cryocooler options for the MWIR sensor. BMDO offered to provide a cryocooler to the UK and they accepted the challenge to make their system work with a tactical cryocooler.

STRV-2 definition and requirements were in constant flux throughout 1993. MWIR cooling requirements evolved from 125 mwatts (at 65K-90K) in January to 400 mwatts (at 80K) in August. Simultaneously, power and weight allocations for the entire STRV-2 payload module started at 40 watts and 45 kg in January 1993 and eventually grew to 150 watts and 250 lbs. by early 1995 as additional experiments were added. Figure 1 depicts the final physical configuration of the STRV-2 Experiment Module. The STRV-2 experiment module is manifested for flight on the Air Force Space Test Program's TSX-5 (Tri-Service eXperiment #5) spacecraft scheduled for initial launch capability in July of 1998. The MWIR and its integrated cooler lie within the module and are isolated from the module (and spacecraft) by a Vibration Isolation and Suppression System (VISS) experiment (not shown) which attaches the MWIR to the STRV-2 module's lower deck. The MWIR looks out a nadir facing aperture; and on the antinadir side, a radiator which is integrally attached to the cooler stares into deep space.

### STRV-2 Cryocooler Surveys

**Initial Cryocooler Survey.** The first cooler survey for STRV-2 was performed in January 1993, and used the SPAS III Tactical Cooler Survey[3] as a departure point. Table 3 lists the candidate Space Coolers which were under consideration. The final column of Table 3

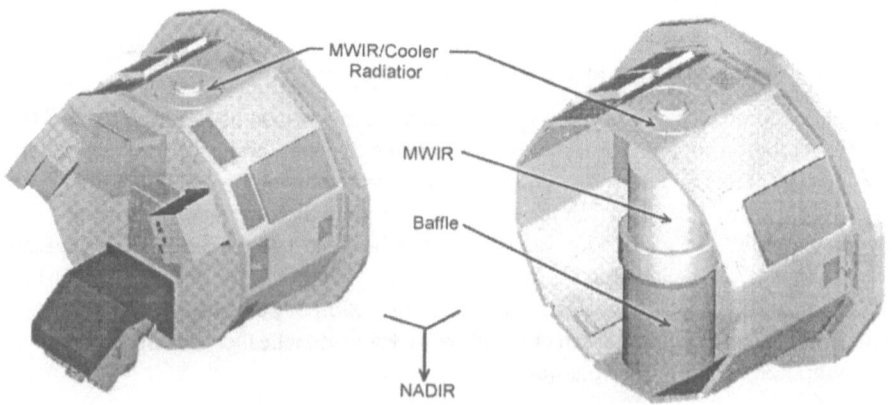

**Figure 1.** STRV-2 Experiment Module

summarizes the STRV-2 requirements which were being targeted by the survey. Table 4 lists the candidate tactical cryocoolers which remained after this initial survey/screening. The STRV-2 requirements shown in Table 3 apply to Table 4 also.

**Second Cryocooler Survey.** By the end of August 1993, the STRV-2 requirements had changed substantially. The content of Tables 5 and 6 is practically identical to that of Tables 3 and 4, respectively; with the exception that the TRW Miniature Pulse Tube was added, some cryocooler data was updated, and the STRV-2 requirements had changed. Boldface-underlined text in Tables 5 and 6 flag cryocooler characteristics which were deemed to be noncompliant with the new STRV-2 requirements. The choice was now between the TRW space coolers and the TI 1.0 Watt tactical cooler.

**Table 3.** Candidate Space Coolers for STRV-2 (Feb. 1993)

| Figure of Merit | TRW Split Stirling | Hughes | Lockheed/ Lucas | STRV-2 Requirement |
|---|---|---|---|---|
| Cooling Load | 0.25 Watt | 2.0 Watts | 2.0 Watts | <125 mWatts |
| Temp. | 65 K | 60 K | 60 K | 65 K- 90 K |
| Life (Hours) | 10 yrs | 10 yrs | 10 yrs | 1000-4000 |
| Weight (lbs) | 3.2 | ? | ? | |
| Input Power | 28 W | 70 W | 70 W | <30 W |
| Lead Time | 1 year | ? | ? | < 1 year |
| ~ Cost ($K) | 400 (2 units) | 1000 | 1000 | Less is better |
| Comments: | Not life tested or qualified No Flight Elect. | In Development | In Development | Desire to minimize mass power & Cost |

**Table 4.** Candidate Tactical Coolers for STRV-2 (Feb. 1993)

| Figure of Merit | Texas Instruments | | Hughes | Magnavox | Ricor | |
|---|---|---|---|---|---|---|
| Cooling Load | AAWS-M 0.15 Watt | Split-Stirling 1.0 Watt | 7004H 1.0 Watt | MX7058 1.0 Watt | K5265 0.5 Watts | ICE MC111B 0.1 Watts |
| Temp. | 80K | 80K | 65K | 80K | 80K | 80K |
| Life (Hours) | 4000 | 4000 | 2000 | 4000 | 1500 | 2000 |
| Weight (lbs) | 0.9 | 4.1 | 4.0 | 4.0 | 2.0 | 0.9 |
| Input Power | 10 W | 32 W | 50 W | 60 W | 30 W | 8 W |
| Lead Time | <4 months | 2 months | 6-8 months | 6 months | 3-4 months | ? |
| ~ Cost ($K) | 15 | 15 | 15 | 11 | 9 | ? |
| Comments: | Developed by ANVL | Developed by ANVL | BETSE Coolers for SPAS III | Stirling TCE | Split Stirling Not Space Qualified | |

**Final Cooler Selection**

The decision to use a tactical cryocooler was the result of qualitative risk considerations which were supported by limited data. Each of the risk area will be discussed, and then a summary risk assessment will be attempted.

**Table 5.** Candidate Space Coolers for STRV-2 (Sept. 1993)

| Figure of Merit | TRW Pulse Tube | TRW Split Stirling | Hughes | Lockheed/ Lucas | STRV-2 Requirement |
|---|---|---|---|---|---|
| Cooling Load | 0.60 Watt | 0.625 Watt | 2.0 Watts | 2.0 Watts | 0.40 watt |
| Temp. | 80K | 80 K | 60 K | 60 K | 80K |
| Life | 10 yrs | 10 yrs | 10 yrs | 10 yrs | 2000 hours |
| Weight | 4.4 | 3.2 | ? | ? | Less is better |
| Input Power | 30 W | 28 W | 70 W | 70 W | <35 W |
| Lead Time | 1 year | 1 year | ? | ? | < 1 year |
| ~ Cost ($K) | ? | 400 (2 units) | 1000 | 1000 | Less is better |
| Notes: | Not life tested or qualified. No Elect. | Not life tested or qualified No Flight Elect. | In Development | In Development | Desire to minimize mass ,power & Cost |

**Table 6.** Candidate Tactical Coolers for STRV-2 (Sept. 1993)

| Figure of Merit | Texas Instruments | | Hughes | Magnavox | Ricor | |
|---|---|---|---|---|---|---|
| Cooling Load | AAWS-M 0.15 Watt | Split-Stirling 1.0 Watt | 7004H 1.0 Watt | MX7058 1.0 Watt | K5265 0.5 Watts | ICE MC111B 0.1 Watts |
| Temp. | 80K | 80K | 65K | 80K | 80K | 80K |
| Life (Hours) | 4000 | 4000 | 2000 | 4000 | 1500 | 2000 |
| Weight (lbs) | 0.9 | 4.1 | 4.0 | 4.0 | 2.0 | 0.9 |
| Input Power | 10 W | 32 W | 50 W | 60 W | 30 W | 8 W |
| Lead Time | <4 months | 2 months | 6-8 months | 6 months | 3-4 months | ? |
| ~ Cost ($K) | 15 | 15 | 15 | 11 | 9 | ? |
| Comments: | Developed by ANVL | Developed by ANVL | BETSE Coolers for SPAS III | Stirling TCE | Split Stirling Not Space Qualified | |

**Cost.** The cost data in Tables 3 through 6 are not a true reflection of the cost of handing a flight-ready cooler for integration with a sensor. The base cost of the tactical cooler must be amplified to account for requisite flight acceptance testing, and for any modifications required to bring it into compliance with the STRV-2 interface requirements (or visa versa). In the end, four TI coolers were procured, and subjected to testing, including a dedicated life-test article, for about $250K. Conversely, in order to secure an experimental flight, two TRW Pulse-Tube were offered to the program for free; but required the development of flight electronics at an estimated cost of $1M.

**Life.** The space coolers have a significant advantage with a design life of 90,000 hours (compared to the tactical's 4000 hours MTTF). However, life test data was nonexistent for the TRW coolers; and the coolers were still in development. TI was, at the time, producing 30 of the

1.0 watt tactical coolers each month and even a derating factor of 2.0 on the 4000 hour MTTF would meet STRV-2 requirements. This family of coolers had received a limited amount of life testing (sample size = 6 coolers) at TI which produced the life probabilities shown in Table 7.[4] Other than the MTTF value, the data in Table 7. were not published until April 1994 and, consequently, were not a factor in the decision under discussion. They are reported here for general interest.

**Input Power.** Comparative values are given in Tables 5 and 6. This was not a significant factor in the decision.

**Induced Vibration.** This is a criterion where the space coolers again had a clear advantage over the tactical coolers. Peak forces were expected to be on the order of 0.2 Newtons as compared to the tactical cooler's 2.0 Newton peak. This was the MWIR principal investigator's greatest concern regarding the tactical cooler. However, this concern was minimized by adding an innovative design approach which added thermal capacitance to the focal plane. This decreases the sensor cooldown rate; but allows collection of mission data after powering down the cooler; thus eliminating cooler vibration altogether. It is worth noting, that while baseline MWIR data collection occurs with the cooler off, data will also be taken with the cooler operating in order to quantify both the effect of cooler induced vibration on sensor performance, and the performance of the Vibration Isolation and Suppression System (VISS) experiment.

The VISS is a complex experiment which demonstrates the space application of advanced adaptive "smart" structures concepts. The VISS will isolate the MWIR from the spacecraft vibration environment, suppress the tactical cryocooler vibration to a level adequate for the MWIR jitter requirements and provide sensor gimballing. The VISS and similar concepts are crucial for providing a mechanism which enables sensitive optical instruments to operate in the presence of jitter sources (such as tactical coolers and spacecraft bus disturbances).

**Cooling Performance.** In less than a year, the nominal MWIR cooling requirement had grown from 125 mwatts to 500 mwatts. This growth coupled with the fact that the MWIR design was still conceptual suggested that a 25% contingency on cooling performance was advisable. All three coolers were capable of meeting this 625 mWatt requirement; but the space coolers only met it marginally while the TI cooler had a 60 percent performance margin to spare. The implication was that the tactical cooler could afford some life-related degradation and still meet performance requirements. Since the typical failure mode for this class of coolers is graceful degradation until performance falls below spec. values, this 60% performance margin (if power is available) can be traded for extended effective life. The tactical cooler had a clear advantage here.

**Schedule.** The TI cooler was in production (at 30 units per month), and there was little if any schedule risk associated with TI's ability to deliver within 2 months of the order as promised. The space coolers, as a development technology, had some level of unquantified schedule risk associated with them. But even neglecting this risk, the near 1-year lead time was excessive to the MWIR experimenter who needed to integrate the cooler with the FPA into a

**Table 7.** Life-Probability of an Individual TI Cryocooler

| Number of Hours | Probability that an Individual Cooler will last longer than the Number of Hours Specified |
|---|---|
| 2000 | 97.13% |
| 3000 | 91.15% |
| 4000 (MTTF) | 79.10% |

dewar as one of the first steps in the MWIR integration. The tactical cooler had an advantage here also.

**Technology Readiness.** This is, perhaps, the most difficult selection criterion to quantify. On first order, it is quite subjective. Tactical coolers, as a production item, are known quantities: the manufacturing processes are well-established and unit-to-unit variability is understood and fairly consistent. Space coolers are development items: the engineering units experience anomalies of the type consistent with development testing, short-term successful performance was yet to be demonstrated, let alone long life testing.

It was determined that the overall technology readiness risks associated with the space coolers was significant and difficult to mitigate, while the concerns regarding the tactical cooler could largely be reduced to a discussion of its flight qualifiability. If and when confidence is established that a tactical cooler is flight qualifiable, the perceived risk of baselining the tactical cooler for a flight program can be deemed acceptable. It is then a matter of verifying cooler flight worthiness by test.

### Tactical Cooler Flight Qualifiability

The question of tactical cooler qualifiability was pursued by: 1) evaluating the tactical design for characteristics which are acceptable in terrestrial application but pose limitations for space operation, and 2) comparing terrestrial environmental qualification levels with space environmental requirements. The following discussion is built around an assessment of the flight qualifiability of the TI 1.0 watt cooler for the STRV-2 program but the approach is applicable to the question of flight qualifiability for tactical coolers in general.

**Texas Instruments 1.0 Watt Tactical Cooler Description.** The TI 1.0 watt cryocooler was developed for high-end use in airborne night vision systems. It is a split stirling cryocooler with the compressor end and expander connected by a customizable transfer tube. The compressor contains dual-opposed pistons for force cancellation, vibration suppression, and noise reduction. Via temperature feedback, the coldhead temperature is maintained within 1.5K of the desired set point. A cooler-inhibit line is available for shutting down the compressor and expander during critical measurements where extremely low vibration levels are desired. This would allow all of the cooler functions, and electronics to remain enabled.

**Custom Modifications to the TI 1.0 Watt Cooler for STRV-2.** There were four minor changes to the production configuration to accommodate STRV-2 requirements: 1) A custom transfer tube geometry was specified in order to coalign the thrust axes of the expander and compressor to control the direction of the primary induced forces. Sensor system analysis suggested that directing the induced forces along the instrument line-of-sight would minimize vibration impact on sensor performance; 2) The coldfinger side of the expander flange received number 16 circular lay surface finish in order to insure a reliable seal with the MWIR FPA dewar; 3) The two opposing compressor coils were wired separately in order to permit each coil to be driven independently. This was done in order to allow the STRV-2 program to insert custom electronics which would have been capable of reducing output vibration via electronic balancing. However, this capability was not fully implemented, and the coils are driven by a single signal; and 4) The drive electronics were specified to have no conformal coating. The reason for this modification are documented later. Figure 2 shows the STRV-2 configured 1.0 watt cooler.

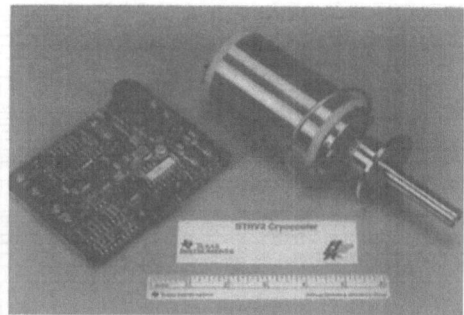

**Figure 2.** TI 1.0 Watt Tactical Cooler & Electronics: STRV-2 Configuration

**Structural Integrity.** The compressor, expander and all tubing have been qualified to four times the Maximum Expected Operating Pressure (MEOP), and individual units can be acceptance tested to two times MEOP as required (the cooler on STRV-1b was proof-tested to 2 X the charge pressure). Over pressure exhibits a leak-before-burst failure mode. Fill pressure can be varied from 200 to 300 psia. The entire assembly is hermetically sealed and leak-rated at less than 1.0 E-07 cc/sec of helium. Each delivered unit is leak tested, and measured values typically range between 1.0E-08 to 1.0 E-09 cc/sec.

**Electronics.** The electronics are designed to operate at 28 volts DC; but will operate properly at input voltages between 17 and 32 volts DC. The electronics consist of surface mounted components which are conformally coated.

**Outgassing and Contamination Control.** The TI electronics' conformal coating consisted of a silicon-based compound that is not on the NASA-approved materials list. This issue was easily addressed by specifying that the STRV-2 electronics cards were not to be conformally coated by Texas Instrument. Upon delivery, they were immediately sent to JPL, where a space-approved conformal coating was applied. The mechanical components are hermetically sealed in stainless steel and pose no outgassing threat.

**Radiation Susceptibility.** The TI cooler and associated electronics have been tested to levels which exceed the total dose and dose rates that they will experience on STRV-2. The exact test levels are classified and cannot be reported here. Suffice it to say that the testing indicated that the coolers are not expected to experience out-of-spec performance induced by the radiation environment during the STRV-2 mission goal of 1-year.

**Random Vibration.** The cooler and electronics are qualification and acceptance tested by TI to the levels shown in Figure 3. The cooler is operational during factory testing. The STRV-2 component acceptance test requirement is shown for comparison and is about a factor of two higher the factory test levels. Because the levels are lower in the factory test, this is deemed to be a risk area; but the test durations and operational condition of the cooler ( in the spaceflight random vibration testing, the cooler and board are unpowered) reduce the perceived risk to acceptable levels.

**Temperature Rating.** The cooler and electronics are rated for operation between -54 C and 71 C and for survival between -62 C and 71 C. Each delivered unit undergoes four thermal cycles

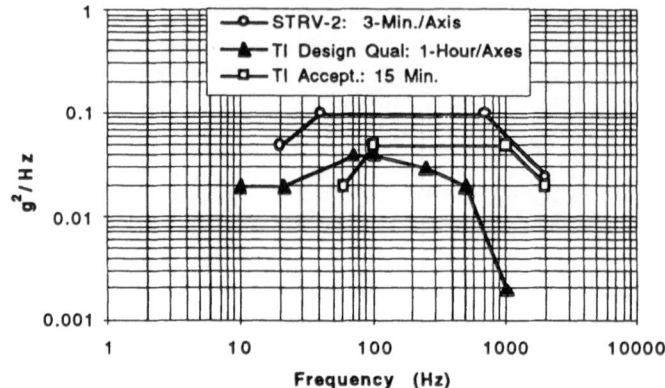

**Figure 3.** TI Qualification and Acceptance Random Vibration Tests

between these operating limits with performance verified during dwells at each extreme. The
electronics are designed for (and rely on) conductive cooling. A thermal plane removes heat from
components, conducts it to and sinks it into the card guide via DESC-DWG-89024-03CT
compliant wedge locks. It is common for terrestrial systems to rely on convection for heat
dissipation, so the TI conductively cooled electronics was a pleasant surprise.

**Life.** Thousands of hours of life testing have been conducted on this family of TI tactical
coolers. Failure typically occurs from seal wear in either the compressor or expander. This is a
graceful failure mode exhibited by a gradual loss in performance. These tests, which are
conducted in thermal chambers cycling between -54 C and 71 C with 10 C/minute ramps between
soaks, have substantiated that the STRV-2 cooler should easily attain the 2000 hour STRV-2
design baseline if operated within its design limits.

**Flight Qualifiability Assessment.**

By January 1994, it was determined that the TI 1.0 Watt met the definition of being flight
qualifiable. The one design characteristic which was deemed noncompliant with space flight
requirements, the silicon based conformal coating on the electronics package, was easily
corrected by specifying that the STRV-2 electronic receive no factory coating; and by applying a
space-approved coating after receipt of the hardware.

**Flight acceptance testing**

The transition from a flight qualifiable cooler to a flight qualified cooler was accomplished by
the series of tests shown in Figure 4. Moser, Roberts and Rawlings[5] give a detailed discussion of
the development, implementation and results of these tests which successfully demonstrated the
flight worthiness of the selected tactical cooler.

**Summary Assessment and the Use of Tactical Coolers in Space.**

Once the tactical cooler was judged to be flight qualifiable, the risk assessment weighed heavily
in its favor. Maturity and cost were the two most significant liabilities associated with the
miniature space coolers. While it is unlikely that these miniature space coolers will ever be
competitive with tactical cooler on cost, they certainly will overcome the questions on maturity.
In spite of this, tactical coolers will continue to be considered for space missions which meet the

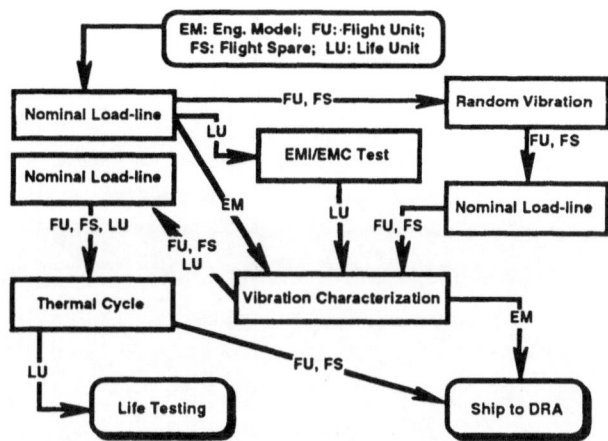

**Figure 4.** STRV-2 cooler flight acceptance test flow.

following three criteria: 1) limited duration, 2) non-continuous operational profiles, and 3) tolerance for the higher vibration output of tactical cooler either through operational work-arounds, vibration suppression technologies, or systems which are relatively insensitive to cryocooler-induced jitter. In the end it was determined that the TI 1.0 watt cooler was the minimum risk option available to STRV-2, and the TI cooler was baselined in January 1994. By the end of January 1996, testing had been completed on the flight article which confirmed the flight qualifiability assessment. The limited life of tactical coolers could potentially be addressed by flying multiple coolers for a given application. As one coolers fails, another could be brought on-line. A reliable, efficient thermal switch would be the enabling technology for this approach.

## ACKNOWLEDGMENTS

Special thanks are extended to Dr. John Stubstad, the BMDO Materials and Structures Program Manager, for his support, genuine interest and leadership in this effort; to the UK/MOD for accepting the challenge to incorporate a tactical cooler into its MWIR design; and to Dick Rawlings and Texas Instruments for their gracious commitment to this effort.

## REFERENCES

1. Glaser, R.J., Garba, J.A., and Obal, LtCol M., "STRV-1B Cryocooler Vibration Suppression", 36th SDM, New Orleans, LA (1995).

2. Glaser, R.J., Ross, R.G. Jr., and Johnson, D.L., "Cryocooler Tip Motion Suppression Using Active Control of Piezoelectric Actuators", Proceedings of the 7th International Cryocooler Conference, Santa Fe, New Mexico (1992).

3. Ludwigsen, J., and Fraser, M., SPAS III Tactical Cooler Survey, Prepared for SDIO/TNS (Mr. M. Harrison), Nichols Research Corporation, Albuquerque (22 Nov. 1991).

4. Rawlings, R.M., *Statistical Analysis of Texas Instruments Linear Cooler Life Test Data, a Report to Phillips Laboratory, PL/VTS*, Texas Instrument, Dallas (26 April 1994).

5. Moser, K.S., Roberts, T.P., and Rawlings, R.M., "Space Qualification Test Plan Development, Implementation and Results for the STRV-2 1.0 Watt Tactical Cryocooler", *Cryocoolers 9*, Plenum Press, New York (1996).

# The Design, Development and Qualification of the SCIAMACHY Radiant Cooler

Ino J.G. Wigbers

Fokker Space B.V.
2303 DB Leiden, The Netherlands

## ABSTRACT

Fokker Space has developed a radiant cryocooler for space applications that provides a cooling capacity of 0.75 W at 123 K and 0.96 W at 175 K. The design is based on a two-stage radiator concept with a parabolic reflector and a cover door and release system. Additionally, a thermal bus unit has been developed to transfer the heat from the source to the radiators. All suspensions are of a glass reinforced plastic (GRFP) type to minimize parasitic heat leaks.

A description of the Fokker Space cryocooler and a summary of its performance and environmental test program are presented.

## INTRODUCTION

The SCanning Imaging Absorption spectroMeter for Atmospheric CartograpHY (SCIAMACHY) instrument is foreseen to fly as an Announcement of Opportunity on the ESA Envisat-1 mission, scheduled for launch in 1999. The primary scientific objective of SCIAMACHY is the global measurement of trace gases in the troposphere and stratosphere. SCIAMACHY uses a total of eight cooled detector-arrays (two in the near ultraviolet, three in the visible, and three in the near-infrared spectrum) to observe the solar-illuminated atmosphere in both the nadir and Earth-limb directions.

The detector-arrays have to be cooled to temperatures in the range of 130-240 K. To obtain this objective Fokker Space started in September 1994 on the design, development and qualification of a passive cooler called the SCIAMACHY Radiant Cooler (SRC). The SRC consists of two parts: the Radiator Reflector Unit (RRU), and a Thermal Bus Unit (TBU)— total mass 31 kg, envelope 900x450x300 mm. The RRU radiates the heat to space and consists of a two-stage radiator and a high reflectivity parabolic reflector. The TBU consists of an aluminum thermal conductor and two cryogenic heatpipes to transfer the heat loads of about 1.7 W total from the detector-arrays.

All technical risks were identified at the beginning of the project. The risks were reduced in parallel with the design and analysis phase by extensive development testing of the thermal low conductance GRFP struts, the high conductance flexible links, the high reflectivity coatings, and the low conductance heatpipe suspensions. Thermal analyses were run based on ESATAN mathematical modeling, and structural analyses were conducted based on NASTRAN/PATRAN finite element modeling. Sensitivity analyses were completed for both.

Cryocoolers 9, Edited by R.G. Ross, Jr.
Plenum Press, New York, 1997

917

The SRC Project is subject to very stringent performance, schedule, and cost constraints. Delivery of the protoflight model of the SRC is scheduled for March 1997.

## SYSTEM DESCRIPTION

The SCIAMACHY radiant cooler, shown in Fig. 1, consists of the radiator reflector unit, the thermal bus unit, and a hold-down and release system. In the launch configuration a cover door of the RRU is closed and locked so as to protect the high reflectivity surfaces against contamination during the launch phase and during the final on-orbit acquisition phase. In the launch configuration the door is also part of the structure and has a load-carrying capability.

Once in orbit and the satellite has stabilized to the correct attitude, the door of the RRU is released by activating a central release unit that is part of the hold down and release system. The door is deployed by torque springs that are located in the door hinges. The function of the door is to provide a shield to protect the RRU from earthshine and albedo effects that might strike the second (cold stage) radiator. Further, it provides protection against heat input coming from other parts of the spacecraft located under the opened door.

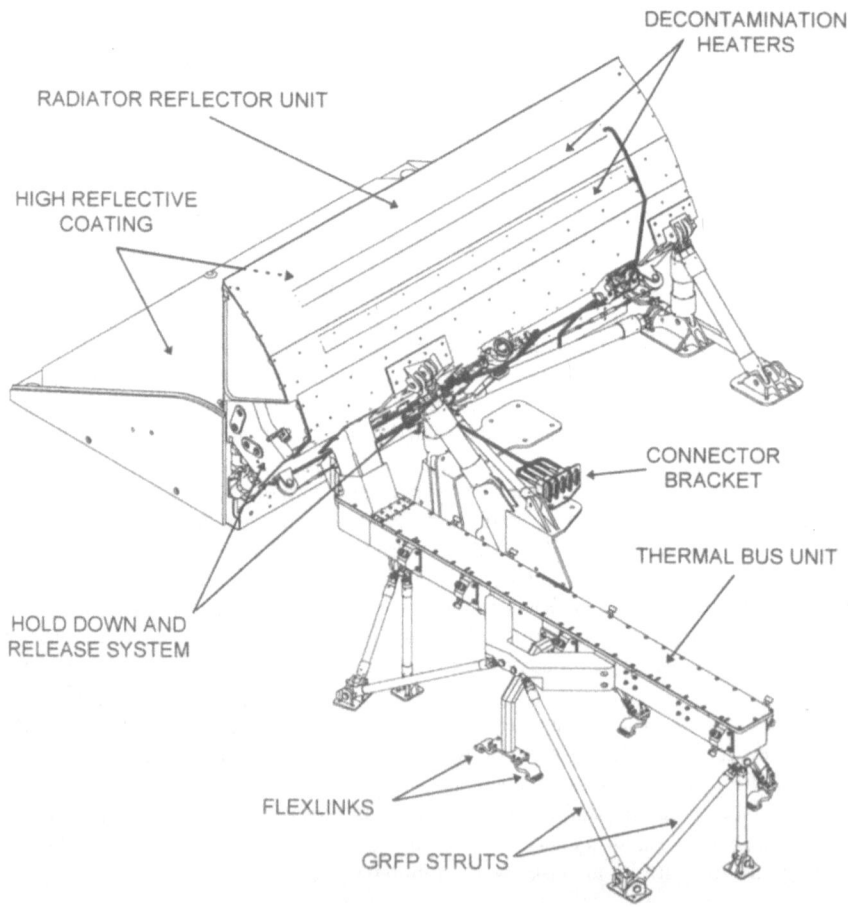

**Figure 1.** Fokker Space Radiant Cooler (SRC)

## Thermal Bus Unit

The Thermal Bus Unit (TBU) is mounted via a support frame on the optical bench module of the SCIAMACHY Instrument. It consists of a shroud with interface bracketry, a cryogenic heatpipe assembly, the GRFP heatpipe suspension, branches to the detectors, GRFP struts, and thermal hardware like multilayer insulation and trim heaters. The function of the TBU is to protect the heatpipe assembly from thermal environmental heat loads. The heatpipe assembly transfers the heat loads from the infrared detectors to the cold stage. Further, the TBU functions as a thermal conductor that transfers the heat loads from the other detectors to the first stage radiator, which is called the intermediate stage of the RRU.

All TBU surfaces are gold plated to avoid radiative coupling to the heatpipe assembly. The heatpipe assembly is a square tube with two separate embedded cryogenic heatpipes running to the infrared detectors. Each heatpipe is totally enclosed and contains high pressure methane gas. The support of the heatpipe assembly within the TBU is by means of GRFP tension wires. The number of wires is determined by a balance between the minimum required structural natural frequency of the TBU, and the heat leak from the shroud to the heatpipe assembly via the GRFP wires.

The branches and the flexlinks function as thermal conductors to transfer the heat loads from the detectors to the shroud. Flexlinks are being used to avoid structural loads from the TBU to the cold-finger of the detector chip; excessive loads would cause erroneous readout of the detector data due to misalignments. The flexlinks, see Fig. 2, consist of a very large number of ultra thin aluminum sheets to provide minimum structural stiffness and maximum thermal conduction.

The suspension of the TBU is statically determinant. The TBU is supported by six filament-wound GRFP struts with tuned cross-sections; the orientation of the fibers guarantees a stiff strut. The cross-section of each strut was selected to obtain an optimum compromise between the structural natural frequency of the bus and the size of the parasitic heat loads. As shown in Fig. 3, metal end fittings, each equipped with a special ball bearing, are bonded to both ends of the struts. The length of the struts varies as a result of the individual strut orientation. The struts have been subjected to an extensive development test program.

## Radiator Reflector Unit

The RRU is located on the top edge of the spacecraft. Due to the orientation of the spacecraft, this part of the spacecraft hardly sees any direct sun, and the SRC can therefore cool down to low temperatures. The RRU consist of a reflector parabolic structure including the intermediate stage, the cold stage, several GRFP struts, the deployment hinge assembly, the hold-down and release mechanism, the door structure including protection side covers, and a stainless steel cable harness with connector bracket.

MULTILAYER ALUMINUM
SHEETS

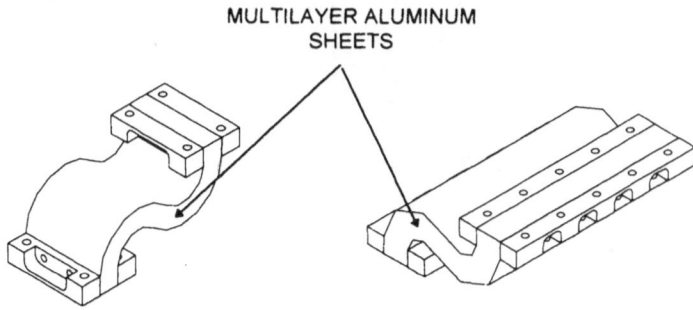

**Figure 2.** High-conductance flexlinks

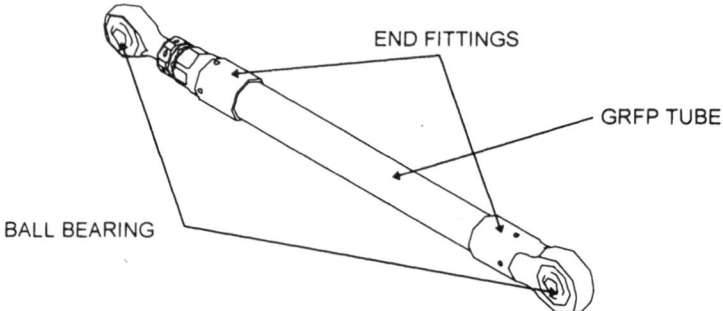

**Figure 3.** Low-conductance GFRP Strut

The curvature of the reflector is selected such that all parasitic reflections from the open door are reflected into deep space. This is also valid for any effects from albedo or earth- and sunshine. The reflector and door facesheet are polished aluminum, which provides a high specularity and a very low emittance. The cold stage and intermediate stage interface with the TBU shroud and heatpipe assembly by means of flexlinks. The GRFP struts thermally decouple the RRU from the spacecraft, and the cold stage is also decoupled from the reflector structure via GRFP struts.

**Hold-Down and Release System**

The hold-down and release system, which is totally integrated in the RRU, supports the door during launch, provides sufficient preload to clamp the door against the RRU, and releases the stowed door to allow deployment by the hinges after launch. The door is held down on both sides in the middle by a locking mechanism that is attached to an aramide cable. Release of the door is performed by cutting the cable by a central release unit that contains two redundant thermal knives. The thermal knife release system has been successfully used in several Fokker Space solar array programs. A cutaway view of the central release unit is shown in Fig. 4.

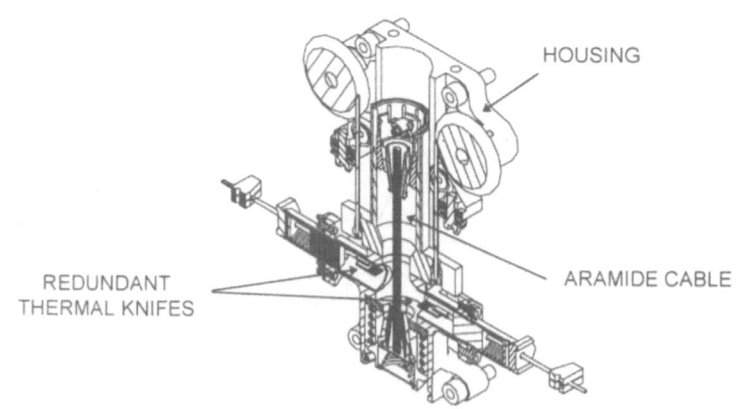

**Figure 4.** Central Release Unit (CRU)

**Table 1.** SCIAMACHY Radiant Cooler Performance Characteristics

| Item | Performance |
|---|---|
| Thermal (cold case, begin-of-life) | 754 mW @ 123 K (Channel 7 & 8) |
| | 959 mW @ 175 K (Channel 1 thru 6) |
| Temperature Stability | 20 mK |
| Decontamination Power | 80W |
| Mass | |
| &bull;    RRU | 23.6 Kg |
| &bull;    TBU | 7.3 Kg |
| Eigenfrequencies | |
| &bull;    RRU | 60 Hz |
| &bull;    TBU | 86 Hz |
| Vibration Levels | |
| &bull;    Sine | 10 g < 60 Hz, 6 g >60 Hz |
| &bull;    Random | 6.4 $g_{rms}$ |

## SYSTEM PERFORMANCE

The required performance of the radiant cooler is divided between requirements for the hot operating case and the cold operating case, and beginning-of-life (BOL) and end-of-life (EOL) conditions; this results from the fact that the heat input to the cooler varies over the season, and the thermo-optical properties of the cooler surfaces degrade over time. By switching on the heaters on the door and reflector, the high reflectivity surfaces can be decontaminated.

In the hot operating EOL case, the temperatures required for channels 1-6 are about 220 K, and for channels 7 and 8, 145 K. The thermal analyses show that channels 1-6 meet their requirements, including a ±8 K uncertainty, with a margin varying between 0 and 4 K. Channels 7 and 8 meet their requirements with no margin.

In the cold operating BOL case the temperatures required for channels 1-6 are about 190 K, and for channels 7 and 8, 130 K. For this case the thermal analyses show that channels 1-6 meet their requirements, including the ±8 K uncertainty, with a margin varying between 2 and 20 K. Channels 7 and 8 meet their requirements and still have a margin of 9 K. In case the detectors get too cold, trim heating will increase the overall temperature level of the detectors.

The thermal stability requirement for the cooler is derived from the thermal stability requirement of the detectors, which is a stability of less than 50 mK per 100 minutes. Almost 60% of this temperature variation is caused by orbit-induced variations. All passive parts of the cooler are considered intrinsically stable except the heatpipe. Therefore, the heatpipe thermal stability shall be less than 20 mK per 100 minutes. The nominal temperature difference over the heatpipe is 2 K, which means that a change of less than 1% has to be verified during heatpipe performance testing. The test results demonstrate compliance with the thermal stability requirement.

Table 1 summarizes the overall performance characteristics of the SCIAMACHY radiant cooler.

## SYSTEM QUALIFICATION

### Development Testing

To reduce the technical risks in the early stages of the program, many development tests were performed. Amongst others, these included vibration testing of the struts, stiffness and conductivity tests on the flexlinks, material conductivity tests, optical property tests of the high reflectivity coatings, and heatpipe suspension tests. Critical parameters were derived from these tests and fed back to the thermal and finite element mathematical models to verify the design.

Input levels SRC RRU

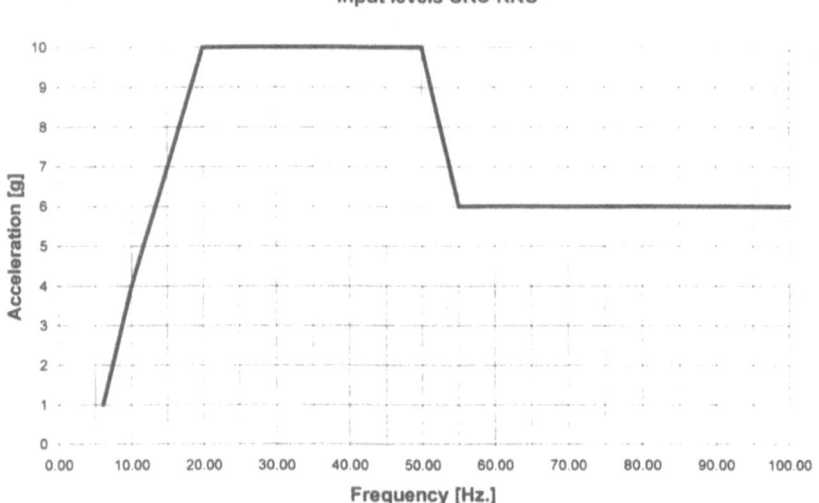

Random input levels SRC RRU

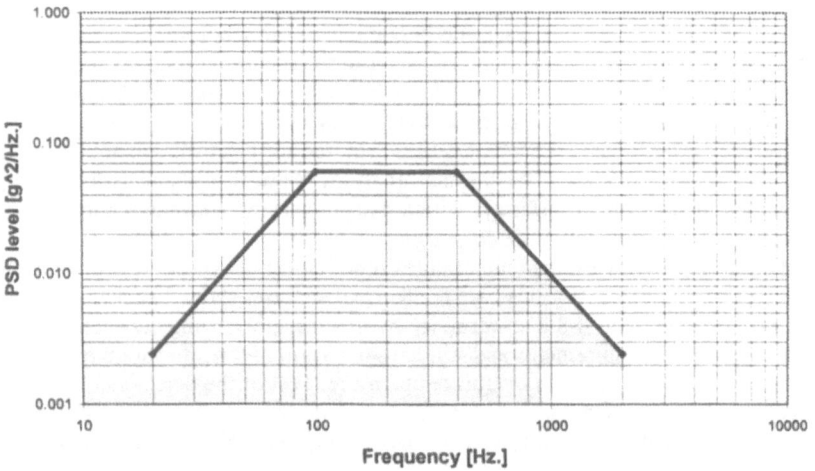

**Figure 5.** Sine and random vibration levels for the SRC RRU

## Vibration Testing

Sine and random environmental testing was performed on structural models of both the TBU and RRU. Levels for the RRU were those required by Envisat and are shown in Fig. 5. The TBU was tested integrally with the structural model of the instrument. All tests conducted were successful. The structural model has been delivered, and is currently integrated with the space-craft structural model.

## Thermal Balance/Vacuum Testing

The TBU and RRU will be subjected to thermal vacuum testing to verify their functionality over the temperature extremes of the mission; their workmanship will be verified by means of cycling under test at the temperature extremes. Thermal balance testing of the TBU and RRU has been performed to verify the mathematical model and the cooler performance.

## SUMMARY/CONCLUSIONS

The successful vibration testing of the SCIAMACHY radiant cooler has proven the structural integrity of the SRC. The structural model has shown that it fulfills its structural requirements in terms of envelope, mass, and natural frequencies.

The thermal analyses have shown that the thermal performance of the cooler will meet its requirements, including the $\pm$ 8 K mathematical model uncertainty. There is still about 4 K margin on channels 1 through 6 in the hot operational mode and end-of-life conditions. The thermal mathematical model has been updated using all critical parameters obtained in the development test program.

Finally, the vibration testing and thermal performance testing of the protoflight model are expected to provide end-to-end characterization and validation of the cooler's ability to function during the Envisat flight for four years.

## ACKNOWLEDGMENT

The author wishes to thank Doriner GmbH, Friedrichshafen, Germany for its contribution of the heatpipe assembly to the radiant cooler.

The work described in this paper was carried out by Fokker Space, The Netherlands, and was funded by the N.I.V.R. (Nederlands Instituut voor Vliegtuigontwikkeling en Ruimtevaart) through an agreement with the European Space Agency. The contribution of many engineers, technicians, and support personnel who were members of the SRC team at Fokker Schiphol Amsterdam, especially Henk Cruijssen, Stef Butterman, Rinus Coesel, Theo Arts, Roel Veldhuis, Louis Koekenberg, Hans Kollen and Rein Goossens are greatly appreciated.

# Integration of HTS SQUIDs with Portable Cooling Devices for the Detection of Materials Defects in Non-Destructive Evaluation

**R. Hohmann**[a]**, M. L. Lucía**[b]**, H. Soltner**[a]**, H.-J. Krause**[a]**, W. Wolf**[a]**, H. Bousack**[a]**,
M. I. Faley**[b]**, G. Spörl**[d]**, and A. Binneberg**[d]

[a]Institut für Schicht- und Ionentechnik (ISI), Forschungszentrum Jülich GmbH,
D-52425 Jülich, Germany
[b]Institut für Festkörperforschung (IFF), Forschungszentrum Jülich GmbH,
D-52425 Jülich, Germany
[c]Departamento de Física Aplicada III, Universidad Complutense de Madrid,
Madrid 28040, Spain
[d]Institut für Luft- und Kältetechnik (ILK), D-01309 Dresden, Germany

## ABSTRACT

We have integrated Superconducting Quantum Interference Devices (SQUIDs) made from High Temperature Superconductors (HTS) with a portable cryostat and a Joule-Thomson cooler. With these magnetic field sensing systems, flaws in metallic test samples were detected using an adapted eddy-current technique. The measurements were performed in the absence of any magnetic shielding by moving the sensors below the sample by means of a scanning table. This publication describes in detail the physical parameters of the two cooling devices, the requirements for the integration of the SQUIDs, and its realization. The measurements of test samples demonstrate the applicability of the eddy current technique to SQUID technology. Furthermore, we discuss steps to improve the systems performances.

## INTRODUCTION

In non-destructive evaluation (NDE), eddy-current techniques are commonly used for the detection of hidden materials defects in metallic structures. Conventionally, one works with an excitation coil generating a field at a distinct frequency. The eddy currents are deviated by materials flaws and the resulting distorted field is sensed by a secondary coil. Because of the law of induction, this technique has its limitations in the low frequency range (because of the inverse proportionality of the skin depth with the root of the excitation frequency), which leads to a decrease of the probability for flaw detection in larger depths. To be able to detect smaller flaws with a higher probability (enhancement of probability of detection -POD), SQUID sensors may be used because their sensitivity is not limited by an induction mechanism and thus at low frequencies their sensitivity is higher compared to induction coils of the same spatial resolution. Furthermore, SQUID electronics are available which endow such systems with a high linearity and a high dynamic range. This is a prerequisite for the quantitative evaluation of magnetic field distributions, i.e the solution of the inverse problem to unambiguously identify the flaw.

The development of liquid nitrogen cooled HTS SQUIDs allows their use as extremely sensitive magnetic field sensors in practical applications because of the less stringent cooling requirements. To insure operation in any orientation, either specially designed cryostats can be applied or machine coolers. The latter may produce problems for SQUID operation due to electromagnetic interference (EMI) by compressor parts or moving parts in the cold end, vibrations which indirectly lead to EMI due to the tilting of the SQUID in the earth magnetic field, or due to instability of the temperature. All these influences have a negative impact on SQUID performance. Therefore, care was taken to choose appropriate cooling devices for these investigations.

In this contribution, we describe two cooling devices. One was a cryostat specifically designed for this purpose, the other was a commercially available closed-cycle Joule Thomson cooler. Because of the operation principles of these devices, we expected only small disturbances which would not prevent the successful operation of a SQUID and the application of the system to nondestructive evaluation. This report will mostly focus on the properties of the system with the Joule Thomson cooler and will use the results on the portable cryostat for comparison.

## SYSTEM COPMPONENTS

### Joule Thomson Cooler

The Joule Thomson Cooler used for these investigation is a commercially available, closed cycle system by APD named KC 100 ("Cryotiger"). Figure 1 shows a cross section of the cold end and the dewar (left) and a photograph. Figure 2 shows a phoptograph of the compressor and a front view of the cold end.

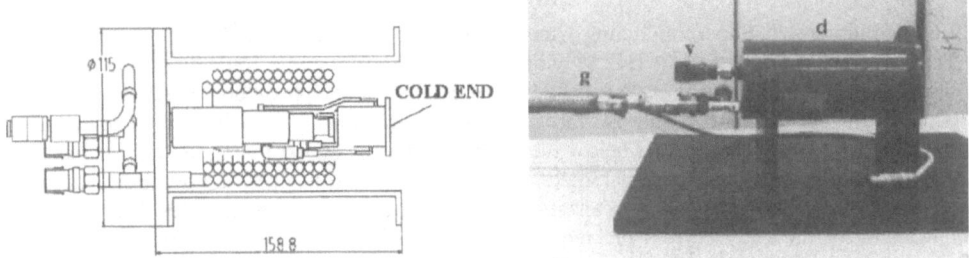

**Figure 1.** Cross section of the dewar (left) and a front view of the cold end (right). The cold end is made from copper and has a diameter of 34 mm. The dewar (d) is connected to the compressor via the gas lines (g). The needle valve (v) is used to adjust the pressure drop and in turn the temperature at the cold end.

**Figure 2.** Compressor (c) (left) and front view of the cold end (right) of the Joule-Thomson cooler by APD . The compressor has the dimensions (w x h x d ) 44.5cm x 35.6 cm x 31.1 cm. Five thread holes (h) serve to screw the probe to the cold head. Also visible are part of the superinsulation (s) and the groove (g) for the o-ring.

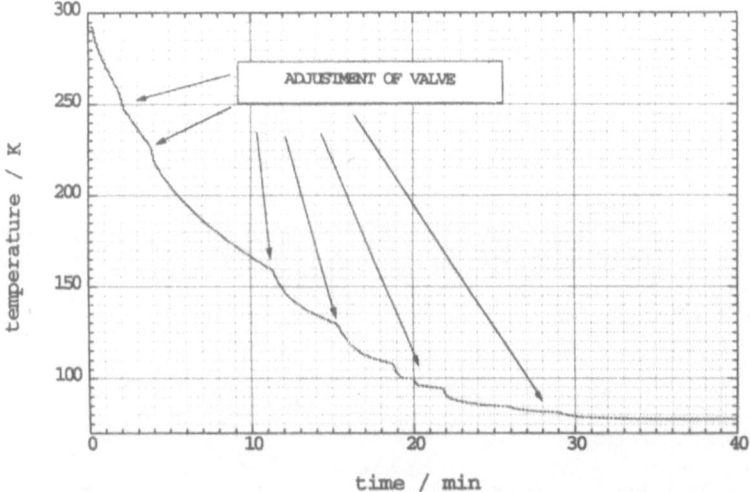

**Figure 3.** Cooling characteristics of the KC 100. The minimum temperature of about 70 K is reached after about 40 minutes. The adjustments of the needle valve to reach this state cause changes in the slope.

The compressor has a power input of about 500 W for 220 V. It is filled with an optimized mixture of nitrogen and hydrocarbons ("PT 13 gas") to ensure -according to the specifications- a minimum temperature of about 73 K at the cold head. The gas lines are about 3 m in lenth. They have a metallic mantle. Their minimum bend radius is about 15 cm. Another system delivered later was satisfactorily operated with gas lines of about 18 m in length. A self constructed end with a flange for pumping was used to close the dewar.

Figure 3 shows the cooling characteristics with a negligible heat load. The minimum temperature is reached after about 30 to 40 minutes. For this purpose, the needle valve has to be successively closed when the slope of the temperature drop becomes too shallow. The system delivered later was equipped with an automatic valve mechanism inside the cold head to facilitate the operation.

Figure 4 shows the temperate of the cold head after the stationary state has been reached. The temperature displays oscillations with a peak-to-peak amplitude of about 0.1 K. A spectral analysis of these oscillations reveals that the major contribution is due to a frequency of about 0.08 Hz and its harmonics. Frequency components higher than 1 Hz were not observed.

Because of the impairment of the SQUIDs sensitivity by temperature variations close to the superconducting transition temperature (90 K for $YBa_2Cu_3O_{7-x}$ used here), we attenuated the temperature oscillations by introducing sheets of PMMA between the cold head and the sensor. (The actual setup for this configuration is displayed in Figure 8.) PMMA was chosen for this purpose because it serves as a very good thermal low pass filter due to its small heat conduction. In Figure 4, the attenuation of these temperature oscillations is represented by the upper curve which was measured with a temperature sensor spaced by 5 mm of PMMA. The dominant temperature oscillations with a frequency of 0.08 Hz could be attenuated to a level lower than the resolution limit of the thermocouples (1 mK). A quantitative analysis of this attenuation behavior based on caculations by ter Brake [1] well described the experimentally observed results.

Figure 5 displays the cooling power of the KC 100, measured with and without superinsulation, and the corresponding specifications by APD. The minimum temperatures obtained are 70 K and 71.3 K. Every 2 K additionally about 1 W of cooling power is gained. This slope is also supported by the measurements by the supplier, whereas the minimum temperaturewas lower than specified.

The cooler was also characterized in terms of its vibrational characteristics, because vibrations at the cold head may lead to tilting of the SQUID magnetometer in the geomagnetic field

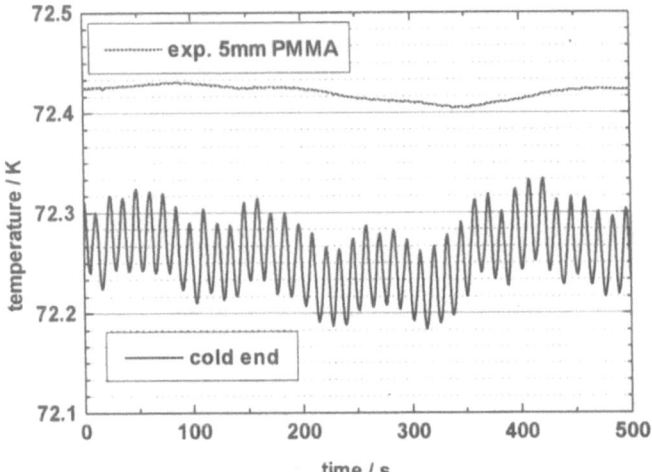

**Figure 4.** Temperature oscillations at the cold end after reaching a stationary state at about 72 K (-- cold end). The resulting temperature for a thermocouple separated by 5 mm of PMMA is represented by the curve above. The temperature oscillations are attenuated by PMMA by a factor of 50.

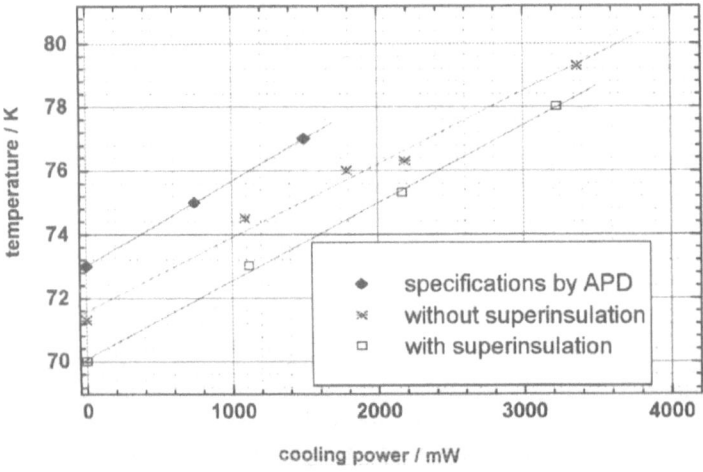

**Figure 5.** Cooling power of the KC 100. Upper curve as specified by APD, middle and lower curve measured without and with superinsulation. Apart from the absolute temperature values, the slopes of the curves consistently show a gain in cooling power of about 1 W every 2 K.

and thus induce a disturbance signal in the SQUID output at the frequency of the vibration. For this measurement, acceleration sensors were attached to the cold end and their response was monitored. Figure 6 shows one of these measurements. The upper trace represents the accelera- tion at the cold end with the compressor switched on. The amplitude of the accelerations are in the range of about 0.6 m/s$^2$. A frequency analysis shows a major contribution of 50 Hz, which is due to the compressor operating at this frequency and transmitting its movements via the gas lines to the cold end. For damping of these vibrations, the gas lines were squeeezed into the grooves of a Trovidur block. This yielded an attenuation of the amplitude of about a factor of five (second trace), whereas the insertion of elbow connectors, giving a 90 degrees turn to the direction of the gas lines, attenuated the acceleration values by a factor of about 2 (third trace).

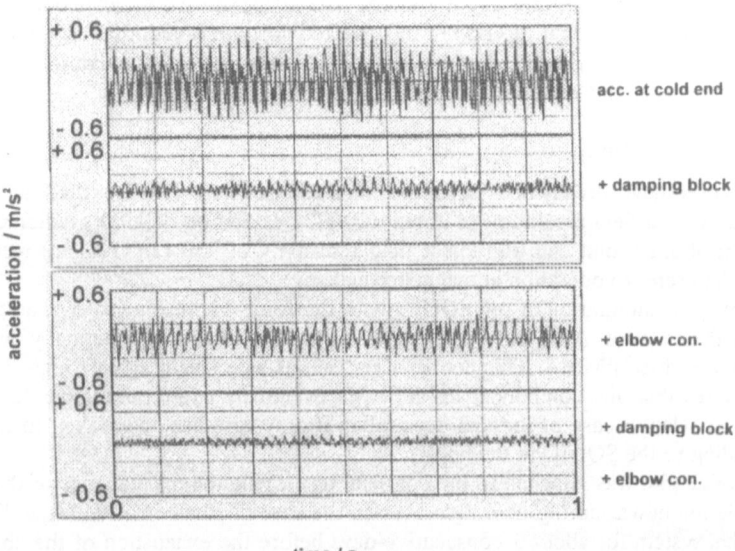

acc. at cold end

+ damping block

+ elbow con.

+ damping block

+ elbow con.

acceleration / m/s²

time / s

**Figure 6.** Accelerations at the cold end with running compressor during a time interval of 1 sec and in four different configurations. First trace - without additional means. Second trace - damping block squeezing the gas lines. Third trace - gas lines equipped with elbow connectors. Fourth trace - damping block and elbow connectors in combination. The damping block yields an attenuation of a factor of 5, the elbow connectors a factor of 2. The combination attenuates by a factor of about 10.

The combination of the block and the elbow connectors yielded an attenuation of 10 (lowest trace), thus suggesting the independence of the two damping methods.

Measurements of the possible electromagnetic disturbance of the SQUID by the compressor, the only active electromagnetic source of this cooler, were carried out with a fluxgate magnetometer. The line frequency disturbance was identified as the major electromagnetic contribution of the compressor and was chosen for evaluation. At a distance of about 2 m from the compressor, the amplitude of the 50 Hz field was 1.5 nT, whereas at 3 m the 50 Hz contribution vanished in the electromagnetic disturbance level of the laboratory.

Another source of electromagnetic disturbance is due to the presence of metallic parts, in particular the copper cold end, close to the SQUID. It was shown by Varpula and Poutanen [2], that an infinite conducting plate with temperature T of thickness d yields a field noise $B_N$ at a distance z

$$B_N = \frac{1}{2}\mu_0\sqrt{\frac{\sigma kT}{2\pi}\frac{d}{z(z+d)}} \quad , \tag{1}$$

where $\sigma$ denotes the conductivity of the material, $\mu_0$ is the permeability of the vacuum, and k is Boltzmann's constant. An estimate for a copper cardboard ($\sigma = 10^8$ Sm$^{-1}$, d = 10 µm, z = 1 mm, T = 300 K) gives a field noise of about 500 fT/$\sqrt{Hz}$. As this value holds up to frequencies in the 100 kHz range, this effect can be one of the major contributions to the impairment of SQUID performance.

### Cryostat independent of orientation

As a reference, similar efforts were undertaken to integrate the SQUIDs with a portable cryostat, designed by ILK, Dresden, for this purpose. Figure 7 shows a cross section of this cryostat. The liquid nitrogen cools the copper end plate in any orientation because the interior walls of the nitrogen reservoir are made from thin copper foil.

The holding time of this cryostat is about 8 hours. The temperature difference obtained when turning the cryostat from the upright position to the inverted position is about 0.5 K. Temperature stability is in the 10 mK range. The cooling power can be measured in a similar way as for the Joule- Thomson cryocooler and was found to be 2.5 W at 80 K.

## SQUIDs and their integration

The SQUIDs used have already been described in detail elsewhere. Briefly, rf washer SQUID magnetometers [3] with a field resolution of about 120 fT/√Hz and dc SQUID gradiometers [4] with a base line of about 6 mm and a gradient field sensitivity of 750 fT/(√Hz cm) were used. These resolution data refer to operation in magnetic shielding.

Figure 7 shows the integration of the SQUID with the Joule-Thomson cooler. The principle is the same as for the cryostat. A copper hood contains the SQUID which is thermally decoupled from the hood by a disk of PMMA. This copper hood serves as a shield against high frequency EMI and also ensures that any component inside it will eventually attain the temperature of the cold end. This holds also in case of thermal decoupling if only care has been taken to thermally sink the wires leading to the SQUID at the hood.

A reservoir of zeolite was attached to the cold end to provide the maintanance of the insulation vacuum inside the dewar during operation even in the case of small external leaks. We were able to operate this system for about 3 consecutive days before the exhaustion of the absorption capacitance of the zeolite.

## Scanning table

To investigate samples in nondestructive evaluation, a scanning table was constructed which allows the displacement in two orthogonal directions of the sample underneath a stationary cryostat with a SQUID inside or the movement of a sensor system (cooler or cryostat) underneath the sample to be investigated.

## Electronics and eddy current excitation

For the operation of the rf SQUID and the dc SQUID, two different types of electronics were used, which have also been described elsewhere [5, 6]. For the generation of eddy currents in the sample, a double D-configuration of induction coils was used [7], which in principle generates a minimized magnetic field at exactly the location of the SQUID. A digital lock-in technique was used to detect signals from the samples at the eddy-current frequency, typically in the range from 20 Hz to 30 Hz.

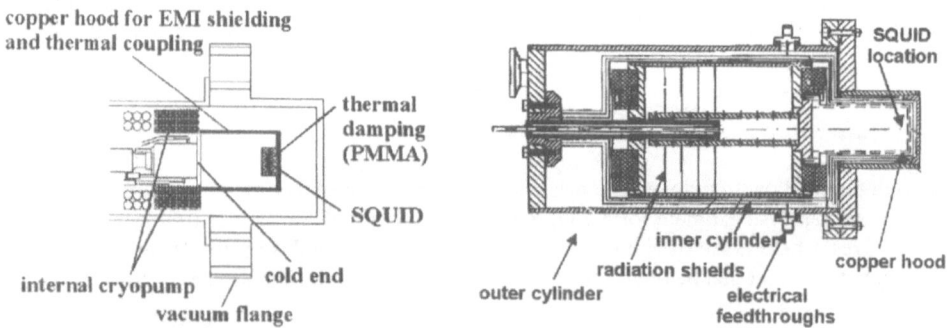

**Figure 7.** Scheme for the integration of the SQUIDs with the KC 100 by APD (left) . The SQUID is located inside a copper hood on top of the cold head. For thermal decoupling, disks of PMMA can be inserted. A reservoir of zeolite (internal cryopump) ensures the maintenance of the vacuum during operation (cf. Figure 1). The superinsulation (not shown here) is wrapped around the copper hood and the internal cryopump. Cross section of the orientation independent cryostat by ILK, Dresden. The SQUID is mounted on the interior side of the copper hood in the vacuum space in a similar way as with the cooler.

## RESULTS OF SQUID INTEGRATION

For comparision with the SQUIDs cooled by the KC 100, noise measurements were performed inside magnetic shielding for an rf washer SQUID inside a conventional conventional cryostat made from fibre reinforced epoxy. The lowest curve in Figure 8 shows the result of this measurement. Here, a field resolution of about 200 fT/√Hz in the white noise region above 3 Hz is obtained, which describes the performance of the electronics used here rather than the field resolution of the SQUID (120 fT/√Hz as mentioned above).

The first experiments after the integration of the SQUID with the KC 100 were also performed inside magnetic shielding and with the damping blocks. After the SQUID had been cooled by the KC 100 to operating temperatures of about 75 K, the compressor was switched off, which did not result in a fast increase in temperature because of the large thermal mass of the zeolite reservoir. During this time the field resolution as depicted in Figure 8 was measured (middle

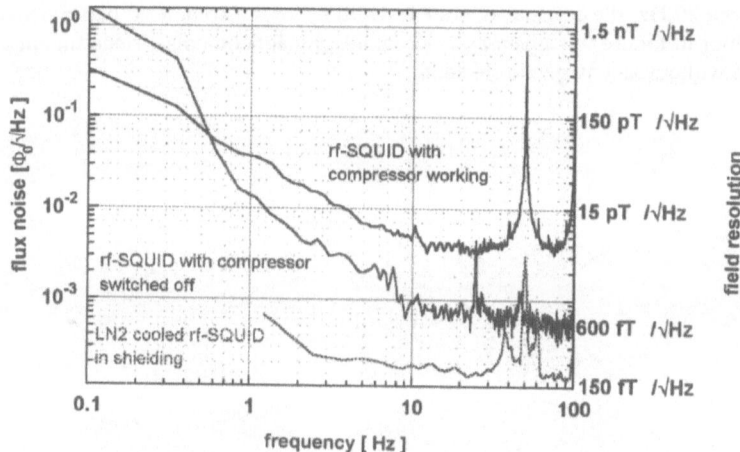

**Figure 8.** Field resolution and flux noise of the rf SQUIDs in magnetic shielding for operation inside a liquid nitrogen dewar (lower curve) and with the Joule-Thomson cooler in the case of the compressor switched off (middle curve) or on (upper curve)

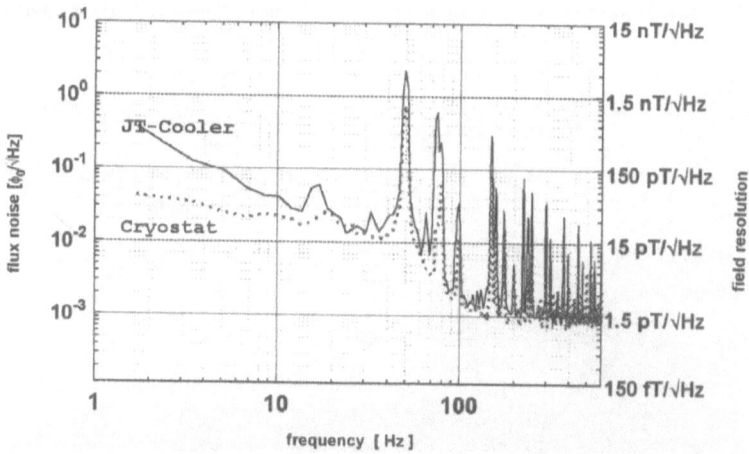

**Figure 9.** Comparison of the field resolutions and flux noise outside magnetic shielding for rf washer SQUIDs operated by the Joule-Thomson cooler (upper curve) and the portable cryostat (lower curve). A significant difference can only be seen below about 20 Hz.

curve). Above about 20 Hz, the field resolution in the white noise region is about 600 fT/√Hz. When the compressor is switched on, a large contribution of 50 Hz is dominant, although from the parameters of the shielding and the electromagnetic disturbances of the compressor this effect should have been much smaller. Also, the white noise region is now elevated to about 6 fT/√Hz, ten times as much as in the case without the compressor operating and 30 times as much as inside the dewar. The removal of the damping block increases the noise further, e.g. by a factor of 4 above 10 Hz (not shown here).

A closer examination of the 50 Hz region for the case without the damping block showed also smaller peaks at 47.7 Hz and 52.3 Hz, which can be explained by frequency mixing of 50 Hz with a compressor vibration of 2.3 Hz. This effect could nearly entirely be suppressed by the damping block. The vibration at 2.3 Hz had been identified by an analysis of the movement of the partially magnetic gas lines by a fluxgate magnetometer.

Figure 9 shows a comparison of the field resolutions outside magnetic shielding of the rf SQUIDs inside the cryostat or with the Joule-Thomson cooler (compressor working). For both situations, a noise floor of about 1.5 pT/√Hz above 100 Hz is obtained. Only in the frequency range below about 20 Hz, the cryostat-operated SQUID shows a somewhat better behavior. Note that the noise floor in Figure 9 is given by environmental magnetic noise, since the measurements were conducted without any magnetic shielding.

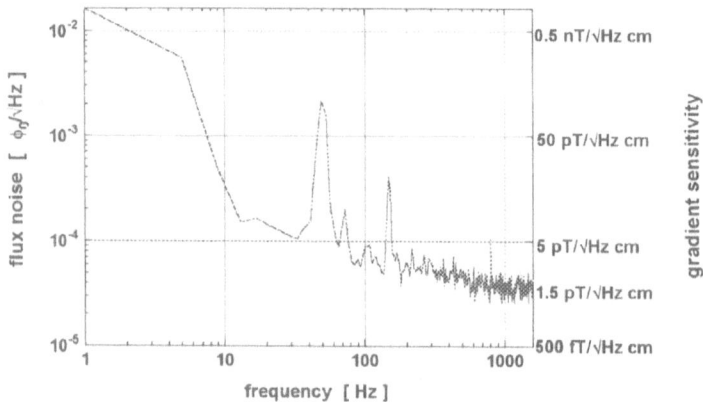

**Figure 10.** Gradient field resolution outside magnetic shielding of the dc SQUID gradiometer cooled by the KC 100.

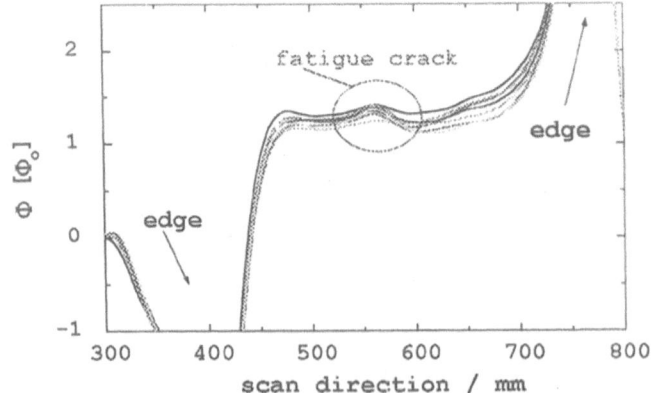

**Figure 11.** Fatigue crack detection in an aluminum plate under four intact plates using an rf SQUID magnetometer cooled by the KC 100. The different traces correspond to different location along the y direction. The signal is dominated by the edge effects due to the finite extension of the aluminum plates.

Figure 10 shows the corresponding gradient field sensitivity outside magnetic shielding of the dc SQUID gradiometer operated with the Joule Thomson cooler (compressor working). The 1/f noise region starts roughly at about 100 Hz. Above 100 Hz, the white noise is characterized by a gradient field resolution of about 1.5 pT/($\sqrt{Hz}$ cm). The 50 Hz contribution is still visible.

With this system it was possible to perform eddy current measurements relevant for NDE. Figure 11 shows the signal of a fatigue crack in an aluminum plate, produced by weakening a specific location by introducing a bore hole and stretching the plate several times. This plate was buried under four intact aluminum plates (all 7.5 mm in thickness) and this stationary stack was scanned with the eddy current excitation setup. The excitation frequency was 20 Hz. Measurements with aluminum plates, which had on some locations been reduced in thickness had also been carried out successfully. All these measurements were performed outside magnetic shielding.

## DISCUSSION

For the successful operation of cooler-based SQUID sensing systems in NDE, the cooler must meet several requirements lest it impairs the sensitivity of the SQUID. The results obtained in the previous section (Figure 11) have shown that SQUIDs can be operated with the portable cryostat and with the Joule Thomson cooler.

The Joule Thomson cooler was chosen for our investigations because of the absence of moving parts in the cold head in contrast to, e.g., Stirling coolers. A cooling power of more than 2 W (at 77 K, see Figure 5) is more than enough for the operation of single HTS SQUIDs, which are usually operated at 77 K and require about 200 mW.

Hence, no electromagnetic disturbances were to be expected from this cooler except for those that were due to the compressor itself. These were supposed to be negligible at a distance of a few meters, i.e the length of the gas lines. This was corroborated by the measurement of the flux gate at a distance of 3 m away from the compressor. Nevertheless, as Figure 8 clearly demonstrates, the operation of the compressor does add additional disturbances particularly at the line frequency, although this measurement was performed inside shielding. We believe that this is due to the transmission of electromagnetic disturbances along the gas lines and thus directly to the cold head. In fact, measurements with the dc-SQUID gradiometer on the scanning table were up to now impossible probably because of these disturbances. We expect that this effect can be avoided when the compressor and the cold head are galvanically decoupled by either using rubber gas lines or at least by inserting insulating ceramic connectors in the gas lines.

Another severe problem for SQUID performance arises from the white noise contribution by the metal parts close to the SQUID ( Eq.(1) ). As can be seen from Figure 8, the white noise floor inside shielding is about 600 fT/$\sqrt{Hz}$ even with the compressor switched off. This noise figure is of the order of the contribution estimated according to Varpula and Putanen [2]. That means that inside magnetic shielding the field resolution will at least be limited by the quantity of metal close to the SQUID. It seems inevitable to use metal parts at least for the Joule-Thomson cooler, but we believe that close to the SQUID, all metal parts should be replaced by sapphire, which at 77 K shows a heat conduction close to that of copper [8]. Also, the detection of 50 Hz disturbances by the gradiometer (see Figure 10) may at least in part be due to the hardly avoidable nonsymmmetric distribution of copper close to the gradiometer. Thus, particularly the copper hood will be subject to corresponding changes in construction.

Outside magnetic shielding, the sensitivity is limited by environmental disturbances rather than by the particular way of how the SQUID is cooled. This is suggested by Figure 9, which shows that the white noise region is practically the same for both the Joule Thomson cooler and the cryostat. Only at low frequencies and at distinct frequencies like 50 Hz, the Joule Thomson cooler leads to somewhat higher noise.

To use rubber gas lines will also be beneficial for the reduction of transmission of vibrations from the compressor to the cold head, because of the higher flexibility of this material. Also, the damping blocks could possibly be more effective because of better coupling to the softer gas lines, thus reducing the second source of electromagnetic disturbances to the SQUID, i.e. periodic tiltings in the geomagnetic field. It was estimated that accelations of the order of 0.6 m/s$^2$ (Figure 6)

-most unfavourably combined in such a way as to lead to a tilting of the cold end- prevent SQUID operation because of unacceptably strong field changes. With the employment of the damping block, we fall short of this limit by a factor of 5 approximately. Due to the distinct frequency of this disturbance, it could be eliminated by filtering. For a planar gradiometer, the tilting effect is even of lesser importance because of the large rejection of common modes.

Temperature oscillations impair the sensitivity of the SQUID at the same frequency. The effect of temperature oscillations on the SQUID can be neglected for NDE measurements, because the oscillations have effectively been suppressed by the use of a thermal low pass filter (PMMA, see Figure 4) and, furthermore, the frequencies of these oscillations are below 1 Hz, i.e much smaller than the eddy current frequencies used here.

## SUMMARY

We have integrated HTS SQUID magnetometers and gradiometers with a Joule-Thomson cryocooler (and a portable cryostat for comparison) for NDE measurements. The HTS SQUIDs could be operated inside and outside magnetic shielding and could be moved in a two dimensional way for scanning metallic samples with flaws. The different contribution of disturbances to SQUID operation have been characterized. Electromagnetic disturbances via gas lines and the abundance of metal parts are the most severe sources for the impairment of SQUID sensitivity, whereas temperature fluctuations at the cold head or vibrations play a minor role. This system will be developed further for the detection of flaws in aircraft structures.

## ACKNOWLEDGMENTS

We acknowledge the support concerning the acceleration measurements by the Staff operating the jolt-ramming machine MAVIS in Jülich, Germany. We would like express our gratitude to the representatives of APD in Aldermaston, England, for their continuous support of this work. This investigation was supported by the German BMBF under contract number 13 N 6682. M.L.Lucía acknowledges financial support by DGICYES, Spain.

## REFERENCES

1.   ter Brake, M., Aarnink, W.A.M., Flokstra, J., Rogalla, H., Proceedings of the HTS-workshop on applications and new materials, Enschede, The Netherlands, 1995.

2.   Varpula, T., and Putanen, T., "Magnetic field fluctuations arising from thermal motion of electric charge in conductors", *J. Appl. Phys.* vol. 55, no. 11 (1984), p. 4015-4021.

3.   Zhang, Y., Mück, M., Herrmann, K., Schubert, J., Zander W., Braginski, A.I., Heiden, C., "Sensitive rf SQUIDs and magnetometer operating at 77 K", *IEEE Trans.Appl.Supercond.*, vol. 3, (1993), pp. 2465-2468.

4.   Faley, M.I., Poppe, U., Urban, K., Hilgenkamp, H., Hemmes, H., Aarnink, W.A.M., Flokstra, J., Rogalla, H., "Noise properties of direct current SQUIDs with quasiplanar YBaCuO Josephson junctions", *Appl. Phys. Lett.*, vol. 67, no. 14 (1995), pp. 2087-2089.

5.   Zimmermann, E., Brandenburg, G., Clemens, U., Halling, H., "Kompensationsregelung für extrem nichtlinearen Sensor mit digitalem Signalprozessor", *DSP Deutschland'95* , Design und Elektronik, Munich (1995).

6.   Technical manual of analog dc-SQUID electronics M301-A, Cryoton, Troitsk, Russia.

7.   Tavrin, Y., Krause, H.-J., Wolf, W., Glyantsev, V., Schubert, J., Zander, W., Bousack, H., "Eddy current technique with high temperature SQUID for non-destructive evaluation of non- magnetic metallic structures", *Cryogenics*, vol. 36, (1996), pp. 83-86.

8.   Pobell, F., *Matter and Methods at low Temperatures*, Springer publishing company, Berlin (1992), p. 51.

# Stirling Cooler Magnetic Interference Measured by a High-Tc SQUID Mounted on the Cold Tip

**H.J.M. ter Brake, H.J. Holland and H. Rogalla**

University of Twente, Faculty of Applied Physics,
P.O.Box 217, 7500 AE Enschede, The Netherlands

## ABSTRACT

Small Stirling-type cryocoolers are available on the market mainly for cooling infrared detectors, but they can also be used to cool high-Tc superconducting devices, like SQUIDs. Because SQUIDs are extremely sensitive magnetic sensors it is questionable whether these devices can be mounted directly on the cold tip of the cooler. To investigate this we attached a high-Tc SQUID to the tip of a representative split Stirling-cycle cryocooler (Signaal Usfa type 7058), and operated it in a magnetically shielded room.

The SQUID that we used for our test experiments was one of the first high-$T_c$ SQUIDs that were manufactured in our group and had a noise level of 0.7 pT/√Hz at about 77 K. With the SQUID attached to the tip of the cold head and the cooler running, the noise appeared to have increased dramatically: at the driving frequency of 50 Hz and at the harmonics by roughly $5.10^5$, and at other frequencies by about a factor $10^3$. The experiments indicated that the noise coupled into the SQUID was dominated by contributions originating from the cold head, and not by compressor interference. The experiments are discussed, and the consequences with respect to the cooling of SQUIDs with standard cryocoolers are briefly considered.

## INTRODUCTION

Superconducting Quantum Interference Devices (SQUIDs) are the most sensitive magnetic flux-to-voltage converting sensors. Although a variety of applications has been investigated, worldwide SQUIDs are mainly applied for biomagnetic measurements. In this field of research multichannel low-$T_c$ dc-SQUID based magnetometers are used that are usually cooled by liquid helium. Because of the higher operating temperature, a much more flexible magnetometer system can be realized with high-$T_c$ SQUIDs. The most widely used material $YBa_2Cu_3O_7$ has a critical temperature of 92 K and allows cooling with liquid nitrogen boiling at 77 K. However, also small-scale turn-key cryocoolers are available that were originally developed for cooling infrared sensors[1]. Cooling of high-$T_c$ SQUIDs by means of such coolers has several advantages compared to a liquid nitrogen bath. The operating temperature can be lower (roughly 50 K), which gives a potentially better SQUID performance[2]. Furthermore, turn-key operation of the magnetometer is possible (replacing liquid nitrogen refills) and also the magnetometer can be operated in all directions without any precautions.

We are currently developing a high-$T_c$ SQUID-based heart scanner that should be cooled with small cryocoolers. An important issue in the design of such a magnetometer is whether a SQUID can be mounted directly on the cold tip of the cryocooler. This was claimed to be possible by Khare and Chaudhari from IBM, who attached a "directly coupled" bicrystal high-$T_c$ SQUID to the tip of a small integral type Stirling cooler (Inframetrics Inc., 0.15 W at 77 K)[3]. Unfortunately, the SQUID that they used had a relatively poor performance: effective sensing area 0.013 mm$^2$, field resolution 75 pT/$\sqrt{Hz}$ at 1 Hz and 10 pT/$\sqrt{Hz}$ at 10 Hz. Furthermore, the SQUID was shielded from the external field but also from the noise of the cryocooler motor by means of μ-metal. Two cans were used: one bigger can surrounded the whole cryocooler, whereas the smaller can surrounded the cold finger and the SQUID only. Apart from these, two plates of μ-metal were applied to shield the cooler from the bottom[3,4]. In this configuration Khare and Chaudhari measured no increase in the field noise of the SQUID (compared to measurements in a liquid nitrogen bath) in the frequency band of 1 Hz to 10 Hz. However, the fact that the cooler apparently did not contribute to the noise can fully be attributed to the high intrinsic noise level of the SQUID and to the μ-metal magnetic shielding. This small-size μ-metal shielding is only applicable if the measuring object can be placed close to the SQUID and inside the shielding (e.g. magnetic measurements on small materials). For the large majority of applications (e.g. biomagnetism, non-destructive testing of relatively large materials, magnetotellurics) this type of small-size shielding cannot be applied.

Therefore, we performed similar experiments with a more sensitive SQUID and without small μ-metal shields shielding the SQUID directly from the cooler. Experiments were performed with and without a μ-metal shield. In the next section of this paper the experimental set-up is described, followed by a presentation of the results. After that these results are discussed in a separate section, and the paper is concluded with some closing remarks.

## EXPERIMENTAL SET-UP

The SQUID that we used also was a directly coupled bicrystal high-$T_c$ SQUID. The specific device that was available for our experiments was one of the first devices that were manufactured in our group[5]. It had an effective area of 0.12 mm$^2$ and a noise level of 0.7 pT/$\sqrt{Hz}$ at 10 Hz at about 77 K, measured inside a magnetically shielded flow cryostat (at higher frequencies the noise level decreased to about 0.5 pT/$\sqrt{Hz}$). The cryocooler that we applied for cooling the SQUID is a split Stirling cooler manufactured by Signaal Usfa (type 7058, 1.5 W at 80 K)[6]. The SQUID was placed in a prototype ceramic package made of 0.7 mm thick $Al_2O_3$ plates. This ceramic package was screwed on to a 1 mm thick square stainless steel plate measuring 32 mm in size. (Calculated thermal noise contribution 0.07 pT/$\sqrt{Hz}$). Between the SQUID package and the stainless steel plate was a 4 mm thick $Al_2O_3$ plate. The steel plate was soldered on the cold-head tip via a small cylinder. A tube of thin kapton foil was placed around the cold finger in order to hold the SQUID leads. Also, a diode was attached to the SQUID package for monitoring the tip temperature. This assembly as depicted in Fig. 1 was covered with 3 layers of superinsulation (aluminized mylar foil) and could be inserted in a stainless steel vacuum can. The bottom of this can was made of glass-fibre reinforced epoxy so that the set-up could also be used for biomagnetic measurements provided the cooler noise contributions were acceptable. A metallic bottom would interfere with such measurements due to eddy-currents and thermal noise.

**Figure 1**. Cold finger with SQUID package and surrounded by kapton cylinder holding the electrical leads.

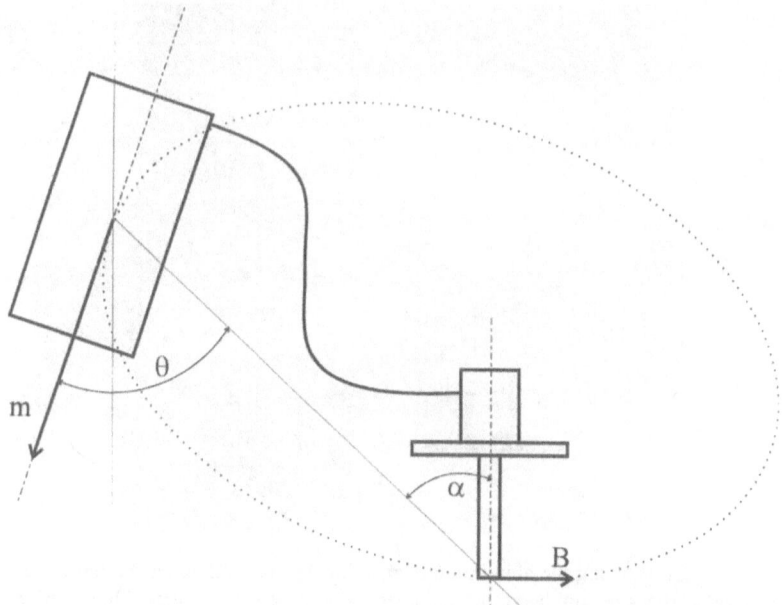

**Figure 2.** Schematic diagram of compressor and cold head configured in one plane.

The cold head holding the SQUID was positioned in such a way that minimum pick-up from the compressor-coil currents could be expected. The magnetic field due to these currents was measured to be dipolar-like with a magnetic dipole moment of 0.5 $Am^2_{RMS}$ along the compressor axis[7]. Consider the configuration that is schematically depicted in Fig. 2. Because the SQUID measures the field component along the cold-head axis, the noise field should be perpendicular to that axis. Referring to Fig. 2, it is simple to derive that this results in the following condition:

$$\tan(\alpha) \cdot \tan(\theta) = 2 \tag{1}$$

We designed an arrangement with $\alpha = 45^{\circ}$ and $\theta = 63.4^{\circ}$, which is depicted in Fig. 3. The cold head as well as the compressor were thermally anchored to an aluminium plate for heat sinking. The SQUID was located at a distance of about 21 cm to the centre of the compressor. As an option a $\mu$-metal shield was available to cover the compressor. In earlier experiments[7] this shield reduced the compressor noise by roughly a factor of 45. This arrangement was installed in the magnetically shielded room of the Biomagnetic Centre Twente in order to measure only the intrinsic noise contributions[8].

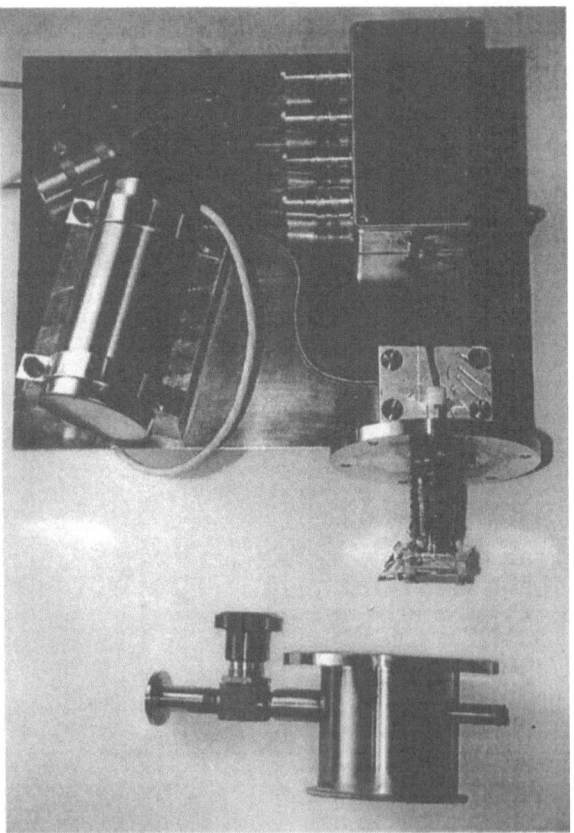

**Figure 3.** Photograph showing the aluminium heat sinking plate holding on the left the compressor and in the lower right corner the cold head. In the upper right corner of the plate a box is placed for the connections to the SQUID electronics. Below the cold finger the cold-head vacuum can is shown.

## RESULTS

After switching on the cooler the SQUID package initially cooled down to 51 K. Because of inadequate heat sinking the compressor and the warm end of the cold head slowly increased in temperature, resulting in a stable SQUID temperature of about 56 K. Noise spectra of the SQUID output were recorded by means of a HP3562A spectrum analyzer. Typical spectra for various configurations are depicted in Fig. 4 and the results are summarized in Table 1. In all the measurements we applied bias-current modulation to reduce the low-frequency noise of the SQUIDs[5]. With the cryocooler running the noise at the SQUID output appeared to have increased dramatically: 0.5 - 1 nT/√Hz at 10 Hz and 0.5 $\mu T_{RMS}$ at 50 Hz. Installation of the μ-metal shield around the compressor did not have any effect on the 50 Hz noise, but it significantly decreased the higher-frequency contributions. This can be explained by considering the measured noise to originate from two sources: firstly, the noise due to the compressor currents, and secondly, the noise due to the displacer moving in the cold head. The first contribution is shielded by the μ-metal and is thus reduced by a factor of about 45 at 50 Hz[7], and possibly more for higher frequencies due to eddy-current effects. The fact that in our experiments the 50 Hz peak did not decrease after installing the μ-metal, confirms that the positioning according to Eq. (1) was succesful. Apparently, the 50 Hz contribution of the compressor was much smaller than that of the cold head. The effect of the positioning of the compressor can also be evaluated from the worst case in which, at 21 cm distance, the compressor-coil currents can yield a 50 Hz noise of 10 $\mu T_{RMS}$, a factor 20 above the measured value. The higher-frequency components behave differently from the 50 Hz component because they mainly result from non-linear effects in the cooler that will not be equally distributed over the two coil sections of the compressor. The equivalent magnetic dipole for higher frequencies is, therefore, not in the centre of the compressor and, as a result, the positioning consideration as applied for 50 Hz does not hold. Comparison of the measured spectra with and without μ-metal shield shows that the contribution from the compressor-coil currents shows much more high-frequency peaks than the displacer contribution. This is because the cooler is a resonant system operated at a frequency of 50 Hz.

In order to investigate the effect of the connecting leads on the noise (50 Hz of the power line with higher harmonics, and RF pick-up), we turned down the power supply so that the cooler stopped operating. Despite the SQUID slowly warming up, the noise could be measured and was about 40 dB lower than with the cooler running, the even harmonics of 50 Hz hardly being present. Now the μ-metal appeared to reduce the remaining noise effectively, at 50 Hz roughly with the expected factor of 45. We conclude that the remaining noise was caused by currents in the compressor coils that are due to 50 Hz noise (and harmonics) in the power supply and pick-up in the leads. The 'white'-noise level was high compared to that measured in the flow cryostat because of temperature drift, and because of RF interference coupled into the shielded room via the power lines[9]. Finally, also the leads were disconnected from the power supply. Then, the noise decreased by a further factor of 5, roughly to the noise level measured in the flow cryostat.

**Table 1.** Noise measured by High-$T_c$ SQUID on cold tip of 7058 cooler

| Configuration | 50 Hz | 10 Hz |
|---|---|---|
| on tip, cooler on (with <u>as well as</u> without μ-metal) | 0.5  $\mu T_{RMS}$ | 0.5 - 1 nT/√Hz |
| on tip, with power connections, cooler off | 5  $nT_{RMS}$ | 5 - 10 pT/√Hz |
| on tip, cooler off, μ-metal around compressor | 0.1  $nT_{RMS}$ | 5 - 10 pT/√Hz |
| on tip, no power connections | 1  $pT_{RMS}$ | 1   pT/√Hz |
| in magnetically shielded flow cryostat | 2  $pT_{RMS}$ | 0.7  pT/√Hz |

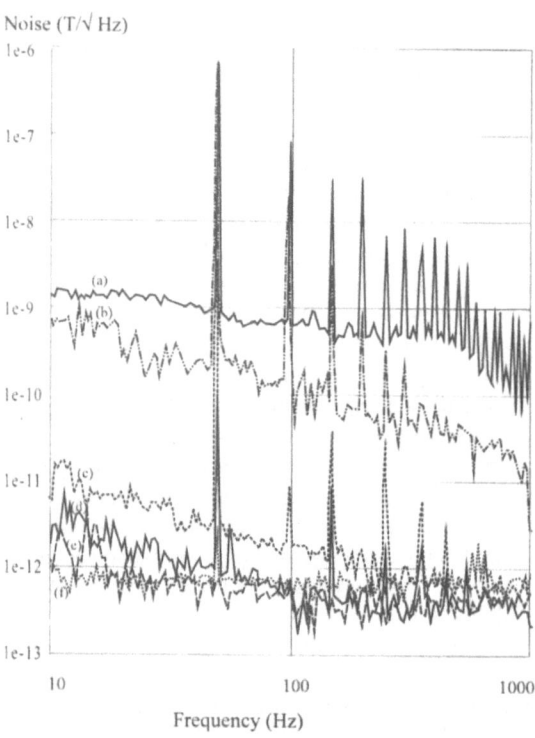

**Figure 4.** Typical noise spectra of a high-$T_c$ SQUID on the cold tip of a Signaal Usfa 7058 cooler in various configurations: (a) cooler on, (b) cooler on (with μ-metal compressor shield), (c) cooler off, (d) cooler off (with μ-metal compressor shield), (e) cooler leads disconnected, (f) flow cryostat at about 77 K.

## DISCUSSION

The results of the noise experiments clearly indicate that for proper compressor positioning the cryocooler noise coupled into the SQUID is dominated by 50 Hz contributions originating from the cold head. At higher harmonics also significant contributions due to the compressor-coil currents arise. These contributions can be reduced by μ-metal shielding. The remaining cold-head contribution is caused by magnetic material in the displacer that oscillates with a 3 mm amplitude more or less close to the SQUID. The overall magnetization was measured with a SQUID and appeared to be dominated by the metallic spring in the warm-end section of the cold head. The equivalent magnetic dipole moment of this spring was $3.10^{-3}$ $Am^2$ pointing in the displacer-axis direction. With the SQUID at about 75 mm distance, and considering the spring to oscillate only partly with an amplitude of 3 mm, a noise contribution of roughly 0.1 $\mu T_{RMS}$ at 50 Hz results. Because the noise measurements yielded 0.5 $\mu T_{RMS}$ we anticipate that also the displacer is slightly magnetic, which is more dramatic because it is much closer to the SQUID (The noise contribution due to an oscillation scales with the distance to the power -4). Some remanent magnetization of the displacer can be expected because CrNi-steel is used as the regenerator material.

We also considered the effect of vibrations of the SQUID on the measured noise. Due to the displacer moving up and down in the cold finger the theoretical acceleration of a freely moving cold head is roughly 5 $ms^{-2}$ implying 50 Hz vibrations of 50 μm in amplitude. Because the cold head is fixed in a rigid construction the vibrations are reduced to at least below 10 μm. Further, the effect of pressure fluctuations in the cold finger results in calculated vibrations of about 2.5 μm in amplitude, which is negligible. We have measured the static field gradient in the centre area of the shielded room by means of a fluxgate magnetometer to be below 5 nT/m. In this field gradient a vibration of 10 μm corresponds to a measured field fluctuation of only 50 fT in amplitude. Due to the magnetic material in the cryocooler the field gradient in our test configuration will have been much larger and thus also the effect of vibrations. In this respect the permanent magnet inside the compressor is expected to give the largest contribution. The magnitude of this field contribution was measured to be $k/r^2$ with k = 0.21 $\mu Tm^2$ and r the distance to the compressor[10]. Although these field recordings were performed at measuring distances ranging from 30 cm to 80 cm, we can estimate the gradient at 21 cm to be roughly 45 μT/m. Vibrations of 10 μm would result in field fluctuations of 0.45 nT, which is 60 dB under the measured noise level. Therefore, we conclude that vibrations of the cold tip, and the resulting movements of the SQUID in a field gradient, did not contribute to the measured noise level in a significant way.


## CONCLUSIONS

Based on our noise experiments with a representative split stirling-cycle cryocooler (Signaal Usfa 7058) we conclude that for biomagnetic applications the SQUID cannot be mounted directly on the cold tip. Shielding of the compressor with μ-metal and proper positioning of the compressor with respect to the SQUID significantly reduces the noise contributions of the compressor-coil currents. But the remaining noise due to remanent magnetization of the displacer assembly is also much too high. In our configuration we measured 0.5 $\mu T_{RMS}$ at 50 Hz; 0.5 nT/√Hz at 10 Hz. To overcome this problem we will apply a flexible copper thermal link between the cold tip and the SQUIDs with a length of 30 cm[11].


## ACKNOWLEDGEMENTS

The authors thank the co-workers of Signaal Usfa for their contribution in the discussions.

# REFERENCES

1    Walker, G., Bingham, E.R., *Low-capacity Cryogenic Refrigeration: Monographs on Cryogenics*, Clarendon Press, Oxford (1994).

2    Rogalla, H., "Superconducting electronics", *Cryogenics*, vol. 34 (ICEC suppl.) pp. 25-30.

3    Khare, N., Chaudhari, P., "Operation of bicrystal junction high-$T_c$ direct current-SQUID in a portable microcooler", *Appl. Phys. Lett.*, vol. 65, no. 18 (1994), pp. 2353-2355.

4    Khare, N., National Physical Laboratory, New Dehli, India, private communication.

5    Hilgenkamp, J.W.M., *High-$T_c$ dc SQUID Magnetometers*, PhD thesis, University of Twente, Enschede (NL), 1995.

6    Verbeek, D., Helmonds, H., Roos, P., "Performance of the Signaal Usfa Stirling cooling engines", *Cryocoolers 7*, Santa Fe (1992), pp. 728-737.

7    Ter Brake, H.J.M., Van den Bosch, P.J., Holland, H.J., "Magnetic noise of small Stirling coolers", *Adv. Cryog. Eng.*, vol. 39B (1994), pp. 1287-1295.

8    Ter Brake, H.J.M., Huonker, R., Rogalla, H., "New results in active noise compensation for magnetically shielded rooms", *Meas. Sci. Technol.*, vol. 4 (1993), pp. 1370-1375.

9    R.H. Koch et al., "Effects of radio frequency radiation on the dc SQUID ", *Appl. Phys. Lett.*, vol. 65, no. 1 (1994), pp. 100-102.

10   Verberne, J.F.C., Bruins, P.C., Van den Bosch, P.J., Ter Brake, H.J.M., "Reduction of the vibrations generated by Stirling cryocoolers for cooling a high $T_c$ SQUID magnetometer", *Cryocoolers 8*, Plenum Press, New York (1995), pp. 465-474.

11   Van den Bosch, P.J., Ter Brake, H.J.M., Holland, H.J., De Boer, H.A., Verberne, J.F.C., Rogalla. H., "Cryogenic design of a high-$T_c$ SQUID-based heart scanner cooled by small Stirling cryocoolers", submitted to *Cryogenics*.

# Cryocooler Integration with High Temperature Superconducting Motors

**B.X. Zhang, D.I. Driscoll, B.A. Shoykhet and A.A. Meyer**

Reliance Electric Company
Cleveland, Ohio, USA 44117

## ABSTRACT

Electric motors with high temperature superconducting (HTS) field windings have been under development for several years[1]. Previous studies concluded that air-core synchronous motors with superconducting field windings offer the best motor topology. A cryocooler is required to provide refrigeration such that the salient-pole superconducting winding coils that reside in the rotor cryostat are kept at their nominal operating temperature of 30 K. Motor efficiency analysis indicates that the compressor input power of the cryocooler is limited to around 10 kW in order to make the motor economically viable. The delivery of refrigeration from the cryocooler in a stationary environment to a rotating frame with minimal loss appears to be the most challenging design issue. This work discusses pertinent issues concerning the integral design of the cryocooler and rotor cryostat of the HTS motor. Various schemes that could be potentially utilized to accomplish such an objective are also presented.

## INTRODUCTION

Large HTS motors offer improved efficiency over conventional motors (approximately 50 % reduction in losses) and a short pay-back period for cost premiums. It was estimated that over the life time of a large HTS motor, the energy savings can entirely offset the initial motor cost. The high field generated by the HTS field windings also substantially reduces the overall motor volume and weight[2].

### HTS Motor Topology

Through the HTS motor development program at Reliance Electric in the past several years, it has been concluded that air-core synchronous motors with superconducting field windings offer the best HTS motor topology. The armature of a HTS motor resembles that of a conventional motor and operates at the ambient temperature (discounting a temperature rise). The field winding coils of a HTS motor reside in the center of the rotor. A cryostat is built around the HTS coils to maintain a cryogenic environment at the coil operating temperature. The rotor cold space is thermally insulated from the ambient using multilayer insulation blankets in the surrounding vacuum jacket. Cryogenic refrigeration is delivered to the rotor cryostat from a cryocooler system through a single or multiple circulation loops.

Cryocoolers 9, Edited by R.G. Ross, Jr.
Plenum Press, New York, 1997

## Development Status

Several HTS motors have been built and tested since the beginning of the HTS motor development program at Reliance Electric in 1987. The latest development milestone was marked by a recent demonstration of an HTS motor. With the field winding coils kept at 27 K, this air-core, synchronous HTS motor delivered 200 horsepower at 1,800 rpm. To date, cryogenic cooling for all the prototype HTS motors developed at Reliance Electric has been limited to the use of open-cycle systems for economic reasons. Liquid nitrogen and gaseous helium have been used to cool the rotor and then exhausted into the atmosphere. Towards the goal of developing commercially viable products (large-scale HTS motors), it is imperative to have a closed-cycle cryocooler system as an integral part of the motor package. An immediate step in the HTS motor development calls for integration of a commercially available cryocooler with a 1,000 horsepower HTS motor system. Preliminary cryogenic process analysis has been completed for a number of system options that meet the rotor cooling requirements, including efficiency and reliability criteria[3].

# CRYOCOOLER SYSTEM REQUIREMENTS

## System Availability

Besides the huge energy savings advantages, a key factor that influences the competitiveness of the HTS motors against conventional motors is its availability. The targeted market sector for the large-scale HTS motors is comprised of applications where continuous operation of industrial equipment, such as pumps, fans and compressors, is required. The HTS motors will be competing with conventional motors of equivalent power rating that are generally very reliable and only require infrequent maintenance at manufacturer recommended intervals. For the conventional motors, unscheduled shut-down for maintenance is rare. The HTS motors, on the other hand, require cryogenic cooling. The maintenance schedule of the cryocooler should match the motor maintenance schedule which is typically 12-18 months.

## Cooling Capacity

**Normal Operation**. The required cooling capacity of the cryocooler depends on the design of the HTS motor. Based on earlier economic analysis, the compressor input power should be limited to around 10 kW to cool HTS motors rated over 1,000 horsepower during normal operation. It is desirable for such an input power to generate a cryogenic refrigeration capacity equivalent to 70 W at 27 K.

**Rotor Cool-down**. During a rotor cool-down from the ambient temperature to a nominal HTS coil operating temperature of 27 K, the total thermal energy to be removed from a typical rotor cold space including the HTS coils is in the order of $10^7$ Joules. While the actual cool-down capacity requirement is mostly limited by allowable thermal stresses in the rotor cold space, it is hardly practical to size the cryocooler system for the cool-down as it significantly boosts the system costs and demands an additional operating mode for the cryocooler system in order to meet its efficiency requirement during normal operation. Alternately, the cryocooler system should provide means for the use of liquid nitrogen to assist the rotor cool-down.

## Additional Requirements for On-board Cryocooler Systems

Transfer losses that occur during the delivery of refrigeration from the cryocooler to the rotor substantially contribute to the overall cryogenic load for the cryocooler. These losses would be greatly reduced if an innovative integral design of a cryocooler into the rotor cryostat could be worked out and turned into reality with part of the cryocooler system such as the cold heads, or even the entire unit built into the rotor cryostat. Additional design requirements for any on-board cryocooler system would be the functionality and reliability of all on-board parts under the rotating environment of the rotor (typically at 1,800 rpm), as well as in depth analysis of the consequential impact on rotor dynamics.

### Transfer Coupling Design Requirements

The transfer coupling connects the cryocooler to the HTS motor and facilitates the transition between the stationary environment of the cryocooler system and the rotating frame of the rotor cryostat. This device is required for most system configurations including that with the cold heads of a cryocooler installed within the rotor cryostat. The only way of avoiding the use of such device is to have the entire cryocooler system including the compressor module installed on board with the rotor.

Depending on the thermal management schemes for the rotor cryostat, there could be one or two cryogen circulation loops with two, three, or four streams flowing through the transfer coupling with the following potential configurations:

a. One supply and one return: cryogen is brought into the rotor cryostat where it is circulated to cool the HTS coils before returning to the compressor suction side. Depending on the process design, the return stream could be either cold or warm. This configuration offers the simplest structural feature of the transfer coupling.

b. One supply and two returns: considering the refrigeration loads at different temperature levels, upon return the cryogen can be split into two streams with one cooling the coils and the other intercepting heat leaks at a higher temperature. It should be noted that although such configuration appears to be more thermally efficient for the rotor cryostat alone, it causes an imbalance in the cryogenic process loop, which could potentially yield a lower overall cryogenic system efficiency. In depth thermal analysis is required for this type of system configuration.

c. Two supplies and two returns: thermodynamically, two separate cooling loops offer a higher system efficiency. The second cooling circuit will be operating at a higher temperature intercepting a large portion of the parasitic heat leaks and therefore reduces the load at the cold end. The drawback of this approach is increased structural complexity in both the rotor cryostat and the transfer coupling.

The most challenging task in transfer coupling design, fabrication and successful operation is probably the development and demonstration of a rotating seal for cryogen containment. Generally, the heat generation at the seal and supporting bearings increases with the motor speed which should be minimized since part of it enters the cryogenic system as heat leak. At the transition point between the rotating and non-rotating halves, thin-walled cylindrical tubes with high aspect ratio (length to diameter) are concentrically telescoped with clearance tolerance are used to reduce parasitic heat leak into the cryogenic system. The use of such parts significantly increases the degree of difficulty in fabrication and assembly of the transfer coupling. The final design will always be a compromise between its thermal performance and manufacturing feasibility.

## CRYOCOOLER SYSTEM DESIGN OPTIONS

### Recuperative Cryocoolers

Incidentally, the cryogenic refrigeration load for large-scale HTS motor applications falls in the capacity gap between commercially available regenerative and recuperative cryocoolers. Strictly from a thermodynamics view point, the preferred cryocooler for the HTS motor applications will probably be a recuperative type that offers higher efficiency. Figure 1 shows a flow diagram of a Brayton cycle (reversed) that could be readily used in such applications.

One of the advantages in adapting the Brayton cycle cryocooler is the convenience in delivery of refrigeration to the rotor cryostat, as the cooling channels in the rotor cryostat are naturally part of the circulation loop of the cryocooler. Variations to the basic Brayton cycle system, such as an addition of a secondary circulation loop shown by the dashed lines in Figure 1 that delivers refrigeration at higher temperatures, may offer increased system efficiency. On the other hand, it is almost certain that an increase in the number of cryogen circulation loops will be penalized by increased structural complexity in both the transfer coupling and rotor cryostat. With regards to the cryogenic refrigeration capacity, recuperative cryocoolers that are currently available as commercial products offer roughly at least twice the capacity that the HTS motors need. These cryocoolers normally carry high price tags (comparable to that of the motor) and therefore can only be used for motor prototype testing purpose. On the low end of the capacity spectrum, a miniaturized recuperative cryocooler has been developed and successfully demonstrated a few years ago[4]. Considering the current recuperative cryocooler technology, it is reasonable to anticipate substantial development at the component level for any cryocooler vendor to come out with a custom designed system that meets the HTS motor application requirements.

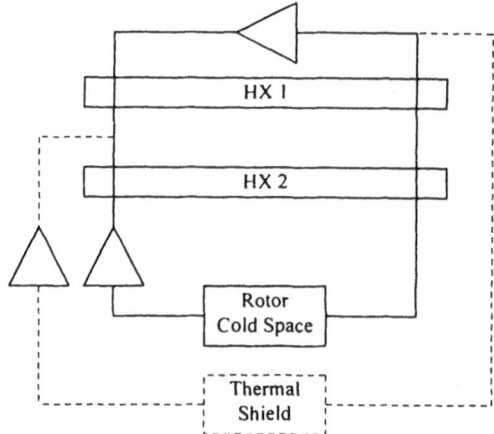

**Figure 1.** Flow diagram of reversed Brayton cycle for HTS motor applications

## Regenerative Cryocoolers

The use of regenerative type cryocoolers for the HTS motor applications appears to be less attractive. Besides the lower efficiencies inherently associated with these cryocoolers, a scheme for refrigeration delivery from the cold heads to the rotor cryostat has to be implemented. As an example, a design option involves the use of a circulation loop as shown in Fig. 2. Such a system

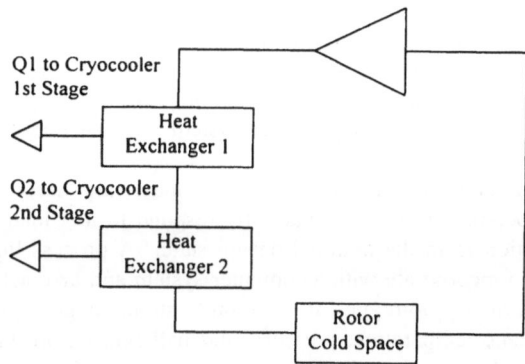

**Figure 2**. Cryogenic circulation loop for HTS motor using a regenerative cryocooler

requires minimal modification to an ordinary regenerative cryocooler. A secondary circulation loop sustained by a pump delivers refrigeration from the cold heads of the cryocooler to the rotor cryostat. The pressure head required for circulation is minimal (a few psi) and therefore little energy, which literally has no effect on the overall efficiency of the HTS motor system, is needed to drive the pump. The circulation stream picks up refrigeration at the heat exchangers that are thermally anchored to the cold heads of the cryocooler and delivers it to the rotor cryostat. Due to transfer losses, the efficiency of such a cooling system would be always lower than that of the regenerative cryocooler itself.

Another option is to have an on-board cold head cooling system configuration. In that case, the cold head would be mounted on the rotor and the refrigeration can be delivered to the HTS coils through a number of ways, such as solid conduction or the use of a heat pipe. At 1,800 rpm, significant centrifugal force results due to the rotation (in the order of $10^3$ g for the size of the motors under development). The impact of such force on the cold head parts, especially the displacer, and the contained helium gas has to be carefully analyzed to ensure reliable system operation. The configuration of the cold head may have to be re-arranged such that the displacer is located along the rotating axis of the rotor to avoid the lateral load due to the centrifugal force. Another disadvantage of such a configuration is that the pressure rating for the transfer coupling has to be increased at least by a factor of ten, which makes it extremely difficult for a workable and reliable seal design. On the other hand, the transfer coupling will be operating at the ambient temperature and therefore will no longer be a cryogenic device, which directly translates into reduced structural complexity.

## SYSTEM INTEGRATION

In addition to the conventional motor system components, a cryocooler and a transfer coupling are required for generation and delivery of cryogenic refrigeration to the rotor cryostat. They are integral parts of the HTS motor system. A system component chart for a HTS motor is shown in Figure 3. Regardless the cryocooler configuration, at least one circulation loop is

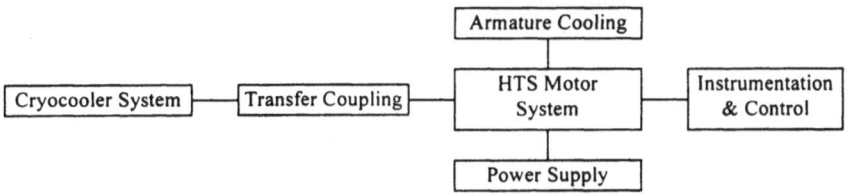

**Figure 3**. System diagram of HTS motor

required for delivering refrigeration from the cryocooler to the rotor cryostat. The transfer coupling is located at the transition point between the stationary environment and the rotating frame of the rotor, where the cryogenic circulation loop enters and exits the rotor cryostat. The transfer coupling is mounted on the non-drive end of the rotor. Ordinary cryogenic bayonets can be used for the connection between the cryocooler system and the non-rotating part of the transfer coupling.

The actual refrigeration load from the rotor cryostat may vary according to the motor operation mode. The cryocooler system will naturally respond to any load shift in a passive way, allowing some deviation from the nominal design state. A process logic control (PLC) unit may be implemented to incorporate with a computer system and take active control during such time periods. In HTS motor prototypes, in addition to strain gauges mounted on the HTS field winding coils, cryogenic temperature, pressure, and hall sensors are installed at various locations within the rotor cold space to monitor the coil operating condition and performance.

## SUMMARY

Cryogenic cooling of HTS motors demands stringent requirements for a cryocooler system and brings out difficult system integration issues. Refrigeration delivery from the cryocooler system to the HTS motor appears to be the most challenging task that requires extensive technology development. System efficiency, reliability and costs are the dominating factors to be properly balanced for successful development of commercially viable products. A recuperative cryocooler system operating in a reversed Brayton cycle offers the highest system efficiency but requires expensive development at the component level. The use of an on-board regenerative cryocooler reduces transfer losses, and can possibly eliminate the need for a cryogenic transfer coupling.

## ACKNOWLEDGMENT

The HTS motor development program has been supported by EPRI, and in part by the U.S. Department of Energy since 1994 under cooperative agreement DE-FC36-93CH10580 with Reliance Electric Company.

## REFERENCES

1. Edmonds, J. S., Sharma, D. K., Jordan, H. E., Edick, J. D. and Schiferl, R. F., "Application of High Temperature Superconductivity to Electric Motor Design", *IEEE Transactions on Energy Conversion*, Vol. 7, No. 2, June 1992, pp. 322 - 329.

2. Schiferl, R, Driscoll, D. and Stein, J., "Current Status of High Temperature Superconducting Motor Development", *American Power Conference*, April, 1994, Chicago, IL, Conference Record, pp. 1351-1355.

3. Zhang, B.X., "Conceptual Design of Cryogenic Refrigeration System for Superconducting Motors", Engineering Report No. 064, Reliance Electric Corporate Research and Development, Jan, 1996.

4. Swift, W.L. and Sixsmith, H.S., "Performance of a Long Life Reverse Brayton Cryocooler", *Proc. 7th Intl. Cryocooler Conf.*, Phillips Laboratory, April 1993, Part 1, pp. 84-97.

# Integration of a Photoconductive Detector with a 4K Cryocooler

**R. S. Bhatia , A. G. Murray , M. J. Griffin and P. A. R. Ade**

Astrophysics Laboratory , Department of Physics
Queen Mary & Westfield College , London , United Kingdom E1 4NS

**T. W. Bradshaw and A. H. Orlowska**

Cryogenics Section , Cryogenics and Materials Engineering Group
Rutherford Appleton Laboratory , Didcot , United Kingdom OX11 OQX

## ABSTRACT

We have characterised the performance of a Ge:Ga photoconductive detector cooled to 4 K by a closed cycle mechanical cryocooler. The tests were performed under a photon background level of around 1 nW, appropriate for Earth observing applications such as rapid-scanning far-infrared Fourier transform spectrophotometry of the stratosphere. Signal frequencies for such applications are generally in the 500 Hz - 5 kHz range, and the main aim of these tests was to determine if the detector performance would be adversely affected by cooler vibration or electromagnetic interference. The cryogenic system was a Joule-Thomson expansion cooler operating at 35 Hz, precooled to 20 K by a two-stage Stirling-cycle mechanical cooler operating at 32 Hz. As a comparison, the detector response was also evaluated in a $^4$He cryostat with and without microphonic excitation. These initial tests have shown that the detector performance is not affected by the cryocooler environment for frequencies greater than a few hundred Hz. At lower frequencies, there are high-Q features at the cooler drive frequency and its harmonics, but the level of white noise is only slightly elevated. With further optimisation of the system, it is expected that an effectively undisturbed spectrum can be achieved. These measurements have demonstrated the feasibility of operating far-infrared photoconductive detectors in conjunction with a 4 K cryocooler for high photon background applications.

## INTRODUCTION

Far-infrared (30 - 200 μm) spectroscopy is a powerful technique for remote sensing of the Earth's atmosphere [1]. Balloon-borne Fourier transform spectrometers (FTSs) have been used in recent years to measure trace species in the stratosphere involved in ozone chemistry [2]. These FTS instruments are operated in rapid scan mode and down-convert the incident radiation frequencies to electrical signals at audio frequencies in the range 500 Hz up to 5 kHz. The Ge:Ga photoconductive detectors which are used at these wavelengths require cooling to 4 K. This is usually achieved using a liquid helium cryostat.

Satellite-borne instruments have the important operational advantages over ground- and balloon-based instruments of repeated near-global coverage and much longer measurement timespans. $^4$He cryostats have been used to cool these satellite instruments in the past, but although the technology for their use is now considered to be mature, this cooling technique does nonetheless have the disadvantages of relatively low lifetime and high launch mass. The use of radiators and closed cycle coolers without a dewar seems to be an attractive option, but the system level integration aspects of this option need to be addressed, in particular the impact of microphonic vibrations on the detectors.

Closed cycle cooler capable of achieving temperatures of 4 K having been developed which use a Joule-Thomson (J-T) expansion cooler precooled to below the helium inversion temperature by a two-stage Stirling-cycle mechanical cooler. Degradation in detector performance may result from a combination of the following causes.

i)      The J-T compressor pistons and the Stirling compressor piston and displacer alternately compress and then expand a working fluid to effect a thermodynamic cycle. The linear reciprocating motion of these moving elements creates the basic momentum imbalance which is of concern here.

ii)     The primary drive signal to the cooler motors is such that the moving elements follow as pure a sine wave as possible, and hence the induced vibration from this motion would only be at the cooler fundamental drive frequency. However, the pistons and displacer move against a non-linear gas spring of the working fluid and the non-linear flexure support springs which maintain alignment of the moving elements. The resultant non-sinusoidal motion induces additional higher harmonics.

iii)    Non-linearities in the practical implementation of the drive electronics also result in non-sinusoidal motion which induces additional higher harmonics.

iv)     Additional frequencies arise from internal resonances of components and subassemblies within the cryocooler system.

v)      The cryocooler cold finger expands and contracts as a result of thermal cycling.

vi)     At the Joule-Thomson cold end, the helium expands rapidly through an orifice, creating acoustic noise.

The cryocooler induced vibration is transmitted to the detector system along the cold finger - focal plane thermal link, and also through the mechanical structure that is used to support the cryocooler system. The vibration can cause mechanical oscillations of the detector, the detector wiring or the detector preamplifier box. This in turn results in microphonic noise from modulation of stray capacitances by mechanical motion or motional emf induced by the movement of wiring in magnetic fields generated by the cooler motors and drive electronics. The vibrations can also cause oscillatory movement of the detector optical feedhorn which modulates the detector field of view, so giving rise to a modulated radiant signal. The magnetic fields themselves vary with time and so can potentially induce pickup even in stationary wiring. In addition, detector performance may be degraded by temperature fluctuations which arise from the cyclic nature of the cooling technique. Adequate temperature stability at the cold stage is important because the detector responsivity and time constants are very dependent on the temperature.

Although much work has been undertaken to minimise the induced vibration levels from cryocooler systems [3,4,5,6,7], the system level integration issues have not yet been fully addressed and detailed characterisation of detector performance within the environment of the cryocooler system has not yet been undertaken. Preliminary investigations into the effects of cryocooler systems on bolometric detectors have identified some of the areas in which further work is necessary [8,9]. Here we report on characterisation of the performance of a Ge:Ga photoconductive detector cooled to 4 K by a closed cycle mechanical cryocooler.

## DETECTOR MODULE

The detector module (Figure 1) comprised a Ge:Ga detector with a cut-off wavelength of 120 µm, viewing an infrared source [10] through a 28 µm low pass edge filter. A useful figure of merit for a photoconductor is the Detective Quantum Efficiency [11] which for this device was greater than 30% from 80 Hz to 3 kHz [12,13]. The detector front end electronics consisted of a dual J-FET pair constituting the first stage of a conventional transimpedance amplifier (Figure 2). A low-noise 5 MΩ resistor was used as the feedback resistor. These components were all located on a common block machined from OFHC copper.

## CRYOSTAT TESTS

The cryostat used for the baseline detector characterisation was an Infrared Laboratories HDL-5 dewar, shown in Figure 3. The detector module was bolted to the cold plate and covered with a 4 K copper shield with aluminium tape wound around the base of this shield to prevent stray light from reaching the detector. Potential wire movement was reduced by fixation of all electrical wires to the work surface with GE 7031 varnish which also provided good thermal anchoring to 4.2 K. A cryogenic accelerometer [14] was mounted next to the detector module to measure the cold plate acceleration levels and a stainless steel low diameter coaxial cable used for output readout. Using a mechanical shaker, the cryostat was microphonically excited at a fundamental frequency of 32 Hz. The warm electronics box was vibrationally isolated from the shaker microphonic excitation. Detector signal and accelerometer spectra were taken using a spectrum analyzer. The tests were repeated with different currents through the infrared source

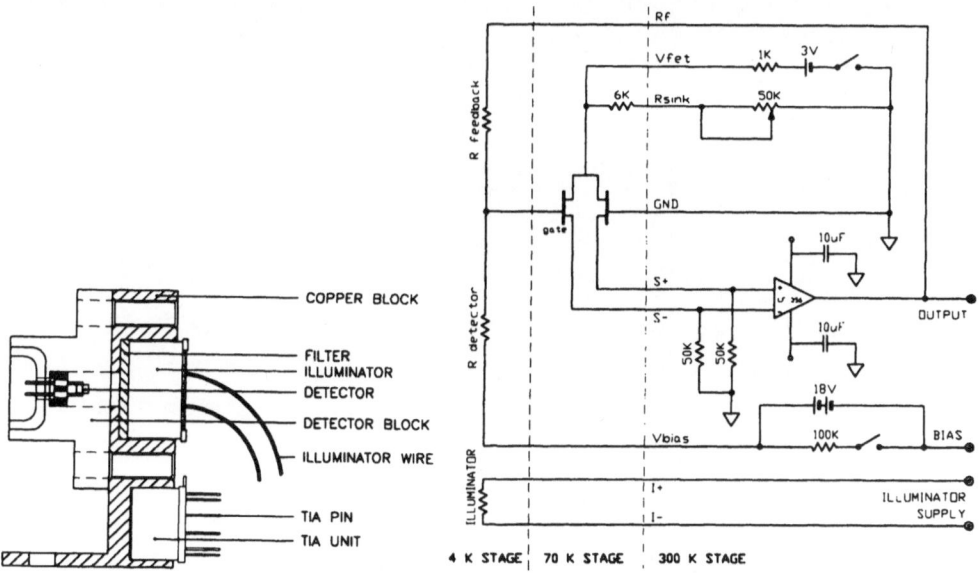

**Figure 1**. Detector Module                    **Figure 2**. Electronics Schematic

giving different levels of illumination on the detector up to a maximum of ~ 1 nW. The Joule heating effect of the maximum current passing through the illuminator was 6.2 mW. These different illumination levels on the detector were indicated by different levels of DC voltage at the preamplifier output. Detector noise spectra were also taken with the shaker orientated in three orthogonal axes but mechanically decoupled from the cryostat, in order to investigate the contribution of EMI pickup. These tests were done with zero illumination, corresponding to the most sensitive case.

## 4 K CRYOCOOLER TESTS

The 4 K cryocooler system consists of a J-T expansion system operating at 35 Hz, precooled by a two-stage Stirling system operating at 32 Hz. Details of the design, development and space qualification of this cryocooler are given elsewhere [15,16]. The cryocooler heatlift capability is 11 mW at 4 K.

The detector module was bolted to an OFHC copper disk which was thermally and mechanically anchored to the J-T 4 K plate via two copper rods. A photograph of the displacer and J-T cold stage is shown in Figure 4. Potential wire movement was reduced by fixation of all wires with GE 7031 varnish firstly to each other and then to each of the cryocooler cold stages. This served the additional purpose of heatsinking the wires to minimise parasitic heat conduction to the 4 K stage. Gold plated radiation shields were fixed to the 4 K, 20 K and 150 K plates. Five layers of MLI comprised of mylar interleaved with dacron netting were placed between the 20 K shield and the 150 K shield, and also between the 150 K shield and the vacuum can at 300 K. The displacer assembly was vibrationally isolated from the Stirling and J-T compressors by floating both the displacer and the compressors pneumatically on separate tables, such that the only connection between the two was through the Stirling cooler transfer pipe and the J-T compressor lines and bypass line. Detector noise spectra and acceleration spectra were taken for different levels of illumination in the same manner as for the cryostat testing. Sufficient time was allowed prior to application of the illuminator current to allow a small quantity of helium to liquefy at the cryocooler 4 K stage. These tests were repeated using a momentum compensator on the displacer driven using open loop control to reduce the cryocooler vibration levels at the cold tip.

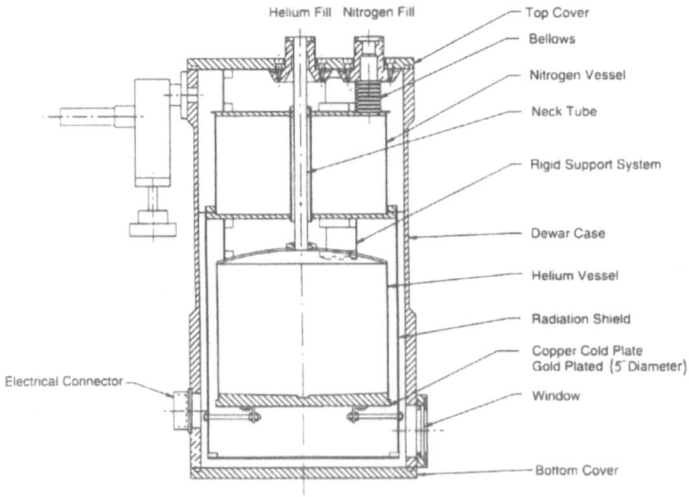

**Figure 3.** Infrared Laboratories HDL-5 Cryostat

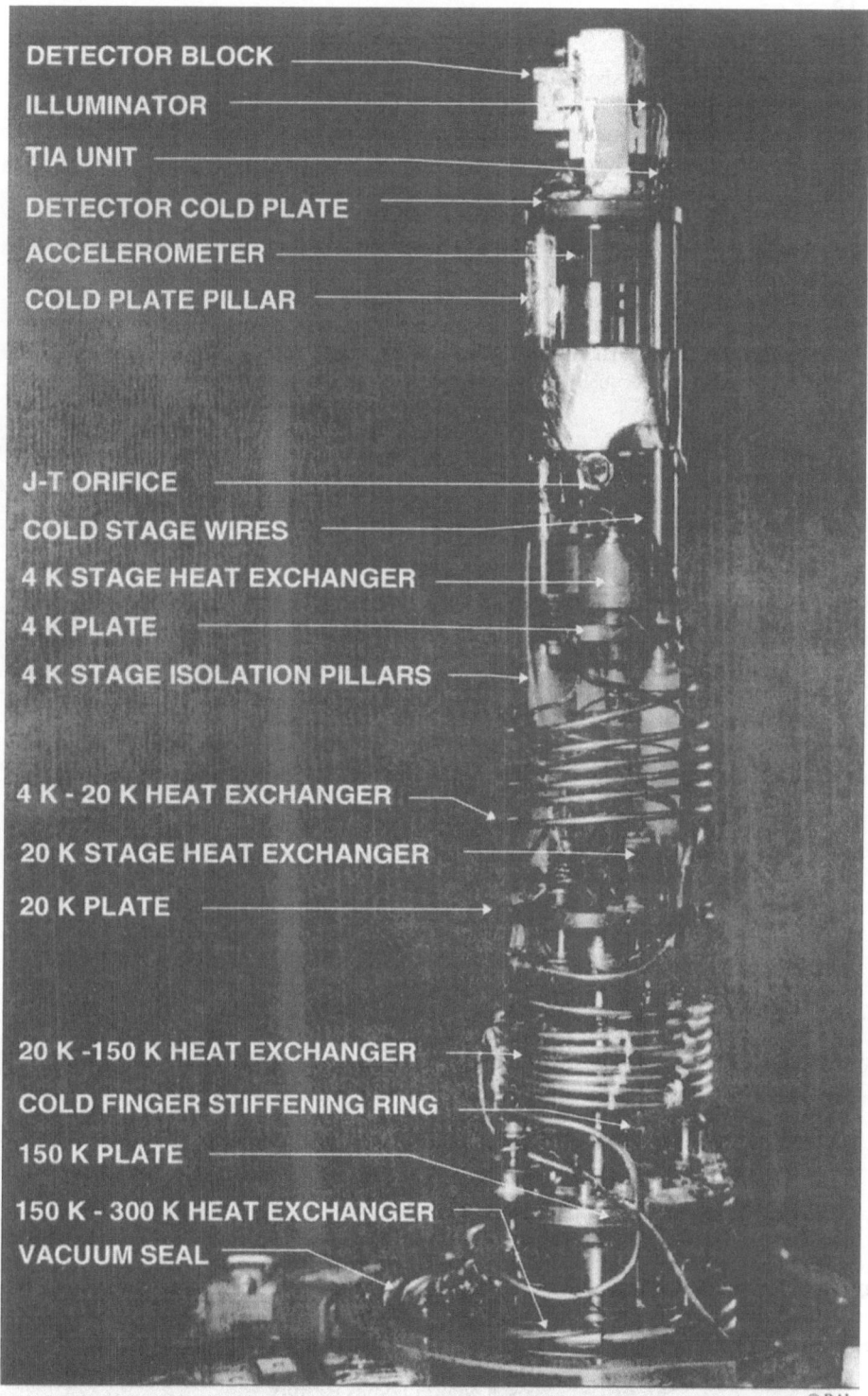

DETECTOR BLOCK

ILLUMINATOR

TIA UNIT

DETECTOR COLD PLATE

ACCELEROMETER

COLD PLATE PILLAR

J-T ORIFICE

COLD STAGE WIRES

4 K STAGE HEAT EXCHANGER

4 K PLATE

4 K STAGE ISOLATION PILLARS

4 K - 20 K HEAT EXCHANGER

20 K STAGE HEAT EXCHANGER

20 K PLATE

20 K -150 K HEAT EXCHANGER

COLD FINGER STIFFENING RING

150 K PLATE

150 K - 300 K HEAT EXCHANGER

VACUUM SEAL

© RAL

**Figure 4 .** 4 K Cryocooler Cold Stages

## RESULTS

### Cryostat Results

The upper panel in Figure 5 shows the longitudinal acceleration levels applied to the cryostat over 0-200 Hz. The external lateral acceleration levels were of the order of 0.3 ms$^{-2}$ rms at the excitation frequencies. There was little mechanical input above ~ 500 Hz in both longitudinal and lateral axes. The lower panels show the resultant detector responses over 0-200 Hz and 0-3200 Hz for 1 nW radiant loading. The detector response to the excitation is clearly evident as high Q features at the excitation frequencies, particularly at the second harmonic at 64 Hz. The mains interference at multiples of 50 Hz is also evident. There is little increase in the baseline noise floor level and no increase in low frequency 1/f noise.

The response with the shaker mechanically decoupled from the cryostat but in close proximity to it showed a noise floor of 0.2 $\mu V_{rms}/\sqrt{Hz}$. This was comparable to the quiescent response. The response at the shaker frequencies was 0.4 $\mu V_{rms}/\sqrt{Hz}$ for all orientations except with the armature positioned vertically, in which instance the detector response at 64 Hz was 4 $\mu V_{rms}/\sqrt{Hz}$.

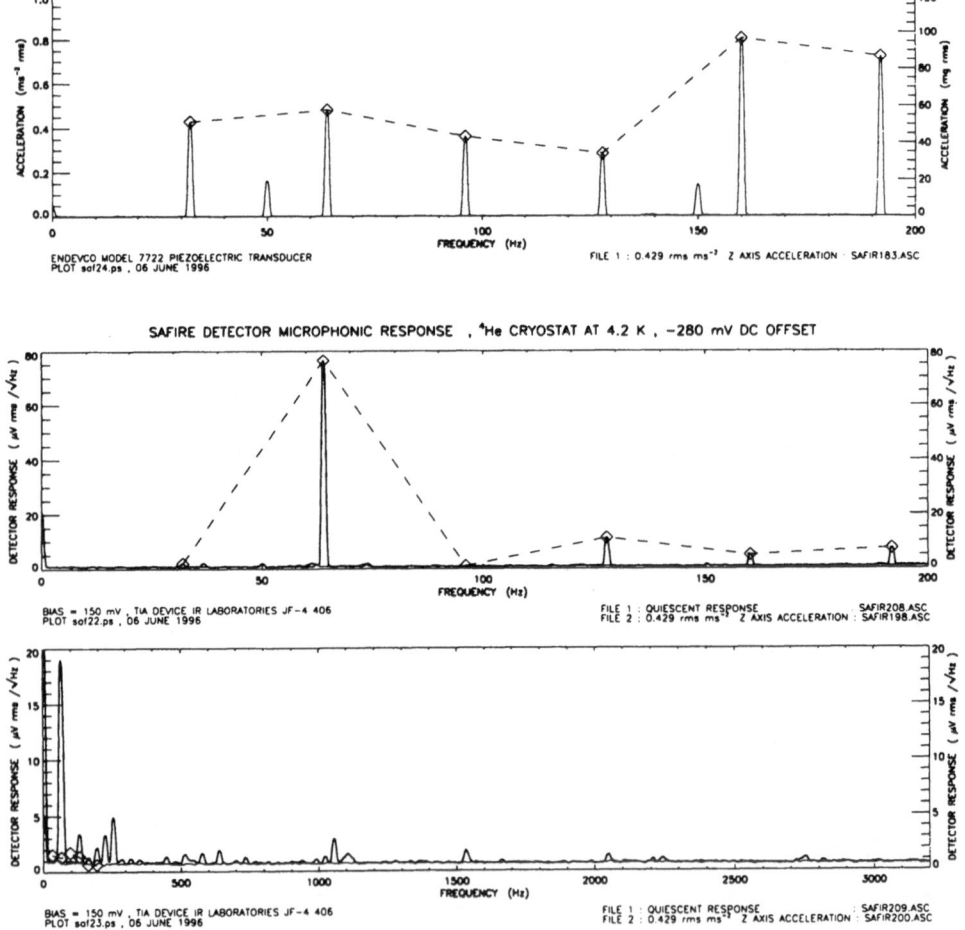

**Figure 5.** Cryostat Microphonics Results

## Cryocooler Results

The upper panel in Figure 6 shows the cryocooler longitudinal acceleration levels at the 4 K plate over 0-400 Hz both with and without displacer momentum compensation. Again, the external lateral acceleration levels were of the order of 0.3 ms$^{-2}$ rms at the excitation frequencies. In all three orthogonal axes there was little mechanical input above ~ 500 Hz.

The lower panels show the detector response (without momentum compensation) plotted in bold over the quiescent cryostat based response, for 1 nW radiant load. From above ~ 30 Hz out to 3200 Hz, the quiescent cryostat and cryocooler responses have comparable noise floor levels of 1 $\mu V_{rms}/\sqrt{Hz}$. Discrete high Q features specific to the cryocooler spectrum are seen, but are of relatively low magnitude except at ~ 400 Hz and ~ 750 Hz; the two spectra are otherwise very similar. Levels of detector response above the noise floor are generally coincident with the cryocooler Stirling vibration frequencies (marked with a ◊) and the cryocooler Joule-Thomson frequencies (marked with a +). Surprisingly, the detector response with displacer momentum compensation was very similar to that obtained without compensation. At different illumination levels the responses were similar to those shown above for both the cryostat and the cryocooler tests, although as expected the infrared photon noise level increased with illumination level.

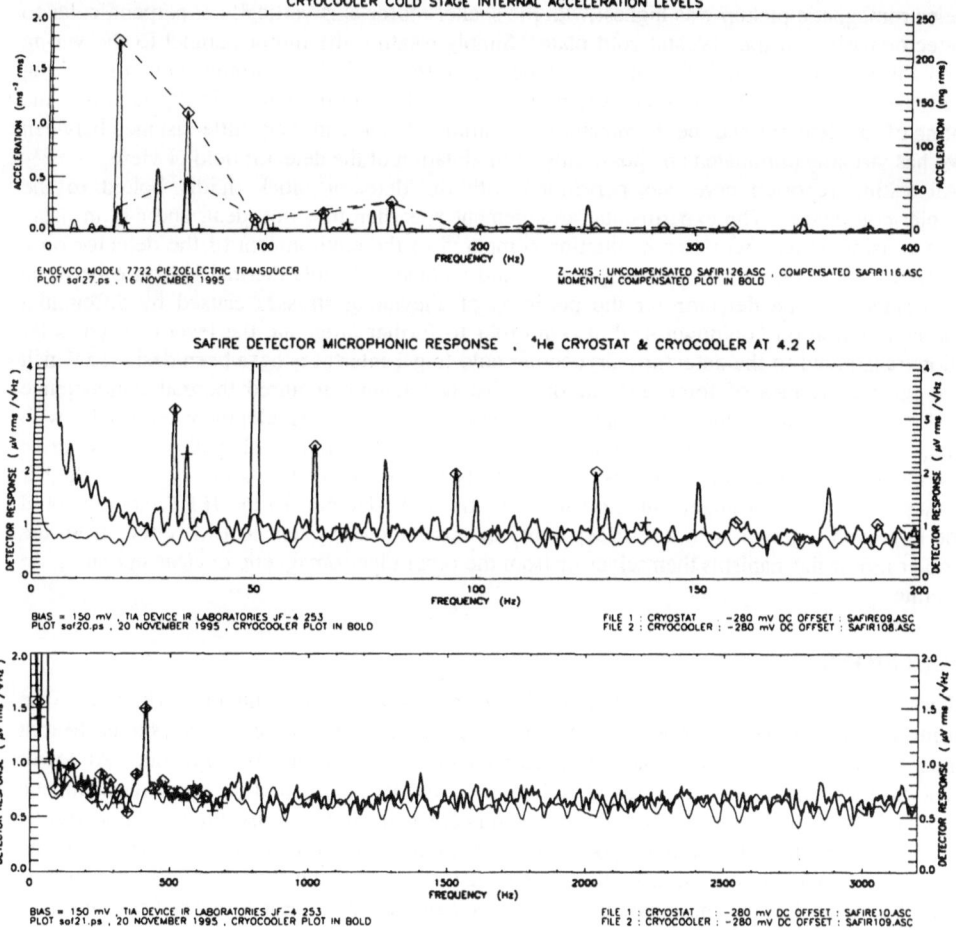

**Figure 6 .** Cryocooler Microphonics Results

## DISCUSSION

Above 500 Hz, the photoconductor response reaches the fundamental limit imposed by photon noise from the infrared source whether the detector is cooled by the microphonically quiescent $^4$He cryostat or by a closed cycle mechanical cryocooler. Below this frequency, detector responses to cryocooler vibrations are evident as features at discrete cooler frequencies together with only a slight increase in the level of white noise. There is little difference in the detector response amplitude depending on whether the displacer is momentum compensated or not, despite this giving an order of magnitude decrease in the fundamental excitation amplitude. This may imply either that a saturation point has already been reached in the vibration of the detector wiring even at the reduced vibration levels with momentum compensation, or that a significant part of the detector pickup is primarily an EMI response to the effects of the cryocooler motors.

The sinusoidal drive from the shaker was less pure than that from the cryocooler motors due to the smaller size of the shaker motor, giving greater levels of acceleration at the higher harmonics for the cryostat tests. The cryostat response exhibited higher susceptibility to microphonic excitation which may be due to the looser wiring of the cryostat.

With the shaker mechanically decoupled from the cryostat, the detector pickup was negligible compared to the cryostat microphonic response for all three orientations of the shaker. The electromagnetic pickup was highest with the shaker motor axis vertical, i.e. perpendicular to the detector wiring on the cryostat cold plate. Simply rotating the motor parallel to the wiring reduced this pickup by an order of magnitude. For the cryocooler arrangement the wiring generally lay in the vertical axis, along which the motor coils were moving. It is expected that the mounting of the detector and the illuminator to a common block with very little distance between the two has virtually eliminated any possibility of modulation of the detector field of view.

All testing reported here was performed with the detector block rigidly bolted to the cryocooler cold stage. The experimental arrangement was therefore not ideal, and can in many ways be considered as a worst case situation compared to the environment of the detector on a flight system. A flight version of the system would include a flexible thermal interface between the cold stage and the detector for the purposes of alleviating stresses caused by differential thermal contraction on cooldown, and also in order to further attenuate the level of cryocooler vibration transmitted to the detector. Cryocooler cold finger interfaces have been designed which achieve several decades of force attenuation whilst maintaining required thermal conductance levels [17]. For these tests, the cryocooler drive electronics was at a development level only, with no capability for precise vibration reduction. Improving the sinusoidal quality of the drive waveforms to the cryocooler motors would implicitly lead to lower levels of cryocooler higher harmonics. The wiring arrangement and harnessing from the detector to the JFET module could also be improved together with the susceptibility of the detector system to EMI either from the cryocooler motor mechanisms themselves or from the other electromagnetic devices operating on the satellite.

## CONCLUSIONS

These initial tests have shown that at frequencies above a few hundred Hz, the noise spectrum of a Ge:Ga photoconductor when cooled to 4 K by a cryocooler is very similar to that obtained when the detector is cooled in a microphonically quiescent $^4$He cryostat. At lower frequencies, there are high-Q features at the cooler drive frequency and its harmonics, but the noise floor is only slightly perturbed above the fundamental photon noise limit. These results were obtained with the detector and cryocooler unoptimised for minimising levels of vibration and EMI emission. With optimisation of the system, it is expected that an effectively undisturbed spectrum can be achieved. These measurements have demonstrated the feasibility of operating far infrared photoconductive detectors in conjunction with a 4 K cryocooler for high photon background applications.

## ACKNOWLEDGEMENTS

The 4 K cryocooler was developed under contract to ESA/ESTEC. R. Bhatia acknowledges a graduate studentship funded by QMW College (University of London) and QMC Instruments Ltd.

## REFERENCES

1. Persky, M. J., "A Review of Spaceborne Infrared Fourier Transform Spectrometers for Remote Sensing" , *Review of Scientific Instruments* Vol. 66 No. 10 (October 1995), pp. 4763-4797

2. Ade, P. *et. al.*, "SAFIRE-A : Spectroscopy of the Atmosphere using Far Infrared Emission – Airborne", *Proposal to The CEC Environment Program (Second Phase 1993-94) for the development of a Fourier transform far-infrared spectrometer operating from high altitude aircraft or balloon platforms for observation of the Earth's atmosphere* (1993)

3. Aubrun, J-N., R. R. Clappier, K. R. Lorell, T. C. Nast and P. J. Reshatoff, Jr., "A High-performance Force-cancellation Control System for Linear-drive Split-cycle Stirling Cryocoolers", *Advances in Cryogenic Engineering*, Vol. 37 Part B, Ed. R. W. Fast, Plenum Press, New York (1992), pp. 1029-1036

4. Bhatia, R. S. and K. P. Rodger, "Reduction of Cooler Vibration" , *Contractor Report No. ESA CRP 3632* , Final Report to The Eurpoean Space Agency, ESA/ESTEC Contract Number 9497/91/NL/FG, (March 1993)

5. Collins, S. A., A. H. von Flotow and J. D. Paduano, "Analogue Adaptive Vibration Cancellation for Stirling Cryocoolers", *Cryogenics* Vol. 34, No. 5 (1994), pp. 399-406

6. Mon, G. R., G. T. Smedley, D. L. Johnson and R. G. Ross, Jr., "Vibration Characteristics of Stirling Cycle Cryocoolers for Space Application", *Proceedings of the 8th International Cryocooler Conference*, Vail, Colorado USA, June 28-30, 1994 Ed. R. G. Ross, Jr., Plenum Press, New York (1995), pp. 13-22

7. Boyle, R., F. Connors, J. Marketon, V. Arillo, E. James and R. Fink, "Non-real Time, Feed Forward Vibration Control System Development & Test Results", *Proceedings of the 7th International Cryocooler Conference*, Santa Fe, NM USA, November 17-19, 1992, Air Force Phillips Laboratory Report PL-CP-93-1001, Kirtland Air Force Base, NM USA (1993), pp. 809-819

8. Bradshaw, T. W., J. Hieatt, M. Griffin and P. Ade, "An Investigation into the Effects of Closed Cycle Coolers on Bolometric Detectors", *European Space Agency Contractor Report* (August 1993)

9. Bhatia, R. S., M. J. Griffin, A. J. Zebedee, P. A. R. Ade and I. D. Hepburn, "Bolometer Microphonic Susceptibility", *Proceedings of a Workshop on Bolometers for Millimetre and Submillimetre Space Projects*, 15-16th June 1995, Orsay, Paris France. Ed. J. M. Lamarre, J. P. Torre and V. Demuyt, Institut d'Astrophysique Spatiale Report Number RS 95-02 (June 1995), pp. 143-150

10. Infrared Laboratories TRS Device, 1808 E. 17th Street, Tucson, AZ 85719 USA

11. Haller, E. E., "Physics and Design of Advanced IR Bolometers and Photoconductors" , *Infrared Physics* Vol. 25, No. 1/2 (1985) pp. 257-266

12. Murray, A. G., M. J. Griffin, P. A. R. Ade, J. Leotin, C. Meny and G. Sirmain, "Optimisation of FIR Photoconductors for Atmospheric Spectroscopy", *Proceedings of a European Symposium on Photon Detectors for Space Instrumentation*, Noordwijk, The Netherlands, 10-12 November 1992, ESA SP-356 (December 1992), pp. 159-163

13 . Lee, C., Astrophysics Group, Department of Physics, Queen Mary and Westfield College, London UK, *Private Communication* (18 October 1995)

14 . Endevco Corporation, 30700 Rancho Viejo Road, San Juan Capistrano, CA 92675-1789, USA

15 . Orlowska, A. H., T. W. Bradshaw and J. Hieatt, "Development Status of a 2.5 K - 4 K Closed-Cycle Cooler Suitable for Space Use", *Proceedings of the 8th International Cryocooler Conference*, Vail, Colorado USA, June 28-30, 1994, Ed. R. G. Ross, Jr., Plenum Press, New York (1995), pp. 517-524

16 . Jones, B. J. and D. W. Ramsay, "Qualification of a  4 K Mechanical Cooler for Space Applications", *Proceedings of the 8th International Cryocooler Conference*, Vail, Colorado USA, June 28-30, 1994, Ed. R. G. Ross, Jr., Plenum Press, New York (1995), pp. 525-535

17 . Sparr, L., R. Boyle, L. Nguyen, H. Frisch, S. Banks, E. James, V. Arillo, "Design and Test of Potential Cryocooler Cold Finger Interfaces" , *Advances in Cryogenic Engineering*, Vol. 39, Ed. P. Kittel, Plenum Press, New York (1994), pp. 1253-1262

# Proceedings Index

This book draws upon the work presented at the 9th International Cryocooler Conference, held in Waterville Valley, New Hampshire, in June 1996. Although this is the ninth meeting of the conference, which has met every two years since 1980, the authors' works have only been made available to the public in hardcover book form since 1994; this book is thus the second hardcover volume. Prior to 1994, proceedings of the International Cryocooler Conference were published as informal reports by the particular government organization sponsoring the conference — typically a different organization for each conference. Most of the previous proceedings were printed in limited quantity and are out of print at this time.

For those attempting to locate references to earlier conference proceedings, the following is a listing of the eight previous proceedings of the International Cryocooler Conference.

1) *Refrigeration for Cryogenic Sensors and Electronic Systems*, Proceedings of a Conference held at the National Bureau of Standards, Boulder, CO, October 6-7, 1980, NBS Special Publication 607, Ed. by J.E. Zimmerman, D.B. Sullivan, and S.E. McCarthy, National Bureau of Standards, Boulder, CO, 1981.

2) *Refrigeration for Cryogenic Sensors*, Proceedings of the Second Biennial Conference on Refrigeration for Cryogenic Sensors and Electronics Systems held at NASA Goddard Space Flight Center, Greenbelt, MD, December 7-8, 1982, NASA Conference Publication 2287, Ed. by M. Gasser, NASA Goddard Space Flight Center, Greenbelt, MD, 1983.

3) *Proceedings of the Third Cryocooler Conference*, National Bureau of Standards, Boulder, CO, September 17-18, 1984, NBS Special Publication 698, Ed. by R. Radebaugh, B. Louie, and S. McCarthy, National Bureau of Standards, Boulder, CO, 1985.

4) *Proceedings of the Fourth International Cryocoolers Conference*, Easton, MD, September 25-26, 1986, Ed. by G. Green, G. Patton, and M. Knox, David Taylor Naval Ship Research and Development Center, Annapolis, MD, 1987.

5) *Proceedings of the International Cryocooler Conference*, Monterey, CA, August 18-19, 1988, Conference Chaired by P. Lindquist, AFWAL/FDSG, Wright-Patterson AFB, OH.

6) *Proceedings of the 6th International Cryocooler Conference*, Vols. 1-2, Plymouth, MA, October 25-26, 1990, David Taylor Research Center Report DTRC-91/001-002, Ed. by G. Green and M. Knox, Bethesda, MD, 1991.

7) *7th International Cryocooler Conference Proceedings*, Vols. 1-4, Santa Fe, NM, November 17-19, 1992, Air Force Phillips Laboratory Report PL-CP-93-1001, Kirtland Air Force Base, NM, 1993.

8) *Cryocoolers 8*, proceedings of the 8th ICC held in Vail, Colorado, June 28-30, 1994, Ed. by R.G. Ross, Jr., Plenum Press, New York, 1995.

# Author Index

# Subject Index